Hal, -a (L) Breathe
Helmin (G) Worm
Hem, -a, -ato, -o (G) Blood
Hepa (G) Liver
Herni, -a (L) Rupture
Hetero (G) Other, different
Hisc (L) Open
Histo (G) Tissue, web
Hom, -eo, -o (G) Like, similar
Hom, -in, -o (L) Man
Hydr, -a (G) Water
Hypo (G) Under, beneath
Hyper (G) Over, excessive, above
Hypno (G) Sleep
Hyster (G) Uterus, womb, lower

Ilio (L) Intestine
Immun, -o (L) Safe, free
Infarct (L) Filled in, stuffed
Infra (L) Below, lower
Infundibul (L) Funnel
Inguin, -o (L) Groin
Innocula (L) Implant
Insul, -a (L) Island
Intro (L) In, within, into
Isch, -o (G) Hold, suppress
Iso (G) Equal
-itis (G) Inflammation

Jacul (L) Throw
Juxta (L) Near to

Kary, -o (G) Nut, nucleus
Kilo (G) Thousand
Kypho, -s (G) Bent

Labi (L) Lip
Lacr (L) Tears, weeping
Lact, -o (L) Milk
Lamin, -a (L) Thin plate, sheet
Lat (L) Broad, wide
Laten, -t (L) Hidden
Later, -al (L) Side
-lemma (G) Husk, peel, shell
Leuco (G) White
Leuk (G) White
Liga, -m (L) Bound, tied
Lingu, -a (L) Tongue
Littor, -a (L) Seashore

Macro (G) Large, long
Mal, -e, -ign (L) Bad, evil
Mandibul (L) Jaw
Mast (G) Breast
Mater (L) Mother
Maxill, -a (L) Jaw
Medi (L) Middle
Medic (L) Heal, healing
Medull, -a (L) Marrow, pith
Melano (G) Black
Mens, -e (L) Month
Menstru (L) Monthly

Metab, -ol (G) Change
Micro (G) Small
Milli (L) One-thousandth
Mini (L) Smallest
Molecul (L) Little mass
Mono (G) Single, one
Morph (G) Form
Mort (L) Death
Muta (L) Change
Myc (G) Fungus
Myo (G) Muscle

Narco (G) Numbness, stupor
Nari (L) Nostril
Naso (L) Nose
Naus (G) Seasickness
Necros (G) Death, deadness
Neo (G) New, recent
Noci, noxi (L) Harmful
Noct (L) Night
Nomen (L) Name
Noxios (L) Harmful
Nutro (L) Feed, nourish

O (G) Egg
Obstetri (L) Midwife
Olfact (L) Smell
-oma (G) Tumor
Omni (L) All
Oo (G) Egg
Opercul (L) Cover, lid
Opti (G) Eye, vision
Orchi, -d (G) Testicles
Oro (L) Mouth
-osis (L) Condition of
-osis (G) Disease
Oss (L) Bone
Ossic (L) Little bone
Ot, -i, -o (G) Ear
Oto (G) Ear
Ov (L) Egg

Palp (L) Touch, feel
Pan (G) All
Papill, -a, -i (L) Nipple
Parasit (G) Near food
Partur (L) Bring forth young
Peri (G) Around
Phall, -o, -us (G) Penis
Phil (G) Love, loving
Phleb (G) Vein
-phobia (G) Fear, dread
Phone (G) Carry, bear
Pili (L) Hair
-plasty (G) Growth, molding
Pleur (G) Side, rib
Pneumo (G) Lungs
Pollut (L) Defiled
Poly (G) Many, much
Pons (L) Bridge
Post (L) Behind, after
Presby (G) Old
Ptery (G) Wing
Ptos, -us (G) Falling

Quadr, -a (L) Four

Rabi, -es (L) Mad, raving
Retro (L) Back, behind
Rheo (G) Flow, current
Rhin, -o (G) Nose
Rupt (L) Broken, burst

Sacchar (G) Sugar
Sangui, -ni (L) Blood
Sapro (G) Rotten, putrid
Scat (G) Dung
Schisto (G) Divided, split
Seb, -i, -um (L) Grease, tallow
Senesc (L) Grow old
Sepsi, -s (G) Putrid, putrefaction
-spire (L) Breathe
-stalsis (G) Constriction, compression
Staphyl, -o (G) Bunch of grapes
Sub (L) Under, below
Supra (L) Above, over, beyond
Symbio (G) Living together
Synap, -s, -sis (G) Joining, union
Syndesm (G) Bond, ligament
Synerg (G) Work together, cooperate
Sys (G) With, together
Systol (G) Contraction

Tach (G) Quickly
Tacti (L) Touch
Tardi (L) Slow
Tax, -i, -is (G) Arrange
Tempor, -a, -o (L) Time
Testi, -s (L) To witness; testes
Tetan, -o, -us (G) Tense, rigid
Thermo (G) Heat
Trache, -a (L) Trachea
Tri (L) Three
Troph, -i, -o (G) Nourish, food
Trunc, -a, -at (L) Cut off
Tympan, -o (G) Drum

Ultr, -a (L) Beyond

Vaccin (L) Of a cow: vaccine
Vagin, -a (L) Sheath
Vener, -a, -ea (L) Pertaining to coitus, intercourse (after Venus)
Vesicul, -a (L) Small bladder
Virul (L) Poisonous
Viscer, -a, -o (L) Organs of the body cavity

Xeno (G) Stranger
Xer, -o (L) Dry
Xyl, -o (G) Wood

Zoo (G) Animal

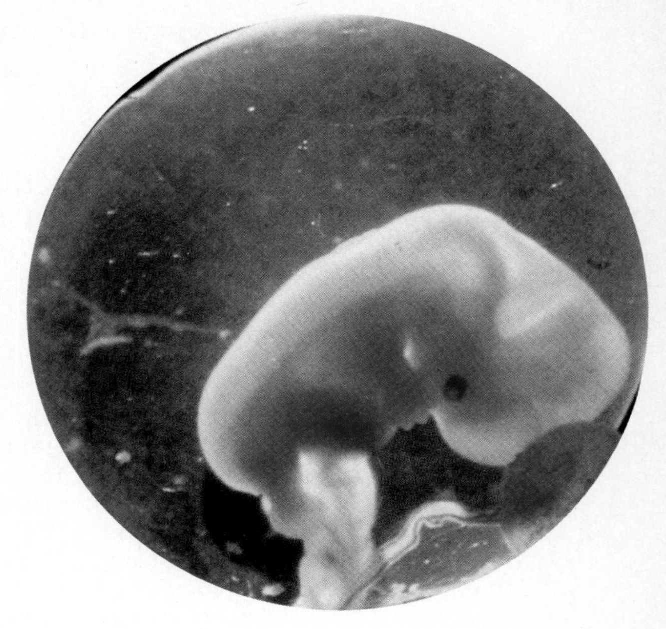

Focus on Human Biology

Carl E. Rischer
Long Beach City College

Thomas A. Easton
Thomas College

HarperCollins*Publishers*

To my gentle wife and feisty friend, Marty!
C.E.R.

For Betty Sue and Joellen
T.A.E.

Sponsoring Editor: Glyn Davies
Development Editor: Barbara Conover
Project Editor: Shuli Traub
Art Director: Teresa J. Delgado
Art Coordinators: Theo Hernandez/Kathy Skultety
Text and Cover Design: Delgado Design, Inc.
Cover Photos: VU/© Michael Webb; Telegraph Colour Library/FPG International; Portrait of Lucrezia Panchiatichi, by A. Bronzino, Galleria Uffizi.
Photo Researcher: Karen Koblik
Production Manager: Willie Lane
Compositor: York Graphic Services, Inc.
Printer and Binder: Arcata Graphics/Kingsport
Cover Printer: The Lehigh Press, Inc.

FOCUS ON HUMAN BIOLOGY

Library of Congress Cataloging-in-Publication Data

Rischer, Carl E.
Focus on human biology / Carl E. Rischer, Thomas A. Easton.
p. cm.
Includes bibliographical references and index.
ISBN 0–06–045416–4 (student ed.) ISBN 0-06-500416-7 (instructor's ed.)
1. Human biology. I. Title.
QP36.R49 1992
612—dc20 91–26058
CIP

92 93 94 95 9 8 7 6 5 4 3 2 1

Contents in Brief

Contents in Detail

Preface

A human biology course can be extremely rewarding for both the teacher and the students. The reward lies in the connection between the subject and the students, for each person has a chance to learn about the intricacies of his or her own body and to build an understanding that will be reinforced and expanded through a lifetime of continuous exposure to new information.

In this course we will discuss the following:

- The structure and methods of basic science
- The minuscule and complex units of life, the cells
- The anatomy and physiology of human body systems
- The causes of pain and suffering, the microscopic agents of disease, and how the body defends itself
- How life probably originated on this ancient planet, Earth
- How humans evolved from simple origins
- Where "life" might be going as we move from the present into the future

We make no attempt to disguise our enthusiasm for our subject. Biology is one of the most enticing areas in the academic curriculum. We discovered that simple truth when we were students. We embraced the study of biology then and pursued it into our graduate work and careers. Now we wish to share some of the things we have learned with the people who will follow us, the students.

Without hesitation or doubt we make each student a rather extravagant promise: If you stay with us and study and work, we will change the way you think about life, your body, and what it means to be human. You will develop an understanding of natural processes that will prepare you for many of the complex issues facing our society—for example, the ethical aspects of genetic engineering and reproductive physiology, abortion, disease and immunity, overpopulation, and the human impact on our planet.

SCOPE OF THE BOOK

Focus on Human Biology discusses the basic principles of biology, cell biology, molecular biology, genetics, human anatomy and physiology, human evolution, ecology, and the human impact on the environment. There is a special emphasis on basic human anatomy and physiology and on disease and immunity. For example, we include a table at the end of each of the chapters that deals with the various body systems and that lists and briefly describes some of the most common diseases affecting these systems. In addition, we have a separate chapter describing the biology and classification of the various pathogens that currently afflict humanity with disease, including the human immunodeficiency virus (HIV).

We also cover the basics of nutrition and metabolism, with special attention to some of the myths about nutrition and weight control.

Focus on Human Biology thus can be used either as a one-year course dealing with the many aspects of human biology or as a semester course dealing with specific topics of human life.

AN EMPHASIS ON TEACHING AND LEARNING

Completing an ambitious project like this one represents the fruition of a career-long dream for us, the authors. As educators with nearly a half century of combined experience in working with college and university students, we have developed a feeling for describing and *teaching* the difficult and complex topics of biology. More important, we have structured this book to facilitate each student's *learning* of human biology as well. As you may well know, teaching and learning are two very different processes.

This book is not another catalogue of information. Our aim is to provide a dynamic teaching instrument that presents the information of human biology in an orderly, smoothly flowing, beautifully illustrated, and exciting way and that makes the information easier for students to absorb. As educators, we have attempted to provide an assortment of carefully structured learning experiences designed to facilitate the students' mastery and understanding of the complex array of topics they will encounter in a class like this one.

To achieve these goals, *Focus on Human Biology* offers the students a number of learning tools built right into the text. Many of these tools are not found in other texts. They are designed to encourage and structure the students' learning efforts in an active, hands-on-way from the start.

The following is a list of some of the most important pedagogical features we have incorporated into the project and of the philosophies behind them.

Our Use of Questions

About 2400 years ago, the great Greek teacher and philosopher Socrates demonstrated that teaching and learning could be made more effective by the judicious use of questions that actively and creatively engage the student. A cliché of pedagogy says that teaching students a fact is a little like giving them a fish; it feeds them for a day. Asking them a question is more like giving them a rod and a reel, line, sinkers, lures, and a boat and a motor; it feeds them for the rest of their lives.

The facts must be learned, but no one should stop with the facts. We use questions in an effort to help students take the next step because we believe questions stimulate interest and creatively engage the students in active (rather than passive) learning. They take the students beyond the facts into the realms of imaginative and critical thinking. Their struggle to answer the questions encourages them to reach beyond simple memorization of facts to the analysis and understanding of relationships and associations. As a result, the process of learning becomes a creative consequence of that higher order of activity and moves away from the boring drills of rote.

As you will see, we apply the use of questions in an aggressive manner. We include the traditional block of end-of-chapter questions, as they are sometimes useful for reviewing the chapter. In this textbook, however, we also make liberal use of in-text questions (including a few in the figure captions) and provide a number of Learning Focus exercises. We have designed questions to encourage the students to pause and consider other dimensions and relationships while they cover the concepts for the first time. We want the students to stop and think critically about the materials in small increments while the topics are still fresh and manageable.

Learning Focus Exercises

This textbook presents a series of hands-on learning and organizing exercises. One or two subjects in each chapter are selected to receive special attention because of their importance or complexity. This attention takes various forms as Learning Focus exercises and responses. In these assignments the students are asked to do a variety of tasks that are designed to facilitate understanding, memory, and critical thinking about the information being presented while the material is still fresh.

The simplest of these exercises help the students master straightforward material by labeling blank diagrams of body systems and by naming the functions of assorted anatomical features of those systems. Other Learning Focus exercises call for the students to extract, organize, and transfer select or conceptually difficult information from the text into tabular form.

The most complex—and surely the most interesting—Learning Focus exercises challenge the students to apply the information from the text to real-life situations and decisions. For instance, we ask the student to design a weight loss program for a person with a specific weight change goal that has both dietary and exercise components using the information they have studied about metabolism and nutrition. First, in the Learning Focus part of the exercise, we show the students how such programs are set up. Then, in the response part of the exercise, we ask that they design a program themselves.

Linking Concepts to Art

Biology is a very visual science. For that reason much of our four-color illustration program has been painstakingly conceived and rendered. Since we firmly believe that it is easier to learn ideas if they can be related to pictures, many of the figures have been carefully laid out and contain extensive explanations of the concepts and principles being taught. In some cases, we use substantial text in the figure to enhance the connection between the abstract concepts or processes and the visual presentation.

The Language of Science

Perhaps the most anxiety-provoking prospect facing students taking a biology course is the tremendous number of scientific terms that have to be learned. It is not inaccurate to say that learning about biology and medicine has a great deal in common with the study of a foreign language. To ease learning of many scientific terms, we have included two types of glossaries. The comprehensive glossary of terms at the end of the book defines virtually all the major terms in the book. Further, to provide instant emphasis and convenient review of selected terms, end-of-chapter glossaries define about 20 to 25 of the most important terms found in each chapter.

ACKNOWLEDGMENTS

We thank the editors, designers, and reviewers at HarperCollins Publishers for their contributions in making this book that you now hold in your hands.

We express our appreciation to the professors and researchers who have reviewed the manuscript and proof of this text in its various stages. Their contributions have been invaluable to the development of the book. They are: William C. Bessler, Mankato State Unvieristy; Thomas W. Collins, Moorhead State University; Roger B. Corzine, Odessa College; John E. Dille, Winthrop College; Frank Einhellig, The University of South Dakota; David Fox, New Orleans, Louisiana; Kenneth W. Gregg, Winthrop College; Madeleine M. Hall, Cleveland State University; Laszlo Hanzely, Northern Illinois University; James Heisinger, The University of South Dakota; J. Robert Hippensteele, Illinois Wesleyan University; Robert J. Huskey, University of Virginia; Elizabeth King, Duke University Medical Center; Jerri K. Lindsey, Tarrant County Junior College District; Larry O. Miller, Moorpark College; Raleigh K. Pettigrew, Denison University; Larry G. Sellers, Louisiana Tech University; William Thieman, Ventura College; Robin M. Tyser, University of Wisconsin—La Crosse; David R. Voth, Metropolitan State College; and Richard Walker, Des Moines Area Community College.

ANCILLARIES

A number of helpful aids have been developed to coordinate with this edition of the text.

Instructor's Manual

An instructor's manual, written by David Fox, includes, for each chapter, an overview, list of key topics, lecture outline, classroom discussion topics and activities, list of readings, list of resources (e.g., software, videos, films), and a section relating the individual themes of each chapter to the book as a whole.

Testbank

A testbank of 1500 questions and answers compiled by John Capeheart and Alfred Avenoso of the University of Houston is available both in hard copy and on Testmaster—computer discs for IBM and Macintosh computers.

Student Study Guide

Prepared especially to accompany the text, this helpful study aid for students is written by John Capeheart and Alfred Avenoso. It not only includes objectives and outlines for each chapter but also provides review summaries of the key concepts and self-tests to help students prepare for examinations. An appendix to the guide contains part-by-part selected readings from generalist technical magazines such as *Scientific American, Science News,* and *Science,* and a comprehensive list of word origins for biology and medicine.

Overhead Transparencies

One hundred twenty-five acetate transparencies—all in full color—are available to adopters. Taken directly from the extensive art program of the text, they illustrate key biological concepts discussed in each of the 26 chapters.

Anatomy Coloring Book, Physiology Coloring Book, Human Brain Coloring Book

With an exciting new approach, the coloring books offer an enjoyable and effective way of learning the fundamentals of biology. Participation by the student, through creative coloring, provides significant learning reinforcement. The supportive explanatory text accompanying each coloring plate leads the reader through the plate in a step-by-step manner. In addition, the finished colored plates provide an excellent review that the student has helped to create.

Writing about Biology

Written by Jan A. Pechenik of Tufts University, this brief but straightforward guide includes sections on writing lab reports, essays, term papers, research proposals, critiques, and summaries and in-class essay examinations. It also includes special sections on effective note taking, how to give oral presentations, and how to prepare applications for summer and permanent jobs in the field of biology. Appendices listing commonly used scientific abbreviations are also featured.

The HarperCollins Biology Encyclopedia Laser Disc

The Biology Encyclopedia Laser Disc, produced in conjunction with Nebraska Interactive Video, Inc., offers the latest in visual technology. It contains transparencies, micrographs, slides, and film and video footage. Over 1500 images were provided by Carolina Biological Supply. The laser disc allows instant access to any image or footage, frame-by-frame or moving, simply by pushing a few buttons on a hand-held remote. The disc enhances the principles of biology covered in the text much more effectively than transparencies or videos.

Harper Dictionary of Biology

Written by W. G. Hale and J. P. Margham, both of the Liverpool Polytechnic Institute, it contains 5600 entries that go far beyond basic definitions to provide in-depth explanations and examples. Diagrams illustrate concepts such as genetic organization, plant structure, and human physiology. The dictionary covers all major subjects (e.g., anatomy, biochemistry, ecology) and also includes biographies of important biologists.

Student Environmental Action Guide

The Earthworks Group and HarperCollins have joined with the Student Environmental Action Coalition to bring students a handbook of the environmental movement on campuses around the country. Through real campus examples, it provides a series of strategies for approaching the administration, the community, political leaders, student leaders, and one's own personal habits to achieve positive change. Examples include population control, transportation, water conservation, and publishing a newsletter. All proceeds are returned to the Student Environmental Coalition.

Carl E. Rischer
Thomas A. Easton

A Note to the Student

We have some suggestions to help you gain the most you can from this book. To start, this book is different from other texts. We want you to get physically involved with your own learning.

That's right, physically involved! We believe that learning complex and detailed material is best achieved when all the senses are involved, not just the mind. Be creative in your studying. Write things down, diagram difficult concepts, define terms, duplicate and then relabel diagrams, and gather supportive information from a variety of sources. This is the way scholars work—and at this point in your career (strange though it may seem), "scholar" is your profession. So think about what you are learning. Connect it to other things you know, to news stories, to events in your life, and to other parts of this textbook.

Actively involving yourself in the learning process is the best way to keep your mind alert and focused. By the time you have completed this course, we hope you will have written all over these pages. Write in the margins and over the figures. Add labels to the figures, sticking in bits and pieces of additional, related information that you have gleaned from the text. Use the figures to correlate as much information from the text as you can.

We have inserted questions at frequent intervals in the text to jar you into organizing the information while learning increments are still small and easy to manage. Usually the answers to questions have just been given in the text, although not necessarily in an explicit form. Occasionally, the answer may not appear until later in the chapter; the reason for this is that we believe that when you are asked these questions, you will have enough information to anticipate the answer.

In each chapter, we have paid special attention to one or two topics that seem especially difficult to learn. To help you master these subjects, we have developed Learning Focus exercises. These exercises take many forms, the most common being a diagram summarizing complex topics, showing graphically what is happening at each step. In some cases, we provide you with a physical structure, for example, a blank table that you will be asked to fill in by gathering information about complicated subjects from the text. This will help *you* organize the information as you proceed and will direct your attention to details your professor is likely to expect you to master.

Other Learning Focus exercises provide material not covered in the rest of the chapter. You should go over each exercise carefully. Once you have mastered the material, possibly using a diagram, you may be directed to a blank version of the diagram (the response) one or two pages later and asked to fill it in with your own labels and notes. Remember, repetition is an important part of the learning process; go over these materials many times as you prepare for your exams. We also recommend you make several photocopies of each blank diagram so that you can practice and quiz yourself more than once as you prepare.

Other Learning Focus exercises ask you to synthesize what you are learning, piecing it together with material from other chapters or with events in your life. Sometimes you are asked questions for which there are no simple answers. In such cases your answers are at least as good and maybe better than the answers we or your professors might propose for the same question. The key point here is that we want you to get used to *thinking* about these subjects and develop confidence in your own judgment.

We invite you to begin your study of human biology browsing through this textbook. Leaf through the book and glimpse the multitude of fascinating things we have to share with you.

Carl E. Rischer
Thomas A. Easton

Chapter 1

Basic Themes in Human Biology

William Shakespeare said it over 300 years ago: "What a piece of work is man." And it's true. No other life form, at least on this planet, even comes close to displaying the complex behaviors that we see in our own species. We are the only species with so many intricate systems for communication—which only begin with innumerable languages and written alphabets, mathematics, and other ways to use symbols to pass and store information. We are the only species practicing anything similar to art, religion, literature, science, and technology. But humans are also a lesson in contradictions. Where else can we find scoundrel and saint, lover and hater, stupid and cunning all wrapped up in the same individual? According to some experts, we are the only species that *sometimes* learns from its own history, stores volumes of information in libraries, and uses its knowledge and ability to solve abstract problems in ways that may improve our futures.

It is obvious that most of what makes humans unique is their brain and the manipulative skills of their hands. In almost all other ways, humans are wonderful and really quite ordinary! For instance, we have such a standard mammalian body, with such a standard internal organization, that it is not unusual for students in a course like this to study laboratory rats to learn about human anatomy (structure). Our physiology (the way our organs function) is so similar to that of other mammals that we test drugs and medical procedures on rats, mice, rabbits, monkeys, and dogs before we try them on people.

Stripped of our technologies, we are not all that exceptional. Humans cannot fly; they are not very fast runners; they can only stay submerged for a minute or two; they are not particularly strong; and they are not armed with any special weapons like fangs and claws. Of course, we are different in several other ways; for instance, we have less hair than most other mammals and have a peculiar habit of walking around on our hind feet.

When you're through with this course, we think you'll look back and say Shakespeare was right!

WHY STUDY HUMAN BIOLOGY?

The news continuously bombards you with a host of complex biological issues that require your understanding and concern. You will have to identify for yourself the real risks attending a promiscuous lifestyle. You will have to decide whether the theory of evolution really poses a threat to your family's religious beliefs and whether it makes any sense to send aid to help famine victims (Figure 1-1). You will have to decide whether abortion is an appropriate form of birth control and whether test tube babies have the same legal rights as other people.

Closer to home, catastrophic illness may strike a loved one and you may be the one asked to consent to complicated, risky medical procedures (Figure 1-2). Or your concerns may be simpler, like designing your own nutrition, weight control, and physical fitness programs (Figure 1-3). Such decisions will be more objective (if not easier) if you understand something about how the human body is structured and how the various systems interact.

On the brighter side, these are the most exciting of all times. With the advent of genetic engineering, we are in the process of taking control over the di-

FIGURE 1-1 A host of complex problems face the world: overpopulation, environmental destruction, and famine. The child in this photo is receiving his daily milk ration at a feeding station in northern Africa during the drought and famine that killed hundreds of thousands in the mid-1980s. International assistance sent millions of dollars for relief until the rains temporarily ended the drought. But in 1991 the famines returned (see Chapter 26).

FIGURE 1-2 Some call acquired immunodeficiency syndrome (AIDS) the disease of the 20th century. In 1981, researchers first recognized that HIV virus was responsible for the spread of this new deadly disease that devastates the human immune system. By 1985, demonstrations like the one in this photo, focusing on the urgent need for extensive government funding for AIDS research, were common. AIDS was identified as the second leading cause of death for college-age men in 1991 (see Chapter 24).

rection of evolution. Our molecular biologists are designing and implanting new genes that promise to eliminate some of our inherited metabolic defects, such as diabetes. They are also experimenting with genetically engineered strains of plants capable of manufacturing their own fertilizer. Such developments will improve agricultural yields and may give us more time to control human overpopulation and help us deal with future episodes of famine. Our biomedical engineers are building artificial organs to replace those worn out or damaged through disease or injury. Our science fiction writers are already dreaming of the day when we will be able to select our childrens' traits from a menu. These are just a few of the subjects we will cover in *Focus on Human Biology*.

FIGURE 1-3 Aerobics class in full swing. Our current cultural embrace of the benefits of physical fitness has made local health clubs popular.

WHAT IS LIFE?

We are fascinated with life. In this country, the dominant view is that *human* life is sacred, and we go to great lengths to protect it. Many religions, however, hold *all* life sacred, making it a sin to step even on ants. Our interest in "life" spreads to all levels: we are concerned about our own bodies, our children, our gardens, agriculture, genetic engineering, disease, world famine and overpopulation, and pollution.

It is an ironic fact that, although we are preoccupied with life in all its facets, there is just no simple way to define what life actually is. The branch of science that focuses on the study of life is called **biology** (*bio(s)* = life; *ology* = study of). At its simplest level, we can say that life is the consequence or by-product of the activities of certain types of physical and chemical systems. Somehow, that is not a very soothing definition. However, in spite of the fact that we cannot define life simply, we still know a great deal about it. Here is a list of characteristics shared by most life forms (Figure 1-4):

Characteristics of Living Systems

1. **Organization.** All life occurs in microscopic, highly organized and compartmentalized units called **cells.** In fact, the cell is the fundamental unit of life (see Chapter 3), and the few exceptions (such as viruses) are still manufactured by cells. Ironically, in spite of the great diversity seen among the various types of living things on this planet, from ferns to alligators to humans,

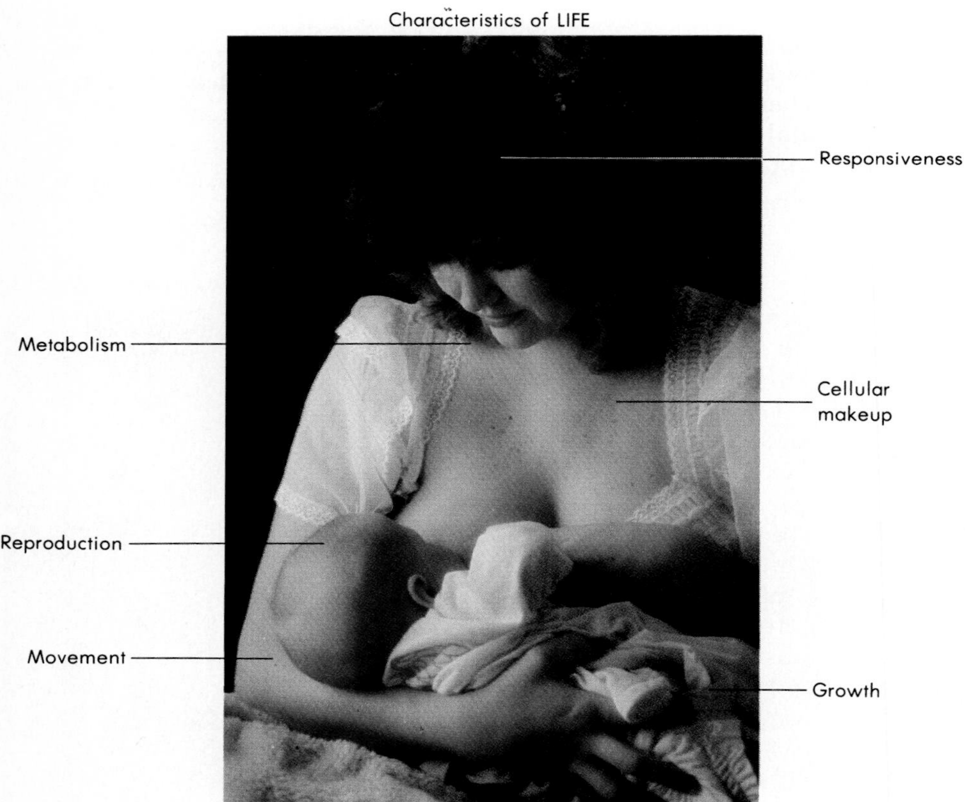

FIGURE 1-4 Living things share many characteristics. Since life requires a continuous supply of energy, living organisms are dynamic chemical systems dedicating substantial effort to procuring fuel (food), processing it, and converting the energy it contains to a form the body can use (the sum total of such chemical activities is called metabolism). In addition, living things reproduce, they grow, they are irritable (they respond to changes in their environment), they move, and virtually all are built on a cellular plan.

they are all made of the same fundamental architectural and structural units, cells. The cells organize the various systems so that the chemical events supporting life occur in orderly and economical ways. The human body is made up of roughly 100 trillion such cellular units.

2. **Selective Chemistry.** One of the ways in which life is organized deserves special notice. Living things select and concentrate only certain types of the various chemical substances from the world around them. In fact, most of each living thing is composed of assorted combinations of only four different chemical elements—carbon, hydrogen, oxygen, and nitrogen. Add two more, phosphorus and sulfur, and you have six of the essential elements that are used in the structures and the various chemical functions needed by most living things. Some students rely on the key word *CHONPS* (not quite "chomps") to help remember these "bio"-elements.
3. **Energy Processing.** Life requires a large and continuous input of energy. We take in energy with the food we eat. We then extract it, using the oxygen we breathe, and convert it to a form that can be used in the cells (see Chapters 11 and 21). The chemical reactions supporting these energy transformations are collectively called **cellular respiration.** In addition, the chemical wastes generated by these fundamental processes must be eliminated from the body (see Chapter 12). The sum total of all these chemical activities is called **metabolism.**
4. **Movement.** Movement is often the easiest way to distinguish the living from the nonliving. Movement is especially well developed in animals, including humans, because they must frequently search for food in order to fill their daily energy quota. The ability to move is much less developed in plants because they can manufacture their own food through a process called photosynthesis. However, all life forms show considerable movement at the level of the cell.
5. **Reproduction.** In order to replace those individuals that have been removed from a population by death or to replace cells that have been damaged by injury or disease, most life forms have evolved several ways to reproduce. Thus, reproduction occurs at the level of the cell (see Chapter 5) and the level of the individual (see Chapters 5 and 17). At the level of the population, reproduction permits species to spread into and colonize new environments (see Chapter 26).
6. **Growth.** As an organism changes size, growth occurs; positive growth occurs as it gets larger, negative growth as it gets smaller (as in weight loss). Living organisms use part of the nutrients and energy they absorb to manufacture more of their own substance. Growth in children occurs as a result of both an increase in cell number as a consequence of cell division and an increase in size of the new cells (see Chapters 5 and 18).
7. **Responsiveness.** Living systems must be able to detect changes in their environment, analyze them, and then respond in a manner that supports their survival. If they are threatened by a noxious chemical or a predator, they must detect it and move to a safer place. On the other hand, if they have an opportunity to feed or mate, they must also perceive that and respond appropriately. We call the ability to detect such changes and activate a response **irritability.** Responsiveness or irritability can occur at the level of the cell, tissue, organ, or individual. Much of human responsiveness is mediated by the high-speed electrochemical activities of the nervous and the sensory systems (see Chapters 14–16).

---

1. List the seven major characteristics shared by living systems.
2. Explain what each of these characteristics contributes to life.

THE SCIENTIFIC METHOD

A nonmajors biology course is only partly a "science" course. It is just as much a history course. It covers what we know about the human body and human biology. It covers what happens in the stomach, what activities are controlled by the various areas of the brain, how hormones affect different organs, and so on. As history, it covers what we have been able to piece together from the researches of the past. It generally spends very little time on how researchers went about systematically discovering all these wonderful things, even though that is what science really is.

Do not feel slighted. Most professional biologists devote almost their entire undergraduate experience just to learning what is already known about biology and relatively little time learning about the information-gathering and testing processes we call **science.** For most, their first real encounters with science are delayed until graduate school. It is important, however, that each educated person know something about how science is actually practiced. This is why we now will tell you about the most powerful of all problem-solving systems, the **scientific method.**

In practice, there are five major components to the scientific method. They include **observation** (measurement), **generalization** (identifying a pattern), stating a **hypothesis** (a tentative extension of the pattern or explanation for why the pattern exists), and **experimentation** (testing that explanation). The results of the tests are then **communicated** to other members of the scientific community, usually by publishing the findings. How each of these components contributes to the scientific method will be discussed next.

Observation

The basic units of science—and the only real facts the scientist knows—are the individual observations. Using them, we look for patterns, suggest explanations, and devise tests for our ideas. Our observations can be casual, as they were among the people who noticed that rabies victims had all been attacked by a rabid animal some weeks before their own symptoms began.

In science, however, our observations are more often numerical measurements of things. We measure how tall people are, how much they weigh, whether they get sick or not after being exposed to some disease, the blood sugar level in a diabetic, the size of a virus, or how fast a given computer chip can complete a set of calculations. A single measurement is called a **datum** (plural *data*). Much of our scientific progress resides in our skill at developing instruments to improve our abilities to measure things.

The actual things we choose to measure are called **variables.** Variables can be classified into two broad categories based on how we go about measuring them: discrete variables and continuous variables. **Discrete variables** are usually counted; they do not come in fractional units. For instance, suppose we want to know how many dollar bills you have in your wallet. It would be simple enough for you to count them, and you would probably come up with a whole number like 3, 4, or 10 (depending on how impoverished you happen to be). It is not likely that you are carrying any bills that are torn into quarters or halves.

On the other hand, **continuous variables** are measured rather than counted; they often come in fractional units. Things like height, weight, and time are often continuous variables. If you were measuring the heights of basketball players, you might find someone to be 6 feet 6 inches tall or 6 feet 6.5 inches or 6 feet 2.875 inches, depending on the accuracy of your measuring devices and your need for precision (Figure 1-5).

(a)

(b)

FIGURE 1-5 Discrete versus continuous variables. (a) Discrete variables are things that can be counted. They are usually measured as whole numbers. (b) Continuous variables consist of measurements. They can occur as fractions.

Generalization

After we have made many measurements, we begin to look to see if there is a discernible pattern among those measurements. For instance, after measuring the height of a great number of males and females, you might eventually notice that males are taller than females on the average. Such statements of patterns are referred to as generalizations. Please keep in mind that these generalizations are only made by examining many single observations and then looking for patterns. Cautious experimenters will often make additional observations to be certain the pattern really holds true. In fact, this necessity to **replicate** the findings is a very important part of the scientific process (Figure 1-6).

The Hypothesis

A tentative explanation suggesting why a particular pattern exists is called a hypothesis. For instance, you might suggest that males are taller than females because they mature later and have two more years of childhood growth before their sex hormones start up, provoking the final growth spurt of adolescence. The mark of a good hypothesis is that it be *testable.* The one just suggested would be troublesome to test since we cannot easily do rigorous experiments on children. But we can set up similar experiments with animal subjects, and we can look for human children with hormonal problems that might shed some light on our hypothetical speculation.

The Experiment

The experiment is the most formal part of the scientific process. There are two major types of experiments; one type is performed to *test a hypothesis;* the other is done to *measure a quantity* (e.g., what is the most effective dose of medication needed to treat a particular medical condition?). The two often overlap because the size of the measurement may tell whether a hypothesis is any good or not.

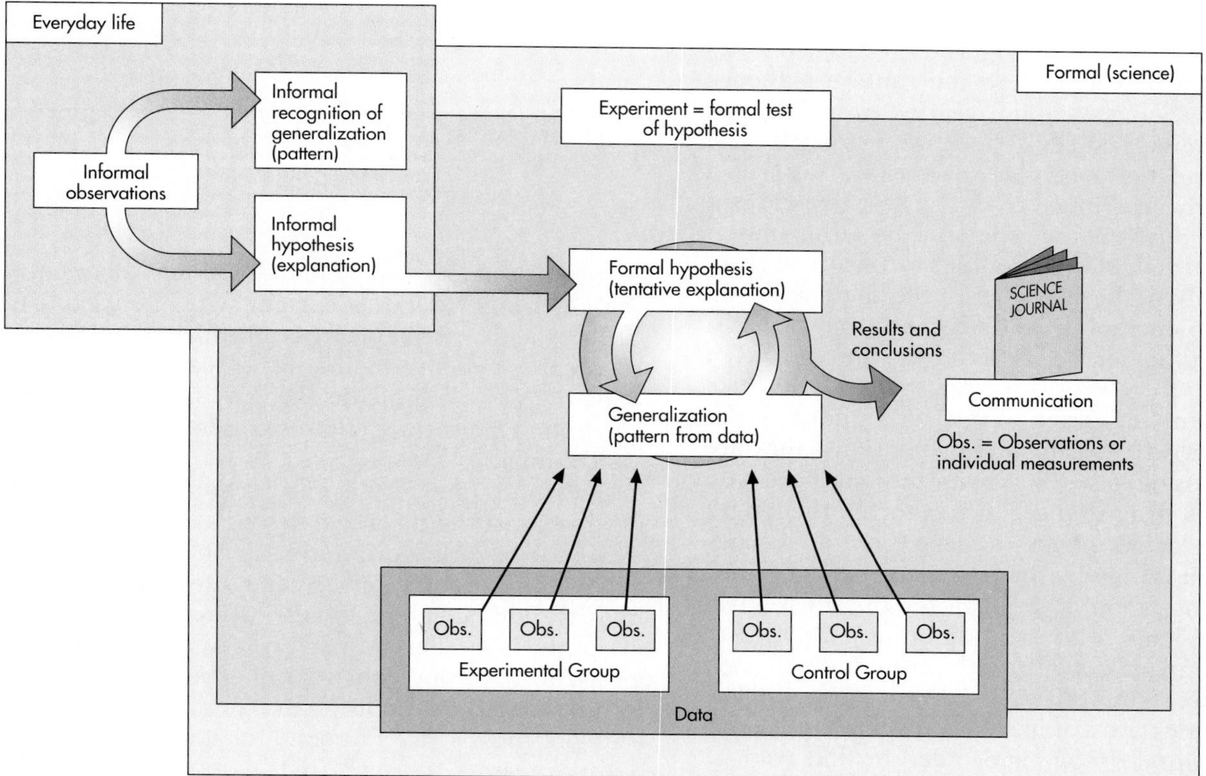

FIGURE 1-6 An overview of the scientific method. The scientific process starts as an everyday, casual observation, or the informal recognition of a pattern among observations. A hypothesis is formed when we offer an explanation of why the pattern exists. Most observations and speculations stop here. We make the transition to science when we formally design, set up, and carry out tests (experiments) to see if our hypothesis is a good one. If the observations fail to support the hypothesis, then the hypothesis must be discarded or revised. Communication enters the picture when we publish our findings.

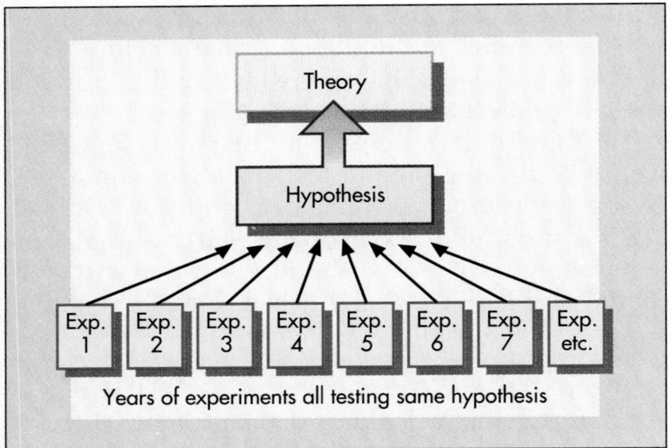

FIGURE 1-7 If a hypothesis can withstand years of experimental testing without being discarded, then it may be elevated to the status of a "theory."

Experiments are usually designed to try to disprove a hypothesis. In fact, *science is the art of disproving,* not proving. If a hypothesis withstands all the tests, then it may be a good explanation of what is going on. If it is good enough to withstand successfully several years of rigorous testing, the hypothesis may be elevated to the rank of a **theory** (Figure 1-7). Science cannot really "prove" things in the mathematical sense because it is physically impossible to perform all the possible tests of a given hypothesis. However, if any test disproves a hypothesis, the hypothesis should be either modified or discarded. Many scientists are ready with an alternative hypothesis, just in case their research shows their current hypothesis is wrong. If that occurs, then they move on to the alternative and so on.

Experiments designed to measure a number are not really directly testing an explanation—they are after something else. For example, suppose you have found a drug that seems to shorten the length of time that lesions of genital herpes remain active. The question is not whether the drug works or how, but what dose should be taken for the most effective results. Thus, a large group of herpes sufferers is divided into subgroups, with individuals in each subgroup receiving different specific dosages. The recovery times are compared between the groups and the optimal dosage is decided in that way.

Communication

Each scientist is obligated to share her or his findings with the rest of the scientific community. Science accumulates and uses information vertically through time, adding knowledge and understanding to the findings of the past, and horizontally by encouraging cooperation and sharing among contemporary researchers. The most common avenue for this sharing is, first, through the "peer review" process, whereby other scientists evaluate one's research reports and recommend them for publication in scientific journals, and, second, through actual publication of the reports. Communication is also supported by press conferences, seminars, and courses. Making scientific findings available allows other researchers to see if they get the same results when they attempt similar experiments. **Repeatability** is a major criterion for scientific truth.

---

1. List the five major components of the scientific method and give the function of each.
2. Explain the difference between a generalization and an experiment.
3. Why do you think repeatability is so crucial as a criterion of scientific truth?

MAD DOGS AND THE SCIENTIFIC METHOD

The power of the scientific method was eloquently demonstrated by Louis Pasteur when he used it to attack the disease rabies, one of the most terrifying medical problems facing the nineteenth-century world. (We will indicate the part of the scientific method in operation in parentheses in the passages that follow.)

Rabies is one of the most feared diseases. Descriptions of the disease date back more than 2300 years to Aristotle (communication). In earlier times it was likely to be spread by the bite of a rabid wolf that wandered into a village, attacking anyone in its path. In modern times, we are likely to be exposed to it through the bite of a pet infected by a rabid skunk or raccoon. The term *rabies* tells something of the disease's ferocity for it comes from the Latin *rabere,* "to rage." Girolamo Fracastor, an Italian physician, wrote this description of the "furious" form of rabies in 1546 (communication):

> **Once the disease takes hold, the patient can neither stand nor lie down; like a madman he flings himself hither and thither, tears his flesh with his hands and feels an intol-**

erable thirst. This is the most distressing symptom, for he shrinks from water and all liquids that he would rather die than drink or be brought near to water; it is then they bite other persons, foam at the mouth, their eyes look twisted, and finally they are exhausted and painfully breathe their last.

The rabies virus (see Chapter 23) eventually invades and destroys the tissues of the nervous system, including the spinal cord and the brain. Viruses are such tiny agents of disease that they can be seen only with the aid of extremely powerful magnifying devices such as the electron microscope. It is a testimony to the power of the scientific method that Pasteur developed an effective treatment for this dreaded disease almost 75 years before we had microscopes powerful enough to show that viruses actually existed. We will trace some of the steps Pasteur went through to illustrate the components of the scientific method.

Pasteur's Work and the Method

Pasteur's work transcended the casual and became science when he decided to pursue formally the cause of rabies and see if he could develop a cure. The popular idea (hypothesis) of the times was that some agent was transferred during the bite from the rabid animal to the new victim. Thus, the logical way to start was to attempt to find this agent in the saliva of rabid dogs. His studies began as he and his staff restrained rabid bulldogs, forced open their jaws, and sucked saliva samples from between their teeth with glass pipettes held up to their own mouths. Each time Pasteur drew a sample and examined it, he was making a single observation (Figure 1-8).

The next step in the process was to attempt to see these agents by looking at each sample through the microscope (observations and experiment). In all the specimens Pasteur examined, he saw nothing unusual (generalization). We now know that viruses are so tiny they were invisible to the microscopes available in Pasteur's day. However, other evidence so strongly suggested the presence of disease agents that Pasteur did not abandon his hypothesis. Instead, he recognized that the limitations probably rested with his equipment (in this case his ability to observe) and designed another test (experiment) for his hypothesis.

Next he injected two groups of dogs with saliva, one group with saliva from a rabid dog (the experi-

FIGURE 1-8 Assisted by his aide, Louis Pasteur took samples of saliva from rabid animals.

mental group), and one group with saliva from a healthy dog (the control group). The **experimental group** in any experiment is given the treatment that tests the hypothesis, while the **control group** is given a "fake" treatment, just to make sure that experimental procedures (such as simply injecting saliva through the skin) are not causing the reactions. With the experimental results came confusion. All four dogs injected with saliva from the healthy dog remained healthy (observations). Of the four dogs that were injected with saliva from the rabid animal, two contracted rabies, but two remained healthy (observations). The experiment was repeated with similar results (generalization).

Thus, Pasteur systematically built up his understanding of what the rabies virus was like. Eventually his experiments showed that rabies viruses are found mostly in the brains and spinal cords of infected animals. He concluded the agent was extremely tiny and that it was sometimes but not always found in the saliva of rabid animals and that is how it is most likely transmitted. Eventually he chased these deadly agents through a series of hypotheses and experiments that led to the development of the first rabies vaccine and saved literally hundreds of thousands of people from one of the most terrifying deaths. And he never got a glimpse of the dreaded enemy.

SCIENTIFIC LOGIC

Scientific reasoning flows along two philosophical channels: inductive reasoning and deductive reasoning. **Inductive logic** occurs when a great number of individual observations suggests a general principle may be true. For instance, suppose you interviewed 25 women on campus concerning their feelings about legalized abortion and every one said she opposed it. From this sample you might be tempted to make the general statement that *all* women are opposed to legalized abortion. Obviously, you are taking a risk when you make such general statements. The smaller the sample of people interviewed or subjects tested, the greater the risk. Science goes to great lengths to reduce the possibility of these inductive errors, or **sampling errors,** by insisting that all scientific studies include as many *replicates* (additional subjects) as time or money allows. For that reason, personal testimony is the worst and most unreliable kind of scientific evidence of all. A single person's opinion or experiences (Like: "Well, I've been taking megadoses of vitamin C for two years now and I haven't had a cold yet!") are usually viewed with great suspicion.

Deductive logic is even riskier. It attempts to use general statements to predict specific situations. For instance, suppose you are operating from the general premise:

> "Malamute huskies are good sled dogs." (General statement)
>
> "Kima is my new malamute puppy." (Fact)
>
> "Therefore, Kima will be a good sled dog when she grows up." (Prediction)

You are using a general statement to predict something about your own dog. It might come true. But it might not. The predictability is only as good as the general statement. As it turns out, Kima came from a highly inbred line of malamutes, and by age two she was horribly crippled with hip dysplasia (her thigh bones did not fit into her hip joints).

Scientists use inductive logic during experiments aimed at revealing general patterns. They use deductive experiments when they are testing the predicting power of their theories or hypotheses.

Yet it is not quite fair to believe, as many people do, that scientists are strictly logical beings. Induction and deduction enter their work mostly after the research has been done, when they are writing up the results for publication in journals and textbooks. The actual process of scientific research has much more in common with the creative arts than most people suspect. The scientist leaps out of bed in the dead of night to cry "Aha!" and rush to the laboratory with a new hypothesis or an experiment or an insight that might make sense of a jumble of data. Dreams and hunches and intuitions all play parts in the process. The history of science is a history of inspiration as much as it is a history of logic.

---

1. Distinguish between inductive and deductive logic. Give your own examples of each type.
2. What are the dangers of deduction and induction in the scientific process and how can you reduce the dangers of making an error with either?

Learning Focus

THE STRUCTURE OF THE SCIENTIFIC METHOD

The following example is based on an actual experiment that we have partially fictionalized for clarity so that it reveals the structure of the scientific method.

A CURE FOR THE COMMON COLD?

Initial Observation: A young patient being treated for leukemia at the University of Texas was given a lozenge of zinc gluconate to correct the zinc deficiency that commonly accompanies that disease. Instead of swallowing the lozenge, she kept it in her mouth and sucked on it. To everyone's surprise, the symptoms of a cold she was developing vanished within an hour.

Hypothesis: The researchers' hypothesis (tentative explanation) was that prolonged contact between the zinc compound and the tissues lining the throat and mouth may somehow shorten the duration of the cold. However, for research purposes, the hypothesis must be reworded to describe the conditions under which no (null) difference appears. That is, because it is impossible to do all the experiments required to prove a positive hypothesis but it takes only one experiment to disprove a negative one, it must become a *null hypothesis.* Since the researchers really expected the zinc lozenges to make a difference, they set up the following testable hypothesis:

Null Hypothesis: A group of patients given zinc lozenges (the experimental group) will have active cold symptoms for the same length of time as another group given lozenges containing no zinc (the control group).

The Experiment: The researchers divided a population of 64 people who were coming down with colds into two groups; 32 people were given the zinc lozenges (experimental treatment) and 32 were given lozenges that contained no zinc (control treatment). As in most modern research on humans, the study was *double blind,* meaning that neither the experimental subjects (the patients) nor the researchers knew what kind of lozenges any particular patient received. This keeps the expectations of the subjects and researchers from influencing the results. (In a *single-blind* or just plain "blind" experiment, only the subjects are kept ignorant.)

Individual Observations: The researchers noted the duration of cold symptoms for each subject. Each measurement of the length of time a subject stayed sick is a single observation.

Looking for the Pattern (Generalization): Much of the creative work in science is involved with attempting to find the pattern (if any) in the data generated by an experiment. Often a simple bar graph (called a histogram) is made by plotting the distribution of the results for each group. In this experiment, the information plotted is the number of days it took each individual to become free of cold symptoms. The recovery patterns for the two groups are then compared to each other to see if any group differences can be seen. In this example, there may be significant differences between the two treatments. In actual practice, a special branch of applied mathematics, called *statistics,* is used to help design experiments and to determine just how "significant" these differences really are. The total number of sick days for each group and the average length of time members from the respective groups were sick were calculated to see if a pattern emerged from the data for the two groups. Their results were strikingly different:

Test Results

GROUP	NUMBER IN GROUP	TOTAL DAYS WITH COLDS	AVERAGE DURATION OF COLD
Experimental Group			
Zinc lozenges	32	128	4.28 days
Control Group			
No zinc	32	352	11.06 days

Conclusion: The results are compared to the null hypothesis to see whether they discredit it or not. In this example, it does seem that taking the zinc lozenge shortens the duration of cold symptoms. The null hypothesis is not supported.

Now go to the Learning Focus Response to break the information from this experiment down into the components of the scientific method and relate the experimental design to the patterns of deductive and inductive reasoning.

Learning Focus Response

DEDUCTIVE VERSUS INDUCTIVE LOGIC

After you have read the material in the preceding Learning Focus and the information about inductive and deductive logic in the text, summarize what you have learned by answering the following questions.

1. What is a null hypothesis?

2. What was the null hypothesis that was tested by the experiment?

3. What was the actual hypothesis from which the researchers were working?

4. Why and how is a null hypothesis used?

5. What is inductive reasoning?

6. Describe deductive reasoning.

7. Would you say the zinc lozenge experiment was designed along inductive or deductive lines? Why?

8. If, as a physician, you prescribed zinc lozenges for a patient after reading that study, would you be proceeding along inductive or deductive lines? Explain.

9. Come up with your own examples of the application of deductive and inductive reasoning.

BASIC AND APPLIED RESEARCH

What scientists do as they apply their methods is called *research* (Figure 1-9). **Basic research** seeks no specific result. It is motivated essentially by curiosity. It is the study of some intriguing aspect of nature for its own sake.

Applied research seeks answers to specific problems. Applied researchers want cures for diseases (like Pasteur), methods for analyzing problems, and ways to control various phenomena. They are mission oriented, and most biologists and other scientists who work for government and industry are applied researchers.

Do you remember Mark Twain's *Tom Sawyer?* Do you recall the time when Tom fed painkiller to the cat (see Figure 1-10)? Tom was only being a small boy, but he had one of the attitudes essential to a basic researcher: he was curious. He wondered what the stuff would do to the cat. If, on the other hand, he had wondered whether the stuff would kill the cat's fleas, he would have been an applied researcher.

FIGURE 1-10 Tom Sawyer feeds painkiller to a cat. Would Tom's behavior come closer to being basic or applied research? Why?

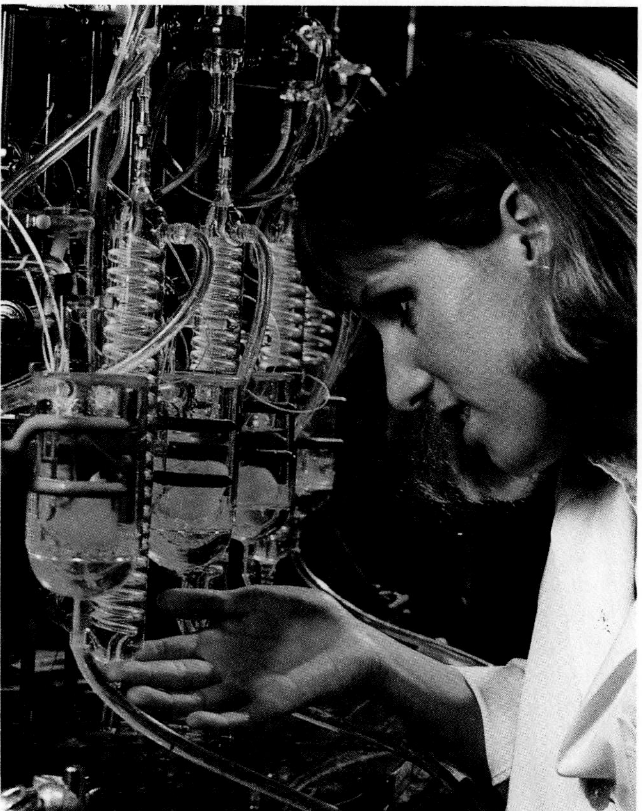

FIGURE 1-9 Research biologist in the lab.

Today, applied research receives far more funding than basic research. The reason is clear, for we have many technical problems that cry for solutions. Yet there is also a need for basic research, for basic research supplies a great many of the observations that applied researchers need to formulate their hypotheses.

THREE BASIC PRINCIPLES OF BIOLOGY

The scientific method, used in both basic and applied research, has allowed us to accumulate a vast amount of knowledge. By itself, however, this knowledge is of little use. Observations, hypotheses, and theories need to be knitted together into concepts that help us make sense of the whole. We call them laws of nature. Some of these concepts seem indisputable; one is the law of gravity.

Others seem to offer more room for debates and occasional apparent exceptions. These are **princi-**

ples, and each field of study has its own. The three principles most important to biology, including human biology, are as follows:

1. **Homeostasis.** Living systems seek chemical and physical stability (see Chapter 12). *Physiology* is one branch of biology to which this principle is central.
2. **Organic Evolution.** Each living thing has arisen from ancestral predecessors unlike itself. *Evolutionary biology* is the branch of biology most directly concerned with this principle, but it enters many other fields as well. See Chapter 25 for a fuller discussion.
3. **Ecology.** The affairs and activities of all living things are interconnected and influenced by other living and physical factors of the environment, and vice versa (see Chapter 26). The branch of biology that focuses most on this principle is the field of *ecology*.

We will discuss each of these three principles briefly in this chapter. Later we will see them in action as we discuss the various aspects of human biology.

HOMEOSTASIS

The basic principle of homeostasis has made it possible to understand a great deal about how the body works. **Homeostasis** is the balance maintained when several systems operate to keep the conditions inside the body roughly constant. That is, the body maintains its temperature and levels of mineral salts, nutrients, wastes, oxygen, and water within the narrow limits that support human life.

Many different diseases can result when various aspects of homeostasis fail. For instance, the human body uses the sugar glucose as its main fuel. Thus, it is extremely important to keep just enough sugar in circulation so that the cells can draw on it to power their own activities; normally that is 90 milligrams (mg) of glucose in each 100 milliliters (ml) of blood (see Appendix A for information about the metric system). If a person's blood sugar falls too low, he or she suffers from *hypoglycemia* (*hypo* = under, beneath; *glyc* = sweet) and lack of energy. If, on the other hand, the blood sugar is too high, the problem is *hyperglycemia* (*hyper* = over, above) or diabetes mellitus (sugar diabetes). Excess blood sugar is excreted in the urine; in severe cases, the body's cells lose water, the brain stops working, and the patient goes into a coma.

Negative Feedback and Homeostatic Control

Controlling Sugar Level in the Blood

How does the body prevent hypo- and hyperglycemia? After a meal, when sugar is being absorbed from the digestive system, specific cells in a gland called the pancreas sense the rising sugar level in the blood and secrete *insulin* (see Chapters 13 and 20). Insulin causes cells in the muscles and elsewhere to quickly absorb this excess sugar from the blood, reducing its level back to the normal 90 mg/100 ml (Figure 1-11).

Between meals, the blood sugar level falls as the cells use it for energy. The pancreas may then secrete another substance called *glucagon,* which signals the liver to release some of the extra sugar stored there. This hormone raises the sugar level back up to the normal level (see Figure 1-11).

This type of control mechanism is called **negative feedback.** Such feedback mechanisms use the level of the substance or the physical condition being controlled as the indicator for turning off or on the homeostatic response. Negative-feedback mechanisms measure the level of whatever is being controlled against some "set point," or specific level determined in advance. When the variable departs from the set point, the mechanism switches on to counteract the change, returning the variable to the set point (Figure 1-12, p. 16).

In contrast, *positive* feedback acts not to diminish the departure of some variable from its starting point, but to *increase* it. These systems, once activated, are often destructive and accelerate the departure from the original starting point. An example is the feedback squeal, of increasing intensity, produced by public address systems when a microphone is placed in front of the speaker. (Why does such a situation produce louder and louder squeals?) Another positive-feedback system develops when an earthen dam is collapsing. At first the water runs out through a tiny crack. But as the water flows through, the crack widens, allowing an ever-increasing gush of water to flow out—collapsing the dam in the process. (Why is it hard to think of any examples of positive-feedback systems existing in a living body?)

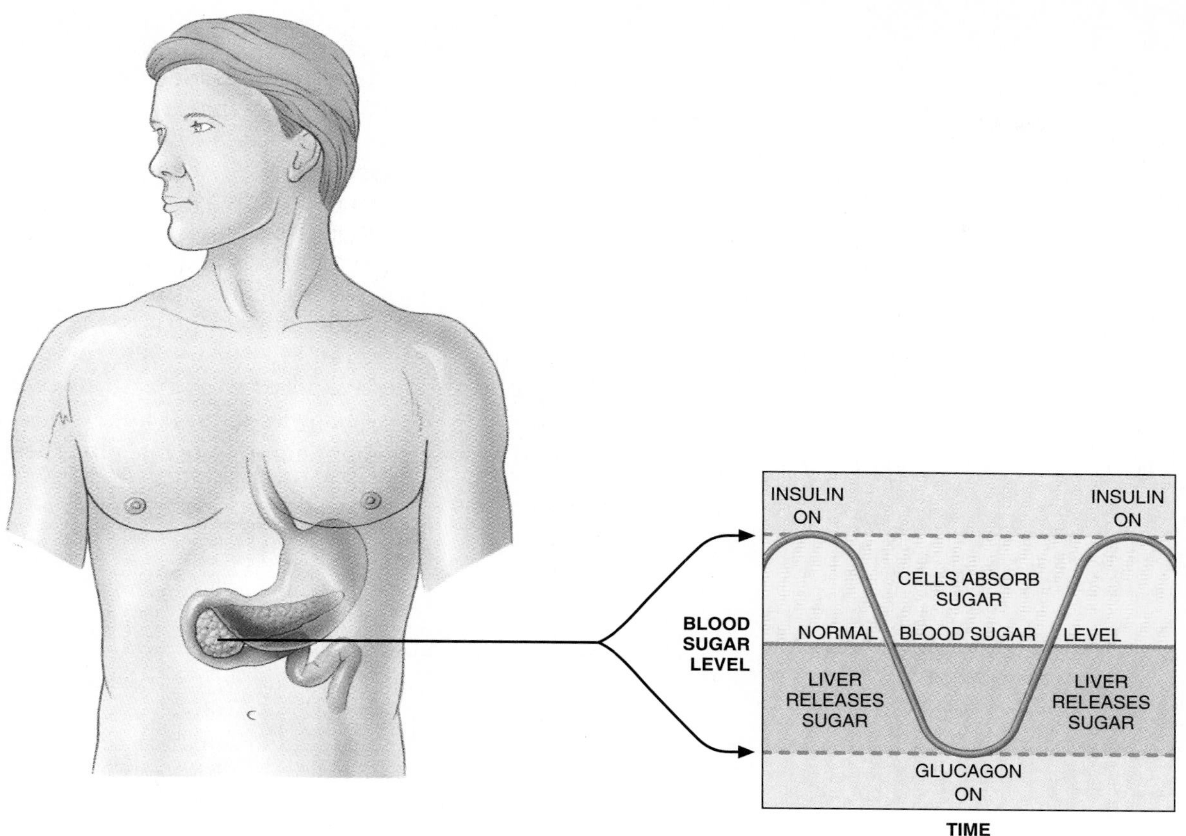

FIGURE 1-11 Homeostatic control of blood sugar level. Many of the body's systems are controlled by feedback systems superficially resembling a thermostat. In this oversimplified diagram, the level of sugar circulating in the blood is regulated by the pancreas. When the sugar level drifts far enough above the normal set point, it stimulates the pancreas to release insulin, which encourages many of the body's cells to take up and use or store the excess sugar, causing the level to decline back toward the set point. When the sugar level drops below the set point, the pancreas releases glucagon, which encourages cells that store sugar to release some of their reserve into the blood, raising the level back toward the set point.

Mechanical Negative-Feedback Controls

A simple example of a negative-feedback mechanism is the thermostat that controls room temperature by turning a heater off and on. As the temperature goes up, a temperature-sensitive switch opens, turning off the heater. As the room temperature cools again, the temperature sensor changes until it turns the heater back on. Thus, by alternately turning the heater off and on, the thermostat keeps the room temperature fairly constant (Figure 1-13).

Body Temperature Control

A similar thermostat works to control human body temperature. Located in the brain, it activates sweating, panting, and skin flushing (reddening of the skin in light-complected people caused by routing the blood to the surface) when the body is overheated. It activates shivering and skin paling, among other measures, when the body is chilled.

Negative Feedback in Other Body Systems

Negative-feedback mechanisms also work to maintain blood pressure, heart rate, fluid volume, and blood levels of oxygen, carbon dioxide, calcium, sodium, and other substances (Figure 1-14, p. 17). Their end result is that the composition of the blood remains constant (within limits) and the

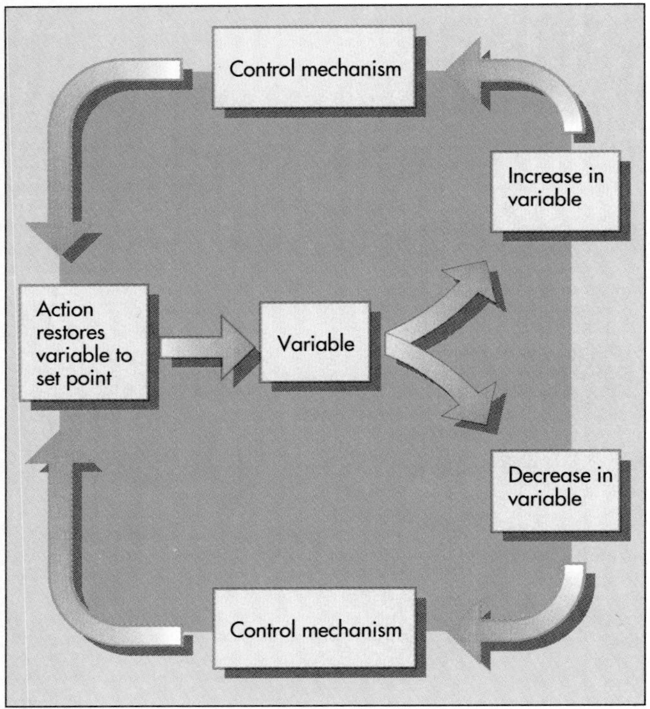

FIGURE 1-12 In negative feedback systems, "effects" feed back on their cause to keep variation within a narrow range around a "set point."

body cells remain in an environment that allows them to survive. Changes in blood composition, the availability of nutrients (food molecules), body temperature, or heart function that are too great can be fatal.

The principle of homeostasis helps us to understand how various aspects of body functions are interrelated. It also tells us that whenever we find a substance or activity in the body that maintains a steady level, we can then expect to find a control mechanism. The concept of homeostasis, therefore, serves both as an organizing principle for our knowledge of the body and as a guide to further research.

1. Explain homeostasis and how it contributes to survival.
2. Diagram and explain how a negative-feedback mechanism operates.

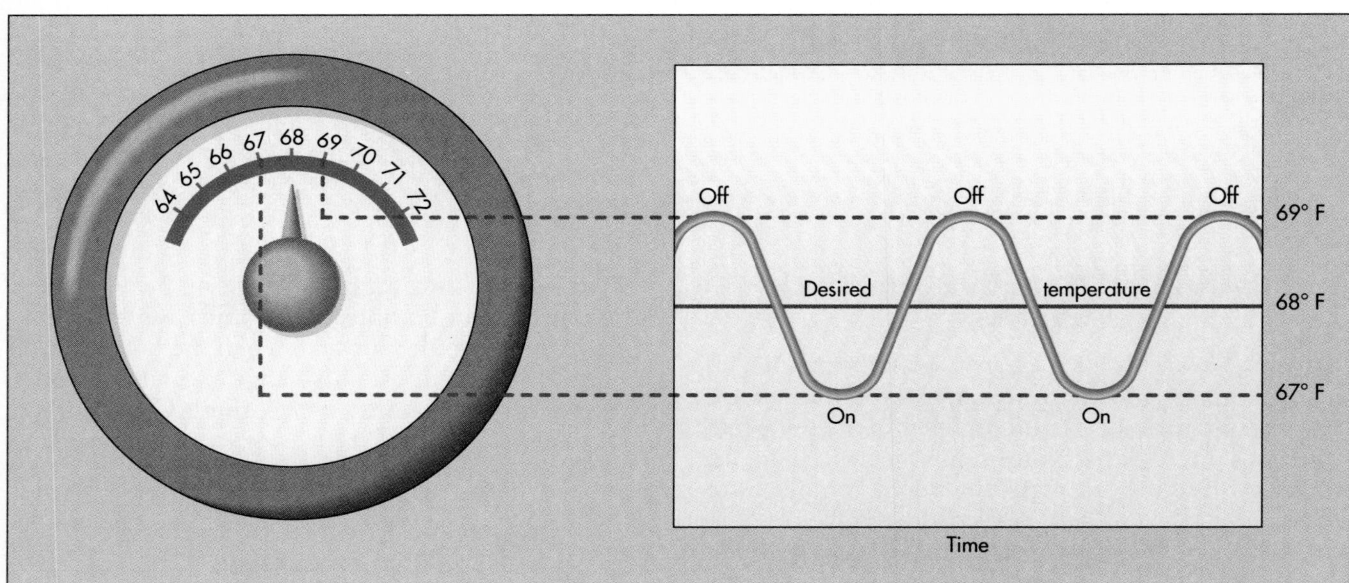

FIGURE 1-13 Negative-feedback control of house temperature. Another example of negative-feedback control in nonliving systems is the thermostat and heater that control room temperature. As the room cools, a temperature-sensitive switch in the thermostat closes, turning on the heater. As the room then warms, the switch opens, turning off the heater. The temperature thus fluctuates within a narrow range.

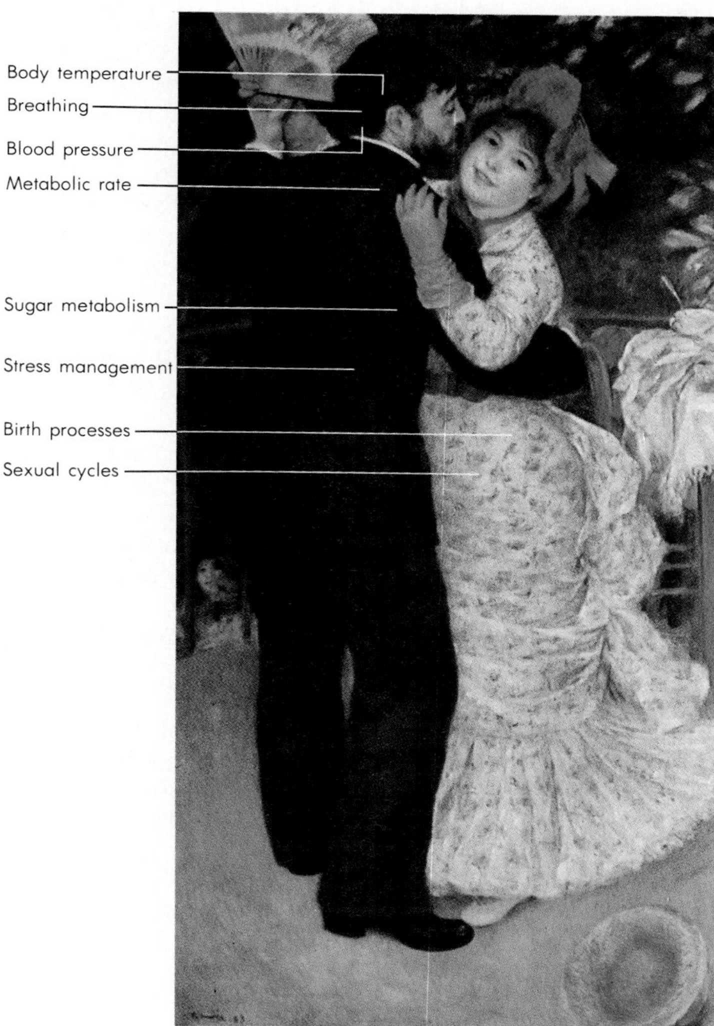

FIGURE 1-14 Summary of several of the body's negative-feedback control systems.

ORGANIC EVOLUTION

In each century, a single idea or principle often surfaces to provide us with a quantum leap in our understanding of the universe. Ironically, it is often the elegantly simple ideas that hold the most far-reaching implications; the theory of evolution is one such concept. Simply and broadly stated, it says that things change. Present forms emerge from past forms. The current conditions of the universe, our planet, and life itself are the consequences of natural processes that obey natural rules.

Inorganic evolution attempts to explain how the nonliving parts of the universe formed and developed. **Organic evolution** focuses on how life has formed and then diversified. Both these processes can be studied in orderly ways, and the second is very much part of the first. There is no need to pull into our explanations such supernatural things as benevolent or angry gods or guiding spirits (that exclusion is the basis for the controversy that seems to exist between the theory of evolution and some fundamentalist religions).

The concept of evolution, both inorganic and organic, extends far beyond our understandable preoccupation with the origins of humankind. The theory of evolution supports, and is supported by, virtually all branches of science, including astronomy, chemistry, biochemistry, geology, physics, and biology. This **unifying theory** explains the origins of our universe and how stars and planets came into being and why they disappear. It tells us how the various types of chemical elements evolve and develop in the interiors of stars and how geological forces shape our planet. And it helps us to make sense out of the tremendous diversity we see among the various life forms that share our world and to understand how that diversity came to be. It also helps us to understand why some life forms succeed, while many others fail or become extinct.

Darwinian Evolution

The theory of organic evolution did not originate with Charles Darwin when he published *On the Origin of Species by Means of Natural Selection* in 1859 (Figure 1-15). Rather, the theory of evolution predated Darwin by 23 centuries. The first recognizable seeds of the theory date all the way back to the Ionian Greek scholars. Darwin's contributions were important, not because he originated the idea, but because he was first to amass and present the data that convincingly supported the theory.

Darwin's key idea is that the environment selects those inherited features of a species that aid survival from generation to generation. Several factors play an important role in this process, and we will discuss them in detail in Chapter 25. For now, let us take a quick look at some of the principles that form the foundations of the theory.

1. **Genetic Variability.** There is a great deal of physical variability among the different individuals of a species (one kind of living thing), and some of the many variations are better able to survive in the species' environment than others. (We say that the environment "selects" those variations that aid survival.) Some of the species' variability is passed from one generation to the next in the genes that are transmitted from parents to offspring during sexual reproduction, and it is this **genetic variability** of the species that is the raw material of evolution. Species (or populations) evolve, but individuals do not. *You* are not evolving, but the human species is.
2. **Competition for Limited Resources.** In order for a species to survive in a given area, its members must be able to cope with all the pressures the environment offers. Since environmental resources are limited, some of those pressures will be in the form of competition among species members for things in short supply, such as nest space, shelter from the cold, and adequate water, sunlight, and food.
3. **"Survival of the Fittest."** Certain individuals in the species will have physical characteristics that make them more competitive in a given environment. Thus, they are more likely to succeed in that environment than are members of the same species lacking their advantageous characteristics. For example, animals with longer fur would be more likely to keep warm in

(a)

(b)

FIGURE 1-15 (a) Charles Darwin and (b) Alfred Russel Wallace coauthored the first widely accepted presentation of the theory of evolution in 1858. Thirteen months later, Darwin published *On the Origin of Species by Means of Natural Selection* and stunned the world.

the face of a coming ice age (there have been four major ice ages in the last million years). Some people refer to this type of advantage as **survival of the fittest.** However, survival, by itself, is not the most important issue. It must be the kind of survival that allows certain individuals to survive long enough to reproduce. It is more accurate to identify the evolutionary rewards reaped by successful individuals as *reproduction of the fittest.* Only those physical advantages that can be passed genetically to the offspring will have anything to do with shaping the species in future generations (Figure 1-16). The accumulation of many such advantages over many generations can lead to the formation of new species.

1. Why is the theory of evolution considered to be the great "unifying" principle in science?
2. Why is the popular term *survival of the fittest* not a good description of what is most important in the evolutionary process?

ECOLOGY

We are not alone. Humans are only a single species. We share this Earth with over a million other animal species, almost a half million species of plants, over a hundred thousand different fungi, and innumerable bacterial, protistan, and blue-green algal species. We have known about many of these for a long time, but only very recently have we realized the importance of their interrelationship. **Ecology** is the study of these relationships.

By the end of the nineteenth century, studies of **taxonomy** (the branch of science that classifies all living organisms into evolutionarily related categories) and evolution had already led some biologists to see how living things depend on each other. Plants and animals supply each other with food and living space, but not always in obvious ways. Biologists also realized that the physical factors (light, temperature, soil, etc.) in the environment determine, to a large extent, the kind and amount of life a given area can support.

But living things also affect their environment. They contribute chemicals to the soil, to the atmosphere, and to rivers, lakes, and oceans. We have also learned a great deal about the impact the activities of one species, humans, has had on the life support systems that keep the **biosphere** (*bios* = life; *sphere* = realm) going. We are now coming to grips with such problems as the local climatic changes that human overpopulation and agricultural and political practices are causing on our planet.

The biologists who study such relationships between the biological and physical elements of the environment are **ecologists.** They tell us that if we destroy a forest, we also destroy the spongelike soil that holds water and prevents flooding and erosion and the formation of wastelands and deserts. They tell us that if we use pesticides carelessly, we destroy not only the insect pests that are our targets, but also insect-eating bugs and birds that help us control these and other such outbreaks. They tell us that if we kill predators such as wolves, we can expect population explosions of deer or rabbits or mice, which will, as they exhaust their food supplies, die in vast numbers, devastating their environments and sometimes spreading disease before they go (see Chapter 26).

We humans are not alone. We are connected to the world of living things in which we live by a web of kinship and interrelationship. What we do affects our world, and our world affects us in ways both subtle and plain. This basic principle does a great deal to shape our modern biological view of the world. It conditions our decisions about population, food supply, resources, land use, and pollution. It affects our present, and it affects our view of the future of human civilization.

THE BIOLOGICAL STORY

Homeostasis, evolution, and ecology are major parts of the story of biology. We will provide more detail on the latter two topics in the last chapters of this book. However, between now and then, we will make plain how heart and brain and stomach, muscle and bone, glands and gonads, all serve the first of our basic principles—homeostasis—and its maintenance of human life.

But first, life is a matter of chemistry and energy, and we must spend the next chapter on the basics of these subjects. Only then can we go on to the elements of human structure—cells and tissues and genes—and to how the body performs this business of being alive.

(a)

(b)

(c)

(d)

FIGURE 1-16 The course of evolution for a species is often dictated by the natural pressures operating in the species' environment. In this cartoon, we see natural selection operating in a diverse population of doglike mammals living in an area where the climate is changing drastically. In (a), the climate is hot and dry, and the short-haired variants are thriving. In (b), the climate becomes colder, the short-haired variants show signs of stress, and the long-haired variants are coming into their own. In (c), continued climatic cooling eventually eliminates the short-haired version from the population. In (d) is shown the most important evidence that evolution has occurred: The reproductive advantages of long hair show in the increase in the number of long-haired offspring.

SUMMARY

Life means taking in and using nutrients and energy, growing, reproducing, and responding to stimuli. All living things are built of cells and cell products. These statements apply to goldfish, ferns, and bacteria as well as to humans. Human biology is the study of human form, functions, and interactions with the world. It thus involves many newsworthy issues, such as birth control and artificial organs, and it raises many ethical and moral questions.

Yet it is the results, not the methods, of science that make news. The methods are less exciting, for they are "merely" observation, generalization, hypothesis, experiment, theory, and communication. The basic criterion of scientific truth is repeatability, and scientific truths are sought in two ways: Basic research is driven by curiosity about nature; applied research seeks to solve specific problems.

Three basic principles of biology serve as cornerstones to our understanding of human biology. The principle of homeostasis says that living things strive to maintain their internal conditions within roughly constant bounds, using various types of negative-feedback mechanisms for controls. Failures of homeostasis lead to many diseases. The principle of evolution says that all living things, including humans, are interrelated and each species is finely tuned to succeed in its environment through the processes of natural selection. The principles of ecology teach us the interdependence of all life forms and the relationships they have with the physical environment.

STUDY QUESTIONS

1. Discuss the ways in which a social organization such as a business or government fits or fails to fit the text's definition of life.
2. Is the scientific method only for scientists? Give an example of how you have already used the sequence of observation, generalization, hypothesis formation, and experimentation to learn something.
3. Distinguish between applied and basic research. Why is more grant money usually available to applied researchers?
4. How do inductive and deductive logic apply to the scientific process?
5. The human body stores calcium in the bones of the skeleton. A homeostatic mechanism keeps the level of calcium in the blood roughly constant by dissolving some of the calcium out of the bone or causing it to be deposited. Design a simple feedback system that would control the calcium level (it does not have to be correct, but it should be logically sound).
6. What do you think accounts for the great diversity of life on Earth?

GLOSSARY

Biology The study of life.

Biosphere Places in the Earth's atmosphere and waters and on the Earth's crust where life can exist.

Cells Tiny, modular, fundamental units in which life occurs.

Communication As part of the scientific method, it usually means publishing a paper about scientific findings in a journal or in some other way sharing them with the scientific community.

Control group The group or population receiving the sham treatment or placebo in an experiment. Results from the control group are compared against those from the experimental group.

Deductive logic A reasoning system that attempts to predict an individual outcome based on a general statement. Starting with the general statement "men are taller than women," you might try to predict (before seeing either of them) that since individual A is a man and individual B is a woman, A will be taller than B.

Ecology The branch of science studying the interrelatedness and interactions between the living and nonliving factors in the environment.

Experimental group The group receiving the treatment that tests the hypothesis.

Experimentation Designing and carrying out formal tests of the hypotheses.

Generalization Identifying a pattern found among many observations.

Homeostasis The process of keeping the chemical and physical environments within the body stable and compatible with life.

Hypothesis A tentative explanation for the patterns seen among the observations.

Inductive logic A reasoning system that makes general statements based on many individual observations. After measuring the heights of a sample of many men and women, you might be inclined to generalize inductively that men are taller than women.

Irritability Responsiveness. The ability of living systems to perceive changes in their environment and respond.

Metabolism The sum total of the body's chemical activities.

Negative feedback Control mechanisms that operate by inhibiting excursions in either direction away from a predetermined set point.

Observation Measurements, either formal or informal, of some physical or biological phenomenon.

Organic evolution The branch of biological thought focusing on how life originated and then diversified to the many forms we see today.

Repeatability An important criterion for scientific acceptance that suggests that experimental findings should be repeatable regardless of who is doing the experimentation.

Science Systematic study of physical, material, and biological phenomena and organization of the information discovered about each.

Scientific method A structured approach to problem solving employed by scientists.

Theory A hypothesis that has successfully withstood every test over a long period of time. It is very likely to be a true explanation for the phenomenon.

Variable The particular thing being measured.

Chapter 2

Elements of Chemistry and Energy

- Atoms, Elements, and Isotopes
- Important Bio-elements
- Chemical Symbols, Molecules, and Compounds
- Chemical Reactions
- Chemical Bonds
- Acids, Bases, and the pH Scale
- Organic Chemistry
- The Nucleotides
- Energy and Life
- The Laws of Thermodynamics
- Learning Focus: Organic Chemistry
- Summary
- Study Questions
- Glossary

A basic tenet of certain Eastern religions is that everything, human and nonhuman, living and nonliving, large and small, matter and energy, is part of the same cosmic universe. Everything is related to everything else. Some arrangements of **matter** have that subtle quality we call "life." Other variations in the organization of this cosmic stuff give us nonliving things, such as planets, stars, rocks, water, and air. Science's view of the universe can be phrased in almost exactly the same words.

Chemistry is the study of the various kinds of matter. It examines their composition, structure, and properties as well as how various substances are likely to react with each other. Since everything consists of chemicals, knowledge of the rudiments of chemistry is necessary to understand almost every branch of science. This is particularly true for biology, whose practitioners feel that many of the undiscovered secrets of life will be unveiled through chemical studies.

ATOMS, ELEMENTS, AND ISOTOPES

Atoms

Chunks of matter cannot be subdivided endlessly. **Atoms** are the smallest bits of what we usually consider matter. These fundamental structural units are so tiny that it would take roughly two hundred million of them to form a line an inch long. The 92 different types of atoms—or **elements**—found in nature (others are synthesized) provide the building blocks from which all the various types of matter are formed (Figure 2-1).

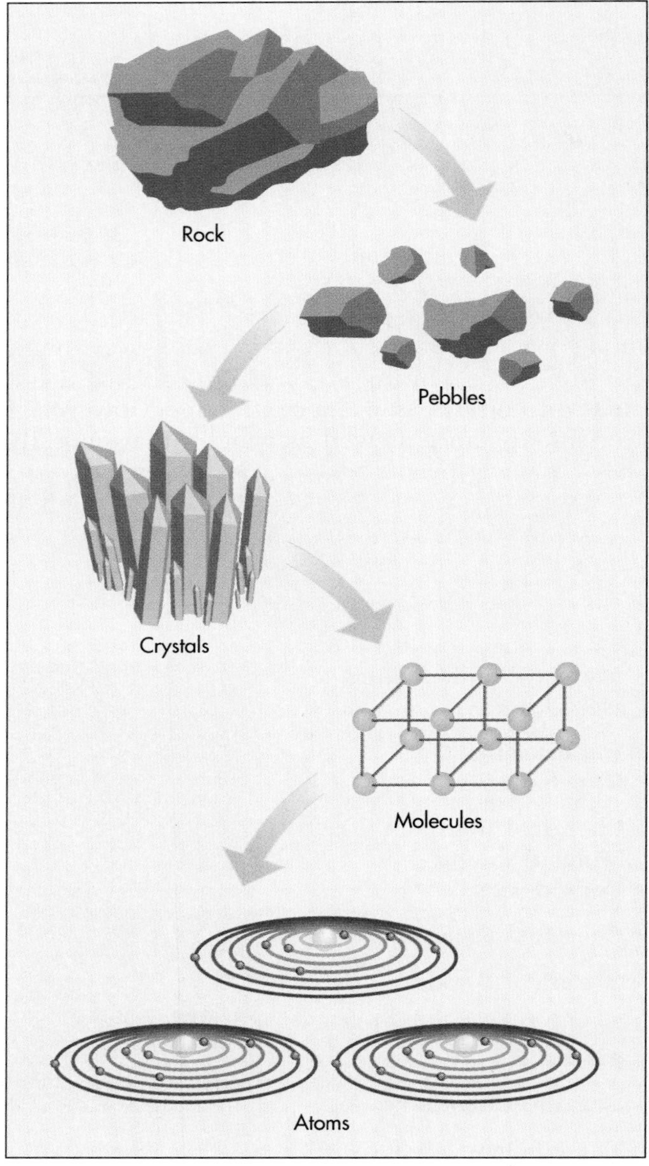

FIGURE 2-1 Pieces of matter, such as rock, cannot be subdivided endlessly. The smallest particle of a substance that remains identifiable as a piece of that substance is an *atom.*

Physicists have learned so much about the structure of atoms that we can predict how the different kinds of atoms are likely to behave. We now know that atoms have three chief components, or subatomic particles. The most common particles are protons, neutrons, and electrons. **Protons** are found in the nucleus (core) of the atom and contain much of its actual mass. In addition, they have a positive electric charge. The nucleus's **neutrons** are of almost equal mass but lack a charge. **Electrons** are extremely tiny negatively charged particles. They orbit the nucleus approximately as planets do a sun; since they do not have specific locations, they have to be drawn as cloudy bands around the nucleus, as in Figure 2-2. The outermost layer of electrons, called the valence electrons, determines the reactions in which an atom is likely to participate.

Each electron's negative charge is of the same intensity as the positive charge of a proton, but electrons are much smaller than either protons or neutrons; a single proton or neutron contains over 1800 times more mass than a single electron (Figure 2-2, p. 26). The protons and neutrons of atomic nuclei contribute virtually all the mass to the various substances in the universe. Atoms are electrically neutral because they have exactly the same numbers of negatively charged electrons as positively charged protons.

The simplest atom, that of the element *hydrogen*, consists of a single proton orbited by a single electron. Larger atoms have more protons and electrons; they also have neutrons. An atom of iron, for instance, contains 26 protons, 26 electrons, and 30 neutrons. It is thus nearly 60 times as heavy as hydrogen. But even such a large atom is one of the tiniest things in the universe. A cubic centimeter of iron weighs only about 8 grams, but it contains about 84,500,000,000,000,000,000,000 atoms.

A carefully scaled model of a hydrogen atom might be compared with a slightly overweight housefly (representing an electron) orbiting a nucleus (proton) the size of a cantaloupe. However, if we scale distance as well as weight, that housefly would have to be orbiting a good 5 miles from the cantaloupe. Thus, at the atomic level everything—including the piece of iron mentioned in the preceding paragraph, the rock of Gibraltar, and your biology professor—consists mostly of open space.

---

1. Why are atoms considered the fundamental building materials of the universe?
2. How many different kinds of atoms are found in nature? What are elements?
3. Characterize protons, neutrons, and electrons and discuss why they are referred to as subatomic particles.

Atomic Number, Atomic Mass, and Isotopes

Atoms are classified in two ways. All atoms with the same number of protons in their nuclei (or **atomic number**) belong to the same element. However, not all atoms of an element need have the same number of neutrons. All carbon atoms have six protons in the nucleus and six orbiting electrons. Thus, all carbon atoms have the same atomic number, 6. However, carbon atoms can differ in the number of neutrons they contain. The most common form of carbon atom has six neutrons. Adding the numbers of neutrons and protons (6 + 6) gives the **atomic mass** (12). Other forms of carbon atoms, called carbon **isotopes,** have the same atomic number (protons and electrons) but different numbers of neutrons; they therefore have different atomic masses. One isotope has eight neutrons in its nucleus instead of six; since isotopes are named for their atomic masses, this one is called carbon-14. Because the configuration of the electron cloud is basically the same, carbon-14 participates in chemical reactions in about the same way as the more common carbon-12. However, it is easily distinguished from carbon-12 because it spontaneously releases fragments of its nucleus. It is radioactive; that is, it is unstable and emits radiation, which makes it easier for a scientist to keep track of carbon-14 atoms during experiments. For that reason, carbon-14 is often used to study the details of biological chemistry, which is very much the chemistry of carbon.

---

1. Define atomic number and atomic mass and discuss how they are related.
2. What is an isotope?

IMPORTANT BIO-ELEMENTS

Of the 92 naturally occurring elements, relatively few are important in living systems. In fact, 95 percent of all **protoplasm** consists of various combinations of only five different types of atoms. The "big five" are carbon, hydrogen, oxygen, nitrogen, and phosphorus (Figure 2-2).

Carbon

In terms of the history of life, carbon is the most important and versatile element in the universe. The arrangement of electrons in its outermost, or valence, shell is such that each carbon atom is capable of bonding chemically to four other atoms, which often include atoms of hydrogen, oxygen, nitrogen, and other elements. Few other atoms can form so many chemical bonds. For that reason, carbon is the ideal material for the complex molecules of living systems.

Over 2 million carbon compounds are known, and thousands more are discovered each year. They show tremendous diversity in form and properties ranging from gases such as methane to crystalline solids. Diamond, among the hardest substances we know, is made of pure carbon atoms arranged in a three-dimensional crystal lattice (Figure 2-3). Some carbon compounds are highly reactive, while others are practically nonreactive. The branch of chemistry dedicated to studying **compounds** of carbon is referred to as organic chemistry. In contrast, the study of compounds that are not built around chains of carbon atoms is called inorganic chemistry.

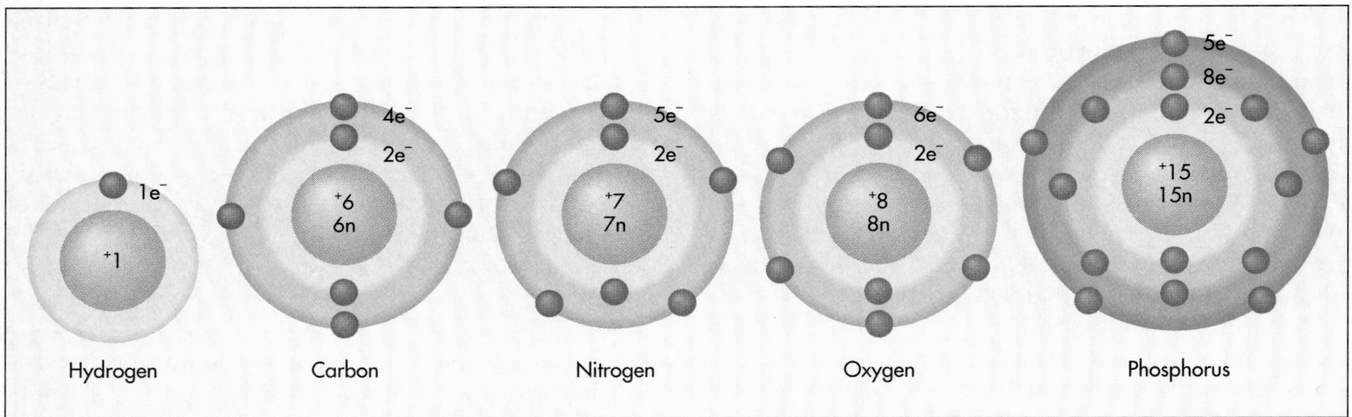

FIGURE 2-2 The five atoms depicted are sometimes referred to as the bio-elements, since different chemical combinations of these five make up about 95 percent of all living substances. Most of the mass of each atom resides in its central nucleus, made up of positively charged protons and uncharged neutrons. Traveling around each nucleus in concentric orbits, called energy levels or shells, are the tiny negatively charged electrons. The number of electrons equals the number of protons in the nucleus.

Carbon can form long sheets, nets, and rings when its atoms are chained together in various ways (Figure 2-4). This ability to form extensive chains with other atoms of the same kind is rare. It makes carbon ideal for forming membranes, muscle fibers, the glues holding cells together, hair, and enzymes, among other things.

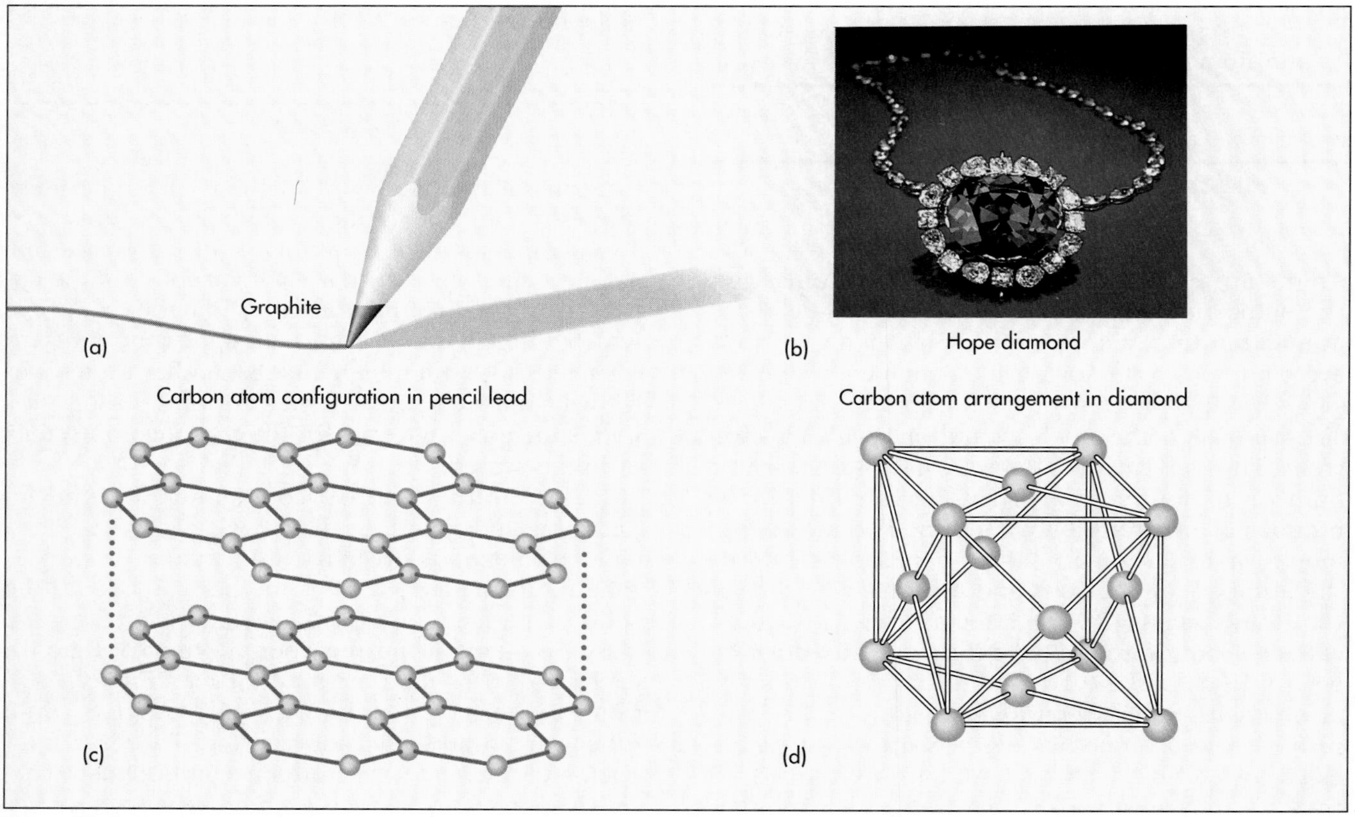

FIGURE 2-3 Pure carbon comes in different forms. In (a) the graphite of a pencil lead, carbon atoms are arranged in (c) removable layers. In (b) the Hope diamond, the world's largest flawless diamond, carbon atoms are arranged in a rigid crystal lattice (d) that makes diamond one of the hardest substances known.

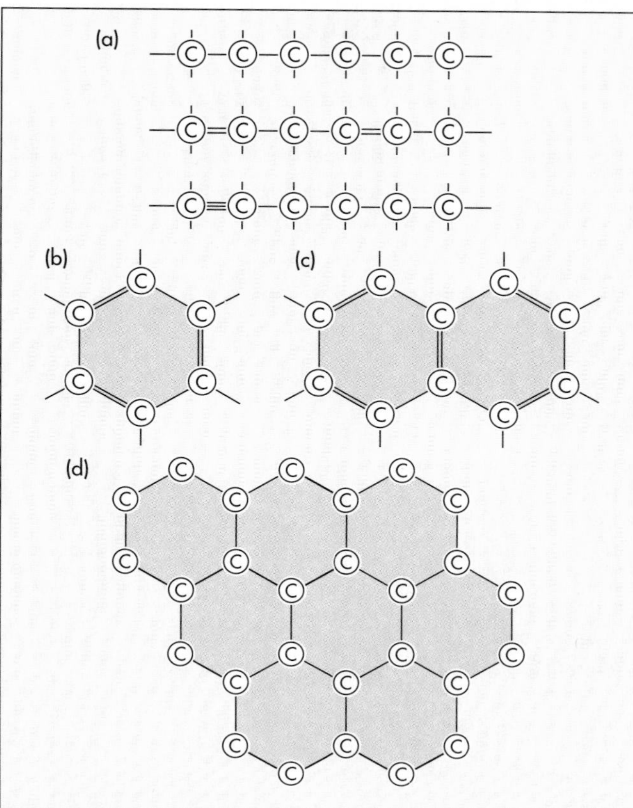

FIGURE 2-4 Because each carbon atom can form four different chemical bonds, it has a structural flexibility that makes it an ideal material for the diverse organic molecules of living systems. It forms (a) straight chains with single bonds, double bonds, and triple bonds connecting adjacent carbon atoms. It also forms (b) single rings, (c) linked rings, and (d) sheets.

Hydrogen

Hydrogen is the simplest atom, consisting of a single proton orbited by a single electron. It is also the most abundant substance in all the universe—over 90 percent of everything that exists consists of hydrogen atoms. For every carbon atom found in the universe there are over 9200 hydrogen atoms. Likewise, for every carbon atom found in the Earth's crust, there are 52 hydrogen atoms.

Hydrogen seems to be the ultimate parent atom. All other types of atoms in the universe were apparently formed in the atomic fusion reactions that begin with the hydrogen atoms in the interiors of stars, including our sun. At the high temperatures and pressures existing inside these stars, other elements are produced by nuclear fusions, releasing vast quantities of energy. It has been calculated that the amount of energy released from the fusion of just one cupful of hydrogen atoms would be more than enough to heat 7 tons of water from room temperature (25°C) to boiling. Some people look to fusion reactions as a future source of unlimited energy for human civilization.

Oxygen

Oxygen is one of the most abundant substances on Earth. Molecular oxygen, an odorless, colorless gas, makes up 21 percent of the planet's atmosphere (see Table 2-1); some oxygen gas also dissolves in water. Combined with hydrogen, oxygen contributes about 90 percent to the weight of water. Combined with another element, silicon, it constitutes about 50 percent of the weight of dry sand.

Atmospheric oxygen is essential to most present forms of life. Living cells depend on molecular oxygen to operate the energy-processing systems that fuel the activities of life. Especially dependent are the very active life forms, such as animals (including humans), which will die within minutes if they are deprived of their oxygen supplies.

In spite of its usefulness, oxygen is a highly reactive and potentially harmful element. Fire is produced when various "fuel" molecules react vigorously with oxygen, releasing heat and light in the process. The higher the oxygen concentration, the more easily such reactions can occur. Many biologists feel that some of the changes of aging are caused by oxygen damage to tissues. The harmful effects of oxygen were not completely understood at the time of some early experiments with the self-contained underwater breathing apparatus (SCUBA). Early experiments involved the use of tanks of pure oxygen, which had **toxic** effects on divers at two atmospheres of pressure, equivalent to a depth of only 9.15 meters (about 30 feet).

Nitrogen

Like oxygen, nitrogen is an odorless and tasteless atmospheric gas. However, unlike oxygen, the molecular gas form of nitrogen is very unreactive and cannot sustain life. Discovered in 1772, it makes up about 78 percent of the air we breathe.

Table 2-1

Composition of Dry Air (in Percent by Volume)

CONSTITUENT	FORMULA	PERCENT
Nitrogen	N_2	78.08
Oxygen	O_2	20.95
Carbon dioxide	CO_2	0.035
Hydrogen	H_2	0.01
Helium	He	0.0005
Other		About 1

Virtually all plants and animals require nitrogen in a combined form to build proteins, vitamins, and genetic materials. Since atmospheric nitrogen is so **inert** (unreactive), it would be unavailable to the biosphere were it not for certain bacteria and algae that can incorporate nitrogen gas into various compounds through a series of reactions collectively called *nitrogen fixation*. For that reason, this is one of the most important chemical processes of all. Another source of usable nitrogen is the manufactured fertilizers we place on our agricultural crops.

Phosphorus

Pure phosphorus is a nonmetallic solid with many uses, such as in match heads, fire bombs, and poison gases. It is also an essential nutrient for living things, as part of the structural material of bones and as part of the genetic material [deoxyribonucleic acid (DNA)]. It is also involved with the multiple energy transfers that take place in each living cell every second as part of a substance called adenosine triphosphate (ATP).

CHEMICAL SYMBOLS, MOLECULES, AND COMPOUNDS

Chemists have a shorthand system for identifying the different elements. The symbol for each element is usually the first one or two letters of the element's name, either in Latin or English (see Table 2-2). For instance, the symbol for oxygen is simply O, for hydrogen H, carbon C, nitrogen N, phosphorus P, sodium Na (Na? Yes, for the symbol comes from its Latin name, *natrium*), and so on (Table 2-2).

While the smallest particles of matter are the atoms, the smallest particle of each type of chemical substance is called a **molecule.** Molecules can be single atoms or they can contain more than one atom of the same type or they can be constructed of two or more different types of atoms that have been chemically combined in a definite ratio. The abbreviation or formula for a molecule is constructed by combining the symbol for the atom with a subscript that indicates the number of atoms in a single molecule. Pure nitrogen, for example, is written as N_2; the formulas for other gases appear in Table 2-3.

Water's formula is written H_2O. The subscript 2 after the H indicates that one molecule of water contains two atoms of hydrogen. The lack of a subscript after the O indicates that there is only one atom of that kind per molecule of water. If it contains any more than one, then the symbol must be followed by a subscript indicating the number.

The formula for the glucose molecule (glucose is a sugar that is used as fuel by the body) is $C_6H_{12}O_6$. Thus, 1 molecule of glucose contains 6 carbon atoms combined with 12 hydrogen atoms and 6

Table 2-2

Some Common Elements

NAME	SYMBOL	ATOMIC NO.	ATOMIC MASS
Arsenic	As	33	74.92
Barium	Ba	56	137.34
Bromine	Br	35	79.90
Calcium	Ca	20	40.08
Carbon	C	6	12.01
Chlorine	Cl	17	35.45
Cobalt	Co	27	58.93
Copper	Cu	29	53.55
Fluorine	F	9	19.00
Gold	Au	79	196.97
Helium	He	2	4.00
Hydrogen	H	1	1.01
Iodine	I	53	126.90
Iron	Fe	26	55.85
Lead	Pb	82	207.20
Lithium	Li	3	6.94
Magnesium	Mg	12	24.31
Manganese	Mn	25	54.94
Mercury	Hg	80	200.59
Molybdenum	Mo	42	95.94
Nickel	Ni	28	58.71
Nitrogen	N	7	14.01
Oxygen	O	8	16.00
Phosphorus	P	15	30.97
Platinum	Pt	78	195.09
Plutonium	Pu	94	242.00
Potassium	K	19	39.10
Radium	Ra	88	226.03
Selenium	Se	34	78.96
Silicon	Si	14	28.09
Silver	Ag	47	107.87
Sodium	Na	11	22.99
Strontium	Sr	38	87.62
Sulfur	S	16	32.06
Tin	Sn	50	118.69
Uranium	U	92	238.03
Zinc	Zn	30	65.37

Notes: Remember the atomic number consists of the number of protons in each atom, and the atomic mass approximates the number of protons and neutrons in each atom. This table contains only 37 of the 92 naturally occurring elements. These are the elements we are likely to have reason to discuss in a biology class.

oxygens. A single molecule is the smallest particle of glucose, and it must contain all of these atoms arranged in a specific sequence if it is to have all the properties of that substance.

1. Identify and describe each of the five elements that collectively make up about 95 percent of the living substance.
2. Why is carbon so important in living systems? What is organic chemistry?
3. What is a molecule? What do the subscripts mean in the formula $C_6H_{12}O_6$?

Table 2-3

Molecules of Some Common Gases

ELEMENT	MOLECULAR SYMBOL
Oxygen	O_2
Nitrogen	N_2
Hydrogen	H_2
Carbon monoxide	CO
Carbon dioxide	CO_2
Chlorine	Cl_2

CHEMICAL REACTIONS

Chemical reactions occur when chemical bonds between atoms form or break. It is not a chemical reaction when a substance simply changes state, as when ice melts to become liquid water. Both ice and liquid water are made up of H_2O molecules; it is only the nature of the arrangement between molecules that changes when ice melts or water freezes. When a reaction occurs, a change takes place within the chemical molecule that causes it to form a new substance.

For example, when you strike a match, you activate two sets of chemical reactions. First, striking the match raises the temperature of the match head and sets off the reaction of the sulfur and phosphorus compounds in the match head with the oxygen in the air. The result is the intense flaring reaction that, in turn, raises the temperature of the match's wood (or paper). The oxygen in the air begins to combine with the chemicals in the match stick, releasing the energy they contain as heat and light and producing end products such as carbon dioxide, carbon monoxide, and the various ingredients of smoke. These reactions involve breaking large molecules and recombining their parts with parts of other molecules.

The events of individual chemical reactions are summarized in **chemical equations.** The symbolism illustrating what is taking place in a reaction is fairly easy to follow. An example is:

$$\underset{\text{(reactants)}}{2H_2 + O_2} \; \underset{\text{(produce)}}{\longrightarrow} \; \underset{\text{(product)}}{2H_2O}$$

This sample equation shows the reaction of hydrogen and oxygen to produce water (although, at this time, we are ignoring the fact that either heat or a catalyst—an agent that speeds up reactions without being consumed—is required to really get this reaction to go). The arrow separates the two sides of the equation and points *to* the reaction's end results, or **products.** It points *from* the substances that are undergoing the reaction, the **reactants.** Read this equation as: two molecules of hydrogen plus a single molecule of oxygen produce two molecules of water. Note that the number preceding a reactant or product tells how many molecules of that substance are needed for the reaction. If no number precedes the formula, then assume *one* molecule.

Another example would be the chemical equation summarizing the events of *cellular respiration* (perhaps the single most important series of chemical processes taking place in your body). Cellular respiration is the process by which your cells transfer the raw energy from food molecules to a chemical form your cells can use directly. A major reason you eat, breathe, and move about is to supply your cells with the raw materials they need for this process. Here is the equation:

$$\underset{\text{(reactants)}}{C_6H_{12}O_6 + 6O_2} \longrightarrow \underset{\text{(products)}}{6CO_2 + 6H_2O +} \;\; \underset{\text{(ATP)}}{\text{Usable energy}}$$

The equation reads: one molecule of glucose plus six molecules of oxygen produce six molecules of carbon dioxide plus six molecules of water and usable energy in the form of ATP, the energy currency molecules of the cell. We will discuss this process in considerable detail in Chapter 20.

CHEMICAL BONDS

Chemical reactions normally result from collisions between molecules. When they hit just right, their outermost layers (orbitals) of electrons—the valence electrons—interact to form chemical bonds. Chemical reactions rarely involve changes in the atomic nuclei or in the electrons orbiting closer to the nucleus. Depending on how the electrons interact between reacting atoms, several different types of chemical bonds can result: ionic bonds, covalent bonds, or hydrogen bonds.

Since each type of atom has its own characteristic number of protons complemented by its orbiting electrons, each element reacts along specific lines. The arrangements in some make them very stable and hence unreactive. The arrangements in other atoms make them aggressively seek electrons from other sources; they are thus highly reactive. An example of such an electron-needy element is oxygen, which readily seizes electrons from other elements as it reacts with them. On the other hand, some elements hold their valence electrons only loosely and readily pass them on to other elements or molecular fragments. Other elements participate in reactions that allow them to share electrons with other atoms, which makes them more stable, or less reactive. The many gradations to this scheme form the basis for the different types of chemical bonds found in nature.

Ionic Bonds

Normally, atoms have equal numbers of protons and electrons. However, under certain circumstances, an atom can gain or lose one or more electrons. It then is no longer electrically neutral. It has a positive electric charge if it has lost electrons and a negative electric charge if it has gained electrons. Such charged atoms are called **ions.**

Ions are often formed when an atom or molecular fragment that would be more stable (less reactive) if it could add one or more electrons to its outermost shell meets one that would be more stable if it could lose the same number of electrons. One reactant then gives up electrons to the other, and both become ions. Because the two ions have opposite electrical charges, they attract each other strongly. That is, they are held together by an **ionic bond** to form an **ionic compound.** An example of such a reaction occurs between sodium (Na) and chlorine (Cl) atoms to form the ionic compound sodium chloride (NaCl, or table salt).

Sodium, in the uncombined form, is a grayish metal containing a loosely held outer electron. Pure chlorine is a greenish, poisonous gas whose atoms need an electron in their outermost orbital, making them very ready to capture electrons from another source. If the collision between the reacting particles occurs with enough force, there will be a rearrangement of the electrons between the sodium and chlorine atoms. Chlorine draws sodium's negatively charged valence electron into orbit around itself to fill its own vacancy. It thus gains an extra negative charge. Since the sodium atom has lost one of its electrons, it has acquired a net positive charge. Electron donors (positive ions) are called cations (pronounced "cat ions"); electron acceptors (negative ions) are called anions ("an ions").

Water molecules are, in a way, partial ions. The chemical bonds that link their hydrogen atoms to their much larger oxygen atoms involve a sharing of electrons (see the discussion of covalent bonds that follows below), but the sharing is uneven. The electrons spend most of their time orbiting the oxygen, thus giving it a small negative charge and leaving the hydrogens with small positive charges. The water molecule thus has opposing electrical charges on its two ends, turning it into a **polar molecule,** or a **dipole** (Figure 2-5). When ionic compounds are dissolved in water, the water dipoles are drawn by the attraction of opposite charges for each other toward the ions. As they surround the ions, they insulate them from the attraction of their ionic partners, weaken their ionic bonds, and pry them free. However, when the ions separate in solution, the displaced electrons remain with their new owners. Consequently, the particles separate as charged ions:

$$\underset{\text{sodium chloride}}{NaCl} \longrightarrow \underset{\text{sodium ion}}{Na^+} + \underset{\text{chloride ion}}{Cl^-}$$

Many important substances in the human body break down into ions when dissolved in water; among them are the various types of minerals. Minerals are usually salts. However, dissolved salts break apart in solutions as ions. Since ionic solutions are capable of conducting electricity, ionic substances are commonly called **electrolytes.** Many of the electrical activities associated with living systems, from nerve signals to the electrochemical activities of a beating heart, are the result of cellular manipulation of the movements of these charged particles. An alphabetical list of many ions likely to be discussed in a biology class is found in Table 2-4. You will note that many single ions contain more than one atom.

Covalent Bonds

Many atoms form a different kind of chemical bond when they react. Rather than simply transfer valence electrons from one to another, they share

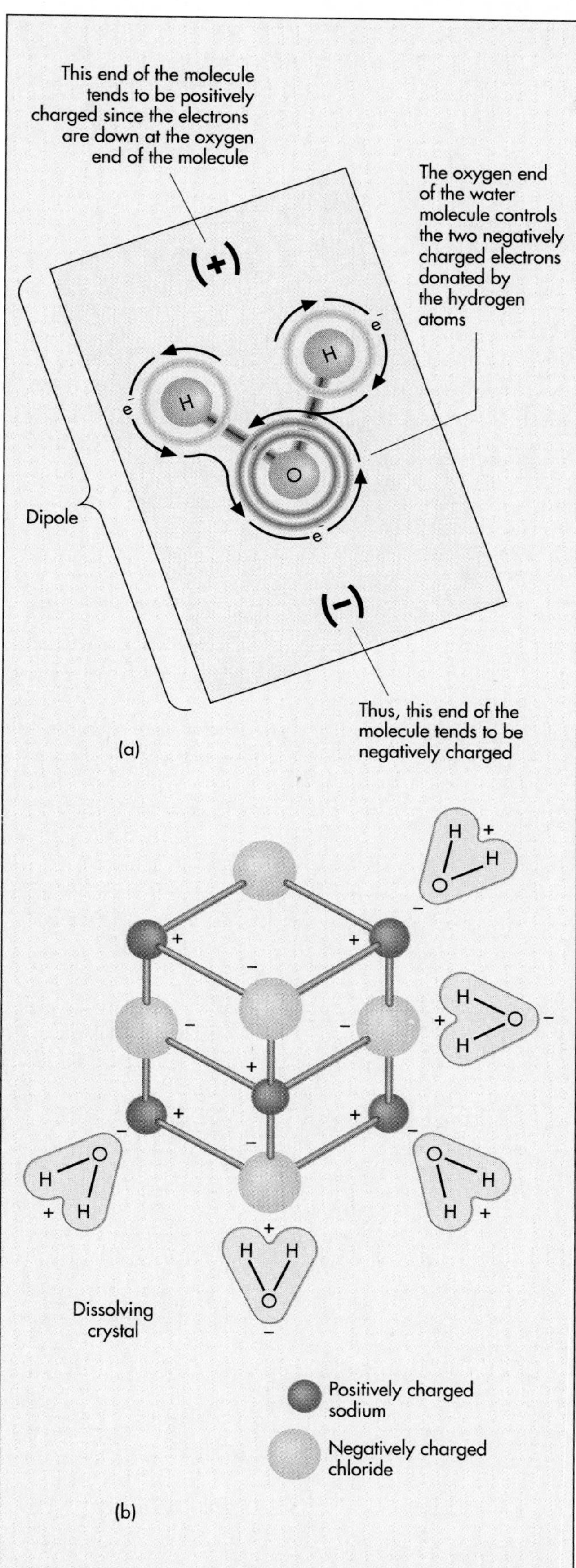

Table 2-4

Common Ions in Biological Systems

NAME	SYMBOL
Ammonium ion	NH_4^+
Bicarbonate ion	HCO_3^-
Calcium ion	Ca^{2+}
Carbonate ion	CO_3^{2-}
Chloride ion	Cl^-
Cyanide ion	CN^-
Ferrous ion	Fe^{2+}
Hydrogen ion	H^+
Hydroxide ion	OH^-
Nitrate ion	NO_3^-
Nitrite ion	NO_2^-
Permanganate ion	MnO_4^-
Phosphate ion	PO_4^{3-}
Sodium ion	Na^+
Sulfate ion	SO_4^{2-}
Sulfite ion	SO_3^{2-}

them. We saw one example in the water molecule. An electron-sharing bond is called a **covalent bond.** Covalent bonds tend to be stronger than ionic bonds, and substances constructed with covalent bonds do not break down into ions when dissolved in water. Covalent bonds are usually formed when a nonmetal reacts with another nonmetal.

A simple example of a covalent bond is that which forms between a hydrogen atom and a fluorine atom. A hydrogen atom is simply one proton orbited by one electron; it would be in a more stable configuration if it had two electrons orbiting its nucleus. A fluorine atom has seven orbiting valence electrons but would be more stable if it had eight. A reaction between hydrogen and fluorine atoms allows them to reduce their instabilities by sharing each other's electrons on a part-time basis (Figure 2-6). Therefore, part of the time the electron from the hydrogen atom becomes the eighth orbiting

FIGURE 2-5 (a) Because of an unequal sharing of the negatively charged electrons between the oxygen atom and the hydrogens in a water molecule, the oxygen end tends to be negatively charged. Conversely, the hydrogen ends of the molecule take on a slight positive charge. Such a polarized arrangement is called a dipole. (b) The dipole nature of water molecules makes water a powerful solvent. For instance, salt crystals dissolve because the negatively charged ends of surrounding water molecules pull on the positively charged sodium ions of the salt crystal. The positively charged ends of the water molecules pull on the negatively charged chloride ions.

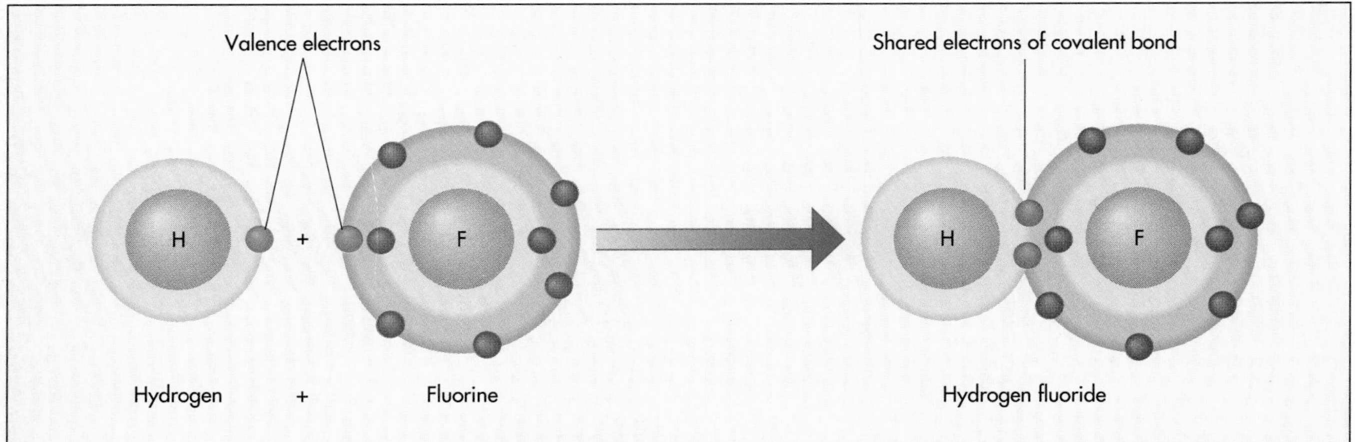

FIGURE 2-6 A covalent bond is formed when reactants come to share a pair of electrons, as do hydrogen and fluorine in hydrogen fluoride. The electrons orbiting each nucleus occupy orbital layers or energy levels. Each energy level contains a specific number of electrons if the atom is in the most stable configuration. The reaction occurs, in part, because the hydrogen atom needs an extra electron to have a total of two in the first energy level and fluorine becomes more stable with eight electrons in its second energy level. This favorable arrangement is approached when the two atoms share their valence electrons.

electron for fluorine and part of the time the hydrogen atom fulfills its need for two valence electrons by drawing from fluorine's supply. Thus, on the average both atoms are more stable than they were before they combined resources.

1. List and characterize the three different types of chemical bonds.
2. How does the polar nature of water molecules make possible the hydrogen bonding that occurs in liquid water?

Hydrogen Bonds

Hydrogen bonds are weak bonds, considerably less than one-tenth as strong as a covalent bond. They often form between different molecules. They can also give shape to a single molecule by binding one portion of that molecule to another part of the same molecule.

Hydrogen bonds resemble ionic bonds in that they depend on the attraction of opposite charges. However, the charges are not those of ions. Instead, they are partial charges such as those we see in water molecules, which offer a good example of hydrogen bonding. In a molecule of water, one oxygen atom bonds covalently to two hydrogen atoms. Because the ends of water molecules bear opposite charges, they can attract each other like tiny bar magnets (Figure 2-7). The positively charged hydrogen of one molecule attracts the negatively charged oxygen of the molecule next to it. This attraction is a hydrogen bond. The hydrogen bonds are weak enough to allow the individual water molecules to flow over one another as the liquid is poured, but collectively they are strong enough to give water the surface tension that allows a needle or other small, dense object to float on it.

ACIDS, BASES, AND THE pH SCALE

Many acids and bases are very reactive chemicals. Some people experience firsthand just how reactive when they spill battery (sulfuric) acid on their clothes. The fabric disintegrates as it reacts with the hydrogen ions. Most people have also seen the graphic television commercials showing the way drain cleaner, a basic solution (often a version of lye), dissolves organic debris from the pipes under the sink. Also, everyone has heard stories about the ability of these substances to burn exposed skin. We understand that these chemicals are potentially reactive enough to deserve considerable respect. But what are they?

In biology, **acids** are substances that release hydrogen ions (H^+) when dissolved in water; **bases** release hydroxyl ions (OH^-). The appropriate chemical equation for hydrochloric acid is

$$\underset{\text{hydrochloric acid (muriatic acid)}}{HCl} \xrightarrow{H_2O} \underset{\text{hydrogen ion}}{H^+} + \underset{\text{chloride ion}}{Cl^-}$$

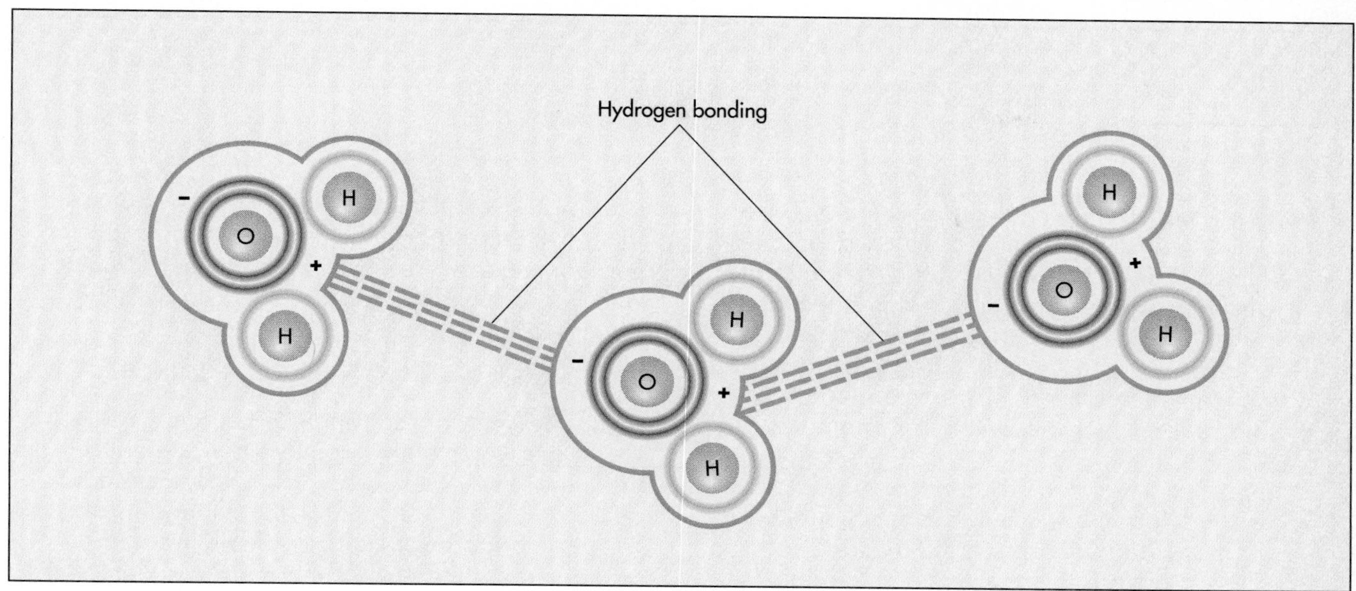

FIGURE 2-7 Hydrogen bonding between water molecules. The hydrogen and oxygen atoms are held together within a water molecule by covalent bonds. However, the oxygen atoms attract and hold the electrons more strongly than the hydrogen atoms do, creating dipoles. The more positively charged hydrogen ends of the water molecules are attracted to the negatively charged ends of adjacent water molecules; this weak attraction is called hydrogen bonding. Hydrogen bonds are collectively responsible for the surface tension of water but are weak enough to allow water to flow as it is poured.

The equation for the release of two hydrogen ions from a sulfuric acid molecule is

$$\underset{\text{sulfuric acid (battery acid)}}{H_2SO_4} \xrightarrow{H_2O} \underset{\text{hydrogen ions}}{2H^+} + \underset{\text{sulfate ion}}{SO_4^{2-}}$$

Equations showing the release of hydroxyl ions from two common bases follow the same lines. The reaction for sodium hydroxide (soda lye) is

$$\underset{\text{sodium hydroxide (soda lye)}}{NaOH} \xrightarrow{H_2O} \underset{\text{sodium ion}}{Na^+} + \underset{\text{hydroxyl ion}}{OH^-}$$

For Milk of Magnesia, generically known as magnesium hydroxide, the reaction is

$$\underset{\text{magnesium hydroxide (Milk of Magnesia)}}{Mg(OH)_2} \xrightarrow{H_2O} \underset{\text{magnesium ion}}{Mg^{2+}} + \underset{\text{hydroxyl ions}}{2OH^-}$$

Water molecules can also break down in this way, or ionize. One water molecule releases both a hydrogen and a hydroxyl ion.

$$\underset{\text{water}}{H_2O} \rightleftharpoons \underset{\text{hydrogen ion}}{H^+} + \underset{\text{hydroxyl ion}}{OH^-}$$

However, the attraction between the hydrogen and hydroxyl ions is so strong that they almost immediately recombine. That is, the reaction proceeds in both directions at once, as indicated by the two arrows. The darker arrow (pointing left) indicates the stronger reaction tendency; thus, molecular water is much more likely to form than ions, and when ionization does occur, it is almost immediately reversed. Water is unlikely to accumulate or build up an excess of either hydrogen or hydroxyl ions; it is therefore neither an acid nor a base.

Strong and Weak Acids and Bases

Some acids and bases are stronger than others. The difference lies in the amounts of reactive ions they release. Strong acids, like hydrochloric (HCl) or sulfuric (H_2SO_4) acids, ionize completely, releasing large quantities of hydrogen ions in solution. Weak acids, like vinegar ($HC_2H_3O_2$, known generically as acetic acid) and carbonic acid (H_2CO_3, normally found in carbonated soft drinks), release very small amounts of hydrogen ions in water. Weak acids are much less reactive than their stronger counterparts.

A similar relationship exists between strong and weak bases. A strong base, like sodium hydroxide

Box 2-1 Spas, Skin Infections, Chlorination, and pH Management

Some of the living things with which we share our world cause problems for us on a very personal scale. They are the bacteria, fungi, and viruses that are responsible for skin infections, bladder infections, colds, and mumps, among other diseases (see Chapter 23). Much to our dismay, these microorganisms often lurk in surprising places, such as swimming pools, whirlpool baths, and spas.

Improperly maintained pools, whirlpool baths, and spas have been responsible, according to many dermatologists, for a "rash" of skin infections because their waters can support infectious bacteria or fungi. *Proper maintenance* means keeping the chemical climate of such waters adjusted to prevent or inhibit the growth of troublesome microorganisms. This is usually done by chlorination. Some very sophisticated systems actually bubble poisonous chlorine gas through the water. Most systems rely on a much safer and less expensive chemical, sodium hypochlorite (NaOCl). Once added to water, sodium hypochlorite releases chlorine and kills microorganisms; the rest of the molecule reacts with the water and is converted to sodium hydroxide (NaOH):

$$2NaOCl + 2H_2O \longrightarrow 2Cl + 2NaOH$$

In time, the concentration of sodium hydroxide in the water builds up. This buildup causes the pH of the water to rise as it becomes more alkaline. To compensate for the change in pH, the pool water must be periodically treated by adding small amounts of muriatic (hydrochloric) acid (HCl). The muriatic acid reacts with the sodium hydroxide to produce sodium chloride (table salt) and water. Since this reaction removes the source of alkalinity—the HCl neutralizes the NaOH—it brings the pH back toward neutral:

$$HCl + NaOH \longrightarrow H_2O + NaCl$$

(NaOH), ionizes completely and releases large quantities of hydroxyl ions, while a weak base, like ammonium hydroxide (NH_4OH), ionizes only partially, releasing substantially fewer hydroxyl ions.

The pH Scale

The presence or absence of hydroxyl and hydrogen ions can influence many of the body's chemical processes and environments. It is thus often important to be able to measure the acidic or basic properties of substances the body is exposed to or of conditions in various parts of the body. A convenient system for measuring and stating the relative strength or weakness of acidic and basic (alkaline) solutions is called the **pH scale.** This scale runs from 0 to 14. The middle of the scale (pH 7) is the neutral point; a solution with a pH of 7 is neither acidic nor basic. A solution with a pH of less than 7 is said to be acidic; it has an excess of H^+ ions. The stronger the acid, the lower the pH (Figure 2-8). Thus, gastric or stomach juice with a pH of about 1.5 is a much stronger acid than milk with a pH of 6.5.

A solution with a pH of more than 7 is by definition a base with an excess of OH^- ions. Once again, the stronger the base, the greater the pH deviation from the neutral reading of 7. Blood plasma (pH 7.4) and lye (pH 13.5) are both basic, but lye is a much stronger base.

1. What is the difference between an acid and a base?
2. Describe the pH scale. Where on the pH scale would you most likely find a strong acid, a weak acid, a neutral solution, a weak base, and a strong base?

ORGANIC CHEMISTRY

The chemistry of living (organic) systems centers, for the most part, around compounds of carbon, which we call **organic compounds.** The study of these carbon compounds is called **organic chemistry.** Because of our interest in life, organic chemistry is one of the most active areas of study in all of

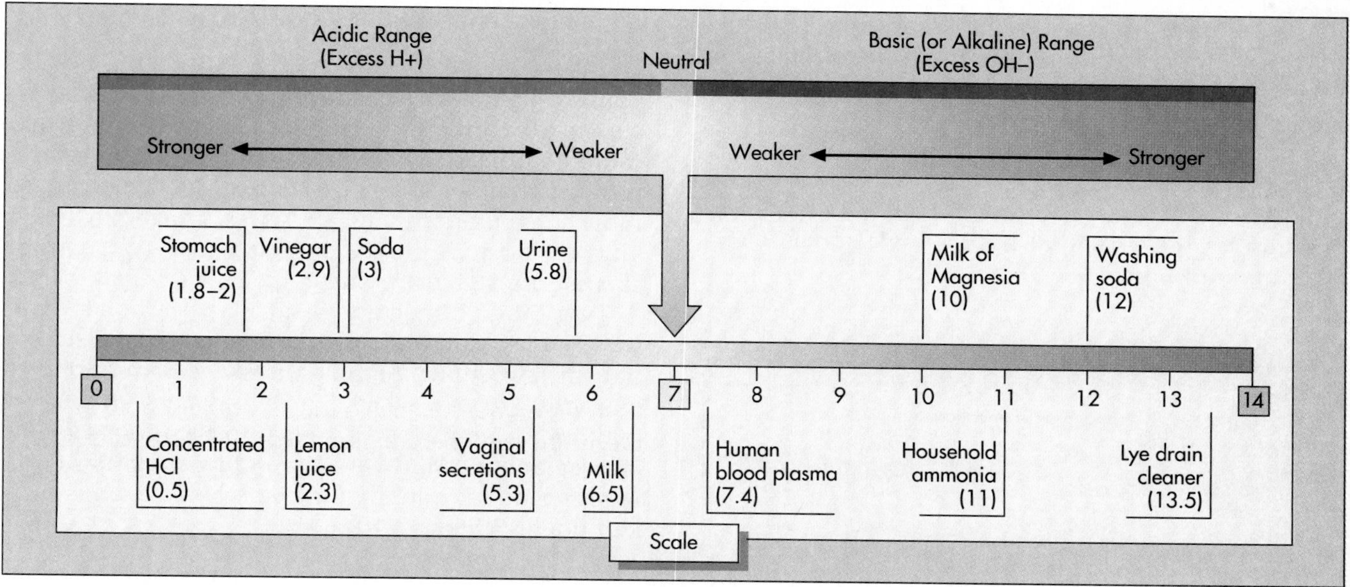

FIGURE 2-8 The pH scale.

science; so far, over 2 million different organic compounds have been identified; in contrast, chemists have formally identified less than 500,000 non-carbon-containing chemicals.

In the chemistry of living systems, there are four dominant organic families: the carbohydrates, the proteins, the lipids, and the nucleotides. We will discuss each of these families in what follows.

Carbohydrates: Sugars and Starches

Carbohydrates are found throughout the living world. Sugars are one of the main sources of energy for cellular activities, just as petroleum products are the main source of energy for automobiles. Some complex carbohydrates, the starches, serve to store energy, especially in plants. Others, such as cellulose, act as important structural materials for plants and contribute roughage or fiber to our diets.

Carbohydrates contain three different elements—carbon, hydrogen, and oxygen—in the approximate ratio of 1:2:1. Unfortunately, when carbohydrates were first named as a group, the chemists were misled by this ratio into thinking that sugar molecules were chains of carbon atoms that had water molecules attached (hydrated) to each carbon. Although they later found that this was simply not the case, by then the name for the group had been firmly established.

Complex carbohydrates are built up as chains of repeating units (or **monomers**). The units are the simpler carbohydrates—the sugars, or **saccharides.** Based on their complexity, carbohydrates are grouped into three major divisions—monosaccharides, disaccharides, and polysaccharides.

Monosaccharides: The Single, or Simple, Sugars

The simplest carbohydrates are the **monosaccharides.** The most abundant and important of the monosaccharides is **glucose** ($C_6H_{12}O_6$), which is not only the main chemical fuel for the human body, but also a chemical connecting link between the biosphere and the solar energy that powers it (Figure 2-9). Glucose is formed when plant photosyn-

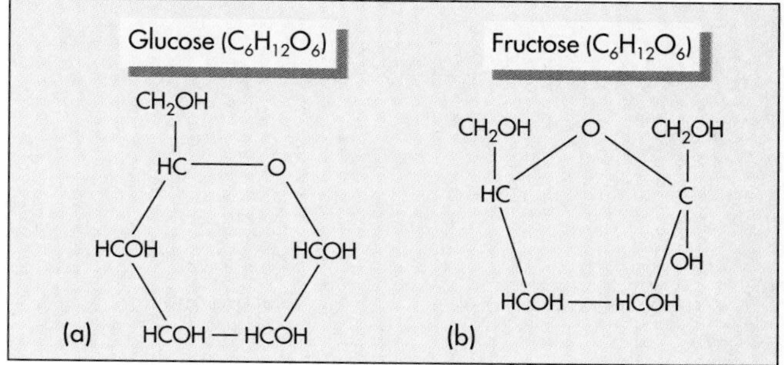

FIGURE 2-9 The chemical structure of the monosaccharides—(a) glucose and (b) fructose. Note that the formula for both glucose and fructose is $C_6H_{12}O_6$ but that the structural arrangements of the components differ markedly, giving each sugar its own unique characteristics. Glucose is used as a monomer for constructing many of the body's complex carbohydrates, the polysaccharides. Fructose is a powerful sweetener found in fruit.

thesis captures the energy of sunlight. It thus makes solar energy available to all life. Glucose is also called **dextrose,** and it is often so identified on food packages.

Another important monosaccharide is **fructose,** or fruit sugar. Like glucose, it also has the formula $C_6H_{12}O_6$, but it has a different internal structure. Fructose is the principal natural sweetener found in fruits and honey.

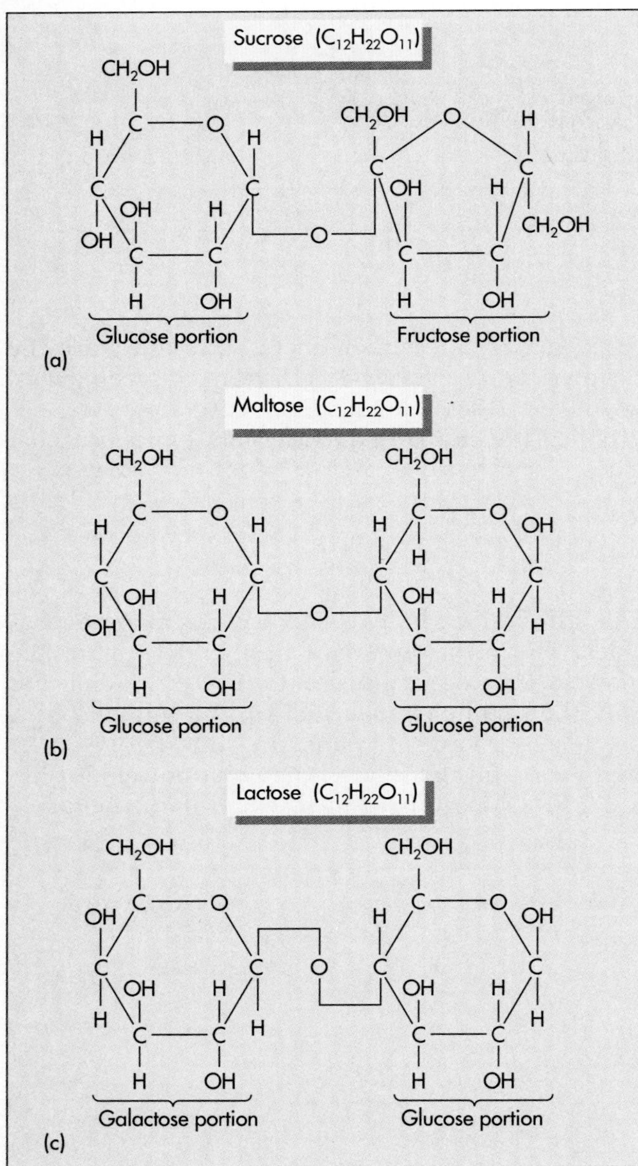

FIGURE 2-10 The disaccharides, or double sugars. One important disaccharide is (a) sucrose, a double sugar formed by chemically linking a glucose molecule to a fructose molecule. Other important disaccharides include (b) maltose (malt sugar), which consists of two linked glucoses, and (c) lactose (milk sugar), which is formed when a glucose monomer is connected to a galactose.

Disaccharides: The Double Sugars

The next category consists of the **disaccharides,** or double sugars. The disaccharides are produced by chemically linking two monosaccharides to form a more complex carbohydrate. One of the most important double sugars is **sucrose** (also known as table sugar). Sucrose ($C_{12}H_{22}O_{11}$) is formed when a glucose sugar is linked to a fructose. We extract sucrose mostly from sugar cane and sugar beets.

Two other important disaccharides are lactose (milk sugar) and maltose (malt sugar). Lactose, the natural sweetener found in milk, is commercially extracted from "whey," a by-product of cheese making. In it, glucose is linked to the monosaccharide galactose. Maltose is two glucose units linked together and is usually formed when starch is broken down (Figure 2-10).

Polysaccharides: The Complex Carbohydrates

The most complex carbohydrates, the **polysaccharides,** are usually constructed by chaining many monosaccharide units together. Each repeating chemical building block is a **monomer,** and a chain of monomers is a **polymer.** Polysaccharides often serve as reservoirs of raw energy, in the form of glucose chains, that can be harvested chemically by releasing one glucose unit after another from the chain. Plant **starch** is the sugar storage molecule found in plants. Animal starch, or **glycogen,** serves the same purpose in animals (Figure 2-11).

Unlike plants, animals rarely store much reserve energy in the form of glycogen. In fact, such energy stores are usually restricted to high-performance tissues that are likely to run out of glucose during vigorous activity; glycogen stores are found in cells of skeletal muscles and in tissues of the liver. However, total storage accounts for only a few hours' reserve (Figure 2-12). Most long-term energy storage in animals occurs in the more concentrated form of fat, which we will discuss later.

Cellulose is another important polysaccharide. It too consists of long chains of monosaccharide (glucose) units. However, the chemical linkages binding these glucose monomers cannot be broken by any of the digestive enzymes possessed by animals. Cellulose is thus one of the most durable chemicals in the biosphere; it may also be the most abundant carbohydrate in nature. It is used as a structural material for plants. It forms the cell walls surrounding most plant cells, is the principal constituent of wood, and forms the indigestible, but *useful,* roughage or fiber found in certain foods (Figure 2-13, p. 38).

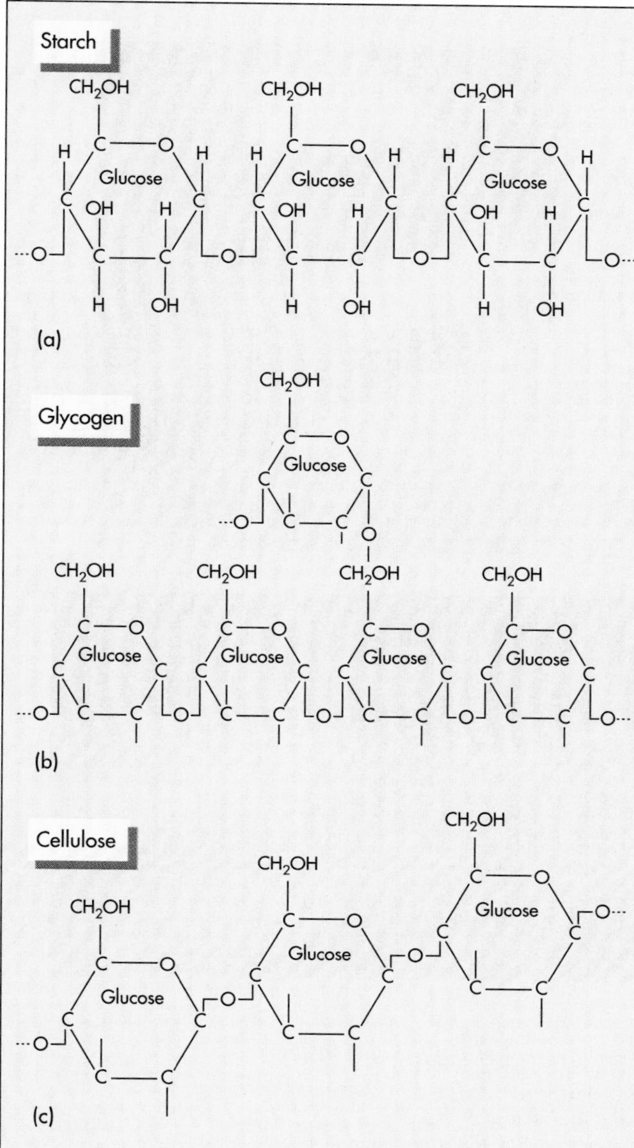

FIGURE 2-11 The three polysaccharides most important to humans are (a) starch (an energy storage component of plants), (b) glycogen (sometimes referred to as animal starch), and (c) cellulose (an important structural component of plants). All three polysaccharides are polymers formed by linking long chains of glucose monomer units together. The differences among them arise because in each a different type of chemical linkage chains the glucose monomers together.

1. List the principal functions for each of the major categories of carbohydrates.
2. Discuss why glucose is so important in the affairs of the biosphere and why it is involved in so many aspects of carbohydrate metabolism.

FIGURE 2-12 Glycogen (animal starch) stores a limited amount of glucose for use by the skeletal muscles and liver. These active organs can draw the glucose out of the glycogen reserves during periods of intense activity.

Proteins: Structural Materials and Catalysts

The human body is largely built on a **protein** framework. The glues that hold your cells together, the network of fibers that underlie your skeleton, the membranes that line your body cavities and compartmentalize your organ systems and cells, the muscle fibers that make skeletal movement possible, and many more bodily elements, including hair, are largely **structural proteins.** A second important category of proteins, the **enzymes,** are the organic **catalysts** that speed up the thousands of chemical reactions in the body under conditions that are compatible with life.

Structure of Proteins

Proteins, like polysaccharides, are complex molecules constructed of chemical building blocks. The blocks are not sugars, however. They are **amino acids** (Figure 2-14, p. 39), and they are chained together through specific chemical linkages called **peptide bonds** (Figure 2-15). Just as we use 26 letters in our alphabet to produce an infinite variety

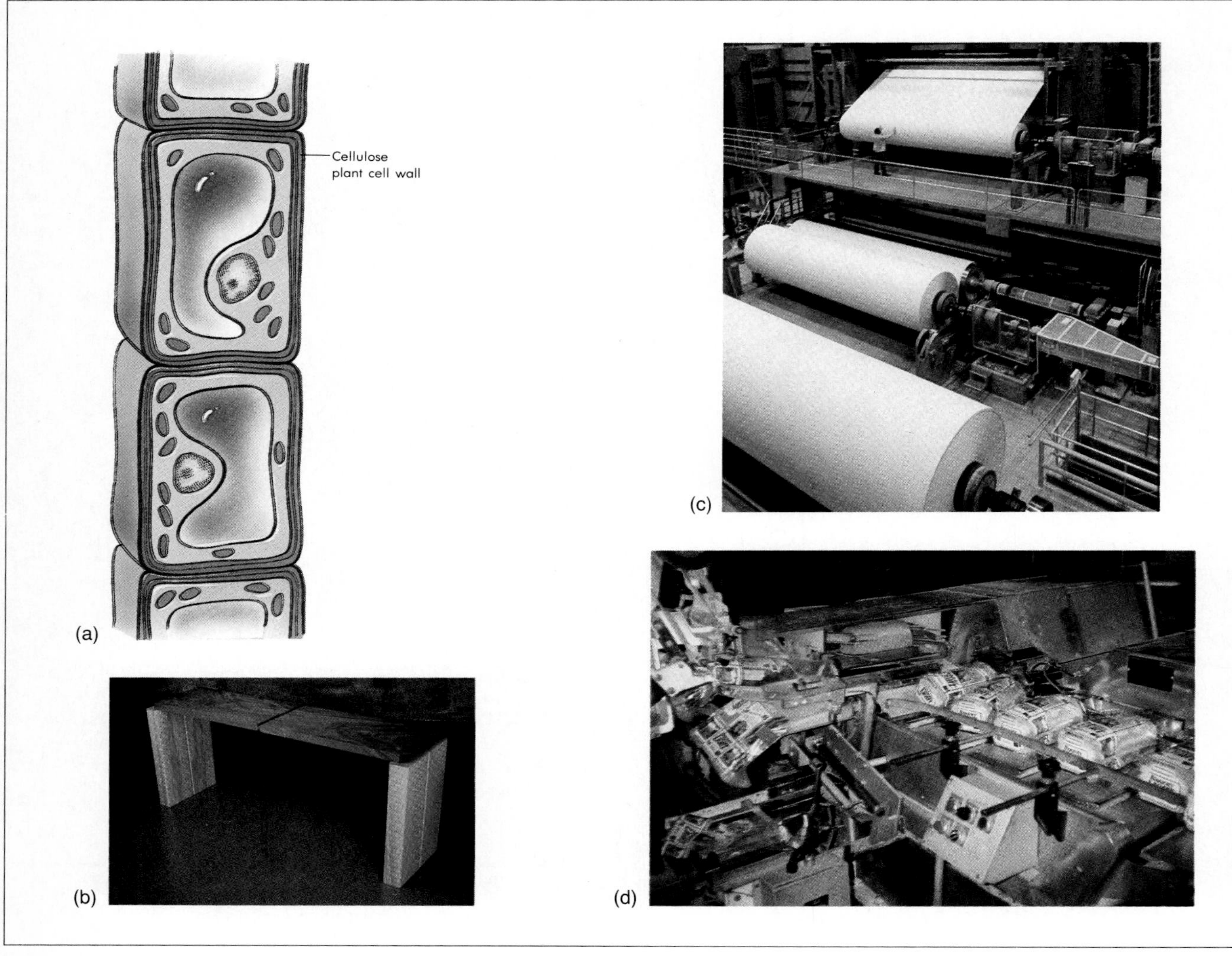

FIGURE 2-13 Some uses of cellulose. (a) Cellulose plant walls strengthen plants and support them against gravity. (b) Woody tissues are thus mostly cellulose. A beautifully sculpted piece of hand-made furniture is made mostly of the cell walls from woody plants. (c) Paper is another cellulose product made from wood. (d) Some types of diet bread work by substituting purified cellulose, obtained from wood pulp, for flour, thus reducing the digestible calories and increasing the amount of roughage in the diet.

of words, human proteins employ a total of just 20 different amino acids in sequences that give the thousands of human proteins their unique properties and shapes.

Chains of up to one hundred amino acids linked by peptide bonds are often referred to as **polypeptides.** Polypeptides with over a hundred amino acids are usually called proteins, although some "proteins"—such as insulin, with 51 amino acids—contain fewer amino acids. It is not uncommon for a specific protein to have hundreds of amino acids in a single molecule. Because of their enormous size, you might suspect proteins to be rangy, ill-defined molecular entities. However, most proteins have compact globular shapes in which the position of each amino acid is specifically defined.

The **primary structure** of a protein is simply the sequence of its amino acids. The **secondary structure** is the coiling of the amino acid chain into an "alpha (α) helix" or some other shape. The pattern of this coiling is determined by the individual shapes of the specific amino acids occupying key positions along the chain.

The **tertiary structure** adds a dimension to the complex folding and coiling of protein strands. It develops when different portions of a protein molecule interact with each other, often as hydrogen bonds form between amino acids located in differ-

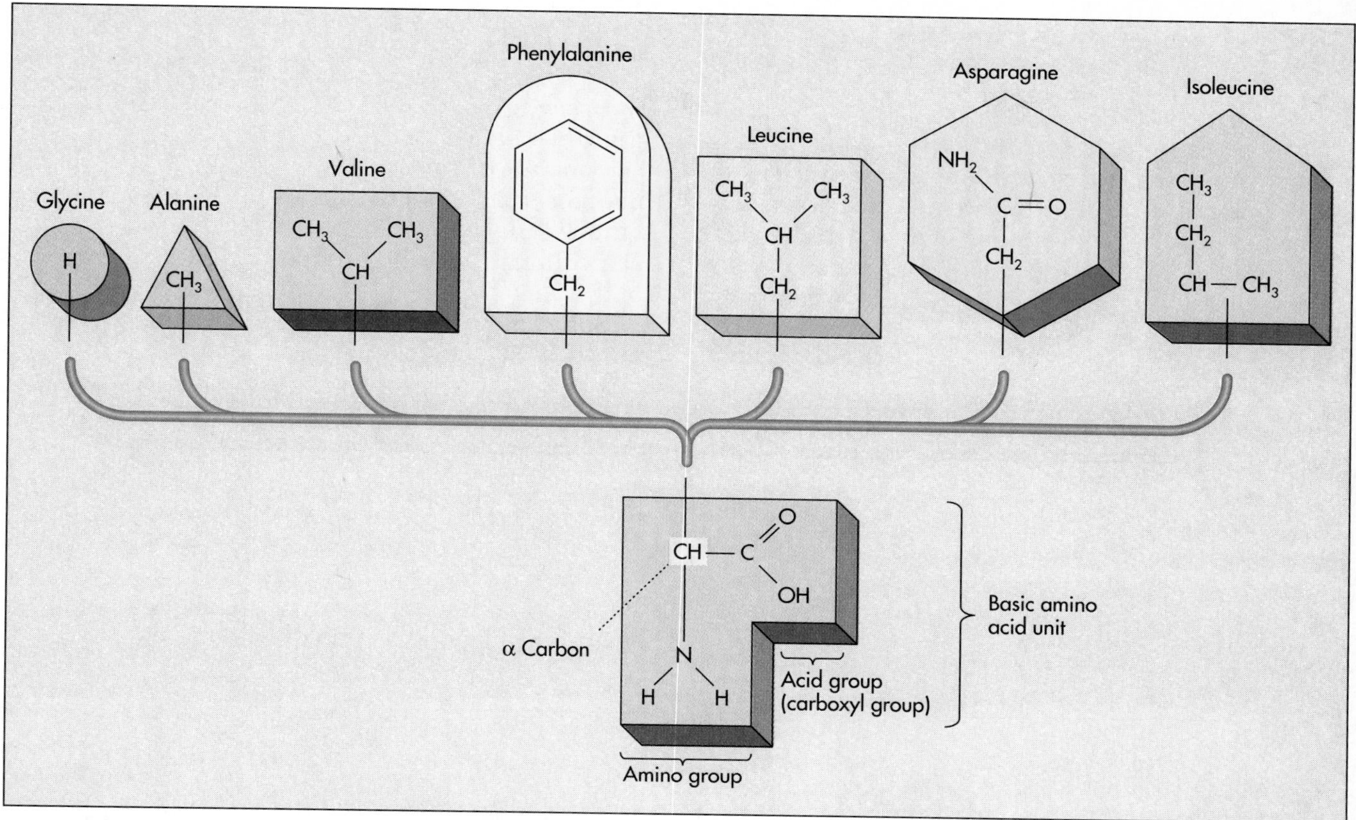

FIGURE 2-14 All amino acids share a two-carbon core. The first carbon bears a carboxyl (organic acid) group. The second carries an amino group (a nitrogen attached to two hydrogens) and is attached to one of the 20 different side chains (called R groups) seen in human proteins.

ent parts of the chain (Figure 2-16, p. 41). In some cases, sulfur-containing amino acids form covalent disulfide bonds. The **quaternary structure** appears when proteins combine with other proteins or other substances, such as vitamins and minerals.

Collagen: The Most Abundant Structural Protein

Perhaps the most abundant of all proteins in the human body is a substance called **collagen.** Some estimates suggest that fully one-third of the protein in the human body consists of collagen (Figure 2-17, p. 41). It is a key constituent of the skin, tendons (the connective tissue straps that tie muscles to bones), ligaments (the tissues that bind the joints of the skeleton together), and cartilage (the flexible supportive tissue of the ear, nose, and throat and the tissue that reduces friction in the joints). It also forms much of the cornea, the transparent outer window allowing light to enter the front of the eye. The individual fibers of collagen have the tensile strength of steel while providing the flexibility and toughness of leather. The body employs collagen in many diverse ways.

Enzymes: The Organic Catalysts

Life is a chemical concert made up of thousands of different chemical instruments, or reactions. These reactions must all "play" at the proper rates and under conditions that are compatible with life. If they play too slowly, death soon puts an end to the concert. One crucial task of life is thus to keep reaction rates fast.

A chemical reaction usually occurs when reacting molecules collide. The energy behind many of these reaction-causing collisions is provided by the amount of heat in the system. The higher the temperature, the faster the molecular movements in the system and the greater the number of collisions. In the lab, chemical reactions are often encouraged or sped up by heating a reaction mixture. However, there are limits to how hot you can make protoplasm (the living substance). Cells thus use another way to speed up reactions—the enzymes.

A Model of Enzyme Function

Molecules have definite shapes and sometimes must be fitted to each other like puzzle pieces before a reaction can occur. Collisions must thus occur with just the right molecular orientations to let the reactive portions of the molecules come together in just the right way (Figure 2-18, p. 42). Enzymes, acting as protein catalysts, speed up virtually all the reactions of the cells by organizing the reactants (called the **substrates**) so that the reaction is likely to take place at the relatively low temperatures found in living systems.

Most enzymes are globular proteins with niches or depressions (called active sites) tailored to fit parts of the specific substrate molecules they help react. According to one model of enzyme function (loosely referred to as the lock-and-key model), an **enzyme-substrate complex** forms when the substrate molecules collide with the enzymes and are drawn into the fitted niches on the enzyme's surface. The substrate molecules are then so positioned that their reactive portions are near each other and the reaction can occur efficiently. Once the reaction occurs, the internal rearrangements that take place as the product is formed cause the enzyme to release the product and be available for another reaction. This whole process occurs so quickly that some individual enzyme molecules can catalyze literally millions of reactions per second.

Some enzymes combine pairs of substrate molecules into larger molecules. Other enzymes split single molecules into two or more products. In general, each type of reaction has its own specific enzyme, shaped to encourage that particular reaction. Cells thus have thousands of different enzymes, one type for each of its thousands of reactions.

Molecular shape and fit are crucial (Figure 2-19, p. 43). Anything that causes an enzyme to change its shape is likely to prevent or inhibit the reaction

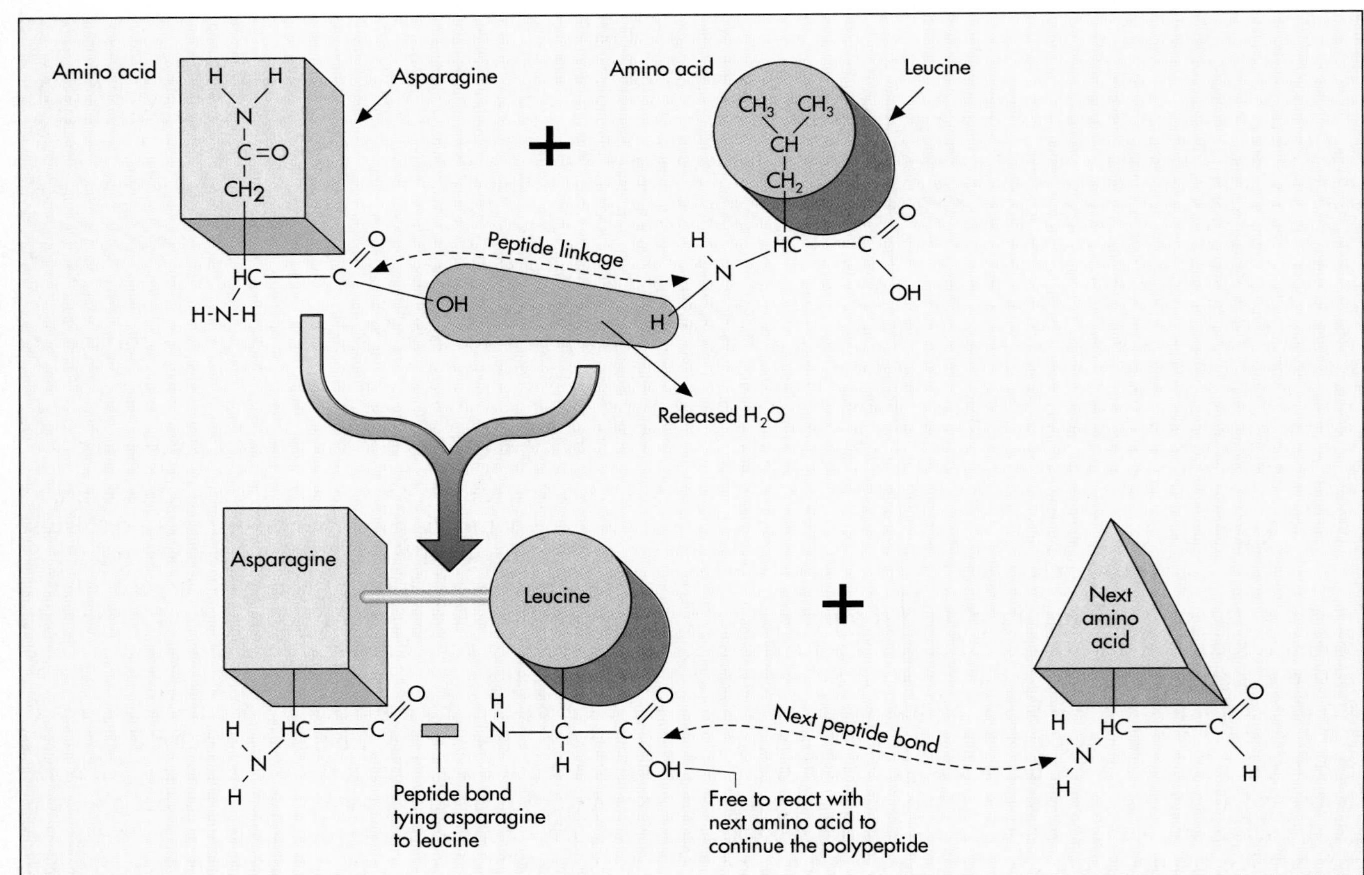

FIGURE 2-15 (a) Amino acids are linked in proteins by a peptide bond when the carboxyl group of one amino acid (here, asparagine) reacts with the amino group on the next amino acid (here, leucine). In the process an OH is removed from the carboxyl group and an H from the amino group (forming water). (b) The peptide bond connects what is left of the carboxyl with what remains of the amino group. The shorthand notation for a peptide bond is —CONH—.

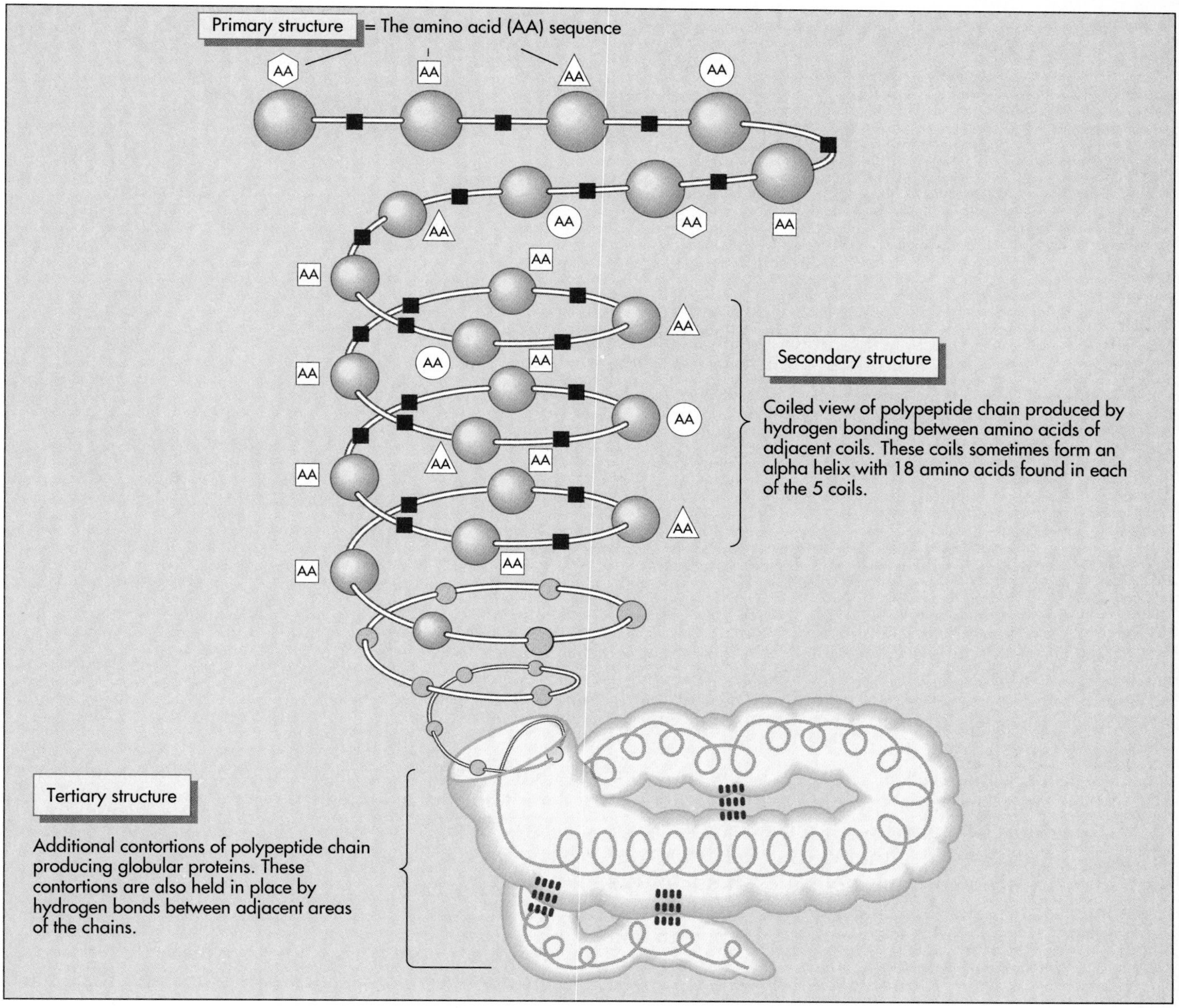

Primary structure = The amino acid (AA) sequence

Secondary structure

Coiled view of polypeptide chain produced by hydrogen bonding between amino acids of adjacent coils. These coils sometimes form an alpha helix with 18 amino acids found in each of the 5 coils.

Tertiary structure

Additional contortions of polypeptide chain producing globular proteins. These contortions are also held in place by hydrogen bonds between adjacent areas of the chains.

FIGURE 2-16 Proteins have three levels of structure: primary, the sequence of amino acids; secondary, the coiling and kinking of the amino acid chain; and tertiary, the tying together of different parts of the molecule.

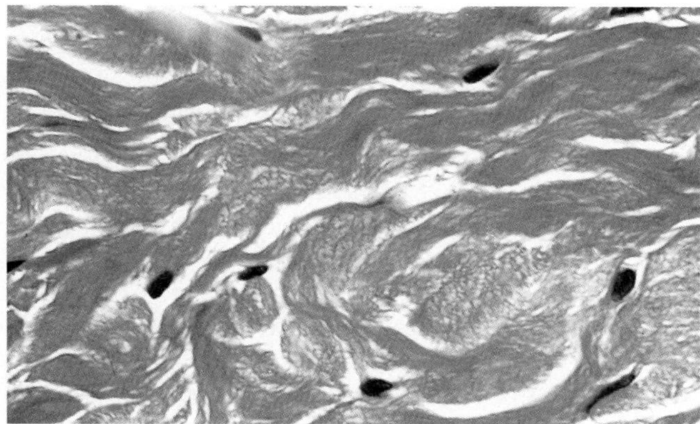

FIGURE 2-17 Collagen fibers are abundant in human connective tissues.

that enzyme catalyzes. When such a protein irreversibly changes its shape, it is said to have been **denatured.** The numerous physical and chemical conditions that can cause enzymes to denature include high temperatures, exposure to heavy metals such as mercury and silver, and changes in the pH of the solutions containing the enzymes.

Naming of Enzymes

By convention, each enzyme is named by adding the suffix *-ase* to the name of its substrate or the type of reaction it catalyzes. Thus, the enzyme catalyzing the breakdown of sucrose (table sugar) is *sucrase.* Lactase breaks down milk sugar, or lac-

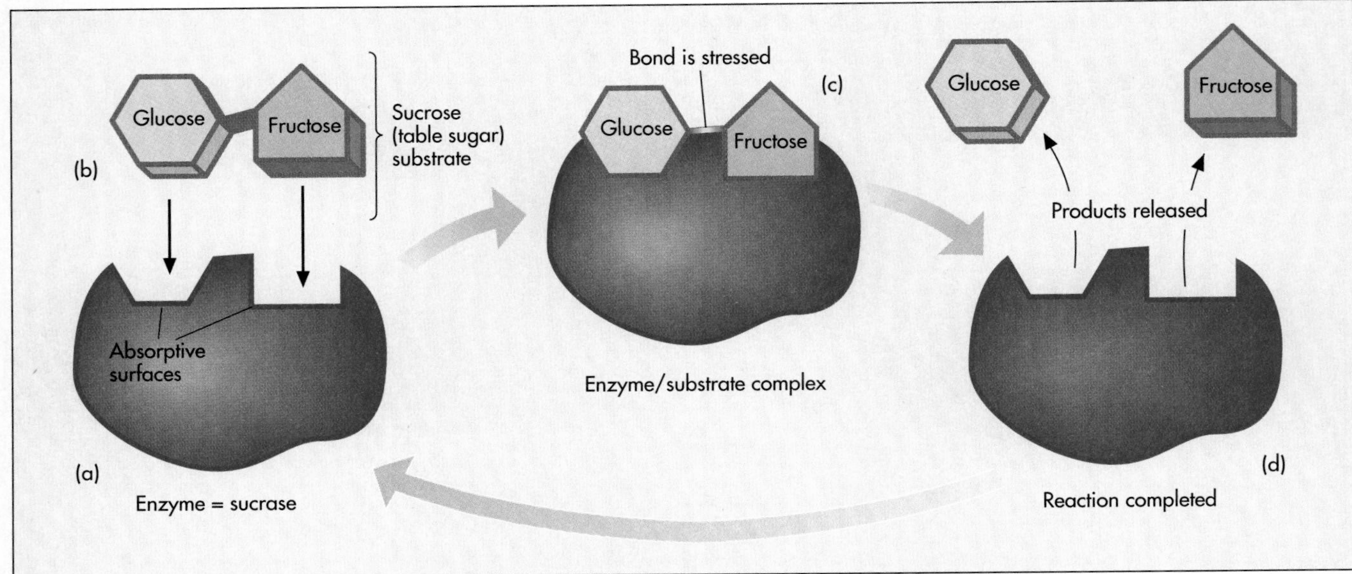

FIGURE 2-18 Enzymes are protein catalysts that speed chemical reactions in cells. They organize the reactions by providing surfaces on which they can occur in an orderly fashion. (a) Enzymes are usually globular proteins with molecular niches on their surfaces; the niches fit the shapes of their specific substrate molecules. (b) The substrate is drawn into an enzyme's niche to form an enzyme-substrate complex. The fit of the substrate to the enzyme used to be compared to the fit of a key in a lock, but recent evidence suggests the enzyme's shape is not rigid but flexible and is actually molded to some extent by the substrate itself. (c) The chemical bond is broken (or formed—depending on the enzyme) and the products are released. (d) The original enzyme is now free to react with new substrate molecules.

tose. Proteases, peptidases, lipases, nucleases, and amylases split proteins, peptides, lipids, nucleic acids, and starches (amyloses), respectively. Dehydrogenases are enzymes that catalyze the removal of hydrogens from certain molecules. Some of the enzymes that were named before the *-ase* suffix was adopted (such as the digestive enzymes pepsin and trypsin) have an *-in* suffix.

1. Distinguish between enzymes and structural proteins in their biological roles.
2. Explain the lock-and-key model of enzyme function. Why is protein shape so important? What is a substrate?

Lipids: Fats, Oils, and Steroids

The **lipids** are a diverse group of energy-rich organic compounds whose main common feature is that they do not dissolve in water. They do, however, dissolve in various organic solvents such as ether, chloroform, and benzene. They include the fats, oils, waxes, and steroids. Their structures vary widely, and they are employed as energy storage molecules (especially in animals), components of cell membranes, insulators for the nervous system, and hormones—just to mention a few. We will focus next on two of the major lipid categories.

Fats and Oils

Pound for pound, fats can store more than twice the energy of carbohydrates. They thus provide a much more economical way for animals, which must be able to move about rapidly, to carry their long-term energy reserves. Stored fat can sustain some people for long periods. For instance, hunger strikers (starting with average body weights) have survived without eating for six to eight weeks by drawing on their fat. Some overweight individuals have been known to fast and survive for over a year by drawing on their more ample fat reserves. Of course, before fat can be used for energy, the body's cells must convert it to a chemical form that can be fed into the enzymatic machinery of cellular respiration (see Chapter 20).

Fats also cushion the internal organs and form an insulating layer beneath the skin that protects the body from excess heat loss. Their distribution under the skin also has sexual implications, for males and females tend to store their fat reserves in different patterns.

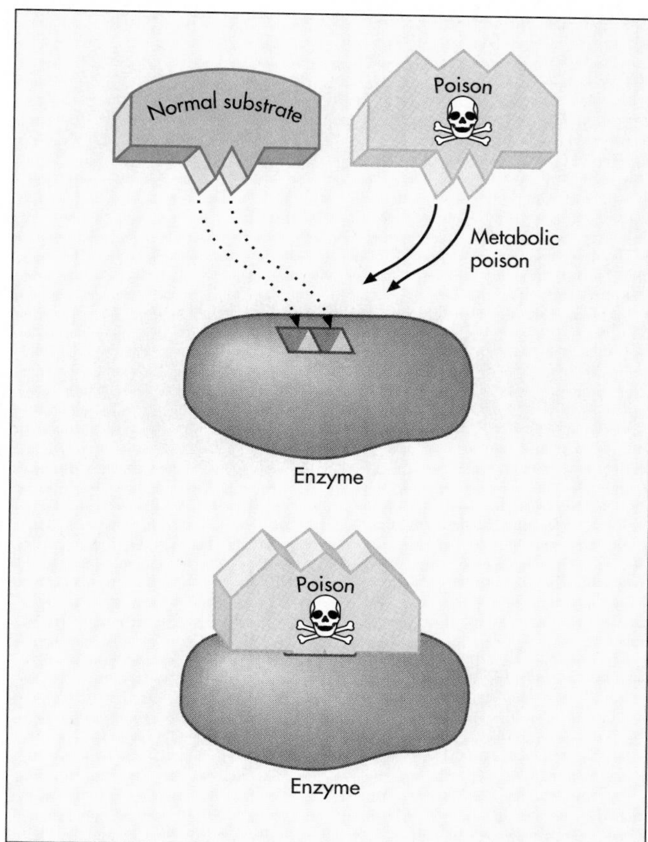

FIGURE 2-19 Competitive inhibition. Some metabolic poisons work by competing with substrate molecules for the active sites on enzyme surfaces. In fact, they may actually fit the active sites on the enzymes better than the normal substrate molecules. By occupying these active sites, they prevent the needed reactions from progressing, sometimes causing death (nerve gas works in this way).

Structure of Fats

Each fat molecule has four chemical components. It is formed when a three-carbon alcohol, called **glycerol,** combines with *three* long-chain molecules called **fatty acids.** Thus

$$\text{Fat} = \text{Glycerol} + 3 \text{ fatty acids}$$

In **saturated fats** (Figure 2-20), the carbons in the fatty acids use two of their four bonds to connect to neighboring carbon atoms. The other two link to hydrogens. **Unsaturated fats** have pairs of neighboring carbons that each link to only one hydrogen, not two. The excess bonds then link the carbons in a special way known as a double bond. These bonds are unsaturated because they have the potential to contain more hydrogen. **Polyunsaturated fats** have many double bonds between their carbons. Unsaturated fats, especially those with shorter fatty acid chains, are usually found as vegetable oils. Most animal fats, including those in the human body, tend to be saturated; they also tend to have a more solid or semisolid consistency.

An important group of chemicals related to fats is the **phospholipids.** As important components of cell membranes, they help control the types of substances that enter the cell. Like fats, they are combinations of fatty acids with glycerol. However, phospholipids use only two fatty acids, substituting a phosphate group for the third (Figure 2-21).

Steroids

Steroids represent another important category of lipids found in the body. They form a crucial family of **hormones** that includes the male and female sex hormones and adrenal hormones, which help the body deal with stress. They also serve as components of cell membranes.

The steroids have a unique chemical configuration consisting of four linked "rings" of carbon atoms (Figure 2-22). The parent molecule for the entire steroid family, **cholesterol,** receives a great deal of attention because studies link high cholesterol levels in the body with an increased risk for heart attacks and strokes, caused when cholesterol accumulates in the linings of the arteries (a condition known as atherosclerosis). As a result, people are advised to avoid foods such as eggs and cream, which contain much cholesterol. They are also admonished to cut down on fatty foods, especially the animal foods, which are heavy in saturated fats, since saturated fats are very easily transformed into cholesterol.

However, many important steroids, such as testosterone, progesterone, and estrogen, are formed in the various tissues of the body using cholesterol as the raw material. Thus, while cholesterol is implicated as a problem causer in certain individuals, it is also an extremely important substance in the normal affairs of the human body. We will discuss the steroid hormones in some detail in Chapter 13.

---

1. What are some of the important functions of lipids in the living system?
2. What peculiar characteristic causes chemicals to be placed in the lipid category?
3. Describe the fats and oils and the steroids.

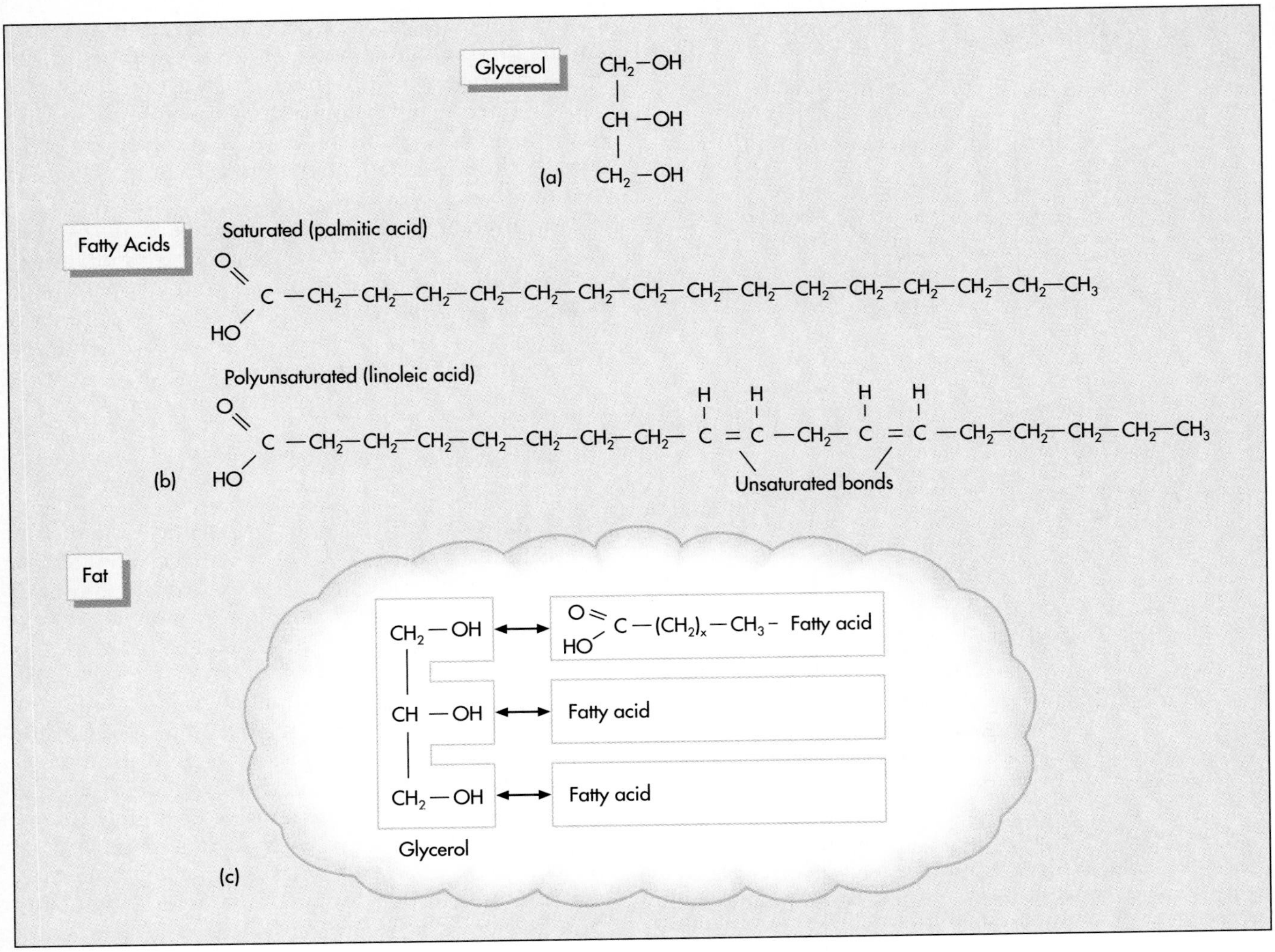

Glycerol

CH_2-OH
|
$CH-OH$
|
(a) CH_2-OH

Fatty Acids

Saturated (palmitic acid)

$HO(O=)C-CH_2-CH_2-CH_2-CH_2-CH_2-CH_2-CH_2-CH_2-CH_2-CH_2-CH_2-CH_2-CH_2-CH_2-CH_3$

Polyunsaturated (linoleic acid)

(b) $HO(O=)C-CH_2-CH_2-CH_2-CH_2-CH_2-CH_2-CH_2-CH=CH-CH_2-CH=CH-CH_2-CH_2-CH_2-CH_2-CH_3$

Unsaturated bonds

Fat

CH_2-OH ⟷ $HO(O=)C-(CH_2)_x-CH_3$ - Fatty acid

$CH-OH$ ⟷ Fatty acid

CH_2-OH ⟷ Fatty acid

Glycerol

(c)

FIGURE 2-20 There are four components to each fat molecule—one glycerol and three fatty acids. (a) Glycerol, a three-carbon alcohol, is bound to (b) the fatty acids. The different types of fats are produced by connecting various fatty acids to the glycerol. Note the presence of double bonds along the unsaturated fatty acid chain. (c) The link between the fatty acids and glycerol resembles that between amino acids in a protein, being formed by the removal of H and OH groups.

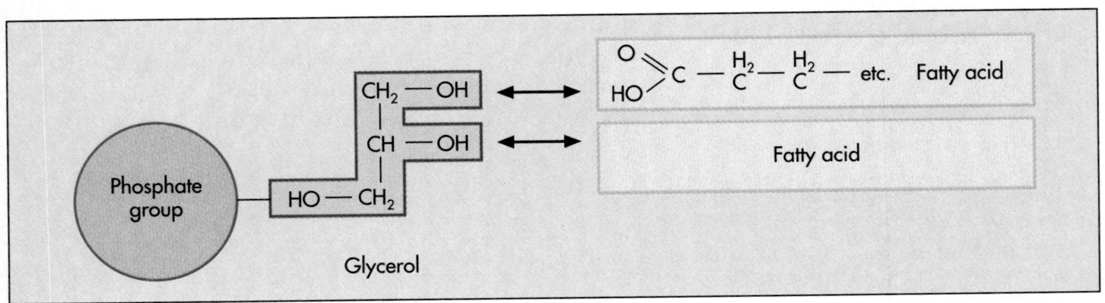

FIGURE 2-21 Phospholipids serve as components of cell membranes. Unlike fats, they use only two fatty acids; the third is replaced by a phosphate group. The phosphate ionizes in water, making that end of the molecule soluble in water, while the fatty acid portion remains insoluble.

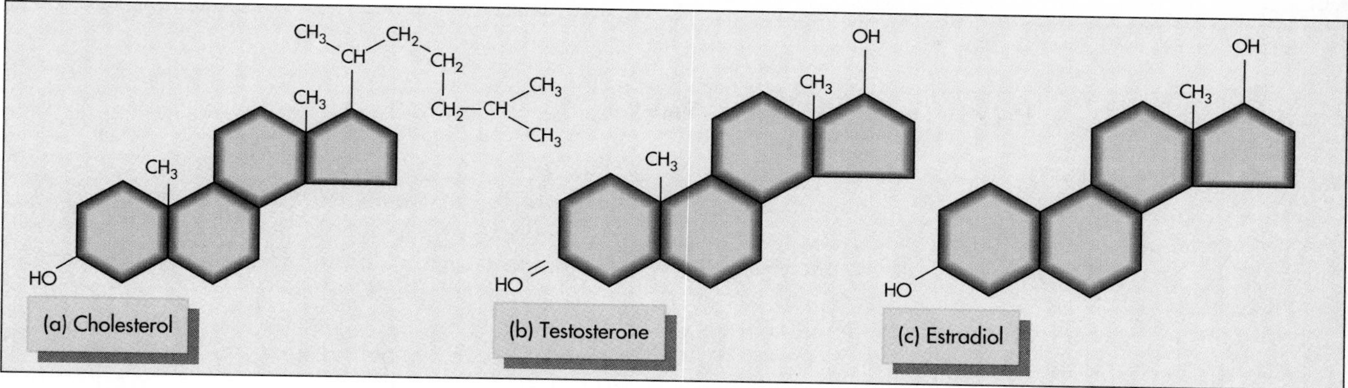

FIGURE 2-22 Steroids include some of the most important chemicals in the human body. The basic steroid configuration is four interconnected rings. The different steroids are formed by attaching different side chains to key locations on the rings. From (a) the parent molecule, cholesterol, are made the sex hormones (b) testosterone (the male sex hormone) and (c) estradiol (the female hormone).

THE NUCLEOTIDES

The **nucleotides** are the chemical building blocks of the genetic material and the energy-rich ATP molecules of cells. A nucleotide has three linked chemical components: a nitrogenous organic base, a five-carbon sugar, and an inorganic phosphate group (Figure 2-23, p. 47).

Cells have several uses for nucleotides. Hundreds and sometimes thousands of nucleotide units may be chained together to form the **nucleic acids,** the most important of which is **deoxyribonucleic acid (DNA),** the substance of the hereditary instructions (genes) that govern cellular activities. A second nucleic acid is **ribonucleic acid (RNA),** which helps synthesize proteins as instructed by the genetic material. These nucleic acids will be covered in some detail in Chapter 4.

One nucleotide is also the basic structural unit of the cell's chief source of usable energy, **adenosine triphosphate (ATP)** (Figure 2-24, p. 48). The energy the body needs for its activities arrives as raw fuels (sugars) with the food. However, food energy cannot be used directly by cells. They must release the energy in a series of reactions known collectively as cellular respiration and then transfer it for storage to the chemical bonds that link two additional phosphates to the nucleotide adenosine monophosphate (AMP). The resulting energy-rich ATP is then employed whenever the cell needs usable energy to perform its chemical tasks. This reaction involves the removal of one or both of the additional phosphates attached to AMP and the release of the energy stored in them. Usually, cells extract energy from ATP by converting it to adenosine diphosphate (ADP) and store energy by converting ADP to ATP, as shown in Figure 2-25 (p. 48).

1. Why are nucleotides important?
2. List three very important nucleotides found in cells.

ENERGY AND LIFE

As in economics and society, "there are no free rides" in living systems. Virtually all life's chemical activities require energy. Just as your automobile requires chemical energy in the form of gasoline to move from one place to another, so too does your body require its own brand of fuel (sugars) to carry on its activities.

We define **energy** as the capacity to do work when we talk about how cells extract it from sugars and other organic materials. We define the units of energy **(calories)** when we talk about the body's use of energy (Chapter 21). For now, let us say a few words about those laws of physics that govern all uses of energy. These are the laws of thermodynamics.

THE LAWS OF THERMODYNAMICS

The three laws of thermodynamics applicable to biology summarize much of what we know about energy relationships and the relationships between

Box 2-2 Isomers: The Chemistry of the Left and Right Glove

Think about a pair of gloves. The left glove and right glove of a pair look basically very similar. They are usually constructed of the same materials—be it calfskin or wool. However, the left and right glove differ from each other in subtle ways; they are constructed as mirror images of each other so that the left hand only fits into the left glove (unless you are inclined to dress in a funny way).

Similar situations exist in organic chemistry. Any time a carbon atom's chemical bonds link to four different chemical groups, there can be two versions, or **isomers,** of the molecule. Because the carbon atom's four bonds point to the four corners of a tetrahedron, any molecule has a definite three-dimensional shape. If a molecule has different corners, then any two corners can be interchanged to produce two structures that cannot be made identical simply by rotating the molecule. One of the structures is left-handed and one is right-handed (Figure 1). If two isomers are simply mirror images of

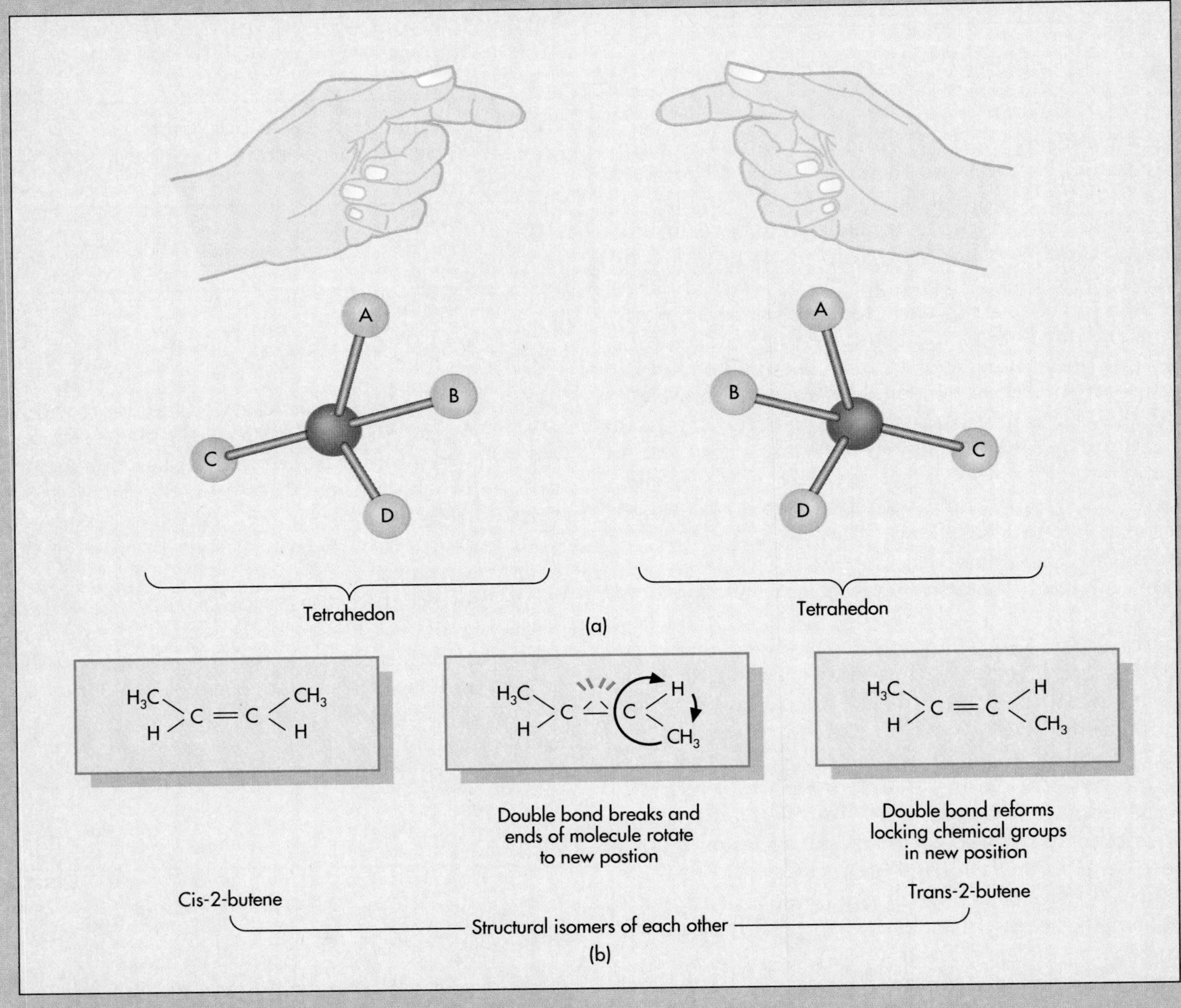

FIGURE 1 (a) Optical isomerism. (b) Structural isomerism.

each other, they are said to be *optical isomers;* their main biological difference is the way each rotates polarized light. However, some isomers differ structurally in other ways. Parts of organic molecules can freely rotate around single (saturated) carbon-carbon bonds. Double (unsaturated) bonds do not allow free rotation of the molecular parts. Thus, it is possible to construct two versions of many molecules that have double bonds.

Both *structural isomerism* and *optical isomerism* can lead to important chemical and physical differences between the isomers. This can be especially important in medications, where one isomer has very beneficial effects but the other has virtually no effect. In fact, drug manufacturers have found that one form of a drug can be over 300 times as biologically active as its isomer even though the two have the same basic formula.

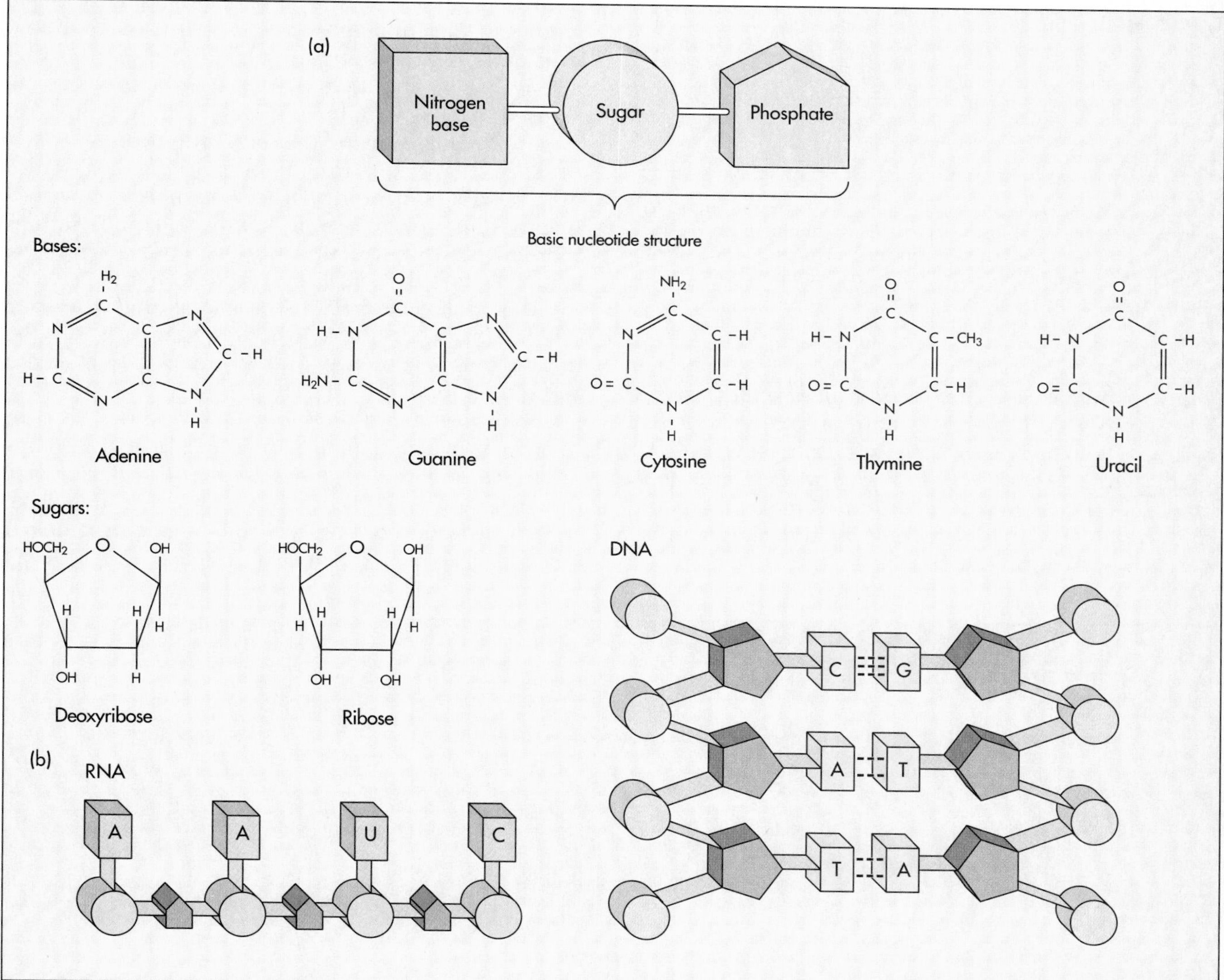

FIGURE 2-23 (a) Nucleotides have three chemical subunits. In each one, a five-carbon sugar (either ribose or deoxyribose) is connected to an inorganic phosphate group on one end and to a nitrogenous base (adenine, thymine, uracil, cytosine, or guanine) on the other. (b) When nucleotides are chained together, they form the nucleic acids (DNA and RNA).

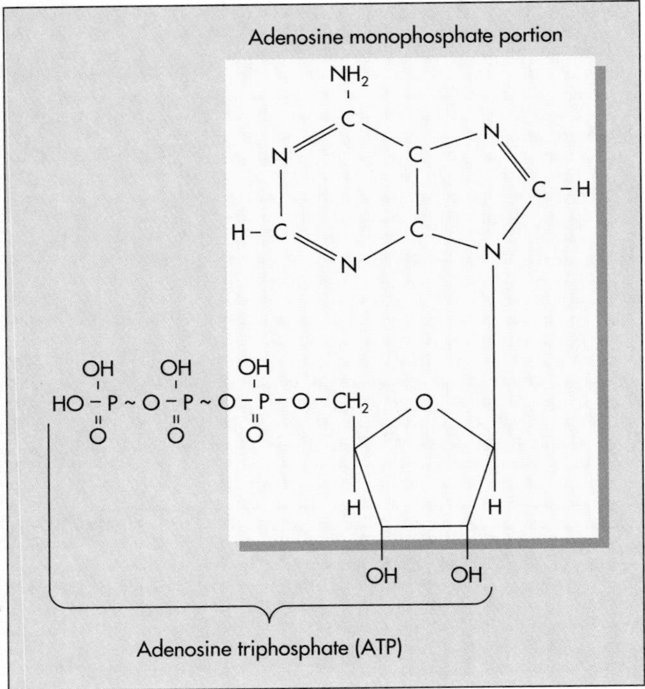

FIGURE 2-24 Adenosine triphosphate (ATP), the molecule used to carry energy within cells, is a specially modified nucleotide. It is the ordinary nucleotide adenosine monophosphate (AMP) with the addition of two extra phosphates by bonds that store unusually large amounts of energy.

matter and energy. The *first law of thermodynamics* says simply that energy is neither created nor destroyed. For that reason, it is sometimes referred to as the *law of conservation of energy.* It means that the amount of energy in the universe is constant and that, as Einstein said, energy and matter must be convertible into each other.

The *second law of thermodynamics* says that whenever energy is converted from one form to another (as when electricity is converted to light in a light bulb), some is lost or wasted, generally as heat (the light bulb gets hot). And if you are wondering how we can say that heat is "wasted" energy, consider that heat is useful only when there is somewhere cooler for it to go. Only then can it be used to warm a house or, indeed, to generate some other form of energy (like that needed to run an engine). Heat is, in a sense, the bottom of the energy hill.

Like everything else that uses energy, your body wastes heat too. A fraction of the energy processed by your body during its normal activities is converted to body heat. This process becomes very clear when you speed up the chemical activities of your muscles during exercise. You usually unload this extra heat by activating cooling responses such as sweating. The heat energy is an actual loss because it is not available to help you move your skeleton, which was your primary intention.

All energy-requiring systems, living and nonliving, need a continuous supply of energy to replace what is lost to the second law of thermodynamics. For our biosphere, that supply comes from our nearest star, the sun. The radiant energy released by the sun is trapped in the sugar molecules produced by the photosynthetic organisms on this planet and becomes available to the rest of the biosphere as food.

The *third law of thermodynamics* states that **entropy** (disorder) naturally increases as time passes. Thus, complex or organized systems will degenerate to simpler or less organized states in time. Likewise, systems containing high concentrations of energy can be expected to release that energy,

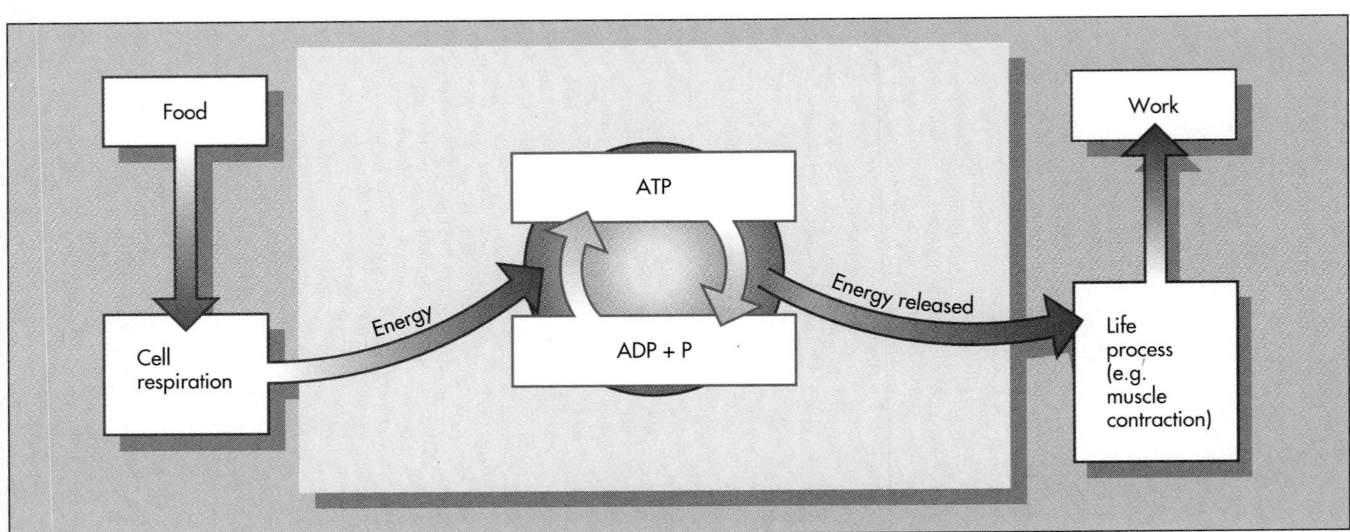

FIGURE 2-25 The ATP-ADP reaction.

Learning Focus

ORGANIC CHEMISTRY

Extract the information from the text dealing with the four categories of organic substances and summarize it in the spaces provided in the following chart.

Summarize Information about Carbohydrates

STRUCTURE	NAME EXAMPLES	FORMULA	COMMENTS
Monosaccharides	(1)		
	(2)		
	(3)		
Disaccharides	(1)		
	(2)		
	(3)		
Polysaccharides	(1)		
	(2)		
	(3)		

Summarize Information about Proteins

STRUCTURE	NAME EXAMPLES	COMMENTS
Amino acids	(1)	
	(2)	
Peptide bonds		
Primary structure		
Secondary structure		
Tertiary structure		
Enzymes		

Diagram and explain enzyme function.

What is competitive inhibition?

Give at least three examples of structural proteins.

(1)

(2)

(3)

Additional examples:

Summarize Information about Lipids

STRUCTURE	NAME EXAMPLES	STRUCTURE	COMMENTS
Glycerol			
Fatty acids	(1) Saturated		
	(2) Unsaturated		
Fats			
Steroids	(1)		
	(2)		
	(3)		

Summarize Information about Nucleotides

STRUCTURE	NAME EXAMPLES

Identify the three basic components.

(1)

(2)

(3)

Nucleic acids: Identify the two major types and describe the functions of each.

Adenosine triphosphate (ATP): What is its role in the cell?

Diagram its structure. Underline the part of the molecule that indicates its nucleotide nature.

eventually degrading it to its least ordered form, heat. To many people, life seems to go against this law, for when a baby or a tree grows, it is clearly becoming *more* ordered in form and energy content. However, life does not violate the third law of thermodynamics; it grows at the expense of energy—and order—obtained from elsewhere. Only when the sun burns out some 5 to 10 billion years from now will the order of life on Earth be forced to end.

Chemical Reactions and Energy Considerations

Chemical reactions normally result from collisions between reactant molecules. A reaction is said to occur whenever a chemical bond is broken, formed, or rearranged. Often energy is transferred, stored, or released as a consequence of these reactions.

Reactions that occur spontaneously are likely to release energy stored in the bonds of the reactants. Such energy-releasing reactions are referred to as **exergonic reactions.** An example of an exergonic reaction occurs when gasoline burns in the cylinders of an automobile engine. The gasoline molecules contain tremendous quantities of stored energy in their bonds. When they are combined with air and ignited with a spark, they release their energy explosively.

Some reactions require energy from outside sources, storing it in the chemical bonds of the reaction products. These energy-trapping reactions are usually called **endergonic reactions.** The most important endergonic (energy-capturing) process in the biosphere is, of course, photosynthesis.

Enzymes, Chemical Reactions, and Activation Energy

You learned earlier that enzymes speed up chemical processes by organizing reactions so they take place more efficiently, even at relatively low body temperatures. However, many reactions need a small amount of energy, called the **activation energy,** to get them started. This is the case even for the combustion of gasoline in an engine, which needs a small spark from the plug to get the reaction going. Once going, however, the energy released is substantial.

Most of the chemicals in cells also have to increase their energy status to a point where they become unstable enough to react. However, in cells, the amount of activation energy required to get the bio-reactions started is reduced by enzymes. By organizing the reactants, the enzymes make it much easier for them to get together with just the right orientation so the process can progress. Therefore, the temperatures inside cells do not have to achieve the high levels needed in nonliving systems to get those same reactions to go.

1. Discuss the definition of energy as the capacity to do work. How many forms of energy (kinds of work) can you think of?
2. State each of the three laws of thermodynamics in your own words and explain what each implies.

SUMMARY

The material universe is made up of fundamental building blocks called atoms. There are 92 naturally occurring different kinds of atoms, which we call elements. Atoms chemically combine to form molecules. Compounds are molecules containing two or more different kinds of atoms. Ninety-five percent of all protoplasm employs only five different elements combined in various ways to form the compounds from which living substance is formed; these five important elements are carbon, hydrogen, oxygen, nitrogen, and phosphorus.

The dominant chemical theme seen in living systems is the remarkable use of carbon frameworks from which the organic chemicals are formed. There are four major categories of organic chemicals. They include the carbohydrates (sugars and starches), the proteins (structural materials and enzymes), lipids (fats, oils, waxes, and steroids), and the nucleotides [nucleic acids and adenosine triphosphate (ATP)].

Life depends on a continuous supply of energy. The rules governing many of the physical and chemical affairs in the universe are the laws of thermodynamics. The first law states that energy is neither created nor destroyed; the second law says that no chemical or physical event occurs without some energy being lost as heat. The third law says that entropy or disorder is increasing in the universe.

STUDY QUESTIONS

1. How are atoms, elements, and molecules related?
2. List and characterize the five most important elements in living systems. Why are these the most important?
3. What is the difference between organic and inorganic chemistry?
4. List the four major groups of organic chemicals found in living systems, and give their major properties and biological roles.
5. Why is "life" said to be ultimately powered by nuclear energy?
6. State the three laws of thermodynamics and what they mean to the affairs of the biosphere.

GLOSSARY

Acids Substances that release hydrogen ions (H^+) in solution. Strong acids release more hydrogen ions than do weak acids.

Adenosine triphosphate (ATP) The principal form of chemical energy used by cells.

Amino acids The building blocks from which proteins are made. Human protein employs 20 different amino acids.

Atoms The tiny fundamental units of matter. They consist of a nucleus made up of positively charged proton(s) and neutron(s) orbited by negatively charged cloud(s) of electron(s). There are 92 different kinds of atoms in nature.

Bases Substances releasing hydroxyl ions (OH^-) in solution. Strong bases release more OH^- than weaker ones.

Calorie A unit of energy. By definition, it is the amount of energy required to raise the temperature of 1 liter of water by 1°C (e.g., from 14.5°C to 15.5°C).

Chemistry The study of the composition, properties, and structure of matter.

Cholesterol The parent material for the steroid lipids.

Compounds Substances made by chemically combining two or more different kinds of atoms.

Covalent bonds Chemical bonds based on the sharing of electrons.

Deoxyribonucleic acid (DNA) The hereditary material. The substance from which the genes are made.

Electrolytes Ionic substances dissolved in water that are capable of conducting electricity.

Electrons The tiny electrically negative particles in orbit around the atomic nucleus.

Elements The various kinds of atoms are called elements.

Energy The ability to do work. Work usually is defined as moving something.

Enzymes The protein catalysts that speed up virtually all the body's chemical reactions under conditions compatible with life.

Hydrogen bonds Weak chemical bonds based on the attraction of opposite small electrical charges (such as those in water dipoles).

Ion An atom that has lost or gained electrons and hence acquired a net electrical charge.

Ionic bonds Chemical bonds based on the attraction between oppositely charged ions.

Isotopes Atoms of an element that differ in the number of neutrons they contain but not in their number of protons or electrons.

Lipids Organic materials that do not dissolve in water but do in organic solvents.

Matter Anything that occupies space and has mass.

Molecule The smallest particle of a particular chemical substance, often containing two or more atoms.

Neutrons The uncharged particles in the nuclei of atoms.

Nucleic acids DNA and RNA.

Nucleotides Complex molecules formed by chaining a nitrogenous base to a sugar and a phosphate; the basic building blocks of nucleic acids and ATP.

pH scale The system, ranging in values from 0–14, for estimating the relative acidic or alkaline properties of various solutions. A neutral solution is one with a pH of 7. An acid by definition has a pH of less than 7 and an alkaline (basic) solution has a pH greater than 7.

Polar molecule A molecule whose ends have opposite electrical charges.

Polymers Large molecules formed by chaining together simpler molecular building blocks called monomers.

Polysaccharides Complex carbohydrates formed by chemically linking more than two monosaccharides in chains or sheets.

Proteins Amino acid polymers with which cells perform many functions.

Protons The positively charged particles found in the nuclei of atoms.

Ribonucleic acid (RNA) Molecules used by cells to help carry out the genetic instructions. They are often employed to help with protein synthesis.

Steroids A chemical family of lipids featuring four interconnected organic rings. Cholesterol is the parent compound for the family; many steroids are powerful hormones.

Chapter 3

Cells: Units of Life

- The Cell Theory
- Prokaryotic and Eukaryotic Cells
- The General Eukaryotic Cell
- How Materials Move in and out of Cells
- Passive Transport
- Active Transport
- Phagocytosis and Pinocytosis
- The Nucleus and the Nucleolus
- Endoplasmic Reticulum and Ribosomes
- Golgi Apparatus
- Mitochondria: Energy Processors
- Plastids
- Learning Focus: The Cell
- Summary
- Study Questions
- Glossary

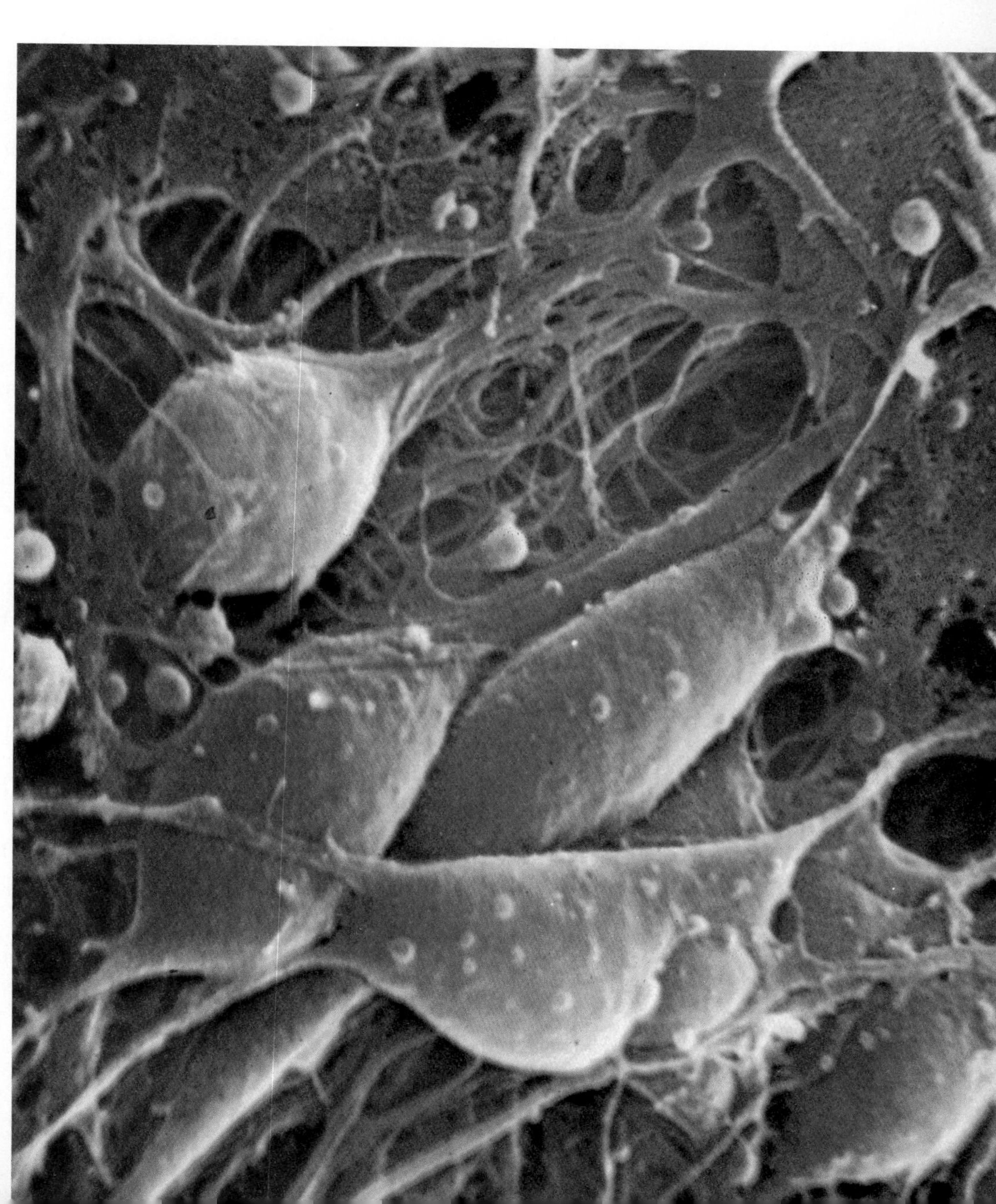

It has taken almost 300 years to piece together one elegantly simple fact: Life occurs in microscopically small, highly compartmentalized specks we call cells. Death occurs when something disrupts or interferes with the activities going on in these tiny cells. This is true whether we are speaking of the result of the predatory dash of a white shark toward an unwary seal or of the systematic destruction of white blood cells that renders an acquired immunodeficiency syndrome (AIDS) sufferer defenseless. It is as true for a microscopic organism whose entire body consists of a single cell as it is for a human, whose body is made up of trillions of cells that work in a coordinated fashion to produce the complexities of human life.

By writing this book, we are attempting to transfer information stored in our network of brain cells to your brain cells. By coding that information into the symbols of language, we can pass it to your brain through the cells of your eyes using the cables of nerve cells that connect your eyes to your brain. As we write these words, nerve cells control the muscle cells that move our fingers over the keyboards of our word processors (Figure 3-1). So, if we are to understand the nature of life, we must start at its simplest level—the single cell.

FIGURE 3-1 Even students and textbook authors communicate at the level of cells. Authors organize and display the information they have stored in their networks of brain cells. Students use various techniques to load that information into their brain cell networks.

THE CELL THEORY

Simply stated, the cell theory says that all life occurs in cells (Figure 3-2) and that all cells come from preexisting cells. That is, the functional unit of life continues through an unbroken chain all the way back to the origins of life, which probably occurred over 3.8 billion years ago. This chain of life is likely to continue into the unforeseeable future, unless our irresponsible actions cause the end of life on this planet.

The recognition that cells are a common denominator shared by all life forms came only slowly. The ancient Greeks recognized that life forms are often constructed of a relatively few repeating units. For instance, a plant is likely to carry thousands of copies of the same basic leaf pattern. A fish's backbone is constructed of repeating vertebral units running down its length. The organ systems found in one group of mammals are likely to resemble organ systems in other species of mammals.

However, we had to wait for the development of glass magnifying lenses before discovering the depth to which nature goes in working with the repeating structural and functional units of cells. In A.D. 1665, an English scientist, Robert Hooke, coined the term *cell* to describe the compartmentalized microscopic organization he observed in a thin section of cork (Figure 3-3, p. 56).

Within 10 years, an eccentric Dutchman, Anton van Leeuwenhoek, had so improved the lenses in his crude hand-held microscopes that his descriptions of what he observed astounded the world. He was the first to see bacteria, sperm cells, and the complex single-celled protozoans that inhabit pond water. His lens was the first window into the realm of microscopic cells; it revealed perhaps the true nature of the demons, spirits, and avenging angels that up to that time had been credited with causing the diseases plaguing humanity (Figure 3-4, p. 56). But it would take other scientists almost 200 years to recognize fully the significance of these early trips down the scale of dimensions into the habitat of microbiology.

The first clue to the cell theory came in 1824 from R.J.H. Dutrochet. He correctly concluded that all tissues, plant and animal, are made up of smaller repeating units—cells. Fifteen years later, M.J. Schleiden (a German botanist) and Theodor Schwann (a German zoologist) were credited with gathering up most of the information known about cells at that time and synthesizing the first fully articulated and integrated cell theory. They con-

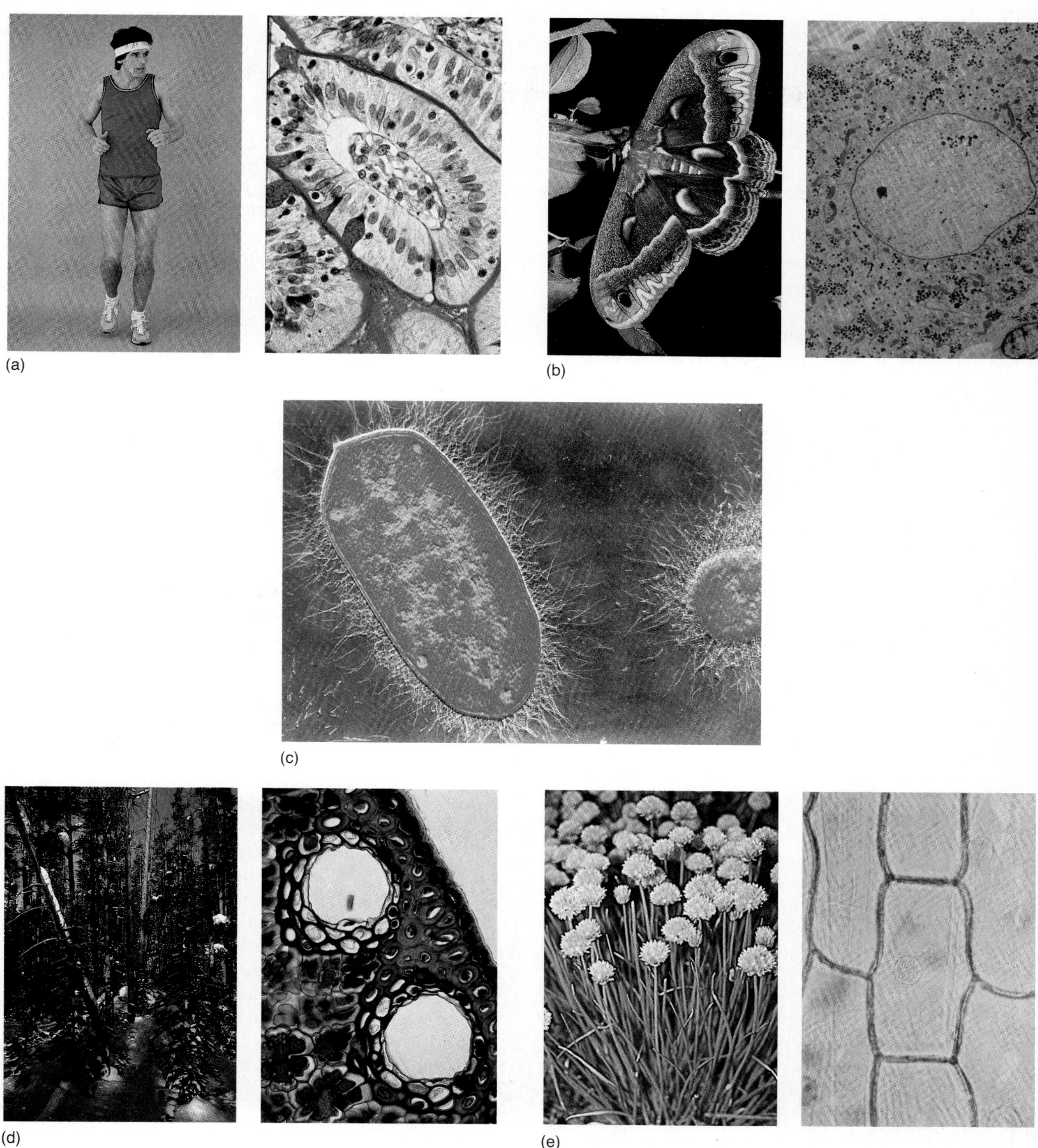

FIGURE 3-2 The cell theory states that all life is constructed of fundamental structural and functional units called cells. The cellular nature of all of life is apparent from the following series of photographs and photomicrographs: (a) Photo of a human; photomicrograph of human cells taken from the digestive system. (b) Photo of a Cecropia moth; photomicrograph of Cecropia moth cells. (c) A falsely colored electromicrograph shows the cellular structure of two *Calymmatobacterium spp.*, similar to those causing sexually transmissible *granuloma inguinale*. The hairlike pili on the cell surfaces help the bacteria adhere to other cells. (d) Photo of Lodgepole pines; photomicrograph showing the cellular makeup of a pine needle. (e) Photo of blooming onion plants; photomicrograph of onion cells.

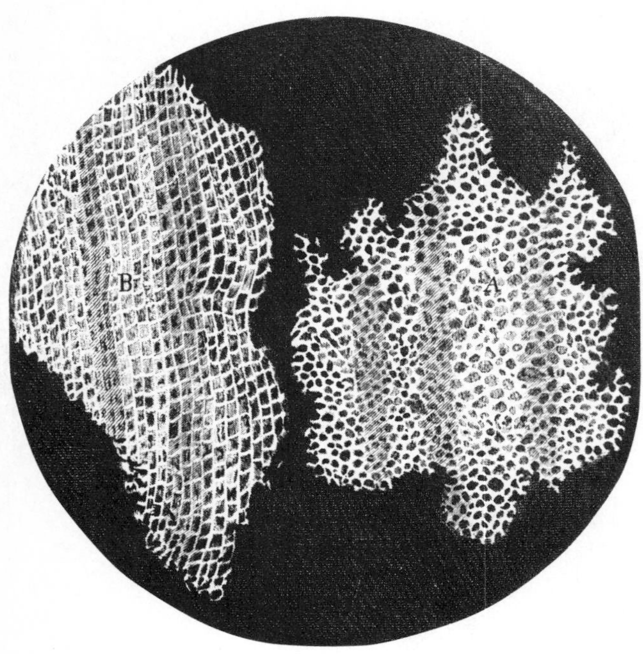

FIGURE 3-3 Three centuries ago, Robert Hooke first coined the word *cells* for the tiny compartments his microscope revealed in thin sections of cork. The word aptly described the compartmentalized nature of most living tissues, and it stuck.

cluded that all tissues are made up of cells and that metabolism and development of tissues are the result of cell activity. In 1858, Rudolf Virchow added the correct suggestion that cells arise only from preexisting cells. He went on to conclude that cells were the fundamental units of life and hence the principal sites of disease.

FIGURE 3-4 Anton van Leeuwenhoek, a Dutch cloth merchant, was one of the earliest and most successful microscopists. One of his most significant achievements was his discovery of bacterial cells in 1675.

---

1. State simply the cell theory.
2. State the roles of Hooke, Leeuwenhoek, Dutrochet, Schleiden and Schwann, and Virchow in developing the cell theory.

Protoplasm: The Living Substance

Long before anyone appreciated the complexities of living systems, someone coined the term **protoplasm** (*proto* = first) to mean simply the "living substance." Since the fundamental unit of life is the cell, protoplasm is by definition found inside cells. Most complex cells, like those found in the human body, are divided into two major compartments: the cytoplasm (*cyto* = cell) and the nucleoplasm. The **cytoplasm** includes everything in the cell except the nucleus. The nucleus is separated from the cytoplasm by the nuclear envelope. The various materials in the nucleus, including the chromosomes, collectively form the **nucleoplasm** (Figure 3-5). Some cells, such as mature red blood cells, have lost their nucleus, making their entire intracellular contents cytoplasm.

Early descriptions of protoplasm called it a semiliquid, grayish substance with various granules suspended in it. Somehow, this mixture was organized in such a way that it had the properties of life. By the eighteenth century, early microscopists had refined their crude microscopes and skills to the point where they could discern the cellular nature common to most living organisms. However, limitations placed on their observations by their crude lenses gave them the impression that protoplasm was a fairly homogeneous material. Scientists did not fully appreciate how limited these observations were until the development of electron microscopes in the twentieth century (Table 3-1) (Figure 3-6, p. 58).

The electron microscope quickly dispelled any belief that protoplasm was homogeneous. It showed the cytoplasm to be highly partitioned and compartmentalized by the numerous organelles that carry out the cell's various chemical routines. We will discuss this division of labor as we describe the "typical" cell.

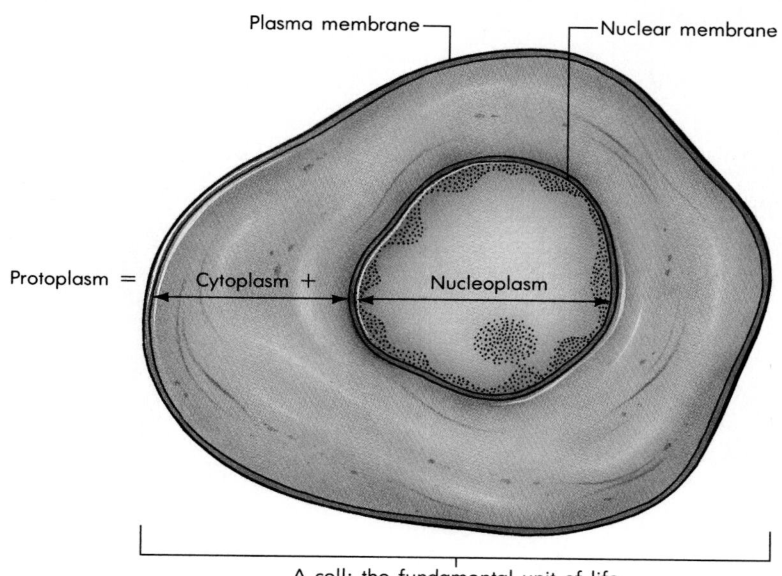

FIGURE 3-5 Protoplasm is the living substance. In eukaryotic cells, it occupies two major cellular compartments. The protoplasm within the cell nucleus is called the nucleoplasm; that between the plasma membrane and the nuclear envelope is called the cytoplasm.

PROKARYOTIC AND EUKARYOTIC CELLS

Cells are divided by history and structure into two broad categories: **prokaryotes** and **eukaryotes** (*pro* = first; *eu* = true; *karyo* = nucleus). The prokaryotic cells—the bacteria and the cyanobacteria, or blue-green algae (Figure 3-7, p. 59)—may structurally resemble some of the earliest life forms on this planet. In fact, members of this group are so distinctly different from the rest of the biosphere that they are set aside in their own kingdom, **Monera,** the most ancient and successful of the five kingdoms of living things. Prokaryotic cells are distinguished by the fact that they lack membrane-bound organelles. Instead of a nucleus, they have a **nucleoid,** a region of their cytoplasm that contains their deoxyribonucleic acid (DNA).

About a billion years ago there appeared the more complex eukaryotes. This cell line eventually gave rise to more modern kingdoms of complex multicellular life forms: the kingdoms of animals (including humans), plants, and fungi (including molds, mildews, and mushrooms), and the most complex cells of all, the single-celled or simple colonial life forms in the kingdom Protista (Figure 3-8, p. 60).

Table 3-1

Relative Resolving Powers for Instruments of Observation

INSTRUMENT	RESOLVING POWER[a]	MAGNIFYING POWER[b]
Unaided eye	0.1 mm (100 μm)	1 ×
Light microscope	0.2 μm (2000 Å) (blue light)	About 1000 ×
Electron microscope	0.0001 to 0.0005 μm (1–5 Å)	100,000–200,000 ×

[a]*Resolving power* is the ability to tell that two objects, lines, dots, or structures that are very close together are really separate. The units given state how close together they can be and still be seen as separate: 1 mm (millimeter) = 1000 μm (micrometers) = 10^7 Å (angstroms).

[b]*Magnifying power* refers to a lens system's ability to enlarge the visual image of the object being observed.

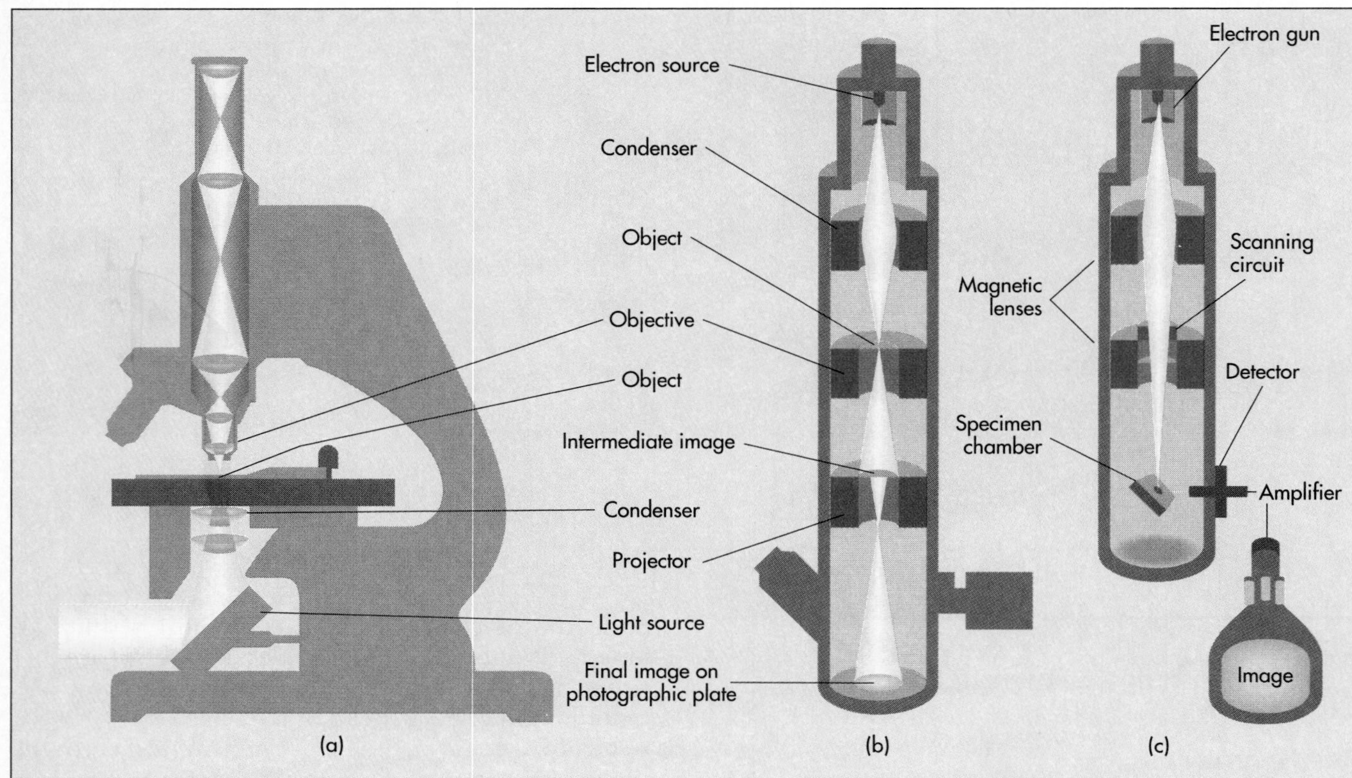

FIGURE 3-6 Modern light microscope and electron microscope. (a) The light microscope passes light through thin preparations of living (or nonliving) cells and tissues and uses glass lenses to produce magnified images. (b) The electron microscope passes a beam of electrons through nonliving specimens. Materials containing heavy atoms restrict the flow of electrons and produce darkened regions on the photographic plate used to record the image. (c) There are also the scanning electron microscope (which uses electrons reflected from the surface of a specimen to build up an image), the scanning tunneling microscope (which uses distance from a specimen's surface to build a three-dimensional map, even of something as small as a DNA molecule), and others.

THE GENERAL EUKARYOTIC CELL

Because of the keen interest in humans, most discussions of cell anatomy and physiology focus on the typical eukaryotic cell with its well-defined nucleus and membrane-bound **organelles.** We too will discuss most of the major organelles that have been studied to date as if they all existed inside a single, typical cell. However, no such single cell containing all these organelles probably exists. There is really no such thing as a typical cell (see Figure 3-17).

The Plasma Membrane

One of the most important and active areas of any cell is the **plasma,** or **cell, membrane** that forms the cell's outer boundary. It separates and protects the cell's chemically controlled interior environment from the extracellular world. It carefully regulates which chemicals are allowed to enter the cell and which are to be eliminated or released. The plasma membrane also contains proteins on its surface and embedded in it. These proteins play a role in the body's defense systems and serve as messengers or receptive sites for chemical communications between cells.

Structure of the Plasma Membrane

The principal chemical components of plasma membranes are lipids and proteins. The lipids are usually phospholipids, formed when glycerol is combined with two long-chain fatty acids and a phosphate group (see Chapter 2). This gives the

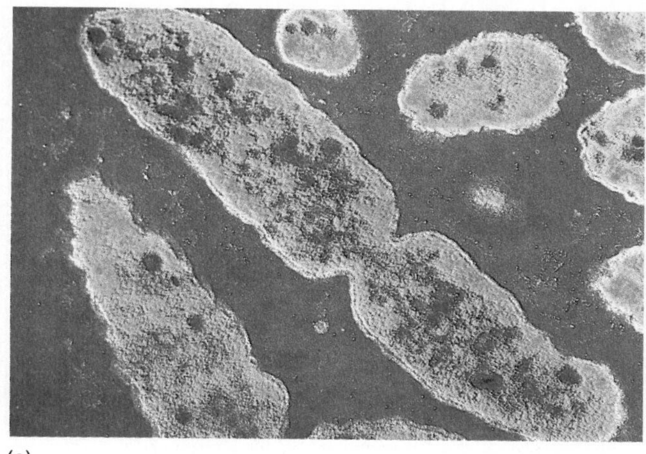

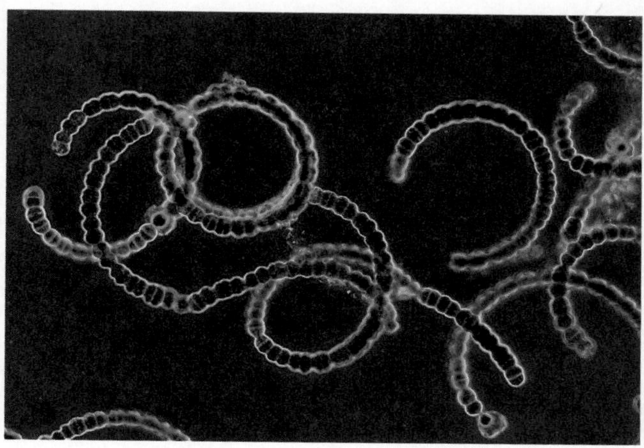

(a)

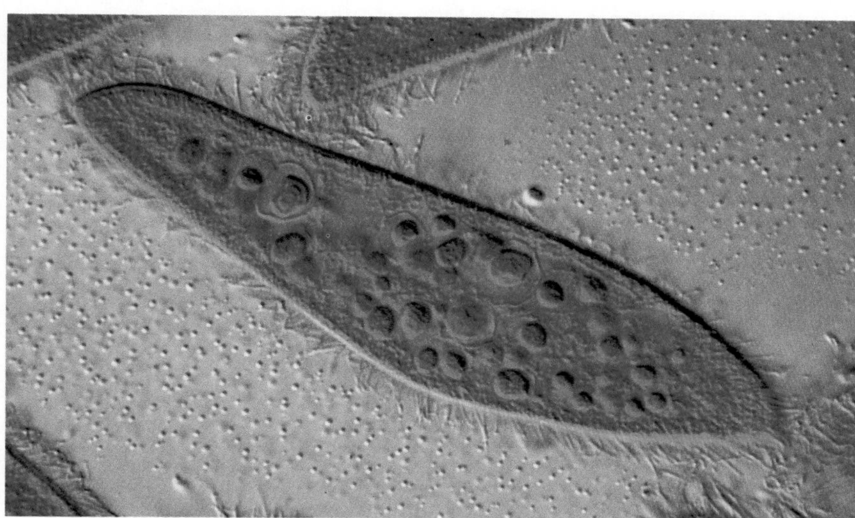

(b)

FIGURE 3-7 (a) Prokaryotic cells (bacteria and cyanobacteria) have no nucleus or membrane-bound organelles. These electron micrographs show the rod-shaped bacterium *Serratia marcescens*, a normal inhabitant of animal (including human) intestines. A second example of prokaryotic cells is seen in this scanning electron micrograph of an ancient line of cells, the filamentous cyanobacteria (blue-green algae). (b) Eukaryotic cells of the kingdom Protista are among the most complex cells in nature. Highly compartmentalized and containing a wide variety of organelles, they perform many activities that more advanced, multicellular forms delegate to separate cells and organs.

phospholipid some unique properties (Figure 3-9, p. 61). The fatty acid portions of these molecules are *not* soluble in water, making them **hydrophobic.** The phosphate portion has an affinity for water, making it **hydrophilic.**

Since cells exist in an **aqueous** environment, their cell membrane phospholipids form a bimolecular layer. The fatty acid portions of the molecules are tucked into the middle of the membrane away from the water, while the water-soluble phosphate ends are found on the exterior and interior surfaces of the membrane closest to the water that makes up most of the extracellular environment and the cytoplasm.

Proteins also form a part of the cell membrane. In the fluid mosaic model of the cell membrane, proteins are embedded in the lipid layer, and because the lipid is a liquid, they are free to move over the surface of the cell rather as icebergs move over the surface of the sea (Figure 3-10, p. 62). Some of these proteins protrude from the exterior surface of the plasma membrane, others from the interior surface of the membrane. Some proteins extend all the way through the lipid bilayer, while others extend only part way through the membrane. In addition, the surface proteins on the outside of the plasma membrane may be complexed with carbohydrates—forming what are known as **glycoproteins.** These glycoproteins are often involved in cell-to-cell bonding and with immunological reactions (see Chapter 24).

---

1. How would you distinguish between a prokaryotic cell and the typical eukaryotic cell found in the human body?
2. How are the phospholipids positioned in the plasma membrane? Where are the proteins found?

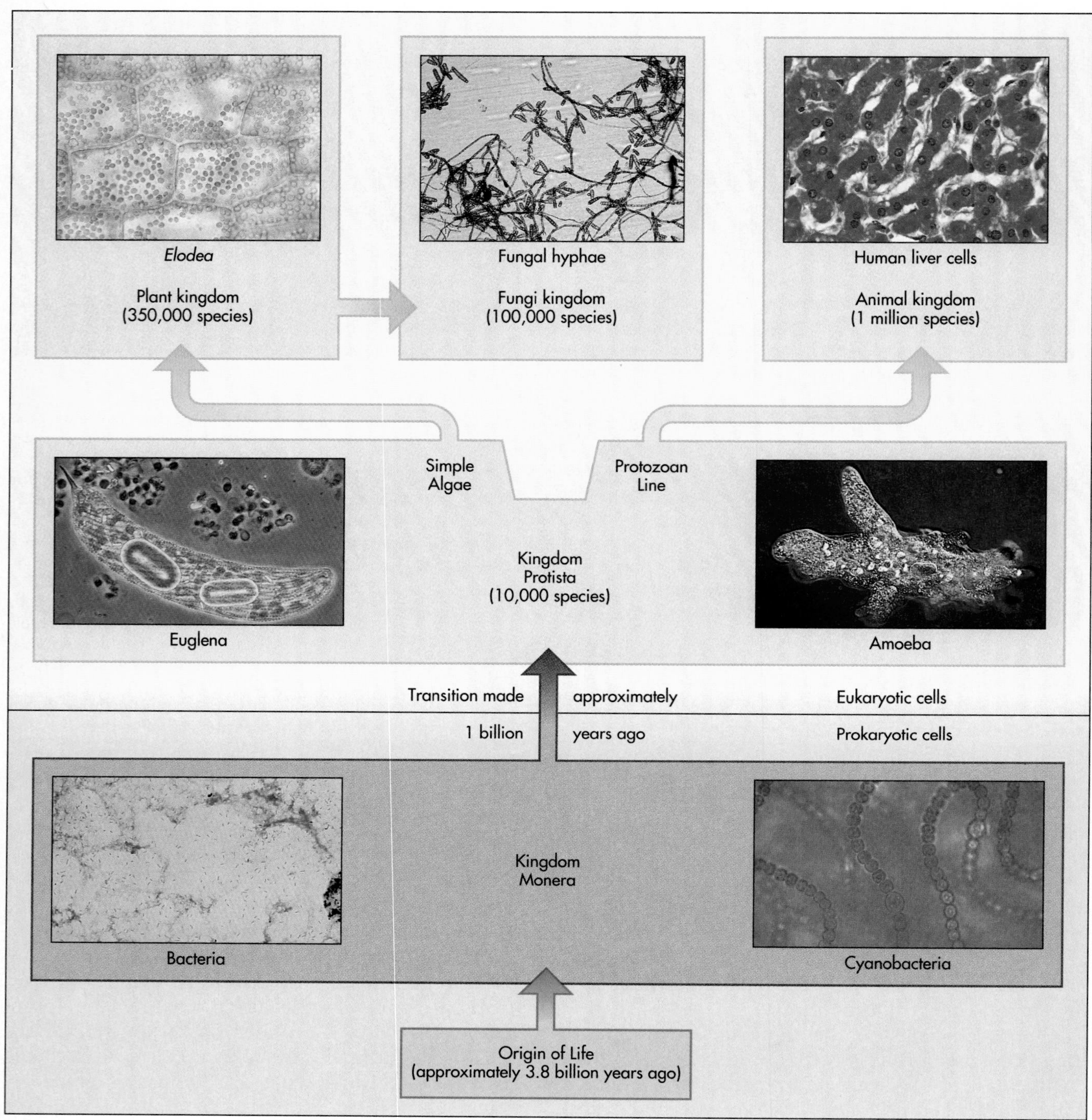

FIGURE 3-8 The five kingdoms of living organisms differ from one another, in part, in cell structure. The prokaryotes of the kingdom Monera first appeared about 4 billion years ago. Examples include (a) the rodlike prokaryotic cells of the bacterium *Hemophilus influenzae* and (b) the photosynthetic cells of the cyanobacteria. About a billion years ago, the monerans underwent major changes to produce the first eukaryotic cells and form the new kingdom Protista. (c) The photosynthetic line of protistans is represented by the free-living *Euglena*. The plant kingdom apparently arose from similar protistans. (d) The animal kingdom apparently arose from the nonphotosynthetic protozoan ("first animals") protistans, represented here by *Amoeba proteus*. (e) The cells of the plant kingdom are represented here by cells from the water plant *Elodea spp*. (f) The cellular makeup of the kingdom Fungi is shown in this scanning electron micrograph of the threadlike fungal hyphae. Fungi apparently branched off from the plant kingdom. (g) Eukaryotic human liver cell with a well-defined nucleus containing a nucleolus.

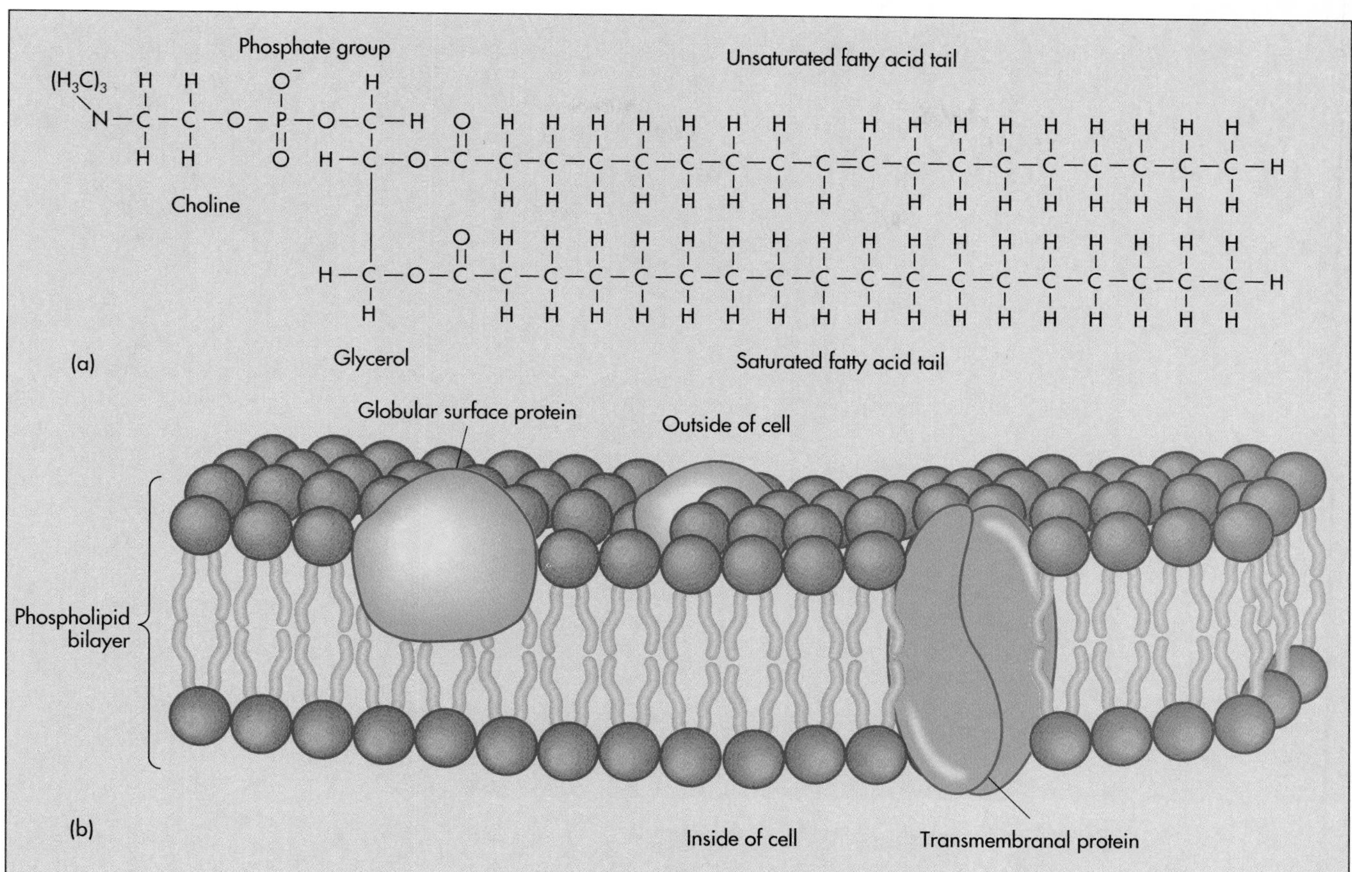

FIGURE 3-9 Structure of phospholipids. (a) The principal constituents of plasma membranes are the phospholipids, formed by hooking two insoluble fatty acids and a phosphate group to a glycerol molecule. The resulting molecule is soluble in water (hydrophilic) at its phosphate end and insoluble (hydrophobic) at the fatty acid end. (b) Since the cellular environment, both inside and out, is mostly water, the phospholipids form a two-layered membrane with the water-loving phosphate ends closest to the watery environment inside and outside the cells. The fatty acids tuck their hydrophobic tails away from the water toward the middle of the membrane.

HOW MATERIALS MOVE IN AND OUT OF CELLS

Cells regulate the molecular traffic necessary to keep their living systems working in several ways. Some methods, like diffusion and osmosis (discussed in the next section), cost the cells nothing; they simply play off the laws of nature and the physical conditions in which life is found. Since the cells do not have to pay the energy expenses for these forms of molecular traffic, they are collectively referred to as **passive** (or physical) **transport.** Other systems of molecular movement require the cells to spend some of their metabolic energy; these are **active transport** systems. We will consider each of these categories in turn.

All molecules are in motion at temperatures above absolute zero (−273°C). The higher the temperature, the more vigorous the movements. Thus, the amount of thermal agitation occurring at temperatures that can support living systems is considerable. At these temperatures, molecules are bouncing around and colliding with each other. It is these intermolecular collisions that are responsible for many chemical reactions and also provide the driving force for much of the molecular traffic so important to cells (Figure 3-11, p. 63).

PASSIVE TRANSPORT

Diffusion

In a sense, living systems are somewhat like businesses. There are certain advantages to be gained

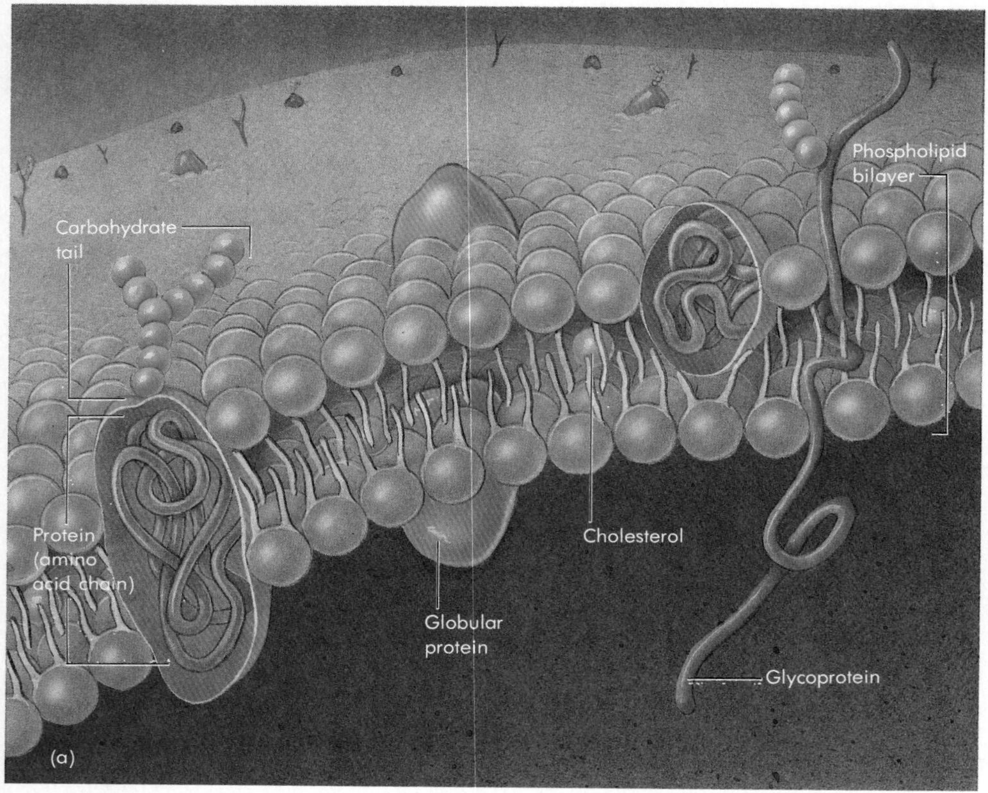

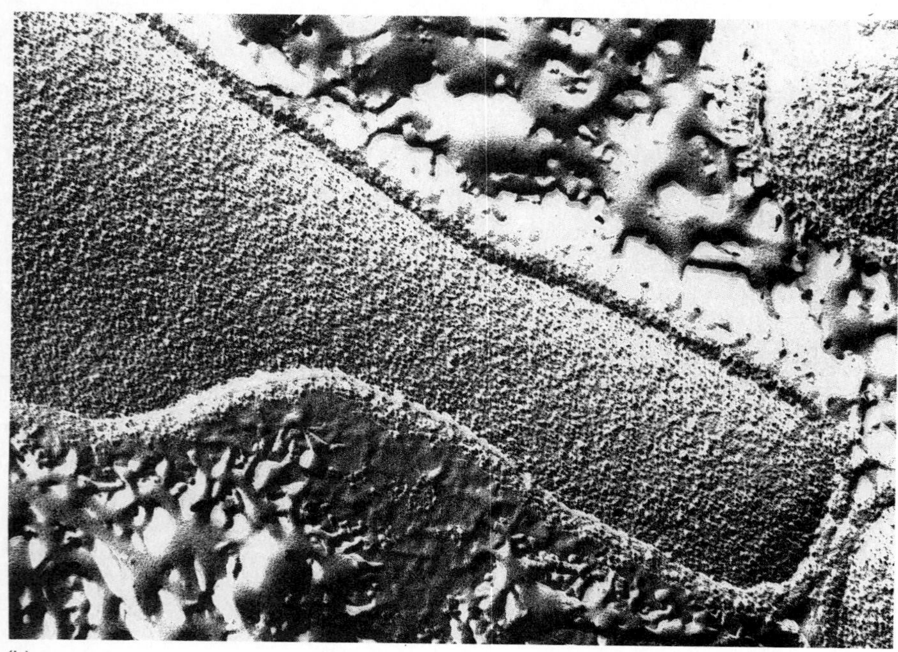

FIGURE 3-10 (a) The fluid mosaic model is the currently favored description of membrane structure. The plasma membrane is seen as a two-layered molecular sea in which proteins float. Some of the proteins extend all the way through both phospholipid layers, while others extend only part way through. These proteins move freely in the lipid. Carbohydrates, attached to some of the outside surface proteins, may serve as receptor sites for intercellular messages. (b) The two layers of a plasma membrane can be separated; the technique called *freeze fracturing* reveals the numerous proteins embedded in the lipid bilayers. The photo shows the plasma membrane of a bacterium, *Bacillus anthraces*, that has been treated with this process (×116,000).

by doing things as cheaply as possible. Diffusion is an example of a very economical way for cells to move chemicals about. Simply defined, **diffusion** is the movement of molecules from an area where they are highly concentrated to an area where they are less concentrated (Figure 3-11). It happens to be one of the most important methods by which materials are moved in and out of cells and from place to place within cells. It is free to the cell because the driving force for these molecular movements is the same as for Brownian movement—namely, that all molecules are in motion. In areas where molecules are highly concentrated, their movements produce collisions causing them to bounce away into areas of lower concentration (and consequently fewer collisions).

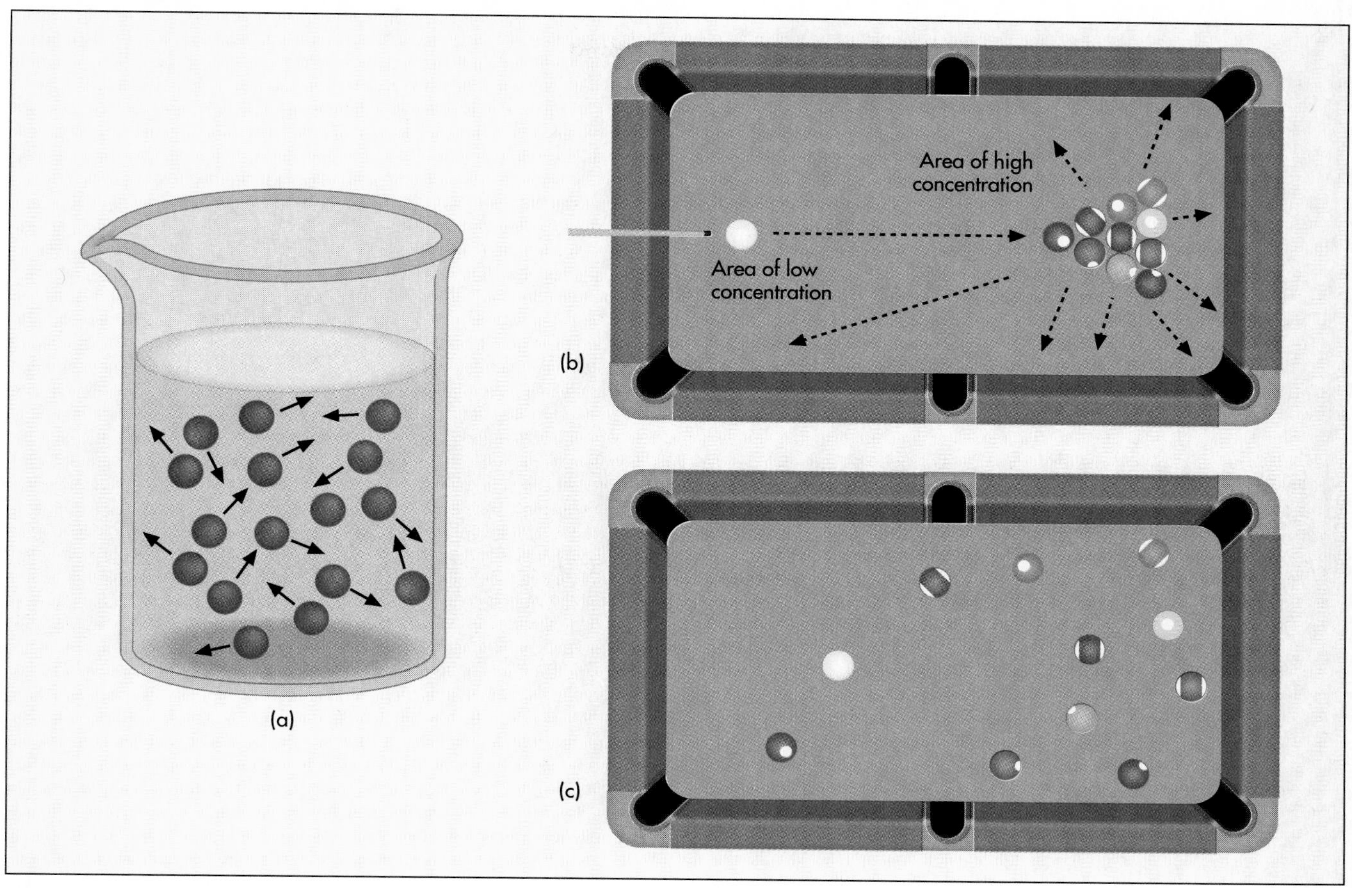

FIGURE 3-11 Brownian movement and diffusion. (a) Brownian movement is the random dance of tiny particles suspended in fluid, such as dye particles in water, caused by collisions between the particles and the molecules of the fluid. These random movements indicate that all molecules are in motion. (b,c) Diffusion is the movement of substances from areas of high concentration to areas of lower concentration. In this example, energy in the form of a moving cue ball powers the "diffusion" of pool balls across a billiard table. The collision between crowded molecules drives them away from highly concentrated areas. Similarly, thermal energy, which causes the movements of atoms and molecules, powers chemical diffusion.

To diffuse, molecules need only a concentration gradient and freedom to move. The steeper the gradient (the greater the difference in concentration for a particular substance between two areas), the faster diffusion occurs. As diffusion takes place, it slows, for the concentration gradient diminishes as the molecules spread out from where they were previously highly concentrated (Figure 3-12). In cells receiving nutrients by diffusion, the concentration gradient of food molecules between inside the cell and outside is maintained because the blood continues to deliver the nutrient molecules to the cells. The food molecules are more concentrated outside the cell, so they diffuse through the cell membrane either through pores or by dissolving in the lipid layer. As the materials move into the cells, they are consumed by the processes they are supplying. Hence, nutrient concentrations remain lower inside the cells than outside, and the inflow continues. Waste products leave the cell by following a similar concentration gradient in the opposite direction. Wastes that have been ejected from the cell are swept away from the tissues by the body fluids, while the metabolic activities inside the cell continue producing more wastes to diffuse out.

In the human, food molecules move from the blood to the tissue fluids and from the tissue fluids into the cells by way of diffusion. Cell wastes diffuse in the opposite direction, and many other materials move from where they are manufactured in cells to where they are needed or stored. Oxygen inhaled into the lungs diffuses through the lung's membranes and into the bloodstream and finally into the cells.

Facilitated Diffusion

Membranes can be classified by how readily they allow various materials to pass through them. A

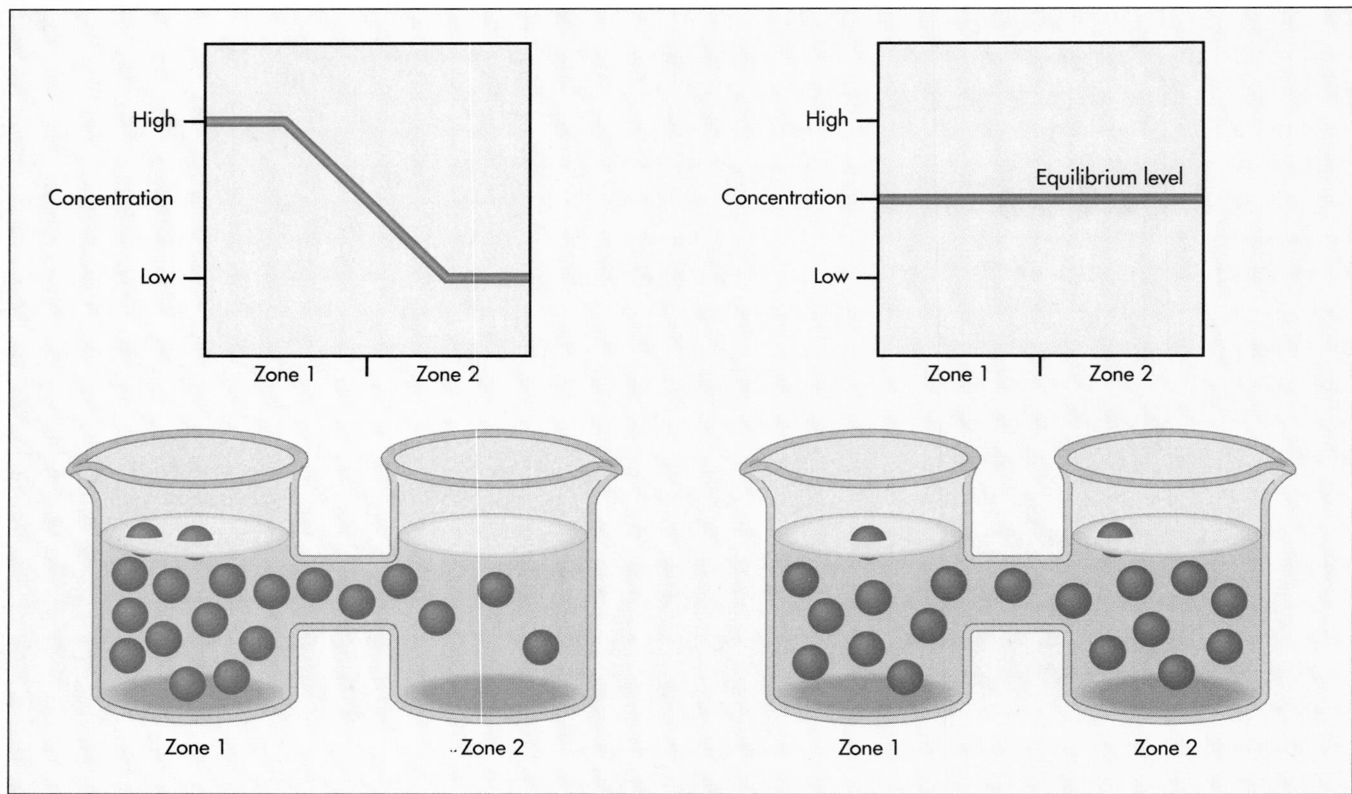

FIGURE 3-12 Diffusion is movement down a concentration gradient. It continues as long as the difference in concentration exists.

membrane whose pores are so large that virtually any molecule can pass through is said to be completely **permeable.** If a membrane prevents everything from passing through, it is **impermeable.** However, most biological membranes are **selectively permeable** (or semipermeable) because they allow some materials (such as water) to pass freely through but control the movements of other materials.

Not all substances required by cells can get through the plasma membrane unaided. For instance, vital molecules like amino acids and glucose are unable to enter cells because they are too large to pass through the membrane pores or are insoluble in the lipids. They require the help of carrier molecules to get them through the membrane. The carrier molecules are located within the membrane and can freely complex with specific molecules they encounter on one side of the membrane, carry them across the membrane, and release them to the other side. Carrier molecules can transport substances in both directions across the membrane, but they tend to combine with more molecules on the side of the membrane where they are the most concentrated and release more on the side where they are less concentrated. Such molecular traffic is called carrier-**facilitated diffusion** (Figure 3-13). The rate of facilitated diffusion is determined by both the differences in concentration across the membrane and the number of available carriers.

Osmosis

Cells are set off from their external environment by membranes that allow some substances, like water, to pass freely in and out of the cell but restrict the movements of many larger, more complex materials like proteins, nucleic acids, and polysaccharides. Thus, the selective permeability of the plasma membrane creates some rather interesting situations for cells. These situations must be carefully considered when injecting people with medications or fluids.

Osmosis is a special case of diffusion; it is the movement of *water* through a selectively permeable membrane from the side where it is more concentrated to the side where it is less concentrated (Figure 3-14, p. 66). Since so many medical conditions, such as profuse sweating, vomiting, burns, and diarrhea, can affect the fluid balance that exists among the blood, tissue fluids, and cells, it is important to understand something of the osmotic relationships that exist between the fluid compartments in the body.

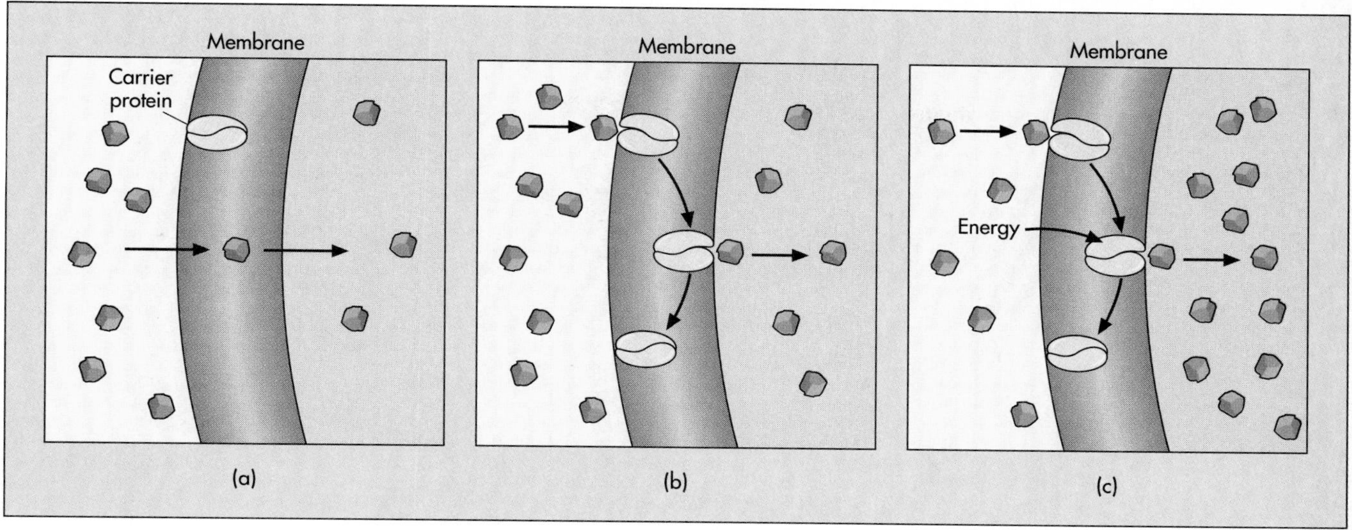

FIGURE 3-13 Two types of diffusion are observed as materials move in and out of cells. (a) Simple diffusion is the movement of materials down a concentration gradient from one side of the cell membrane to the other. (b) In facilitated diffusion, carrier molecules help move certain materials across the membrane down a concentration gradient. (c) Active transport also uses carrier molecules, but it moves the carried material *against* a concentration gradient. The cell must pay for such movements by using some of its metabolic energy, that is, ATP.

In Chapter 2 we said that a **solution** is the mixture formed when something is dissolved in a liquid. The dissolved substance is the **solute,** and the liquid in which it is dissolved is called the **solvent.** Thus, a cup of instant coffee is a solution, with the coffee crystals serving as the solute and hot water acting as the solvent. Protoplasm, in a sense, is an extremely complex solution with various salts and organic materials dissolved in water. The membranes of cells are selectively permeable; since water readily moves in any direction, osmotic changes can occur between the cells and their surroundings fairly quickly.

Osmosis and Hypertonic, Isotonic, and Hypotonic Solutions

Suppose a dehydrated patient is mistakenly given an injection of pure, distilled water directly into his or her veins. What is likely to happen? Simply put, the blood cells exposed to pure water are likely to swell up and explode. The reason is fairly straightforward: Red blood cells are encased in selectively permeable plasma membranes that freely allow the passage of water molecules in either direction. The cytoplasm of red blood cells normally contains some water, but also millions of organic molecules that *cannot* pass through the membrane and leave the cells. When the individual red blood cells are exposed to pure water, there is thus an immediate change in the concentration gradient; the extracellular fluid will have a much higher water concentration than the cytoplasm. Water then enters the cells, causing them to swell and eventually rupture, spilling their contents and killing them. Such exploding of red blood cells is called **hemolysis.**

Three terms are usually employed to describe a fluid's osmotic activity relative to the cells of the body; they are hypotonic, isotonic, and hypertonic (Figure 3-15, p. 67), and they refer to the *amount of solute* found in the solution. The solution in the preceding example was **hypotonic** (*hypo* = too low) because it had too little solute or *too much water.* Be careful, because the terminology seems backward here; it is the water movement (and concentration) that causes the osmotic problems, yet the terms refer to the amount of solute in the solution.

The problems could have been avoided simply by lowering the water concentration of the inflowing fluid. The easiest method is to add enough solute, like sugar or salt, to the injected fluid so that its water concentration just matches the water concentration normally found in the cytoplasm of red blood cells. The extracellular fluid is now **isotonic** (*iso* = equal) to the cells. Such solutions will promote no net movement of water either into or out of the cells. An isotonic solution for human blood contains the equivalent of 0.9 grams of salt per 100 grams of blood (0.9 percent NaCl). Such injectible solutions are called **physiological,** or **normal, saline.**

It is also possible to expose the cells to solutions

FIGURE 3-14 Osmosis is the diffusion of water through a selectively permeable membrane. In the cartoon, the perforated wall represents the plasma membrane (with pores), side A the cell interior, and side B the extracellular environment. The dark balls are water molecules; the white balls are larger intracellular molecules such as proteins. All the balls are in motion, but only the dark ones can pass through the pores. Since there are more dark balls on side B, more of them pass through the barrier into the cell than out of the cell. Osmosis is the net movement of the water molecules across the barrier and down the water concentration gradient.

containing too much dissolved material and to a water concentration that is too low; such solutions are said to be **hypertonic** (*hyper* = over, above, too much). When the cells are exposed to hypertonic solutions, they contain higher water concentrations than does the surrounding medium. Thus, their water flows out and they shrivel. Such shriveling in blood cells is called **crenation** (see Table 3-2).

1. How are diffusion, facilitated diffusion, and osmosis related?
2. How are they different from one another?

ACTIVE TRANSPORT

Some substances are so important to cellular metabolism that cells transport them across their cell membranes (in or out) against concentration gradients. In order to go against a concentration gradient, the cell must use some of its metabolic energy to pay for the movement. Such energetically expensive forms of molecular transport are called active transport. Since energy is in limited supply for living systems, active transport mechanisms are gen-

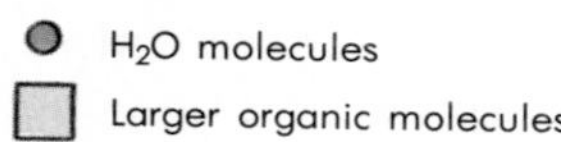

(a) ISOTONIC SOLUTION

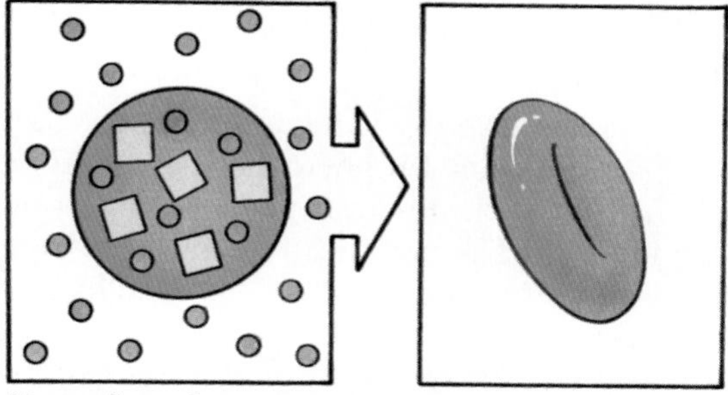

No net flow of water in either direction; therefore cells are not osmotically challenged by the solution

(b) HYPERTONIC SOLUTION

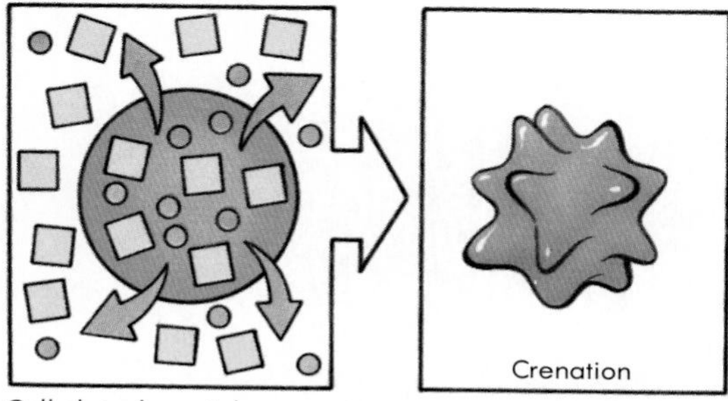

Cell shrivels as it loses water:

(c) HYPOTONIC SOLUTION

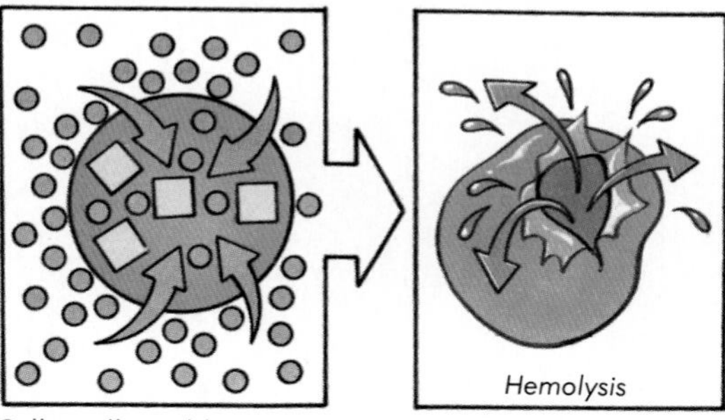

Cell swells and bursts as water flows in, producing (in blood cells)

FIGURE 3-15 (a) Human blood in an isotonic (0.9 percent) solution. Since the osmotic pressure is the same inside and outside, the blood cells remain one size. (b) Human blood cells in a hypertonic solution (1.6 percent). Note the crenated (shriveled) look to the cells. (c) Swollen blood cells in a hypotonic solution (0.2 percent).

erally reserved for moving substances with very high metabolic priorities.

Active transport is often used to move ions such as Na^+ and K^+ through the cell membrane and to develop substantial concentration gradients for these materials across the membrane. For instance, many cells maintain lower concentrations of Na^+ *inside* the cell than are found in the surrounding fluid. These same cells also maintain higher concentrations of K^+ than are found in the tissue fluid. When these positive ions are actively pumped across the membranes, a disproportionate number of negative ions, often chloride (Cl^-), is left behind since they are not transported through. This results in one side of the membrane becoming more positively charged, while the other side takes on a negative charge. Since the ions are electrically charged, the concentration gradient is also an electrical gradient; the electrical gradient is what makes possible such things as nerve impulses, coordination of the heart muscle during contraction, brain activity, sensory perception, communication, and defense (like that seen in electric eels).

Like carrier-facilitated diffusion, active transport depends on carrier molecules that pick up the material being transported on one side of the membrane, carry it across to the other, and release it. The action of the carrier molecules, against the natural tendency of molecules to move down concentration gradients, is powered by energy released from ATP molecules.

Table 3-2

Hypertonic, Isotonic, and Hypotonic Osmotic Solutions

SOLUTION TYPE	SOLUTE CONCENTRATION	SOLVENT (H_2O) CONCENTRATION	WATER MOVEMENT
Isotonic	Equal to cell's	Equal to cell's	No movement
Hypertonic	More than cell's	Less than cell's	Out of cells; cells shrivel
Hypotonic	Less than cell's	More than cell's	Into cells; cells swell (burst?)

PHAGOCYTOSIS AND PINOCYTOSIS

Some materials enter the cell by mechanisms other than simple diffusion, osmosis, or active transport. For example, **phagocytosis** is a form of cellular "eating" employed by amoebae, one-celled protistans that live in pond water, and by certain types of blood cells (neutrophils and macrophages; see Chapter 10). During phagocytosis, a blood cell or amoeba detects an edible object (such as another cell), flows to surround it with cytoplasmic arms called **pseudopods,** and captures it by forming a pocket of cell membrane that pinches off into the cytoplasm (Figure 3-16). Such a pocket is a **food vacuole,** or endocytic vesicle. It often fuses with other organelles, the lysosomes, which supply digestive enzymes for breaking down the ingested item and releasing nutritive molecules that can cross the food vacuole's membrane by diffusion or active transport.

Phagocytosis usually consumes solids such as cells or parts of cells. **Pinocytosis** resembles phagocytosis in that it draws in a pocket of membrane, capturing, or "drinking," fluid from the extracellular environment in the process. It is used by the cell to consume droplets of solution, often containing small amounts of protein or some other useful material, plus whatever proteins may have been attached to the membrane that folded in to form the pocket. **Exocytosis** is essentially the reverse of pi-

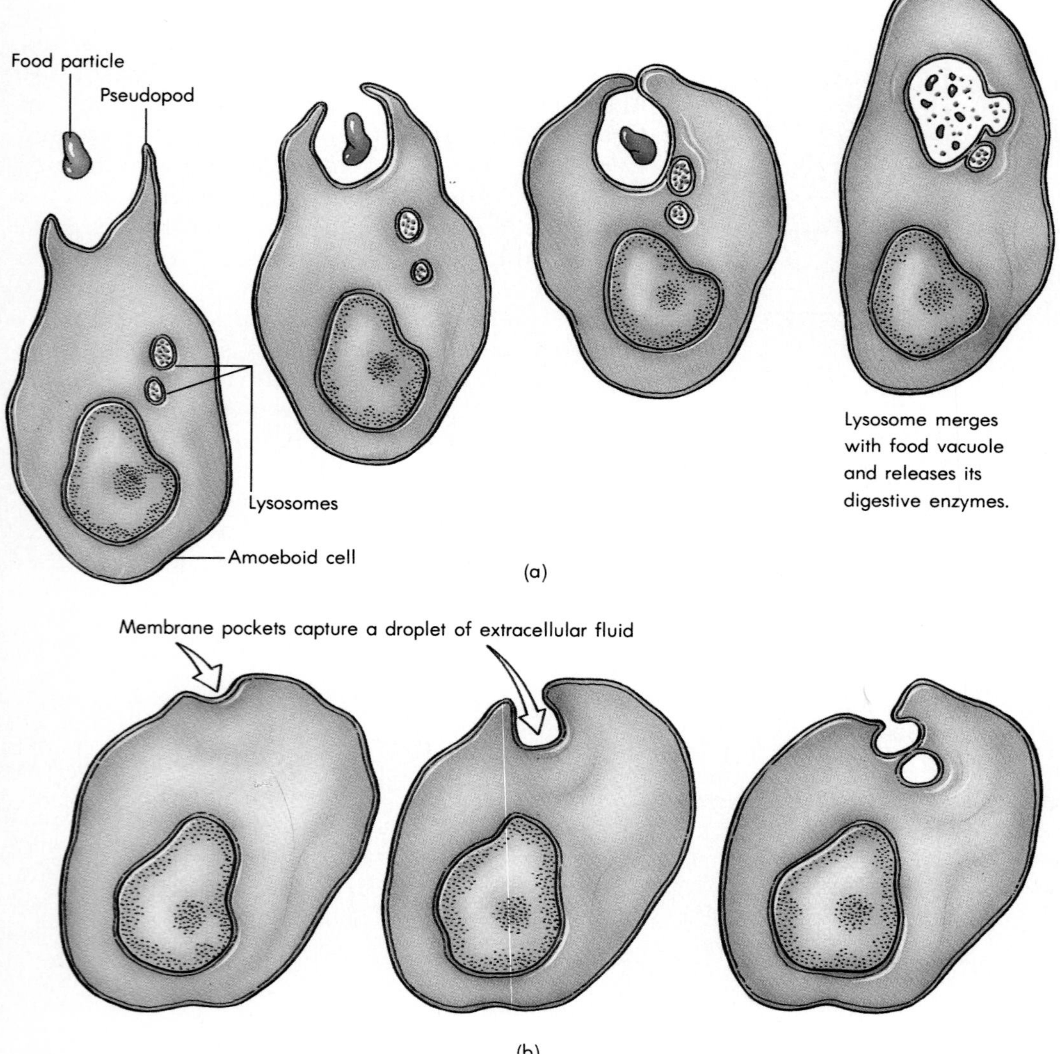

FIGURE 3-16 (a) Phagocytosis (cellular eating) usually involves the ingestion of particles such as bacteria by cells. The particles are enclosed in food vacuoles (or endocytic vesicles) that merge with lysosomes. (b) Pinocytosis (cellular drinking) usually involves the ingestion of droplets of extracellular fluid and proteins attached to the cell membrane.

nocytosis. It moves a droplet of materials made in the cell, wrapped in newly synthesized membrane, to the interior surface of the cell membrane. This droplet then fuses with the cell membrane, dumping its contents outside the cell. Exocytosis thus serves as a mechanism of secretion and excretion. It also serves to add new membrane to some growing cells. Since there is an expenditure of cellular energy involved with both exocytosis and endocytosis, some biologists consider them to be separate variations on the active transport theme.

1. How does active transport differ from passive transport?
2. Describe how active transport works.
3. What is the difference between phagocytosis and pinocytosis?

THE NUCLEUS AND THE NUCLEOLUS

A conspicuous feature of most eukaryotic cells is the **nucleus,** usually located near the center of the cell (Figure 3-17). It is often the largest organelle; it is certainly one of the most important. Most eu-

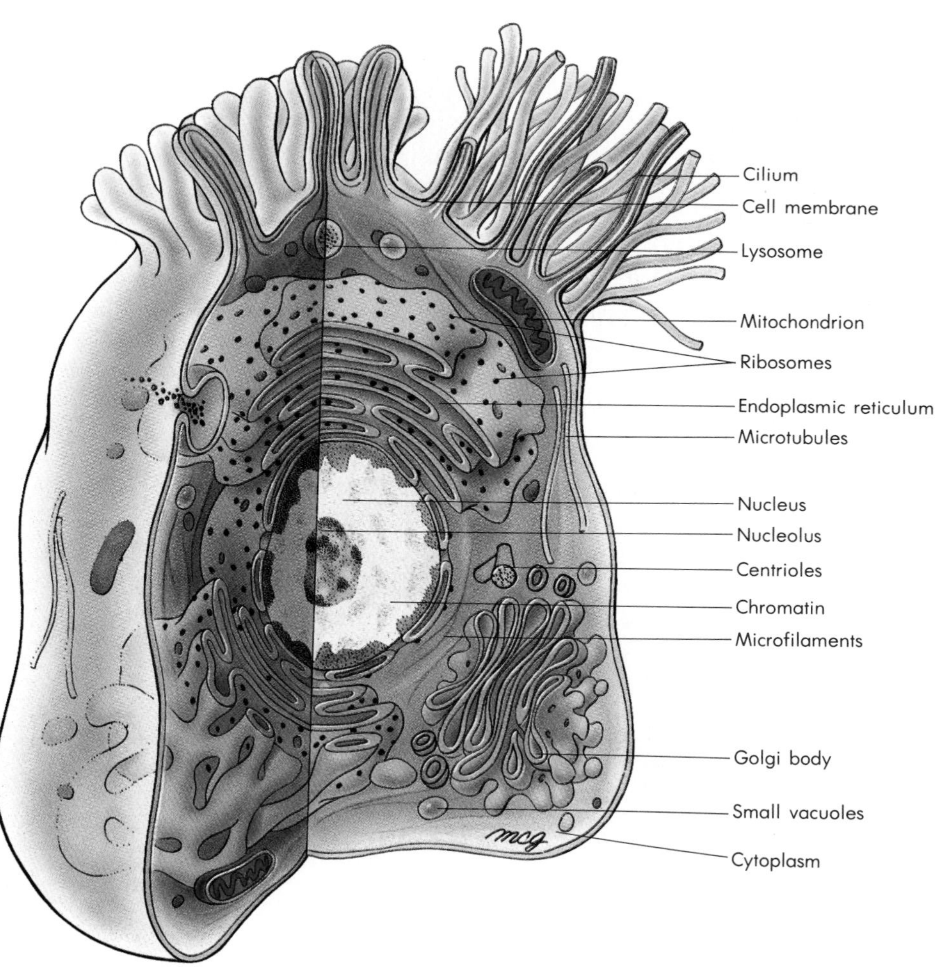

FIGURE 3-17 The "typical" eukaryotic cell illustrated here is not very typical. For convenience, we have included many organelles you would see only in different human cells, not in any single cell. For instance, the fingerlike folds in the upper right portion of the cell membrane are found in cells that line cavities in which absorption is taking place, such as the small intestine. The cilia on the upper left can be seen in cells that line tubes such as the trachea and Fallopian tubes; their waving action sets up moving currents.

karyotic cells have one nucleus at some stage in their lives, but some, like red blood cells, lose them as they mature. The nucleus acts as a control center for the cell, for it contains the DNA (or genes), in the form of chromosomes, that directs the synthesis of enzymes and other proteins.

The nucleus is set off from the rest of the cytoplasm by the **nuclear envelope.** Continuous with the endoplasmic reticulum, this double membrane shields the chromosomes from the chemical reactions taking place in the cytoplasm. However, it does not isolate the nucleus. The nuclear envelope contains pores through which various substances can readily move to and from the cytoplasm.

Eukaryotic chromosomes are long, threadlike structures made of the nucleic acid DNA and protein, or *chromatin,* so named because it stains vividly with certain dyes. If we can compare a cell to a computer, we can say the chromosomes are the equivalent of a computer program that governs the activities of the cell, while the organelles in the cytoplasm are equivalent to the output devices, such as the printer and modem, that are controlled (see Chapter 4).

The **nucleolus** is a dense, granulär or fibrous region visible within the nucleus of many cells. The number of nucleoli varies between different kinds of cells and from time to time within the same cell. They usually disappear completely in cells undergoing cell division, reappearing after the division is complete. The nucleolus appears to form around sections of certain chromosomes. It uses the information in these chromosomal segments to manufacture ribosomal RNA (another nucleic acid); later, the cell uses the ribosomal RNA to make ribosomes (see what follows), which are used in the manufacture of protein.

ENDOPLASMIC RETICULUM AND RIBOSOMES

Early studies of cells using light microscopes did not reveal much complex structure in the cytoplasm. Electron microscopes show it contains an elaborate network of membranous channels, tubes, and flattened sacs called the **endoplasmic reticulum (ER)** (Figure 3-17). Portions of the ER expand as flattened sacs to form the double-walled nuclear envelope. The rest forms a pipeline that interconnects many of the cytoplasmic organelles. It conducts recently manufactured proteins and other substances from where they are manufactured to where they are stored or receive additional processing or are used. It also provides a tremendous surface area on which certain chemical reactions can occur.

Embedded in some sections of the ER are tiny granules made up of protein and ribosomal RNA. These granules are called **ribosomes;** they are the principal sites of protein synthesis in the cell (Figure 3-17). Their presence gives the ER a bumpy or roughened look and the name **rough ER.** The ribosomes in these areas are thought to be manufacturing *exportable proteins,* which will move through the rough ER to storage sites and be released later, to the outside. Other ribosomes, floating free in the cytoplasm, unattached to the ER, apparently manufacture proteins that will remain in the cytoplasm.

Smooth ER contains no attached ribosomes. These areas of the ER may be involved in the synthesis of lipids or in chemical detoxification of the intracellular environment. In some cells these areas may also be involved with the absorption and release of calcium ions.

GOLGI APPARATUS

The **Golgi apparatus** represents the storage and packaging center for the cell. It consists of three to seven (or more) hollow chambers, called cisternae, that resemble a stack of dinner plates. The chamber on one side of the stack receives membrane-bound bubbles or vesicles containing proteins recently synthesized by the ribosomes and pinched off from the endoplasmic reticulum. Enzymes in the cisternae then modify the new proteins, often by adding sugar molecules. Successive steps in the processing are performed in successive chambers in the stack. Material moves from chamber to chamber as vesicles pinch off from one and move to the next. When they reach the next chamber, they fuse with it. After processing, the last chamber in the stack pinches off vesicles that serve to store waste materials and intracellular enzymes and to package material for secretion by exocytosis (Figure 3-18).

Lysosomes

Some of the vesicles pinched off the Golgi apparatus contain powerful digestive enzymes that will be retained by the cell rather than released. These vesicles are the **lysosomes.** They are frequently used to digest materials in the food vacuoles of phagocytic cells. A lysosome fuses with a vacuole and adds its contents to the captured material (Figure 3-19).

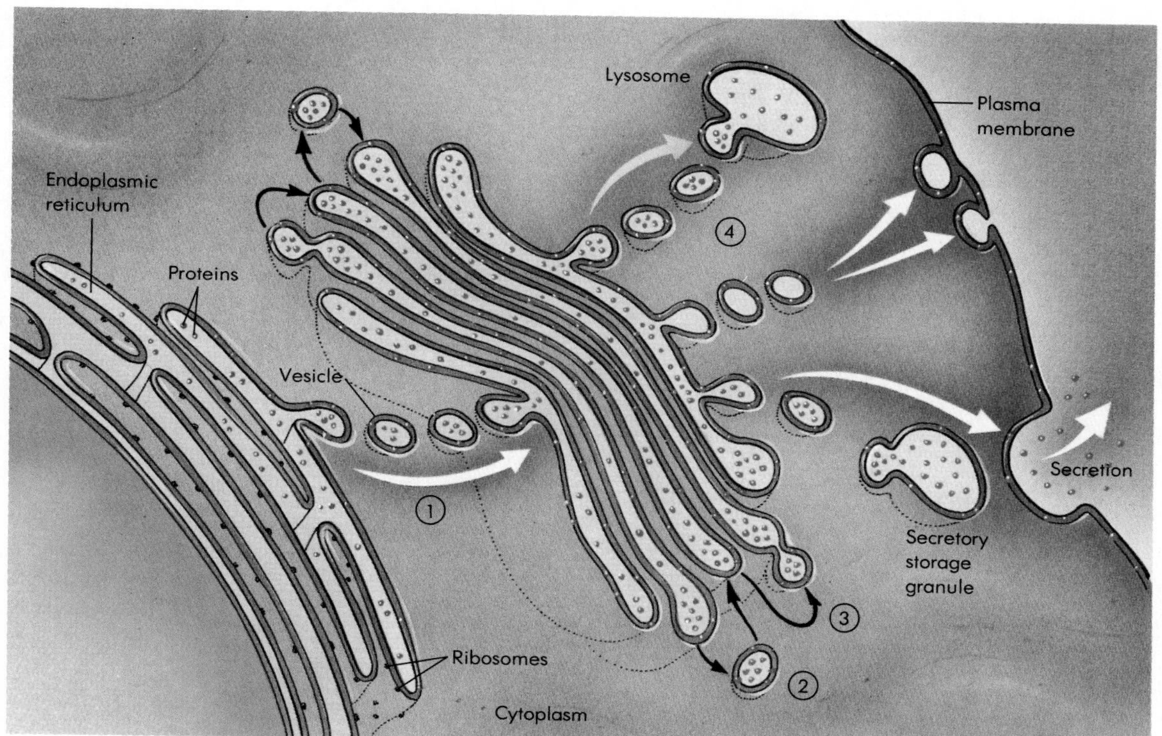

FIGURE 3-18 Golgi apparatus. (1) Proteins manufactured by the ribosomes of the rough endoplasmic reticulum usually migrate in vesicles to the face of the Golgi chambers closest to the endoplasmic reticulum. (Note: In rare electron micrographs some branches of the endoplasmic reticulum appear to be directly connected to the Golgi chambers.) (2) After their enzymes have performed the first processing steps on the proteins, more vesicles bud off, travel to, and merge with the chambers in the center of the Golgi apparatus. (3) Different enzymes then modify the proteins further, and a new batch of vesicles carries the proteins to the last of the Golgi chambers. These chambers sort the proteins and package them for delivery in vesicles that (4) migrate away from the Golgi apparatus. Some proteins go into lysosomes (vacuoles containing degradative enzymes); others go into secretory vesicles that will migrate to the edge of the cell, fuse with the plasma membrane, and release their contents into the extracellular environment. Others will simply migrate to and become incorporated with the plasma membrane.

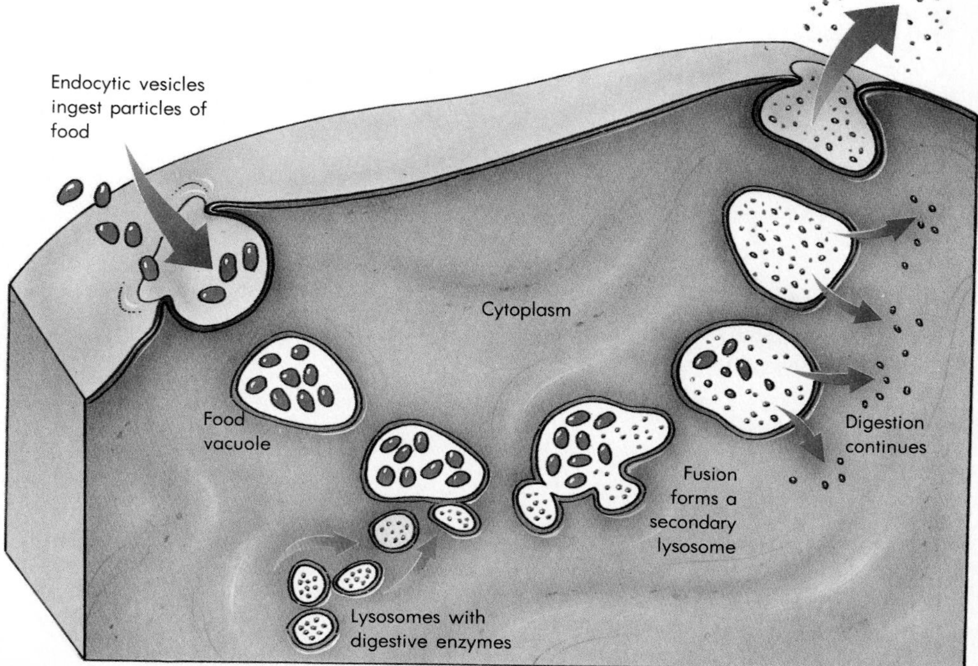

FIGURE 3-19 Intracellular digestive system. The lysosomes containing the cell's digestive enzymes fuse with endocytic vesicles (food vacuoles) containing ingested food particles. The food materials are then broken down into small molecules that leave the vesicles for use by the cell. The indigestible end products and wastes are carried back to the cell membrane and jettisoned through the process of exocytosis.

Lysosomes are also used to break down the body's own damaged or worn cells so they can be replaced with new cells. Evidently, cellular damage increases the likelihood that lysosomes will burst and digest the faulty cell that contains them. The skin peeling that occurs after a severe sunburn is provoked by ultraviolet damage to the cells of the skin, which causes the lysosomes to rupture. Similarly, a sudden reduction in the oxygen supply to the brain (as when someone has nearly drowned) can cause brain damage because low oxygen levels also cause lysosomes to burst, thereby killing brain cells.

Microtubules and Microfilaments

Early observations of the spindle fibers attached to the chromosomes of dividing cells hinted that there was a network of fibers and tubules in the cytoplasm. Current research using electron microscopes, lasers, fluorescent markers, and phase-contrast microscopy have indicated that these tiny tubules and filaments function as a cell skeleton and are responsible for many cell movements. They also seem to affect cell metabolism by providing a three-dimensional setting for the chemical processes of the cell.

The **microtubules** are the most studied of the cytoskeletal elements. They are hollow tubes formed by combining many spherical protein (tubulin) molecules (Figure 3-20). The microtubules often radiate out from an organizing center called the centriole near the cell's nucleus (see the next section). They form the basic cytoskeletal elements of the nondividing cell. However, as a cell prepares to divide, this network of microtubules breaks down and reforms to become the spindle fibers responsible for the separation of chromosomes during cell division (see Chapter 5). Microtubules also form the **flagella** and **cilia** used in the locomotion of highly motile cells, like sperm cells and protozoans (Figure 3-21). Flagella and cilia are cylindrical projections from the cytoplasm covered by an extension of the cell membrane. Their 20 microtubules are arranged as a central pair surrounded by nine additional pairs. Movements of flagella or cilia result from interactions between pairs of microtubules of the outer ring and the pair at the center. Microtubules are also used in amoeboid movement.

Microfilaments are smaller in diameter, nontubular in structure, and constructed of different proteins (actin and myosin). They are often found in parallel bundles of stress fibers just inside the cell membrane, and they play roles in muscle contraction, cytoplasmic streaming (circulating movements of the cytoplasm), and changes in cell shape. They are most abundant in flattened cells

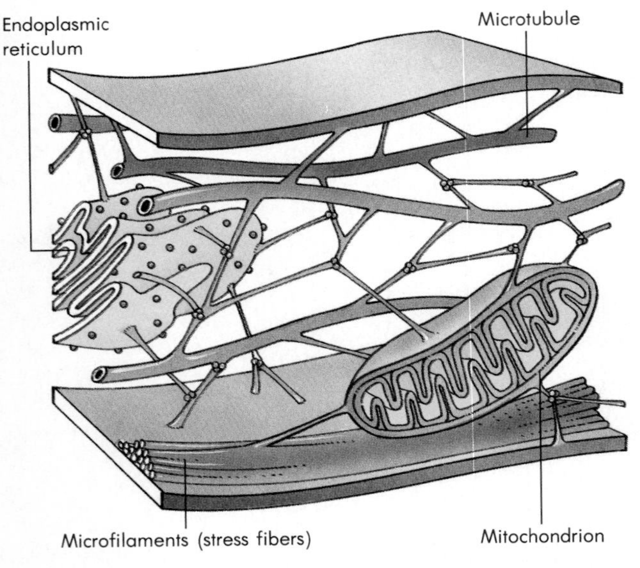

(a)

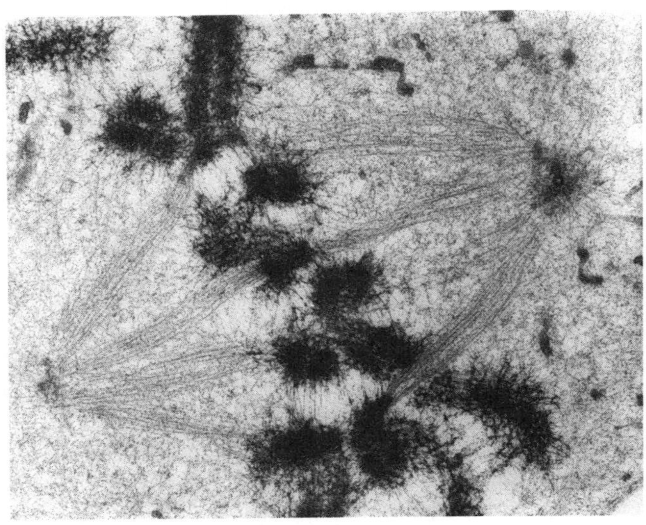

(b)

FIGURE 3-20 (a) Microfilaments are elements of the cytoskeleton. They give the cell some of its three-dimensional organization and structural support. These filaments are primarily constructed of the protein actin and are often used in certain types of cellular movement. (b) During cell division, the microtubules disassemble and reform to make the spindle fibers that move the chromosomes, as seen in this preparation made from a vertebrate epithelial cell. Each spindle pole contains centrioles and other materials. The microtubules also form the core materials of cilia and flagella.

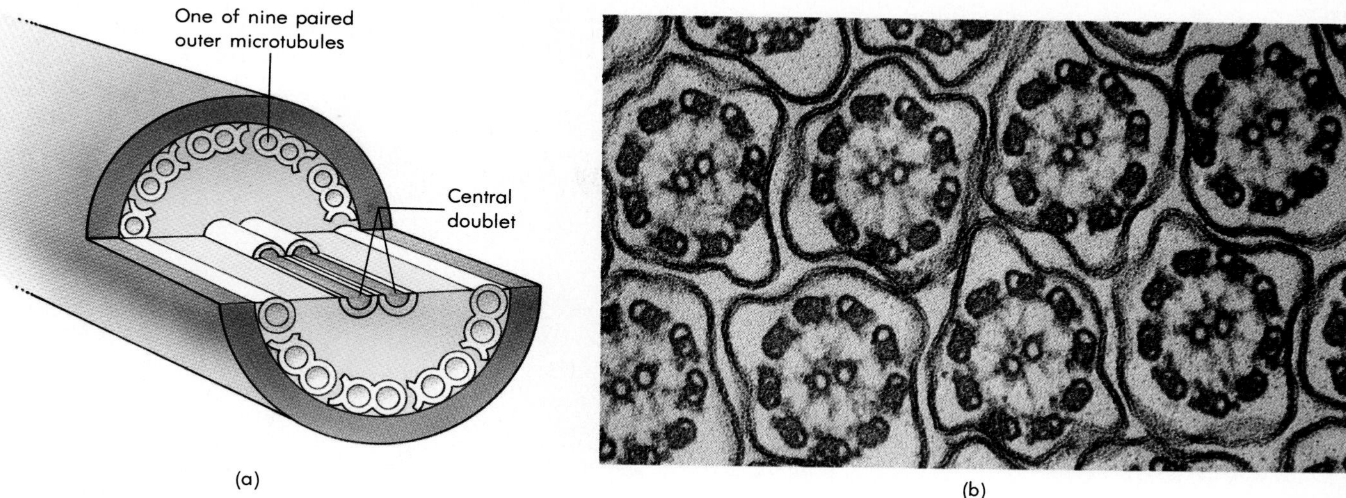

FIGURE 3-21 (a) A longitudinal section through a group of *Tetrahymena* cilia showing their microtubular structure. (b) A cross section of the cilia shows their 20 microtubules—a central pair surrounded by nine additional pairs. The outer covering of each cilium is an extension of the plasma (cell) membrane. The cilia are anchored at their bases by basal bodies just within the cell membrane.

that spread out over a fairly large area or are resting on a membrane or some other supporting surface.

Centrioles

Animal cells possess a pair of tiny **centrioles** located just outside the nucleus in nondividing cells. They resemble short segments of cilia held at right angles to each other without the enveloping membrane and without the pair of tubules located in the center. They serve as organizing centers and anchoring points for the microtubules that form the spindle fibers during cell division (Figure 3-22) (see Chapter 5).

Vacuoles

Vacuoles are membrane-bound compartments found in various eukaryotic cells. In most simplified cell diagrams, they resemble fluid-filled cytoplasmic bubbles. They are used for a number of purposes. *Storage vacuoles* hold materials being saved, such as fat and starch, for future use and keep them separate from the cytoplasm. *Food vacuoles* hold particles of food such as those obtained by phagocytosis. *Contractile vacuoles,* seen in free-living single-celled protozoans, hold liquid wastes until they can be excreted. Plant cells often use a large, rather conspicuous *central vacuole,* occupying much of the cell's center, to store wastes and

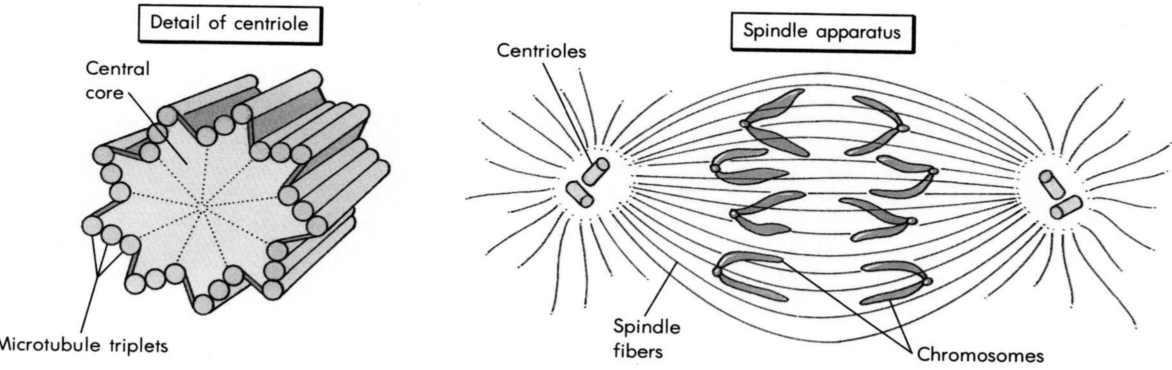

FIGURE 3-22 Centrioles are also formed of microtubules, but they are arranged as nine triplets of microtubules in a circle around a central core. Each nondividing cell has two centrioles arranged at right angles to each other, usually just outside the nuclear envelope. Prior to cell division, the centrioles duplicate themselves. The resulting two pairs of centrioles separate and then migrate to opposite sides of the cell to serve as organizing centers for the spindle fibers. Note that the arrangement of microtubules in the basal bodies of cilia and flagella is the same as in the centrioles (see Figure 3-21).

Box 3-1 Origins of Life

Many biologists believe that the earliest life on Earth had to occur in forms resembling simple cells and that it had to emerge from the primordial seas, but even their agreement is speculation. We cannot know the actual details of what happened to produce the dawn of life. It was simply too long ago, and its conditions were too unlike those we know today.

THE EARLY EARTH

When researchers first addressed the question of how life may have begun, they worked from the little they knew about the conditions that probably existed on the early Earth. They knew, for instance, that the early Earth had water, carbon monoxide, carbon dioxide, ammonia, silicon dioxide, and many other inorganic materials, all of which were present in the cloud of gas and dust that condensed to become the solar system. There may also have been many simple organic compounds, including methane, ethane, acetylene, ethylene, formaldehyde, formic acid, and methanol, in spite of the fact that the most important sources of these organic materials—living systems—had not yet made their appearance.

There was very little free oxygen in the atmosphere because plants either had not yet appeared or had not yet evolved ways of chemically releasing it from water molecules (Figure 1). The planet's rocks were thus shades of black and green, not the reds and browns (produced by exposure to oxygen) we are used to seeing. There were no sedimentary rocks nor coal nor oil nor fossils. The torrential rains that fell to fill the seas were acidic because of the volcanic fumes in the atmosphere.

The primordial Earth was bombarded with energy from many sources. There were shock waves from abundant meteors, electrical discharges from lightning, background radiation triple today's level, and heat from molten lava. More energy reached the planet's surface from the sun in the form of heat, light, and ultraviolet radiation (which was more intense then than now). And all these energies reacted with the organic and inorganic chemicals in the early Earth's atmosphere, on its surface, and in the seas, changing them in very specific ways that led to the formation of the protocells that eventually gave rise to "life."

FIGURE 1 The atmosphere of the early Earth.

ORGANIC CHEMISTRY AND THE EARLY EARTH

Research into the origins of life began with efforts to see how the components of modern cells might fit together to make those cells. Since this approach used highly evolved cells, however, it could say little about the first cells, which had to be much simpler. Once researchers realized this, they tried to duplicate the conditions they thought had existed on the primitive Earth, seeking events that might have led to the formation of the first living cells.

This synthetic approach began in 1953 with the work of Stanley Miller and Harold Urey. They put a mixture of the gases methane, ammonia, water vapor, and hydrogen—a "primitive" atmosphere—in a glass vessel, exposed it to electric discharges (simulating lightning), and found that many important organic molecules, such as amino acids and nitrogenous bases, formed spontaneously. Their experiment showed that molecules essential to the formation of life could easily have been produced in the conditions of the primitive Earth (Figure 2).

Other researchers soon showed that all 20 amino acids common to modern protein, porphyrins (molecules related to vitamins, chlorophyll, and hemoglobin), sugars, lipids, and even nucleotides can form under similar conditions. Some of the experiments producing large quantities of organic products used only mild heat as the energy source. However, showing that the chemicals essential to life can be produced easily in the absence of life and under conditions resembling those on the primitive Earth is still a long way from showing the origins of a living cell.

IN SEARCH OF THE FIRST CELLS

Some researchers have found that certain clays can encourage amino acids to string together to form proteins. Perhaps more encouraging, the experiments of Sidney Fox showed that a heated mixture of amino acids can form proteinlike **proteinoids.** Proteinoids are not quite true proteins because they are highly branched in structure. However, they do have many of the properties of true proteins. For instance, proteinoids resemble enzymes in that they have limited powers to speed up certain chemical reactions. They may thus have served as the starting point for the evolution of more powerful and specific enzymes. They may also have paved the way for the development of the first cells.

The first cell, or the **protocell** that foreshadowed it, may not even have been alive in the technical sense. It would have existed in seas full of molecules identical to those within it. Thus, it would have been surrounded by an environment full of "spare parts." It could absorb what it needed from its surroundings. In fact, the first cell needed only three things:

1. A way of separating itself from its surroundings, for example, by means of a simple membrane.
2. The ability to absorb materials and grow.
3. The ability to reproduce.

Only later, as a consequence of evolution, need it acquire the other characteristics we use to define life (such as energy acquisition and processing and the ability to respond to stimuli).

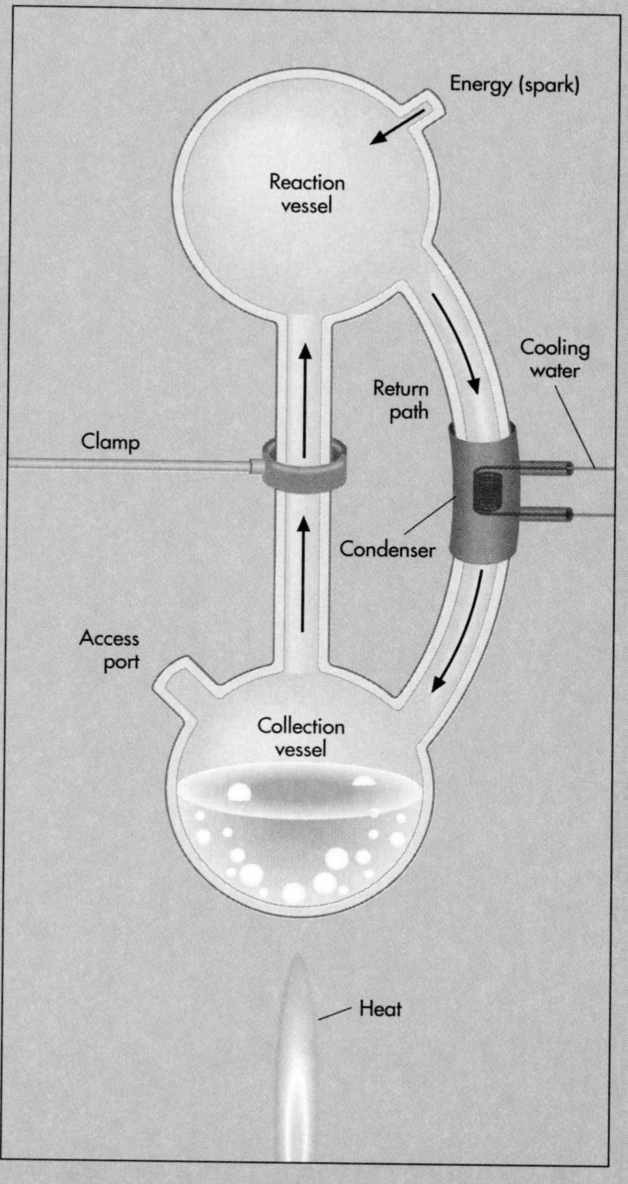

FIGURE 2 Early experiments attempted to show that conditions on the primordial Earth could produce the organic molecules necessary for the origins of life. Miller and Urey, for example, filled the upper chamber of a closed container with a mixture of gases they believed would mimic the early atmosphere. The gases were exposed to electrical discharges, radiation, ultraviolet light, and/or heat. As the gases reacted with each other, many organic molecules used by living things collected in the water at the bottom of the container.

(Continued on p. 76)

RISE OF THE PROTOCELL

There are several theories about how the first protocells came into being. Each of these theories might be contributing part of the actual explanation of how cells came into existence. Many types of "protocells" might have experimented at the edges of life several times independently. One theory suggests that proteinoids or clusters of proteins coagulated into *coacervate droplets* (globs of protein suspended in water). Such droplets can be made in the laboratory; artifically loaded with enzymes taken from modern cells, they can then behave something like cells, carrying out internal chemical reactions and occasionally even dividing. However, they lack any structure (such as a cell membrane) that effectively separates them from their surroundings.

The best theory of the origin of protocells is Sidney Fox's *microsphere theory* because it starts with a rudimentary membrane. In the 1960s, Fox discovered that small quantities of proteinoid dissolved in water spontaneously form billions of small, hollow spheres (Figure 3). These *microspheres* would readily have formed in prebiotic times, every time waves washed proteinoids from the shore into the water.

Microspheres are about the same size as bacterial cells. They have layered surface membranes that may surround an interior containing additional proteinoids and other substances as well. They are also cell-like in their abilities to separate charges across the membrane and to reproduce by budding off smaller versions of themselves. Some microspheres have even been seen to move about; one was so active (as revealed by time lapse photography) that the people in the lab named it "Motile Marvin." Also, they can grow by absorbing more proteinoid from the surrounding solution. In addition, they tend to preserve the enzymatic abilities of their constituent proteinoids, making them good candidates for protocells. In time, some would incorporate bits of nucleic acids and protein particles with slightly more efficient enzymatic properties. Some would even pick up bits of pigment that would make them slightly more efficient in absorbing energy. From such humble beginnings, perhaps, began the history of life.

Eventually, the substances that used to be available for the taking from the sea were all locked away in protocells. The resulting shortage of resources began the competition that eventually became natural selection. Some of the early experiments at the edges of life were surely rewarded with survival and refined by the intensifying competition. Most disappeared without a trace.

FIGURE 3 Microspheres resemble protocells in that they are separated from the external environment by a rudimentary membrane, they divide or bud off small versions of themselves, and they grow.

pigments and to help maintain the pressure in the hydraulic system that acts as a supporting system in herbaceous plants.

MITOCHONDRIA: ENERGY PROCESSORS

Energy reaches cells in the form of nutrients. Before cells can use it, however, they must release this nutrient energy and store it in a form the cell's energy-requiring systems can use. The energy is stored in ATP molecules through a complex series of reactions known as **cellular respiration.** The most efficient reactions of cellular respiration occur in the 1000–10,000 **mitochondria** found in almost every cell. Most mitochondria lie in or near parts of the cell that use the most energy. Some people thus call mitochondria *cellular powerhouses.*

The usually oval mitochondrion is composed of two membranes. The outer membrane is smooth, but the inner membrane is folded into characteristic shelflike **cristae** (Figure 3-23) whose surfaces bear numerous small particles and molecules. The latter, called **cytochromes,** play a vital role in the movement of electrons required for the energy-transferring reactions of cellular respiration (see Chapter 20). The space between the cristae is filled with a gelatinous material, the matrix, containing some ribosomes, enzymes that support the set of energy-transferring chemical reactions known as the Krebs cycle, and DNA.

The presence of DNA in mitochondria (and chloroplasts) has prompted the idea that these organelles originated a billion years or more ago as moneran parasites or symbiotes of primitive eukaryotes.

PLASTIDS

The **plastids** are a class of organelles found in plant cells, distinguished by the type of substances they contain. The leucoplasts are usually colorless storage depots for starches. The chromoplasts contain various colored pigments that often give coloring to fruits, vegetables, and other plant parts. One type of chromoplast—the **chloroplast**—is extremely important because of its impact on virtually all life forms on this planet. Plants use their chloroplasts to trap and incorporate the energy from sunlight into the sugars and other organic molecules that serve animals as food. As a by-product, they release oxygen. The process is called **photosynthesis,** and it is one of the most important chemical events on this planet (Figure 3-24, p. 79).

1. Why might we describe the events taking place in chloroplasts as *energy-capturing* processes and those taking place in mitochondria as *energy-releasing* processes?

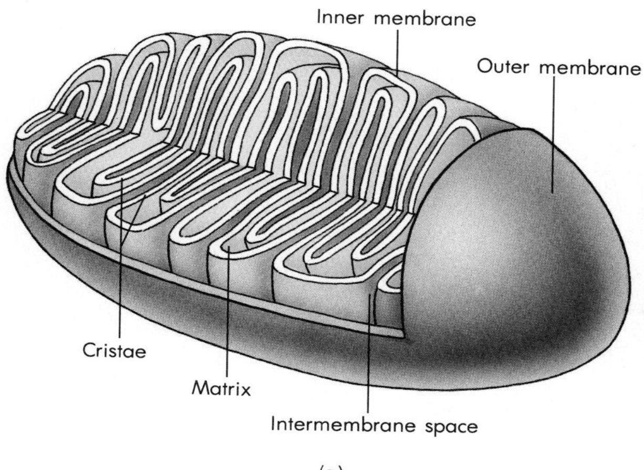

(a)

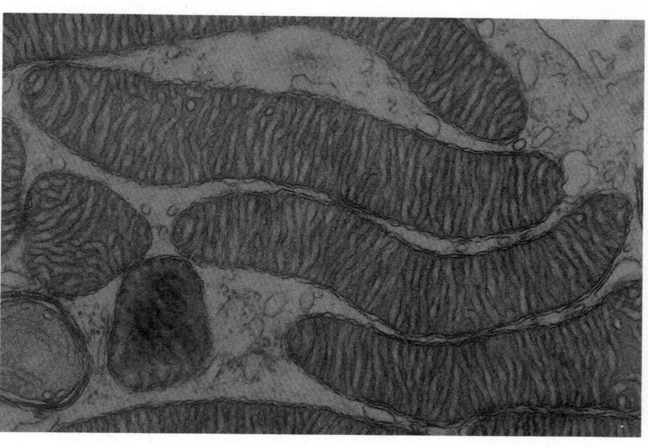

(b)

FIGURE 3-23 The mitochondria are the organelles responsible for converting nutrient energy into a form the cell can use. That complex process is called cellular respiration. (a) Each mitochondrion is an elongated vesicle with a double membrane surrounding an interior matrix. The inner membrane is folded into flattened shelves called cristae. Embedded in the cristae are the cytochromes used in cellular respiration. (b) In the micrograph, the elaborate array of cristae define the mitochondria.

Learning Focus

THE CELL

Since the fundamental unit of life is the cell, it is important that you can understand and organize the information you have read about cells. For that reason, we are providing you with an unlabeled diagram of the generalized eukaryotic cell. We recommend that you make several photocopies of this blank diagram to use as a study aid.

Review the anatomy of the typical cell shown in Figure 3-17 of this chapter. Once you have mastered it, return to the unlabeled diagram and identify the organelles from memory. At the same time, organize what you know about the organelles by writing notes about each structure right on this figure next to the label. Check your performance by referring back to the original figure. Then go over the description of each organelle, extract any useful information from the text, and note it on this page. Go over your lecture notes and record any additional information about each organelle on the blank diagram as well.

Repeat this exercise as many times as you need to master the subject.

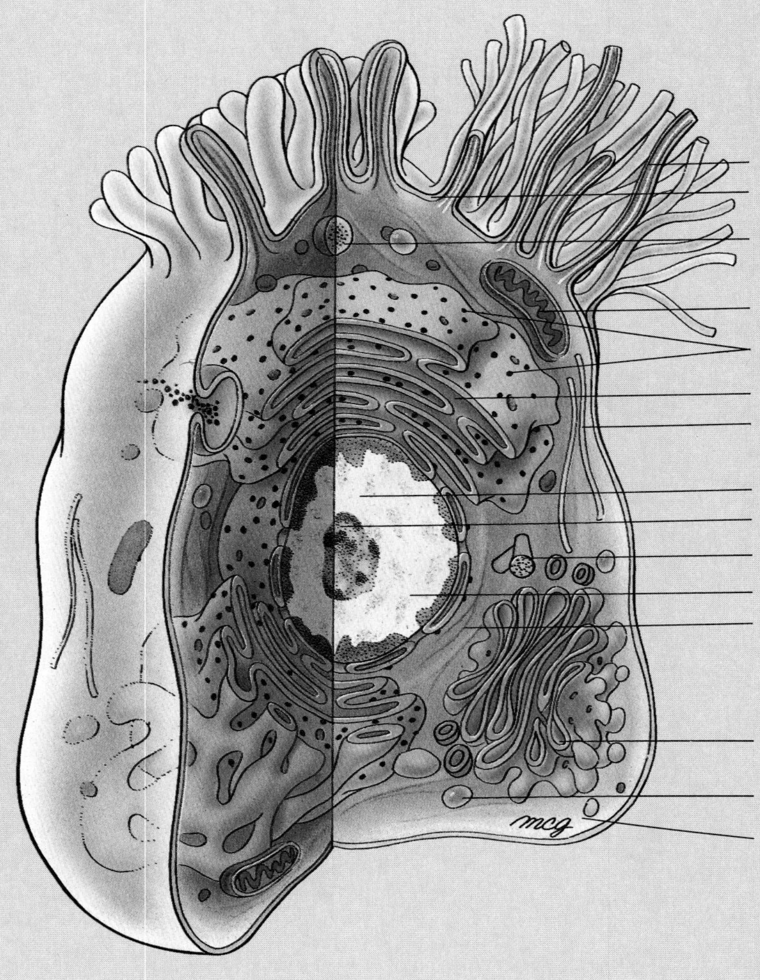

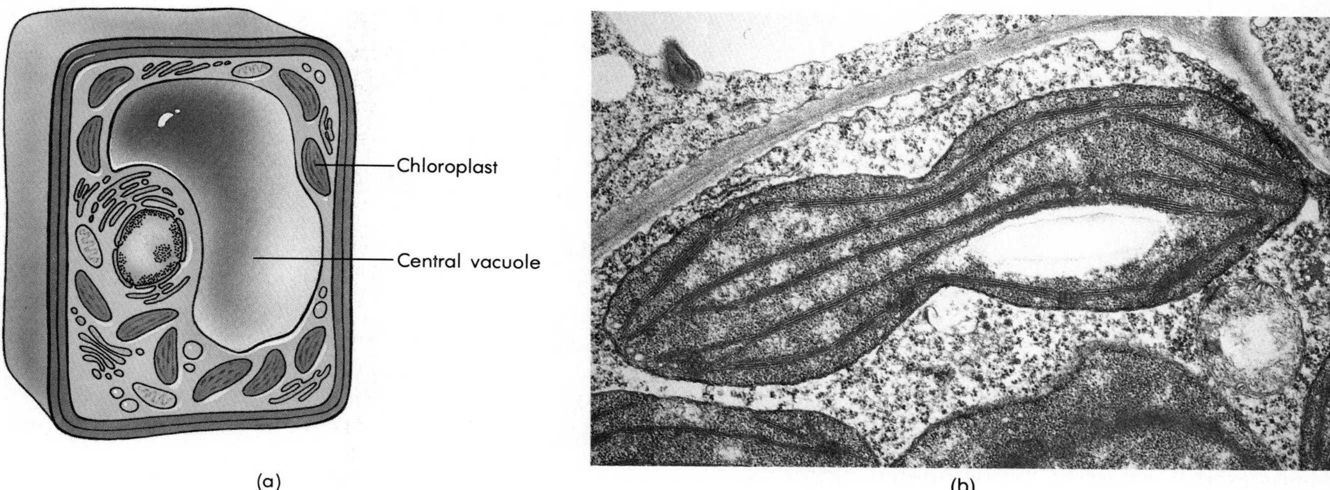

FIGURE 3-24 (a) Chloroplasts are the green energy-capturing organelles found in photosynthetic plant cells. They capture a portion of the energy in light and use it to manufacture the sugar (glucose) that powers the biosphere. (b) False-color transmission electron micrograph of a dividing chloroplast in a leaf cell of the pea plant.

SUMMARY

The cell theory states that all life takes place in cells, the fundamental units of life, and that all cells come from preexisting cells (except the first cells).

Cells come in many shapes and sizes. Most are microscopic, and all are encased in a cell membrane that controls the chemical climate within. Prokaryotic cells (bacteria and cyanobacteria) are simpler and more primitive in structure than the more modern and complex eukaryotic cells from which the human body is made. Eukaryotic cells are more highly compartmentalized; they confine their genetic material within a membrane-bound nucleus, and they have many other membranous organelles such as mitochondria (for energy processing), chloroplasts (for photosynthesis), endoplasmic reticulum (intracellular pipeline), Golgi apparatus (packaging center), and vacuoles (food processing, waste storage and excretion, and fat storage). Cellular appendages, the cilia and flagella, serve as organs of locomotion or create currents or movements in extracellular fluids. Both eukaryotic and prokaryotic cells use organelles called ribosomes as manufacturing centers for their proteins. Table 3-3 summarizes many differences between prokaryotes and both plant and animal eukaryotes.

Table 3-3

A Comparison of Cell Structures in Different Cell Types

Organelle	Bacteria (Prokaryote)	Animal (Eukaryote)	Plant (Eukaryote)
Cell membrane	+	+	+
Nucleus	−	+	+
Nucleolus	−	+	+
Nucleoid	+	−	−
Ribosome	+	+	+
Endoplasmic reticulum	−	+	+
Golgi apparatus	−	+	+
Lysosome	−	+	+
Mitochondrion	−	+	+
Chloroplast	−	−	+
Vacuole	−	+	+
Centriole	−	+	−
Cell wall	+	−	+
Flagellum	+	+	+

Molecules move in and out of cells both passively and actively. Passive transport requires no expenditure of energy, for it carries molecules only "downhill," from zones of greater concentration to zones of lesser concentration. Its mechanisms are diffusion, facilitated diffusion, and osmosis. Active transport uses energy and can carry molecules up concentration gradients. Active transport often relies on "pump" molecules embedded in the cell membrane. In phagocytosis and pinocytosis, the cell membrane deforms to admit bubbles of exterior fluid into the cell. In phagocytosis, the bubbles contain solid materials like cells, parts of cells, and debris. In pinocytosis, only liquid is ingested.

The physical and chemical conditions existing on the

early Earth aided the origins of life. The organic materials necessary for life to form were already present over 3.8 billion years ago, and they presumably led to the formation of proteinoids, microspheres, and protocells. Time and competition then refined the protocells into rudimentary cells. From these early, simple cells came the prokaryotes and eventually the eukaryotes, setting the stage for the evolution of the various forms of multicellular life such as plants, fungi, and animals, including humans.

STUDY QUESTIONS

1. Discuss why the cell is considered the fundamental unit of life.
2. Explain the structural and historical differences between prokaryotic and eukaryotic cells.
3. Give the functions of each of the following cell parts: nucleus, nucleolus, endoplasmic reticulum, Golgi apparatus, mitochondria, chloroplasts, plasma membrane, centrioles, microtubules and filaments, cilia and flagella, vacuoles, ribosomes, and lysosomes.
4. Why is the microsphere model for the origin of protocells favored over the coacervate theory?

GLOSSARY

Active transport The use of energy to move substances into and out of cells.

Centrioles Tiny structures made of microtubules; organizing centers for the spindle apparatus.

Cilia Locomotor organelles built of microtubules.

Cytochromes Proteins that make ATP in cellular respiration.

Cytoplasm That portion of a eukaryote's protoplasm between the plasma membrane and the nuclear envelope, or all the cytoplasm of a prokaryote.

Diffusion The movement of molecules from where they are highly concentrated to where they are less concentrated.

Endoplasmic reticulum (ER) The intracellular pipeline. A series of membranous channels linking the various areas of the cytoplasm; the principal route for molecular traffic around the cell.

Eukaryotes Complex modern cells, including those of humans, that have a well-defined nucleus and other membrane-bound organelles.

Exocytosis Fusion of a membrane-wrapped bubble of material made in the cell with the cell membrane; mechanism of secretion and membrane growth.

Facilitated diffusion The transport of materials across a membrane, down a diffusion gradient, assisted by carrier molecules in the membrane.

Flagella Locomotor organelles built of microtubules; longer than cilia.

Golgi apparatus A stack of three to seven (or more) flattened chambers that package proteins and lipids made in the cell.

Hypertonic solution Solution containing more solute (less water) than the cells of the body.

Hypotonic solution Solution containing less solute (more water) than the cells of the body.

Isotonic solution Solution containing the same amount of solute (and water) as the cells of the body.

Microfilaments Thin, nontubular fibrils responsible for changes in cell shape.

Microtubules Hollow fibers that serve as skeletal and motor structures in the cell.

Mitochondria The sites of cellular respiration in eukaryotes.

Nuclear envelope The double membrane separating the nuclear compartment from the cytoplasm.

Nucleolus The region of the nucleus where ribosomal RNA is manufactured.

Nucleus A conspicuous organelle near the center of most eukaryotic cells; contains the chromosomes and nucleolus and serves as the control center for the cell.

Organelles The tiny structures and compartments that act as the cell's "organs."

Osmosis The diffusion of water through a selectively permeable membrane.

Phagocytosis Cellular eating by surrounding an external object with cell membrane.

Photosynthesis The trapping of light energy and its use in the synthesis of glucose in plant cells.

Pinocytosis Cellular drinking by drawing in a bubble of cell membrane.

Plasma (cell) membrane The active outer covering of cells. It controls which materials will be allowed to enter or leave cells.

Prokaryotes Primitive cells that lack a nucleus and other membrane-bound organelles (bacteria, cyanobacteria).

Ribosomes Tiny granular organelles that are the principal sites of protein synthesis in the cytoplasm.

Selectively permeable membranes Membranes that allow certain substances to pass through but prevent, inhibit, or otherwise control the movements of other materials.

Vacuole Membrane-bound compartment in a eukaryotic cell; contains food, fluid, or stored material.

Chapter 4

Control of the Cell: Molecular Genetics

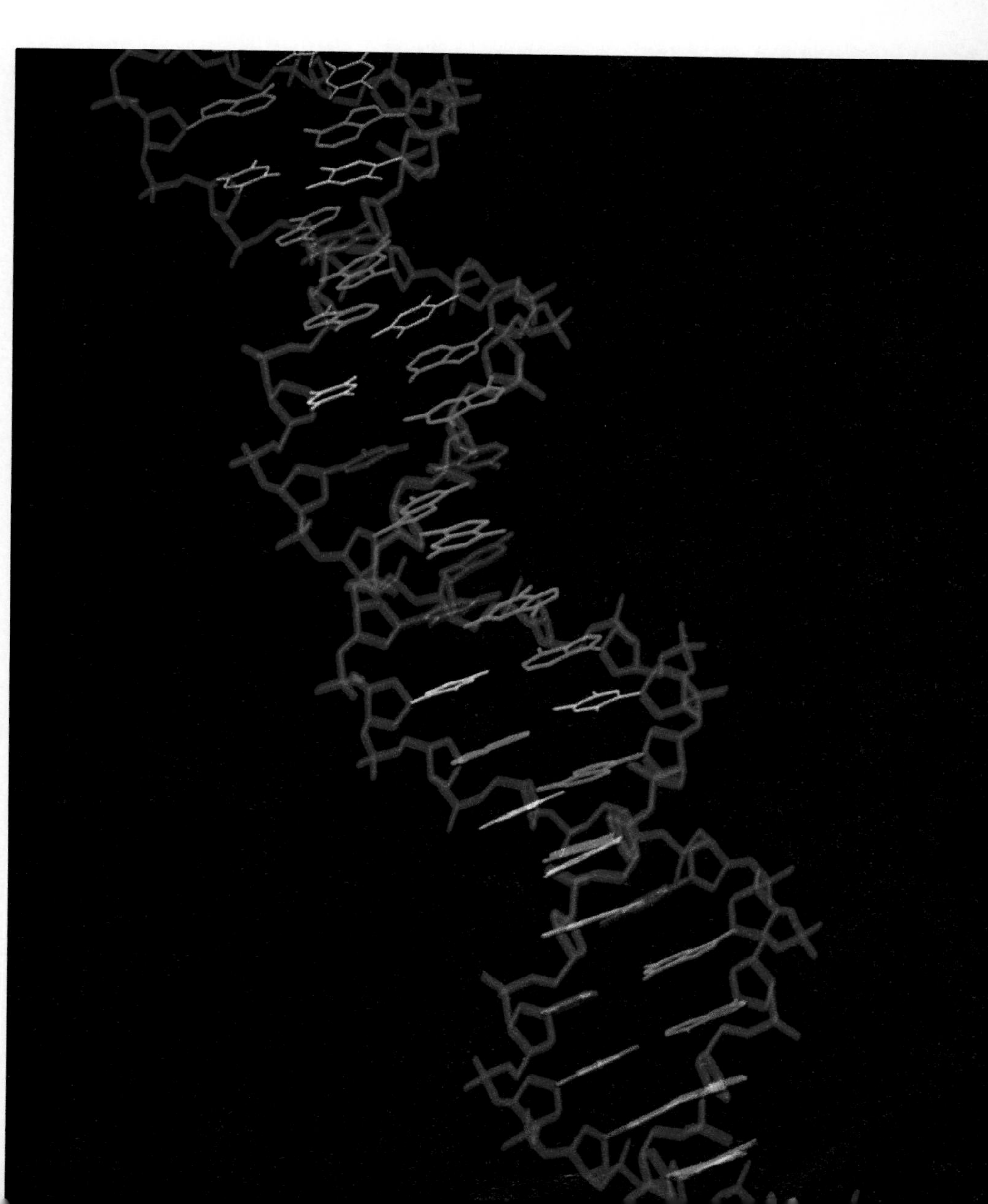

The astounding complexity of the tiny cell has become obvious only in the past few decades. However, the cell seemed complex enough a century ago to baffle biologists. How, they wondered, did the cell manage to generate so many structures? How did it control so many activities? How could one cell—a fertilized egg—give rise to so many other kinds of cells—muscle, liver, brain, skin, bone, and so on?

There must, they thought, be some mysterious substance within each cell that encoded the vast amount of necessary information. Furthermore, that substance must be able to pass the information with relatively little change from generation to generation. Only in this way could the similarities of parent and child be explained. Only in that manner could the workings of evolution have a firm basis.

But what was that substance? How did it work? No one knew, although some tried to guess.

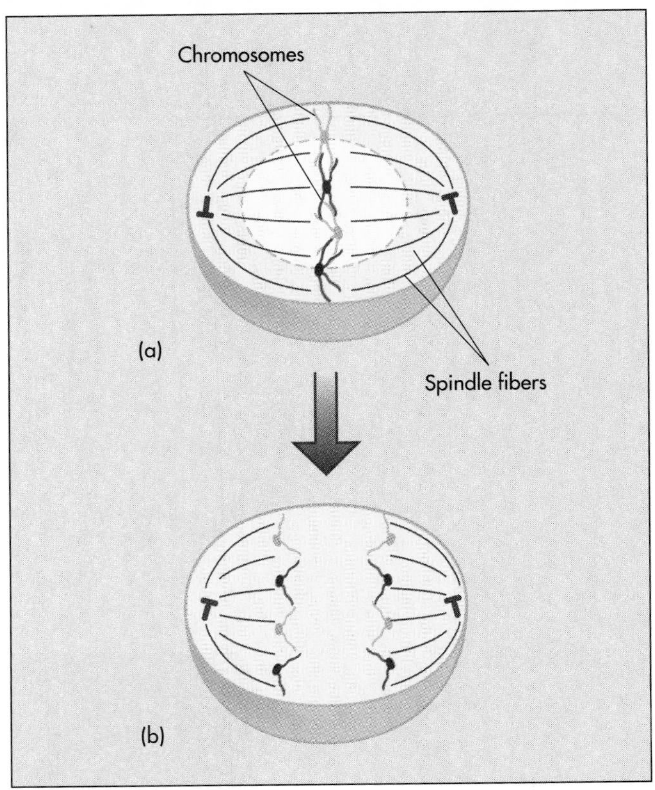

FIGURE 4-1 A dividing cell. (a) The X-shaped chromosomes (each consisting of two identical "sister" chromatids) of a dividing cell in late metaphase with spindle fibers attached. (b) The spindle fibers contract, pulling the chromatids apart and dragging one complete set of genes to each side of the cell. When the cell pinches in half, each daughter cell will have a complete single-stranded set of genetic instructions.

THE CHROMOSOMES

By the beginning of this century, biologists recognized that within the nucleus of each eukaryotic cell were minute X-shaped threadlike structures, the **chromosomes.** As cells prepared to divide, these structures condensed out of the nucleoplasm, becoming readily visible with the aid of a microscope. During cell division, the connected strands that formed the two sides of the X moved apart, so that when the dividing cell pinched in half, each of the daughter cells thus formed received a single strand from each of the chromosomes (Figure 4-1). Because the strands (or bodies: *somes*) absorbed large amounts of certain dyes (or colors: *chromo*), they were given the name *chromosome.* A more detailed look at chromosome structure can be found in Chapter 5.

The fact that each newly formed daughter cell received a part of each chromosome was consistent with the idea that chromosomes might be carrying the information of heredity. The questions that remained were how they carried that information and how the cell used that information to control its structure and activities. Unfortunately, it took many years to find the answers.

THE DISCOVERY OF THE GENE

Early in this century, it was also clear that the information of heredity came in discrete units, the **genes,** with one gene for each of an organism's many characteristics or traits. In Chapter 19, you will discover the ways these traits are inherited. For now, let us note simply that many of these hereditary units pass from parent to offspring as if each one belonged to a separate chromosome. However, we now know there are over 50,000 human genes (some estimates suggest the actual number may be double that) spread over the 46 chromosomes that normally make up the complete set.

It was Thomas Hunt Morgan and his co-workers at Columbia University who, in 1911, made the leap to the idea that the hereditary units were lined up on the chromosomes like beads threaded on a string (Figure 4-2). Today we consider this descrip-

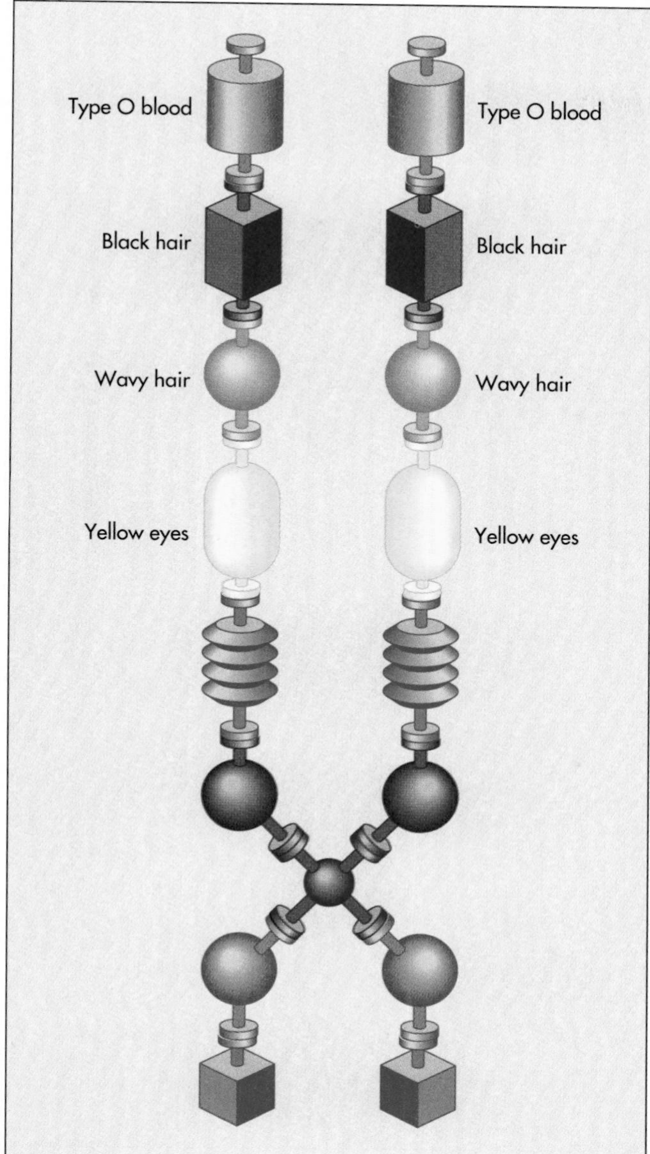

FIGURE 4-2 A simple model paralleling the relationship between a chromosome and the genes it carries is a child's string of plastic pop-together beads. Each bead is shown as a gene. A string of such genetic beads is a chromosome.

tion of genes along a chromosome simplistic, but it remains a useful analogy.

THE COMPOSITION OF THE GENE

Less than 50 years ago, no one understood how a gene might carry information. Biologists had figured out that genes must carry the patterns or recipes for assembling the proteins, especially enzymes, that perform the many activities of the cell. Chemists had shown that chromosomes were made of protein and nucleic acid. But opinions differed strongly about whether the protein or the nucleic acid carried the hereditary information. Proteins seemed to have the advantage because they were complex molecules made of 20 different units, the amino acids. If the genes were made of proteins, the cell needed no more than a way to copy them.

Nucleic acids, on the other hand, were each made of only four different units, the nucleotides (see Chapter 2). To many investigators, the nucleic acids seemed poorly suited to coding the vast amounts of information necessary to spell out the various cellular proteins. The reason is that nucleic acids used an "alphabet" of only 4 chemical letters to specify the 20 different amino acids normally found in proteins.

Still, some researchers did favor nucleic acids as the genetic material. They thought it significant that the amount of nucleic acid in a cell's nucleus always doubled before the cell divided, as if the cell were making a copy of its information for its offspring.

1. Why should the number of letters in an alphabet affect the amount of information that alphabet can be used to carry?
2. A cell contains a great deal of protein in and out of the nucleus. Why might this protein make it difficult to detect a predivision doubling of "hereditary" protein?

Genetic Transformation

The first step toward resolving the argument came in the 1920s, when Fred Griffith of the British Ministry of Health discovered two forms, S and R, of the bacterium responsible for pneumonia. The S form had a gelatinous (polysaccharide) coating, grew in smooth-surfaced colonies in laboratory dishes, and caused fatal disease in mice. The R form lacked the coating, grew in rough-surfaced colonies, and was not lethal in mice. The difference was hereditary and therefore due to a gene.

Griffith later found that when he injected dead S (lethal) cells into a mouse, the mouse survived. But if he injected both dead S cells and live R cells, some of the R cells turned into S cells, killing the mouse (Figure 4-3). The conversion of R cells into S cells by some substance apparently released from dead S cells is called **transformation.** It requires the transfer of a bit of hereditary material—one or

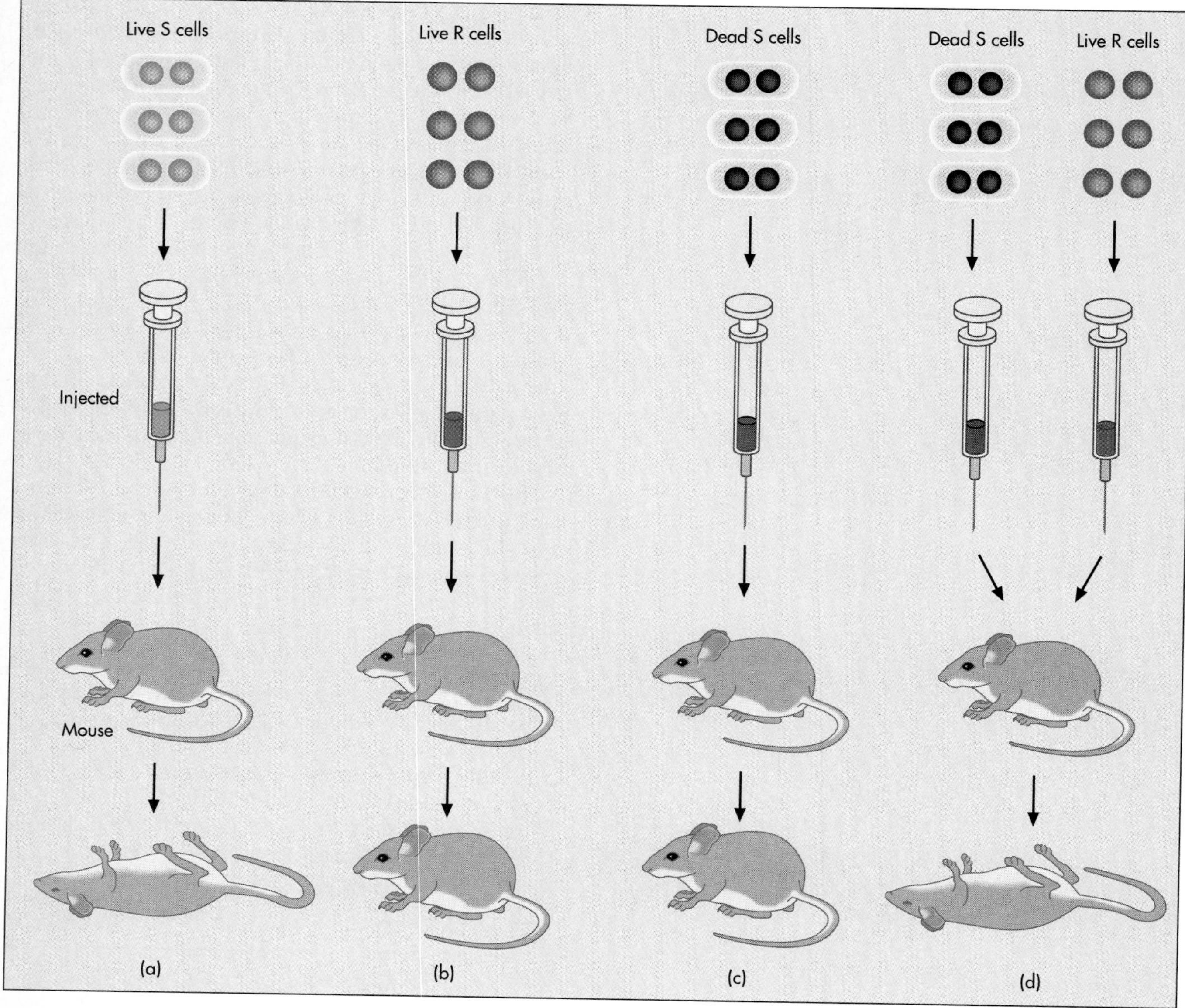

FIGURE 4-3 (a) Injections of live S-type pneumonia bacteria kill mice by causing a lethal pneumonia. (b) When mice are injected with the R strain of bacteria, they do not die. (c) Nor does injecting them with S cells that have been previously killed cause them to die. (d) Injecting mice with a mixture of dead S cells and live R cells, however, kills. The nonlethal R cells are somehow transformed to lethal S cells by exposure to the dead S-type bacteria.

more genes—from the dead S cells to the living R cells.

Griffith never figured out his "transforming principle." In the 1940s, however, Oswald T. Avery and his co-workers attacked the problem. They separated the S form of the pneumonia bacterium into all its chemical components—lipids, proteins, polysaccharides, and nucleic acids—and mixed them, one by one, with R bacteria. They found that only the nucleic acids—specifically the form of nucleic acid found in the chromosomes, **deoxyribonucleic acid,** or **DNA**—could transform R cells into S cells. They had identified the substance of the gene. Despite the shortness of its alphabet, DNA encoded all the information that made each living thing what it was.

Unfortunately, it was another decade before this discovery was universally accepted. In 1952, A.D. Hershey and M. Chase were studying viruses (bacteriophages) that infect bacteria. Like chromosomes, these viruses were known to be made of both protein and DNA. They were also known to reproduce by inserting their hereditary material into bacterial cells to force those cells to manufacture more viruses. To identify the hereditary material, Hershey and Chase grew two batches of virus, one in nutrients labeled with radioactive sulfur, which became incorporated in their protein component, and one in nutrients labeled with radioactive phosphorus, which became incorporated in their DNA. They then allowed each batch to infect bacteria and looked at the bacteria to see where the radioactivity went. They found that only the radioactive phosphorus, and hence the DNA, entered the bacterial cells (Figure 4-4). The case for DNA as the carrier of the hereditary information was now strong enough to convince everyone.

We now know that DNA is not *always* the substance of heredity. Some viruses use the related substance **ribonucleic acid,** or **RNA.** But many viruses, all bacteria, and all plants and animals use DNA to carry their hereditary information. Even the mitochondria and chloroplasts within plant and animal cells use DNA as the substance of their few genes (which suggests their origins, billions of years ago, as parasitic bacteria). However, RNA does serve all living things, as we will see later in this chapter, as an information carrier, an intermediary between the gene and the trait the gene controls.

THE STRUCTURE OF DNA

As we mentioned in Chapter 2, DNA has four different subunits, the **nucleotides.** Each nucleotide has

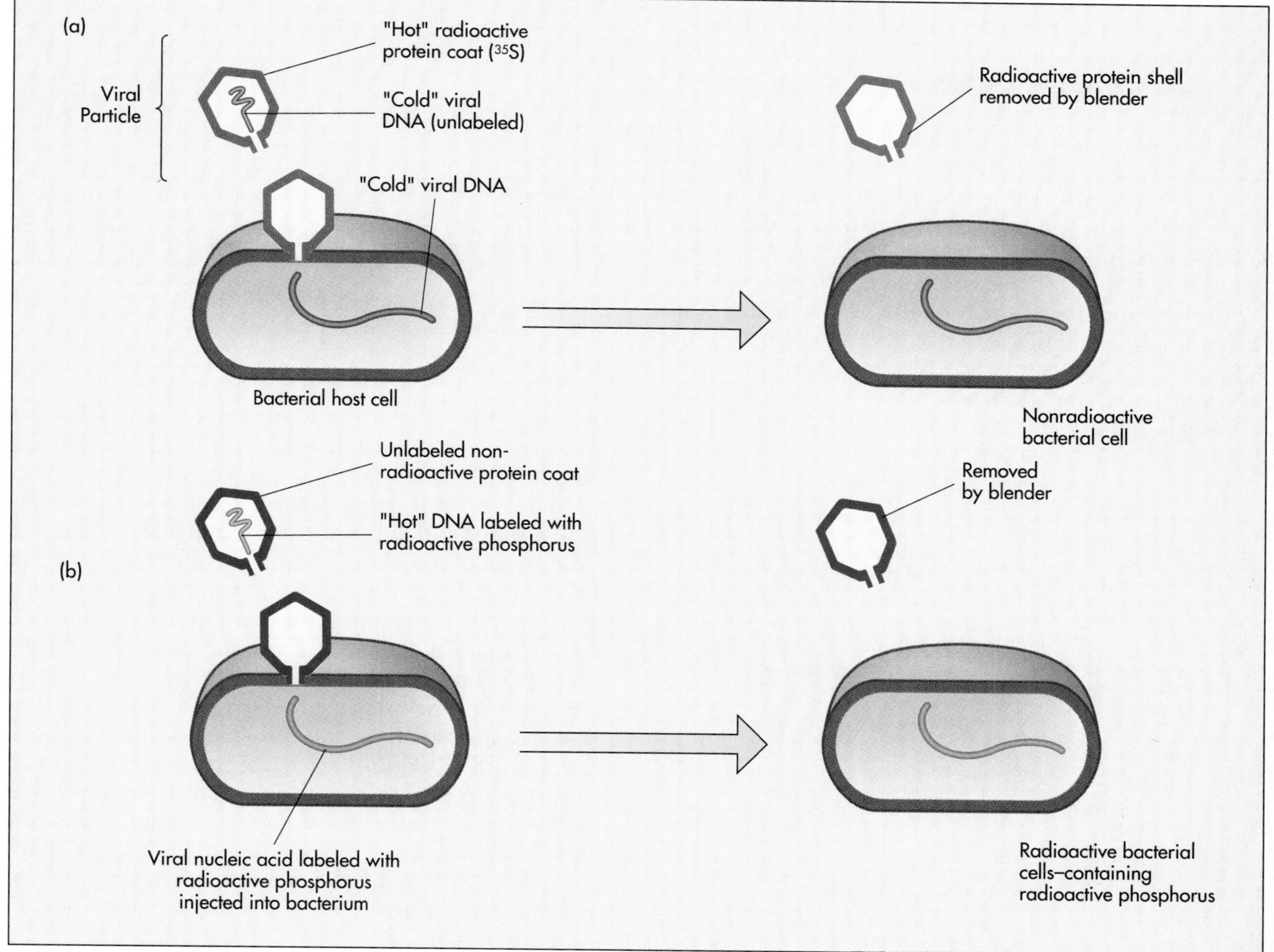

FIGURE 4-4 The Hershey-Chase experiment showed the genetic material of viruses and bacteria was DNA, not protein. (a) Bacteriophages constructed with either radioactively labeled protein shells (made with ^{35}S) or (b) radioactively labeled DNA (using ^{32}P) were used to infect bacterial cells. Only the radioactive viral DNA was later found inside the infected bacteria. This discovery significantly advanced the case for DNA as the genetic material.

three parts: a phosphate group on one end, a central sugar (deoxyribose), and a nitrogen-containing base (adenine, thymine, cytosine, or guanine) (Figure 4-5).

By 1953, it was known that DNA always contained exactly as much of the base adenine (A) as of the base thymine (T) and as much cytosine (C) as guanine (G), that for each base there was a molecule of the sugar deoxyribose and a phosphate group, and that the DNA molecule had a long linear structure. In 1953, James Watson and Francis Crick put together all the available evidence and figured out the structure of the DNA molecule; in 1962, in recognition of their insight, they won a Nobel Prize.

The DNA molecule is called a **double helix** because it takes the shape of a twisted ladder. Each side rail of the ladder is formed by a single strand of nucleotides (Figure 4-6). In the side rail, the phosphate of one nucleotide binds to the sugar of the next, so that the side rail is a chain of alternating sugars and phosphate groups. The ladder's rungs are formed by the bases that project from the side rails.

The two sides of the double helix are held together by hydrogen bonds between compatible bases. The base adenine binds to thymine and cytosine to guanine, and this **base pairing,** or **base complementarity,** explains why DNA contains equal amounts of A and T and of C and G (Figure 4-7). The complementarity of the chains also means that each chain specifies exactly the structure of the other—only one sequence of bases can match up with any existing DNA chain.

You might think that the structure of the DNA molecule was little more than a challenging puzzle whose solution offered the same kind of satisfaction as solving a difficult crossword puzzle. But it was much more than that. As soon as Watson and Crick recognized that DNA was a sequence of nucleotides, they realized that the sequence could store a great deal of information, just as does the sequence of 1's and 0's in a modern computer; having only four letters in the alphabet was not necessarily a handicap at all. When they understood the double-stranded nature of the molecule and the role of base pairing in holding the strands together, they also recognized how the information in the molecule could be duplicated (for inheritance), read, and used to control a cell. For duplication, all that was necessary were enzymes that could separate the DNA strands, line up new nucleotides on each strand according to the matching pattern of base pairing, and join the new nucleotides together, as indicated in Figure 4-8. For information retrieval, a similar mechanism could copy the base sequence of a gene into another molecule that could then be used to specify events elsewhere in the cell. And these steps of duplication and retrieval, as we shall see, are exactly what happen.

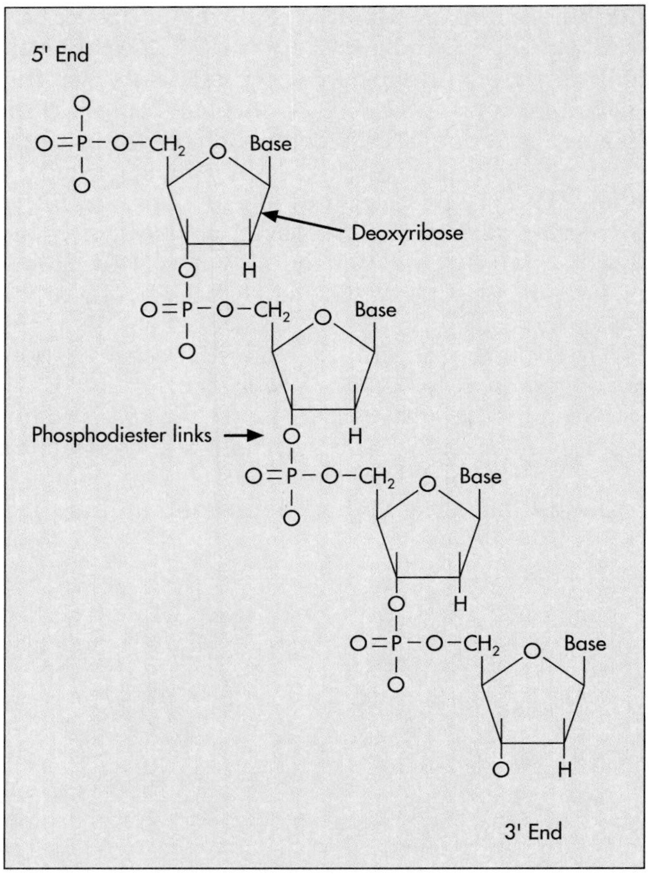

FIGURE 4-6 One strand of a DNA molecule shows the phosphodiester bonds that link its component nucleotides.

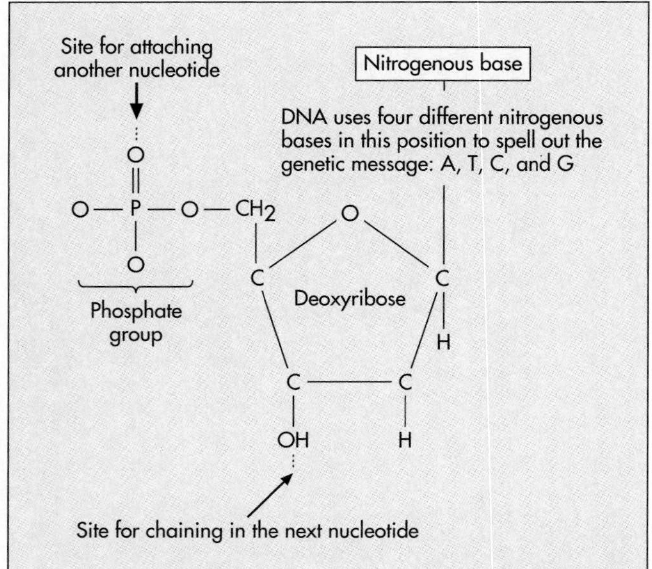

FIGURE 4-5 The basic nucleotide structure.

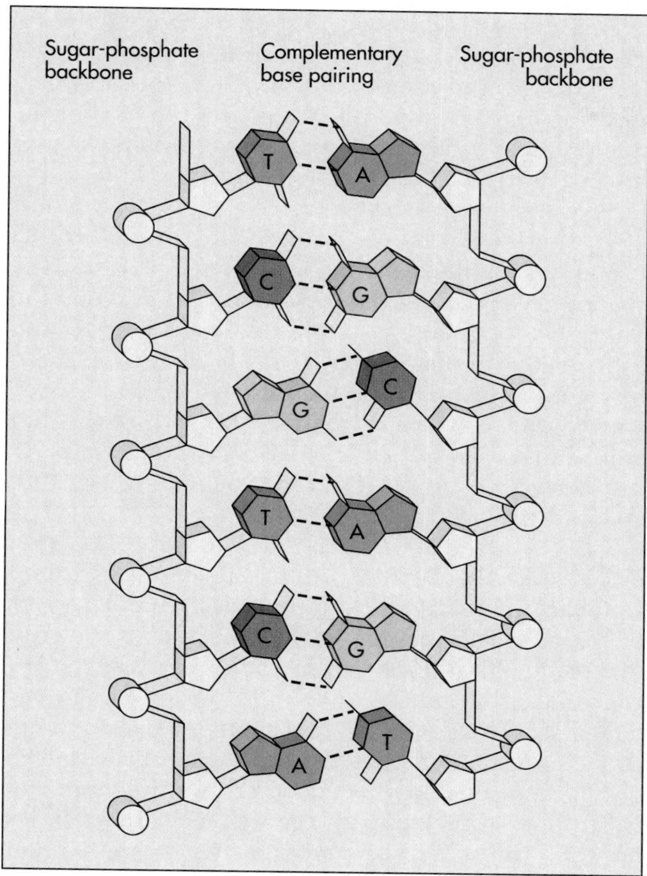

FIGURE 4-7 The paired strands of a DNA molecule (untwisted) showing the base pairing.

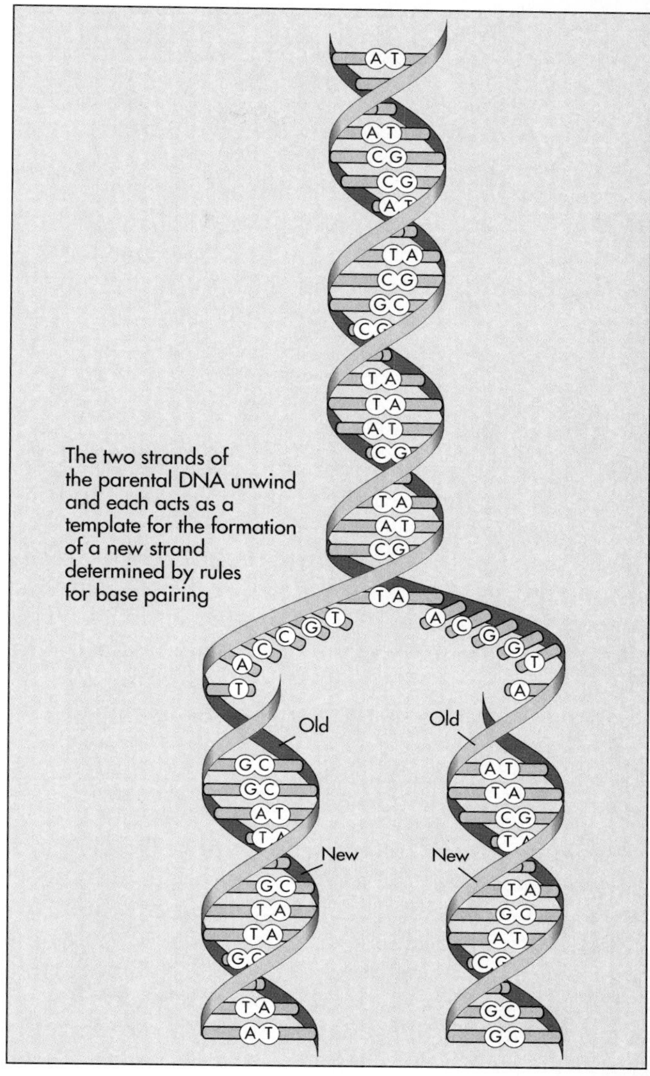

FIGURE 4-8 The DNA double helix is formed by two complementary nucleotide strands that coil around each other along the length of the molecule. In the bottom half of the figure, the original strands have separated, with each single strand being used as a pattern to form additional copies of the DNA during replication.

THE OTHER NUCLEIC ACID: RNA

Cells do not contain only one nucleic acid. Besides DNA, they also contain RNA, which differs from DNA in several important respects:

1. Like DNA molecules, RNA molecules are chains of nucleotides. However, RNA nucleotides are built around the sugar *ribose,* not deoxyribose. Ribose contains one more oxygen atom than deoxyribose (which explains the latter's name).
2. DNA nucleotides use the bases adenine, thymine, cytosine, and guanine (A, T, C, and G). The RNA nucleotides do not use thymine, replacing it with uracil (U). Like thymine, uracil acts as a complement to adenine.
3. Whereas DNA molecules are double stranded, RNA molecules have but one strand of nucleotides. They may have double-stranded sections, but only because the single strand folds back on itself to let complementary segments bind to each other.
4. Deoxyribonucleic acid is found only in the nucleus of the cell (except for the small amount in mitochondria and chloroplasts). Ribonucleic acid is found there too, for that is where it is made, but it is also found in the cytoplasm, where it performs its various functions.
5. There is only one kind of DNA, but there are several kinds of RNA. The three most important to understanding how genes work are messenger RNA (mRNA), transfer RNA (tRNA), and ribosomal RNA (rRNA).

DNA, RNA, AND PROTEIN SYNTHESIS

When Watson and Crick announced the structure of DNA, they opened up a highly fruitful field of research. Immediately, many researchers embraced the study of the gene (**molecular genetics**). Before long, they had confirmed and greatly extended Watson's and Crick's initial glimpse of how DNA must work, and within two decades they had learned how to snip genes apart and rebuild them, founding the technology of genetic engineering.

In essence, the path from DNA to a cell's traits is simple. Using the information encoded in the DNA requires **transcription,** copying the nucleotide sequence of a gene into the complementary nucleotide sequence of an RNA molecule that can leave the nucleus. In the cytoplasm, the process of **translation** converts the information in the RNA into a sequence of amino acids in a protein molecule. The protein may be an enzyme or a structural protein; either way, it is responsible for some physical or functional feature of the cell, a trait (Figure 4-9).

The Genetic Code

Complementarity makes transcription fairly easy to understand. All the enzymes in the nucleus need to do is find the RNA nucleotides, line them up with their complements on the DNA strand, and then chain them together. The newly constructed strand of RNA, called **messenger RNA (mRNA),** acts as a copy of the gene's protein pattern or template that can move from the nucleus to the cytoplasm, where it directs the synthesis of a protein.

But how can 4 different DNA nucleotides transcribed to 4 different RNA nucleotides specify 20 different amino acids? Molecular biologists quickly realized that they cannot *if* there is to be a one-to-one correspondence in the coding between single nucleotides and single amino acids; the numbers just do not match up. The answer had to be that the nucleotides formed code words, or **codons,** two or more at a time. The question was then how many letters each code word had.

Could two letters be enough? The first letter in a two-letter coding system could be A, T, C, or G. The second letter could also be any of the four, giving only 16 (4^2) different two-letter code words, and that was not enough. But what if each code word had four letters? That would give 256 (4^4) code words, far too many. What about three? That would give 64 (4^3), which was still more than 20, but it was the smallest number of code words that covered the need for 20.

Only a few years after Watson and Crick had announced the DNA structure, researchers found that the cell really did use three-letter code words, or triplets. They also matched the 64 possible triplets to the 20 different amino acids of proteins. We reproduce this **genetic code** in Table 4-1. So far, the code has proved to be the same in every living thing—animals, plants, bacteria, and viruses. The only exceptions have been found in mitochondria. In human mitochondria, for instance, UGA codes for the amino acid tryptophan, not a stop signal, and AUA codes for methionine, not isoleucine.

1. The code words in Table 4-1 include U's but no T's. Why does this tell you that these code words are the code words of mRNA, not DNA?
2. Using the code words of Table 4-1 and the complementarity rules, prepare a list of the code words as they would appear in the cell's DNA.

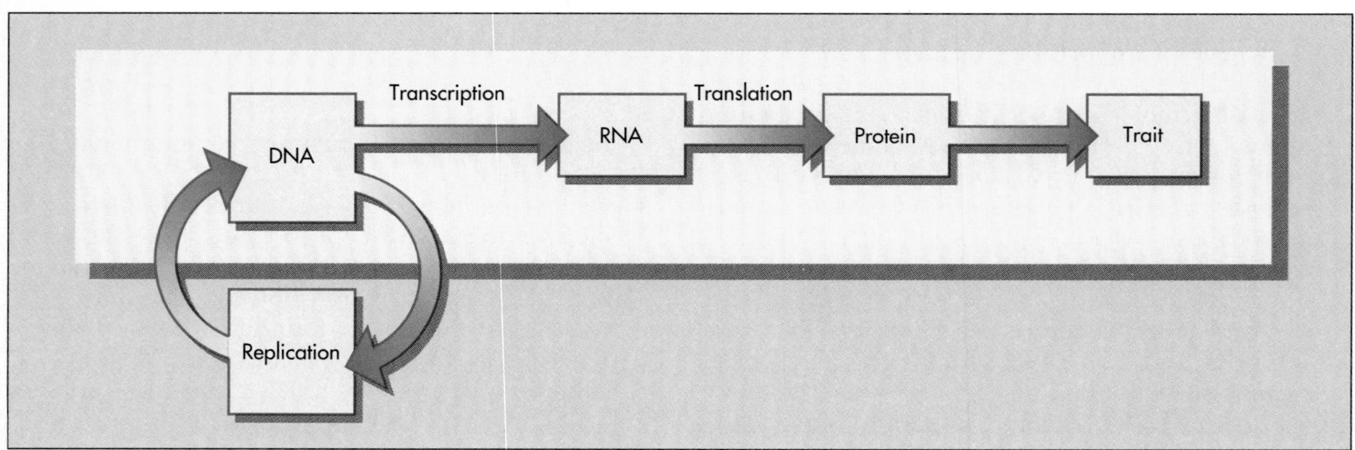

FIGURE 4-9 The path from DNA to trait.

Table 4-1

The Genetic Code

FIRST BASE	SECOND BASE	THIRD BASE	AMINO ACID	FIRST BASE	SECOND BASE	THIRD BASE	AMINO ACID
U	U	U	Phenylalanine	A	U	U	Isoleucine
		C	Phenylalanine			C	Isoleucine
		A	Leucine			A	Isoleucine
		G	Leucine			G	Methionine, Start
U	C	U	Serine	A	C	U	Threonine
		C	Serine			C	Threonine
		A	Serine			A	Threonine
		G	Serine			G	Threonine
U	A	U	Tyrosine	A	A	U	Asparagine
		C	Tyrosine			C	Asparagine
		A	Stop			A	Lysine
		G	Stop			G	Lysine
U	G	U	Cysteine	A	G	U	Serine
		C	Cysteine			C	Serine
		A	Stop			A	Arginine
		G	Tryptophan			G	Arginine
C	U	U	Leucine	G	U	U	Valine
		C	Leucine			C	Valine
		A	Leucine			A	Valine
		G	Leucine			G	Valine
C	C	U	Proline	G	C	U	Alanine
		C	Proline			C	Alanine
		A	Proline			A	Alanine
		G	Proline			G	Alanine
C	A	U	Histidine	G	A	U	Aspartic acid
		C	Histidine			C	Aspartic acid
		A	Glutamine			A	Glutamic acid
		G	Glutamine			G	Glutamic acid
C	G	U	Arginine	G	G	U	Glycine
		C	Arginine			C	Glycine
		A	Arginine			A	Glycine
		G	Arginine			G	Glycine

Note that there may be several different code words for a single amino acid. That is, the genetic code is *redundant*. It thus provides some insurance against changes in the DNA (mutations; see a later section), for many such changes simply alter a code word to another code word for the same amino acid.

Three code words act as "stop" signals, marking where the apparatus of translation stops "reading" an mRNA molecule and making a protein. They are the triplets UAA, UAG, and UGA. One triplet, AUG, marks where the translation apparatus starts reading an mRNA. When this triplet appears within the recipe for a protein, rather than at the beginning, it codes for the amino acid methionine.

Replication

DNA **replication** begins when an enzyme breaks the hydrogen bonds between bases that hold the two nucleotide strands of a DNA molecule together. Because the enzyme seems to unzip the molecule, it is sometimes called an "unzipase." As it works, the strands separate and the twist of the molecule jams up toward the parts of the molecule whose strands are still joined. A second enzyme, **DNA gyrase,** breaks the strands, lets the jammed twist or "supercoil" unwind, and then rejoins the strands.

Once the DNA strands have been separated for some distance along the DNA molecule, an enzyme lines a few RNA nucleotides up against the DNA strands and hooks them together, or polymerizes them. This enzyme is an **RNA polymerase,** and the short bits of complementary RNA it attaches to the DNA are primers essential to making more DNA. Once the primers are in place, a **DNA polymerase** can begin to lay down DNA nucleotides and hook them together to make complementary DNA strands. These DNA strands are attached to the RNA primers.

DNA replication begins in several places on the DNA molecule, as indicated in Figure 4-10. As it proceeds, the replication zones approach each other. When the lengths of new DNA meet, a second

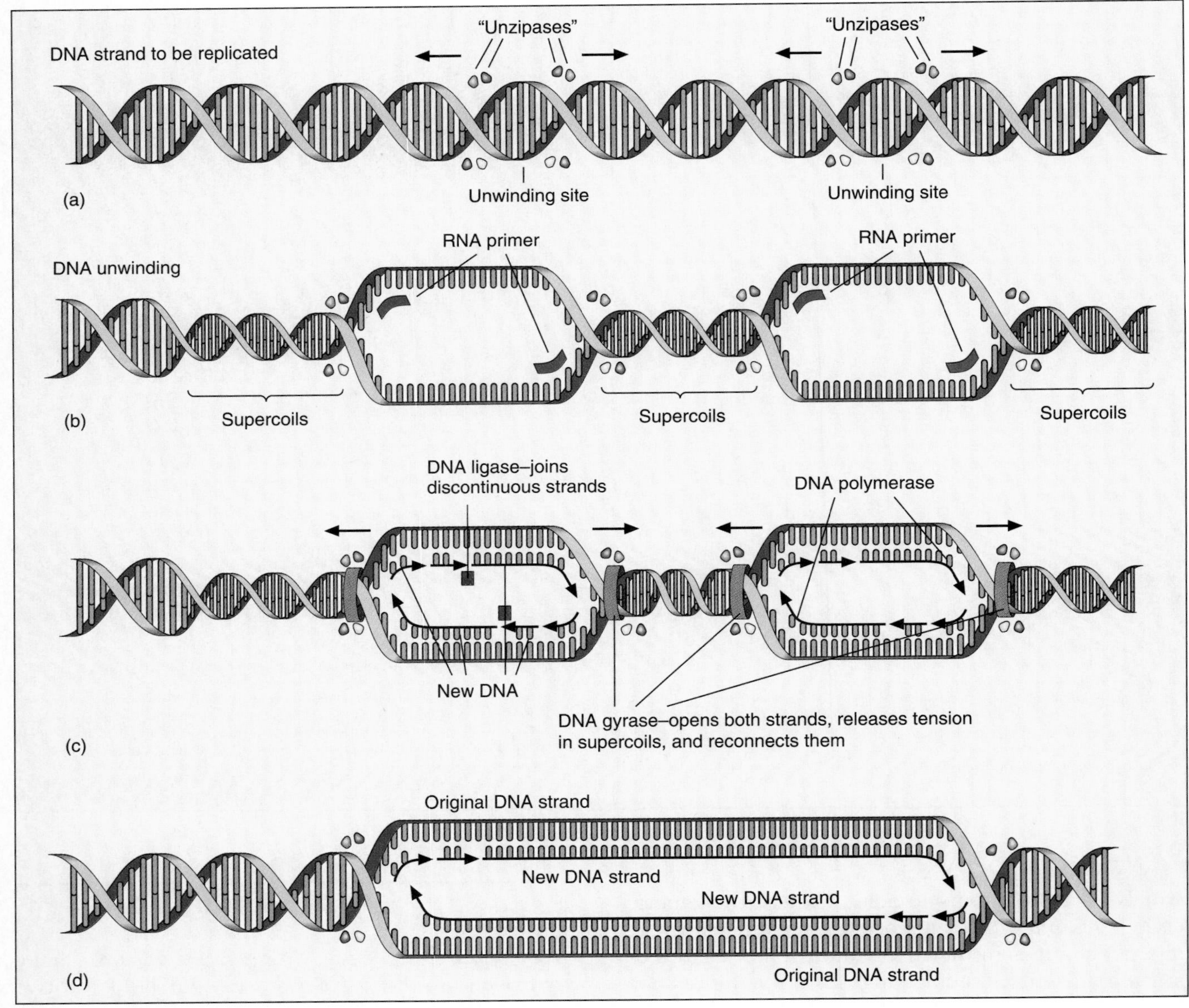

FIGURE 4-10 A model for DNA replication. (a) Replication begins when unwinding proteins ("unzipases") undo the DNA helix at several specific sites along the length of the DNA molecule. (b) As the unwinding progresses, the opening areas converge on each other and produce areas of tightly wound supercoils. Using RNA polymerase, short sections of RNA primer are synthesized in the areas where DNA synthesis begins. (c) Another enzyme, DNA gyrase, uses energy from ATP to break both DNA strands, thus releasing the tension on the supercoils, and then to reconnect them. DNA polymerase promotes the synthesis of new complementary strands of DNA along the length of the original strands. The synthesis begins at several places simultaneously along the DNA. As DNA replication continues, the individual short sections being synthesized are joined into a continuous strand by another enzyme, DNA ligase. (d) The process results in replication of the original DNA, forming two identical double-stranded DNA molecules.

DNA polymerase replaces the RNA primers with DNA, and another enzyme, **DNA ligase,** joins the DNA lengths together. The result is the formation of two identical DNA molecules where before there was only one. Note that each of the two molecules contains one strand of nucleotides from the original molecule and one strand that has just been synthesized. When a cell divides, the daughter cells thus each receive half the parent's DNA; the other half is newly made.

Transcription: Copying the Genetic Pattern

Transcription resembles replication in that it involves an initial separation of the DNA strands. It differs in that the separation is only local, confined to the region of a single gene. It also differs in that it does not require DNA polymerase, for the object of the process is not a duplicate of an entire DNA molecule, but an RNA copy of the DNA in one gene.

Once the DNA strands have separated, an RNA polymerase binds to just one of the DNA strands (the "sense" strand). It then assembles complementary RNA nucleotides into an RNA molecule, ending the process when it encounters a sequence of DNA bases that marks the end of the gene with a "stop here" message. Finally, it releases the RNA molecule.

In eukaryotes, there are four different RNA polymerases. One makes the RNA primers necessary for replication. Another makes the mRNA that carries protein templates or recipes into the cytoplasm. Two more make the RNA tools the cell needs to use the information carried by the mRNA. These tools are **transfer RNA (tRNA),** which delivers amino acids to the ribosome for incorporation in a protein molecule, and **ribosomal RNA (rRNA),** which forms an essential part of the ribosome. Most genes specify mRNA molecules; relatively few genes specify rRNA and tRNA, for the cell, like a cook, needs many fewer tools than recipes.

The genes for rRNA are concentrated near the nucleolus, in an area called the *nucleolar organizing region;* in the nucleolus, the rRNA and proteins synthesized in the cytoplasm are used to make ribosomes (Figure 4-11).

Transfer RNA is unlike other RNAs in that after its construction on the template of its DNA gene, a number of its component nucleotide bases are chemically modified. It also comes in many forms, one for each of the code words or base triplets that specifies the position of a particular amino acid in a protein. Because of the redundancy in the genetic code, there are several different tRNAs for most of the 20 amino acids found in proteins. In form, a tRNA molecule resembles a clover leaf (Figure 4-12). Its single strand of nucleotides folds back on itself, and the loops are held in place by complementarity binding between different parts of the molecule. The tip of the clover leaf's stem is the part that binds to an amino acid. The tip of the

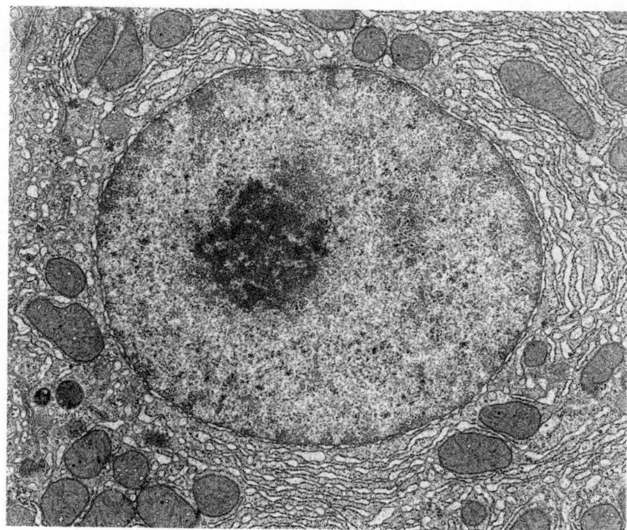

FIGURE 4-11 Photomicrograph of the nucleolar organizing region and nucleolus of a human liver cell (hepatocyte).

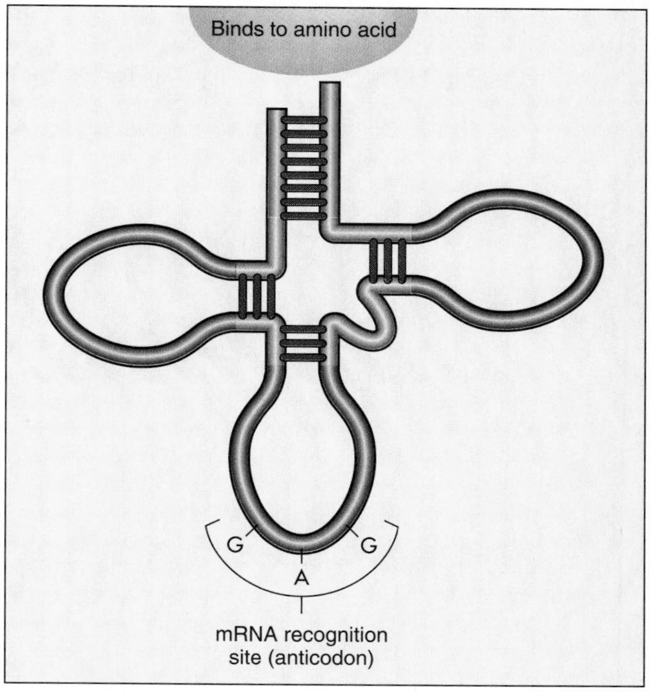

FIGURE 4-12 The tRNA molecule aligns amino acids along the length of a mRNA template in genetically directed protein synthesis. The stem binds with the specific amino acid; the central anticodon is the mRNA recognition site.

central leaflet contains a three-nucleotide sequence, an **anticodon,** complementary to an mRNA codon. The anticodon is the key to how the translation apparatus lines up amino acids to match the codon sequence of mRNA and make a protein.

Exons and Introns: Sense and Nonsense Gene Segments

In eukaryotes, mRNA must be processed before it can leave the nucleus to do its job of directing protein synthesis. Eukaryotic genes, unlike the genes of prokaryotes (bacteria), contain nonsense nucleotide sequences. These sequences, or **introns,** encode no information and are embedded among the informational parts of the gene, or **exons.** They must be removed from the mRNA if the mRNA is to be useful.

One way to remove the introns is to keep them out of the mRNA in the first place. This can be accomplished if, in the DNA, the introns somehow avoid transcription. Another way is to remove the intron segments from the mRNA and splice the

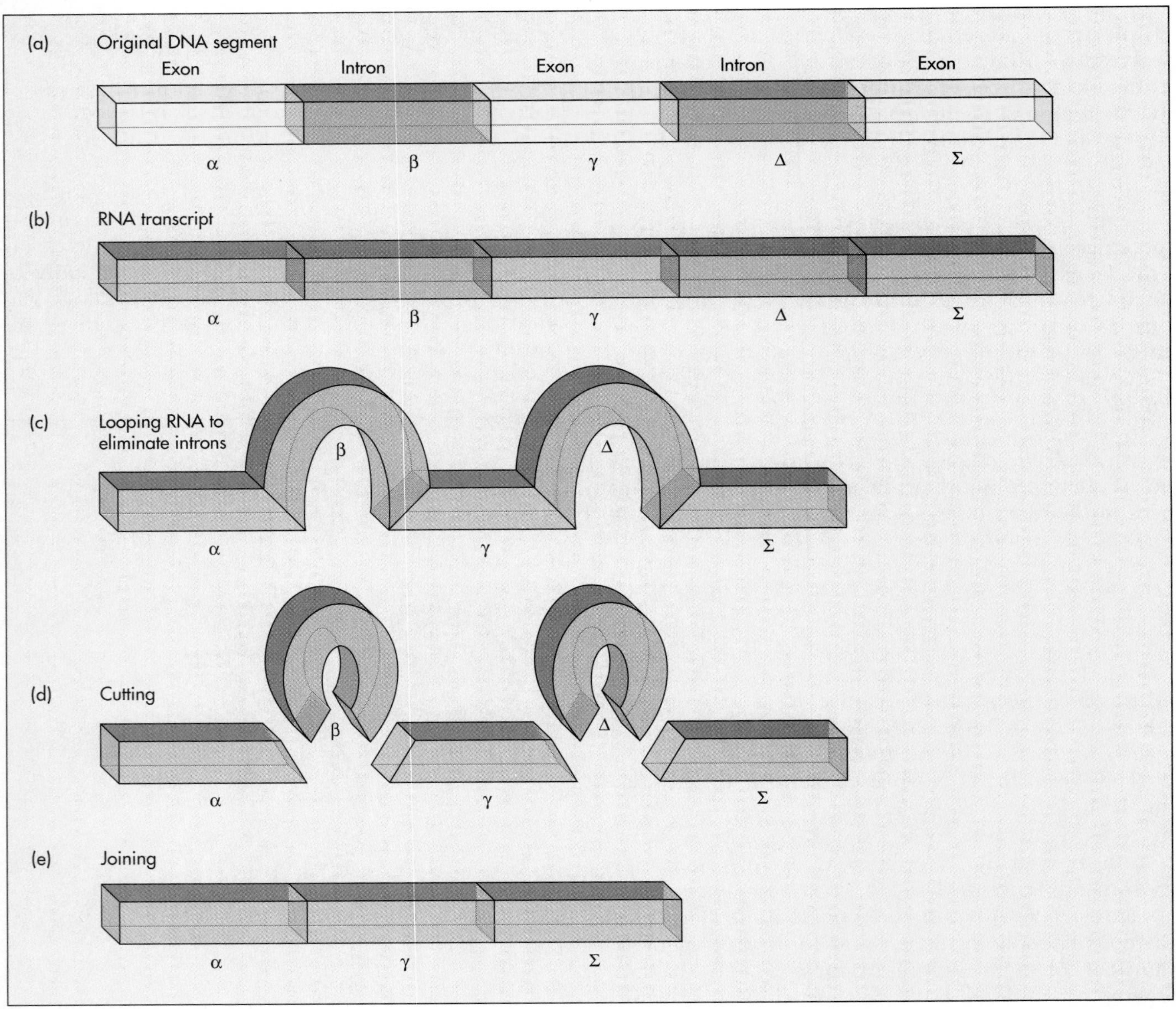

FIGURE 4-13 The removal of introns from DNA transcripts. (a) The original DNA gene segment complete with introns (β,Δ) and exons (α,γ,Σ). (b) The transcribed mRNA with copies of both the intron and exon sections. (c) Looping the RNA to isolate the introns. (d) Cutting out the intron loops. (e) Joining the exon transcriptions to assemble the genetically active mRNA.

functional informational segments together into a single molecule, as shown in Figure 4-13. Intriguingly, some RNAs seem able to act like enzymes and catalyze their own snipping and splicing operations.

Recent work seems to indicate that the exons each code for a single functional part of a protein. Many modern eukaryotic proteins are known to have similar segments serving similar functions. Proteins may have evolved as strings of several simpler, more primitive proteins, each one of which was able to perform one of the functions of the modern protein. The way the strings were pieced together shows in the genes. Each exon codes for one of the pieces; the introns are the stitches between them.

Translation: Genetically Directed Protein Synthesis

Translation of the nucleotide sequence of a mRNA into the amino acid sequence of a protein is performed by the ribosomes. Each ribosome is composed of two roughly spherical subunits of unequal size that are joined together. The larger unit contains about 50 proteins, a 5000-nucleotide rRNA, and two smaller rRNAs of about 100 nucleotides. The smaller subunit is built of about 30 protein molecules and a single molecule of rRNA that is 2000 nucleotides long; within its structure are two notchlike sites, the A and P sites, into which tRNAs can fit (Figure 4-14).

Translation also requires the various tRNAs to carry in the different amino acids that will eventually be assembled into a protein. Before they can be used, each kind of tRNA must be attached to the specific type of amino acid it can carry. Before the tRNAs can bind to their respective amino acids, enzymes in the cytoplasm must bind to the amino acids that are going to be picked up and to the adenosine monophosphate (AMP) formed by the removal of two phosphate groups from ATP. The amino acid–AMP–enzyme complex can then react with the particular tRNA specific for that amino acid, transferring the amino acid to the stem of the tRNA molecule and releasing the AMP and enzyme for recycling (Figure 4-15). The enzyme lines up the amino acid so it can complex with its tRNA. There is a particular enzyme for each of the possible amino acid–tRNA combinations.

Once mRNA, ribosomes, and amino acid–tRNA pairs are all available, translation can begin. First a ribosome binds to the leading end of the mRNA. As shown in Figure 4-16 (p. 95), the ribosome covers two of the mRNA's codons. It immediately guides into place the two tRNAs whose anticodons match the codons, and an enzyme in the ribosome forges a peptide bond between the amino acids carried by the tRNAs. Then the tRNA in the P site releases its amino acid, leaving the tRNA in the A site attached to two joined amino acids, the beginning of a poly-

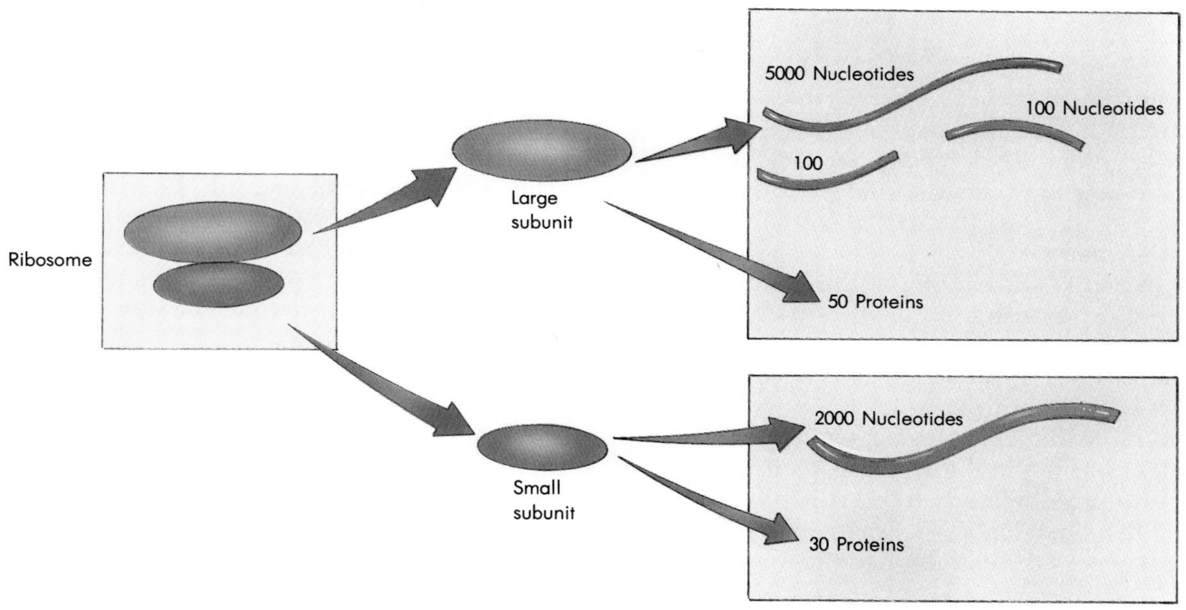

FIGURE 4-14 Ribosome structure. Each ribosome is composed of two roughly spherical subunits. The larger unit contains about 50 proteins, a 5000-nucleotide rRNA, and two smaller rRNAs of about 100 nucleotides. The smaller unit contains 30 proteins and a single molecule of rRNA 2000 nucleotide units long.

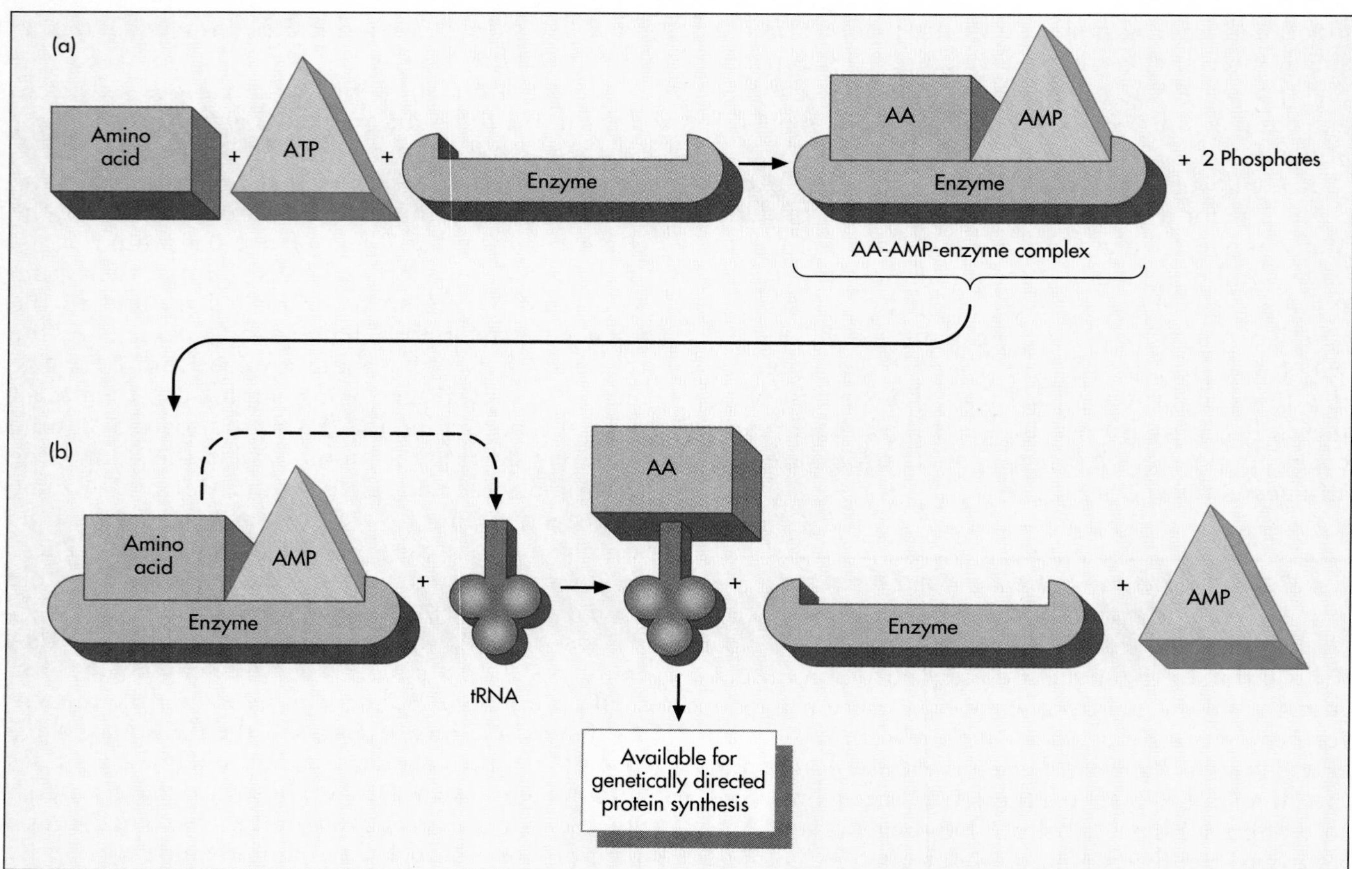

FIGURE 4-15 Preparation of amino acids for protein synthesis. (a) An amino acid (AA) plus an energy source, such as ATP, and the proper enzyme react to produce an AA-AMP-enzyme complex plus two phosphates. (b) The AA-AMP-enzyme complex reacts with the proper tRNA to produce an AA-tRNA and releases the enzyme to be reused and the AMP to be recycled back to ATP. The AA-tRNA is now available for use in the translation process.

peptide or protein, and unhooks from the ribosome. When the P site is clear, the ribosome moves one codon length along the mRNA. The tRNA attached to the amino acid pair now occupies the P site, and the A site is vacant, ready for the arrival of another amino acid–tRNA pair.

The process repeats, adding one amino acid to the growing protein at each step of the ribosome along the mRNA until the ribosome comes to the end of the mRNA and the stop signal. It then releases the mRNA and the protein. Its job is done, and the information that began as a sequence of DNA nucleotides is in its final form, that of an enzyme or structural protein, ready to play its role in the life of the cell—dictating shape, pigmentation, or one of the many possible cellular chemical reactions.

The linear nature of the mRNA molecule allows a single mRNA to support the synthesis of many identical protein molecules simultaneously. As soon as the first ribosome to attach to the mRNA has moved down from the mRNA's leading end, a second ribosome can attach. As it moves along the mRNA, it makes room for a third, and a fourth, and so on. The combination of one mRNA with many attached ribosomes is called a **polyribosome.** Under the electron microscope, a polyribosome and the multiple protein copies it is manufacturing, each one slightly longer than the next, resemble a feather (Figure 4-17, p. 96).

---

1. Explain why the multiple protein copies manufactured by a polyribosome differ so systematically in length.
2. Magnesium ions must be present if ribosomes are to bind to mRNA. A deficiency of magnesium in the diet is very unlikely, for magnesium ions are present in virtually every food, but what would you expect to be a major symptom of such a deficiency?

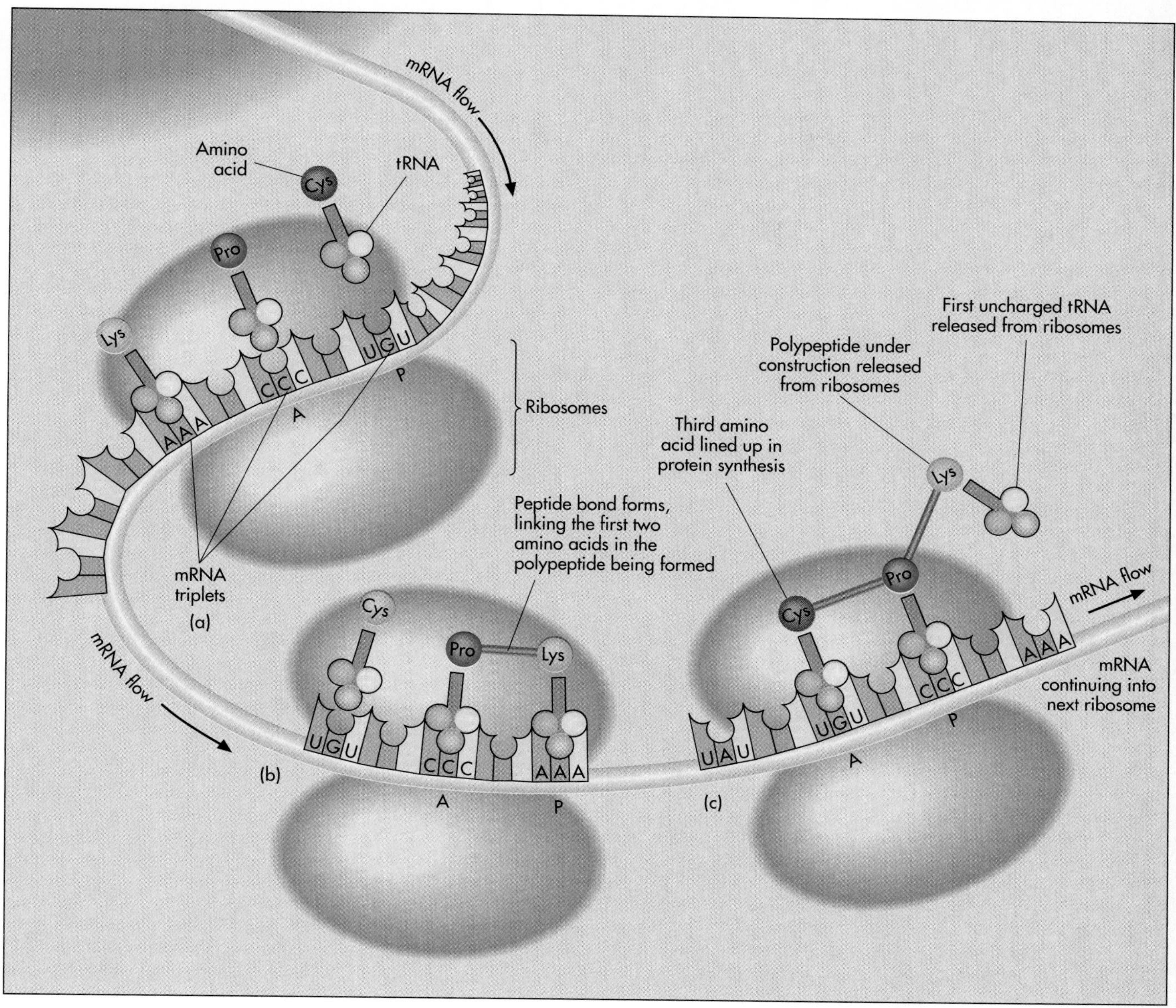

FIGURE 4-16 The role of the ribosome in protein synthesis. (a) Messenger RNA, carrying a transcribed genetic template from DNA, enters the larger subunit of the ribosome. The first two mRNA triplets move into positions A and P in the ribosome. The mRNA triplets determine the amino acid sequence in a protein by attracting tRNA amino acid carriers to their specific position. (b) Enzymes link adjacent amino acids with peptide bonds. (c) As the polypeptide forms, each amino acid is released from its specific tRNA carrier and the tRNA molecule is ejected from the ribosome. The next mRNA triplet moves up into slot A of the ribosomes to position the third amino acid.

CONTROLLING THE GENE

A fertilized human egg contains all the genes needed to make all the proteins found in all the cells of all the tissues of an adult human body. But no single cell makes all those proteins. No cell transcribes all its genes at once, and a major question of molecular genetics is "Why?" How does the cell control or select which genes to use?

So far, researchers have found only partial answers for eukaryotic cells such as our own. They have had better luck with the simpler, more primitive cells of prokaryotes (bacteria) and with viruses. They too fail to transcribe all their genes at once, but they have many fewer genes and the basic scheme of genetic control has been easier to work out.

Learning Focus

ONE GENE–ONE ENZYME, OR . . . ?

Ten years before Watson and Crick announced the structure of DNA and while Avery and his colleagues were using transformation to identify DNA as the substance of the gene, George Beadle and Edward Tatum were learning that a gene—whatever it was—was the pattern or template for a protein. They were studying the bread mold, *Neurospora crassa,* which can normally synthesize all the vitamins, amino acids, and other organic substances it needs for survival, growth, and reproduction. When they irradiated the mold's spore-forming organs with x-rays, however, they found that many of the spores gave rise to organisms that could *not* make all the vitamins, amino acids, and so on, they needed; each of the many x-ray-produced variants, or *mutants,* could survive, grow, and reproduce only if Beadle and Tatum supplied a special nutrient. Different mutants required different nutrients. (We will discuss mutants in more detail later in the chapter.)

Careful research showed that the reason why the mutants could not make their own nutrients was that the enzymes the "wild-type" mold used to make them were defective. Cross-breeding experiments then revealed that the various defects were inherited as separate traits and that their inheritance obeyed the same laws as other traits (such as blue eyes in humans). Each defect therefore had to be caused by a change in a single gene. Furthermore, Beadle and Tatum found mutants that had defective genes for virtually every trait known to exist in *Neurospora.* Their conclusion was inevitable: If damaging a gene with x-rays leads to a defective enzyme and if damaging *different* genes leads to *different* defective enzymes, then each gene has to carry the recipe for a different enzyme. This insight was the birth of the **one gene–one enzyme theory.**

We know now that a gene does not always carry the recipe for an enzyme. It can also carry the recipe for a nonenzymatic structural protein such as collagen. Therefore, Beadle and Tatum's theory is better rephrased as the **one gene–one protein theory.**

Or is it? Use the following Learning Focus Response to review what you have learned about transcription and translation. Try to see how some genes fail to fit even the one gene–one protein theory.

The first hint of the control scheme for bacteria came with the discovery of **inducible** enzymes, enzymes the cell makes only when it needs them. These are enzymes such as beta-galactosidase, which breaks the nutrient disaccharide lactose into its components, glucose and galactose. A bacterial

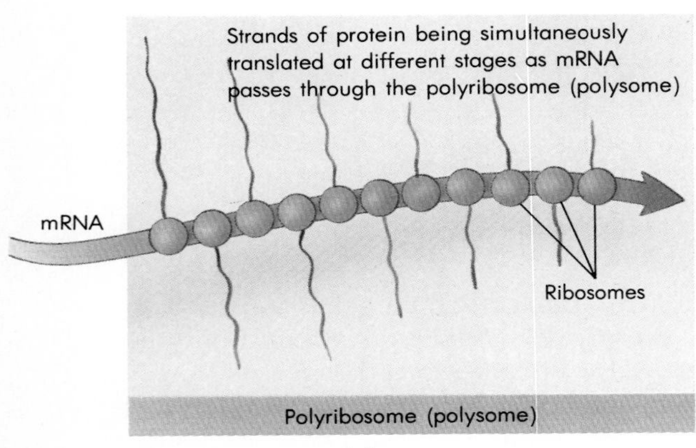

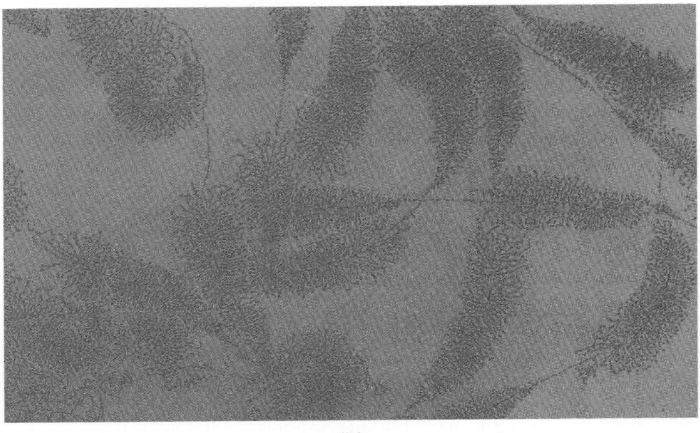

FIGURE 4-17 (a) Polyribosomes are produced when one mRNA molecule is supporting the synthesis of identical proteins in several different ribosomes at the same time. (b) Different ribosomes along the length of the polyribosome synthesize the proteins for gradually increasing intervals of time, giving the overall structure a Christmas tree look.

Learning Focus Response

SUMMARIZE WHAT YOU KNOW ABOUT MOLECULAR GENETICS

On the accompanying diagram, identify mRNA, tRNA, rRNA, amino acids, ribosomal proteins, and ribosomes. Then answer the questions.

1. Where does transcription occur?

2. What are the products of transcription?

3. What parts of the drawing represent the process of translation?

4. What is the product of translation?

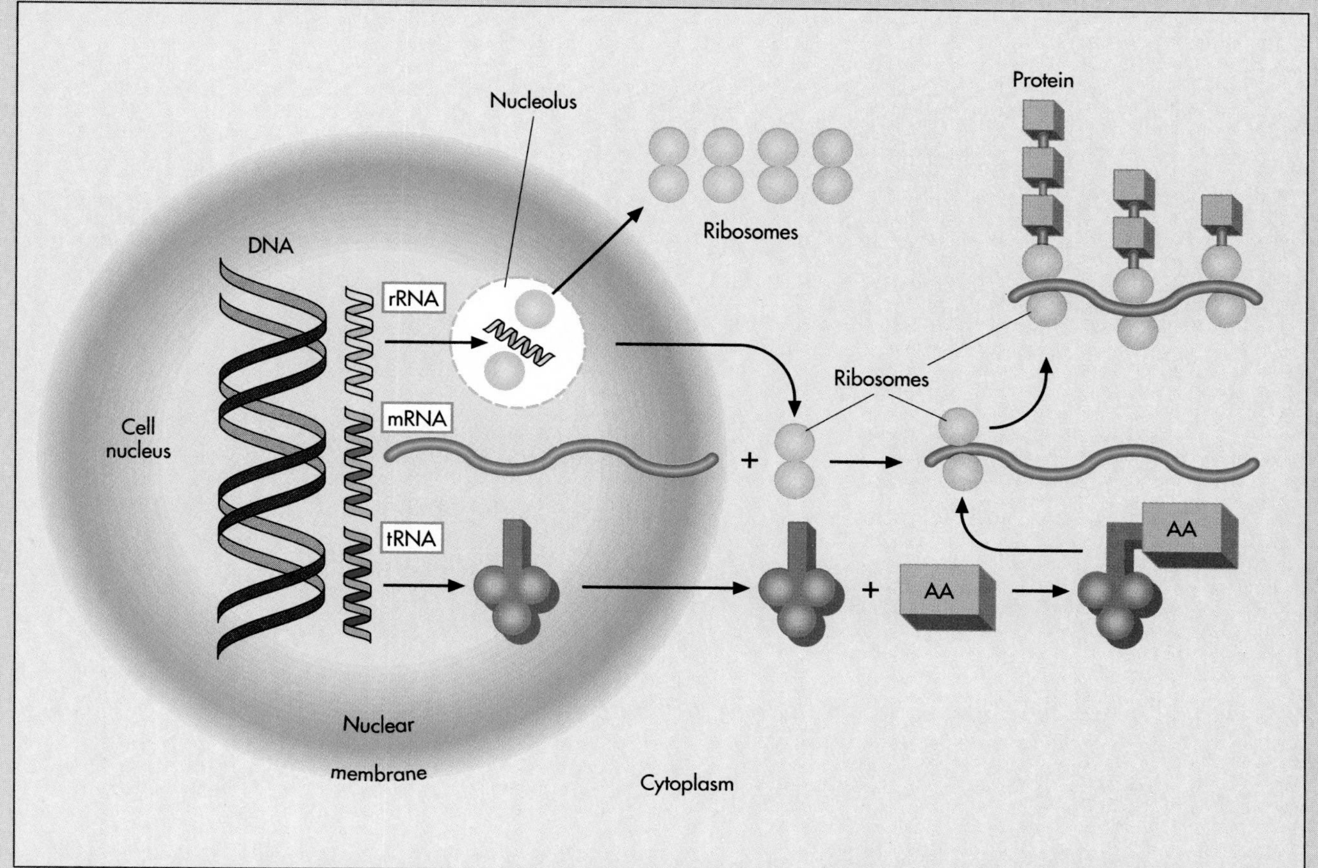

5. Describe how amino acids are prepared for use in protein synthesis.

6. Describe how ribosomes and mRNA interact to synthesize proteins.

7. What is the difference between a codon and an anticodon?

8. The mutants that Beadle and Tatum studied were produced when x-rays damaged DNA and altered the recipes for proteins. What carries these recipes from the DNA to the protein-synthesizing apparatus?

9. Genes are lengths of DNA that directly specify the construction of molecules of one substance. What is that substance?

10. Why, then, is the Beadle-Tatum theory only *almost* right?

11. How do your answers to questions 9 and 10 suggest that we should rephrase the Beadle-Tatum theory?

12. How might x-rays produce changes in proteins *without* interfering with protein recipes?

13. Do you think such changes would be more or less common than the kind of changes Beadle and Tatum studied? Why?

cell does not make the enzyme unless lactose is available. When lactose becomes available, it quickly synthesizes the enzyme and begins to process the sugar for energy. When the cell has used up the lactose in its environment, it stops making the enzyme. We say that the sugar, the substrate for the enzyme, *induces* the enzyme. When the sugar is used up, the enzyme is **repressed.**

We now know how induction and repression work in bacteria. Research has shown that the bacterial chromosome contains several kinds of genes. **Structural genes** are the ones that code for proteins such as beta-galactosidase. Attached to them are **operator** and **promoter** genes or regions. Together, the structural, operator, and promoter genes comprise an **operon,** a functional genetic unit. **Repressor** genes may be located nearby or further away on the chromosome.

All structural genes seem to have promoter regions. They are the sites to which RNA polymerase binds before it begins to transcribe mRNA. In the lactose operon, the operator is located between the promoter and the structural gene, as shown in Figure 4-18. The repressor gene codes for a protein that binds to the operator and blocks the movement of the RNA polymerase from the promoter to the structural gene and hence the synthesis of mRNA and protein.

When lactose is present, it binds to the repressor molecule and alters its shape so that it no longer fits the operator. The RNA polymerase is then free to transcribe the structural gene, producing the mRNA that eventually makes it possible for the ribosomes to make the enzyme. Once the cell has exhausted the lactose supply, the repressor is no longer inactivated, it binds to the operator, and enzyme synthesis stops.

---

1. Together induction and repression give the cell a sensitive way to adjust its production of enzymes to circumstances. What advantage does the cell gain by making enzymes only when it needs them?
2. Not all enzymes process nutrients. Some make substances the cell needs, such as amino acids, but only until the cell has enough of the substances. How might these substances control enzyme production?

Many other bacterial genes are controlled in a similar way, although there are plenty of variations. Sometimes the repressor is the end product of a chemical reaction governed by the induced enzyme, so that the cell stops the reaction when it has enough of the end product. Sometimes the repressor gene can itself be repressed or induced. Often, the structural gene controlled by a single operator and promoter is actually several structural genes whose enzyme products are all involved in a single chemical process (the lactose operon actually includes three structural genes).

Gene control is more complex in eukaryotic cells, and it is much less well understood. However, we do know that eukaryotic genes too have promoters to which RNA polymerase must bind. In at least some cases, there are also operators and repressor genes and proteins. In addition, there may be special "sensor" genes that, in response to external signals such as hormones or nutrients, produce "activator" molecules of RNA or protein that can initiate transcription of structural genes.

One important way of controlling eukaryotic genes may be based on the way eukaryotic DNA is complexed with numerous proteins to form the **chromatin** substance of eukaryotic chromosomes, as will be discussed in Chapter 5. These proteins are of two types, **histones** and **nonhistones,** and the latter seem to dictate which genes are transcribed. If DNA and histones from a person's thymus gland are mixed in a test tube with nonhistones from bone marrow cells and with the necessary enzymes for transcription, only RNA typical of bone marrow cells is transcribed.

MUTATIONS

Mutations are inheritable changes in the DNA of the genes. They can be of two general types: **chromosomal mutations** and **point mutations.**

Chromosomal Mutations

Chromosomal mutations are changes in structure of one or more chromosomes. They include

1. **Deletions,** in which a piece of a chromosome is lost, together with whatever genes it carried;
2. **Duplications,** in which a segment of a chromosome is doubled, together with its genes;
3. **Inversions,** in which a piece of chromosome is snipped out of the chromosome and replaced, after being flipped head for tail; and
4. **Translocations,** in which a piece of a chromosome is deleted from one chromosome and moved to another (it may be added to one end of the receiving chromosome or inserted in its middle).

Such changes in chromosome structure can have serious effects, for they can change how genes are

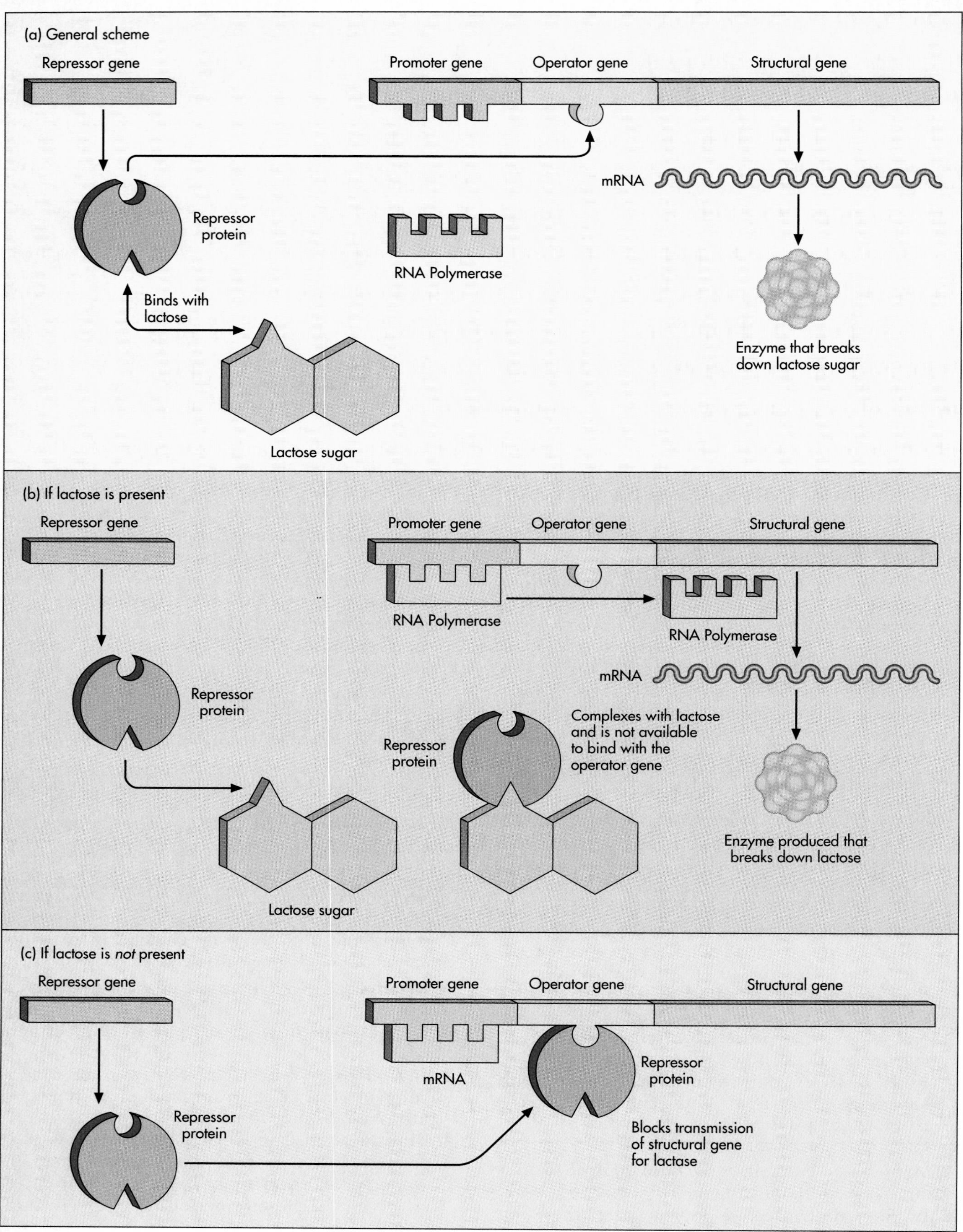

FIGURE 4-18 The lactose operon. (a) A simplified overview of the lactose operon. (b) If lactose sugar is present, it reacts with the repressor protein, which then lets go of the operator and allows the structural gene to synthesize the enzyme, lactase. (c) If lactose is not present, the repressor protein attaches to the operator and blocks the synthesis of lactase.

controlled. For instance, consider the translocation of a structural gene from one operon, where it is controlled by one repressor or sensor gene, to another. The gene may then be turned on at a different stage in the cell's life cycle or in response to different environmental events. The change may be beneficial, but far more often it makes the organism less able to survive. In humans, for instance, translocations have been linked with cancer (Figure 4-19).

Duplications of all sizes, from single genes to whole sets of chromosomes, may be more likely to be beneficial, for they supply extra copies of genes. These extra copies can then mutate freely, accumulating both beneficial and harmful changes, while the original gene continues to serve its original function. Eventually, the changes can add up to turn the copy into a new and useful gene, coding for a protein the organism could never have previously made. This process, in fact, seems to be responsible for the great increase in gene number that we see in evolution, from a bacterium's 2000 or so genes to a human's 50,000–100,000. It is thus also responsible for the accompanying increase in the complexity of organisms.

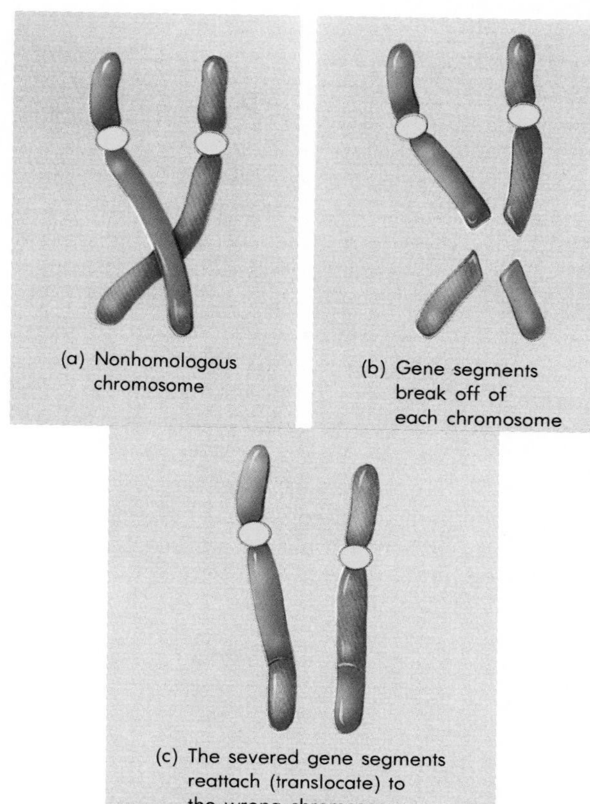

FIGURE 4-19 Chromosomal translocations, in which part of one chromosome has moved to another chromosome, have been linked to cancer in humans. One translocation is responsible for the *cri du chat* (cat cry) birth defect; affected babies sound like crying cats because of an improperly formed voice box; they are also severely retarded. (a) Nonhomologous chromosomes. (b) Segments break off each chromosome. (c) The severed segments reattach (translocate) to the wrong chromosome.

Point Mutations

Point mutations are changes in individual nucleotides or bases. They can change one nucleotide into another. They can also remove nucleotides entirely, in which case they cause *frame shift mutations.*

1. Consider the DNA base sequence ATA GGC GCA CCT. What will the effect on the resulting protein be if the first T is changed to a C?
2. What will happen if that first T is removed? Remember that the DNA must still be read three bases at a time. Why is this kind of mutation called a frame shift mutation?
3. Try thinking of an English sentence made up of run-together "triplet" words in the same sort of way (so that you read the sentence by taking it in three-letter chunks, just as the cell reads DNA, not by looking for spaces). Here is such a sentence:

   THECATATETHERATANDSAWONEDOG.

   What kind of damage can chromosomal-type inversions, deletions, and insertions do to the meaning of this sentence? What kind of damage can point mutations do to the meaning of this sentence? What does a frame shift mutation do to the sentence?

Because most genes by far serve as templates for mRNA and code for proteins, most point mutations alter the structures of proteins. Most such changes stop the proteins from doing what they are supposed to do, as Beadle and Tatum found. Only occasionally will a mutation improve a protein's efficiency or even make it able to do something else useful.

Causes of Mutations

Mutations have several possible causes. Beadle and Tatum used x-rays, a form of radiation that can damage both whole chromosomes and individual DNA nucleotides. Other kinds of radiation—ultraviolet light, cosmic rays, and nuclear radiation—can do the same thing.

Viruses can disrupt chromosomal structure, for they often plug their own genes into a chromosome. If they plug into the right—or wrong—spot, they can interfere with normal control of genes. A virus can also unplug itself, and when it does, it can carry a cellular gene with it. It can thus be responsible for deletions; if it or its progeny later plug into another cell's chromosomes, it can then be responsible for a gene duplication.

Chemicals can also damage the DNA, by replacing a base with a variant base that does not replicate faithfully (Figure 4-20a), by reacting with a base to change its structure and hence its base-pairing characteristics (Figure 4-20b), or by breaking the DNA strand itself, thereby enabling inversions, deletions, and translocations.

Mutations can also occur spontaneously, for replication is not a perfectly reliable process. Structural shifts like that shown in Figure 4-20a for 5-bromouracil can occur for any of the normal bases, though they are rare.

Some kinds of mutations are so common that cells have evolved ways to repair them. A common effect of ultraviolet light is to cause adjacent thymines in a DNA strand to link together, or dimerize. "Proofreader" enzymes constantly scan the DNA double helix, looking for such TT dimers. When they find them, they break the links or snip out the dimer and replace it with bases complementary to those on the other DNA strand. When people lack these enzymes, they suffer from diseases such as *xeroderma pigmentosum*, in which exposure to sunlight quickly triggers skin cancers; the ultraviolet component of sunlight causes gene damage, which causes skin cells to lose control of their growth and reproduction.

Even in normal individuals, the cell's repair processes do miss mutations, and every organism shows a typical *spontaneous mutation rate*. Bacterial genes accumulate one permanent DNA change for every 100 million cell divisions. Human genes acquire one mutation for every 100,000–1,000,000

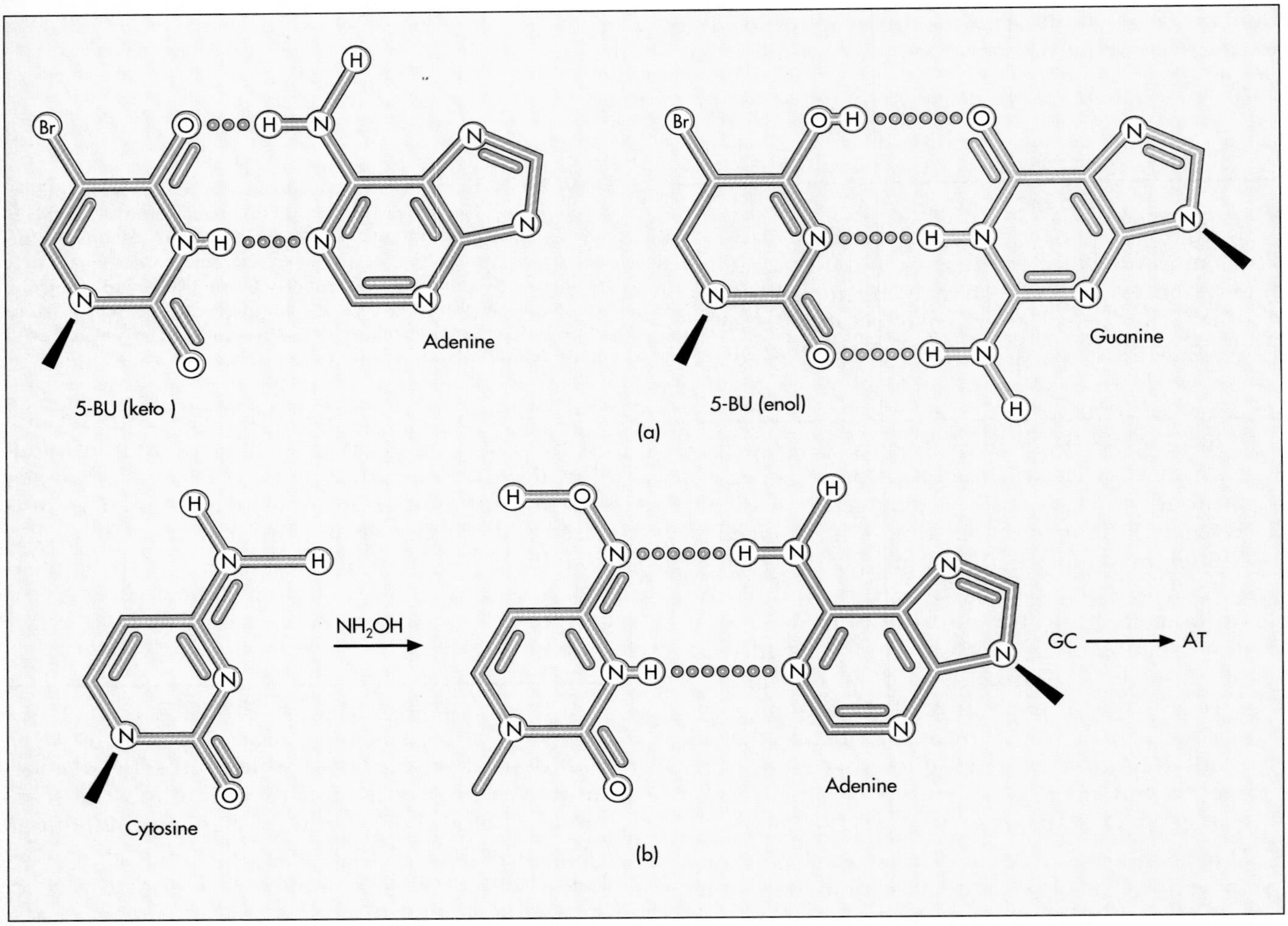

FIGURE 4-20 (a) 5-Bromouracil can pair with either adenine or guanine. (b) Hydroxylamine converts cytosine to a form that can base pair with adenine.

gametes (reproductive cells: sperm and eggs) formed. Other organisms vary greatly in their spontaneous mutation rates, with mice holding the record at one mutation per 10,000–100,000 gametes.

Mutations and Time

Because organisms accumulate changes in their DNA base sequences, and hence in the amino acid sequences of their proteins, it is possible to tell how closely related two species are by comparing these sequences. Researchers get their best results along these lines by examining DNA base sequences.

1. The first interspecies comparisons of mutations considered only the amino acid sequences of proteins. Why would this miss many mutations?
2. Why would comparing base sequences in DNA be more informative than comparing base sequences in mRNA?

Differences in the DNA base sequences of two species can tell us how closely related the species are. Consider that *any* two species must have had a common ancestor at some point in the history of their evolution. Any mutations acquired by that common ancestor will show in the DNA of both modern species (Figure 4-21). Mutations acquired after the split will tend to be different in the two species, and the longer the time since the split, the more different mutations the two species will accumulate. Therefore, the number of base differences between two species is a measure of how much time has passed since the common ancestor. Unfortunately, mutation differences do not reveal time in years or even generations; they say only "more" or "less."

Yet "more" or "less" can be very informative. When we compare the DNA base sequences of humans and apes, we learn that humans are most closely related to chimpanzees. Table 4-2 shows the DNA base differences between humans and chimpanzees, gorillas, orangutans, and gibbons for part of the gene for one mitochondrial enzyme [nicotinamide adenine dinucleotide (NAD) dehydrogenase 5]. Similar patterns emerge for comparisons of the whole enzyme, of other enzymes, and of larger samples of DNA.

We can translate protein- and DNA-based measures of relatedness into rough measures of time by using estimates of spontaneous mutation rates. Spontaneous mutation rates appear to remain more or less constant over time for any particular kind of organism. In particular, the human muta-

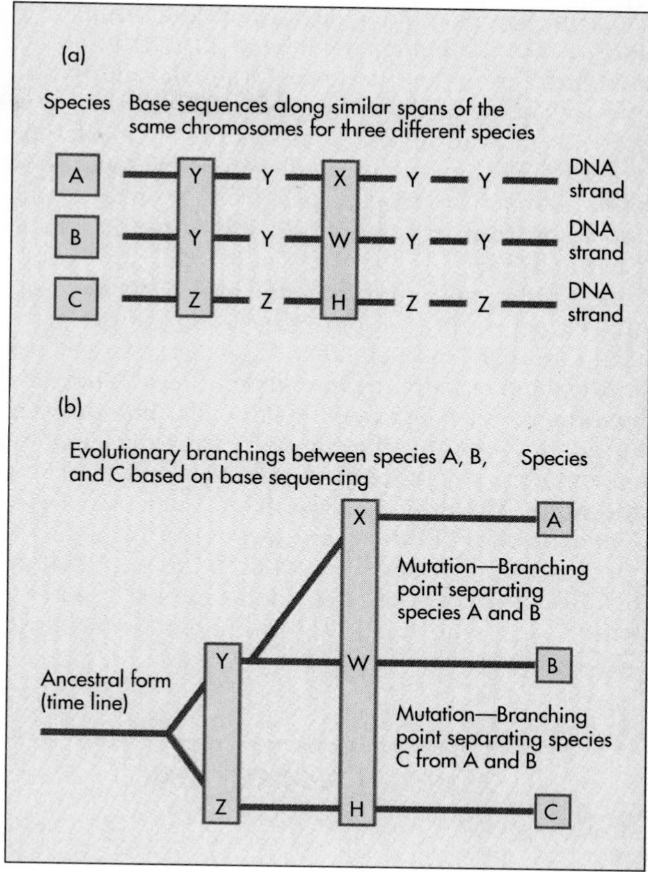

FIGURE 4-21 (a) Comparisons of the DNA base sequences for the same genes in three species show how mutation counts reveal relatedness. Here, sequences A and B differ at two points, while sequences B and C and A and C differ at six. (b) This variation implies that A and B had a common ancestor more recently than did B and C or A and C.

Table 4-2

DNA Base Differences Between Humans and Apes

	DNA BASE DIFFERENCES WITH HUMANS
Chimpanzee	22
Gorilla	28
Orangutan	59
Gibbon	56

Source: A.C. Wilson, "The Molecular Basis of Evolution," *Scientific American*, October 1985, p. 169.

tion rate seems to have been much the same for as long as human beings have been on Earth and not much different for the ancestors of humanity, all the way back to the primitive primates that gave rise not only to humans but also to monkeys and apes. Using this rate gives the times for separation of humans from apes, apes from monkeys, and monkeys from still earlier stock that we discuss in Chapter 25.

A similar and controversial approach suggests that every living human shares an umpty-great grandmother some 200,000 years ago. We get our mitochondria only from our mothers. They are present in both egg and sperm cells, but those of sperm do not enter the egg during fertilization; all an embryo's mitochondria come from the egg. Furthermore, the mitochondria have their own DNA, which also accumulates mutations. And analysis of these mutations suggests that all human mitochondria had a single source, in one woman, a hypothetical "Eve," who lived 200 millennia in the past.

GENETIC ENGINEERING

Taking place right now is a scientific revolution that will forever alter the study of biology, the course of evolution, and the practice of medicine. Called *genetic engineering,* it exploits the very nature of DNA to transfer gene segments from one cell to another and from one life form to another. As a new division of science, it may be the most important scientific development that will occur during your lifetime.

Genetic engineering was born with the discovery of two enzymes, reverse transcriptase and restriction endonuclease. **Reverse transcriptase** is an enzyme used by RNA viruses to make a DNA copy of their RNA genes. Genetic engineers use it to make DNA copies (cDNA) of mRNA extracted from cells and thus to get "pseudogenes" corresponding to the mRNA's target proteins.

Restriction endonucleases are bacterial enzymes that break DNA strands at specific sequences of bases, such as GGATCC. When such an enzyme encounters its specific sequence, it cuts it, as in Figure 4-22, leaving single-stranded projections from the two sides of the break. The projections are complementary in base sequence and can therefore pair up again and be rejoined. Projections obtained from different cuts by the same enzyme in a cell's DNA, or from cuts by the same enzyme in the DNA of different cells or organisms, can also pair up and be joined.

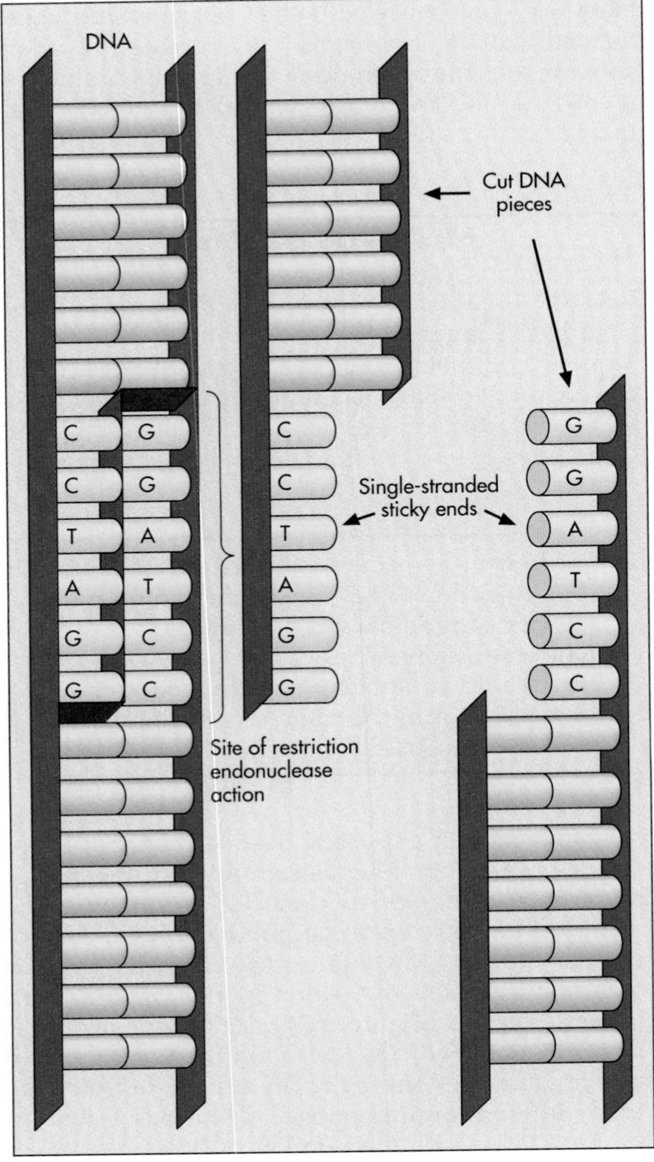

FIGURE 4-22 When restriction endonucleases cut the DNA double helix, they leave single-stranded projections from each side of the break. Since the projections are complementary in sequence, they can pair up again. Restriction endonucleases are therefore said to leave "sticky ends" after they cut the DNA.

Treating a cell's DNA with a restriction endonuclease breaks the DNA into numerous pieces of many lengths, depending only on how the enzyme's recognition sites are scattered in the DNA. Some of the pieces are much smaller than a gene. Some include several genes. Some include just one gene.

The next step is to treat the DNA from a virus or a *plasmid* (a tiny loop of DNA that serves a bacterium as an extra chromosome) with the same restriction endonuclease, mix the resulting DNA fragments with those obtained from the cellular DNA in the first step, and add DNA ligase to join the sticky ends together. The resulting mixture of rejoined fragments includes some viruses or plasmids that have incorporated a bit of cellular DNA, as shown in Figure 4-23. The fragments are said to be recombined, which gives the technique its name of **recombinant DNA.**

Recombinant viruses can then be allowed to infect bacterial (or other) cells. They will then either multiply within the cells or merge their DNA with that of the cells and multiply as the cells multiply. Recombinant plasmids can be mixed with bacteria and suitable chemicals that encourage the bacteria to take the plasmids into themselves. Then they too multiply as the cells multiply. Either way, the result is thousands of copies—or *clones*—of the recombinant DNA. If the recombinant DNA is attached to suitable promoters, operators, and other regulator genes, the host cell can even make proteins it could never make before.

Because the recombinant DNA technique produces so many copies of the genes trapped in virus or plasmid DNA, it allows researchers to identify and study genes with greater ease than ever before. It has thus vastly increased our knowledge of many genes. But it can do more as well. Once researchers have isolated a gene from recombinant DNA or made a DNA copy of mRNA with reverse transcriptase, they can deliberately transplant that gene into bacteria, yeast, and even plant and animal cells, giving the host cells the ability to make new proteins. And they have done so. Bacteria carrying various human genes now produce—in quantity and cheaply—hormones for treating diabetes, pituitary dwarfism, and other ailments. Viruses carrying the genes for proteins native to other viruses and bacteria are being used to make safer, more effective vaccines.

Many people expect that as the techniques of genetic engineering improve, we will see the replacement of the genes responsible for hereditary diseases such as hemophilia (failure of the blood to clot). We will see crop plants that make their own fertilizer. We may even see people designed for greater intelligence, strength, health, and longevity. In fact, in 1990, the first human gene transplants were performed in an effort to cure an immune system deficiency. The idea was to replace a defective gene with a gene that worked as it should.

The promise of genetic engineering is tremendous. Nevertheless, some people are worried that the technology may—deliberately or accidentally—lead to the creation of new disease organisms. They fear that adding foreign genes to bacteria and viruses may increase their ability to cause disease in people or in crop (or wild) plants and animals, in-

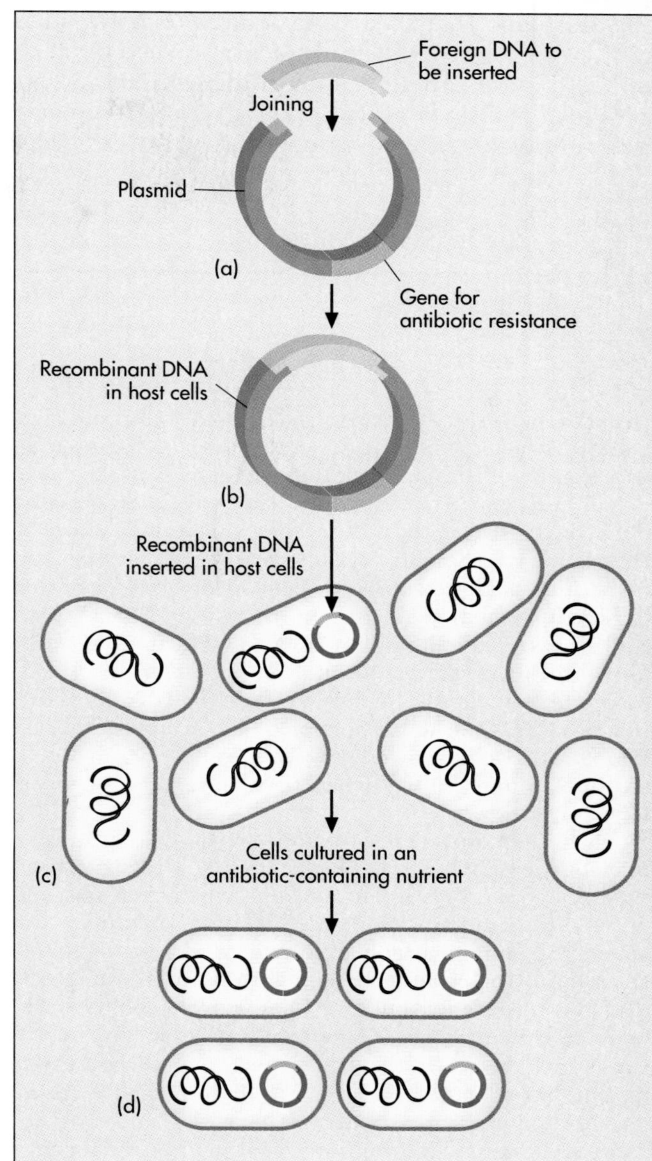

FIGURE 4-23 Insertion of cellular DNA into a virus or plasmid and subsequent gene cloning. (a) The foreign DNA, plus a gene section that confers antibiotic resistance, is inserted into a bacterial plasmid (minichromosome). (b) The resulting blend of nucleic acids is called recombinant DNA. (c) The recombinant DNA is inserted into a bacterial cell. The bacterium is grown in a nutrient medium containing low levels of antibiotics. The surviving cells in such an environment are those containing the genes for antibiotic resistance, the same cells that contain the foreign genes being tested. (d) If those genes produce insulin, then the colony may produce economically useful quantities of that genetically controlled product.

crease drug resistance, or make diseases more often fatal or more easily spread. At worst, such a development could threaten human life on Earth.

When genetic engineering was new in the early and mid-1970s, the researchers themselves recognized the hazards and suspended their research until safeguards could be devised and agreed to. Since then, there have been no signs of danger, and the safeguards have been relaxed. But the potential for danger—and the need for caution—remains.

SUMMARY

Decades before they recognized that deoxyribonucleic acid (DNA) was the substance of the gene, biologists realized that hereditary information was carried as strings of genes, each one of which controlled some characteristic of the organism, on the chromosomes. The chromosomes, they knew, were made of nucleic acid and protein, but just which of these substances encoded the information of heredity remained a mystery until the 1940s. At that time, researchers found that DNA could be used to transfer a trait from one bacterium to another. In the 1950s, further experiments clinched the case for DNA by showing that it was the active portion of bacterial viruses, and Watson and Crick showed how DNA could work in heredity and cellular control when they discovered its structure.

Each DNA molecule is composed of two strands of nucleotides. Each nucleotide is a sugar (deoxyribose) molecule with a phosphate group attached to one end and a nitrogen-containing base (adenine, thymine, cytosine, or guanine) attached to the other. In one DNA strand, the nucleotides are joined sugar to phosphate so that the bases stick out from a sugar-phosphate backbone. The two strands of a molecule are linked by hydrogen bonds between the bases; base complementarity means that adenine bonds only to thymine and cytosine only to guanine. The sequence of bases in a strand encodes the genetic information; most genes code for proteins, and each triplet of bases is a code word that stands for a single amino acid in a protein.

The key to transmitting the information in the DNA from one generation of cells or organisms to the next generation lies in base complementarity. Enzymes unzip the hydrogen bonds between the DNA strands, line up complementary nucleotides, and fasten them together to create new strands of DNA. The result of this replication process is two identical double-stranded DNA molecules, and in each molecule, one strand is old and one is new.

The key to translating the DNA information into protein is similar, but it depends on an intermediary substance—the other nucleic acid, known as ribonucleic acid (RNA). RNA differs from DNA in that its nucleotides use the sugar ribose, not deoxyribose, and the base uracil instead of thymine. In addition, it exists as single-stranded molecules, and there are several types of RNA. Ribosomal RNA (rRNA) becomes part of the ribosomes. Transfer RNA (tRNA) binds to amino acids and delivers them to the ribosomes for incorporation in proteins. Messenger RNA (mRNA) carries protein recipes from the DNA to the ribosomes.

Transcription occurs when enzymes copy the DNA message by making a complementary RNA version. Some genes are transcribed into rRNA or tRNA. Most are transcribed into mRNA. Translation occurs when the mRNA leaves the nucleus and is read like a tape by the ribosomes, which match tRNA molecules to appropriate base triplets (codons) in the mRNA and chain the amino acids carried by the tRNAs together to build a protein.

Genes are not transcribed and translated continuously. They exist as units (operons in prokaryotes), in which one or more structural (protein) genes are attached to a promoter, to which the enzyme that copies DNA as RNA attaches, and an operator, to which repressor substances (proteins, or end products of enzyme-controlled chemical reactions) can bind to stop transcription. Repressor genes code for repressor proteins and have their own operators and promoters. There are also other regulatory genes, which conspire to ensure genes are active only when the cell needs their products.

Mutations are inheritable changes in DNA structure caused by chemicals, radiation, and errors in replication. Point mutations affect single nucleotides, losing them or replacing them with other nucleotides; they change code words and can change the amino acid makeup of a protein. Chromosomal mutations are the deletion, duplication, or inversion of a piece of a chromosome (sometimes including many genes) or the transfer of a chromosome piece from one chromosome to another (translocation). Mutations are random and occur at a typical rate in any species; they can thus be used to measure the degree of relatedness of two species and to estimate the time since two species last shared a common ancestor.

Genetic engineering (recombinant DNA) involves transplanting genes from DNA of one species, such as humans, to DNA of another, such as bacteria. It requires the restriction endonuclease enzymes to cut DNA in such a way as to leave "sticky ends," complementary single-stranded edges to the cuts that can be rejoined. Using the same enzyme on the DNAs of two species produces a mixture of DNA fragments; rejoining the fragments then merges the DNAs of the two species. In this way, the human gene for, say, insulin can be inserted into the DNA of a bacterial or yeast cell, which will then transcribe the gene and make the human protein. Because gene transplants may produce new disease organisms, genetic engineering techniques must be used with caution.

STUDY QUESTIONS

1. Why did biologists first think heredity had to involve the chromosomes?
2. What is wrong with the idea that each chromosome *is* a gene?
3. In the 1940s, a crucial experiment demonstrated than one component of the chromosomes could transfer genetic information from one organism to another. Identify the component and describe the experiment.
4. Is DNA always the substance of heredity? Why or why not?
5. Describe the structure of the DNA molecule.
6. Why does base pairing or base complementarity make it easy to see how DNA must replicate itself?
7. Name five differences between DNA and RNA.
8. Describe how the 4 DNA nucleotides can code for the 20 amino acids in the cell's proteins.
9. How are replication and transcription alike? How are they different?
10. Describe the role of tRNA in translation.
11. What is the difference between a ribosome and a polyribosome?
12. How does the operon permit the processes of induction and repression?
13. Why might a mutation that substituted one DNA base for another have no effect on protein structure?
14. What kind of mutation could turn an entire gene into nonsense?
15. What major difference must there be between a cDNA "pseudogene" and a real gene? Why?
16. How does transformation resemble recombinant DNA genetic engineering?

GLOSSARY

Anticodon The base triplet on a tRNA that is complementary to an mRNA codon.
Base complementarity Base pairing.
Base pairing The matching, via hydrogen bonds, of adenine to thymine or uracil and of cytosine to guanine along the length of DNA and RNA molecules.
Chromatin The complex of DNA and protein that makes up a eukaryotic chromosome.
Chromosomes Threadlike structures of DNA and protein, found in the nucleus; their DNA carries the information of heredity.
Codon A DNA or mRNA code word, or base triplet, standing for an amino acid.
Deoxyribonucleic acid (DNA) The substance of heredity; the DNA molecule is a double-stranded helix; each strand is a chain of alternating deoxyriboses and phosphate groups. From each strand projects a series of nitrogen-containing bases—adenine, guanine, thymine, and cytosine—that encodes the genetic instructions. The bases pair to link the two DNA strands.
DNA polymerase An enzyme that links DNA nucleotides together on a DNA template.
Exon A gene segment that codes for protein.
Genes Segments of chromosomes whose DNA encodes specific characteristics of the cell or organism, usually in the form of structural proteins or enzymes.
Genetic code The list of codon–amino acid equivalences.
Histones Structural chromatin proteins.
Induction Activation of genetic transcription of a gene.
Intron A gene segment that does not code for protein.
Messenger RNA (mRNA) The form of RNA that carries genetic information from the nucleus to the protein-synthesizing apparatus (ribosomes) in the cytoplasm.
Mutation An inheritable change in the DNA.
Nucleotides The component units of nucleic acids; composed of a sugar, a phosphate group, and a nitrogen-containing base.
One gene–one protein theory The idea that each gene specifies the structure of one type of protein molecule.
Operator The site to which repressors bind to block transcription of a gene.
Operon A functional genetic unit, regulated as a whole, consisting of one or more structural genes, an operator, and a promoter.
Promoter The site to which RNA polymerase must bind before it can transcribe a gene.
Recombinant DNA DNA formed by the fusion of genes from two or more organisms; the technique for producing such DNA.
Replication The duplication of a cell's DNA.
Repression Inactivation of genetic transcription of a gene.
Ribonucleic acid (RNA) A nucleic acid differing from DNA in that its sugar is ribose, thymine is replaced by uracil, and its structure is single stranded; it is essential to translating the DNA information into protein.
Ribosomal RNA (rRNA) The RNA that forms part of the ribosome.
RNA polymerase An enzyme that links RNA nucleotides together on a DNA template.
Transcription The making of an RNA copy of the information in the DNA.
Transfer RNA (tRNA) The RNA that attaches to amino acids and delivers them to the ribosome for use in protein synthesis.
Translation The use of the information encoded in RNA to make protein.

Chapter 5

Reproduction at the Level of the Cell

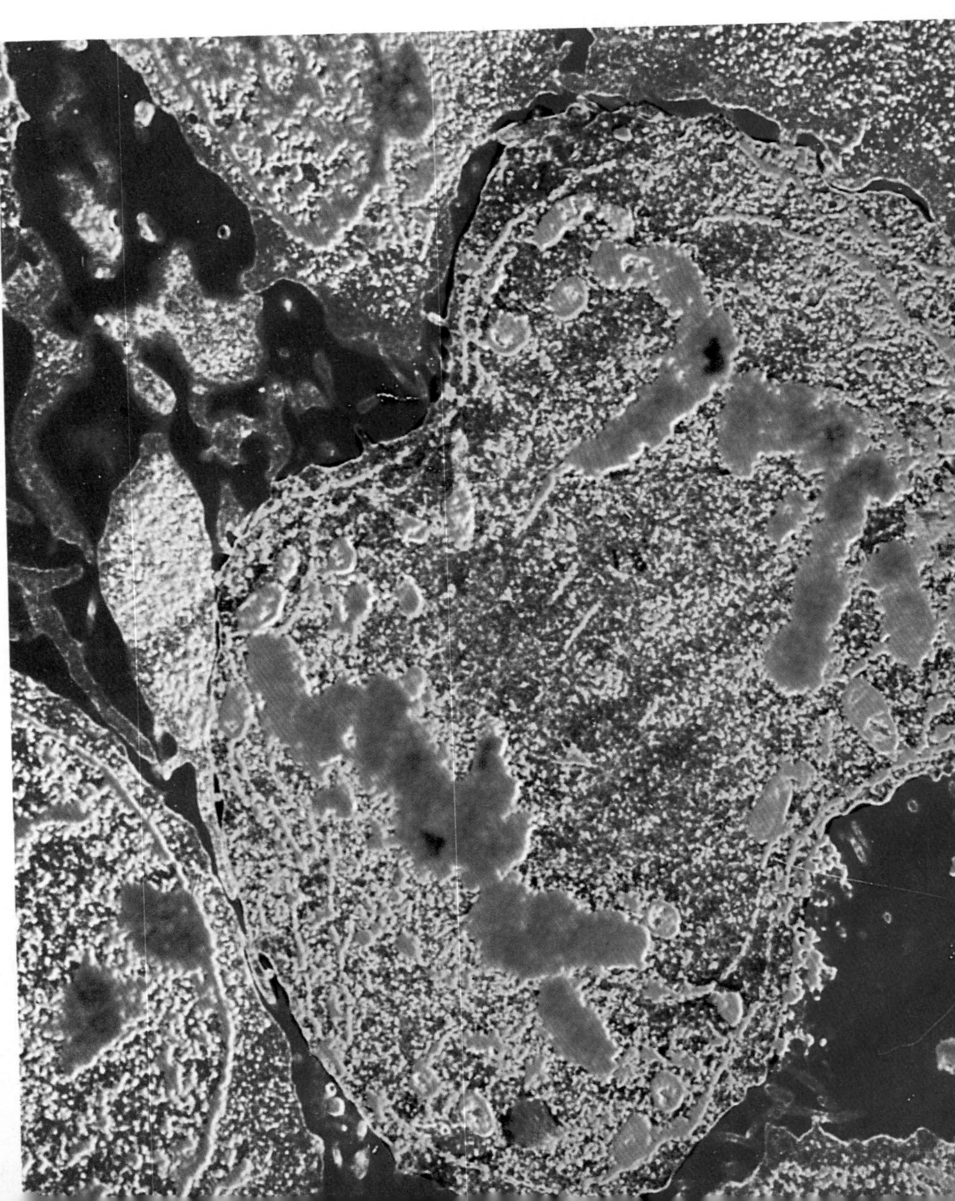

Death is inevitable. Living things wear out. They are eaten by other creatures. They fall victim to accidents, storms, earthquakes, and floods. It is thus something of a miracle that life continues to exist on the planet Earth. It is even more amazing that living things continually increase in number and variety.

The key to these miracles is the ability of life to renew itself by reproducing. Each living thing strives to beat death to the punch by generating at least one duplicate or near duplicate of itself before it must leave the scene. Barring complications from overpopulation, the most successful individuals and species have many such offspring before they die, and their success is what accounts for how the number and variety of living things have increased over the eons.

For most living things, offspring are literally genetic duplicates of their parents, with the same genes. Many plants reproduce by **cloning,** sending out runners, dropping plantlets from the edges of their leaves, or growing multiple tubers, all of which can grow into new plants. Yeast cells and some primitive animals grow their offspring as **buds.** Bacteria increase slightly in size before they split in two in a process called **fission** (Figure 5-1).

We call reproduction that duplicates a single parent **asexual reproduction. Sexual reproduction** requires two parents, each of which contributes

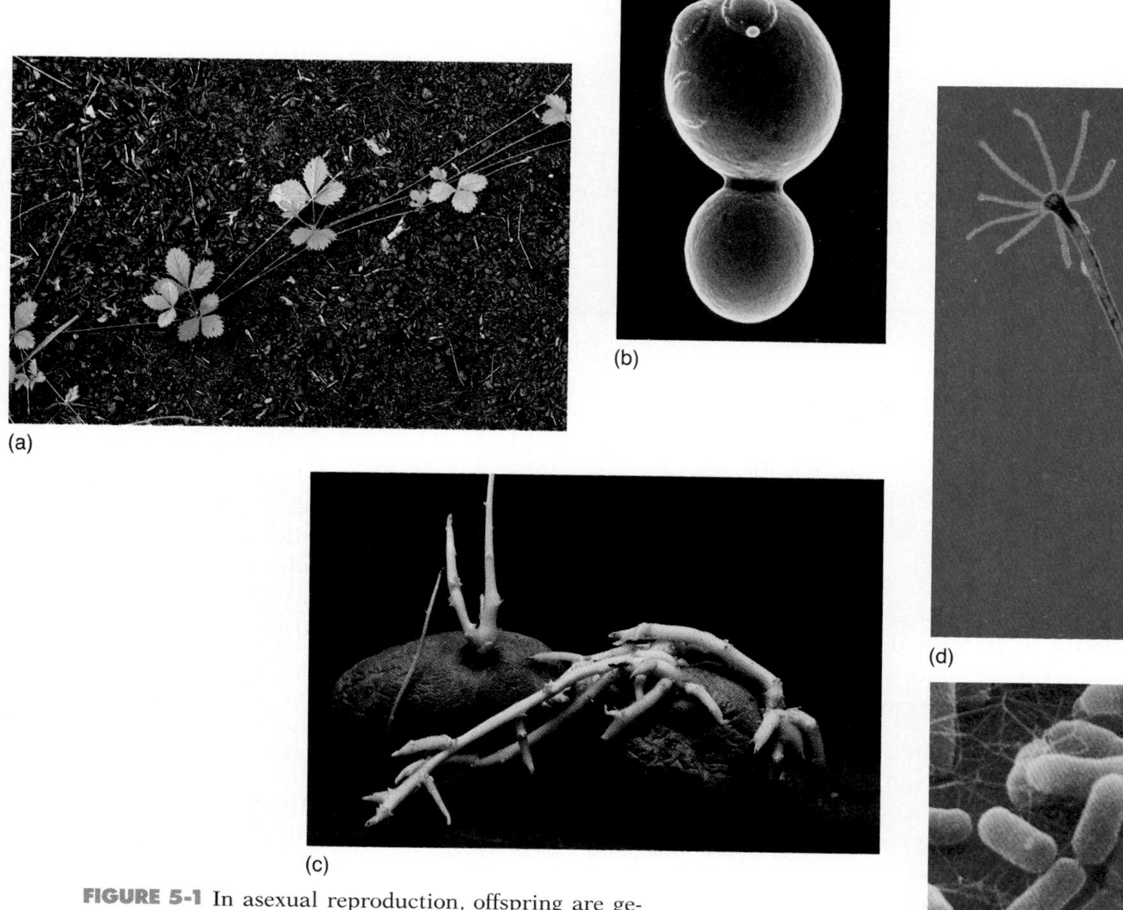

(a) (b) (c) (d) (e)

FIGURE 5-1 In asexual reproduction, offspring are genetic duplicates or clones of the parents. (a) Strawberry plants can grow from runners. (b) Brewer's yeast cells reproduce by budding. (c) Potato tubers contain stored energy and buds that sexually produce new potato plants. (d) The tiny aquatic hydra can reproduce by budding. (e) Bacteria such as *Escherichia coli* (from the human intestine) reproduce by fission.

copies of half its genes to its offspring. To achieve this, each parent produces specialized cells, called sex cells or **gametes,** that fuse to produce the unique blend that forms the genetic basis of the new individual. A special kind of cell reproduction, called meiosis, is used to shuffle each individual's genetic deck so that copies of half its genes can be dealt into its sex cells. The gametes thus formed are the sperm and ova for the higher animals and lower plants and are the pollen and eggs for the higher plants (Figure 5-2). In general, the male gamete (sperm) is much smaller than the female gamete (egg).

The advantage of sexual reproduction is that when the gametes fuse, during fertilization, the **sexual recombination** of genes from two separate parents can create genetically different organisms that may be better adapted to their environment.

Sexual reproduction is the human mode. But so is asexual reproduction. People do not bud like yeast or split down the middle like bacteria (although very small embryos may subdivide to form identical twins or triplets). But our individual component cells do split. Our cells divide in two, enlarge for a time, and divide again. In this way, our bodies grow in size and complexity, from one cell—an egg—to trillions, from a speck to a child to an adult.

---

1. The kinds of asexual reproduction permit no reshuffling of genes and hence yield no genetic variety by recombination. But variety is an essential prerequisite for evolution by natural selection. How then can organisms that reproduce asexually evolve? That is, where do they get the necessary genetic variety? (*Hint:* See Chapter 4.)
2. Why do you think it would be difficult or impossible for humans to reproduce asexually? If we *could* do it, do you think fission or budding would work better? Why?

In every case, sexual or asexual, reproduction requires one thing: the preparation of a duplicate set of the parent's genes. Without such duplication, the offspring would have no blueprints to tell them how to grow, function, and reproduce in their turn.

Preparing the duplicate begins with deoxyribonucleic acid (DNA) replication, and in bacteria, it requires little more. A bacterial cell has one chromosome, in the form of a single molecule of DNA whose ends are joined to form a circle. The DNA molecule is bound at one point to the inside surface of the cell membrane. Once the molecule has replicated, two molecules are bound to the membrane side by side. The cell then manufactures new cell membrane between the attachment points. As the membrane grows, the attachment points separate, pulling the DNA molecules with them (Figure 5-3, p. 112). Once the DNA molecules are entirely separate, the cell pinches in between them, forming two daughter cells, each with its own chromosome and gene set.

The process is more complicated in eukaryotic cells than in bacteria for four main reasons. First, because eukaryotic chromosomes are not attached to the cell membrane, separating them is more difficult. Second, there are many chromosomes, and the cell must somehow keep them straight so that the daughter cells get one and only one copy of each chromosome. Third, a nuclear envelope restricts chromosome movement during cell division. Fourth, the chromosomes are not just DNA molecules, but complexes of DNA and protein.

We will begin our discussion of eukaryotic cell division by looking at the structure of chromosomes. Then we will consider how eukaryotic cells sort out their chromosomes for asexual reproduction (mitosis) and for sexual reproduction (meiosis). We will also discuss the difference between nuclear division (karyokinesis) and cell division (cytokinesis).

EUKARYOTIC CHROMOSOMES

If the DNA in a human cell were a single molecule, it would contain 6 billion base pairs and be 2 meters (6 feet) long. Yet all that length is crammed into the tiny space of the cell nucleus. Even more marvelous, all that length remains untangled. The cell has no trouble finding the parts it needs for transcription and, after replication, separating the copies for its offspring.

Ploidy

We simplify the mystery only a little when we realize that the cell's DNA is not in the form of a single molecule. Rather, there are 46 DNA molecules in each human nucleus, each molecule a chromosome. The cells of other kinds of organisms have different numbers of chromosomes, but the numbers are usually even. They are even because sexually reproducing life forms receive two complete sets of chromosomes, one set from each parent. We humans have two sets of 23 for a total of 46 chromosomes.

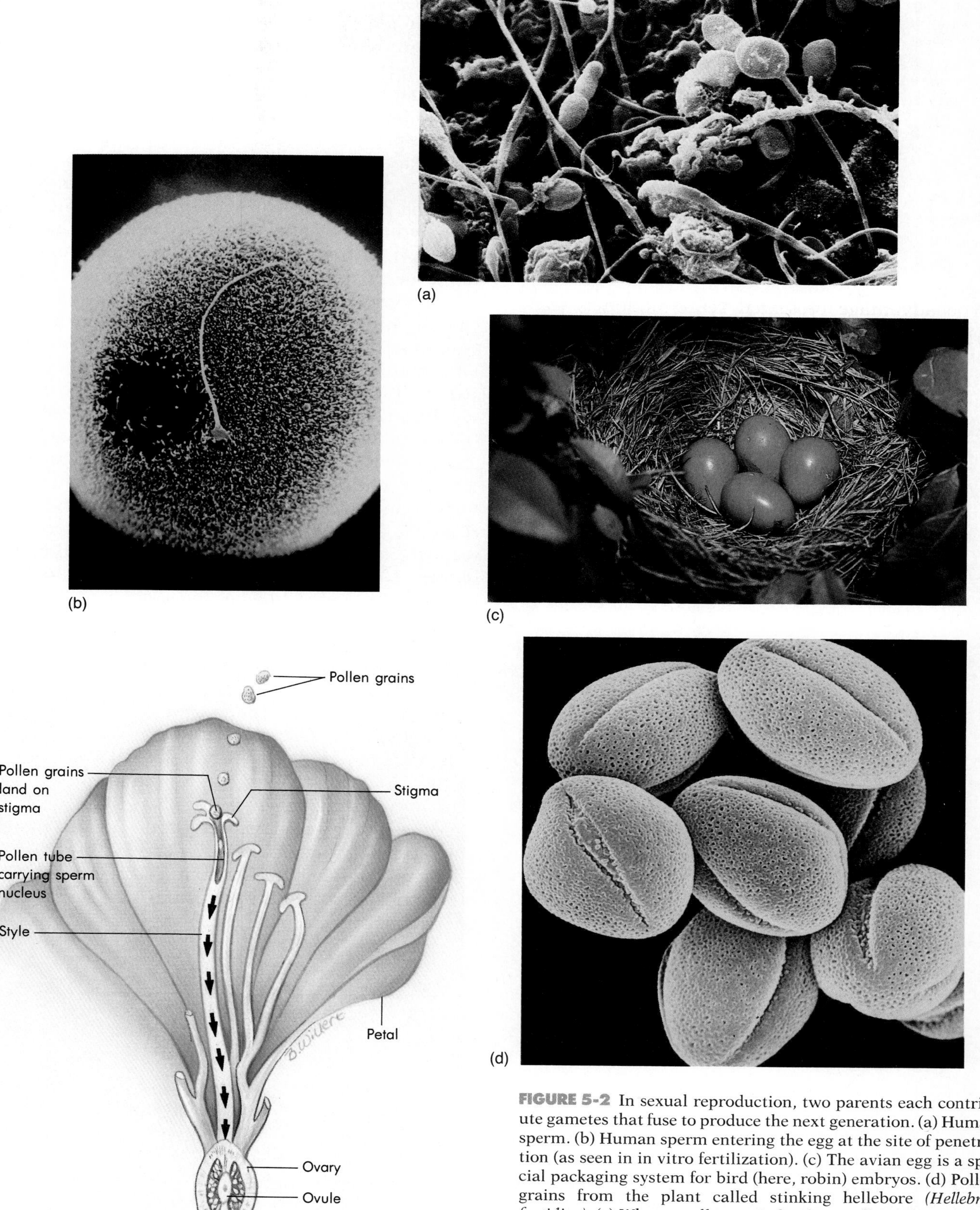

FIGURE 5-2 In sexual reproduction, two parents each contribute gametes that fuse to produce the next generation. (a) Human sperm. (b) Human sperm entering the egg at the site of penetration (as seen in in vitro fertilization). (c) The avian egg is a special packaging system for bird (here, robin) embryos. (d) Pollen grains from the plant called stinking hellebore *(Hellebros foetidius)*. (e) When a pollen grain lands on a flower's stigma, it sprouts a "pollen tube," which then grows through the style's tissue, carrying the sperm nucleus, to reach an egg in the floral ovary.

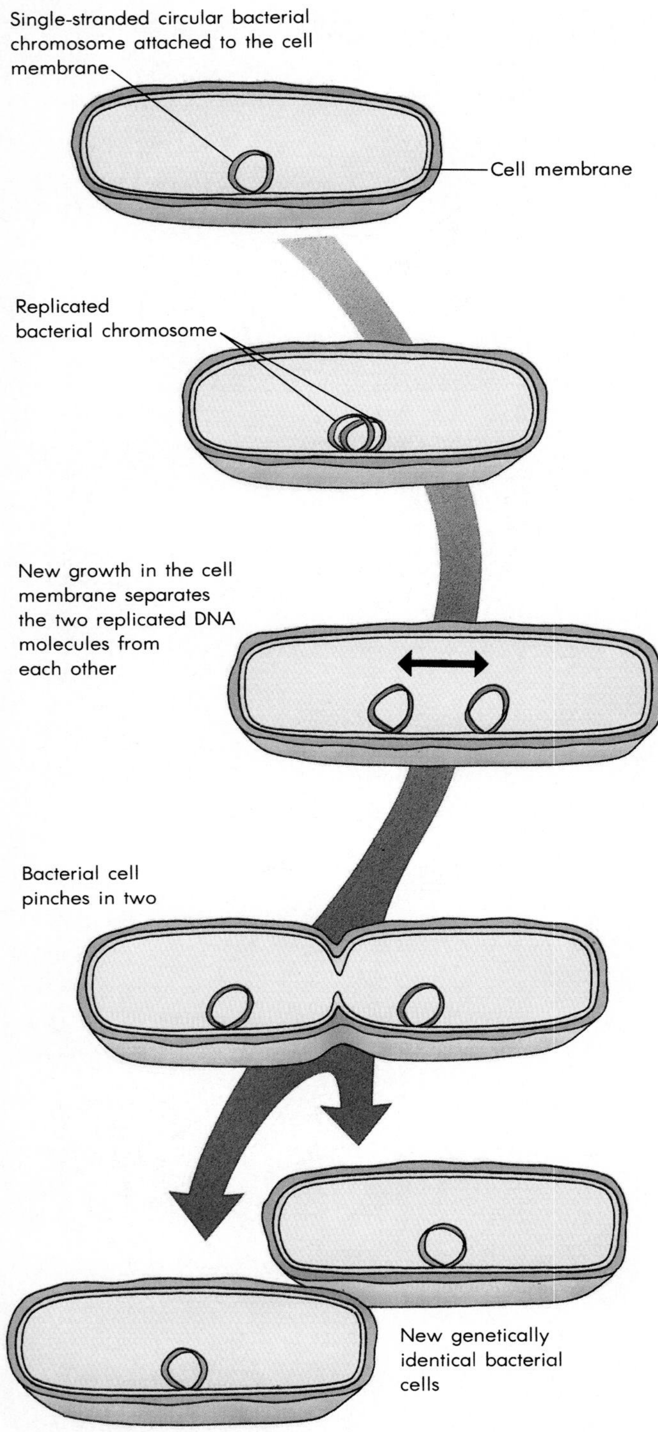

FIGURE 5-3 Bacterial fission.

The chromosomes in an individual's cells form homologous pairs, with one member of each pair coming from each parent. The two **homologous chromosomes** in a given pair are physically similar to each other, and both contain genes governing the same physical traits. Each parent thus has something genetic to say about each of the offspring's body characteristics.

When an individual produces its gametes, its cells divide in a special way **(meiosis)** that reduces the chromosome number from 46 to 23. As it produces the gametes, it cuts the number of chromosomes in half, giving each sperm or ovum only one of the homologous chromosomes from each pair. When sperm and egg fuse to give rise to offspring, each contributes a single set of homologues, reconstituting the complement of 46 chromosomes. We call the number of chromosomes in a gamete, or the number of *pairs* in a body cell, the **haploid number;** for humans, it is 23. The number of chromosomes in a body cell (twice the haploid number) is the **diploid number.**

One way biologists study chromosomes is by staining the cells to make the chromosomes visible and photographing them (Figure 5-4a). They then cut out the pictures of individual chromosomes and arrange them in order of size and other features; this arrangement is a **karyotype** (Figure 5-4b). Homologous chromosomes are arranged side by side, for they are almost identical except in their origin and in some of the genes they carry. Note that in karyotypes each chromosome looks X shaped, as if it were constructed by connecting two chromosomes together. In reality, you are looking at two identical strands of the same chromosome, the results of DNA replication; they are called **sister chromatids.**

Biologists give most of the chromosome pairs in a human karyotype numbers, 1–22; these are the **autosomes.** The remaining pair is called the **sex chromosomes,** which consist of the two X chromosomes (in females) or an X paired with a Y chromosome (in males). The sex chromosomes carry (among others) genes that primarily determine maleness or femaleness. Body cells contain two of each type of autosome and two sex chromosomes. Gametes contain one of each autosome and one sex chromosome; eggs always contain an X chromosome; sperm can have either an X or a Y.

Chromatin and Nucleosomes

The chromosomes are not just DNA; rather they are a complex of DNA and protein, called **chromatin.** The protein is of two types, **histone** and **nonhistone** proteins. As we noted in Chapter 4, the nonhistone proteins seem to be involved in regulating gene activity. The histones are structural proteins. They provide a framework on which the DNA is wound to reduce the length of the chromosome while preventing tangles.

There are five histone proteins. Four of these proteins form tiny beads consisting of two copies of

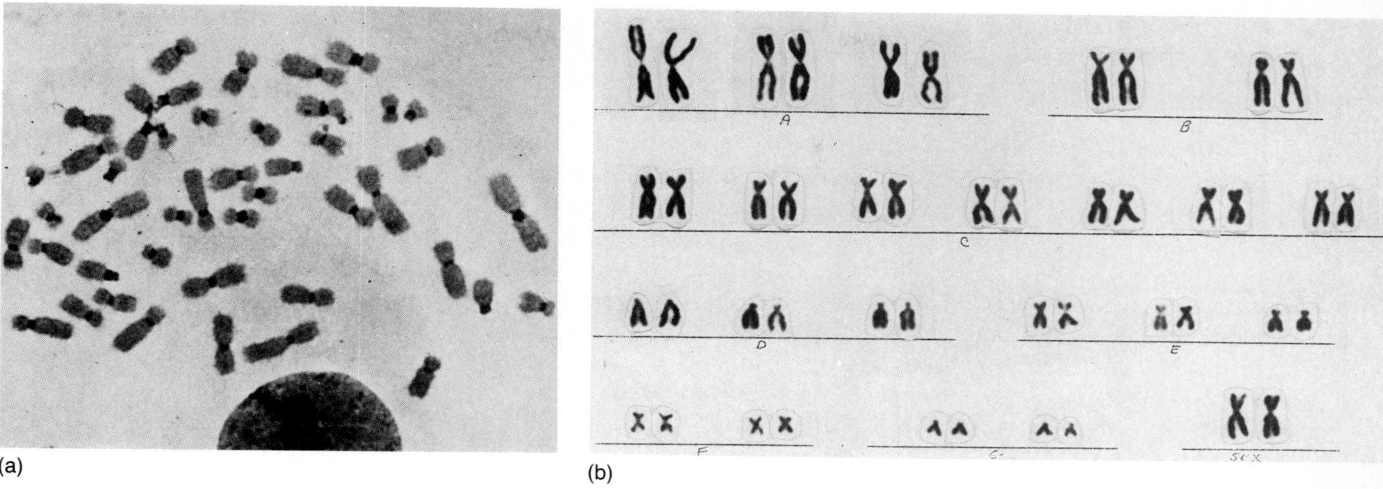

FIGURE 5-4 (a) Mitotic cell prepared to show spread-out chromosomes. (b) Chromosomes arranged in a karyotype.

each of the four (see Figure 5-5). The DNA winds twice around each bead, so that a chromosome looks like a string of pearls. The pearls, the protein beads with the DNA wrapped around them, are called **nucleosomes.** The fifth histone attaches to the DNA between the nucleosomes, acting like a spacer.

When the cell is not preparing for division, its chromosomes exist as loose strings of nucleosomes. The DNA is accessible to enzymes for transcription

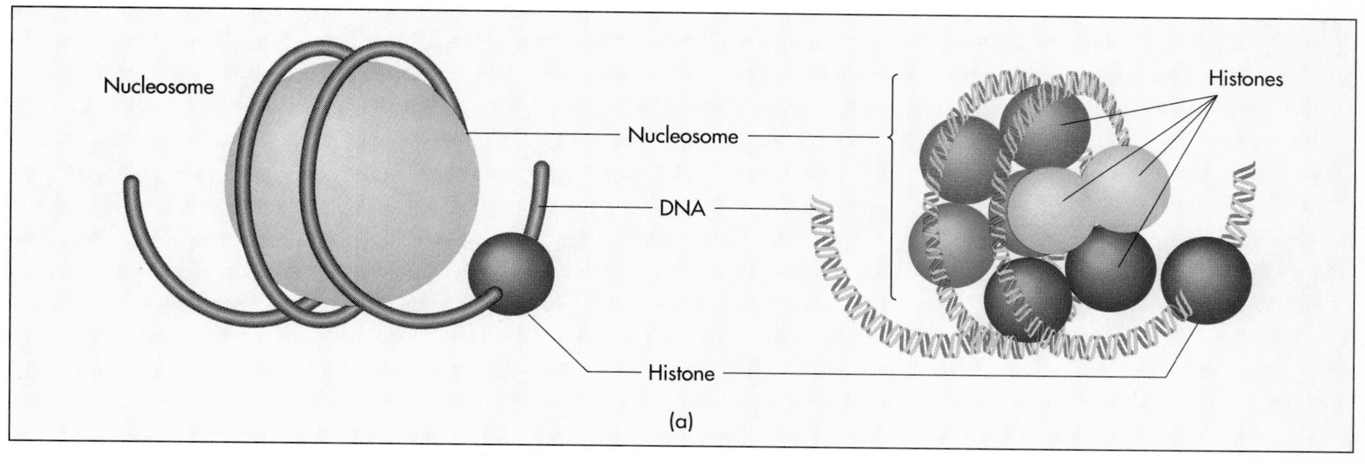

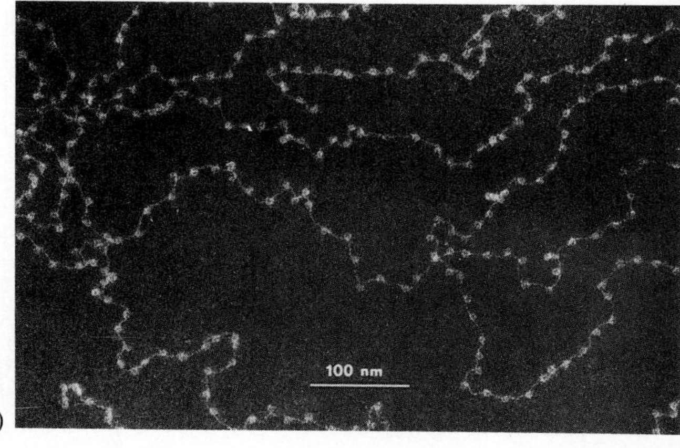

FIGURE 5-5 (a) In chromosomes, strands of DNA are coiled around beads made of histone proteins. (b) The resulting histone-DNA complexes, visible in this photomicrograph, are known as nucleosomes.

and replication, but it is also spread throughout the nucleus. To allow the separation of DNA copies in cell division, each chromosome must somehow be condensed into a much more restricted space. In fact, it must be shortened 10,000-fold.

When the process begins, each chromosome has already been replicated. The two strands of each chromosome are joined to each other in a region known as the **centromere** (see Figure 5-6), and they remain attached until cell division is nearly done.

The condensation of a chromosome begins with the coiling of its string of nucleosomes into a *supercoiled* strand of chromatin. This strand then folds back on itself numerous times. The resulting structure looks solid under the light microscope, but the electron microscope reveals loops of chromatin strands that give it a fuzzy appearance.

The single chromosome in Figure 5-6 looks like two chromosomes joined at a centromere. The two "half chromosomes"—duplicate structures formed in the process of replication—are called sister chromatids. The joined sister chromatids separate during cell division, one of each pair going to each daughter cell.

Heterochromatin

The chromosomes exist in two physical forms. Loosely spread throughout the nucleus when the cell is not dividing, they change by condensing as they prepare for cell division. However, the whole truth is not so simple. Some parts of each chromosome remain condensed even between cell divisions, and their chromatin is even more condensed than that in pre–cell division chromosomes. We call it **heterochromatin.** Normal chromatin is **euchromatin.**

Some heterochromatin is permanent; it contains "nonsense" DNA that is not arranged as genes. Some is temporary; it contains inactive or repressed genes, genes that are not being used by that particular cell. Heterochromatin thus seems to be a device used by the cell to keep useless or unused parts of the chromosomes out of the way.

Heterochromatin remains identifiable even in condensed chromosomes. Suitable chemical treatments reveal it in a pattern of dark bands that is unique for each chromosome (Figure 5-7). The

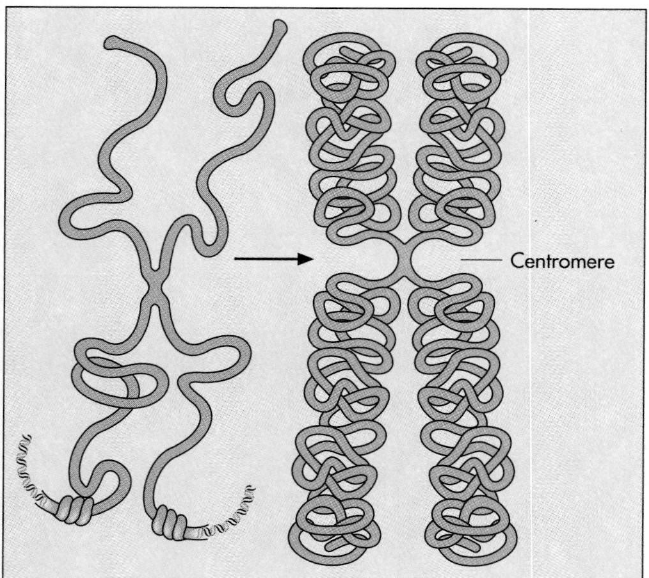

FIGURE 5-6 The coiling and folding of chromatin to form a pre–cell division chromosome.

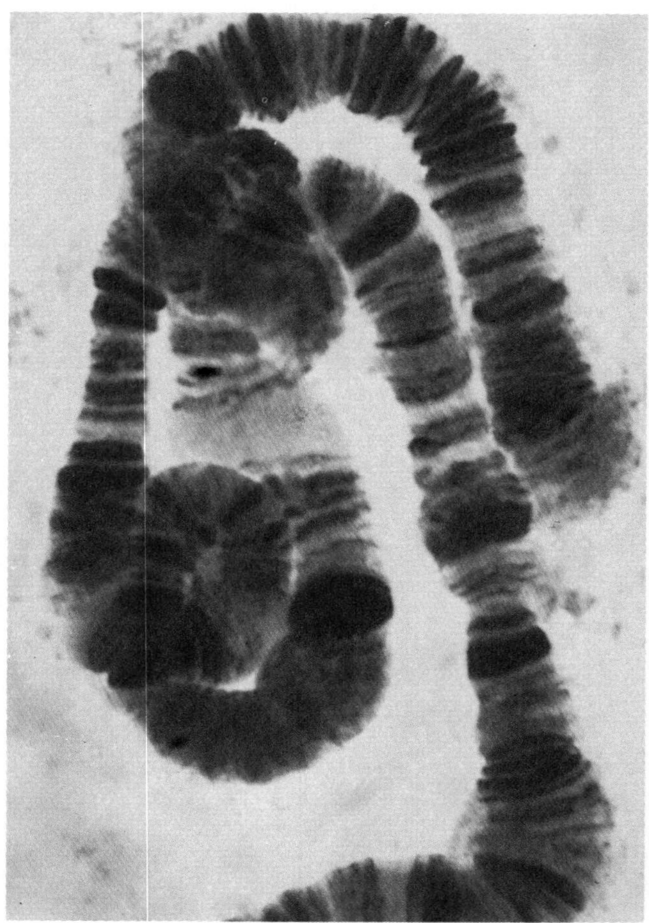

FIGURE 5-7 Chromosome banding patterns are sometimes used to detect changes or abnormalities in chromosomes, like those caused by mutations. The darkly stained bands consist of less genetically active, denser heterochromatin, and this may be a way for the cell to keep inactive chromosomal regions, or non-genes, out of the way.

banding pattern can be used to detect and study chromosomal mutations or changes.

1. How would you expect a chromosomal deletion to show up in a chromosome's banding pattern?
2. How would you expect a chromosomal translocation to show in banding patterns? *Hint:* See Figure 4-19.

Similar bands are visible in the special chromosomes found in the salivary glands of the fruit fly. These **polytene chromosomes** are formed when the DNA is replicated many times and the resulting DNA strands remain connected together in a single bundle; such multiplication of DNA provides many copies of the cell's genes and allows much greater gene transcription and protein synthesis. The active genes are those *between* the dark bands in Figure 5-8, where the DNA seems to take the form of **Z-DNA.** Z-DNA is a double helix, but it twists in the direction opposite that of the more common "B-DNA." Apparently as one result of the change in structure, the DNA strands between the bands puff out from the polytene chromosomes and are more accessible to RNA polymerase.

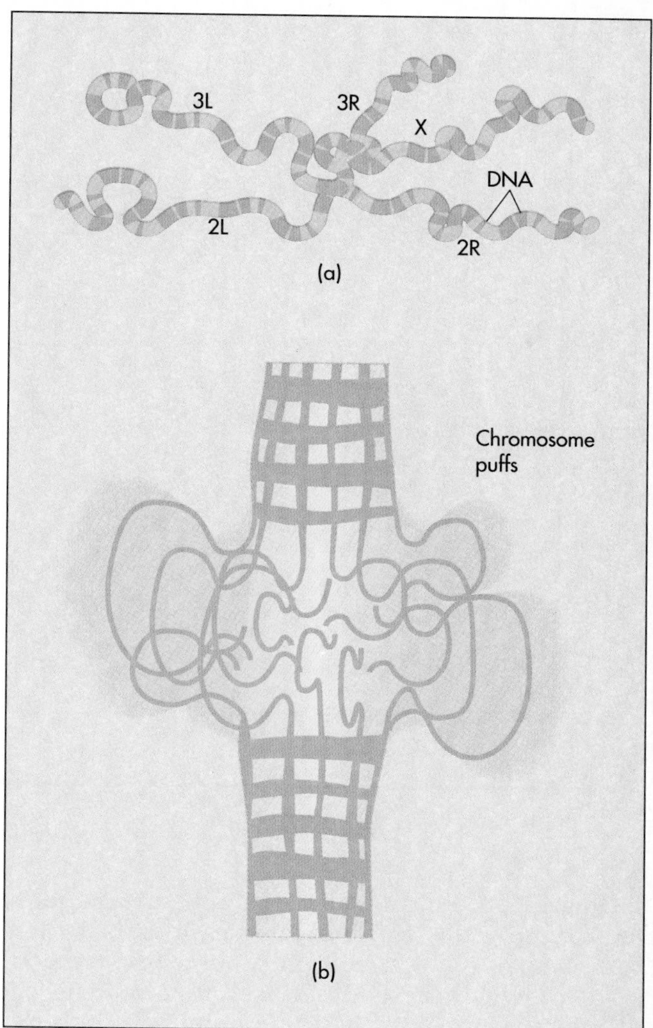

FIGURE 5-8 Polytene chromosomes are formed when DNA is replicated many times and the resulting DNA strands remain together. (a) Drawing of a polytene chromosome for the fruit fly, *Drosophila melanogaster.* (b) Chromosome puffs are areas of active genes, produced where the DNA uncoils and becomes more accessible to RNA polymerase.

THE CELL CYCLE

Like organisms, cells have life cycles. They are "born," they grow and develop, and they reproduce. They even grow old and die, but oddly, only cells that have lost their ability to divide, such as nerve and muscle cells, seem to age in the same sense as individuals.

The **cell cycle** is the sequence of events through which a cell passes from cell division to cell division (Figure 5-9). It is usually described as having two main phases. Cell division itself is one phase, the **M** (for mitosis) **phase.** Between cell divisions comes **interphase.** Interphase has three subphases. Immediately after cell division is a time of growth and development for the cell, the $\mathbf{G_1}$ (first gap) **stage.** This is followed by the **S** (synthesis) **phase,** when the cell replicates its DNA in preparation for the next cell division. After the S phase is the $\mathbf{G_2}$ (second gap) **stage,** a second time of growth and development.

The phases of the cell cycle vary in length from cell type to cell type. In many human cells, cell division takes less than an hour. The G_1 stage then lasts for about six hours, the S phase for about seven hours, and the G_2 stage for about two. However, these times can vary greatly, and cells can suspend indefinitely their progress through the cell cycle in either G_1 or G_2, as they do when they develop into specialized nerve, muscle, or other cells. Some specialized cells may resume their progress when damaged tissue needs replacement; nerve cells generally cannot.

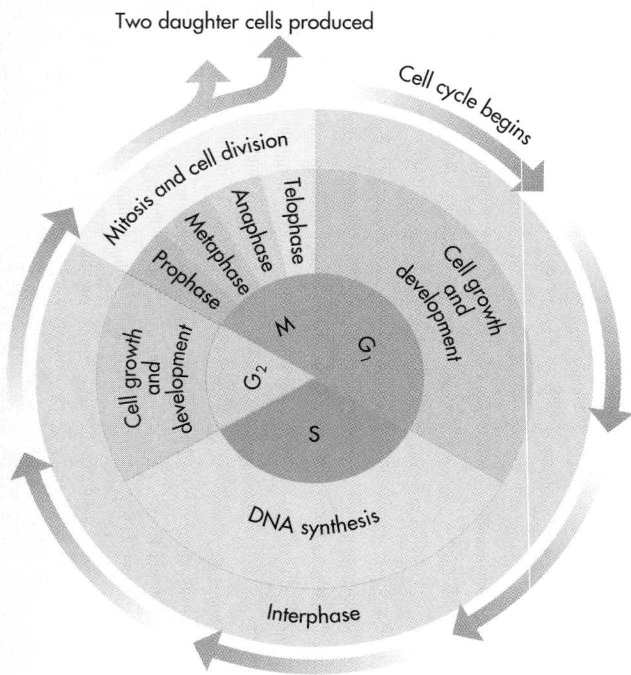

FIGURE 5-9 The cell cycle for a typical cell.

MITOSIS

Mitosis is the kind of cell division that is unique to eukaryotic cells. In multicellular organisms, it is the means by which cells multiply within the body and by which the body grows. After mitosis, the daughter cells have the same number of chromosomes as their parent cell; they are clones.

Mitosis seems complicated, for during it many things happen within the cell and to the chromosomes. But it is really very simple: All the steps of the mitotic process are designed to accomplish the same thing, the separation of sister chromatids from their double-stranded arrangement and their assignment to separate daughter cells. In essence, all mitosis does is line up the chromosomes and pull them apart.

There are two main events in mitosis. **Karyokinesis** is the separation of the chromosomes and the division of the nucleus. **Cytokinesis** is the division of the cell itself.

Karyokinesis

Figure 5-10 shows the stages of mitosis. The first stage is **prophase.** As it begins, the two centrioles just outside the nucleus separate and move toward opposite sides of the cell. The centrioles are small barrel-shaped structures built of microtubules. As they separate, each one duplicates itself, growing a new centriole at right angles to its own structure. As a result of this duplication, once cell division is completed, each daughter cell begins life with the pair of centrioles it will need when its turn comes to divide.

Centrioles seem to serve as organizing centers for an array of microtubules known as the **mitotic spindle.** As the centrioles move apart and the spindle begins to form, the nuclear membrane breaks down, the nucleolus vanishes from the nucleus, and the chromosomes condense. When the spindle is mature, spindle fibers radiate as a starlike *aster* from each centriole. In mitosis, the centromere of each chromosome is attached to fibers from both centrioles, coming from opposite sides of the cell. Some of the spindle fibers extend across the cell, connecting centriole to centriole without coming into contact with the chromosomes. By the end of prophase, each chromosome is visible as a pair of sister chromatids joined at the centromere.

In the second stage of mitosis, **metaphase,** some spindle fibers attach to the centromeres of the chromosomes. These fibers adjust their lengths, pushing and pulling on the centromeres, until the chromosomes lie in the central plane or equator of the cell. At the end of metaphase, the chromosomes are so arranged, with spindle fibers extending from each centromere toward both centrioles.

In **anaphase,** the third stage of mitosis, the spindle fibers linking the centromeres to the centrioles shorten while those linking the centrioles lengthen. Apparently, the cell removes protein subunits from the microtubules connected to the centromeres of the chromosomes and adds them to the microtubules that connect the two centrioles. The result is that the centrioles move further apart while the centromeres are pulled until they split. Once the centromeres have split, the sister chromatids move apart, drawn toward opposite ends of the cell by the shortening spindle fibers. Note that this process ensures that each end of the cell receives one sister chromatid from each pair and hence a complete set of single-stranded chromosomes.

Telophase is the end of karyokinesis, or nuclear division. During this phase of mitosis, the spindle disappears, the nuclear envelope reforms, the chromosomes reverse their condensation, and the nucleolus reappears.

Cytokinesis

During telophase for most cells, the cell divides into two daughter cells. This cytokinesis depends on the formation of a belt of microfilaments around

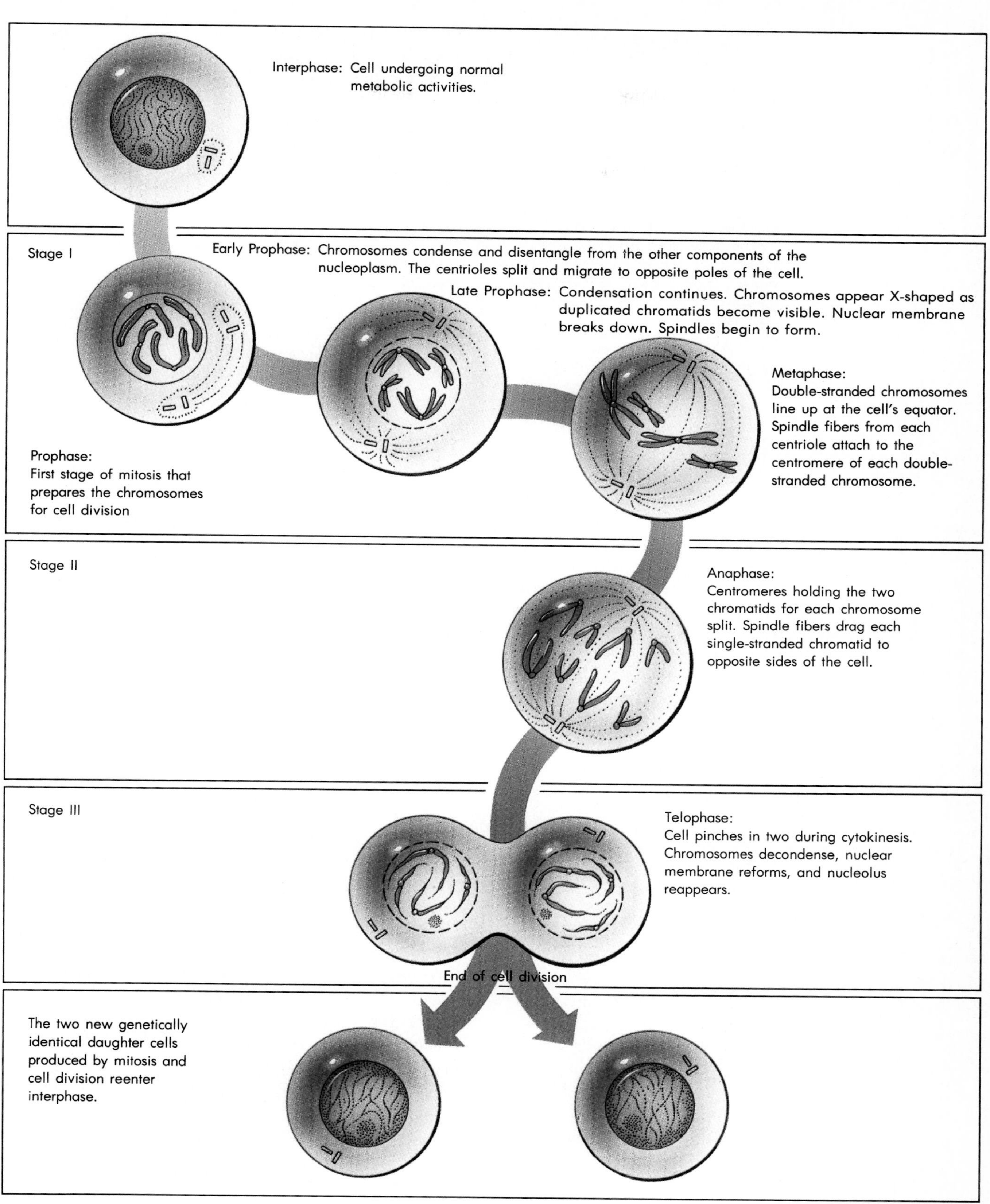

FIGURE 5-10 The stages of mitosis, diagrammed for a four-chromosome (two homologous pair) animal cell.

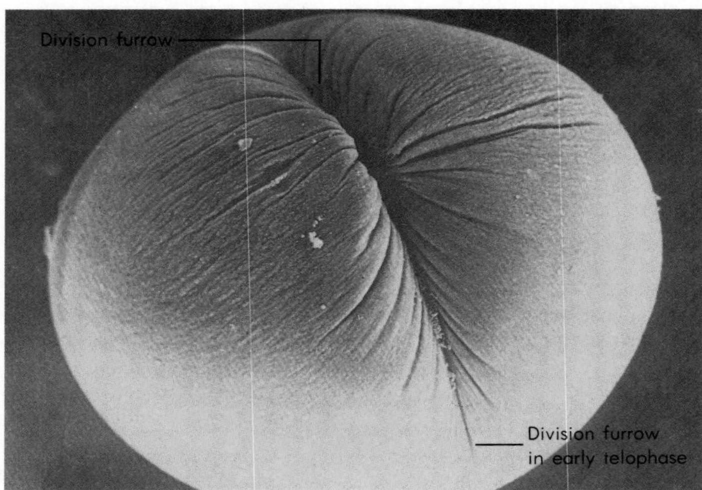

FIGURE 5-11 Photomicrograph showing the microfilament belt that helps pinch the cell in half during cytokinesis.

the cell's equator, just beneath the cell membrane (Figure 5-11). Once karyokinesis has taken place, these microfilaments contract and pinch the cell's equator inward, just as if someone had tied a string around it. However, not all cells perform cytokinesis during telophase. Some delay it for a time. Still others never get around to it; in such cases, repeated mitoses produce single cells or syncytia (such as slime molds) that accumulate many nuclei.

The cells of higher plants handle cytokinesis differently. In them, globules of material merge on the cell's equator to form a *cell plate,* which becomes a new cell membrane and cell wall.

Controlling Mitosis

We know that mitosis must be controlled in the body, for our organs stop growing when they reach a certain size. Yet they can start growing again (to a small extent) when injuries need healing. The control is exercised in two ways. Many cells secrete *growth factors* that stimulate mitosis in neighbor-

Learning Focus

MITOSIS REVIEW

Inter means between. *Pro* means first. *Telo* means last. These definitions are enough to let you remember where interphase, prophase, and telophase belong in the cell cycle. But *meta* means among. Does metaphase come before or after anaphase? Remember that while *ana* can mean without, here it means again or double—the second set of chromosomes is formed when the double-stranded chromosomes pull apart. It is from these two sets of single-stranded chromosomes that the two double-stranded chromosomes will eventually arise again. It has to come just before telophase, leaving metaphase to follow prophase.

But the inter-pro-meta-ana-telo (IPMAT) sequence is the least of the things you should remember about mitosis. When does replication occur? What is the difference between a chromosome and a chromatid? What role do the centrioles play? What is the function of the centromeres? When do they split, and why? What is the function of the spindle and its fibers? When (and why) does the nuclear envelope disappear and reappear? What is the difference between karyokinesis and cytokinesis? What does mitosis *do*?

Answer the above questions, referring to the text where necessary, as you complete the accompanying drawing. First, however, make several copies of the drawing to use when you are reviewing for exams. Then draw a single pair of homologous chromosomes alongside the first cell. Show what happens when the chromosomes replicate. Sketch what happens to the chromosome in prophase, metaphase, anaphase, and telophase. Include in your drawings the changes in the centrioles and spindle fibers.

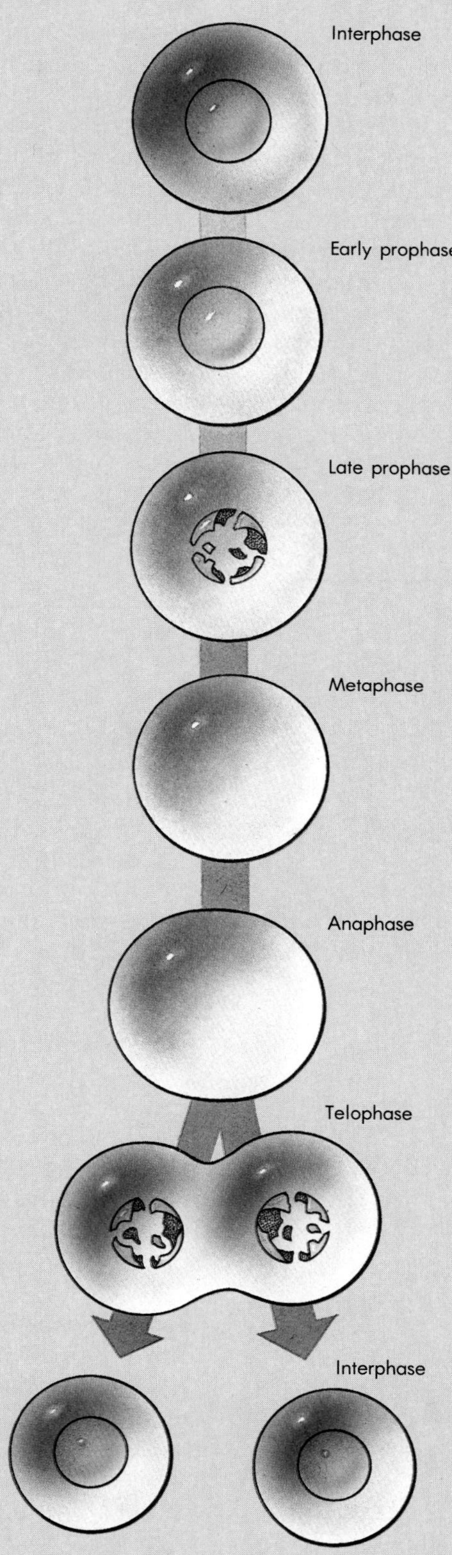
Interphase
Early prophase
Late prophase
Metaphase
Anaphase
Telophase
Interphase

ing cells. *Platelet-derived growth factor* is released by the blood platelets that play a key role in blood clotting (see Chapter 10). It stimulates mitosis in connective tissue cells and the cells that line blood vessels to repair breaks in blood vessels. Other growth factors stimulate the growth of blood vessels into areas that need a blood supply, the multiplication of skin cells, and the growth of nerves.

There are also inhibitory factors. Many cells produce **chalones,** proteins or glycoproteins that inhibit their own mitosis, halting them in the G_1 stage of their cycle. They may produce these substances in response to signals from neighboring cells that say, in effect, "This organ is big enough. Stop now." We can see the effect of such inhibition in the way cells grown in a laboratory dish (cell culture) stop growing when they begin to touch each other. In cell culture, normal cells form a layer one cell deep, and no more. This is the phenomenon of **contact inhibition** (Figure 5-12).

Cancer

The control mechanisms stop working when cells become cancerous. As discussed in more detail in Chapter 23, *carcinogenesis* is the process of cancer initiation. It has two steps, initiation and promotion, that appear to be triggered by mutations, caused by chemicals or radiation, in separate genes. They may also be triggered by *oncogenes,* "cancer genes," carried into the cells by viruses or activated by chemicals or radiation. Whatever the trigger, the result is a failure of the affected cell to respond to mitotic inhibitors and other growth controls. In cell culture, the changed cells show no sign of contact inhibition; they continue to multiply even after they touch each other, and the bottom of the laboratory dish is covered with mounds of cells. In the body, their freedom from inhibition can lead to *tumors,* masses of cells that interfere with various bodily functions. Cancer cells can undergo further changes as well. One such change makes them able to spread, a process called *metastasis,* which makes cancer hard to cure.

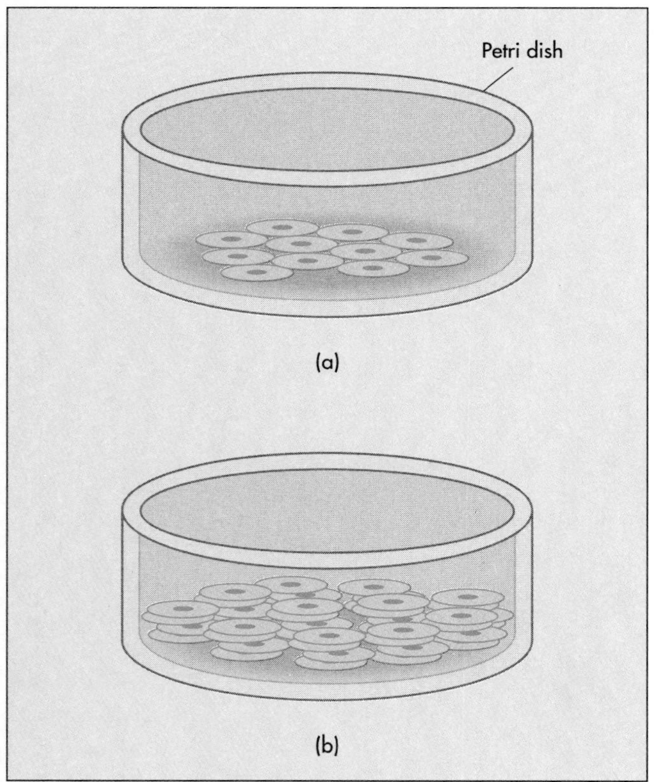

FIGURE 5-12 (a) Normal cells respond to contact inhibitors when they cover the bottom of the dish by ceasing to increase in number. (b) Transformed cells keep multiplying even when they are in contact with each other.

MEIOSIS

The function of the kind of cell division we call **meiosis** is to produce daughter cells that each have *half* the number of chromosomes of the parent cell. *Diploid* parent cells produce daughter cells, and as they do so, they convert their daughter cells to the *haploid* condition. In general, the cells produced by meiosis become gametes or sex cells, that is, the sperm and ova of animals. Their destiny is to fuse during sexual reproduction to produce new diploid cells, the fertilized eggs or zygotes that can then grow by mitosis into new individuals.

Meiosis resembles mitosis in that it separates pairs of sister chromatids. It differs in that it also separates pairs of homologous chromosomes, and it takes two cell divisions to do it. That is, when meiosis begins, a human cell contains 2 homologues and 4 chromatids for each of the 23 chromosomes. When it ends, there are 4 daughter cells, not 2, and each one contains 1 chromatid for each chromosome; each one contains 22 autosomes and 1 sex chromosome (Figure 5-13).

To reach the end state of meiosis, the cell first replicates its chromosomes in the S phase of interphase. Then, as the cell prepares for the first division, the two homologous chromosomes (each one a pair of chromatids joined at a centromere) come to lie next to each other in the cell's central plane. Thus, in human cells, there are 23 pairs of chromosomes lined up across the middle of the cell before the first division.

As the chromosomes take their positions, spindle fibers extend between the two centrioles, just as in mitosis, and from each centriole to the nearest centromere or chromatid pair. The fibers then contract, dragging the members of each pair of homologous chromosomes to opposite sides of the cell. *The centromeres do not split yet.* At this point, there are 23 double-stranded chromosomes resting on different sides of the cell. The cell now divides, forming two cells, each one with only 23 chromosomes (or chromatid pairs). Since the chromosome count was reduced from 46 in the parent cell to 23 in each daughter cell, this first division is often called the reduction division. Note that the chromatids are not separated in this first division, but homologous chromosome pairs are.

The second division is similar to mitosis, except that the cells undergoing it are really haploid, not diploid, when it begins. In this case, spindle fibers from both spindles attach to each of the centromeres of the 23 double-stranded chromosomes. *This time the centromeres split, just as they do in mitosis.* The chromosomes are pulled apart, with a single strand or chromatid going to opposite sides of the cell. Each of the two daughter cells from the first division divides—producing a total of four haploid daughter cells with single-stranded chromosomes. It is these cells that are eventually modified into gametes.

The two divisions are called divisions I and II, and each division—as in mitosis—has the stages of prophase, metaphase, anaphase, and telophase.

Division I: The Reduction Division

In many ways, the first division is the more important. Besides reducing the number of chromosomes and ensuring that each homologous pair will be represented in the gametes, it is also responsible for much of the genetic mixing that occurs in sexual reproduction. This genetic mixing is responsible for much of the variability that enables a species to cope with changing environmental conditions.

Figure 5-13 shows the stages of meiosis. The first meiotic division begins with prophase I. As in mitotic prophase, the chromosomes condense as joined pairs of sister chromatids. A difference from mitosis quickly appears, however. As the chromosomes condense, homologous chromosomes move toward each other and form bundles of four chromatids, or **tetrads.** Within each tetrad, nonsister chromatids then break and exchange segments in the process of **crossing over.** The crossing over links gene combinations from different family lines together on the same chromosome.

Crossing over thus enhances the genetic differences between parents and their offspring by moving gene variants from one homologous chromosome to the other. Since the homologous chromosomes carry different sets of variants, crossing over reshuffles the genetic deck by moving about blocks of genes (Figure 5-14, p. 123). The resulting new combinations of genes can make significant contributions to the survival of the species.

1. Crossing over means that the members of each generation thus carry genes from more original sources. To see how this is so, consider Sam. As a sexually produced individual, he has pairs of chromosomes, with one member of each pair coming from his mother (Annie) and one from his father (Mike). Determine why the chromosomes Sam contributes to the next generation each carry a mix of Annie's and Mike's genes. If Sam's mate is Jo, with parents Jan and Alec, then meiosis in Sam's and Jo's child, Albert, will generate chromosomes with genes from Annie, Mike, Jan, and Alec. Why is this true?
2. Let's say that there is no such thing as crossing over. Now, what can you say about the sources of Albert's genes? Will one chromosome carry genes from all four grandparents? Or from only one? Why?

Toward the end of prophase I, the nuclear envelope breaks down, the spindle forms, and the spindle fibers attach to the centromeres of the chromosomes. However, each centromere is linked to only *one* centriole, not both. In metaphase I, the tetrads move to the cell's equator. In anaphase I, the shortening spindle fibers draw each tetrad's pairs of sister chromatids, still joined by their centromeres, toward opposite ends of the cell. This step further shuffles the cell's genes, for though the cell originally acquired its two sets of homologous chromosomes from its parents, one set from each, anaphase I separates the pairs of homologous chromosomes randomly. Note that at this point each *chromosome* consists of a pair of *sister chromatids* joined at a centromere.

The first meiotic division may end with a distinct telophase, including reformation of the nuclear envelope. Often, however, cytokinesis divides the original cell without waiting for nuclei to reform. Nor will there be any replication of the chro-

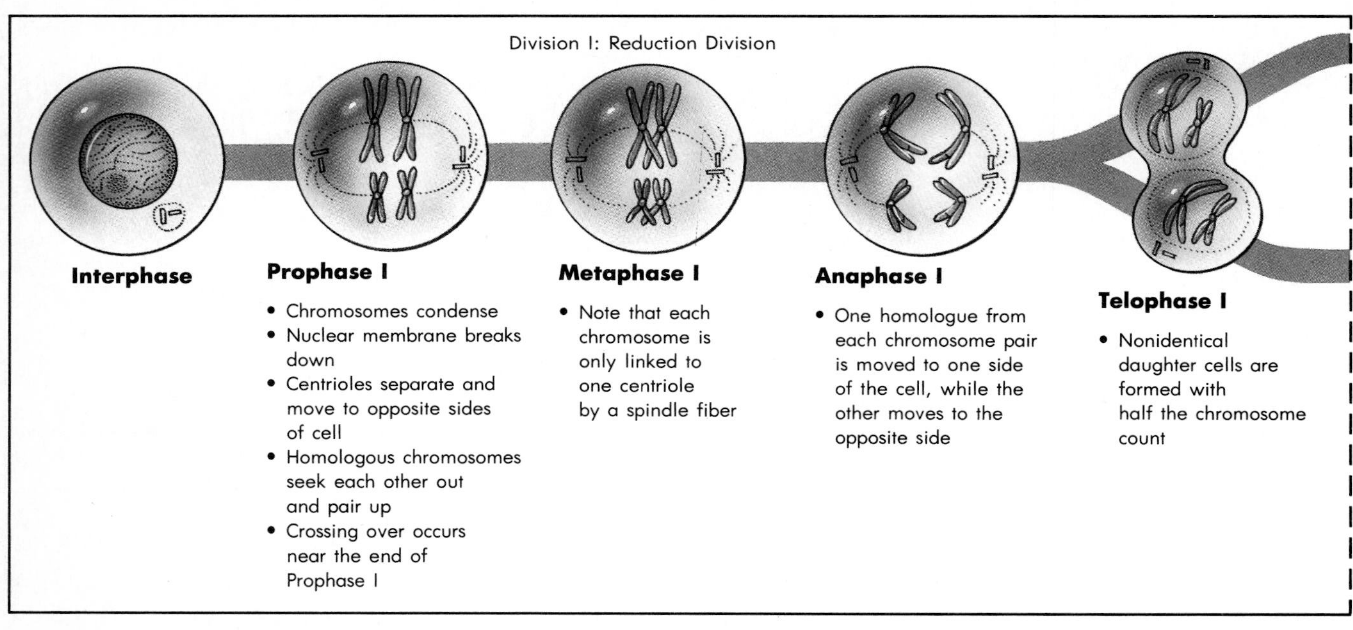

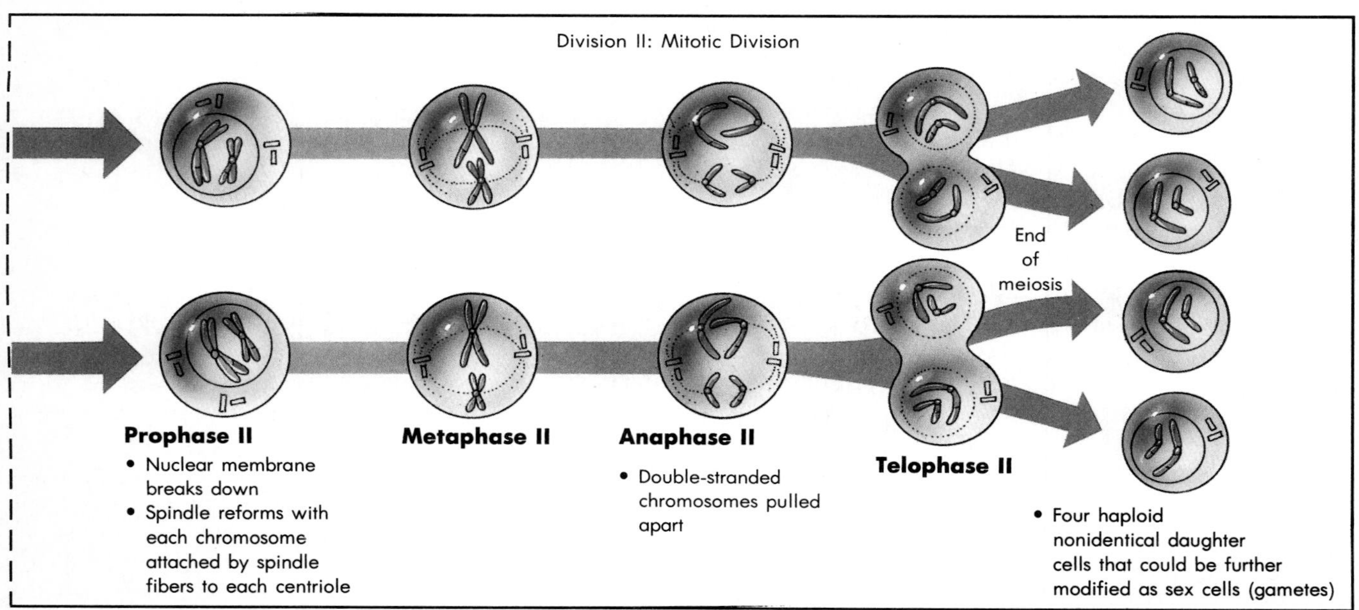

FIGURE 5-13 The stages of meiosis.

mosomes before the second division, and the chromosomes need not lose their condensed state.

Division II: The Mitotic Division of Meiosis

The second meiotic division thus begins with the chromosomes already condensed, as pairs of sister chromatids joined by centromeres. Prophase II involves only the dissolution of the nuclear envelope (when it is present) and the reorganization of the spindle. During metaphase II, the chromatid pairs line up along the cell's equator. Anaphase II then draws them apart, breaking the centromeres. Telophase II reforms the nuclear envelope, and cytokinesis splits the cells once more.

The end result of meiosis is *four* daughter cells, not 2, and each of these cells is haploid, not diploid. That is, it contains 1 single-stranded chromosome representing each of the original homologous pairs. There are 4 daughter cells because the meiotic pro-

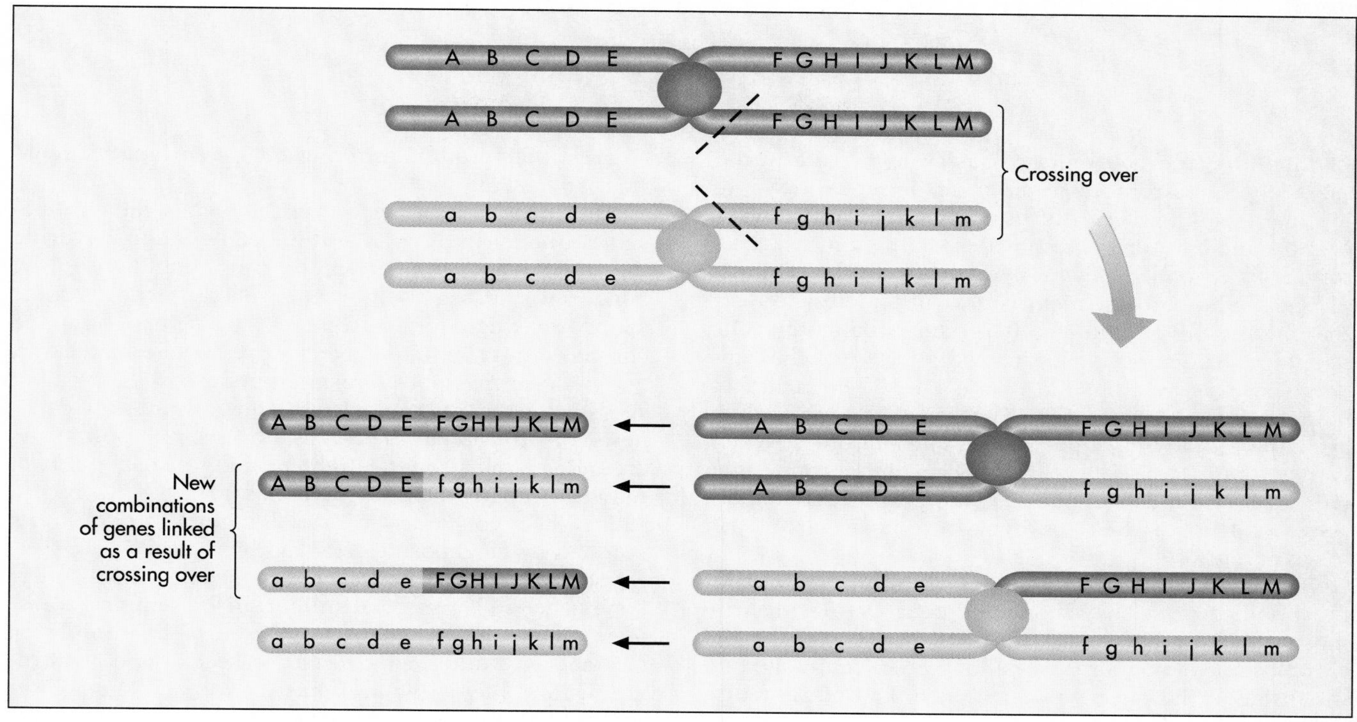

FIGURE 5-14 The mixing effect of crossing over on the set of gene variants carried by homologous chromosomes.

cess involves two sequential cell divisions. The first division produces 2 haploid daughter cells, each one containing 23 double-stranded chromosomes. Each of these daughter cells divides in turn, each producing 2 more daughter cells—yielding a total of 4 haploid cells.

In males, the four daughter cells are identical in size (although each cell contains its own unique set of genes). Each cell receives an equal share of the parent cell's cytoplasm and other materials, and each one will become a sperm cell. The cellular events leading to the production of sperm cells are collectively referred to as **spermatogenesis.**

In females, the outcome is somewhat different. Female gametes are egg cells, which require all the stored nutrients and cellular equipment they can acquire to support the eventual growth of the embryo that results from the fusion of sperm and egg. In ovum-producing meiosis, called **oogenesis,** the daughter cells are not identical. The first division splits the parent cell into one large cell, which will eventually become the egg, and a tiny "polar body," containing almost nothing other than half the parent cell's chromosomes. In the second division, the polar body divides evenly, forming two even tinier polar bodies, but the large daughter cell repeats this unequal division of resources, so that the egg winds up with virtually all the parent's cytoplasm and stored nutrients. The polar bodies are the packages in which the surplus chromosomes are discarded.

Errors of Meiosis

Meiosis works perfectly most of the time, but not always. Sometimes pairs of homologous chromosomes or sister chromatids fail to separate during meiosis. Such *nondisjunctions* result in gametes that have more or less than their proper number of chromosomes. The same is true of any embryos to which these gametes may later contribute. Occasionally, a child is born with only one or three or even more of a particular chromosome, not two. And the results can be tragic. Down syndrome, a condition marked by mental retardation and other handicaps, occurs when a child is born with three copies of chromosome number 21. The babies that result from many other nondisjunctions die before or soon after birth. When they survive, they are generally seriously deformed or retarded.

SUMMARY

Living things reproduce in two basic ways: sexually and asexually. Cells reproduce only asexually, by dividing into pairs of daughter cells, but they too have two ways of reproducing: Mitosis duplicates the parent cell, preserving its chromosome number. Mitotic divisions multiply cells while an organism is growing. Meiosis halves the chromosome count, for it produces daughter cells specialized as gametes, sperm and eggs. Cells that undergo mitosis are diploid; that is, they have pairs of homologous chromosomes. The products of meiosis are haploid; they have only single chromosomes. When gametes fuse to produce an embryo, the chromosome pairs are reestablished; one chromosome of each homologous pair comes from one parent, and one from the other parent.

Before a cell can divide by either mitosis or meiosis, it must replicate its deoxyribonucleic acid (DNA) and chromosomes. For a bacterium, this is one step, as its single chromosome is no more than a loop of DNA attached to the inside of the cell membrane; once the DNA is replicated, two loops are attached to the membrane, and the membrane between them grows to separate them. For eukaryotic cells, the process is more complicated since they have many chromosomes made of chromatin, a complex arrangement of DNA on a frame of histone proteins. The histones form beads around which the DNA strand is wrapped.

The cell cycle runs from division to division. Division itself is the M phase of the cycle. It is followed by the G_1 growth stage, the S phase (when the DNA and chromosomes are replicated), and the G_2 growth stage. When the cell is preparing for division, the chains of nucleosomes are coiled and condensed into thick cords, which are then folded many times.

During the first part of the M phase, prophase, the chromosomes become visible as rods of condensed heterochromatin. Each chromosome appears as a pair of duplicate chromosomes, or sister chromatids, joined at a centromere. Also during prophase, the cell's nuclear envelope dissolves, and its two centrioles duplicate themselves and move toward opposite ends of the cell, where they serve as organizing centers for the mitotic spindle, an array of microtubules that helps the chromatids separate later in cell division.

After prophase comes metaphase, when the chromosomes line up in the plane of the cell's equator. During the following anaphase, spindle microtubules shorten and draw the sister chromatids of each chromosome apart, breaking their centromeres. During telophase, the chromosomes clump together at opposite ends of the cell, and nuclear envelopes reform around them. Karyokinesis, the division of the nucleus, is now complete. Cytokinesis, the division of the cell as a belt of microfilaments around the cell's equator tightens and pinches it in two, may accompany or follow telophase.

Mitosis is controlled by growth factors and inhibitors secreted by neighboring cells and by the cell itself. Cancer cells differ from normal cells in that they have undergone genetic changes that immortalize them and transform them, or free them from control by inhibitors. Contact with other cells no longer keeps them from dividing, and they grow into massive tumors. Further genetic changes make them able to travel throughout the body and establish "colony" tumors elsewhere, or metastasize.

Meiosis differs from mitosis in that it involves one DNA replication and two divisions. During the prophase of the first division, the four chromatids of each pair of homologous chromosomes bundle together to form a tetrad, in which the homologous chromatids exchange segments or cross over. The tetrads then line up on the cell's equator (metaphase I), and the spindle fibers draw apart the homologous chromosomes, each a pair of sister chromatids still joined by a centromere (anaphase I). The second division separates the sister chromatids, breaking the centromeres, and results in daughter cells that have but one copy, not two, of each chromosome. Errors in meiosis can cause failure of homologous chromosomes or chromatids to separate properly; such nondisjunctions result in cells with not one but zero, two, three, or four copies of a chromosome. The babies that result from the fusion of such faulty gametes often do not survive to or past birth; if they do, they may be seriously defective, mentally or physically.

STUDY QUESTIONS

1. What is the advantage of sexual reproduction over asexual reproduction?
2. What is the prerequisite for reproduction? Why?
3. What is the difference between karyokinesis and cytokinesis?
4. What is the difference between homologous chromosomes and sister chromatids?
5. Distinguish between diploid and haploid.
6. What are nucleosomes? What is their function?
7. When does DNA replication occur?
8. Describe briefly the events of mitosis.
9. Describe briefly the events of meiosis.
10. What is the main difference between mitosis and meiosis?
11. Describe crossing over. When does it occur?
12. How do cancerous cells differ from normal cells?

GLOSSARY

Anaphase The third stage of mitosis, in which the spindle fibers draw sister chromatids apart.

Asexual reproduction Reproduction of a single parent, without mating; produces clones.

Autosomes Non–sex chromosomes; homologous autosomes look identical.

Cell cycle The sequence of events from cell division to cell division.

Centromere Region linking sister chromatids.

Chromatin The material (DNA plus protein) of eukaryotic chromosomes.

Contact inhibition Cessation of cell division when cells touch each other in cell culture.

Crossing over Exchange of segments between homologous chromatids in a tetrad.

Cytokinesis Division of the cell.

Diploid number The number of chromosomes in a body cell; 46 in humans.

Gamete Specialized reproductive or sex cells; male and female gametes (sperm and ova) fuse to produce offspring.

Haploid number The number of chromosomes in a gamete, or the number of homologous pairs of chromosomes in a body cell; 23 in humans.

Histones Structural proteins of the chromosomes; comprise the nucleosomes.

Homologous chromosomes Matching pairs of chromosomes. Each parent contributes one of the homologues (chromosomes) to each pair. There are 23 pairs of homologous chromosomes in human cells.

Interphase The stage of the cell cycle between cell divisions.

Karyokinesis Division of the nucleus.

Meiosis The type of cell division (actually two sequential divisions) employed to produce four haploid daughter cells (often the gametes) from a diploid parent cell.

Metaphase The second stage of mitosis, marked by alignment of the chromosomes on the cell's equator.

Mitotic spindle An array of microtubules, centered on the centrioles, that helps the chromosomes separate in karyokinesis.

Nucleosomes Histone beads around which DNA wraps to shorten its length.

Oogenesis Process by which ova (eggs) are formed.

Prophase The first stage of mitosis, marked by condensation of the chromosomes, disappearance of the nuclear envelope, separation and duplication of the centrioles, and formation of the spindle.

Sex chromosomes X and Y chromosomes; females have two X's; males have an X and a Y.

Sexual recombination Mixing of genes from two individuals resulting from the fusion of gametes.

Sexual reproduction Reproduction with mating; involves the fusion of gametes from two different parents.

Sister chromatids Postreplication duplicates of a chromosome, joined at a centromere.

Spermatogenesis Process by which sperm are formed.

Telophase The fourth and last stage of mitosis, during which the nuclear envelope reforms.

Tetrad A bundle of four chromatids (two homologous pairs of sister chromatids) formed during prophase I of meiosis; site of crossing over.

Chapter 6

Human Tissues and the Basic Body Plan

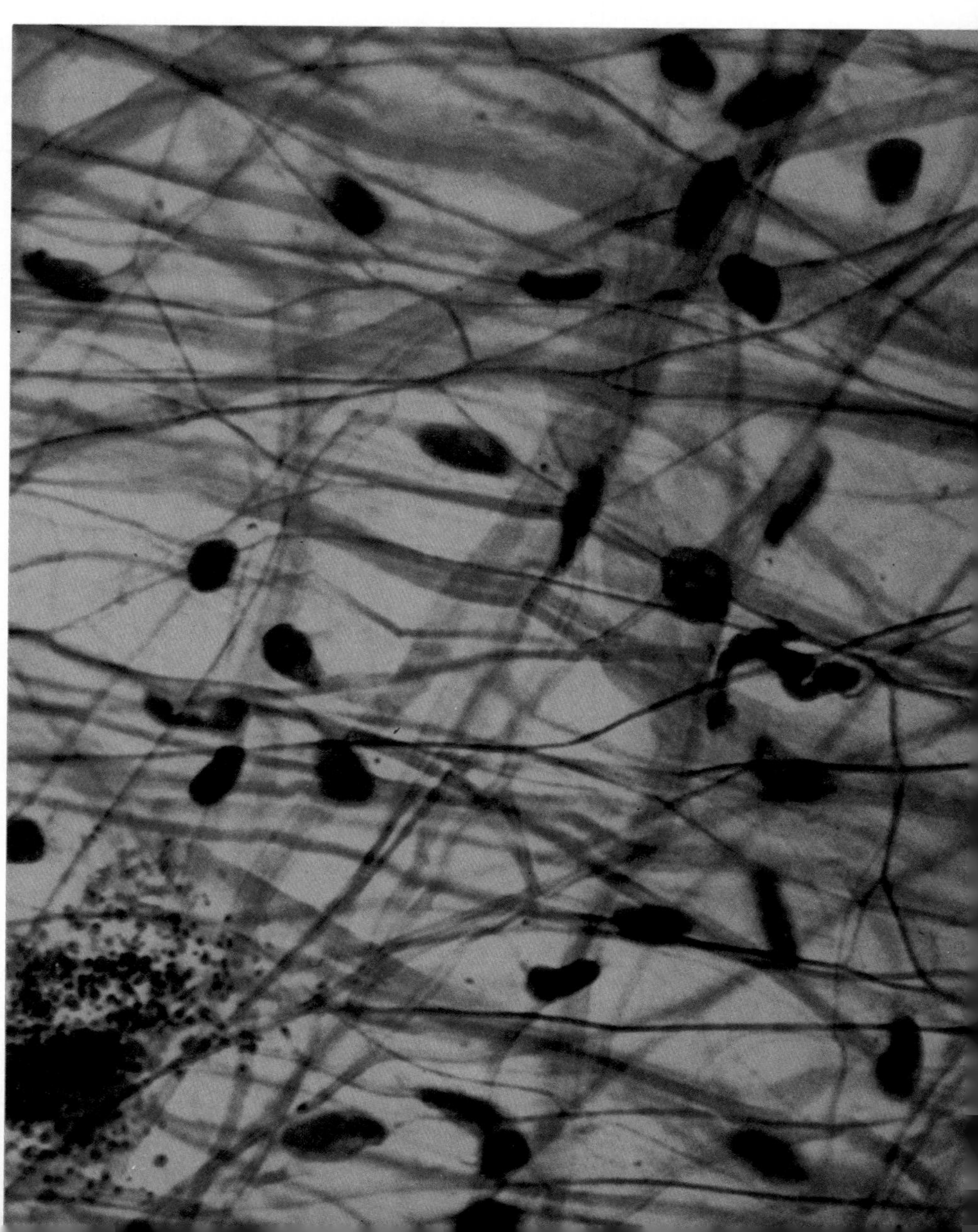

The most complicated single cells of all are found among the protozoans—microscopic unicellular creatures that inhabit ponds, rain gutters, and standing water in flower pots. Their single cell is so structurally complex they can do all the things they need to survive just using the battery of organelles each possesses. They find and process their own food, sense and respond to changes in the environment around them, and reproduce—and they do it all without help from other cells. However, when cells began to group together in multicellular organisms, such as humans and pine trees, they could no longer be so self-sufficient. There had to be a shift in cellular strategy. For one thing, only a few of the cells in such an organism can be exposed to the external environment. Cells on the interior of the organism needed help in getting food and in sensing external events. In addition, among other things, the organism could no longer reproduce as a whole.

The answer was division of labor. Some cells specialized to form a shield against the outside world for the rest. Others specialized to process food. Still others specialized to carry the processed nutrients to the rest of the body, to sense outside events, to coordinate the body's responses, to support the body against the pull of gravity, and to move the body around. A few even specialized to handle the task of reproduction on behalf of the whole.

In complex life forms such as ourselves, we find that many groups of cells work together to serve the needs of *all* the body's cells. These groups are what we call **tissues.** Each tissue provides the body with a different service.

Tissues consist of aggregations of similar cells, the intercellular matrix (which can be watery, jelly-like, gristly, or hard) between the cells, and the cell products such as fibers, salts, or organic materials distributed throughout the matrix. The cells of tissues are held together in sheets, clumps, and strips varying in size from mere dozens of cells to millions, all working together to make a collective contribution to the well-being of the body.

Groups of different tissues combine to form the major structures and compartments of the body, the **organs.** The organs are linked together into **organ systems,** which handle particular categories of biological functions like processing food, coordinating quickly or slowly changing systems of the body, or keeping track of the outside world. Some of the categories of tissues, types of organs, and the names of organ systems are listed in Table 6-1. The various human major organ systems are shown in Figure 6-1.

BETWEEN THE CELLS

The body's cells are held together in several ways. Many cells are attached to their neighbors by proteins embedded in their membranes. Many others are held in their places within the body by two substances that serve as intercellular glues. These substances are **hyaluronic acid** and **chondroitin sulfate.** Hyaluronic acid is a carbohydrate that combines with a protein to form a viscous gelatinous material. Among other functions, it serves as a lubricant in the joints of the skeleton. Chondroitin

Table 6-1

Levels of Organization in the Body

TISSUES	ORGANS	ORGAN SYSTEMS	ORGANISMS
Epithelial	Stomach	Digestive	Humans
Nervous	Brain	Nervous	Dogs
Connective	Bone	Skeletal	Snakes
Muscle	Biceps	Muscular	Elephants
	Heart	Cardiovascular	
	Thyroid gland	Endocrine	
	Eyes	Sensory	
	Kidney	Excretory	
	Testes or ovaries	Reproductive	
	Lungs	Respiratory	
	Skin	Integumentary	

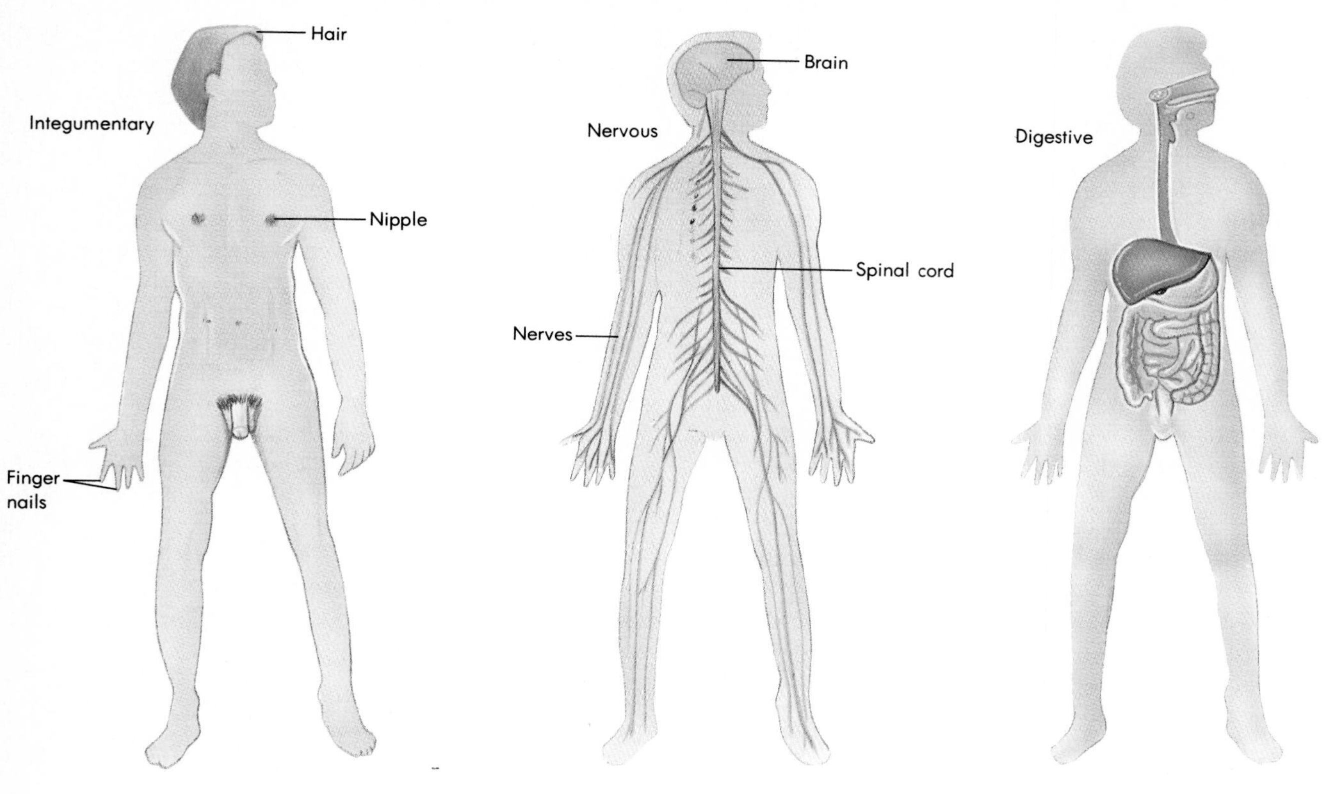

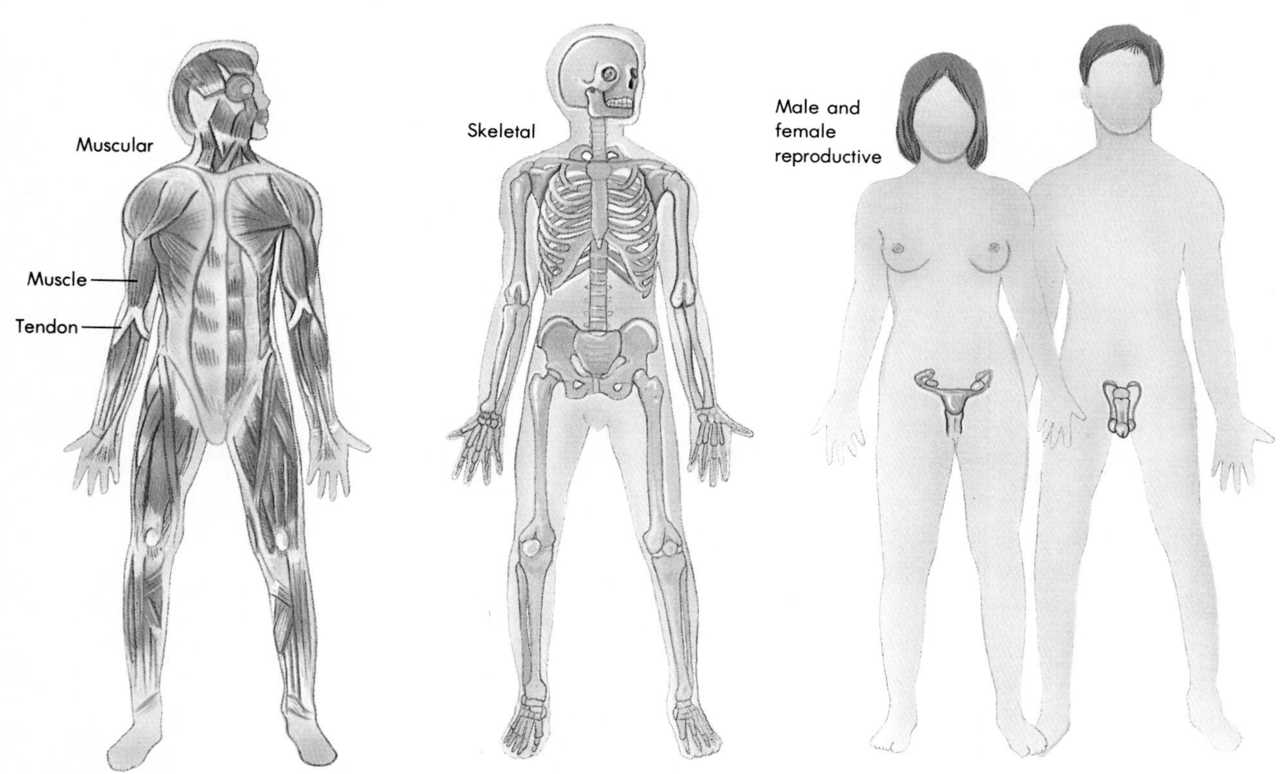

FIGURE 6-1 Various systems of the human body.

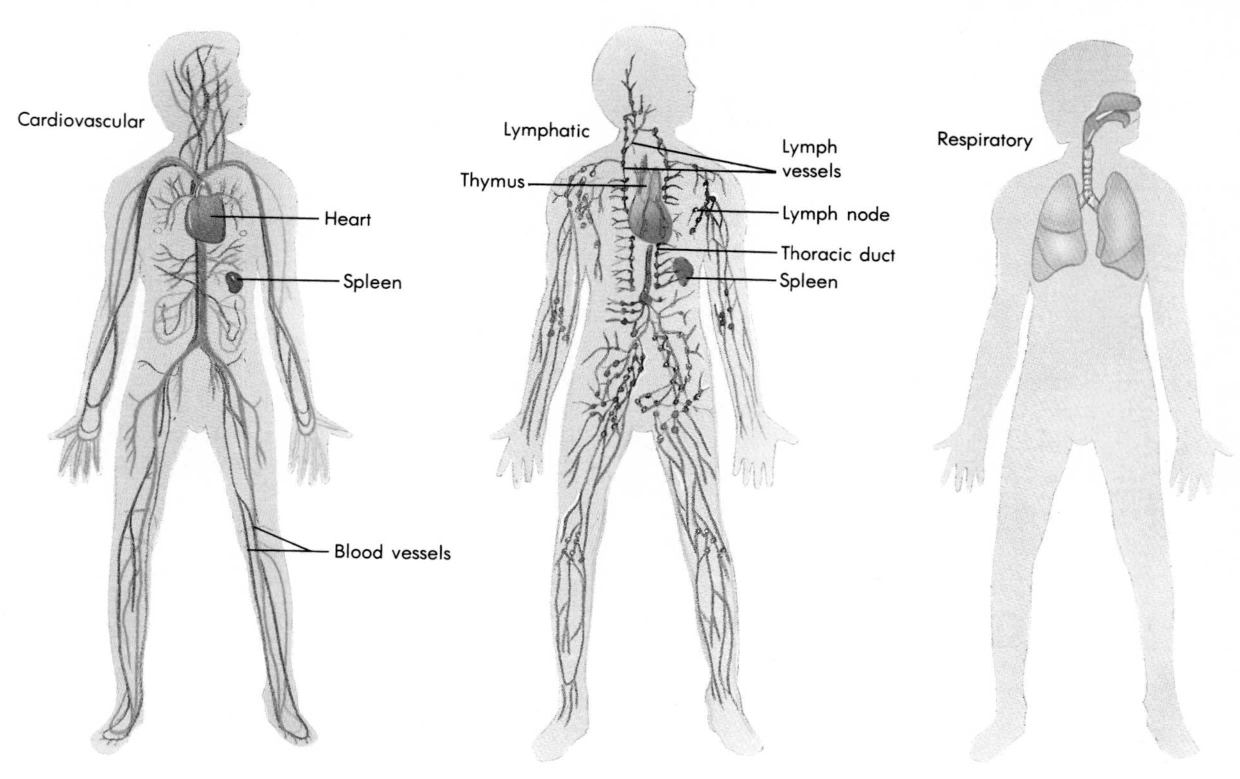
Cardiovascular
Heart
Spleen
Blood vessels
Lymphatic
Lymph vessels
Thymus
Lymph node
Thoracic duct
Spleen
Respiratory

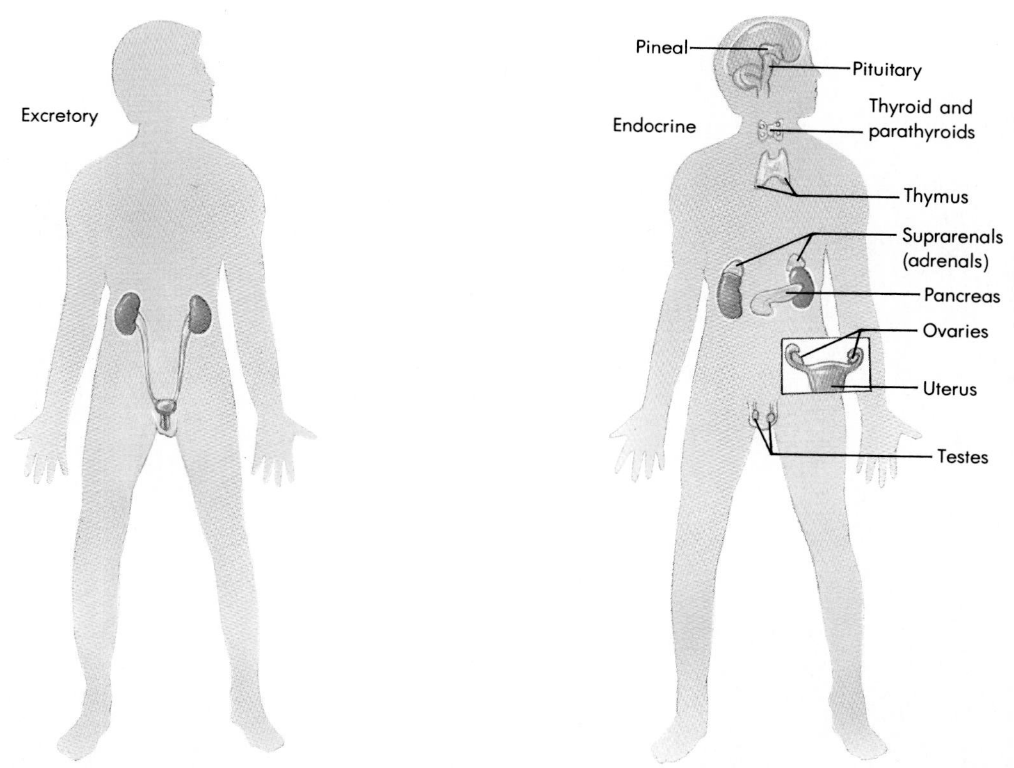
Excretory
Pineal
Pituitary
Endocrine
Thyroid and parathyroids
Thymus
Suprarenals (adrenals)
Pancreas
Ovaries
Uterus
Testes

sulfate also combines with a protein to form a thick gel. It is produced by cartilage cells and forms the intercellular matrix of many connective tissues.

TISSUES

Animal tissues are classified into four broad categories; they include the epithelial, connective, muscle, and nerve tissues. Each category is made up of cells and intercellular materials that help the body cope with related kinds of problems. For instance, epithelial tissues form many of the body's coverings, and they include several types of membranes that line the free surfaces of various compartments of the body, the skin, and the inner linings of blood vessels. Muscle tissues are composed of cells that can change shape by contracting, and they are employed wherever some movement is required or where the body needs to regulate the rate of flow through a tube by changing its diameter. Nerve tissues are made up of cells that can carry electrochemical signals from one area to another.

Each major category of tissue can be subdivided into several groups of related tissues. Discussion of the major tissue types and their subdivisions is in the form of a table because it makes it easier to see the overall organization of the various categories (Table 6-2).

(Text continues on p. 136.)

Table 6-2

Categories of Animal Tissues

TISSUE TYPE	FUNCTION
Epithelial Tissues Tissues covering many of the body's external and internal surfaces (Figure 6-2). Examples include the skin and the linings of blood vessels and body cavities. This category also includes glandular (secretory) tissues and tissues involved in absorption, such as the lining of the digestive system. Most epithelial tissues are closely packed with very little space between the cells. The epithelial tissues include:	
Squamous Epithelium 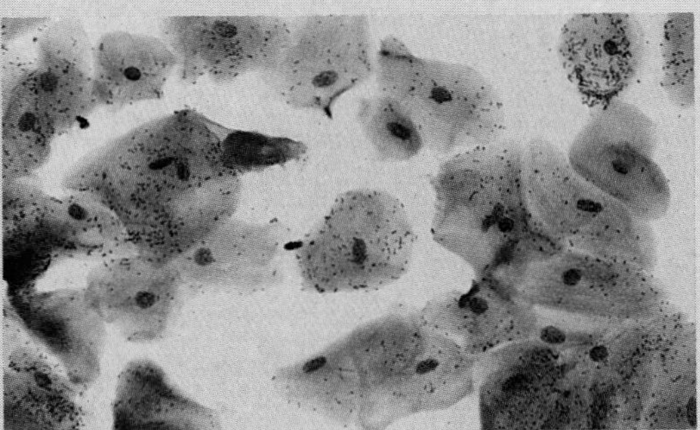	Characterized by flattened cells resembling "fried eggs." Simple squamous epithelium lines the blood vessels, body cavities, and lymph vessels.
Stratified Squamous Epithelium 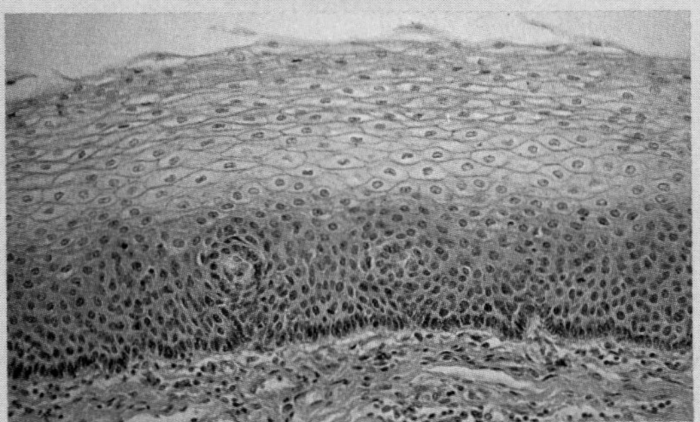	Stratified squamous (multilayered) epithelium forms outer layers of the skin.

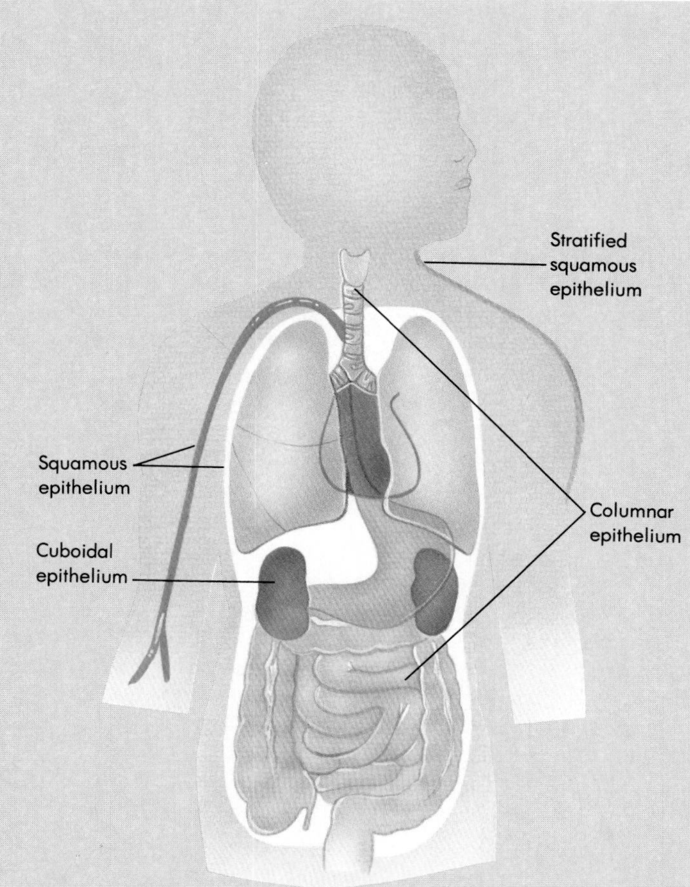

FIGURE 6-2 Epithelial tissues cover the body's various free surfaces. Stratified squamous epithelium forms the skin. Other forms of squamous epithelium line the blood vessels and the body cavities. Columnar epithelium lines the respiratory and the digestive tubes. Cuboidal epithelium lines the kidney tubules and forms the secretory cells in glands.

TISSUE TYPE	FUNCTION
Columnar Epithelium 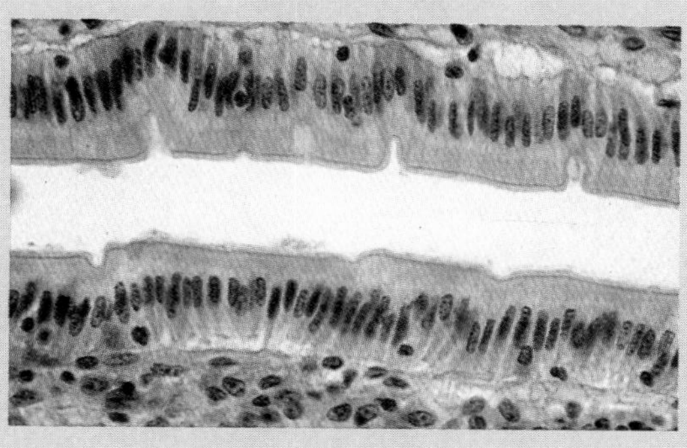	Elongated column-shaped cells that line the absorptive surfaces of the digestive system and various ducts. Those lining the respiratory system are ciliated for removing inhaled particles.
Cuboidal Epithelium 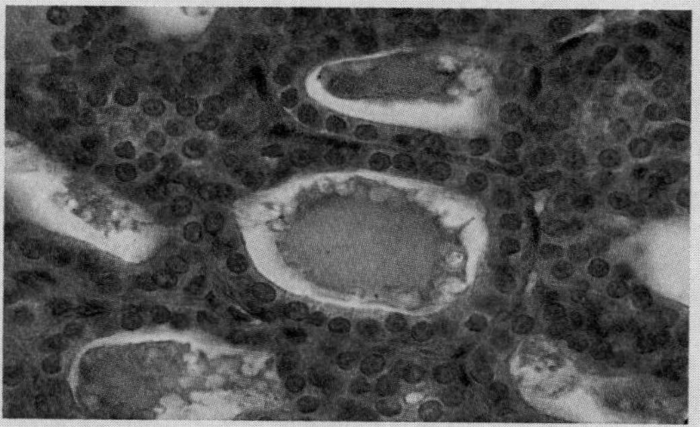	Cube-shaped cells that form small ducts, such as the tubules of the kidneys, covering the ovaries, and as secretory cells in some glands.

(Continued on p. 132)

TISSUE TYPE	FUNCTION

Connective Tissues

Widely distributed tissues for binding structures together, filling in spaces, storing fat, providing support and protection, and distributing materials from one part of the body to another. Connective tissue cells are generally spaced farther apart than those of epithelial tissues. Connective tissues are often characterized by the nature of the extracellular material (matrix) found in the spaces between the cells.

Osseous Tissue

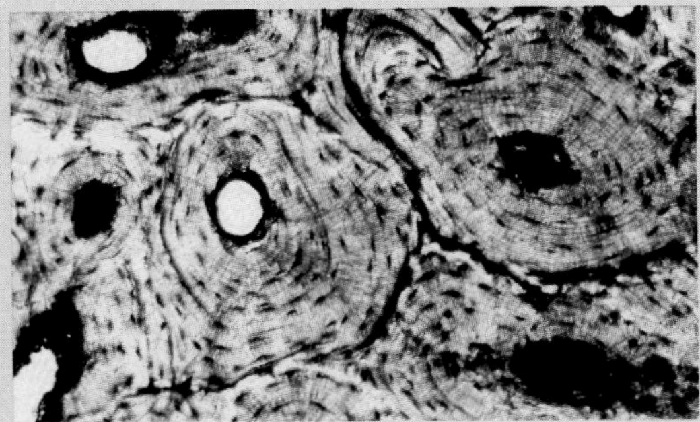

Osseous tissues are the hardest of all connective tissues; they form the bones of the skeletal system. The extracellular matrix consists of calcium salts laced with collagenous fibers. The functional cells (osteocytes) are organized into Haversian systems (see Chapter 8).

Fibrous Connective Tissue

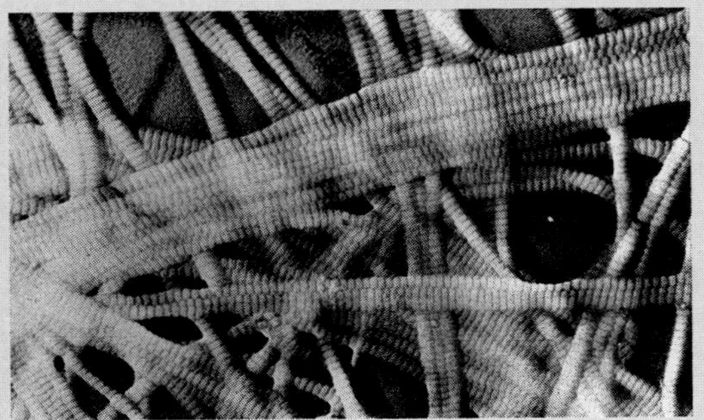

Contains large amounts of collagenous fibers. These fibers are strong and flexible but not very elastic. They form the tendons, which bind the skeletal muscles to the bones, and ligaments, which bind different bones together across a joint.

Loose Connective Tissue

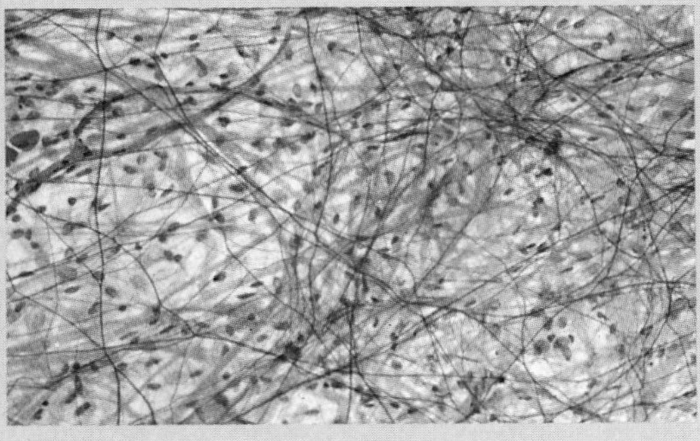

Forms delicate membranes throughout the body characterized by jellylike matrix laced with numerous fibers. Areolar connective tissue is found under the skin and other epithelial tissues and around many internal organs.

Adipose Tissue

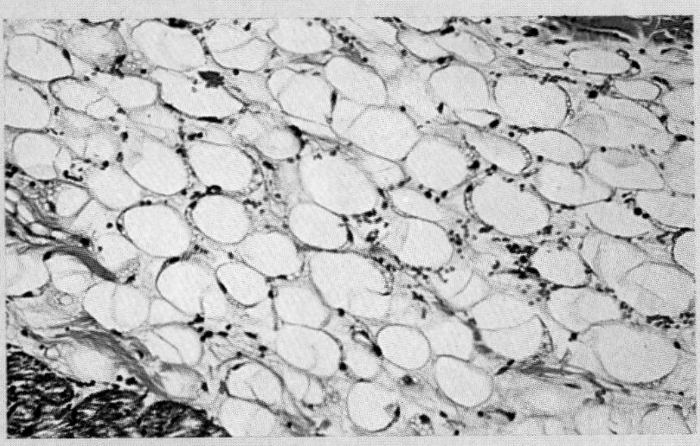

Fatty tissues whose cells store substantial amounts of fat in vacuoles that are the dominant feature of the cells. Fat deposits are located under the skin, around the heart and kidneys, in back of the eye sockets, and in some abdominal membranes.

TISSUE TYPE	FUNCTION
Cartilage	Strong tissues that provide support while retaining flexibility in key areas of the body (Figure 6-3). It is often referred to as "gristle." It has a dense but flexible matrix secreted by chondrocytes. Hyaline cartilage is found at the end of bones, supporting the nose, and holding open the windpipe or trachea.

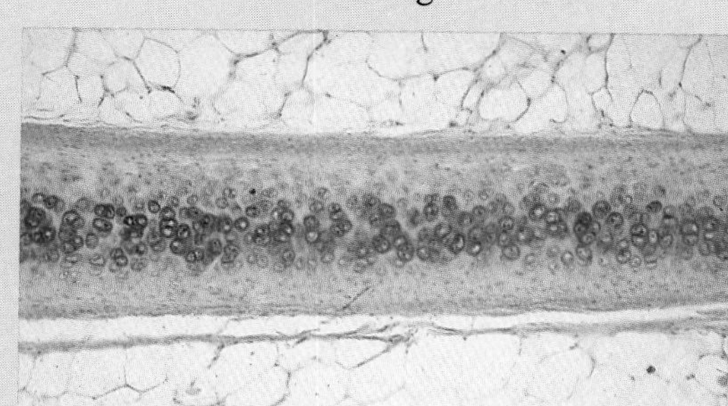

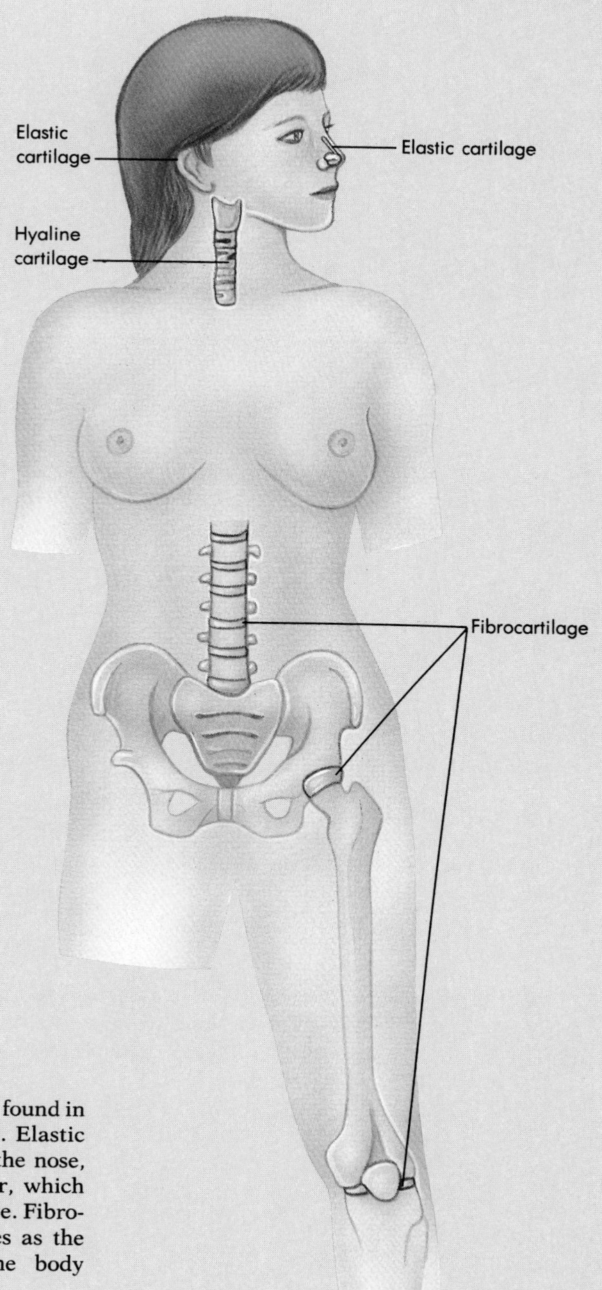

FIGURE 6-3 Hyaline cartilage is found in joints, where it reduces friction. Elastic cartilage is found in the tip of the nose, the larynx, and the external ear, which need to be flexible and supportive. Fibrocartilage is found in such places as the intervertebral disks, where the body needs tough, cushioning pads.

(Continued on p. 134)

TISSUE TYPE	FUNCTION
Elastic Cartilage 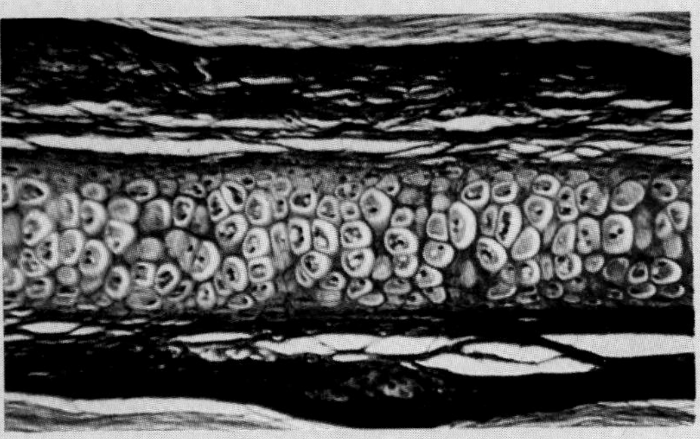	Elastic cartilage is the framework for part of the ears and structural elements of the voice box (larynx).
Fibrocartilage 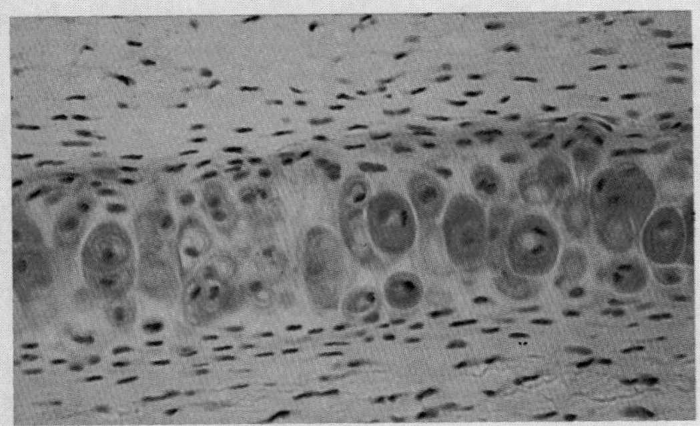	Fibrocartilage acts as shock-absorbing cushions (disks) between the vertebrae; also found in the knee joints and hips.
Blood 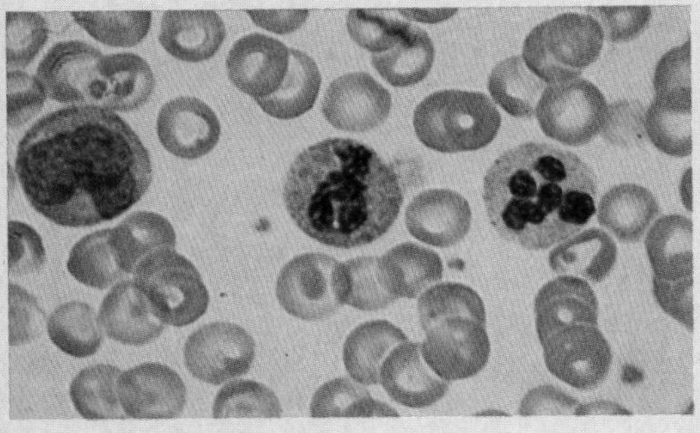	Made up of the red blood cells, the white blood cells, platelets, and plasma. Forms the most important transport and exchange medium for the body. It carries oxygen, wastes, hormones, nutrients, and various other materials through the cardiovascular system of the body.

Muscle Tissues

Tissues made of cells specialized for changing shape by contraction of microfilaments. They are the principal tissues employed for moving the bones and thus changing the body's posture or location, propelling the blood through the cardiovascular network, and regulating the flow of materials through the tubes of the body.

TISSUE TYPE	FUNCTION
Skeletal Muscle 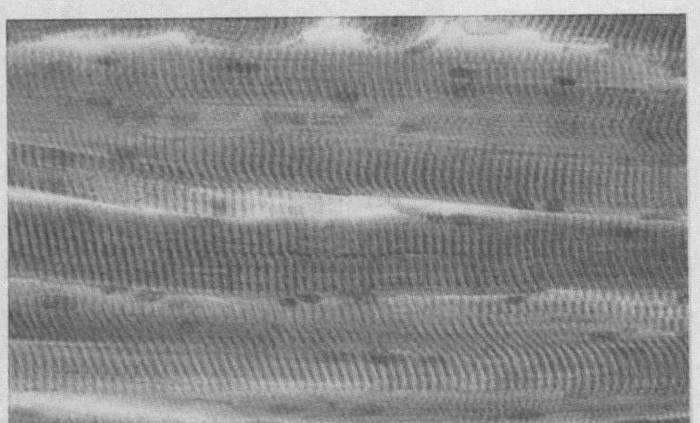	Complex cells attached to the skeletal system and responsible for skeletal movement. They are formed of complex, multinucleate cells identifiable by distinctive striations (stripes).

TISSUE TYPE	FUNCTION
Smooth Muscle 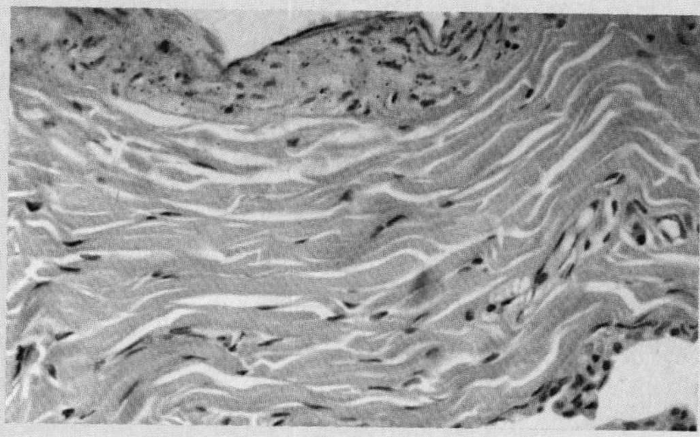	Spindle-shaped cells found in layers in the walls of tubular organs (the uterus, digestive system, blood vessels, etc.). They regulate the diameter of the lumens of these tubes by contracting or relaxing.
Cardiac Muscle 	The special type of muscle from which the heart is formed. Its fibers are branched.

Nerve Tissues

Complex cells that specialize in transmitting electrochemical signals from one part of the body to another. They form a high-speed network of cells responsible for regulating and controlling many bodily functions.

Neurons 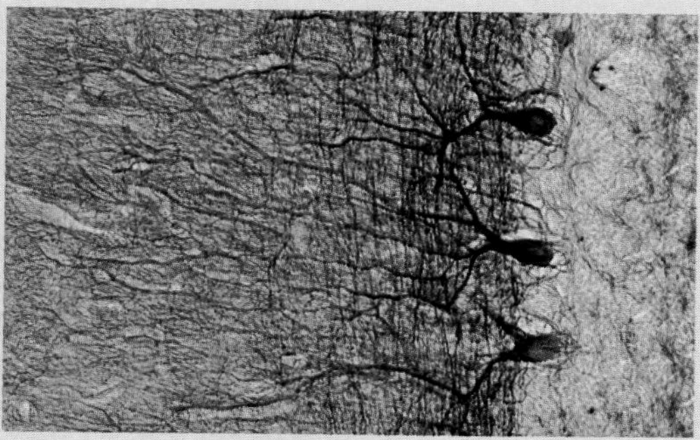	Very complex cells with elongated conducting processes called axons and dendrites that conduct the action potential (electrochemical signal) from one place to another.
Neuroglial Cells 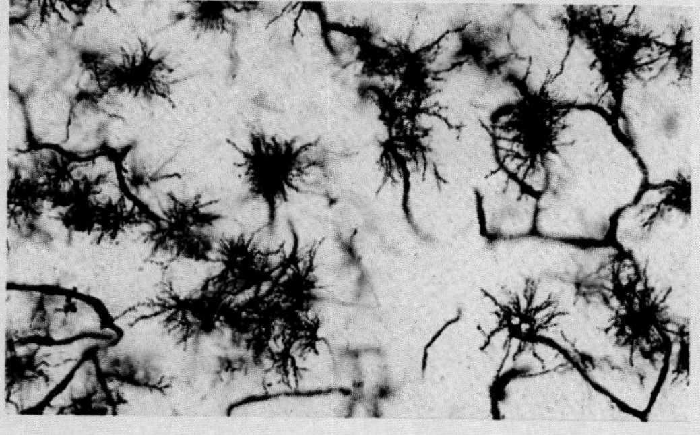	Cells that support the neurons and hold the nervous tissues together. They also help supply nutrients to the neurons.

Learning Focus

Human Tissue Types

In the following pages you will find photomicrographs of the various types of human tissues. Examine them, and based on the general appearance of each tissue, the structure of the individual cells, and the nature of the matrix between the cells, identify each tissue type. Also use these figures to summarize what you know about tissues and as a study aid. Gather information about each tissue from the book and any other sources, such as lecture notes or a visit to the library, and write it in the space below the photograph.

Tissue Type ______________________

Diagnostic Feature ______________________

____

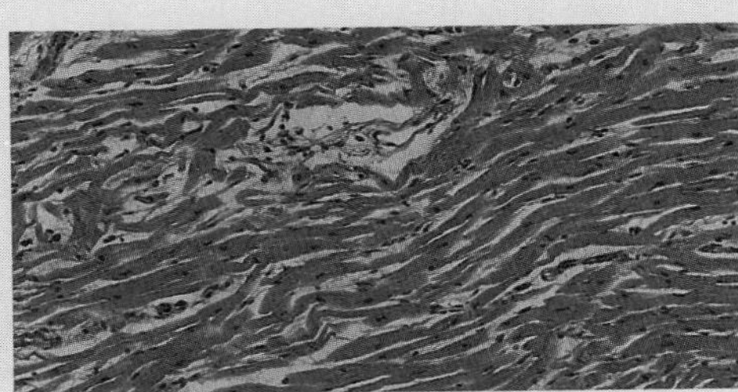

Tissue Type ______________________

Diagnostic Feature ______________________

____

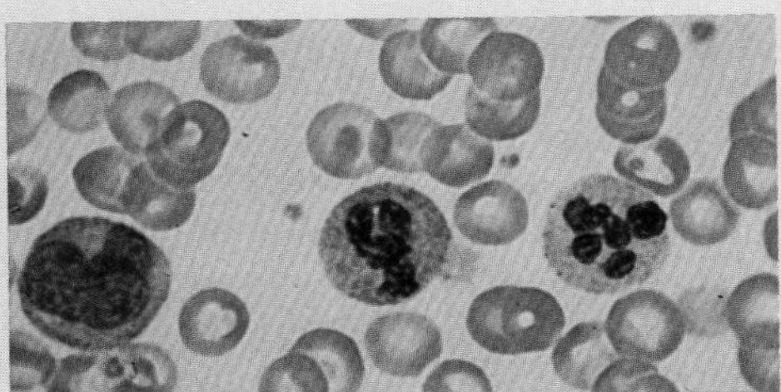

Tissue Type ______________________

Diagnostic Feature ______________________

____

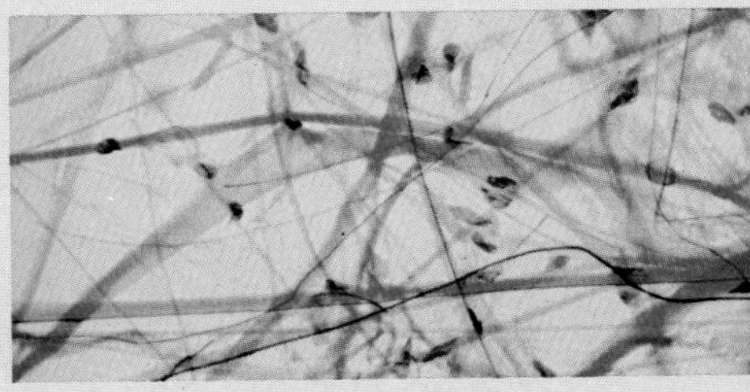

Tissue Type ______________________

Diagnostic Feature ______________________

____

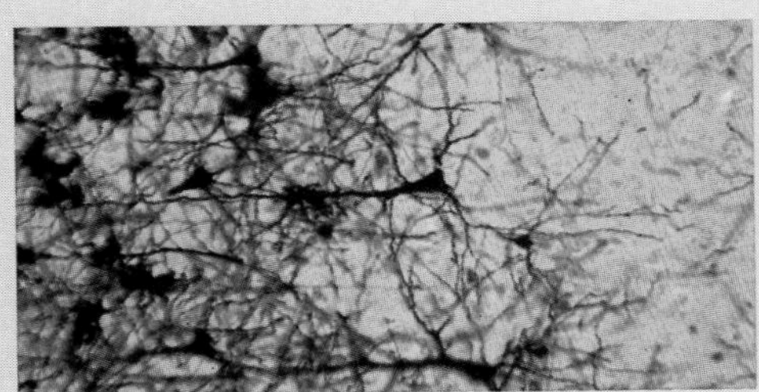

Tissue Type ______________________

Diagnostic Feature ______________________

____

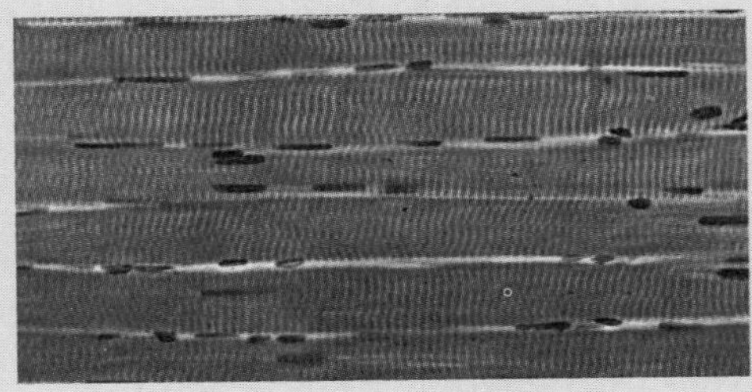

Tissue Type ______________________

Diagnostic Feature ______________________

____

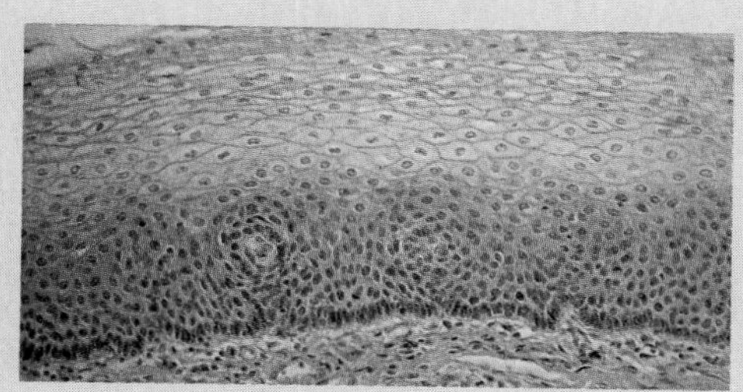

Tissue Type ______________________________

Diagnostic Feature ______________________________

____

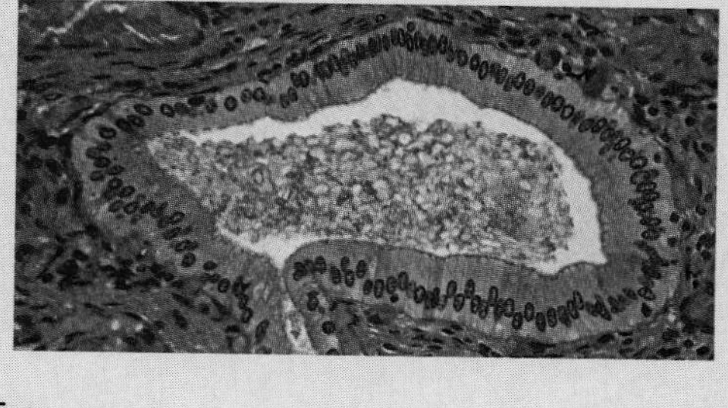

Tissue Type ______________________________

Diagnostic Feature ______________________________

____

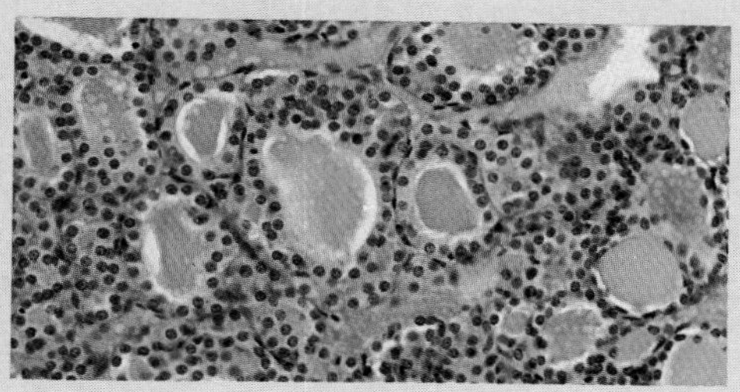

Tissue Type ______________________________

Diagnostic Feature ______________________________

____

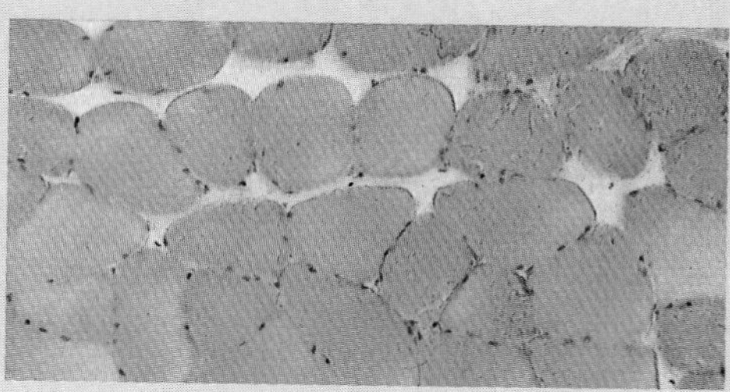

Tissue Type ______________________________

Diagnostic Feature ______________________________

____

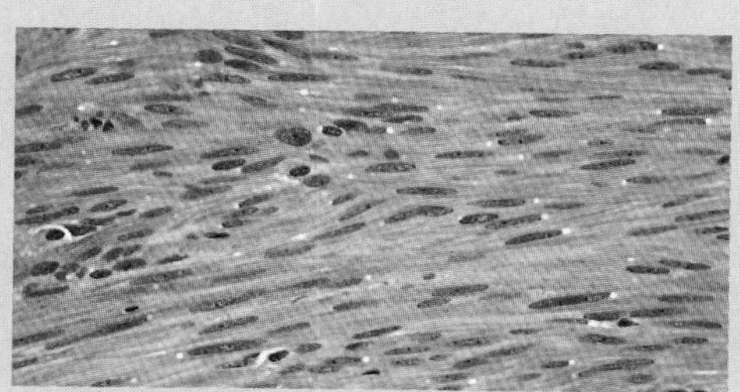

Tissue Type ______________________________

Diagnostic Feature ______________________________

____

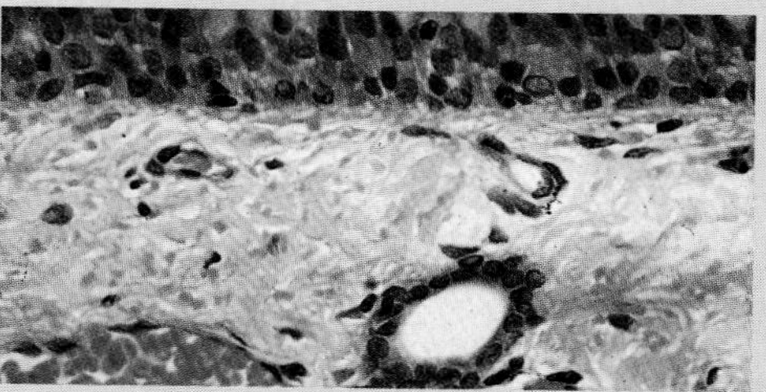

Tissue Type ______________________________

Diagnostic Feature ______________________________

____

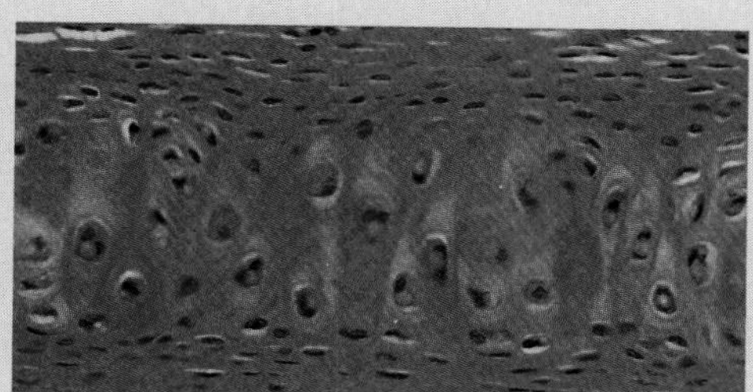

Tissue Type ______________________________

Diagnostic Feature ______________________________

____

AN OVERVIEW OF THE HUMAN BODY PLAN

As we progress more deeply into our studies of human biology, we need a few terms to help us describe the locations and relative positions of the organs and systems of the human body.

Anatomical Position and Common Terms

All written descriptions of the body and its parts refer to the *anatomical position,* a body standing upright, arms held with the palms forward, and the feet facing forward. Pairs of terms then describe the positions of different parts of the body with respect to each other (see Figure 6-4 and Table 6-3).

Pairs of directional terms are used to describe the relative positions of various structures and organs on and in the human body. **Superior** means above another reference organ (the nose is superior to the navel), while **inferior** means below (the knee is inferior to the hip). Using the body's midline as a reference, one structure is said to be **lateral** to another if it is farther from the midline. Conversely, one structure is **medial** to another if it is closer to the midline. The nipples are superior and lateral to the navel (for most people).

The terms proximal and distal usually describe the relative positions of structures located on the limbs. The part of a limb that is closer to where the appendage attaches to the body (or closer to the midline) is said to be **proximal** to an area that is farther away. If the area under discussion is farther from the limb's point of attachment, it is said to be **distal.** The elbow is proximal to the wrist, while the wrist is distal to the elbow.

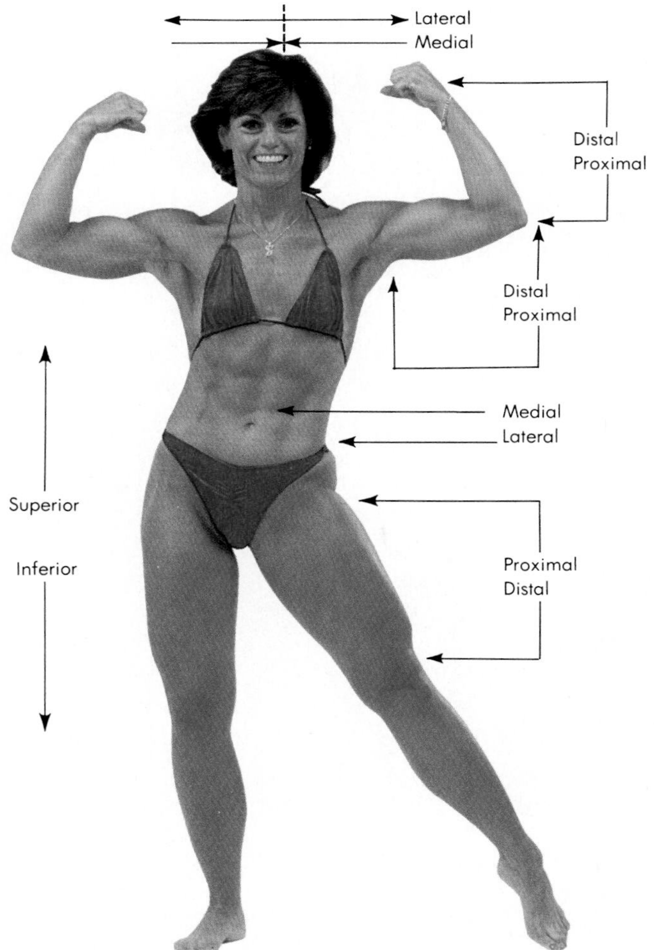

FIGURE 6-4 In formal anatomical descriptions of the body, the subject is always described as if it is standing upright, feet slightly separated, toes pointing forward, and the palms of the hands facing forward with the thumbs rotated out. How does the position in this figure differ from the formal anatomical position? The pairs of positional terms employed to describe the relative positions of the various components of the body include: *superior* (above) and *inferior* (below); *medial* (closer to the midline) and *lateral* (further from the midline); *proximal* (closer to a point of reference, such as where a limb attaches to the body) and *distal* (further from the point of reference). Thus the elbow is said to be proximal to the wrist.

Table 6-3

Pairs of Anatomical Terms

TERM	MEANING
Superior	Above or higher than another part of the body; e.g., the head is superior to the shoulders.
Inferior	Below or lower than another part of the body; e.g., the feet are inferior to the knees.
Medial	Closer to the midline of the body; e.g., the eyes are medial to the ears.
Lateral	Farther from the midline of the body (more to the side); e.g., the ears are lateral to the eyes.
Proximal	Closer to the point of attachment. Sometimes refers to the midline; e.g., the elbow is proximal to the wrist.
Distal	Farther from the point of attachment; e.g., the hand is distal to the shoulder.
Anterior	In front or on the front side; e.g., the tip of the nose is anterior to the eyes.
Posterior	Behind, in back of; e.g., the ears are posterior to the eyes.
Ventral	Toward the belly side; e.g., the navel is ventral to the small of the back.
Dorsal	Toward the backside; e.g., the small of the back is dorsal to the navel.

The front of the body is said to be **anterior.** The rear of the body is **posterior.** Two other terms are very similar in meaning: ventral and dorsal. **Ventral** means the belly side of the organism. Since the human anatomical position is standing up, the belly and the front are on the same side. **Dorsal** refers to the back—actually the side away from the belly (nearest the backbone in vertebrates). Like anterior and ventral, posterior and dorsal are often used interchangeably, even though the meanings are somewhat different (see Figure 6-4).

Body Cavities

Many of the body's organs and organ systems lie in a series of body cavities, the largest of which are called simply the dorsal and ventral cavities. The **dorsal cavity** is located on the dorsal side of the body; it has two subdivisions. The **cranial cavity** inside the skull contains the brain; the **vertebral cavity,** a channel passing through the bony elements of the backbone, contains the spinal cord, which connects the brain to the rest of the body (Figure 6-5).

The **ventral cavity** is the large cavity system in the body's trunk; it has two main subdivisions, the thoracic and abdominopelvic cavities. The **thoracic cavity** is that portion of the ventral cavity found within the rib cage. The lower margin of the thoracic cavity is marked by a muscular sheet, the diaphragm, at approximately the lower edge of the rib cage. The diaphragm aids in the respiratory movements associated with breathing. The thoracic region is subdivided into four additional cavities. Each lung lies in its own **pleural cavity,** bounded by the pleural membranes. The heart lies in the **pericardial cavity,** bounded by the pericardial membranes. Everything else found in the thoracic

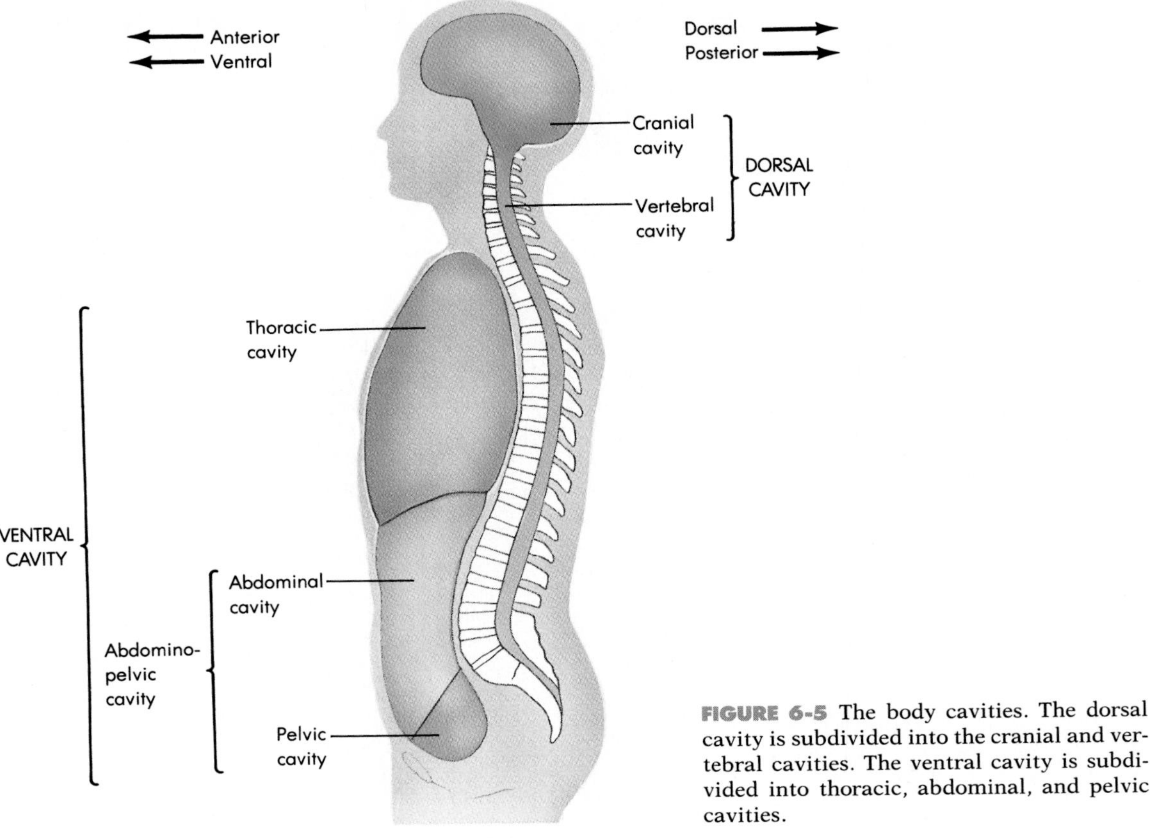

FIGURE 6-5 The body cavities. The dorsal cavity is subdivided into the cranial and vertebral cavities. The ventral cavity is subdivided into thoracic, abdominal, and pelvic cavities.

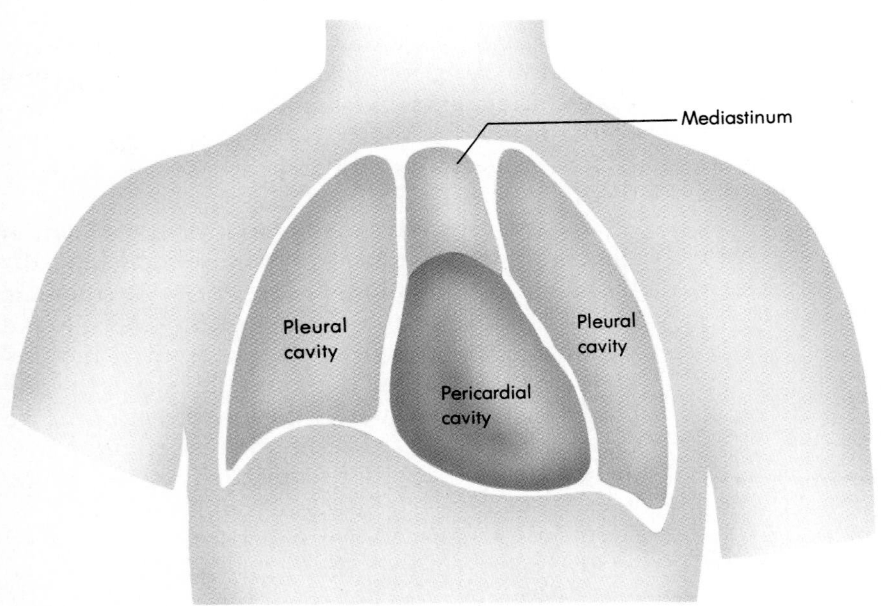

FIGURE 6-6 Subdivisions of the thoracic cavity. Each lung is housed in one of the two pleural cavities; the heart is located in the pericardial cavity, and almost everything else, including the major vessels of the chest cavity, the esophagus, and the windpipe or trachea, is said to lie in the mediastinum.

Table 6-4

The Ventral Cavity System

MAJOR CAVITY	SUBDIVISION	SPECIFIC ORGANS WITHIN CAVITY
Thoracic	Pleural cavities	Both lungs
	Pericardial cavity	Heart
	Mediastinum	Major blood vessels connecting to heart; windpipe or trachea; esophagus
Abdominopelvic	Abdominal cavity	Liver, stomach, pancreas; upper regions of small intestine; kidneys
Pelvic		Bladder; reproductive apparatus; portions of colon and rectum

cavity other than the heart and lungs is in the **mediastinum.** This includes the major blood vessels connected to the heart, part of the windpipe or trachea, and the esophagus (the gullet) (Figure 6-6 and Table 6-4).

Below the diaphragm, the ventral cavity consists of the **abdominopelvic cavity.** It has two subdivisions. The abdominal portion begins just below the diaphragm and continues down to include everything above a line drawn across the top of the pelvis. Below the line lies the **pelvic cavity.**

Most of the organs found in the abdominopelvic cavity are covered with the peritoneal membrane that supports and protects them. The stomach, liver, gallbladder, kidneys, spleen, pancreas, and parts of the small intestine and colon are found in the abdominal region. Lower in the pelvic cavity lie the internal organs of the reproductive system, the bladder and some of the tubes connected to it, portions of the small intestine, the cecum, the appendix, and the sigmoid colon and rectum. We will discuss the functions of most of these organs later, in the chapters where we deal with the various systems of the body.

ANATOMICAL SECTIONS

The diagrams used for illustrating the anatomy of the human body are made from specimens that have been cut ("sectioned") along a number of specific planes. Sections that slice the body into left and right portions are said to be in a **sagittal plane.** A cut that divides the body right down the midline into left and right halves is called a **midsagittal section.** Cuts that cross the body at right angles to the long axis are said to be **transverse sections.** Sections that divide the body into front and back portions are called **coronal sections** (Figure 6-7).

1. What kind of a section do you use when you slice a banana for a banana split?
2. What kind of a section do you use when you slice a banana to put on your breakfast cereal?
3. When a man shaves, he occasionally nicks the tip of his chin. What kind of a section is this?

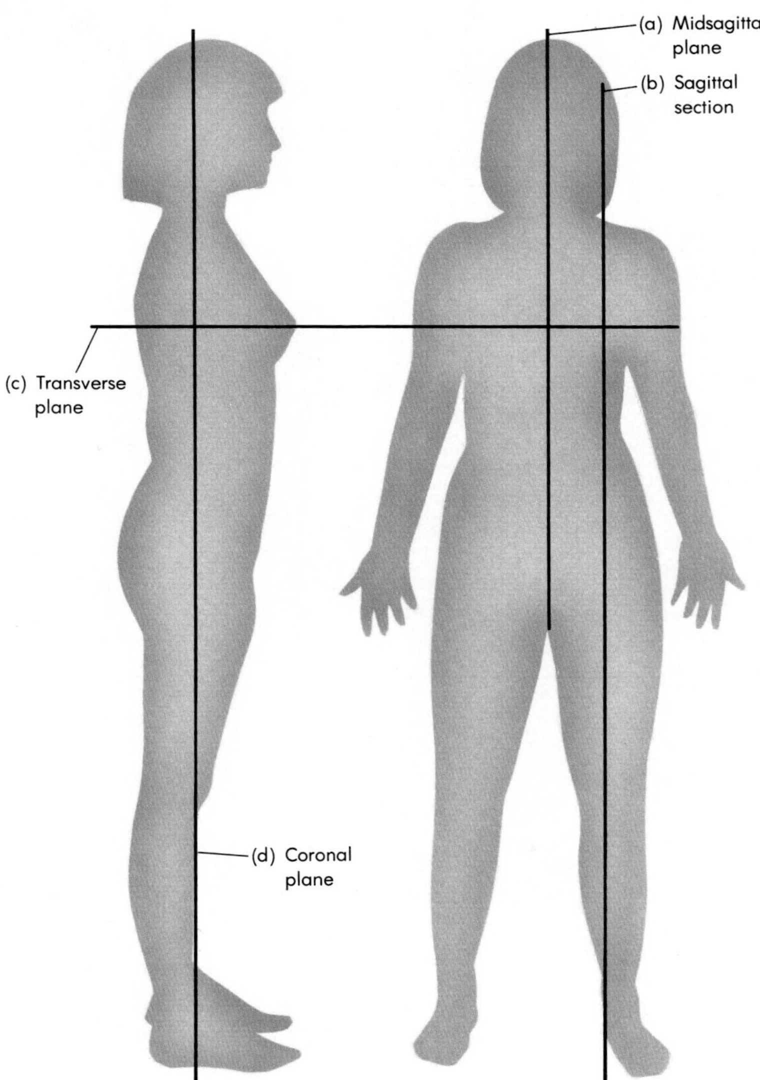

FIGURE 6-7 Traditional planes of cuts used for preparing the anatomical sections of the human body: (a) midsagittal section (divides body into left and right halves); (b) sagittal section (divides the body into left and right portions); (c) transverse section (cuts perpendicular to the long axis of the body); (d) coronal section (divides the body into front and back portions).

SUMMARY

In multicellular life forms such as higher plants, animals, and humans, the cells are arranged in aggregations of physically similar cells that support a common function or service, called tissues. In humans, there are four major categories of tissues: epithelial tissues, which line the free surfaces of the body, absorb, and secrete; connective tissues, which hold the body together, provide flexibility and strength, and carry nutrients from one area to another; muscle tissues, which perform movements; and nervous tissues, which convey rapid electrochemical signals for intercellular communication.

Descriptions of human anatomy always assume the subject being described is in the anatomical position: standing upright, palms rotated forward, with feet pointing ahead. Pairs of terms describe the position of one organ or part of the body with respect to another. There are two major systems of cavities in the body: the dorsal cavity, containing the brain and spinal cord; and the ventral cavity, in the trunk. The portion of the ventral cavity found within the rib cage is the thoracic cavity. The thoracic cavity is subdivided by the membranes surrounding the heart and lungs. Below the diaphragm, the abdominopelvic cavity contains many of the digestive, excretory, and reproductive organs.

STUDY QUESTIONS

1. Define *tissues* and describe and give the functions of the four major types of animal tissues.
2. What is meant by the term *anatomical position?*
3. How are the terms *anterior, posterior, ventral,* and *dorsal* related in descriptions of anatomical position? Also define the terms *distal/proximal, superior/inferior,* and *lateral/medial.*
4. List and describe the various types of connective tissues. How are they similar and how do they differ from one another?
5. Identify the three types of muscle tissues and what each type contributes to the normal body functions.
6. Describe the major types of epithelial tissues and what kind of body functions each type supports.
7. Distinguish between the dorsal and ventral cavities of the human body. Indicate their boundaries and identify the major subdivisions of each of these cavities.

GLOSSARY

Abdominopelvic cavity Portion of ventral cavity in the belly and pelvis.
Anterior Toward the front.
Cranial cavity Portion of dorsal cavity containing the brain.
Distal Further from a limb's point of attachment.
Dorsal Toward the back or spine.
Dorsal cavity Body cavity containing the central nervous system.
Inferior Below.
Lateral Farther from the midline.
Medial Closer to the midline.
Organ Major structures or compartments of the body made by combining different tissues.
Organ systems Groups of organs that cooperatively provide the body with certain services.
Posterior Toward the rear.
Proximal Closer to a limb's point of attachment.
Sagittal section An anatomical section cut parallel to the long axis of the body that divides the body into left and right portions.
Superior Above.
Thoracic cavity Portion of ventral cavity in the chest.
Tissues Groups of similar cells that have a common function or provide a similar service for the body.
Transverse section A section cut perpendicular to the long axis of the body.
Ventral Toward the belly.
Ventral cavity Body cavity within the trunk.
Vertebral cavity Portion of dorsal cavity containing the spinal cord.

Chapter 7
The Integument

The human body is not simply a collection of tissues or even of organs and organ systems. It is *organized* into a coherent whole. There is a framework, the skeleton. There are the muscles, ligaments, tendons, and nerves that hold the skeleton together and make it move. There is the circulatory system that carries nutrients from the digestive tract and oxygen from the lungs to the rest of the body and carries wastes to the lungs and kidneys for disposal. There is the nervous system that receives information from the outside world and coordinates responses. Each of these body systems must operate in a carefully controlled physical and chemical environment. The need for control requires the body to have a boundary between itself and the outside world.

That boundary is called the skin, or **integumentary system.** It encloses the rest of the body with a covering that has a surface area of almost 2 square meters (3000 square inches), the equivalent in area of a small bedspread. Its functions include softening blows, preventing microinvasions of bacteria (both physically and chemically), preventing dehydration, blocking the sun's radiation, and communicating. It supports a sensory network that keeps the brain informed of surrounding conditions, and it helps the body control its temperature. It is even the site for the synthesis of one essential nutrient, vitamin D. Numerous skin structures and organs help it carry out its various functions. They include hair, nails, and teeth, the various types of skin glands that secrete fluids onto its surface, and buried in it, numerous sensory nerve endings, blood vessels, and muscles.

THE STRUCTURE OF THE SKIN

The skin is thickest on those parts of the body that receive the most wear, such as the palms of the hands and the soles of the feet, where it may be 3 millimeters (about $\frac{1}{8}$ inch) thick. Elsewhere, it is usually thicker on the dorsal sides of body parts; at its thinnest, it is only $\frac{1}{2}$ millimeter thick.

Epidermis

Whatever its thickness, the skin has two main layers, the **epidermis** and the **dermis** (Figure 7-1). The epidermis is the surface layer. It consists of stratified squamous epithelium generated by a single **basal layer** of cells. These cells are the only epidermal cells that can divide, and they do so continuously to replace the surface cells lost to wear and tear. They increase their rate of multiplication under the influence of pressure and friction. They may also increase their multiplication under the influence of certain viruses, resulting in small, raised thickenings of the skin, or **warts.** Most warts eventually disappear spontaneously, but the viruses responsible for them can be spread by contact. Genital warts—the most distressing kind—can thus be spread by sexual contact.

1. A **callus** is a patch of thickened, hardened skin that develops at the site of prolonged pressure or friction, usually on the hands or feet. Explain how a callus forms.
2. The skin of a newborn baby's palms and soles is nearly as thin as the rest of its skin. It gains most of its extra thickness only later. Why?

When a basal layer cell divides, one daughter cell remains in the basal layer. The other is forced nearer the surface, where its nucleus degenerates, the cytoplasm fills with granules of a tough waterproofing protein, **keratin,** and the cell dies and shrinks to become a flat plate. This process of change is complete by the time the cell has been pushed all the way to the surface, where it eventually wears off. *Dandruff* consists of flattened patches of epidermal cells that have flaked off from the dry scalp. Typical epidermal cells last about six weeks from the time they originate until they are lost from the body; in the course of a lifetime, the average human sheds about 40 pounds of these cells.

With a microscope, we can see up to five layers in the epidermis. At the base is the single layer of basal cells (stratum basale). Next comes the *spiny layer,* (stratum spinosum), 8–10 rows of cells that may have a prickly appearance (due to the way the piece of skin is prepared for study). In the *granular layer* (stratum granulosum), 3–5 rows thick, the keratin granules first become visible and the cells begin to flatten. In the *clear layer* (stratum lucidum), missing in hairy skin such as that of the scalp and prominent in thick skin such as that of the palms, the cells are filled with a translucent form of keratin called *eleiden.* Finally, the *horny layer* (stratum corneum) is 25–30 rows of flat, dead, keratin-filled surface cells. It effectively blocks water loss from the inside and water gain from the outside and the penetration of bacteria, many

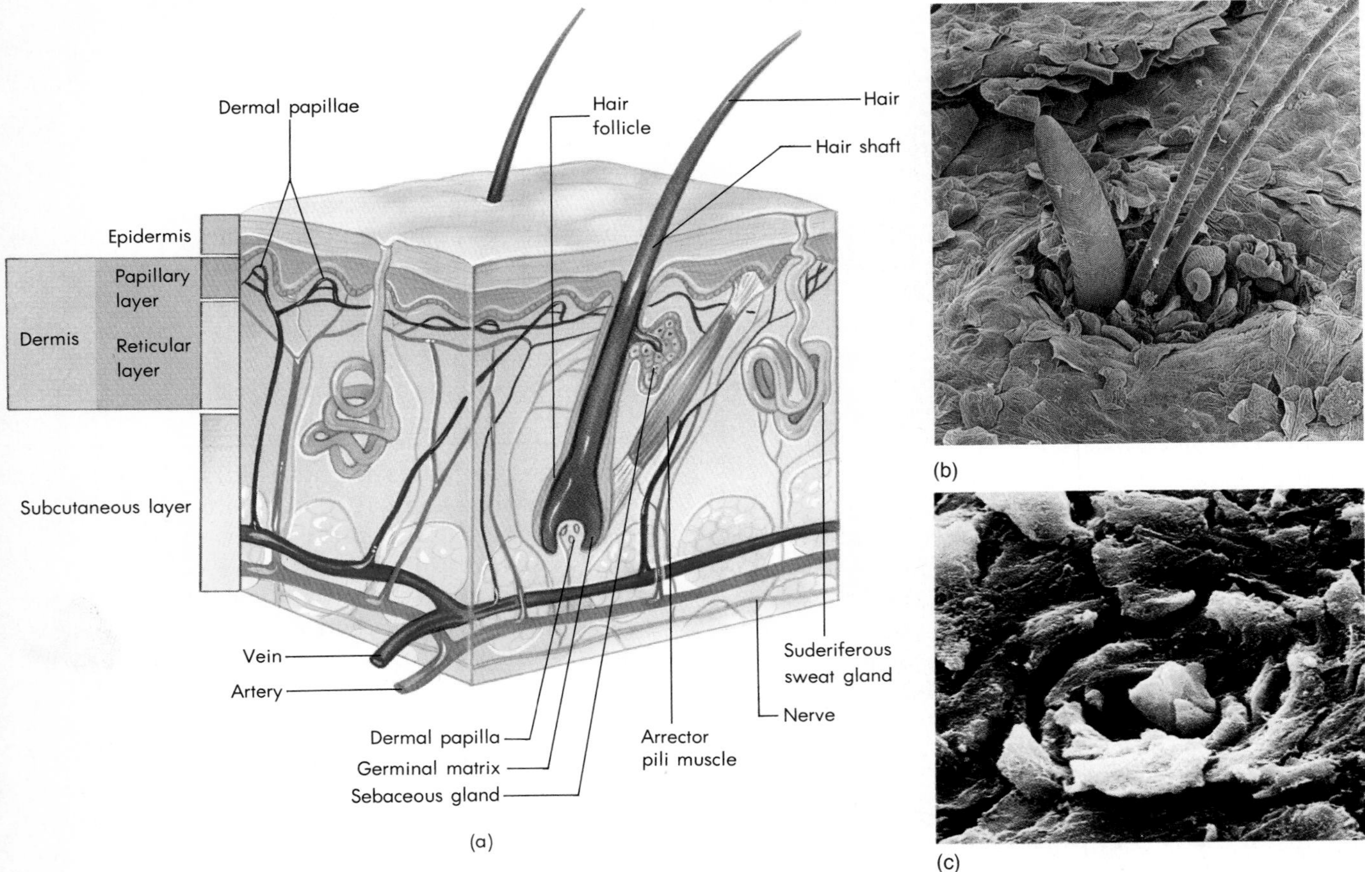

FIGURE 7-1 Structure of the skin. (a) Cross section through the skin. (b) A new hair emerging from the human scalp. The scalelike keratinized epidermal cells, surrounding the hair shafts, are continuously shed throughout life, waterproof the skin, and make it impermeable to many disease organisms. (c) A scanning electron micrograph looking down at a sweat pore in human skin.

chemicals, light, and heat. It even resists minor scratches, nicks, and cuts.

The epidermis also contains **Langerhans cells,** phagocytes recently recognized to play an important role in helping certain cells of the immune system recognize and attack foreign materials on the skin. More obvious in their function are the **melanocytes,** which provide most of the pigment present in human skin (see the Learning Focus on pigments of the skin).

Where the body needs traction, as on the tips of the fingers and toes and on the palms and soles, the epidermis is folded to form **epidermal ridges.** On the fingertips, these ridges form the lines and loops of the fingerprints (Figure 7-2, p. 149); because they are determined by heredity, they are unique to each individual and can be used as a means of identification. Since the ridges deposit a matching pattern of skin oils on any smooth surface, they can also be used to tell who last touched an object and have made fingerprinting an essential tool of law enforcement.

The epidermis is also folded elsewhere on the body, though the folds are not so close together. Look at the back of your hand; here the folds form **epidermal grooves** that mark off diamond-shaped areas. They become larger near joints, where they may resemble the folds of an accordion; here their function is to let the skin stretch to accommodate joint movement (Figure 7-3, p. 149).

Dermis

Do you have a tattoo? Do you know someone who has one? Then you know that tattoos are for keeps, and now that you have read about how the epidermis constantly sheds its cells, you may be wondering why people don't shed tattoos with their skin. The answer is that tattoos are made by injecting tiny droplets of ink beneath the epidermis, into the dermis, and the dermis does not wear away.

And yes, getting a tattoo does hurt. The epider-

Learning Focus

The Pigments of the Skin

There are no truly "white" people in the world. However, some very pale individuals, who are technically pink skinned, are found in all races. Inaccurately described as being white, they are white because they lack the genetic ability to make the skin pigment, melanin, and the color of their blood shows through. They are **albinos,** and they usually do have white (or pale yellowish) hair and even unpigmented pink irises in their eyes.

The rest of us have skin pigment, in varying amounts. It is synthesized from the amino acid tyrosine in the pigment-forming cells of the skin, called *melanocytes*. These cells sit among the cells of the basal layer of the epidermis and, like so many octopuses, extend long cytoplasmic arms among the cells of the spiny layer. At the tips of its arms, each melanocyte passes packets of the pigment **melanin** (melanosomes) to the epidermal cells. In the skin cells of Caucasians (whites), much of the pigment is broken down before it accumulates. In Negroids (blacks), the melanosomes remain intact, and the pigment accumulates in the skin. Orientals also have melanin in their skin, but they gain a yellowish hue from another pigment, called **carotene.** This pigment is found in both their epidermis and their dermis.

Melanocytes respond to the ultraviolet component of sunlight by increasing their melanin production. The melanin absorbs the ultraviolet, darkening somewhat in response, and prevents it from reaching deeper tissues, where it could cause damage. The buildup of melanin in the skin from exposure to sunlight is called **tanning.** Unfortunately, this buildup is not instantaneous. It takes time; thus it is important for light-skinned people to expose their skin to the sun only briefly until the melanin has begun to accumulate. If the exposure is too intense, the ultraviolet will cause the cell damage, inflammation, and skin peeling we call sunburn.

Some ultraviolet is necessary, for our skin cells use it to make vitamin D, required for proper absorption of calcium and phosphorus from the small intestine and the proper formation of bones (see Chapter 22). Yet too much vitamin D can be toxic. Some biologists therefore believe that human skin colors are the result of natural selection. Tropical, dark-skinned peoples have evolved to have enough pigment to absorb most ultraviolet light, letting through just enough to give them the amount of vitamin D they need. More northerly peoples, who are exposed to less intense sunlight, have evolved to have less pigment, letting them too get just the amount of vitamin D they need.

Sadly, too much exposure to the sun, even with a tan or a heavy-pigment birthright, is not good for you. Ultraviolet that reaches the eyes can cause blindness due to cataracts (clouding of the eye's lens). In the skin, melanin lets enough ultraviolet through to damage connective tissue and hasten the development of wrinkles and even to increase the chances of developing skin cancer due to damage to the deoxyribonucleic acid (DNA) of skin cells. It is thus wise for everyone to use a sunscreen lotion when outdoors for long periods; the best lotions contain *para*-aminobenzoic acid (PABA), which binds to the epidermis's horny layer and absorbs ultraviolet light.

Makers of tanning booths advertise that their products avoid damaging the skin because they use a nondamaging component of ultraviolet light. However, recent research suggests that even "safe" ultraviolet may increase the risk of skin cancer.

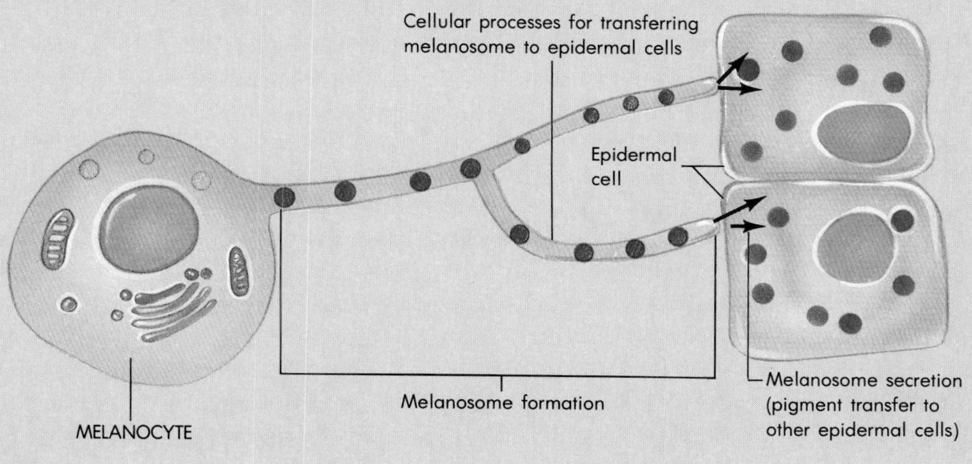

In the basal epidermis, melanocytes pass melanosomes (membrane-bound sacs containing dark pigments) to adjacent epidermal cells. In fair-skinned Caucasians, melanosomes in the skin cells are broken down by organelles similar to lysosomes. In dark-skinned races, the melanosomes do not break down and the skin remains darkly pigmented.

Learning Focus Response

THE PIGMENTS OF THE SKIN

Using the information about skin pigmentation in the preceding Learning Focus, answer the following questions:

1. No tan lasts forever. Why not?

2. In northern parts of the United States, people do not get much sun in the winter. If they take their last sunbath on Labor Day, by what date would you expect them to have lost the last of their summer tan?

3. Children (and adults) require vitamin D for strong bones. Too much vitamin D, however, can be toxic. Use these two facts to come up with an explanation for why people from tropical areas tend to have more heavily pigmented skin.

4. Many people are concerned because chemicals released into the atmosphere by human activities, such as the chlorofluorocarbons, destroy the ozone in the upper levels of the atmosphere. Ozone absorbs much of the ultraviolet light that comes from the sun, and destroying it will apparently expose us to levels of ultraviolet light beyond the ability of our tanning mechanism to cope. What consequences would you expect us to suffer from the destruction of the Earth's ozone layer?

5. What group(s) of people would you expect to suffer least from the destruction of the Earth's ozone layer? Most? Why?

6. The human races differ in skin color because of the pigment in their epidermis. Name and explain the exception to this statement.

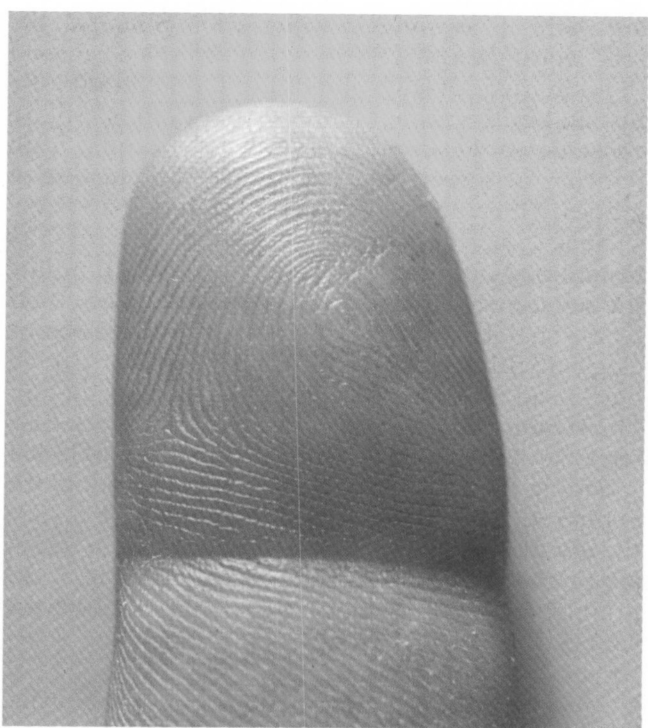

FIGURE 7-2 The epidermal ridges, like those in this fingertip, increase the traction of the gripping surfaces.

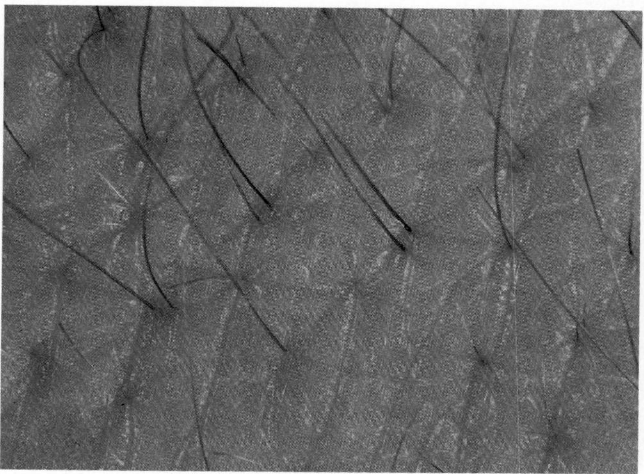

FIGURE 7-3 Epidermal grooves create patterns of lines and diamonds in various regions of the integument; they are often most conspicuous on the skin covering the major skeletal joints. Compare the pattern of epidermal ridges seen on the hand in the figure to that seen on the back of your own hand.

mis and the hair and nails that grow from it contain no nerve endings or blood vessels. The outer layer of the skin consists, for the most part, of nonliving materials like proteins and layers or plates made of dead epidermal cells. But the skin is nevertheless very sensitive to touch, temperature, and pain, for the dermis, the layer of the skin below the epidermis, contains an immense variety of nerve endings and receptors (Table 7-1). The dermis also contains the skin's blood vessels and muscles as well as the glands and hair follicles formed when the epidermis's basal layer dips away from the surface.

Table 7-1 shows that the most abundant skin structures are those involved in self-preservation. For instance, the most numerous are the pain sensors that set off an alarm when a part of the body has been damaged. Heat receptors are, likewise, more numerous than cold sensors. A wide variety of touch receptors keep the brain informed about the materials that are in contact with the skin. The numbers of sweat glands indicate the importance of the skin's contribution to controlling body temperature. The sebaceous glands secrete oils that condition the skin and keep it in good shape.

The dermis is composed of tough connective tissue. Its surface layer, nearest the epidermis, is loose connective tissue with many fine elastic fibers; it is folded into numerous **dermal papillae** that give it its name, the *papillary layer.* As the epidermis conforms to these folds, it produces its own epidermal ridges and grooves. Many of the papillae contain nerve endings and capillaries. The deeper *reticular layer* of the dermis is dense connective tissue, in which bundles of coarse collagenous and elastic fibers form a network (or reticulum). This region contains more nerve endings, capillaries, and muscles.

The collagenous and elastic fibers of the dermis make the skin both strong and stretchy. The strength is apparent when you realize that the dermis is the part of the skin (of animals) that tanners turn into leather for shoes, belts, jackets, and other articles. The stretchiness becomes obvious during pregnancy, sudden weight gain (when the skin has no time to adjust by growing), and swelling (as with infections).

It is the cells of the dermis that are responsible for making vitamin D. They contain a precursor of the vitamin that is activated by the ultraviolet component of sunlight. The process is so efficient that in the winter, when sunlight is weakest, as little as an hour's walk outdoors each day with the face exposed satisfies the body's needs for most fair-skinned people.

Since skin pigment blocks the penetration of solar ultraviolet, people with dark skins can make enough vitamin D in their skins only in those parts of the world (the tropics) where the sun is brightest. Since vitamin D is necessary for the body's proper management of calcium, dark-skinned children

Table 7-1

Calculated Totals for Skin Structures of the Human Body

STRUCTURE	CALCULATED TOTALS[a]
Hairs	195,000
Sebaceous glands	300,000
Sweat glands	1,950,000
Nerves	8400
Cold receptors	39,000
Heat receptors	234,000
Pressure sensors	495,000
Pain sensors	3,900,000

[a]Calculated assuming the individual possesses 3000 square inches of skin surface and each square inch is made up of average skin.

who live in nontropical, less sunny areas are more frequently burdened with the skeletal deformities seen in rickets. For that reason, we now routinely put vitamin D in milk. Because light-skinned peoples tend to come from climatic zones with more clouds and less sun and dark-skinned peoples tend to come from areas with more sun, many biologists believe that racial variations in human skin pigmentation evolved, in part, to ensure that members of each group got enough vitamin D (without overdosing).

Beneath the dermis lies a *subcutaneous layer* that links the skin to the underlying organs. A prominent component of this layer is subcutaneous fat, which provides protective padding for the deeper tissues, some insulation from the cold, and a reserve supply of energy. Females generally have more subcutaneous fat than males, which accounts in part for their less angular contours.

The Glands of the Skin

The skin has four kinds of glands: the sebaceous glands (oil glands), two types of sweat glands (eccrine and apocrine), and the ceruminous (wax-secreting) glands. The **sebaceous glands** are structurally complex glands that secrete **sebum,** an oily mixture of fats, cholesterol, proteins, and salts. Most sebaceous glands release their secretions into a hair follicle, from which they spread up the hair shaft to reach the surface of the skin. These oily secretions keep the hair from becoming brittle, aid the surrounding skin in staying moist and supple, and help in keeping it waterproof. Some of the sebaceous glands open directly onto the surface of the skin. Beds of such glands are found around the lips and in the skin of the nose and eyelids. All are constructed of branching tubules leading from bunches of secretory cells, resembling clusters of grapes. An example is visible in Figure 7-1.

Acne sometimes complicates puberty, especially in males. At puberty, the sex hormones stimulate the sebaceous glands to enlarge and increase their secretion of sebum. The glands may become clogged with excess sebum, which discolors into a blackhead as it combines with oxygen in the air. The sebum provides a fertile environment for bacteria, and the glands can become infected. Responding to the infection, the tissue around the gland reddens and swells, white blood cells invade the gland to form pus, and the blackhead becomes a whitehead or larger pimple. In severe cases, the pimples become still larger cysts, surrounded by connective tissue capsules, and can leave scars when they have healed.

The **sudoriferous (sweat) glands** take two forms. The **eccrine sweat glands** are coiled tubules that secrete a mixture of water, salts, urea, uric acid, ammonia, and other substances onto the surface of the skin. They play a minor role in the disposal of wastes (urea, uric acid, ammonia), but their main role is temperature regulation (Figure 7-4). They are especially active when the body is overheated; evaporation of sweat from the skin then provides cooling. They also serve to moisten the soles and palms and hence improve traction. They are most numerous—3000 per square inch—on the palms.

The **apocrine glands** are branched tubules that secrete a thick sweat rich in organic materials into hair follicles in the armpits, groin, anal region, and pigmented regions surrounding the nipples. The secretions are responsible for an individual's personal body odor—the musk-related scent—not the bad odor associated with uncleanliness. The apocrine glands begin secreting at puberty. If the apocrine secretions are not washed frequently enough from the skin and the clothing, the resident bacteria will metabolize the secretions and produce the foul-smelling compounds associated with bad "body odor," or "B.O."

The **mammary glands** of the breasts (discussed in Chapter 17) are modified, enlarged apocrine sweat glands that secrete milk to feed infants. The **ceruminous glands** of the ear canal are modified eccrine sweat glands that secrete a waxy material. Ear wax, or *cerumen,* is a mixture of the substances produced by these glands and the oil of the sebaceous glands. Your genes determine whether your cerumen is moist and waxy or dry and crumbly.

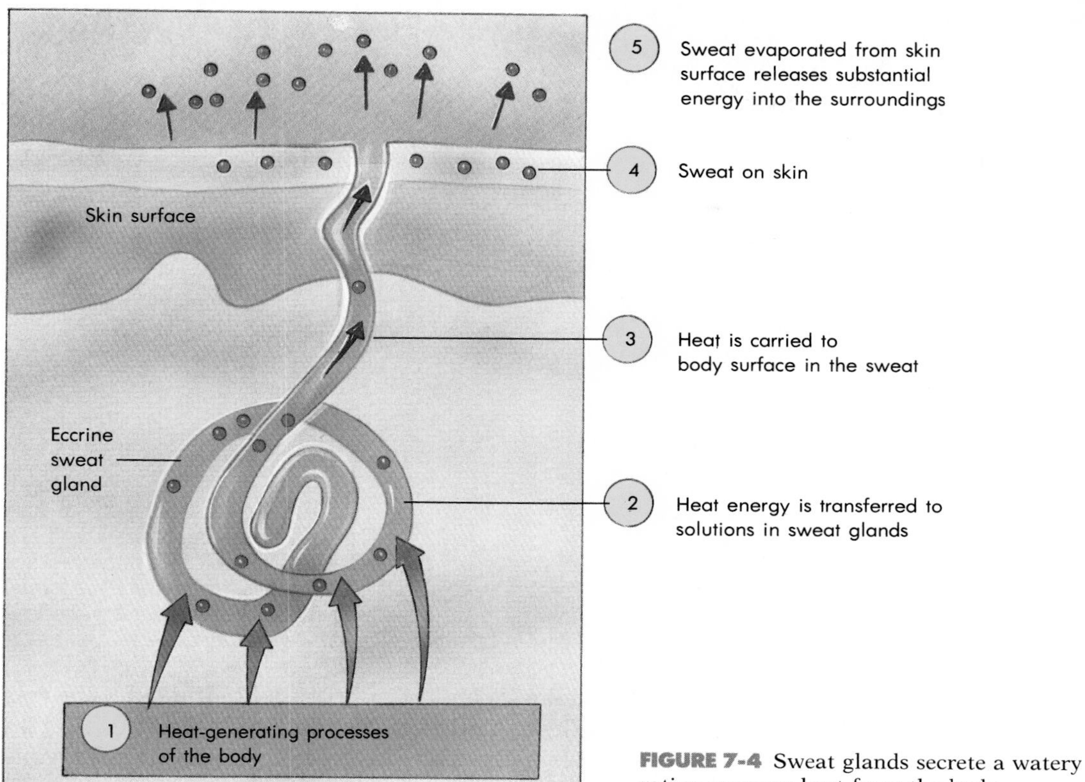

FIGURE 7-4 Sweat glands secrete a watery fluid whose evaporation removes heat from the body.

DERIVATIVES OF THE SKIN

The skin is not simply a covering sheet. The epidermis dips deep into the dermis—and beyond—to form the glands. As a result, the sebaceous and apocrine sweat glands, though they are made of epidermal cells, lie deeply embedded in the dermis. Eccrine sweat glands lie in the subcutaneous layer, closer to the metabolic heat sources they have to regulate.

Three other skin-derived structures project above the epidermis, though they have their roots in deeper layers of the skin. These structures are teeth, hair, and nails.

Teeth

Surprisingly enough, teeth are skin structures, not simply outgrowths of the bones of the jaws. Early in embryonic development, the epithelium (epidermis) of the jaws gives rise to a series of **enamel organs,** one for each future tooth. As it develops, the enamel organ stimulates the underlying connective tissue to form a *dental papilla,* which will form the bulk of the tooth. The enamel organ then secretes the hard enamel that covers the surface of the tooth (Figure 7-5).

The abilities of epithelium to form enamel organs and of connective tissue to respond to these organs by forming dental papillae are controlled by different genes. And the fact that both genes (or sets of genes) are necessary for the formation of teeth was proved by research on chickens. One scientist found that if he transplanted the enamel organs of embryonic mice into the jaws of embryonic chicks, the chicks' connective tissue would respond to the enamel organs by developing dental papillae. We can conclude that chickens, who have a strong reptilian evolutionary heritage, still have the genes for growing teeth, but they lack the genes for developing enamel organs. Supply them with enamel organs from mice, and they can grow the proverbially scarce "hen's teeth."

Hair

The males of many animal species, such as African lions, have well-developed manes of hair on their heads, necks, and shoulders. In humans, both sexes are also considered to have well-developed manes.

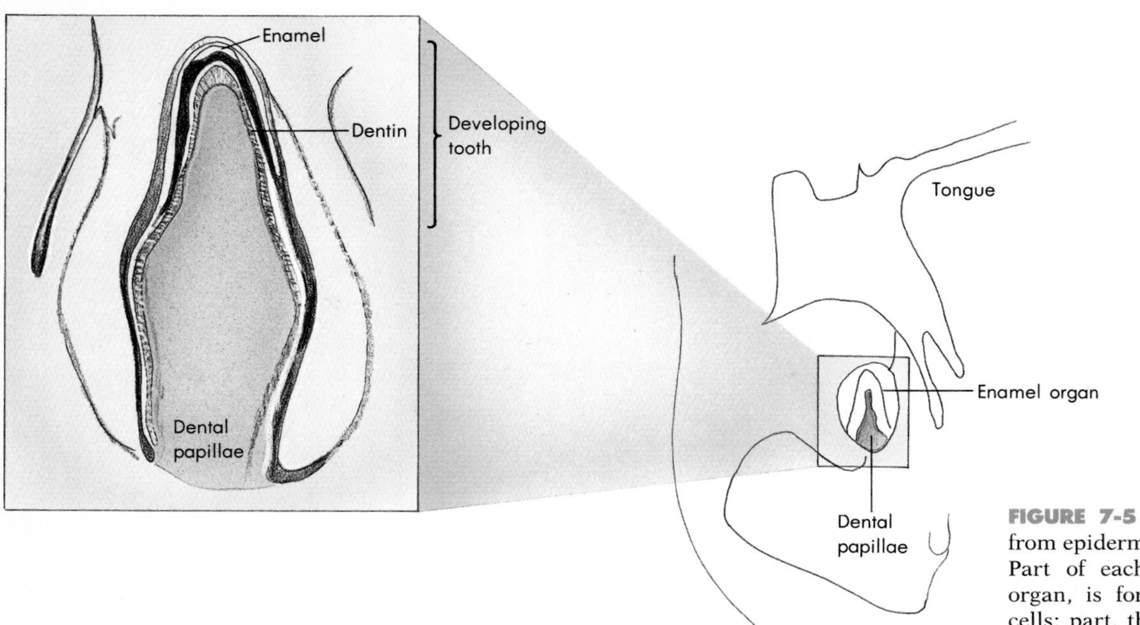

FIGURE 7-5 The teeth develop from epidermal buds in the jaws. Part of each bud, the enamel organ, is formed by epidermal cells; part, the dental papilla, is dermal in origin.

The hair on the heads of both males and females grows continuously and conspicuously throughout life. However, as in other species, the mane in human males includes conspicuous facial and neck hair, making it fuller than it is in females. Currently, our cultural standards of beauty have declared male facial hair out of style—so most males scrape it off daily.

Hair covers every part of our bodies—including the interior of the nose and the ear canal—except the lips, palms, and soles. It is even present on the forehead and the tip of the nose, although it is very fine and short in these places. It is longest on the scalp.

A hair is a column of dead epidermal cells covered with a layer of flattened, scalelike cells, the **cuticle** (Figure 7-6). Beneath the cuticle lie the hair's **cortex** and **medulla;** their cells are unflattened and contain eleiden granules, melanin pigment, and air spaces. The air spaces increase in number and size with age, changing not only the color of hair but also its texture. Because of the way the air spaces scatter light, they cause hair to gray as you age. Hair tonics and other preparations restore or darken hair color because they fill the spaces and allow the pigment to show better. The hair continues to lighten with age, since the natural production of hair pigment also slows as a person gets older, continuing the transition from gray to white hair.

The part of the hair that shows above the skin's surface is the hair's **shaft.** The rest of the hair, the **root,** is embedded in the skin, in the **hair follicle.** Like the skin glands, the hair follicle is an extension of the epidermis, a tube of cells, the **external root sheath,** extending into the dermis. At its tip, the tube forms a **bulb** into which a papilla of dermal connective tissue protrudes. A layer of epidermal cells (the germinal matrix) covers the papilla; as they divide, these cells give rise to the cells of the hair itself and the **internal root sheath** (Figure 7-7).

The epidermal cells covering the papilla do not divide continuously. They alternate periods of active growth, during which they produce about a hundredth of an inch of hair per day (1 millimeter in 3 days), with periods of nongrowth. At the end of the quiet period, the hair is sloughed off—you lose 70 to 100 hairs per day from your scalp. The next growth period replaces it. The length of the growth

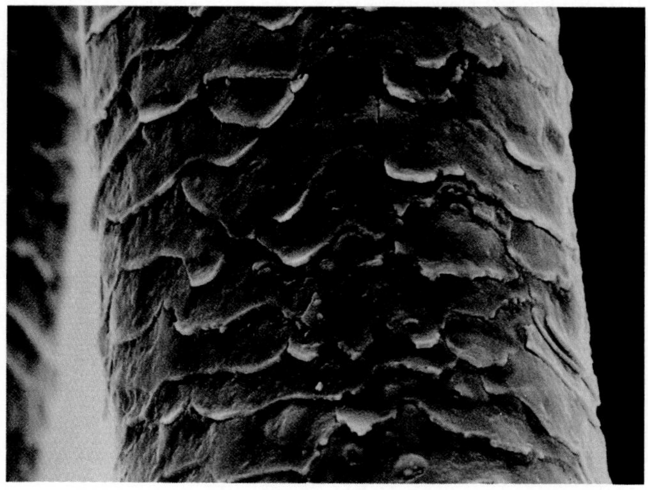

FIGURE 7-6 The surface of a hair carries distinctive cuticle scales.

THE FUNCTIONS OF THE SKIN

We have already mentioned several functions of the skin—communicating, sensing, waterproofing, repairing cuts, preventing infections, filtering ultraviolet, alerting the immune system, and producing vitamin D. We have also mentioned the role of the sweat glands in regulating temperature, but sweating is not our only way to lose heat.

Temperature Regulation and the Skin

Heat can be transferred from one object to another in four ways (Figure 7-10). *Evaporation* takes advantage of the fact that converting a liquid such as water to a vapor consumes large amounts of heat energy. *Radiation* refers to the way a warm object emits infrared (heat) waves to its cooler surroundings; we feel this radiation when we sunbathe or pass near a heater. About 60 percent of the heat lost from the human body leaves as infrared radiation, at normal room temperature. *Conduction* is the transfer of heat from a warm object to a cool one by contact, as when we warm a chair seat. *Convection* is the movement of a warmed fluid or gas (like the atmosphere) away from a heat source, as when warmed air rises from our bodies or when we chill in an ocean or lake. Conduction accounts for about 3 percent and convection for about 15 percent of our room temperature heat loss.

When we are overheated because of hot weather, temperature receptors in the skin signal the brain, which then activates the sweat glands. As they release sweat onto the surface of the skin, it evaporates, keeping the body cool. The process works best when the humidity is low, for high humidity interferes with evaporation. It also requires an ample supply of water and salts to replace what we lose in our sweat. Under these conditions, sweating is so efficient that we can sit for hours in a room heated to 145°F (63°C) and remain alive and even fairly comfortable (if the humidity is low). At the same time, a steak on the bench beside us will cook to a turn. We enhance evaporation, as well as conduction and convection, when we seek breezes, fan ourselves, and lie spread-eagle to expose more surface area to the air.

We can also grow overheated as the result of exertion, when we generate heat in our muscles faster than it can leave the body by radiation, conduction, and convection. In this case, heat sensors in the

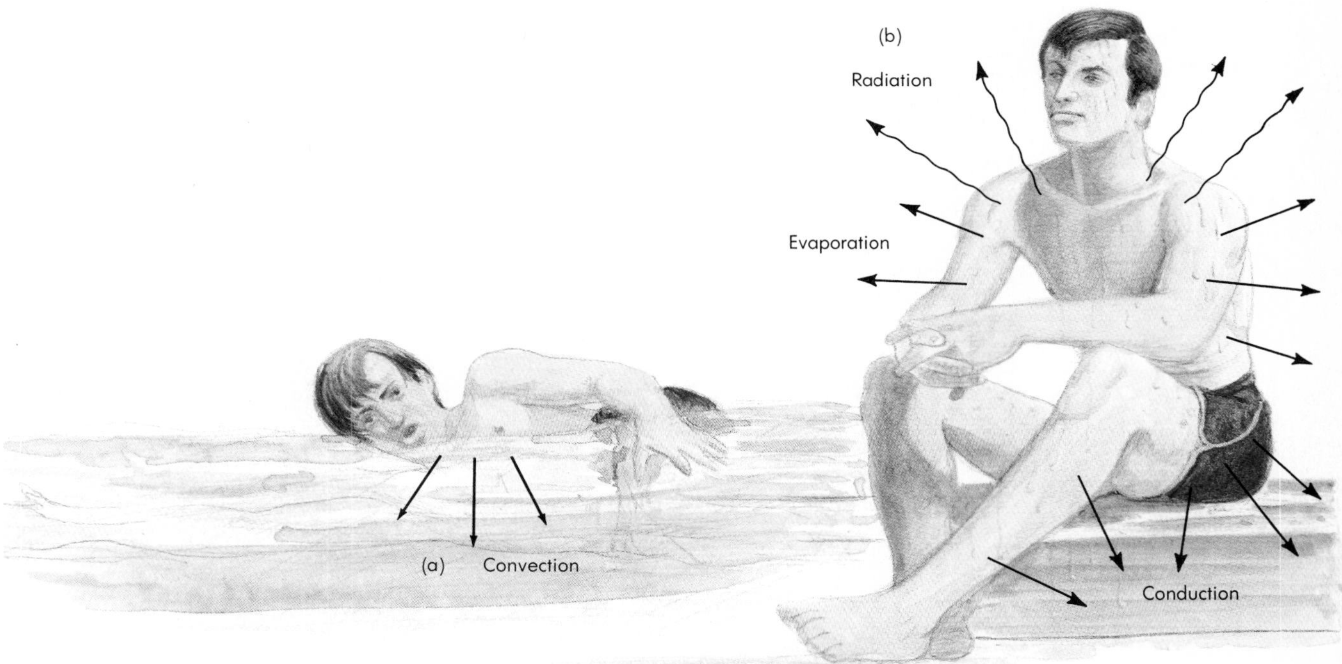

FIGURE 7-10 Mechanisms of heat transfer. (a) In convection, heat is removed as a gas or liquid flows away from the body. A swimmer loses heat by warming the water right next to him or her. (b) In radiation, heat is released from a warm object, such as the body, to its surroundings; in evaporation, heat is carried away from the body's surface as water—as sweat—evaporates from the skin; in conduction, heat moves from a warm object to a cooler object where the two are in contact.

brain react to the higher temperature of the blood and activate a complex array of physiological avenues for heat loss. One of these mechanisms is sweating. Another takes advantage of the numerous blood vessels in the dermis. To lose heat, the brain relaxes muscle fibers around these vessels, allowing them to fill with blood. As the hot blood flows near the body's surface, we flush or turn pink, the skin warms, and the excess heat radiates away just as if we were a radiator.

The same mechanisms work in reverse too. Heat loss can be reduced by turning the sweat glands off, by reducing blood flow to the skin, by avoiding moving air, and by huddling. In addition, animals can fluff their fur to increase insulation. Humans enjoy some insulating benefit from head and facial hair, but they rely much more on a wide variety of insulating garments.

Communication and the Skin

We generally think of communication as a matter of voice and gesture, but the skin is an important communicative organ. Animals communicate their moods and sometimes make threats by fluffing their fur or causing it to rise on the backs of their neck and shoulders; humans cannot do that. However, the patterns of human hair distribution and color do signal sex and age. We can also—involuntarily—change blood flow to the skin and blush to indicate embarassment, turn red with rage, and go pale with shock. In addition, we have considerable voluntary control over subcutaneous muscles in the face and neck, many of which attach to the skin to produce the stretchings and wrinklings of facial expressions (Figure 7-11).

(a)

(b)

(c)

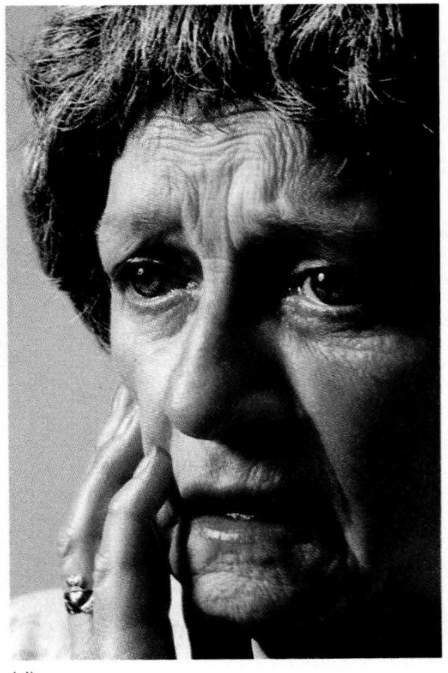

(d)

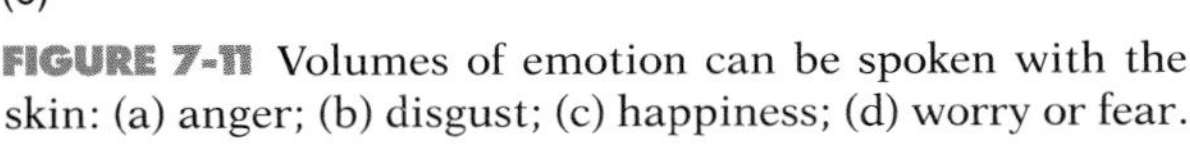

FIGURE 7-11 Volumes of emotion can be spoken with the skin: (a) anger; (b) disgust; (c) happiness; (d) worry or fear.

An additional communicative role is served by the skin's apocrine sweat glands, particularly in adults. The substances they secrete have strong odors, especially after they have been worked on by the bacteria that dwell on the skin, and these glands are found in areas, such as the armpits, groin, and anal region, that have wicklike tufts of hair that can spread the odor onto the air. Emotionally charged situations are likely to stimulate apocrine secretions, which may have played a role in our evolutionary past, signaling that a certain level of sexual readiness (or emotional intensity) had been attained.

SUMMARY

The skin is the boundary between the body and the outside world. It provides a barrier resistant to penetration by physical objects, infectious agents, heat, and sunlight, and it prevents excessive water loss and gain. In addition, it contains many sensory nerve endings, synthesizes vitamin D, controls heat gain and loss, and serves a communicative function via changes in blood flow, changes in configuration due to muscle action, and odorous secretions.

The skin has two major layers, the surface epidermis and the underlying dermis. The epidermis is a sheet of cells generated by a single basal layer of dividing cells; as they are pushed toward the body's surface, they fill with the tough, waterproof protein keratin. They also acquire the pigment melanin from melanocytes; melanin production and accumulation increase with exposure to ultraviolet light.

The dermis, a tough layer of connective tissue, contains all the skin's nerve endings and blood vessels. Embedded in it are the sweat and sebaceous glands and the hair follicles, all of which are derived from the epidermis. Sebaceous glands secrete an oily substance that helps keep the skin waterproof and supple. Eccrine sweat glands secrete a watery fluid whose evaporation helps cool the body. Apocrine sweat glands secrete a more viscous fluid that is largely responsible for bodily odors, which may serve important communicative functions. Mammary glands are modified apocrine glands; the ceruminous or ear wax glands are modified eccrine glands.

Skin derivatives include teeth, hair, and nails. Teeth are formed when the embryonic epidermis overlying the jaw bones produces enamel organs that cause the deeper connective tissue to develop into dental papillae, which form the bulk of the teeth. Hairs are formed in hair follicles, tubes of epidermal cells that dip into the dermis; at the base of the follicle, specialized basal layer cells generate the column of cells that is a hair. Attached to each follicle is a muscle whose contractions raise the hair in response to cold or fear and produce goosebumps. Finger and toe nails are formed by sheets of specialized epidermis on the tips of the fingers and toes. These nail beds produce thick layers of hard keratinous cells that slowly move toward the tips of the fingers and toes.

The body controls its temperature only in part with the skin's sweat glands. It can also regulate heat loss by controlling the flow of blood into the dermal blood vessels.

Table 7-2

Clinical Summary: Selected Diseases of the Skin

DISORDER	DESCRIPTION
Acne	Common inflammation (infection) of the sebaceous glands. Severe cases (cysts, not pimples) can leave scars. Generally begins in puberty when sebum production increases.
Melanoma	Cancer of the melanocytes, caused chiefly by exposure to solar ultraviolet light. Most deadly of skin cancers, with about 14,000 new cases per year, mainly among people aged 35–50. Affects six times as many whites as blacks or Asians.
Psoriasis	Skin condition marked by raised, scaly, red patches produced by excess mitosis of epidermal cells. The patches are often localized and temporary but may cover large areas for long periods. Cause is unknown, but stress and infections are known to trigger outbreaks.

STUDY QUESTIONS

1. List the functions of the skin.
2. How does the epidermis respond to wear?
3. Describe the formation of the horny layer of the epidermis.
4. What is the biological function of epidermal ridges? The social function?
5. How does the dermis serve a nutritive function?
6. Name and describe four kinds of skin glands.
7. What is the function of sebaceous glands?
8. In what sense are teeth *not* a derivative of skin?
9. How are the formation of nails and the formation of hair similar? Different?
10. Describe four ways in which the skin helps us communicate with each other.

GLOSSARY

Apocrine glands Modified sweat glands producing a rich organic sweat associated with personal odor (the musky odor of sex or intense emotion).

Arrector pili The smooth muscle attached to the outside of a hair follicle.

Basal layer The layer of actively dividing cells at the base of the epidermis.

Ceruminous glands Modified eccrine sweat glands found in the ear canal; they secrete cerumen, or ear wax.

Dermis The deeper, connective tissue layer of the skin, containing nerve endings, blood vessels, muscles, glands, and hair follicles.

Eccrine glands Sweat glands producing a watery sweat primarily involved in evaporative cooling.

Enamel organ An embryonic epidermal structure that produces the enamel of a tooth; it also stimulates deeper tissues to produce the rest of the tooth.

Epidermis The surface, epithelial layer of the skin.

Hair follicle The tube of epidermal cells that dips into the dermis and surrounds the hair root.

Keratin The tough, waterproof protein that fills epidermal, hair, and nail cells.

Mammary glands Modified, milk-secreting apocrine sweat glands.

Melanin The dark (brown or black) pigment of the skin.

Nails Sheets of hard, keratinized cells covering the tips of the fingers and toes.

Sebaceous glands The oil glands of the skin; they secrete a lipid-rich waterproofing and softening substance into hair follicles and onto the skin.

Sebum The mixture of fats, cholesterol, proteins, and salts secreted by the sebaceous glands.

Sudoriferous glands Sweat glands; they secrete a watery fluid whose evaporation aids heat loss.

Chapter 8

The Skeletal System

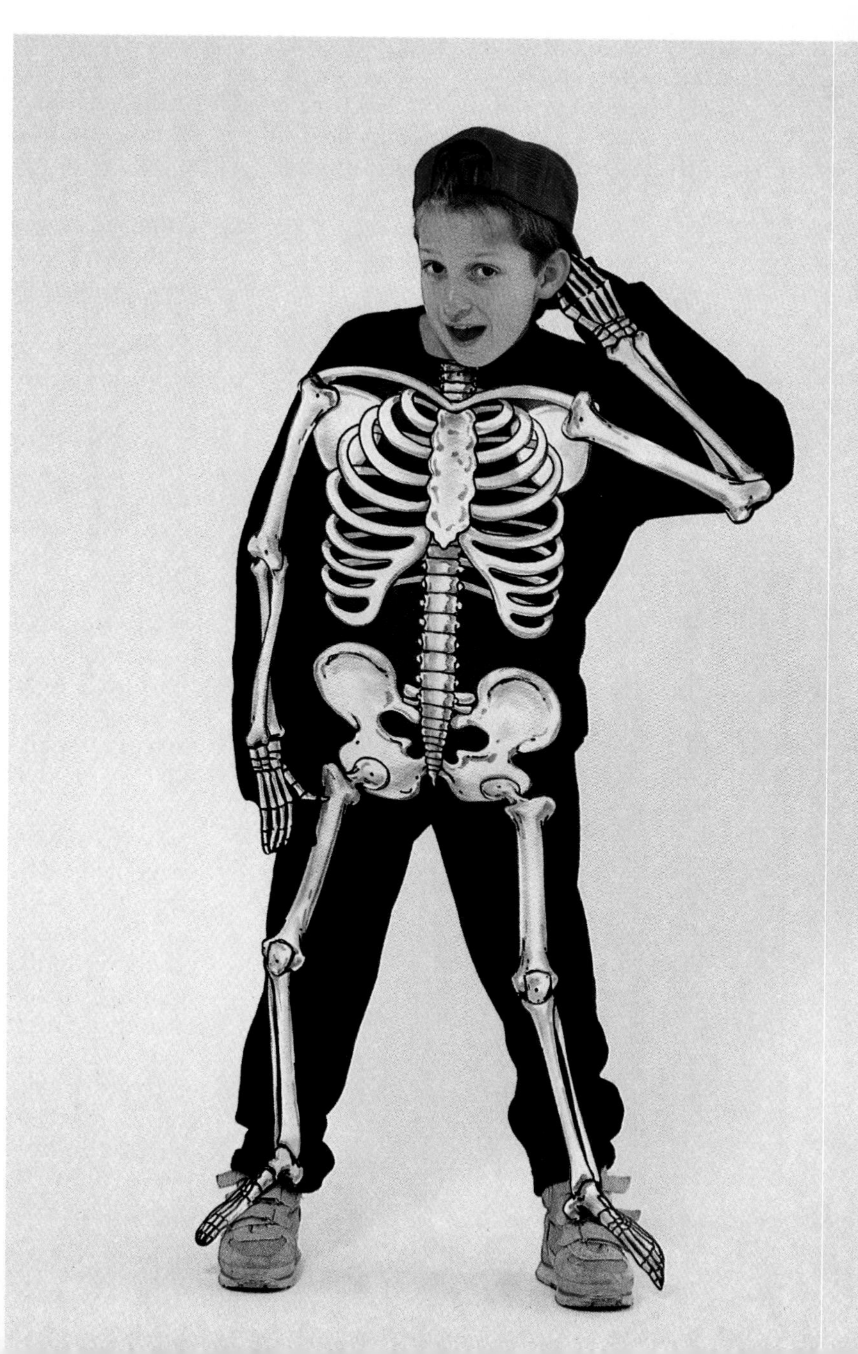

People have long known how to build those costumed dolls, called marionettes, that move like humans when one pulls on strings attached to their limbs. The limbs are each made of three wooden rods. The rods are hinged at shoulder or hip, elbow or knee, and wrist or ankle. The marionette's strings are tied to the head, hands, and feet in such a way that pulling on the strings moves the marionette in a more or less lifelike way (Figure 8-1).

What we are looking at is the marionette's skeleton, the structural framework that holds its clothes in place. It also allows the puppet to move, for it is a system of hinged levers pulled to and fro by the puppeteer's hand on the strings.

The human **skeleton** is also a structural framework and a system of hinged levers, but it is much more complicated. The marionette's skeleton has only 14 pieces. The human skeleton has 206 pieces, the **bones,** and it is moved not by a few strings but by hundreds of muscles (discussed in Chapter 9) controlled by the central nervous system (the brain and spinal cord, discussed in Chapter 15). Yet there remain two great similarities: It is on the skeleton that we wear our "clothes"—our muscles, organs, and skin—and the muscles, like the marionette's strings, act by pulling on the skeleton and swiveling the bones around the hinges or joints that link them to their neighbors.

FIGURE 8-1 A marionette, like a human, has a skeleton. In addition, its strings serve the same function as the human's muscles.

THE SKELETON

The skeleton is divided into two major sections: the axial and the appendicular skeletons (Figure 8-2). The **axial skeleton** consists of all the bony elements that form the body's central axis, including the skull, vertebral column, and rib cage. The **appendicular skeleton** consists of the bones of both the upper and lower appendages and the bony elements used to attach them to the axial skeleton. Thus, the appendicular skeleton includes the bones of the arms, wrists, hands, and shoulder girdle. It also includes the bones of the legs, feet, and pelvic girdle.

Bones can be classified by shape as well as by location in the body (Figure 8-3, p. 162). **Flat bones** are thin plates. They serve as armor for the delicate brain and chest organs, and they provide broad areas for the attachment of muscles. They include the bones of the skull, the scapulae (shoulder blades), the ribs, the sternum (breastbone), and the pelvis.

Long bones, designed to bear great strain, are much longer than they are wide and have a shaft, or **diaphysis,** and two ends, or **epiphyses.** They are the bones of the arms and legs.

Short bones are more blocky in shape and, in fact, they include the "knucklebones" used in times past as dice. They are the bones of the wrists and ankles.

Irregular bones have shapes that fail to fit these categories. The best examples are the bones of the vertebral column, the vertebrae, which add wings and buttresses to a basic block shape. These elaborate projections increase the surface area for attaching muscles, tendons, and ligaments and for strengthening the connections between adjacent bony elements.

In addition, there are two bone types that vary in number and location. *Wormian,* or *sutural, bones* are small flat bones that sometimes lie between the major bony plates of the skull. The number of sutural bones varies from individual to individual. *Sesamoid bones* are small rounded bones usually

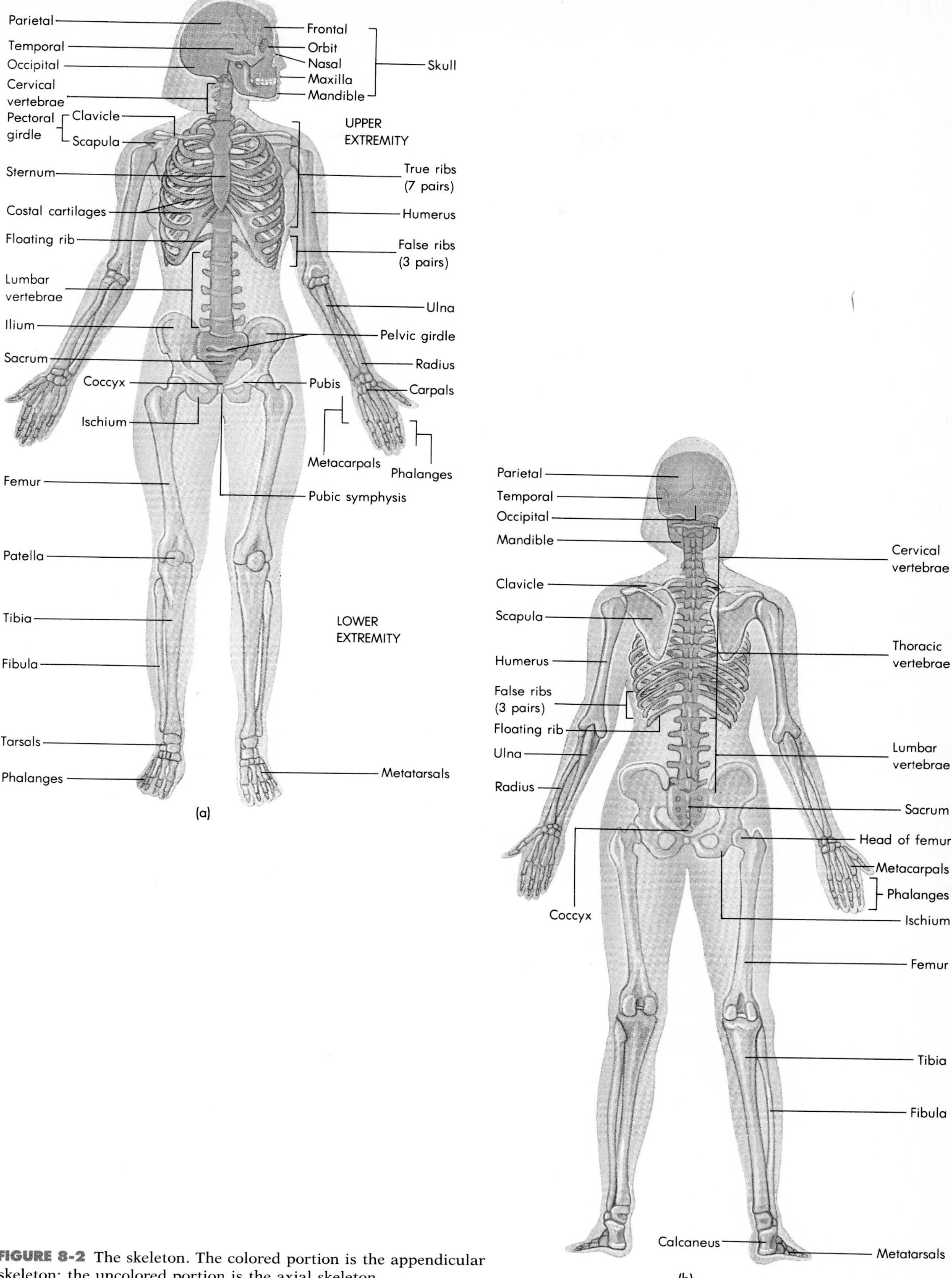

FIGURE 8-2 The skeleton. The colored portion is the appendicular skeleton; the uncolored portion is the axial skeleton.

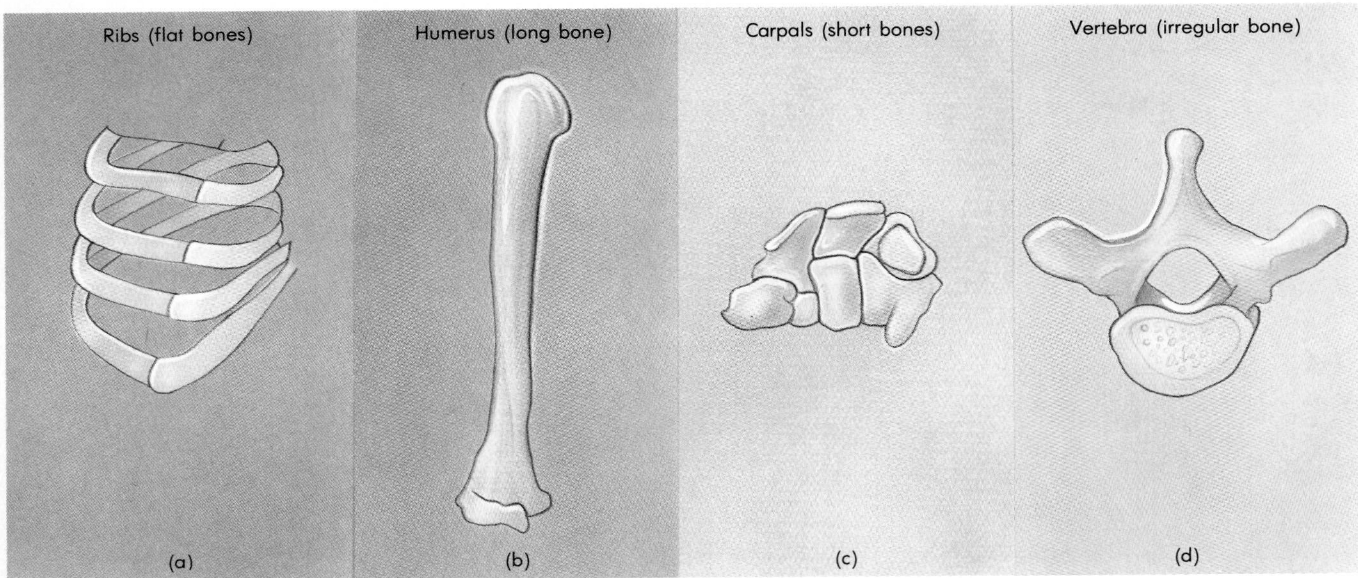

FIGURE 8-3 There are several categories of bones based on their individual shapes.

found in joints, often embedded among the ligaments and tendons around them. Intriguingly, sesamoid bones may spontaneously appear and disappear at different times in a person's life. Their function seems to be to support joint movement along specific lines of motion. When the lines of motion change, as when you change your activities, they too change, disappearing where they are no longer needed and appearing where they are. There is one set of sesamoid bones that virtually everyone is born with; they are the **patellae,** or kneecaps.

Together, these bones compose the skeleton. Most of the flat and irregular bones are found in the central axial skeleton. All the long bones, most of the short bones, and some irregular bones are found in the appendicular skeleton of the limbs and the bones that tie the limbs to the axial skeleton.

The Axial Skeleton

The 80 bones of the axial skeleton include those of the **skull, vertebral column** (or **spine**), and **rib cage.** Collectively, the bones of the axial skeleton form the central superstructure (or "axis") of the human body. We will discuss each of the major axial elements in turn.

The Skull

The skull has two major regions: the cranial vault and the facial region. Tightly fused, curved bony plates form the very strong, almost spherical **cranial vault (cranium),** which houses and protects the delicate brain. The bones of the face support several of the sensory systems of the head including the visual, olfactory, and gustatory. They position these organs to facilitate their information-gathering activities and to make eating efficient (Figure 8-4, p. 165).

The cranium is formed by fusing eight (sometimes more) bony plates together with jagged immovable seams called sutures. The roof of the cranium is formed by three bones. The single **frontal bone** forms the forehead and the anterior roof of the cranial vault (Figure 8-4). The lower margin of the frontal bone flares out slightly to form the brow ridges above each eye socket. The posterior roof of the cranium is formed by an arching, flattened pair of **parietal bones** joined at the midline of the head by a suture. The posterior cranial floor is formed by a single **occipital bone.** The occipital bone contains the **foramen magnum,** the large hole through which the spinal cord passes. Slightly lateral to the anterior edge of the foramen magnum lie two rounded projections, the **occipital condyles,** where the base of the skull meets the top of the vertebral column.

The side walls and part of the internal cranial floor are formed by the two **temporal bones.** Tubular openings conduct sound into the skull, through the **external auditory meatus** into the apparatus of the inner ear, embedded deep within the temporal bones. The butterfly-shaped **sphenoid** forms the anterior internal floor of the cranial vault. In the well-protected center of the head, a saddle-shaped depression in the sphenoid, the **sella turcica** (literally, the "Turkish saddle"), holds the pituitary gland (the "master" gland of the hormone system, discussed in Chapter 13). The smallest of the cra-

Learning Focus

THE BONES OF THE SKELETON

Study Figure 8-2. Then make several copies of the drawings that follow. On one copy, identify and label as many bones as you can. Using colored pencils, color the bones of the axial skeleton in one color and the bones of the upper and lower appendicular skeleton in another. In the left margin, write all the bones of the axial skeleton; in the right margin, write the bones of the appendicular skeleton. When you are done, check your list against the labels in Figure 8-2. Repeat this exercise every time you are studying for a quiz or an exam.

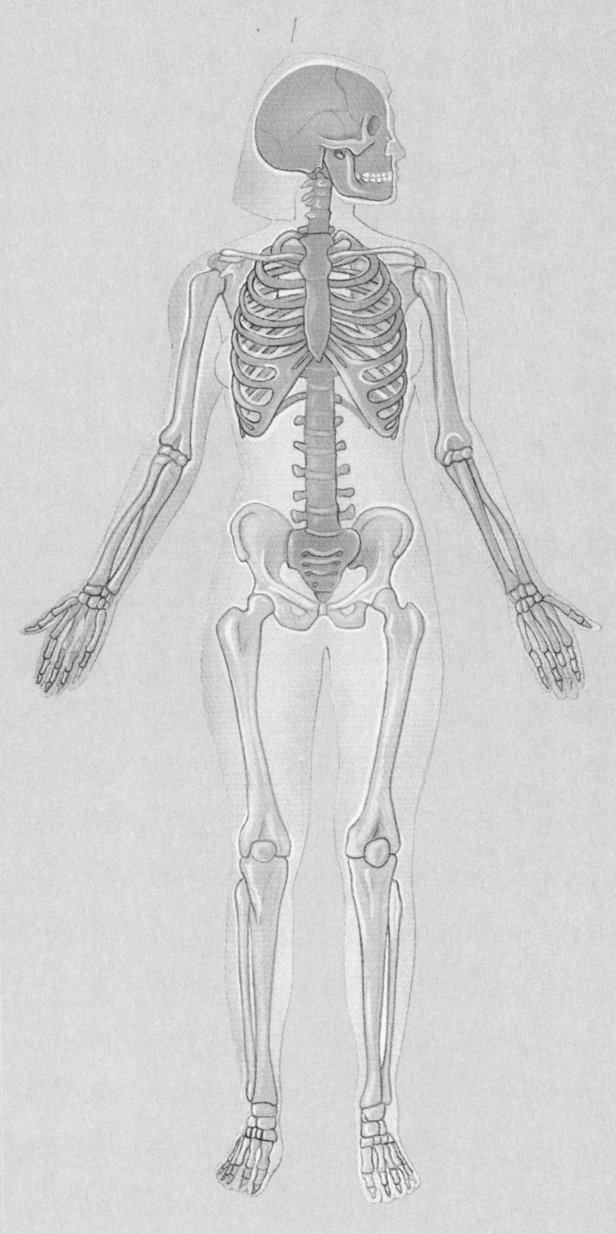

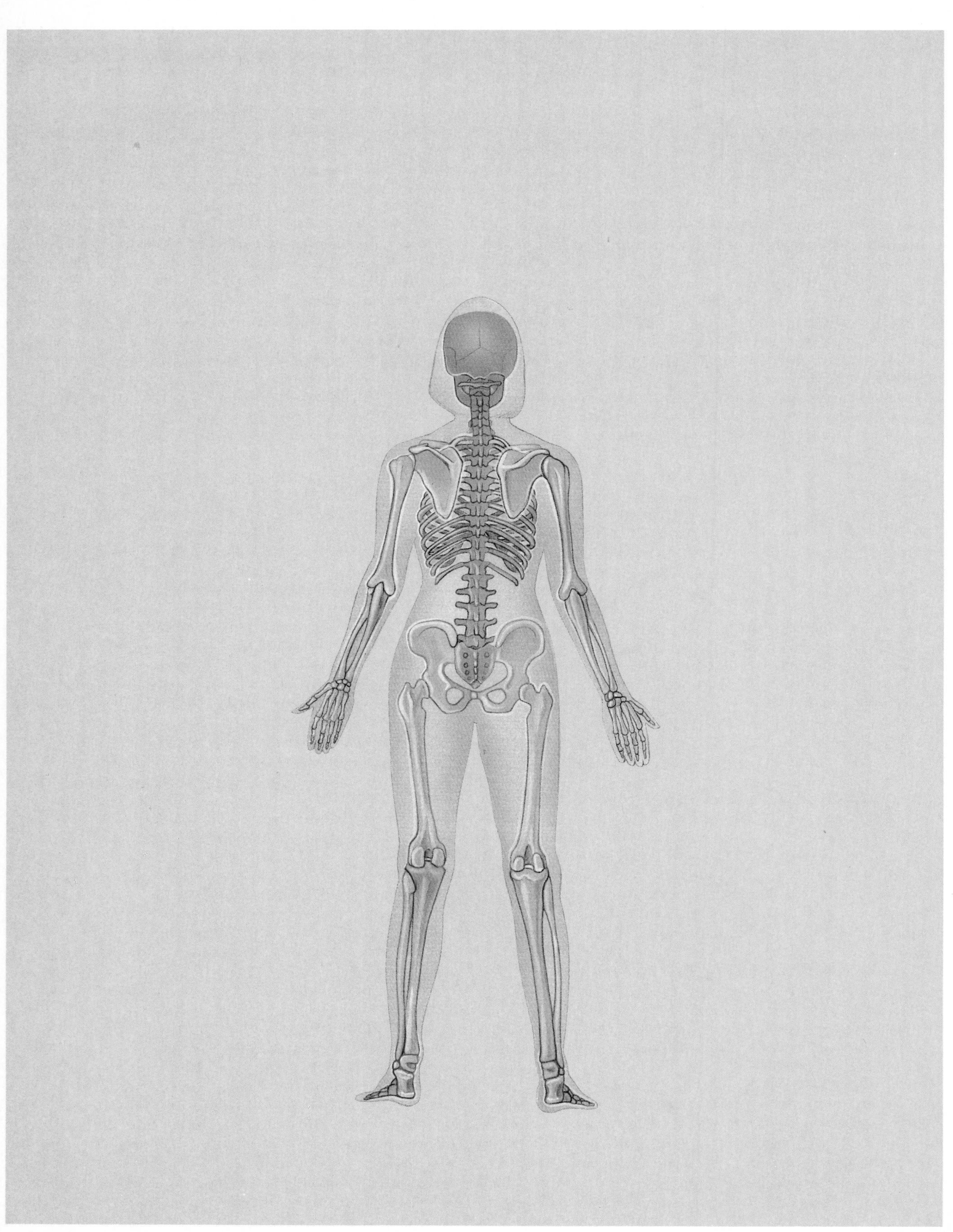

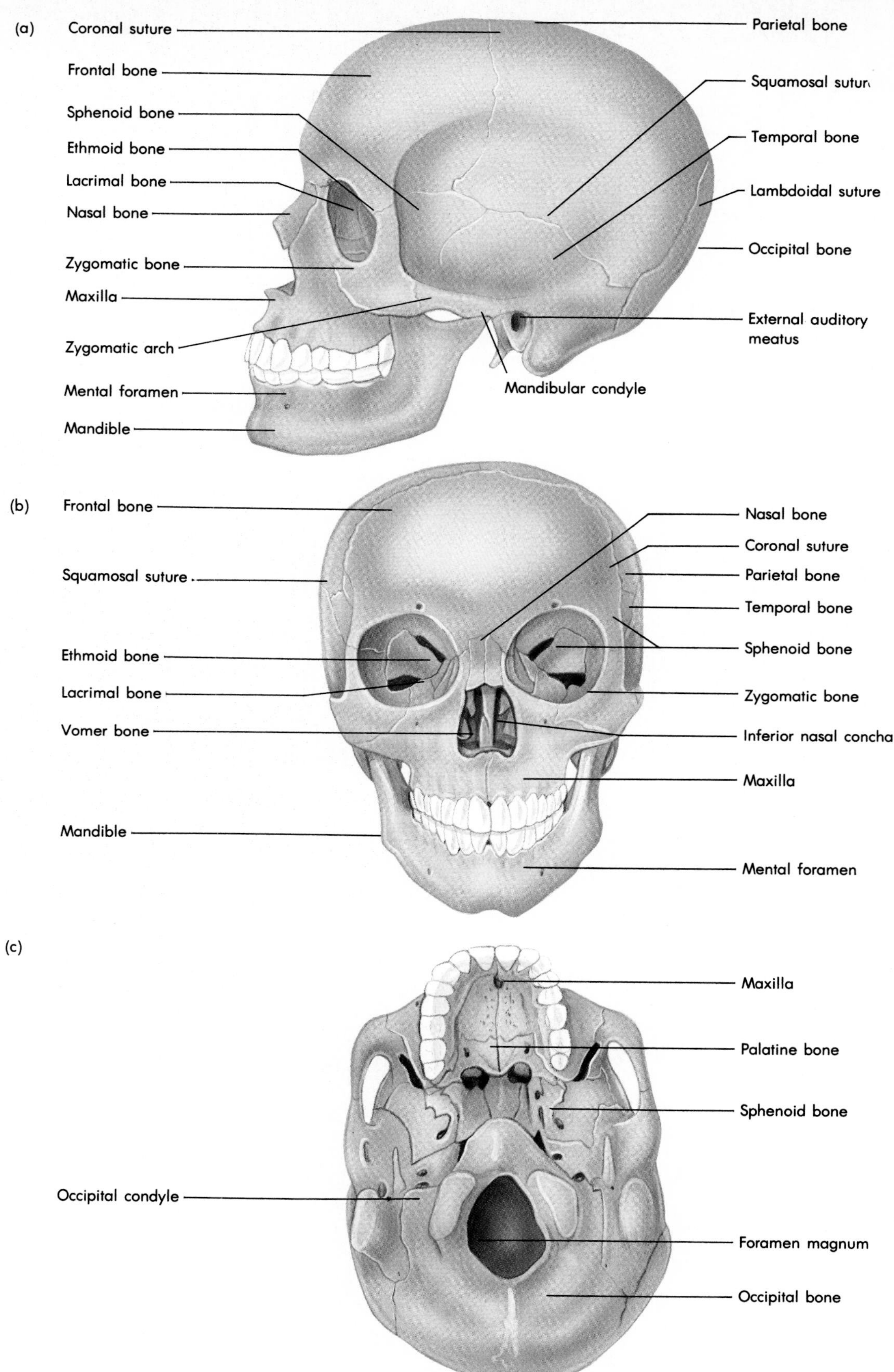

FIGURE 8-4 The skull: (a) lateral view; (b) anterior view; (c) inferior view. Do not be put off by the many Latin terms in the labels on the drawing. A *meatus* is a tubular passage running through a bone; the *external auditory meatus* is where the ear canal penetrates the temporal bone to reach the middle ear. A *fossa* is a depression. A *foramen* is an opening; the spinal cord leaves the brain case for the vertebral column at the *foramen magnum.* Other structures labeled here are discussed in the text.

nial bones, the **ethmoid,** separates the nasal cavity from the cranial cavity. Perforations in the **cribriform plate** of the ethmoid serve as nose-to-brain channels for olfactory nerve fibers.

In the fetal skull, the bones of the cranium have not fused at their suture lines. The spaces between the bony plates are membranous areas called **fontanels** (or soft spots) (Figure 8-5). This incomplete development of the skull at the time of birth allows the bony plates of the cranium to shift, changing the skull shape to facilitate its passage through the birth canal. These soft areas of the skull eventually ossify, and all are closed by the second birthday.

Bones of the Face

Fourteen bones form the superstructure of the human face. They support the sensory apparatus of the head and the muscles responsible for the wide range of human facial expressions. They also provide attachments for the muscles of chewing.

The bridge of the nose is formed by a pair of **nasal bones** (2–3 centimeters long) that are fused at the midline. A flexible, cartilaginous "bumper" extends from the end of the nasal bones out to the tip of the nose. The nasal cavity is divided into a left and right chamber by a vertical partition, the

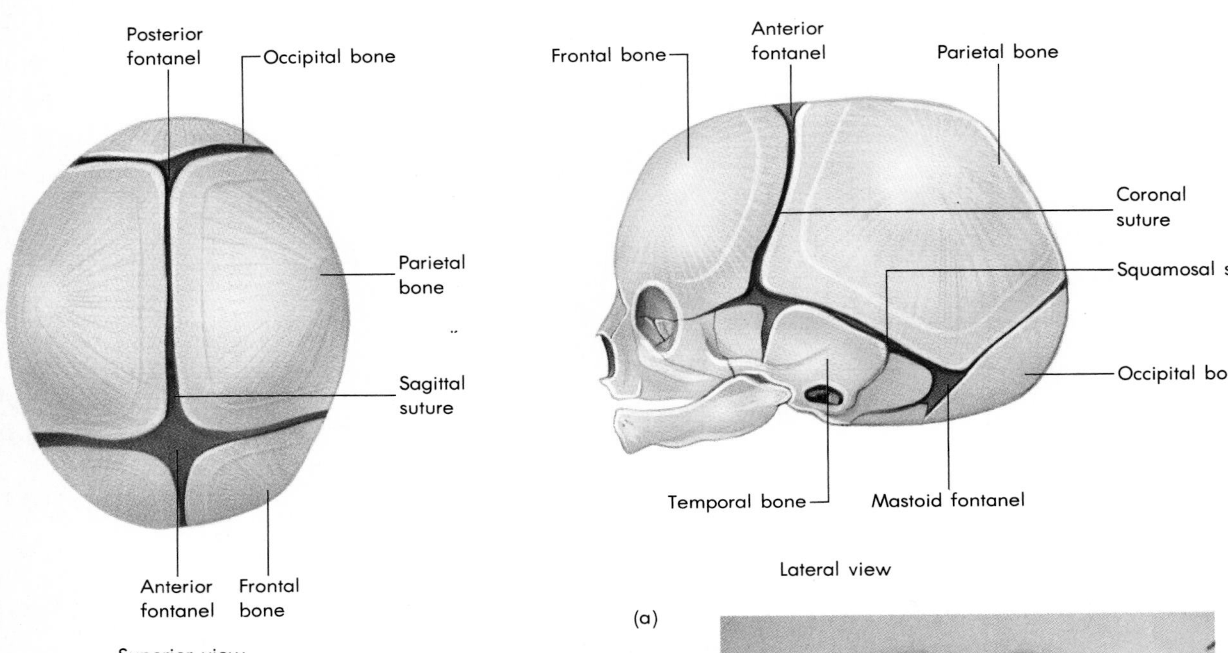

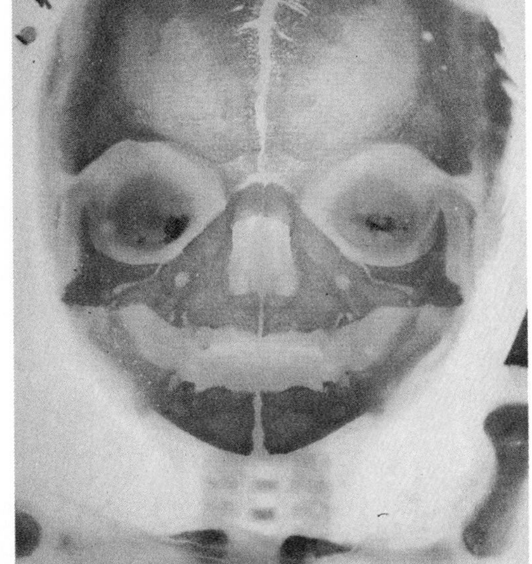

FIGURE 8-5 The fetal skull. (a) Major fontanels in the fetal skull. (b) X-ray view of the human fetal skull (frontal view) showing the extent of ossification. The fontanels allow the cranium to change shape during birth.

nasal septum (or **vomer**). The lateral walls of the nasal cavity are formed by downwardly curved, shelflike plates, the **inferior conchae.** In life, the conchae and the vomer are covered with mucous membranes that trap some airborne dust as inhaled air swirls over them. The membranes also warm very cold air to body temperature and humidify dry air.

The two cheekbones, the **zygomatic bones,** support the face immediately below, and including, the lower margins of the eye sockets. An extension of each zygomatic bone curves back to join a similar forward-reaching projection from the temporal bone and form the zygomatic arch. The tiniest facial bones, the **lacrimal bones,** are located near the corners of the eyes, just inside the medial margin of the eye socket. Each lacrimal has a groove (a tear duct) through which tears drain from the eye into the nasal cavity.

The upper jaw is formed by two bones, the **maxillae,** that are fused at the midline. In a sense, the maxillae form the hub of the facial skeleton, since most other facial bones (except the lower jaw) are joined to them by way of sutures. They fold back as the anterior portion of the bony plate—the hard palate—that forms the roof of the mouth. The posterior region of the bony palate is formed by the two **palatine bones.**

In some babies, the maxillae fail to fuse at the midline. The result is a **cleft palate,** in which the oral cavity is not fully separated from the nasal cavity. If the failure to close extends from the front of the mouth up to the nasal cavity, the result is a **harelip.** Today, these conditions are rarely seen in adults because the techniques for their surgical repair are so successful.

The lower jaw is the **mandible.** It is connected to the rest of the skull by way of the hingelike temporomandibular joint. The muscles for chewing extend from the sides of the skull to this bone.

The Vertebral Column

The vertebral column, backbone, or spine extends downward from the base of the skull (the foramen magnum). It consists of a stack of **vertebrae** through which the spinal cord passes, protecting it from injury. Occasionally, people are born with vertebrae that have not fused properly on their dorsal side, leaving them open toward the back. Such a condition is called *spina bifida.* The opening may even extend through the skin, leaving the spinal cord exposed to the air and infection. Even in cases where the vertebral development is incomplete but the skin is intact, the cord is still vulnerable to physical injury.

As shown in Figure 8-6, the vertebral column is composed of 26 bones, 7 **cervical vertebrae** in the neck, 12 **thoracic vertebrae** in the chest region, 5 **lumbar vertebrae** in the small of the back, 1 **sacrum** at the back of the pelvis (formed by the fusion of 5 *sacral vertebrae*), and 1 **coccyx** (formed by the fusion of 4 *coccygeal vertebrae*). The sacrum joins the backbone to the pelvis. Because of the great stresses produced when the leg muscles contract, the elements of the sacrum are fused into a block to reduce their vulnerability. The coccyx is the vestige of a tail that we carry tucked into our pelvic cavity and for that reason is often referred to as the tailbone.

Between the unfused vertebrae above the sacrum are the **intervertebral disks,** pads of fibrous cartilage that absorb shock and help make the vertebral column flexible. The combined thicknesses of the disks account for about a quarter of the length of the backbone. Over a lifetime, the disks naturally compact; that is one of the reasons your parents seem to be getting shorter (the other is that you got taller!). Each day there is also a settling of the disks. You are usually taller in the mornings than you are at night. Periods of rest or sleep, when you take the pressure off your back, allow the disks to rebound, making you taller by the next morning.

Excessive strain, especially in the lumbar region, can make the disks bulge out of shape as *herniated* or *slipped disks.* The condition can be very painful, for the bulge can press on the spinal nerves where they leave the vertebral column through the **intervertebral foramina.** If the damaged or out-of-place disk presses upon the sciatic nerve, the pain (referred to as *sciatica*) can extend down the leg as well. The displaced disk can also press on the cord itself, destroying some of its tissue and hence some of its ability to transmit nerve signals carrying sensory information or motor commands. Treatment stresses bedrest and, after recovery, exercise. Surgery is a last resort.

---

1. Why might a protein-digesting enzyme help relieve the pain of a slipped disk?
2. Why do you imagine the five sacral vertebrae are fused?

Viewed from the side, the curved nature of the vertebral column becomes obvious. It curves forward in the cervical and lumbar areas and backward in the thoracic and sacral regions (Figure 8-6). The curves add strength and flexibility and

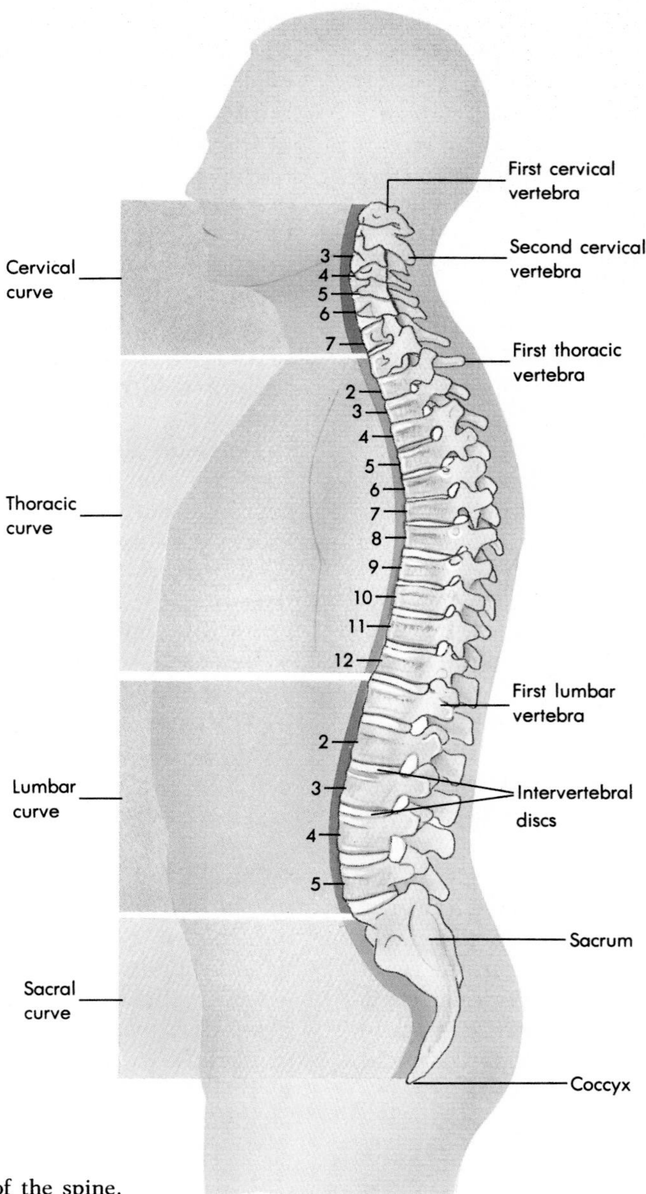

FIGURE 8-6 Normal curvature of the spine.

absorb shocks transmitted to the backbone during walking and running. Poor posture and diseases such as osteoporosis, polio, rickets (vitamin D deficiency), and tuberculosis can alter the curvatures of the spine. The most common abnormal curvature is **scoliosis,** a sideways spinal curve, often due to imbalanced muscular tensions on the two sides of the spine. Treatment may require corrective exercises, braces, or surgery. An experimental treatment involves correcting the imbalanced muscle tensions by electrically stimulating the back muscles. **Kyphosis** (hunchback) is an exaggeration of the thoracic curve. **Lordosis** (swayback) is an exaggeration of the lumbar curve.

Vertebral Structure

The vertebrae vary in shape and size along the length of the spine, but they all share certain features (Figure 8-7). Each one has a *body,* a weight-bearing cylinder of bone cushioned between the disks of adjacent vertebrae. The bodies of the vertebrae line up in the vertebral column and, along with the cushioning intervertebral disks, can withstand compressions many times greater than the body weight. As a result, the backbone rarely suffers serious damage as a consequence of the jumps and collisions associated with our normal athletic pursuits.

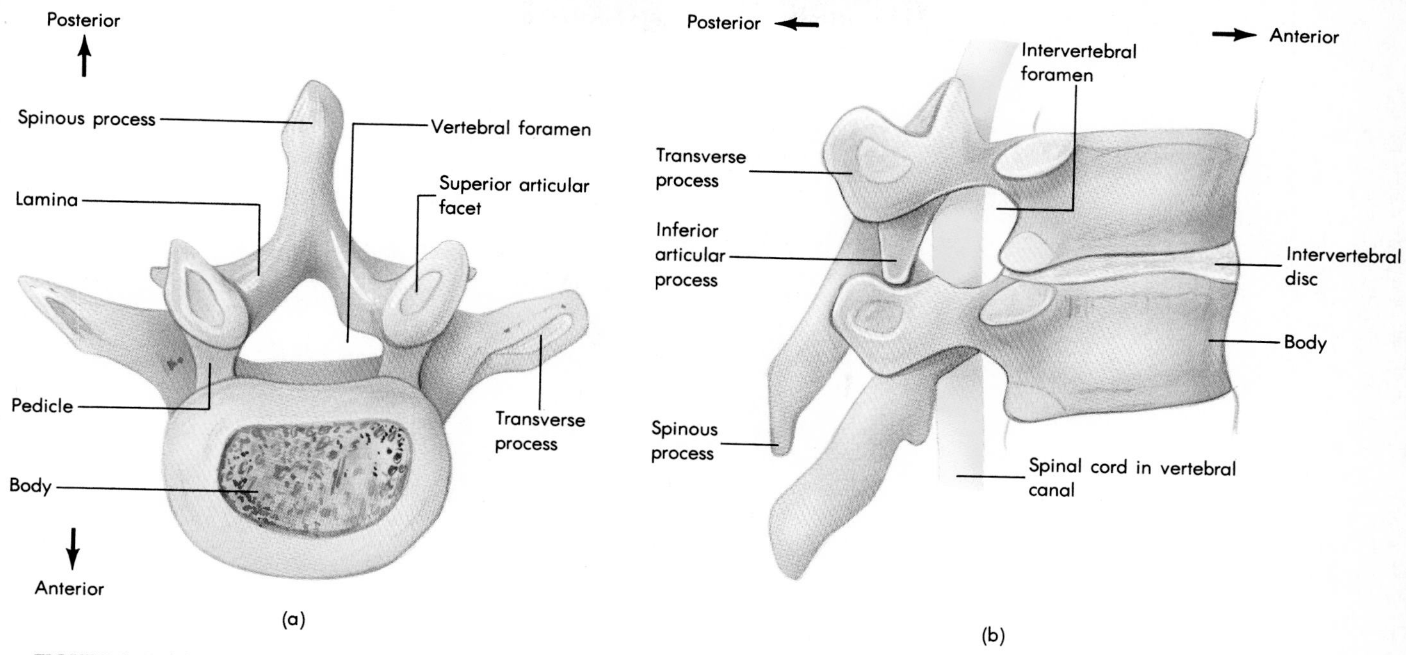

FIGURE 8-7 (a) Superior view of a thoracic vertebra. (b) Lateral view of two thoracic vertebrae as seen in the spinal column.

Extending behind the body of each vertebra is the *vertebral arch,* enclosing and sheltering the spinal cord as it runs through the *vertebral foramen.* The bony outer walls of the vertebral arch are formed by the *lamina* and the *pedicles.* The lamina forms the "roof" of the arch projecting over the dorsal surface of the spinal cord. It extends from the *spinous process* near the midline to the *transverse processes* on each side. The pedicles protect the sides of the spinal cord and connect the vertebral body to the transverse processes. From the arch, *articular processes* extend forward and backward to provide the pivots between the vertebrae. The articular, spinous, and transverse processes are the sites of attachment for many muscles of the back as well as for connective tissue bands, or ligaments, that help hold the vertebral column together.

The 12 thoracic vertebrae have additional features, including the *facets* on the tips of the transverse processes and on the sides of each vertebral body. These facets are where the ends of the 12 pairs of ribs meet the vertebrae and pivot when we breathe (see Chapter 11).

The Rib Cage

The ribs form a protective cage around the vital organs of the chest by connecting to the thoracic vertebrae in back and curving around the sides of the chest to connect to the **sternum,** or breastbone, in the front (Figure 8-8). The sternum has three recognizable parts; the uppermost, shield-shaped portion is called the *manubrium;* the dagger-shaped *body* of the sternum receives attachments from most of the ribs. The lowermost area of the sternum, the *xiphoid process,* is a flexible cartilaginous flap in children; it eventually ossifies in adults.

The upper seven pairs of ribs, the **true ribs,** are linked directly to the sternum by strips of hyaline **costal cartilage.** These cartilaginous attachments provide a flexibility important for breathing, act as shock absorbers, and reduce the chances of ribs breaking when the chest wall receives a blow. The next five pairs, the so-called **false ribs,** are not tied directly to the sternum. Ribs 8–10 have costal cartilages, but the cartilages merge with that of the seventh rib. Ribs 11 and 12 have no costal cartilage at all and do not attach to the sternum—they are the **floating ribs.**

The Appendicular Skeleton

Certain animals possess only an axial skeleton. Most snakes, for instance, have only a skull, spine, and ribs. Most animals, including humans, tack on two pairs of additional appendages—the forelimbs (arms, in our case) and the hindlimbs (legs). The bones of the arms and legs and the bony shoulder and pelvic girdles that attach them to the axial skeleton form the appendicular skeleton.

Arms and legs share a basic design. Each limb is

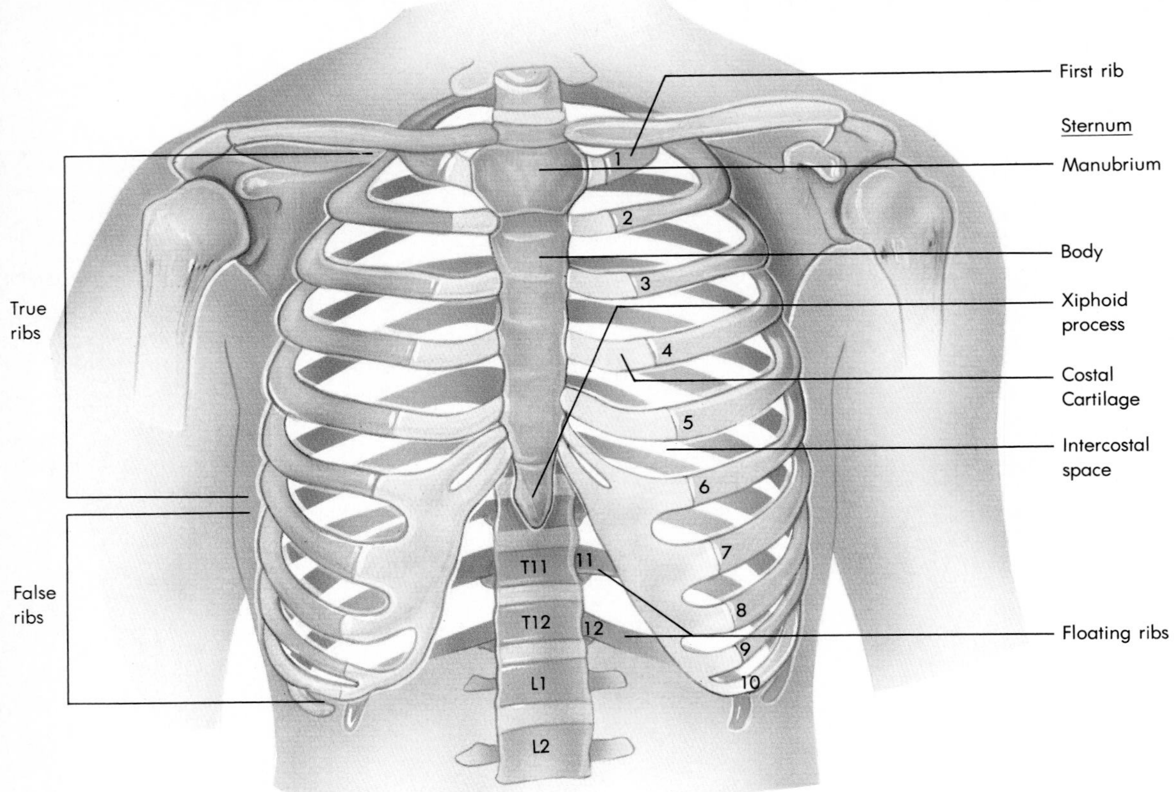

FIGURE 8-8 The rib cage (anterior view).

connected to the axial skeleton by a **girdle** of bones; the pectoral girdle connects the arms to the rib cage, and the pelvic girdle attaches the legs to the sacrum at the lower end of the spine. In addition, the upper elements of the arms and legs consist of a single bone. There are two bones in the lower part of each limb. Distally, the arm bones attach to the flexible joints of the wrist, while those of the leg join with the ankle elements. At the end of each appendage are the phalanges—the bones forming the fingers and toes.

The Pectoral Girdle and Arm

The **pectoral girdle** consists of the **scapulae** (shoulder blades) and **clavicles** (collarbones) (Figure 8-9). Each clavicle is a curved rod that braces a scapula against the top of the sternum. The triangular scapula does not directly touch any part of the axial skeleton. Rather it is tied to the axial skeleton by muscles and ligaments, which give it a great deal of freedom of movement. Some four-legged animals, such as antelopes, that require still more freedom of foreleg movement than humans have eliminated the clavicle entirely and rely on muscles and ligaments alone to support the scapula.

The narrow (shoulder) end of each scapula forms a socket for the upper end of the upper arm bone, the **humerus.** At the elbow, the humerus joins with the two bones of the lower arm, the **radius** and **ulna.** At the junction between the arm and the hand, these bones join, in turn, with the eight **carpals** of the wrist. The small size and the complex array of joints formed by the carpals is responsible for the great flexibility of the wrist. The five bones covered by the fleshy portion of the hand are the **metacarpals.** The elements forming the fingers are the **phalanges.** There are two phalanges in each thumb and three in each of the remaining fingers. (The pattern repeats in the toes.)

The Pelvic Girdle

The arrangement of the elements of the pectoral girdle suggests that a solid attachment of the appendicular skeleton to the axial skeleton is not essential. However, the **pelvic girdle** is much more solidly attached. It consists of two **pelvic bones** (the

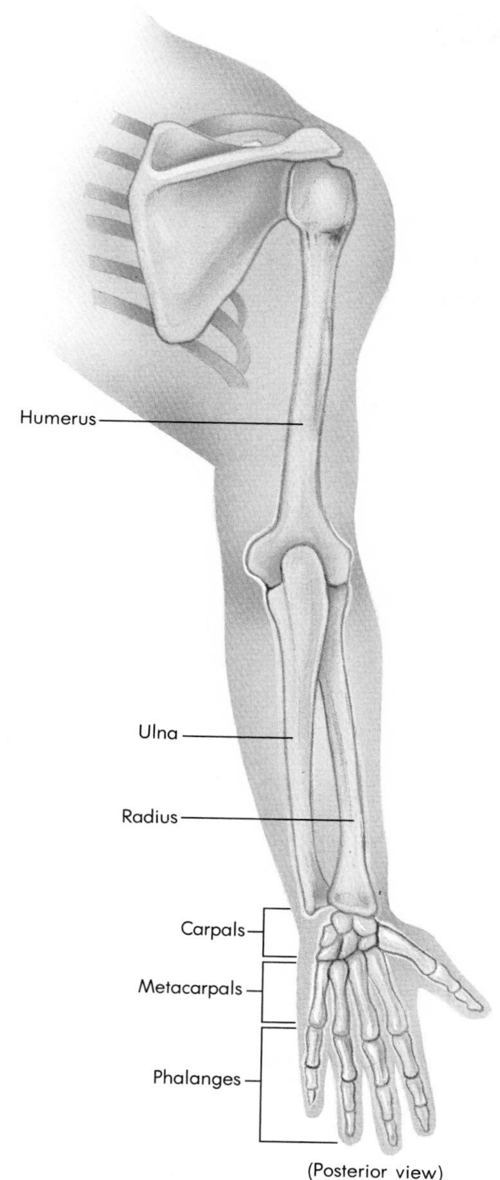

FIGURE 8-9 The pectoral girdle, arm, and hand (posterior view).

os coxae), which form a basin that supports the organs of the lower abdomen (Figure 8-10). The pelvic bones are firmly connected to the sacrum by fibrous connective tissue. They arch around and meet in front at the **pubic symphysis.**

Each pelvic bone is formed in the fetus by the joining of three bones, the *ilium*, the *ischium*, and the *pubis*. The three sections of each pelvic bone meet in the cuplike socket called the **acetabulum,** which receives the rounded head of the thighbone, or femur. By adolescence the fusion of these pelvic elements is complete. The upper flaring region of the pelvis is the ilium. The two ischia support our weight when we are sitting. The left and right pubes fuse at the midline, forming the pubic symphysis, completing the front of the pelvic girdle.

The elements of the pelvic girdle are more rigidly joined in the male than in the female. This rigid configuration translates to a more efficient transfer of energy during athletic activities, like running and jumping. However, the greater flexibility of the female pelvis supports the processes of childbirth. During labor and delivery, the ligaments that hold the pelvic elements in place soften and allow the pelvis to spread, facilitating passage of the fetus through the birth canal.

In the male, the angle the pelvic bones form below the pubic symphysis is less than 90 degrees; in the female, the angle is more than 90 degrees (Figures 8-10 and 8-11). The flare of the male ilium is usually narrower, and the female pelvis is usually lighter in weight and density. These differences, and others, allow archeologists, anthropologists, and police detectives to distinguish skeletons of males from those of females (sometimes).

The Leg

The **femur** (thighbone) of the upper leg is perhaps the largest and strongest bone in the body. The muscles of the thigh can exert tremendous forces on the skeleton when they contract. Just standing up from a squat causes the thigh muscles to exert over 2000 pounds of force. The leg bones thus have to be strong.

At the knee, the femur meets the bones of the lower leg, the **tibia** (the shinbone) and **fibula** (the slender bone running down the outside of the leg). A sesamoid bone, the **patella** (kneecap), covers the knee joint and transmits the pull of some of the muscles of the thigh and lower leg across the joint. At the ankle, we find the **tarsal** bones, which also form the proximal portion of the foot. The bones located in the fleshy area of the anterior foot are the **metatarsals.** The toes, once more, are phalanges.

Because the tarsals are larger than the carpals and the metatarsals are more firmly knitted together by ligaments than the metacarpals, the foot is less flexible than the hand. Its distinguishing functional feature is its springiness, given it by the way its bones are arranged in two arches, the **longitudinal arch** and the **transverse arch** (Figure 8-12, p. 173). "Fallen arches" result when the ligaments and tendons that hold the foot bones in position are weakened by repeated and prolonged strain. As the foot no longer absorbs the repeated shocks of footfalls very well, chronic foot, leg, and back pain can result.

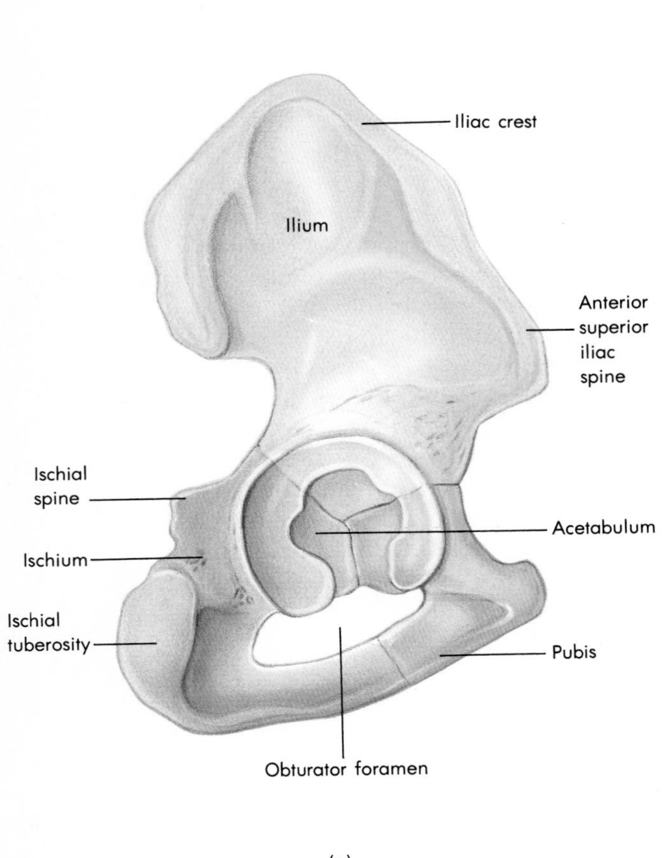

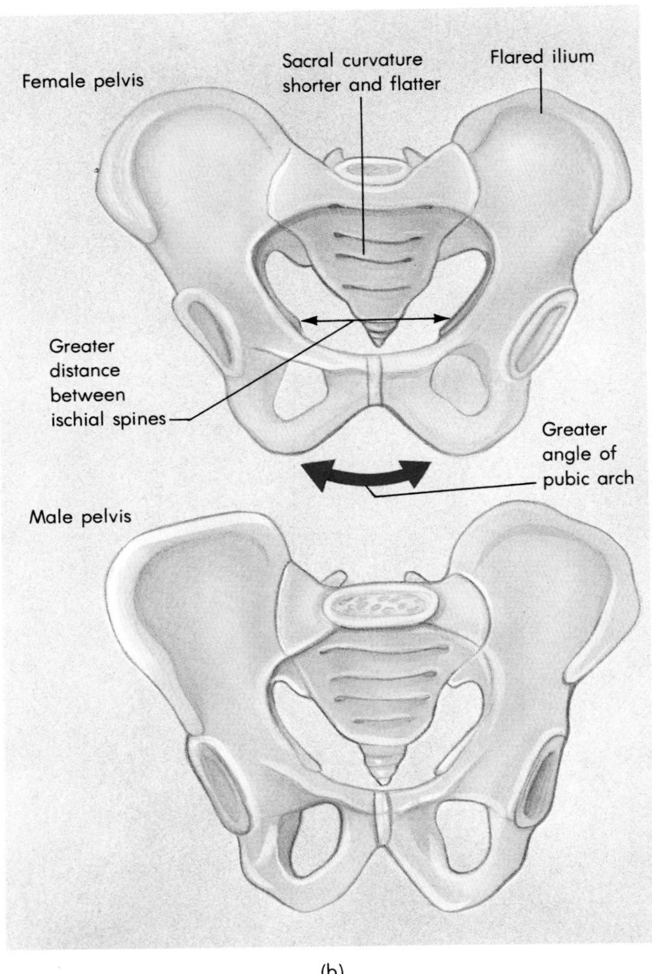

FIGURE 8-10 (a) The pelvis. (b) The differences between the male and female pelves.

1. What activities would you expect to strain the supporting tissues of the arches? What occupations would you expect to be subject to fallen arches?
2. How do the skeletons of the arms and legs resemble each other in structure?

BONE AS A LIVING TISSUE

The most obvious functions of bone are to support the body's weight against gravity, to shield certain body parts against blows, and to provide solid levers against which the muscles can pull. To do all these things, bone must be hard and rigid. Yet it cannot be so rigid that it is brittle; it must be flexible enough to bend before it breaks. It must also be adaptable, able to modify its own structure to meet changing demands on its strength.

Bone gains both its strength and its flexibility from its chemical composition. The main material of bone is a mixture of the mineral salts calcium phosphate and a calcium carbonate called hydroxyapatite. The mineral crystals are embedded in a matrix of collagen (protein) fibers that accounts for one-third of bone's weight. It is the mineral that makes bones hard and strong; the protein fibers let it bend under strain without breaking.

1. You can see the difference between the roles of mineral and protein in bone if, the next time you have chicken for dinner, you save two drumstick bones. Put one in a fire. When it has turned white, let it cool. Then pinch it between your fingers. What happens? Why?
2. Put the other bone in a jar of vinegar for a week. Then take it out and pinch it. Try to bend it. What happens? Why? (Remember that acid dissolves calcium phosphate and carbonate.)

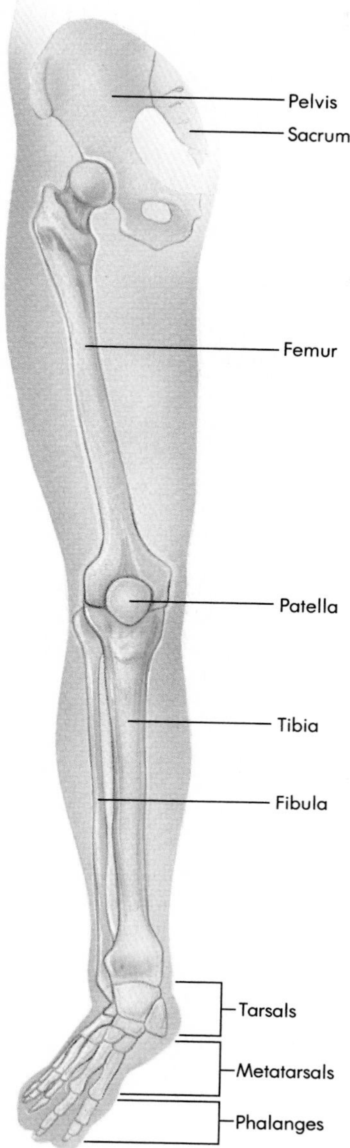

FIGURE 8-11 The pelvic girdle, leg, and foot (anterior view).

Bone is adaptable because it is not dead. It is as much a living tissue as muscle or liver. Like them, it is permeated by blood vessels. It has cells too, the **osteocytes,** which are embedded in tiny spaces, or **lacunae,** in the substance of the bone.

Because it is alive, bone can heal breaks and remodel its internal structure to meet changing stresses. It can also serve two additional functions. Bone is the source of blood cells, for in the interior of many bones (the sternum, vertebrae, and some others in adults) lies the **red bone marrow,** which generates red blood cells, some white blood cells, and blood platelets (discussed in Chapter 10). Bone is also the body's storage depot for calcium and phosphate, depositing these substances when they are plentiful in the diet and releasing them again when the body needs more than the diet can provide. The mineral of bone is thus replaced, bit by bit, many times in a lifetime.

The Structure of a Long Bone

Figure 8-13 shows the humerus as an example of a typical long bone. Its long, straight portion is the diaphysis, or shaft. Its two ends are the epiphyses. In growing bones, in the areas where the epiphyses are joined to the diaphysis, there is an actively growing plate of cartilage, the **epiphysial plate,** where most long-bone growth occurs. Growth occurs when these active sites generate additional cartilage, forcing the ends of the bones farther apart and lengthening the bone. The new cartilage is eventually replaced by bone. The outer ends of the epiphyses, where they meet other bones in joints, are covered with a thin layer of hyaline car-

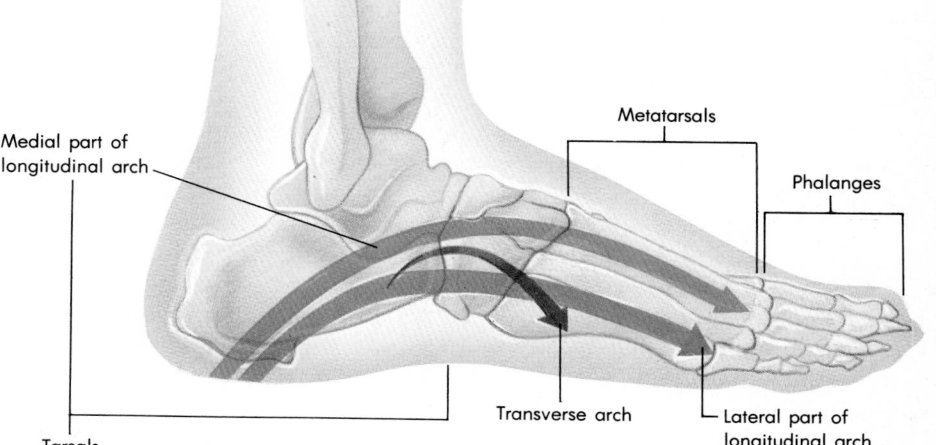

FIGURE 8-12 The arches of the foot.

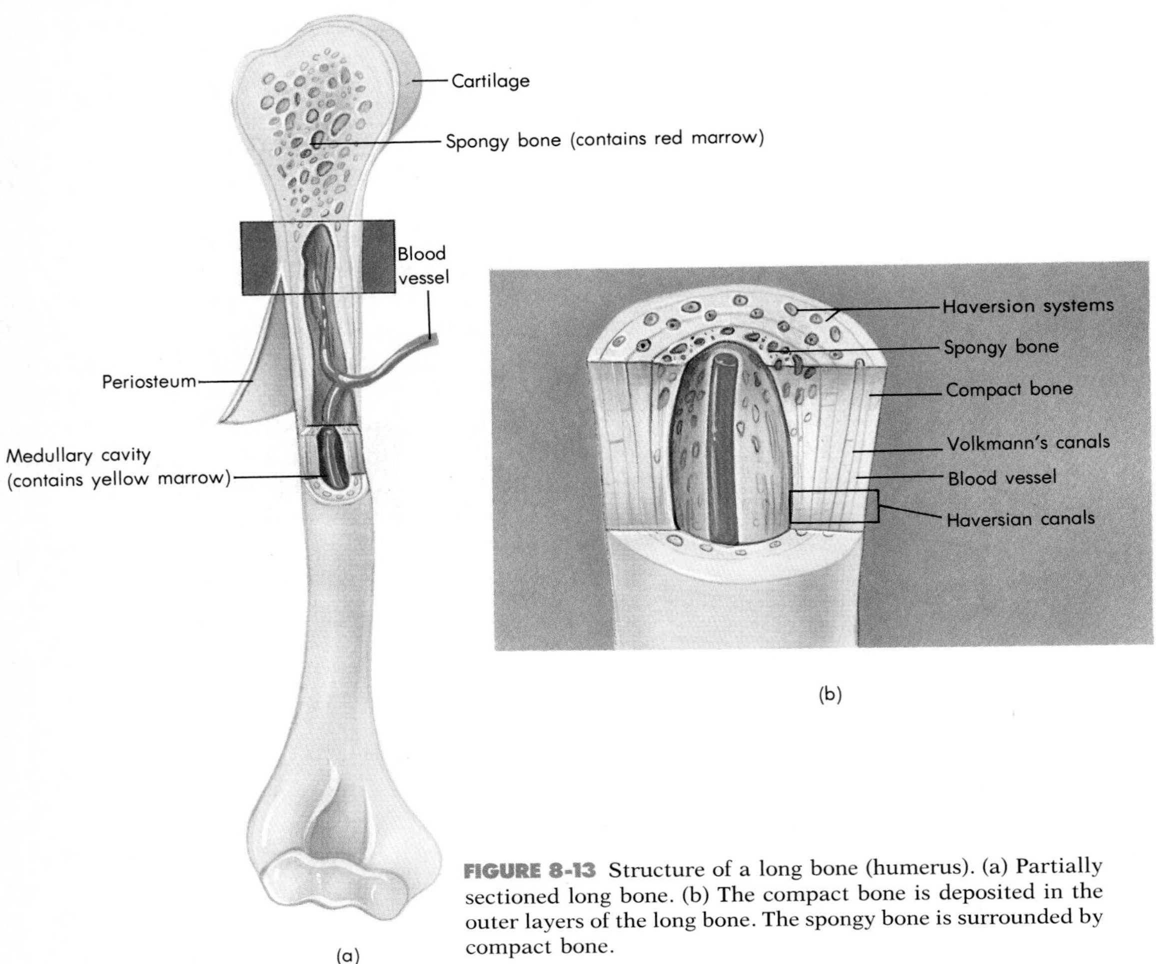

FIGURE 8-13 Structure of a long bone (humerus). (a) Partially sectioned long bone. (b) The compact bone is deposited in the outer layers of the long bone. The spongy bone is surrounded by compact bone.

tilage; this **articular cartilage** provides a virtually frictionless bearing surface for the joints. The surface of the bone, wherever it lacks a cartilage coat, is covered with a connective tissue membrane, the **periosteum.**

The diaphysis of a long bone is hollow. Its outer wall is composed of **compact bone,** a dense, hard layer of bone. The interior of a long bone contains a **marrow cavity,** which in children contains red bone marrow and is one of the sites where blood cells are produced. In adults, the red marrow has been replaced by *yellow marrow,* a fatty tissue that has lost its blood-cell-producing abilities. In some cases, the yellow marrow can be converted back to red marrow.

The epiphyses are also covered with compact bone, but more thinly. In them, and in the ends of the diaphysis, the interior is filled with **spongy bone.** Spongy bone consists of a network of strengthening, girderlike, bony plates and rods called **trabeculae** (Figure 8-13). Embedded in the network are numerous spaces that, in many bones, contain red marrow. Spongy bone also occupies the interior of flat, short, and irregular bones.

Microscopic Structure of Compact Bone

Under the microscope, compact bone proves to have an intricate structure dominated by concentric rings of bony tissue (Figure 8-14). Just beneath the outer surface, we can see several rings, the **circumferential lamellae,** that encircle the entire bone. Deeper in the bone, we see cylindrical structural units, the **Haversian systems.** In the center of each Haversian system is a *Haversian canal,* carrying nerves and the blood vessels that supply the bone cells with oxygen and nutrients and remove wastes. Transverse *Volkmann's canals* bring the blood vessels and nerves into the bone from the periosteum and link the Haversian canals to each other.

Around each Haversian canal lie the **Haversian lamellae.** The lacunae containing the osteocytes,

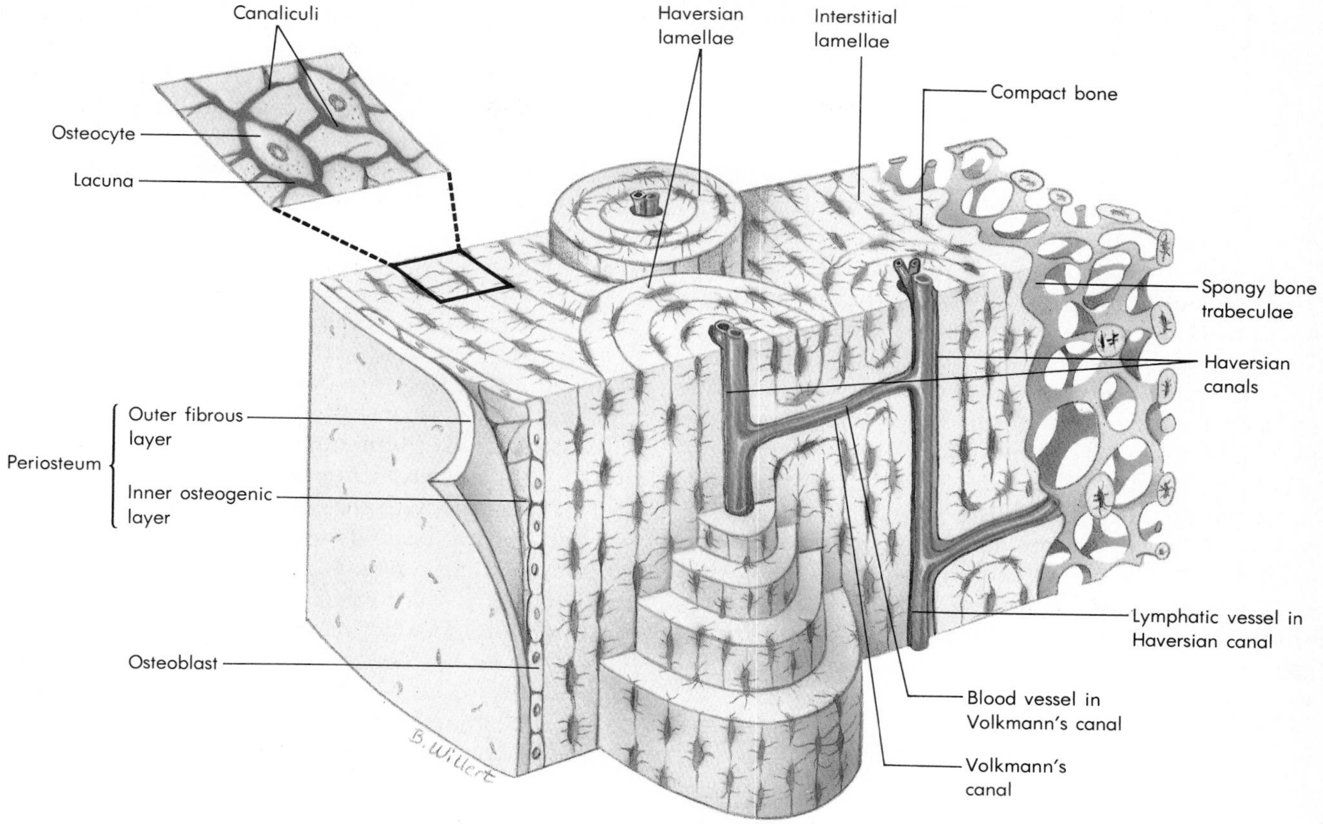

Canaliculi
Haversian lamellae
Interstitial lamellae
Compact bone
Osteocyte
Lacuna
Spongy bone trabeculae
Haversian canals
Periosteum
Outer fibrous layer
Inner osteogenic layer
Osteoblast
Lymphatic vessel in Haversian canal
Blood vessel in Volkmann's canal
Volkmann's canal
B. Willert

(a)

(b)

FIGURE 8-14 Structure of compact bone. (a) Drawing showing lacunae, osteocytes, Haversian canals, Haversian lamellae, and canaliculi. (b) Photomicrograph of compact bone.

inactive bone-forming cells trapped when they laid down the bone, are between the lamellae. Fine *canaliculi*, containing threadlike extensions of the osteocytes, link each lacuna to its neighbors and to the Haversian canal and provide a route for transport of nutrients and wastes.

Since the Haversian systems are cylindrical, they leave triangular spaces where they meet. These spaces are filled with **interstitial lamellae.** The layers of bone often look as if they were fragments of Haversian systems, and in fact they are. The Haversian systems are continually being destroyed and rebuilt as the body draws upon and replaces its stores of calcium and phosphate and as the bones adapt to changing stresses.

The periosteum, the sheet of connective tissue surrounding the bone, is the source of the osteocytes. Its inner *osteogenic* layer gives rise to the **osteoblasts.** These cells rest on the surface of the bone, where they secrete the protein matrix and minerals of new circumferential lamellae. When they are trapped by the bone they form, they become osteocytes.

The Formation of Bone

By the sixth week of a human embryo's existence, the skeleton exists as connective tissue. Where the long bones will later be are rods of hyaline cartilage in the shape of bones. The flat bones of the skull (and elsewhere) exist as membranes of fibrous connective tissue. By the time of birth, much of each rod and membrane will be bone, but the process of **ossification** (bone formation) will be far from complete. Much of the skeleton will still be cartilage and membrane, and the fetus will thus be flexible. This flexibility lets the fetus deform enough to fit through the birth canal. The incompleteness of the skeleton also permits rapid growth, especially of the brain, in the months and years immediately after birth.

Because the precursors of the bones take two forms—cartilage and membrane—there are two methods of ossification. **Intramembranous ossification** begins when osteoblasts migrate into the membranes that will become flat bones. These cells form clusters called *ossification centers,* where they secrete the collagenous bone matrix. Calcium salts promptly crystallize in the matrix, and the ossification centers become surrounded by bone. Bony regions appear in the prebone membrane, lengthen, and merge to form the trabeculae of spongy bone. Red bone marrow develops in the spaces, and the surrounding connective tissue becomes periosteum. Eventually, osteoblasts from the periosteum deposit a layer of compact bone on the surface, and the bone is complete.

At birth, the bones of the cranium are far from complete. Around the edges of each one remains a region of unossified membrane. The region is largest where several bones meet, as where the frontal bone (formed by the fusion of two bony plates) meets the parietal bones. Such a region is called a "soft spot," or fontanel. The fontanels begin to harden about two months after birth and are gone by the baby's second birthday.

Ossification of long bones follows a different pattern, that of **endochondral ossification** (Figure 8-15). It begins with the formation of a small model of the future bone from hyaline cartilage, surrounded by a membrane, the *perichondrium.* When the model is complete, a blood vessel grows through the perichondrium and into the cartilage near its center. The presence of this blood vessel stimulates some of the perichondrial cells to become osteoblasts, at which point we call the membrane periosteum. The osteoblasts begin to deposit compact bone as a collar around the cartilage bone model. At the same time, the cartilage cells in the center of the model, at the bone's **primary ossification center,** swell and burst. When they burst, they raise the pH of the extracellular material, which triggers the deposition of calcium salts in the cartilage.

However, the interior of the cartilage bone model does not become bone. The initial calcification of the cartilage cuts off the flow of nutrients to cartilage cells. The cells die, and large cavities develop in the cartilage. In time, the cavities merge to form the marrow cavity of the bone.

At the same time, the collar of compact bone is thickening and lengthening, and the cartilage model is growing at its ends, so that a zone of cartilage constantly remains beyond the reach of the primary ossification center. In time, blood vessels enter the ends (epiphyses) of the bone and provoke the formation of **secondary ossification centers** there. They act to fill the epiphyses with spongy bone.

The three areas of newly formed bone grow toward each other, but they do not meet (Figure 8-16, p. 178). The zone of cartilage narrows until it forms an epiphyseal plate, but cartilage continues to be formed between the primary and secondary ossification centers and the bone continues to lengthen. The process ends only after puberty; hormones released at that time shut down cartilage growth and

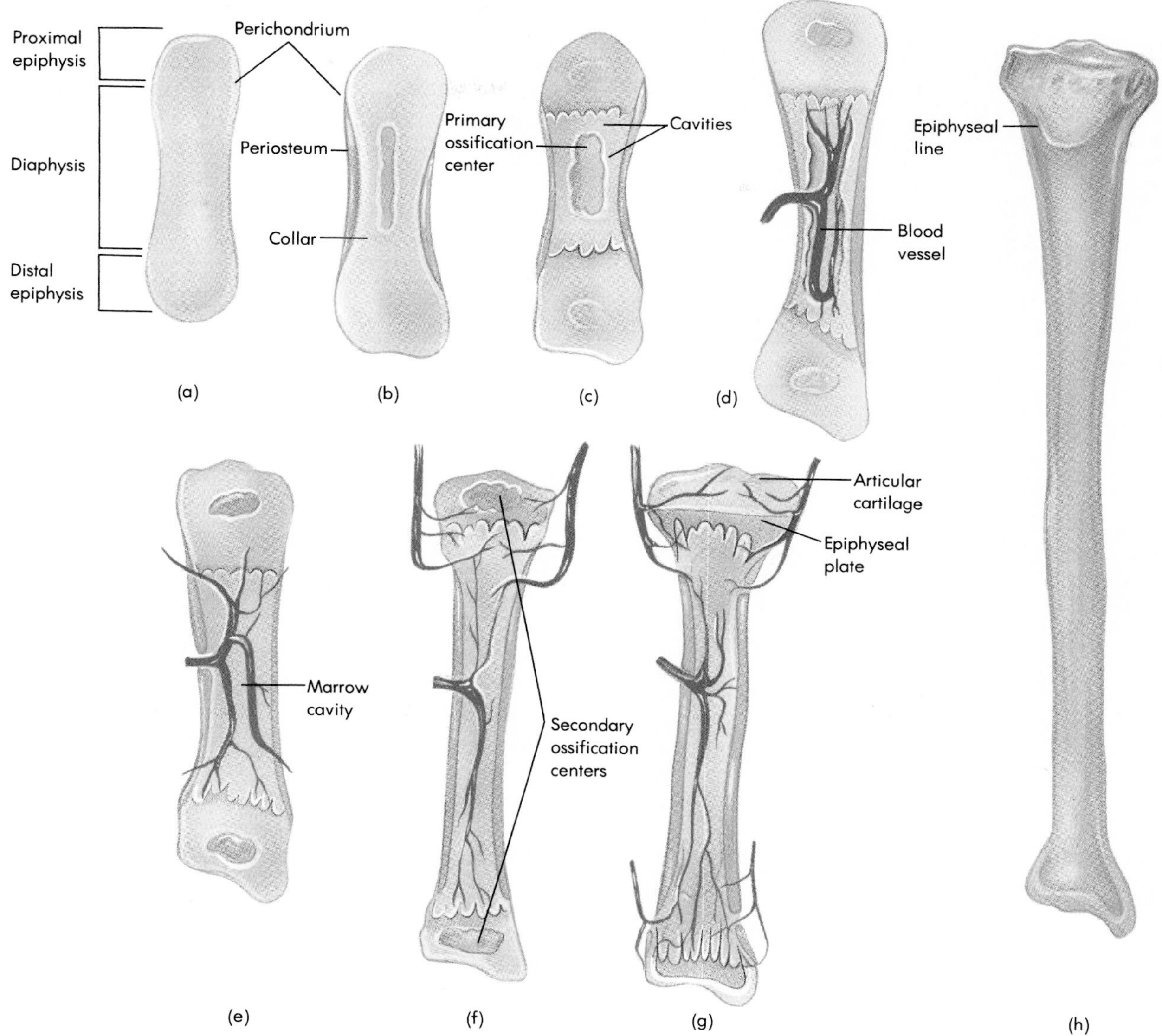

FIGURE 8-15 Endochondral ossification of the tibia (shinbone). (a) Cartilage forerunner to bone forms. (b) Collar develops. (c, d) Primary ossification centers develop and blood supply expands. (e) Marrow cavity develops. (f, g) Secondary ossification centers form, collar lengthens and thickens. (h) Epiphyseal line forms. See text for details.

allow the ossification centers to catch up with and overrun the epiphyseal plate.

Bone Remodeling

Bone formation does not end with the closure of the epiphyses and the fontanels. It continues throughout life, controlled by the interaction of osteoblasts and **osteoclasts,** large multinucleated cells formed from certain white blood cells (monocytes).

Because osteoclasts secrete substances that dissolve bone (Figure 8-17, p. 179), they are the mechanism by which the body withdraws calcium and phosphate from its skeletal reserves as needed (see Chapter 13). Normally, osteoclasts and osteoblasts match their paces of bone resorption and deposition, respectively, so that there is no net change in the body's amount of bone. In older people, however, resorption may come to exceed deposition in the condition we call **osteoporosis.** As the bones lose calcium, they weaken, often so much that they break easily. Many physicians believe that the best

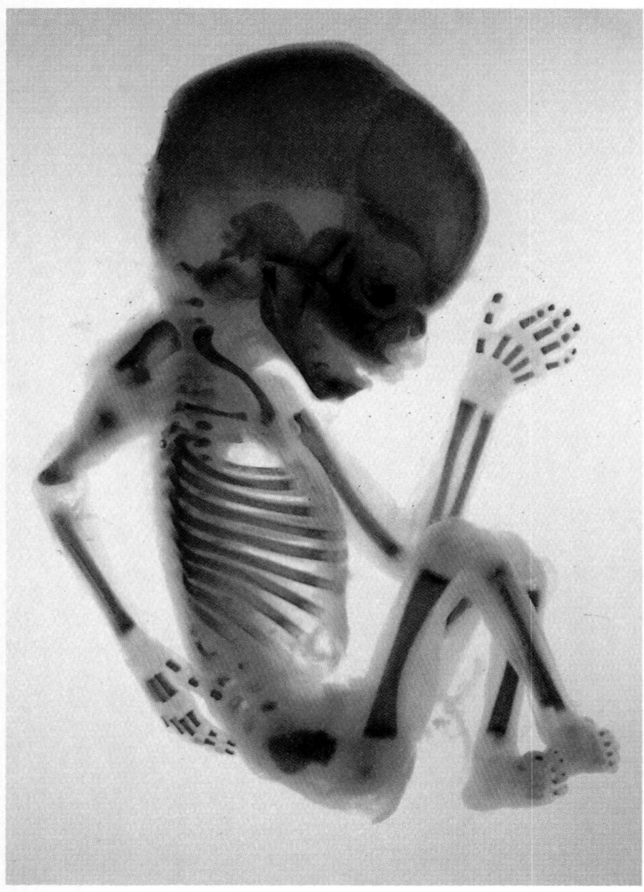

FIGURE 8-16 Fetal skeleton cleared and stained to show the extent of ossification. The darker areas show regions of ossification. The lighter areas near the joints are mainly cartilage and are very active sites of bone growth.

way to prevent or treat osteoporosis is to increase calcium intake by eating more dairy products or taking calcium supplements. However, recent research suggests that while increasing calcium intake increases calcium deposition in bone in children and adolescents, and hence increases bone density, it does not do so in adults.

The resorptive action of osteoclasts, coupled with the deposition of bone by osteoblasts, helps the cranium enlarge to accommodate the growing brain and helps long bones enlarge their marrow cavities. Osteoclasts remove bone from the brain side of the cranium's flat bones and from the wall of the marrow cavity in long bones, while osteoblasts lay new bone down on the outside.

The interaction of osteoblasts and osteoclasts also helps spongy bone develop strength in exactly the right direction to meet external strains. Under stress, bone crystals appear to produce small electrical currents that stimulate osteoblasts to deposit new bone on stressed trabeculae. Where the currents are absent, the osteoblasts are not stimulated, and osteoclasts are free to reduce the size of unstressed trabeculae. As a result, when someone takes up weight lifting (for instance), the trabeculae in the epiphyses of the long bones will change their predominant orientations and strengthen the bones to cope with the new pattern of stress.

---

1. Astronauts tend to lose calcium from their bones while they are in orbit, where they do not experience the pull of gravity. Explain this phenomenon in terms of the interplay of osteoblasts and osteoclasts.
2. In Chapter 13, you will learn how the body controls bone growth with hormones. Try now to anticipate that chapter: What must the body do in order to stop the growth of long bones and become a dwarf? What must it do to become a giant?

The interaction of osteoblasts and osteoclasts is also responsible for the formation of Haversian systems. Throughout life, clusters of osteoclasts eat into compact bone from the surface, forming long **resorption canals.** Blood vessels follow them, and so does a wave of cells that give rise to osteoblasts. Behind the osteoclasts, osteoblasts lay down layers of new bone to form a new Haversian system within the resorption canal. Since the resorption canal cuts across preexisting Haversian systems, so does the new Haversian system. The remnants of the old systems thus come to occupy the spaces between new systems as the interstitial lamellae.

Bone Disorders

Table 8-1 (p. 180) summarizes several disorders of the skeletal system. There are also a number of disorders of bone formation, most due to problems with the hormonal mechanisms that regulate the deposition and resorption of bone. They include giantism, acromegaly, pituitary dwarfism, and osteoporosis. Rickets is a bone weakness that results from a shortage of vitamin D, which is essential to the absorption of calcium and phosphorus from the small intestine and to the deposition of these substances in bone. We will discuss these disorders in more detail in Chapters 13 (on the endocrine system) and 22 (on nutrition).

(Text continues on p. 182.)

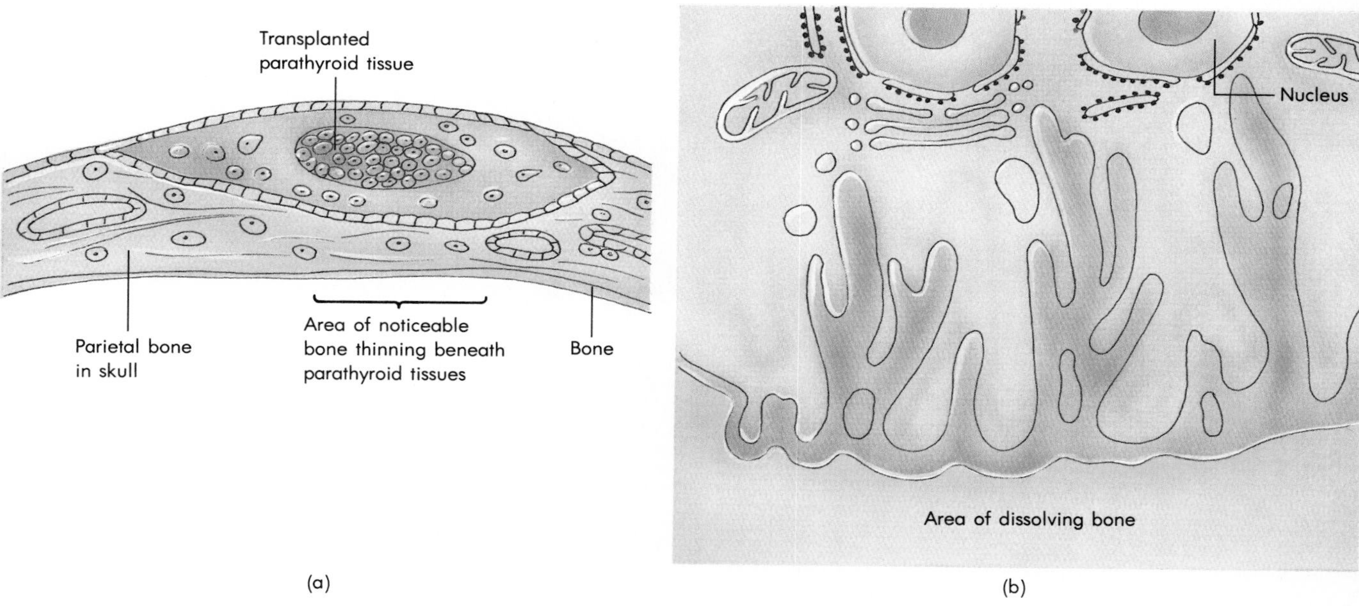

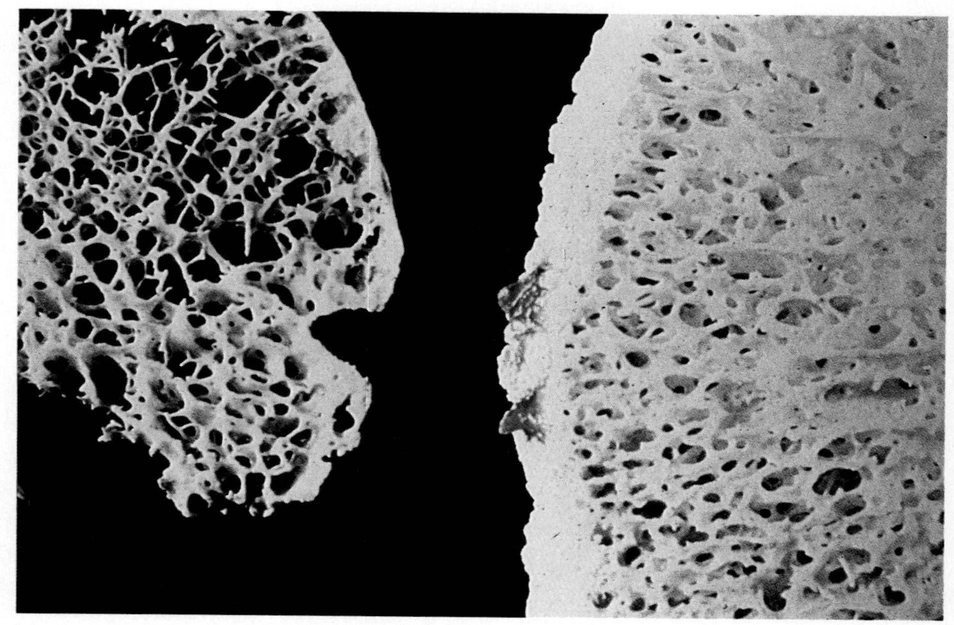

FIGURE 8-17 Osteoclast dissolving bone. (a) When tissues of the parathyroid are transplanted over the parietal bone in rats, there is a noticeable thinning of the bone just under the transplant, caused by the hormonal stimulation of the osteoclasts. (b) Drawing of a multinucleate osteoclast dissolving bone. (c) There is a marked loss of calcium from the bone (on the left) in this sample taken from a person suffering from osteoporosis. The bone becomes not only more hollow, but also more brittle. Normal bone (on the right) contrasts sharply in calcium content. Loss of bone calcium is usually more striking in mature women (about 25% of women will develop osteoporosis) than in men and is caused by the decline in estrogen that accompanies menopause. Recent work suggests that the stage may be set for the later development of osteoporosis in women in their thirties and forties, as a result of hormonal irregularities during the menstrual cycle.

Table 8-1

Clinical Summary: Selected Disorders of the Skeletal System

DISORDER	DESCRIPTION
Arthritis	Inflammations of the joints due to damage, deposits of uric acid crystals (gout), or attacks by the body's immune system. Symptoms include joint pain and swelling and inability to use the joint. In severe cases, the joint may be so damaged that it must be replaced.
Cleft palate	A birth defect in which the maxillae and/or the bones of the palate fail to unite, making the oral and nasal cavities a single chamber. Generally repairable by surgery.
Herniated (slipped) disk	Rupture of the intervertebral disk, with ensuing pressure on spinal nerve roots. Effects can include severe pain, numbness, and paralysis. Usually occurs in lower back (lumbar) region and affects the leg.
Osteoporosis	Loss of bone mineral in older people, resulting in hunchback, fractures, pain, and height loss. Related to decline in sex hormone levels with age. Results when osteoclasts become more active than osteoblasts.
Paget's disease	Accelerated activity of both osteoclasts and osteoblasts, resulting in lumpy, softened bones. Rarely affects people below age 50. Cause unknown.
Scoliosis, kyphosis, lordosis	Aberrant curvatures of the spine. Scoliosis is a lateral curving; kyphosis (hunchback) exaggerates the thoracic curve, leading to a condition known as hunchback; lordosis (swayback) exaggerates the lumbar curve. Can be a birth defect or the result of polio and even poor posture.

Learning Focus

HEALING BROKEN BONES

Bones are hard, strong, and flexible. They stand up under the weight of the body and all the impacts of walking, jumping, falling, and so on. But they are not unbreakable. A bone can easily be stressed beyond its strength or its ability to flex. When this happens, the bone breaks. We call a break in a bone a **fracture.** *Partial* fractures fail to go all the way across a bone; they include cracks and *greenstick* fractures, in which one side of a bone breaks while the other bends. *Complete* fractures break the bone all the way; they can be *simple* fractures, in which the broken ends of the bone do not poke through the skin, or *compound* fractures, in which the ends do reach the light of day. *Spiral* fractures result when a bone is twisted. *Transverse* fractures cross the bone at right angles to its long axis; they result from bends. *Comminuted* fractures can follow heavy blows to bones; in them, the bone is smashed into many small fragments. In *impacted* fractures, broken fragments are driven into each other, as when one jumps from the roof of a five-story building onto pavement.

Since bone is alive, fractures can heal, although the process may take months. It is aided by immobilizing the broken bone, as in a cast or sling. Healing begins as the periosteum in the region of the break generates numerous osteoblasts. These cells produce a mass of new bone tissue, a **callus,** that forms a bridge between the broken ends of the bone.

At first, the callus consists of trabeculae like those of spongy bone. Once it is formed, osteoclasts resorb the dead portions of the original, broken bone and the trabeculae themselves. Compact bone forms over the surface of the callus, and remodeling restores the original bone contour. Eventually, the bone is just as it was before, except that the break shows a "fracture line" that is detectable in x-rays. There may also remain some thickening of the bone.

Occasionally, especially in older people, bone fractures may be slow to heal. In such cases, physicians have found, it is possible to speed up the healing process. Coils of wire hooked to batteries are installed in the cast that immobilizes the bone while it heals. As electrical current flows through the coils, it generates electromagnetic fields. When the fields are made to pulsate, the osteoblasts in the region of the fracture are stimulated, and the pace of ossification is increased.

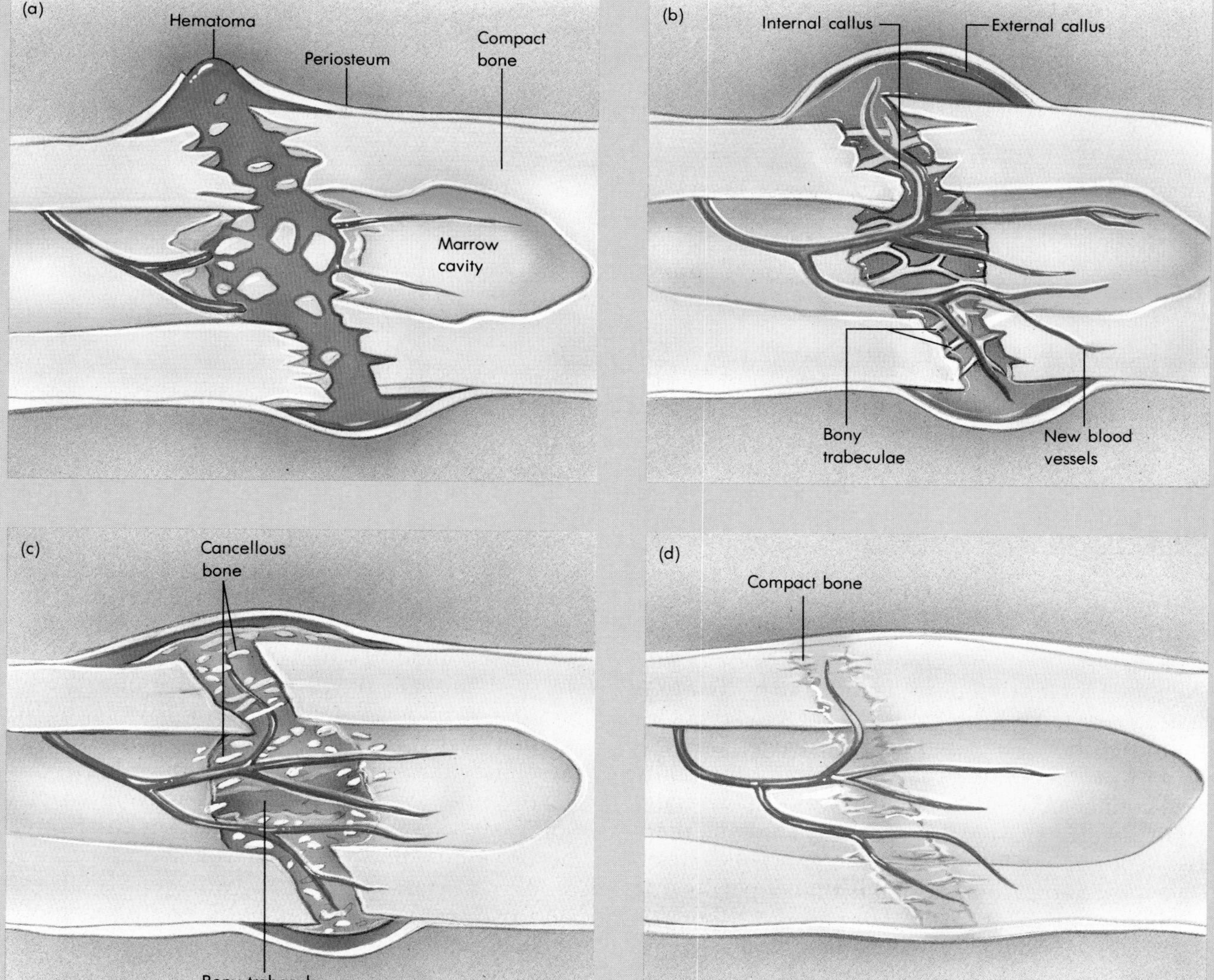

Healing of a bone fracture. (a) Fracture hematoma forms at the site of the break. (b) Callus forms at the site of the injury and acts as a bridge to connect the fractured ends of the bone. (c) At first the callus resembles spongy bone, which is later covered over with a growth of compact bone. (d) The compact bone is eventually remodeled to resemble the original bone.

Learning Focus Response

HEALING BROKEN BONES

1. Compare the repair of a fracture with the formation of new bone.

2. Compare the use of electromagnetic fields to speed fracture healing with how spongy bone controls its own remodeling of trabeculae.

3. How might electromagnetic fields be used to solve the bone loss problem of astronauts?

4. Do you think a healed fracture is likely to be stronger or weaker than the original unbroken bone? Why?

CONNECTIONS

The skeleton is both a framework for the body and a system of levers that permits movement. The first function might be satisfied if the skeleton were all one unified bone, as rigid as a birdcage. The second, however, requires that the skeleton be many bones connected to each other in ways that usually permit relatively free movement. That is, there must be hingelike **joints** wherever two bones meet.

One problem of joint design is that of permitting the bones that meet in the joint to move but not allowing them to separate. The body handles this problem in several ways. In joints that allow no movement (such as the sutures of the skull), the bones are actually interlocked like the pieces of a jigsaw puzzle. In all joints, the bones are tied together by connective tissue links. The links may be very short and composed of fibrous connective tissue, as in the sutures. They may also be more flexible inserts of cartilage, as between the ribs and the sternum. In joints that must allow more movement, like those of the spine and limbs, the links are bands of dense, flexible connective tissue, **ligaments,** that connect bone to bone across the joint area and, like ropes, tie the bones together. Tension in the ligaments helps to limit movement of the joint. Muscles and their tendons, which also reach across joints, play a similar role.

We can classify joints in two ways, by function and by structure. Functionally, there are immovable joints (*synarthroses*), slightly movable joints (*amphiarthroses*), and freely movable joints (*diarthroses*). Structurally, there are **synovial** joints, in which a fluid-filled sac, the joint cavity, lies between the bones, and **fibrous** and **cartilaginous** joints, which lack joint cavities.

Fibrous Joints

In fibrous joints, the bones are linked tightly together by fibrous connective tissue. In **sutures,** which we see between the bones of the skull, the edges of the bones that meet form interlocking projections. The narrow gap between the bones is crossed by short connective tissue fibers. In adulthood, many sutures are filled in by bone. In a **syndesmosis,** two long bones (such as the tibia and fibula) are more loosely linked along their length by a band or membrane of fibrous connective tissue. In a **gomphosis,** a peg is anchored into a socket by fibrous connective tissue; gomphoses are found where our teeth meet our jawbones.

Cartilaginous Joints

You can actually feel one cartilaginous joint with your fingers: It is the band of hyaline cartilage that joins the end of a rib to the sternum. Defined by the cartilage insert between two bones, it is a **synchondrosis.** Another example exists in the epiphyseal plate, before the long bones have completely ossified.

A **symphysis** contains, not hyaline cartilage, but fibrous cartilage, in the form of a disk or pad. One symphysis is found where the halves of the pelvis join in front, in the pubic symphysis. Another is found in the intervertebral discs.

Synovial Joints

Synovial joints are designed to allow much more movement. They occur wherever limb bones meet each other, as in the elbow, knee, wrist, ankle, fingers, and toes or their girdles in the shoulder and hip, as well as where the lower jaw meets the skull, the ribs meet the vertebrae, and the pelvis meets the sacrum. In each case, the surfaces of the two bones that meet in the joint are covered with a smooth, friction-reducing layer of hyaline *articular cartilage.* Pressure on the cartilage, such as develops when we put our weight on a leg, causes it to release fluid that helps to lubricate the joint. The cartilage reabsorbs the fluid when the pressure ends.

---

1. Five types of synovial joints are ball-and-socket, hinge, gliding, pivot, and saddle joints. Examine your own body, a plastic classroom demonstration skeleton, or the pictures of the skeleton in this chapter, and try to find joints whose shapes and permitted motions fit these names.
2. Artificial knee and hip joints replace the natural bone and articular cartilage with stainless steel. What problems might you expect to arise from the loss of the articular cartilage?

Each synovial joint is enclosed in an **articular capsule** (Figure 8-18). The outer layer of the capsule is a sleeve of dense connective tissue (ligament) that merges with the periosteums of the two bones. It is lined with **synovial membrane,** a layer of loose, elastic connective tissue and adipose tissue that secretes a **synovial fluid** rich in hyaluronic acid (a major component of the extracellular material of connective tissue). It also contains phagocytes that remove bacteria and the debris produced by wear in the joint.

The synovial fluid serves to nourish the articular cartilage and to lubricate the joint. When it is not under pressure, it is about as viscous (thick) as egg white; under pressure, it becomes less viscous, partly because of the fluid that leaves the articular cartilage. It also becomes more fluid when the joint is moving. Synovial fluid is also found in **bursae,** sacs that cushion the movement of tendons, muscles, and ligaments over bones; *bursitis* is an inflammation of a bursa.

In many synovial joints, the articular capsule is surrounded by strong *extracapsular ligaments* that strengthen the joint. When a joint is sprained, as by a twisting fall, it is these ligaments, as well as the joint capsule, that are stretched or partially torn, with tearing of blood vessels, inflammation, and swelling. A dislocation involves more severe damage, tearing ligaments, tendons, and the capsule and even pulling bones out of line with each other.

Intracapsular ligaments such as the cruciate ligaments of the knee joint may link the bones within the capsule, although folds of the synovial membrane keep them out of the joint cavity itself. The knee joint also contains pads of fibrous cartilage, the *articular discs,* or *menisci,* that match the contours of the femur to those of the tibia (Figure 8-19, p. 185). A common athletic injury is "torn cartilage" or menisci in the knee; the damage is currently repaired by *arthroscopy,* in which the surgeon makes small incisions in the skin and ligaments of the knee and uses pencil-thin instruments to whittle and reshape the damaged cartilage and suck out the shavings from his or her work.

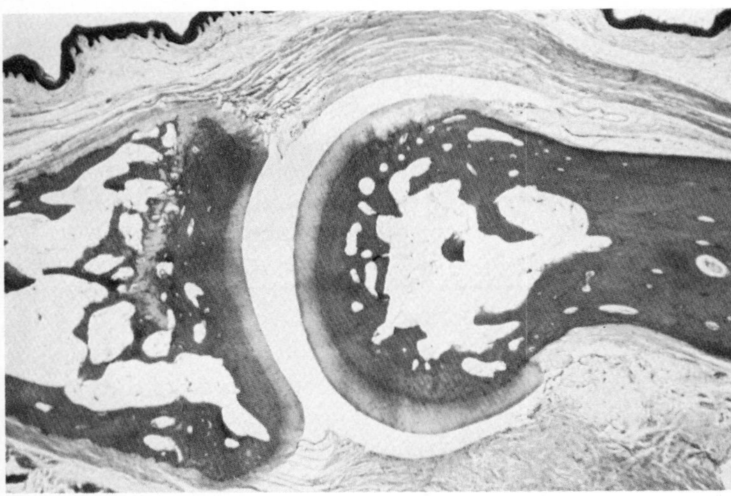

(a)

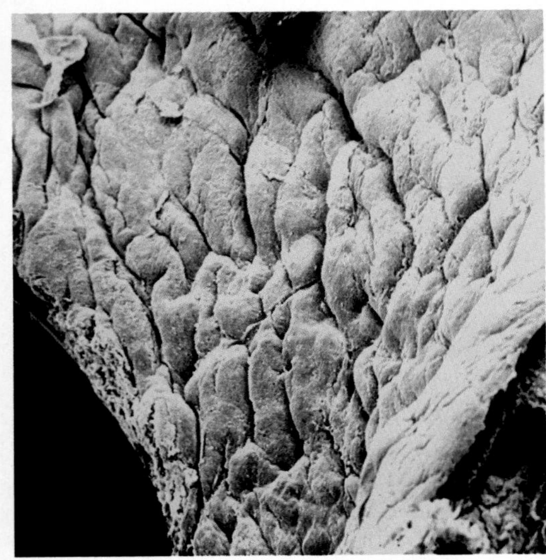

(b)

FIGURE 8-18 (a) Synovial joint in the human finger. Note the articular cartilage covering the ends of each bone and the membrane-lined joint capsule. (b) Scanning electron micrograph of the synovial membrane from a knee joint. It lines the joint cavity and produces the synovial fluid, which lubricates the joint and nourishes the articular cartilage.

Arthritis refers to an irritation or inflammation of the joints, but the term encompasses 25 or more different diseases. One of the most common is *rheumatoid arthritis,* an inflammation of the synovial membrane that leads to excessive accumulation of synovial fluid, swelling, pain, and joint stiffness. It can lead to erosion of the articular cartilage and its replacement with fibrous connective tissue, which then ossifies and fuses the joint. *Gouty arthritis* occurs when the body deposits crystals of uric acid in the soft tissues of the joint. The crystals irritate the cartilage and can lead to fusion of the bones. This condition, but not rheumatoid arthritis, responds well to treatment with drugs that inhibit the production of uric acid or hasten its removal from the body in the urine.

Severe arthritis can so badly damage the hip and knee joints—which suffer the most wear and tear under the body's weight—that they must be rebuilt, using stainless steel replacements for the appropriate end of the femur (Figure 8-20). The surgeon removes the damaged end of the bone and anchors the long shaft of the steel replacement in the marrow cavity with a special cement. Unfortunately, the resulting "artificial joint" does not work as well or hold up as long as the natural joint. It lacks the lubricating qualities of the articular cartilage and the intact synovial membrane, and it cannot repair wear and tear. In addition, stainless steel is a foreign material, and the bone in which it is anchored often withdraws from it, making the joint too loose to stay in place properly.

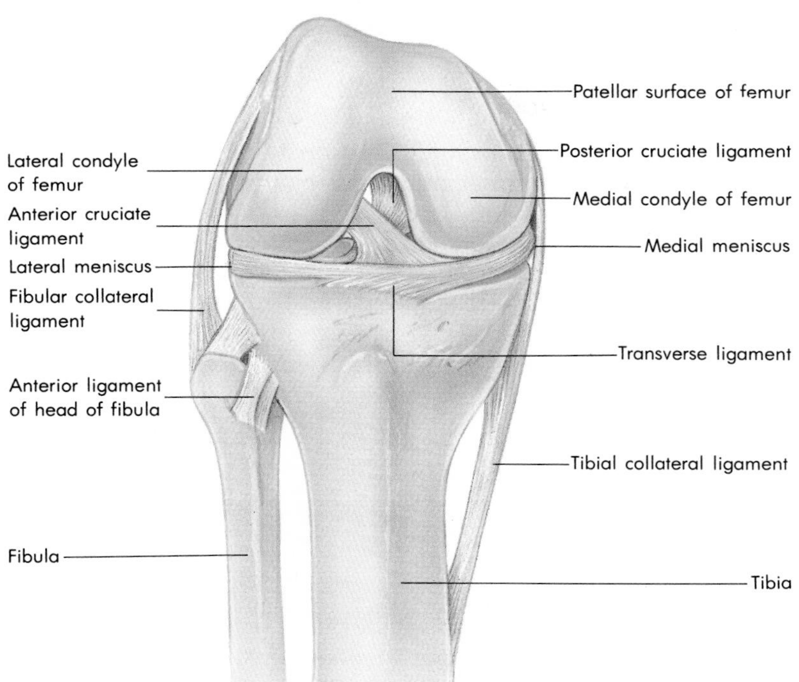

FIGURE 8-19 The knee joint (anterior view with the leg bent in sitting position) showing lateral and medial menisci, cruciate ligaments, and joint capsule. These ligaments are often damaged in sports knee injuries. *Note:* the kneecap has been removed for this drawing.

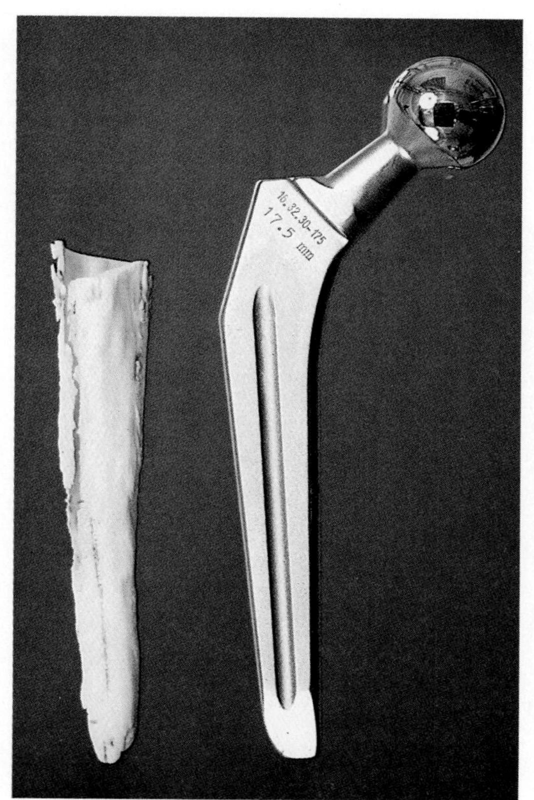

(a)

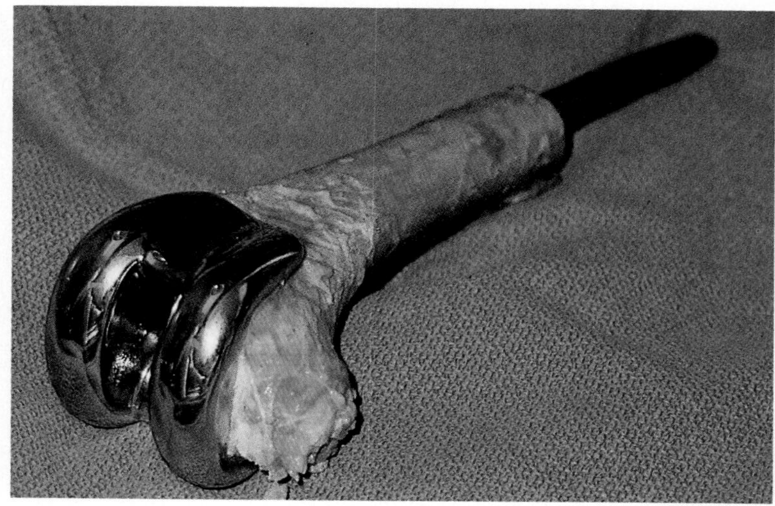

(b)

FIGURE 8-20 Artificial joints. (a) Artificial hip joint with the corresponding shaft of bone from the femur. (b) Knee replacement joint.

Joint Types and Movements

Synovial joints are not all built to the same pattern. The details of their structure depend very much on the kinds of movements they must permit. *Ball-and-socket* joints, for instance, are designed to permit rotation, flexion, extension, adduction, and abduction (Figure 8-21). *Gliding* joints have flat articular surfaces and permit movements in two planes (forward and backward, side to side) (Figure 8-22a). They are found between the carpals and tarsals and between the clavicle and the sternum and scapula. *Hinge* joints permit movement in only one plane (forward and backward), as in the knee, elbow, and fingers (Figure 8-22b). *Pivot,* or *trochoid,* joints set a bony point in a ring of bone and ligament and permit rotations only, as where the radius and ulna meet just below the elbow (Figure 8-22c). In an *ellipsoidal* joint, the oval end of one bone meets a matching depression in another, as where the radius meets the carpals of the wrist; it permits both forward-and-backward and side-to-side movements. *Saddle* joints such as that at the base of the thumb also permit movement in two directions.

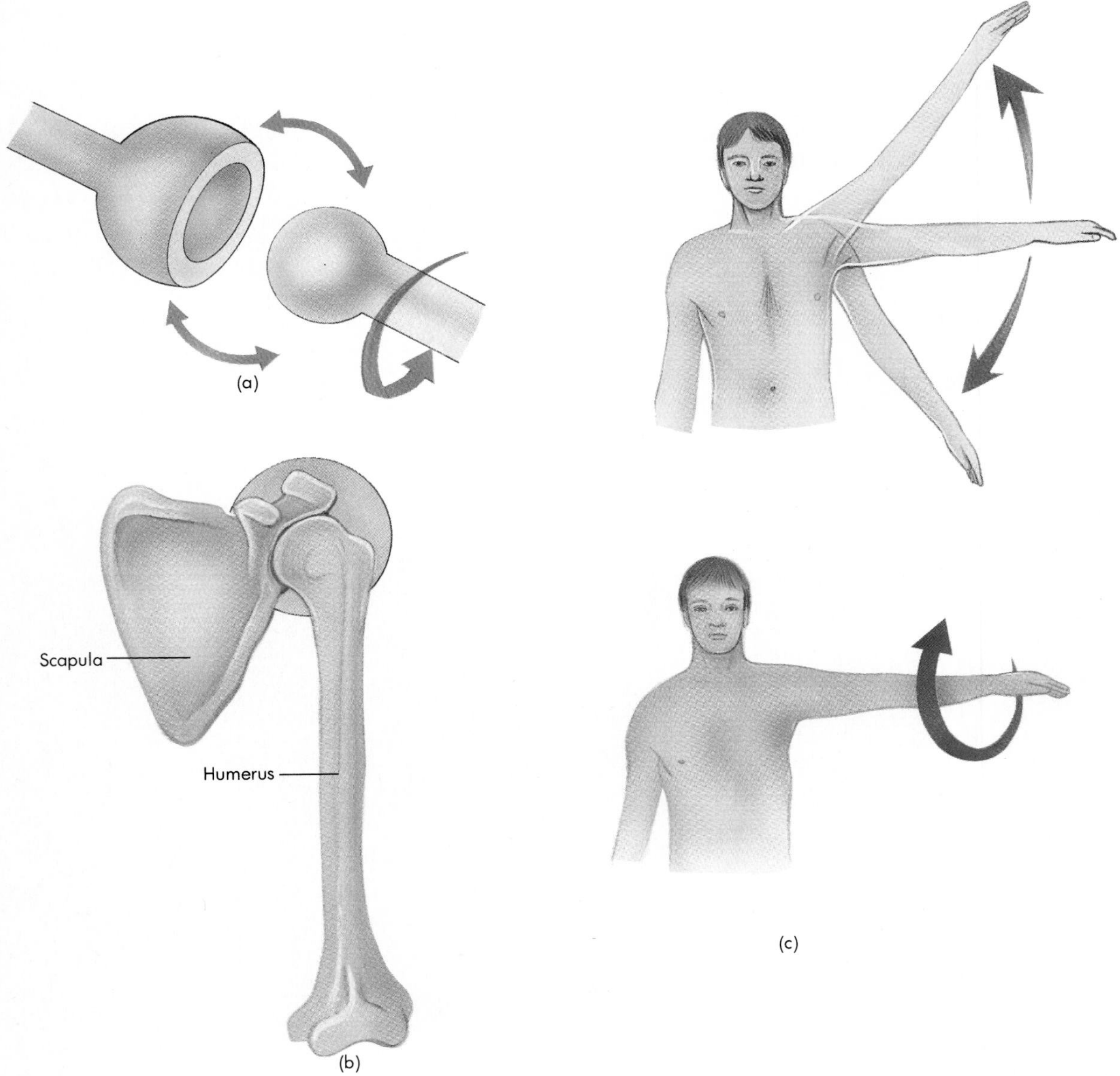

FIGURE 8-21 (a) A mechanical ball-and-socket joint. (b) The shoulder joint. (c) Shoulder movements.

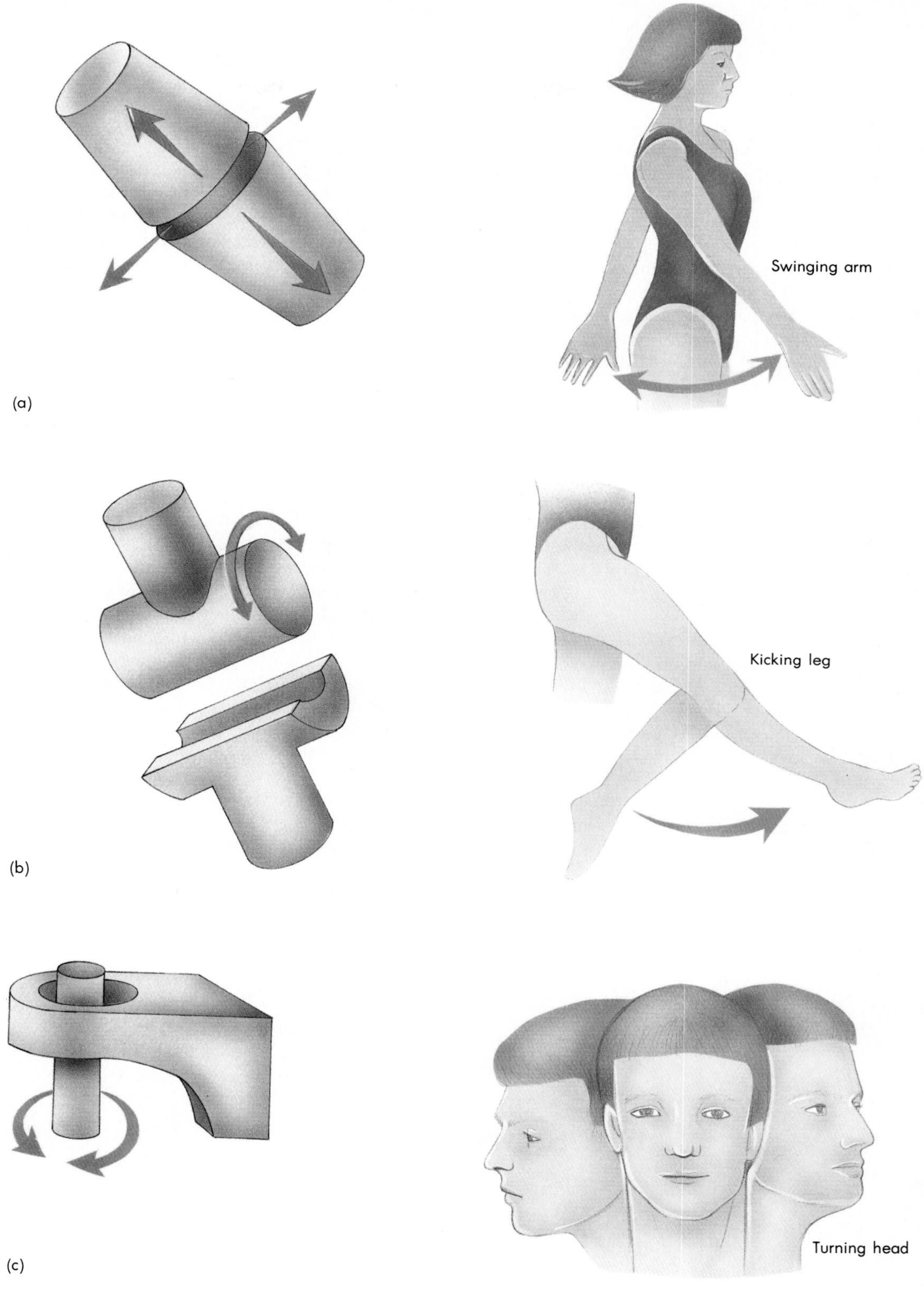

FIGURE 8-22 Mechanical representation and human range of movement for a number of joints: (a) gliding joint (swinging arm); (b) hinge joint (kicking leg); (c) pivot joint (turning head).

SUMMARY

The skeleton is the body's structural framework. Its component units, the various types of bones—flat, long, short, and irregular—pivot on each other at the joints and provide a system of levers that lets the body move. It is divided into two major segments: the central axial skeleton (the skull, ribs, sternum, and vertebral column) and the appendicular skeleton (the bones of the limbs and the pectoral and pelvic girdles).

Because bone is a living tissue, it can heal breaks and adapt to changing stresses. It also contains the red bone marrow, which produces many blood cells. It gains its hardness from calcium phosphate and calcium carbonate, and it serves as a reservoir for calcium and phosphate. It gains its flexibility from the matrix of collagen fibers in which these minerals are deposited.

A typical long bone consists of a shaft, or diaphysis, and two ends, or epiphyses. The shaft is hollow, containing a marrow cavity; its wall is composed of dense, compact bone. The epiphysis has a shell of compact bone surrounding spongy bone, a network of thin bony plates and splinters, or trabeculae.

The surface of compact bone is built of several circumferential lamellae, layers of bone surrounding the bone. Deeper lie the cylindrical Haversian systems, consisting of many concentric rings of bone surrounding a central canal containing blood vessels. Between the rings lie lacunae containing the bone cells, or osteocytes. The cells that form bone, the osteoblasts, arise from the connective tissue membrane, the periosteum, surrounding the bone.

Bone formation, or ossification, starts long before birth and follows two patterns. Intramembranous ossification begins when osteoblasts migrate into the membranes that will become flat bones, secrete the collagenous bone matrix, and trigger the deposition of calcium salts. Endochondral ossification begins when the membrane surrounding the cartilage model of a future long bone begins to produce osteoblasts, which secrete a collar of compact bone around the middle of the cartilage model. As blood vessels then invade the cartilage, the cartilage breaks down and is replaced by bone. Later, ossification begins in the epiphyses, but the zones of ossification do not meet until adolescence. Until then, the cartilage grows just fast enough to maintain a zone of cartilage (the epiphyseal plate) between the zones of ossification.

Once formed, bone is continuously remodeled. Osteoclasts derived from blood cells dissolve bone, driving long resorption canals into compact bone. Osteoblasts then fill the canals with new Haversian systems. Osteoclasts also weaken unstressed trabeculae in spongy bone, while osteoblasts strengthen those under strain.

Where the bones meet in joints, they are tied together by connective tissue in ways that allow more or less movement. In fibrous joints such as the sutures between skull bones, the bones interdigitate and are locked tightly into place by fibrous connective tissue. In cartilaginous joints such as the link between the ribs and the sternum (breastbone), the link is a more flexible band of cartilage. In synovial joints, the most movable joints of all, a lubricating capsule or joint sac encloses the space between the bones. The bones themselves are covered with a smooth, friction-reducing layer of hyaline cartilage, and strong ligaments keep the bones in position. Because synovial joints vary in shape, they also vary in the movements they permit.

STUDY QUESTIONS

1. Describe the functions of the skeleton.
2. Distinguish between the axial and the appendicular skeletons.
3. Sketch the curvature of the spine. What are scoliosis, kyphosis, and lordosis?
4. What is the difference between a true rib and a false rib?
5. What are the respective roles of mineral and protein in bone?
6. What are the differences among circumferential lamellae, Haversian lamellae, and interstitial lamellae?
7. What is an ossification center?
8. Describe the formation of a Haversian system.
9. What is the difference between a syndesmosis and a synchondrosis?
10. What is the function of synovial fluid?

GLOSSARY

Appendicular skeleton The bones of the limbs and the pectoral and pelvic girdles.
Articular cartilage The layer of hyaline cartilage covering the surfaces of bones that meet in synovial joints.
Axial skeleton The skull, vertebral column, ribs, and sternum.
Cranium The portion of the skull enclosing the brain.
Diaphysis The hollow shaft of a long bone.
Epiphysis The end of a long bone.
Fontanels The membranous areas between the bony plates of the fetal cranium.
Haversian system The cylindrical structural unit of compact bone.
Intervertebral discs Pads of fibrous cartilage between the vertebrae.
Intervertebral foramina Openings between the vertebrae for passage of the spinal nerves.
Joint A junction between two bones.
Ligament A connective tissue strap that links bones across a joint.
Ossification Bone formation.
Osteoblasts Bone-forming cells; derived from periosteum.
Osteoclasts Bone-destroying cells; derived from monocytes.
Osteocyte A bone cell.
Pectoral girdle The scapula and clavicle.
Pelvic girdle The pelvis.
Periosteum The connective tissue membrane covering bone; gives rise to osteoblasts.
Red bone marrow The blood-cell-producing tissue found in the marrow cavities of long bones in the young and in spongy bone in adults.
Sacrum The triangular element of the backbone formed of five fused vertebrae; attaches the backbone to the pelvis.
Sutures The seamlike joints between the bones of the skull.
Vertebrae The bones of the vertebral column, or backbone.
Vertebral column The stack of bones (vertebrae) enclosing the spinal cord; the backbone.

Chapter 9

The Muscular System

Unlike a marionette, the human body comes with no strings attached. But it still moves, for instead of strings it has **muscles,** masses of specialized cells that—like strings—pull on the bones and make them pivot on the joints between them.

Here lies the key to the first wobbly steps of an infant, the gliding grace of a ballet dancer, the power spin of a discus thrower, all parts of a concert played on the physiological instruments of muscle, sinew, bone, and nerve. It is the specialized cells of muscle tissues that propel the body through its wide range of movements. In fact, whenever some degree of movement is required, whether it involves a leap, driving the blood through the arteries and veins, or regulating the flow of food through the digestive system, muscle tissues are somehow involved.

The muscles never push, for the secret of their action lies in filaments of the proteins actin and myosin (and others) that fill each muscle cell. These filaments use the energy of adenosine triphosphate (ATP) to slide over each other, thus shortening (contracting) the cells containing them and the muscles of which the cells are part. This shortening pulls the ends of the muscle, and the bones to which the ends are attached, closer together, forcing the bones to pivot around their joint.

There are at least two muscles for every movable joint in the body (Figure 9-1). Whenever a muscle contracts, it pulls the joint in one direction; an example is the biceps, which flexes the arm at the elbow. When the muscle on the other side of the joint contracts, it pulls the bone in the opposite direction; the triceps straightens or extends the arm at the elbow. Since their actions oppose each other, they are referred to as *antagonistic* pairs of muscles.

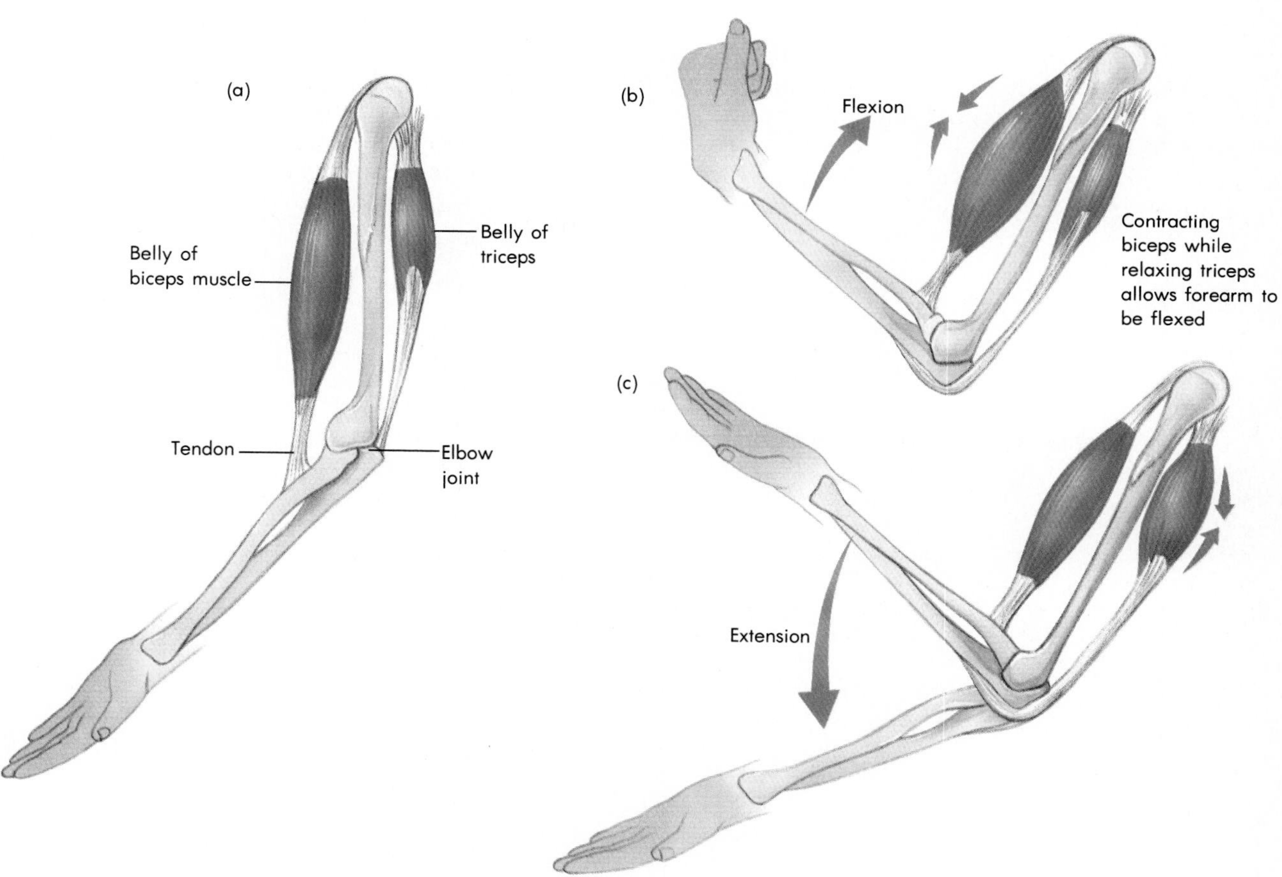

FIGURE 9-1 Muscles work primarily by contracting (shortening). To achieve the wide range of possible skeletal movements, muscles often work in antagonistic pairs or groups. (a) The biceps on the front of the upper arm and the triceps on the back of the arm are antagonists to each other. (b) When the biceps contracts while the triceps relaxes, the arm flexes across the elbow joint. (c) Triceps contraction with the biceps relaxed produces extension (straightening) of the arm.

TYPES OF MUSCLE TISSUES

There are three kinds of muscle tissue (Table 9-1). Known as skeletal, cardiac, and smooth muscle, they differ in microscopic appearance, in the way they are controlled, and in the roles they play in the body. However, they share four essential characteristics:

1. All muscle is **contractile;** that is, it acts by shortening.
2. All muscle is **excitable;** that is, it contracts in response to stimuli such as nerve signals, mechanical and chemical irritation, and electrical shocks.
3. All muscle is **extensible;** that is, it can stretch, as indeed each member of an antagonistic pair must when the other is contracting.
4. All muscle is **elastic;** that is, it can return to its original length after contracting or stretching.

The main function of all muscle is to produce movement, but only skeletal muscle does it mainly by pulling on bones. The other two kinds of muscle work by squeezing; they propel blood from the heart, food through the intestines, and secretions from glands; they also act like valves to control the flows of fluids in many of the body's various tubes.

Skeletal Muscle

Skeletal muscle is the major topic of this chapter. It is the kind of muscle that attaches to the bones and produces the various kinds of skeletal movements. Because it is under conscious or voluntary control, it is also called *voluntary muscle.* The numerous contractile protein filaments within its cells are bundled in such a way that under the microscope they make the cells seem striped or striated; this feature accounts for its third name of *striated muscle* (Figure 9-2). The individual cells of skeletal muscle are so much longer than they are wide that we call them **muscle fibers.** Each skeletal muscle cell contains several nuclei and many mitochondria to supply the energy needed for contraction.

Cardiac Muscle

Cardiac muscle is also striated, but because it is not under conscious control, it is not classified as a type of voluntary muscle. It is the muscle of the heart, and it can contract either spontaneously (and rhythmically) or when stimulated by neighboring cardiac muscle cells or nerves (as discussed in Chapter 10). Like skeletal muscle cells, the cells

Table 9-1

Muscle Types and Characteristics

MUSCLE TYPE	CONTROL	LOCATION	FEATURES
Skeletal	Voluntary	Skeletal muscles	Striated, multinucleate, long, unbranched cells; rapid-onset, brief contractions
Cardiac	Involuntary, spontaneous, rhythmic	Heart	Striated, one-nucleus, branched cells
Smooth	Involuntary	Tube walls	Unstriated, one-nucleus, tapered cells; slow-onset, prolonged contractions

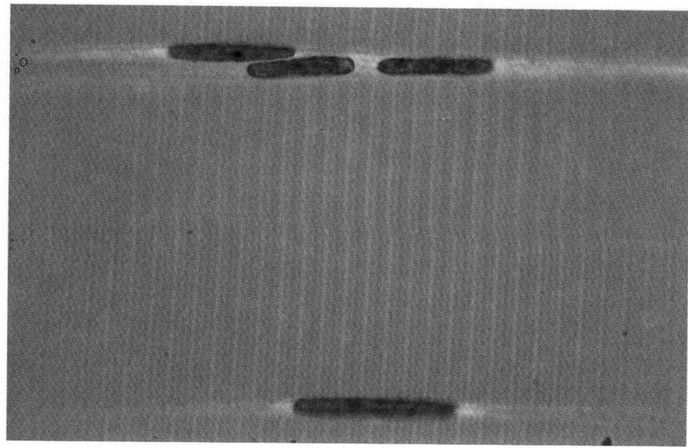

FIGURE 9-2 The well-developed pattern of striations (cross bands) of skeletal muscle tissue is formed by overlapping contractile filaments.

of cardiac muscle have many mitochondria, but they each usually have only one nucleus. The cells are also shorter and branching. The cells butt up against each other at their ends, where their deeply folded cell membranes interdigitate to form the **intercalated disks** that are characteristic of cardiac muscle (Figure 9-3). These disks connect the cells of the heart to each other and facilitate the transmission of signals for contraction from cell to cell.

Smooth Muscle

Smooth muscle is found in many places in the body. Most is found in walls of various tubes of the body, such as the blood vessels, the digestive system, and the drainage ducts for glands like the salivary glands and the pancreas. Changes in the diameter of these vessels regulate the flow of blood to various areas, propel food through the digestive system, and propel the secretions from the different glands. The smooth muscle in the wall of the uterus also propels babies into the world. Tiny bands of smooth muscle fibers, the arrector pili, are attached to each hair follicle—pulling the hair shaft erect in times of stress or coldness. Other smooth muscles control the iris of the eye.

Like cardiac muscle, smooth muscle is classified as involuntary. Unlike both cardiac and skeletal muscle, smooth muscle is not striated (Figure 9-4). Smooth muscle also differs in that it contracts more slowly and holds its contraction longer than cardiac or skeletal muscle. The cells are relatively short and tapered at the ends; they each contain a single nucleus.

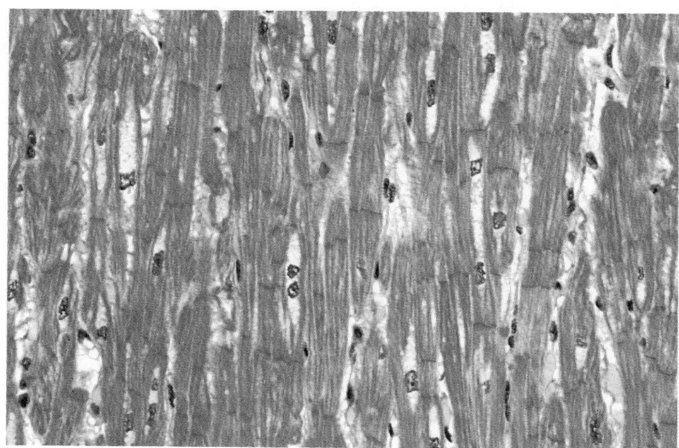

FIGURE 9-3 Cardiac muscle tissue is a little less conspicuously striated than is skeletal muscle. Its distinguishing features include the branching of its cells or fibers and the intercalated disks where the ends of cells meet.

THE MUSCULAR SYSTEM

Normally, the **muscular system** means the body's set of skeletal muscles. We use these voluntary muscles whenever we walk, talk, chew, run, write, or move our skeleton in any way. Skeletal muscles are also regulated, in part, by certain automatic control systems such as those essential to main-

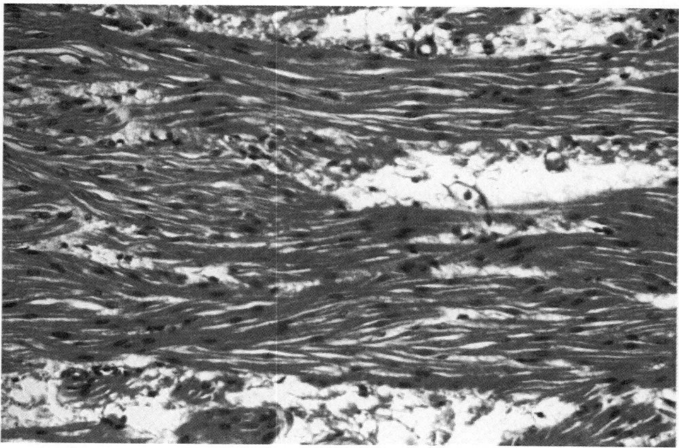

FIGURE 9-4 Smooth muscle cells are structurally very different from both skeletal and cardiac muscle cells. The cells are not striated, and they tend to be tapered at each end. They contract more slowly and less forcefully and are often used by the body to change the diameter of tubular organs such as blood vessels and intestines.

taining both posture and a baseline degree of muscle tone. Voluntary muscle control acts as a willed override of involuntary postural control.

Muscle Action

As we noted earlier, two or more muscles cross each of the body's joints and control its range of movements. When a muscle is contracting to move a joint, it is called the **agonist** muscle; the biceps is the agonist for flexing the arm at the elbow. The opposing muscle, whose contractions move the joint in the opposite direction, must relax when the agonist is contracting; it is the **antagonist** muscle. When the biceps is the agonist, the triceps at the back of the upper arm is the antagonist. **Synergists** contract to provide a steady base for the agonist's action; their function is like that of the hydraulic jacks that balance and support a hook-and-ladder fire truck when it is working. For the biceps, the synergists include the shoulder (deltoid) and chest (pectoralis major) muscles as well as those muscles of the trunk and legs that maintain the body's balance against changes in the body's center of gravity caused by the arm motion. Antagonists relax and synergists contract automatically when you use any muscle to make a movement.

1. Consider what you do when you flex your arm at the elbow. Just what are you controlling voluntarily?
2. What part of the action is controlled involuntarily?

Muscle Structure

The individual muscles are all built along the same lines. Each muscle is surrounded by a sheath of fibrous connective tissue, the **epimysium.** Enclosed by this sheath are numerous **fascicles** (bundles) of muscle fibers. Folds of the epimysium dip into the muscle to wrap each fascicle in **perimysium.** Folds of perimysium enclose the muscle fibers themselves in **endomysium** (Figure 9-5).

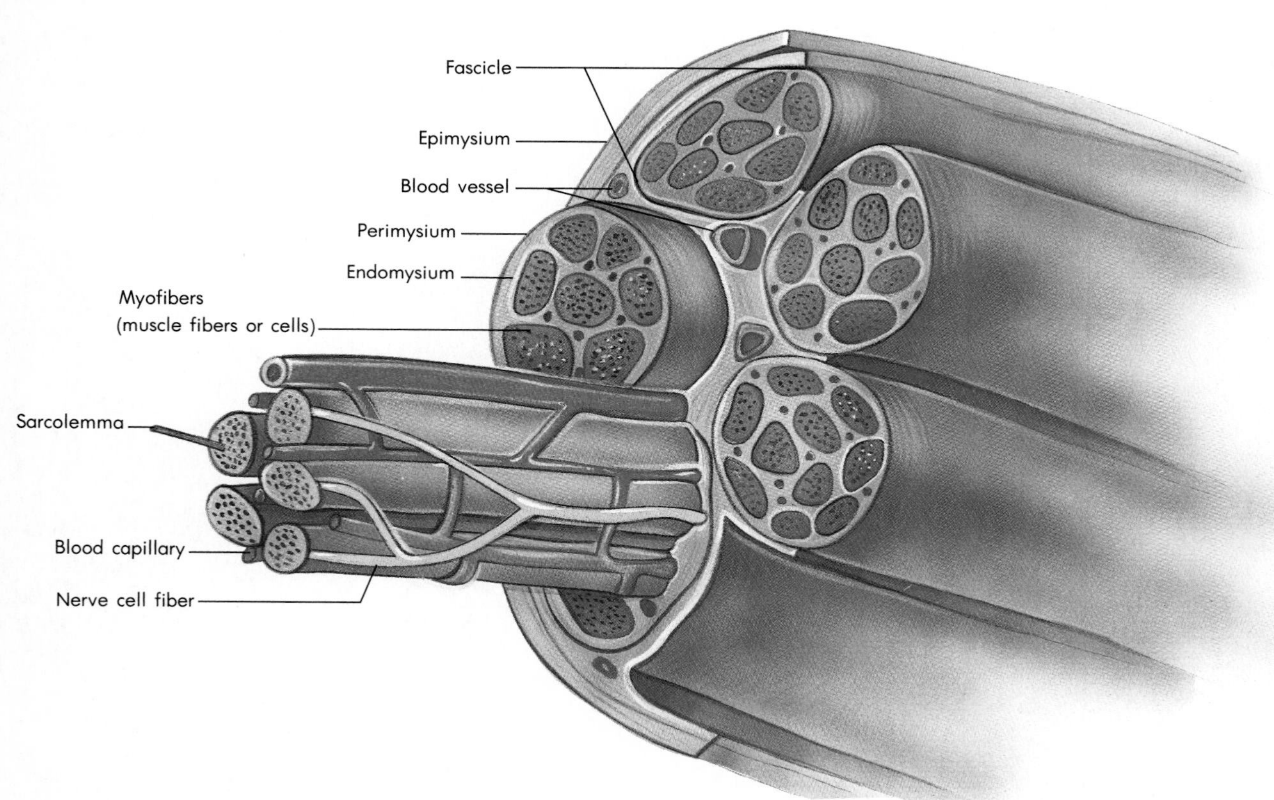

FIGURE 9-5 The connective tissue sheaths of muscle.

The three connective tissue sheaths merge at the ends of the muscle to form a **tendon** (see Figure 9-1). A tendon is a cord of tough, fibrous connective tissue that attaches the muscle to the periosteum of a bone. It transmits the muscle's pull to the bone. Together with the connective tissue sheaths of the muscle, it also accounts for a large part of the muscle's elasticity.

Some tendons are flat, sheetlike structures called **aponeuroses.** They link muscle to bone, muscle to muscle, or muscle to **fascia** (a sheet of connective tissue between the muscles and the skin). The cordlike tendons of the wrists, ankles, and some other joints are enclosed in tubular **tendon sheaths** lined by synovial membranes; the sheaths serve to reduce friction. Injury or excessive exercise can cause the sheaths to become inflamed and painful, thus restricting movement; the condition, known as **tendinitis,** resembles bursitis.

When muscles contract, they pull on the bones or other structures that are attached to their tendons. The biceps muscle of the upper arm, for instance, pulls on the shoulder blade at one end and on the radius and forearm fascia at the other. Typically, one bone remains relatively immobile, and the end of the muscle that attaches to it is called the **origin** of the muscle. The biceps originates on the scapula (shoulder blade). The other bone (or bones, in the case of the forearm) moves freely, and that more movable attachment of the muscle is called the **insertion** of the muscle. The biceps inserts on the radius and forearm fascia. (See Figure 9-6 for another example of an origin and an insertion.)

The Muscles

Figure 9-7 shows all the muscles that are visible on the surface of the human body, just beneath the skin and subcutaneous fat (many more muscles lie underneath these). These are the muscles that pad the skeleton and give the body its distinctive contours. For instance, the bulge of the calf is due to the gastrocnemius muscle; that of the buttocks is largely due to the gluteus maximus; and that of the chest is due to the pectoralis major. The muscle at the back of the neck that gets so tense when we are tired or anxious is the trapezius. The one that bulges in our cheek when we chew or grit our teeth is the masseter.

We list some of these muscles in Table 9-2 (p. 198), together with their origins, insertions, and actions. Note that each muscle has a particular effect and that several muscles can have similar effects. The exact action of a muscle depends on which other muscles are contracting at the same time. For instance, by itself the serratus anterior rotates the scapula laterally; when other muscles prevent the scapula from moving, it serves to elevate the ribs for deep breathing.

The names of the muscles may seem formidable, but they are actually quite straightforward—if you know Latin. Some of the components of the names indicate the direction in which the muscle fascicles run in relation to the body's midline (*rectus* means straight up and down; *transversus* means across or at right angles to the long axis of the limb or body). Some indicate position (*anterior, posterior, medi-*

(continued on p. 199)

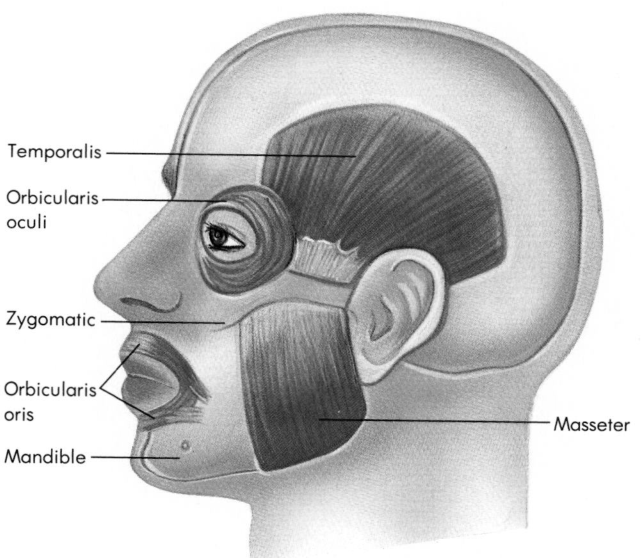

FIGURE 9-6 Some of the major muscles of the side of the head. The masseter is the powerful muscle that moves the jaw when we chew. Determine which attachment moves most and which moves least as the masseter contracts. Then identify the muscle's origin and insertion.

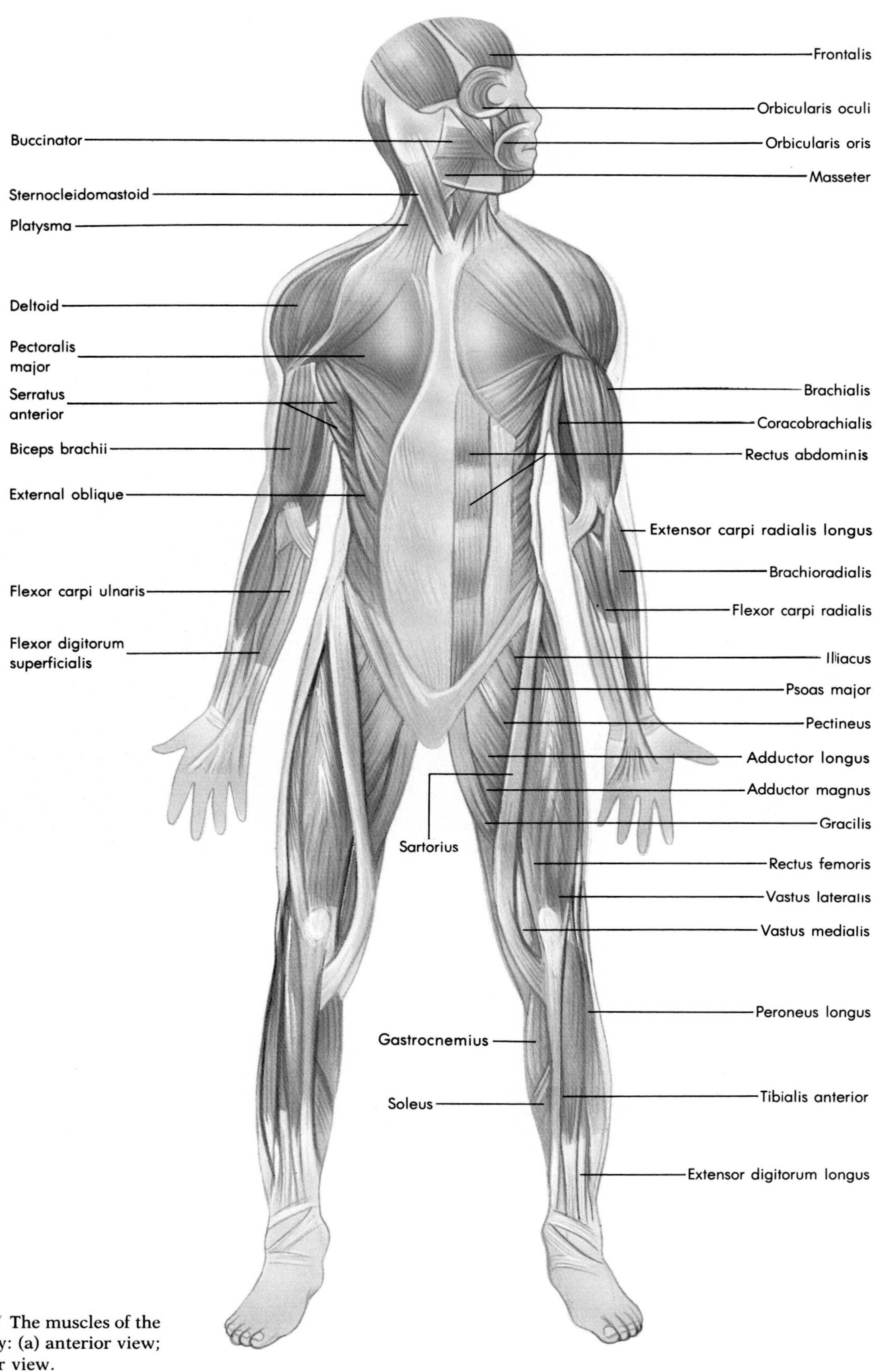

FIGURE 9-7 The muscles of the human body: (a) anterior view; (b) posterior view.

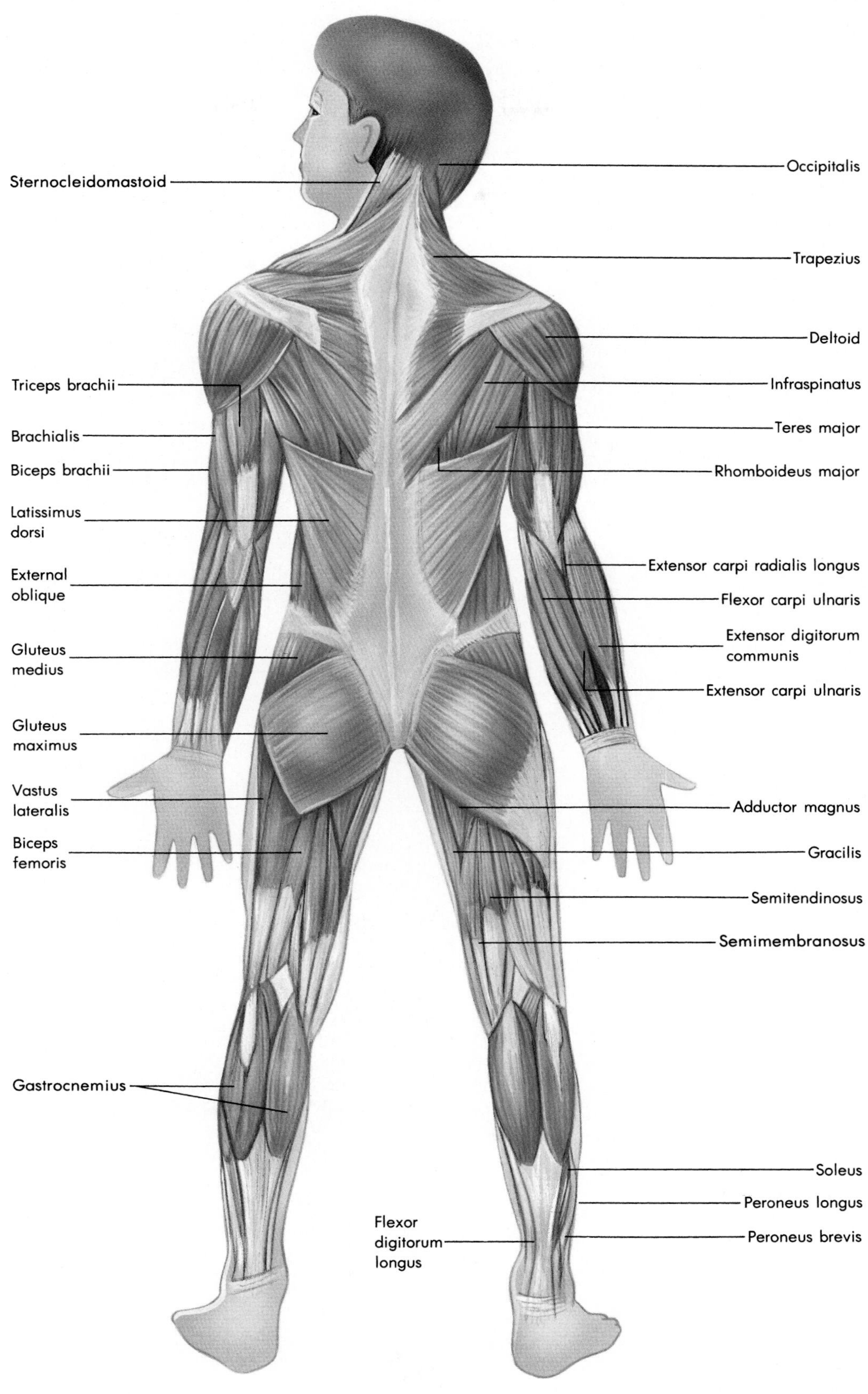
Sternocleidomastoid
Occipitalis
Trapezius
Deltoid
Triceps brachii
Infraspinatus
Brachialis
Teres major
Biceps brachii
Rhomboideus major
Latissimus dorsi
External oblique
Extensor carpi radialis longus
Flexor carpi ulnaris
Gluteus medius
Extensor digitorum communis
Extensor carpi ulnaris
Gluteus maximus
Vastus lateralis
Adductor magnus
Biceps femoris
Gracilis
Semitendinosus
Semimembranosus
Gastrocnemius
Soleus
Peroneus longus
Flexor digitorum longus
Peroneus brevis

(b)

Table 9-2

Selected Major Muscles of the Human Body

MUSCLE	ACTION
Head and Neck	
Masseter	Raises, protrudes mandible
Temporalis	Raises, retracts mandible
Orbicularis oculi	Closes eye
Orbicularis oris	Closes, shapes lips
Buccinator	Compresses cheek
Sternocleidomastoid	Bends neck; turns head
Trapezius	Elevates shoulder; extends head
Platysma	Lowers mandible; pulls lower lip down and back
Shoulder, Back, Chest, and Arm	
Deltoid	Pulls arm outward
Teres major	Extends, pulls down arm; helps rotate arm
Teres minor	Rotates arm
Pectoralis major	Flexes, pulls in, rotates arm
Latissimus dorsi	Extends, rotates arm; lowers shoulder
Serratus anterior	Rotates scapula; raises ribs
Rhomboideus major	Raises shoulder
Biceps brachii	Flexes, supinates arm (palm up)
Triceps brachii	Extends elbow
Brachioradialis	Flexes elbow
Extensor carpi radialis longus	Extends wrist
Pronator teres	Pronates wrist (palm down)
Extensor digitorum communis	Extends fingers
Brachialis	Flexes arm
Flexor carpi radialis	Flexes wrist
Flexor digitorum superficialis	Flexes fingers
Abdomen	
External oblique	Compresses abdomen; bends spine to side
Rectus abdominis	Flexes spine; tightens abdomen
Hip and Leg	
Iliacus	Flexes hip; rotates thigh
Psoas major	Flexes hip and spine; rotates thigh
Pectineus	Flexes hip; rotates thigh
Adductor longus	Adducts, flexes hip; rotates thigh
Gluteus medius	Abducts, rotates thigh
Gluteus maximus	Extends hip; rotates thigh
Biceps femoris	Flexes knee; extends hip
Semitendinosus	Flexes knee; extends hip
Sartorius	Flexes knee and hip; rotates thigh
Adductor magnus	Adducts, flexes, extends hip
Gracilis	Flexes knee; adducts hip
Quadriceps femoris (includes rectus femoris, vastus lateralis, vastus medialis, vastus intermedius)	Extends knee
Gastrocnemius	Extends ankle
Soleus	Extends ankle
Flexor digitorum	Flexes toes
Tibialis posterior	Extends ankle; inverts foot
Tibialis anterior	Flexes ankle; inverts foot
Extensor digitorum	Extends toes; flexes ankle; everts foot.

alis, and *lateralis* are clear; the *orbi*cularis oris *orbits,* or surrounds, the mouth). Some indicate size (*maximus, longus*) or shape (*deltoid* means triangular; *trapezius,* trapezoidal; *serratus,* toothed). *Biceps, triceps,* and *quadriceps* indicate how many origins or heads (*-ceps*) a muscle has (two, three, or four). Some are named for their points of origin and insertion (the *brachioradialis* links the upper arm, or *brachium,* and the radius) or for their action (an *extensor digitorum* extends the digits—fingers or toes).

MUSCLE CONTRACTION

The most important feature of muscle is its ability to contract when stimulated by a nerve signal. It owes this ability to the structure of its component cells, the muscle fibers. Contraction itself is produced by many contractile filaments within each fiber. The ability to respond to stimulation by contracting is due to the way these filaments work.

The Structure of Muscle Fibers

Each muscle fiber is a single cylindrical cell between 10 and 100 micrometers thick and up to 30 centimeters (12 inches) long. The many nuclei found in a single muscle fiber (or cell) reflect its origin in the fusion of many cells in the embryo. Like other cells, each muscle fiber has a cell membrane, cytoplasm, and an endoplasmic reticulum, but these components differ in several ways from those of more ordinary cells and have their own names. The names all include the root *sarco-,* meaning flesh (which is what muscle is). The cell membrane of a muscle fiber is called the **sarcolemma.** Its cytoplasm is the **sarcoplasm.** Its endoplasmic reticulum is the **sarcoplasmic reticulum.**

Sarcoplasm differs from ordinary cytoplasm in that it is almost completely filled with contractile protein filaments, the **myofilaments** (*myo* = muscle), organized into bundles called **myofibrils.** The sarcoplasmic reticulum forms a network of tubules surrounding the myofibrils in each muscle cell. Extending inward from the sarcolemma is a second set of tubules, the **transverse (T) tubules.** The T tubules are flanked by larger sarcoplasmic reticulum tubules to form **triads.** The triads are linked by smaller sarcoplasmic reticulum tubules (Figure 9-8).

There are two kinds of contractile elements (myofilaments) in the sarcoplasm. The *thick myofilaments* are composed of the protein **myosin.** The *thin myofilaments* are composed mainly of the protein **actin.** Together these myofilaments form the contractile units of a muscle fiber, the **sarcomeres.** A single sarcomere segment is only 2.6 micrometers long, so that a 30-centimeter-long muscle fiber may contain over 100,000 of them.

Through the microscope, we see that each sarcomere is bounded by two dark lines (Figure 9-9, p. 202). The thin myofilaments run parallel to the long axis of the fiber, away from each of these **Z lines.** They do not, however, reach all the way to the next Z line. They end somewhat short of the midpoint, leaving a gap. The thick myofilaments of myosin fill the gap and overlap with the thin myofilaments of actin. When the muscle contracts, the different types of myofilaments slide over each other, pulling each end of the sarcomere toward the middle, shortening it in the process. The net effect increases the amount of overlap between the actin and myosin filaments. The clear zone near the Z

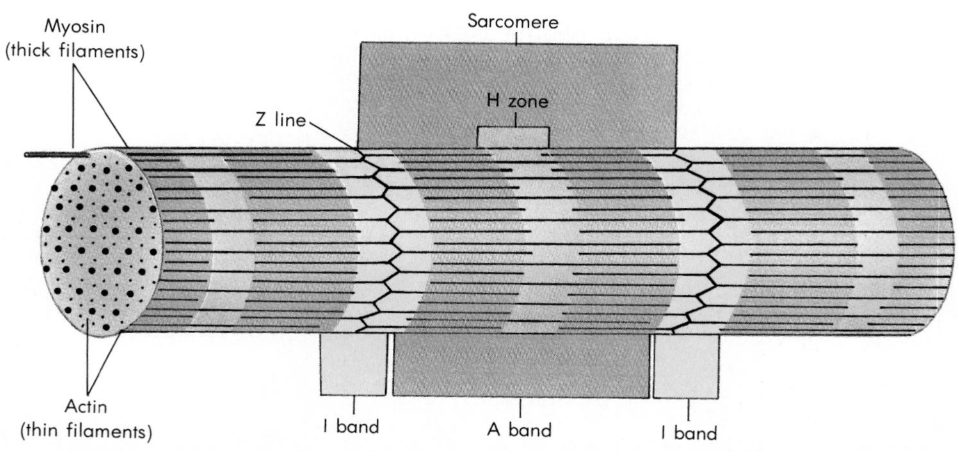

FIGURE 9-8 Skeletal muscle fiber showing some of the components of the contractile system.

Learning Focus

The Major Muscles of the Human Body

Label the following diagrams of the human muscles. Draw an arrow to each of the major muscles, write the name and the function of each muscle right on the diagram. You might want to make several copies of the diagrams (before you label them) so you can use them several times for review before you take your exam.

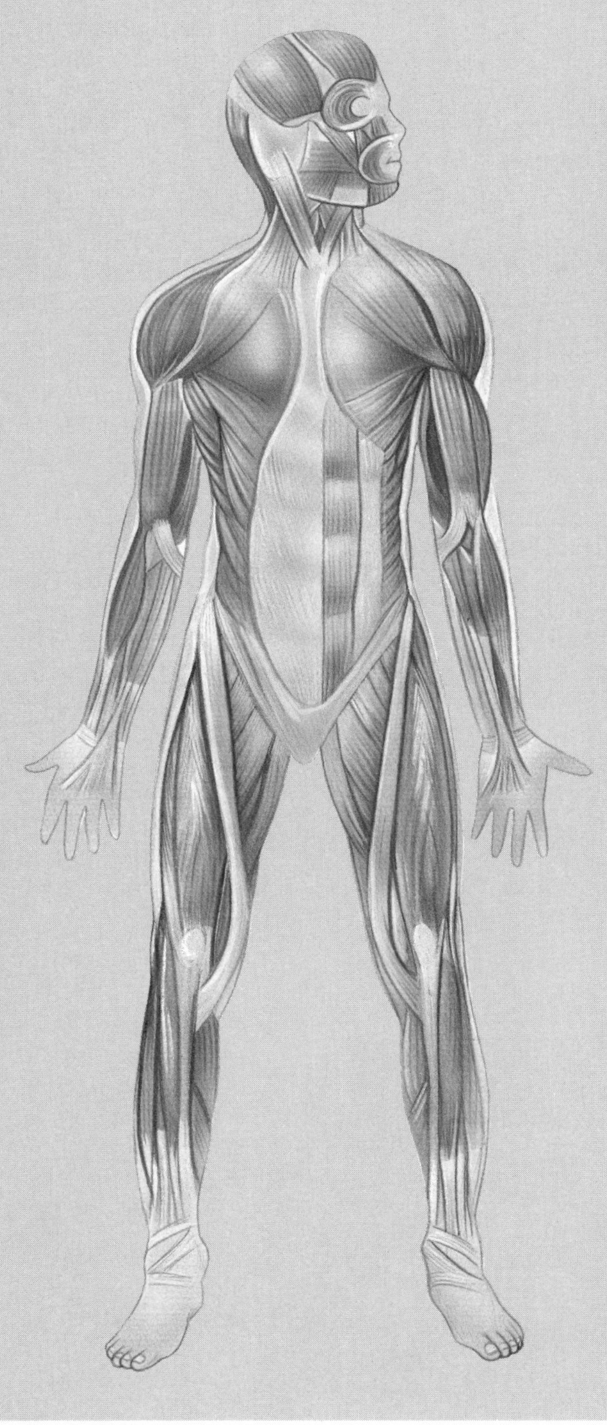

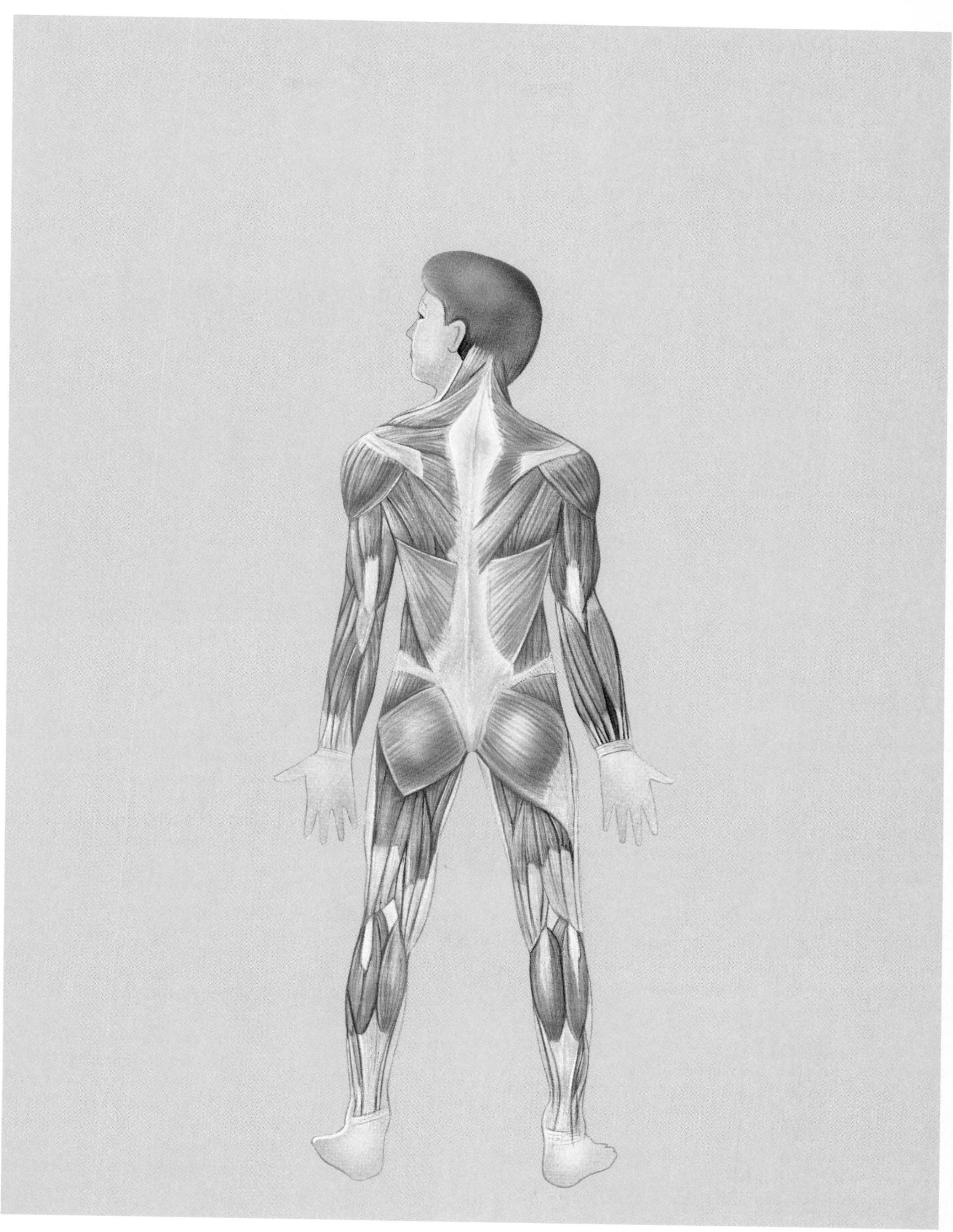

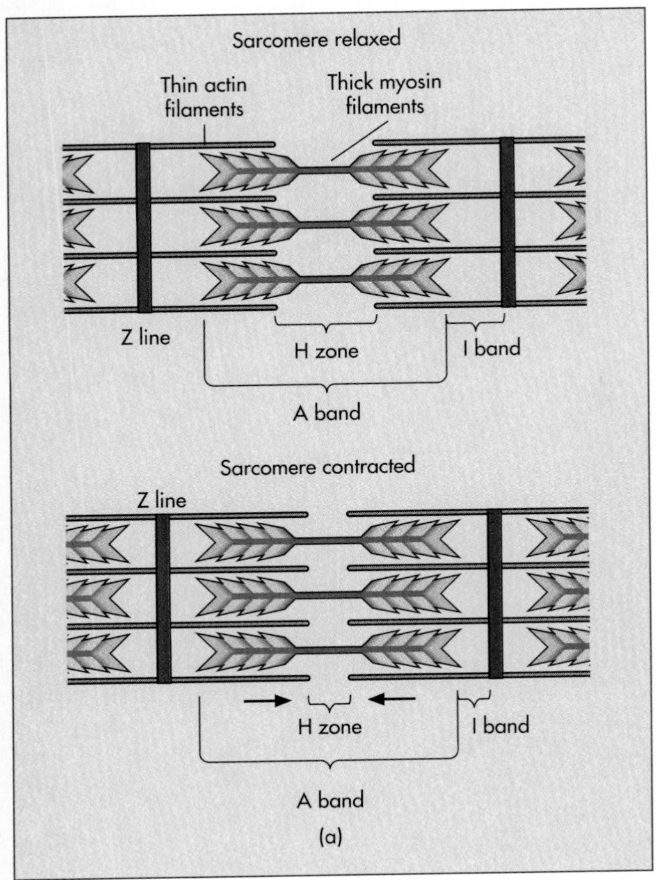

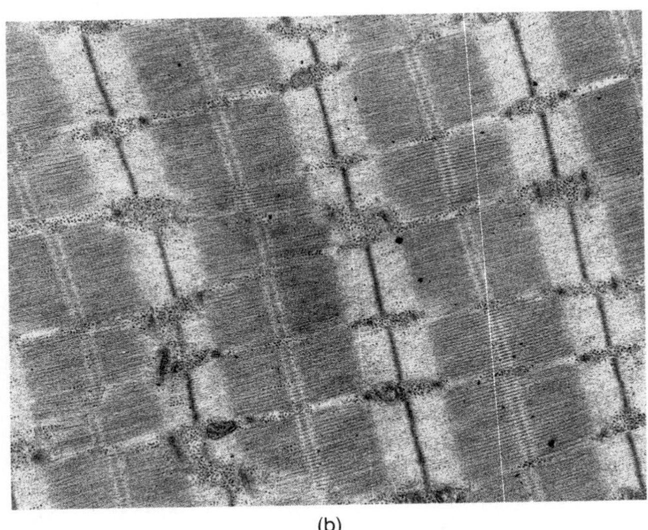

FIGURE 9-9 (a) Diagram of the sarcomere. (b) Electron micrograph of sarcomeres found in rat skeletal muscle cells.

line, occupied only by the thin actin myofilaments, is called the **isotropic (I) band.** The gap in the middle of the sarcomere, occupied only by thick myosin myofilaments, is the **H zone.** The **anisotropic (A) band** is occupied by the thick myofilaments; it includes the central H zone. The triads of sarcoplasmic reticulum and T tubules, two for each sarcomere, lie at the ends of the A band.

The Sliding Filament Theory of Muscle Contraction

Electron micrographs reveal that crossbridges link the thick and thin myofilaments where they overlap. These crossbridges are the heads of the myosin molecules. The thick myosin myofilaments are formed as bundles of the molecules, their shafts together and their heads jutting out to the sides (Figure 9-10).

The thin myofilaments are composed of actin, whose globular molecules are arranged in a double coil. Threadlike molecules of **tropomyosin** lie in the grooves of the coil. **Troponin** molecules sit near the ends of the tropomyosins (Figure 9-11). In the presence of calcium ions, the troponin and tropomyosin molecules seem to shift their positions slightly. The shift allows the myosin heads to bind to the actins of the thin myofilament. The calcium ions also stimulate the myosin heads to break down ATP and use the energy released to bend on their shafts (Figure 9-12, p. 204). Each myosin head then releases the actin it is bound to, snaps forward, and binds to another actin a little further down the thin myofilament. It then repeats the process as long as calcium ions remain available, and the thick myofilament "runs" along the thin myofilaments that surround it, increasing the zone of overlap. As it does so, the H zone and I band shrink, and each sarcomere shortens by about a micrometer. As each myosin head binds and lets go, it also emits tiny crackling sounds; the millions of such sounds that accompany a contraction add up to make creaks, groans, and thumps that are sometimes loud enough to hear.

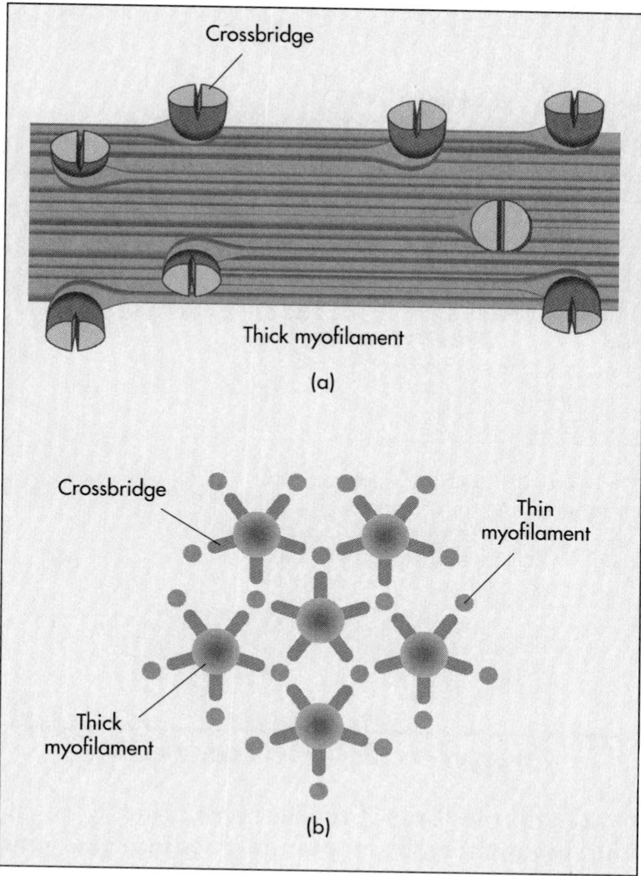

FIGURE 9-10 (a) In the thick myofilaments, the myosin molecules bundle together to form heavy strands with the knobby heads, or crossbridge units, sticking out at intervals along the length of the strands. These heads bind to actin during muscle contraction. (b) Cross-sectional view showing the relationships between several thick and thin myofilaments.

---

1. During muscle contraction, a sarcomere shortens by about a micrometer but the muscle itself may shorten by many centimeters. Why?
2. A stretched muscle, in which the zone of overlap between the thick and thin myofilaments is reduced, cannot contract as strongly as a relaxed muscle or a partially contracted muscle. Why?
3. Why do you think the triads are located near the ends of the A band instead of, say, at the Z lines?

When calcium ions are not available, the myosin heads release their grip on the actins and the muscle relaxes. However, this process also requires ATP. If no ATP is available, the myosin heads remain locked in position, which accounts for the rigidity (*rigor mortis*) that develops soon after death as ATP disappears from the body's cells.

Calcium Ions and Control

Clearly, calcium ions offer a convenient way to control muscle contractions. In fact, they are so used, and this is the function of the sarcoplasmic reticulum and T tubules. In a relaxed, noncontracting muscle, the sarcoplasmic reticulum absorbs and stores calcium ions from both the extracellular fluid and the sarcoplasm.

When a nerve signal (discussed in Chapter 14) reaches the sarcolemma of a muscle fiber, it triggers the movement of sodium and other ions across the sarcolemma. The resulting movements of ions produce electrical fields that can be detected from outside the body. A recording of these fields is an **electromyogram** (Figure 9-13). It is used to study the way muscles contract under different conditions and for different purposes.

The ion movements and electrical activity expand like a wave across the sarcolemma and down into the transverse tubules. Once there, the signal stimulates the large tubules of the sarcoplasmic reticulum nearest the T tubules to release calcium ions into the sarcoplasm, where they trigger contraction. When the signal ends, the membrane of the sarcoplasmic reticulum pumps the calcium ions back into their reservoir and contraction ends.

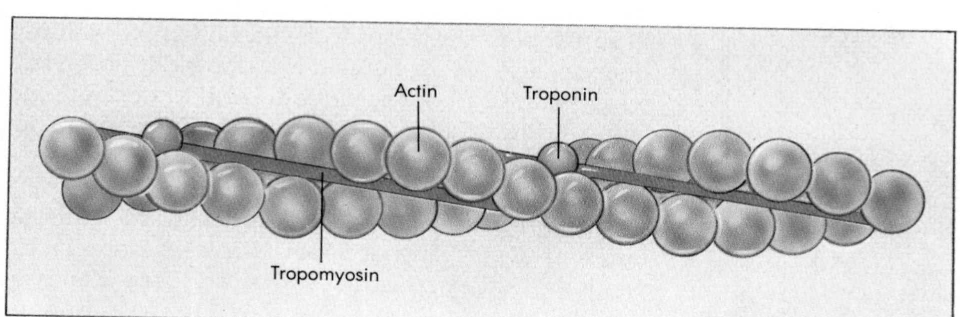

FIGURE 9-11 The thin myofilament consists of tiny globular actin molecules arranged in a double coil. The tropomyosin threads lie in the grooves between the actin molecules. At the end of each tropomyosin is found a troponin molecule.

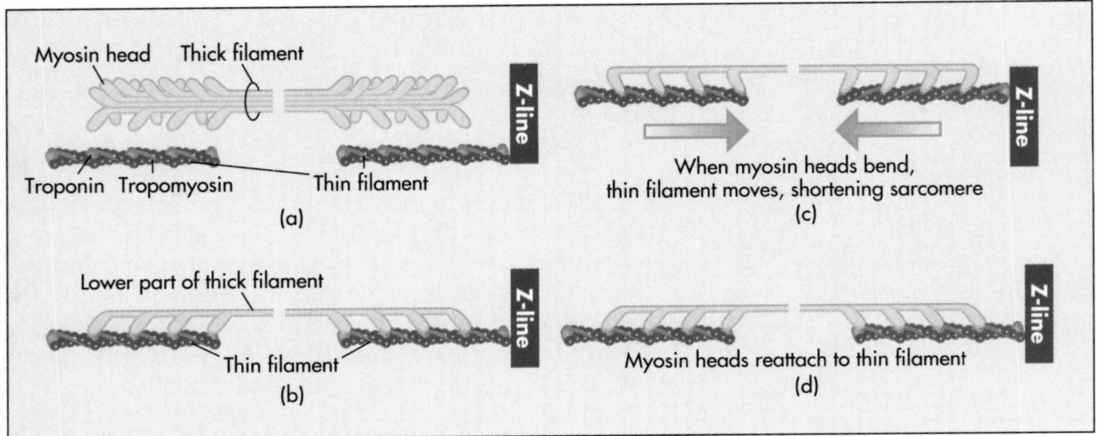

FIGURE 9-12 The "sliding filament" mechanism of contraction. (a) The relationship of the thick and thin filaments in a relaxed muscle fiber. (b) Myosin heads bind to the actin units on the thin filaments. (c) The myosin heads bend (using energy from ATP), and ratchet the two sliding actin filaments toward the center of the sarcomere, shortening it in the process. (d) The myosin heads extend the contraction by releasing and reattaching to other actin molecules farther along the filament and then bending again. To prevent back-sliding, not all myosin heads release and reattach at the same time.

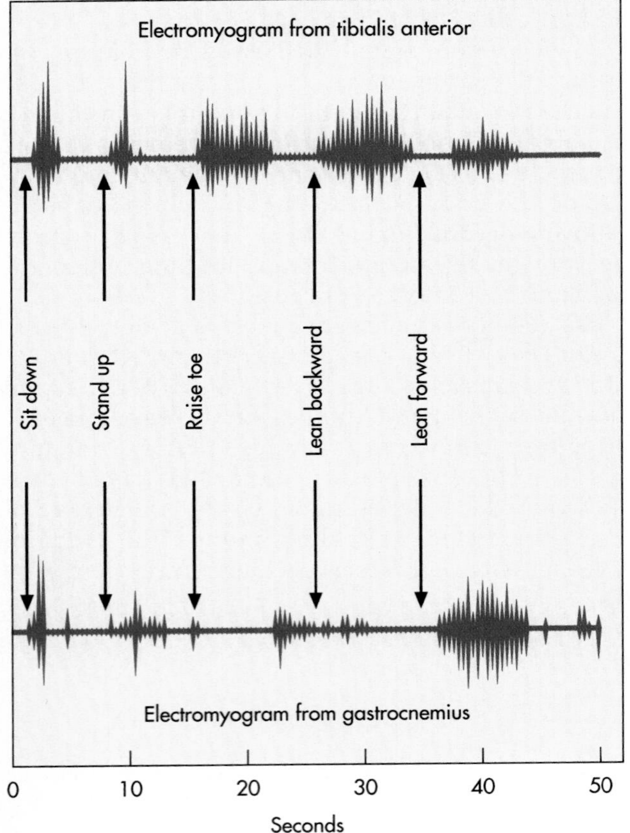

FIGURE 9-13 Electromyograms are graphic pictures made of the electrical activities generated by muscle contractions.

Energy for Muscle Contraction

Powering the "steps" of the myosin heads on the thin myofilaments and pumping sodium and other ions across the sarcolemma and calcium into the sarcoplasmic reticulum after a muscle contraction requires energy. Muscle fibers obtain this energy from ATP, but the ATP stored in a single muscle fiber is enough for only two or three seconds of muscle activity. To sustain more prolonged activity, the ATP must be continuously regenerated from glycogen (animal starch) stored in the muscle fibers, from glucose delivered by the blood (the mechanics of ATP generation are discussed in Chapter 21), and from the energy storage substance creatine phosphate.

Muscle activity is limited not by the amount of energy available, but by the supply of oxygen and by the blood supply, which delivers oxygen and glucose and carries away by-products of energy generation such as lactic acid. Without oxygen, the amount of ATP generated from glucose is severely limited and lactic acid production increases tremendously. The lactic acid interferes with both calcium release from the sarcoplasmic reticulum and the binding of calcium to troponin. Its buildup accounts for the loss of strength we experience as we grow fatigued, and its removal is an important part of the postexercise recovery process. We call the amount of oxygen necessary to remove this lactic acid the "oxygen debt" incurred by the exercise.

Learning Focus

Muscle Contraction

The process of muscle contraction can be hard to grasp if you do not take the time to follow it through all its steps. Let us therefore slow down for a moment and work through the process once more. Begin by making several copies of this Focus box so that you can repeat this review exercise when you are preparing for exams.

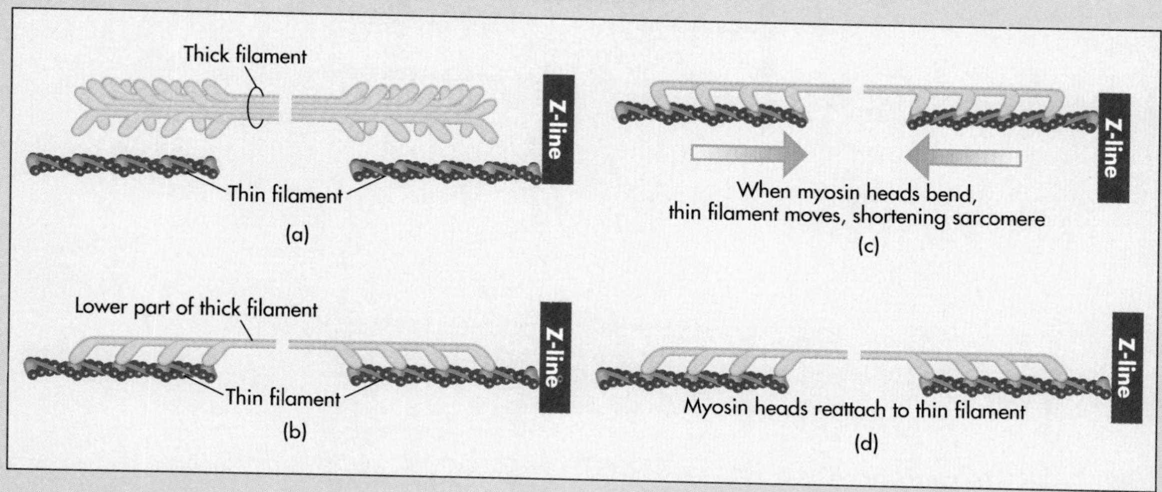

1. On the diagram, label the myosin myofilaments and label the myosin heads.

2. Identify the actin filaments. Show the position of troponin and tropomyosin on the thin filaments.

3. How do the myosin heads make the thick and thin myofilaments increase their overlap?

4. Where do the myosin heads get the energy they need to power the contraction process?

5. From the text, describe how calcium ions promote muscle contraction.

6. Draw on the diagrams where ATP is likely to be involved in the contraction process (see text).

7. What happens when ATP is not available? Why?

(continued on p. 206)

Now label the sarcoplasmic reticulum, T tubules, Z line, A and I bands, H zone, and thick and thin myofilaments in the following diagram of a sarcomere:

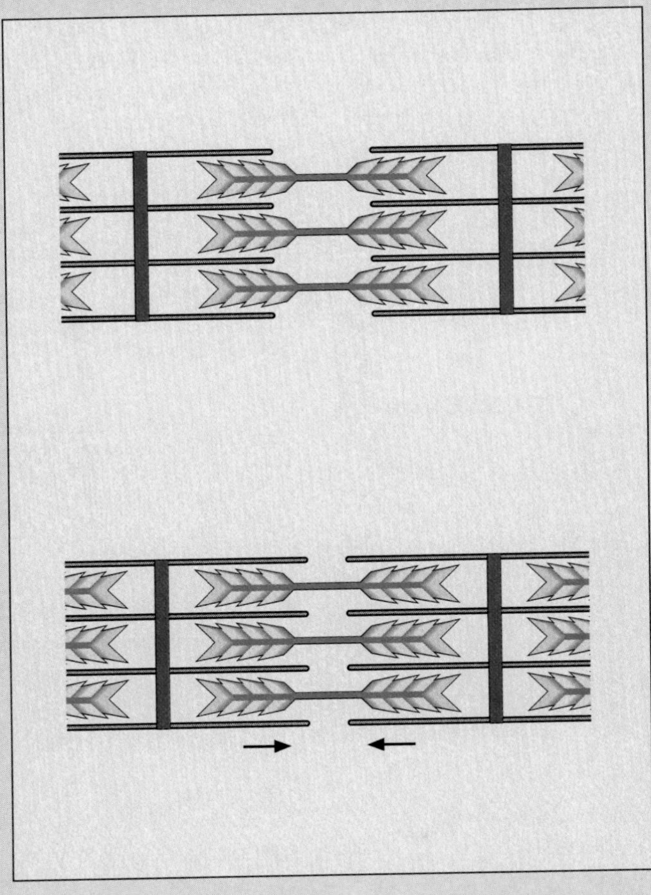

1. Do the thick or thin myofilaments themselves shorten during contraction? Why or why not?

2. What is the role of the T tubules?

3. What is the function of the sarcoplasmic reticulum?

4. What happens to the A band during contraction?

5. What happens to the I band during contraction? Why?

6. What happens to the H zone during contraction? Why?

7. Where are the enzymes that break down ATP for the energy a sarcomere needs to contract?

8. What determines the limit on how much a sarcomere can contract?

9. What determines how much a muscle fiber can contract?

Chapter 10

The Blood and the Cardiovascular System

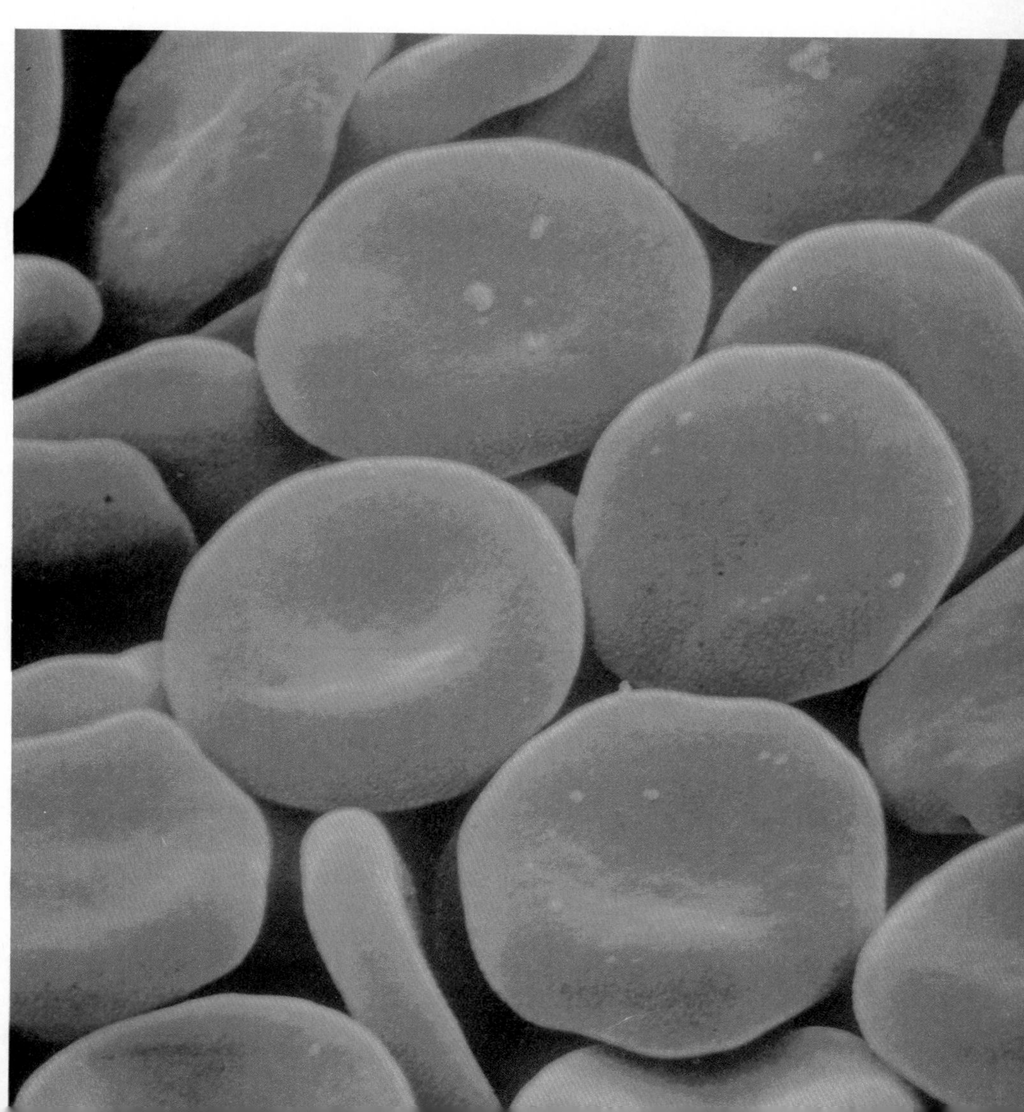

Fast and Slow Muscles

There are two kinds of skeletal muscles in the muscular system. One contracts often and powerfully. As an adaptation to this high level of activity, it has more capillaries to supply oxygen and glucose and remove lactic acid, more mitochondria, and more of the protein called **myoglobin,** which stores oxygen within muscle fibers. Because the myoglobin and the mitochondrial ATP-generating enzymes are reddish in color and because of the increased blood supply, these muscles are known as **red muscles.** They are also called **slow muscles,** for their contractions, while powerful, are relatively slow.

Fast muscles are **white muscles.** They have fewer capillaries and mitochondria and less myoglobin than do slow muscles, but what they have is enough for their needs. Their contractions tend to be fast twitches, as in the eye muscles.

In humans, many muscles have both red and white fibers (Figure 9-14). The different fibers are activated by the nervous system for different purposes, and they respond differently to training. Endurance athletes (such as marathon runners) develop more capillaries in the muscles they use and more mitochondria and myoglobin in both their white and red fibers, so that both fiber types become redder. "Impulse" athletes such as sprinters, shot putters, and weight lifters, who need strength and speed more than endurance, tend not to develop more mitochondria, myoglobin, or capillaries. For them, training increases the number of myofilaments in and the size of muscle fibers, especially the white ones, and the amount of stored ATP, creatine phosphate, and glycogen.

1. Relate the differences in the muscle changes in marathoners and sprinters to their energy needs.
2. Why do impulse athletes develop bulkier muscles than endurance athletes?

CONTROLLING THE MUSCLES

Muscles are controlled by nerve signals generated in the brain and spinal cord. These signals are transmitted to a muscle by way of long nerve fibers called axons. As the axons near the muscle, they branch, and the end of each branch forms a *neuromuscular junction* on a muscle fiber (Figure 9-15). When a nerve signal reaches the junction, the end of the axon releases small packets of the chemical acetylcholine into the gap between the axon's cell membrane and the sarcolemma. The acetylcholine diffuses across the gap between the membranes and binds to protein receptors on the surface of the sarcolemma, where it initiates the wave of ion movements that triggers contraction. An enzyme in the gap promptly destroys the acetylcholine to keep the signal brief. Anything that interferes with the enzyme (such as the poison strychnine) prolongs the signal and sustains the contractions causing convulsions. Anything that interferes with the receptors in the sarcolemma and prevents them from binding to the acetylcholine (such as curare) causes paralysis. **Myasthenia gravis,** a progressive

FIGURE 9-14 Red and white fibers in human muscle. The red (slow-twitch) fibers are darker and contain more mitochondria and large amounts of myoglobin. They contract more slowly and have greater endurance than the fast-twitch white (light) fibers. Light fibers fatigue easily, but they have a very fast contraction velocity.

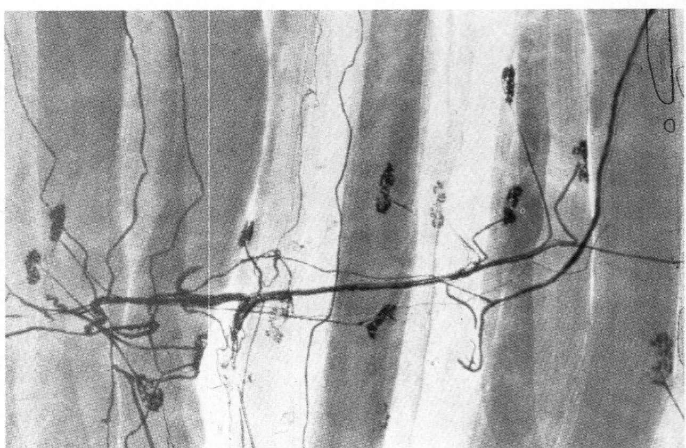

FIGURE 9-15 Neuromotor end plates, where nerve fibers release chemical transmitters on skeletal muscle cells that cause the cells to contract.

Box 9-1 Making Muscles Stronger: Exercise Versus Anabolic Steroids

Young people often wish for larger, stronger, sleeker, sexier bodies. Some of the change they crave comes simply with the transition from childhood to adulthood. More can come if they exercise their muscles with weights, swimming, running, and sports (among other activities). Impulse exercises (e.g., weight lifting) cause the cells of fast, white-fibered muscles to gain myofilaments and thus increase in size; they also increase the amount of energy they store. Endurance exercise (distance running, swimming) causes both fast, white-fibered and slow, red-fibered muscles to change by adding capillaries, mitochondria, and energy storage.

Exercise has other benefits: It can help control weight and keep cholesterol level down. But many people think it has serious drawbacks as well: It takes time, sweat, and tears.

There is an alternative. It cannot help with weight or cholesterol. In fact, it might even make the latter worse. But it will make your muscles bigger, partly by enhancing their response to exercise. It is the use of *anabolic steroids.*

Surveys have indicated that about 500,000 U.S. teens use anabolic steroids to enhance muscle size and strength. Almost half use steroids to improve athletic performance, as do many professional athletes, even though such use has been banned as giving "improved" athletes an unfair advantage. Over a quarter of teen users, on the other hand, take anabolic steroids—as pills or injections—to improve their appearance.

Anabolic steroids are related to the male sex hormone testosterone (see Chapter 17), which among its several effects accelerates protein synthesis in muscle cells and causes muscles to increase in size and strength. Unfortunately, these muscle-building drugs have side effects more serious than just disqualifying users who win competitions. Depending on the type, dose, and quality of the steroid, males may develop male pattern baldness, testicular atrophy (shriveling), impaired fertility, breast enlargement, and/or acne. Females may develop a deeper voice, enlarged clitoris, decreased breast size, and increased facial hair. More serious, both sexes may develop high blood pressure and increased levels of the kind of blood lipids linked to cardiovascular disease and heart attacks.

People who take large doses of anabolic steroids also risk serious liver damage, ranging from jaundice to liver cancer. Young teens may see their limb bones stop growing, leaving them stunted. Since steroids stimulate the growth of muscle but not a matching increase in strength of connective tissue ligaments and tendons, there is a risk of joint injuries. Anabolic steroids may also lead to behavioral changes, including increased aggressiveness, disturbed sleep patterns, and dramatic mood swings (from excessive aggressiveness and hostility when on the drugs to apathy, listlessness, and poor self-esteem when off the drugs).

Want to look good? Work up a sweat. Exercise has fewer unfortunate side effects. Even better, the muscles you strengthen and enlarge by sheer exercise are all yours. They are muscles of which you can truly be proud.

weakening of the muscles, appears to happen when the body's immune system attacks and destroys the receptors.

TYPES OF MUSCLE CONTRACTIONS

We often speak of **isometric** and **isotonic** muscle contractions. When a muscle contracts isometrically, it remains the same length, although the strength of the pull it exerts on its attachments to the skeleton may vary widely. It does not actually contract or change length because something is preventing the joint it crosses from moving. An example is attempting to lift an object that is too heavy. Another example is doing isometric exercises, such as pushing the arms outward against a doorframe.

In isotonic contractions, the muscle is free to change length. The load is free to move, and as soon

as the muscle develops sufficient strength, it does move. Thereafter, the strength of the contraction remains roughly constant, which is what *isotonic* means.

Studied in the laboratory, isotonic contractions reveal several interesting features of muscle action. The researcher anesthetizes an animal such as a frog or rat and frees one end of a muscle from its attachment to the skeleton, leaving intact its blood vessels and nerve. He or she then attaches the free end of the muscle to a device that can record the pull exerted by a contraction and stimulates the muscle or its nerve electrically (Figure 9-16).

The response to a single brief stimulus is a **twitch.** The twitch does not begin until two or three milliseconds after the stimulus; it is during this **latent period** that the sarcoplasmic reticulum is releasing calcium ions and the myosin heads are beginning to pull on the thin myofilaments. After the latent period, the muscle begins to contract. Within a few more milliseconds, the contraction ends and the muscle relaxes. When the muscle has returned to its initial length, the twitch is over (Figure 9-17).

Twitches vary in strength with the strength of the stimulus, but not because of any change in the response of individual muscle fibers. The fibers obey the **all-or-none principle;** that is, if a stimulus is strong enough to make them contract, each fiber contracts all the way. The variation in the strength of muscle twitches arises because stronger stimuli activate more motor units.

When a second identical stimulus follows the first quickly enough, we see the phenomenon known as **temporal summation.** The twitches the stimuli evoke merge, and the total contraction is stronger than that for a single stimulus. The gain in strength arises because the second twitch begins before the first has finished its relaxation phase. The second twitch therefore need not overcome the muscle's elasticity (slack) before beginning its contraction. The same thing happens with a third stimulus, a fourth, and so on (Figure 9-17).

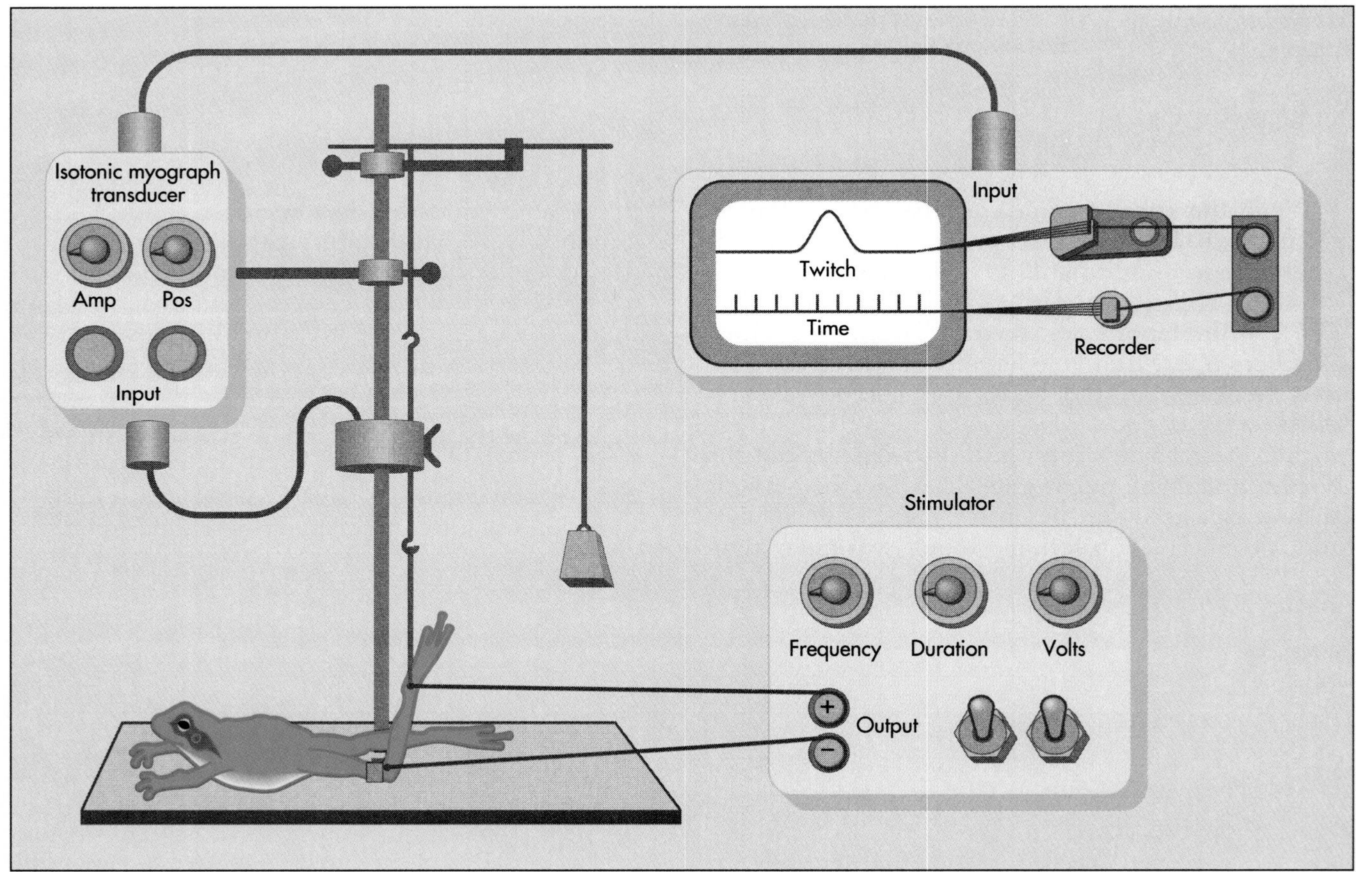

FIGURE 9-16 Isotonic contractions of the gastrocnemius muscle can be measured graphically when the muscle is electrically stimulated. When the frog's muscle contracts, it pulls the hook, lever, and weight and activates the transducer that measures the extent and duration of the contraction. The recorder plots both the muscle "twitch" and indications of the stimulus and of elapsed time.

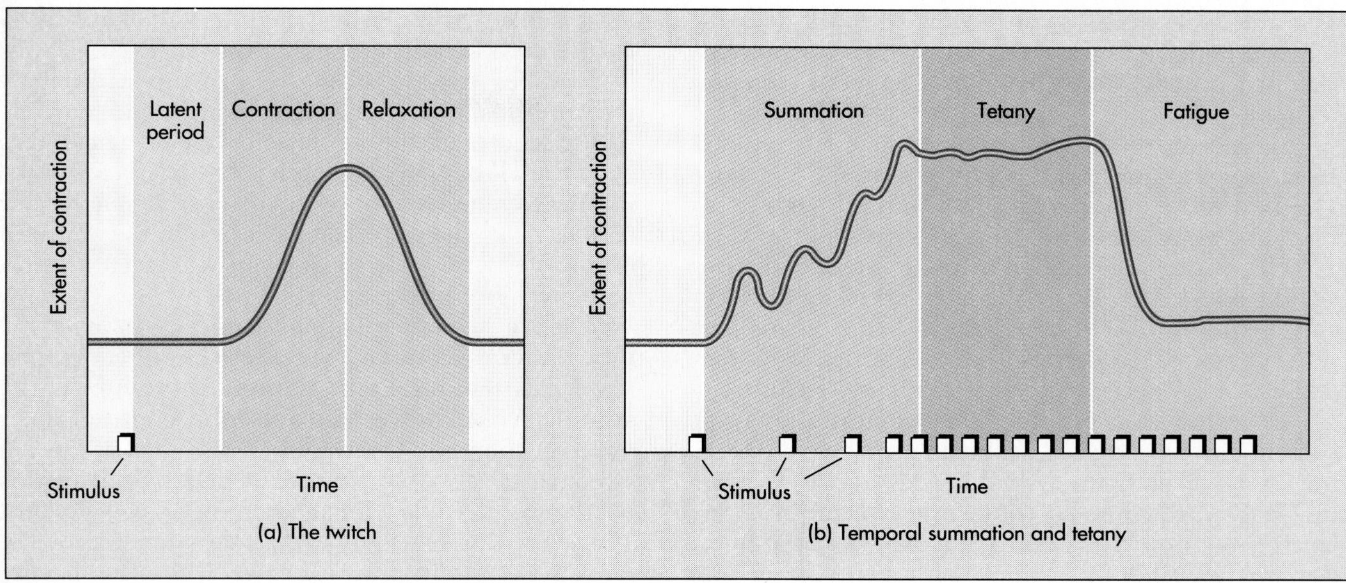

FIGURE 9-17 (a) The stages of muscle twitch produced by a single electrical stimulus to a muscle. (b) Muscle contraction. Temporal summation occurs when a muscle is stimulated by a fast series of electrical impulses delivered by way of a nerve (23.5 times per second). The muscle does not have enough time between stimuli to return to its relaxed state. As a result, each contraction occurs on top of the previous contraction, and the level of muscle tension continuously increases. Sustained contractions, tetany, are produced at high rates of stimulation (greater than 114 per second).

When the stimuli come rapidly enough, the resulting twitches merge to produce a smooth contraction curve and a still stronger maximum contraction. This condition is called **tetanus** (Figure 9-17). In the laboratory, the energy stored in muscle fibers is eventually exhausted, and the contraction weakens. In the body, the nervous system shifts activity from one group of muscle fibers within the muscle to another, thus sharing out the burden and delaying fatigue. However, the nervous system *can* activate the entire muscle all at once and produce contractions much stronger than usual, as in cases of "hysterical strength," for example, when a parent lifts a car off a young child or during epileptic convulsions. Unfortunately, such excessive contractions can tear tendons loose from their attachments and even break bones.

1. Explain the phenomenon of temporal summation.
2. How can a muscle have a "reserve strength" beyond what can normally be called upon?

SUMMARY

Muscles are masses of cells specialized for the ability to contract. They come in three types: the cardiac muscle of the heart, the smooth muscle of the walls of tubular organs and a few other places, and the skeletal muscle. Like cardiac muscle, skeletal muscle is striated because of the way its component protein filaments line up. Unlike both cardiac and smooth muscle, it is under voluntary control.

Table 9-3

Clinical Summary: Selected Diseases of the Muscular System

DISORDER	DESCRIPTION
Tendinitis	Inflammation of tendon sheaths or synovial membranes of some joints, usually in wrists, elbows, shoulders, and ankles. May involve severe pain and swelling. Often results from exercise (as in tennis elbow).
Muscular dystrophy	Progressive degeneration of muscle fibers leading to muscle atrophy and death in the teens. Cause is a defective gene for the protein dystrophin that leads to leakage of calcium ions from storage sacs into the cell. The most common muscular dystrophy (Duchenne type) affects one male in 25,000.
Myasthenia gravis	Muscular weakness due to destruction by the immune system of acetylcholine receptors in the neuromuscular junction. More common in women aged 20–50. Most often affects muscles of the head and neck. Death can result (rarely) if the disease affects the respiratory muscles.

Skeletal muscles stretch across the joints of the skeleton, attaching by way of tendons to the periosteum of the bones or to connective tissue sheets and bands. Muscle contractions cause the bones to pivot on their joints. When muscles contract, the bone to which one end attaches remains relatively immobile; this is the muscle's origin. The more mobile point of attachment is called the insertion. Between the muscle and its attachment is a cordlike or sheetlike tendon, an extension of the layers of connective tissue that surround the muscle itself, the bundles of cells or muscle fibers within it, and the individual cells.

Skeletal muscles often come in pairs of antagonistic muscles, which pivot the joint in opposite directions when they contract. The contracting muscle of a pair is the agonist; the other is the antagonist. Muscles that contract to stabilize the body against movements are synergists.

All muscle is contractile, excitable (responsive to stimulation), extensible (stretchable), and elastic (it recovers after contraction or stretching). The elasticity is largely due to a muscle's connective tissue sheaths. The contractility is due to its protein myofilaments. The myofilaments are arranged in bundles, the myofibrils, within each muscle fiber. Within the myofibrils, the myofilaments show a regular pattern within repeating contractile units, the sarcomeres.

Thick myofilaments are composed of myosin molecules, whose heads (the crossbridges) project from the sides of the filament to interact with the thin myofilaments. Thin myofilaments consist of actin, to which the myosin heads bind. A nerve signal and the presence of calcium cause the myosin and the actin to slide over each other, shortening the sarcomere (and the muscle) during contraction.

To contract, muscles use the energy stored in adenosine triphosphate (ATP) and replenished from glycogen, creatine phosphate (stored in the muscle cells), and glucose (delivered by the blood). Muscles with ample blood supply are called red or slow muscles; they contract often and powerfully, and they are strengthened by endurance training. Muscles that have less blood supply are called white or fast muscles; they are designed to contract for shorter periods, but more quickly, and their forte is strength and speed, not endurance.

Muscles are controlled by signals delivered to neuromuscular junctions by long extensions of nerve cells in the brain and spinal cord. In isometric muscle contractions, the muscle develops tension but remains the same length. In isotonic contractions, the muscle maintains a constant tension while it shortens. Studies of the latter in the laboratory reveal that the smallest, simplest muscle contraction, in response to a single electrical shock to a nerve, is the twitch.

STUDY QUESTIONS

1. Muscle is contractile, excitable, extensible, and elastic. Why is each of these features important to muscle function?
2. Name four differences between cardiac and skeletal muscle.
3. Name four differences between smooth and skeletal muscle.
4. Distinguish among epimysium, perimysium, and endomysium.
5. How can you tell which end of a muscle is its origin?
6. Draw a sarcomere and label its parts.
7. Explain the role of calcium ions in initiating muscle contraction.
8. Why is creatine phosphate an essential energy source for muscle contraction?
9. Explain the different roles of red and white muscles.
10. Why do you think motor units contain only red or only white muscle fibers?
11. Why do muscle twitches vary in strength?
12. Why does a rapid sequence of twitches produce a stronger total muscle contraction than a single twitch?

GLOSSARY

Actin The major protein of the thin myofilaments.

Agonist Of the two or more muscles crossing a joint, the muscle that is contracting during a movement.

All-or-none principle The idea that a muscle fiber contracts either fully in response to a stimulus or not at all.

Antagonist The muscle that, because its action opposes an agonist, must relax to permit motion.

Cardiac muscle The striated, involuntary muscle of the heart.

Electromyogram A recording of a muscle's electrical activity.

Insertion The attachment to bone of the more mobile end of a muscle.

Latent period The delay between a stimulus and a contraction.

Muscle fiber A muscle cell.

Muscular system The body's skeletal muscles.

Myofilaments The contractile protein filaments within a muscle fiber.

Myosin The protein of the thick myofilaments.

Origin The attachment to bone of the relatively immobile end of a muscle.

Red muscle A muscle with a rich blood and myoglobin supply; specialized for endurance.

Sarcomere The functional unit of muscle fiber contraction; a repeating unit of myofibril structure.

Skeletal muscle Striated, voluntary muscle; responsible for moving the skeleton.

Smooth muscle The nonstriated, involuntary muscle of the walls of the body's tubular organs, the iris of the eye, and the arrector pili (hair erectors).

Synergist A muscle whose contraction steadies the agonist's origin.

Temporal summation The merging of successive twitches to produce a stronger, longer contraction.

Tendon A cord or band of connective tissue formed by the merging of epi-, peri-, and endomysium at the end of a muscle, which links the muscle to periosteum or fascia.

Transverse (T) tubule An invagination of the sarcolemma that delivers signals to the myofibrils.

Tropomyosin A control protein in the thin myofilaments.

Troponin A control protein in the thin myofilaments.

White muscle A muscle with less myoglobin and fewer capillaries than red muscle; specialized for strength and speed.

Whether you are studying emergency room procedures in medical school or first aid with the American Red Cross, you learn that the most serious kinds of injury are usually those involving severe blood loss or an interruption of blood flow. When you are treating a victim, you first control profuse bleeding; once that is done, other problems like broken bones and shock can be handled successfully.

What is it about the blood and the cardiovascular system (the pump and tubes channeling its flow) that make them so important? The human body is an interconnected community of cells in which each cell contributes to the survival of the whole body but in which each cell is also dependent on a wide variety of exchange systems for its own survival. The blood picks up oxygen from the respiratory system and distributes it, so that all cells receive their required share. The blood also picks up nutrients from the digestive system and distributes them. In addition, it manages to carry the various chemical wastes to their appropriate disposal sites, participates in body defense, delivers hormonal messages from one group of tissues to another, and redistributes body heat.

The blood and the cardiovascular system are thus the body's pickup and delivery service. And this service does not dawdle. The body's cells need their oxygen and nutrients *now*. To meet this rush demand to supply the large amounts of energy the body needs to maintain its high temperature and activity levels and growth rates, the heart pumps the blood at fairly high pressure, rapidly. A leak can therefore be disastrous.

The **cardiovascular,** or **circulatory, system**—the arteries, arterioles, veins, venules, capillaries, and heart—usually contains about 5 liters of whole blood. You can lose up to a tenth of this quantity—1 pint, the volume of a blood donation—and hardly notice the loss. You can recover from a loss of 30 percent or more of your blood (about 1.5 liters) following an accident. Lose much more and you die.

THE BLOOD

Blood is not simply a liquid, but a complex mixture. About 45 percent of its volume consists of solid components, the blood cells (both white and red) and platelets that we call the **formed elements.** The other 55 percent is **plasma,** a solution of water in which are dissolved a host of materials such as proteins, lipids, nutrients, and minerals.

The blood's main function is to transport materials within the body. It delivers nutrients, oxygen, and chemical messages to the body's cells. It carries away wastes. It also regulates the body's pH (acidity) and water content, helps control body temperature, and is a main site of the body's immune system (discussed in Chapter 24).

The blood does not bathe the body's cells directly. Its flow is confined within a network of tubes, the blood vessels, whose narrowest elements, the capillaries, pass very close to virtually all the cells of the body. Nutrients, wastes, and other substances pass between the blood and the tissues mostly by diffusion. The materials enter and leave the blood through the walls of the capillaries. They pass into the **interstitial fluids** surrounding the tissue cells and then across plasma membranes into and out of the cells.

Blood Plasma

The liquid fraction of the blood is called the plasma. It consists mostly of water, as shown in Table 10-1. The fluid character of plasma makes it the ideal medium for transporting the blood cells and the dissolved materials it contains throughout the body. Dissolved in the plasma are numerous proteins, over half of which are albumens; they bind to and transport hormones, drugs, and other substances and, because they contribute greatly to the osmotic pressure of the blood (its tendency to draw water from more dilute solutions), keep water from leaving the plasma for the tissues. The other plasma proteins include fibrinogen and prothrombin, both essential for blood clotting (discussed in what follows), and the globulins, or antibodies, produced by immune system cells in the blood. The albumens, prothrombin, and fibrinogen are produced by the liver.

The plasma also contains dissolved minerals—calcium, sodium, potassium, magnesium, chloride, phosphate, sulfate, bicarbonate, and other ions—oxygen, carbon dioxide, nutrients, and hormones (chemical messengers discussed in Chapter 13). The minerals are involved, like the albumens, in regulating the water content of the blood, but they also help maintain normal pH and meet the mineral needs of the tissues.

The plasma also carries and distributes nutrients absorbed from the digestive system or drawn from the body's reserves. Sugars are used as raw fuels by the cells, and *glucose* (also called blood sugar) is the primary way it is carried in the

Table 10-1

Plasma Components

COMPONENT	REMARKS
Water	91.5%; solvent and carrier
Proteins	
Albumens	Smallest plasma proteins; carriers, osmotic regulators
Fibrinogen	Essential for blood clotting
Prothrombin	Essential for blood clotting
Globulins	Antibodies; attack foreign materials
Nutrients	Absorbed from the digestive system
Glucose	Blood sugar
Amino acids	Building blocks for proteins
Fats	Used in cell membranes and energy storage
Wastes	Include CO_2, urea, creatinine, ammonia
Oxygen	Minor; most is carried in red blood cells
Minerals	Na^+, K^+, Ca^{2+}, Mg^{2+}, Cl^-, PO_4^{3-}, SO_4^{2-}, HCO_3^-; help control pH, osmotic pressure, serve tissue needs
Hormones	Intercellular chemical messages produced and released by cells of the endocrine glands. Regulate slow changes in the body such as growth or sexual cycling.

plasma. Amino acids likewise are distributed in the plasma as raw materials the individual cells can draw on to manufacture their own proteins. Fats usually travel as fats.

Serum is the amber fluid that is squeezed out of clotted blood. It really consists of plasma—*minus* the blood cells and the protein (fibrin) net that produces the clot. Serum is the liquid left behind when the soluble **fibrinogen** is converted to insoluble **fibrin,** which forms a network of fibers to trap blood cells and seal small breaks in blood vessels.

Serum is rich in antibodies (globulins), which the body uses to destroy disease organisms and toxins. One animal can therefore be temporarily protected against exposure to such dangerous materials using serum produced by another animal. For instance, blood is taken from horses that have been exposed to snake venom and developed antibodies against the venom. Once the blood has clotted, the resulting serum, now called antivenin, can be injected into a human snakebite victim and the antibodies it contains can help destroy the venom's toxicity.

Red Blood Cells

The most numerous blood cells are the **erythrocytes,** or **red blood cells.** Each cubic millimeter of blood contains about five million of these tiny dished-in (bi-concave) disks (Figure 10-1). Mature red blood cells lack nuclei and live in the blood for only about four months. By that time, the plasma membrane of the aging cells has deteriorated. White blood cells in the liver, spleen, and bone marrow destroy them, making way for their replacement by fresh new cells. Each red blood cell contains relatively large amounts of the protein **hemoglobin,** which is responsible for their red color.

The primary function of the red blood cells is to carry oxygen. Each red blood cell carries a supply of hemoglobin that binds to the oxygen absorbed from the lungs. The blood cells transport the oxygen to the various body tissues where the oxyhemoglobin complex breaks down and releases the oxygen. The oxygen then diffuses out of the blood to be absorbed by the tissue cells.

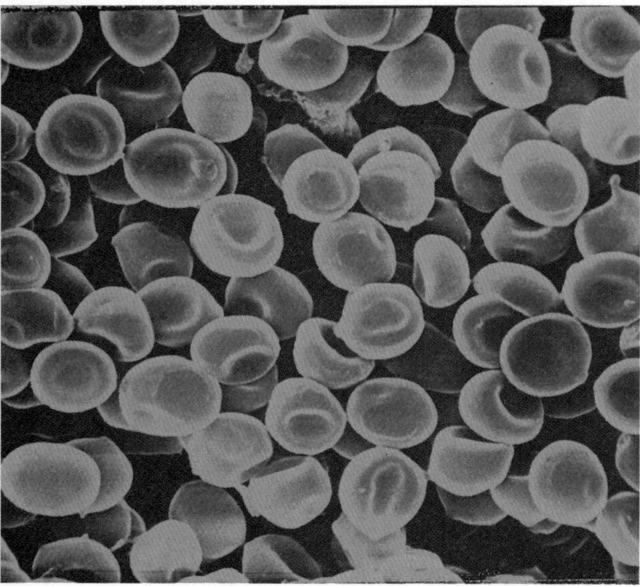

FIGURE 10-1 Scanning electron micrograph (color enhanced) of human red blood cells (3000 ×). Note their lack of a nucleus and the biconcave shape of the cells.

Hemoglobin is a complex multiple-unit protein (Figure 10-2). It contains four polypeptide **globins,** two *alpha* globins and two *beta* globins. Each globin is bound to a pigment molecule, a **heme** group, containing an iron ion. It is the iron that binds to oxygen in the lungs and releases it in the tissues.

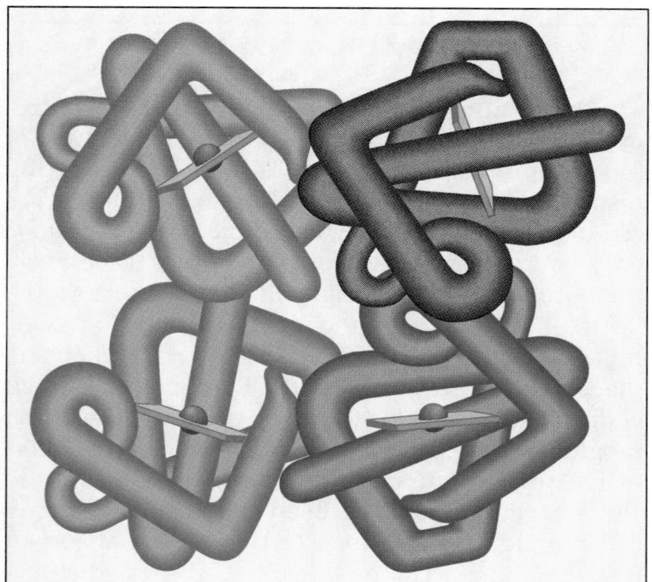

FIGURE 10-2 The hemoglobin molecule consists of four clustered polypeptides (two alpha and two beta globins), each one wrapped around a heme group. When hemoglobin carries oxygen molecules, the oxygens are bound to iron ions in the heme groups.

However, the globins play an important role. Their shape controls the iron's access to oxygen, and small changes in their amino acid sequence can drastically alter hemoglobin's ability to carry oxygen. They also bind to carbon dioxide in the tissues and release it in the lungs, carrying roughly a quarter of the carbon dioxide in the blood in this way (the rest is carried in bicarbonate ions dissolved in the plasma).

1. Carbon monoxide binds to hemoglobin's iron ions 200 times as powerfully as oxygen. Why then can breathing air containing even small amounts of carbon monoxide suffocate you?
2. The globins of fetal hemoglobin differ in structure from those of adult hemoglobin. The result is that fetal hemoglobin can bind more oxygen when less is available. Explain why this should help the fetus survive.

Erythropoiesis: Red Blood Cell Production

The formation of red blood cells is known as **erythropoiesis.** In adults, red blood cell formation occurs in the red bone marrow in the spongy interiors of the cranial bones, the sternum, the vertebrae, and the proximal ends of the humerus and femur. There a blood-cell-producing parent cell, the *hemocytoblast,* divides to generate cells that can give rise to all the different kinds of blood cells. One of these cells is the *rubriblast,* whose daughter cells mature into red blood cells. These red blood cells synthesize hemoglobin, expel their nuclei, and leave the marrow to circulate in the blood (Figure 10-3).

Erythropoiesis is controlled by the hormone (chemical messenger) **erythropoietin,** produced in the liver and elsewhere. Red blood cell production increases when blood oxygen levels fall, as can happen when one moves from sea level to high altitude, where the air is less dense and therefore contains less oxygen. In addition, medical conditions can reduce the oxygen-carrying capacity of the blood. These conditions, caused by low red cell counts or reduced levels of hemoglobin, are referred to as **anemias.** Severe blood loss can have similar effects.

In all such cases, the kidney responds to low oxygen levels by releasing **renal erythropoietic factor** into the blood, where an enzyme converts it into erythropoietin. When the hormone reaches the bone marrow, it stimulates the hemocytoblasts to produce more red blood cells. It works best when the need for more red blood cells and hemoglobin

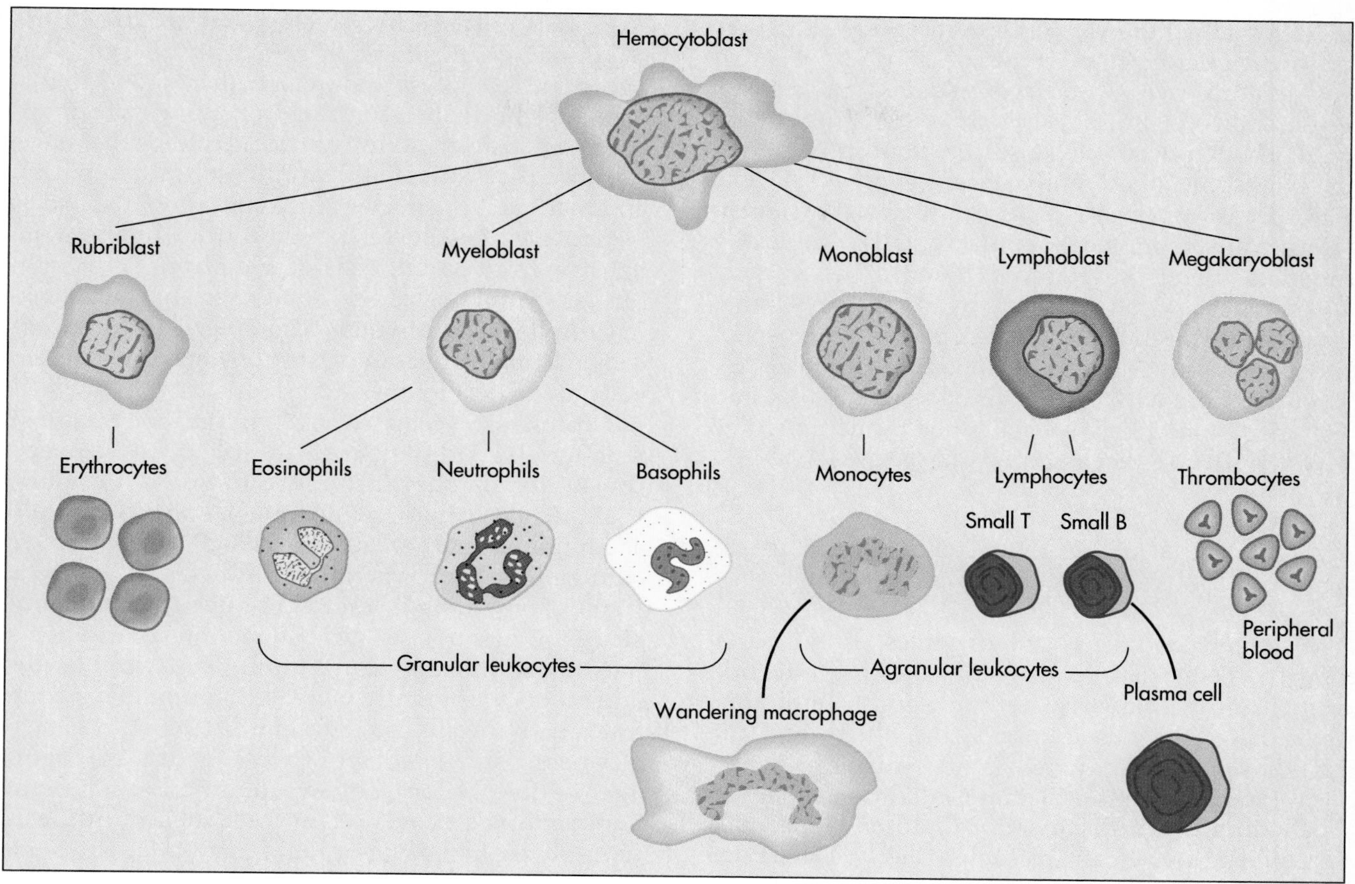

FIGURE 10-3 Hematopoiesis begins in the bone marrow, where a single type of cell, the hemocytoblast, gives rise to all the different kinds of blood cells.

has such causes as altitude changes. When hemoglobin is low because of a lack of dietary nutrients—iron, vitamin B_{12}, or amino acids—essential to its production, the hormone can have little effect.

Most of the disorders related to the red blood cells are anemias. **Pernicious anemia** is due to the body's inability to produce *intrinsic factor,* essential to the absorption of vitamin B_{12} in the small intestine. **Hemorrhagic anemia** is due to rapid blood loss; prolonged slow bleeding produces **chronic anemia,** a chief symptom of which is fatigue. **Hemolytic anemia** occurs when inherited defects in hemoglobin structure, parasites (such as malaria), toxins, or other factors cause red blood cells to self-destruct in the blood. In the genetic disorder **sickle cell anemia** (discussed in Chapter 19) the hemoglobin differs from normal in just one amino acid and crystallizes when oxygen level is low; the result is that the red blood cells change to a shape resembling a sickle (a long curved knife used for harvesting agricultural crops by hand). A sickling episode may cause the cells to plug small blood vessels and burst. In **aplastic anemia,** the red blood cell shortage is due to destruction of the bone marrow, as by the radiation used to treat cancer. It was also a serious threat to people exposed to radiation released by the 1986 nuclear reactor failure in Chernobyl, Russia.

The opposite of anemia is **polycythemia.** This condition is marked by abnormally large numbers of red blood cells in the blood. The blood becomes viscous or syrupy (resistant to flow), the blood pressure rises, and blood clots form more easily.

Blood Types

One important kind of anemia, **erythroblastosis fetalis,** appears when a mother's immune system destroys her unborn infant's red blood cells, as we discuss in Chapter 19. This problem can arise because red blood (and other) cells possess chemical "markers," called antigens, in their cell membranes. These antigens are used by the body's immune system to identify the various cell types and to identify foreign cells that might be causing an infection (see Chapter 24). With red blood cells, the markers define the various **blood types.** With other cells, they define **tissue types.**

There are over 30 different sets of markers for red blood cells, but only two sets are of great concern. The Rh (Rhesus) causes the condition known as erythroblastosis fetalis (see Chapter 19). The ABO set is important to successful blood transfusions (transfusion means replacing all or some of a person's blood with stored blood). The various blood types are distinguished by the antigens found on a person's red blood cells. A person with type A blood possesses A antigens; type B blood has B antigens; people with type AB blood have both A and B; type O blood has neither A nor B antigens. The blood types are discussed in Chapter 19.

White Blood Cells

The white blood cells, or **leukocytes,** do not carry oxygen. Their task is primarily fighting infection and cleaning up debris (see Table 10-2). To do this, they travel the tissue spaces, the blood, and the glands removing any debris and attacking whatever foreign ("nonself") material they encounter. The **lymphocytes** (discussed in Chapter 24) produce antibodies and destroy foreign cells. **Monocytes** phagocytize bacteria and the remains of dead cells; they become the large phagocytes, or macrophages, found in infected tissues. The **granular leukocytes** contain granules. They include the *eosinophils, basophils,* and *neutrophils.* Neutrophils, the most numerous of the leukocytes, are the main phagocytes in inflamed areas, to which they are drawn by chemicals released by platelets and damaged cells (Figure 10-4). Eosinophils help fight allergies and parasites. Basophils and mast cells respond to foreign substances by secreting histamine (which antihistamine drugs counteract) and other substances to cause the redness and swelling of inflammation and to attract phagocytes into contaminated areas.

Granular leukocytes are produced in the bone marrow from parent cells derived from hemocytoblasts. Lymphocytes and monocytes are produced in lymph nodes, thymus, and spleen from parent cells that migrated out of the bone marrow to complete their development.

Lymphocytes can survive in the body almost indefinitely. Other leukocytes must be replaced constantly, for they last only a few days (or hours, when they are fighting infections). The blood rarely accumulates more than 5000 to 9000 leukocytes per cubic millimeter. When the "white count" exceeds 10,000, something is wrong. In **infectious mononucleosis** ("mono"), a viral infection that affects mostly children and young adults, the lymphocytes and monocytes greatly increase in number, causing fever, sore throat, cough, and discomfort; it generally passes in a few days or weeks, but it can damage the liver, heart, kidneys, and nervous system. **Leukemia** is a cancer, or uncontrolled multiplication, of the leukocyte-producing cells in the bone marrow. It can kill because rubriblasts are crowded out of the marrow, causing anemia and bleeding; because the leukocytes fail to mature properly, allowing infections to rage unchecked; and because the massive numbers of leukocytes produced can damage essential organs such as the liver.

Table 10-2

The Blood Cells

CELL TYPE	NUMBER (PER MM³)	SIZE (μM)	LIFE SPAN	ROLE
Erythrocytes (red blood cells)	4.8 million	7.7	120 days	Transport of oxygen and carbon dioxide
Leukocytes (white blood cells)	5000–9000		Hours to days	
Neutrophils	60–70%	10–12		Phagocytosis
Eosinophils	2–4%	10–12		Combat allergies, parasites
Basophils	0.5–1%	8–10		Release histamine, attract phagocytes
Lymphocytes	20–25%	7–15		Immunity
Monocytes	3–8%	14–19		Phagocytosis

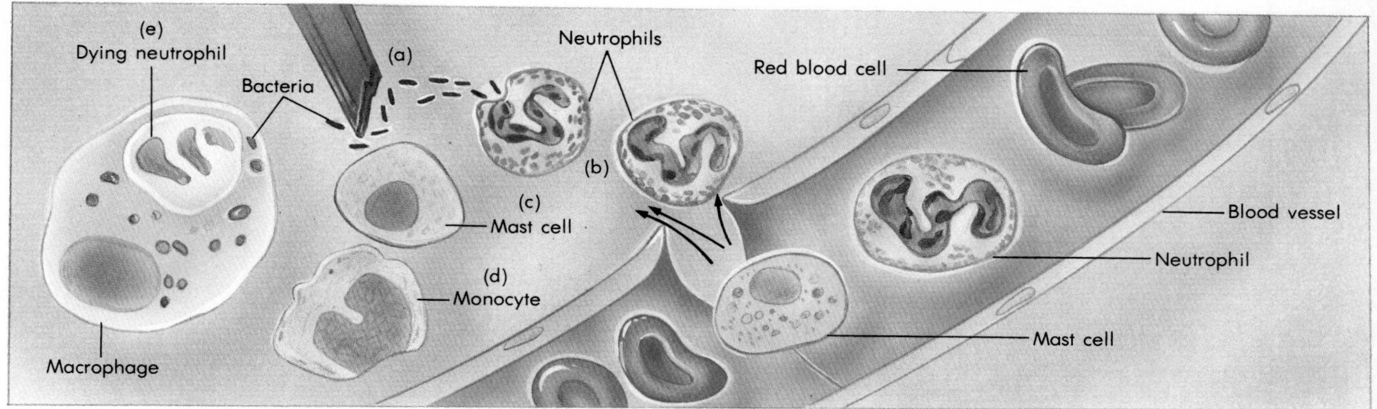

FIGURE 10-4 Infection and defense. (a) An infection starts when bacteria penetrate the skin, as on splinters and other sharp objects. (b) Near the beginning of the defense process, neutrophils migrate through capillary walls, move to the invasion site, and begin ingesting and digesting the bacteria. (c,d) Monocytes and other slow-moving phagocytes arrive later and ingest more of the bacteria. Mast cells enhance the local inflammation reaction by releasing histamines, which increase blood flow through the area. (e) Monocytes and other tissue phagocytes grow larger as they ingest bacteria and dying white blood cells. These huge phagocytes become the macrophages.

Blood Clotting

Hemostasis is the blood's ability to seal leaks—cuts and breaks—in the blood vessels and thus to limit blood loss. The secret lies in the plasma proteins prothrombin and fibrinogen, which we mentioned previously. An injury to a blood vessel sets off a series of chemical reactions that ultimately convert fibrinogen to fibrin, whose fibers knit across and plug the leak with a **blood clot.**

But fibrin is not all there is to clotting. Accidents can occur at any time, and the blood must be ready to clot on a moment's notice. Yet clotting must be activated only when it is needed and quickly stopped when the need is over. To ensure the necessary responsiveness, the clotting process therefore involves several steps. Each of these steps triggers another in an orderly sequence that ultimately leads to the formation of the fibrin that stops the blood loss.

1. What would you expect to happen if the blood clotted too easily?
2. What would you expect to happen if the blood clotted too slowly?

There are two sequences of events that lead to the formation of blood clots (Figure 10-5). The **intrinsic pathway** begins with **platelets,** or **thrombocytes.** They are one of the blood's formed elements, but they are not cells. Rather, they are cellular fragments, bits of cytoplasm enclosed in cell membrane, that form by pinching off from **megakaryocytes** in the bone marrow. They last five to nine days, and each cubic millimeter of blood contains 250,000–400,000 of them.

When a blood vessel is damaged by a cut or some other injury, platelets contact damaged cells and the collagen of the vessel wall. They then enlarge, stick to the exposed collagen, and secrete substances that make other platelets stick to them. The result is a plug of platelets in the break. The plug grows stronger as a clot forms around it.

Other substances released by the platelets begin the actual clotting process by leading to the formation of *intrinsic thromboplastin,* a lipoprotein. Intrinsic thromboplastin in turn acts as an enzyme to catalyze the conversion of the plasma protein *prothrombin* to *thrombin.* Finally, thrombin catalyzes the conversion of fibrinogen to fibrin to form a meshwork of fibers that traps red blood cells, binds the platelets together, and links the edges of the break (Figure 10-6).

The **extrinsic pathway** to a clot begins when damaged cells release *tissue thromboplastin,* which plasma "coagulation factors" promptly convert to *extrinsic thromboplastin.* Like intrinsic thromboplastin, this enzyme then converts prothrombin to thrombin, which catalyzes the formation of fibrin.

One of the plasma coagulation factors is simply calcium ions. The others are proteins, and one of them is known as antihemophilic factor. Hemophilia, an inherited inability of the blood to clot, usually results when the liver cannot make this factor in a working form because of a defective gene. Other coagulation factors are also essential to clotting, and rarer forms of hemophilia result from de-

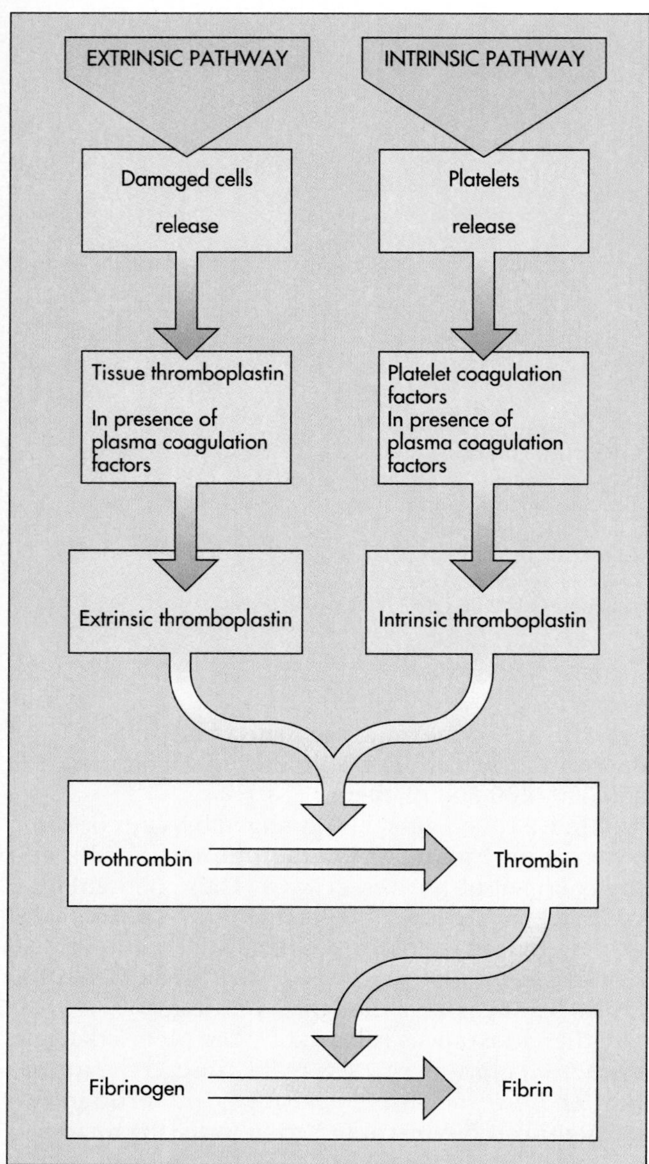

FIGURE 10-5 The intrinsic and extrinsic clotting pathways.

fects in the genes responsible for their synthesis. Treatment for hemophilia calls either for blood transfusions to replace the blood lost through bleeding or for injection of the missing coagulation factor. Antihemophilic factor used to be extracted from human blood collected by blood banks, and it was expensive, rare, and sometimes contaminated by hazardous viruses. Today, it is produced cheaply, plentifully, and cleanly by genetically engineered bacteria.

Once a clot has formed, the fibers of fibrin gradually contract. This contraction squeezes fluid from the clot and draws the edges of the broken blood vessel closer together to facilitate healing. At the same time, a new process begins; it will become dominant only when healing is complete. Even as

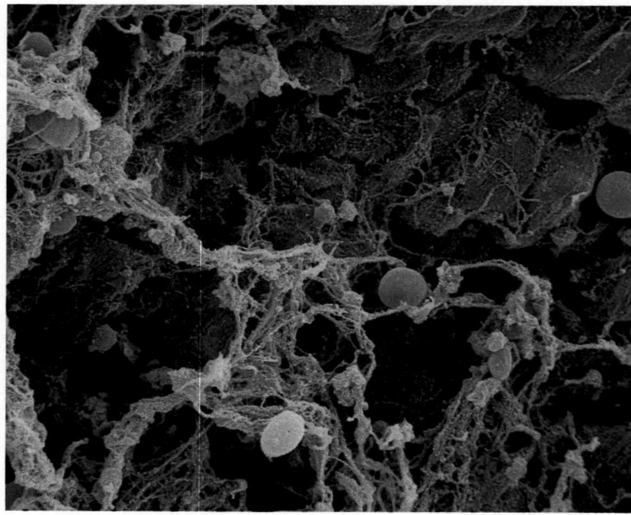

FIGURE 10-6 Scanning electron micrograph of red blood cells ensnared by fibrin strands during clot formation.

the damaged cells were releasing tissue thromboplastin, they were also releasing an enzyme that converts some of the plasma protein into *plasmin.* Plasmin breaks down fibrin, but until healing is done, new fibrin forms faster than plasmin can remove it. Once fibrin formation stops, plasmin destroys the clot.

Occasionally, a clot will form inappropriately in an unbroken blood vessel as platelets stick to a rough spot in the vessel wall. Such a clot forming in an intact vessel is called a **thrombus,** and it can be very dangerous. If the clot, or a portion of it, breaks loose from the thrombus and floats away in the blood, it is called an **embolus.** A moving clot becomes an **embolism** when it lodges in a small blood vessel, blocking blood flow into the lungs, heart, or brain, where it may cause pain, damage, or even death of the tissues thus deprived of their blood supply. Embolisms are a major cause of **strokes** (brain damage) and heart attacks.

THE BLOOD VESSELS

The blood travels through the body in a system of tubes, the blood vessels. Thick-walled **arteries** leave the heart, the muscular pump that keeps the blood moving, and branch into smaller and smaller vessels—of which the smallest are the **arterioles.** The arterioles pass the blood into an extensive network of delicate channels, the **capillaries,** where the exchanges between the blood and the tissues take place. The blood is collected from the capillaries by **venules,** which merge to form the **veins** that return blood to the heart (Figure 10-7). See Table

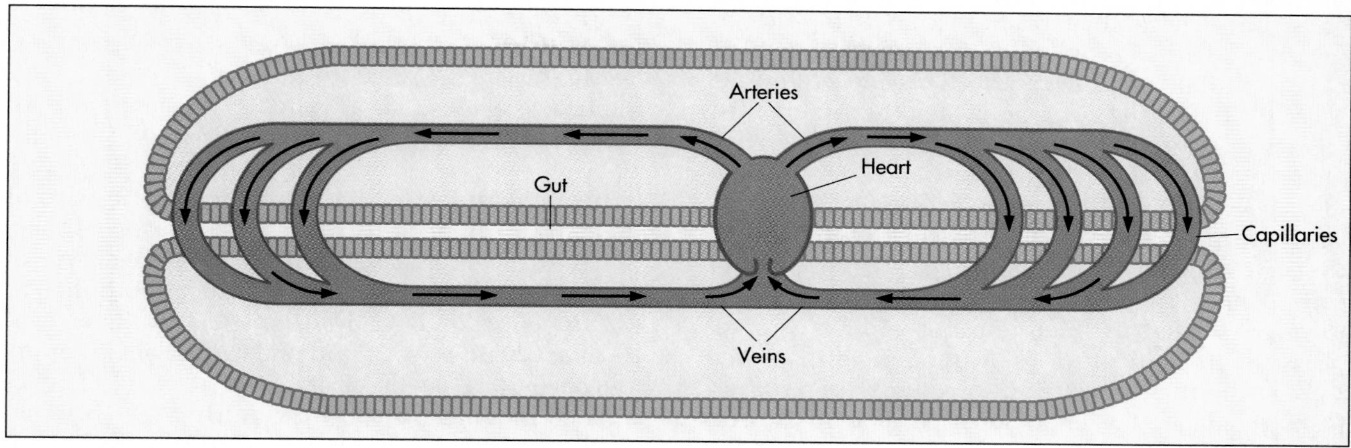

FIGURE 10-7 In the closed circulatory systems found in complex animals, such as humans, the blood flows continuously through the closed tubing of arteries, arterioles, capillaries, venules, and veins.

Table 10-3

The Blood Vessels

VESSEL TYPE	MAJOR FEATURES
Arteries	Carry blood *from* the heart; have thick, elastic, muscular walls
Veins	Carry blood *toward* the heart; have thinner, less elastic, less muscular walls; serve as blood reservoir; have valves
Capillaries	Site of all tissue-blood transfers; narrow; wall is endothelium only

10-3 for a summary of the major features of blood vessels.

Capillaries

All blood vessels are lined by a thin layer of flattened epithelial cells, the **endothelium.** For capillaries, this lining is all there is. They have the thinnest possible walls, so that water, nutrients, oxygen, and wastes can pass easily between the blood and the tissues. They are also very narrow, in some cases not much wider than is necessary to allow one red blood cell to pass. They thus maximize the surface area of the blood, which speeds the transfer of substances to and from the blood.

Networks of capillaries link the finest of the arteries (arterioles) and the finest of the veins (venules) (Figure 10-8), but they are not merely passive channels. Some capillaries, the **metarterioles,** pass directly from arteriole to venule; smooth muscle cells wrapped around their endothelium let them constrict or dilate to control blood flow. **True capillaries** branch extensively and lack smooth muscle

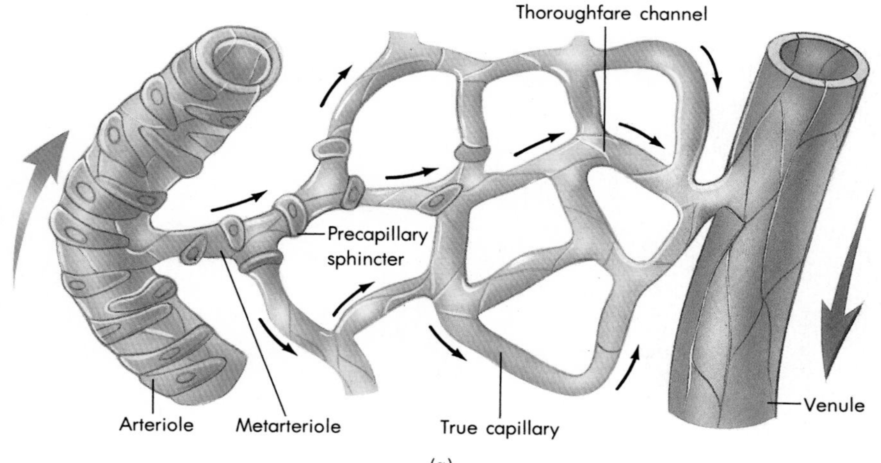

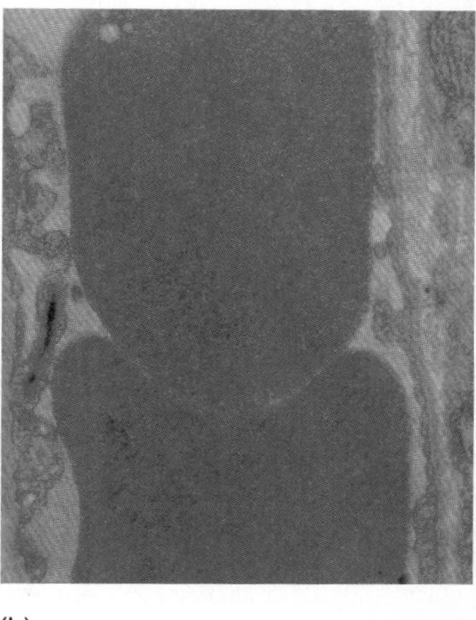

FIGURE 10-8 (a) A capillary network. (b) Red blood cells passing through a mammalian capillary.

cells, but where they branch from arterioles or metarterioles, rings or sphincters of smooth muscle play a similar controlling role. The smooth muscle is controlled by the autonomic nervous system (discussed in Chapter 15).

Many substances, like oxygen and carbon dioxide, cross the capillary wall by simple diffusion. Others are driven by the pressure of the blood through the tiny gaps between the endothelial cells. These gaps are too small to let proteins through, so that the blood retains enough osmotic pressure to hold onto its water and draw some water from the tissues. Some proteins are removed from the blood by endocytosis (pinocytosis) and encased in bubbles of endothelial cell membrane, which cross the cells and are released into the tissue fluid by exocytosis (see Chapter 3 and Figure 10-9).

Arteries

Arteries have much thicker walls than do capillaries. Immediately surrounding their endothelial lining is a layer of elastic connective tissue; together, the endothelium and elastic layer compose the artery's **tunica interna** (inner coat). The much thicker **tunica media** (middle coat) consists of elastic connective tissue and smooth muscle. The outer coat, the **tunica externa,** or **adventitia,** is a layer of both elastic and collagenous connective tissue (Figure 10-10).

It is the tunica media that gives arteries two of their most important features. The elastic connective tissue lets them expand to receive the charge of blood pumped by each beat of the heart and then to use the elastic rebound of the vessel walls to continue squeezing that blood and help propel it on its journey.

The amount of blood flowing through the various blood vessels is largely controlled by the **autonomic nervous system** (a division of the nervous system, not under conscious control, that regulates many important body systems). The autonomic nervous system uses arterial smooth muscle to constrict and dilate the arteries and decrease and increase the amount of blood flowing to the skin (for temperature control), muscles (to support activity), and other tissues. It can also use smooth muscle to shut down small arteries entirely to control bleeding after a wound.

The largest arteries, such as the aorta that leaves the heart (see Figure 10-11), have the greatest

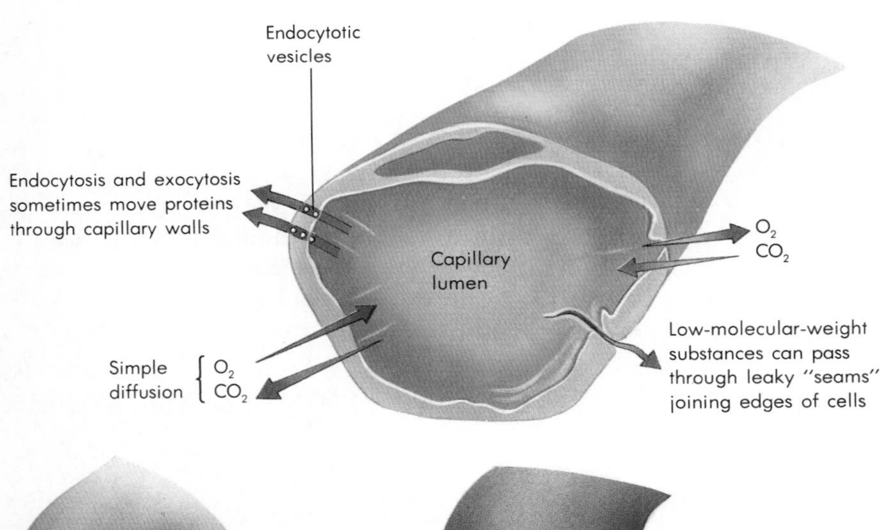

FIGURE 10-9 Movement of materials through capillary walls.

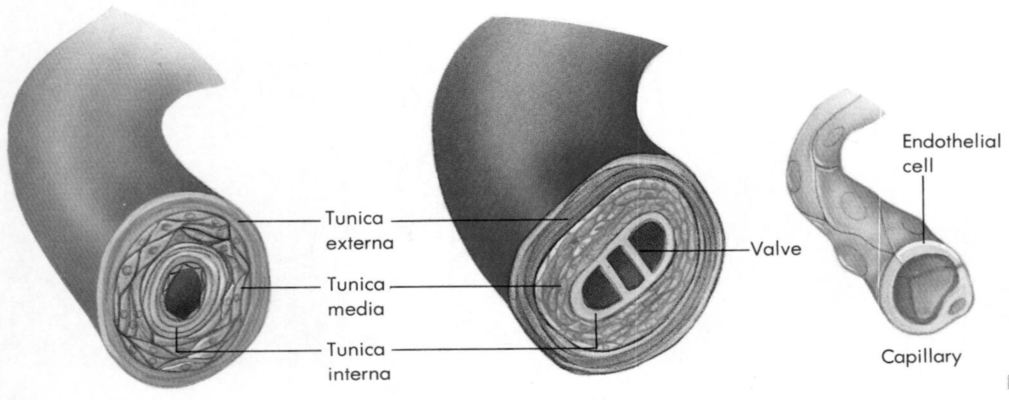

FIGURE 10-10 The structures of an artery, a vein, and a capillary.

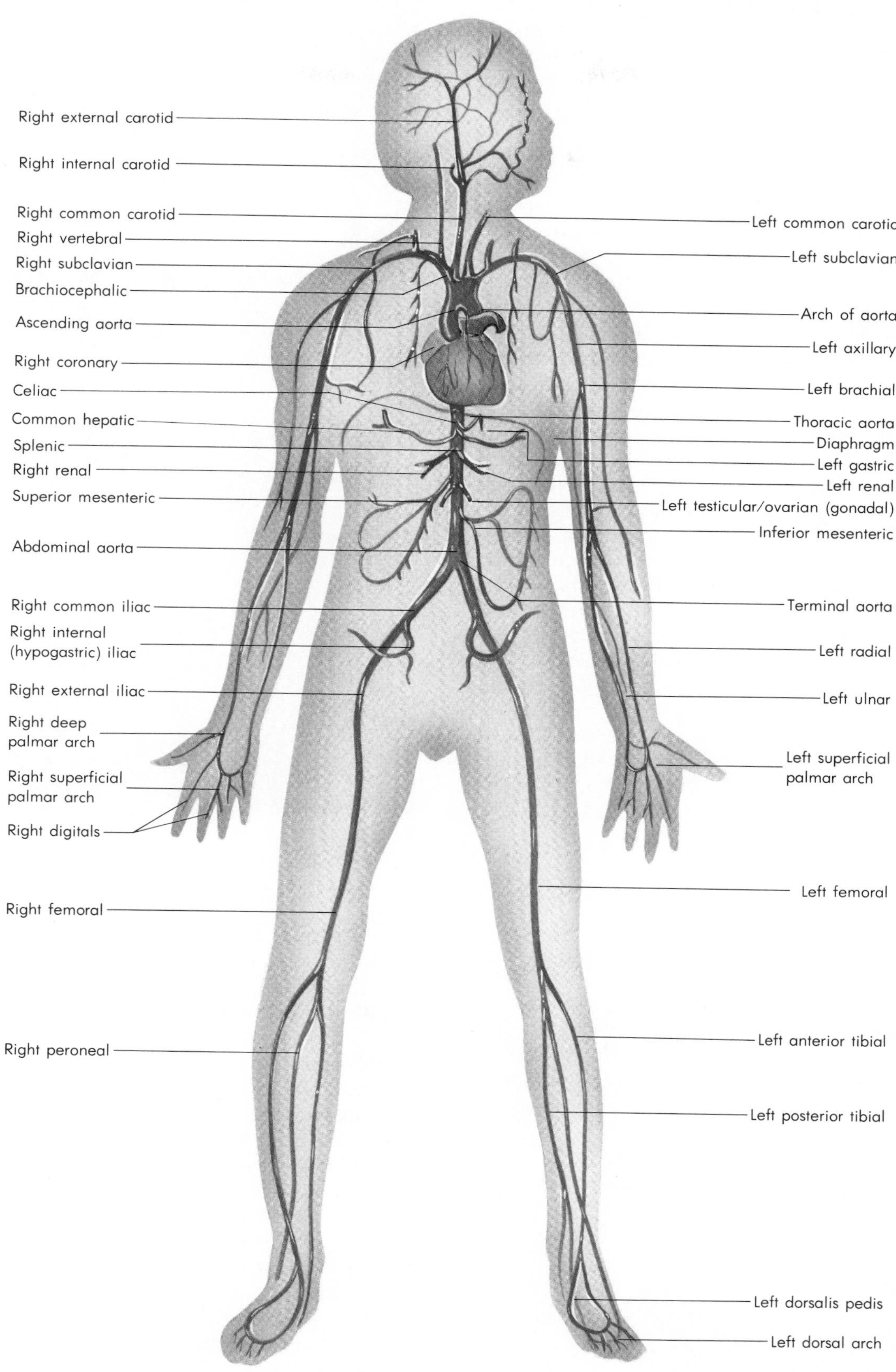

FIGURE 10-11 The arteries.

amounts of elastic tissue. Smaller arteries, which distribute blood to the body's parts, are dominated by the smooth muscle component of their walls. Many of these arteries are linked by arterial branches, or **anastomoses.** Thanks to these interconnections, blockage of one artery by a blood clot or wound does not deprive other body areas of their entire blood supply.

The walls of the smallest arteries, the arterioles, consist of little more than endothelium and smooth muscle (Figure 10-12). From these vessels branch the capillaries that deliver blood to the tissues; their smooth muscle provides the major control over blood flow to the tissues.

Arterial disorders are of two main types. In one, a weak spot in the arterial wall balloons out under the force of the blood pressure to form an **aneurysm.** If an aortic aneurysm bursts, the ensuing massive blood loss can be quickly fatal. If an aneurysm of a brain artery bursts, the loss of blood supply and the sudden drop in blood pressure can cause the death of brain tissue, producing a stroke. The second type of disorder is a blockage, caused by a thrombus (clot), or by **atherosclerosis,** a condition marked by deposits of cholesterol in the arterial lining and the proliferation of smooth muscle cells. These deposits (Figure 10-13) appear to form when the endothelium is damaged by chemicals such as those in tobacco smoke, by high blood pressure, and by large amounts of cholesterol in the blood. They are responsible for most heart attacks, for they often form in the coronary arteries that serve the heart.

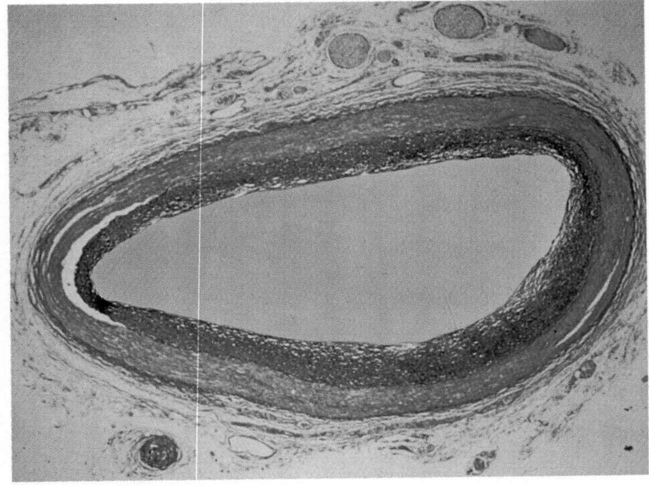

(a)

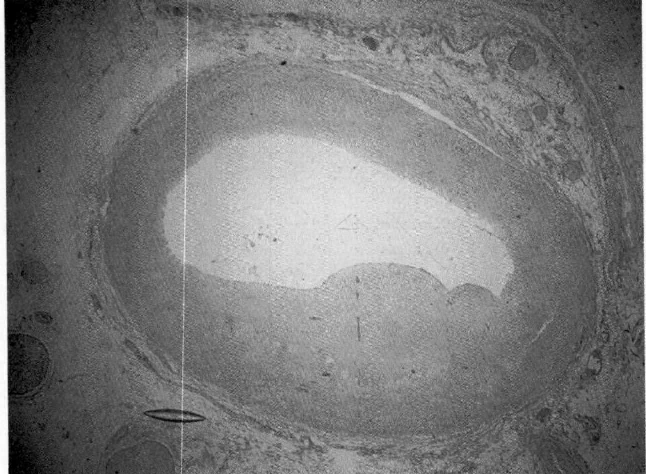

(b)

FIGURE 10-13 (a) Cross section of a normal artery. (b) Cross section of an artery partially blocked by cholesterol deposits.

Veins

By the time the blood has passed through the arteries, arterioles, and capillaries, most of the blood pressure has been exhausted in overcoming frictional resistance. For that reason, the blood pressure in the veins is much lower, and the walls of the veins are thinner than those of arteries of equivalent capacity. Veins do have the same layers of smooth muscle and connective tissue as those of arteries, but much of the elastic tissue is replaced by fibrous connective tissue (Figure 10-10). The smallest veins, the venules, have no smooth muscle at all.

Veins drain blood from all the tissues (Figure 10-14) and must carry just as much blood as the arteries. Because the pressure in them is low, they must therefore be larger than the arteries that deliver blood to the tissues they drain. At any moment, the veins contain about 60 percent of the body's blood. They thus serve as storage reservoirs for blood. During exercise and after blood loss, the autonomic nervous system contracts the smooth muscle of the veins, forcing more blood into the rest of the circulatory system.

The residual blood pressure in veins carrying blood from areas below the heart is too low to move the blood back up to the heart. Veins coming from

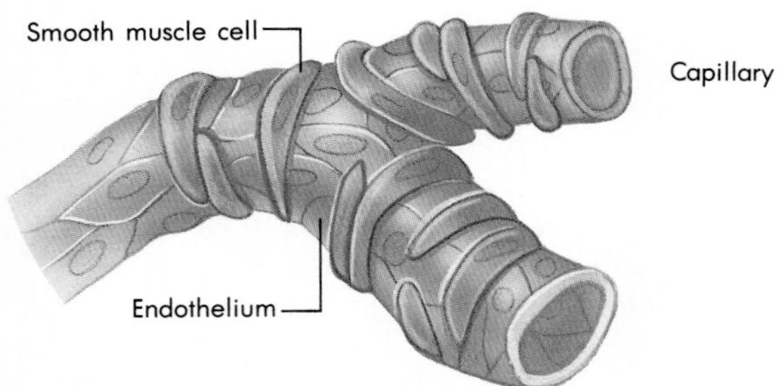

FIGURE 10-12 Structure of an arteriole.

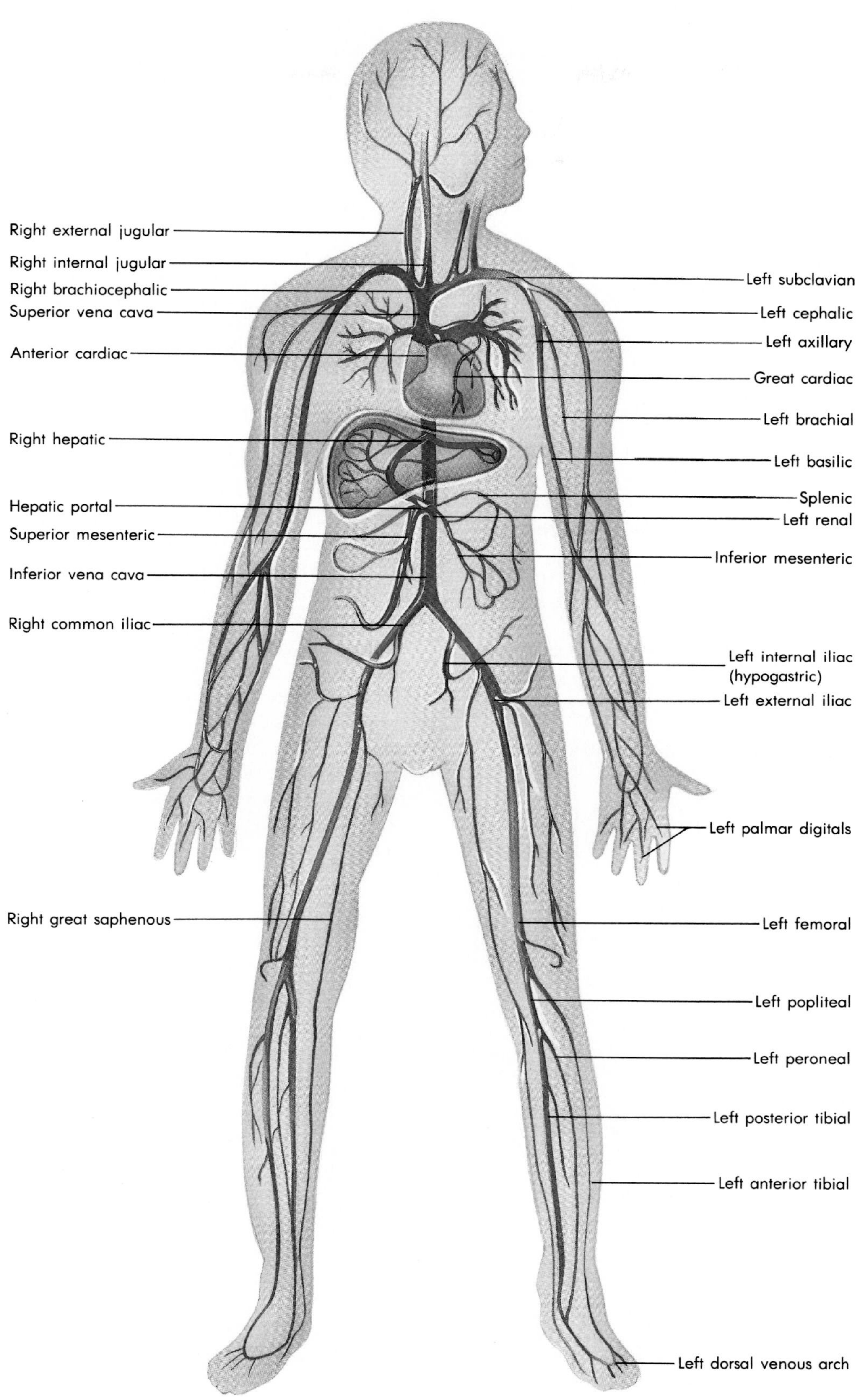

FIGURE 10-14 The veins.

these areas contain one-way *valves* that prevent the blood from dropping back down, away from the heart, because of gravity. Contractions of skeletal muscles surrounding the veins propel the blood upward, valve by valve, as if it were climbing a ladder (Figure 10-15). In a sense, the skeletal muscles function as auxiliary hearts or pumps in helping the blood move up the body. Thus long-distance runners often continue running for some distance after the race is over in order to provide muscle contractions that ensure the blood will not pool in the veins and will reach other body parts, such as the brain. Pressure changes in the chest cavity, caused by breathing, also help propel blood in the veins.

1. Novice soldiers standing at rigid, motionless attention for long periods sometimes faint. Experienced soldiers do not, for they know better than to remain literally motionless. Why? What sorts of inconspicuous motions do you think they make?
2. Massage helps bed-ridden patients by improving their circulation and easing the load on their hearts. Why?

THE LYMPHATIC SYSTEM

As blood flows through the capillaries, some of its fluid escapes through the pores in the capillary walls into the tissues and does not return. If there were no way for the blood to reclaim most of this lost fluid, the tissues would bloat, the blood would thicken and become more viscous, and death would follow. Fortunately, the **lymphatic system** collects the lost fluid and returns it to the circulation.

The lymphatic system begins as an array of narrow, thin-walled dead-end tubules, the **lymph capillaries,** among the tissue cells (Figure 10-16a). They collect excess fluid, or **lymph,** from the spaces among the cells and convey it to larger lymph vessels, the **lymphatics.** Eventually, the lymphatics return the fluid to the blood through one-way valves that open into large veins near the heart (Figure 10-16b).

Like veins, the lymphatics contain valves to ensure one-way flow under the massaging effect of skeletal muscles. In the groin, neck, armpits, and abdomen, the lymph vessels converge on **lymph nodes** (Figure 10-17, p. 228). These nodes are filtering systems that remove cellular debris, bacteria, and viruses from the lymph collected from the various regions of the body. Each node contains phagocytic macrophages and the parent cells of lymphocytes. They are thus an important part of the immune system. Bacteria and viruses collected from the tissues are carried in the lymph to the nodes, where macrophages can remove them and lymphocytes can be activated to combat them elsewhere in the body.

The lymph nodes often swell and grow tender to the touch during infections because as the nodes attempt to filter, trap, and destroy the infectious agents, macrophages and lymphocytes in the area

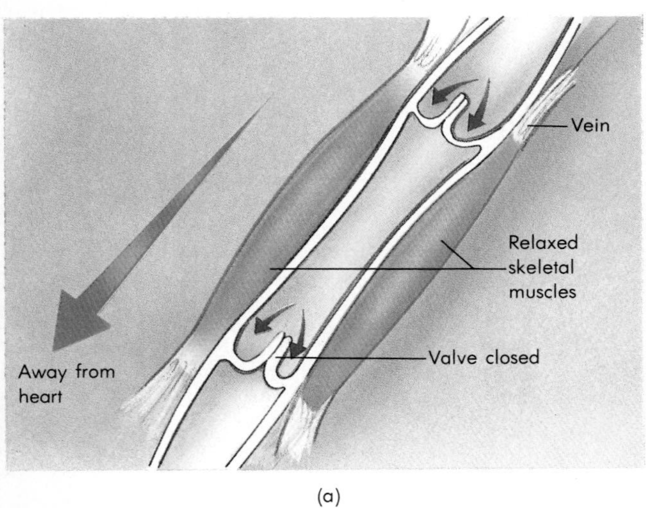

(a)

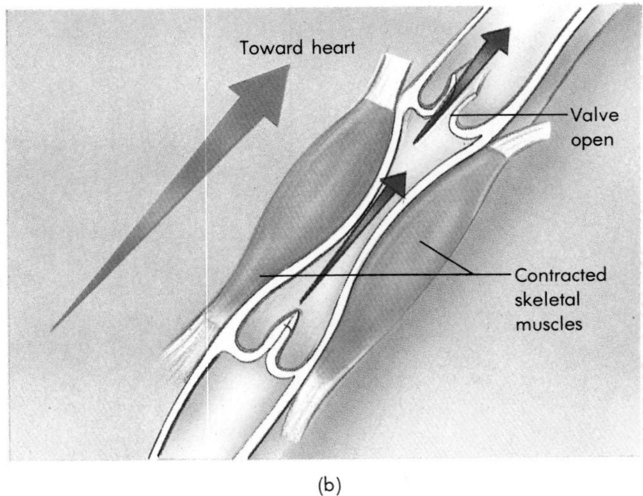

(b)

FIGURE 10-15 Vein valves. Skeletal muscles act as auxiliary pumps to facilitate the flow of blood through the veins. (a) Backflow of blood is prevented by one-way valves inside the veins. (b) Pressure exerted on the walls of the veins during skeletal muscle contractions propels the blood from valve to valve on its way back to the heart.

FIGURE 10-16 (a) Lymph capillaries begin as blind pouches collecting fluids that have leaked from the cardiovascular capillaries. (b) Overview of the body's lymphatic drainage system. Lymph is filtered by lymph nodes before it is returned to the blood as plasma. Notice the asymmetric nature of the lymph return. How would you describe it?

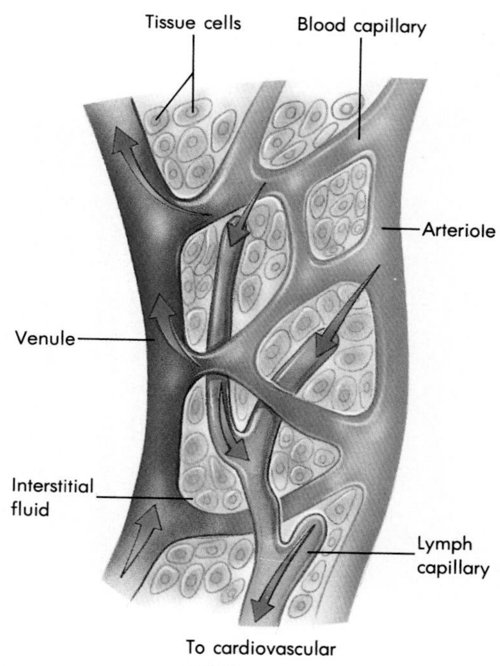

(a)

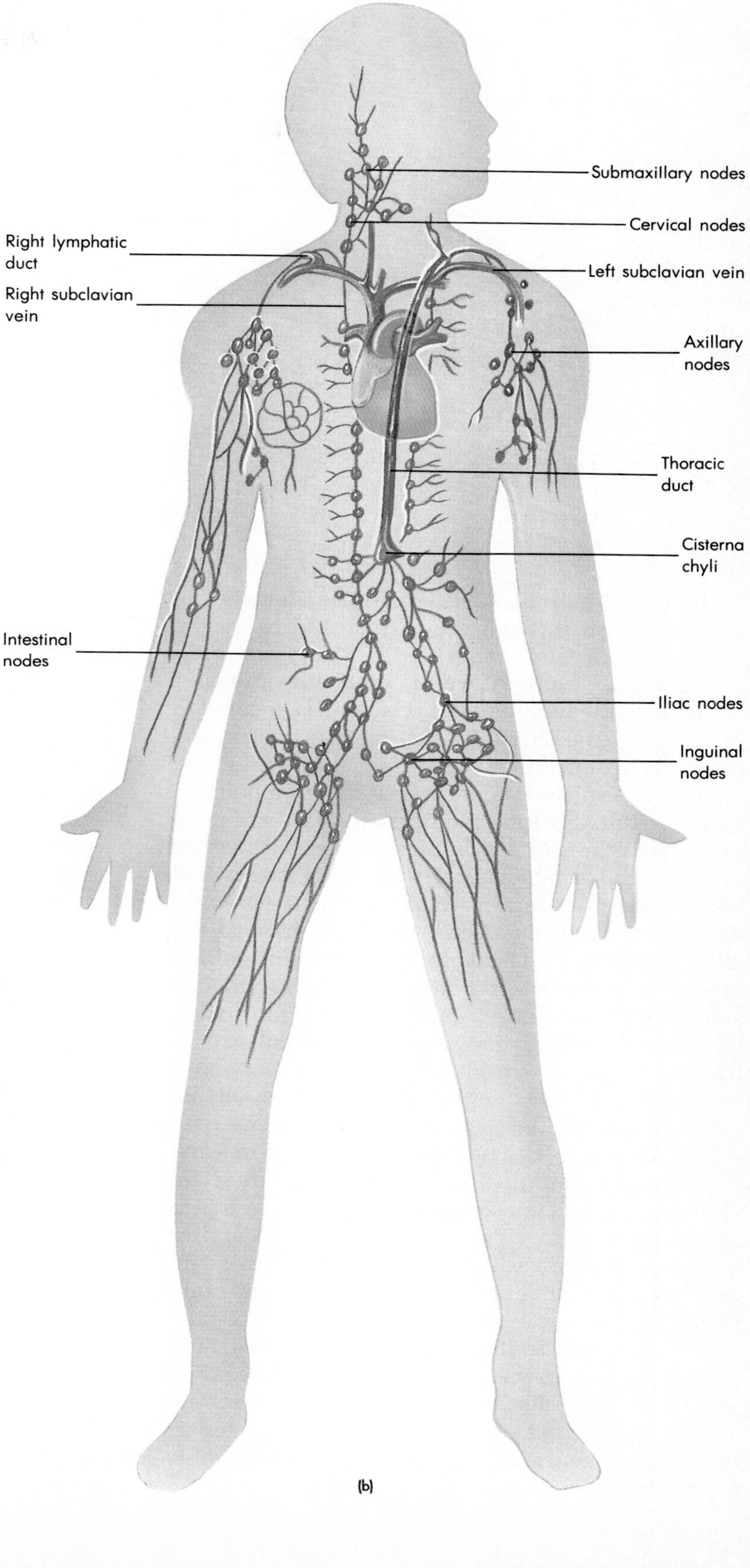

(b)

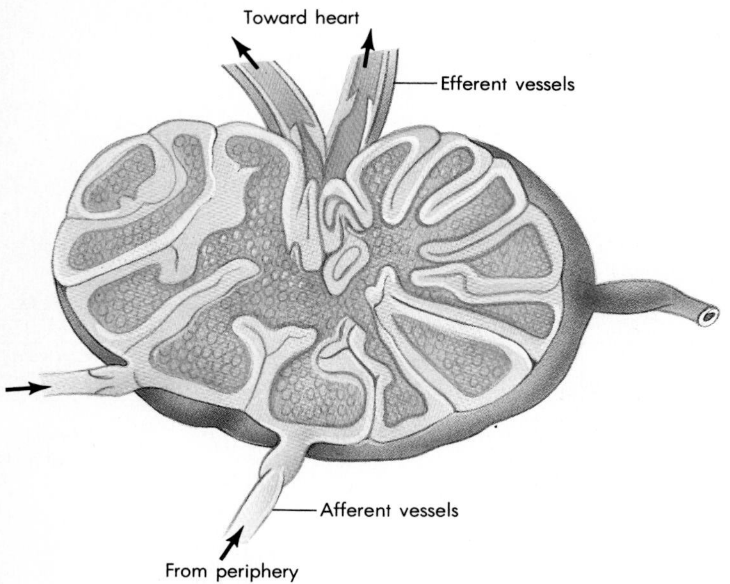

FIGURE 10-17 Lymph flows into the lymph nodes by way of the afferent lymphatic vessels. After filtering, it is carried on its way through efferent lymphatic vessels.

proliferate. Their increased numbers impede the flow of lymph through the system and cause the nodes to enlarge.

The body contains several **lymphoid organs** in addition to the lymph nodes. The **tonsils,** located in the sides of the throat at the back of the tongue, respond to infectious microbes that enter the mouth. The **spleen,** the largest of the lymphoid organs, is located behind the stomach and serves two functions. The *white pulp* of the spleen generates and stores lymphocytes. Its *red pulp* stores red blood cells that can be released when the blood needs to enhance its oxygen-carrying capacity—as after blood loss. The spleen returns the stored blood cells to the circulation by contracting its smooth muscles. The red pulp also contains macrophages and lymphocytes that filter blood just as they do lymph in lymph nodes; they also remove worn-out red blood cells. The **thymus gland,** located behind the breastbone, is where some lymphocytes (T cells, discussed in Chapter 24) mature.

There are several diseases that strike the lymphatic system. Some parasites, such as the filaria roundworm (carried by tropical mosquitoes), live in the lymph nodes and provoke the growth of connective tissue. Eventually, lymph flow is completely blocked from a leg, arm, or other body part. The tissue swells as fluid accumulates (*edema*), and a leg may grow to the size of an elephant's leg (Figure 10-18). The disease is known as *filariasis;* its advanced stage, where the limb has grown to enormous proportions, is appropriately referred to as *elephantiasis.*

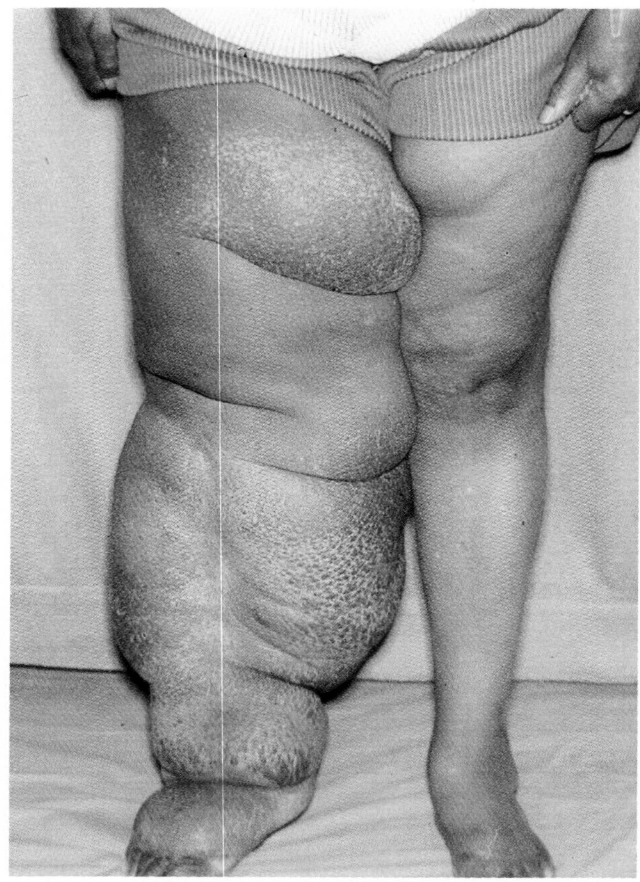

FIGURE 10-18 Elephantiasis can cause severe deformities of the leg. The parasite causing it physically obstructs the flow of lymph through the nodes, thus trapping fluid in the lower extremities.

1. Edema is not only a symptom of elephantiasis. It accompanies bruises, sprains, and infections as well, where inflammation has increased the loss of fluid from the capillaries. What treatments would you expect to reduce edema in such cases? Why?
2. Why are "swollen glands" associated with colds and other infections? What are the "glands" in question?

People can also develop **lymphomas,** cancers of the lymphoid tissues. *Hodgkin's disease,* now often curable with radiation therapy, is a cancer of the structural tissue in lymph nodes. Lymphomas can also involve lymphocytes.

THE HEART

The **heart** is a fist-sized muscular pump located near the center of the thoracic cavity between the lungs. Its pumping action powers the circulation of blood through the vascular channels distributed throughout the body, eventually driving it through the various exchange systems. When the pump fails, for whatever reason, death quickly follows.

The heart's performance is so impressively successful—every nine months, it moves enough blood to fill an Olympic-sized swimming pool—that we are only beginning to attempt to match it with manufactured artificial hearts. So far, however, artificial hearts have met with only limited success, while creating horrendous ethical problems for the people involved. Powered by compressed air from outside the body, these machines can keep working for a year or more, but most implant patients die within months from blood clots, bleeding, or infection. The artificial heart is a long way from being a permanent replacement for the real thing.

The bulk of the heart wall, the **myocardium,** is a mass of specialized muscle tissue. This **cardiac muscle** is striated, like skeletal muscle, but its mitochondria (energy generators) are larger and more numerous and it is not under voluntary control. It contracts spontaneously with its own natural rhythm, about 70 times per minute. The resting heart rate can be speeded up or slowed down by the autonomic nervous system.

Technically, three distinct tissue layers make up the heart wall. The chambers inside the heart are lined with the **endocardium,** a layer of endothelium that merges with the endothelium of the blood vessels. The bulk of the heart wall—the middle layer—consists of the myocardium. The outer surface of the heart is covered with a membrane, the **visceral pericardium,** or **epicardium.** Surrounding this membrane is the **pericardial sac,** composed of the **parietal pericardium.**

The outer portion of the parietal pericardium is fibrous. The inner is serous tissue, as is the visceral pericardium. These serous tissues secrete a layer of lubricating fluid that reduces the friction the heart encounters as it rubs against other organs in the chest during its own contractions and the movements associated with breathing (Figure 10-19).

The interior of the heart is divided into four chambers: two atria and two ventricles. The two upper thin-walled chambers, the **atria,** receive blood from four primary sources: the superior vena cava, the inferior vena cava, the coronary sinus, and the pulmonary veins. The **superior vena cava** delivers blood to the right atrium from the upper trunk, the head and neck, the shoulders, and the arms. The **inferior vena cava** returns blood from below the heart. In addition, the blood returning from the heart's own capillary system returns to the right atrium by way of the *coronary sinus.* The left atrium receives its blood from the lungs by way of four **pulmonary veins.**

The thicker walled **ventricles** are primarily pumping chambers and are separated from each

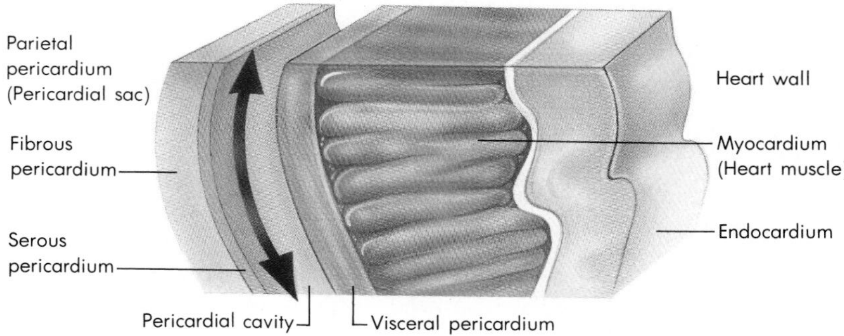

FIGURE 10-19 The heart wall.

other by the *interventricular septum.* The ventricles eject the blood into the arteries of both the pulmonary and the systemic circuits. The right ventricle pumps blood out to the lungs through the **pulmonary artery** and its tributaries. The left ventricle pumps blood into the **aorta,** whose branches deliver it to all the body's organs (Figure 10-20). The coronary arteries branch from the base of the aorta, carrying blood to the capillary beds located in the muscular walls of the heart.

Heart valves primarily promote a one-way blood flow through the chambers of the heart. There are two categories of valves—the two *cuspid valves* prevent blood from being forced from the ventricles back into the atria; the two **semilunar valves** prevent blood from flowing from the arteries back into the relaxing ventricles. The **tricuspid valve** guards the opening between the right atrium and ventricle. When the ventricle contracts, the pressure it exerts closes the valve's flaps, which are anchored to the ventricular wall by cords of tissue, the *chordae tendineae* (sometimes referred to as the heart strings). The blood can then move out of the right ventricle only through the pulmonary artery. The **bicuspid valve** (mitral valve) guards the passageway between the left atrium and ventricle and works similarly. The valves at the bases of the pulmonary trunk and the aorta each contain cuplike flaps of tissue, forming the *aortic semilunar valve* and the *pulmonary semilunar valve,* that fold back against the arterial wall while blood is flowing from the ventricles. When the ventricles relax and the elasticity of the arteries begins to push the blood backward, the flaps fill with blood and close, blocking the return flow (Figure 10-21).

The valves can be damaged by bacterial infections. Once the infection is cured, residual scar tissue may interfere with the valves. If the valve flaps cannot seal properly, the result is a leaky valve that produces a "whooshing" with each heart contraction—such a condition is called a **heart murmur.** Most murmurs represent no hazard, but severe ones can indicate that the heart's efficiency is seriously impaired. If as a consequence of the buildup of scar tissue the passageway through the valve has narrowed, a **stenosis** results. Stenotic valves make the heart work harder. Treatment can involve replacing the defective valves with artificial valves made of metal and plastic (Figure 10-22).

The Circulatory System

We call the system of arteries, veins, and capillaries in which the blood flows the circulatory system. The blood flow is powered by the muscular con-

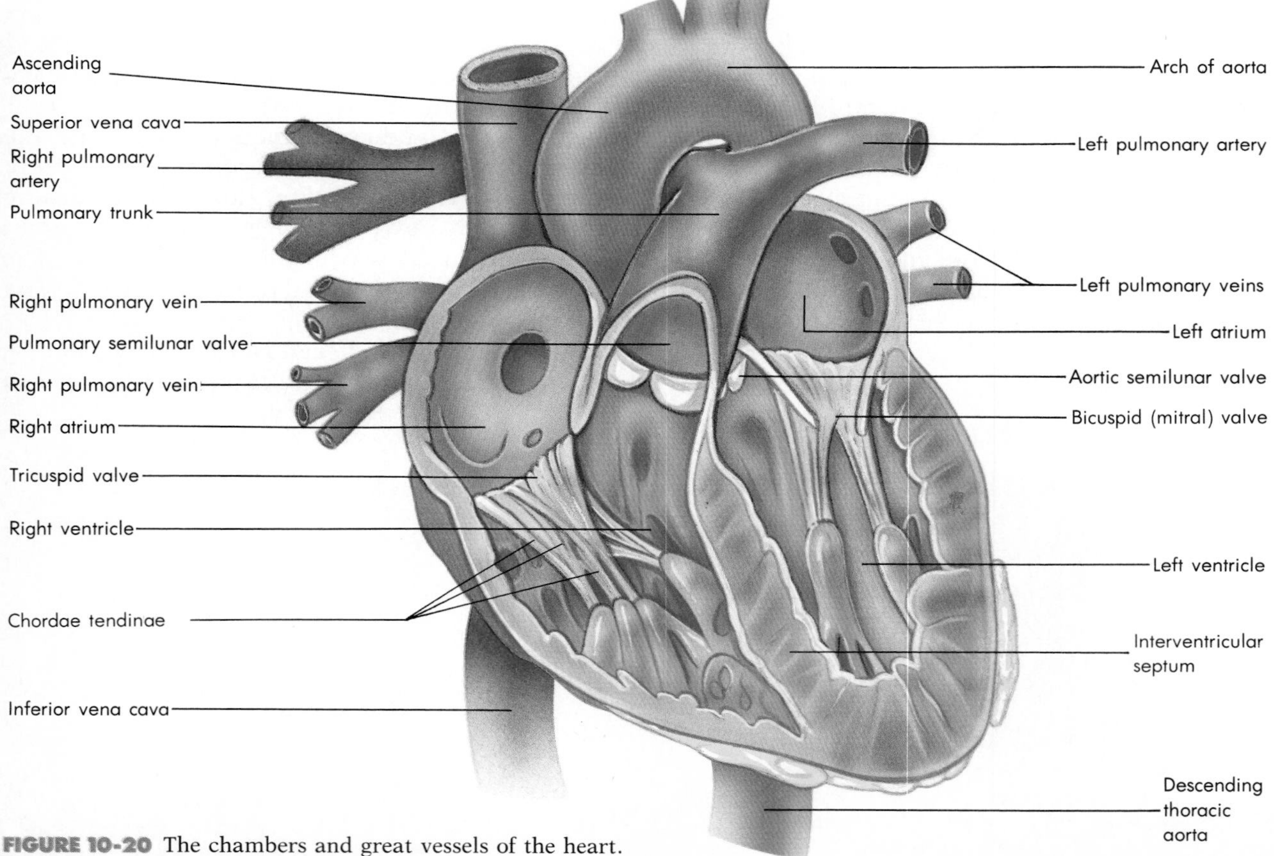

FIGURE 10-20 The chambers and great vessels of the heart.

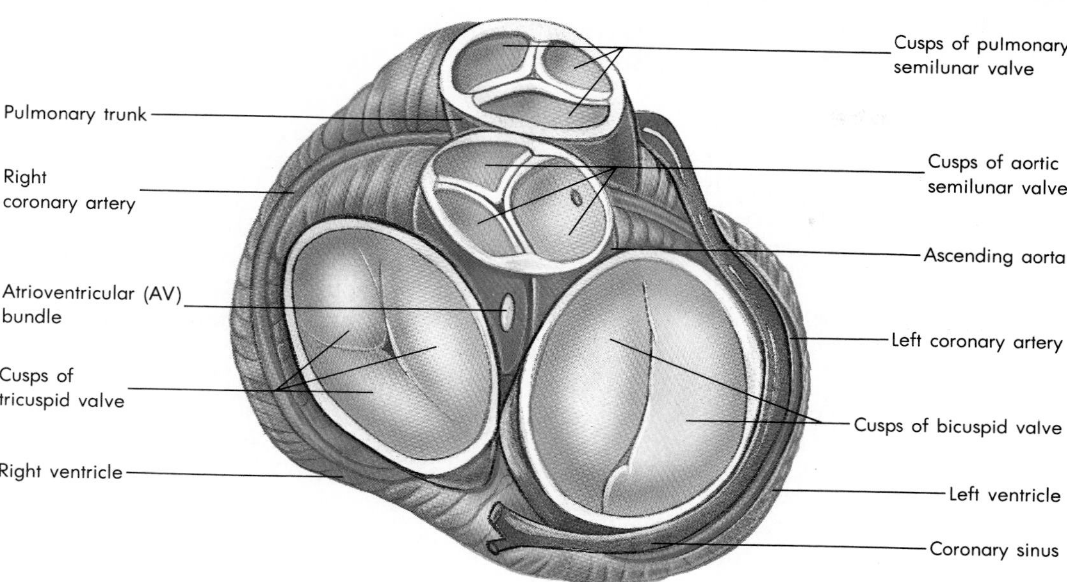

FIGURE 10-21 In this view of the heart, the atria and the pulmonary and aortic arteries have been removed to show the structure of the cuspid valves (separating the atria from the ventricles) and semilunar valves (between the major arteries and the ventricles).

tractions of the heart, the elastic rebound of the arterial walls in the interval between heartbeats, and the contractions of skeletal muscles surrounding the veins.

The blood flows through two major circuits. The left side of the heart collects oxygenated blood returning from the lungs and drives it through the arteries to the capillaries of the various systems of the body. This blood supports the muscular, nervous, sensory, digestive, excretory, and various other systems of the body. For that reason, blood from the left side of the heart is said to be flowing through the **systemic loop.** Deoxygenated blood from this systemic loop is collected by veins, transported to the right side of the heart, and pumped out through the relatively short **pulmonary loop** to the lungs, where it releases its load of carbon dioxide and recharges the hemoglobin of the blood cells with oxygen (Figure 10-23).

FIGURE 10-22 Replacement valve for a faulty aortic semilunar valve. It is made by modifying a valve from a pig's heart.

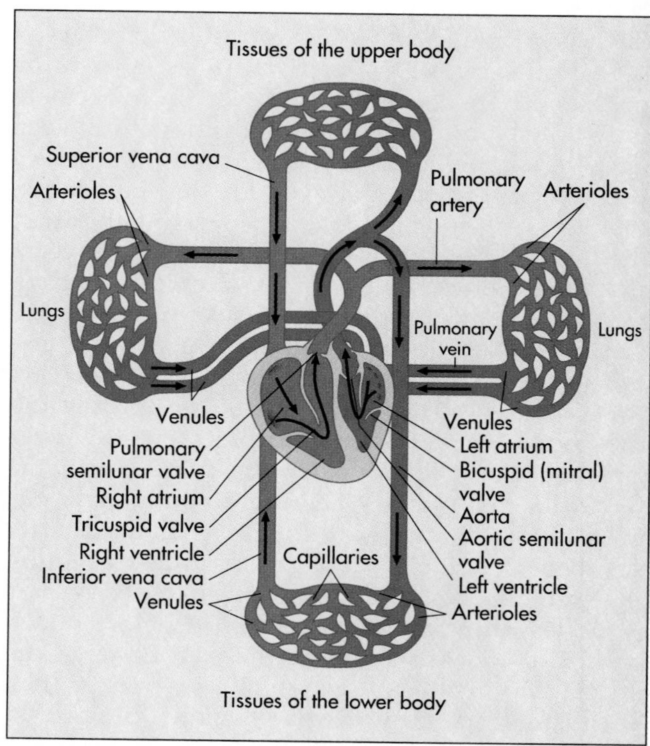

FIGURE 10-23 The systemic and pulmonary circulations.

As the blood leaves the right side of the heart for the lungs, in the pulmonary artery, it is poor in oxygen (and hence a dark, purplish red) and rich in carbon dioxide, for it has already been through the systemic circulation, where it delivered oxygen to the tissues and picked up waste carbon dioxide. In the capillaries of the lungs, it unloads the carbon dioxide, recharges its hemoglobin with oxygen, and becomes bright red. It then returns to the heart in the pulmonary veins. The heart then pumps the oxygen-rich, carbon-dioxide-poor blood back into the systemic circulation.

The heart is really two pumps, side by side, that happen to be fused into the same organ. The two upper chambers, the atria, are the receiving chambers for blood returning to the heart from each circuit. The two lower chambers on both sides, the ventricles, are the major pumping chambers that drive the blood they receive into either the systemic or pulmonary loops. The right ventricle drives the blood it receives, from the right atrium, into the pulmonary loop. The blood flowing from the lungs collects in turn in the left atrium, passes to the left ventricle, and from there is sent out into the systemic circuit and back to the right atrium. The two pumps of the heart thus shunt blood from one circuit to the other.

Fetal Circulation

Normally, intact partitions—the interatrial and interventricular septa—separate the chambers of the left and right atria and ventricles, preventing the blood from one side of the heart from accidentally flowing into the other. These partitions keep oxygen-rich blood from being contaminated by oxygen-poor blood. However, in the fetal heart, there is an opening in the wall separating the atria, the *foramen ovale.* This opening allows the oxygen-rich blood coming in from the placenta to transfer directly across to the systemic side to be pumped out to the fetus's body. In addition, a shunt artery, the *ductus arteriosus,* passes any blood that enters the pulmonary artery to the aorta. The opening and the shunt are necessary because the fetal lungs have never been functional. At the time of birth, the first inhaled air opens the lungs to pulmonary circulation. The foramen ovale and the ductus arteriosus then usually close up tight, promptly isolating the systemic from the pulmonary circuits.

Occasionally, the opening fails to close and a baby is born with a hole in the wall between the atria. The blood in the systemic arteries is then purplish, which gives the baby's skin a bluish cast and earns it the term "blue baby." Fortunately, such birth defects can be repaired with open-heart surgery.

The Conduction System

If the various areas of the heart fail to contract in a coordinated manner, the heart cannot pump blood efficiently. The heart achieves its coordination and efficiency thanks to its own regulatory system, the **conduction system.** All heart muscle can contract spontaneously and rhythmically. However, a small region of specialized cells, the **sinoatrial (SA) node,** or **pacemaker,** is located in the wall of the right atrium near where the superior vena cava attaches. This pacemaker has a slightly faster rhythm than the rest of the cardiac muscle. When it contracts, the electrical activity it generates spreads through the atrial myocardium and sets off atrial contractions—slightly ahead of the ventricles. The electrical activity quickly reaches a secondary pacemaker, the **atrioventricular (AV) node,** located in the interatrial septum. From there, after a brief delay, it travels quickly through muscle fibers specialized for conduction, the **atrioventricular bundle,** or **Bundle of His** (pronounced "hiss") and its branches, to the walls of the ventricles (Figure 10-24, p. 234).

When the conduction system is damaged, as by a heart attack, the heart may no longer pump blood efficiently, and the patient may suffer from debilitating or even life-threatening arrhythmias. In such situations, it is sometimes advisable to use an artificial pacemaker to restore efficient cardiac rhythms (Figure 10-25, p. 234).

The Electrocardiogram (EKG)

The heart's electrical activity carried by the conduction system can be detected with suitable instruments at the surface of the body. The record of this activity is called an **electrocardiogram (EKG)** (Figure 10-26, p. 236). It typically shows three upward spikes, the P, R, and T waves, and two downward spikes, the Q and S waves, in the sequence PQRST. The P wave represents the impulse spreading across the atrial walls; it is followed by atrial contraction. The QRS complex represents the spread of the impulse across the ventricles. The T wave represents the recovery of the ventricles in preparation for the next contraction.

Abnormalities in the EKG are very useful for diagnosing problems with the heart. A flat T wave may suggest the heart is receiving too little oxygen, as when the coronary arteries are blocked by ather-

Learning Focus

THE CIRCULATORY SYSTEM

Review the materials on the heart and circulatory system. Then study the accompanying drawing. It simplifies the circulatory system. On it, label the systemic arteries, veins, and capillaries; the pulmonary arteries, veins, and capillaries; and the chambers of the heart. Add arrows to show the direction of blood flow. Bear in mind that the drawing shows the heart as if it were in the body of a person lying on his or her back facing you; that is, its right is on your left, and its left is on your right.

Let's say that the path of blood through the heart begins as the blood leaves the systemic circulation. What, then, is the *first* heart chamber the blood encounters?

List in sequence the heart chambers through which the blood flows.

Read the part of this chapter on the heart and list in sequence the *valves* through which the blood passes.

List in sequence the arteries and veins the blood passes through as it enters and leaves the heart.

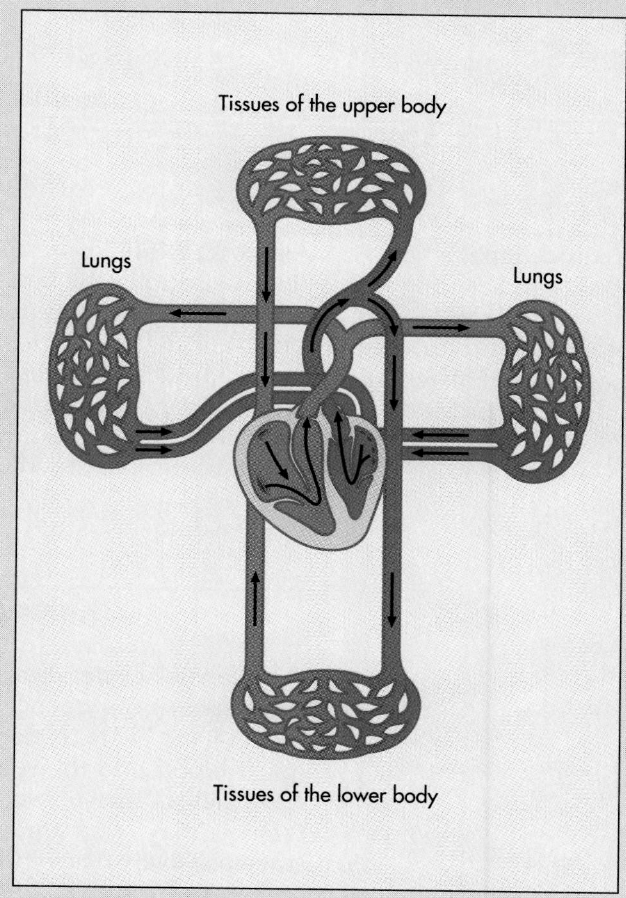

The systemic and pulmonary circulations.

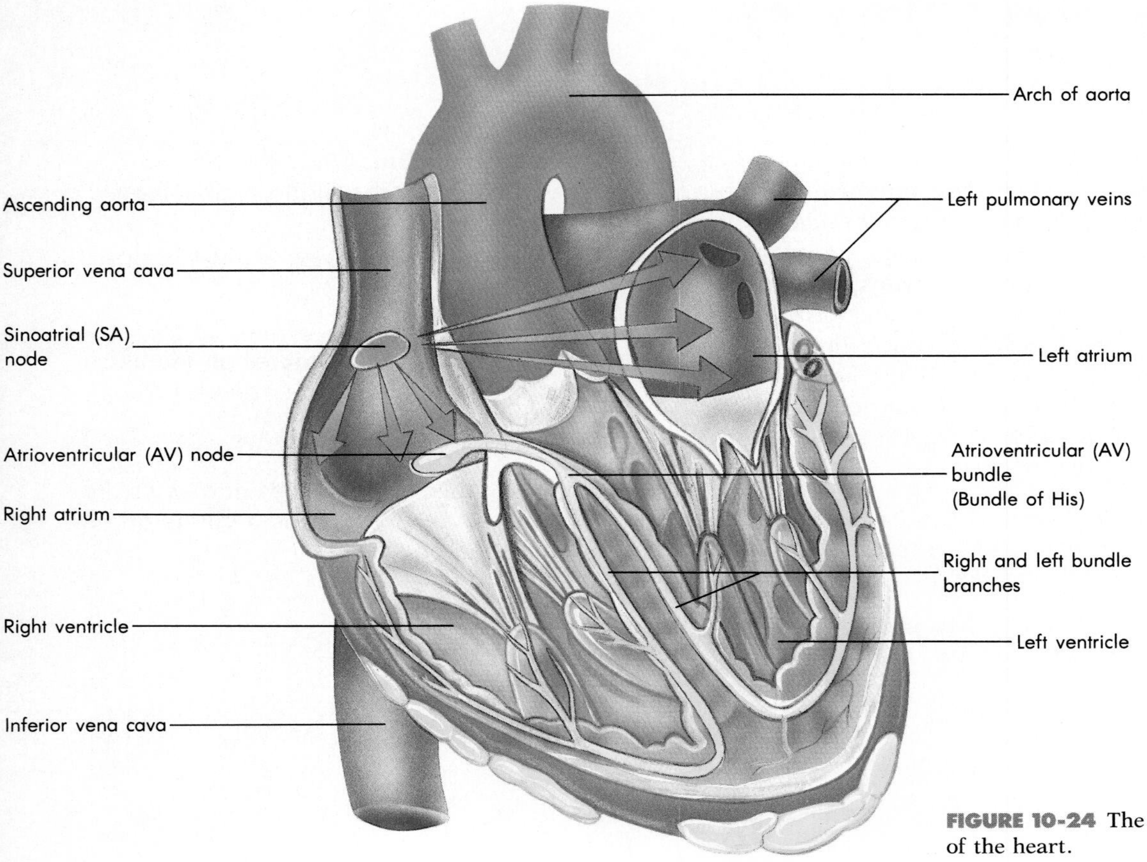

Arch of aorta
Ascending aorta
Left pulmonary veins
Superior vena cava
Sinoatrial (SA) node
Left atrium
Atrioventricular (AV) node
Atrioventricular (AV) bundle (Bundle of His)
Right atrium
Right and left bundle branches
Right ventricle
Left ventricle
Inferior vena cava

FIGURE 10-24 The conduction system of the heart.

osclerosis. An enlarged Q spike could signal a heart attack. A long P-R interval may mean scarring or inflammation from infection.

Some abnormalities are more dramatic. In *ventricular fibrillation,* for instance, the ventricular contractions are weak, rapid, and sometimes chaotic; they fail to pump blood. The cause may be interruption in the blood supply or damage to part of the conduction system by a heart attack. Treatment, or *defibrillation,* requires a substantial electrical shock to the chest wall to "reset" the heart. Ventricular fibrillation is common enough that most ambulances and all emergency rooms have defibrillators on hand.

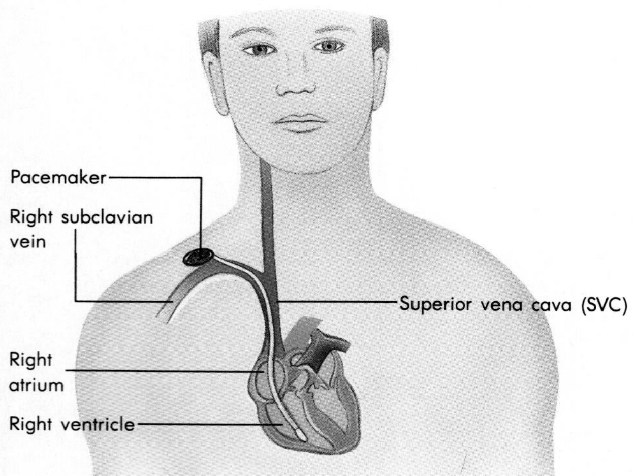

FIGURE 10-25 Diagram of an internal pacemaker's position when it is implanted in a patient. The wires run through the right subclavian vein and superior vena cava and pass through the right atrium and valve to the right ventricle where they stimulate rhythmic contractions.

The Cardiac Cycle

The **cardiac cycle,** shown in Figure 10-27 (p. 237), is the repeating sequence of events that marks each beat of the heart. It begins as the atria contract to push blood into the ventricles. The ventricles then contract to drive blood into the arteries. Once empty, they relax and refill again. As they expand, they passively draw blood from the atria. About 70 percent of ventricular filling occurs as blood flows through the relaxed atria into the waiting ventricles. The weak atrial contractions that start each cardiac cycle pump only an additional 30 percent into the ventricular blood volume.

When the various regions of the heart are contracting, they are said to be undergoing **systole.** A

Box 10-1 Heart Attacks

One of every five people who reach the age of 60 will have a heart attack. That is, their hearts will stop pumping blood. One major form of attack involves an interruption of the blood supply to the heart that damages or kills part of the heart's muscle or the heart's conduction system. The cause can be a blood clot. More often, it is a blockage of the coronary arteries by fatty deposits (atherosclerosis).

Researchers have found that such attacks run in families. But with or without a genetic predisposition to heart attacks, they are most common among people who share the "risk factors" of high blood cholesterol levels, high blood pressure, cigarette smoking, obesity, lack of exercise, and diabetes mellitus. Smoking, for instance, puts into the blood chemicals that can damage the endothelial lining of the coronary arteries and provide sites for the growth of atherosclerotic plaques. Obesity, high cholesterol levels, and diabetes provide the raw material—cholesterol—from which to build the plaques.

Fortunately, you needn't be among the 20 percent. Most of the risk factors are controllable. Don't smoke. Exercise. Keep your weight down. Eat a diet low in cholesterol and saturated fats (from which the body can make cholesterol). Even diabetes can often be prevented, for it can be caused by obesity.

memory trick is to think of it as forcing the blood to flow through the heart's valves with a "Sssss!" The relaxation phase is called **diastole.**

Blood Pressure

The blood is forced into the arteries under substantial pressure each time the ventricles contract. As the blood rushes out of the heart, it causes the arterial walls to distend; the rebound of these walls keeps the blood pressure up and the blood moving forward between heart contractions. In a healthy young adult, the blood pressure is said to be 120 over 80 (written 120/80). Those figures really represent the range of pressures the arterial blood goes through in one cardiac cycle. When the ventricles contract, the arterial pressure is equivalent to 120 millimeters of mercury (mm Hg). Between contractions, the pressure drops to 80 mm Hg. The higher figure (120) is thus said to be the systolic pressure, while the lower figure is called the diastolic pressure. The systolic pressure states the higher arterial pressure that occurs during ventricular contraction; the diastolic represents the lower blood pressure between heartbeats.

1. Study Figure 10-27 carefully. Note that it shows systole and diastole separately for the atria and ventricles, but it also shows pressure changes in the aorta and pulmonary arteries, the atria, and the ventricles. Relate the atrial and ventricular pressure changes to their contractions.
2. Why does the arterial pressure rise only some time *after* the ventricles begin to contract?
3. Why is there a peak in the arterial pressure after the semilunar valves close?
4. Sketch the EKG in the blank space provided. What must the P wave and QRS complex accompany?

Cardiac Output

The **cardiac output** is the amount of blood pumped by either the left or right ventricle per minute (Figure 10-28, p. 237). It depends on several things, including the heart rate (beats per minute) and the *stroke volume* (the amount of blood pumped by each heartbeat). The stroke volume in turn depends on the amount of blood left in the ventricles after their contraction, which depends on the force of the contraction and on the blood pressure in the arteries that resists the heart's efforts. High blood pressure, **hypertension,** reduces cardiac output.

The stroke volume also depends on the amount of blood that enters the ventricles before their contraction, which depends on the blood pressure in the veins. To a large extent, the more blood that enters the ventricles, the more the muscle of the ventricular walls is stretched and the stronger is the subsequent contraction. The heart increases its output during exercise partly because exercise increases the rate of venous return.

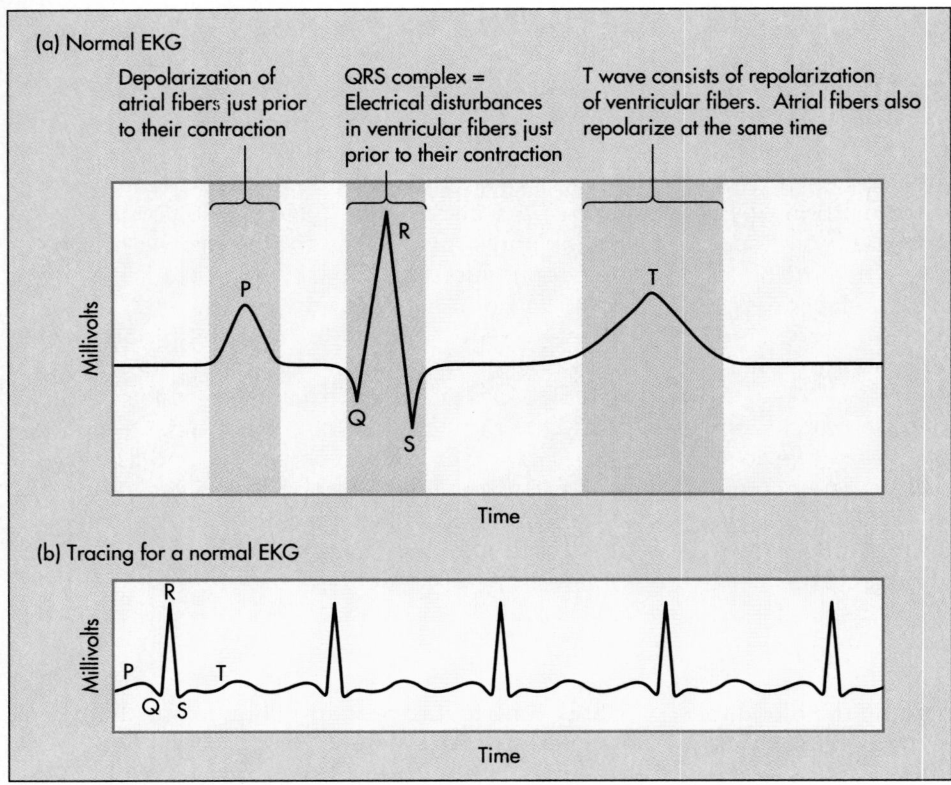

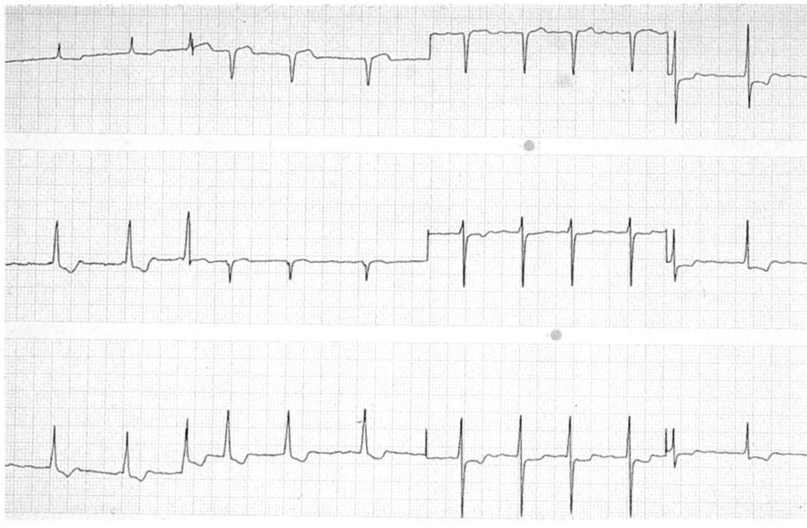

FIGURE 10-26 (a) A single cycle of a normal EKG identifying and naming the electrical disturbances (waves) that occur in the heart tissues during a normal cardiac cycle. (b) The repeating cycles as they would appear in a patient's normal EKG. (c) An EKG showing atrial fibrillation similar to that experienced by President Bush during the spring of 1991. Notice the loss of the coordination between the small, upwardly rounded p-waves and the rest of the electrical events of the cycle. Why would ventricular fibrillation be more serious than atrial fibrillation?

Heart rate, contraction force, and blood pressure can all be controlled by the body. In the medulla at the base of the brain (discussed in Chapter 15) lie two groups of nerve cells. One acts through the autonomic nervous system to speed up and strengthen the heart beat. The other slows it down and weakens it. Both respond to blood pressure detectors located in the carotid artery (in the neck), in the base of the aorta, and in the right atrium. The carotid and aortic sensors respond to arterial blood pressure; when it is high, their signals decrease cardiac output. The atrial sensor responds to venous blood pressure; when it is high, its signals increase cardiac output (to keep up with the inflow of blood).

The body can also control blood pressure di-

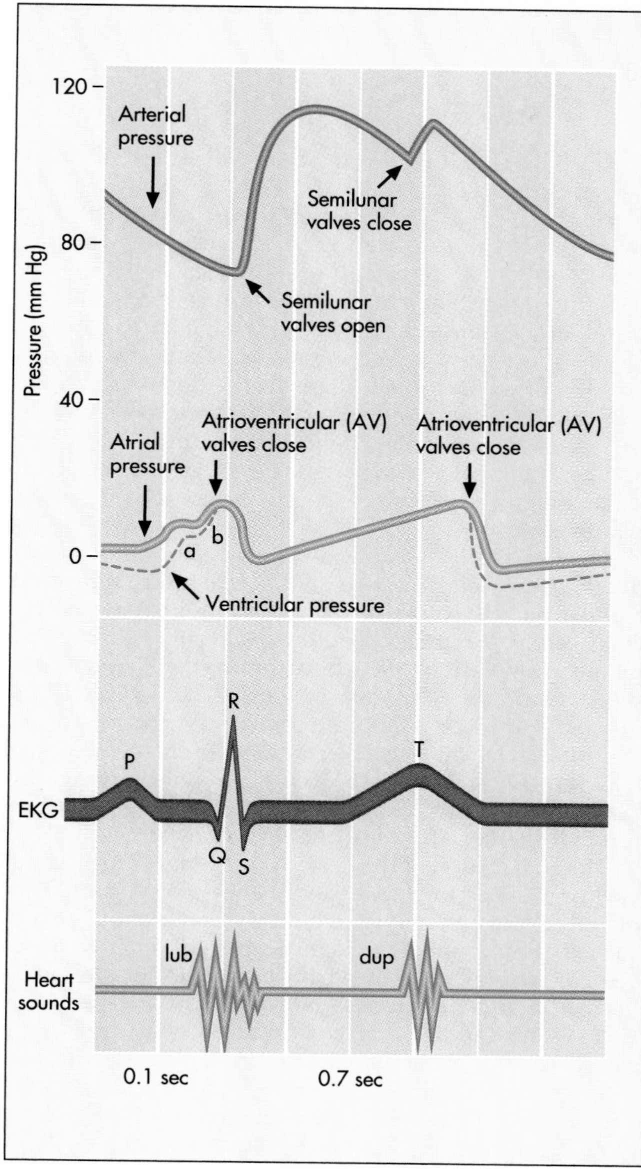

FIGURE 10-27 The cardiac cycle.

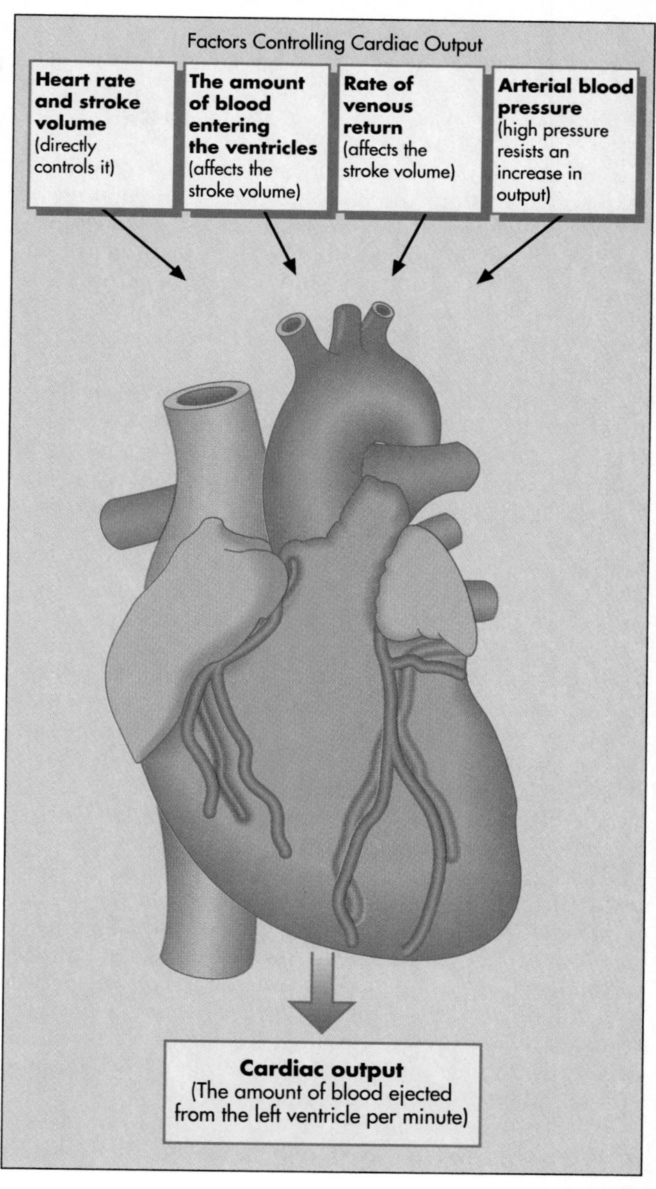

FIGURE 10-28 The major factors controlling cardiac output.

rectly. As we discuss in Chapter 13, both the kidney and the brain's hypothalamus respond to decreases in blood pressure due to loss of fluid (as from blood loss or dehydration) by secreting substances that cause the kidney to conserve water. The autonomic nervous system can cause the smooth muscle of arteries and arterioles to contract, thus decreasing the space available for the blood and raising blood pressure.

The control of cardiac output and blood pressure is a negative-feedback system. Changes in cardiac output affect blood pressure, and blood pressure affects cardiac output. Under normal conditions, they keep each other roughly constant. However, the system does not always work. When massive bleeding reduces venous blood pressure, the atrial sensors tell the heart to decrease cardiac output, which reduces blood pressure and slows the heart still more. Because of the blood loss, there is not enough blood in the system for the mechanisms that increase cardiac output in response to low arterial pressure to work. The blood supply to the heart, brain, and other organs declines, and the resulting **circulatory shock** can easily end in death.

SUMMARY

The cardiovascular or circulatory system is the blood vessels that contain the body's 5 liters of blood and the heart that pumps the blood through the vessels. The blood is 55 percent plasma (water and dissolved protein, salts, and other substances) and 45 percent formed elements (the red and white blood cells and platelets). The formed elements originate from parent cells in the bone marrow.

Red blood cells lack nuclei and are full of hemoglobin, which carries oxygen from the lungs to the tissues and carries some carbon dioxide from the tissues back to the lungs. Red blood cell formation is controlled by the hormone erythropoietin, generated in part when the kidney responds to low oxygen levels. On their cell membranes, red blood cells bear chemical markers called antigens, responsible for the ABO and Rh (and other) blood types.

White blood cells, or leukocytes, are the blood's defense and cleanup squad. Lymphocytes produce antibodies and attack foreign cells such as bacteria. Monocytes (which can become macrophages) and neutrophils phagocytize bacteria and cellular debris. Eosinophils fight allergies and parasites. Basophils and mast cells are involved in the body's inflammation response.

The blood can seal breaks in blood vessels by clotting. Initially, platelets (cellular fragments) stick to the edges of the break and release coagulation factors, as do damaged cells. The coagulation factors trigger the conversion of plasma proteins into enzymes that convert fibrinogen in the plasma into insoluble fibrin. As the fibrin fibers form, they trap red blood cells and more platelets to form a blood clot and plug the break. Later the fibrin fibers contract to squeeze out liquid and draw the edges of the break together and thus speed healing. Still later, the enzyme plasmin breaks the clot down.

All blood vessels are lined with endothelium. Blood travels away from the heart in thick-walled, elastic arteries that branch into smaller arteries. The smallest arteries are the arterioles. They branch into capillaries, whose thin walls, composed only of endothelium, permit oxygen, nutrients, and other substances to pass into the tissues by diffusion. The endothelial cells also move some substances from the blood by endocytosis and exocytosis. As they leave the tissues, the capillaries merge into venules, which merge into veins, which carry blood toward the heart. Veins have thinner walls and larger diameters than arteries; they contain about 60 percent of the blood at any moment. Limb veins have valves to aid blood movement.

Fluid escaping from the capillaries (or lymph) is recovered and drained from the tissues through the lymphatic system. Lymph capillaries collect the lymph and pass it to larger lymphatics, which return the fluid to the blood, in the veins near the heart. Lymph nodes filter the lymph and expose it to lymphocytes and macrophages, which remove bacteria and debris; they also contain parent cells for lymphocytes. The thymus, a lymphoid organ, is where T cell lymphocytes mature. The spleen, another lymphoid organ, removes worn-out red blood cells, stores red blood cells, and generates and stores lymphocytes.

The heart is a specialized muscular pumping system divided into four chambers. Its right atrium receives blood from the veins of the systemic circulation. The right ventricle pumps blood into the arteries of the pulmonary circulation. The left atrium receives oxygenated blood from the pulmonary veins. The left ventricle pumps it into the main systemic artery, the aorta. The ventricles have much thicker walls than the atria. Valves between the atria and the ventricles and between the ventricles and the arteries ensure that blood flows through the heart in only one direction.

The contraction of the heart is coordinated by specialized portions of the heart muscle that generate and conduct the contraction signals. This conduction system consists of the sinoatrial node, or pacemaker; the atrioventricular node; and the atrioventricular bundle and its branches. The electrical signals carried by the conduction system can be detected and recorded at the body's surface as the electrocardiogram (EKG). Heart attacks occur when the heart is damaged by interruptions in the heart's own circulation. The damage can occur to both the heart muscle and the conduction system.

The speed and strength of the heart's contractions are controlled by negative feedback from blood pressure detectors in the carotid artery, right atrium, and aorta as well as by signals from the brain that reflect emotions and by body temperature, age, and sex. The control system can, however, fail if there is massive bleeding. The result is circulatory shock, which can quickly lead to death.

Table 10-4 summarizes several common diseases of the blood and the circulatory system.

Table 10-4

Clinical Summary: Selected Diseases of the Blood and the Cardiovascular System

DISORDER	DESCRIPTION
Anemia	Deficiency of red blood cells or of hemoglobin, and thus of the ability to supply the body with oxygen, due to failure to absorb vitamin B_{12} (pernicious anemia) or iron, destruction of the bone marrow (aplastic anemia), or blood loss (hemolytic anemia), among other causes.
Atherosclerosis	Blockage of arteries by fatty (cholesterol) deposits. Can cause heart damage or failure when coronary arteries are blocked. Can begin in childhood; risk factors include smoking, drinking, lack of exercise, obesity, hypertension, high blood cholesterol levels, diabetes mellitus, and a family history of the condition.
Erythroblastosis fetalis	Destruction of a newborn Rh^+ infant's red blood cells by the immune system of an Rh^- mother. Can be treated by transfusing the infant, destroying the mother's anti-Rh antibodies, or destroying the Rh antigens that reach the mother's blood. See Chapter 19.
Hemophilia	Inherited failure of the blood to clot due to lack of an essential coagulation factor. Usually treated today with injections of *antihemophilic factor* synthesized by genetically engineered bacteria.
Hypertension	High blood pressure (greater than 140/90). May be due to chronic stress, dietary factors (excess table salt can stimulate water retention), blood vessel blockage, or other causes. Affects about 20 percent of Americans. Reduces cardiac output and may make strokes more likely. Can be treated with diuretics, antistress medications, and vasodilators.
Leukemia	Cancer of the bone marrow cells that produce white blood cells. Interferes with red blood cell production and immune function. Damages organs. Can be treated like other cancers with radiation and drug therapy.
Lymphoma	Cancer of the lymphoid organs (spleen, thymus, lymph nodes). One variety, Hodgkin's disease, is often curable with radiation therapy.
Sickle cell anemia	Genetically determined defect in hemoglobin structure. The hemoglobin of sickle cell victims crystallizes when oxygen levels are low. Affected red blood cells deform and burst. Affects mainly blacks and others whose ancestors dwelt in malarial regions.
Stroke	Brain damage due to lack of oxygen. Can be caused by blockage of an artery by atherosclerosis or an embolism or by bursting of an aneurysm. Hypertension is a major risk factor.

STUDY QUESTIONS

1. Why do we say the blood contains "formed elements" instead of "cells?"
2. What is the difference between blood and plasma? Between plasma and serum?
3. People who live at high altitudes have more red blood cells per cubic millimeter of blood than people who live at sea level. Why?
4. Which of the white blood cells are phagocytes? What are their functions?
5. Cancer is uncontrolled multiplication of cells. White blood cells fight infections. Explain why, in leukemia, despite the multiplication of white blood cells, infections are not stopped more quickly than ever.
6. Outline the intrinsic and extrinsic clotting pathways.
7. What are the differences between arteries and veins? What are the differences between capillaries and veins?
8. How does the body prevent fluid from accumulating in its tissues?
9. Describe the path of blood through the heart.
10. What is the function of the heart's sinoatrial node?
11. Your blood pressure should be somewhere near 120/80 if you are in your teens or twenties. How would a blood pressure of 300/180 affect your heart function? Your likelihood of a heart attack?

GLOSSARY

Aorta Largest artery in the body; branches deliver blood to all tissues.

Arteriole One of the smallest arteries; subdivides to form capillaries.

Artery A blood vessel carrying blood away from the heart.

Atherosclerotic plaque Deposit of fatty substances (cholesterol and triglycerides) and smooth muscle on artery wall.

Atria The chambers of the heart that receive venous blood.

Bicuspid valve The two-flapped valve between the left atrium and the left ventricle.

Blood types (or groups) Blood classification system that is based on markers in red blood cell membranes.

Capillary Smallest of blood vessels; wall is endothelium only; site of oxygen and nutrient transfer to tissues and waste transfer to blood.

Cardiac cycle The sequence of events marking each contraction of the heart.

Cardiovascular system The heart and blood vessels.

Circulatory system The cardiovascular system.

Clotting Formation of a platelet and fibrin plug in damaged blood vessels.

Conduction system That part of the heart specialized for the generation and conduction of electrochemical signals that causes the heart to contract in an orderly and efficient manner.

Diastole Relaxation of the heart.

Electrocardiogram (EKG) A graphic record of the electrical activity generated in the heart's conduction system; the term is abbreviated *EKG* because in German *cardio* is *kardio*.

Endothelium Lining of blood vessels and the heart.

Fibrin Insoluble, fibrous protein formed from fibrinogen when blood clots.

Fibrinogen Soluble plasma protein converted to insoluble fibrin during the clotting process.

Heart The muscular organ that pumps the blood.

Hemoglobin Iron-containing oxygen-binding protein found in red blood cells.

Interstitial fluid Liquid in the spaces between the cells of the body's tissues.

Leukocytes White blood cells.

Lymphatic system Drainage and filtering system for excess tissue fluid.

Lymph nodes Lymph filters, containing lymphocytes and their parent cells, that remove bacteria, viruses, and cellular debris from the lymph.

Lymphocytes Leukocytes that secrete antibodies and attack foreign cells.

Macrophages Large phagocytes found in infected tissue.

Myocardium The muscle of the heart wall.

Pacemaker The SA node.

Plasma Blood minus its formed elements; a solution of water, proteins, salts, nutrients, and other substances.

Platelets Membrane-enclosed cellular fragments pinched off from megakaryocytes in the bone marrow; essential to blood clotting.

Red blood cells Blood cells lacking nuclei and filled with hemoglobin; biconcave disks.

Semilunar valves The valves at the entrances to the aorta and pulmonary artery that prevent the backflow of blood into the ventricles; each one is composed of three cups of valvular tissue.

Systole Contraction of the heart.

Tricuspid valve The three-flapped valve between the right atrium and the right ventricle.

Vein A blood vessel carrying blood toward the heart.

Ventricles The chambers of the heart that pump blood into the arteries.

Venule One of the smallest veins; formed when capillaries merge.

Chapter 11

The Respiratory System

- The Respiratory Organs
- Gas Exchange
- Oxygen Transport
- Carbon Dioxide Transport
- Control of the Respiratory Apparatus
- Learning Focus: The Transport of Respiratory Gases
- Summary
- Study Questions
- Glossary

Unlike fish, humans live in an ocean of air, not water. Like fish, we too have special anatomic structures for exchanging gases between our external environment and our cells. Our cells need a continuous supply of oxygen to help release the energy contained in food molecules so it can be used to operate our bodies. Whereas fish use gills to absorb oxygen, we use a series of baffles and passageways to guide air deep into special chambers, our lungs. There the oxygen in the air goes into solution and eventually passes through membranes into our blood. We also unload potentially harmful carbon dioxide gas, produced in our tissues, by running the system simultaneously in the other direction. These transfers of gases are collectively handled by our *respiratory system*.

What most of us call breathing, biologists call **pulmonary ventilation.** By this, they mean simply drawing air in through the nose, mouth, and windpipe (or trachea) to the lungs and then expelling it. They reserve the term *respiration* for what happens next. **External respiration** takes place when inhaled oxygen moves through our lung membranes into the blood and carbon dioxide passes in the opposite direction. **Internal respiration** is the movement of oxygen out of the blood and into tissues such as muscles and the movement of carbon dioxide from the tissues, where it is being produced, into the blood (Figure 11-1).

Cellular respiration refers to the use of oxygen inside cells as the cells break down their nutrients (especially glucose) to obtain energy (Chapter 21). *Anaerobic* cellular respiration is the part of that process that does not require oxygen. *Aerobic* cellular respiration does.

THE RESPIRATORY ORGANS

The respiratory organs are those organs through which air flows during pulmonary ventilation. They include the nose, oral cavity, pharynx, larynx,

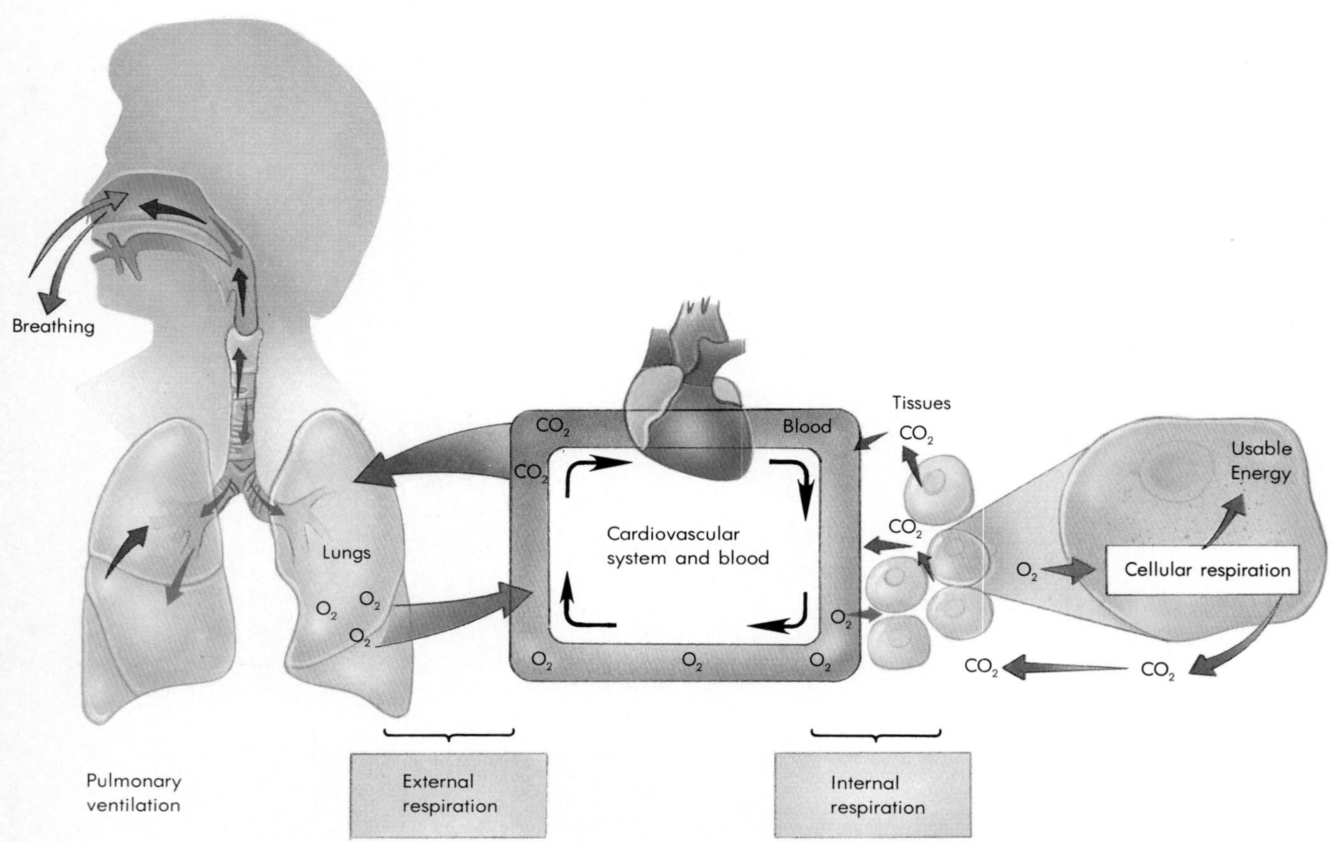

FIGURE 11-1 The types of respiration.

trachea, bronchi, bronchioles, and lungs (Figure 11-2). We can breathe through the mouth, but the nose is the primary entrance and exit for air. The mouth is a backup, essential only when the nasal passage is blocked, as by a cold, or when we are struggling to get in as much oxygen as possible, as during vigorous exercise. Normally, air flows out the mouth only when we are talking.

1. How does cellular respiration differ from external and internal respiration?
2. What is the difference between internal and external respiration?

The Nose

The nose conditions the air we breathe by filtering, warming, and humidifying it as it passes across the membranes of the nasal cavity. The two openings into the nose, the nostrils, are technically called the **external nares.** The air passes from the nasal cavity to the throat, or pharynx. The wall between the nostrils, the *nasal septum,* divides the nasal cavity into left and right chambers. Three thin plates of bone covered with mucous membrane, the *nasal conchae,* extend out a short distance from the lateral walls of the nasal passage, curving downward and forming baffles that swirl the air as it passes through.

At the top of the nasal cavity, above the conchae, lies the part of the nose responsible for the sense of smell. The mucous membrane in that area contains the cell bodies of numerous nerve cells; their fibers pass into the brain through the cribriform plate that forms the bony floor to the cranial vault above. When airborne molecules are inhaled, they go into solution in the moisture coating the membranes and stimulate receptors on the nerve cells; the nerve cells then send nerve impulses to the area of the brain responsible for the sense of smell. A duct from the eye also opens into the nasal cavity; this

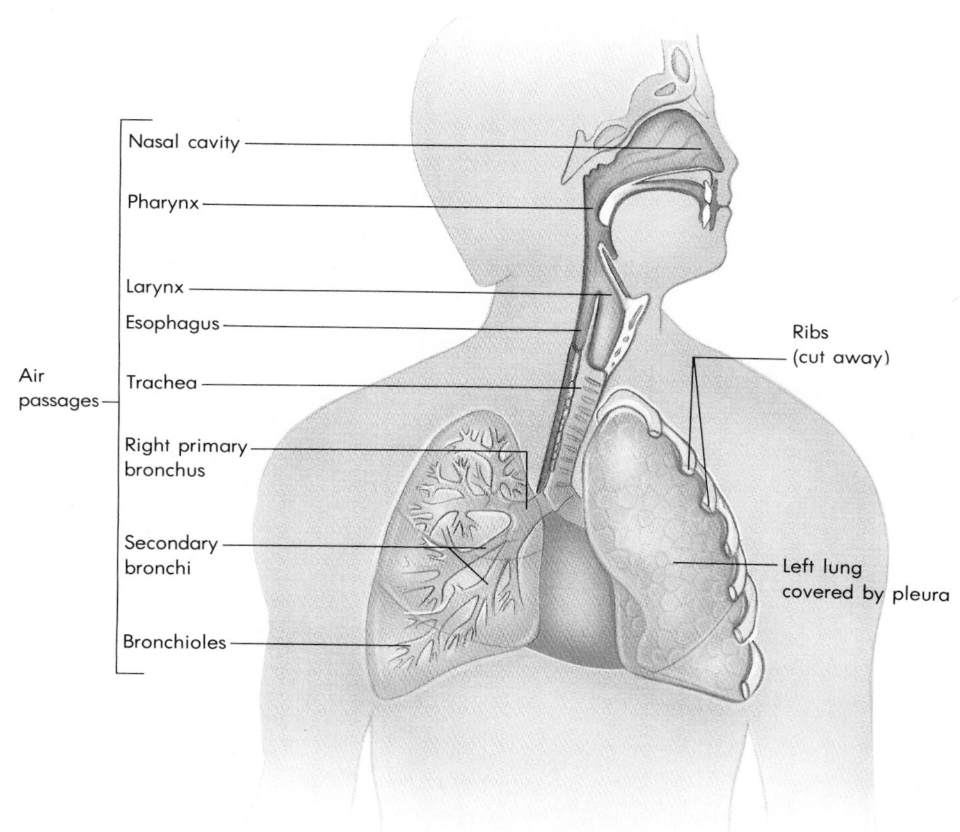

FIGURE 11-2 The respiratory organs.

nasolacrimal duct drains tears and explains why our noses run when we cry (Figure 11-3).

Hairs just within the external nares (nostrils) filter dust and other particles from the air and cause moving air to swirl and strike the mucous membranes covering the conchae and lining the nasal cavity. The nasal mucus traps whatever particles the hairs miss. The nasal conchae also help control the temperature and humidity of the air passing through the respiratory passages.

Temperature control is especially important for people living in very cold climates, like the northern states, where the air just outside the nostrils can be 30°C below zero. By the time that same air has flowed through the upper respiratory passages to the lungs, it has been warmed to 37°C above zero—to body temperature. This prevents ice crystals from forming in the delicate tissues of the lungs.

In many animals, including people, the conchae help to conserve heat. As cold air flows into the nose, it draws warmth from and cools the lining of the nose. On the way out, carrying that warmth and more from the body's depths, it passes its warmth to the previously cooled mucous membrane. The air that leaves the nose may be little warmer than the air outside.

The conchae can also help conserve moisture. When dry air enters the nose, it draws moisture from the nasal mucus, drying the nasal lining instead of the delicate tissues of the lungs. On the way out, the moisture moves from the flowing air back to the mucus. In that way, the body reduces the potential water loss.

---

1. Do you think the nasal air conditioning system can work in reverse? How would it work with hot (above body temperature) air?
2. Why would nasal cooling work more efficiently with hot, dry air?

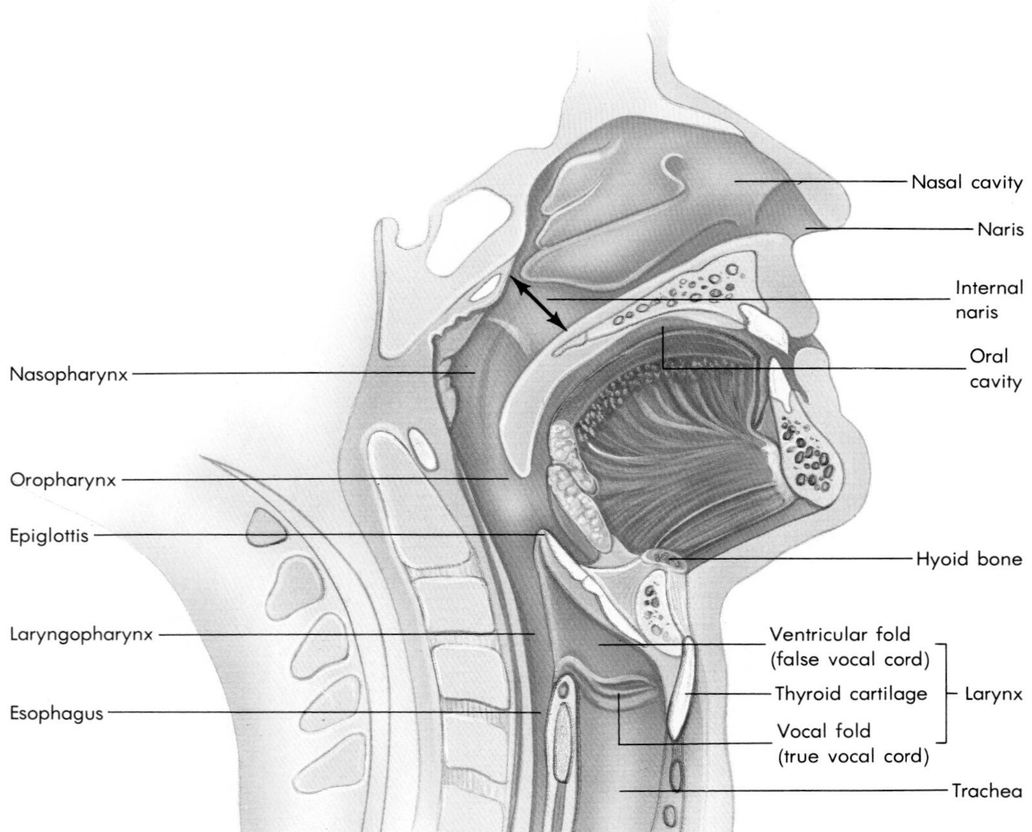

FIGURE 11-3 Sagittal section of nose, mouth, larynx, and upper trachea.

The Pharynx

Behind the oral and nasal cavities and in the vicinity of the voice box is the **pharynx.** The pharynx is divided into three zones: behind the nasal cavity it is called the nasopharynx, behind the mouth it is the oropharynx, and around the top of the voice box, the laryngopharynx. It is the latter two that we generally call the throat. The **internal nares** are the channels at the back of the nasal cavity that open into the **nasopharynx** (see Figure 11-3). The nasopharynx extends from the top of the pharynx down to the border of the soft palate on the roof of the mouth. Below that point and behind the mouth, it is called the **oropharynx.** The tag of flesh hanging from the posterior edge of the soft palate, the **uvula,** marks the border between the mouth and oropharynx.

The oropharynx belongs to both the respiratory and the digestive systems, for air, liquid, and food pass through it. It ends at the level of the hyoid bone and epiglottis, above the voice box, where the **laryngopharynx** begins. In the laryngopharynx, the digestive and respiratory passages separate. The esophagus carries food to the stomach, while the **trachea,** or windpipe, carries air to the lungs. The **epiglottis** folds down to cover the passageway into the larynx during swallowing, thus preventing food or liquid from entering the windpipe.

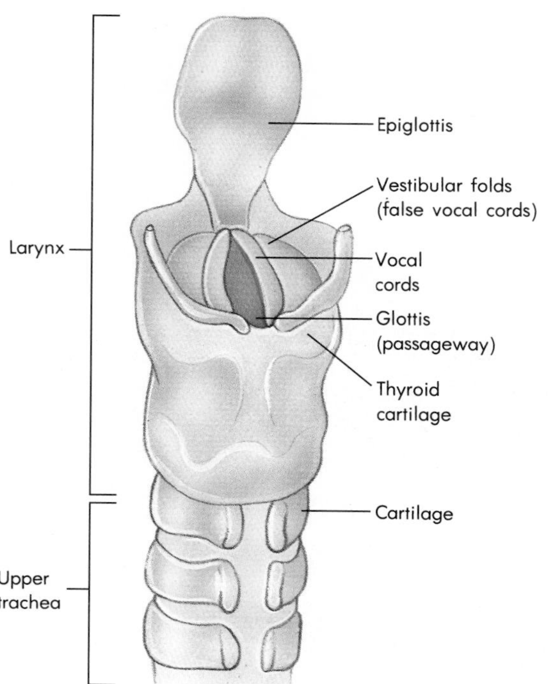

FIGURE 11-4 Posterior view of the larynx, showing the vocal folds, glottis, and epiglottis.

The Larynx

The **larynx,** or voicebox (Figure 11-4), sits atop the trachea. It is a relatively rigid box of cartilaginous plates. Its front is formed by the **thyroid cartilage,** a projection of which is the Adam's apple. Other cartilages form the sides and back of the box.

The mucous membrane lining the larynx forms two pairs of folds (Figure 11-4). Skeletal muscles attached to the larynx can stretch these folds to block the larynx partially or completely. The upper pair of folds, the **ventricular folds,** or *false vocal cords,* serves to close off the airway, as when we are lifting a heavy weight (see Figure 11-13). The lower pair, the **vocal folds,** or *true vocal cords,* is used to generate sounds. The actual passageway into the larynx, through which air flows, is called the **glottis.**

The laryngeal muscles stretch the vocal cords taut and narrow the passageway between them. A controlled stream of exhaled air sets the cords to vibrating like a reed in a clarinet or saxophone, thus producing voice sounds. The more tightly the cords are stretched, the higher the pitch of the sound. The faster the air flows, the louder the sound. Because the larynx is larger in males, their vocal cords are generally thicker and longer than in females; their cords therefore vibrate more slowly and produce sounds of lower pitch.

Once generated, the sounds resonate in the mouth, pharynx, nasal cavity, and **sinuses.** Variations in the sizes and shapes of these cavities account for the characteristic differences in voices. Swollen membranes, like those caused by colds, change the resonating properties of these chambers to produce noticeable differences in voice quality. Variations produced by movements of the tongue and lips yield the sounds of speech.

When the larynx is removed for treatment of cancer, people can still use their respiratory cavities, tongue, and lips to speak, but they need a different sound generator. Some fill this need by learning to inflate their stomachs with air and generate sounds with prolonged belches (teenagers often practice a version of this art). Others use a buzzer about the size and shape of an electric shaver held against the side of the neck to produce their "voice" sounds.

The Trachea, Bronchi, and Bronchioles

Food or liquid forces the esophagus open during swallowing; between swallows, this tube collapses. The trachea cannot collapse, for air is not pushed into this tube. It is sucked in, and the tube must stay open at all times. Its wall is therefore braced and stiffened with C-shaped rings of cartilage, as shown at the bottom of Figure 11-4. The open ends of each ring are linked by smooth muscle and elastic connective tissue that forms the back wall of the trachea.

The trachea extends from the larynx to just above the level of the heart, where it divides to form the primary **bronchi;** the left primary bronchus carries air to the left lung and the right primary bronchus carries it to the right. In the lungs, the primary bronchi branch into secondary bronchi, which carry air into each of the lung lobes. There, the secondary bronchi branch into tertiary (terminal) bronchi; these in turn branch into the **bronchioles.** Further branchings yield the **terminal bronchioles,** which deliver air to the actual tissue of the lungs (Figure 11-5).

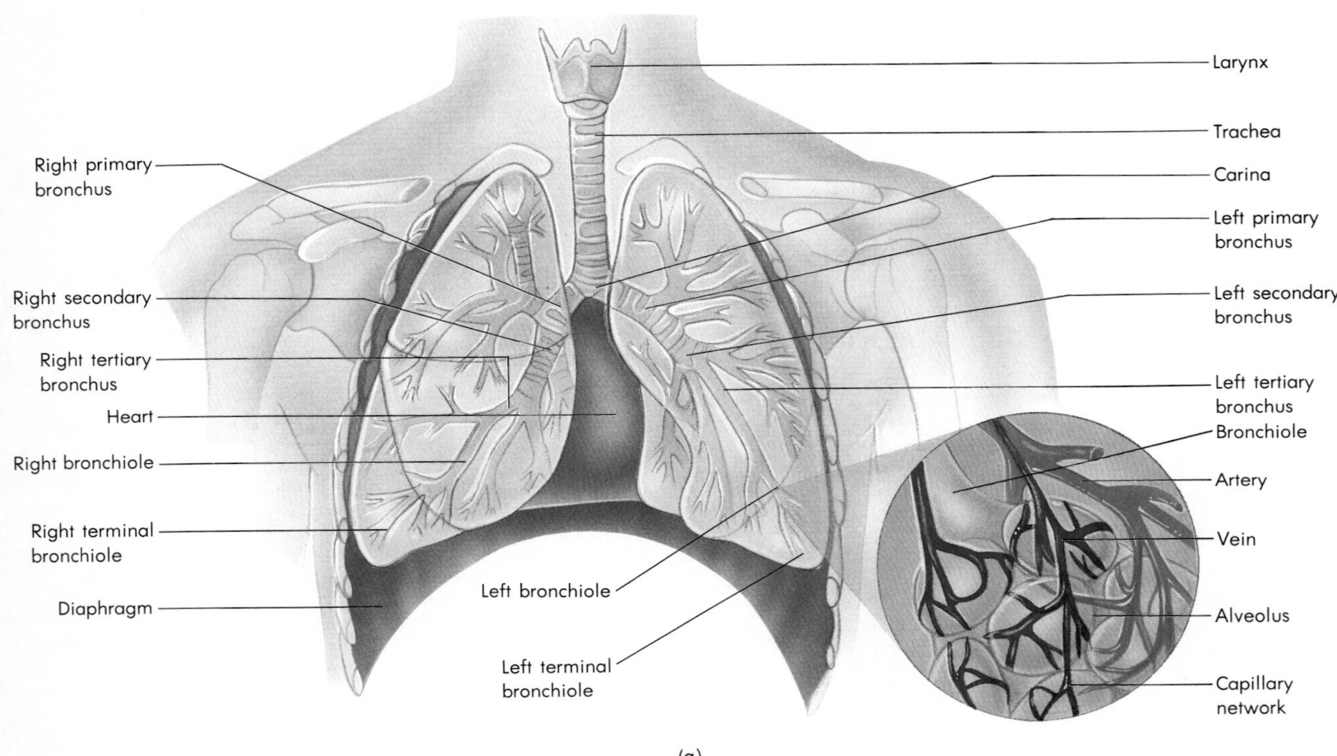

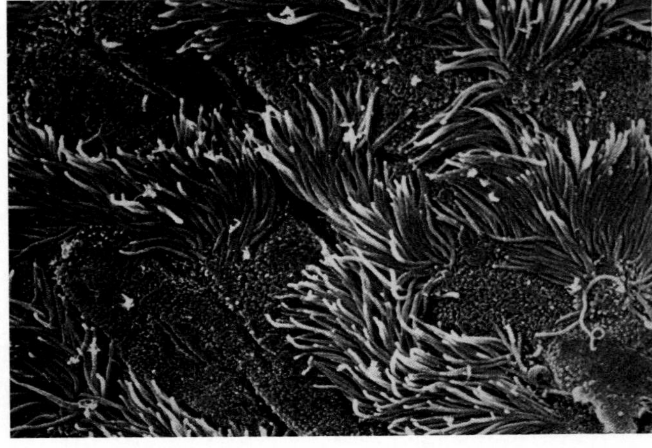

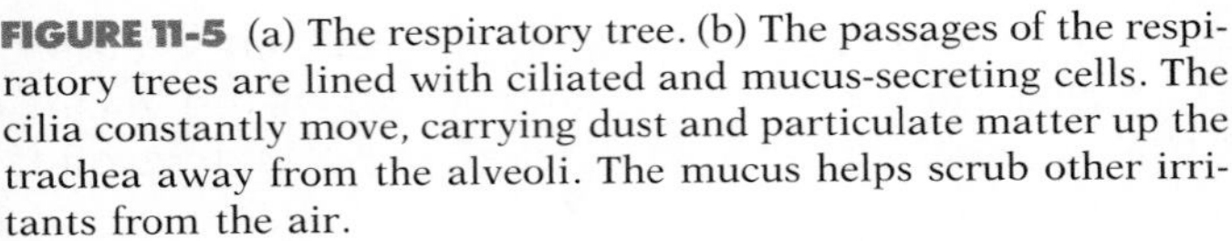

FIGURE 11-5 (a) The respiratory tree. (b) The passages of the respiratory trees are lined with ciliated and mucus-secreting cells. The cilia constantly move, carrying dust and particulate matter up the trachea away from the alveoli. The mucus helps scrub other irritants from the air.

Like the trachea, the bronchi are stiffened and kept open by rings of cartilage. As the branching of the bronchi proceeds, the rings are replaced by plates of cartilage. They disappear entirely in the bronchioles. As the amount of cartilage diminishes, the relative amount of smooth muscle in the walls of the respiratory tubes increases. Irritation of the lining of the trachea, bronchi, and bronchioles, as by an allergy or infection, can cause spasms of the smooth muscle that narrow or close the airway. Such spasms, coupled with the increased secretion of mucus that accompanies an allergic response, account for the wheezing and respiratory difficulty of **asthma.**

The trachea, bronchi, and bronchioles (but not the terminal bronchioles) are lined with ciliated epithelium coated with a layer of mucus. The mucus traps dust in inhaled air. The cilia then propel the dust-laden mucus upward to the pharynx, where it can be swallowed. Tobacco smoke contains substances that both cause the epithelial cells to lose their cilia and inhibit the action of remaining cilia. In smokers, mucus transport is thus impaired. Yet the mucus must still be removed from the airway. Smokers accomplish this task with the hacking cough for which they are so infamous.

The Lungs

The lungs are two masses of spongy tissue on either side of the heart within the rib cage. They are surrounded by two **pleural membranes. Visceral pleura** is attached directly to the outer surface of the lungs. **Parietal pleura** lines the chest cavity. Between them is a layer of lubricating fluid, just as there is between the pericardial membranes of the heart (Figure 11-6; see Chapter 10).

The lungs themselves are subdivided into the **lobes** that receive the secondary bronchi. The lobes in turn are subdivided into **lobules** enclosed in elastic connective tissue and receiving a lymphatic vessel, an arteriole, a venule, and a terminal bronchiole. The terminal bronchiole branches into two or more **respiratory bronchioles** that subdivide into the **alveolar ducts,** which feed into the air sacs of the lungs, the **alveolar sacs** (see Figure 11-5).

Gas exchanges between the blood and lungs take place in alveolar sacs, which resemble a bunch of tiny grapes or a cluster of bubbles. Each grape or bubble is an **alveolus,** and it is through the walls of each alveolus that the gas transfer takes place. The alveolus is lined by epithelium and surrounded by capillaries, so that the air in the alveoli is separated

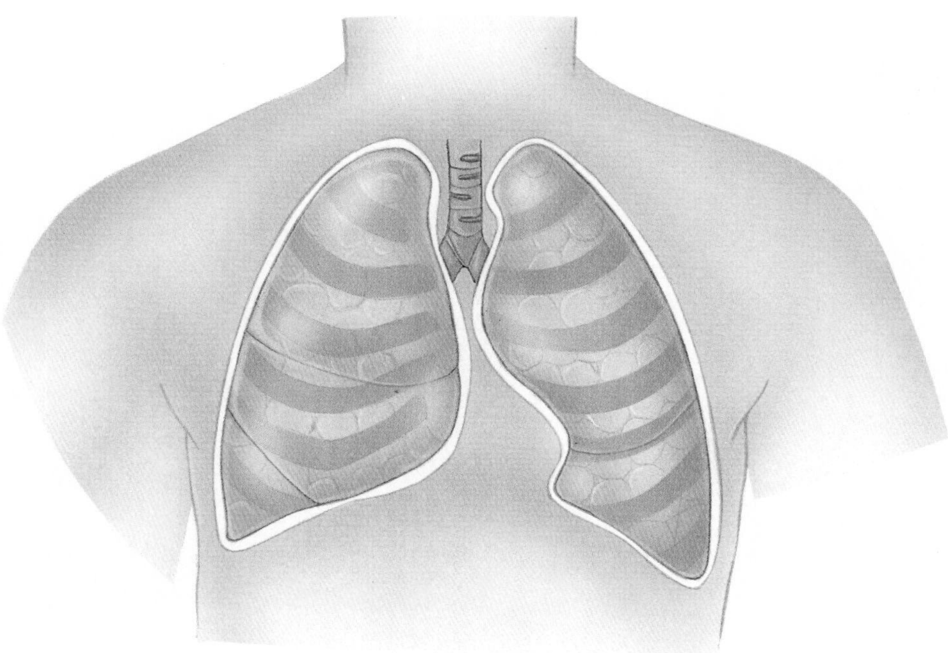

FIGURE 11-6 The pleural membranes and lobes of the lungs.

from the blood by a thin wall just two cells thick (Figure 11-7). This wall is only 0.5 micrometers (μm) thick. The lungs contain some 300 million alveoli, with a combined surface area of 70 square meters (750 square feet).

Pneumonia can cause death primarily because the fluids produced by infected respiratory tissues physically block the gas exchanges within the alveoli and thus reduce the supply of oxygen to the brain and rest of the body. Pneumonia can have numerous causes, ranging from bacterial and viral infections to irritations provoked by inhaling dangerous gases that damage and inflame the linings of the lungs. In response to such irritations, the alveoli fill with fluid and dead white blood cells. Bacterial pneumonia now rarely kills, for we can use antibiotics to end the infection in its early stages. Also, a vaccine has recently been developed that protects high-risk patients against the most common bacterial form of pneumonia. Viral pneumonia can still be fatal (antibiotics have no effect on viruses).

Certain cells in the alveolar epithelium secrete *surfactant*, a substance that helps keep the alveoli from collapsing when they lose air and shrink during exhaling. Babies—often premature babies—that are born with an inadequate supply of surfactant cannot keep their lungs inflated. The result is *infant respiratory distress syndrome;* physicians can now supply as an aerosol spray surfactant produced by genetically engineered bacteria, and death is less likely than it used to be.

The alveolar wall also contains **dust cells** (macrophages) that phagocytize and remove dust particles and debris, which otherwise would collect deep within the lungs. They remove the dead white cells after a bout with pneumonia. They also remove the tars deposited in the lungs by tobacco smoke, allowing smokers who stop smoking to improve significantly the condition of their tar-

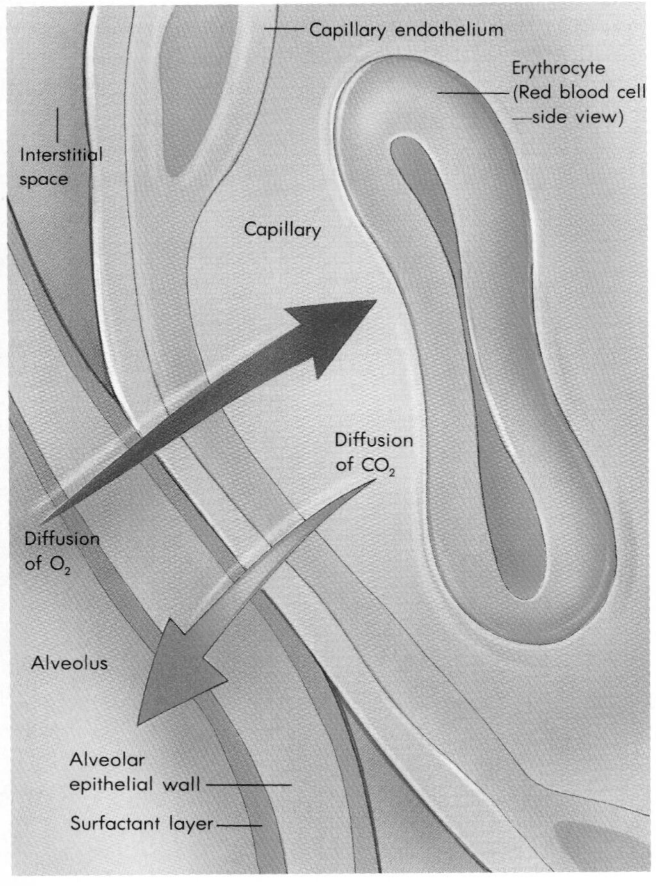

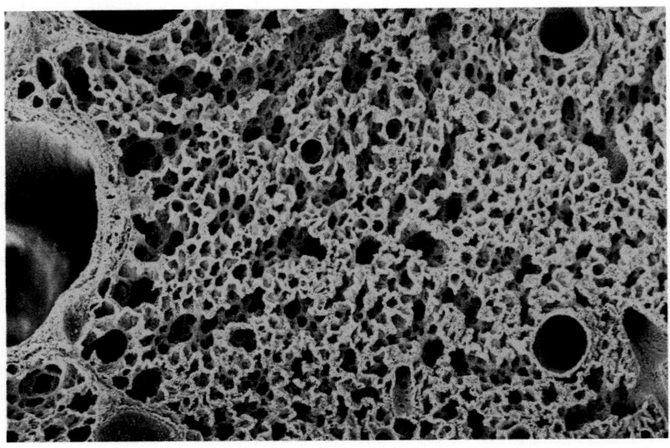

FIGURE 11-7 (a) The alveolus-capillary membrane. (b) A section of human lung showing the network of alveolar ducts and alveoli. Gas is exchanged between the alveoli and the pulmonary capillaries in the walls surrounding each tiny chamber.

blackened lungs (though the improvement may take many years) (Figure 11-8).

1. List the mechanisms that keep the respiratory organs free of dust.
2. What is the benefit of the lungs' immense interior surface area?

Lung cancer is not generally a cancer of the alveolar cells. Rather, it develops when chemicals such as those in tobacco smoke damage bronchial cells enough to cause them to become malignant. The resulting **bronchogenic carcinoma** can escape from the bronchial walls to spread in the actual lung tissue. The growth of cells blocks alveoli and bronchial tubes and prevents oxygen from reaching the blood. The cells also *metastasize,* or break loose from the tumor and ride the blood or lymph to other parts of the body, including the brain. After metastasis, even complete replacement of the lungs with a transplant often cannot cure lung cancer, and lung cancer is rarely detected before metastasis has occurred. It is among the most frequently fatal of cancers.

The Mechanism of Ventilation

We have two ways to draw air into our lungs, but both work on the same principle: During *inspiration* (inhaling), the chest cavity is enlarged through contractions of muscles of the rib cage and the diaphragm. The resulting change in the volume of the thoracic cavity lowers the air pressure within the lungs. Air then flows from the zone of high pressure, outside the body, through the nose, pharynx, trachea, and bronchi, and into the lungs. During *expiration* (exhaling), the muscles for breathing relax, lowering the ribs and allowing the abdominal viscera to push the diaphragm back up into the thorax. This action decreases the size of the chest cavity, increasing the pressure in the lungs and forcing air to flow in the opposite direction. At the same time, the elasticity of the lungs helps them contract.

1. Can you think of other cases where creation of a low-pressure zone forces the movement of air (or liquid)?
2. When Ronald Reagan was shot in an assassination attempt, the bullet passed into the space between the two layers of pleural membranes, allowing air to enter the chest. Why would this condition—known as *pneumothorax*—make breathing difficult or impossible?

The principle is the same as the one we see in action when we use a bicycle tire pump. In that case, pulling on the handle expands an internal air chamber and air flows in. Pushing on the handle shrinks the air chamber and air flows out.

We exploit this principle with our ribs, or the muscles between them, and our diaphragm. During normal, quiet breathing, we use mostly our rib

(a)

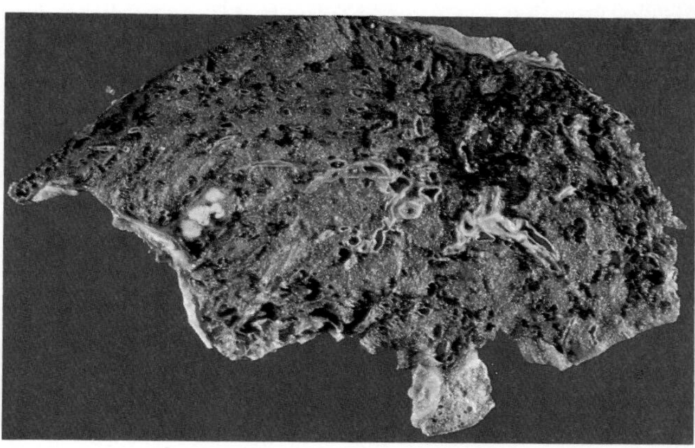
(b)

FIGURE 11-8 (a) Cross section of a normal nonsmoker's lung. (b) Cross section of a blackened smoker's lung. The white areas to the left are lung cancer. The sooty stains are the marks of smoking.

Box 11-1 Smoking: More Treacherous than Just Cancer

The more you smoke, the shorter your life expectancy. The average 25-year-old who smokes two packs of cigarettes a day will live 8.3 years less than a nonsmoker. If you smoke more than that, you can expect to lose even more years, often to lung cancer.

To what else can you lose those years? Smokers are 2 to 17 times as likely as nonsmokers to develop cancer of the lips, mouth, larynx, esophagus, bladder, and pancreas. They are 4 to 25 times as likely to develop **emphysema,** a destruction of the alveoli caused by exposure to irritants such as air pollutants (see Chapter 26 for a discussion of pollution) and tobacco smoke; the symptoms include shortness of breath, panting, and coughing; the heart has to work harder to pump blood through the lungs, and even then it cannot deliver sufficient oxygen to the body's cells, especially to those of the brain.

Smoking is also a major risk factor in heart disease. Smoking mothers are more likely to miscarry and to have low-birthweight babies that are vulnerable to infection. Even seven years after their birth, children whose mothers smoke are likely to be physically and socially underdeveloped.

Smoking also affects the health of those who share offices, homes, and classrooms with smokers. Research has indicated that even those who inhale only second-hand smoke are at greater risk of cancer, heart disease, and other ills. For this reason, more and more cities and states are banning smoking in government buildings, restaurants, and other public places.

Many people find second-hand smoke offensive as well as dangerous. For both reasons, considerate smokers take their cigarettes, pipes, and cigars outdoors. The most considerate smokers stop smoking, for their risks of cancer, heart disease, and other ills immediately begin to decline. Ten to 15 years after quitting, their life expectancy is very close to that of the lifetime nonsmoker.

muscles. As the external intercostal muscles draw the ribs closer together, they swing forward and upward because of the way they are hinged on the vertebrae and because of their flexible attachments to the sternum (breastbone). This movement expands the rib cage and lungs. During *forced inspiration,* additional, *accessory* muscles enhance the expansion. During *forced expiration,* the internal intercostal and abdominal muscles pull the ribs down and shrink the cavity, forcing air out of the lungs (Figure 11-9).

The **diaphragm** is a sheet of muscle separating the base of the thoracic cavity from the top of the abdominal cavity. Anchored on the lowermost ribs, it domes up beneath the lungs and heart (see Figure 11-5). When it contracts, it flattens, pushing the abdominal organs downward and expanding the chest cavity. When it relaxes, it domes up again, shrinking the chest cavity and pushing air out of the lungs.

You can watch the two ventilating mechanisms in action if you lie flat on your back on the floor with one book on your chest and another on your abdomen. When you breathe with your diaphragm, the first book does not move, but the second rises and falls. When you breathe with your ribs, the second book is motionless, but the first one rises and falls.

Pulmonary Volumes

The maximum amount of air the lungs of the average person can hold—the **total lung capacity**—is about 6 liters, or 6.5 quarts. Research with a *spirometer,* a device that measures the amount of air moved in breathing, shows that during normal, quiet breathing (about 12 breaths per minute) you move about 500 milliliters (1 pint) of air in and out of your lungs with each breath. That amounts to about 1 cup (250 milliliters [ml]) of air moved for each lung. Of this **tidal volume,** about 350 ml reaches the alveoli; the other 150 ml occupies the **dead air volume** of the nose, pharynx, trachea, bronchi, and bronchioles. A very deep breath can draw in an additional 3100 ml, the **inspiratory reserve volume.** A forced exhale, beyond a normal exhalation, can blow out an extra 1200 ml, the **expiratory reserve volume.** Even then, the lungs con-

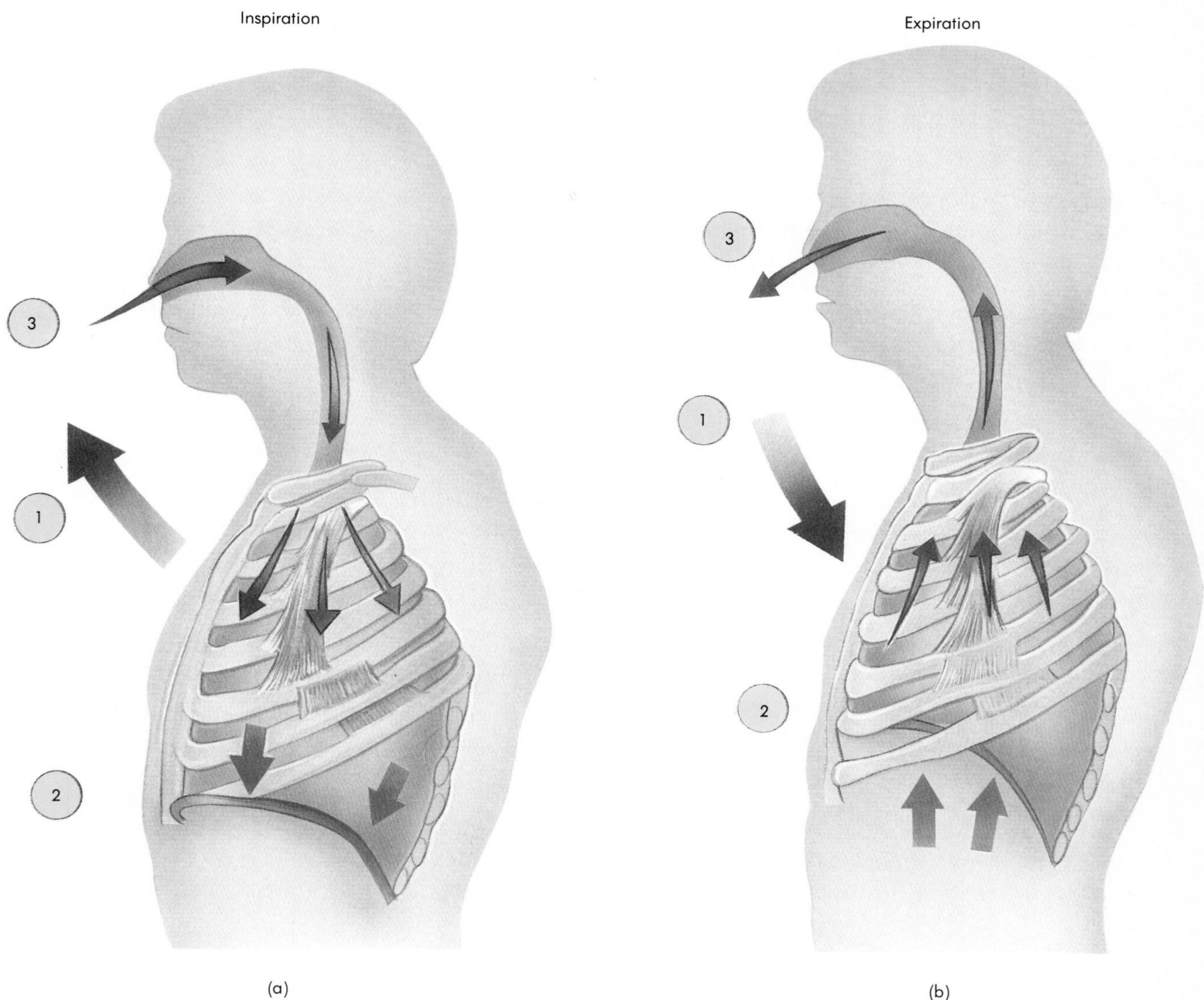

FIGURE 11-9 Ventilation. (a) *Inspiration.* (1) Intercostal muscles contract, ribs rise, chest cavity expands—lung pressure drops. (2) Diaphragm contracts. (3) Air flows into the respiratory system as lung pressure drops. (b) *Expiration.* (1) Intercostals relax, lowering ribs and increasing lung pressure. (2) Diaphragm relaxes—viscera push up, increasing lung pressure. (3) Air flows out as lung pressure increases.

tain a **residual volume** of 1200 ml that you cannot blow out. The total lung capacity minus the residual volume is the **vital capacity,** which consists of all the lung volume available for deep breathing. Your vital capacity is determined by measuring the volume of air you can blow out in a maximum exhalation after taking in as deep a breath as possible (a maximum inspiration) (Figure 11-10, p. 253).

The residual volume is what makes the **Heimlich maneuver** possible. Sometimes people get a piece of food caught in their throat. They choke, trying desperately to cough it out. Unfortunately, it refuses to budge, they exhale all the air they can (their expiratory reserve volume), they cannot inhale, and they suffocate. The solution is that someone comes up behind them, positions a fist (clutched by the other hand) just below the choker's breastbone (the sternum), and firmly pulls back and slightly up under the ribs. This is the Heimlich maneuver, and it has saved many lives by compressing the lungs enough to force residual air out hard enough to blow out the obstruction. A choking victim who happens to be alone can produce much the same effect by leaning over the back of a chair and forcing it up under the rib cage. Remember, a person with an obstructed airway often is not coughing because he or she cannot get any air out. Sometimes your only warning indicating that the person is asphyxiating is a panicked look in the eyes and a gesture toward the mouth.

Box 11-2 Cardiopulmonary Resuscitation

Heart attacks kill by stopping the circulation of blood. Interruptions of pulmonary ventilation kill by halting the flow of oxygen to the blood. Both of these problems can be treated by ambulance and emergency room crews, but by the time the medics reach the victim it is often too late. The brain can survive only minutes without a continual supply of oxygen.

Fortunately, there is something bystanders can—and should—do immediately and on the spot. It is called **cardiopulmonary resuscitation,** or **CPR,** and courses in the technique are available nationwide. Take one, and you may save the life of a victim of heart or respiratory failure due to drowning, poisoning, drug overdose, electrocution, or heart attack.

With accident victims, you should be sure to move them as little as possible, for movement can severely damage the spinal cord if the neck or back is broken, resulting in paralysis or death. In victims without spinal injury, you begin CPR on a nonbreathing person by making sure the airway is clear. Push down on the chin to open the airway to its maximum. This may be enough to start the victim breathing spontaneously. If it is not, keep the chin down and pinch the nostrils shut. Take a deep breath, cover the victim's mouth with your own, and exhale. Blow in about twice as much air as a normal breath. Remove your mouth to let the victim's lungs deflate. Repeat about 12 times per minute. Stop when the victim starts breathing spontaneously, when medics arrive to take over, when you can continue no longer (because of exhaustion), or when a physician declares the victim dead.

If the heart has stopped, you must perform *external cardiac compression.* You put the heels of your hands over the victim's sternum (breastbone) and press down firmly and smoothly 60 times per minute. This action compresses the heart and propels blood into the arteries, it cannot match a working heart, but it can achieve a circulation about a third as efficient as normal.

If the victim's heartbeat and breathing have both stopped, you must alternate mouth-to-mouth resuscitation and external cardiac compression. Give three or four breaths, and then a dozen compressions. Repeat.

Warning: Do not attempt CPR if you have had no training in the technique. Get help instead. This might be a good time to sign up for a certified CPR course so that you do not get caught being unable to help.

What if you are the victim? Although you cannot do mouth-to-mouth resuscitation on yourself, you can do a kind of cardiac compression. If your heart stops, and you notice it before you lose consciousness, start coughing—*hard!*—at a rate of 60 coughs per minute. Coughs raise pressure in the chest, squeeze the heart, and propel blood into the arteries enough to maintain consciousness, at least for awhile. One physician whose heart stopped drove himself to the hospital, coughing all the way.

GAS EXCHANGE

Pulmonary ventilation puts a thin layer of air as close as possible to a thin layer of blood. Only then can *external respiration,* the exchange of oxygen and carbon dioxide between air in the lungs and the blood, begin.

The thinness of the layers of air and blood maximizes the efficiency of the exchange. The narrowness of their separation by the alveolus-capillary membrane (just 0.5 μm) and the large surface area offered by all the alveoli of the lungs (70 square meters) maximize the speed of exchange. When the surface area is decreased, as in emphysema and pneumonia, gas exchange diminishes greatly.

We generally discuss gas exchanges between the air, the blood, and the tissues in terms of partial pressures. Scientists established this system after they learned that the atmosphere at sea level exerted a pressure that would push mercury (Hg) 760 millimeters (mm) up a tube "filled" with a vacuum. Since atmospheric air contains 21 percent oxygen and 0.04 percent carbon dioxide (most of the rest is nitrogen), we say that oxygen contributes roughly 21 percent of the 760 mm Hg pressure (21 percent $\times$ 760 mm Hg), or a *partial pressure* (pO_2) of

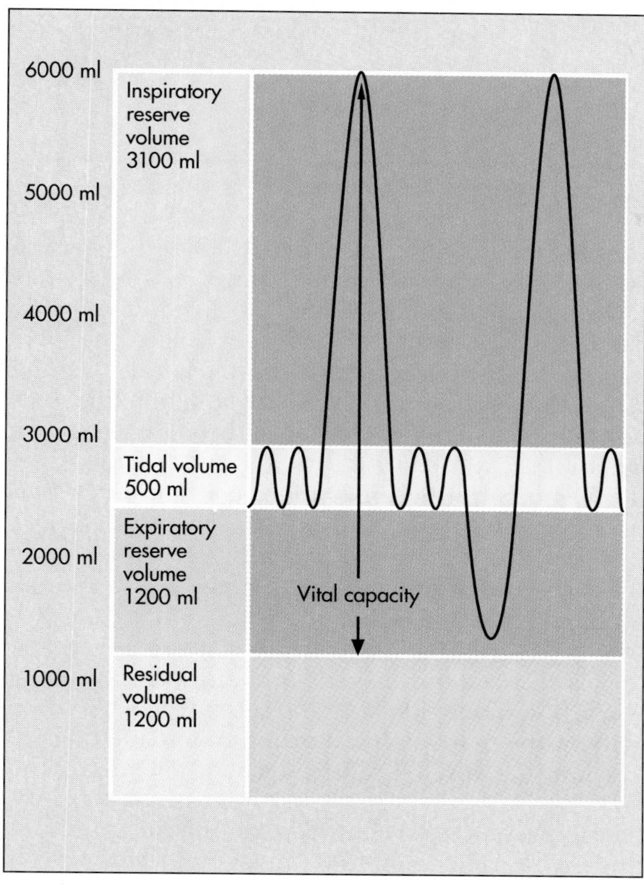

FIGURE 11-10 Lung volumes.

160 mm Hg, and that carbon dioxide has a partial pressure (pCO_2) of 0.3 mm Hg. In the alveoli, this fresh air mixes with oxygen-poor and carbon-dioxide-rich dead air; the mixture has a pO_2 of 105 mm Hg (14 percent) and a pCO_2 of 40 mm Hg (5.5 percent).

The blood that reaches the alveolar capillaries has a pO_2 of 40 mm Hg and a pCO_2 of 45 mm Hg. It thus contains less oxygen and more carbon dioxide than the alveolar air, and there is a concentration gradient across the alveolus-capillary membrane for each gas. Oxygen diffuses into the blood and carbon dioxide diffuses out of it. When the blood leaves the alveoli, it has a pO_2 of 105 mm Hg and a pCO_2 of 40 mm Hg (Table 11-1).

During *internal respiration,* the exchange of gases takes place between the blood and the cells of the body's tissues, which use oxygen to generate energy in cellular respiration and produce carbon dioxide as a waste product. In the cells, the pO_2 is 40 mm Hg and the pCO_2 is 45 mm Hg. Oxygen thus diffuses from the blood into the cells, and carbon dioxide diffuses from the cells into the blood. The venous blood that leaves the tissues has the pO_2 and pCO_2 characteristics of the blood that arrives at the alveoli.

OXYGEN TRANSPORT

The blood's pO_2 depends only on the oxygen dissolved in the plasma, but this is only 3 percent of the oxygen carried in the blood that leaves the lungs, and it is entirely insufficient to meet the body's need for oxygen. The body needs another oxygen transport mechanism, and it has it in the red blood cells and the hemoglobin they contain. Hemoglobin's heme groups and iron ions allow each molecule of hemoglobin (Hb) to bind to four molecules of oxygen (O_2) to form *oxyhemoglobin* (HbO_2):

$$Hb + O_2 \longrightarrow HbO_2$$

Hemoglobin has such a high affinity for oxygen that when it is exposed to a pO_2 of 105 mm Hg, as it

Table 11-1

Gas Concentrations in Respiration[a]

	ATMOSPHERE	ALVEOLI	ARRIVING BLOOD (TO LUNGS)	DEPARTING BLOOD (TO TISSUES)
pO_2	160 21%	105 14%	40	105
pCO_2	0.3 0.04%	40 5.5%	45	40

[a]Concentrations expressed in mm Hg.

is in the lungs, almost every available heme-iron binding site grabs an oxygen molecule. The hemoglobin becomes 100 percent *saturated* with oxygen. Iron is such an important component of hemoglobin that it has been estimated that about 65 percent of ingested iron will go for hemoglobin synthesis. The body contains a total of about 4 grams of iron in the total blood volume.

In the tissues, the reaction goes in the opposite direction:

$$HbO_2 \longrightarrow Hb + O_2$$

Here the pO_2 is much lower, only 40 mm Hg. In addition, the temperature is higher and the blood is more acidic because of higher levels of carbon dioxide and the presence of such metabolic waste products as lactic acid. Under these conditions, hemoglobin has a much lower affinity for oxygen, and it releases about a third of its oxygen load (Figure 11-11). The effects of increased temperature, acidity, and other factors on oxygen release help explain why it is helpful to warm up before exercising.

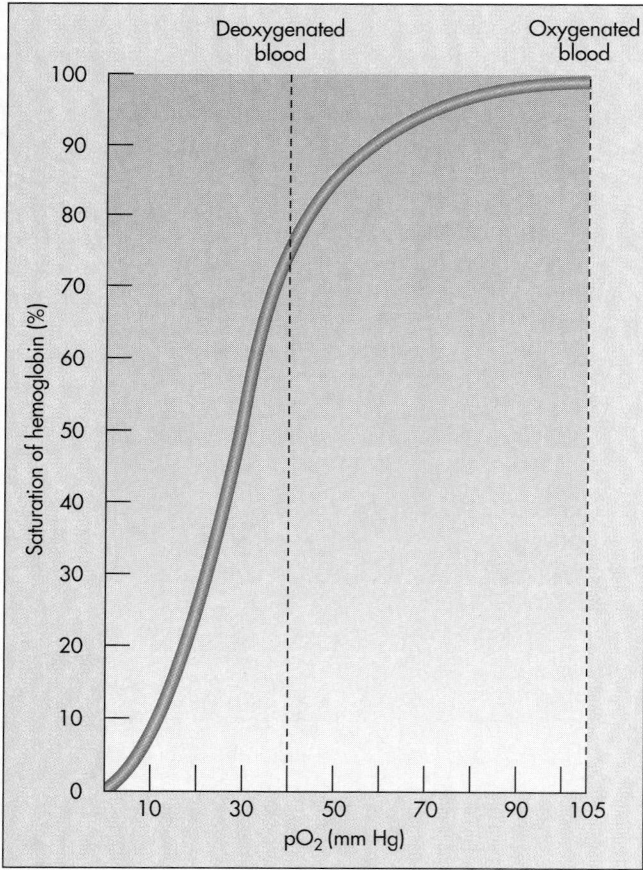

FIGURE 11-11 Oxygen-hemoglobin dissociation curve showing the relationship between hemoglobin saturation and pO_2. When the pO_2 goes up, more oxygen complexes with the hemoglobin.

Carbon Monoxide Poisoning

Carbon monoxide binds to the heme-iron part of hemoglobin more than 200 times as strongly as oxygen. Even a small amount of carbon monoxide can therefore prevent oxygen from binding by locking on to the binding site first. Because of the strength of the binding, the carbon monoxide does not let go easily. On each pass through the lungs, where it is exposed to additional carbon monoxide, the hemoglobin becomes more contaminated and its oxygen-carrying capacity is reduced. Before long, too little oxygen is reaching the tissues, and the body dies of suffocation. In fair-skinned people, a characteristic feature of carbon monoxide poisoning is bright pink skin, for carbon monoxide makes the blood even redder than does oxygen.

Carbon monoxide *does* infrequently let go of its grip on the hemoglobin. Carbon monoxide poisoning can therefore be reversed slowly by having the victim breathe pure oxygen at higher than atmospheric pressures. When hemoglobin is exposed in this way to much higher than normal levels of oxygen, the added oxygen can grab freed binding sites and block out the poison, which can then leave the body in exhaled air. Additional carbon dioxide is often added to the "therapeutic" gas mixture to stimulate an increase in the rate of breathing, so the released carbon monoxide can be flushed more quickly from the body.

CARBON DIOXIDE TRANSPORT

As we mentioned in Chapter 10, about a quarter (23 percent) of the carbon dioxide in the blood is bound to the globin (protein) portion of the hemoglobin molecule. A little more (7 percent) is dissolved in the plasma and accounts for the blood's pCO_2. The rest, 70 percent, is handled very differently. In the tissues, where it is produced, it diffuses into the blood. From there it diffuses into the red blood cells, where an enzyme, *carbonic anhydrase*, combines it with water to form carbonic acid. The carbonic acid promptly releases a hydrogen ion (H^+), which binds to hemoglobin. The remainder of the carbonic acid molecule is bicarbonate ion (HCO_3^-); it diffuses out of the red blood cells into the plasma, where it combines with a sodium ion to form sodium bicarbonate ($NaHCO_3$):

$$\underset{\text{carbon dioxide}}{CO_2} + \underset{\text{water}}{H_2O} \xrightarrow[\text{carbonic anhydrase}]{} \underset{\text{carbonic acid}}{H_2CO_3}$$

Then:

$$\underset{\text{carbonic acid}}{H_2CO_3} \longrightarrow \underset{\text{hydrogen ion}}{H^+} + \underset{\text{bicarbonate ion}}{HCO_3^-}$$

This reaction is driven by the high pCO_2 in the tissue capillaries. In the lungs, where the pCO_2 is lower, it reverses, releasing carbon dioxide. As carbon dioxide dissolved in plasma is lost to the alveolar air, it is replaced by carbon dioxide released from hemoglobin and generated by the conversion of bicarbonate back to carbonic acid, water, and carbon dioxide. The end result is restoration of the blood's ability to pick up and carry carbon dioxide on its next trip through the tissues.

Carbonic acid and bicarbonate provide one of the **buffer systems** that control blood pH or acidity. Carbonic acid is a weak acid; that is, it releases relatively few hydrogen ions and has a small impact on pH. Bicarbonate is its salt. Together, this weak acid and its salt resist changes in blood pH from the addition of stronger acids. Such strong acids release many hydrogen ions; those excess ions, however, are consumed by the reaction that converts bicarbonate back to carbonic acid. Since carbonic acid is a weak acid (which releases few hydrogen ions), its buildup in the blood affects the pH much less than the presence of an unbuffered strong acid. The buffer system also resists pH changes from the addition of bases, which release hydroxyl (OH^-) ions. These ions then react with carbonic acid to make bicarbonate.

CONTROL OF THE RESPIRATORY APPARATUS

The relation of carbon dioxide to blood pH is intimately connected to the control of breathing. Cells on the surface of the medulla at the base of the brain are sensitive to the acidity of the fluid (the cerebrospinal fluid) that bathes the brain, which in turn depends on the acidity of the blood. These cells respond directly to the concentration of hydrogen ions, and hence of carbonic acid, or indirectly to carbon dioxide. When this concentration is high, they signal nerve centers in the medulla to speed up the pace of pulmonary ventilation. As a result, more carbon dioxide is removed from the blood and the blood pH rises. When it is low, they signal the nerve centers to slow the pace, less carbon dioxide is removed from the blood, and the blood pH falls.

This control mechanism promotes *homeostasis* in two ways. First, it ensures that pulmonary ventilation keeps up with the body's need for oxygen, for high carbon dioxide levels generally imply low oxygen levels. Second, it helps maintain the blood pH at a nearly constant level (Figure 11-12). Doubling the normal breathing rate can shift blood pH from its norm of 7.4 to 7.63. Cutting the rate to a quarter of normal can drop blood pH from 7.4 to 7.0. And such changes take only 1 to 3 minutes.

Sensors in the carotid arteries and the aorta also play a role in respiratory control. Some respond to blood pH, and hence carbon dioxide levels, signaling the respiratory control centers in the brain to speed up or slow down breathing as necessary. A few respond to blood oxygen levels, but they are much less sensitive. They respond only when arterial oxygen levels fall from their norm of

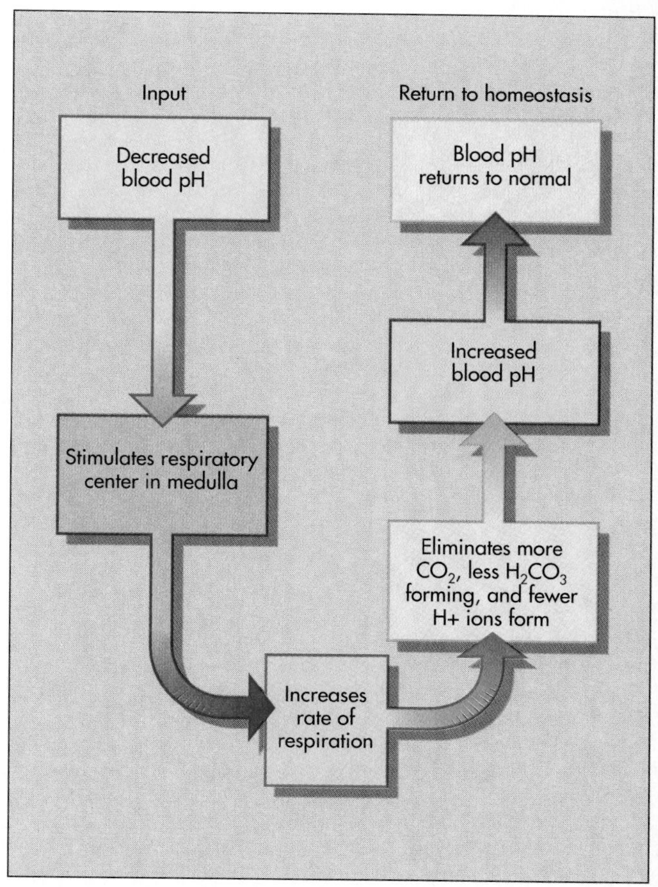

FIGURE 11-12 Respiratory control of blood pH.

Learning Focus

THE TRANSPORT OF RESPIRATORY GASES

Make a duplicate of the following figure, which is a sketch of the pathway between the lungs and the tissues. On it, write the pO_2 and pCO_2 values for the air outside the body, the alveoli, the blood entering the lungs, the blood leaving the lungs, the blood entering the tissue capillaries, and the blood leaving the tissue capillaries. Add arrows to mark the direction of diffusion of oxygen and carbon dioxide in the alveoli and the tissues.

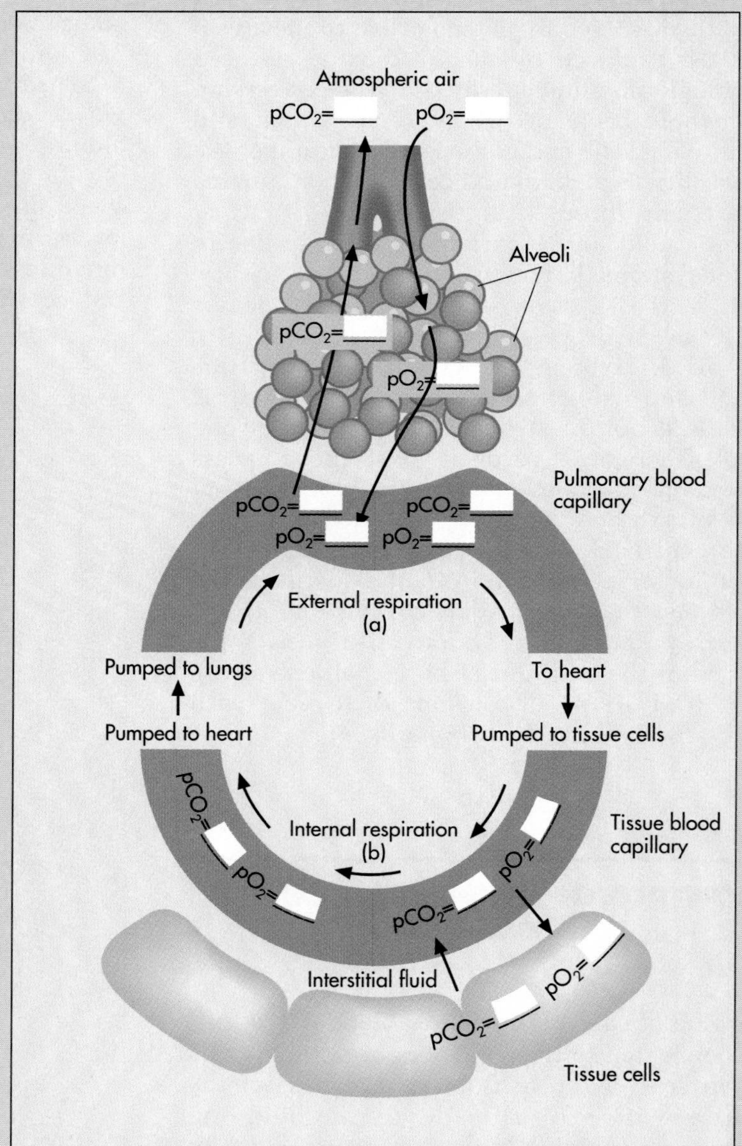

Partial pressures involved in respiration. Using the information from the text, indicate on the diagram the partial pressure for each of the gases in various parts of the system. For instance, you have learned that the partial pressure of oxygen in the atmosphere is 160 mm Hg. However, because inhaled air is mixed with returning air in the alveoli, the pO_2 in the alveoli is only 105 mm Hg.

The sketch that follows also shows the pathway between the lungs and the tissues. Indicate on it how oxygen and carbon dioxide are carried in the blood. Include the percentages for the various modes of transport.

What determines where oxygen is released?

What determines where oxygen is picked up?

What determines where carbon dioxide is released?

What determines where carbon dioxide is picked up?

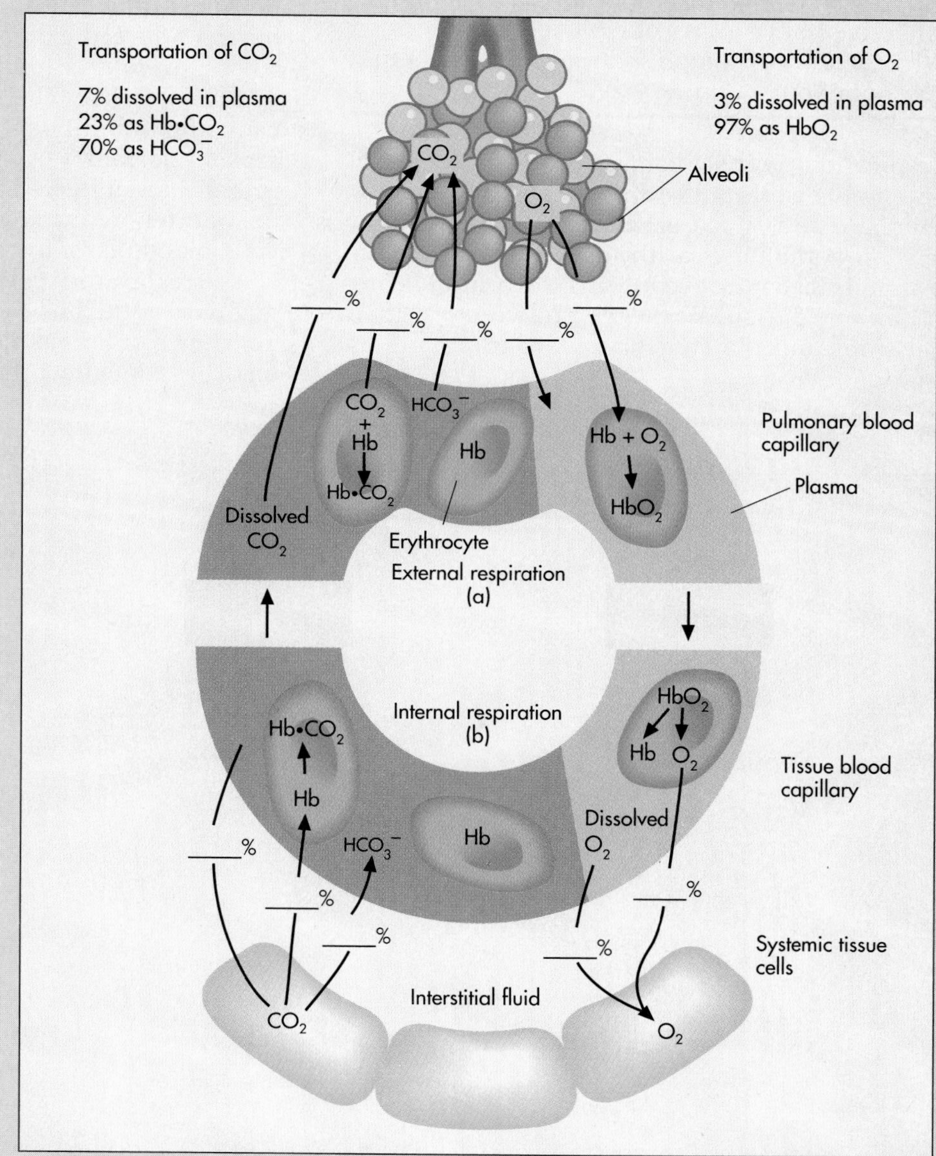

Transport of respiratory gases. Use this diagram to summarize text information about how the respiratory gases are carried in various parts of the system. For instance, what fraction of oxygen passing through the alveoli to the blood is carried in the red blood cells and what fraction is carried in the plasma? How is the carbon dioxide picked up from the tissues carried? Give the percentages for each type.

105 mm Hg to 70 mm Hg. Most chemical respiratory control is related indirectly to carbon dioxide concentration, which is what makes *hyperventilation* dangerous. People hyperventilate when they take several deep breaths in rapid succession, as a swimmer might in preparation for a long underwater swim. Swimmers often think they are charging their systems with oxygen when they do this, but they are really "blowing off" carbon dioxide. This indeed prolongs the time before they feel the need to take a breath. However, they can run out of oxygen before that time arrives. When that happens, they can drown.

1. Why does hyperventilation *not* increase the amount of oxygen in your blood?
2. Why does "blowing off" carbon dioxide help you hold your breath?

Respiration is also controlled by mechanical sensors in the walls of the bronchi and bronchioles. When these *stretch receptors* are activated by the expansion of the lungs at the end of a deep breath, they signal the control centers in the brain to command an exhalation. When the stretch ends, their signal stops and the control centers command an inhalation.

There are four control centers for respiration in the brain (see Chapter 15). During normal quiet breathing, the **inspiratory center** in the medulla is active for 2 seconds and inactive for 3 seconds. When it is active, it commands the diaphragm and external intercostal muscles to contract, producing an inspiration (inhalation). When it is inactive, these muscles relax, producing an expiration. The center's rhythm is spontaneous. Its activity is limited by the **pneumotaxic center** and stimulated by the **apneustic center,** both of which lie in the pons anterior to and above the medulla (Figure 11-13).

During fast breathing, at the end of each cycle of activity the inspiratory center signals the nearby **expiratory center** to command a contraction of the internal intercostal and abdominal muscles, producing a forced exhalation (Figure 11-14).

The stretch receptors work by inhibiting the inspiratory center. The pH sensors stimulate it. You can either inhibit or stimulate it at will, but the pH sensors normally override any effort to hold your breath indefinitely. They fail only when you have deliberately reduced the carbon dioxide content of your blood, as in hyperventilation.

The respiratory control centers are as reliable as any part of the body and more reliable than some. Yet they can fail, as they do when people overdose on drugs that affect the functions of the central ner-

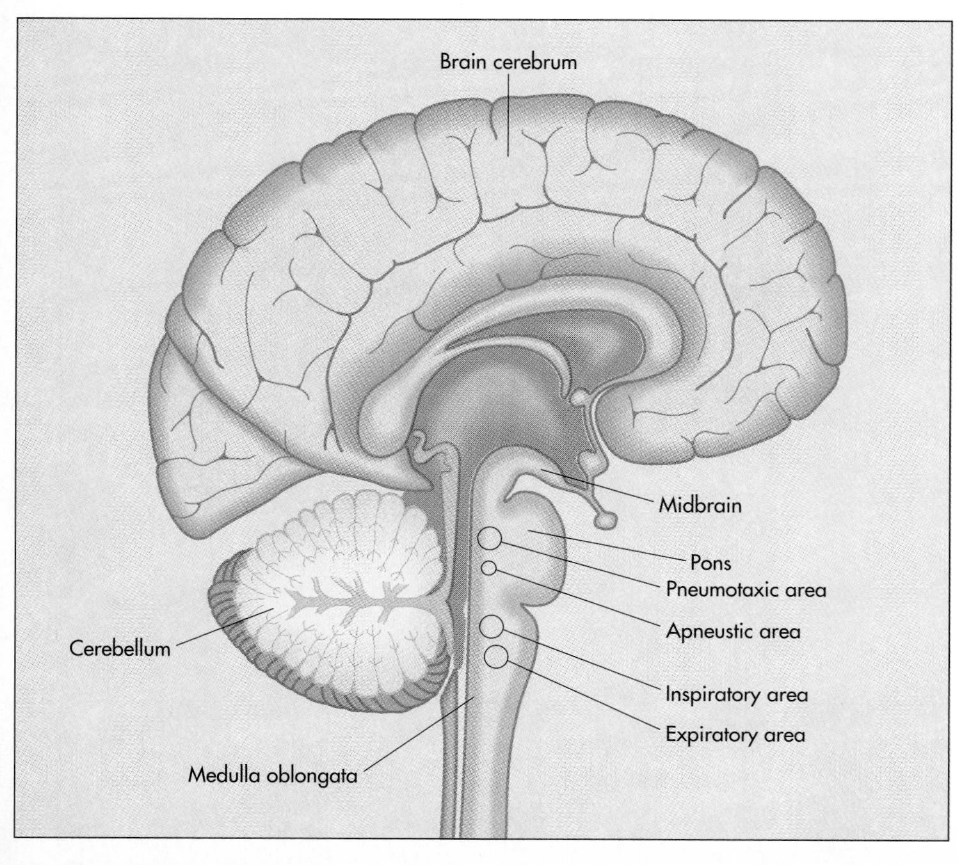

FIGURE 11-13 The respiratory control centers in the brain.

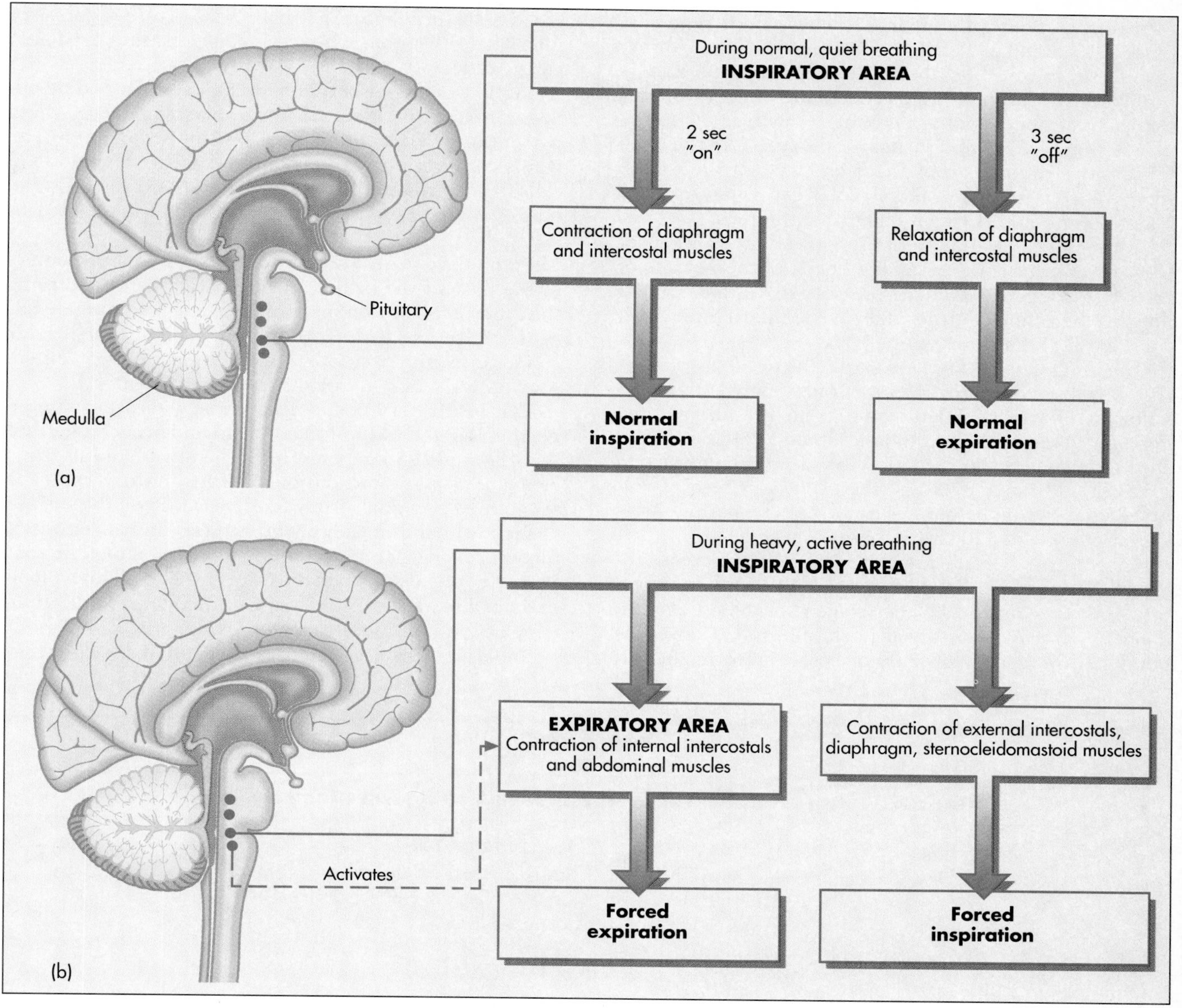

FIGURE 11-14 Nervous control of respiration: (a) During normal quiet breathing; (b) During heavy active breathing.

vous system. Overdoses of cocaine, heroin (and other opiates), and alcohol can all kill by interfering with normal respiratory control. They also depress heart function.

The respiratory control centers can also fail in **sudden infant death syndrome (SIDS),** also called crib death. This condition kills 10,000 U.S. babies per year and is the leading killer of infants between one week and one year of age. The victims die quietly, in their sleep, apparently because they simply stop breathing. Half have had colds in the previous two weeks, but the rest seemed healthy. The cause of SIDS is therefore a mystery, though many possibilities have been suggested, from botulism bacteria in the intestine, viral infections, and allergies to malfunctions of the respiratory control centers.

SUMMARY

The respiratory organs serve the function of pulmonary ventilation. That is, they move air in and out of the lungs, where oxygen and carbon dioxide can diffuse between the air and the blood in the process of external respiration. The respiratory organs are the nose, which serves as a passageway, a filter, and a temperature and humidity control; the mouth; the pharynx, whose three segments (naso-, oro-, and laryngopharynx) are only a passageway; the larynx, or voicebox; the trachea, or windpipe, and its branches, the bronchi and bronchioles; and the lungs themselves. The terminal bronchioles deliver air to the alveoli, bubblelike subdivisions of the lungs. In the alveoli, air is separated from the blood by a 0.5-μm-thick membrane consisting of the alveolar cells, the capillary endothelium, and their basement membranes.

Air is moved in and out of the lungs by the actions of the ribs and diaphragm. As they expand the volume of the chest cavity, external air pressure drives air into the lungs. When they contract the cavity, internal air pressure drives the air out again.

The lungs can hold 6 liters of air (the total lung capacity), but a normal breath moves only 500 ml (the tidal volume), of which 150 ml remains in the airway (the dead air volume). A deep inhalation adds 3100 ml of inspiratory reserve volume. A forcible deep exhalation can drive an additional 1200 ml (the expiratory reserve volume) from the lungs beyond a normal exhalation. A residual volume of 1200 ml remains unusable.

Concentration gradients drive oxygen from the alveolar air into the blood and carbon dioxide from the blood into the alveolar air. Of the oxygen the blood carries, about 3 percent is dissolved in the plasma. The rest is bound to the heme-iron part of hemoglobin in the red blood cells. It is released in the tissues, in the process of internal respiration, where the concentration gradients go the other way.

In the tissues, carbon dioxide diffuses into the blood. Seven percent is dissolved in the plasma. Twenty-three percent attaches to the protein part of hemoglobin. The rest is converted by carbonic anhydrase in the red blood cells to bicarbonate ion. The conversion is reversed in the lungs, where carbon dioxide is released for removal in exhaled air.

Pulmonary ventilation is controlled by nerve centers in the base of the brain. An inspiratory center controls the cycle of inhalation (inspiration) and exhalation (expiration) with a spontaneous rhythm. An expiratory center comes into play for forced exhalations. Other centers modulate the activity of the inspiratory center.

The inspiratory center receives signals from stretch receptors in the walls of the bronchi and bronchioles that report when the lungs are full. The signals inhibit this center and allow exhalation. The inspiratory center is stimulated by signals from acidity sensors on the surface of the brain and in the carotid arteries and aorta. Since blood acidity is largely a function of carbon dioxide content, they are also carbon dioxide sensors, and high blood carbon dioxide levels increase pulmonary ventilation, which blows off more carbon dioxide.

Table 11-2 lists selected diseases of the respiratory system.

Table 11-2

Clinical Summary: Selected Diseases of the Respiratory System

DISORDER	DESCRIPTION
Atelectasis	Collapse of the lung, usually caused by air entering the space between the pleural membranes. Causes can be punctures of chest wall (as from bullet or knife wounds) or internal tear of lung.
Bronchial asthma	Paroxysmal narrowing of bronchial airways, wheezing, and difficulty in breathing, caused by extreme irritability of bronchial smooth muscle. Narrowing may be reversed quickly by certain medications.
Chronic bronchitis	Defined as bronchial inflammation present in any patient with a persistent cough with abnormal sputum production for at least three months in two consecutive years. Often caused by tobacco smoke and other irritants and by bacterial infections. Can precede cancer.
Emphysema	Abnormal, permanent enlargements of the air spaces in the lungs, caused by destruction of alveolar walls. Provoked by chemical irritants in tobacco smoke and other air pollutants.
Pneumonia	Inflammation of the respiratory tree and alveoli by viruses, bacteria, or chemical irritants. Causes mucous membranes to secrete fluids that can fill the alveoli and asphyxiate the victim. A leading cause of death.
Rhinitis	A cold. Chronic or acute inflammation of the mucous membranes lining the nasal passages.

STUDY QUESTIONS

1. Through what respiratory organs does air pass on its way to the lungs?
2. What is the difference between pulmonary ventilation and respiration?
3. Discuss how the nose serves as an air conditioner.
4. Why is the subdivision of the lungs into alveoli crucial to external respiration?
5. Singers "breathe from the diaphragm." Why?
6. When do you use your lungs' residual volume?
7. Describe the concentration gradients for oxygen and carbon dioxide in the lungs and tissues. How do these gradients influence the movement of oxygen and carbon dioxide in and out of the blood?
8. Why does hemoglobin release oxygen in the tissues, despite its high affinity for this gas?
9. Describe how the blood carries carbon dioxide.
10. Explain how changes in carbon dioxide levels in the blood influence the control of pulmonary ventilation.

GLOSSARY

Alveolar duct The branch of a terminal bronchiole delivering air to an alveolar sac.

Alveolus Sites of gas exchange in the lungs. Bubble-like swellings in the walls of the alveolar sacs; site of external respiration.

Apneustic center Respiratory control center in the brain (pons); stimulates inspiratory center.

Bronchi Air tubes branching from the trachea and major bronchi; stiffened by cartilage rings and plates.

Bronchioles Branches of the smallest bronchi; not stiffened by cartilage.

Cellular respiration The oxygen-dependent breakdown of nutrients to obtain energy.

Diaphragm Dome-shaped muscle sheet at bottom of chest cavity; expands chest cavity for inhalation on contraction.

Expiratory center Respiratory control center in the brain (medulla); stimulates forced exhalation.

Expiratory reserve volume The amount of air exhalable in addition to the tidal volume; 1200 ml.

External respiration The exchange of gases between air in the lungs and the blood.

Inspiratory center Respiratory control center in the brain (medulla); stimulates inhalation with a spontaneous rhythm.

Inspiratory reserve volume The amount of air inhalable in addition to the tidal volume; 3100 ml.

Internal respiration The exchange of gases between blood and tissues.

Larynx The voicebox.

Pulmonary ventilation The flow of air in and out of the lungs.

Tidal volume The air moved in an ordinary, quiet breath; 500 ml.

Trachea The windpipe; the tube carrying air from the pharynx to the lungs; stiffened by U-shaped cartilaginous supports.

Vital capacity Total lung capacity minus residual volume; the lung volume that is available for deep breathing.

Chapter 12

The Excretory System

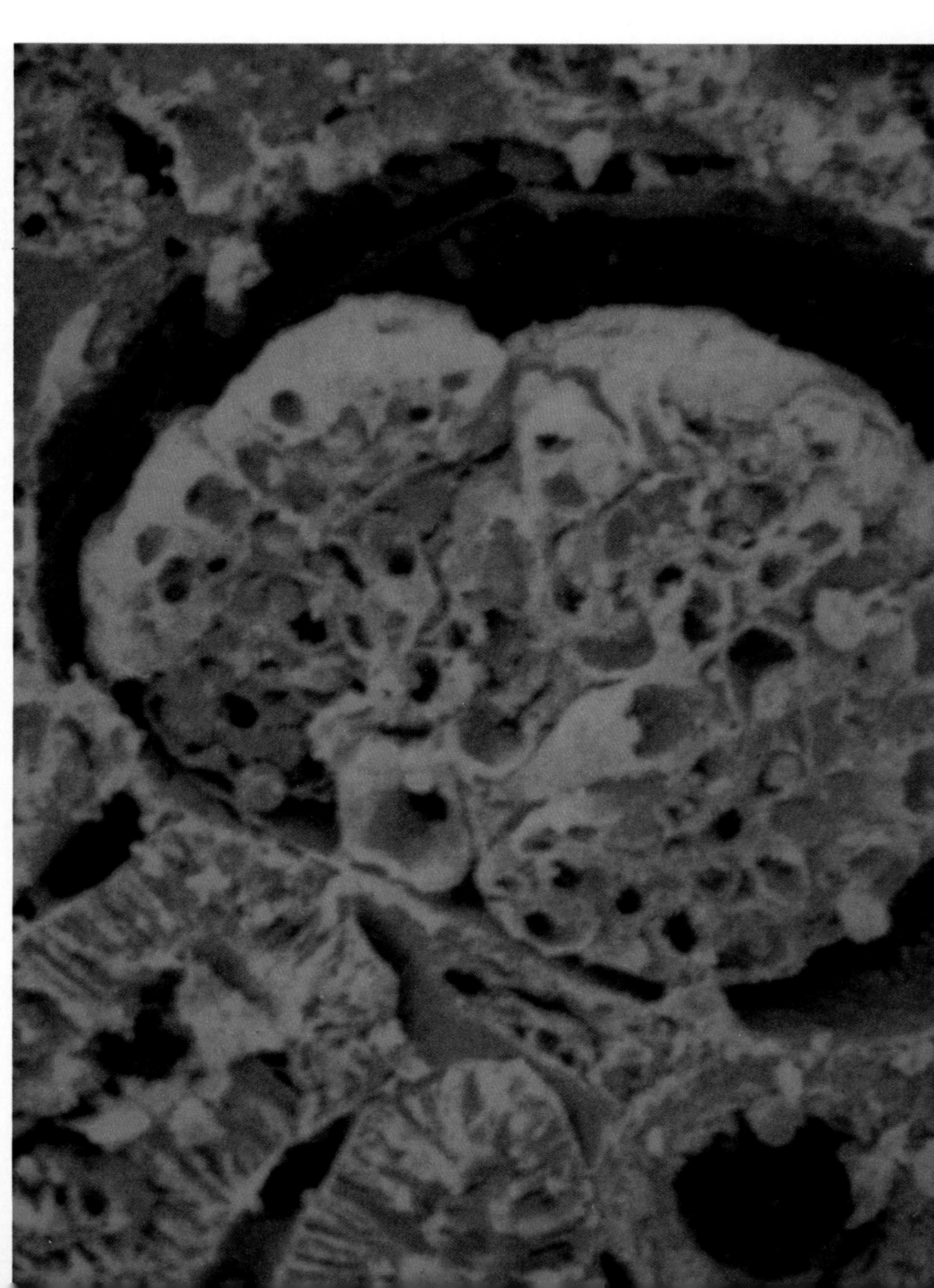

Like a factory, your body produces its equivalent of smoke, sewage, and junk that it must eliminate as waste products of its activities (Figure 12-1). As the body uses oxygen to generate energy, it also produces carbon dioxide, which it then must dispose of in exhaled air. The digestive system must get rid of the solid wastes it produces, for not all the food we eat can be absorbed into the body and used. This residue, plus such things as bile pigments and the bacteria that flourish in the digestive system, passes through the intestines and leaves the body as feces.

The body produces many other wastes as well. Included among them are excess salts, urea and ammonia (by-products of protein breakdown), creatinine (a waste product of muscle action), and other substances. These wastes collect as solutes in the blood plasma and must be filtered from the blood and eliminated from the body.

The body accomplishes these tasks in two ways. The skin's sweat glands extract a portion of the soluble wastes from the blood, which includes some salts and a little urea, and eliminate them as sweat. However, sweat glands play only a minor role in waste disposal. Their main function is temperature control.

The major waste-filtering system for the blood is the kidneys. They remove excess water and dissolved wastes in bulk, transfer them to the bladder

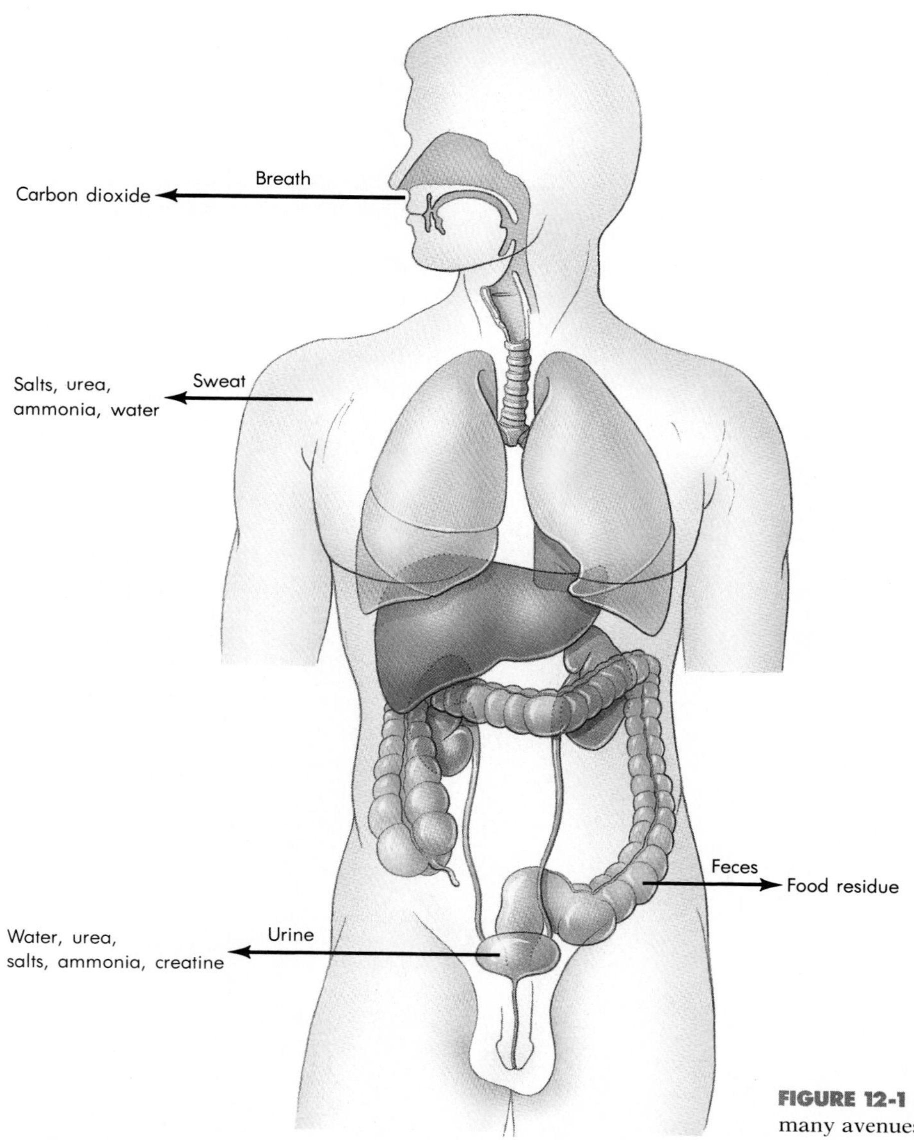

FIGURE 12-1 Routes of excretion. The body has many avenues for eliminating its various wastes.

for storage, and pass them out of the body as urine. Without them, the wastes accumulate in the blood, eventually rendering it literally toxic to the body's cells. The sweat glands are no substitute and no lifesaver.

THE URINARY SYSTEM

The **urinary system** consists of the organs that produce urine, store it, and pass it from the body. The key organs are the two **kidneys,** which sit against the back wall of the abdominal cavity, behind the stomach and liver (Figure 12-2). Unlike the other abdominal organs, they lie between the lining of the abdominal cavity (the peritoneum) and the inner body wall. They receive a massive flow of blood—1.2 liters per minute, a quarter of the cardiac output. Thus, about 25 percent of the blood flowing through the aorta is diverted through the renal arteries to the kidneys for cleansing of chemical wastes.

1. Study Figure 12-2. Does the location of the kidneys in relation to the rib cage suggest that they are as important to life as the heart and lungs? Why?
2. A "kidney punch" to the small of the back, below the ribs, cannot usually damage the kidneys. What would happen if it did?

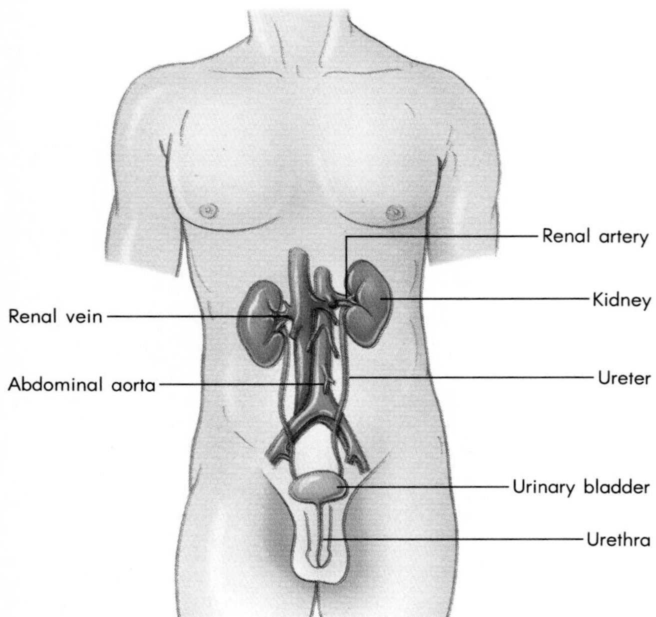

FIGURE 12-2 Organs of the urinary system.

The urine formed as the kidneys filter the blood consists mostly of excess water and a mixture of various salts. Once formed, the urine flows from each kidney through a **ureter** to the **urinary bladder** for storage. When the bladder is full, it empties through the **urethra** to the outside (Figure 12-3).

The Ureters

The ureters are lined with mucous membrane. The mucus this tissue secretes protects the ureters' cells from the acid urine. The wall of each ureter also contains a layer of smooth muscle, whose contractions help propel urine toward the bladder. The ureters pass under and attach to the backside of the bladder near their entrance points. The pressure of a full bladder pushes on the ureters, sealing them off and discouraging the backflow of urine. When the sealing fails to work, bladder infections (a common cause of cystitis, or inflammation of the bladder) can spread up the system to the kidneys, with potentially serious results.

The Urinary Bladder

The wall of the urinary bladder is constructed of tissues that allow it to expand. Its lining is both folded and coated with *transitional epithelium* (Figure 12-4). As it fills, the bladder's folds first smooth out. As it continues to fill, the rounded, piled-up cells of the epithelium stretch and slide to form a single layer of flattened cells.

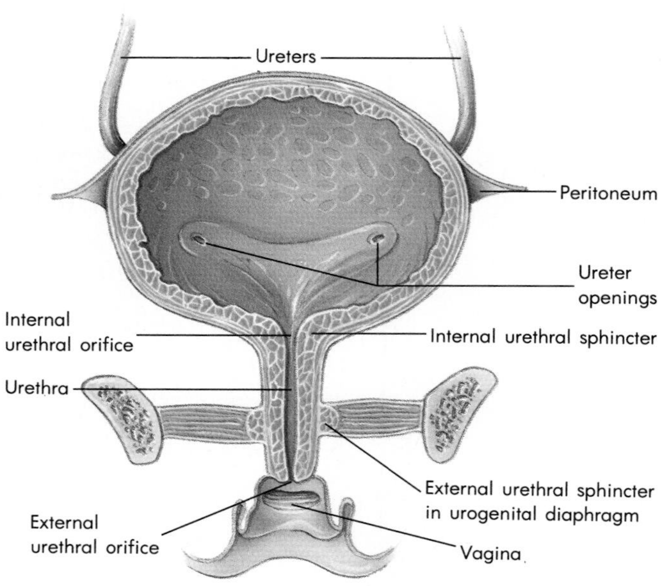

FIGURE 12-3 Ureters, urinary bladder, and the urethra in a female.

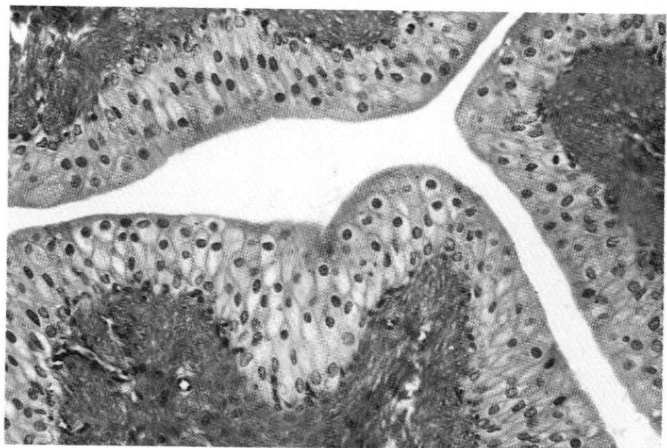

FIGURE 12-4 Transitional epithelium of bladder wall.

When it is full, stretch sensors in the wall of the urinary bladder notify the spinal cord, prompt the conscious urge to urinate, or *micturate,* and activate an involuntary *micturition reflex.* Nerve impulses from the spinal cord, via the parasympathetic branch of the autonomic nervous system (discussed in Chapter 15), stimulate the smooth muscle of the bladder wall to contract and relax the bladder's **internal sphincter,** a ring of smooth muscle surrounding the base of the urethra. Micturition is not automatic, however. It awaits "permission" from the conscious levels of the nervous system, in the brain, which relax the skeletal muscle of the bladder's **external sphincter,** just beyond the internal sphincter.

The Urethra

The urethra leads from the bladder through the internal and external sphincters. In males, it passes through the penis, at the base of which it is joined by the ejaculatory duct, which delivers semen for ejaculation. In the female, the urethra ends just below the clitoris (see Chapter 17); because of its shorter length (3.8 cm [1.5 in.] in females; 18–20 cm [7–8 in.] in males), women are more prone than men to bladder infections.

In women, but not in men, the wall of the urethra has a layer of circularly arranged smooth muscle; its contractions may, by expelling urine, make it more difficult for microorganisms to reach the bladder and cause infections. However, it is important for women to minimize the chances of infection by good hygiene, that is, not only frequent use of soap and water but also such practices as always wiping *from* the vulva *toward* the anus. This keeps the bacteria in fecal material from getting anywhere near the urethra and vagina (where they can also cause infections).

In men, but not in women, the urethra is surrounded near its base by a gland (the prostate gland) that secretes part of the semen. When this gland enlarges, as it often does in elderly men, urination may become difficult and painful.

THE KIDNEY

The kidney is a marvelously intricate structure. It is enclosed in a three-layered capsule. The outermost layer, the *renal fascia,* anchors the kidney to its surrounding tissues. The middle *adipose capsule* provides a cushioning layer of fat. The innermost *renal capsule* covers the kidney itself with a barrier to infection and injury.

The ureter and the renal artery and vein all meet the kidney at the **hilus,** an indented region on its medial side, that gives the kidney a distinct resemblance to a "kidney" bean and accounts for the bean's name. If we slice a kidney in half as if splitting a bean into its two halves, we see that the ureter is an extension of a funnellike cavity within the kidney, the **renal pelvis.** Cuplike extensions of this cavity form 2 or 3 **major calyces** and up to 18 **minor calyces.** Each minor calyx embraces a **renal papilla,** a domelike protrusion of the kidney's substance, and collects urine from that part of the kidney (Figure 12-5).

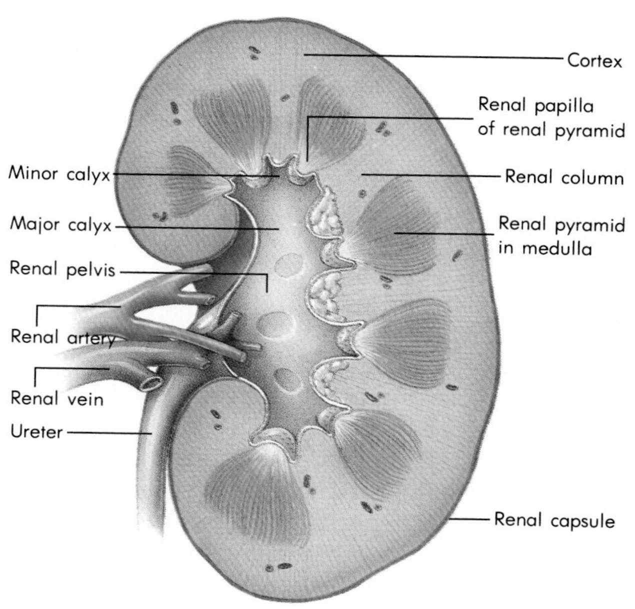

FIGURE 12-5 The interior of a kidney.

Box 12-1 Polycystic Disease

Each kidney normally weighs about 0.5 lb (225 g). However, it can reach weights of as much as 30 lb (14 kg) in **polycystic disease.**

A 30-lb kidney is highly unusual, even though polycystic disease is the most common inherited kidney defect. It results in deformed nephrons, whose tubules become studded with balloonlike swellings or cysts. Generally, the cysts are small, perhaps the size of a pinhead. Though they increase in number with age, they usually do not cause kidney failure until after the age of 40, and often after 60. However, some people get cysts as big as hen's eggs; when these cysts increase in number, the kidneys swell to many times their normal size and fail sooner. Fortunately, the disease's progress can be controlled by drugs and diet.

The tissue of the kidney is divided into two zones, an outer **cortex** (rind or bark) and an inner **medulla** (core). The medulla in turn is divided into up to 18 **renal pyramids.** Each pyramid is tipped by a renal papilla; there are one pyramid and one papilla for each minor calyx. The cortex extends as the **renal columns** into the spaces between the pyramids.

The Nephron

The kidney's working units that actually cleanse the blood of wastes are the **nephrons.** Each kidney contains 1.2 million of these tiny structures, each producing less than one drop of urine per day. Together, they produce 1–2 liters of urine per day.

The nephron begins in the renal cortex, where the tiny **afferent arteriole** forms a tuft or ball of specialized capillaries, the **glomerulus** (Figure 12-6). The tuft is enclosed in a funnellike double-walled **glomerular (Bowman's) capsule,** which opens into the tubule that produces urine (Figure 12-7).

The endothelium of the glomerular capillaries is pierced by numerous openings, which permit much of the plasma of the blood flowing through them to escape. Plasma proteins are kept from escaping by the basement membrane of the endothelium. Since the plasma proteins maintain a high osmotic pressure in the glomerular capillaries, much of the water that escapes with the plasma promptly returns to the capillaries.

The fluid that leaves the glomerular capillaries—known as the **glomerular filtrate**—amounts to 180 liters per day. The nephron processes this filtrate to make urine, and the processing begins immediately. The filtrate flows from the glomerular capillaries into the capsular space and then into the tubule of the nephron.

The nephron consists of the glomerulus, the glomerular capsule, and the tubule that leaves the capsule. The part of the tubule near the capsule bends back and forth to form the **proximal** (near) **convoluted tubule.** It then dips into the renal medulla as the **loop of Henle,** returning to the cortex to form the **distal** (distant) **convoluted tubule** (Figure 12-7). As the glomerular filtrate enters the proximal convoluted tubule, it is isotonic (equal in concentration) to the plasma. As filtrate travels through the convoluted tubules of the nephron, it loses water and nonwaste substances such as glucose, which are passed back to the blood, and its waste content becomes concentrated (Table 12-1).

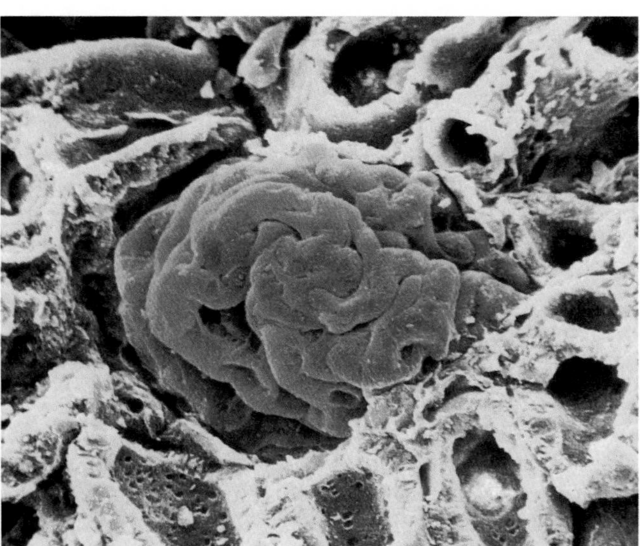

FIGURE 12-6 False-colored scanning electron micrograph (SEM) showing the glomerulus (orange) and Bowman's (glomerular) capsule (gold), which together form the renal corpuscle. Parts of Bowman's (glomerular) capsule have been removed to show the tightly packed capillaries of the glomerulus. The renal corpuscle is the actual filter and trap involved in the filtration of plasma and the separation of chemical wastes from the blood.

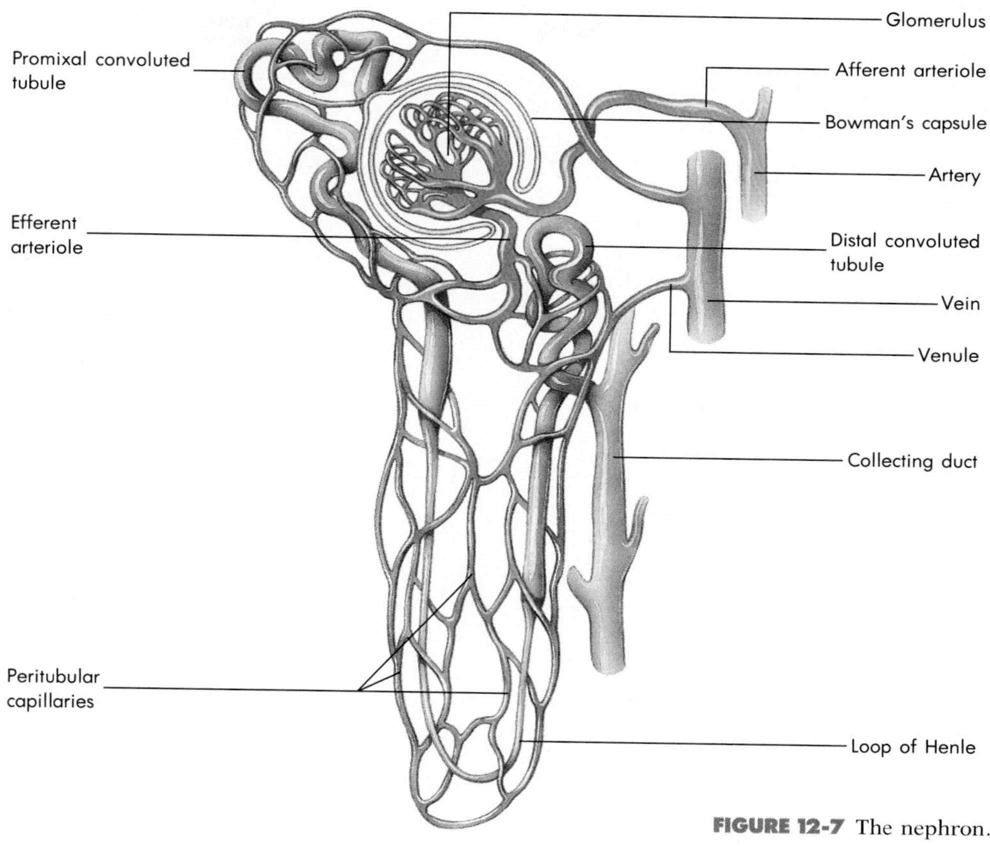

FIGURE 12-7 The nephron.

Table 12-1

Relative Compositions of Plasma, Filtrate, and Urine (as excreted per day)[a]

COMPONENT	PLASMA	FILTRATE	URINE
Water	180	180	1–2
Protein	7–9	0.01–0.02	0.0
Chloride	0.630	0.630	0.005
Sodium	0.540	0.540	0.003
Bicarbonate	0.300	0.300	0.0003
Glucose	0.180	0.180	0.0
Urea	0.053	0.053	0.025
Potassium	0.028	0.028	0.004
Uric acid	0.0085	0.0085	0.0008
Creatinine	0.0015	0.0015	0.0015

[a]Compositions for water expressed in liters. All other compositions expressed in grams.

By the time the filtrate has traveled to the end of the distal convoluted tubule, it has become finished (but still dilute) urine. It is now ready to flow into the **collecting ducts** that empty through the renal papillae into the minor calyces, renal pelvis, ureter, and eventually the urinary bladder.

1. As an exercise, calculate the change in concentration from plasma to urine for the substances in Table 12-1. For example, if 180 liters of plasma and 2 liters of urine each contain 1 g of substance *X*, the formation of urine has concentrated substance *X* 90-fold.
2. What do you think would happen to you if your kidneys failed to reabsorb most of the water in the glomerular filtrate?

Urine Formation

Filtration occurs when the afferent arteriole passes blood, under pressure, through the glomerulus. The glomerular filtrate is collected by the glomerular capsule. The nephron tubule converts glomerular filtrate into urine with the aid of three processes: osmosis, reabsorption, and secretion. Osmosis is responsible for the removal of water from the filtrate, and it begins in the glomerular capsule. It continues in the nephron's tubule, where the kidney exploits the fact that the glomerular filtrate is more dilute than the blood leaving the glomerulus, largely because proteins cannot leave the capillaries.

After leaving the glomerulus, the capillaries merge to form not a venule, but an **efferent arteriole.** This arteriole soon gives rise to a second capillary network surrounding the nephron's proximal and distal convoluted tubules and the loop of Henle. Since the blood in these *peritubular capillaries* is more concentrated, water is drawn back into the blood through the walls of the proximal tubule, from the filtrate (Figure 12-8). At the same time, reabsorption moves sodium and other ions, glucose, vitamins, amino acids, and other substances out of the filtrate, returning them to the blood. Drawing these solutes out of the filtrate makes the filtrate more dilute again—which indirectly raises the water concentration in the tubules, causing still more water to move by osmosis back into the blood. The result is that 80 percent of the water in the filtrate is returned to the blood through the walls of the proximal convoluted tubules in a process known as *obligatory water reabsorption.* The remainder of the water is reabsorbed in a process called *facultative water reabsorption* in the distal convoluted tubule and collecting duct under the control of the hormones aldosterone and antidiuretic hormone (ADH).

Most reabsorption of nonwater substances from the glomerular filtrate also occurs in the proximal convoluted tubule. The surfaces of the tubule cells that face the interior of the tubule are folded into

Box 12-2 Portal Systems

Blood usually passes through only one capillary network as it travels from the heart through arteries and arterioles to the capillaries and then into venules and veins and back to the heart. However, there are exceptions. Sometimes, blood leaves one capillary net not for the heart, but for a *second* capillary net. The two capillary nets, and the blood vessels that link them, compose a **portal system.** The name refers simply to the way the system trans*ports* blood from one exchange site directly to another.

The blood vessel between the two sets of capillaries may be an arteriole, as it is between the capillaries of the nephron's glomerulus and the peritubular capillaries farther down the line. It may be a vein like the hepatic portal vein that links the capillaries of the small intestine with those of the liver or the hypophysial portal veins that link the capillary nets of the brain's hypothalamus and the pituitary gland, both of which will be covered in later chapters.

The difference between a portal arteriole and a portal vein, as you might expect, depends on the vessel's structure, which in turn depends on the pressure within the vessel. The efferent arterioles in the kidney carry blood under much higher pressure than ordinary venules, for the pressure in the glomerular capillaries is also higher. It is this pressure that drives the filtering movement of plasma from the capillaries into the glomerular capsule of the nephron.

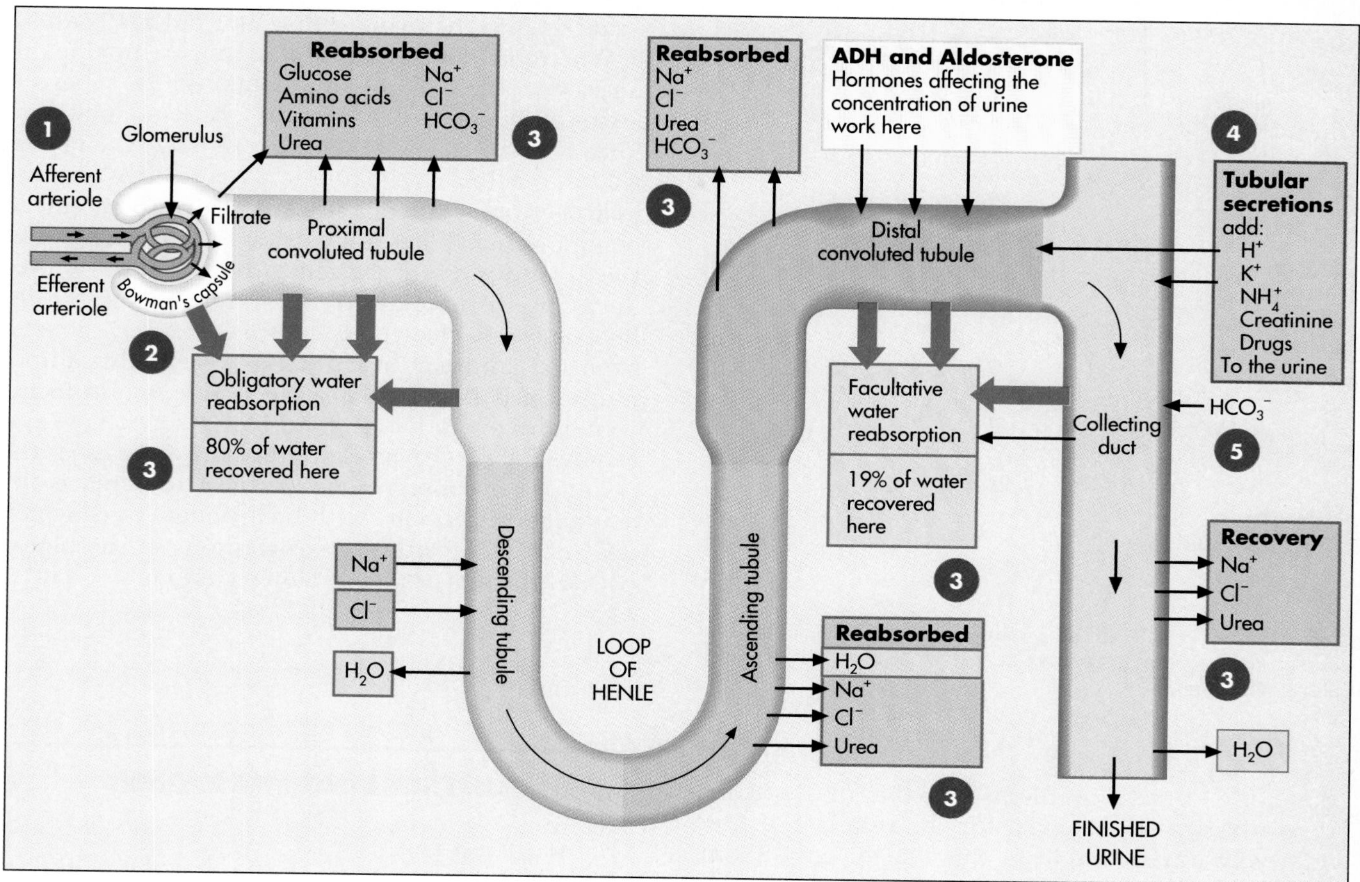

FIGURE 12-8 Each kidney produces urine by continuously running a fraction of the blood through a million filtering units called nephrons. (1) Blood, under pressure, enters each glomerulus (the actual filter) from its afferent arteriole. (2) Plasma filters through the glomerular walls and is collected as filtrate by the glomerular (Bowman's) capsule. From there, the filtrate passes through the proximal convoluted tubule, the loop of Henle, the distal convoluted tubule, and the collecting duct. (3) Water and other useful materials are reabsorbed by the peritubular capillaries surrounding each nephron and returned to the blood. (4) Additional wastes are secreted into the filtrate by the cells lining the tubules. The places in the nephron where each of the major components moves into and out of the blood or filtrate are shown. (5) The output from many nephrons flows into each collecting duct, where it is modified slightly as it becomes finished urine.

numerous *microvilli* to increase surface area (Figure 12-9). The cell membranes contain carrier molecules that, powered by adenosine triphosphate (ATP), bind to sodium, amino acids, and glucose in the fluid in the tubule and bring them into the tubule cells. Amino acids and glucose then diffuse out of the cells to the interstitial fluid; from there, they enter the blood in the peritubular capillaries. Pinocytosis removes protein from the filtrate; once broken down by lysosomes, its amino acids also diffuse out of the cells to the blood.

Active transport moves sodium, the main positively charged ion in the filtrate, along the same path. Negatively charged ions such as chloride and bicarbonate then follow it passively. In the ascending limb of the loop of Henle, it is chloride that is actively absorbed; here, sodium is the passive follower. Sodium is once more actively reabsorbed in the distal convoluted tubule. As the various ions are reabsorbed, they change the relative concentrations of filtrate and blood and draw water with them by osmosis (see the Learning Focus that follows). The end result is the conservation of nonwaste materials and the increased concentration of the glomerular filtrate as water is withdrawn from it to form urine (Table 12-2).

Wastes are handled somewhat differently. They are present in the glomerular filtrate, and though they are not actively reabsorbed, most (except creatinine) do diffuse passively, following concentra-

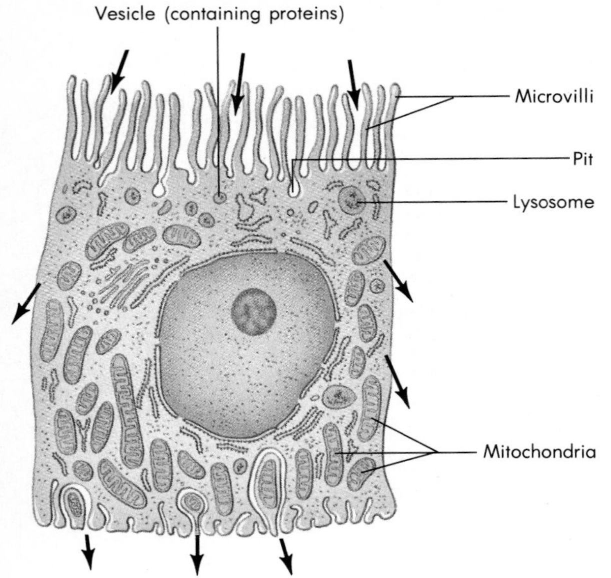

FIGURE 12-9 A cell of the proximal convoluted tubule. The numerous mitochondria reflect the cell's need for ATP to power active transport.

tion and electrical gradients into the tubular cells. Urea diffuses only slowly, for the tubule cells are relatively impermeable to this waste; only about half the urea in the filtrate returns to the blood. Other wastes or surplus materials, including ammonia, creatinine, and potassium and hydrogen ions, are actually secreted from the blood into the distal convoluted tubule. Foreign molecules, including drugs such as penicillin, are also secreted.

The secretion of hydrogen ions and ammonia serves to regulate the pH of the blood by pumping acid out into the urine, whose normal pH is about 6. When the blood is too acidic, carbon dioxide diffuses into the cells of the distal tubule, where it forms carbonic acid. As the carbonic acid ionizes to form H^+ and HCO_3^-, the cells exchange the H^+ for Na^+ in the fluid in the tubule. That is, they pass the hydrogen ions out of the cells and into the urine being produced, and they draw sodium ions out of the filtrate and into the cells. Ammonia (NH_3), produced by the breakdown of amino acids in the tubule cells and elsewhere, combines with H^+ ions to form the ammonium ion (NH_4^+). Once the ammonium ion is secreted into the tubule, again in exchange for sodium ions, it both raises blood pH and removes a toxic waste product. In each case, the result is to conserve sodium and to replace the acidic hydrogen ions with bicarbonate ions, which can then buffer other hydrogen ions in the blood. (Recall the discussion of buffer systems in Chapter 11.)

CONTROLLING THE KIDNEY

The kidney is controlled in two ways. The smooth muscle of its arterioles receives nerve signals from the autonomic nervous system. By regulating the degree of constriction of these blood vessels, the autonomic nervous system can thus control blood flow through the kidney and hence the loss of fluid and other substances. This control can be important in cases of massive blood loss.

Table 12-2

Reabsorption from Glomerular Filtrate

COMPONENT	PERCENTAGE REABSORBED
Useful materials: normally recovered	
Water	99.2
Protein	100
Chloride	99.2
Sodium	99.4
Bicarbonate	99.9
Glucose	100
Waste materials: at least partially excreted	
Urea	52.8
Potassium	85.7
Uric acid	90.6
Creatinine	0

Learning Focus

CONCENTRATING THE URINE

By the time urine leaves the kidney, it is four times as concentrated as the blood plasma. You may think that this is impossible, for osmosis should ensure that water moves between the forming urine and the plasma until the two have equal osmotic pressures or concentrations of solutes. The kidney counteracts this natural tendency with the aid of a **countercurrent multiplier mechanism** (see the accompanying diagram). The same mechanism also helps the kidney conserve the body's supply of sodium and chloride ions. The result is a very efficient process for maintaining the body's *ionic balance,* an important part of homeostasis.

In many of the kidney's nephrons, the loop of Henle dips deeply into the renal medulla (in some nephrons, the loop barely leaves the renal cortex). As the glomerular filtrate enters the proximal convoluted tubule, it is isotonic to the plasma. Measured in terms of its osmotic pressure, its concentration is about 300 milliosmoles (mOsm). The interstitial fluid outside the tubule, in the renal cortex, has the same concentration. As the filtrate proceeds through the loop of Henle, its concentration increases, to 1200 mOsm, and so does the concentration of the interstitial fluid.

The concentration mechanism depends on the active transport of chloride ions from the filtrate in the *ascending* limb of the loop of Henle. Sodium ions follow passively, so that as the filtrate rises out of the medulla toward the distal convoluted tubule, it becomes more dilute, ending at a concentration of about 100 mOsm. At the same time, sodium chloride (NaCl) accumulates in the interstitial fluid of the medulla, ac-

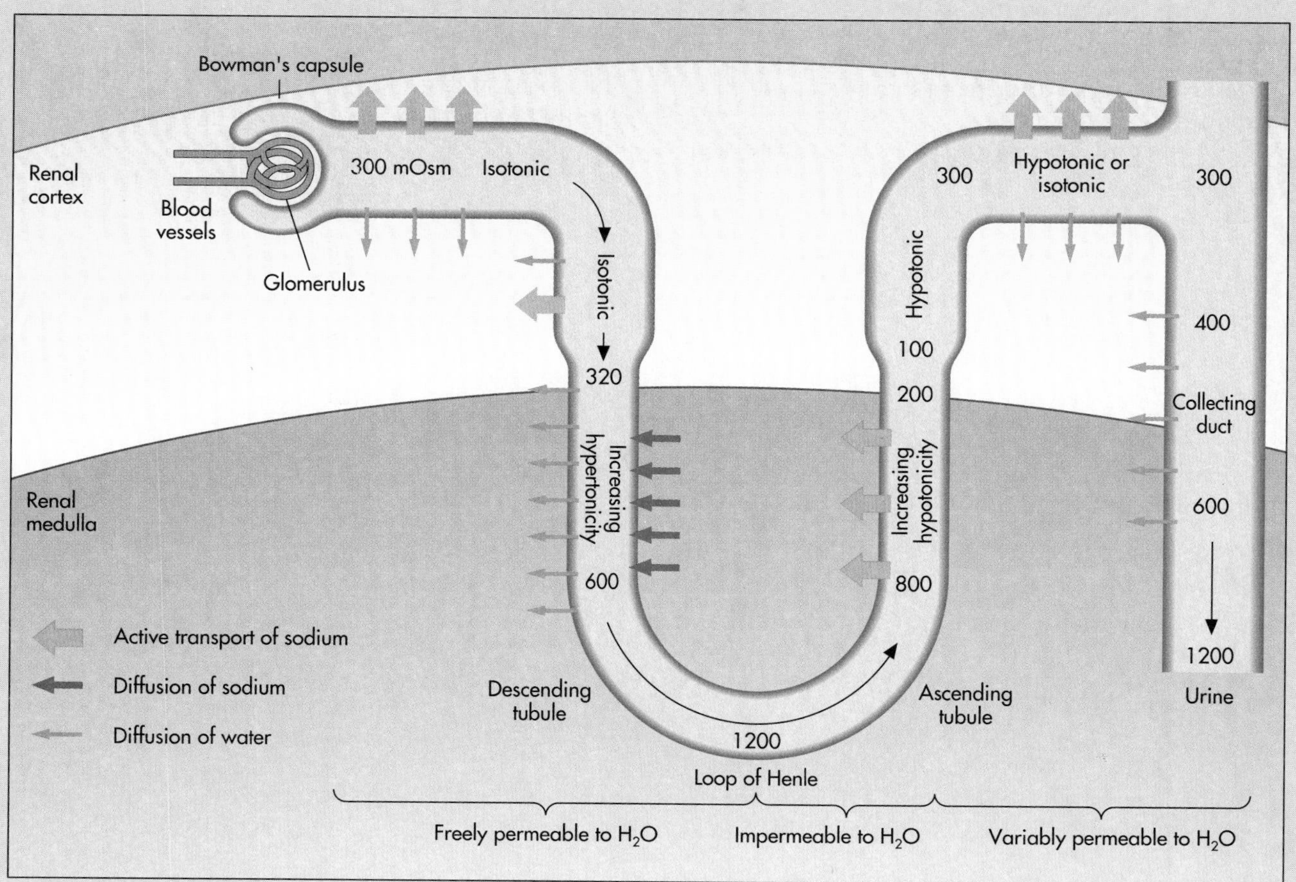

The countercurrent multiplier mechanism.

(Continued on p. 272)

counting for its high concentration there.

The *descending* limb of the loop of Henle is very permeable to entering NaCl. The sodium and chloride ions immediately diffuse into the filtrate as it descends into the medulla, raising the filtrate's concentration. Since these ions are removed again in the ascending limb, they accumulate in the medulla, producing an intensely hyperosmotic environment. The two limbs of the loop provide two flows of filtrate in opposite directions, the countercurrent. The movements of sodium and chloride ions provide the multiplier.

If the blood vessels in the medulla passed straight through, sodium and chloride ions would diffuse into the blood and be carried away. However, the renal blood vessels in these areas help to preserve the high levels of these ions in the renal medulla. As they descend into the medulla, they pick up sodium and chloride ions, and the blood reaches a concentration of 1200 mOsm. As the renal blood vessels rise toward the cortex, the concentration gradients reverse and the blood loses sodium and chloride ions. Most of these ions remain in the medulla. Only enough remain in the blood to leave the medullary concentrations constant and to remove excess water from the medullary interstitial fluid by osmosis.

Only a small amount of water leaves the filtrate in the loop of Henle. In the upper descending limb, the entry of sodium and chloride ions keeps the filtrate isotonic to the interstitial fluid. However, as the descending tubule dips deeper into the medulla, the filtrate becomes increasingly hypertonic by taking on more Na^+ and Cl^- and losing some of its water.

In the ascending limb, the tubule's cells are not very permeable to water. The water leaves the filtrate by osmosis in the distal convoluted tubule, where 100-mOsm filtrate is exposed to a 300-mOsm interstitial environment, and in the collecting duct, which descends again through the medulla, exposing the filtrate to ever more concentrated levels of sodium chloride. When the filtrate leaves the medulla at the renal papilla, it is in equilibrium with the 1200 mOsm concentration of the interstitial fluid at the bottom of the medulla. It is **urine,** four times as concentrated as plasma.

The degree of concentration achieved by this mechanism can vary. The hormone ADH opens pores in the cell membranes of the cells of the distal convoluted tubules and collecting ducts, thus permitting maximum osmosis and concentration of the urine. When ADH is not present, the pores close and the kidneys produce a highly dilute urine.

Learning Focus Response

CONCENTRATING THE URINE

Use the unlabeled diagram to organize visually your information about what kinds of things happen in the various parts of the nephron.

1. What happens in the glomerulus and glomerular (Bowman's) capsule?

2. What blood components do not appear in the glomerular filtrate?

3. What components of the glomerular filtrate are reabsorbed in the proximal convoluted tubule? In the distal convoluted tubule?

The nephron.

4. What substances does the nephron secrete into the filtrate, and what part of the nephron handles the secretion?

5. Describe the role of the loop of Henle.

6. What is the role of the collecting duct in the formation of urine?

7. How does the water that is removed from the urine in the distal convoluted tubule and collecting duct return to the circulation?

8. Chapter 11 described another countercurrent mechanism in the nose. Compare it to that of the loop of Henle.

9. Still another countercurrent mechanism conserves body heat in cold weather and explains why your hands and feet are colder than the rest of you. How must this mechanism work?

The other control mechanism for the kidney is chemical. When the concentration of solutes in the blood rises above a certain point, the pituitary gland releases **antidiuretic hormone (ADH).** (A **diuretic** is a drug or chemical contained in such things as asparagus and coffee that increases urine output and dilution.) This hormone increases the permeability of the nephron's distal convoluted tubule and the collecting ducts to water, allowing osmosis to make the urine more concentrated and conserving water. When the blood is more dilute, the pituitary does not release ADH; the urine is then less concentrated and more voluminous, and the blood loses water. The end result is maintenance of the concentration of the blood at a roughly constant level (Figure 12-10).

We see the effects of ADH when we get up in the morning. Over the night, we have not had anything to drink. To conserve water (and perhaps to let us sleep all night), the pituitary has been releasing ADH, and the first urine of the day is dark yellow due to its high concentration of *urochrome,* a pigment derived from bile. It may also have a strong odor.

Color and odor both vary with diet. Asparagus gives urine the characteristic smell of methyl mercaptan, and beets can turn it reddish. Indeed, practical jokers have long used dyes to turn their victims' urine alarming colors.

Within an hour after we drink two or three cups of coffee or glasses of beer, our pituitary reduces its release of ADH. With the loss of ADH's antidiuretic effect, the kidneys increase their rate of urine production from as little as 50 to 500 milliliters per hour. This urine is very dilute. It consists almost entirely of excess water being dumped from the blood, and it is virtually colorless and odorless. When ADH production shuts down entirely, as in the disease **diabetes insipidus,** the urine is even more dilute, and its volume can be as much as 15 liters per day. Victims of this disease suffer from constant thirst and lack of sleep (for obvious reasons).

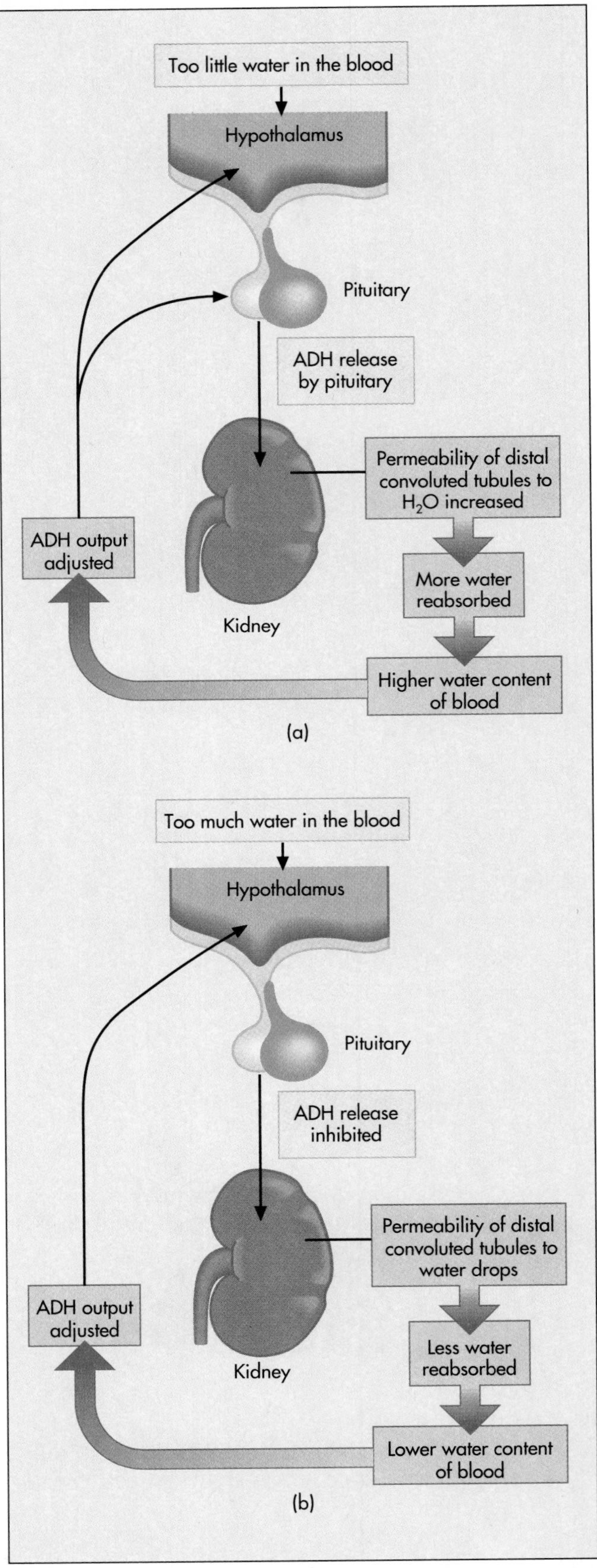

FIGURE 12-10 The control of water reabsorption.

THE KIDNEY AND BLOOD PRESSURE CONTROL

The functioning of the kidney depends on the blood pressure. When the blood pressure falls, the rate of glomerular filtration declines. If the blood pressure falls too low, too little glomerular filtrate may enter the kidneys' nephrons for the kidneys to cleanse the blood adequately. Not surprisingly, the kidney has a mechanism to ensure that blood pressure remains high enough for adequate kidney function.

This mechanism begins with the **juxtaglomerular apparatus.** One of the bends of the distal convoluted tubule touches the afferent arteriole just before that blood vessel divides to form the glomerular capillaries. At the point of contact, the smooth muscle cells of the arteriole wall change shape and replace their myofibrils with secretory granules containing the enzyme **renin;** they become *juxtaglomerular cells.* The cells of the distal tubule nearest the arteriole narrow to form the *macula densa.* Together, the juxtaglomerular cells and macula densa form the juxtaglomerular apparatus (Figure 12-11).

No one is quite sure how the juxtaglomerular apparatus works. However, we do know that the juxtaglomerular cells receive sympathetic (autonomic) nerve signals from stretch receptors that respond to declines in blood volume and pressure. The juxtaglomerular cells may also respond to other signals associated with loss of fluid or pressure. Their response is the release of renin into the blood. There the renin converts the plasma protein **angiotensinogen** into **angiotensin I,** which in turn is converted by enzymes in the lungs to **angiotensin II.**

Angiotensin II has two effects (Figure 12-12). First, it stimulates contraction of the smooth muscle in arterioles throughout the body, thus reducing the space available for the blood and raising the blood pressure. The increase in blood pressure promotes an increase in the blood filtration by the kidney, ensuring that the wastes will be efficiently removed from the blood. Second, it stimulates the adrenal cortex (an endocrine gland, discussed in Chapter 13) to release **aldosterone.** Aldosterone stimulates the cells of the distal convoluted tubule to reabsorb more sodium ions and water, thereby preventing further loss of blood volume and pressure.

THE KIDNEY AND REGULATION OF BLOOD COMPOSITION

The kidney regulates more than just the pressure of the blood. It also regulates the *composition* of the blood. It removes wastes and controls pH. And it removes many of the smaller and simpler chemicals that have become too highly concentrated in the blood.

An example is glucose, or blood sugar. Normally, the interplay of the hormones insulin and glucagon, secreted by the islets of Langerhans in the pancreas, keeps the concentration of glucose in the

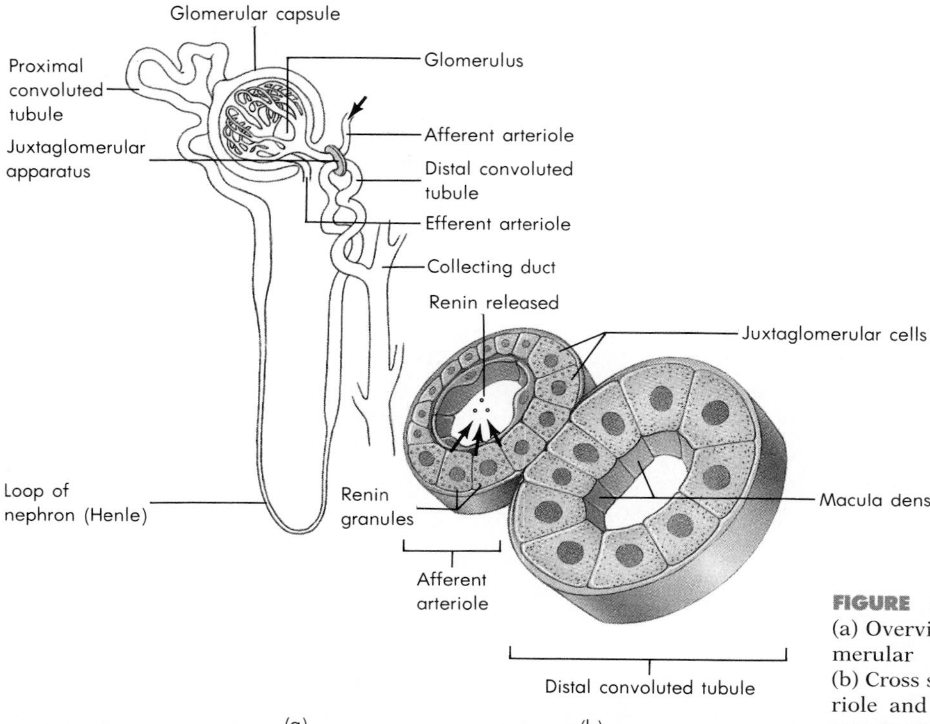

FIGURE 12-11 The juxtaglomerular apparatus. (a) Overview showing the position of the juxtaglomerular apparatus relative to the nephron. (b) Cross section illustrating how the afferent arteriole and the distal convoluted tubule form the apparatus.

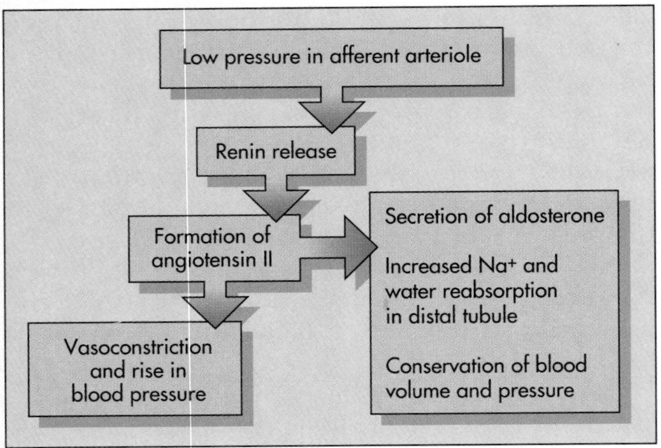

FIGURE 12-12 The role of the juxtaglomerular apparatus.

blood very near 90 milligrams per 100 milliliters of blood (see Chapter 13). In the kidney, glucose is filtered from the blood with the glomerular filtrate, but normally it is entirely reabsorbed in the proximal convoluted tubules of the nephrons, and none appears in the urine. However, when too little insulin is secreted or when the body's cells fail to respond to it, the level of glucose in the blood—and hence in the glomerular filtrate—can rise very high. Then the nephron does not reabsorb all the glucose; the blood leaving the kidney has a nearly normal blood glucose level, with the excess showing up in the urine. Persistent episodes of sugar in the urine is the definitive symptom of **diabetes mellitus** (sugar diabetes).

---

1. The presence of glucose increases the osmotic pressure of the glomerular filtrate. What, then, must be an additional symptom of diabetes mellitus?
2. What would the effect of this symptom be on thirst?

Because diabetics have difficulty using glucose to generate energy, they must metabolize fat instead. In the process, they generate an unusual waste product, **ketone bodies** (such as acetone). Ketone bodies give the diabetic's breath and urine a characteristic sweetish odor reminiscent of nail-polish remover. Ketone bodies also appear during starvation and sometimes when the diet contains too little carbohydrate.

Still another kind of waste that is removed from the blood by the kidneys is the normal breakdown products of hemoglobin. Hemoglobin from obsolete red blood cells is turned by the liver into the bile pigment *bilirubin,* which is secreted into the intestines and converted by bacteria into *urobilinogen.* Some bilirubin and urobilinogen normally appear in the urine; higher than normal levels indicate liver disease or excessive breakdown of red blood cells (as in some kinds of anemia). Excess urobilinogen also accompanies congestive heart failure and infectious mononucleosis.

The kidneys also remove excess salts from the blood, for the nephrons reabsorb only enough sodium, chloride, calcium, and other ions to keep blood levels constant. Excesses are excreted. Unfortunately, this process can lead to trouble, as when the bones release large amounts of calcium and phosphate in response to hormonal disorders, prolonged bed rest, or exposure to microgravity (a problem for orbiting astronauts). In the glomerular filtrate, the calcium and phosphate ions combine to form calcium phosphate, and when the filtrate contains large amounts of this relatively insoluble compound, it can crystallize to form kidney stones, or *renal calculi.* Large stones, which can be extremely painful, used to require surgical removal. Today, physicians immerse a patient in a water bath and aim ultrasonic vibrations at the kidneys to pulverize the stones; the resulting powder exits the body with the urine.

Kidney stones can be composed of substances other than calcium phosphate. One common alternative is uric acid. Uric acid is the major nitrogen-containing waste in birds and reptiles, but it is formed in large amounts by relatively few people, all of whom have a particular gene. Others have a hereditary inability to excrete the small amounts of uric acid normally produced. Both groups of people tend to develop **gout,** a painful condition marked by the deposition of uric acid crystals in the joints and in the kidney tissue. Gout is aggravated by eating foods high in nucleic acids (such as sardines), whose breakdown produces large amounts of uric acid. It is also aggravated by dehydration, starvation, and the use of diuretics.

Neither the glomerular filtrate nor the urine normally contains protein, hemoglobin, or blood cells. When these items show up in the urine, they indicate disease. Protein in the urine indicates a breakdown in the basement membrane of the glomerular capsule due to injury, high blood pressure, or poisoning by bacterial toxins, heavy metals (such as lead), or organic solvents (such as ether). The presence of red blood cells indicates bleeding due to injury, infection, or kidney stones. White blood cells—and microorganisms—in the urine are a sign of infection.

Box 12-3 The Artificial Kidney

Injuries, infections, poisons, cancer—all can destroy the kidneys. Without these cleansing organs, the blood accumulates waste materials, which are themselves toxic to the body's other cells. The body cannot effectively control blood pH. Nor can it get rid of excess salts, and the blood and tissue fluid accumulate water. In response, the tissues swell, the heart experiences added strain, and heart failure can result.

Would a sauna help? Certainly sweating can help rid the body of water and salts, and even some waste materials, but the basic problem remains. For some people the answer is a kidney transplant. Physicians take a kidney from a healthy donor, preferably a close relative whose tissue type is as similar as possible to the patient's tissue type (see Chapter 10). They can also take a kidney from someone who has just died in an accident or from a disease that left the kidney healthy; again they try to match tissue types. Failure to match tissue types can lead to rejection of the transplant by the patient's immune system. Since the tissue type match is rarely perfect, the patient must take drugs such as cyclosporine, which suppress the immune system and rejection, for the rest of his or her life.

Transplants work, but there are many fewer kidneys available for transplanting than there are people who need them. Too few healthy people realize that they carry a spare. That is, one kidney is enough to meet the body's needs. (Bear this in mind if someone you know ever needs a kidney.)

The shortage of kidneys for transplant means that most victims of kidney failure must rely on the *artificial kidney machine*. This device works on the principle of **dialysis,** by which two solutions separated by a membrane exchange diffusible solutes.

To use the artificial kidney machine, a physician sticks a needle into the patient's radial (forearm) artery and lets the blood flow between cellophane sheets (Figure 1). The cellophane sheets are the membranes across which dialysis occurs. They sit in a tank of fluid designed to mimic normal blood plasma. The fluid contains no proteins or waste materials, but it does have normal plasma concentrations of the various ions (salts), glucose, amino acids, and some other substances.

Each cellophane sheet in the artificial kidney machine has blood on one side and dialysis fluid on the other. As the blood flows past it, glucose and amino acids may diffuse into the blood, which makes it possible to use dialysis to supplement a patient's nutrition. At the same time, wastes such as urea and excess water and salts diffuse down their concentration gradients across the membrane into the dialysis fluid. Some studies have suggested that dialysis also removes from the blood of schizophrenics something that seems to cause their psychotic symptoms; at least, after dialysis, the symptoms are sometimes reduced.

The blood does not lose cells or protein in the artificial kidney because, just as in a healthy

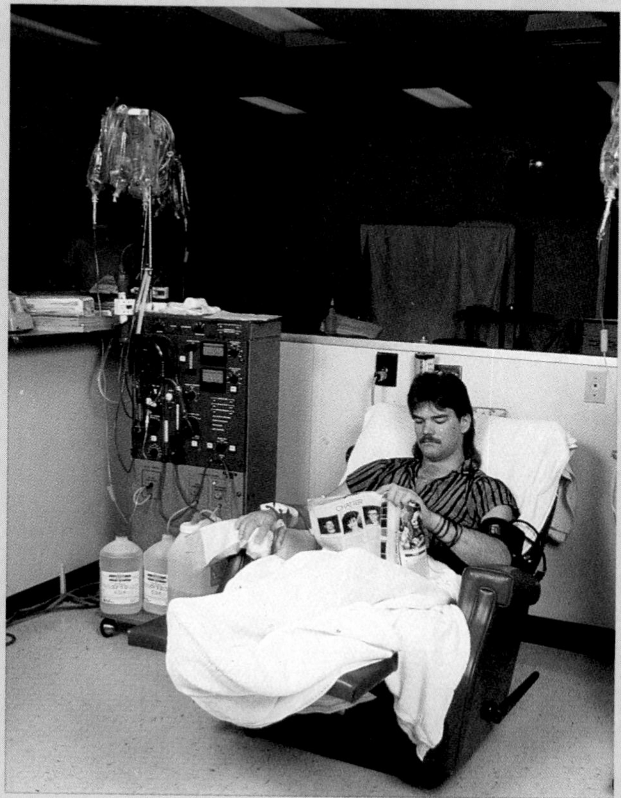

FIGURE 1 A patient undergoing dialysis treatment to remove chemical wastes from the blood because of kidney failure.

(Continued on p. 278)

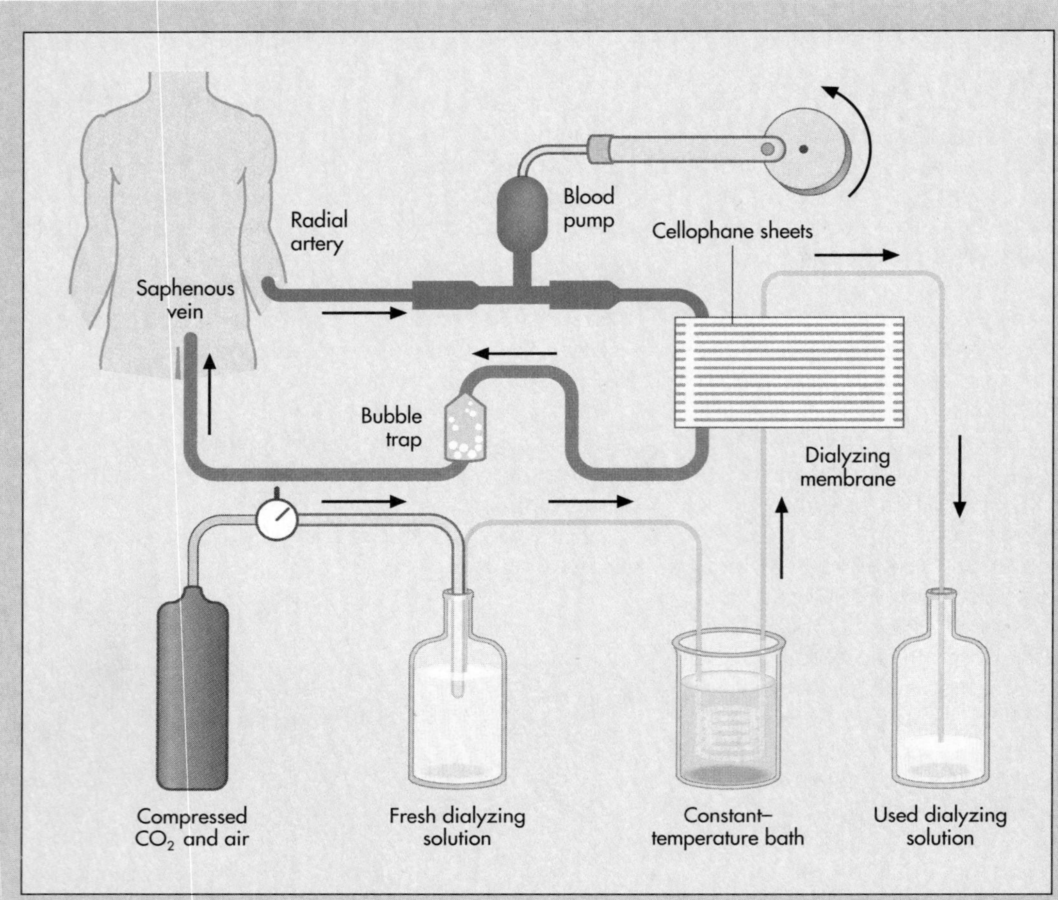

FIGURE 2 Overview of the dialysis treatment scheme.

kidney, they will not pass the membrane. Once it has been cleansed of wastes and its water and salt content have been brought to equilibrium with those of the dialysis fluid, the blood is returned to a vein in the patient's groin.

A variant of the artificial kidney is *continuous ambulatory peritoneal dialysis* (Figure 2), in which the dialysis fluid is made to flow into and out of the abdominal cavity, with dialysis occurring across the peritoneal membrane lining the cavity. Unfortunately, neither kidney substitute works as well as a genuine kidney. They fail to contribute to pH control or to remove large drug and toxin molecules, and they do not operate continuously. Patients may undergo dialysis two or three times a week, but a working kidney never quits, and it keeps blood levels of wastes and toxins much lower.

SUMMARY

The body disposes of wastes in several ways. The lungs vent carbon dioxide. The digestive system excretes the solid residue of what we eat as well as some cholesterol and products of hemoglobin breakdown contained in the liver's bile. The sweat glands, though their main function is temperature control, also remove excess salts, water, and waste materials from the blood. But most excess water, salts, and soluble wastes are removed from the blood by the kidneys in the form of urine. The kidneys also remove excess levels of other substances, including glucose and hydrogen ions. Their function is not just the removal of wastes but also the regulation of the blood's composition.

The kidneys, ureters, urinary bladder, and urethra compose the urinary or excretory system. A quarter of the cardiac output passes through the kidneys, where the nephrons produce 1 or 2 liters of urine per day. The urine drips into the renal pelvis and flows through the ureter, to be stored in the urinary bladder. The bladder empties through the urethra.

The process of urine formation begins in the kidney's outer zone or cortex. Arterioles there give rise to tufts of capillaries, the glomeruli, enclosed by double-walled glomerular (Bowman's) capsules. Each capsule merges with a tubule, and the glomerulus, capsule, and tubule form a nephron. The tubule has three zones: the proximal convoluted tubule; the loop of Henle, which dips into the renal medulla or core; and the distal convoluted tubule. The distal tubule connects to a collecting duct that delivers urine to the renal pelvis.

As blood passes through the glomeruli, much of the plasma, minus its proteins, passes through pores in the capillary endothelium and into the space between the capsule walls to become the glomerular filtrate. Some of the water in the filtrate immediately returns to the blood by osmosis. More returns as the filtrate flows into the nephron's proximal convoluted tubule, where diffusion and active transport also recover salts, glucose, amino acids, and other substances. The loop of Henle serves to concentrate sodium and chloride ions in the renal medulla. As the filtrate leaves the loop for the distal tubule, it is hypotonic to the cortical interstitial fluid and can lose more water by osmosis. In the distal tubule, such wastes as creatinine and some drugs are secreted into the filtrate and sodium ions are reabsorbed. Hydrogen and ammonium ions are also secreted there, making the kidney an important controller of blood pH.

When the filtrate passes through the medulla's hypertonic environment in the collecting duct, it can lose still more water, until the final product, urine, may be four times as concentrated as plasma. Reabsorption in the distal tubule and collecting duct is controlled by antidiuretic hormone (ADH), which is secreted by the pituitary gland.

To maintain blood flow through the kidney, the kidney has a way to control blood pressure. The juxtaglomerular apparatus, formed of cells belonging to the distal convoluted tubule and the arteriole leaving the glomerulus, responds to low blood pressure by secreting the enzyme renin. Renin leads to the formation in the plasma of angiotensin II, which stimulates vasoconstriction (which raises blood pressure) and adrenal secretion of aldosterone, a hormone that acts like ADH to make the kidney conserve water.

Kidney failure, due to injuries, infections, or toxins, leads promptly to death if not treated. Treatment may necessitate use of the artificial kidney machine, by means of which dialysis (diffusion across a membrane) regulates the blood's content of wastes, salts, water, and other substances. However, no artificial kidney made so far can work as continuously or as effectively as a real kidney. A far better treatment for kidney failure is a kidney transplant. Unfortunately, too few kidneys are available from live and dead donors to meet the need.

Table 12-3 provides a summary of selected diseases of the excretory system.

Table 12-3

Clinical Summary: Selected Diseases of the Excretory System

DISORDER	DESCRIPTION
Cystitis	Inflammation of the bladder. Usually caused by ascending bacterial infection of the urinary tract. Can sometimes involve other structures such as the urethra or prostate. Primary symptoms are frequent and painful urination.
Glomerulonephritis ("Bright's disease")	Inflammation of the kidneys. Involves damage to the glomeruli. Symptoms include hypertension, blood in the urine, and edema. In severe cases, there can be painful breathing and neurologic problems leading from delirium through convulsions to coma. Cause is unknown, but it frequently follows respiratory infection. Perhaps caused by allergic tissue sensitization.
Polycystic disease	Hereditary defect responsible for about 6% of human renal failures. Large fluid-filled cysts develop in kidneys from malformed nephron tubules. Affects about 1 adult in 500 and is the reason for 5–8% of kidney transplants and dialysis. Onset is most likely in one's thirties or forties, but it can strike in childhood.

(Continued on p. 280)

DISORDER	DESCRIPTION
Urinary calculi	Stones formed in the urinary tract, usually in the kidneys but often in the bladder. The stones may consist of uric acid, calcium carbonate, phosphate, or oxalate, or even the amino acid cysteine. Stone formation, especially in the bladder, often follows an infection. Diet may also be a factor.
UTI	Urinary tract infection; usually bacterial. Symptoms include discomfort or pain during urination, frequent urination, and sometimes low back pain.

STUDY QUESTIONS

1. List the body's ways of removing wastes.
2. How do the urinary systems of men and women differ?
3. The micturition reflex is involuntary. How then can you hold your urine until you get to the bathroom?
4. Describe the flow of blood through the kidney.
5. What is the role of the renal corpuscle in urine formation?
6. What is the role of the nephron's proximal convoluted tubule?
7. What is the role of the loop of Henle?
8. What is the role of the distal convoluted tubule?
9. Why is there normally no protein in the urine?
10. Describe the role of ADH.
11. What is the function of the juxtaglomerular apparatus?
12. What functions of a genuine kidney does the artificial kidney fail to duplicate?

GLOSSARY

Aldosterone An adrenal hormone that stimulates sodium reabsorption, thus enhancing water recovery in the distal convoluted tubule.

Angiotensin I The product of renin's action on angiotensinogen; becomes angiotensin II in the lungs.

Angiotensin II The end product of the renin blood pressure control mechanism; stimulates vasoconstriction and aldosterone secretion.

Antidiuretic hormone (ADH) A pituitary hormone that stimulates water and sodium reabsorption in the distal convoluted tubule and collecting duct.

Collecting duct A tubule that receives urine from the distal convoluted tubules of several nephrons and delivers it to the renal pelvis.

Countercurrent multiplier mechanism An arrangement of opposing fluid flows that permits the magnification of concentrations.

Dialysis The exchange of solutes between two solutions separated by a membrane.

Distal convoluted tubule The coiled segment of the nephron's tubule farthest from the renal corpuscle.

Glomerular (Bowman's) capsule The double-walled sac enclosing the glomerulus; the input end of the nephron.

Glomerular filtrate The fluid that moves from the blood, through the glomerulus, and into the glomerular capsule; plasma minus most of its proteins.

Glomerulus A tuft or ball of capillaries branching from the afferent arteriole; the actual filter.

Juxtaglomerular apparatus A small organ, composed of cells of the efferent arteriole and the distal convoluted tubule, that responds to low blood pressure by secreting renin.

Kidneys The blood-filtering organs located between the peritoneum and the back wall of the abdominal cavity; they control blood composition and pH.

Loop of Henle A hairpin-shaped loop in the nephron's tubule extending into the renal medulla; builds high salt concentrations in the medulla.

Nephron The functional, urine-producing unit of the kidney; consists of the renal corpuscle, the proximal convoluted tubule, the loop of Henle, and the distal convoluted tubule.

Proximal convoluted tubule The coiled segment of the nephron's tubule nearest the renal corpuscle.

Renin An enzyme that catalyzes the conversion of angiotensinogen to angiotensin I and hence increases blood pressure and the conservation of water and salt.

Ureter The tube that carries urine from the kidney to the urinary bladder.

Urethra The tube that carries urine from the urinary bladder outside the body.

Urinary bladder The storage organ for urine.

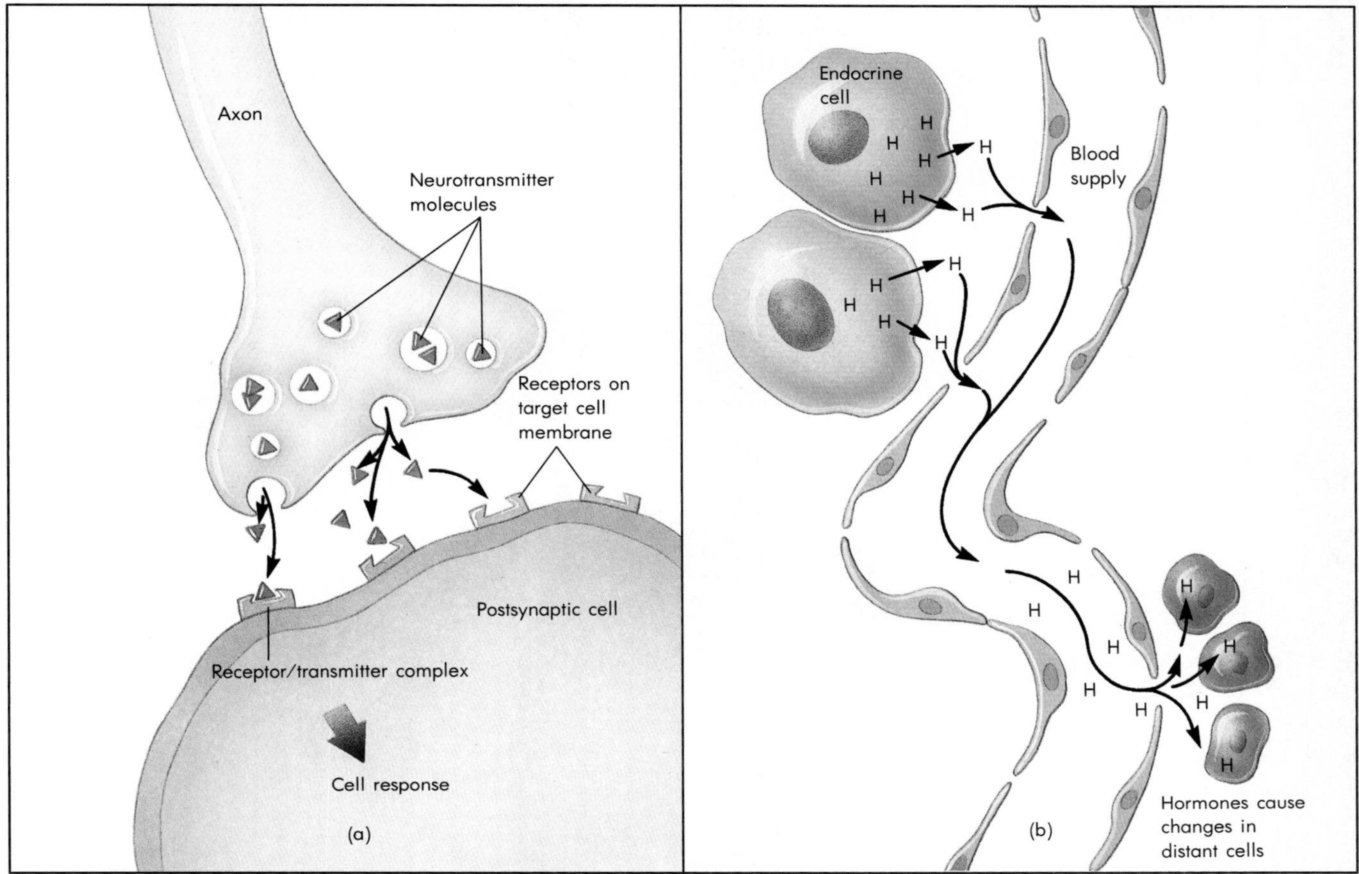

FIGURE 13-2 (a) When nerve cells release neurotransmitter, the transmitter molecules move directly into the space (synaptic cleft) between the communicating cells. The transmitter diffuses the short distance separating the cells, binds to the receptors on the cell membrane of the postsynaptic cell, and provokes a response in that cell. (b) Endocrine cells manufacture their hormones (H) and secrete them into the tissue fluids surrounding them. The hormones then diffuse into the nearest capillary, whence the blood carries them throughout the body. Each hormone provokes responses from the target cells carrying receptors for that particular hormone.

THE ENDOCRINE GLANDS

Researchers have found that cells in the heart, lungs, stomach, small intestine, uterus, and even brain all produce hormones. Because these organs all have other main functions, we can say that even though their endocrine functions are important to the body, they are only secondarily hormone producers.

Glands with a primary purpose of producing hormones are called **endocrine glands** (Figure 13-3). Endocrine glands are sometimes referred to as "ductless glands" because their secretions are not delivered through conducting channels to specific areas of the body, the way they are in such organs as the salivary, sweat, and digestive glands. Glands possessing ducts that deliver their secretions to specific compartments or surfaces of the body are collectively called *exocrine* glands. Endocrine glands simply release their secretions directly into their extracellular surroundings; from there, the secretions diffuse into the bloodstream for dispersal throughout the body.

The organs with secondary endocrine functions tend to control their own production of hormones. They secrete hormones in direct response to environmental signals. The small intestine, for instance, responds to the presence of food by secreting hormones that control the emptying of the stomach, the flow of bile and digestive enzymes, and the movement of food through the small intestine (the hormones of the digestive system will be

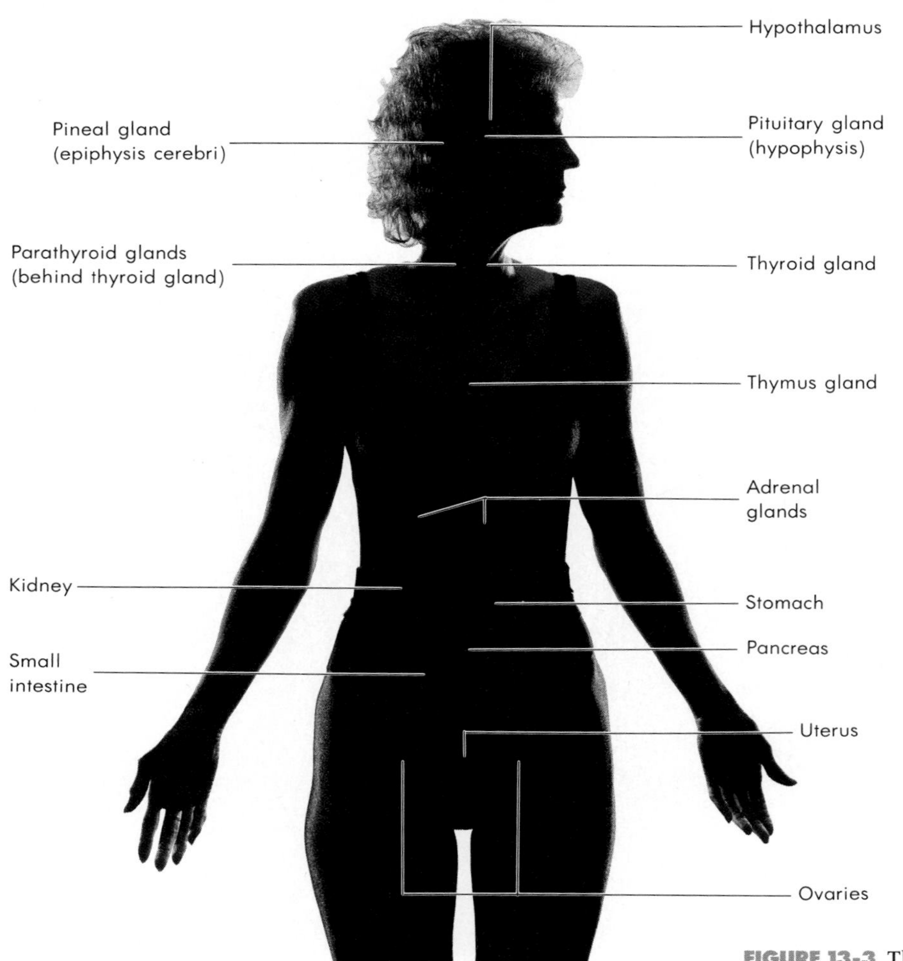

FIGURE 13-3 The endocrine glands.

covered in Chapter 20). The heart responds to changes in blood volume with a hormone that influences blood pressure.

The **thymus gland** at the base of the neck is considered part of the lymphatic and immune systems. Though it differs in structure from other lymphoid organs (such as the spleen), it is essential to the development of those lymphocytes (white blood cells) known at T cells. (In fact, the *T* stands for thymus.) Lymphocytes arise in the bone marrow; those that are destined to become T cells migrate to the thymus, where they mature under the influence of several thymic hormones (including *thymosin, thymic humoral factor, thymic factor,* and *thymopoietin*).

Some people consider renin to be a hormone produced by the kidney. However, renin is actually an enzyme that converts a protein in blood plasma into a real hormone, angiotensin II.

Renal erythropoietic factor is also an enzyme. Produced when the kidney is short of oxygen, as might happen at high altitude or after serious blood loss, it too acts on a plasma protein, but the end result is **erythropoietin,** a hormone that stimulates the production of red blood cells in the bone marrow, thereby raising the blood's oxygen-carrying capacity.

Most endocrine glands are controlled by a "master gland," the pituitary. The pituitary, in turn, is controlled by that portion of the brain called the hypothalamus. We will consider how these control mechanisms work after we have described the individual glands and their hormones.

THE PINEAL

Let us go through the endocrine system, except for the pituitary, from the head downward, beginning with the **pineal gland.** Unfortunately, this small structure near the center of the brain is not well understood. We know only that it develops calcium deposits (*brain sand*) after puberty and that it produces the hormone **melatonin,** which seems to inhibit the reproductive system in many animals, even—maybe—in humans. The target for melatonin action in mammals may be the brain's hypothalamus (Chapter 15), which controls the hormones being released by the pituitary gland. Melatonin may act by inhibiting the hypothalamic releasing factors that normally stimulate the pituitary to release luteinizing hormone (LH), which in turn triggers ovulation in human females.

Some researchers believe the pineal gland may be the site of a "biological clock" in humans and other animals. Secretion of melatonin is controlled at least in part by nerve signals from the eyes. Daily periods of light may provide some of the time-keeping signals that "inform" the pineal of the passage of time; the duration of the light periods may contain information about the season of the year for other species. The pineal seems to use this information to prepare an animal for its mating season, cutting down on melatonin release as the season approaches. High levels of melatonin may block the premature onset of puberty in children (both sexes). As the time of puberty approaches, melatonin secretion diminishes, allowing the reproductive hormones to work their "magic."

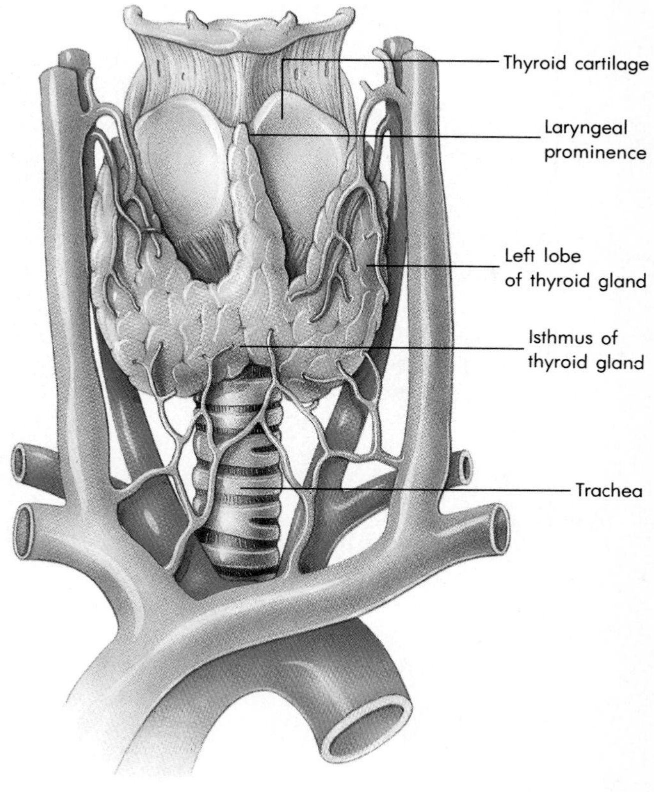

(a)

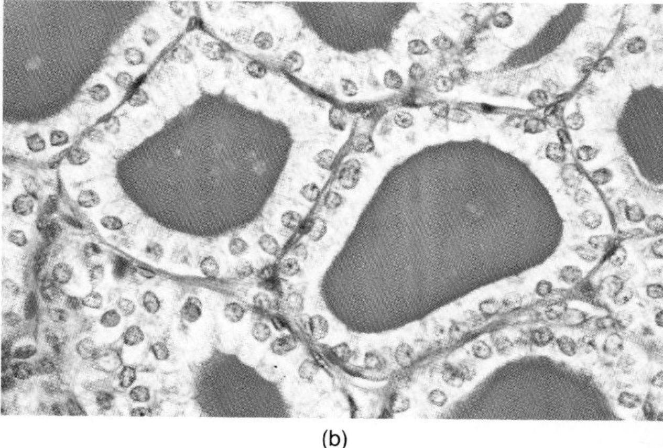

(b)

FIGURE 13-4 (a) The thyroid gland and its blood supply. (b) Thyroid follicles are chambers, surrounded by cells, in which the hormone thyroxine is stored.

THE THYROID

The **thyroid gland** is located in the neck, just below the larynx and in front of the trachea (Figure 13-4a). It is shaped roughly like a W or a butterfly, and it is composed of numerous small chambers, the **follicles.** The one-cell-thick walls of the follicles contain two types of cells (Figure 13-4b). Those cells that extend all the way through the wall are **follicular cells;** they secrete the thyroid hormone **thyroxine.** Those that do not extend all the way through the wall are the **parafollicular cells;** they secrete the hormone **calcitonin.**

Thyroid Hormones

The follicles of the thyroid gland accumulate iodine from the blood to use in the synthesis of thyroxine. The ability of the thyroid to concentrate iodine is impressive, since its concentration is 50–300 times higher in the gland than in the blood from which the iodine is drawn. The most potent form of thyroxine consists of two linked molecules of the amino acid tyrosine and four attached iodine atoms.

As we will see later in this chapter, the release of thyroid hormone is controlled largely by hormonal signals from the pituitary gland and the hypothalamus. The hypothalamus calls for increased thyroxine release when its chemical sensors detect a fall in the level of thyroxine in the blood as well as when the body's temperature drops. In addition, pregnancy and high altitude (both of which boost

Box 13-1 Jet Lag and the Hormonal Clock

Life has its own rhythms. From flowering plants to bacteria to humans, most life forms are controlled, in part, by built-in cellular rhythms that regulate the timing of many activities. This biological clock dictates when each cell is active, when it manufactures enzymes, when it secretes, when it divides, and when it rests. The basic rhythm pulses with a near-24-hour, or circadian ("about a day"), cadence. But there are also seasonal aspects to these rhythms that may coordinate long-term changes like the timing of sexual activity and reproduction. Recent research suggests that these biorhythms in humans are coordinated, in part, by the actions of the hypothalamus, the pineal gland, and the pineal hormone melatonin.

The inherent rhythm tends to drift out of synch with local day-night conditions and has to be reset each day to the local conditions, usually by exposure to sunlight or other forms of bright light that are at least 10 times brighter than room light. But even with the absence of synchronizing factors, the basic rhythms continue. For instance, human research subjects housed in almost complete isolation for extended periods—with no clocks or other external time cues like sunlight, television, or simply the sounds of other people's activities—still maintain their normal sleeping and waking cycle in a period slightly longer than a 24-hour day (for most humans, 24.5–25.5 hours). These natural, circadian rhythms control a wide range of physiologic activities including the daily cycles of alertness and sleepiness, circulating hormone levels, and body temperature. In most people, the body temperature is lowest at 6:30 a.m. It also plays a role in the timing of the reproductive cycle. Recently, it has been discovered that there is even a rhythmic nature to catastrophe, since most strokes and heart attacks occur around 9:00 a.m.

For a time, it was believed that the most powerful resynchronizing cue for humans was social contact. It seems that other mammalian groups were much more dependent on bright light as a synchronizing signal than humans. We now think differently—bright light has a very strong impact on human behavior and attitudes.

Rapid, extended travel to places where the day-night cycle is completely out of synch with one's natural biologic rhythms can cause a form of psychologic and physiologic confusion called "jet lag." This has a dramatic impact on a person's ability to function, at least until his or her body is resynchronized with local day-night patterns. A traveler to Europe, for instance, can take several days to come into synchrony with the new schedule of daily activities. This is especially important for business travelers, since they may be at a disadvantage if they have to negotiate deals during hours when they would usually be sleeping. For that reason, business people sometimes arrive at their destination several days in advance, to resynchronize to local conditions before they discuss an important deal.

The pineal gland may help reset the biologic clock by its secretion of melatonin. Strong light somehow suppresses the secretion of melatonin in the pineal gland. Melatonin secretion acts as a sedative in humans, promoting sleepiness or feelings of fatigue or disinterest. During long periods of dim light, like the stormy months of winter, melatonin secretions run high, causing "winter depression"—an annual psychologic complaint for some people. Psychiatrists have found that they can frequently treat winter depression successfully by exposing its sufferers to 2–4 hours of high-intensity light per day.

The use of light therapy has also been suggested to quickly reset the biologic clock. Some researchers believe that exposure to an hour of high-intensity light the evening of arrival could reduce jet lag in travelers who had journeyed across six time zones (or less) from east to west. If they traveled in the other direction, their exposure to light should come during the local morning. They are now testing whether exposure to bright light can help reset the clocks of shift workers who must periodically change from working days to working evenings or nights. About a third of working men and a quarter of working women are subject to such dramatic changes in their working patterns (and lifestyles).

the body's need for energy) stimulate the thyroid, whereas high levels of the sex hormones testosterone and estrogen can inhibit it.

Functions of Thyroxine

Thyroxine affects the body's metabolic rate (the pace of the body's chemical reactions) by stimulating the breakdown of carbohydrate for energy. A portion of the energy contained in carbohydrates is released as heat, especially from the electron transport system in the mitochondria (Chapter 21). The body can adapt to cold, in part, by increasing its production of thyroxine, and hence of heat. Thyroxine seems to be actively involved in 20 different enzyme systems. In addition, it increases the rate of insulin secretion, which promotes carbohydrate metabolism.

Hyperthyroid people, who have excess levels of thyroxine, have slightly higher than normal body temperatures. Since thyroxine also increases the reactivity of the nervous system, these people are "nervous" too. They have increased blood flow and blood pressure, more rapid heart rates, and increased intestinal motility. In contrast, hypothyroid individuals are lethargic and have low blood flow and pressure, slow heart rates, and decreased intestinal motility causing constipation in some.

Adult hyperthyroidism is more common in women than in men. It is marked by an enlarged thyroid gland, or **goiter,** and swelling of the tissues behind the eyes, making the eyes protrude. Together, these symptoms give the condition its name of **exophthalmic goiter.** Treatment requires either drugs to slow down production of thyroxine or surgical removal of part of the thyroid gland.

Hypothyroid people have underactive thyroids that can cause a number of problems. Thyroxine, together with growth hormone from the pituitary gland, also stimulates growth and development, especially of the nervous system in children. By aiding cell respiration, thyroxine makes energy readily available for physiologically expensive processes, like growth. Childhood hypothyroidism results in the condition known as **cretinism,** marked by dwarfism and severe mental retardation. Lacking adequate amounts of thyroxine, the child simply does not have enough energy available for the brain to develop properly and the body to grow. Unless such a condition is rapidly diagnosed and the child is given replacement doses of artificial thyroxine, the damage will be permanent. Cretins are also slow to develop sexually and are marked by fat deposits causing rounded faces and protruding abdomens.

Adult hypothyroidism, or **myxedema,** is less serious than cretinism, is more common in women than in men, and can usually be easily corrected with daily doses of artificial thyroxine. Myxedema is marked by facial swelling, difficulty in concentrating (a temporary form of mental retardation), lethargy, weight gain, and sometimes lowered resistance to infection. It is less serious because the onset occurs after the brain and other body systems have already formed properly. Problems resulting from myxedema are related more to not having enough energy to conduct daily business than to not having enough energy to properly support the development of the body's systems.

---

1. Use what you have learned about the thyroid to say why doses of radioactive iodine, often used to treat thyroid cancer, expose mostly the thyroid to toxic levels of radiation.
2. Explain how doses of nonradioactive iodine can protect the normal thyroid against exposure to radioactive iodine, such as may be emitted in a nuclear power plant breakdown like the one that happened in Chernobyl (USSR) in 1986.

Simple goiter, which results from a dietary deficiency of iodine, was once very common, especially in inland areas where iodine-rich seafood was scarce. As blood levels of thyroxine fell, the hypothalamus and pituitary would secrete their hormones to stimulate the thyroid gland. The thyroid, in its effort to obey, would increase in size in order to intercept more blood flow and filter more iodine from the blood. Lacking iodine, however, the process would not end. The resulting goiters could be impressive—some goiters weighed over 500 grams (Figure 13-5). Today, with iodine added to most table salt, such goiters are rare.

Calcitonin

For a time, it was mistakenly thought that the hormone calcitonin was secreted by the four tiny glands, the parathyroids, found on the backside of the thyroid. We now know it is secreted by the thyroid gland's parafollicular cells, and it helps regulate the level of calcium and phosphate in the blood. It inhibits the breakdown of bone, stimulating the deposition, not the removal, of calcium and

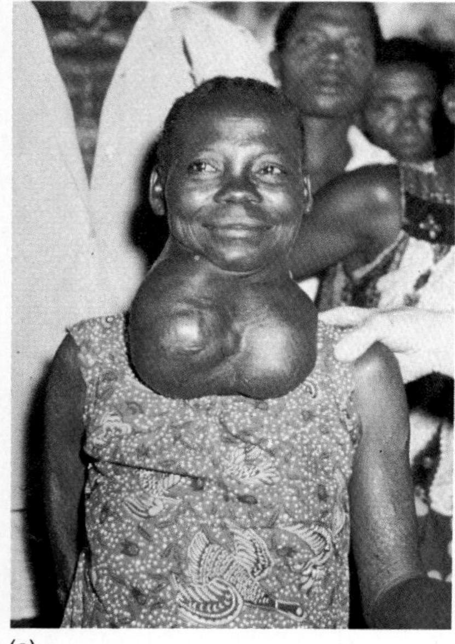
(a)

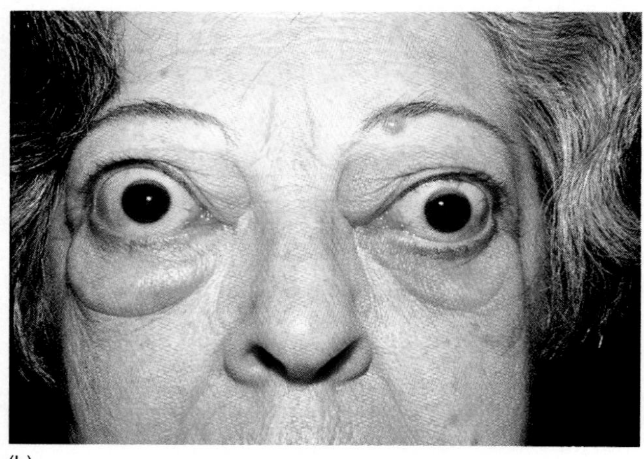
(b)

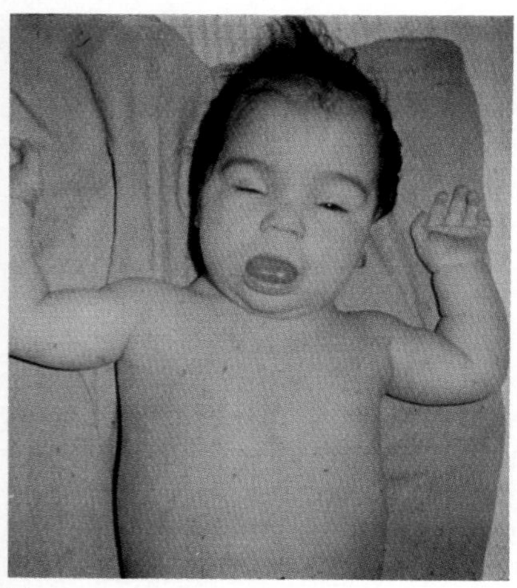
(c)

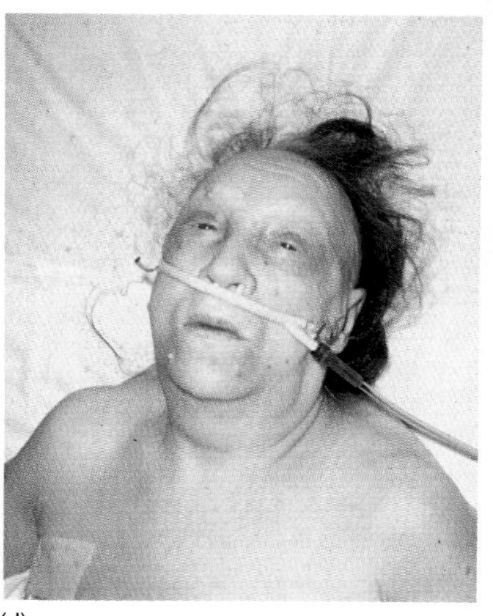
(d)

FIGURE 13-5 (a) A rare double goiter is an overgrowth of thyroid tissue in the neck, seen in people from geographic regions where iodine is in short supply. Their thyroids are unable to produce enough thyroxine to shut off (by negative feedback) the pituitary hormone (TSH) that commands its release. (b) Exophthalmia (protruding eyes) is frequently seen in people suffering from hyperthyroidism. (c) Cretinism is seen in babies and small children suffering from hypothyroidism. These children fail to grow and their nervous systems do not develop properly. (d) Myxedema is the result of hypothyroidism in adults. Its sufferers become sluggish and concentrate poorly. Because they retain fluids, their bodies and faces take on a puffy or swollen look.

phosphate. It seems to work by inhibiting osteoclasts (cells that dissolve bone) and inhibiting the production of the parathyroid hormone parathormone. It is released when blood levels of calcium are high. Too little calcitonin (or too much parathormone) can lead to kidney stones and calcium deposits in normally soft tissues. Too much calcitonin can lower blood calcium levels and cause muscle spasms.

THE PARATHYROIDS

Attached to the back of the thyroid gland are four small, pinkish **parathyroid glands** (Figure 13-6). When blood levels of calcium fall, these glands secrete **parathormone (PTH),** which raises the calcium and phosphate in the body fluids back to normal levels. Parathormone achieves this balance in a number of ways. It stimulates the osteoclasts to release calcium and phosphate from the bones. Parathormone also stimulates the kidney to recover more calcium from the urine and to excrete more phosphate into the urine. In addition, it helps activate vitamin D and, when enough of this vitamin is present, increases the absorption of calcium, phosphate, and some magnesium from the intestine.

High levels of blood calcium or calcitonin cause the parathyroid to cut back on its secretion of parathormone. Thus, the control of parathormone secretion is directly regulated, in part, by the mate-

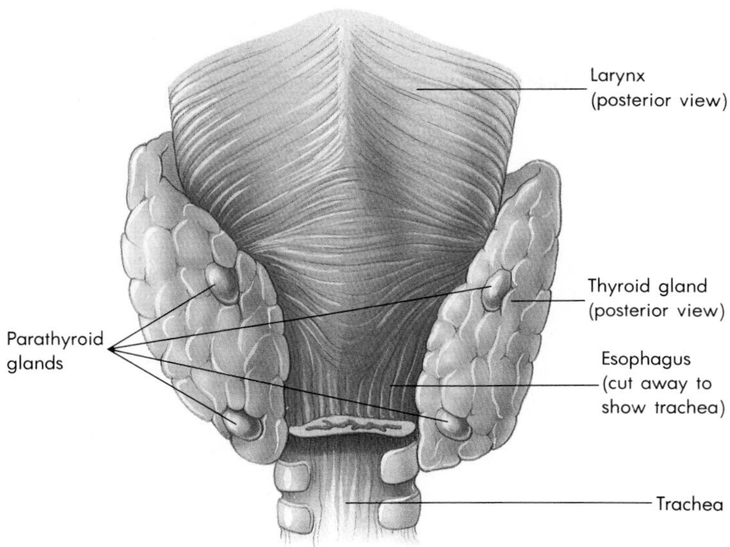

FIGURE 13-6 The four tiny parathyroids are located on the posterior surface of the thyroid.

rial that it controls—namely, calcium. Less parathormone is secreted when blood levels of calcium are high and calcitonin is present. However, the regulatory mechanisms can fail. Tumors of the parathyroids, or surplus parathyroid glands, can put more parathormone into the blood than the body needs. The result is too high calcium levels in the body fluids, calcification of soft tissues, skeletal fragility as the bones become demineralized, and kidney stones. Surgery is required; it necessitates removing one or more of the parathyroids as well as part of the thyroid.

Although, unlike most parts of the body, the thyroid can regenerate, the parathyroids cannot. Thus, when thyroid surgery removes one or more of the parathyroids, the remaining glands may not be able to secrete enough parathormone. The result is too low a calcium level, just as when the thyroid produces too much calcitonin. A lowered calcium level causes increased irritability of the nerve tissues, which results in spasms in the skeletal muscles. These spasms lead to a sustained state of contraction—called tetany—the results of which can be fatal (Figure 13-7).

THE PANCREAS

The **pancreas** is an extremely versatile organ lying just below the stomach. As an exocrine gland, it supplies digestive enzymes to the small intestine through the pancreatic duct. Within it, however, are small clumps of cells, the **islets of Langerhans,** which do not release their secretions into the gland's ductwork. They secrete hormones instead, making the pancreas both an endocrine and an exocrine gland.

The islets contain two major cell types. The **alpha cells** secrete the hormone **glucagon.** The **beta cells** secrete **insulin.** Both hormones are polypeptides (Figure 13-8).

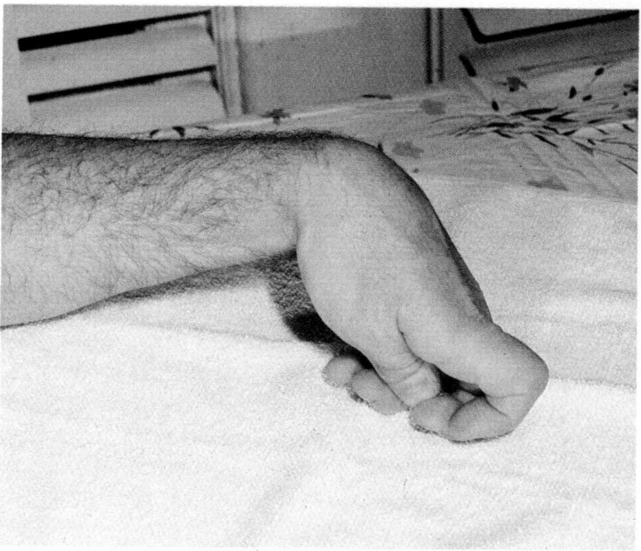

FIGURE 13-7 The level of calcium in the body fluids is controlled, in part, by the hormone parathormone. Low levels of parathormone cause a decrease in calcium levels in the body fluids. This causes increased nerve irritability, which in turn leads to sustained muscle contractions, called tetany, seen here in the hands.

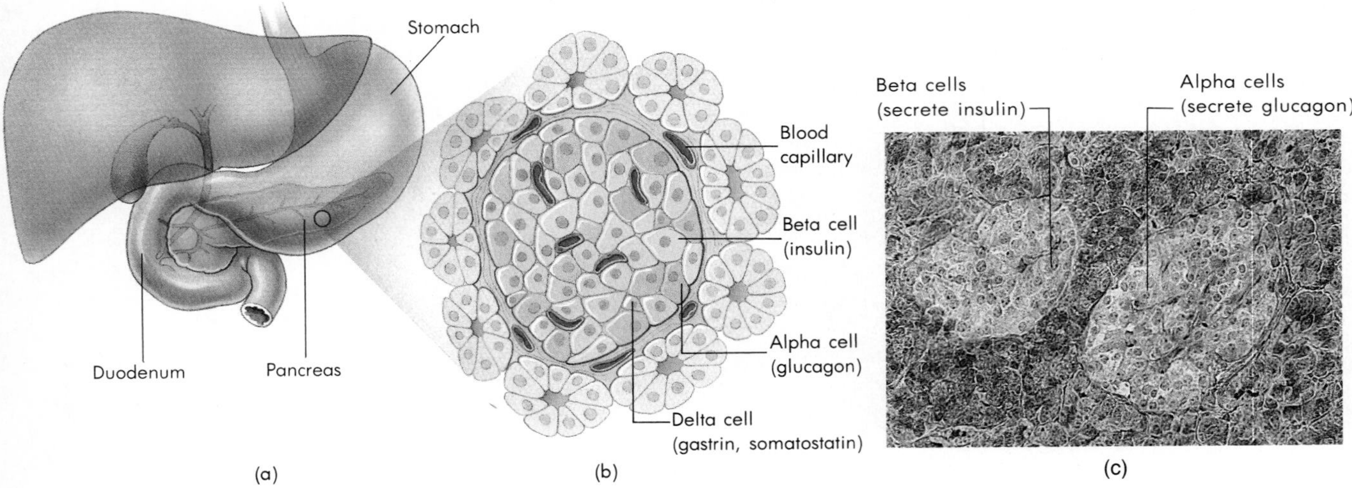

FIGURE 13-8 (a) Location of the pancreas. (b) Section through the pancreas showing an islet of Langerhans. (c) Photomicrograph of an islet of Langerhans, showing alpha and beta cells. The alpha cells secrete glucagon; the beta cells secrete insulin.

Insulin and Glucagon

Insulin and glucagon are the main hormones secreted by the pancreas. Together they control the blood levels of glucose. When glucose levels are high, as after a meal, the pancreas releases insulin. This hormone acts to lower blood glucose levels by stimulating skeletal muscle (and other) cells to absorb glucose, thereby removing it from the bloodstream and rapidly converting it to glycogen—a glucose storage molecule, sometimes called animal starch. It also stimulates fat cells to absorb the sugar and convert it to fat. In addition, it stimulates protein synthesis by encouraging cells to absorb amino acids and use them to synthesize peptides. This helps keep the sugar level down by keeping cells from using newly acquired amino acids to make glucose instead.

The pancreas secretes glucagon when glucose levels are low, as between meals. Exercise, which increases the demands for glucose to fuel the muscles, has a similar effect. Glucagon stimulates liver cells to convert glycogen back into glucose and release it to the blood. It also stimulates liver cells to convert amino acids, glycerol, and lactic acid into glucose. This has the net effect of elevating the level of glucose in the blood and making more fuel available to the various tissues of the body.

Negative Feedback

Together, insulin and glucagon act to maintain the blood sugar at a nearly constant level of about 90 milligrams (mg) per 100 milliliters (ml) of blood (normal range is 80–120 mg glucose/100 ml blood). They are thus a crucial part of the body's *homeostasis* and provide an excellent example of *negative feedback*. That is, their effects control their secretion. The alpha cells respond to a blood sugar level below 90 mg/100 ml by secreting glucagon. This hormone raises the sugar level until it exceeds 90 mg/100 ml, at which point the alpha cells shut down. Beta cells respond to blood sugar levels above 90 mg/100 ml by secreting insulin, which lowers the level of glucose in circulation. When the level falls below 90 mg/100 ml, the beta cells shut down. As a result, the blood sugar level under normal conditions never departs very far from the 90-mg/100-ml range (Figure 13-9).

1. Growth hormone, among its other effects, increases blood sugar levels. What effect would you expect it to have on insulin secretion?
2. Prepare a diagram like that in Figure 13-9 to show how negative feedback works in the control of calcitonin and parathormone secretion.

Diabetes

There are a number of ways in which the secretion of glucagon and insulin can go awry. Tumors of the islets of Langerhans can increase glucagon secretion, causing **hyperglycemia** (too high levels of blood glucose), or increase insulin secretion, causing **hypoglycemia** (too low blood glucose). Hypo-

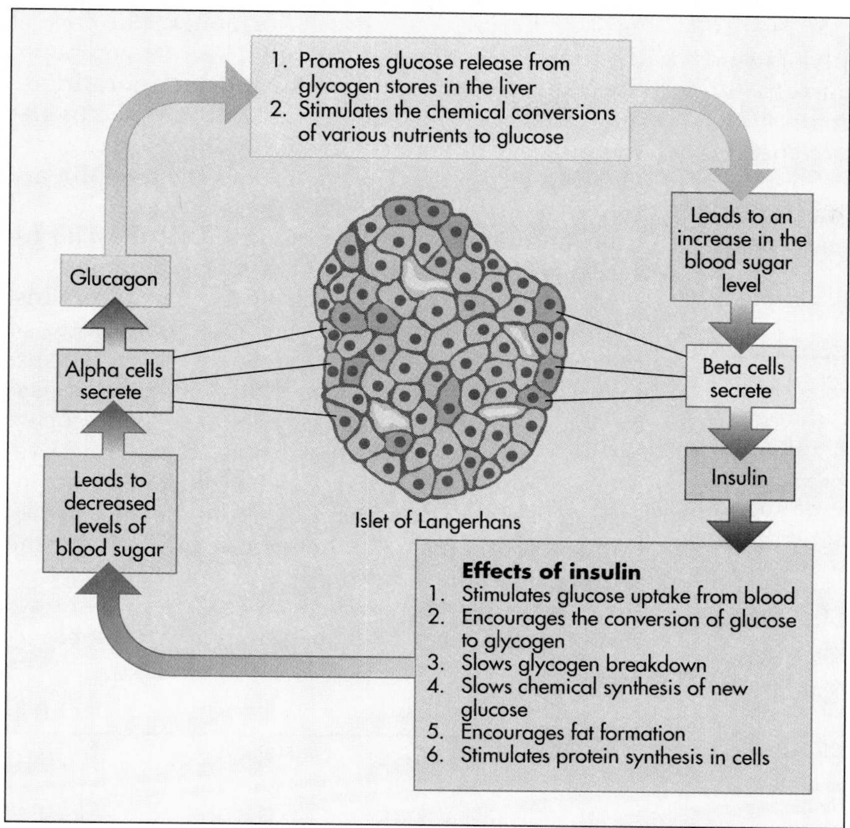

FIGURE 13-9 Regulation of insulin and glucagon secretion.

glycemia starves the body's cells, especially in the brain, which normally uses only glucose for fuel. The symptoms include weakness, tremors, convulsions, mental disorientation, unconsciousness, and even death. Severe hypoglycemia, when caused by overdosing an insulin injection, causes the blood glucose level to plummet, producing what is known as *insulin shock.*

Diabetes mellitus is marked by hyperglycemia (a diabetic's blood sugar level can be as high as 300–1200 mg of glucose/100 ml of blood), excretion of glucose in the urine (glycosuria), immense volume of urine, and thirst. In *juvenile-onset* or *insulin-dependent diabetes,* childhood viral infection (such as mumps or measles) can sometimes lead to the destruction of beta cells as part of a delayed reaction to those diseases. In other cases, heredity may also play a major role. Lacking functional beta cells, the abnormal pancreas cannot secrete adequate amounts of insulin. Without enough insulin, the body's cells cannot absorb glucose. The body therefore draws upon fat as an alternative fuel, causing the production and release into the blood of acidic *ketone bodies.* These compounds give the breath an odor like nail-polish remover and can raise the blood acid level enough (acidosis) to cause unconsciousness (*diabetic coma*) and death. In addition, changes in the way the body metabolizes fat also lead to atherosclerosis and the numerous cardiovascular problems it causes—including higher risks for heart attacks, strokes, and impaired circulation. The body also breaks down protein, loses weight, and grows weak. Lack of fuel to the brain can induce mental confusion, coma, and eventually death.

In some cases, the pancreas contains enough beta cells so that "oral insulins," which stimulate insulin secretion, can effectively control diabetes. More often, diabetics require injections of human insulin obtained from genetically engineered bacteria. Unfortunately, injections cannot match the flow of insulin to changes in blood sugar level in the way a functioning pancreas can. Insulin pumps, implanted under the skin to release insulin into the blood, do a better job. Still better results may be possible in the future, for researchers are developing an artificial pancreas that can be implanted in the body to release insulin as needed. Other researchers are experimenting with the injection of beta cells into the abdominal cavity, where they may survive to play the role of a true pancreas.

Adult-onset diabetes most often strikes people

who are over 40 and obese. Apparently, it results when the body's cells diminish their supply of insulin receptors. At any rate, the cells cannot respond well to insulin. Blood glucose levels remain high, even though the blood may contain more insulin than in healthy people. This kind of diabetes clearly cannot respond to insulin injections. Treatment involves controlling the diet and losing weight.

1. Insulin is a protein or polypeptide. Why do you think it must be administered to diabetics by injection? Why can't someone just take an insulin pill?
2. Diabetics need to control their intake of sugar. Nevertheless, many diabetics carry with them at all times a little candy. What do you think is the reason?

THE ADRENALS

One **adrenal gland** sits atop each kidney embedded in a fatty covering, where it secretes two basic categories of hormones. The outer layer of the gland, the **adrenal cortex** (cortex means "rind" or "bark"), secretes the *adrenal steroids,* derived from cholesterol and related to the sex hormones. The core of the gland, the **adrenal medulla,** secretes **epinephrine** (adrenaline) and **norepinephrine** (noradrenaline), which are also neurotransmitters used by the sympathetic division of the autonomic nervous system (Figure 13-10). The cortex and the medulla are actually two different glands. They are derived from different embryonic tissues, and they lie together more or less by accident of development. In some animals, they are actually separate organs.

In spite of their separate origins, both areas of the adrenal glands help the body cope with the various aspects of stresses—both physiologic and psychologic. Generally speaking, the adrenal cortex releases hormones to help the body cope with physiologic stresses—such as those provoked by an injury (blood loss) or specific metabolic difficulties (kidney problems). The hormones released by the adrenal medulla help the body cope with psychologic stresses; they trigger the "fight-or-flight" response to alarming or frightening situations.

The Adrenal Cortex

Each of the three layers in the adrenal cortex is responsible for producing specific hormones or a family of related hormones. The outer layer of the adrenal cortex secretes the **mineralocorticoids,** chiefly **aldosterone.** These hormones help control the body's water and mineral (salts) contents. Aldosterone causes the tubules of the kidney's nephrons

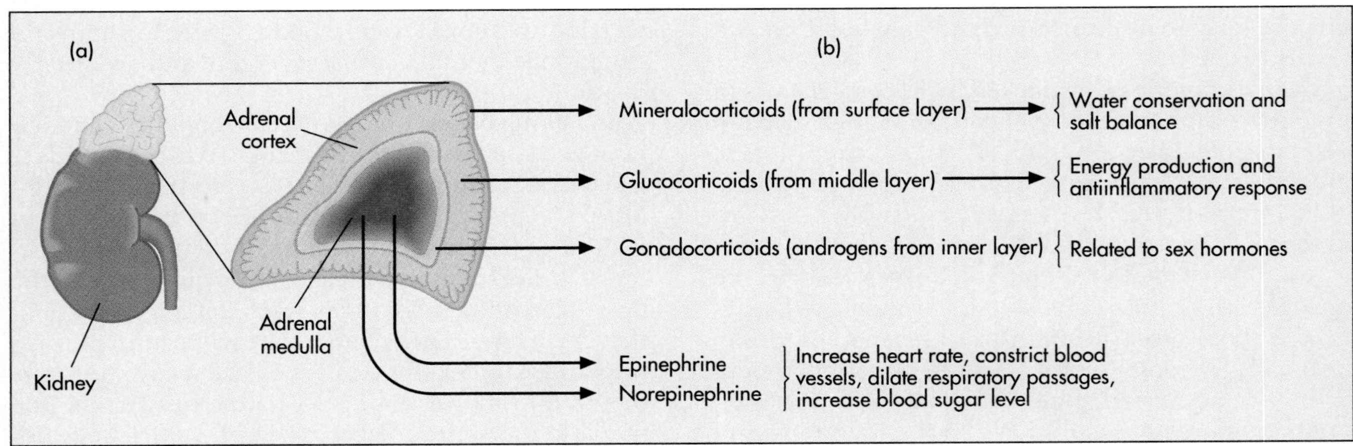

FIGURE 13-10 The adrenal glands. (a) The adrenal glands are embedded in a body of fat on top of each kidney. (b) The adrenal cortex (outer layer) secretes three groups of hormones: the mineralocorticoids, the glucocorticoids, and the gonadocorticoids. The adrenal medulla (or core) releases the two hormones most likely released as a consequence of fright—epinephrine and norepinephrine.

to decrease their reabsorption of potassium and increase their reabsorption of water and sodium. In the process, the tubule cells secrete hydrogen ions into the urine, thus reducing the acidity of the blood.

Aldosterone is part of the body's mechanism for controlling blood pressure, as shown in Figure 13-11. Cells in the (1) kidney's juxtaglomerular apparatus (Chapter 12) respond to low blood pressure by (2) secreting *renin,* an enzyme that (3) converts *angiotensinogen,* a blood plasma protein made in the liver, to *angiotensin I.* Angiotensin I encounters a second enzyme in the lungs that (4) converts it further to *angiotensin II.* This protein (5) constricts arterioles and hence (8) helps raise blood pressure; it also (6) stimulates aldosterone secretion. Because aldosterone (7) promotes water conservation in various ways in the body, it too (8) helps raise the blood pressure.

The middle layer of the adrenal cortex secretes **glucocorticoids,** which include **cortisol (hydrocortisone), corticosterone,** and **cortisone** (Figure 13-12). Of these three hormones, cortisol accounts for about 95 percent of all glucocorticoid activity. These hormones increase energy production by stimulating protein breakdown and conversion of amino and fatty acids to glucose. An increase in the supply of glucose makes more energy available to help the body handle stress or to promote healing of damaged tissues. These hormones also have anti-inflammatory effects, which makes them useful as medications.

Glucocorticoid secretion is controlled by negative feedback. The adrenal cells respond to the pituitary hormone adrenocorticotropic hormone (ACTH), which is released when hypothalamic cells detect either low levels of glucocorticoids in the blood or stress. The hypothalamus secretes corticotropin releasing factor as its command to the pituitary.

The inner layer of the adrenal cortex secretes small amounts of **gonadocorticoids,** or sex hormones (Figure 13-12). In both men and women, the major product is **androgen,** which resembles the male sex hormone, testosterone. Its effects on the body are generally unnoticeable, except when a tumor of the adrenal gland or pituitary leads to the production of large amounts of androgen. In male children, the result can be early puberty. In female children and women, it can be the appearance of male secondary sexual characteristics such as a beard (Figure 13-13, p. 295). Occasionally, an adrenal tumor can load a male body with female sex hormones and cause breast development.

The Adrenal Medulla

The adrenal medulla secretes epinephrine (adrenaline) and norepinephrine (noradrenaline) in response to signals from the sympathetic branch of the autonomic nervous system (Figure 13-14, p. 295). The sympathetic branch is the neurologic

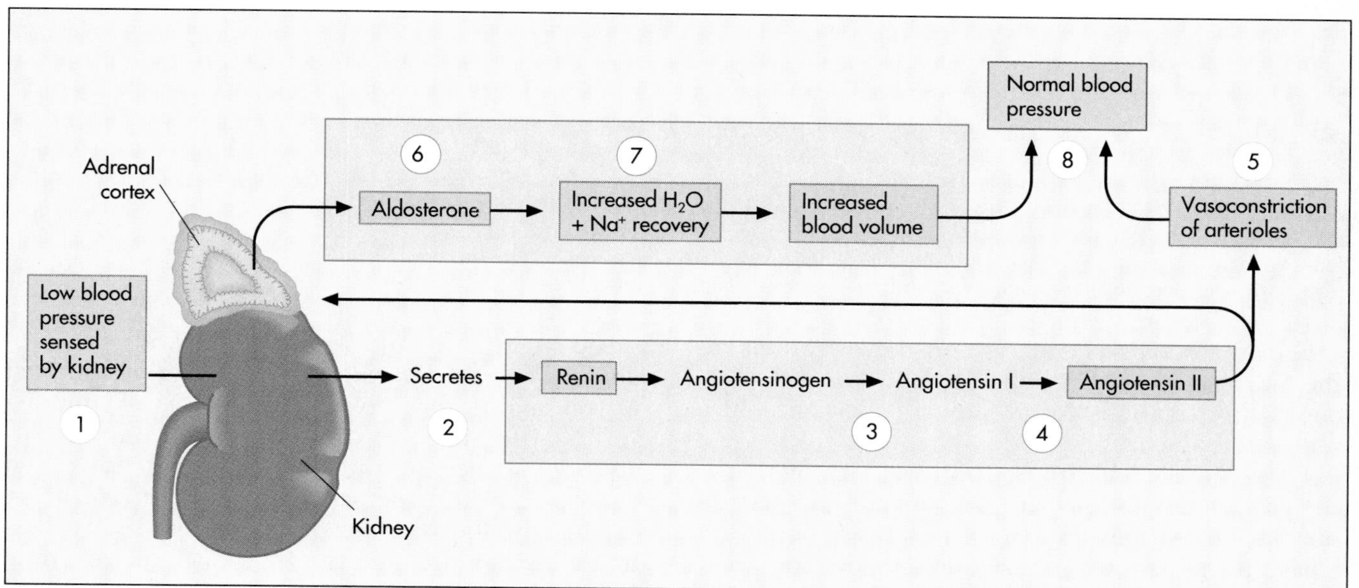

FIGURE 13-11 Renin-angiotensin-aldosterone blood pressure control.

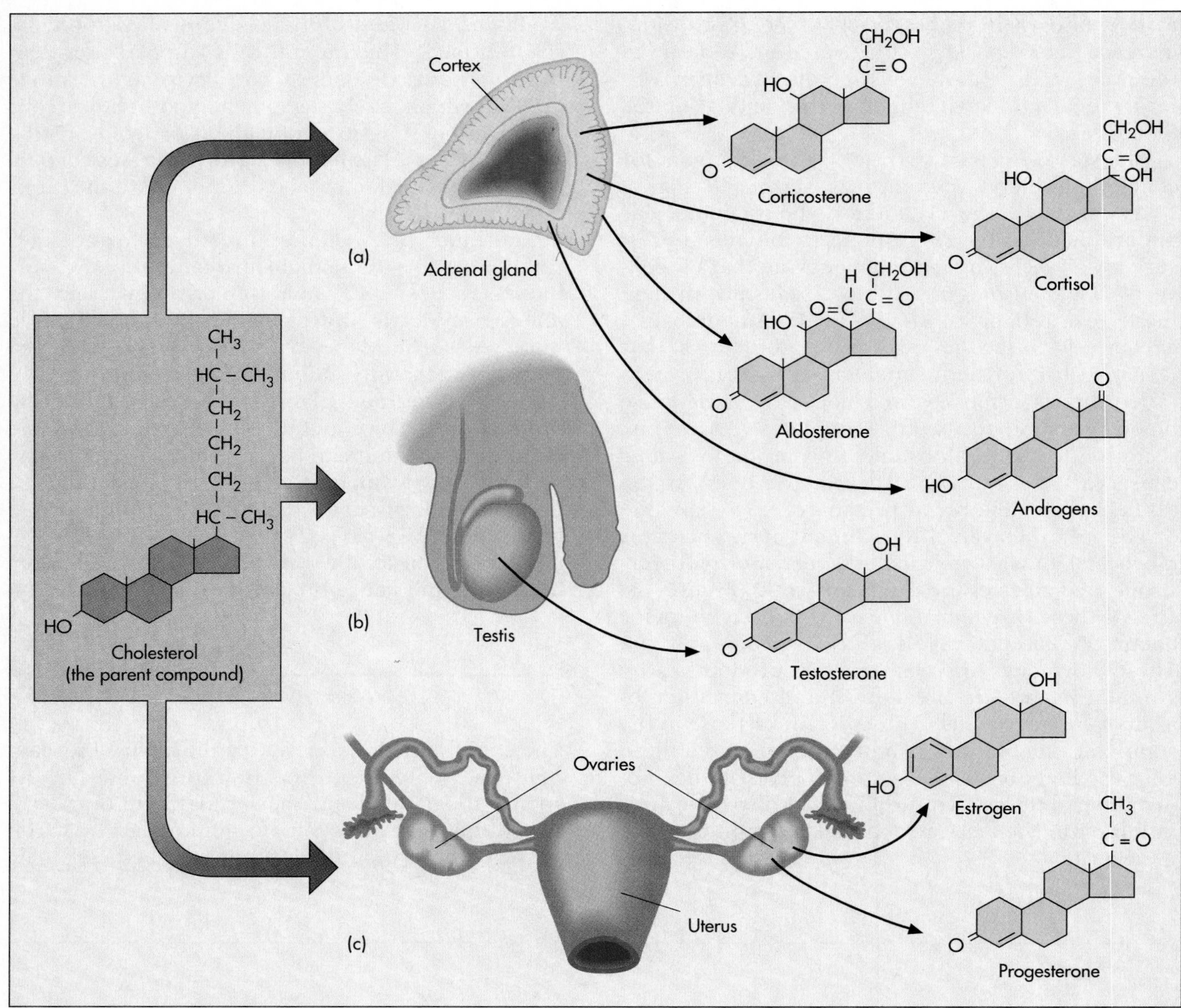

FIGURE 13-12 Structures of cholesterol and the adrenal steroid hormones. (a) The adrenal cortex uses cholesterol as the raw material for synthesizing its various hormones. The mineralocorticoids (aldosterone) are produced by the outermost layer of the adrenal cortex. The glucocorticoids (corticosterone and cortisol) are released from the middle layer of the cortex. The androgens, which are most closely related to the sex hormones, are released from the innermost cortical layer. (b) The testes secrete one of the most powerful steroids, testosterone. (c) The ovaries produce two female sex hormones, estrogen and progesterone.

side of the body's alarm system; the secretions of the adrenal medulla enhance and prolong the body's preparations for whatever emergency has set off the alarm.

Epinephrine, the more potent hormone, accounts for about 80 percent of the medullary secretion. Both epinephrine and norepinephrine increase heart rate and muscle efficiency, constrict the blood vessels, accelerate breathing, dilate the respiratory passages, slow digestion, increase the blood sugar level, and stimulate the body's metabolism. They thus prepare the body for action, just as you might expect from your memory of your last "adrenaline rush." It came when you were frightened or alarmed, in traffic or at a horror movie, and its function was to get you ready either to run away or to fight for your life. That adrenaline rush is understandably called the fight-or-flight reaction. It is the body's response to sudden, urgent, psychologic stress—a real or imagined threat to your life.

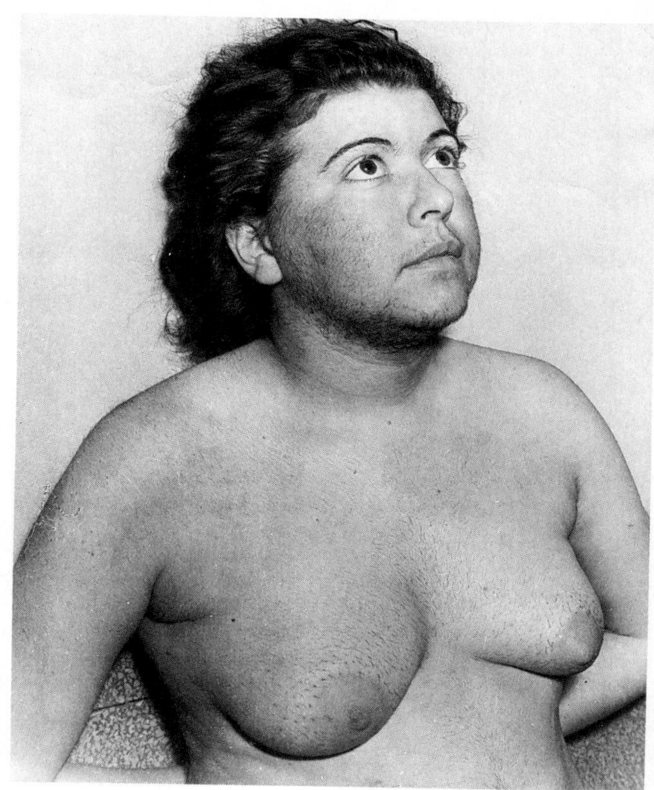

FIGURE 13-13 Adrenal tumors in females sometimes produce male hormones that result in the adrenogenital syndrome (masculinization of the female). A conspicuous symptom is an increase in facial and body hair.

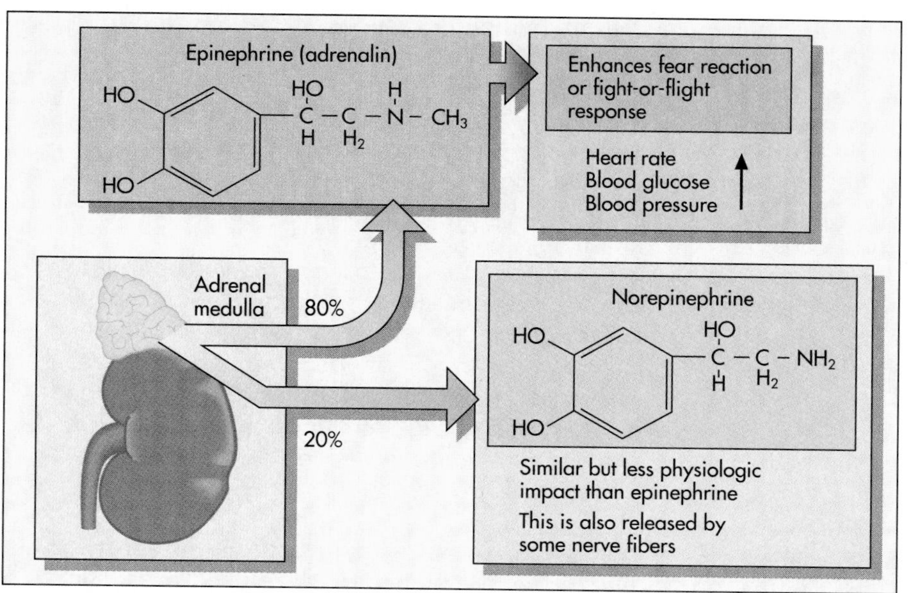

FIGURE 13-14 Structures of epinephrine and norepinephrine.

Learning Focus

STRESS AND THE GENERAL ADAPTATION SYNDROME

Stress occurs when the body is damaged or when there is even a threat of such damage. Stress consists of wounds, including surgery. Stress is produced by toxins in food or by infections. It can be caused by severe heat or cold or pain. Strong emotions—extreme anger, fear, sadness, and even joy and love—can cause stress. It can be provoked by anything that upsets the brain's interpretive power or threatens to upset the body's chemical and physical equilibrium. And to each stress, physical or psychologic, the body responds in much the same way.

The body's response begins in the hypothalamus, which detects stress in nerve signals and in sudden changes in blood pressure, chemistry, or temperature. It then sends nerve signals to the autonomic nervous system, which we will discuss in Chapter 15, and chemical signals (hormones) to the pituitary gland. The pituitary then releases ACTH, growth hormone, and thyroid-stimulating hormone (TSH).

The autonomic nervous system stimulates the adrenal medulla to release epinephrine and norepinephrine. The ACTH stimulates the adrenal cortex to release mineralocorticoids and glucocorticoids. Growth hormone stimulates the breakdown of fats and the release of glucose in the liver. The TSH stimulates the release of thyroxine. The overall effect is an increase in the supply of energy for the body to use in coping with the stress. There is also an increase in the heart and respiration rates, constriction of blood vessels in the digestive organs and skin, sweating, dilation of the respiratory passages, and a decrease in the production of digestive enzymes and urine. In addition, pain sensitivity diminishes and the spleen contracts, squeezing into the blood a surge of white blood cells, ready to combat the bacteria that may enter the body through wounds, and red blood cells, ready to replace those lost to wounds. The body is ready to exert itself in fight or flight and to survive whatever damage it may incur in a fight (see the diagram that follows).

Hans Selye was the first to study the body's response to stress in any detail. In the 1940s, he gave the name of **general adaptation syndrome (GAS)** to the body's response and noted that the GAS is indeed helpful in many cases. It preserves the body's homeostasis in the face of blood loss, infection, cold, heat, and many other assaults. However, he also noted that it can become a hazard in its own right if it persists too long.

Selye called the body's initial response to stress the **alarm reaction.** This is the same as the fight-or-flight response. It is mediated by the sympathetic branch of the autonomic nervous system and by the adrenal medulla, and it immediately boosts the supply of blood, fuel, and oxygen to the brain and muscles. A severe alarm reaction can lead to shock, unconsciousness, and even death.

The second stage of the GAS is the **resistance reaction.** It is mediated by the pituitary, adrenal cortical, and thyroid hormones and growth hormone. It is thus slower to develop and longer lasting than the alarm reaction. Its point is to adapt the body to longer lasting stress by ensuring a plentiful supply of glucose, while returning many of the alarm reaction's effects on the body to normal. Unfortunately, blood pressure remains high in this stage, and many people believe the long-lasting stresses of modern life—from fear of urban crime to the frustrations of commuter traffic—account for at least some of the high incidence of high blood pressure. Also, unfortunately, the glucocorticoids released in the resistance reaction impair the inflammatory response to wounds and hence may cause slow healing.

When the resistance stage continues for too long—as it may when stresses are not escapable—the GAS enters its **exhaustion** stage. Mineralocorticoids have been stimulating the kidney to retain water and sodium but excrete potassium and hydrogen ions. When the body has lost too much potassium, its cells begin to die. At the same time, the heart, blood vessels, and adrenal cortex, which have been under prolonged strain, may fail. Depletion of glucocorticoids may cause blood sugar levels to suddenly fall, starving cells throughout the entire body. The result can be weakness, general collapse, and even heart failure and hence death.

Stresses do not usually last long, at least for humans who live unhurried, relaxed lives, such

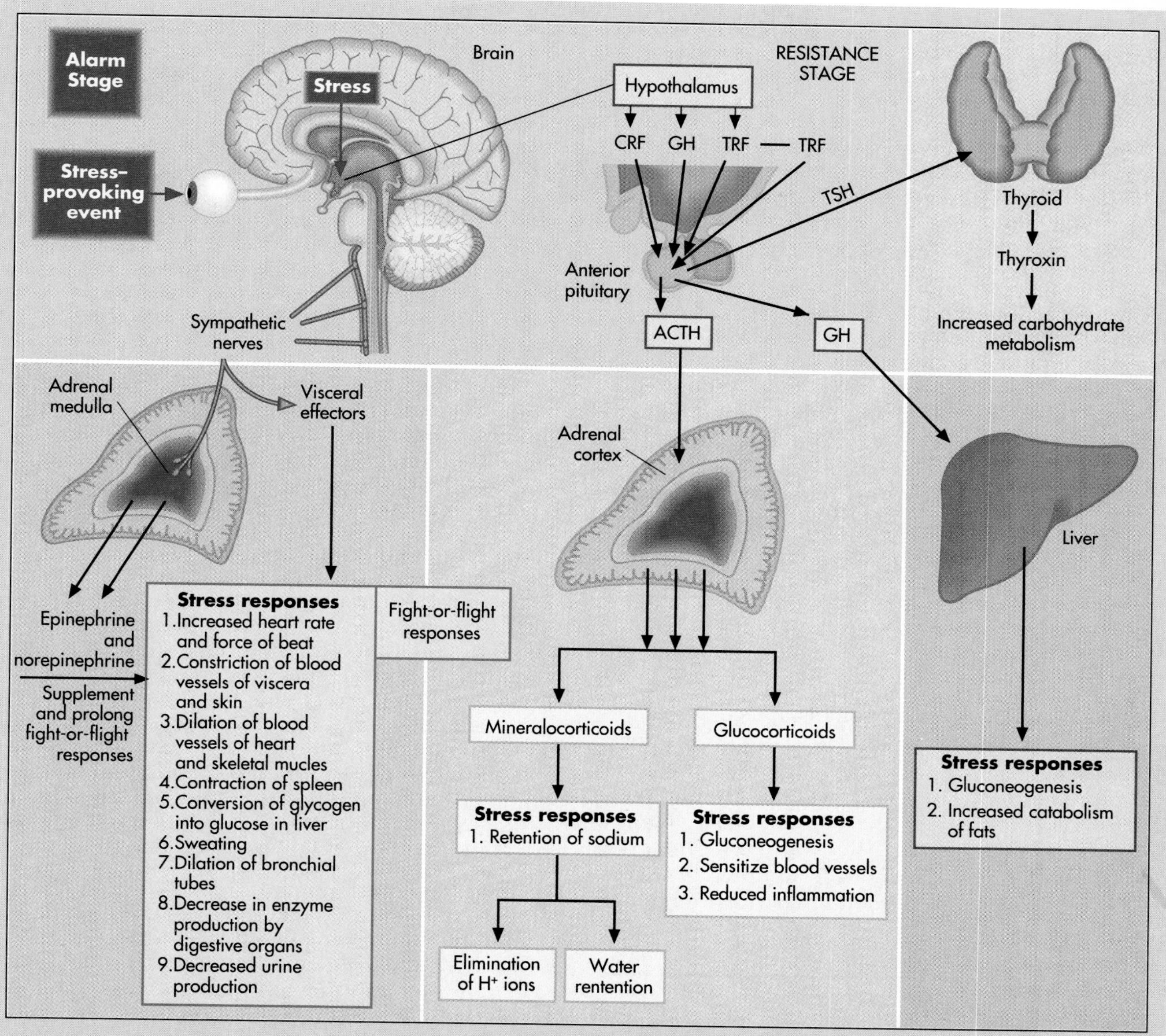

The hormonal response of the general adaptation syndrome (GAS): CRF, corticotropin releasing factor; ACTH, adrenocorticotropic hormone; GH, growth hormone; TRF, thyrotropin releasing factor; TSH, thyroid-stimulating hormone.

as many rural people. Stresses are brief, the alarm reaction is often enough to see people through their emergencies, and the capacities of the resistance reaction are rarely exceeded. However, when people choose to live more frantic lives—as many urban people do—they may spend much of their time in the resistance reaction. If they are well nourished, with plenty of potassium and calories, they may never run into the exhaustion stage. But if they encounter additional stress—an accident, an illness, loss of a loved one, change in their profession, or a divorce—they may weaken or even die. Regular exercise and an adequate, balanced diet can help to ward off the effects of stress.

Carefully analyze these stages of the stress reactions and understand what is going on in each stage. Then go on to the Learning Focus Response exercise and dissect a truly stressful event.

Learning Focus Response

WHAT HAPPENS WHEN . . .

You are on an airplane flying over the Mediterranean. Suddenly, a terrorist leaps out of his seat and begins to wave a bomb over his head. The plane, he says, is to land. The local government must release several of his compatriots from jail and put them aboard. The plane will then fly to their home base. If the local authorities refuse to cooperate. . . . Well, you can guess what will happen then.

But what is happening right now? What is going on inside your body? What hormones are flooding your system? What are their effects?

The plane lands, and it sits there for hours, for days, while the terrorist repeats his demands and the local authorities delay and delay and delay some more. You know they are hoping to outlast the terrorist, waiting for him to collapse from exhaustion. They will then storm the plane, arrest him, and free you. But he is on amphetamines. He doesn't seem to need sleep.

What is happening inside your body now?

Finally, the local government gives in. The other terrorists arrive. The plane takes off and lands again, somewhere in Europe. But you and your fellow passengers are not freed. You are hostages, prisoners, fed on bread and water. Now what can you expect to happen inside your body if your ordeal does not end soon?

THE GONADS

Normal humans have two **gonads** apiece, **testes** for men and **ovaries** for women. These organs produce the gametes, the sperm and ova, necessary for sexual reproduction. They also produce the sex hormones, which regulate the reproductive process and those features of the body that go with it. We discuss these organs in detail in Chapter 17; here, we focus only on some of their endocrine functions.

The Testes

At puberty, the **interstitial cells** between the testes' sperm-producing tubules begin to produce the male sex hormone, **testosterone.** This hormone stimulates the development of those features—facial hair, additional body and pubic hair, deep voice, musculature—that mark the adult male. It also stimulates enlargement of the penis and the maturation of the sperm-producing tubules and seems to enhance aggressive and sexual behavior.

1. It was once common for European choirmasters to have their best boy sopranos castrated. The resulting *castrati* retained their high voices for the rest of their lives. Why?
2. The practice of castration is no longer in fashion. However, castration *is* sometimes proposed as a way to "rehabilitate" rapists and murderers. Why might it work? Do you approve or disapprove?

Testosterone secretion begins when a hypothalamic hormone signals the pituitary to secrete the gonadotropins, luteinizing hormone (LH) and follicle-stimulating hormone (FSH). One gonadotropin, LH, stimulates testosterone production. The other, FSH, stimulates sperm production.

Within the testes' tubules, certain cells (Sertoli cells) secrete the hormone **inhibin** when enough sperm are on hand for reproductive purposes. Inhibin inhibits gonadotropin secretion, both in the hypothalamus and in the pituitary.

The Ovaries

The female sex hormones are **progesterone** and **estrogen.** At the onset of puberty, when the sex hormones are "switched on," estrogen causes the female sex organs to mature and stimulates the female body to acquire its distinctive characteristics. It stimulates the enlargement and development of the breasts, the broadening of the pelvis, and the thickening of subcutaneous fat layers. It also stimulates a final period of rapid skeletal growth, causing the young girl suddenly to become taller. However, this growth period is fairly brief, for it causes the epiphyseal growth zones in the

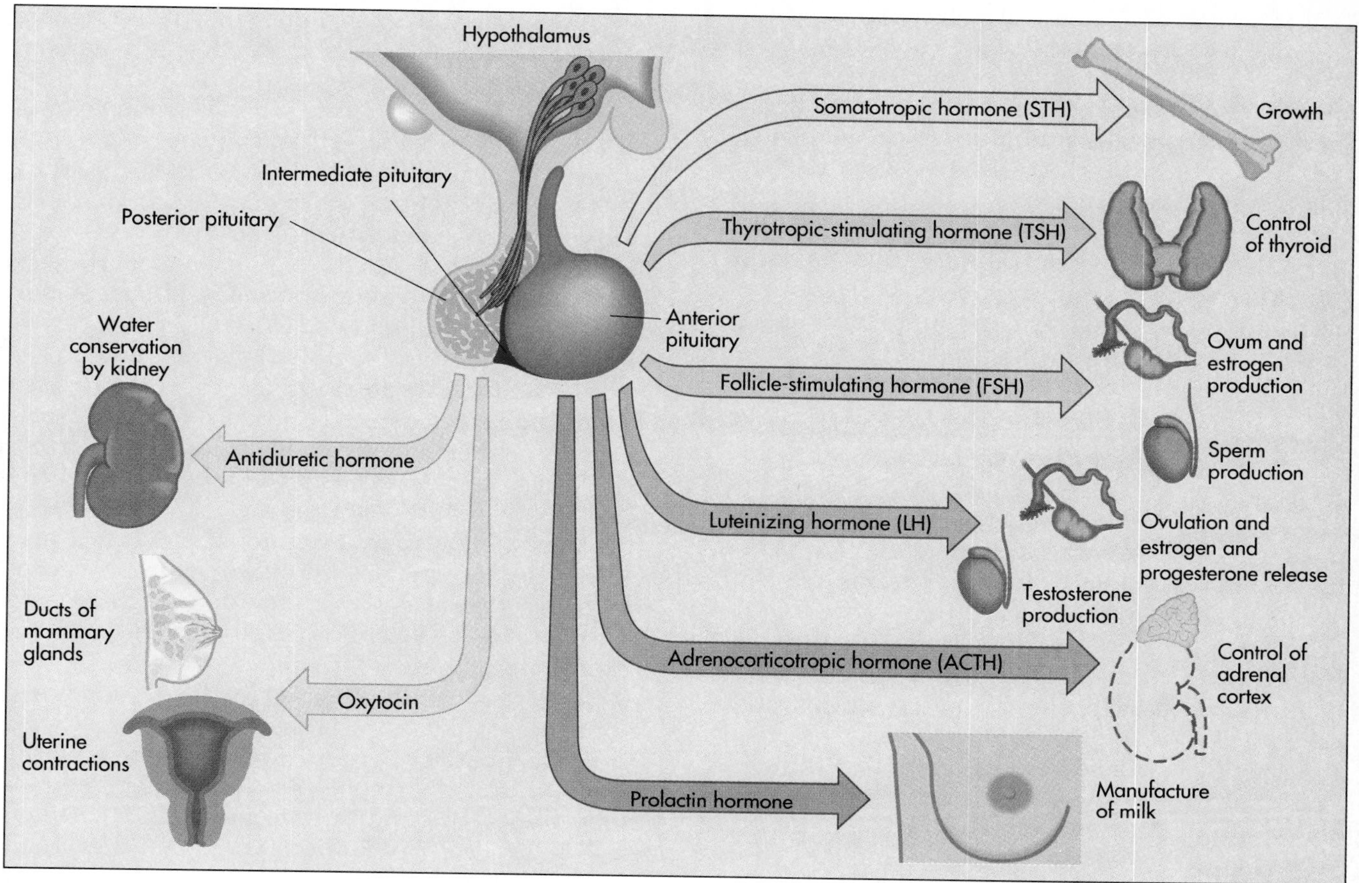

FIGURE 13-15 The lobes and hormones of the pituitary.

limb bones to shut down or ossify, preventing her from continuing to become taller. Males experience a similar growth spurt when they mature. But because males mature about two years later than females, they have an additional two years of childhood growth to add to their stature before their own hormones set off their final phase of skeletal growth. Adult males thus tend to be taller than adult females. The estrogens also stimulate the growth of pubic hair, but the face remains bare and the voice remains high; those features change only under the influence of testosterone (or adrenal androgen).

We discuss in Chapter 17 the way estrogen and progesterone influence the female's monthly reproductive or menstrual cycle. In Chapter 18, we consider the role of **human chorionic gonadotropin (HCG),** which is produced by the developing placenta until this organ can take over the production of estrogen and progesterone from the ovaries. Late in pregnancy, the ovaries produce an additional hormone, **relaxin.** Relaxin appears near the time of childbirth, and its function is to soften the connective tissues of the pelvis and birth canal to ease the passage of the infant from the uterus.

THE PITUITARY

The **pituitary,** or **hypophysis** (Figure 13-15), is a pea-sized organ located in one of the most protected areas of the human body. It is found in the cranial vault, roughly in the center of the head, above the roof of the mouth. It is suspended below the brain by a short stalk called the **infundibulum.** There are two main parts to the pituitary, each responsible for releasing its own set of hormones. The smaller **posterior lobe** is a distribution point for hormones actually produced in the hypothalamus of the brain, just above the pituitary's stalk. The larger **anterior lobe** is the main endocrine portion of the pituitary; many of its hormones control the activities of other endocrine glands (see Table 13-1 in the summary to this chapter). The anterior lobe's own activity is controlled by hormones secreted by the hypothalamus into a special network of blood vessels, the *hypophysial portal system.* In addition, a small zone, sometimes called the **intermediate pituitary,** partially separates the anterior from the posterior pituitary.

The Intermediate Pituitary

In lower animals, the intermediate pituitary secretes **melanocyte-stimulating hormone (MSH),** which promotes the synthesis of *melanin* (dark pigments coloring various skin structures) in the skin's pigment cells. It also causes stored pigment to spread out in the pigment cells and thus to darken the skin. Some researchers think MSH does not even exist in humans; certainly it is not crucial to health, though it may be involved in tanning and in the darkening of the skin during pregnancy.

The Anterior Lobe

The pituitary's anterior lobe secretes six major hormones that are essential to normal development or functioning of the body. It also contains other materials, such as endorphins and lipotropins, but it is not known if these other materials are technically operating as hormones. We will highlight each of the major hormones released by the pituitary (Figure 13-15).

Growth hormone (*somatotropin*) stimulates overall growth of the body. It causes cell division and multiplication, protein synthesis, the use of fat instead of carbohydrate for energy, and the conversion of liver glycogen to glucose and hence hyperglycemia. It seems to work by causing the liver to produce **growth factors,** including the *somatomedins*. The secretion of growth hormone is controlled by a hypothalamic "releasing factor"; it also controls its own levels by its own effects on the body (negative feedback). When control breaks down, the result is a growth disorder. Too much growth hormone in childhood overstimulates the mechanisms of growth, leading to *gigantism* (or giantism). Some giants have exceeded 8 feet 6 inches in height. Unfortunately, people thus afflicted are "lumbering" giants, not athletic ones, and they often have a host of medical problems that are likely to shorten their lives.

Growth hormone continues to be secreted after puberty, when it stimulates protein synthesis in muscle and the use of fats for energy. It does not cause the long bones to lengthen further, for their epiphyses have ossified. However, excess secretion of growth hormone can cause bones to thicken, leading to acromegaly, a condition marked by enlarged hands, feet, jaws, and joints and thickened eyebrow ridges (Figure 13-16).

Too little growth hormone during childhood produces *pituitary dwarfism*. In pituitary dwarfism, the skeletal elements (head, limbs, and trunk) remain well proportioned but are significantly smaller than average. Fortunately, pituitary dwarfism can be treated with injections of growth hormone, most of which is now obtained from genetically engineered bacteria (Figure 13-16).

Thyroid-stimulating hormone (TSH; thyrotropin) controls the activities of the thyroid gland. It stimulates growth of the thyroid gland and the production and release of its hormones. It is itself controlled by a hypothalamic releasing factor as well as by blood levels of thyroid hormone. Since TSH exerts its effects through its impact on the thyroid gland, malfunctions of the system express themselves as either hyperthyroidism or hypothyroidism, depending on whether too much or too little TSH is released.

Adrenocorticotropic hormone ACTH; (also called **adrenocorticotropin** or **corticotropin**) regulates the adrenal cortex's secretion of glucocorticoids and mineralocorticoids in response to another hypothalamic releasing factor.

The **gonadotropins** released by the pituitary regulate the activities of the ovaries and the testes. They affect both the production of gametes and the secretion of sex hormones by the gonads. They include **follicle-stimulating hormone (FSH),** which encourages the production of gametes in both sexes. In females, FSH promotes development of the ovum and the follicle supporting it. It also causes the ovarian follicle to secrete estrogens early in the monthly cycle. In the male, FSH stimulates sperm production in the seminiferous tubules. The other gonadotropin, **luteinizing hormone (LH),** stimulates sex hormone production and secretion in both sexes. It also causes release of the mature ovum in females. In females, it causes the ovaries to continue estrogen secretion and stimulates the progesterone release that completes uterine preparations in case pregnancy occurs. It also helps prepare the breasts for milk production. In males, LH causes the interstitial cells to produce and release testosterone.

Prolactin earned its name because its main effect is on the breasts. During pregnancy, it stimulates their growth and development in preparation for nursing a baby. Prolactin requires estrogen, progesterone, insulin, thyroxine, growth hormone, and adrenal cortical steroids for its full effect. Its production is limited by hypothalamic **prolactin inhibiting factor (PIF)** until the high levels of estrogen and progesterone typical of pregnancy in turn inhibit PIF production. At that time a hypothalamic releasing factor further boosts prolactin production. In addition, mechanical stimulation of the nipple by a nursing baby stimulates secretion.

The Posterior Lobe

The posterior lobe of the pituitary gland is not itself a true endocrine gland. It neither manufactures nor secretes its own hormones. Instead, it serves as a release point for hormones produced by cells in the brain's hypothalamus, just above the pituitary's stalk. These cells have long extensions, or *axons*, that reach from the hypothalamus, through the infundibulum, into the posterior lobe, where they end close to capillaries. Hormones migrate down the axons and accumulate in their tips. Under the right circumstances, the hypothalamic cells send nerve impulses down the axons; when they too reach the tips, they trigger the release of the hormones into the blood.

Currently, only two hormones are known to be released from the posterior lobe of the pituitary. One is **antidiuretic hormone (ADH), or vasopressin.**

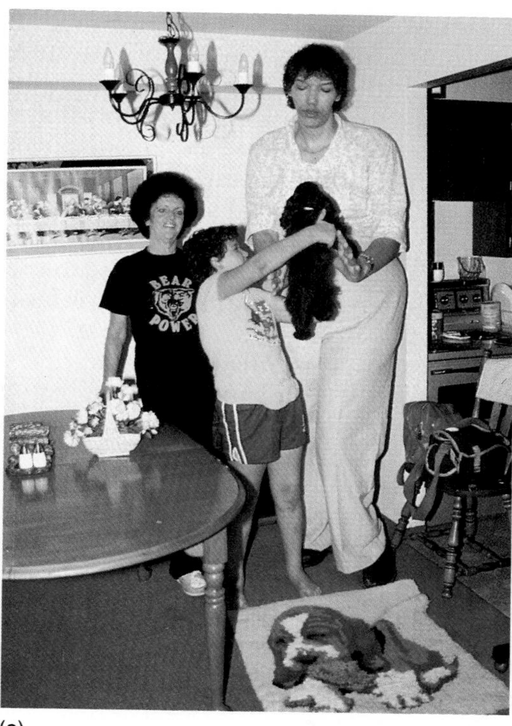

(a)

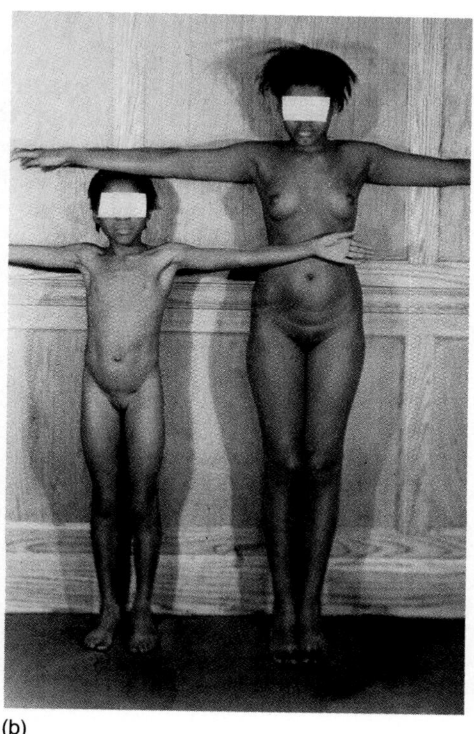

(b)

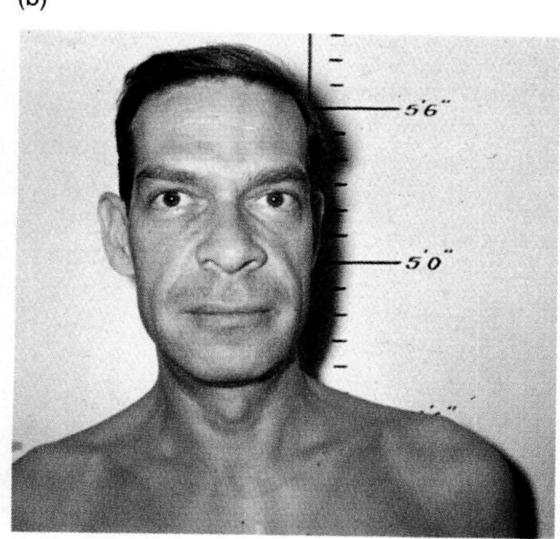

(c)

FIGURE 13-16 (a) Gigantism results from overproduction of growth hormone in childhood. The woman in the photograph is 7 feet 7.5 inches tall and weighs 480 pounds. Shown with her are her normal-sized mother and her 12-year-old brother. (b) Insufficient amounts of growth hormone during childhood result in pituitary dwarfism. The symptoms include a well-proportioned but childlike body. The two girls in the photo are both $12\frac{1}{2}$ years old. The one on the right is normally developed for her age. The one on the left remains childlike and tiny. (c) Acromegaly occurs when too much growth hormone is released after long-bone growth has stopped at the end of childhood. It results in the thickening of bones and joints, especially in the hands, feet, and head; notice also the increased thickness in the eyebrow ridges and the jaw length in the x-rayed skull.

It is secreted when the hypothalamic cells detect a decrease in the water concentration in the blood plasma. The hypothalamus activates neurons with axons terminating in the posterior pituitary, stimulating them to release ADH to the nearby capillaries. Once in the blood, ADH sets off a variety of physiologic mechanisms that collectively encourage the body to conserve water. The ADH decreases urine volume by enhancing water recovery from the nephrons. It also cuts back on the water lost via the sweat glands. In addition, it can raise blood pressure by constricting arterioles. All these effects can be important in the event of blood loss. If the blood is too dilute—as after drinking a soft drink—ADH production shuts down and urine volume increases greatly. When ADH production shuts down permanently (as after damage to the hypothalamus), the result is **diabetes insipidus.** Like diabetes mellitus, this disease is marked by the loss of immense quantities of urine (as much as 10 times the normal output); however, the urine does not contain sugar so it is not sweet (it is tasteless or "insipid"). The condition is treated effectively by injections of ADH.

The posterior lobe's second hormone is **oxytocin.** It is secreted at the end of pregnancy, during labor, in response to the pressure of the infant against the opening of the womb (the cervix). As the pressure increases, nerve impulses from the cervix signal the hypothalamus to release oxytocin. The oxytocin then stimulates contractions of the uterine muscles, which further increase the pressure on the cervix. Oxytocin thus provides us with one of the few examples of a *positive-feedback* system within the human body.

1. Antidiuretic hormone production varies with the time of day. Contemplating the patterns of your own life, determine when your ADH production is greatest. When is it least?
2. Both diabetes mellitus and diabetes insipidus are marked by increased production of urine. How might you—*without laboratory tests*—distinguish between the two conditions? (Would you believe that this is precisely what old-time physicians used to do?)
3. In positive feedback, a cause exerts an effect that increases (not decreases) the cause. But the process cannot go on forever. (Why?) What event must halt the ever-increasing release of oxytocin during labor? Why?

Oxytocin also affects the release of milk from the breasts. A nursing baby stimulates nerve signals from the nipple that tell the hypothalamus to release the hormone. It then causes smooth muscle cells in the milk glands to contract and squeeze milk into the ducts leading to the nipple.

HYPOTHALAMIC CONTROL

The hypothalamus controls the secretions of the anterior lobe of the pituitary somewhat less directly than it does those of the posterior lobe. It uses the releasing and inhibitory factors we have mentioned. It secretes these hormones into the local transport system, the hypophysial portal system, which links the hypothalamus and anterior lobe. It does so when its cells detect changes in the blood, such as a decline of blood sugar, thyroxine, or sex hormones. To reverse these changes and maintain homeostasis through negative feedback, the hypothalamus then secretes appropriate **releasing** or **inhibiting factors.**

Growth hormone is controlled by both *growth hormone releasing factor (GRF)* and *growth hormone inhibiting factor (GIF)*. Also known as somatostatin, GIF is secreted in the islets of Langerhans in the pancreas as well as in the hypothalamus.

Thyroid-stimulating hormone (thyrotropin) is controlled by *thyrotropin releasing factor (TRF)*. The ACTH is controlled by *corticotropin releasing factor (CRF)*. The gonadotropins LH and FSH are controlled by *gonadotropin releasing factor (GnRF)*. Prolactin is controlled by both *prolactin releasing factor (PRF)* and *prolactin inhibiting factor (PIF)*.

This system of hypothalamic control makes many of the body's hormones directly controllable by the brain. The hypothalamus receives nerve signals from other parts of the brain, including those parts responsible for thought or stress. It is thus possible for the brain to affect the body in many ways once thought impossible. For instance, a child deprived of human interaction suffers from *marasmus,* or failure to thrive. It does not grow, perhaps because it is not producing growth hormone. Similarly, adult women who strongly desire to have children sometimes experience *pseudocyesis,* or false pregnancy, perhaps because their brains are skewing the secretion of GnRF and gonadotropin. Brain control of hormone secretion may also be involved in many other *psychosomatic* conditions. It is definitely involved in the body's physical responses to psychologic stress (see the preceding Learning Focus).

HORMONELIKE SUBSTANCES

The body produces a number of substances that resemble hormones in their effects but are not secreted by glands. **Prostaglandins,** related to the fatty acids in structure, are like the hypothalamic hormones in that they act very near the site of their release. They are *local* or *tissue hormones,* and they are produced by all or most of the body's tissues.

There are a great many prostaglandins, and they have many effects in the body. They modify the effects of true hormones; stimulate smooth muscle contraction; raise and lower blood pressure; affect immune responses, respiration, and reproduction; and cause inflammation, fever, pain, and sedation. They have thus evoked a great deal of medical interest. Prostaglandins themselves have been used to induce labor, and drugs that block their synthesis have proved useful for controlling menstrual cramps. Aspirin is an effective pain killer because it inhibits prostaglandin synthesis. Researchers expect that the prostaglandins and prostaglandin synthesis inhibitors will also be useful in treating arthritis, kidney disease, glaucoma, and many other diseases.

Other hormonelike intercellular messengers include the **kinins,** substances released by damaged cells to attract white blood cells and aid the development of inflammation; **somatomedins,** growth factors that aid the working of growth hormone; **interleukins** and **leukotrienes,** which modulate the actions of the immune system; and the **endorphins** and **enkephalins,** one of whose main roles seems to be to modify pain sensitivity. These substances too should prove useful in medicine, either directly or through drugs that block their effects.

HOW HORMONES WORK

At the beginning of this chapter we mentioned that hormones work by interacting with receptors in the membranes of their target cells. This method is true for basically all hormones except the steroids, that is, the sex hormones and the adrenal cortical hormones. These substances—which are lipids—dissolve in the lipid of the cell membrane and pass into the cell's interior. They then bind to protein receptors *within* the cytoplasm, and the protein-hormone complex moves by diffusion into the cell's nucleus. The hormone exerts its effect when the complex interacts with the deoxyribonucleic acid (DNA) to activate specific genes. The active genes are transcribed into messenger ribonucleic acid (RNA), which is translated into new protein. The end result is that the hormone causes the production of specific enzymes, which then carry out the "intended" action of the hormone (Figure 13-17). Sex hormones, for instance, cause cell growth, and glucocorticoids cause the production and release of glucose in this manner.

The peptide hormones (insulin, releasing factors, etc.) are peptides (small proteins). They, as well as epinephrine, norepinephrine, and the prostaglandins, work by binding to membrane receptors. Each hormone affects only cells with the appropriate receptors. The receptors thus serve to define the "target" cells; they, not the hormones, provide the hormones' specificity of action.

Many receptors are closely linked to enzymes in the cell membrane. When a hormone binds to such a receptor, the complex activates the enzyme. The enzyme then produces a "second messenger" within the cell, which activates other enzymes that carry out the "first messenger's," the hormone's, ultimate effect (Figure 13-18). Most often, the enzyme converts adenosine triphosphate (ATP) into **cyclic adenosine monophosphate (cAMP).** It may also create other second messengers, such as cyclic guanosine monophosphate. Prostaglandins affect hormone function by influencing the formation of second messengers such as cAMP.

Some hormones use both cAMP and another second messenger, calcium ions. The cAMP or the activated receptor opens membrane pores that let calcium ions enter the cell. The ions then bind to **calmodulin,** a specialized protein that, when activated by calcium, can in turn activate intracellular enzymes.

In some cases, calcium ions may serve as the only second messenger. In other cases, peptide hormones may actually enter cells and bind to intracellular receptors, much like the steroid hormones. There is evidence that insulin, growth hormone, and prolactin can do this, though they also bind to membrane receptors. Thyroxine, on the other hand, has a unique feature. It is a peptide, but it enters cells and binds to receptors in their nuclei as well as to receptors in the cytoplasm. It can thus directly activate genes.

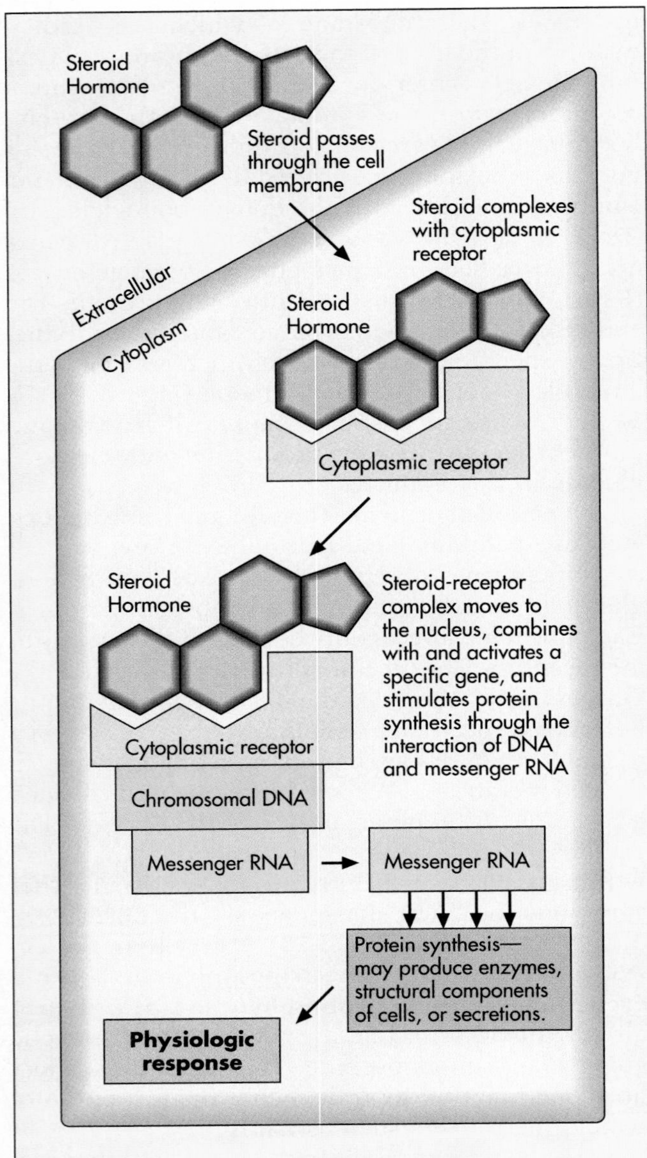

FIGURE 13-17 Mechanism of action of steroid hormones.

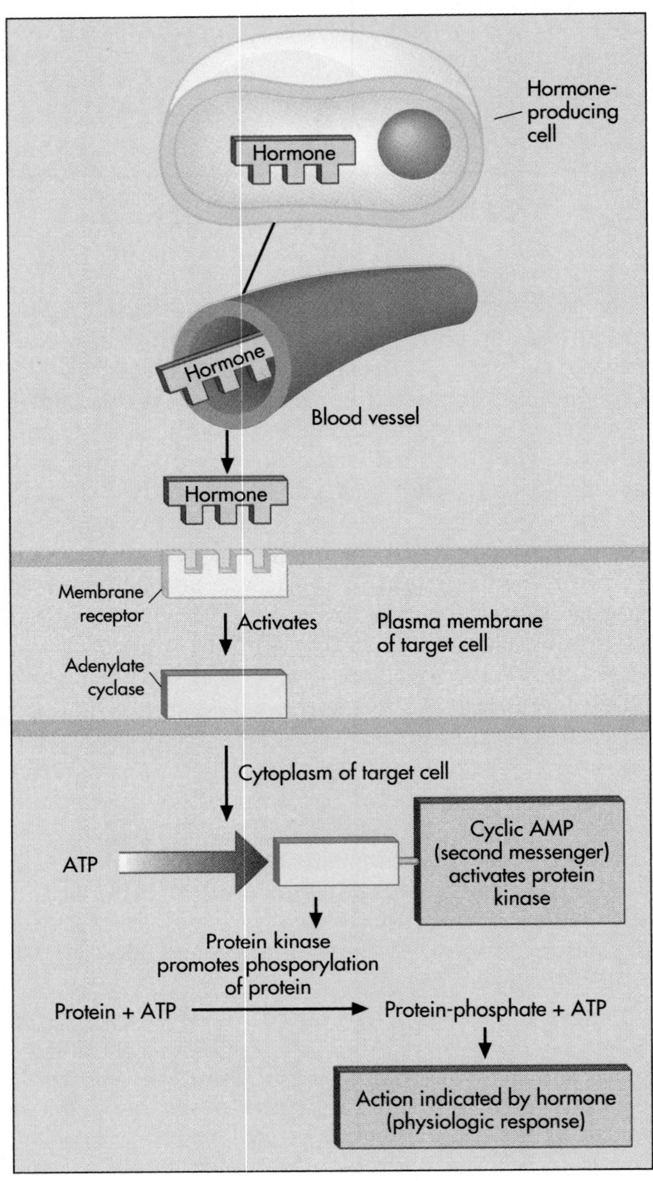

FIGURE 13-18 Mechanism of hormone action via cAMP.

SUMMARY

The body's cells coordinate their efforts by communicating with each other through two different, but related, systems. One is the nervous system. The other is the endocrine system, whose endocrine glands put chemical messengers—the hormones—into the blood. The blood then carries the hormones to target organs and cells throughout the body. When hormones reach their targets, they bind to receptors through which they exert their effects. These effects include growth and metabolism, mineral balance, reproduction, blood pressure, heart rate, digestion, stress, and the secretion of other hormones.

The endocrine glands are the pineal, thyroid, parathyroids, thymus, pancreatic islets of Langerhans, adrenal cortex and medulla, gonads (testes and ovaries), pituitary, and hypothalamus. In addition, several other organs also secrete hormones; they include the placenta, heart, kidney, and small intestine. We discuss the hormones of the heart and small intestine elsewhere. The others are summarized in Table 13-1.

Table 13-1

A Roster of Hormones

HORMONE	SOURCE	EFFECT ON	CONTROLLED BY
Melatonin	Pineal gland	Pigment, reproductive timing	Light
Thyroxine	Thyroid	Metabolic rate, development	Pituitary TSH
Calcitonin	Thyroid	Calcium and phosphate deposition	Calcium level
Parathormone	Parathyroids	Calcium and phosphate release, absorption	Calcium level
Thymosin	Thymus	T cell maturation	Unknown
Insulin	Pancreas, beta cells	Glucose absorption and use	Glucose level
Glucagon	Pancreas, alpha cells	Glucose release	Glucose level
Aldosterone	Adrenal cortex	Mineral balance, blood pressure	Angiotensin II, renin, blood pressure
Glucocorticoids	Adrenal cortex	Energy production	ACTH
Androgen	Adrenal cortex	Masculine features	Unknown
Renin	Kidney	Blood pressure	Blood pressure
Renal erythropoietic factor	Kidney	Red cell production	Hypoxia
Epinephrine, norepinephrine	Adrenal medulla	Stress response	Autonomic nervous system
Testosterone	Testes	Masculine features, maturation	LH
Inhibin	Testes	Sperm-production inhibition	Sperm inventory
Estrogens	Ovaries, placenta	Feminine features, maturation, menstrual cycle	FSH
Progesterone	Ovaries, placenta	Menstrual cycle, pregnancy	LH
Relaxin	Ovaries	Birth canal (softening)	Timing (end of pregnancy)
HCG	Placenta	Estrogen, progesterone	Pregnancy

(Continued on p. 306)

HORMONE	SOURCE	EFFECT ON	CONTROLLED BY
MSH	Pituitary intermediate lobe	Pigmentation	Light? Pregnancy?
ADH	Pituitary posterior lobe	Water excretion	Blood dilution
Oxytocin	Pituitary posterior lobe	Uterine contraction, milk release	Pressure on cervix, suckling
Growth hormone	Pituitary anterior lobe	Growth	GRF
TSH	Pituitary anterior lobe	Thyroid	TRF
ACTH	Pituitary anterior lobe	Adrenal cortex	CRF
FSH and LH	Pituitary anterior lobe	Gonads	GnRF
Prolactin	Pituitary anterior lobe	Milk production	PIF, PRF, suckling
GRF, GIF, TRF, CRF, GnRF, PRF, PIF	Hypothalamus	Anterior pituitary hormones	Negative feedback from target hormones and their targets

Most hormones are controlled by negative-feedback mechanisms; that is, their effects on the body produce stimuli that reduce their secretion. For instance, the pancreas secretes insulin when blood glucose levels are high. Insulin then causes muscle, fat, and other cells to remove glucose from the blood. As the blood glucose level falls, the pancreas secretes less insulin. The only exception to this pattern is oxytocin, whose secretion is stimulated by pressure on the cervix during childbirth (labor). It causes uterine muscles to contract and thus to increase the pressure; hence its own secretion is an example of positive feedback. Secretion ends only when the pressure ends, after the child is born.

Many hormones, such as thyroxine and the sex hormones, are controlled by hormones secreted by the anterior lobe of the pituitary, which are in turn controlled by releasing factors, or hormones, released by the hypothalamus. In such cases, feedback mechanisms work through both the pituitary and the hypothalamus.

Through hypothalamic control the brain—and the "mind"—has a way to control hormone secretion. This brain-gland link explains many psychosomatic conditions as well as the body's response to stress. When the body is damaged or anticipating damage, the hypothalamus emits nerve signals that tell the adrenal medulla to secrete epinephrine and norepinephrine. Hypothalamic hormones stimulate the release of growth hormone, adrenocorticotropic hormone (ACTH), and thyroid-stimulating hormone (TSH). The overall effect is to prepare the body for action, in fighting or fleeing the damaging agent, and to cope with blood loss and infection.

The body's response to stress is called the general adaptation syndrome (GAS). It has three stages: the initial alarm reaction, the longer resistance reaction, and the final exhaustion stage, which appears when the body has used up its resources. The GAS effects account for many stress diseases such as ulcers and heart attacks.

Most hormones work by binding to receptors in target cell membranes. The receptor-hormone complex then activates an enzyme that creates within the cell a second messenger such as cyclic adenosine monophosphate (cAMP). The cAMP then activates other enzymes that carry out the hormone's effect.

The steroid hormones and thyroxine pass through the cell membrane and bind to receptors within the cell. The receptor-hormone complexes then interact with deoxyribonucleic acid (DNA) to stimulate the synthesis of specific enzymes.

Table 13-2 provides a summary of selected diseases of the endocrine system.

Table 13-2

Clinical Summary: Selected Diseases of the Endocrine System

DISORDER	DESCRIPTION
Acromegaly	Result of excess secretion of growth hormone by anterior pituitary after closure of epiphyseal plates in long bones. Symptoms include striking enlargements of hands, feet, jaw, and various joints; heavy eyebrow ridges; abnormal curvature of spine; and coarse, thickened, deeply furrowed skin.
Addison's disease	Chronic failure of the adrenal glands to secrete adrenal corticoids (usually caused by destruction of adrenals). Symptoms include progressive weakness; pigmentation of the skin (especially around nipples, genitals, face, and hands); very low blood pressure, gastrointestinal involvement, including nausea, vomiting, and diarrhea.
Adrenal virilism, "adrenogenital syndrome"	Result of excess release of masculinizing hormones (androgens) from the adrenal cortex. Can cause precocial sexual development in children (early appearance of adult genitals in very young boys and development of facial and body hair in both sexes). Can also enhance muscular development.
Cretinism	Hypothyroidism in the fetus and infant, often caused by failure of thyroid to develop and secrete thyroxine. Skeletal system, brain, and sexual organs do not develop properly. Cretins are often severely retarded and have rough, dry skin, very coarse hair, and a large, protruding tongue.
Cushing's syndrome	Result of overproduction of glucocorticoids by adrenal cortex. Confusing array of symptoms, including "full moon face;" fat deposits on head, neck, and trunk; excess body hair in females; sexual atrophy; and muscular weakness. Similar symptoms may accompany prolonged therapeutic use of ACTH or cortisone (as for treating arthritis).
Diabetes mellitus, "sugar diabetes"	Result of the pancreas's inability to produce insulin or body cells' inability to respond to insulin. Symptoms include polyuria (dramatic increase in urine volume); glycosuria (sugar in the urine); hypertension; excessive hunger and thirst; marked weakness and weight loss; inability to metabolize fat; acidosis; gallstones; diabetic gangrene; coma; and if untreated, death.
Giantism, gigantism, or hyperpituitarism	Usually caused by excess secretion of growth hormone by anterior pituitary during years of skeletal growth (prior to closure of epiphyseal plates in long bones). Generally caused by adenoma (tumor) of the pituitary. "Giants" have exceeded 8 feet 6 inches in height.
Graves' disease	Result of hyperactive thyroid, more often in women than in men. Most conspicuous symptom is exophthalmic goiter (protrusion of eyeballs). Neck goiters are also common. Patients seem very tense or nervous and have hand tremor, rapid pulse, moist skin, and very high metabolic rate.
Myxedema	Hypothyroidism in adults and older children. Caused by dietary iodine deficiency or surgical removal or degeneration of thyroid. Symptoms include puffiness of hands and face; slow speech; inability to concentrate, mental apathy; drowsiness; sensitivity to cold; and sometimes loss of hair.
Pituitary dwarfism, or hypopituitarism	Result of too little secretion of growth hormone during childhood growth years. Patient remains bright, well proportioned, and graceful but sexually undeveloped and small.

STUDY QUESTIONS

1. Discuss the body's need for intercellular communication. How does it resemble the human use of speech to communicate? How does it differ from speech?
2. Why does a hormone such as FSH communicate only with cells of the gonads?
3. Why does a hormone such as thyroxine or growth hormone communicate with cells throughout the body?
4. What is the difference between an endocrine and an exocrine gland?
5. Explain how negative feedback works as a control mechanism. Use either glucose or calcium metabolism as an example.
6. Explain how positive feedback works. Give an example.
7. Outline the renin-angiotensin-aldosterone mechanism of blood pressure control.
8. Describe the general adaptation syndrome.
9. How can the brain, via the hypothalamus, cause high blood pressure?
10. Why is the pituitary's posterior lobe not a true endocrine gland?
11. The hypothalamus is part of the brain, but it is also an endocrine gland. Why?
12. Explain the concept of the second messenger.

GLOSSARY

Adrenal cortex Outer layer of adrenal gland; source of steroid hormones.

Adrenal medulla Inner portion of adrenal gland; source of epinephrine (adrenalin) and norepinephrine (noradrenalin).

Alpha cells Pancreatic cells secreting glucagon.

Beta cells Pancreatic cells secreting insulin.

Cyclic adenosine monophosphate (cAMP) A chief intercellular carrier of hormone signals; a chief second messenger.

Endocrine glands Organs that secrete hormones into the blood; ductless glands.

Endorphins Modulators of pain sensitivity.

Enkephalins Modulators of pain sensitivity.

General adaptation syndrome (GAS) The body's response to stress.

Hormones Peptides, steroids, and other chemicals carried in the blood as intercellular messengers.

Inhibiting factors Hypothalamic hormones inhibiting release of anterior lobe hormones.

Islets of Langerhans Endocrine portions of pancreas.

Ovaries Female gonads.

Parafollicular cells Cells in wall of thyroid follicle that secrete calcitonin.

Parathyroid glands Small glands on the back of the thyroid; secretors of parathormone.

Pituitary gland (hypophysis) Endocrine gland suspended below the brain and above roof of mouth; many of its hormones control other endocrine glands.

Prostaglandins Fatty-acid-like substances that act as local, short-range hormones. Sometimes described as "tissue hormones."

Receptors Proteins and glycoproteins in cell membranes, cytoplasm, and nucleus that bind to hormones.

Releasing factors Hormones produced in the hypothalamus that stimulate the release of hormones from the anterior lobe of the pituitary.

Stress Damage or threat of damage to the body (physical or psychologic).

Testes Male gonads.

Thymus gland Located at the base of the neck; involved in maturation of immune system T cells.

Thyroid gland Secretor of thyroxine; located in the neck below the larynx.

Chapter 14

High-Speed Regulation: The Nervous System

It was almost as if it were happening in slow motion. I was watching my son, who had lost his balance, fall backward off a precipitous cliff into a deep glacial valley in the eastern Sierras. His eyes grew larger with realization. His arms began waving in graceful arcs as his face retreated from me. There was no sound. Everything was muted—only his face and shoulders were crystal clear. I was surprised to see my arms slowly entering the picture, my hands reaching out carefully, almost finger by finger, grasping the shoulder straps on his backpack. He was feather-light when I literally yanked him upright the split second before he could fall beyond my reach. The rest of my feelings you can guess.

This event is just one illustration of why the body needs a high-speed response system. These high-speed responses are mediated by electrochemical signals distributed by the conductive cells of the **nervous system.** Some of these nerve impulses can travel 130 meters per second (m/sec; 291 mph) from one area of the body to another. Even though the events just described seemed to be happening in slow motion, the situation was perceived and analyzed, and the responses were activated, so quickly that the feeling of "slowness" was merely an illusion.

In the last chapter we learned of our slow-speed response system, the endocrine system. Hormones are slow because they must travel in the blood, limited by the speed of blood flow, to their sites of action. Many hormones work by causing their target cells to make more of certain enzymes, and this process takes time as well. An additional delay arises because the hypothalamus and pituitary must respond to the internal events that signal the need for more hormone. It can take hours for the body to respond in this way. Obviously a faster system is also needed to handle problems when the speed of response is important.

The nervous system responds to events outside (and inside) the body in seconds at most. With simple events, such as stepping on a thorn or touching a hot pot, it produces the appropriate response—yelling "Ouch!" and lifting the foot or dropping the pot—in about a quarter of a second, an interval known as the human *reaction time.*

It can respond so quickly because the nervous system's cells are specialized for processing information and sending messages from one part of the body to another. Each cell has long, tubular extensions of its cell body that use a combination of electrical and chemical activity to carry signals from one end of the nerve cell to the other. The signals "leap" from nerve cell to target cell when a signaling cell secretes *neurotransmitters* that, like hormones, bind to receptors in the target cell's membrane. Typical target cells activated by the nervous system include other nerve cells, muscle cells, and gland cells (both endocrine and exocrine). The nervous system is thus responsible for its own activity, for visible actions such as running and swimming, and even for many actions of the endocrine system.

DIVISIONS OF THE NERVOUS SYSTEM

We will discuss the structure and function of nerve cells later in this chapter. For now, let us concentrate on the nervous system as a whole, beginning with its major divisions. These divisions are more or less arbitrary, since the nervous system functions as an integrated whole.

You already know some of these parts: You have a brain, a spinal cord, nerves, and sense organs such as the eye and ear. To understand how these parts fit together, think of the brain as a compact central computer and of the nerves as input and output lines. Attached to the brain is the spinal cord, which is both a somewhat less sophisticated computer and a distribution channel for signals to and from the brain and the other organs of the body. Both the brain and the spinal cord functionally occupy central positions in the body—the brain in the head and the spinal cord down the middle of the back. They thus make up the **central nervous system (CNS).** Their importance to the body is suggested by the way they are shielded by the bony skull and vertebrae.

It is in the CNS that most "nervous activity" takes place. The CNS receives sensory information, integrates the data from one sense with those from others and with "experience," and calculates and activates the appropriate responses. Responses may be internal (a thought or feeling or the formation of a memory) or external (signals from the brain to the body's muscles and glands). Consider what happens when, walking down a dark street, you see a shadowy figure lurking, partially hidden, in a doorway. You have a thought, form a memory, activate your adrenal medulla (for the hormonal fight-or-flight response), and, at the very least, are likely to change your course.

Nerves are cables that carry signals to and away from the brain and spinal cord. They are bundles of the thin extensions or conducting processes (nerve fibers) of neurons (nerve cells). Since they conduct signals between the CNS and the body's periphery, they collectively form the **peripheral nervous system (PNS).**

The PNS has two sections. The **afferent system** conducts sensory signals from the eyes, the ears, and other sense organs *toward* the CNS (see Chapter 16). Many afferent, or sensory, nerve cells have their cell bodies in small clusters of cells, or **ganglia,** lying just outside the CNS. In contrast, the **efferent system** carries motor signals *away from* the CNS to muscles and glands throughout the body. Most nerves carry both sensory (afferent) and motor (efferent) fibers and for that reason are called *mixed nerves*.

The efferent division of the PNS also has two segments. One serves the body's voluntary muscles and is under conscious control; it is the **somatic** (body) **nervous system.** The second controls glands, digestive organs, the heart and blood vessels, and other organs over which we normally have no conscious control; it is the **autonomic nervous system (ANS).** The **sympathetic** and **parasympathetic divisions** of the ANS generally oppose each other's effects on organs; one stimulates activity, the other inhibits. In general, the sympathetic division activates the various mechanisms of body defense and is employed when there is some kind of perceived danger or threat. On the other hand, the parasympathetic division promotes a state of physiologic well-being or homeostasis. Its actions encourage the body to save energy, slow down, promote digestion, and eliminate wastes. We will discuss the ANS in more detail in Chapter 15.

Figure 14-1 sketches the CNS and the PNS as they might appear if we could see them within the body. Figure 14-2 outlines the divisions of the nervous system in more schematic form in order to show the relationships of the afferent, efferent, somatic, and autonomic divisions of the PNS.

CELLS OF THE NERVOUS SYSTEM

Like the rest of the body, the nervous system is made up of various types of cells. However, its cells are of only two main types, the nerve cells themselves, the **neurons,** and the **neuroglia.**

The neuroglia are cells that support the conducting cells of the nervous system, the neurons, in many different ways. They are 5–10 times as numerous as neurons, smaller, and simpler in structure. Their relative simplicity allows them to divide and replace themselves throughout an

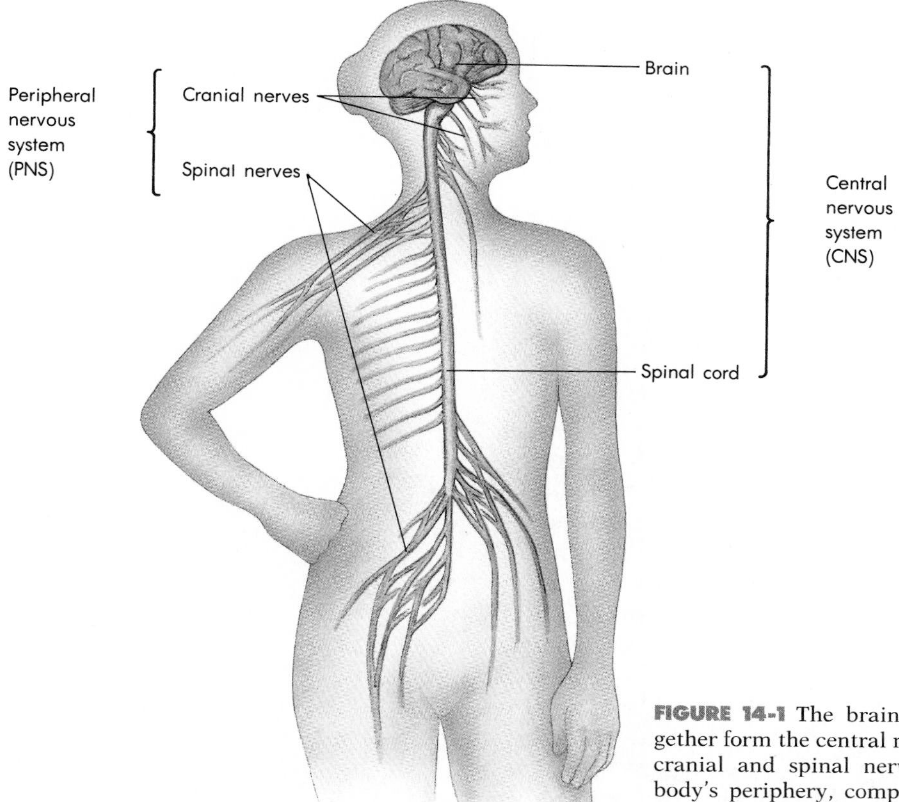

FIGURE 14-1 The brain and the spinal cord together form the central nervous system (CNS). The cranial and spinal nerves link the CNS to the body's periphery, composing the peripheral nervous system (PNS).

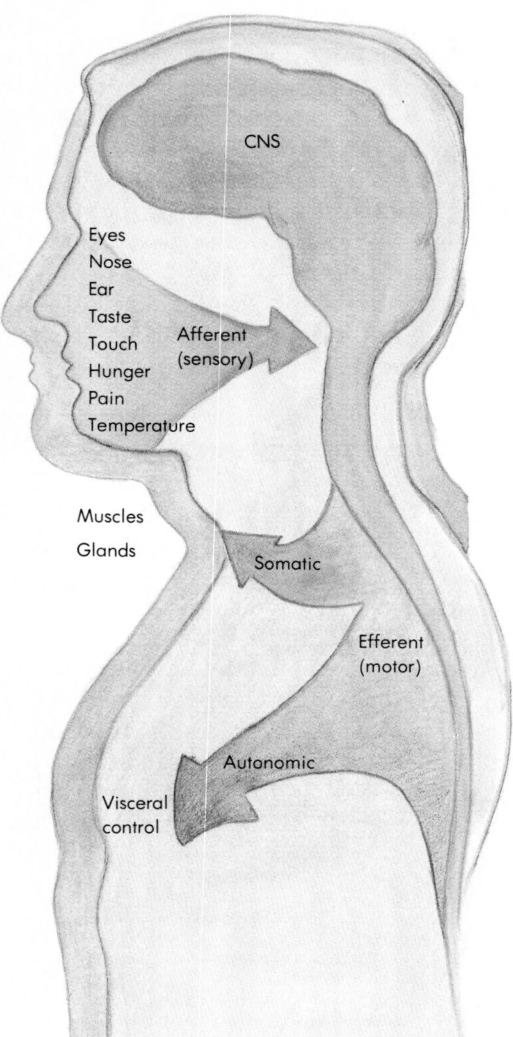

FIGURE 14-2 Afferent signals carry sensory information toward the CNS and from lower levels of the CNS (such as the spinal cord) to higher levels (such as the brain). Efferent signals travel from the CNS toward the body's glands and muscles.

individual's life. It is this continuing ability to divide that makes neuroglia the major causal elements of brain tumors; about 45 percent of intracranial tumors are caused by neuroglia.

The structural complexity of neurons prevents them from dividing continuously, and the maximum number of neurons in the body is attained within the first year or two of life. Thereafter, individual neurons can grow and increase the complexity of their interconnections, but they cannot usually multiply or replace themselves if destroyed. After about the age of 20, each of us loses about 100,000 neurons per day. Fortunately, we have some 10^{12} (one trillion) neurons. At that rate, it would take about 2500 years to lose a tenth of our supply, and surgical experience has shown that removal of a tenth of the brain can sometimes have only minor effects.

Neurons

The neurons are the functional units of the nervous system. They are the generators and processors of nerve signals and the conveyors of commands to the body's muscles and glands. They come in many shapes and sizes, each one serving a different function in the nervous system, but all share certain basic features.

Every neuron has a cell body containing the nucleus and the bulk of the cytoplasm. From the cell body extend two or more narrow, tubular extensions of the cell. One of these extensions conducts nerve signals *away from* the cell body; it is the cell's **axon.** The other (or others) conducts nerve signals *toward* the cell body; it is a **dendrite** (Figure 14-3).

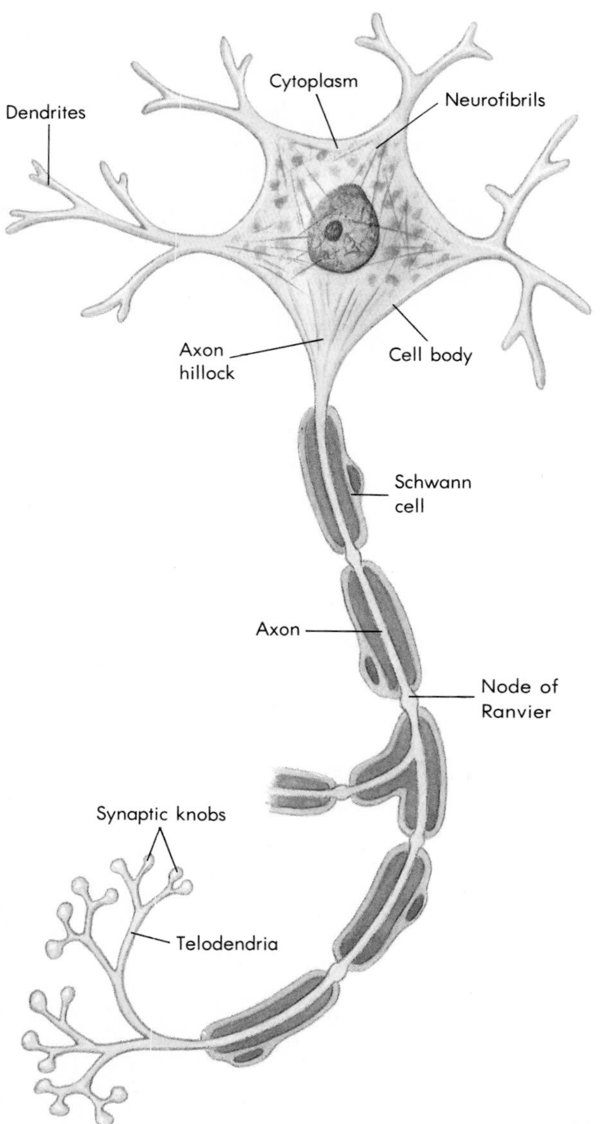

FIGURE 14-3 Diagram of neuron structure.

Both axons and dendrites may be enclosed in an insulating layer of myelin (see below). In the PNS, myelin is produced by Schwann cells.

Axons generally have relatively few branches until near their tips, where they may split into many fine filaments known as *telodendria*. Each telodendrion ends in a swelling, a **terminal button,** or **synaptic knob,** which contains numerous membrane-enclosed *synaptic vesicles*. It is at the synaptic knobs that nerve signals pass from neurons to their target cells.

When the target cell is another neuron, the synaptic knobs lie very close to the membrane of the target cell's body or dendrites, where they form *synapses* (see what follows). Dendrites generally branch profusely in order to provide a large surface area for many synapses and for receiving signals from many other neurons. The resulting convergence of many nerve signals on a single neuron allows each neuron to serve as a single integrator or processor of a great deal of information.

---

1. We said that the adult brain loses 100,000 neurons per day. Why is this not likely to make any significant difference during a normal human lifespan? What practical limits would it set on physical immortality?
2. Neurons do not simply pass on the nerve signals they receive. In fact, many generate nerve signals only when they receive two or more signals within a short period. How might this feature help a neuron process information? Why would a neuron's information-processing ability be improved if some of the signals it received canceled out other signals?

If we look at neurons under a light microscope, we would find it hard to tell dendrites and axons apart, for they resemble each other closely. Both are long, tubular extensions continuous with the cell body, and both contain *neurofibrils*, arrays of microtubules that help move newly synthesized proteins and other substances from the cell body toward the tips of the axons and dendrites. However, axons tend to be longer than dendrites (although dendrites on sensory neurons can also be quite long); the longest axons stretch from the brain to the lower end of the spinal cord and can be 1 meter (39.4 in.) long. Axons also have synaptic knobs at their ends and branch less profusely than dendrites.

We can see these differences well if we glance briefly at a few of the many kinds of neurons in the nervous system (Figure 14-4). Figure 14-4 shows a

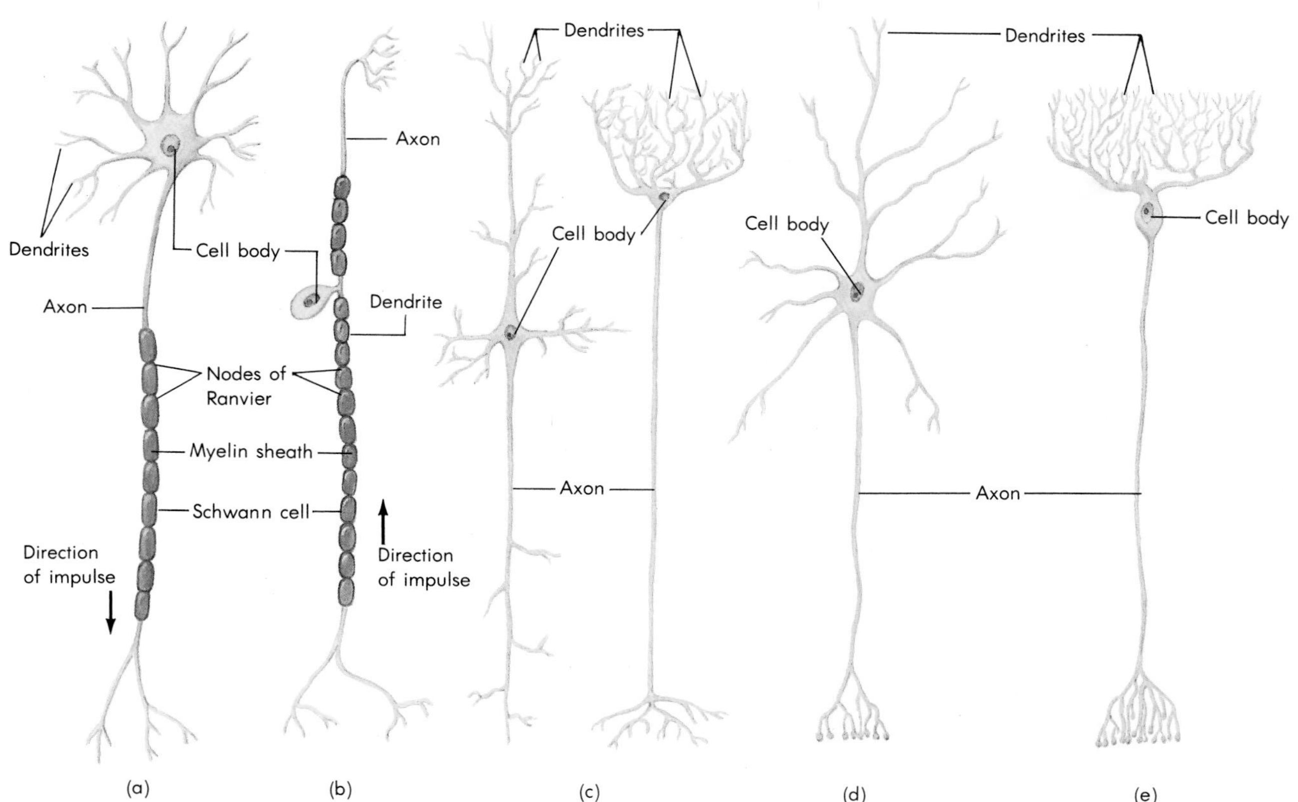

FIGURE 14-4 Types of neurons: (a) spinal motor neuron; (b) sensory neuron; (c) spinal interneuron; (d) pyramidal cell of cerebral cortex; (e) cerebellar Purkinje cell.

spinal motor neuron; its cell body and dendrites are in the spinal cord; its axon extends to a muscle. The figure also shows a sensory neuron; its cell body is in a small ganglion just outside the spinal cord; it has a single root from which both dendrite and axon branch, the dendrite extending to sense organs in the skin or elsewhere and the axon carrying sensory information into the spinal cord. Figure 14-4 shows a typical *interneuron*, or *association neuron*, in the spinal cord; both axon and dendrites are short. In addition, it shows a pyramidal cell from the surface of the brain. Figure 14-4 also shows a Purkinje cell from the cerebellum. Both pyramidal and Purkinje cells are brain association neurons.

Neuroglia

The word *neuroglia* means nerve glue. Although the cells that bear this name greatly outnumber the neurons, they play no known role in information processing or signal transmission. Rather, they surround, support, and protect the neurons. There are four types of neuroglia: *astrocytes*, *microglia*, *ependymal cells*, and *oligodendrocytes;* they are described in Table 14-1.

Some neuroglia help to form the **blood–brain barrier,** which keeps many foreign substances—toxins, viruses, and even some drugs—away from the brain's neurons. The barrier consists of three components: brain capillaries, basement membrane, and astrocytes. The wall cells of brain capillaries are closely packed; brain capillaries are surrounded by a relatively impermeable basement membrane (made of proteins secreted by connective tissue cells), and the brain capillaries themselves are surrounded by numerous astrocytes. Only small molecules, such as those of nutrients, can cross the barrier unaided. Others cross only by active transport, which is controlled by the astrocytes.

The microglia are the phagocytes of the CNS; they remove bacteria and the remnants of dead cells. The ependymal cells line the cavities within the brain and spinal cord; in certain regions, they are specialized to secrete *cerebrospinal fluid,* the fluid within the cavities. The oligodendrocytes form a sheath of **myelin** around the axons of CNS neurons. This sheath serves to insulate the axons against electrical "crosstalk" from neighboring axons. Its main function, however, is to speed the transmission of nerve impulses.

Myelinization

Myelin was first observed as a fatty sheath surrounding nerve fibers (axons and dendrites) in the nerves of the PNS. There it is formed by **Schwann cells.** Each Schwann cell anchors a portion of its cell membrane to the surface of the nerve fiber. It then wraps around and around the nerve fiber, leaving a spiral trail of cell membrane behind it. As it withdraws most of its cytoplasm from between the layers of membrane, it forms one segment of the *myelin sheath* that surrounds the entire length of the nerve fiber (Figures 14-3 and 14-5).

Table 14-1

Types of Neuroglia

TYPE	DESCRIPTION	ROLE
Astrocytes	Star (astro) shaped; numerous dendritelike processes	Support neurons in brain and spinal cord; link neurons to capillaries
Microglia	Small; few processes; derived from blood monocytes	Are the phagocytes of the CNS
Ependymal cells	Epithelial cells arranged in layers	Line cavities of brain and spinal cord; secrete cerebrospinal fluid
Oligodendrocytes	Like astrocytes, but fewer processes	Support; help form myelin in the CNS

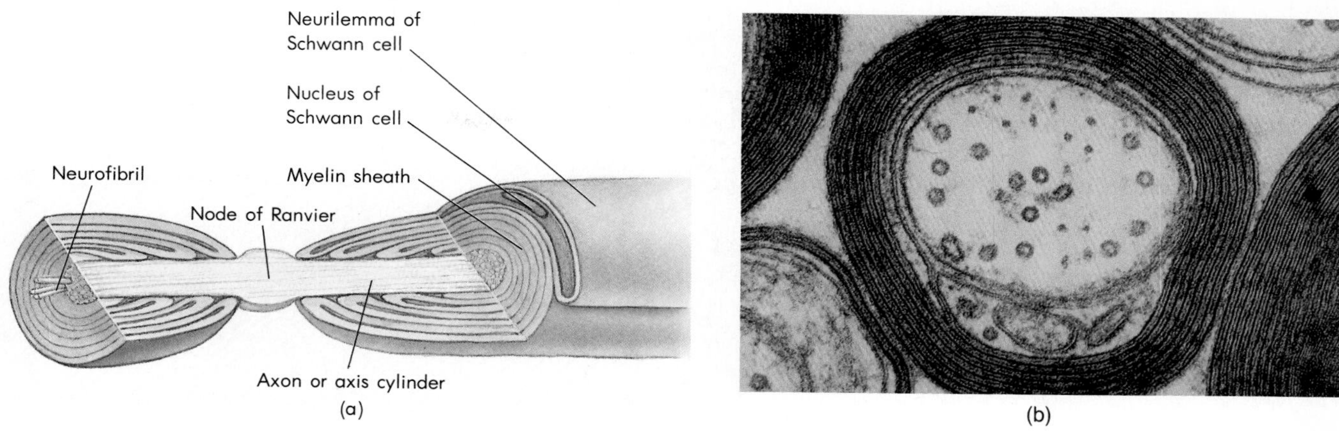

FIGURE 14-5 Formation of myelin. (a) In the PNS, Schwann cells produce the myelin sheath by growing around axons and dendrites. (b) Electron micrograph of a cross section of a myelinated axon in a sensory nerve.

Not all nerve fibers are myelinated in this way. However, all peripheral nerve fibers are surrounded by Schwann cells. Unmyelinated fibers are embedded in Schwann cells that do not coil around them (Figure 14-6).

Myelinated nerve fibers are not entirely covered by myelin. Each Schwann cell is very small, and it takes many of them to cover the length of a nerve fiber with myelin, each one covering a small segment of the fiber. Between each pair of Schwann cells, there is a small gap in the sheath known as a **node of Ranvier** (Figure 14-3). These nodes are crucial to the way myelin accelerates the conduction of nerve signals. In unmyelinated fibers, the signal must travel across every bit of the nerve cell membrane; in myelinated fibers, the signal leaps from node to node of Ranvier and travels much more quickly, up to 130 m/sec (290 mph). Unmyelinated fibers have conduction speeds of about 0.5 m/sec (1.13 mph).

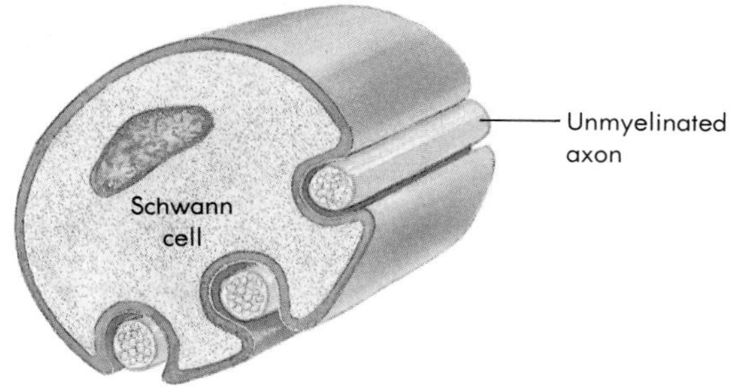

FIGURE 14-6 Unmyelinated axons embedded in recesses of a Schwann cell.

In the CNS, myelin also speeds conduction, but it is not formed by Schwann cells. Instead, oligodendrocytes wrap the ends of their tentaclelike processes around axons to form the myelin. Each oligodendrocyte myelinates several axons, and they are essential to normal function, as we can see from the effects of certain diseases. One of these diseases is *multiple sclerosis,* which destroys myelin in the CNS and interferes greatly with the ability to process information and control our bodies. The loss of myelin slows down the transmission of nerve signals and disrupts the timing of neural events on which many aspects of coordination depend. It also leads to some short-circuiting, as signals in one nerve fiber cross unwanted into other fibers. This can cause both sensory and motor disturbances. The sensory disturbances include numbness and tingling of the hands and feet. Motor problems include stiff clumsy walking, fleeting paralysis, tremors in the hands, and twitching movements of the eyes (nystagmus).

In the PNS, Schwann cells provide an additional benefit to speed. When a nerve is accidentally cut, the nerve fibers isolated from their cell's nucleus die and degenerate, but the Schwann cells remain and multiply, forming a tube along the line followed by the vanished nerve fiber. As a new nerve fiber grows out from the undamaged portion of the nerve cell, it can follow the tube to its end and reestablish connections with muscles and sense organs. When a nerve is badly damaged, the damaged portion can be cut out and replaced with a piece of a different nerve to aid regeneration.

In the CNS, the myelin-forming oligodendroglia do not form tubes after axons are cut. Axons of damaged cells of the CNS cannot regenerate. However, researchers have recently found that if a piece of peripheral nerve, with its Schwann cell tubes, is inserted to cross a damaged area, the CNS axons can follow the Schwann cell tubes and reestablish their former connections. There is thus some hope that one day we will be able to undo the permanent damage of spinal cord and brain injuries that sever axons and break nerve pathways to muscles and from sense organs.

Nerves

Peripheral nerves are bundles of myelinated and unmyelinated nerve fibers connecting the CNS—the brain and spinal cord—with muscles, glands, and sense organs throughout the body. But they are not simply bundles. Within each nerve, the nerve fibers are grouped into smaller bundles wrapped in membranous (connective tissue) coverings, as shown in Figure 14-7. Individual nerve fibers are surrounded by **endoneurium.** Groups or fascicles of fibers are wrapped in **perineurium.** Several fascicles form the whole nerve, which is sheathed in **epineurium.** Together the membranes make the nerve tough enough to weather the stretching and bending it must undergo as the body moves. You can feel that toughness by fingering the large nerve—your "funny bone"—that crosses your elbow.

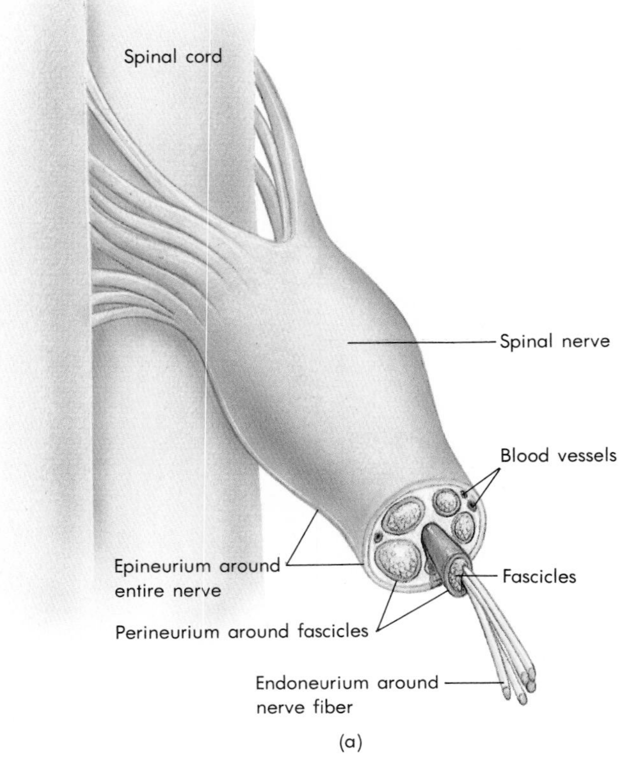

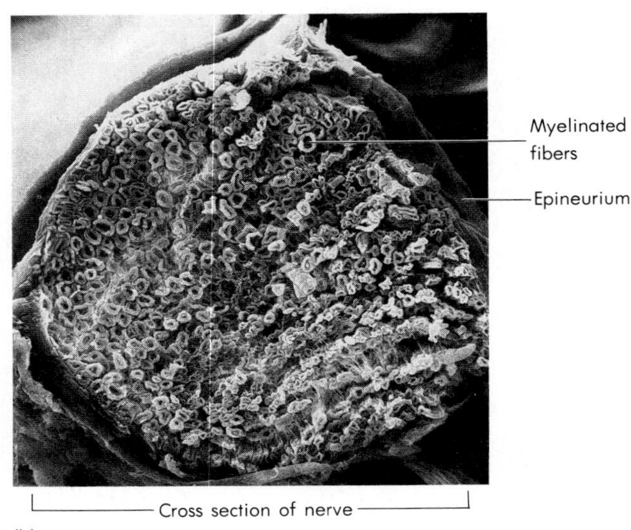

FIGURE 14-7 Nerve structure. (a) Diagram of nerve cross section. (b) Transected bullfrog auditory nerve shows the cablelike makeup of nerves (c × 1600). The outer membrane wrapping the bundles of nerve fibers is the epineurium. Each of the thousands of nerve fibers is encased in its own myelin sheath. Note that this nonmammalian nerve does not have its fibers arranged in individual bundles or fascicles.

ACTION POTENTIAL: THE NERVE SIGNAL

People often speak of nerve signals as if they were purely electrical. They are not, for they are not produced as flows of electrons down the length of a wire. Instead, they are produced as flows of electrically charged ions that move back and forth across a neuron's cell membrane. They are thus both electrical and chemical, or *electrochemical.* Fortunately, their mechanism is very simple in principle (though the details are complex).

Every cell has in its membrane protein "pumps" for moving various substances in and out of the cell. One of these pumps, plentiful in nerve and muscle cell membranes, is the **sodium-potassium pump.** Using adenosine triphosphate (ATP) as an energy source, it binds to a sodium ion (Na^+) inside the cell and a potassium ion (K^+) outside the cell and carries each ion across the cell membrane. For each sodium ion it transfers out of the cell, it transfers a potassium ion into the cell (Figure 14-8).

Potassium ions then diffuse out of the cell through protein pores, or *ion channels,* moving down their concentration gradient, some 100 times faster than sodium ions diffuse back in. Positive electrical charges thus accumulate outside the cell, and though some small negatively charged ions dif-

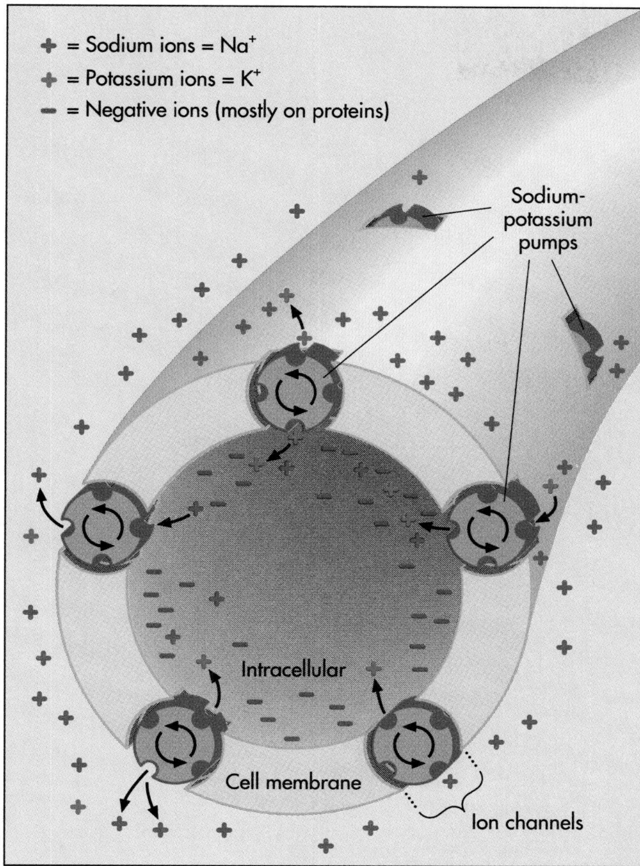

FIGURE 14-8 Formation of the membrane potential by the sodium-potassium pump. A neuron's cell membrane contains ion channels or pores. The flow of ions into and out of the cell through these ion channels is regulated by specific proteins—the sodium-potassium pumps—embedded in the channels. Powered by the cell's ATP, the sodium-potassium pumps selectively move sodium ions out of the cell while simultaneously moving extracellular potassium ions inside the cell. Since there are more negatively charged ions (mostly protein molecules that are too large to diffuse out of the cell through ion channels) inside the cell than there are positively charged potassiums, the inside of the membrane takes on a net negative charge. The resulting *potential difference* between the number of positive charges outside the cell and the number of negative charges inside the cell is the resting membrane potential (or voltage).

fuse out of the cell to balance the positive charges, most of the cell's negative charges are on proteins and other large molecules that cannot cross the membrane. The result is the development of a positive electrical charge outside the cell membrane and a negative charge inside. That is, there is a charge or voltage difference across the cell membrane. This voltage is called the membrane's **resting potential.** It is about −70 millivolts (mV) (the minus sign indicates that the inside of the cell membrane has the negative charge).

A nerve cell membrane with a fully developed resting potential is said to be *polarized.* A nerve signal involves *depolarizing* the membrane. The trigger or stimulus for depolarization can be an electrical shock, a chemical (such as a neurotransmitter), or a mechanical disturbance (such as a pinch). In each case, the effect is to open *sodium channels* in the cell membrane. Positively charged sodium ions flood into the negatively charged interior of the cell, canceling out the resting potential and depolarizing the membrane. The inside of the cell actually becomes positive with respect to the outside; the membrane potential reaches +30 mV (Figure 14-9).

Within about 1 millisecond (msec) of the initial stimulus, the potassium channels open wider, letting potassium ions flow out from the cell and begin to restore the original distribution of electrical charges. The sodium channels close. The sodium-potassium pump removes the sodium ions once more. And the membrane *repolarizes,* reestablishing its resting potential within just 1 or 2 msec. The voltage spike represented by depolarization and repolarization (Figure 14-10) is called an **action potential.**

The membrane cannot produce a second action potential until after it has repolarized. That is, the membrane goes through a **refractory period** while it repolarizes. The length of the refractory period determines how many nerve signals a nerve fiber can transmit per second. In large nerve fibers, it may last as little as 0.4 msec; such fibers can transmit up to 2500 signals per second in the laboratory, although in the body they rarely transmit more than 500 signals per second. In small fibers, the refractory period may last 4 msec; such fibers can transmit at most 250 signals per second.

The action potential is the nerve signal. It travels along the nerve cell membrane and along nerve fibers because the flows of ions that depolarize and repolarize the neuron's membrane themselves act as stimuli for neighboring patches of membrane. The action potential triggers itself, and it moves—or *propagates*—across the membrane much the way a ripple moves across the surface of a pond. It does not fade out the way a ripple does because it obeys the **all-or-none principle.** Any stimulus strong enough to trigger the initial opening of the sodium channels—a *threshold* stimulus—triggers an entire action potential. Stronger stimuli do not produce larger action potentials. Weaker (subthreshold) stimuli produce no action potential at all, although several subthreshold stimuli may together do the job.

Each action potential or nerve signal spreads out in all directions from its point of origin. When it begins on the cell body, it spreads as a ring of activity until it reaches the roots of the cell's dendrites and axons. It then travels along their lengths, as

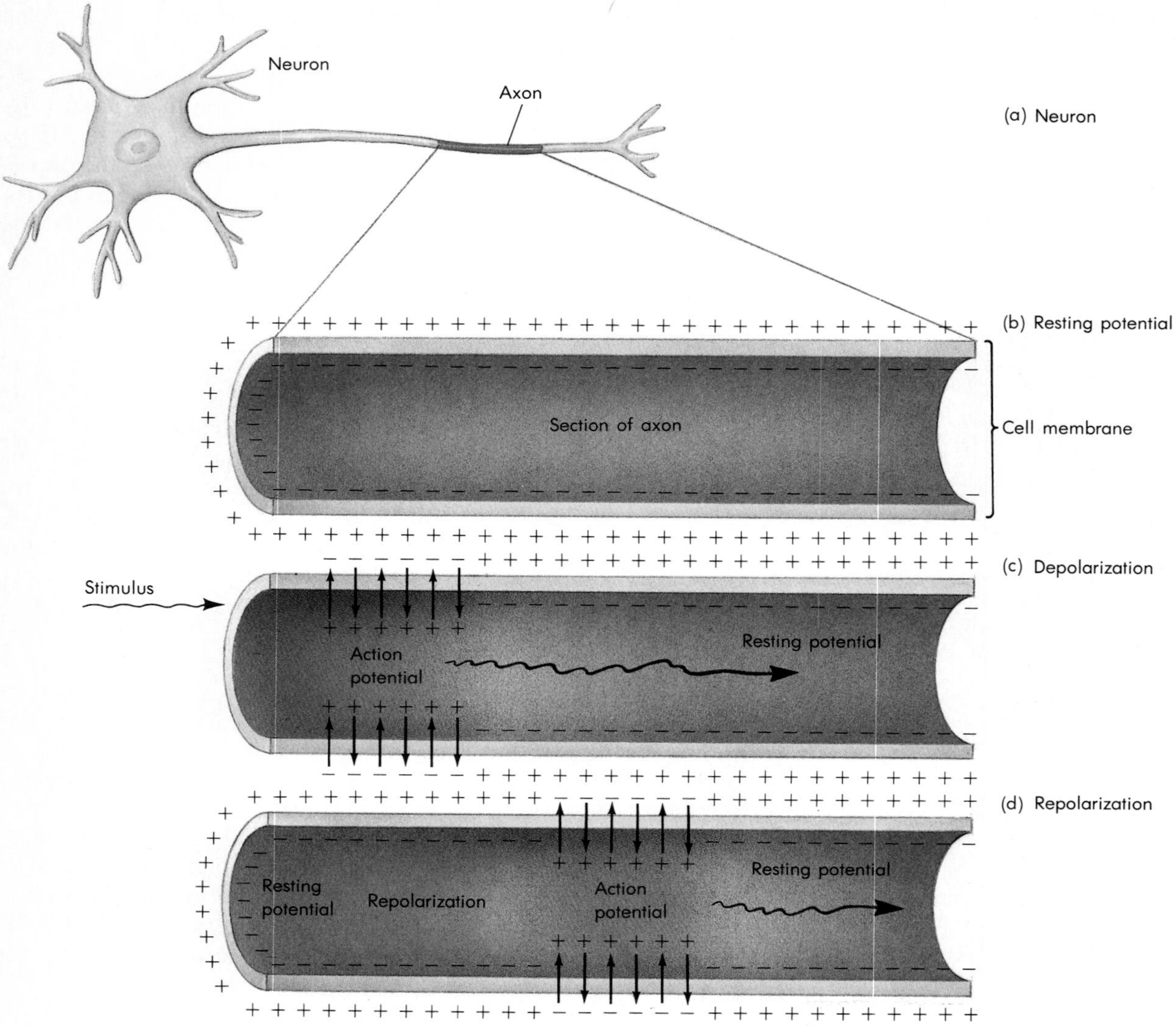

FIGURE 14-9 Depolarization and the action potential in an unmyelinated fiber. (a) Neuron. (b) Resting potential across the neuronal membrane. (c) Depolarization changes the neuronal membrane, allowing Na^+ ions to rush in. The flow of charged particles creates the electrochemical aspect of the nerve signal. (d) Repolarization occurs behind the action potential as it moves along the neuron. The electrochemical events of the action potential cause the depolarization of nearby portions of membrane as the nerve signal moves along.

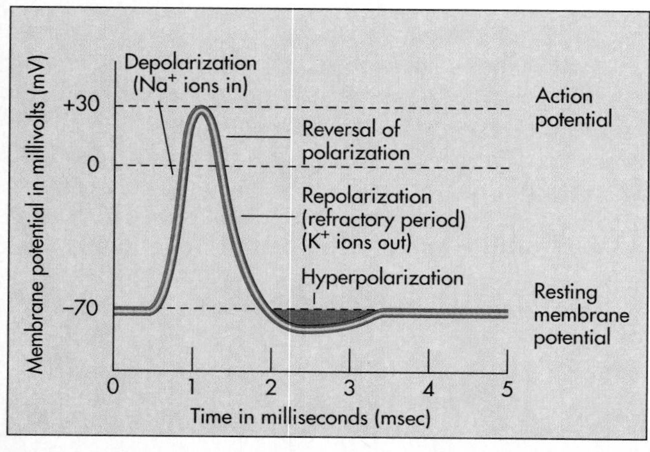

shown in Figure 14-11. When it reaches the tips of the dendrites, it stops dead, unable to go further. When it reaches the tip of the axon, however, it stimulates the synaptic knobs, and a synaptic knob is part of the structure we call a synapse (see the next section). Its importance is that it permits the action potential to pass on to another cell.

FIGURE 14-10 The graph shows a negative resting membrane potential of −70 mV. Stimulation of the nerve cell allows sodium ions to rush in, raising the potential to +30 mV. Within 1.5 msec, the sodium-potassium pump begins working to repolarize the cell membrane. During this period of repolarization, known as the refractory period, the cell cannot respond to a second stimulus.

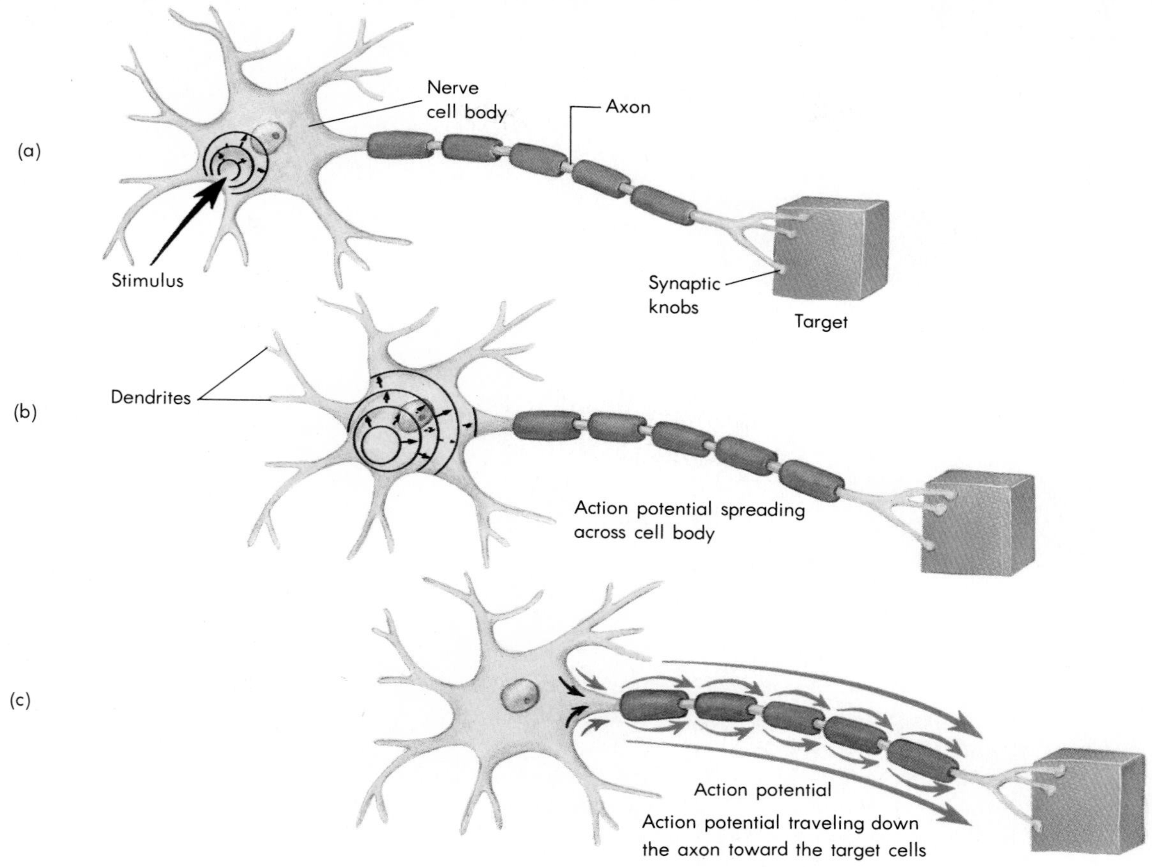

FIGURE 14-11 (a) The cell is stimulated. The action potential spreads as electrical waves moving (b) from the site of stimulation across the cell and (c) along the axon to the target cell.

Learning Focus

THE ACTION POTENTIAL

Action potentials are essential to the operation of the nervous system, for they are nerve signals. They are the basic events behind thought, movement, muscle contraction, and other basic processes of human life. However, because they are not easy to visualize, they are harder than something like neuron structure to understand thoroughly.

Consider a simple feature of the action potential: its shape. Figure 14-10 shows in graph form how the membrane potential changes with time after a stimulus. There is an initial steep rise followed by a somewhat slower return to the resting level. The graph reflects the potential changes at the site of the stimulus.

But the action potential moves, propagating across the cell membrane like a ripple in water. How then would an observer or a measuring device (such as a tiny electrode) positioned at one spot on the membrane "see" the action potential? Which side of the graph would sweep by the observer's position first? Use the accompanying graph to draw the action potential as it would appear when moving across the cell membrane.

The area of membrane that actually supports the action potential is depolarized. That is, the inside of the membrane is dominated by positive ions (Na^+ and K^+), while the outside is negatively charged. The area of membrane toward which the propagating action potential is moving shows just the opposite pattern: the outside is positively charged with respect to the inside. As a result, there are flows of ions between the two areas, and these flows (or *local currents*) serve as stimuli to provoke a new

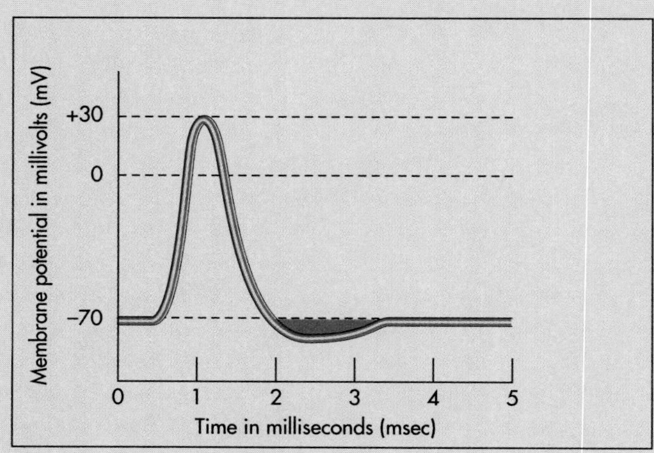

action potential in the previously inactive membrane area. Their first effect is to cause a flow of positive ions into the cell. How do they do this?

Restoring the membrane potential to the resting level requires moving positive ions out of the cell. How does the membrane begin this process?

When a propagating action potential reaches a neuron's axon, it may or may not encounter myelin. The myelin serves to block local currents and keep them from stimulating the next patch of cell membrane. However, the currents can extend far enough through the fluid outside the axon and through the cytoplasm inside to reach the first node of Ranvier. There they stimulate a new action potential, which in turn sets up local currents that stimulate the next node in the series (see the figure that follows). The action potential thus leaps from node to node until it reaches the end of the axon. This mode of conduction is called **saltatory** (leaping) **conduction.** Note that the action potential does not exist at areas of axonal membrane covered with myelin.

Local currents weaken with distance. Does this characteristic suggest to you a reason why nodes of Ranvier cannot be very far apart? Think of what we said about threshold stimuli and the all-or-none principle.

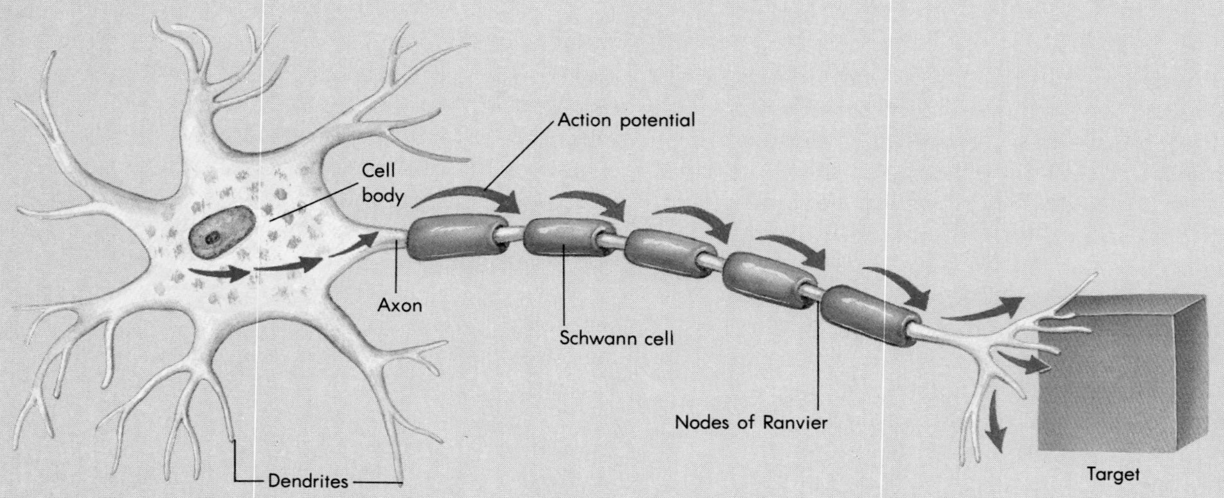

Since myelin is an insulator, action potentials traveling down myelinated axons or dendrites cannot move gradually along the length of the fiber. Instead, the action potential must leap from node to node of Ranvier in a manner known as saltatory conduction. This type of conduction is actually faster than the continuous conduction seen in unmyelinated fibers.

THE SYNAPSE

In a very few cases, nerve cells pass signals from one to another electrically. Portions of their membranes lie close to each other, and membrane proteins form small pores (gap junctions) that permit ions to flow from cell to cell. However, in most cases by far, nerve signals pass from one neuron to another, or to a gland or muscle cell, at **synapses.** A synapse consists of three portions, the presynaptic membrane, found on the end of the signaling cell, the postsynaptic membrane, belonging to the receiving cell, and the **synaptic cleft** between them. The cleft is only about 0.02 micrometer across. The presynaptic membrane generally belongs to a synaptic knob at the end of an axon branch. The postsynaptic membrane, when we are talking about neuron-to-neuron synapses, is usually on a dendrite or cell body, but it can be on an axon; it can even be on a synaptic knob (Figure 14-12).

NEUROTRANSMITTERS

Synaptic knobs contain numerous membrane-enclosed **synaptic vesicles,** which contain supplies of the activating chemicals, the **neurotransmitters.** When an action potential reaches a knob, it stimulates the movement of several vesicles to the interior surface of the knob's membrane. The vesicles then fuse with the membrane (exocytosis) and spill their contents into the synaptic cleft. The neurotransmitters diffuse across the cleft and bind to receptors in the postsynaptic membrane (Figure 14-13). Once bound, they open sodium and other ion channels in the postsynaptic membrane and trigger a rush of charges that can contribute to setting off the action potential in the target cell. Some neurotransmitters, like some hormones, work by first activating the enzyme adenylate cyclase, which makes cyclic adenosine monophosphate (cAMP) to serve as a "second messenger."

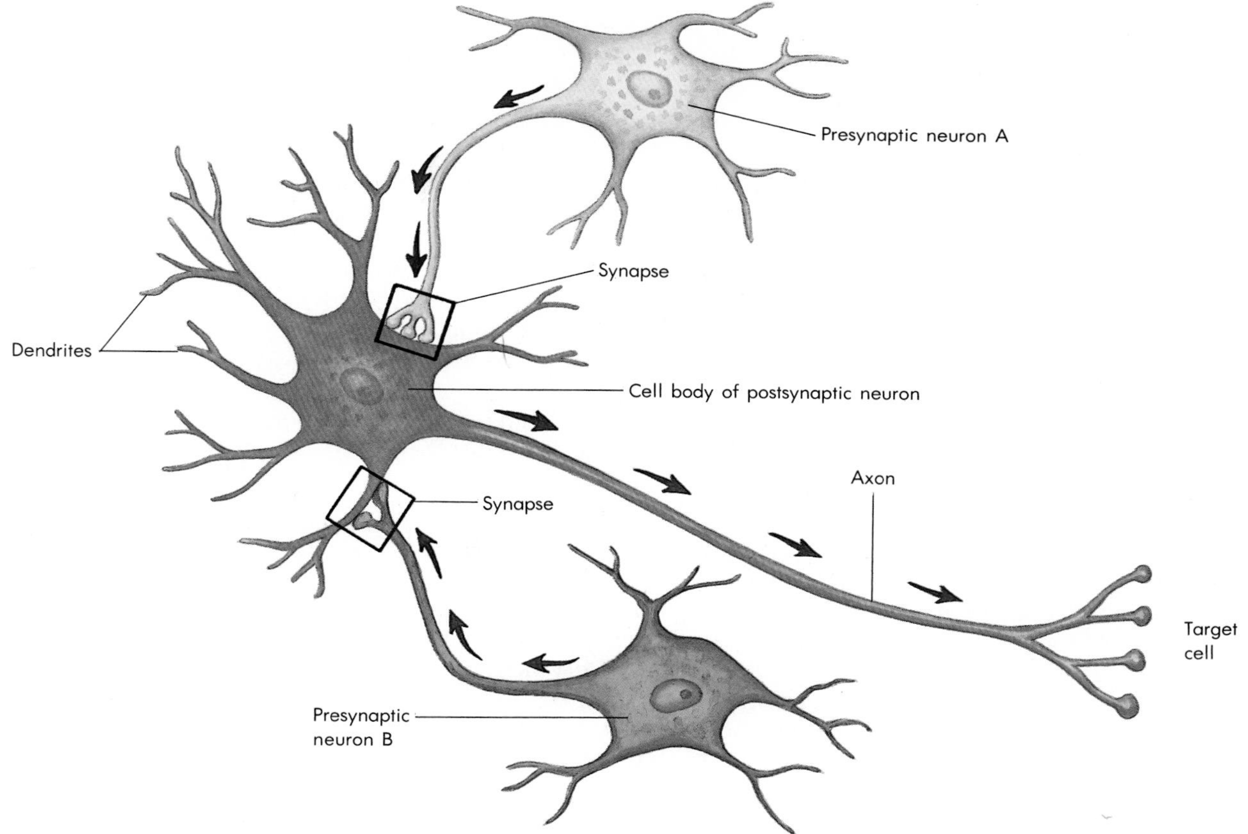

FIGURE 14-12 Synapses are one-way conducting junctions through which one neuron stimulates another. A single neuron can participate in many synapses with numerous other neurons. In this example, a single postsynaptic neuron synapses with two presynaptic neurons. The synaptic knobs of a presynaptic neuron can stimulate either the cell body or the dendrites of the postsynaptic neuron by releasing neurotransmitters.

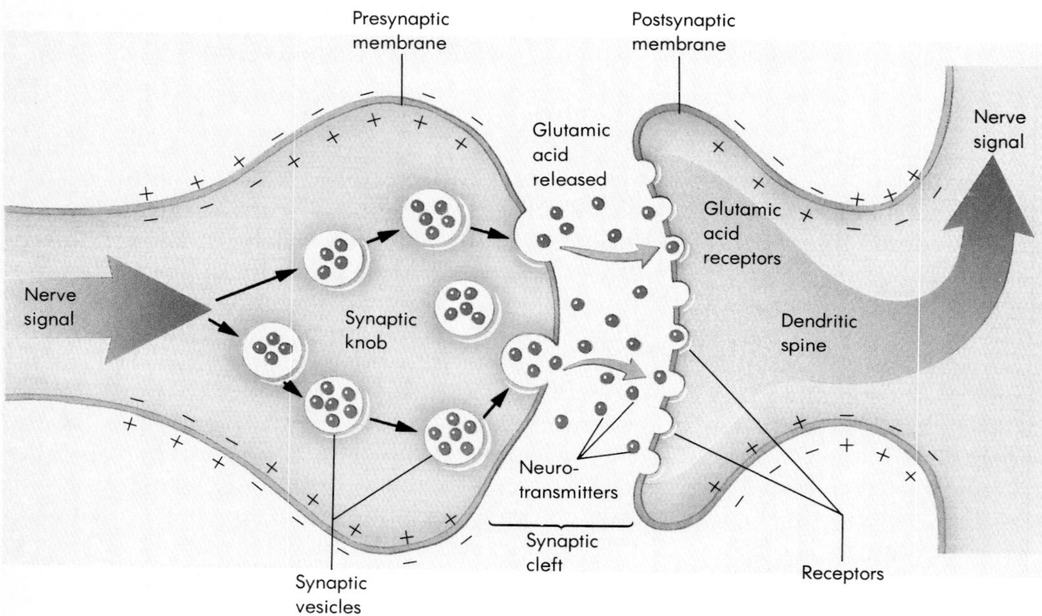

FIGURE 14-13 The action potential (nerve signal) travels down the axon of the presynaptic neuron to the synaptic knob, where it provokes the vesicles to release their neurotransmitters into the synaptic cleft. The transmitters diffuse across the cleft, complex with receptors on the postsynaptic cell membrane, and set off an action potential in the second neuron.

There are over 40 known neurotransmitters (see Table 14-2), some of them identical to various hormones, and researchers believe that the number will eventually pass 100. Some are common. Acetylcholine, for instance, is used in 5–10 percent of the brain's synapses as well as in all nerve-muscle synapses. The amino acid neurotransmitters are used in 25–40 percent of the brain's synapses. Norepinephrine, dopamine, and serotonin are each used in only one synapse in 200.

Many neurotransmitters are associated with specific regions of the brain or with specific functions. Serotonin is linked to temperature regulation, sleep, and emotion. Dopamine is involved in motor coordination, and destruction of dopamine-using brain cells causes the tremors and rigidity of Parkinson's disease. Cholecystokinin (a hormone) seems to be involved in appetite control. Substance P is used in spinal sensory pathways. The enkephalins and endorphins inhibit the transmission of pain signals and, with adrenocorticotropic hormone (ACTH), melanocyte-stimulating hormone (MSH), antidiuretic hormone (ADH), the hormone oxytocin, and some others, seem to be involved in memory formation.

The involvement of neurotransmitters in thought shows up in some of the accompanying signs of mental illness. Schizophrenics tend to release more dopamine, and perhaps more endorphins, than usual. Serotonin has been linked to depression.

Their involvement shows up again in the way psychoactive drugs such as heroin, marijuana, cocaine, angel dust, and lysergic acid diethylamide (LSD) work. Many of these drugs mimic or inhibit the effects of specific neurotransmitters and therefore strengthen or weaken certain kinds of processes within the brain. Unfortunately, the effects sometimes last weeks, months, and even years, as when LSD leaves one prone to repeated hallucinations or cocaine damages the brain's ability to feel pleasure.

Not all neurotransmitters provoke action potentials. Those that do are called *excitatory* neurotransmitters, and it takes the arrival of signals on several excitatory synapses to trigger an action potential in a target neuron. But there are also *inhibitory* neurotransmitters that open different ion channels and *hyperpolarize* target neurons. A neuron that has received an inhibitory signal must receive more excitatory signals if it is to produce an action potential.

In general, each neuron produces and uses only one neurotransmitter. It receives signals using many different neurotransmitters. However, some neurons actually use more than one, changing what they synthesize or release according to the hormonal and nervous signals they receive. They may

Table 14-2

Known or Suspected Neurotransmitters

PEPTIDES	AMINO ACIDS	OTHER
Thyrotropin releasing factor	Gamma-aminobutyric acid (GABA)	Acetylcholine
Gonadotropin releasing factor	Glycine	Norepinephrine
Corticotropin releasing factor	Glutamic acid	Epinephrine
Growth-hormone releasing factor	Aspartic acid	Dopamine
Vasopressin (ADH)		Serotonin
Oxytocin		Histamine
Melanocyte-stimulating hormone		Adenosine
Neurophysin		
Enkephalins		
Endorphins		
ACTH		
Prolactin		
Luteinizing hormone		
Growth hormone		
Thyrotropin		
Vasoactive intestinal polypeptide		
Cholecystokinin		
Gastrin		
Substance P		
Neurotensin		
Pancreatic polypeptide		
Insulin		
Glucagon		
Bombesin		
Secretin		
Motilin		
Angiotensin II		
Bradykinin		
Carnosine		
Sleep peptide		
Calcitonin		
Neuropeptide Y		

even use different neurotransmitters in separate synaptic knobs. The result is enormous flexibility; neurons can respond in many different ways to many different stimuli, a characteristic that gives them their capacity for processing information.

Neurotransmitters do not linger in the synaptic cleft or on the postsynaptic receptors. If they did, their effects would linger too. The postsynaptic cell would continue to generate action potentials in response to the original release of neurotransmitter, and it would be unable to respond to a new signal. In some synapses, such as those that use dopamine, the presynaptic cell reabsorbs the intact neurotransmitter and repackages it for reuse. In others, the synaptic cleft contains an enzyme that breaks down the neurotransmitter and inactivates it; the presynaptic cell then absorbs the fragments of the neurotransmitter molecule and uses them to recharge its supply of neurotransmitter. One such enzyme is acetylcholinesterase, which breaks down acetylcholine, a common neurotransmitter in the brain and the neurotransmitter used in nerve-muscle synapses (or *neuromuscular junctions*) (Figure 14-14). Poisons that inactivate this enzyme, such as some insecticides and tetanus toxin, cause muscles to go into permanent contraction. Other poisons, such as curare, block neurotransmitter receptors and prevent signal transmission across synapses; they cause paralysis.

In the brain, synapses appear to be intimately involved in the development of proper function and in the formation of memories. When a child is born, the brain contains many more neurons, dendrites, and axon branches than it needs. As it matures, some of these "surplus" neurons die and disappear and the surviving cells lose dendrites and axon

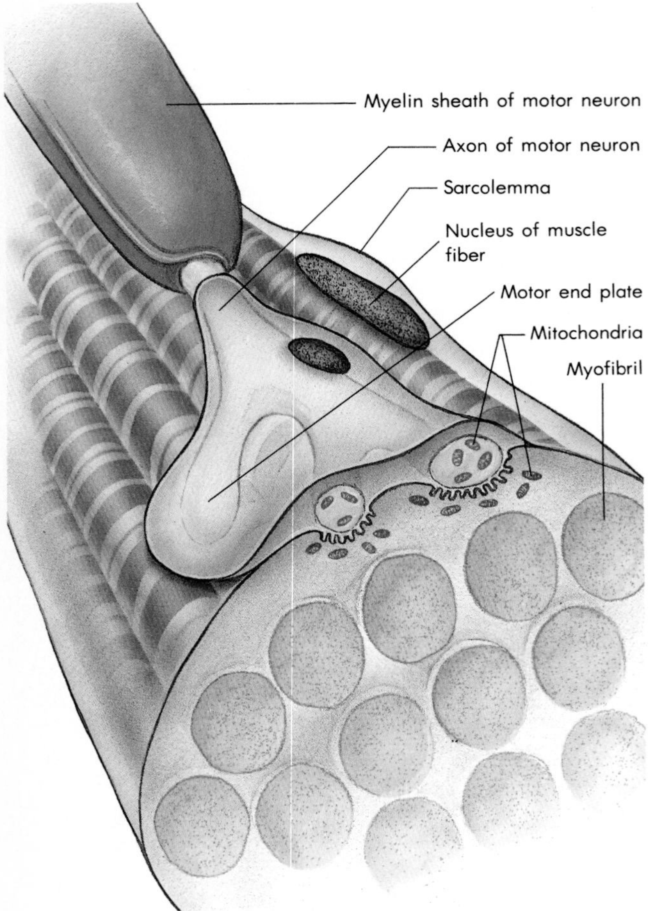

FIGURE 14-14 The neuromuscular junction is a special kind of synapse. Through it, nerve cells stimulate and control muscle contraction. The action potential traveling down the axon of the motor neuron triggers the release of the neurotransmitter acetylcholine right at the motor end plate, causing the muscle cell to contract.

branches, apparently keeping those that are actually used in sensing, processing information, and generating commands to muscles. At the same time, the number and distribution of synapses change, and many synapses enlarge or shrink according to how intensely they are used. Observations of such phenomena have long made researchers think that memories might be formed by changes in synapse size, and recent research has actually revealed mechanisms by which repeated activity at a synapse can cause increases in the number of postsynaptic receptors for neurotransmitter and changes in size and shape of dendritic spines. Such changes can make a synapse's effect on a postsynaptic cell more powerful and hence make it easier to generate particular signals. Memories may be simply signals (or groups of signals) that are easily triggered.

---

1. When you study, you repeatedly read a term, such as *synapse,* and its definition. In your brain, you thus generate, over and over, a particular sequence of nerve signals that corresponds to that visual stimulus. How would synapse strengthening help explain why the exam question "Define synapse" makes you think of the definition?
2. Some of the drugs used to treat mental illness work by interfering with the production, release, or reabsorption of specific neurotransmitters. Why do you think they work? What does the concept of synapse strengthening suggest to you about why, even after the drugs have controlled an excess of neurotransmitters, the mentally ill may continue to think "crazy" thoughts?

NERVE ACTION

You might think that every time you activate a particular neuron, the same thing should happen: The action potential should go to all the same synapses, provoke further action potentials in all the same neurons, and cause the same thoughts, feelings, or movements. To some extent, you would be right. At the simplest level, that of the built-in reflexes (see Chapter 15), a stimulus (such as a pin prick) will always have the same result (a twitch).

However, in general, a specific stimulus will not always have the same effect, either in one person or in different people. The CNS generates perceptions, thoughts, emotions, and actions in endless and unpredictable variety. There is nothing rigid about the nervous system as a whole, no matter how cut-and-dried its operations may seem at the level of the single cell or synapse.

There are three reasons why the CNS generates such variety. First is sheer number: The CNS contains about a trillion neurons and hundreds of trillions of synapses, and they can interact in countless ways. The possibilities inherent in the CNS are too numerous for us to know them all, much less predict them. Second is statistics: When the excitatory and inhibitory synaptic signals converging on a particular neuron add up to a total close to the threshold for triggering an action potential, sometimes they fail to trigger that action potential. That is, even when we know everything that is happening, we cannot always predict the result. The third reason is that everything that we can sense (sounds, sights, weather, our clothes, the state of our health) or have sensed (memories and even past thoughts) produces nerve signals that, through the effects of many synapses on each neuron, affect every neuron's readiness to produce action potentials. Thus, two inputs that we think are the same are never really quite the same.

Neural Circuits

We cannot go into all the details of how the brain generates thoughts, emotions, and actions here. Instead, we will briefly outline a few of the patterns neurons form as they hook together to process information. These patterns are called **neural circuits.**

One of the simplest neural circuits is the **divergent circuit.** We encounter it when we examine sensory input to the brain, for a sensory signal enters the brain on a single axon, but it must then go to many different parts of the CNS for integration, memory formation, and decision making. The circuit diverges because the input axon forms synapses on several CNS neurons, each of which in turn has synapses on several more (Figure 14-15a).

Another simple circuit is the **convergent circuit,** in which several neurons have synapses on the same target cell (Figure 14-15b). This allows several information sources (different senses, memory, signals generated during information processing) to be integrated during decision making before activating a particular nerve pathway.

In the CNS, convergent and divergent circuits are often combined. One combination allows a single action potential to be turned into several in sequence, as in the *parallel after-discharge* circuit in Figure 14-15c, where each neuron introduces a time delay. Another combination, shown in the *reverberating* circuit of Figure 14-15d, both multiplies a signal and preserves it (by making the signal do "laps" around the closed loop); such circuits were once thought to be a possible basis for memory.

These circuits and others are the basic elements of the organic computer that controls your body.

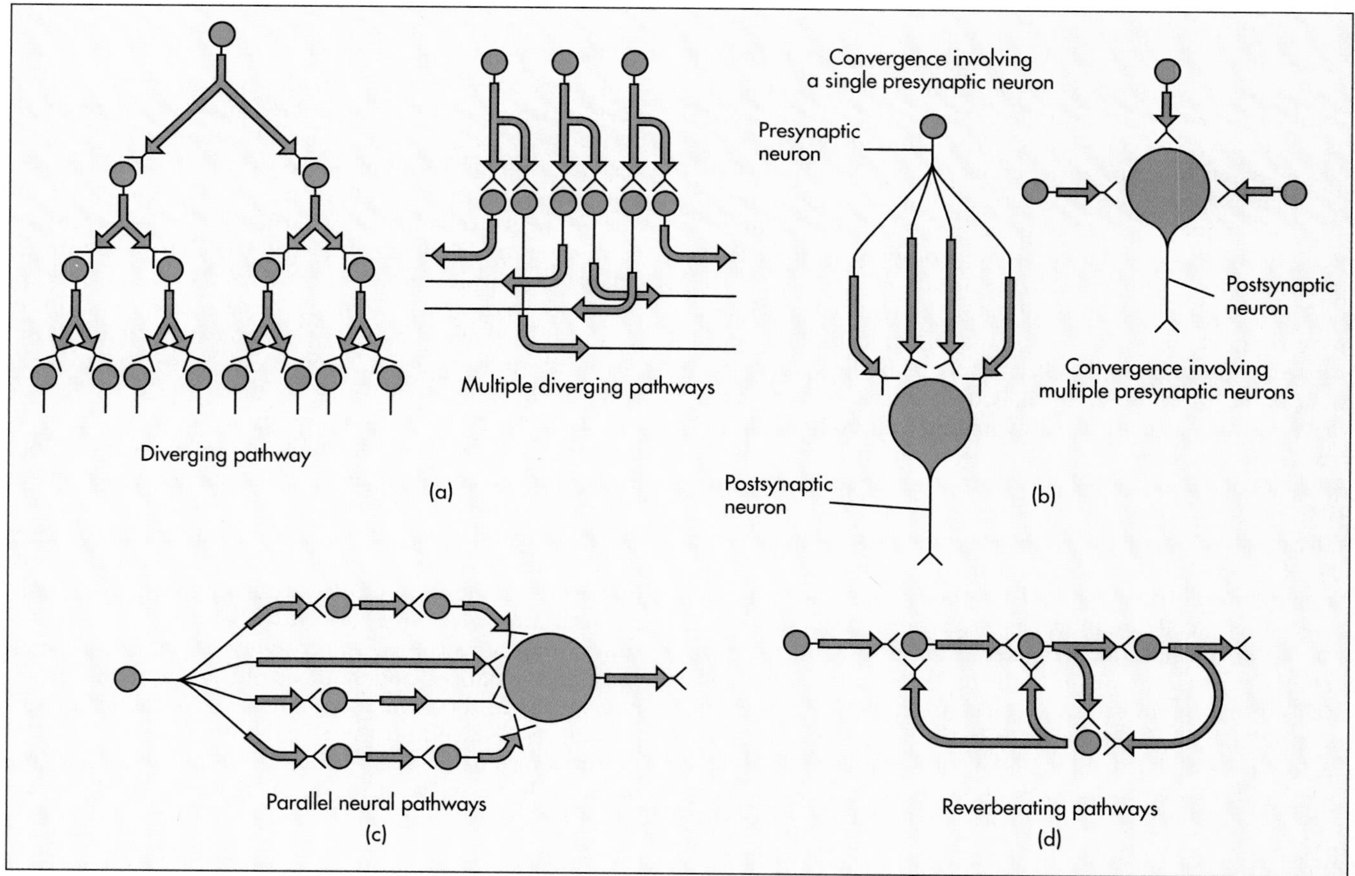

FIGURE 14-15 Neural circuits: (a) divergent; (b) convergent; (c) parallel after-discharge; (d) reverberating.

However, the actual circuits in your brain and spinal cord are far more complex than these simple examples. They involve many more neurons and synapses. They also involve both excitatory and inhibitory signals, so that circuit elements can, in effect, be turned both on and off. You can recognize the inhibition at work if you think of how you respond when you step barefoot on a stone: Normally, a reflex forces you to jerk your foot off the ground, but if you are in a situation where balance is crucial, as on a narrow ledge in the mountains, your foot remains still; you inhibit the reflex.

SUMMARY

The nervous system is the body's high-speed control system. Where the endocrine system communicates at long distance, via the blood stream, the nervous system communicates across fractions of a micrometer at synapses. Each system produces chemicals—hormones or neurotransmitters—that exert their effects by binding to receptors in the cell membranes of target cells.

We divide the nervous system into several parts. Most of its cells lie in the central nervous system (CNS), the brain and spinal cord. They communicate with the body's organs via long extensions or nerve fibers bundled into the nerves of the peripheral nervous system (PNS). The PNS has two components. The afferent system carries information from sense organs to the CNS. The efferent system, which conveys signals from the CNS to muscles and glands, has a voluntary segment, the somatic nervous system, and an automatic component, the autonomic nervous system (ANS).

The CNS has two kinds of cells, the neuroglia, which play support roles, and the actual nerve cells, the neurons. Neurons, like all cells, have cell bodies; they are unique in their long, branched extensions—the dendrites, which carry signals toward the cell body, and the axons, which carry signals away from the cell body. Many axons and some dendrites bear a sheath of lipid-rich myelin, produced by Schwann cells in the PNS and oligodendrocytes in the CNS. The myelin serves to insulate the nerve fibers electrically and to accelerate the transmission of nerve signals in them.

In their cell membranes, neurons have sodium-potassium pumps that transfer sodium ions out of the cell and potassium ions into the cell. The potassium ions leak out again, and positive electrical charges build up outside the cell membrane, establishing the neuron's resting potential. When the membrane is disturbed by at least a threshold amount, sodium channels open and sodium ions rush into the cell, reversing the potential from −70 to +30 mV (depolarizing the membrane). The sodium channels then close, potassium channels open wider, and the sodium-potassium pump restores the resting potential. The result is a brief (1–2-msec-long) electrical spike, the action potential. Since the action potential itself disturbs neighboring patches of membrane, it triggers its own propagation across the membrane. Action potentials come in only one size, according to the all-or-none principle.

At the end of each axon lie the synaptic knobs, which form synapses on target cell dendrites, cell bodies, and even axons. Within each knob are a number of synaptic vesicles, which contain neurotransmitter substances. When an action potential reaches the knob, the vesicles release neurotransmitter. The neurotransmitter molecules then diffuse across the synaptic cleft to the target cell (postsynaptic) membrane, where they bind to receptors and trigger new signals; some neurotransmitters inhibit the production of signals. Soon after their release, neurotransmitters are destroyed or reabsorbed to limit their effect and permit new signals to pass. Synapses may change size, shape, and receptor number in the formation of memories. Surplus synapses, nerve fiber branches, and even neurons are lost during development from the embryo to the adult.

The key to the way the CNS processes information lies in the circuits neurons form in their interconnections. Some of the simpler circuits are divergent, convergent, parallel after-discharge, and reverberating circuits. Actual circuits are complex, involving a great many neurons and synapses and both excitatory and inhibitory neurotransmitters. The sheer numbers of these elements, as well as chance and the continually changing flux of input from outside and inside the CNS, account for the never-ending variety of responses the CNS can produce.

STUDY QUESTIONS

1. Why do we call the nervous system the body's high-speed control system?
2. What is the difference between the central and the peripheral nervous systems?
3. What is the difference between a dendrite and an axon?
4. What is the role of astrocytes in the blood–brain barrier?
5. Describe the formation of myelin in the peripheral nervous system. In the central nervous system.
6. How does myelin serve to speed up the conduction of nerve signals?
7. How does a neuron maintain its resting potential?
8. Describe what goes on in a neuron's cell membrane as it produces an action potential.
9. How does an action potential propagate itself across a neuronal membrane?
10. Describe how an action potential crosses a synaptic cleft.
11. How are inhibitory neurotransmitters useful in information processing?
12. Both anesthesia and low body temperatures stop neural activity, but they do not interfere with long-term memory. Why does this suggest that reverberating circuits cannot be the basis of long-term memory?

GLOSSARY

Action potential Reversal of the membrane potential from −70 to +30 mV, and recovery, on disturbance of the membrane; the nerve signal.

All-or-none principle Above a set threshold, all stimuli produce action potentials of the same size; subthreshold stimuli do not produce action potentials.

Autonomic nervous system (ANS) The automatic, involuntary branch of the efferent PNS; controls the viscera.

Axon Nerve fiber carrying signals away from a neuron's cell body.

Blood–brain barrier Arrangement of capillary walls, basement membrane, and astrocytes that keeps many molecules carried in the blood from reaching the brain's neurons.

Central nervous system (CNS) Brain and spinal cord.

Dendrite Nerve fiber carrying signals toward a neuron's cell body.

Myelin Lipid-rich sheath around many axons and dendrites; electrical insulator and transmission accelerator.

Nerve A bundle of nerve fibers carrying signals to or from the CNS.

Nervous system The network of nerve cells and fibers that receives and processes electrochemical information and coordinates the body's activities.

Neural circuit Arrangement of neurons; analogous to an electrical circuit.

Neuroglia Supporting cells of the CNS.

Neuron Nerve cell.

Neurotransmitter Substance released from synaptic vesicles into synaptic clefts; on binding to postsynaptic receptors, it initiates (or inhibits) an action potential in the target cell.

Parasympathetic division One branch of the ANS; generally opposes the sympathetic division. The branch of the ANS that encourages physiologic well-being or homeostasis.

Peripheral nervous system (PNS) The nerves that carry impulses from the peripheral areas of the body to and from the central nervous system.

Refractory period Period during and immediately after an action potential when a nerve cell membrane cannot generate a second action potential.

Resting potential The electrical charge (voltage) difference across a neuron's cell membrane; −70 mV (inside of cell is negatively charged with respect to the outside).

Saltatory conduction Leaping of action potential from node to node of Ranvier.

Sodium-potassium pump Membrane protein that uses ATP to move sodium ions out of the cell and potassium ions into the cell.

Sympathetic division One branch of the ANS. The alarm, or fight-or-flight division of the ANS.

Synapse Point at which nerve signals pass from neuron to neuron.

Chapter 15

The Brain and the Spinal Cord

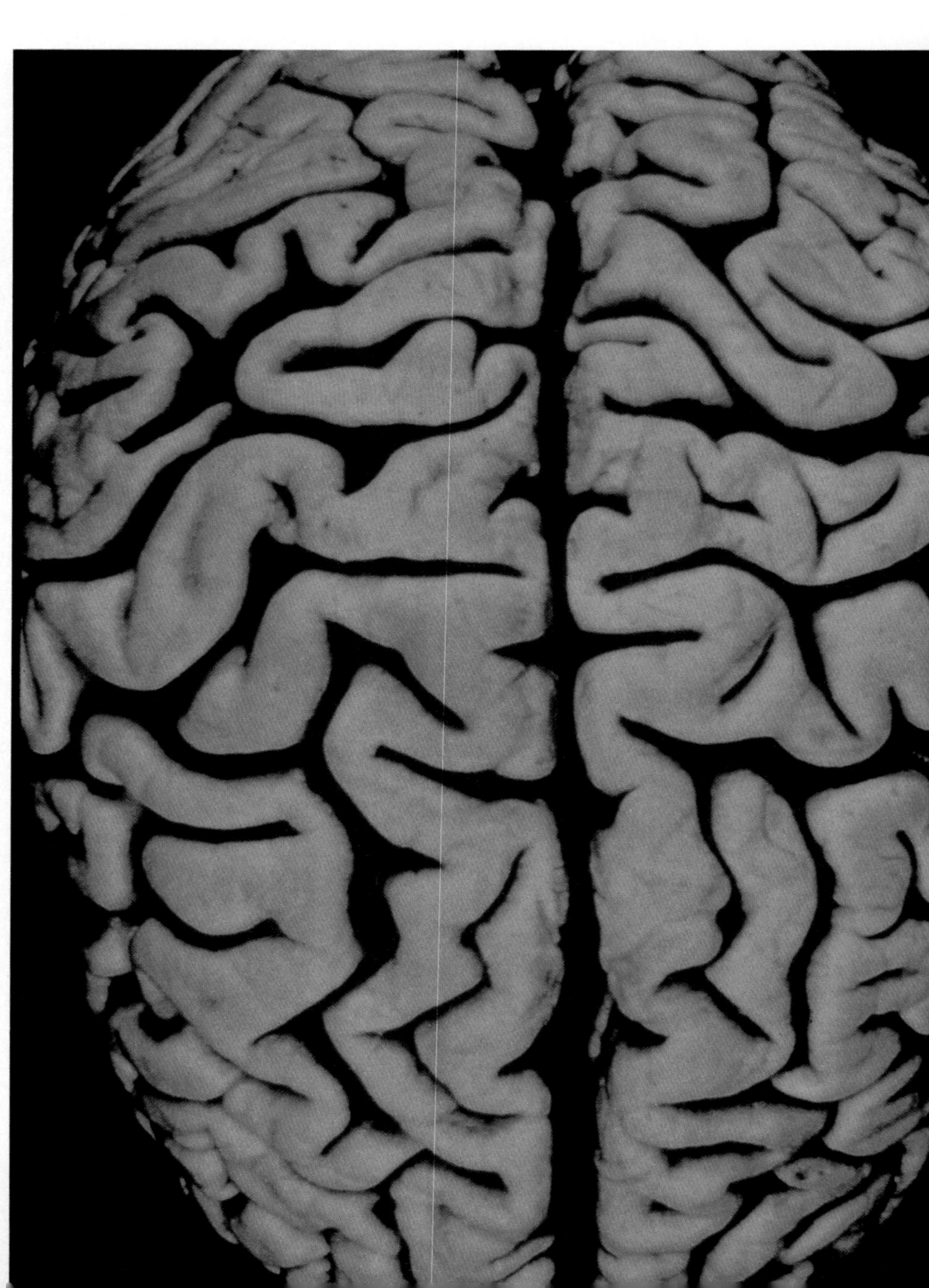

The human brain is extremely demanding, with a voracious appetite for numerous forms of entertainment. Many of our common activities are simply different ways we have of entertaining our brains. They include reading, problem solving, exercising, traveling, doing hobbies, and even studying. We also entertain the brain by giving it new tasks to work on, like controlling our bodies as we learn a new dance step or a piece on the piano. And if that is not enough, when things are "slow," we are likely to position the brain and its input devices in front of a television set.

It is the complexity of the human brain that makes it so fascinating. It receives a perpetual array of sensory reports that keep it informed of the changing patterns of light (vision), air vibrations (hearing), body position, temperature, pain, and an assortment of other conditions. It stores vast amounts of information and uses it to generate many different responses to various events. It gives us consciousness, intelligence, and speech and our abilities to learn and to create. It gives us our talents for art, music, science, and technology. It gives us some degree of mastery of the world around us, and it enables us to recognize at least some of the problems that come with such mastery. We owe religion and politics, love and hate, war and peace, and art and science to the complexity of the human brain and to its need to entertain itself. The brain is what makes us distinctively human.

THE BRAIN

In the last chapter, we compared the brain to a computer. It is not, of course, a computer in the same sense as a personal computer. However, like a personal computer, it does come in an attractive container (some more attractive than others) complete with input and output devices. This container is the head. Instead of a keyboard, it has eyes and ears (and other senses) as input channels. Instead of a screen as an output device, it often uses the mouth, and via the spinal cord and nerves, it accesses the body's muscles and glands (*effectors*) to produce changes in facial expression and posture and to exercise other behavioral options (Figure 15-1).

Just as a computer has a protective casing, the brain has the skull, or cranium. Within the skull is a cushioning suspension system of membranes (the

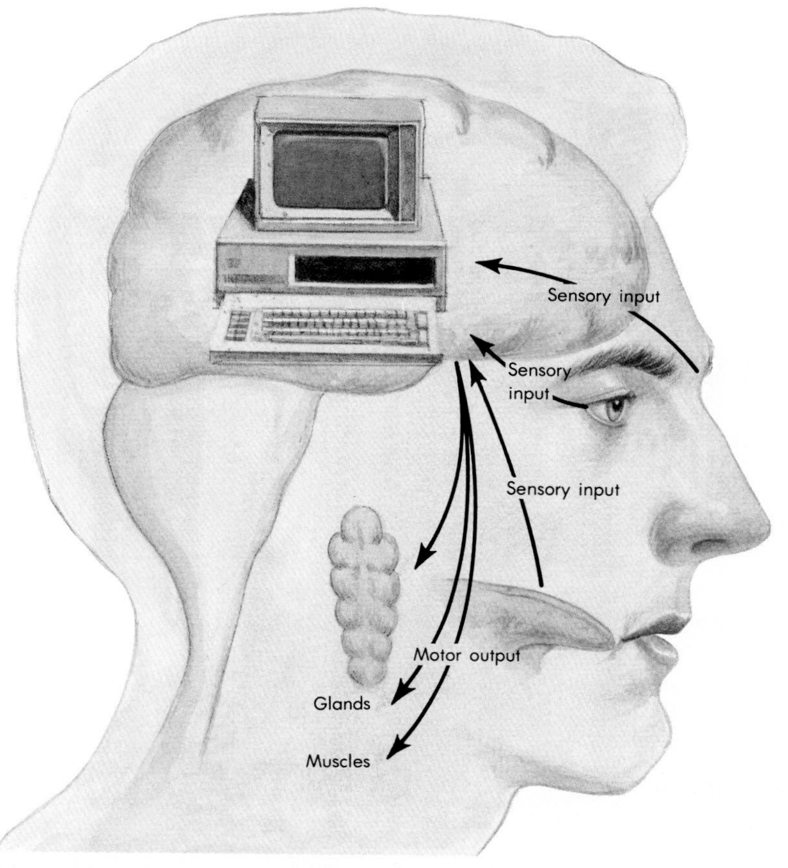

FIGURE 15-1 What are the various components of the nervous system *really* like? One view describes the brain as a central computer and the nerves as the cables that connect the brain to its various input (sensory) and output (muscles and glands) devices.

meninges) and fluid. The most conspicuous features of the brain are the two large hemispheres of the cerebrum that fold over the top of the brain. They form the outer regions of the **forebrain** (Figure 15-2). Within the cerebral hemispheres are found the other divisions of the forebrain, the basal ganglia, thalamus, and hypothalamus. The **brainstem** consists of the midbrain, pons, and medulla oblongata. Behind the brainstem and below the cerebrum lies the **cerebellum.** The medulla is continuous with the spinal cord that runs down the back. Peripheral nerves arise from both the brain and the spinal cord.

The Cerebral Cortex

A student in a nonmajors biology class once described the brain as "a two-and-a-half-pound walnut without its shell." This is a colorful answer to an exam question, but it could stand some elaboration. The surface of the cerebrum is indeed conspicuously grooved and folded in arrangements referred to as *convolutions,* which increase surface area without increasing the brain's overall volume. The upfolds, or ridges, are called **gyri** (singular *gyrus*) and the grooves between them are the **sulci** (singular *sulcus*) (Figure 15-3). Specific sulci, or

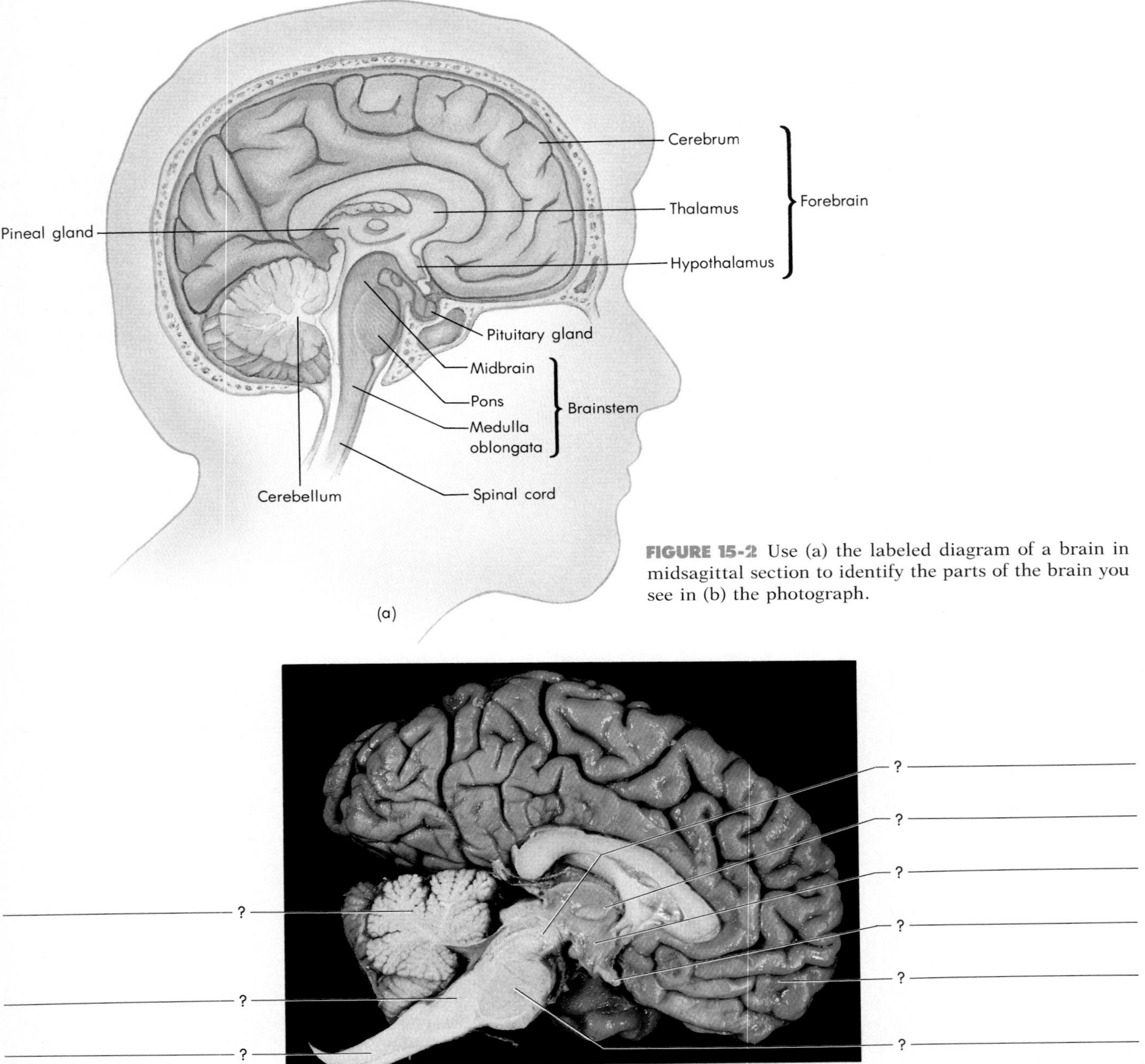

FIGURE 15-2 Use (a) the labeled diagram of a brain in midsagittal section to identify the parts of the brain you see in (b) the photograph.

occipital lobe. It contains stored information of past visual experiences that can be used to recognize previously seen patterns and analyze new visual experiences. A significant amount of the input to the visual association area comes from the thalamus.

Sources of Information about Brain Functions

People sometimes ask a rather pointed question about the functions controlled by specific regions of the brain, namely, "How do we know?" What we know comes from three sources: In laboratory experiments, researchers record the arrival of nerve signals from sense organs in the sensory areas of the cortex and watch muscles twitch when they stimulate motor areas electrically. In "experiments of nature," wounds, strokes, and tumors damage specific cortical areas and impair matching functions. Some of the functions of the frontal lobe were first revealed in 1848, when Phineas Gage, a Vermont construction foreman, was using a crowbar to tamp gunpowder into a hole in rock (for blasting). The powder blew, and the crowbar destroyed much of his frontal lobes and left him impulsive, profane, and rude. The side effects of surgery can also be revealing. Follow-up studies of patients recovering from neurosurgery often indicate which functions were regulated by the areas that were operated on.

1. If we deliver a small electrical shock to the right part of the primary motor area, we can make an arm move. This experiment suggests that control of the arm is affected by that bit of cortex. What result would you expect if a stroke destroyed that bit of cortex?
2. A brain tumor in the occipital lobe can prevent cortical cells from working properly. What would you expect the symptoms of such a tumor to include?

Brain Waves

Because nerve signals depend on flows of electrically charged ions, they produce electric fields that can be detected with suitable apparatus. Since the cerebral cortex contains a great many neurons signaling simultaneously, these fields add up and are strong enough to be detected at the surface of the scalp as **electroencephalographic** waves. The apparatus that detects them is an electroencephalograph, and the record they leave on a paper chart is an electroencephalogram (EEG).

The cells of the cerebral cortex work together, and their joint activity produces EEGs of various types. *Delta waves,* with a frequency of 1–5 hertz (Hz) (cycles per second), occur during sleep. *Theta waves* (5–8 Hz) are common in children and during stress. *Beta waves* (15–60 Hz) mark periods of attentiveness and mental activity. *Alpha waves* (10–12 Hz) occur when people are awake and resting, or mentally quiet (Figure 15-7).

Because physicians can now keep the heart beating and the lungs breathing long after there is any chance that a patient will revive, the lack of brain waves—a "flat" EEG—is today often taken as the definitive sign of death. The premise is that since it is the brain's activity that accounts for memory, thought, personality, and identity, if that activity has stopped, the person no longer exists. The brain-dead "person" is dead in the only way that truly matters, and the body may be unplugged from the machinery or disassembled for organ transplants.

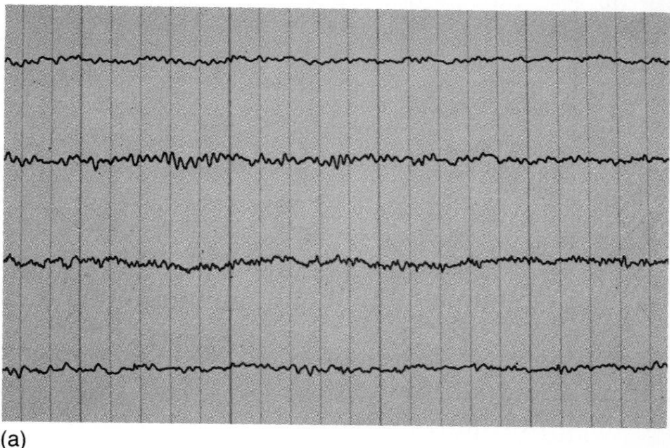

(a)

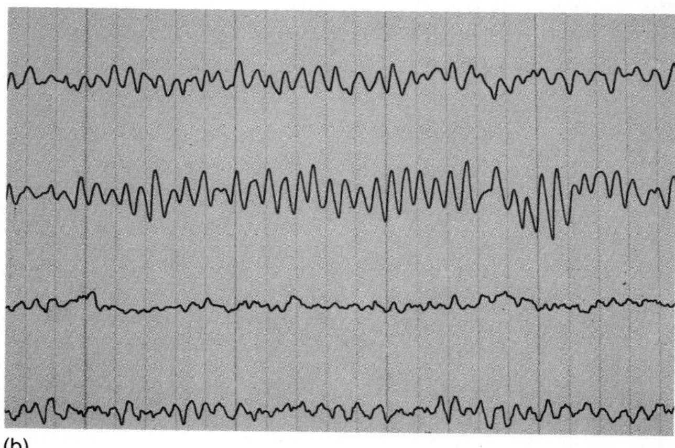

(b)

FIGURE 15-7 The electroencephalogram varies according to what the brain is doing. Beta waves (a) accompany mental activity. Alpha waves (b) occur during rest or mental quiescence.

1. If you accept that brain death is truly death, what can you say about the status of an embryo whose brain has not yet developed enough to show an EEG? Is it alive? Or is it dead?
2. Might this line of thought affect the debate over whether—and when—abortion is murder? In what way?

Electroencephalograms are useless for mind reading. They reflect the activity of too many brain regions to allow us to interpret what is going on in any one region. At best, abnormal EEG patterns are helpful in diagnosing epilepsy, brain tumors, and other disorders that interfere with brain function. However, it is possible to use more sensitive apparatus to study the electrical activity of smaller parts of the brain. When researchers do this, they find that specific electrical waveforms accompany decision making, precede motor acts, and even accompany the receipt of specific words by the brain. There is even evidence that electrical fields outside the brain can cause it to produce specific patterns of activity and hence—perhaps—to think specific thoughts.

1. A part of the cerebral cortex translates thoughts into speech. Since a particular pattern of neural (electrical) activity must be generated to stimulate the muscles that produce each word, "reading" or monitoring this region of cortex with an EEG-like machine may someday allow us to tell what someone is about to say. Name the region of cortex.
2. Would this "reading" of cortical activity be true mind reading? What kinds of thoughts would it miss?
3. To use the method would require building up a vast dictionary of electrical patterns for each person whose mind you wished to read (since the patterns would surely differ from person to person). Under what circumstances might the method nevertheless be useful?

The Cerebral Hemispheres

Tracts of axons that conduct neural information between the left and right sides of the brain, from one hemisphere to the other, are called *commissures*. Several such tracts interconnect the two sides of the brain. The largest of these commissures is the **corpus callosum** (visible in Figure 15-8). It transfers information between the hemispheres and permits "crosstalk" between them and coordination of their activities. When the corpus callosum is cut, isolating the hemispheres from each other (as may be done to treat severe epilepsy), it is apparent that the two hemispheres have somewhat different functions.

For some reason, the left side of the brain monitors sensory information from and controls the effectors on the right side of the body, and vice versa. It is thus not too surprising that in right-handed people, the left parietal lobe is slightly larger than the right (in lefties, the right parietal lobe is only *sometimes* larger). More surprising is that in most people, whatever their handedness, the left hemisphere dominates in such functions as speaking and writing, arithmetic, and reasoning (it is considered by some to be the "linear thinking hemisphere), whereas the right hemisphere handles music, art, pattern perception, and imagination (according to some people, it is the "intuitive" hemisphere). The differences show up when the hemispheres are prevented from communicating with each other by cutting the corpus callosum (Figure 15-8). Then, if the left hand is given a key to

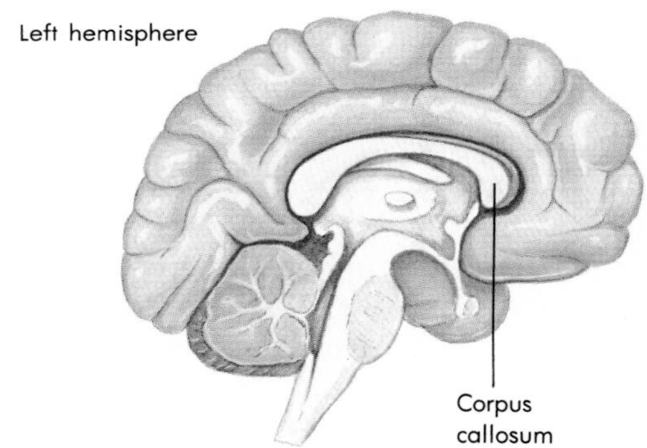

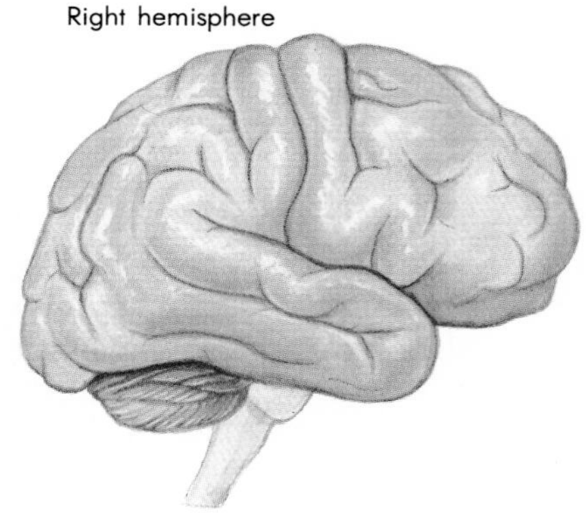

FIGURE 15-8 Split-brain drawing showing the corpus callosum.

feel, so that sensory cues go only to the right hemisphere, the person cannot name it but can match it to a picture. If the key is put in the right hand, the person can name it but cannot match it to a picture.

Since the right and left cerebral hemispheres are not isolated from each other in normal people and the two sides of the brain communicate extensively with each other on most decisions and tasks, it is now considered to be inaccurate and misleading to refer to people that are "creative" and "artistic" as "right-brain" people and those good with numbers as being "left-brain" people.

Injuries to the brain that seem identical except in whether they affect the right or left side can thus have very different effects. Damage to the motor cortex on one side of the body paralyzes the other side of the body. Damage to other parts of the cortex can destroy other functions, such as speech, reading, writing, or musical ability. When such damage strikes older people, as after a stroke, the loss can be permanent. When it strikes children, the lost abilities can often be regained. Other parts of the cortex can pick up the slack. Sometimes, the other hemisphere can even broaden its functions to take over abilities such as speech that previously belonged to the damaged hemisphere.

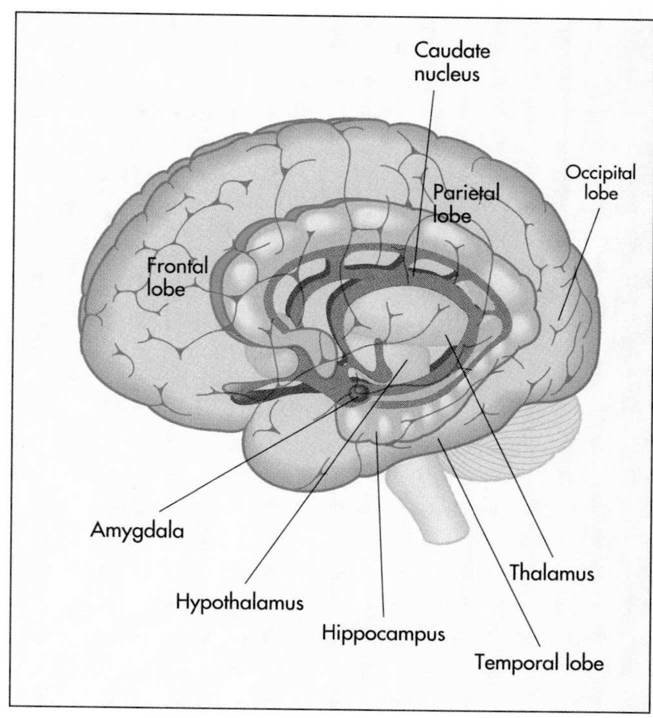

FIGURE 15-9 The limbic system, showing the amygdala at the tail of the caudate nucleus, the hippocampus, and parts of the hypothalamus and thalamus.

The Rest of the Forebrain

Embedded deep within the white matter of each cerebral hemisphere are a number of nuclei, or clusters of nerve cells. They include the **basal ganglia,** some of which play important roles in regulating muscle tone and actions such as walking, which involve large muscles. The **limbic system** is intimately involved with the control of emotions. Besides the amygdaloid nucleus, the limbic system also includes portions of the basal ganglia, the **hippocampus,** and portions of the hypothalamus and thalamus (Figure 15-9).

The hypothalamic portion of the limbic system is also the seat of the "pleasure center." A rat fitted with a "rat-controlled" self-stimulating electrode implanted in this group of nerve cells can trigger sensations so pleasurable that it will ignore its own need for food and starve to death before giving up its artificial high. It is this pleasure center that "burns out" after prolonged cocaine use, leaving cocaine addicts incapable of feeling pleasure, a condition known as *anhedonia*.

The limbic system is also crucial to forming long-term memories. Surgical damage to the hippocampus has left a few people able to remember only their lives before the damage. They cannot form new memories, and when they go outside to mow the lawn, for example, they must be repeatedly reminded of what they are doing. They can, however, form some new motor patterns and habits, though with difficulty.

The **thalamus** consists of two chestnut-sized oval masses of neurons flanking the third ventricle. It receives signals from the body's various senses (except olfaction) by way of the brainstem or from the cerebellum and then relays them to specific regions of the brain for analysis. If an incoming signal deserves conscious evaluation, it will be routed to the appropriate region of the cerebral cortex.

One of the most important areas of the brain, in spite of its small size, is the **hypothalamus,** which lies just below the thalamus where it forms the lower lateral walls and the floor of the third ventricle. It contains many crucial centers (or nuclei) that regulate some of the body's most important physiologic systems. The hypothalamus plays important roles in governing the endocrine system, the secretions of glands (both endocrine and exocrine), the emotions, and many "maintenance" functions such as water balance and thirst, body temperature, and

Box 15-1 Alzheimer's Disease

We used to think that senility, marked by failing memory and impaired thought processes, was a natural consequence of aging. However, not all old people become senile, and today we recognize that a leading cause of senility is the condition known as *Alzheimer's disease*. Inspection of the brains of deceased victims of senility has revealed tangled masses of neurofibrils in many brain neurons and collections of unusual protein (amyloid) in brain blood vessels and among the brain's cells, as shown in Figure 1. There is also severe loss of neurons, particularly those that use the neurotransmitter acetylcholine, in regions of the brain concerned with memory and thought.

Some evidence suggests that Alzheimer's disease may be hereditary, for it seems to run in families. There are also signs that it may be due to a virus or a prion (discussed in Chapter 23) or to some toxin, such as aluminum, from the environment. Other theories blame the disease on poor blood flow or energy metabolism. Unfortunately, none of the theories has yet proved helpful in preventing or curing the disease.

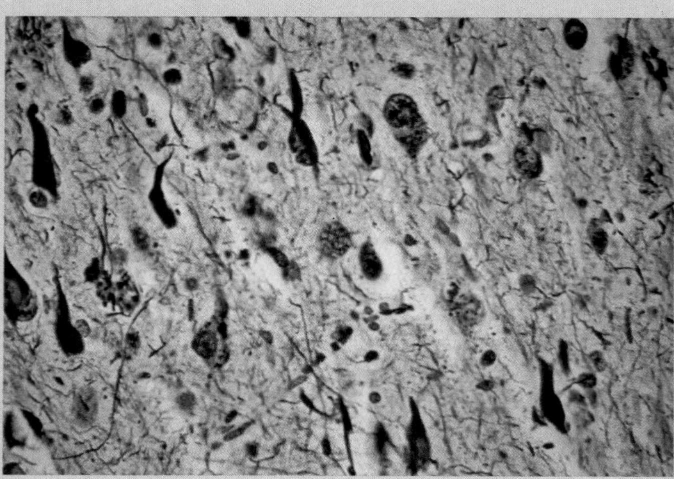

FIGURE 1 Alzheimer's neurofibrillary tangles and neuritic plaques in hippocampus.

hunger and satiety. It also integrates much of the incoming sensory information from the viscera.

Many of the functions controlled by the hypothalamus are those regulated by the parasympathetic division of the autonomic nervous system. The hypothalamus is thus one of the primary regulators of homeostasis. It regulates heart rate, digestion, and responses to stress. It also plays a role in keeping the brain in an alert, or "awake," mode.

THE BRAINSTEM

The **brainstem** bears its name because it resembles the stem of a flower or mushroom. It is a narrow structure atop which rests the great expansion of the forebrain and cerebral cortex. Through the brainstem flow nerve signals connecting the various areas of the brain to the body as well as sensory information from the body to the forebrain. The brainstem consists of the midbrain, the pons, and the medulla oblongata (Figure 15-10).

The **midbrain** is a relatively short area (about 2.5 centimeters long) of the brainstem that contains not only the bundles of axons that convey nerve signals up and down the central nervous system but also a number of nuclei (groups of nerve cells dedicated to regulating specific functions). Some of these nuclei are concerned with the control of movement. Visual information receives its first processing in the **superior colliculi,** two small protuberances on the dorsal side of the midbrain. The superior colliculi also generate reflex responses to sudden changes in the visual input, such as the flick of the eyes or turn of the head that brings the change into better view. The **inferior colliculi,** another pair of dorsal nuclei, serve a similar function

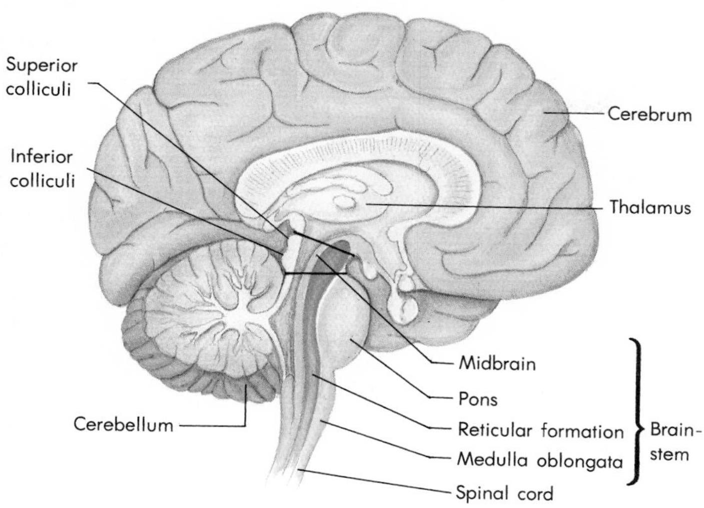

FIGURE 15-10 The brainstem and the reticular formation.

for hearing. It is believed that some circuitry interconnects the superior with the inferior colliculi and relates visual images to sound patterns. In fact, the feelings of discomfort we get when watching an out-of-phase "lip-syncher" may be due, in part, to interactions between these two areas of the midbrain.

Running through the midbrain, extending from the medulla through the pons, hypothalamus, and thalamus, is a cone-shaped and rather elaborate network of fibers, the **reticular formation,** interconnecting various areas of the brain. It serves as the arousal or alertness center that helps to keep an individual awake and to maintain muscle tone by sending nerve signals to the muscles via the spinal cord. Damage to the reticular formation can result in permanent sleep, or coma.

The **pons** ("bridge") is so named because it lies on the side of the brainstem opposite the cerebellum, just above the medulla, and looks like a bridge spanning the medulla. From it arise the nerve bundles that serve the cerebellum. In it are a number of nuclei, including two that help control respiration.

The **medulla oblongata,** as the lowest, or hindmost, part of the brain, is one of the most important areas of the brain. The medulla contains nuclei regulating many of the body's vital functions, including ones that control breathing, heart rate, vasoconstriction, vomiting, coughing, swallowing, sneezing, and hiccoughing. Additional nuclei are associated with the workings of the cerebellum and the sense of balance. The medulla contains bundles of axons carrying motor and sensory information, and it is in the medulla that many of the axons carrying motor commands from the cerebral cortex to the spinal cord cross from one side of the body to the other.

THE CEREBELLUM

The **cerebellum,** roughly an eighth of the brain, lies below the cerebrum and behind the pons and medulla (Figure 15-11). Deep within the cerebellum lie several nuclei that pass nerve signals into and out of the cerebellum. These nuclei now seem to play some role in the acquisition of learned (conditioned) reflexes. As a whole, the cerebellum plays an important role in coordinating and timing voluntary movements, maintaining muscle tone, and preserving balance. It also keeps track of the positions of the limbs as they are moved about during specific tasks. The cerebellum instantaneously adjusts outgoing motor commands, taking into account the rapidly changing positions of the muscles, so that the intentions of the cerebrum are honored.

"On-the-spot" sobriety tests given to suspected drunk drivers actually check the state of the cerebellum. In the tests, subjects are instructed to close their eyes, extend their arms out to their sides, and then bring their fingertips together in front of them with their arms fully extended. In a sober person the cerebellum senses the changes in position in the arms and adjusts the muscular tension so the fingertips will come smoothly and directly together. In someone "under the influence," the cerebellum fails to do this and the subject misses the mark.

The cerebellum is also involved in "simple" movement sequences, like reaching out and picking up a soda can. As the hand approaches the target, the cerebellum, using continuously changing information from the position sensors in the muscles, tendons, ligaments, and skin, inhibits some of the outgoing motor commands and facilitates others to bring the hand to the target.

Table 15-1 summarizes the functions of each of the major areas of the brain.

THE MENINGES AND VENTRICLES

Between the bones of the skull and the brain lie three membranes, or **meninges** (Figure 15-11). The outermost is the sturdy, protective **dura mater** ("tough mother"). The innermost is the **pia mater** ("tender mother"). Between them lies the delicate **arachnoid membrane** that has the consistency of a spider's web (*arachnoid* means spidery). The three form a triple shield around the brain, extending to surround the spinal cord as well. **Cerebrospinal fluid** flows out of the ventricles of the brain, where

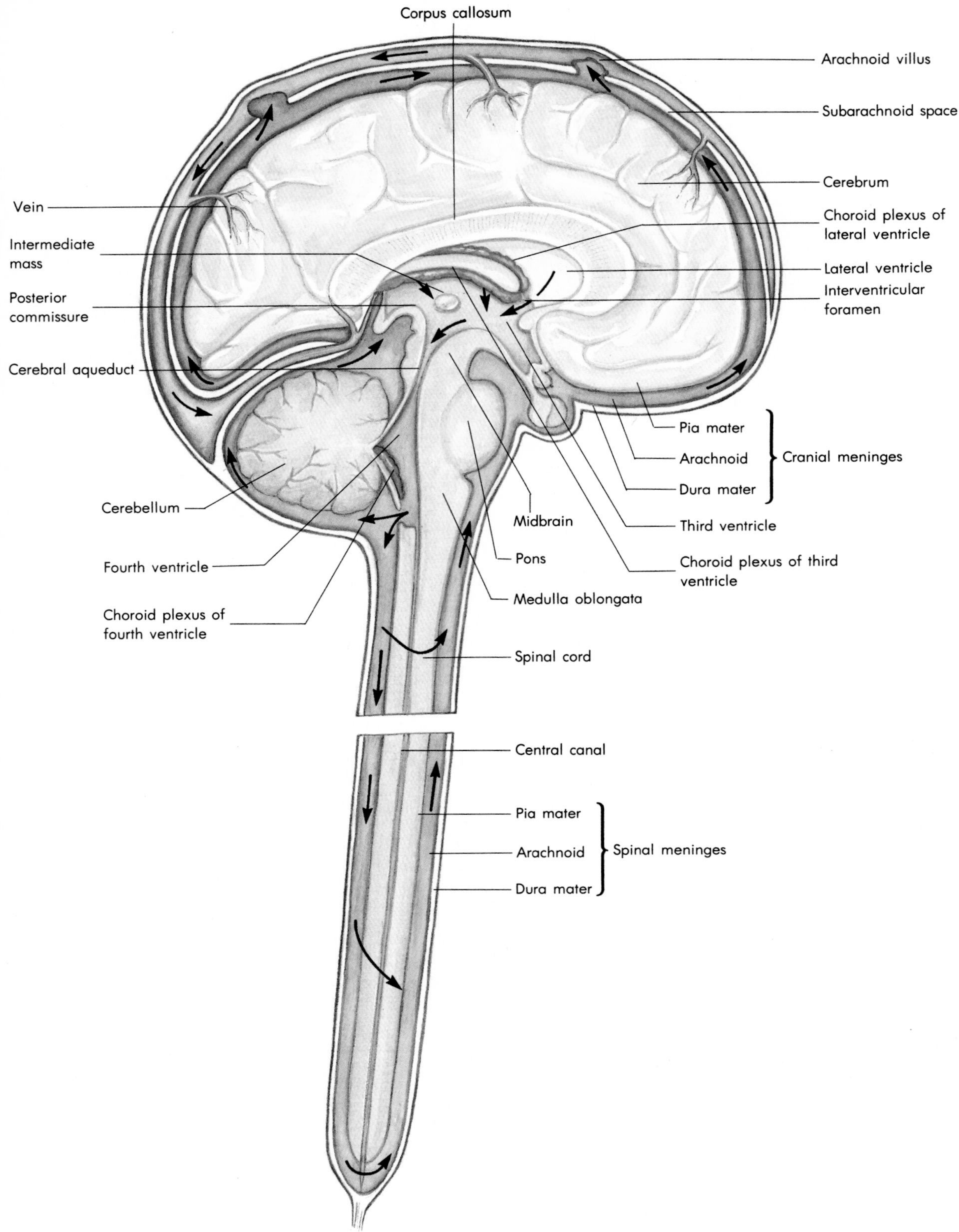

FIGURE 15-11 Sagittal section through the brain, showing meninges, ventricles, and the flow of cerebrospinal fluid.

Table 15-1

Brain Regions and Their Major Functions

REGION	FUNCTIONS
Cerebral hemispheres	
Frontal lobe	Initiates and controls voluntary movement; premotor association; controls speech muscles; intellectual processing (abstract thought, planning, studying); partly controls social behavior and inhibitions
Parietal lobe	Taste and body senses; analyzes data from skin, muscles, ligaments, and tendons for information about body position and condition; correlates data from different senses
Temporal lobe	Hearing; voice and sound records; long-term auditory memory
Occipital lobe	Vision; processes retinal data; long-term visual memory (e.g., faces)
Corpus callosum	Thick, flattened band of white matter (nerve fibers) interconnecting the cerebral hemispheres
Thalamus	Sensory relay station; routes afferent signals to appropriate brain areas for processing
Hypothalamus	A chief regulator of body functions; controls thirst, water balance, appetite, digestion, body temperature, endocrine system, and reproductive cycling
Basal ganglia (including limbic system)	A filter for neural noise that interferes with control of skeletal muscles; essential to formation of long-term memory; limbic portion partially controls emotions
Brainstem	
Midbrain	Passage for ascending and descending fibers carrying signals between CNS areas; controls part of righting reflex activated by loss of balance; contains superior and inferior colliculi for early processing of sight and sound data
Pons	Passage for ascending and descending fibers; routes some to and from the cerebellum
Medulla oblongata	Second major regulator of body's vital physiologic systems; controls heart rate, breathing rate, and blood pressure
Cerebellum	A major controller of posture, muscle tone, and balance; also helps coordinate many complex voluntary movements

it is produced, and then circulates around the outside of the brain and spinal cord in the space between the arachnoid and the pia mater, in what is called the subarachnoid space.

The cerebrospinal fluid is secreted by capillary networks (*choroid plexi*) in the walls of cavities within the brain. There are four of these **ventricles,** one *lateral ventricle* in each of the brain's halves, or hemispheres, a *third ventricle* in the center of the brain (surrounded by the thalamus and hypothala-

Learning Focus

ORGANIZING YOUR INFORMATION ABOUT THE BRAIN

Gather all the information you can about the various areas of the brain indicated on the diagram. Name each area of the brain and list the functions that area controls. Use all of your available sources of information.

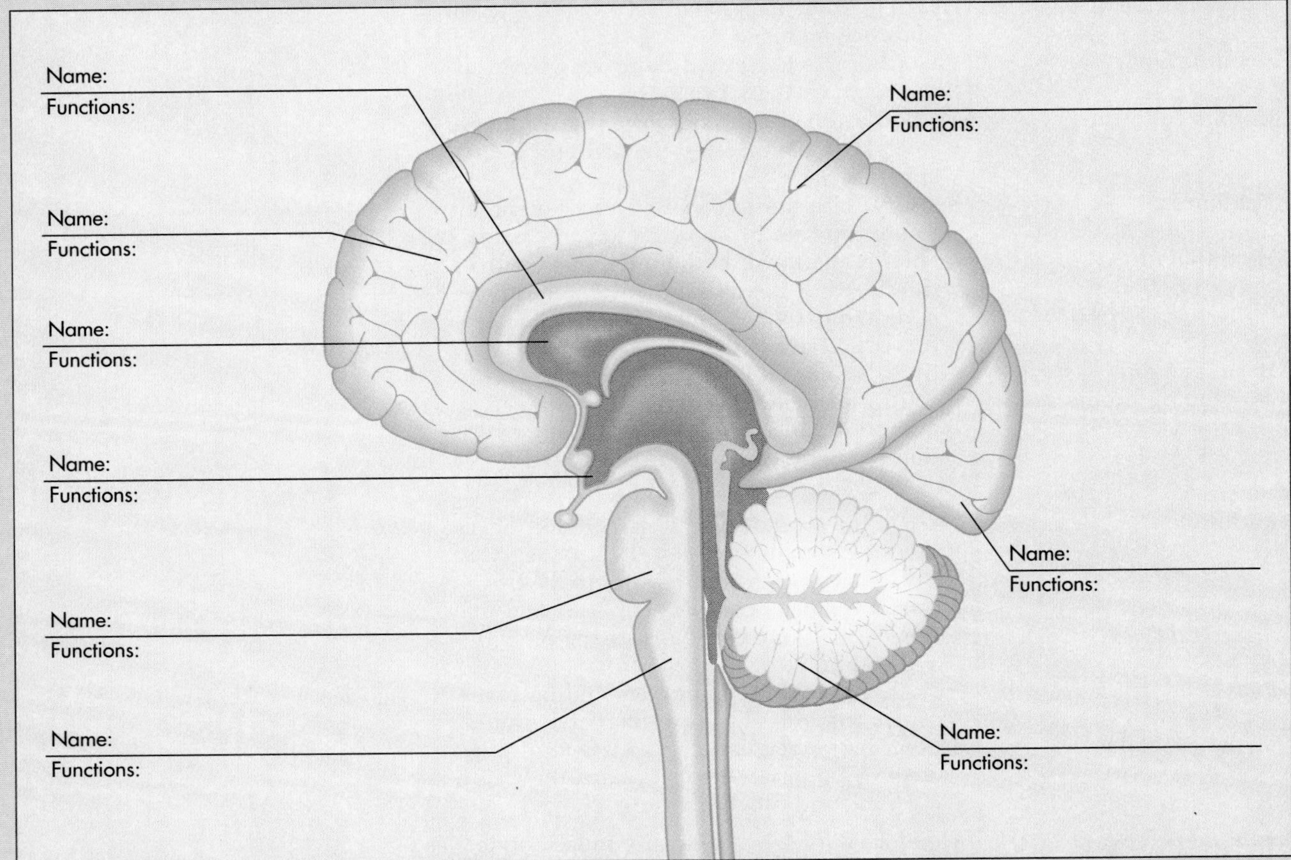

mus), and a *fourth ventricle* ventral to the cerebellum. They are linked by ducts and connected to the *central canal* of the spinal cord. The fluid flows from the choroid plexi through the ventricles to an opening behind the cerebellum. There it emerges to bathe the exterior of the brain and cord. Fingerlike projections of the arachnoid membrane, the *arachnoid villi,* protrude through the dura mater into venous sinuses, where they pass excess cerebrospinal fluid back to the blood.

Production and removal of cerebrospinal fluid normally keep pace with each other. However, sometimes production exceeds removal. In this case, in a fetus or a newborn baby, pressure can build up within the brain. The brain may actually be squeezed into a thin layer just within the skull. Since the sutures binding the bony plates of the skull together have not yet fused, the increasing intracranial pressure causes the head to enlarge dramatically (Figure 15-12). This condition, called *hydrocephaly* (water on the brain), can be treated even in the womb by installing a drainage tube to shunt excess fluid directly into a vein or the heart. Without treatment, mental retardation and even death are likely.

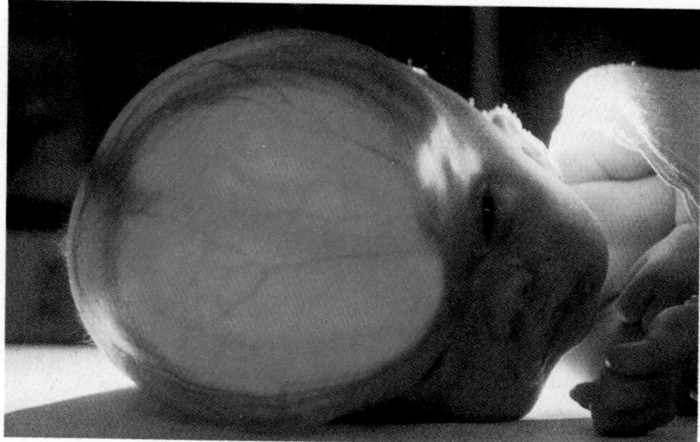

(a)

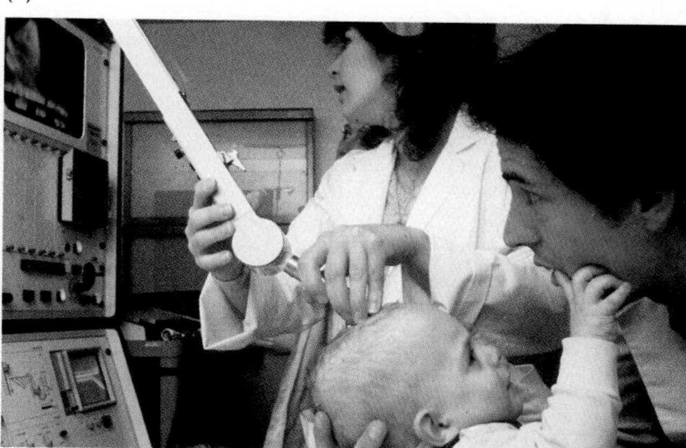

(b)

FIGURE 15-12 (a) Hydrocephaly can produce severe cranial distortions because restricted flow of cerebral fluid through the ventricles of the brain causes pressure buildup inside the cranium. This infant's head was strongly transilluminated to show the detail. (b) This 16-month-old hydrocephalic child was originally treated in the womb; his progress is being checked with ultrasound.

THE CRANIAL NERVES

The **cranial nerves** are bundles of nerve fibers directly connected to the brain. They bring afferent information into the brain from the various sensors in the head and neck or carry outgoing motor commands to muscles, glands, or organs located in the head and neck. One pair of cranial nerves, the vagus, carries signals into the chest and abdomen as well. Most of the 12 pairs of cranial nerves arise from nuclei in the medulla, pons, and midbrain. The nerves are visible in Figure 15-13.

Among the cranial nerves, the olfactory, optic, and vestibulocochlear nerves are purely sensory. They deliver sensory information, respectively, from the nose, eye, and inner ear. The rest of the cranial nerves carry both sensory and motor signals. One, the vagus (the name means the "wanderer"), is the only cranial nerve that leaves the immediate area of the head and neck; it is a major nerve of the autonomic nervous system, distributing as much as 80 percent of the fibers of the parasympathetic division to the organs of the thoracic and abdominal cavities.

THE SPINAL CORD

Continuing where the medulla ends, the **spinal cord** runs through the vertebral cavity protected by the vertebrae. It runs as a single cable, with the diameter of a finger, to just below the rib cage, where it breaks up into a bundle of many small nerves that resembles a horse's tail and is thus called the *cauda equina* (horse tail). Just before it loses its cablelike structure, it enlarges to contain the neurons that give rise to (and receive) the nerves of the legs. Another enlargement in the neck (cervical) area serves the arms. From each side of

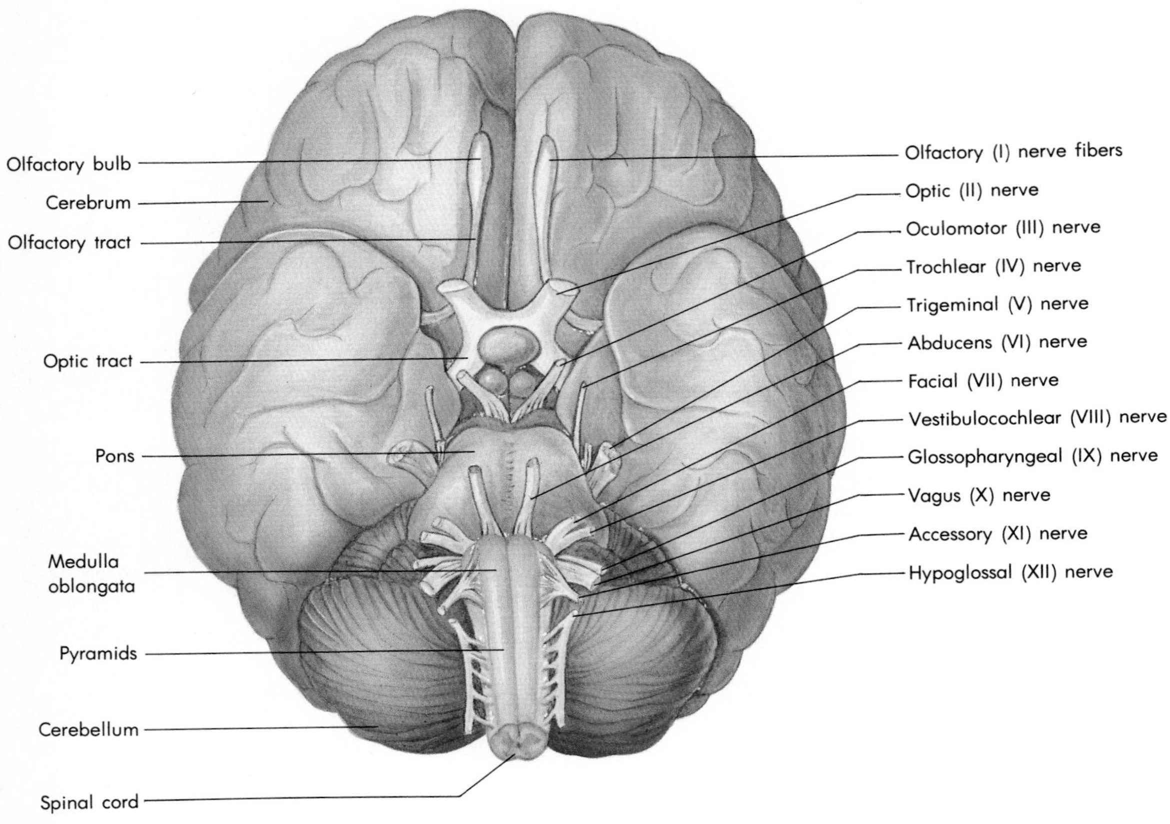

FIGURE 15-13 The underside of the brain, showing all 12 pairs of cranial nerves.

the cord arise 31 **spinal nerves,** 8 cervical, 12 thoracic (chest), 5 lumbar, 5 sacral, and 1 coccygeal (Figure 15-14).

Like the brain, the spinal cord is surrounded by the meninges, including the pia mater, arachnoid, and dura mater. These are the sites of the infections that cause spinal meningitis. The space between the arachnoid and the pia mater is the target for *spinal taps,* a procedure in which a physician inserts a hypodermic needle between the vertebrae to remove a sample of cerebrospinal fluid for diagnosis or to inject antibiotics, anesthetics, or other substances.

Like the brain, the spinal cord contains both gray matter and white matter. If we slice across the cord, we see how these masses of cells and nerve fibers are arranged (Figure 15-15). In the very center of the cord is the **central canal,** which is continuous with the fourth ventricle of the brain. Surrounding it is a butterfly-shaped mass of gray matter, made up of the cell bodies of many of the neurons of the spinal cord. The outer white matter of the cord consists of bundles of myelinated axons transporting nerve signals to and from the brain and other areas of the cord.

The **anterior** (or ventral) **horn** of the spinal gray matter contains the cell bodies of the motor neurons. The axons from these motor neurons leave the cord through numerous **spinal rootlets** that soon merge to form the **anterior** (or ventral) **roots** of the spinal nerves (Figure 15-15). The rest of the gray matter contains only interneurons (association neurons) that receive, process, and pass on signals from sense organs and the brain. The sensory neurons have their cell bodies in ganglia on the **posterior** (or dorsal) **roots** of the spinal nerves. Their dendrites extend all the way from sense organs in the skin, muscles, and other organs. Their axons pass into the posterior (dorsal) horn of the gray matter through the posterior (dorsal) spinal rootlets to form synapses on interneurons and motor neurons.

The dorsal (posterior) roots of the spinal nerves are purely sensory. The anterior roots are purely motor. The spinal nerves formed by their merger are mixed sensory and motor.

In the regions of the neck, shoulders, and hips, many spinal nerves merge to form a complex nerve net called a *plexus* (plural *plexi*). The major plexi include the **cervical, brachial, lumbar,** and **sacral plexi.** From these plexi emerge nerves bearing components from several spinal nerves to serve structures such as the neck, arm, leg, and respiratory muscles.

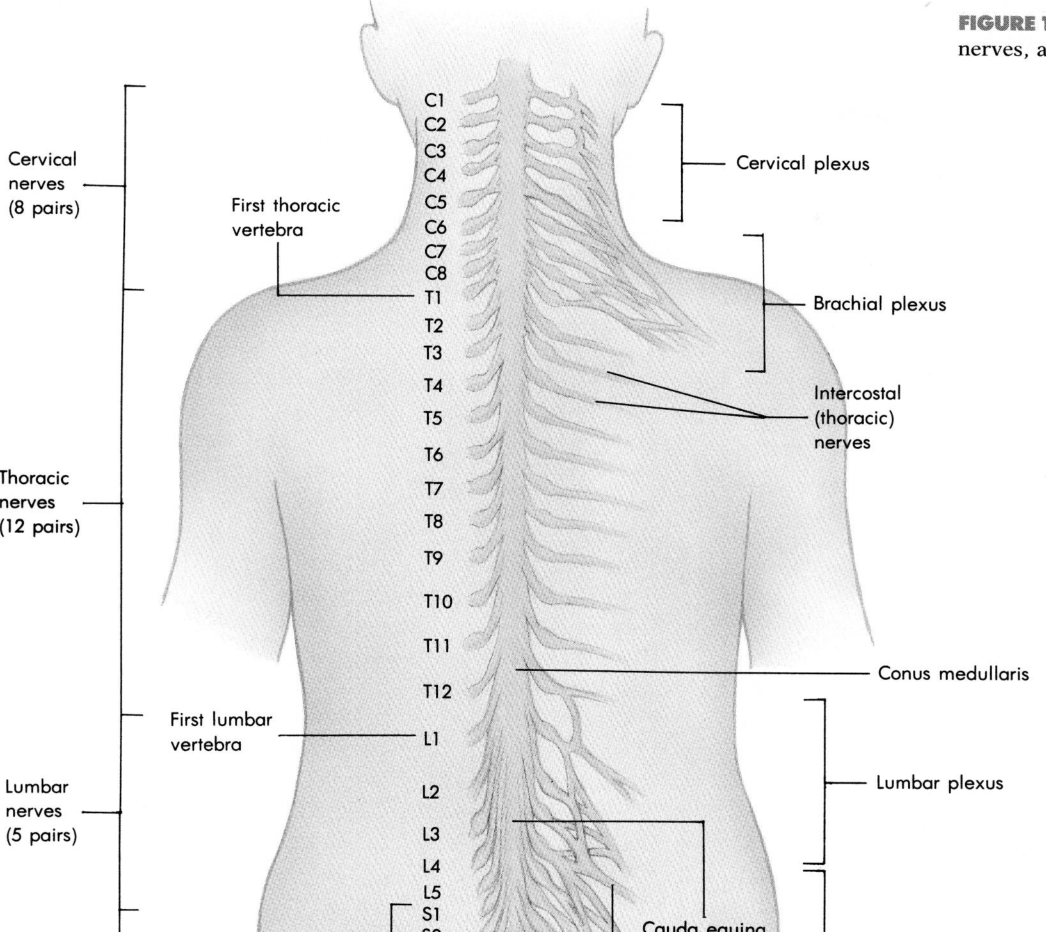

FIGURE 15-14 The spinal cord, spinal nerves, and nerve plexi.

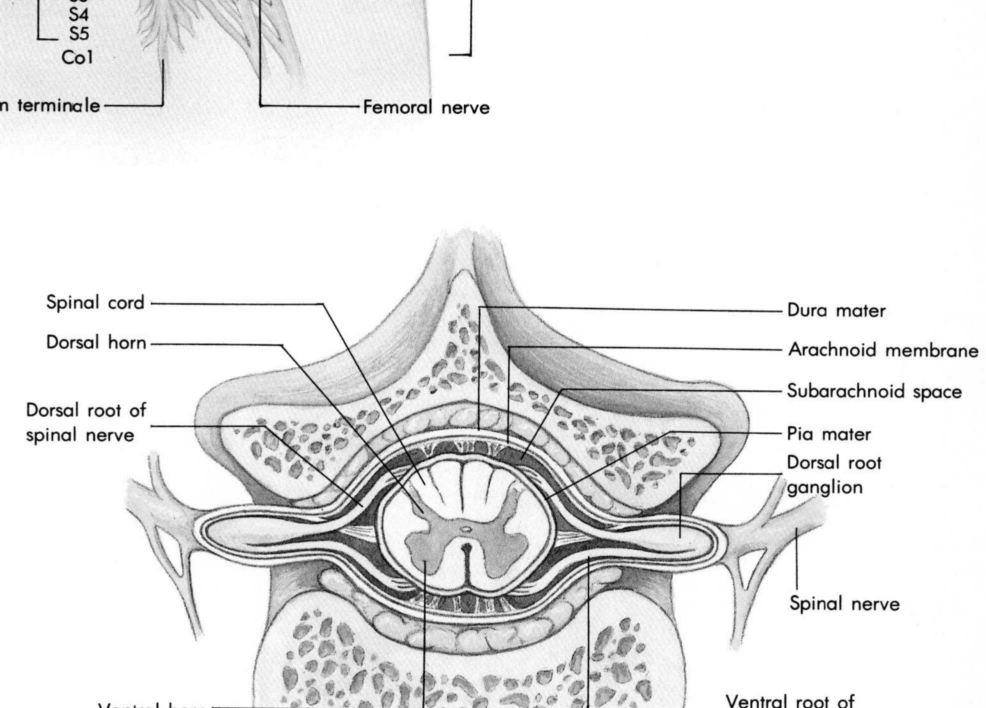

FIGURE 15-15 Cross section of the spinal cord, showing the spinal roots for the spinal nerves.

SPINAL REFLEXES

We have referred several times to reflexes; they are automatic, built-in nerve circuits that activate responses to certain stimuli. We see a reflex whenever we jerk our foot in response to the doctor's tap on the tendon just beneath the knee or twitch on being stuck with a pin. But how do reflexes work? What does *built in* mean?

The simplest reflex is the **stretch reflex.** This is the reflex that causes the knee jerk reaction to the doctor's rubber hammer. The tap on the tendon stretches the muscles of the thigh and activates stretch sensors (muscle spindles). A sensory nerve fiber carries the nerve signal from the sensors to the spinal cord, passing through the posterior (dorsal) root into the posterior horn of the spinal gray matter. The fiber passes through the posterior horn into the anterior horn to form a synapse on a motor neuron (Figure 15-16). The sensory signal thus goes directly to the motor neuron and sets off a motor nerve signal that activates the contractile elements of the same muscle. When the muscle contracts, it counteracts the stretch.

The function of the stretch reflex is to keep the length of a muscle constant. It can do this because when the central nervous system (CNS) commands a muscle to contract, it also commands the muscle spindles (which are modified muscle fibers) to shorten. When the muscle is stretched, so is the spindle, and nerve endings on the spindle generate the nerve signal that initiates the reflex. The stretch reflex is essential for holding position and balancing the body.

A more complex reflex is the **flexion reflex,** an automatic response to stimuli such as a pin prick. The sensory nerve fiber carries the pain signal to the posterior horn of the spinal cord, where it synapses with spinal interneurons. One of these interneurons sends an axon branch to a motor neuron that activates the muscle that will withdraw the pricked skin to safety. Other axons carry the pain signal to the brain and to other spinal motor neurons (Figure 15-17).

The complexity of the flexion reflex shows best when we consider how it works for a leg. Imagine stepping barefoot on a tack or a sharp stone. The reflex immediately makes you flex your leg and lift your foot. However, if that were all it did, you would promptly fall over. Fortunately, the reflex also activates motor neurons on the opposite side of the spinal cord to contract the extensor muscles in the other leg, providing additional support for the sudden change in body position. In addition, it activates other muscles involved in balance.

When the brain recognizes that a sudden lifting of a foot could be hazardous, as when you walk along a narrow ledge, it can inhibit the interneuron responsible for the flexion reflex. This interneuron thus represents a way to turn on or off a large group of muscles, and the action they perform, all at once. Not surprisingly, axons from the brain and from other parts of the spinal cord can also activate the interneuron. There is evidence that precisely such activations are involved in the control of walking, running, and other activities; that is, the CNS does not put movements together muscle by muscle but in terms of muscle groups like those activated by reflexes.

1. The longer the pathway a nerve signal must travel, the longer it takes. Do your muscles respond to a painful stimulus before or after you feel the pain? Why?
2. Strictly speaking, the flexion reflex flexes the pained leg. It is the *crossed extensor reflex* that stiffens the opposite leg. Compare the combined reflex to the natural movements of walking.

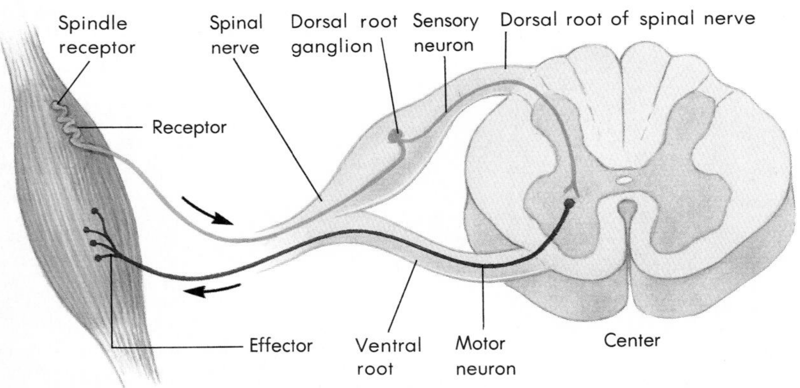

FIGURE 15-16 The stretch reflex begins when the muscle spindle or stretch receptor is activated, sending an action potential into the spinal cord by way of the sensory fiber. In the cord the sensory axon directly stimulates the motor neuron, which responds by generating an outgoing action potential that activates the motor end-plate and provokes the muscle to contract.

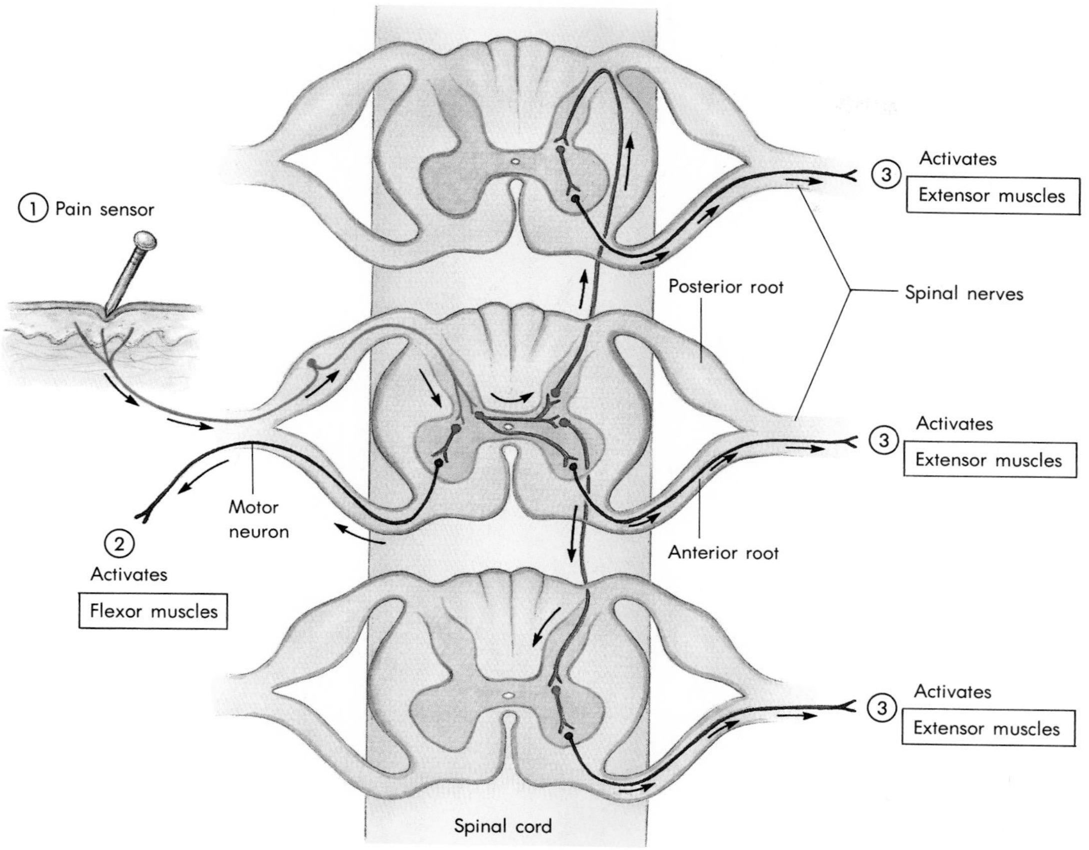

FIGURE 15-17 Complex reflex. In this flexor/crossed extensor reflex, the incoming sensory signal activates responses in both sides of the body almost simultaneously. (1) Stepping on a pin sets off a pain sensor, which sends a signal via interneurons to both sides of the spinal cord. (2) One branch of the signal path loops back to activate the flexor muscles of the leg and pull the injured foot away from the painful stimulus. (3) At nearly the same time, the extensor muscles on the other side of the body are activated to support the body during the weight shift. Note that the reflex activates neural responses at three different levels of the spinal cord.

Learning Focus

THE SEQUENTIAL PROCESSING OF LANGUAGE

Strokes happen when a blood clot lodges in a blood vessel in the brain or when a brain blood vessel bursts. During a stroke a portion of the brain loses its supplies of oxygen and nutrient and dies (or is seriously impaired). Damage from stroke and other causes can affect the gray matter of the cortex and subcortical nuclei and the white matter connecting patches of cortex or nuclei. The effect of the damage depends on the part of the brain it strikes, but any damage costs the brain some of its capabilities.

Some of the most interesting strokes are those that interfere with language processes. Damage to Broca's area in the frontal lobe results in difficulty in speaking (but not, intriguingly, in singing). The stroke victim's speech (and writing) is ungrammatical, slow, and labored, though the person has the words he or she needs.

Damage to Wernicke's area in the temporal

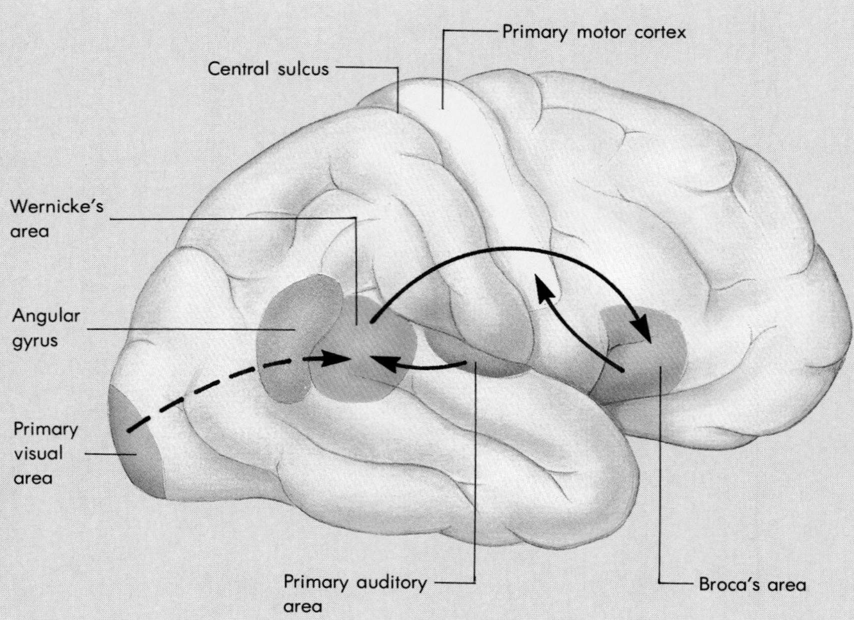

Diagram of information pathways in spoken (——) and written (_ _ _) words.

lobe has a very different effect. The victim loses vocabulary, though grammar remains. In addition, he or she has a great deal of trouble understanding spoken or written language.

Researchers have interpreted these symptoms (and other data) to mean that when one repeats a word someone else says, the information travels from the ear to the primary auditory area in the temporal lobe (see the accompanying figure). From there, the information goes to Wernicke's area, where the meaning of the word is extracted. The meaning then goes to Broca's area, where neurons generate the appropriate commands to the motor cortex, which controls the speech muscles.

When one reads a written word aloud, the visual information passes from the eyes to the primary visual area in the occipital lobe, through the angular gyrus in the parietal lobe, and then to Wernicke's and Broca's areas and the motor cortex. Writing a spoken word calls for transmitting information from the auditory cortex to Wernicke's area to the angular gyrus and then to Broca's area and the motor cortex. Speaking and writing internally generated (spontaneous) words use similar pathways, though—especially for writing—they are not as well understood.

Aphasia is the word for an inability to speak. *Agraphia* means an inability to write. Aphasia and agraphia often go together. Damage to what parts of the brain would you expect to produce such a dual defect?

Aphasia *can* occur without agraphia. Where must the damage be to produce this defect?

Agraphia can also occur without aphasia. Where must the damage be to produce this defect?

Clearly, the effects of a stroke, in terms of lost abilities, can tell a knowledgeable physician a lot about the location of a stroke. They can reveal this location even more precisely than our simple discussion of speech defects suggests, for there are many different kinds of aphasias and agraphias. To hint at the complexity concealed by the terms, let us just mention that aphasiacs can lose not simply the ability to speak but particular parts of their speech ability as well. Aphasics can lose nouns, verbs, colors, or numbers. They can even lose one entire language but not another—and a one-language person who loses it all can nevertheless learn to speak again *in another language*.

THE AUTONOMIC NERVOUS SYSTEM

The **autonomic nervous system (ANS)** controls many of the body's major physiologic systems. It works by regulating the activity of smooth muscle in blood vessels, respiratory tubes, the stomach and intestines, the digestive glands, the bladder, the genitals, and the eye. It also controls the heart rate and the adrenal medulla, sweating, salivation, and many other body functions. And it does it all automatically, without involving the conscious will. However, though the ANS is not under conscious control, you can learn, with the aid of *biofeedback* training, to affect it in many ways.

The ANS is purely efferent. That is, it has no sensory component of its own. The sensory signals it responds to are the same ones that flow into the CNS on the sensory fibers of the peripheral nervous system. Its own neurons are all interneurons and motor neurons, and it carries information only from the CNS to the organs it regulates.

The ANS has two anatomic and functional subdivisions. The **parasympathetic division** originates in the brain (the brainstem) and the lower (sacral) end of the spinal cord; it promotes relaxation and energy conservation, or *homeostasis*. The neurons of the parasympathetic division send axons through the cranial and sacral nerves to the organs they control. These axons do not, however, form synapses on the target organ's own cells. Instead, they first synapse on groups, or ganglia, of other autonomic neurons in or near the target organs. The fibers of the parasympathetic division use acetylcholine as their neurotransmitter. The tenth cranial nerve, the vagus, is the principal distributor of about 80 percent of the fibers of the parasympathetic system. The remaining 20 percent of the fibers are distributed from the sacral end of the spinal cord.

In contrast to the parasympathetic division, the **sympathetic division** of the ANS is an alarm network that simultaneously prepares a number of body systems for fight or flight in response to real or perceived threats. It is primarily involved with promoting survival and therefore often directly opposes the energy-conserving, homeostatic effects of the parasympathetic division. In fact, the two divisions of the ANS send fibers to many of the body's organs and often provoke opposite reactions in the same organs, as indicated in Table 15-2.

Table 15-2

Autonomic Effects

TARGET ORGAN	SYMPATHETIC EFFECT	PARASYMPATHETIC EFFECT
Pupil of eye	Dilation	[a]
Sweat glands	Secretion	[a]
Tear glands	[a]	Secretion
Salivary glands	Inhibition	Secretion
Digestive glands	Inhibition	Secretion
Adrenal medulla	Secretion	[a]
Adrenal cortex	Glucocorticoid secretion	[a]
Bronchi	Dilation	Constriction
Heart	Stronger contraction; accelerated rate	Weaker contraction; slowed rate
Blood vessels, heart and skeletal muscle	Dilation	Constriction
Blood vessels, skin and viscera	Constriction	Dilation
Liver	Stimulation of glycogen breakdown; inhibition of bile release	Stimulation of glycogen formation; increase in bile release
Gallbladder	Relaxation	Contraction
Stomach	Decreased motility	Increased motility
Intestines	Decreased motility	Increased motility
Kidney blood vessels	Constriction	[a]
Pancreas	Inhibition of secretion	Stimulation of secretion
Urinary bladder	Relaxation	Contraction
Sex organs	In male: constriction of ducts; ejaculation	In male: erection; in female: secretion; stimulation of reverse peristalsis of uterus

[a]No known effect.

As shown in Figure 15-18, the fibers of the sympathetic division originate in the thoracic and lumbar regions of the spinal cord. Some of the neurons also use acetylcholine as a neurotransmitter. Many use norepinephrine. The axons of the sympathetic division leave the cord through the anterior roots of the spinal nerves (like most other motor neurons). A short distance from the cord, the fibers leave the spinal nerve and form two chains of sympathetic ganglia that parallel the spinal cord. Most sympa-

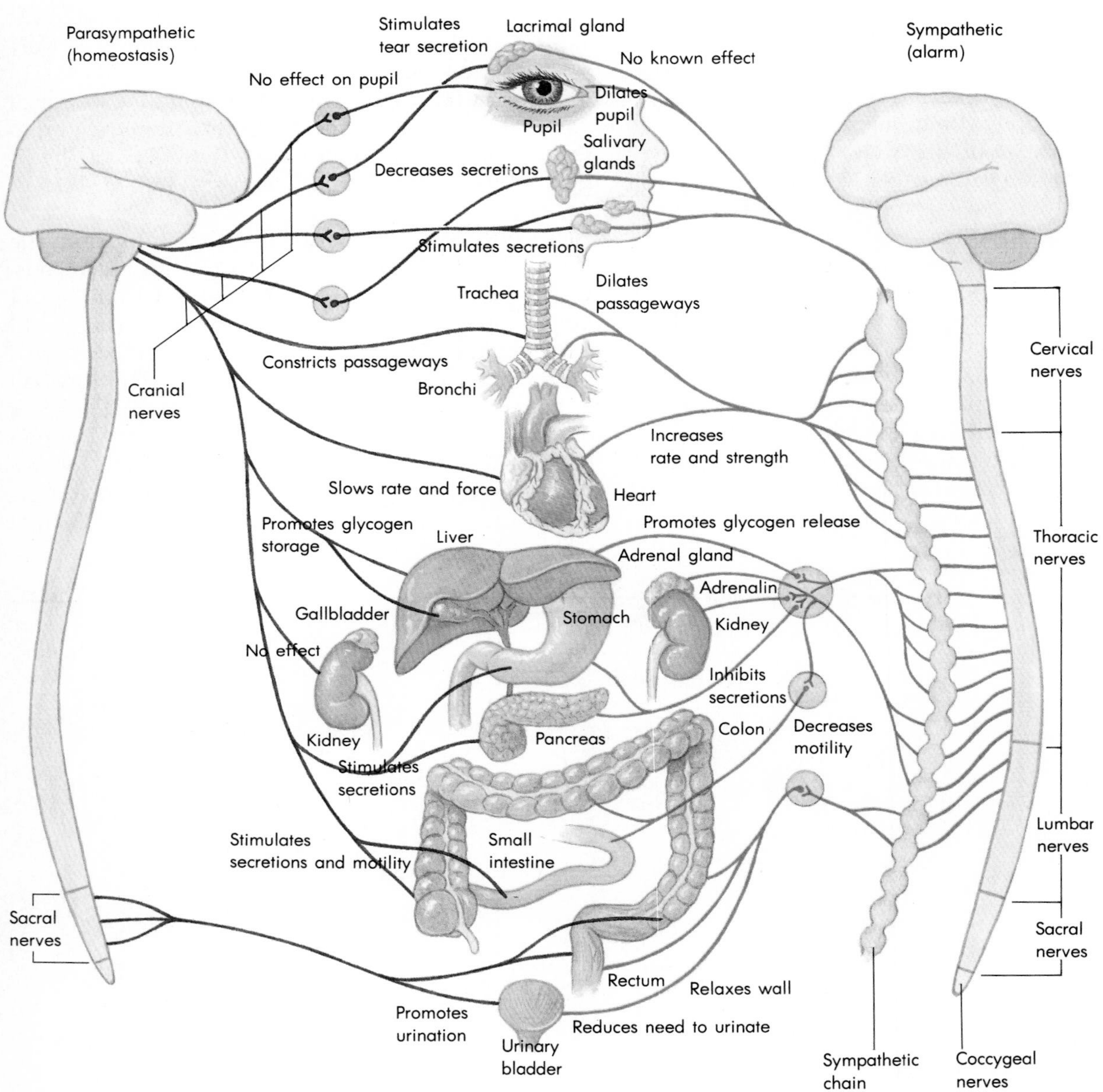

FIGURE 15-18 The fibers of the parasympathetic and sympathetic divisions of the ANS are not identically distributed. Parasympathetic fibers come from four of the cranial nerves. The vagus nerve distributes about 80 percent of the parasympathetic fibers and is the only cranial nerve that sends fibers to the organs of the thoracic and abdominal cavities. The lower portion of the parasympathetic division exits the CNS from the sacral plexus in the pelvic cavity. Sympathetic fibers leave the CNS via two chains of ganglia that parallel the spinal cord. Many organs of the body receive fibers from both ANS divisions, which generally oppose each other's actions. In general, parasympathetic fibers encourage a physiologic quieting of the body's systems. Sympathetic fibers activate changes that prepare the systems for coping with real or imagined threats.

thetic neurons synapse in these ganglia with other neurons whose axons then rejoin the spinal nerves and travel with them to the organs they regulate.

The modes of operation for the parasympathetic and sympathetic divisions of the ANS are different from one another. Because the sympathetic division prepares many areas and systems of the body to cope with potentially life-threatening events, it simultaneously activates many different response systems. However, the parasympathetic nervous system dominates during quiet times by individually adjusting the organs it controls.

1. How do the sympathetic effects listed in Table 15-2 prepare the body for sudden expenditures of energy, as in fight or flight?
2. What does the sympathetically induced dilation of the eye's pupil have to do with the need to fight or flee?
3. How do the parasympathetic effects listed in Table 15-2 support the body's activities when it is at rest, as after a meal?

BIOFEEDBACK

Biofeedback refers to one common method of learning to control some autonomic functions. In essence, it gives you a sense of (feedback on) what your body is doing at the autonomic level. For example, a biofeedback trainer who wishes to teach you to control your heart rate may fit you with an electronic heart rate monitor. A beeper then sounds whenever your heart rate increases (or slows) by 10 percent. Your goal is then to think of as many things as you can, trying on various states of mind until you find one that makes the beeper beep. Often the states of mind that successfully induce a particular physiologic response have no apparent relationship to the condition you desire to change. Indeed, it is very unlikely that visualizing a slowly beating heart will cause your heart rate to slow. Once you do find a thought pattern that induces the desired change, you must learn to reproduce it, and the beep, at will. Some people have learned to control heart rate, blood pressure, and even brain waves in this way.

SUMMARY

The central nervous system (CNS) consists of the brain and the spinal cord. The brain, encased within the skull, receives and processes information from the senses and issues commands to the body's muscles and glands. The spinal cord conveys information to and from the brain and is responsible for many reflexes, automatic matchings of stimulus and response.

The brain has three main parts: the forebrain, consisting of the cerebral hemispheres, basal ganglia, thalamus, and hypothalamus; the brainstem, consisting of the midbrain, pons, and medulla oblongata; and the cerebellum. As a whole the brain and spinal cord are surrounded by three membranes, or meninges: the dura mater, the arachnoid, and the pia mater. The cerebrospinal fluid, secreted by capillary-rich choroid plexuses in the brain's cavities or ventricles, circulates through the ventricles and then outside the brain and spinal cord between the arachnoid and the pia mater; it is absorbed into the bloodstream by the arachnoid villi.

The surface of the brain, the cerebral cortex, is marked by numerous folds (gyri) and grooves (sulci), some of which form boundaries between the frontal, parietal, temporal, and occipital lobes of the cerebral cortex. The central sulcus separates the frontal motor cortex from the parietal sensory cortex. The longitudinal fissure separates the right and left cerebral hemispheres. Each hemisphere registers sensation from and issues commands to the opposite side of the body. The left hemisphere is specialized for speech and logical, linear thinking. The right deals with artistic, intuitive thought.

The cerebral cortex consists of gray matter, numerous neurons, arranged in layers. The electrical activity of these neurons is responsible for electroencephalographic "brain waves." Beneath the gray matter lie bands of myelinated axons, or white matter, that link different parts of the cortex, connect the two hemispheres, and convey information to the rest of the brain and the spinal cord. Embedded in the white matter are several groups of neurons, or nuclei, the basal ganglia, thalamus, and hypothalamus. The basal ganglia are involved in muscle control and with the limbic system, which deals with emotion and memory. The thalamus is the routing center for sensory input destined for the cerebrum. The hypothalamus controls emotion, the endocrine system, and many body functions through the autonomic nervous system (ANS).

The brainstem links the forebrain and spinal cord, and through it pass many axons in both directions. In it also are numerous nuclei whose cells process sensory input and motor commands. It contains the reticular formation, which supports muscle tone and serves to filter out irrelevant sensory data. It also contains the nuclei from which arise most of the cranial nerves. The attached cerebellum plays a crucial role in coordinating motor activity; some of its cells even seem to be involved in acquiring conditioned (learned) reflexes.

The spinal cord's gray matter occupies an H-shaped region around the central canal. The anterior horn of the spinal gray matter contains the motor neurons that give rise to the anterior roots of the 31 spinal nerves. Sensory neurons have their cell bodies in the sensory ganglia on the posterior roots; they often synapse with interneurons in the posterior horn. The spinal white matter contains axons conveying nerve signals to and from the brain and other levels of the spinal cord.

Spinal reflexes occur when a sensory axon activates a spinal motor neuron, thus eliciting an automatic response to the sensory stimulus. In the simplest reflex, the stretch reflex, the sensory axon synapses with the motor neuron directly. In other reflexes, such as the flexion reflex, it synapses first on interneurons. These interneurons allow commands from the brain or from elsewhere in the spinal cord to activate the reflexes as parts of more complex actions.

The ANS regulates smooth muscle, glands, and the heart. It differs from the rest of the nervous system in that its motor neurons lie in ganglia outside the spinal cord; these neurons are thus called postganglionic neurons. Preganglionic neurons link the brain and spinal cord to the ganglia. In the sympathetic division of the ANS, most of the ganglia lie alongside the spinal cord, in the sympathetic chain; a few ganglia lie near the abdominal and pelvic organs they serve. In the parasympathetic division, whose preganglionic nerves arise from the brain and the sacral end of the spinal cord, the ganglia all lie in or near the target organs.

In those organs that receive both sympathetic and parasympathetic postganglionic axons, the effects of the two divisions of the ANS generally oppose each other. The sympathetic division prepares the body for action. The parasympathetic division is dominant in quieter times and encourages homeostasis. Both divisions are controlled ultimately by the hypothalamus, and neither is consciously controllable directly. However, the cerebral cortex does have links to the hypothalamus, and one can learn through biofeedback training to use some of these links deliberately, controlling heart rate and blood pressure, among other things.

Table 15-3 provides a summary of selected diseases of the nervous system.

Table 15-3

Clinical Summary: Selected Diseases of the Nervous System

DISORDER	DESCRIPTION
Alzheimer's disease	Failing memory and impaired thought processes linked to neurofibril masses inside neurons, protein deposits outside neurons, and loss of acetylcholine-using neurons. Cause unknown.
Brain abscess	Pocket of infection in brain. Likely to form as a secondary complication of bacterial infection elsewhere in body, often in the middle ear, sinuses, or lungs.
Hydrocephaly	Extreme enlargement of head caused by blockage of drainage of cerebrospinal fluid through ventricles in newborns and infants; resulting pressure damages the brain and often leads to severe mental retardation.
Meningitis	Inflammation of the membranes covering the brain and spinal cord. Usually caused by bacterial or viral infection. Severe cases may lead to brain and nerve damage and sometimes death.
Multiple sclerosis	Slowly progressing, chronic CNS disease marked by destruction of myelin, resulting in tremors, nystagmus (eye twitching), speech disorders, numbness and tingling of hands and feet, difficulty in motor coordination, and other sensory and motor impairments. The overall deterioration of the disease is relieved by periods of remission or improvement. The cause is unknown, but some suspect it is related to a failure of the immune system.
Neuralgia	Severe, sharp pain in region served by a sensory nerve. Possible causes include physical pressure on the nerve, nutritional deficiencies, and exposure to certain toxins.

DISORDER	DESCRIPTION
Parkinson's disease	Sometimes called the "shaking palsy." A chronic, nonfatal, degenerative brain disease. Symptoms include hand tremors and weakness and extreme rigidity of voluntary muscles, frozen facial expression, and distinctive gait.
Poliomyelitis	Potentially fatal or paralyzing viral infection that attacks the CNS, especially the spinal motor neurons. Incidence may be sporadic or epidemic. It is readily prevented by oral vaccines, and all of us should keep our polio vaccinations current.
Shingles	Painful infection of a sensory nerve by the herpes zoster (chicken pox) virus, which may lie dormant for long periods. Eruptions are often marked by skin blisters and discoloration along the line of the infected nerve (usually in the area of the ribs).
Stroke	Rapid, severe reduction in blood flow to parts of the brain, resulting in unconsciousness, paralysis, loss of sensation, and sometimes death. An ischemic stroke is caused by blockage of an artery by a blood clot or atherosclerosis (cholesterol deposits); clots are responsible for about half of all strokes. A hemorrhagic stroke results from bleeding in the brain, as after severe injury or a ruptured blood vessel (aneurysm).
Tabes dorsalis	Damage to the posterior columns of the spinal cord, resulting in inability to sense limb position, caused by syphilis infection. As disease progresses, the ability to coordinate leg movements is also lost.

STUDY QUESTIONS

1. Describe the flow of the cerebrospinal fluid. Where does it originate? Where does it go?
2. Sketch a side view of the cerebrum. Name the lobes and briefly outline their functions.
3. What is the difference between the central sulcus and the longitudinal fissure?
4. How might electrical shocks to the cerebral cortex help reveal the functions of different cortical areas?
5. How can brain damage (strokes, tumors) help reveal the functions of brain areas?
6. What are brain waves? Why are they useless for mind reading?
7. Under what circumstances might blind people retain a kind of vision?
8. How might the brain use the flexion-crossed extension spinal reflex to generate the movements of walking?
9. Do the posterior roots of the spinal nerves contain preganglionic or postganglionic autonomic nerve fibers?
10. Explain how you can learn to control some of your autonomic functions.

GLOSSARY

Anterior roots Bundles of motor axons formed from merger of anterior spinal rootlets.

Autonomic nervous system (ANS) Portion of nervous system controlling smooth and cardiac muscle and glands; not normally consciously controllable.

Basal ganglia Nuclei within forebrain devoted to motor control, memory formation, and emotion.

Brainstem Midbrain, pons, and medulla oblongata; contains nuclei for most cranial nerves and control centers for automatic functions such as breathing.

Cerebellum The "little brain" located just behind and below the cerebrum. It controls muscle tone and posture and adjusts the complex movement patterns initiated by the cerebrum in a manner that brings the movement to target in a coordinated manner.

Cerebral cortex Surface layer of cerebrum.

Cerebral hemispheres Two halves of the cerebrum; each one controls the opposite side of the body.

Cerebrospinal fluid Fluid secreted in brain's ventricles; circulates through ventricles and between arachnoid and pia mater.

Corpus callosum Band of millions of nerve fibers linking the two cerebral hemispheres.

Electroencephalogram Record of the electrical activity of brain neurons, recorded by electrodes on the scalp.

Forebrain Cerebrum, basal ganglia, thalamus, and hypothalamus; highest level of the brain.

Gyrus Fold of cerebral cortex.

Hypothalamus Located below thalamus; contains nuclei controlling endocrine system, autonomic system, and emotion.

Longitudinal fissure Deep cleft between the two cerebral hemispheres.

Meninges (singular *meninx*) Membranes that enclose the brain and spinal cord; dura mater, arachnoid, and pia mater.

Midbrain Anterior portion of brainstem; contains nuclei for processing visual and auditory input and for some cranial nerves.

Parasympathetic division Portion of autonomic system arising in head and sacral cord; promotes energy conservation and a state of physiologic well-being.

Posterior roots Bundles of sensory nerve fibers entering the posterior or dorsal spinal cord.

Primary motor cortex Strip of cerebral cortex on frontal side of central sulcus; generates voluntary motor commands.

Primary sensory cortex Strip of cerebral cortex on parietal side of central sulcus; receives sensory data from body.

Reticular formation Network of interconnected neurons in brainstem; acts as a sensory filter and to maintain muscle tone and alertness.

Spinal cord A cablelike portion of CNS protected by vertebrae; carries signals between brain and the spinal nerves; its neurons handle many reflexes.

Spinal nerves Thirty-one nerves entering and leaving the spinal cord.

Stretch reflex Wired-in link between sensory neurons that detect muscle stretch and motor neurons; keep muscle length constant.

Sulcus Groove or cleft between adjacent gyri.

Sympathetic division Portion of autonomic system arising in thoracic and lumbar cord; coordinates the body for action.

Thalamus Portion of brain in wall of third (central) ventricle. It relays incoming sensory signals to the appropriate area of the cerebrum.

Ventricles Cavities within the brain.

Chapter 16

Monitoring the Universe: The Sensory Organs

- The Problem of Transduction
- Sense Organs and Receptors
- The Survival Senses
- The Muscle Senses
- The Sense of Pain
- The Skin Senses
- Taste and Smell
- Learning Focus: Receptive Fields
- Vision
- The Senses of the Ear
- Hearing
- The Senses of Balance
- Learning Focus: How Many Senses Do You Have?
- Learning Focus Response: How Many Senses Do You Have?
- Summary
- Study Questions
- Glossary

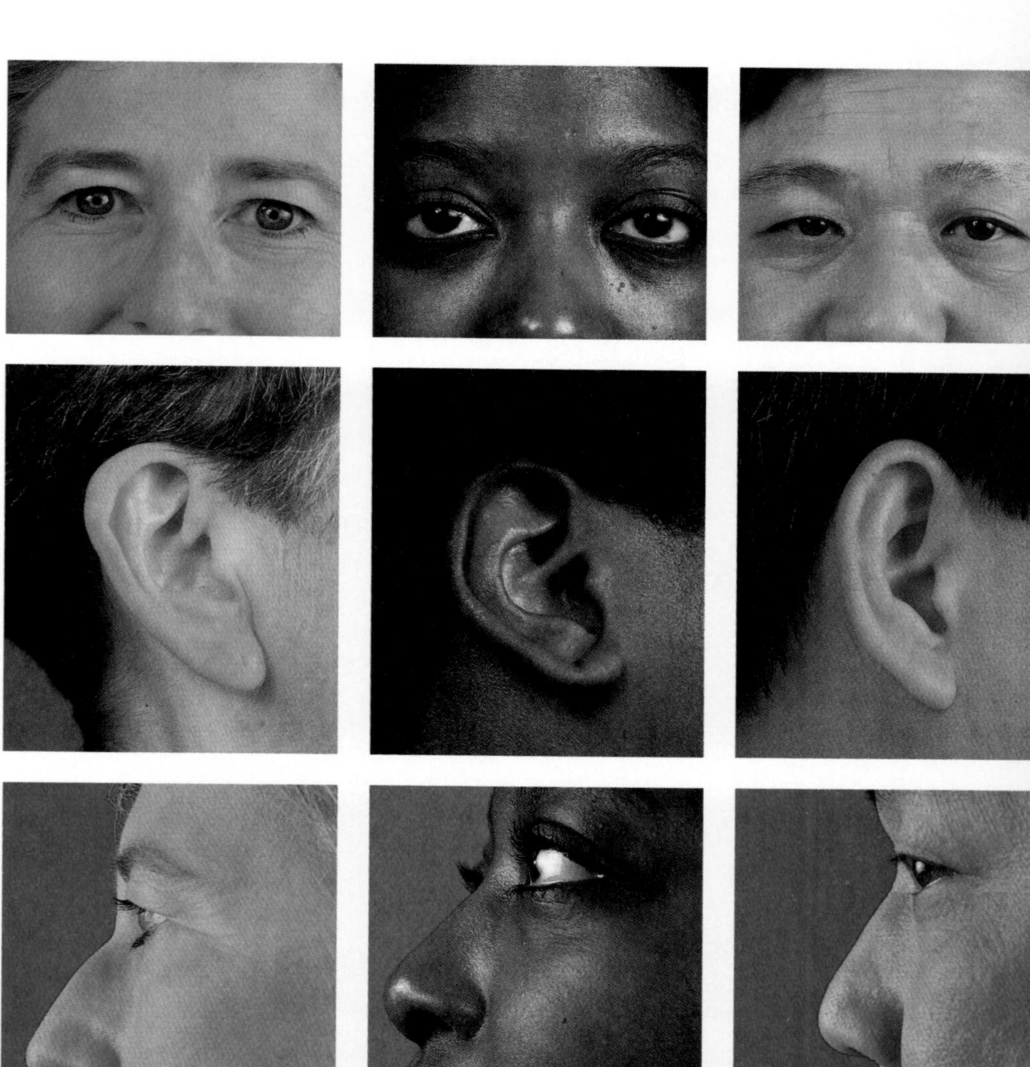

The central nervous system (CNS) cannot function properly all by itself. For proof, consider what happens in sensory isolation experiments: The researchers put the subject in a head-to-toe suit resembling a scuba diver's drysuit, adding a blindfold and earplugs. They then immerse the subject in a tank full of body temperature water, supplying odorless air through a hose (Figure 16-1). Soon after immersion, the subject gets used to the pressure of the suit on his or her skin and stops feeling it. No light reaches the eyes. No sound reaches the ears. As far as the brain is concerned, the universe has vanished.

For a while, some people enjoy the experience, finding it very restful. But this state does not last. The sense of time goes awry—moments seem like hours and hours like days or weeks. The brain grows hungry for sensation of any kind, and before long it is generating hallucinations as substitutes for the sensations it lacks. Eventually an artificially induced psychosis very much like schizophrenia develops. Fortunately, the madness passes when the isolation ends.

Less extreme sensory isolation, such as that experienced by a prisoner in solitary confinement or a shipwrecked sailor, has similar effects, though it takes longer to develop. The similarity has led some people to think that laboratory-type isolation could be used in the same ways as solitary confinement—for instance, as an aid to brainwashing. Victims of such treatment can be so grateful to those who return sensation to them that they will reveal all their secrets, swear loyalty, or turn over their fortunes.

THE PROBLEM OF TRANSDUCTION

The brain needs input. Yet the brain's neurons have no way of interacting directly with the outside world. They generate nerve signals only when stimulated by neurotransmitters released by other neurons. The question we must discuss therefore becomes how the brain gets the input it needs.

When we discussed the nerve signal in Chapter 14, we said that the action potential appears when the neuron's cell membrane is disturbed mechanically, chemically, or electrically. This is the key.

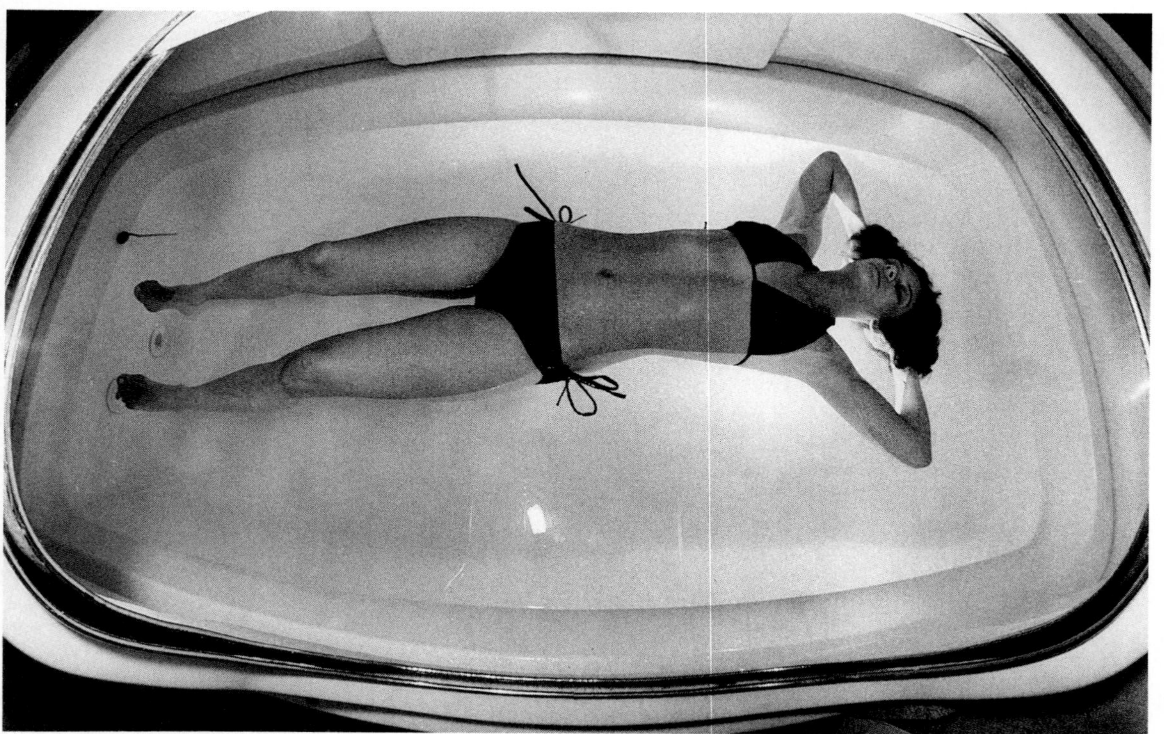

FIGURE 16-1 Sensory deprivation tanks isolate a subject's brain from as much sensory input as possible. After the initial period of getting used to the new environment—which lacks visual input, sound, temperature, and even pressure changes—the subject's brain becomes very hungry for sensation, sometimes with bizarre effects.

The nervous system's various sensory cells are equipped with specialized structures that magnify the side effects of external (and internal) events enough to disturb their cell membranes and generate action potentials. Some are sensitive to light. Others are sensitive to sound, vibrations, pressure, chemicals, stretch, temperature, and damage.

Transduction is the conversion of an event into a signal that we can interpret. The conversion is very simple in the case of a thermometer. A mercury thermometer depends on the way mercury expands or shrinks with changes in temperature. We turn mercury into a temperature transducer by confining it in a thin glass tube so that as it expands or shrinks, the top of the mercury column moves along a printed scale. We will see many examples of transducers as we discuss the body's senses. Some are as simple as a thermometer. Others are more complex. But all convert events outside the CNS into signals the CNS can process and interpret. However, the signals do not differ from sense to sense. An action potential is an action potential. The CNS determines which signals correspond to which senses according to what nerves they arrive on and where they arrive in the brain and spinal cord.

---

1. Go back to the discussion of the cerebral cortex in Chapter 15 and list the destinations in the cortex of several senses.
2. *Synaesthesia* means a confusion of the senses, as in seeing sounds or hearing touch. It can accompany drug-caused hallucinations, insanity, and other disorders. Explain how synaesthesia might develop.

SENSE ORGANS AND RECEPTORS

The transducers for the body's senses are the **sense organs.** Some sense organs, like the eye and ear, are multicellular, multitissue organs. Within them are the specialized neurons that respond to specific stimuli and actually generate the nerve signals that go to the CNS. We call these neurons **sensory receptors.** For other senses, like touch, the receptors are the tips of the dendrites of sensory neurons, and the sense organs are not much bigger, consisting of tiny knots of connective tissue surrounding the dendrite tips. For the sense of pain, mediated by naked and branching dendrites, the sense organs and the receptors are the same structures.

THE SURVIVAL SENSES

We divide the senses into two groups. The **exteroceptive senses** respond to events outside the body; they are the traditional five senses of vision, hearing, smell, taste, and touch. The **interoceptive senses** respond to events inside the body; they include the senses of muscle stretch, joint angle, gut distension, pain, and body temperature.

Although all the senses are essential for survival, some are less essential than others. People without vision or hearing are handicapped, but with support from family, friends, and professionals they can, to a large extent, use their other senses as substitutes. People without a sense of smell or taste may be a little more likely to occasionally ingest spoiled food, but their eyes usually protect them from making such mistakes. People without a sense of pain, however, die easily, for they do not know when they are being damaged. People who lack a sense of muscle stretch or tension cannot move properly and are therefore prone to accidents, damage, and death. People who lack senses of body temperature, blood sugar or carbon dioxide levels, and other internal conditions die most easily of all. Pain and the interoceptive senses may thus be the most important ones for survival.

We generally reserve the term *sense* for those cells and groups of cells that turn nonneural events into nerve signals, but we have previously mentioned several interoceptive senses that produce hormonal signals, either alone or in combination with neural signals. They include pancreatic sensors for blood sugar level; parathyroid sensors for blood calcium; hypothalamic sensors for hormones, body temperature, and blood volume; and kidney sensors for blood volume and pressure. Many of these senses are single cells that rely on cell membrane receptors to register the presence of specific molecules—such as glucose and hormones—in the extracellular fluid. They are crucial to survival because of their role in the body's homeostatic feedback mechanisms.

THE MUSCLE SENSES

Although we might describe many interoceptive senses in detail, let us focus on those that help us control our muscles. We are rarely as aware of these **proprioceptive** (self-feeling) **senses** as we are of hearing or vision, but they are working all the time, whenever we make a movement. They are also essential, for the CNS requires feedback on the results of its motor commands; if a researcher cuts the sensory nerves from the body's muscles, movement becomes poorly coordinated.

Motor coordination relies on three senses. In the ligaments surrounding the joints, there are nerve endings that are sensitive to joint movement; the CNS uses their data to tell where the body's parts are and calculate how to move them to specific places. In the tendons that join each muscle to its bones there are **tension receptors** (Golgi tendon organs) that report the load on the muscle; the CNS uses their data to limit the strength of contraction and prevent damage to muscles and tendons. In the muscles themselves are the **muscle spindles.**

Each muscle spindle consists of several modified muscle fibers enclosed in a fibrous sheath or capsule (Figure 16-2). These fibers receive commands to contract at the same time as normal muscle fibers, but unlike normal muscle fibers, they play a sensory role. The tips of sensory dendrites wrap around the spindle fibers and spread over their surfaces. When a sudden load, such as a book dropped into an outstretched hand, stretches the muscle beyond its commanded length, the spindle fibers also stretch, as do the nerve endings or receptors on them. This stretch disturbs the receptors' membranes mechanically and initiates action potentials, which inform the CNS of the stretch. The CNS then corrects the stretch with additional commands to the muscle (the stretch reflex).

1. We said the muscle and other interoceptive senses are essential to survival. Consider a person who lacks receptors for joint movement (or angle). What would you expect this lack to do to his pattern of movement? Would he be able to swat a fly, play ping-pong or handball, or even drive a car? Why or why not?
2. Consider a person without muscle spindles. What would happen if you handed her a book? If she lost her balance?

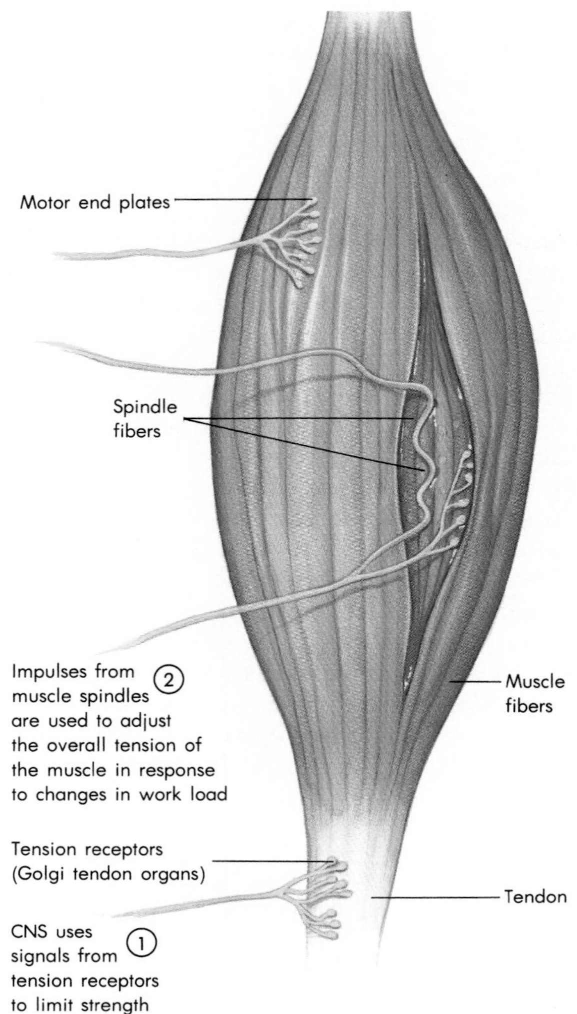

FIGURE 16-2 Two kinds of tension receptors are associated with muscles. The CNS uses signals from Golgi tendon organs (1) to limit the strength of muscle contractions. Muscle spindles are adjustable tension receptors in the muscles. When a sudden load stretches the spindles beyond a preset length, they send off sensory signals (2) that the CNS uses to send compensating signals to the muscle. The result is that the length of the muscle remains constant.

THE SENSE OF PAIN

Pain receptors are found everywhere in the body except in the CNS. It is thus possible to do brain surgery while the patient is still conscious using only a local anesthetic to deaden sensations from the scalp and skull. Once the surgeon is into the brain itself, the patient does not feel pain.

The pain receptors are the terminal branches of the dendrites of certain sensory neurons (Figure 16-3). Those receptors in skin, muscles, tendons, and joints give rise to **somatic** (body) **pain.** Those in the organs found in the body cavities produce **visceral pain.** Often, visceral pain seems to come from the surface of the body even though it is the heart, liver, or other internal organ that is hurting. We then call the pain *referred pain.*

Typically, pain receptors respond to any kind of stimulus if it is strong enough to cause or threaten damage to the affected cells. We all know that excess heat (a burn) and pressure (a blow) are painful, but so are excess cold, light, and sound and excessive distension of the stomach and intestines. Pain receptors also respond when a part of the body re-

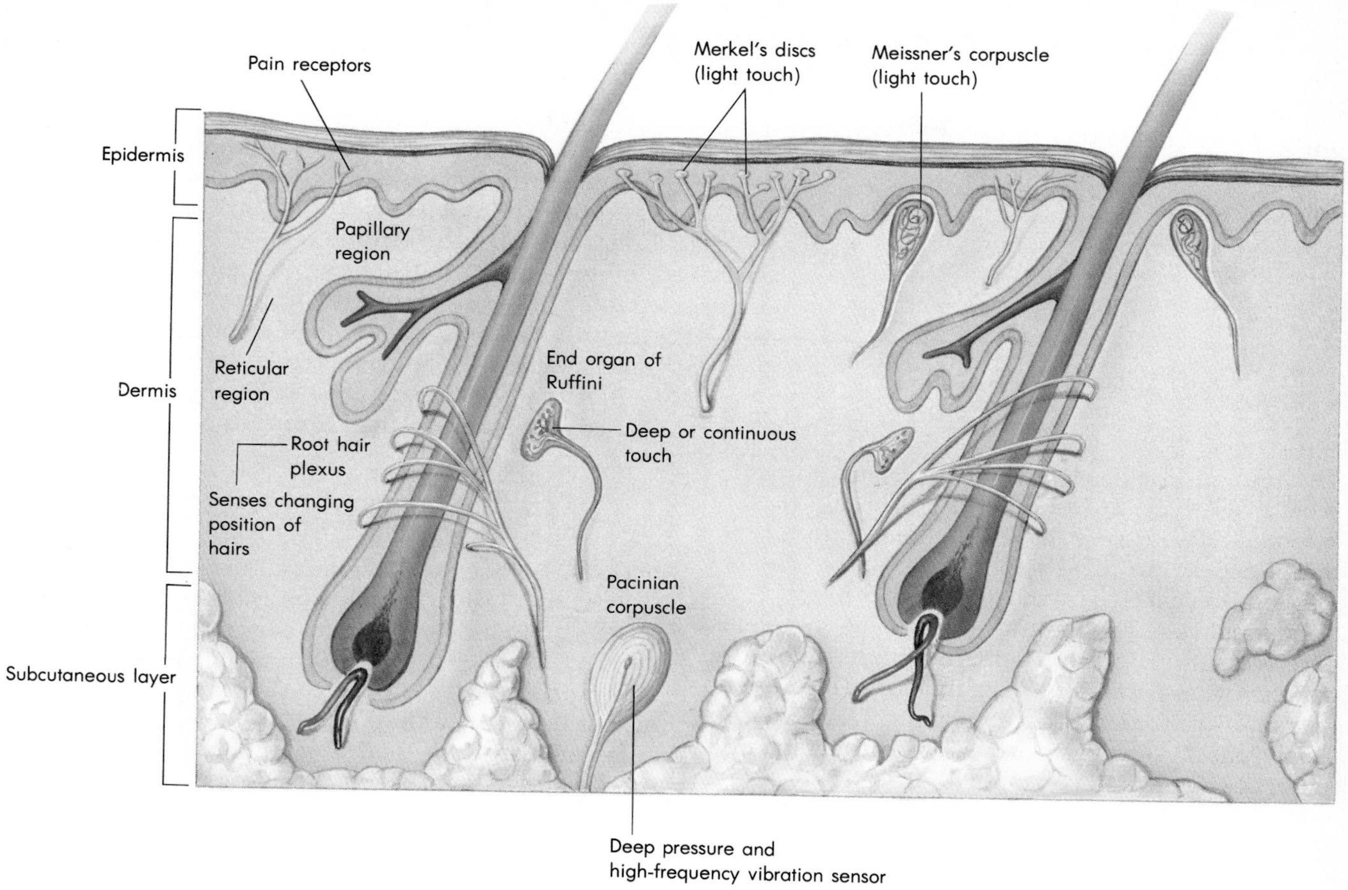

Pain receptors
Merkel's discs (light touch)
Meissner's corpuscle (light touch)
Epidermis
Papillary region
Reticular region
Dermis
End organ of Ruffini
Deep or continuous touch
Root hair plexus
Senses changing position of hairs
Pacinian corpuscle
Subcutaneous layer
Deep pressure and high-frequency vibration sensor

FIGURE 16-3 The skin contains many different kinds of nerve endings.

ceives too little blood flow, and hence oxygen. This sensitivity may account for the pain of prolonged muscle contractions and muscle cramps. The pain of wounds is due to the sensitivity of pain receptors to substances released by damaged cells.

The chains of neurons that convey signals from pain receptors through the spinal cord to the brain use some special neurotransmitters, including substance P. The brain can inhibit the transmission of pain signals with its own opiatelike substances, the endorphins and enkephalins. The brain can also inhibit its own reception of pain signals, via endorphins, enkephalins, and inhibitory synapses in the reticular formation and spinal cord. And strange though it may seem, pain can be inhibited by some forms of competing sensory or neural activity. Dentists often suppress pain by giving their patients music through earphones, and people who are intent on a task—especially in stressful emergencies—have been known to ignore even severe pain.

How important is the sense of pain? Leprosy, or Hansen's disease, was once thought to make the victim's fingers, toes, and nose fall off. We now know that the infection stimulates the resorption of bone and destroys nerve endings, impairing the sense of pain. The bone resorption causes fingers and toes to shorten, but it was the insensitivity to pain that cost lepers their parts, for they failed to notice when rats were chewing on their bodies (Figure 16-4). Today, with better sanitation and rodent control, lepers remain whole.

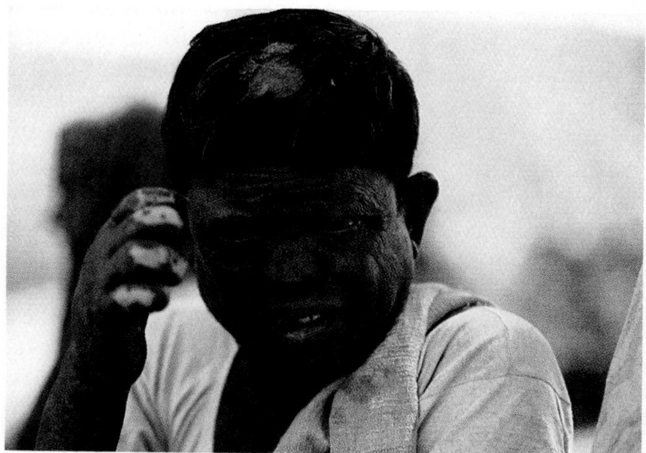

FIGURE 16-4 Leprosy, or Hansen's disease, is an infection that causes resorption of bone and insensitivity to pain. In many cases the latter caused victims of the disease to lose their body parts by not noticing when the afflicted parts were damaged in some way.

A few people lack a sense of pain from birth. They find it difficult to avoid burns and other injuries and used to die young from infections. Modern medical care helps them live long enough to learn the caution so essential to their survival.

THE SKIN SENSES

The human body meets the world surrounding it most extensively at that boundary we call the skin. Sensory receptors in the skin collect data on how the environment affects the skin, and we call the information carried by those receptors *touch*. However, touch is not a single sensation. Our skin registers pressure, vibration and roughness (vibration caused by moving skin over an object), temperature, hair movement, and even pain. There are separate sensory receptors for each of these components of touch (Figure 16-3).

We have already mentioned pain receptors. The others include networks of dendrite endings surrounding the roots of skin hairs. They generate signals when the hairs move and are responsible for the sensations we feel when a light breeze blows, an insect creeps across our skin, or a horror movie raises the hair on our necks. Just under the epidermis lie **Merkel's discs** and **Meissner's corpuscles.** Both are receptors for light touch, and the latter are most common in the fingertips, palms, soles of the feet, lips, tip of the tongue, nipples, and genitals. Meissner's corpuscles also respond to low-frequency vibration. **End organs of Ruffini** lie deeper in the dermis and respond to heavier and continuous touches. **Pacinian corpuscles** lie still deeper and respond to pressure and high-frequency vibration. They are also in the membranes surrounding muscles, in the viscera, and in the joints, where they may serve as movement detectors. **Free nerve endings,** resembling the pain receptors, may be responsible for our sensitivity to heat and cold.

Free nerve endings are both sense organs and sensory receptors. The other skin receptors make a distinction between the two. The Pacinian corpuscle, for instance, consists of a nerve ending, or receptor, enclosed in an onionlike structure. The onion or corpuscle consists of many layers of connective tissue; when the corpuscle is distorted by pressure, the layers slide against each other and magnify the pressure to disturb the receptor's membrane.

TASTE AND SMELL

Like touch, the sense of taste reports on events at the boundary between the body and the world. It responds to molecules of food and other substances dissolved in saliva in the mouth. Smell is a longer range sense, for it detects molecules in air. These molecules can evaporate from substances in the mouth and rise into the back of the nose, modifying our perception of how food tastes. But they can also enter the nose on the air we breathe and inform us of events much further from the body.

Taste

The sense of taste is the more limited of the two. Its sense organs are the **taste buds** of the tongue (a few are also found on the sides of the throat). They are found on the sides of the tongue's *papillae,* the bumps that give the tongue its rough appearance. There each taste bud is embedded in the tongue's epidermis as an oval cluster of **gustatory (taste) cells** surrounded by a capsule of *supporting cells* (Figure 16-5, p. 362). Taste cells are not nerve cells, but *neuroepithelial* cells generated by actively dividing parent cells at the base of the taste bud. They live for about 10 days before being replaced. While they live, they extend fine **gustatory hairs** through a pore at the surface of the tongue. The cell membrane of the hairs bears receptor molecules that can bind to molecules of specific shape in the saliva. There are four such receptor molecules, one each for sweet, sour, bitter, and salty substances. The taste buds of the tip of the tongue bear all four of these receptors, though they have more of those for sweet and salty substances. The sides of the tongue are most sensitive to sour and the rear to bitter.

When suitable molecules combine with the receptors on the taste cells' hairs, the cells generate action potentials. They cannot pass these signals directly to the brain, however, for they lack axons. Instead, their cell bodies form synapses on the tips of dendrites, which then carry the signals to the brain through four of the cranial nerves. Eventually, the signals reach the taste area in the parietal lobe.

Smell

We do not usually think of taste as limited to salt, sweet, sour, and bitter because in the brain it combines with smell to give us our distinctive aware-

Learning Focus

RECEPTIVE FIELDS

Each skin sensory receptor does not have its own nerve fiber to the CNS. Rather, each nerve fiber branches many times in the skin, and each branch ends with a receptor of the same kind. The branches tend to stay close to each other, however, so that each nerve fiber carries one kind of sensory data from a single patch of skin, known as a **receptive field.** Because receptive fields overlap, we can accurately locate stimuli such as pin pricks, but the accuracy is best in areas where there are many receptors and the receptive fields are small. Accuracy is worst in places that contain relatively few receptors and large receptive fields.

To examine the distribution of receptive fields and receptors, obtain an ordinary artist's compass. Ask a friend to help you in this exercise by closing his or her eyes. Then choose a patch of skin on your friend, spread the points of the compass, and gently touch the points to the skin. Ask your friend whether he or she feels one touch or two. If the answer is one, spread the points further apart and try again. Repeat this procedure until your friend reports feeling both compass points at once. The points are now in separate receptive fields. Now bring the points slightly closer together and touch the skin again. Repeat, until your friend feels only one prick. You have now found the *two-point discrimination threshold,* a measure of the width of single receptive fields and the closeness or density of receptors.

Now measure the two-point discrimination threshold in different areas of the body. Determine how large it is on the following:

The thigh.

The back.

The lips.

The tongue.

The fingertips.

The back of the neck.

Where would you expect to be able to locate skin stimuli most accurately?

Where would you expect to be able to locate skin stimuli least accurately?

How do your results match up with what you already know about the tactile (touch) sensitivity of the parts?

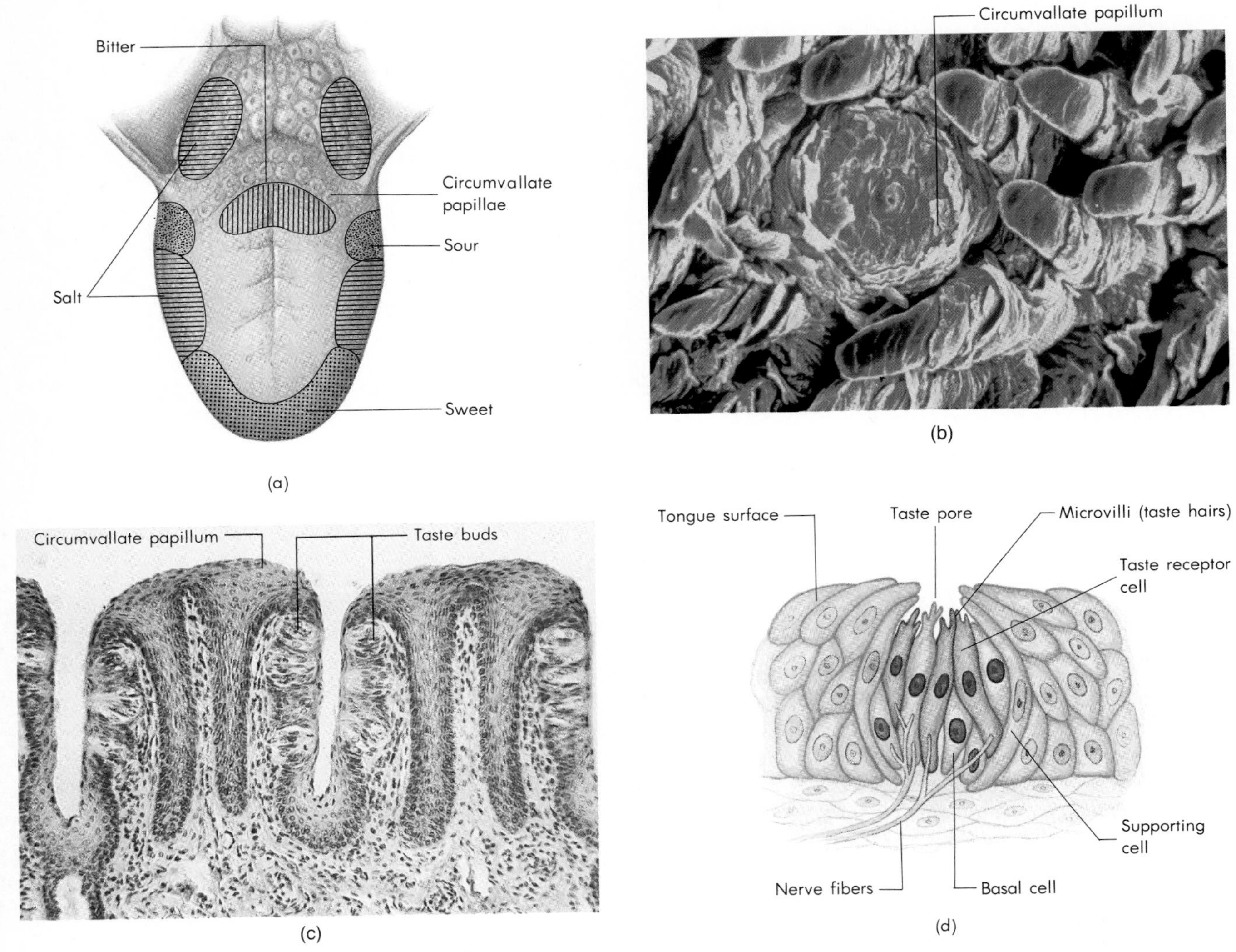

FIGURE 16-5 (a) Drawing of the tongue's surface showing the circumvallate papillae and the taste zones. (b) Scanning electron micrograph of the surface of the tongue. (c) Section showing the locations of the taste buds on the lower margins of the circumvallate papillae. (d) Diagram of the structure of a taste bud.

ness of cider, steak, green peppers, and other foods. Yet smell has its own sense organ, the nose, and its own sensory receptors, the **olfactory cells** embedded in the mucosa of the roof of the nose, just below the frontal lobe of the brain (Figure 16-6).

The olfactory cells are actually neurons. Their stubby dendrites lie on the surface of the nasal mucosa, where odor molecules dissolved in the nasal mucus can reach them and combine with membrane receptors to stimulate action potentials. The action potentials travel on the olfactory cells' axons through holes in the portion of the skull just below the frontal lobe (the cribriform plate of the ethmoid bone) to cells in the **olfactory lobe,** at the tip of the olfactory tract. The olfactory tract is generally considered the first of the cranial nerves; however, the olfactory lobe is really a cellular part of the brain, and the olfactory tract resembles a strip of white matter linking two parts of the brain.

One of the great mysteries of sensation has long been how we can recognize thousands of odors with only a limited number of olfactory cells. The answer emerged with the discovery of many different types of olfactory cells whose dendrites bear receptors that bind to molecules of different general shapes, corresponding to floral, musky, pepperminty, camphorlike, etherlike, pungent, and putrid odors. If each odor molecule binds to two or more of these receptors, differing in which receptor-bearing cells they activate and how strongly they activate them, then thousands of different odor molecules could be recognized.

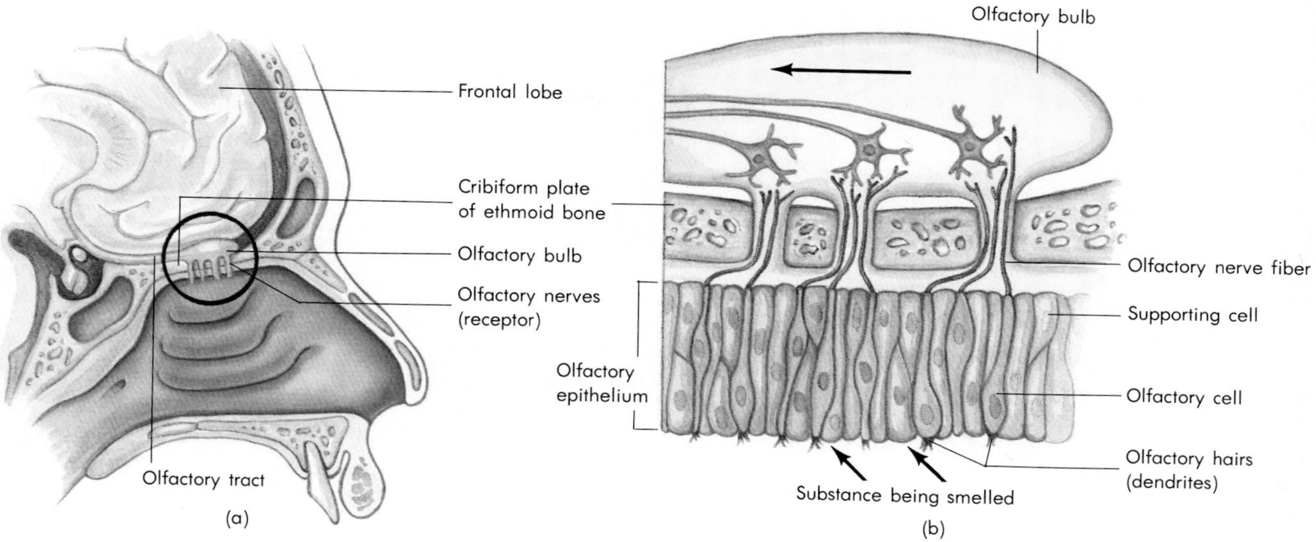

FIGURE 16-6 The olfactory apparatus: (a) olfactory receptors in the nasal cavity; (b) enlarged view of the olfactory receptors and the olfactory bulb of the brain.

Sensory Adaptation

You may have noticed that taste and smell are always sharpest when the source is fresh. That is, as exposure to a flavor or odor continues, you become less aware of it. Sensory **adaptation** occurs as your sensory receptors stop responding. As a result, you do not notice the load of garlic on your own bad breath or the stench of polluted air. The same thing occurs with touch, and it explains why you do not notice the constant rub of your clothing against your skin. It also happens with other senses. It happens least with pain, which ends only when we do something about the cause.

1. Why does a cold interfere with your sense of "taste"?
2. Why is it easier to get close to a garlic-eating loved one if you eat garlic yourself?

VISION

Vision is the most long-range sense of all, for it can detect events—starlight—trillions of miles away, though we normally use it at more reasonable distances. It is the sense of light, and its sense organ, the **eye,** is arranged (like a camera) to focus light reflected from or generated by objects outside the body onto a sheet of receptor cells, the **retina,** in the back of the eye. The focused light forms images that the brain can interpret as highway traffic, a person or an animal, or a page of print.

Many people are sensitive to light in a less precise way. They go outdoors, and if the sun is bright and it strikes their face at just the right angle, they sneeze violently. The reason seems to be that certain sensory endings that should be in the mucosa lining the nose, where stimulation by dust or pepper would evoke a sneeze, have wound up in the skin near the eye. The result, understandably, can be disconcerting, but it can also be useful. Some blind people can use this sensitivity to tell light from dark.

The Eye

The eye is a roughly spherical structure. To its tough outer wall, the white **sclera,** are attached the muscles that move the eyeball within its bony socket. At the front of the eyeball, the sclera is replaced by the transparent **cornea,** through which light enters the eye.

As the reflected light enters the eye, its image is partially focused by the transparent cornea and the **lens,** built of concentric layers of transparent protein. The lens's job is to focus light on the retina. A ring of smooth muscle, the **ciliary body,** changes the shape and curvature of the lens by contracting and relaxing. The ciliary body is attached to the edge of the lens by way of the **suspensory ligament.**

When the smooth muscles of the ciliary body contract, they release tension on the edge of the lens, allowing it to relax into a more spherical shape. Increasing the curvature of the lens in this manner bends, or **refracts,** incoming light more strongly in order to focus the images of nearer objects on the retina. Relaxing the ciliary body increases the tension in the suspensory ligament and pulls the lens into a flatter configuration that is better for viewing objects at greater distances. Such changes in lens shape, used for viewing objects at different distances, are collectively referred to as **accommodation.**

Just in front of the lens lies the **iris,** a membranous disk containing the pigments that give color to the eyes. The diameter of the opening in the center of the iris, the **pupil,** is controlled by autonomic reflexes changing the tension in the smooth muscle of the iris. By opening and closing, the pupil regulates the amount of light entering the eye just as a diaphragm does for a camera. In dim light, the pupils reflexively dilate, allowing more light in. In bright light the iris constricts the pupil, reducing the light allowed to enter.

The iris and lens are suspended in the eye's interior. In front of the lens is the **anterior cavity,** divided by the iris into a *posterior* and an *anterior chamber* filled with liquid *aqueous humor*. Behind the lens, occupying the bulk of the eyeball, is the **posterior cavity;** it is filled with the more jellylike *vitreous humor* (Figure 16-7).

The Retina

The retina lines most of the eyeball behind the lens. It consists of numerous receptor cells of two basic types, rods and cones, and several layers of neurons that perform the initial processing of the receptors' data (Figure 16-8). The receptors are closest to the sclera, and their receptive tips point toward the *outside* of the eye. The processing neurons lie between them and the lens, and light must pass through these neurons in order to strike the receptors. Fortunately, the neurons, like the cornea, lens, and humors, are transparent. Most of the light that enters the eye reaches the receptors. The light that goes past them is absorbed by the **choroid,** a black pigmented layer between the retina and the sclera

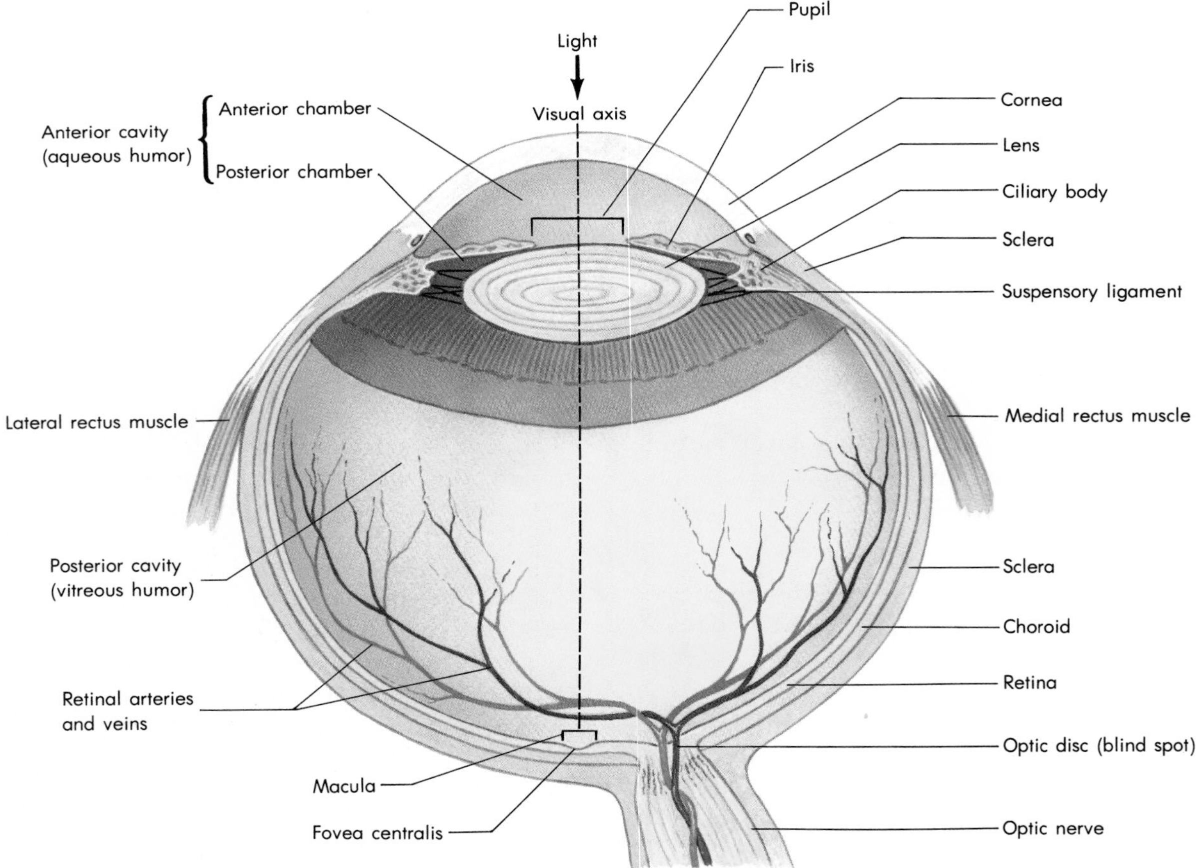

FIGURE 16-7 Anatomy of the eye.

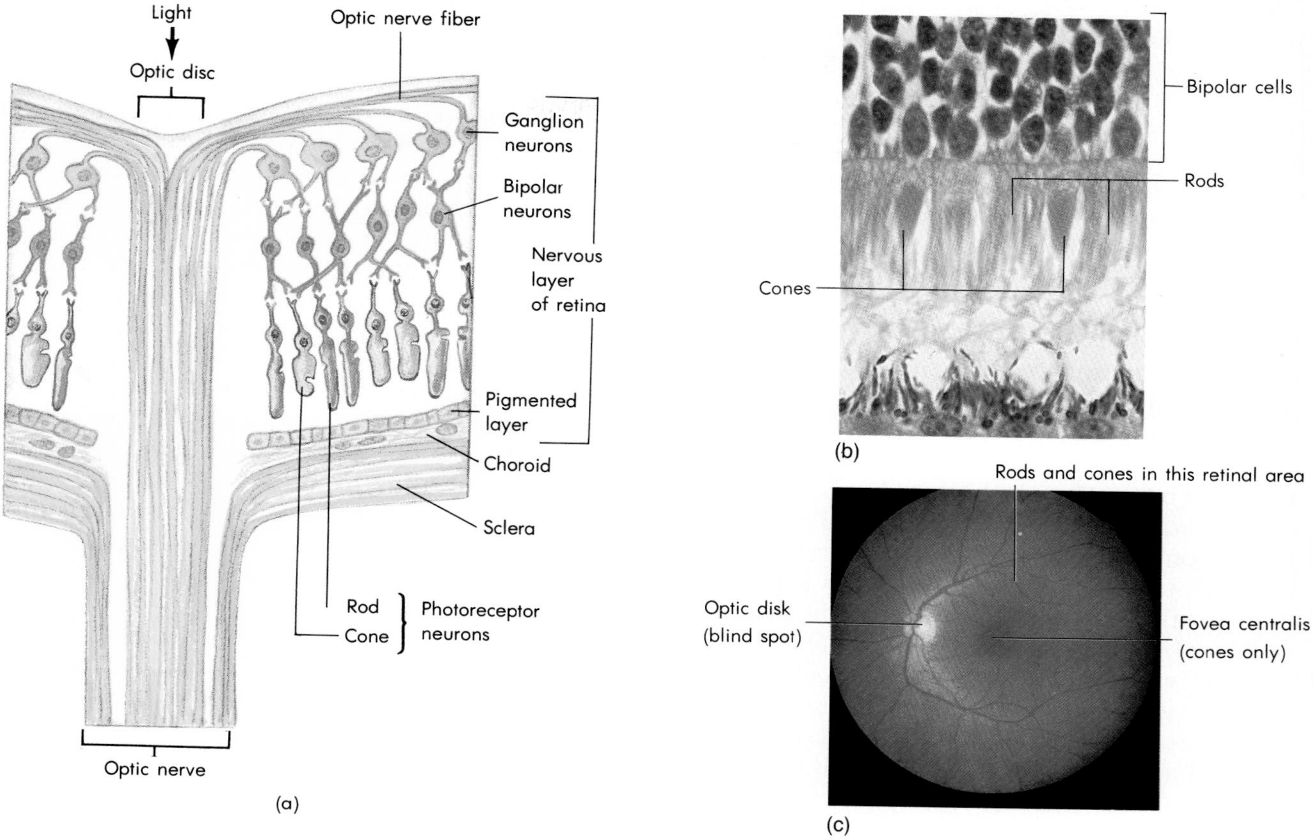

FIGURE 16-8 (a) The arrangement of the visual cells (rods and cones), bipolar neurons, and ganglion neurons in the retina. (b) Retinal rods, cones, and bipolar neurons (×800). (c) The interior of the eyeball as seen through an ophthalmoscope (looking in through the pupil). The optic disk is the site where the optic nerve enters the eye. Blood vessels also enter the eye through the disk and can be seen radiating outward. Because the optic disk contains no visual cells, it is sometimes referred to as the "blind spot." The fovea centralis is the area of detailed vision. It contains only densely packed cones that give the area great resolving power. As you read this page, you are using the fovea to register the letters of each word.

that prevents light from reflecting around the eye's interior and washing out the image. Many blood vessels enter the back of the eye along with the optic nerve and then radiate out through the choroid layer to the retina.

The retina's 140 million **rods** operate only at low light levels and do not distinguish between the different colors of light. Discs in the rod tips (Figure 16-9) contain the purplish pigment **rhodopsin,** composed of a protein (scotopsin) and a derivative of vitamin A, retinene. When the pigment absorbs light, it breaks down and in the process provokes an action potential in the rod. The rods regenerate the pigment quickly, but they are most sensitive when their supply of pigment is at its maximum, as in the dark. Because of the relationship of the pigment to vitamin A, vitamin-A-containing foods such as carrots are said to prevent night blindness.

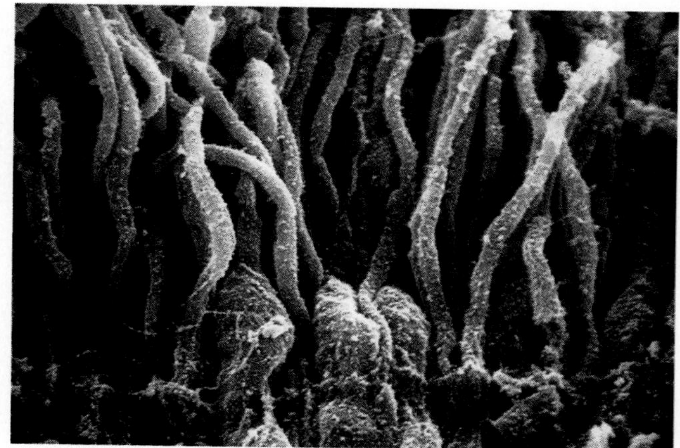

FIGURE 16-9 Scanning electron micrograph showing cones (foreground) and rods (background).

In bright light, **cones** give us our sensitivity to color. There are three types of cones, each one containing a different pigment. All three pigments are related in structure to rhodopsin and to each other, and all three include retinene, but one absorbs mainly red light, the second green light, and the third blue light. The brain interprets specific colors according to how light activates all three receptors; the mixing principle is similar to that for smell. Each eye contains 6–7 million cones; they are concentrated especially in the retina's **fovea centralis** and the surrounding **macula,** or "yellow spot," which have no rods. It is upon the fovea centralis that we focus images when we want to see details. When we wish to see a very dim object, such as a star at night or someone lurking in the shadows, we must look to one side of the target to make its image fall on a part of the retina with rods.

The receptors respond to an image as a pattern of dots, but the "picture" that reaches the brain is more organized. The layer of processing neurons nearest the receptors consists of **bipolar cells;** these cells collect signals from many receptors. The next layer is that of the **ganglion cells,** which collect signals from many bipolar cells. Together, the bipolar and ganglion cells extract simple patterns such as edges and spots from the dot picture received by the receptors. These patterns are what go to the brain on the axons of the ganglion cells. The axons converge from all parts of the retina on the eye's **blind spot,** where there are no receptors of any kind (Figure 16-10). The axons then leave the eyeball as the optic nerve. Since each optic nerve contains about one million axons, there is roughly a 150-fold concentration of information before the signals from the rods and cones even leave the eye.

The blood flow through the choroid layer of the eye can be restricted, with devastating effects, by an increase in intraocular pressure. This condition, called **glaucoma,** is caused by impaired drainage of the aqueous humor in the eye's anterior cavity. The increased pressure within the eye counteracts the blood pressure driving the blood flow through the choroid, starving the retinal cells and leading to blindness. Glaucoma accounts for 5500 cases of blindness per year in the United States (out of 34,000); the blindness strikes most often the elderly. Fortunately, when glaucoma is diagnosed early, it can be treated with drugs that lower the intraocular pressure. Severe cases may require surgery or the use of lasers to open the canals that normally drain fluid from the anterior cavity. Unfortunately, glaucoma has no early symptoms such as pain or fuzzy vision; diagnosis requires a pressure check by an ophthalmologist.

The Retinal Image

In the normal eye, the lens focuses the image precisely on the fovea of the retina. In nearsightedness **(myopia),** either the eyeball is too long or the lens's ability to focus the light is not strong enough and the image falls short (Figure 16-11). In farsightedness **(hyperopia),** the eyeball is too short or the lens focuses the image beyond the retina. In both cases, the retina gets only a fuzzy image, unless artificial lenses (glasses or contacts) are used to bring the image onto the retina properly. An alternative, and currently experimental, treatment is surgical modification of the cornea's shape.

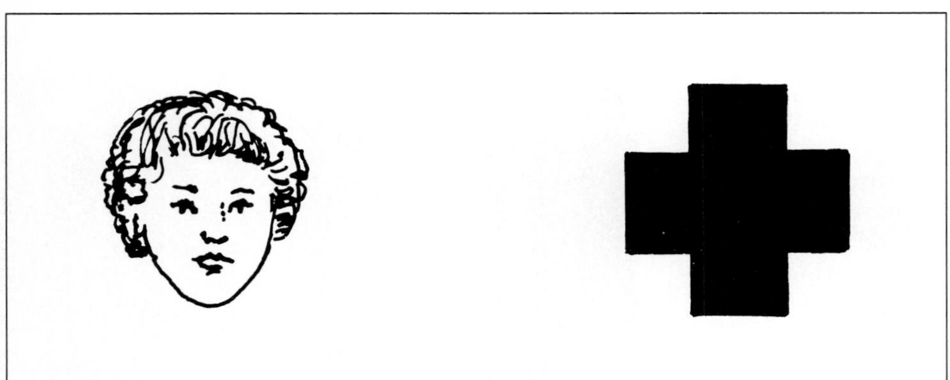

FIGURE 16-10 To demonstrate the position of the blind spot on your own retina, close your right eye while looking directly at the cross and holding the book nearly an arm's length away. Move the page closer to your eye while continuing to focus on the cross. When the page is 14–16 inches from your eye, the face will seem to disappear from the page. As you continue moving the page closer, the face will reappear. The image of the face disappeared as it moved across your blind spot, where there are no rods or cones.

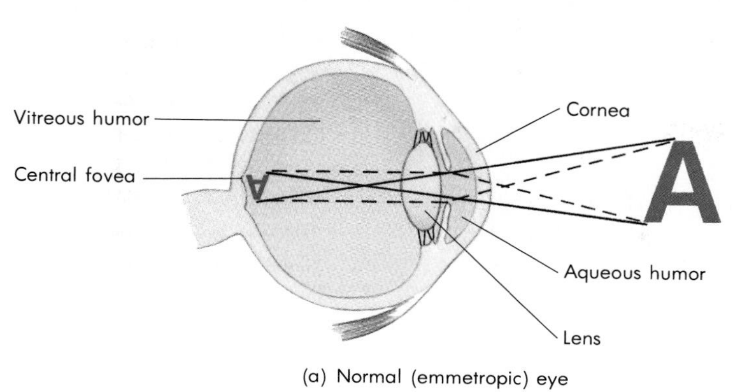

(a) Normal (emmetropic) eye

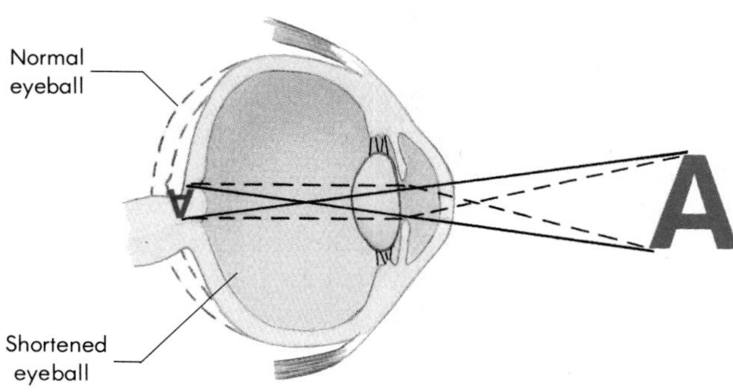

(b) Farsighted (hypermetropic) eye, uncorrected

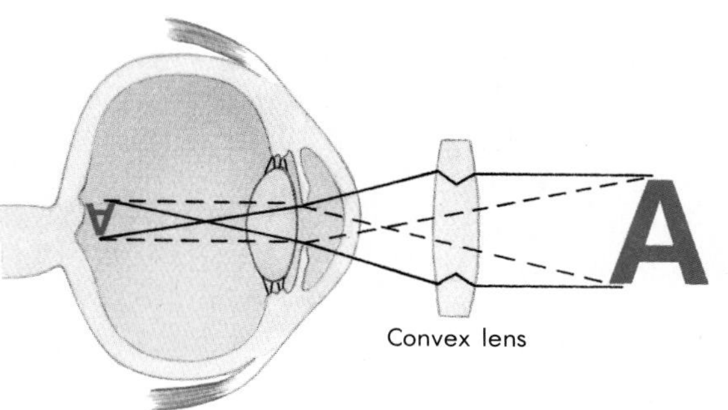

Farsighted (hypermetropic) eye, corrected

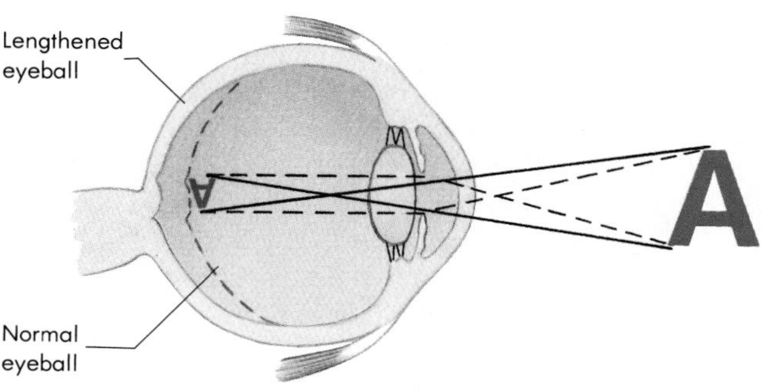

(c) Nearsighted (myopic) eye, uncorrected

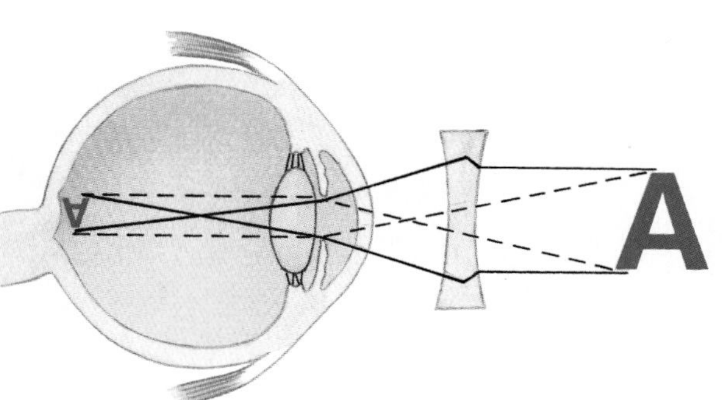

Nearsighted (myopic) eye, corrected

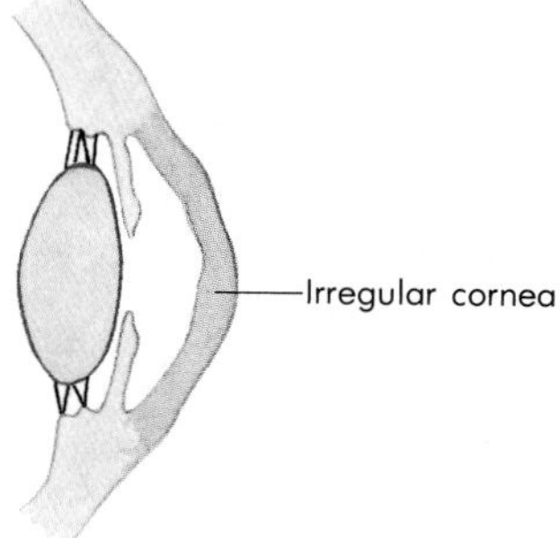

(d) Astigmatism from an irregular cornea

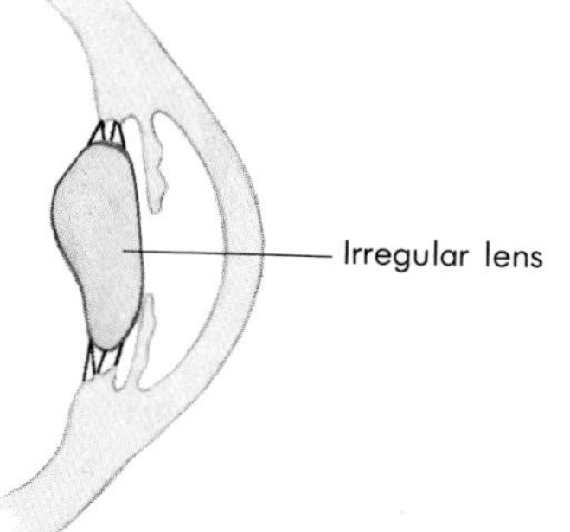

Astigmatism from an irregular lens

FIGURE 16-11 (a) Image refraction in normal vision. (b) Farsightedness, or hyperopia. (c) Nearsightedness, or myopia. (d) Astigmatism.

In **astigmatism,** irregularities in the curvature of the lens or cornea put ripples or flat spots in the retinal image. Corrective lenses for this condition must have compensating irregularities in their curvature.

Lenses are useless for some visual problems, however. In some people the cornea or the lens clouds over and becomes opaque; the result is **cataracts.** These opacities may develop as a consequence of aging, exposure to radiation, or other causes; when severe, they block vision entirely. Treatment depends on the nature of the cataract. If it is caused by opacity of the lens, it may be necessary to surgically remove the lens and replace it with an artificial lens, either externally (glasses or contacts) or internally. If the cataract is caused by a cloudy cornea, a corneal transplant is the appropriate option.

It is strange but true that the brain must turn the world upside down in order to see it right-side up. The image the lens focuses on the retina is always inverted, up for down and right for left. That is, when we look at an object that is in the upper right of our **visual field,** the lens projects its image on the lower left quadrant of the retina (Figure 16-12).

The Optic Pathway

The inversion of visual images has a profound effect on how visual data reach the brain. The optic nerves from the two eyes meet in the **optic chiasma,** just anterior to the pituitary gland. In the chiasma, optic nerve fibers from the left half of each retina join to form the left **optic tract.** Optic nerve fibers from the right half of each retina join to form the right optic tract. The optic tracts end in the left and right *lateral geniculate nuclei* of the thalamus, where they synapse with neurons that carry the visual information to the occipital lobes of the cerebral cortex's corresponding hemispheres. There the separate parts of the visual field project in sequence, just as sensory information projects onto the sensorimotor cortex of the parietal lobe.

Because of inversion and the way the optic nerves split and rejoin, the left occipital lobe "sees" only the right half of an image. The right occipital lobe sees the left half. We see images as wholes because the two halves of the brain communicate with each other and merge their fields of view.

1. **Dark adaptation** is the increase in sensitivity of the eye that occurs after one has been in the dark for awhile. Explain this phenomenon in terms of regeneration of rhodopsin.
2. In Chapter 15, we mentioned studies of people whose corpus callosums have been cut, saying that it is possible to show an object to only one cerebral hemisphere. To what part of the visual field of the right eye would you have to aim an image of a flower, say, in order to get that image to the right hemisphere? To the left hemisphere?

THE SENSES OF THE EAR

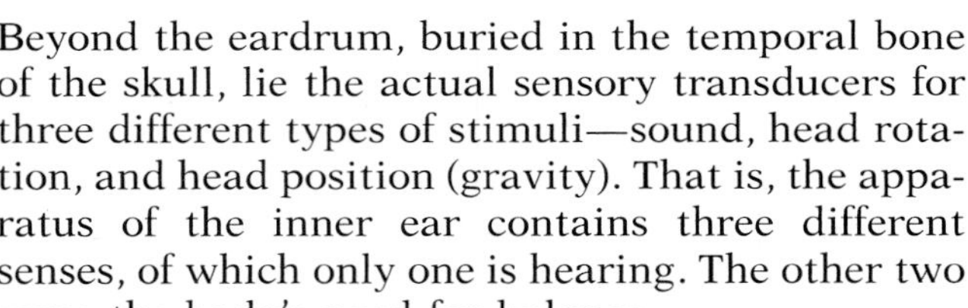

Beyond the eardrum, buried in the temporal bone of the skull, lie the actual sensory transducers for three different types of stimuli—sound, head rotation, and head position (gravity). That is, the apparatus of the inner ear contains three different senses, of which only one is hearing. The other two serve the body's need for balance.

HEARING

Like vision, hearing is a long-distance sense. It detects sound, vibrations of the air set up by events at some distance from the body (usually). We also rely on it for communication and for awareness of what we cannot see, including hidden dangers.

The hearing sense is extremely complex. We are taught from infancy to code information into patterns of vibrating air and transmit it by way of the spoken words we send to people around us. The ear is the receiver of the vibrations, which it converts to nerve signals. The temporal lobe of the brain, in turn, analyzes the patterns of nerve signals and extracts from them the information they contain—including emotions for which we have no words. We call this process hearing. With practice we can listen to the blend of vibrations produced by an orchestra playing Mozart and from it we can isolate the lilting phrases produced by a single flute.

The External Ear

What we generally call the "ears" are those structures attached to the sides of the head, the **pinnas** or **auricles.** They are shaped and positioned to capture vibrating air waves and direct them into the channels leading into the head, where the appara-

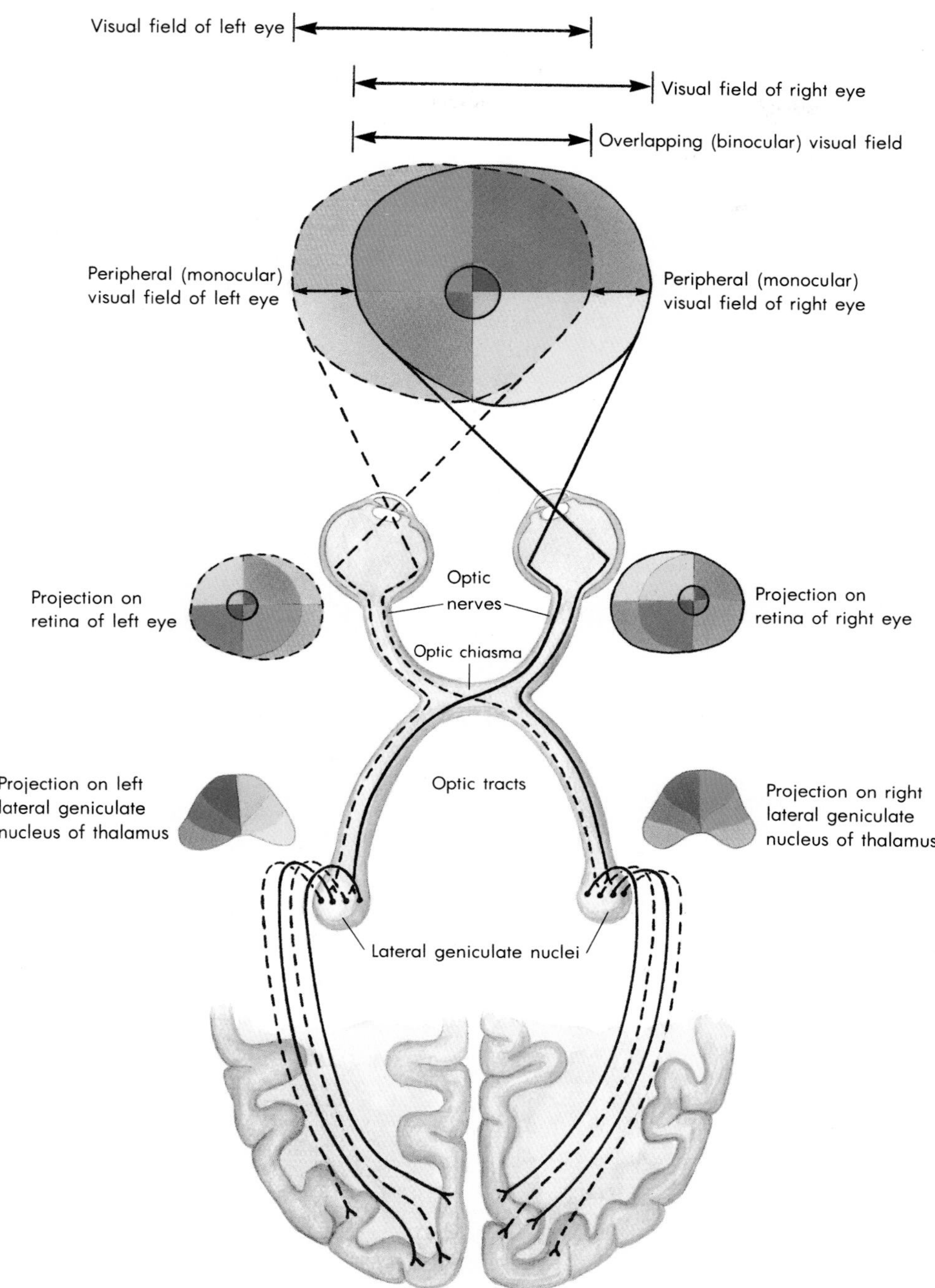

FIGURE 16-12 The overlapping visual fields of both eyes and the optic pathways. Even though the two eyes are both mounted on the front of the head, they have slightly different views of the world. Portions of their visual fields overlap; other portions of the fields are unique. The dark area at the center of each visual field is the area of detail vision examined by the fovea. The incoming visual messages from each eye are split as the signals pass through the optic chiasm. A portion of each eye's visual information is sent, by way of the lateral geniculate nuclei, to the visual cortex of the left and right occipital lobes of the brain.

tus of the inner ear eventually converts them into nerve impulses that the temporal lobe translates as sounds. The vibrations are channeled down the **external auditory canal** to the **tympanic membrane,** or eardrum (Figure 16-13). Everything outside the eardrum, including the pinna, is considered to be part of the **external ear.**

The Middle Ear

At the end of the inch-long ear canal (external auditory canal), sound waves make the concave eardrum (tympanum) vibrate. On the inner side of the eardrum is the air-filled cavity of the **middle ear.** The vibrations of the eardrum cross the middle ear on a chain of three small bones, the **malleus** (hammer), **incus** (anvil), and **stapes** (stirrup), all named for their shapes or functional relationships (Figure 16-13b). The malleus is embedded in the eardrum. When sound activates the eardrum, the malleus vibrates against the incus, which in turn makes the stapes vibrate against the membranous oval window. This window allows the vibrations to enter the inner ear, where they will eventually be converted to nerve signals.

Some cases of deafness are due to problems with

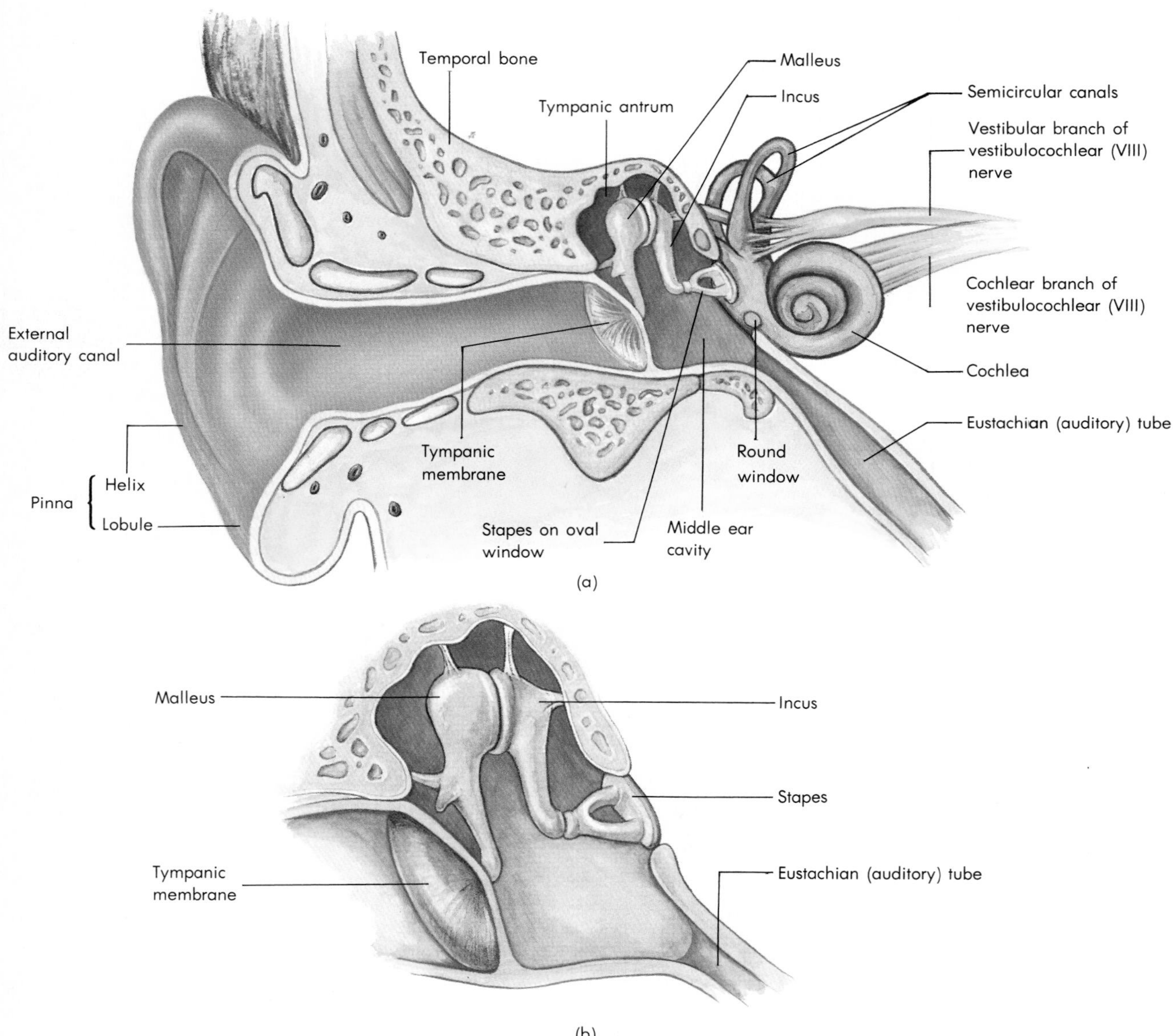

FIGURE 16-13 (a) The ear consists of the external ear, the ear canal, the eardrum, the middle ear, and the inner ear, containing the cochlea. (b) Ear ossicles in the middle ear.

these three small bones. They may, for instance, fuse together. If they remain able to move at least a little, an ordinary hearing aid can amplify quiet sounds to a level that will reach the inner ear. If they become entirely rigid, they may need surgery or even replacement with artificial substitutes.

When the outside air pressure changes, as when we drive up or down mountains, dive to the bottom of a swimming pool, or take a trip in an unpressurized small airplane, the air pressure difference across the eardrum between the outside of the head and the middle ear (which is filled with air) can become so pronounced that it can stretch the eardrum enough to produce severe pain or even break it. Fortunately, there is a passageway, the **Eustachian (auditory) tube,** that connects the middle ear to the back of the throat. It allows us to relieve pressure changes in the middle ear, often by swallowing or by holding the nose and blowing air into the Eustachian tube.

Because the middle ear is connected to the back of the nasal region by the Eustachian tube, it is a common site of ear infections in children—for two reasons. First, they frequently have colds. Second, because their heads are smaller than adults', their Eustachian tubes are shorter, making it easier for bacteria-laden mucus to move from the nasal cavity into the middle ear. Once infected, the mucous membranes swell painfully and the middle ear fills with fluid, exerting pressures on the tympanum that occasionally cause the eardrum to burst. Repeated infections can sometimes lead to scarring of the eardrum and hearing loss.

The Inner Ear

The flat base of the stapes fits against a small opening, the **oval window,** in the bone that surrounds the **inner ear,** where the ear's sensory receptors are located. Through this opening, the stapes transfers its vibrations to the fluid-filled **cochlea,** a long coiled tube resembling a snail. The vibrations travel as waves of pressure through the cochlear tube until they reach the smaller *round window.* This window allows the vibrations to leave the inner ear instead of echoing inside it for several seconds; it therefore protects the delicate tissues of the cochlea proper from damage.

The cochlea is coiled through about 2.5 turns and is divided by membranes into three subtubes, the **cochlear duct,** the **scala tympani,** and the **scala vestibuli.** The oval window marks the beginning of the scala vestibuli.

As the fluid in the scala vestibuli vibrates in response to sound, it sets the membrane separating the scala vestibuli from the cochlear duct to vibrating as well. The fluid in the cochlear duct then vibrates too, as does the **basilar membrane** that separates the duct from the scala tympani (Figure 16-14). It is the vibration of this membrane that activates the hearing receptors.

The hearing receptors are the **hair cells** (Figure 16-15) of the **spiral organ,** or **organ of Corti.** These cells form a row extending the entire length of the basilar membrane. From each cell's tip projects a cluster of microvilli and a cilium. These "hairs" brush an overlying flap of tissue, the **tectorial** (roof) **membrane.** When the basilar membrane vibrates, it bends the hairs against the tectorial membrane. As the hairs bend, they disturb the hair cells' membranes, which gives rise to nerve impulses. The nerve impulses are immediately passed by synapses on the hair cells to the dendrites of the cochlear branch of the vestibulocochlear cranial nerve and then to the temporal lobe of the cerebral cortex.

The basilar membrane does not vibrate as a whole. It is narrower near its base, near the oval window, and wider near its tip. Because of the change in the width along its length, specific regions of the basilar membrane vibrate in response to the different frequencies of the sounds being processed. It thus vibrates in response to high-frequency sounds near its base and to low-frequency sounds near its tip, and the various hair cells—and fibers of the cochlear nerve—respond accordingly to sound frequencies.

This feature of the cochlea has made it possible to design and install effective hearing aids for people who are deaf because of damage to their hair cells. A microphone passes sound to an antenna mounted under the skin near the outer ear. From the antenna, a thin cable snakes along the scala tympani. Electrodes, one for each of several frequency ranges, then provide an electrical substitute for the action of the hair cells in stimulating the cochlear nerve fibers (Figure 16-16, p. 373).

---

1. Excessive sound can vibrate the hair cells severely enough to damage or destroy them. Would you expect a sound dominated by high frequencies to damage all the hair cells of the cochlea? Why or why not?
2. Which hair cells *would* such a sound damage?

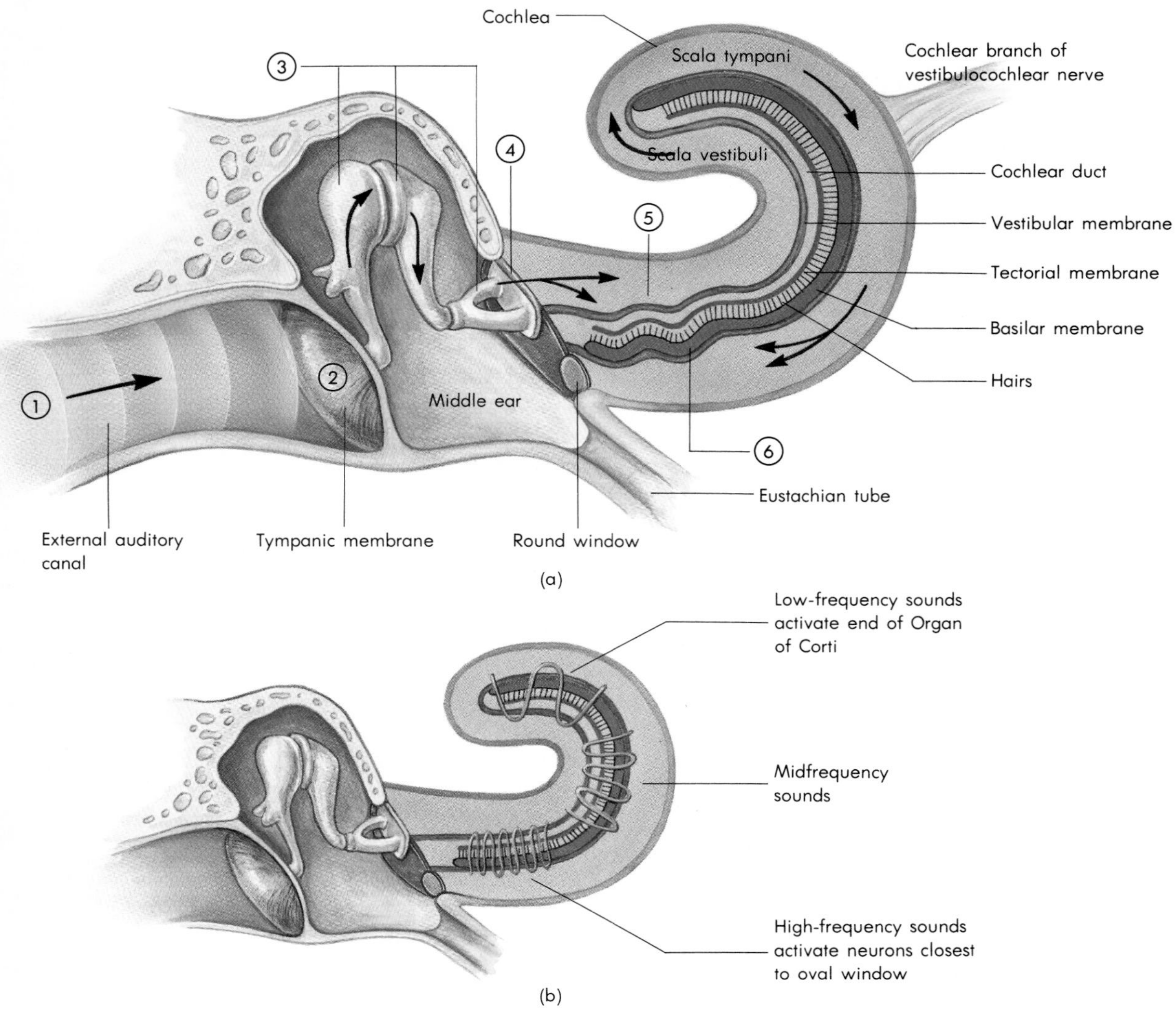

FIGURE 16-14 (a) The external auditory canal (1) channels sound into the ear, where it makes the tympanic membrane (2) vibrate. The ossicles (3) pass the vibration to the oval window (4) and the fluid within the cochlea (5), which in turn vibrates the basilar membrane (6) in the organ of Corti, pushing hair cells against the tectorial membrane, causing them to initiate nerve signals. (b) The structure of the organ of Corti is such that the portion of it nearest the oval window responds best to high-frequency sounds. That portion furthest from the oval window responds best to low-frequency sounds.

FIGURE 16-15 Electron micrograph showing rows of hair cells in the monkey cochlea.

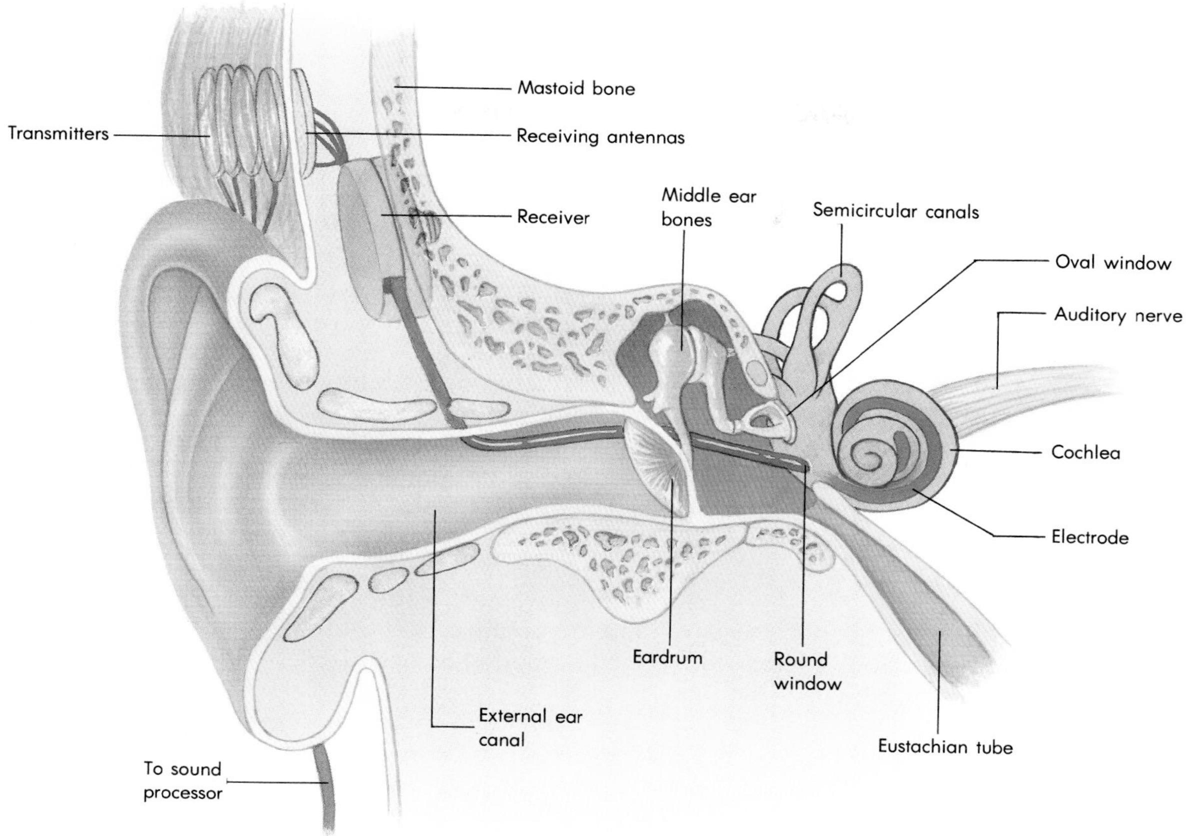

FIGURE 16-16 Cochlear implant hearing prosthesis. This tiny implanted electronic device is threaded into the cochlea, where it converts processed sound to eight frequencies that can be used to activate the patient's auditory nerve endings.

THE SENSES OF BALANCE

The cochlea does not end at the oval window. Rather, it connects to the series of membranous sacs, the **saccule,** the **utricle,** and the **semicircular canals,** that contain the apparatus for the inner ear's other two senses, balance and position sense for the head (Figure 16-17).

The saccule and utricle contain patches of hair cells whose hairs are embedded in a jellylike mass. On the surface of the jelly lie many small crystals of calcium carbonate, the **otoliths.** Gravity pulls the otoliths downward, and changes in head position change the direction of their pull on the jelly and on the hairs of the hair cells. Sudden changes in straight-line movement, as in braking a car, also cause the otoliths to move. In either case, the hair cells give rise to patterns of nerve impulses that tell the brain what is happening. The nerve impulses also trigger reflexes of the neck and limbs that adjust the body's posture for balance, or *static equilibrium*.

Each of the three semicircular canals is a tubular hoop. In the body's anatomic position, one hoop is horizontal, one lies in the right-left plane, and one lies in the anterior-posterior plane. When the head rotates in one of these planes, the fluid in the corresponding canal tends to stay still due to its inertia. That is, the fluid seems to move in the direction opposite that of rotation, *relative to* the walls of the canal.

At the base of each canal is a bulbous swelling of the canal, the **ampulla.** In the ampulla is a patch of hair cells, whose hairs are imbedded in a mass of jelly, the **cupula,** that nearly blocks the ampulla (Figure 16-17). When the fluid in the canal moves, it pushes on the cupula. As the cupula sways, it bends the hair cells' hairs. The hair cells then generate nerve signals that tell the brain of the head movement and trigger reflexes that balance the body against movement; they are responsible for *dynamic equilibrium*. The nerve signals also trigger eye movements that keep the visual field steady despite the rotation of the head.

Learning Focus

HOW MANY SENSES DO YOU HAVE?

We are used to thinking that we monitor the universe with only five senses—vision, hearing, taste, smell, and touch. Occasionally we speak of a sixth sense: a sense of danger, or prescience, or telepathy.

There is no evidence for such a sixth sense. There are only anecdotes and wishful interpretations of sometimes eerie coincidences. However, researchers have found bacteria that contain rows of tiny crystals of magnetic iron oxide, lodestone (see the figure). These bacteria seem to use their natural magnets as compasses, following the lines of the Earth's magnetic field downward into the seabottom ooze in which they live.

Similar magnetic deposits have been found in the abdomens of bees and the brains of homing pigeons and porpoises, all of which would find a compass-based sense of direction useful for navigation. Indeed, homing pigeons actually do seem to use their built-in compasses for this purpose. When a researcher makes these animals wear small magnets on their heads, where the magnets' fields can interfere with the compass sense, they have difficulty navigating.

Do humans have a magnetic sense? Some researchers say they have found lodestone in human brains, but there is no evidence that if we indeed are sensitive to magnetic fields, we use that sensitivity in any way.

Even if we do have and use a magnetic sense of direction, that sense is no sixth sense. Consider that the ear is sensitive not just to sound, but to movement and position as well. Touch includes the senses of pressure, vibration, hair movement, heat, cold, and pain (although pain is really a chemical sense indicating that cell damage has occurred). In addition, there are all the senses that monitor the features of our internal universe—muscle stretch and tension, carbon dioxide, glucose, blood volume and pressure, body temperature, and so on.

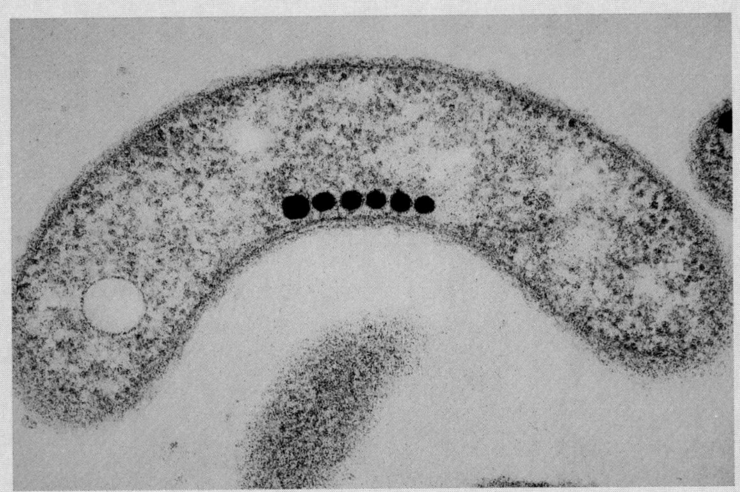

Electron micrograph of bacteria with magnetite grains that seem to be used for position sensing (×64,000).

1. If you stood and spun around six times, the following scenario would be played out in your vestibular apparatus. The horizontal rotation of your head would make the fluid in your horizontal semicircular canals move in the direction opposite that of the rotation. How does the fluid respond when the rotation stops? What signal must then go to brain? How does this match up with your sensations when you stop spinning?
2. Why are you likely to stagger and fall when you stop spinning?

Learning Focus Response

How Many Senses Do You Have?

How many senses do you have? This question gives you a good opportunity to review this chapter. We have provided a blank table with headings for the senses of the skin, eye, ear, nose, mouth, and body. For each heading, write down all the senses you can (one is provided as an example under skin). Then add brief descriptions of the appropriate sense organs, state what they sense, and record their specific location (e.g., retina, not eye).

The Senses of the Body

SENSES	SENSE ORGANS	LOCATION	WHAT IS SENSED?
Skin: (e.g., hair movement)	Free nerve endings	Around hair follicles	Air movement, insects walking
Eye:			
Ear:			
Nose:			
Mouth:			
Interoceptors:			
Muscle:			
Gut:			
Blood:			

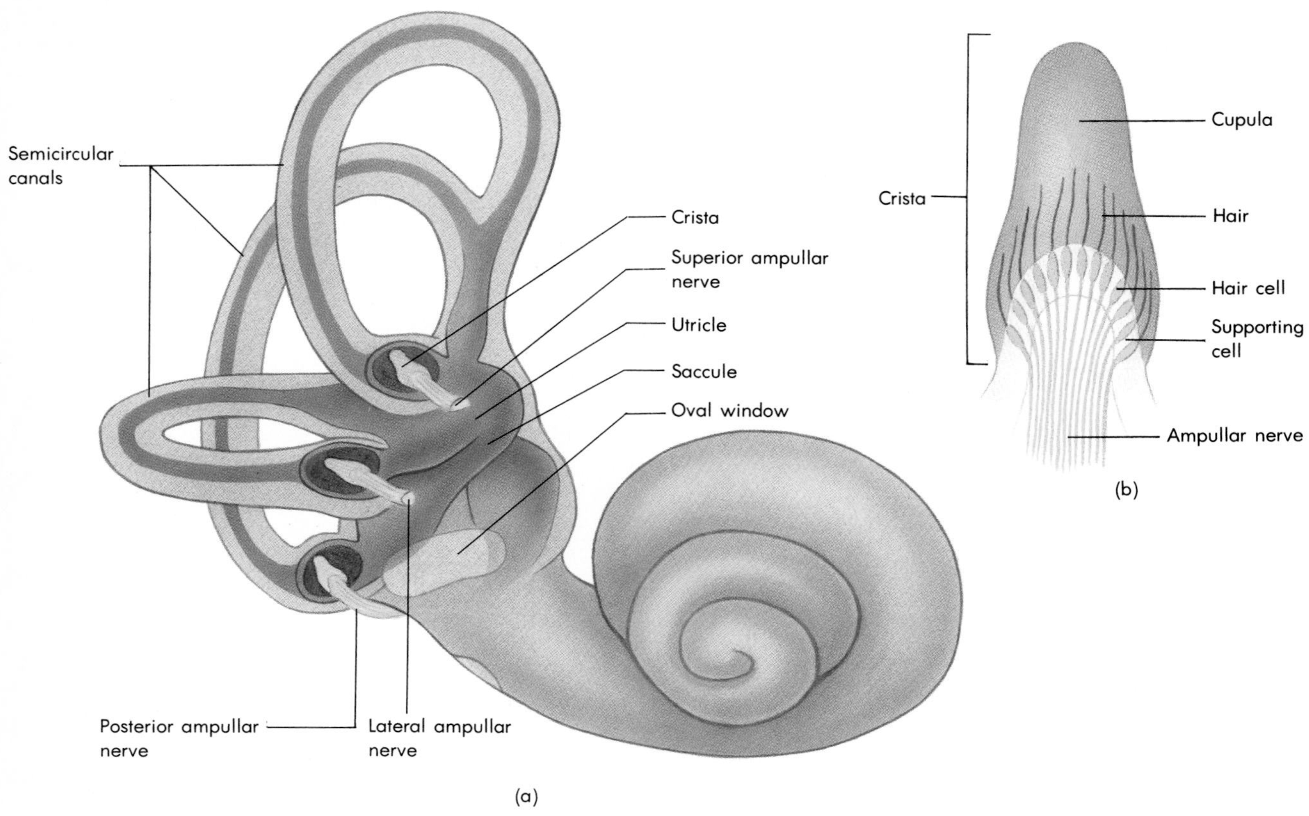

FIGURE 16-17 (a) The saccule, utricle, and semicircular canals. (b) The structure of a crista in an ampulla.

SUMMARY

The body's sense organs provide the input without which the brain cannot function. Each one translates (or transduces) some event outside or inside the body into a stimulus for a sensory receptor, which generates the nerve signal that goes to the central nervous system (CNS). The place at which the signal arrives in the CNS identifies it as visual, auditory, tactile, or some other sensation.

The exteroceptive senses respond to events that originate outside the body. They include vision, hearing, touch, taste, and smell. The interoceptive senses respond to muscle stretch and tension, blood pressure, volume, and carbon dioxide content, body temperature, gut distension, pain, and other internal events. The muscle senses (proprioception) are essential to motor control. Other interoceptive senses are crucial to homeostasis.

Pain is both an interoceptive and an exteroceptive sense, depending on whether the pain receptors are located in internal organs or in the skin. The receptors respond to excessive stimulation of all kinds and to substances released by damaged cells. Pain signals reach the brain on pathways using special neurotransmitters, and they can be inhibited by the brain.

The sense of touch is actually several distinct senses, each one with its own sense organs and sensory receptors and all located in the dermis of the skin. There are receptors for temperature, pressure, vibration, and hair movement as well as pain. Each sensory dendrite supplies many branches to a small region of skin, and each branch of a single dendrite ends in the same kind of sensory receptor and sense organ. The skin area from which one dendrite collects one kind of sensory information is known as a receptive field. Receptive fields are smaller in more sensitive parts of the body, such as the fingertips and lips, and bigger in less sensitive areas, such as the back of the neck.

The sense organs for the sense of taste are the tongue's taste buds. The four types of taste cells have in their membranes protein receptors for molecules dissolved in saliva; they respond to salt, sweet, sour, and bitter. The sense of smell, whose receptor cells lie in the back of the nose, works similarly, but their membrane proteins can bind to many more different molecular shapes. Taste cells are not neurons; they are neuroepithelial cells that generate nerve signals and pass them to sensory dendrites. Olfactory cells are neurons; their axons enter the olfactory lobe of the brain. Both taste and smell, along with other senses, illustrate the phenomenon of adaptation, whereby the sensory response to a continuing stimulus fades.

The eye is responsible for vision. Light reflected from or generated by objects in the universe outside our bodies enters the transparent cornea of the eye. The lens then focuses it, inverted, on a sheet of receptor cells lining the back interior wall of the eye, the retina. There are two types of receptors. Rods are most sensitive in dim light and give little information about color. They are used primarily at night. Cones of three types—those sensitive to red, to green, and to blue light—give us color vision. Cones also give us detail vision, for they are densely packed in the eye's fovea, in the center of the visual field. The fovea lacks rods, which is why when we want to see a very dim star at night, we must look to one side of our target.

Both rods and cones contain pigments (consisting of a protein plus retinene, a derivative of vitamin A). When a pigment molecule absorbs a photon of light, it breaks down and initiates a nerve impulse. Vision thus begins as a pattern of single nerve signals, or activated receptors, that matches the light pattern in the image focused on the retina. However, this pattern is a pattern of dots, and it is not what the brain "sees." The rods and cones pass their signals to other neurons in the retina that combine the dots into larger features of the image, such as edges. The retina thus concentrates the raw data derived from 140 million rods and 6–7 million cones into 1 million optic nerve fibers, which leave the eye as the optic nerve at the eye's blind spot, which lacks receptors of any kind.

The two optic nerves meet at the optic chiasma just anterior to the pituitary gland. There the nerve fibers from the left half of each retina join to form the left optic tract, which delivers nerve signals to the left lateral geniculate nucleus of the thalamus. The nerve fibers from the right half of each retina form the right optic tract. From the thalamus, the visual signals go to the occipital lobes of the cerebrum.

The ear gives us our sensitivity to sound, head position, and head rotation. Sound is funneled by the external ear into the ear canal, where it sets the eardrum to vibrating. The three small bones of the middle ear carry the vibrations to the inner ear and the fluid of the coiled cochlea. In the cochlea, the vibrations are picked up by the basilar membrane, on which rests the spiral organ of Corti, a row of hair cells positioned beneath a rooflike tectorial membrane. When the basilar membrane vibrates, the hairs of the hair cells bend against the tectorial membrane, disturb the hair cells' cell membranes, and initiate nerve signals. Different parts of the basilar membrane vibrate in response to sounds of different frequencies so that each sound stimulates its own set of hair cells and hence specific nerve fibers to the brain.

The cochlea is linked to the membranous sacs that contain the sense organs for head position and rotation. The saccule and utricle contain hair cells whose hairs are embedded in gelatinous masses weighted down with calcium carbonate otoliths. Changes in head position shift the pull of gravity on the otoliths and hence their pull on the hairs and the nerve signals that go to the CNS. The semicircular canals are fluid-filled tubules that lie in the three possible planes of rotation. When the head rotates, inertia causes the fluid in a canal to move in the opposite direction. The fluid then pushes on the gelatinous cupula in the ampulla at the base of the canal. As the cupula sways, it bends the hairs of hair cells and provokes nerve signals to the brain.

Table 16-1 provides a summary of selected diseases of the senses.

Table 16-1

Clinical Summary: Selected Diseases of the Senses

DISORDER	DESCRIPTION
Anosmia	Loss of the sense of smell, caused by damage to the nasal mucosa (including the sensory receptors in the upper nasal cavity), the olfactory nerve, or the olfactory bulb or tract.
Astigmatism	Inability of the eyes to focus properly due to irregularities in the curvatures of the lens and/or the cornea. Causes are unknown, although a genetic basis is suspected.

(Continued on p. 378)

DISORDER	DESCRIPTION
Cataract	Loss of vision due to clouding or developing opacity of the lens or cornea. Commonly accompanies aging. May also result from injury or exposure to ultraviolet light and radiation.
Conduction deafness	Deafness caused by physical disruption of the transmission of vibrations from outside the ear through the middle ear, thus preventing activation of auditory receptors in the inner ear. Possible causes include excess ear wax in the ear canal and fusion (ankylosis) of the ear ossicles.
Glaucoma	Increases in intraocular pressure due to disturbed drainage of anterior cavity fluid. Can result in retinal and optic nerve damage and blindness.
Hyperopia	Farsightedness. The ability to discern objects better from a distance than close up. Occurs when the lens and cornea focus the visual image behind the retina.
Labyrinthitis	Inflammation of the inner ear, with symptoms of vertigo, nausea, and nystagmus. May follow spread of an infection from the middle ear. Often seen as a complication in flu.
Myopia	Nearsightedness. The ability to discern objects better close up than from a distance. Occurs when the lens and cornea focus the visual image in front of the retina. Probably genetic in origin.
Nystagmus	Continuous involuntary twitching of the eyes. Causes include congenital, occupational, and certain neurologic disorders and disturbances of the inner ear.
Photophobia	Extreme sensitivity of the eyes to bright light. Accompanies measles, meningitis, eye inflammation, certain vitamin deficiencies, and reactions to certain drugs.
Presbyopia	"Old eyes." As the lenses lose their elasticity with age, the eyes become hyperopic. Effect is usually noticeable after 40 years of age.
Tinnitis	Unrelenting ringing, tinkling, or clicking sounds in the ears. Causes range widely, from impacted wax and middle-ear infections to labyrinthitis and reactions to certain drugs.

STUDY QUESTIONS

1. Explain the concept of transduction. Besides thermometers what other transducers are familiar to you?
2. What is the difference between a sense organ and a sensory receptor?
3. Name the three interoceptive senses associated with muscle action. What roles do they play?
4. What is the function of pain? Of being able to suppress pain?
5. Why do we say that "touch" is not one sense but several? Where must it *become* one sense?
6. Discuss the *differences* among the taste buds of the various parts of the tongue.
7. What part does the sense of smell play in "taste"?
8. Discuss the difference between *adaptation* and *dark adaptation*.
9. List the parts of the eye through which light must pass on its way to the visual receptors.
10. Diagram the path of sound from a radio's speaker to your brain.
11. Most people who are blind or deaf say they would rather be blind. How about you? Why?
12. Which part of the inner ear tells you which way is up? How?
13. You are sitting, blindfolded, in a car. Which part of the inner ear lets you track the turns the car makes (be specific)? How?

GLOSSARY

Accommodation Changes in the shape of a lens that adapt it for viewing objects at different distances.

Adaptation Fading of a sensation with continued exposure to the stimulus.

Ampulla Swelling at the base of a semicircular canal; contains the cupula and hair cells.

Basilar membrane The wall of the cochlear duct that holds the hearing receptors.

Ciliary body The ring of smooth muscle that controls the shape and curvature of the lens in the eye.

Cochlea The coiled tube containing the receptors for hearing.

Cones Retinal light receptors sensitive to color and used for detailed vision and vision in bright light; each type of cone is sensitive to either red, blue, or green light.

Cupula Gelatinous mass blocking ampulla; it sways when the head rotates, bending the hair cell hairs embedded in it and provoking nerve signals.

Dark adaptation The increase in visual sensitivity that comes after being in dim light or darkness for awhile.

External ear The pinna, ear canal, and eardrum.

Exteroceptor Sense organ for events outside the body.

Fovea centralis Retinal region with greatest density of cones and no rods; used for detailed, color, and bright-light vision.

Gustatory (taste) cells Sensory cells in taste buds.

Hair cells The receptors for hearing and balance.

Inner ear The region containing the ear's sensory apparatus; embedded in the temporal bone of the skull.

Interoceptor Sense organ for events inside the body.

Lens The mass of transparent protein behind the cornea that focuses light on the retina.

Middle ear The air-filled chamber between the eardrum and the oval window, containing the malleus, incus, and stapes.

Olfactory cell Neuron embedded in nasal mucosa; its dendrites carry protein receptors for odor molecules.

Optic tracts The bundles of axons emerging from the optic chiasma; each tract contains axons from one side of both retinas.

Organ of Corti The spiral organ.

Otoliths Crystals of calcium carbonate on the surface of the gelatinous masses holding the hairs of the hair cells in the utricle and saccule.

Proprioceptor Sense organ for muscle stretch or tension or for joint angle.

Receptive field Skin zone in which all the receptors of one kind belong to a single sensory neuron.

Refraction The bending of light rays and images as light passes from one transparent medium to another.

Retina The sheet of light-sensitive receptor cells at the back of the eye.

Rhodopsin The light-sensitive pigment in rods.

Rods Retinal light receptors responsible for dim-light and night vision. They do not respond to the color of light.

Saccule With the utricle, the location of the receptors for head position.

Semicircular canals Hooplike tubes, one lying in each of the three planes of head rotation; when the head rotates, the fluid in the appropriate canal pushes on the cupula to provoke nerve signals.

Sense organ Bodily structure that converts an event or stimulus into a form that can activate an action potential in a nerve ending or neuron.

Sensory receptor Nerve ending or neuron in a sense organ that generates an action potential.

Suspensory ligament The ligament that attaches the edge of the lens to the ciliary body of the eye.

Tectorial membrane The tissue flap overlying the hair cells.

Utricle With the saccule, the location of the receptors for head position.

Visual field That portion of the potential field of view whose image falls on the retina.

Chapter 17

Sexual Reproduction

- Principles of Sexual Reproduction
- Primary and Secondary Sexual Characteristics
- The Human Male Reproductive System
- The Human Female Reproductive System
- Hormonal Control
- Fertilization
- Implantation
- Birth Control
- Learning Focus: Human Reproduction
- Summary
- Study Questions
- Glossary

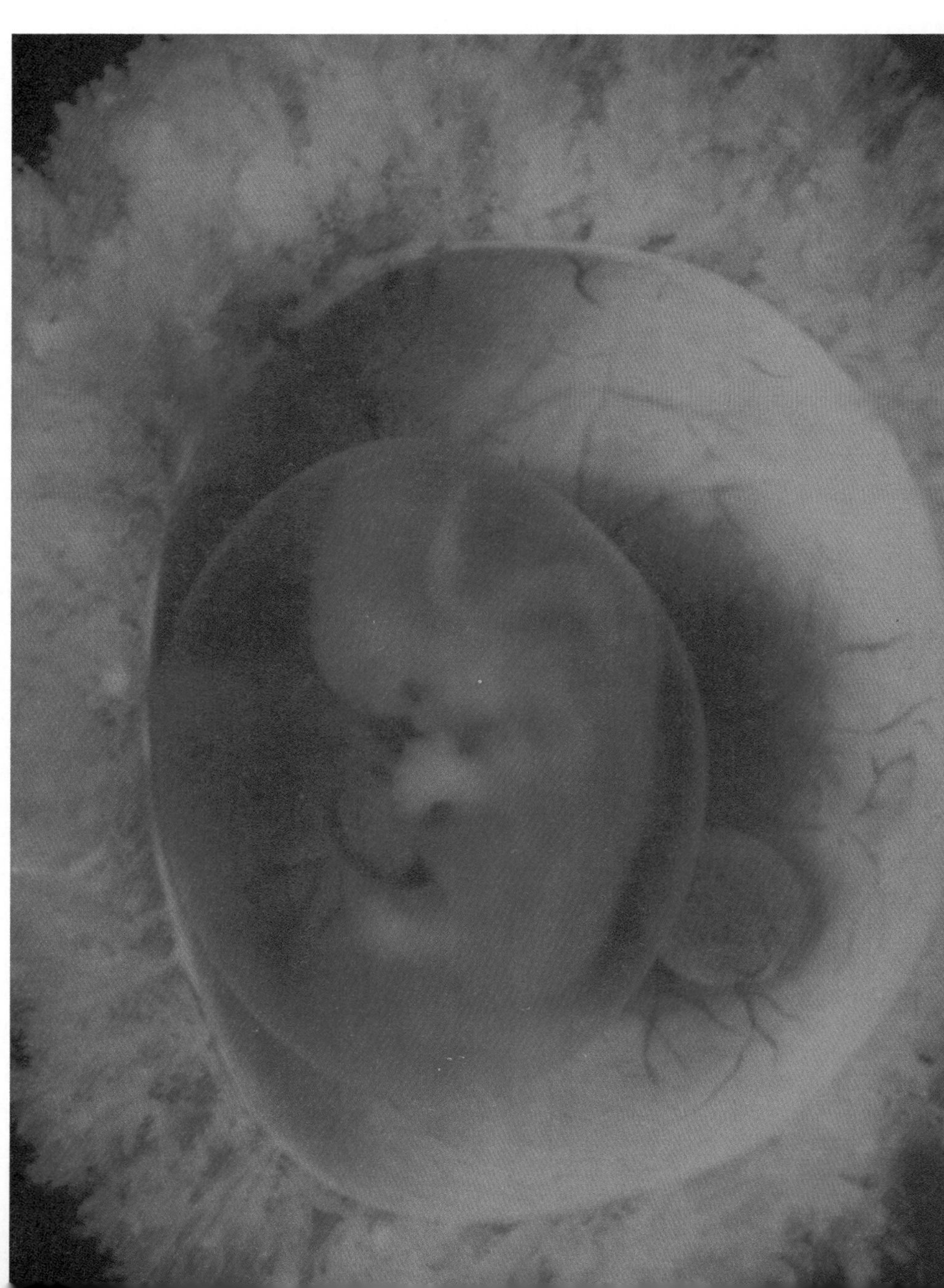

Among the most exciting and rewarding of all human activities are those involved in the process of reproduction. They are also fascinating, for behind desire, sex, pregnancy, and family life lies an intricate tapestry of physiology and anatomy. Hormones released at puberty cause our bodies to ripen and stimulate complicated feelings and urges. We grow interested in the opposite sex in new ways, and as opportunity offers, we naturally explore how our bodies can produce immense pleasures. We fall "in love," and ideally, when we are mature enough to accept responsibility for another life, we form long-term partnerships, using our bodies and their pleasures to create new individuals—our children.

These events provoke such strong feelings in us that they serve as the reservoir from which flows much of our poetry, music, and drama. From these feelings come our love songs, the masterpieces of Shakespeare, and the forgettable clichés of today's "soap operas." For biologists, however, the intricacies of sexual reproduction possess their own fascinations and generate a different kind of poetry—the intricate blend of anatomy and physiology.

PRINCIPLES OF SEXUAL REPRODUCTION

If someone asked you to define reproduction, how would you answer? It seems a simple question, but the more you think about it, the more complex it becomes, for there are two forms of reproduction in humans. **Asexual reproduction** occurs continuously at the level of the individual cells (see Chapter 5) and accounts for most of the body's growth, development, and repair. A single cell simply divides by mitosis to produce two smaller daughter cells. Those cells, in turn, may grow and then divide again.

We normally say that humans reproduce sexually, but that is really an oversimplification. Most events of human reproduction—for example, the formation and development of a new baby—occur as the consequence of asexual processes. It is more accurate to say that the offspring for each generation *are started* by **sexual reproduction.** Technically speaking, sexual reproduction occurs when two individuals each contribute half of the genetic material to a child by combining specialized reproductive cells called **gametes** (sperm and ova). Each parent has specialized organs, the **gonads** (testes in males and ovaries in females), that have the responsibility of producing these gametes. The gonads also manufacture and release the various sex hormones that cause us to develop either male or female body characteristics.

However, once those gametes have fused and the fertilized egg is converted to a zygote (the first-generation cell from which the embryo and the other structures and membranes may form), the processes of growth and development shift back to asexual reproduction as the new line of cells increases and diversifies through processes of mitosis and differentiation. In a sense, this embryonic cell line then assumes the genetic responsibility for promoting and controlling its own development (see Chapter 18). At the simplest level, the body of the mother provides the nutrients, oxygen, and filtering systems for supporting the development of the baby. Biologically speaking, the father's role is finished, although he can help by supporting and protecting the mother and their developing baby.

Why bother with sexual reproduction at all? Sexual reproduction is a much more complicated and less efficient process than asexual reproduction. However, sexual reproduction leads to tremendous variability among the offspring, and because of that variability, some of the offspring may be better equipped to win in the evolutionary struggle for survival and reproduction or to cope in a continuously changing environment. Sexual reproduction produces new combinations of genes and speeds the production of variations that can be tested for value by the processes of natural selection. Although the offspring of asexual reproducers can be produced more rapidly, they are usually little more than genetic duplicates of their parents (Figure 17-1). In asexual reproduction, variation arises only from mutation, for the genes in asexual forms are not reshuffled with every generation.

The key to sexual reproduction is that version of cell division called **meiosis,** which we discussed in Chapter 5. During meiosis *crossing over* interchanges bits of homologous chromosomes, producing new combinations of linked genes. In addition, during meiosis the germ cells in the gonads reduce their normal diploid number of chromosomes (46 for humans) in half, to 23 (Figure 17-2, p. 383). The resulting haploid cells may become the *gametes.* (Some people use *germ cells* or *sex cells* to refer to the gametes; we prefer to use germ cells for those cells that give rise to the gametes.) In males, the germ cells lie in the male gonads, the **testes,** and the gametes are called **sperm,** or **spermatozoa.** In females, the germ cells lie in the female gonads, the **ovaries,** and the gametes are **ova,** or egg cells.

Sexual reproduction begins with sexual intercourse, or copulation. The male inserts his erect penis into the female's vagina, and as a conse-

(a)

(b)

(c)

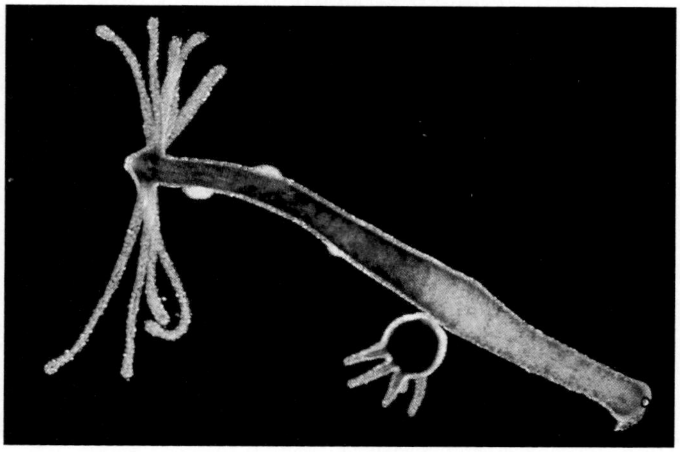

(d)

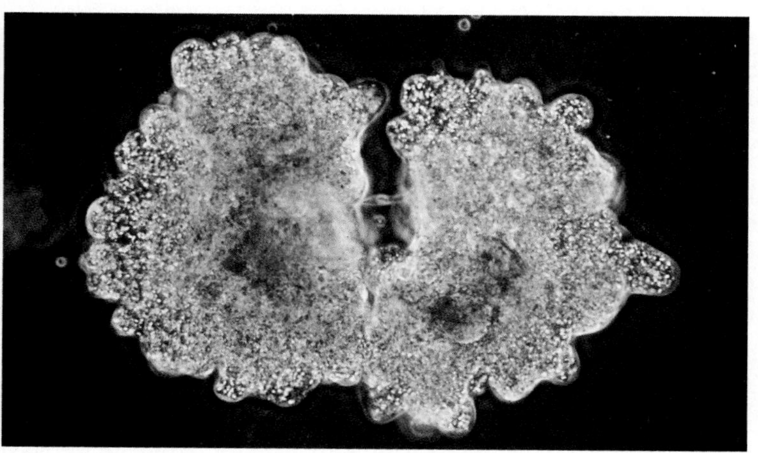

(e)

FIGURE 17-1 Several common modes of asexual reproduction. (a) Potatoes (tubers) can sprout new individuals from their "eyes." (b) grass runners (stolons) grow along the ground, sending out roots and new stems with new leaves at intervals along their length. If a stolon is accidentally severed, each plantlet can survive independently. (c) Common puffballs release clouds of asexually produced spores that can grow into new fungi if they land in a suitable situation. (d) Hydra reproduce asexually by budding tiny duplicates of themselves from their flanks. (e) Single-celled amoebae reproduce asexually by fission, first moving half of the cell's components into separate halves of the cell and then pinching in two.

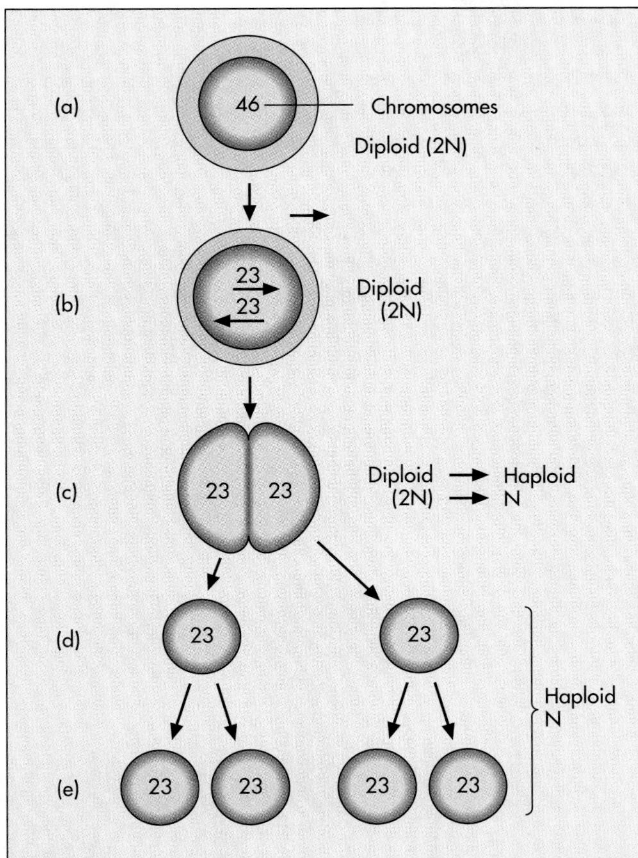

FIGURE 17-2 This oversimplified version of meiosis begins (a) with a diploid cell containing 46 chromosomes. The homologous pairs separate (b,c) to produce two haploid cells each containing 23 double-stranded chromosomes. (d). The haploid cells undergo a second division (e) to form four cells, each of which contains a haploid set of single-stranded chromosomes. In males, all four of these cells become gametes, the sperm. In females, only one becomes a gamete, the ovum; the other three are discarded.

quence of the sensations produced, reflexively releases the whitish sperm-containing **semen.** Besides sperm, semen contains mucus and glandular secretions that support and nourish the sperm and neutralize the acid environment of the female vagina. The sperm move through the female reproductive apparatus, in part under their own locomotive power and in part by actions taking place in the female system. Only a fraction of the sperm released by the male reach the vicinity of the ovum. Of that fraction, one sperm fuses with the ovum in the process of **fertilization.** The two haploid nuclei of the ovum and sperm also fuse, constructing a new diploid nucleus whose genes were contributed equally by each parent. The fertilized ovum becomes the zygote, the first cell from which all the tissues and organs of the new individual, including the membranes that surround the fetus in the womb, will be derived. The zygote soon begins the series of mitoses (cell divisions) that will provide the cells for growth, development, and body maintenance for the embryo, child, and eventual adult. Figure 17-3 summarizes these events.

Sexual reproduction is simple in outline. It becomes complex only when we begin to look at all the organs and functions the body needs to make it work. Fortunately, most people are familiar with many of these organs and functions, at least to some extent, long before they take a biology course. Let us now begin to fill in the details by examining the apparatus of reproduction.

---

1. How does sexual reproduction produce new combinations of genes?
2. Explain why children produced by sexual reproduction tend to resemble their parents.

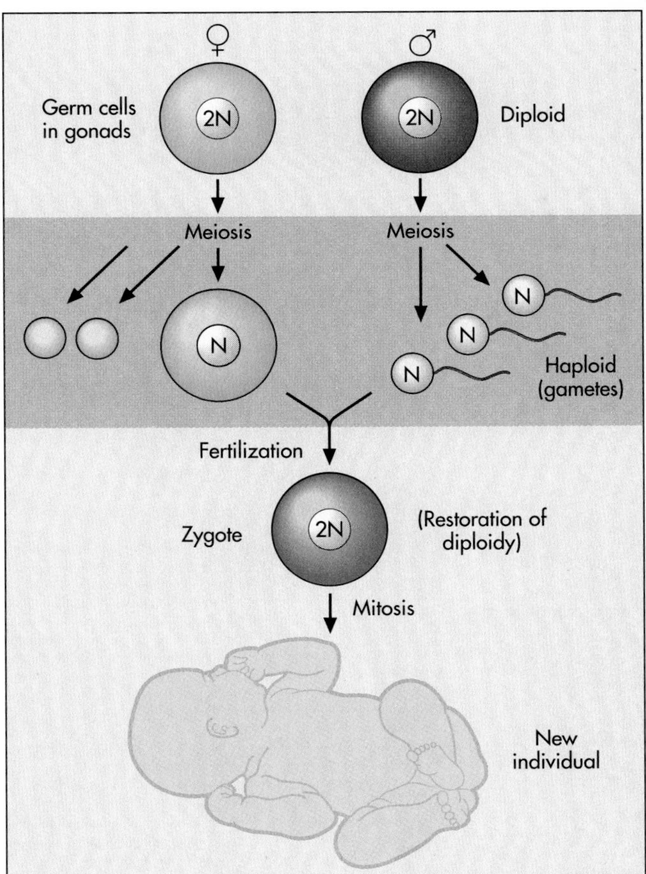

FIGURE 17-3 An outline of sexual reproduction. Diploid germ cells in the gonads divide via meiosis to give rise to haploid gametes. Fertilization restores diploidy in the zygote, which eventually gives rise to a new individual via mitosis.

PRIMARY AND SECONDARY SEXUAL CHARACTERISTICS

Most people are very aware of the major differences between females and males. However, many of these differences influence the way we look and are not directly related to reproduction. Such features are called **secondary sexual characteristics,** and they account for those differences between the sexes that may be apparent even when we are dressed. Male secondary sexual characteristics include the deeper voice, more conspicuous facial and body hair, bone configuration, and greater muscle development and hence strength. Female secondary sexual characteristics include the higher voice, less facial and body hair, wider pelvis, sleeker musculature, more skeletal flexibility, larger breasts, and a thicker layer of subcutaneous fat. In both sexes, these features develop after puberty triggers the flow of sex hormones, and they depend for their maintenance on the continual production of these hormones (Figure 17-4).

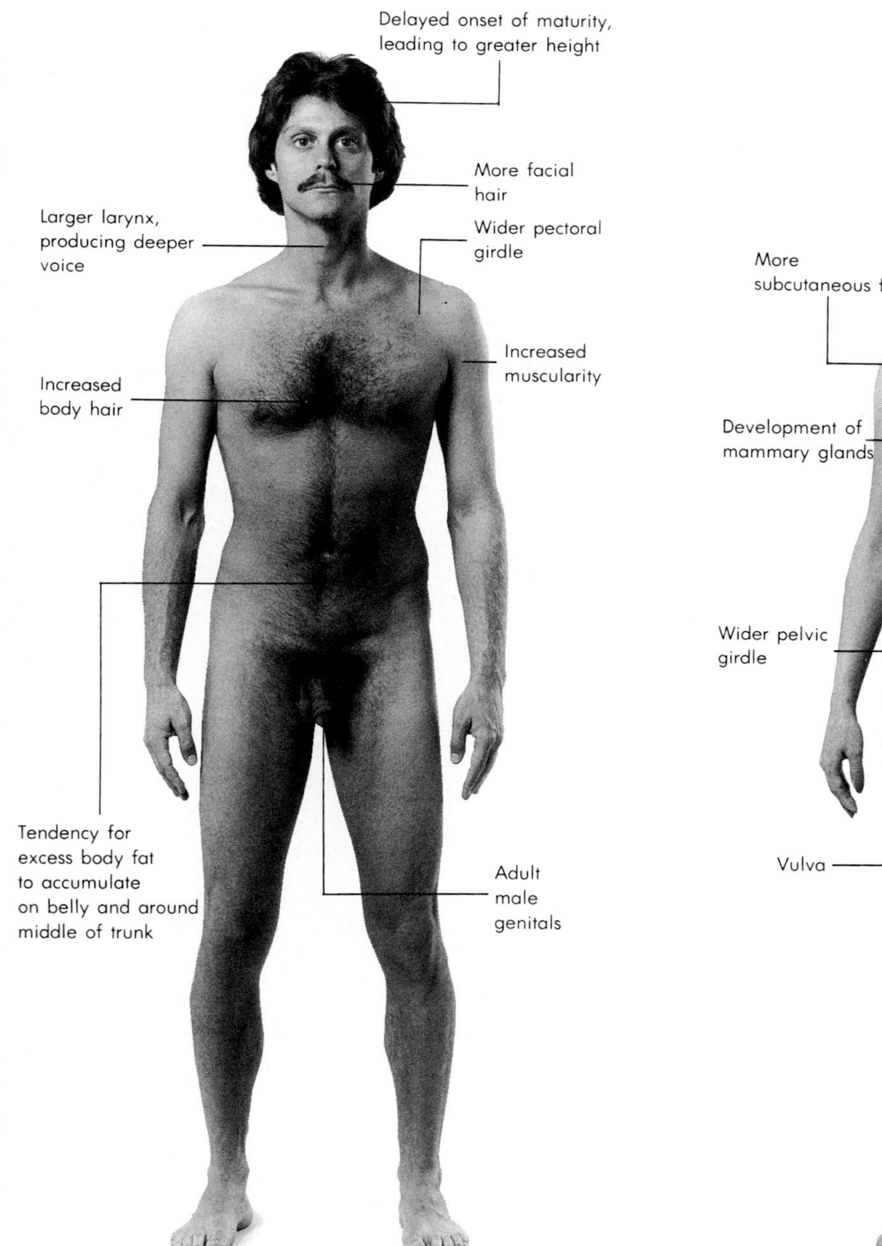

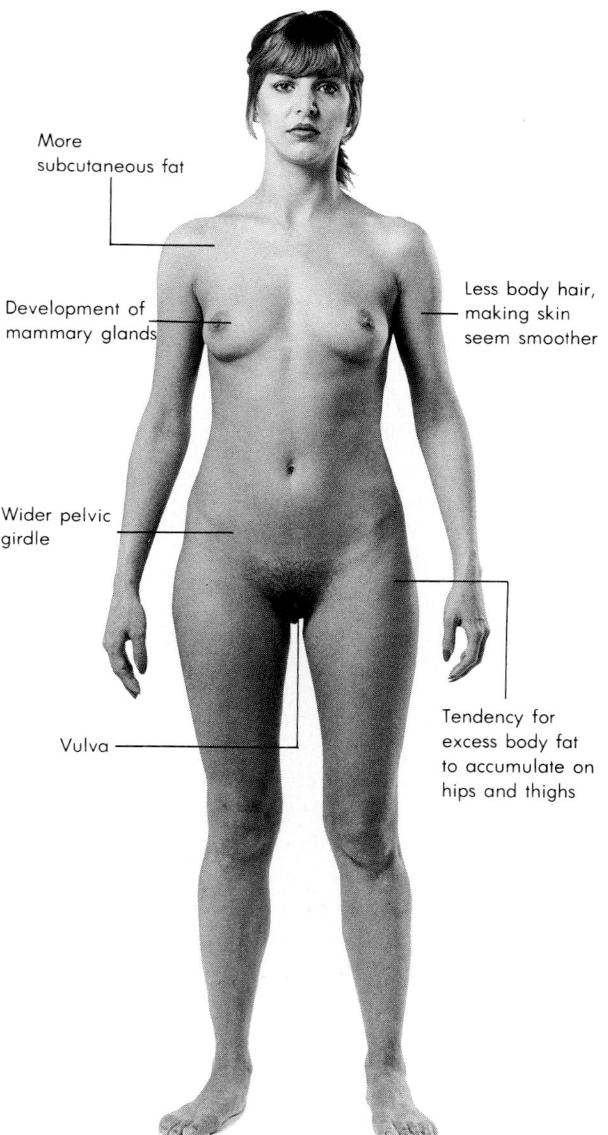

FIGURE 17-4 Comparison of male and female secondary sex characteristics.

Men and women also differ in several less obvious—though often controversial—ways. Women tend to live longer, for instance. Men seem to be more aggressive, perhaps because of the influence of the male sex hormone on the brain. There are also differences in the size of some brain regions. However, these differences are only indirectly related to reproduction, and we will ignore them here.

The **primary sexual characteristics,** on the other hand, are directly related to sex, and they include the various components of the genitals. Since the gonads are responsible for actually manufacturing the gametes, they are considered the **primary sex organs.** The other parts of the genitals, namely the tubes, ducts, and glands of the male and female reproductive systems, when connected during intercourse, form a continuous conduit between the testes of the male and the ovaries of the female. As such, these are the **accessory sex organs** that provide a channel through which the gametes can move and in which they eventually combine. The accessory organs in the female also provide a suitable environment for the developing baby. Technically, then, the penis, vagina, and uterus are accessory sex organs.

THE HUMAN MALE REPRODUCTIVE SYSTEM

The primary *biological* purpose of a male is to produce sperm and fertilize ova. That done, the female can handle the rest of the reproductive process alone. She can even raise the resulting offspring, finding food and shelter as necessary.

Other species push this limited usefulness of males to extremes. For example, female spiders sometimes kill and consume their mates during the sex act. The male's body thus contributes to the nourishment of his own offspring. In another version, the female angler fish wears the male as a tiny ornament on her brow; he fastens there in childhood, sharing his mate's bloodstream and supplying sperm as needed. In many species, like termites, ants, aphids, and bees, virtually all the adults are female; males are found primarily during specific seasons or periods of drought or cold.

The often prolonged partnership we see between human mates and in some species of birds and other mammals does have its benefits. In species where there is an extensive period of parental care for the young, the mated pair shares the burdens of "child rearing" and makes the survival of their offspring much more likely.

The Sperm Cells

A sperm cell is primarily a delivery system for a set of chromosomes. It consists of a dense haploid nucleus and very little cytoplasm. The cell body is an oval structure capped by an **acrosome,** which contains enzymes needed to break through the outer coverings surrounding an ovum. The other end of the cell is marked by a long **flagellum,** a whiplike propulsive organelle that partially powers the movement of the sperm through the fluids of the female reproductive system. Where the cell body and flagellum join, there is a collar or midpiece filled with mitochondria that provide the flagellum with the energy (ATP) it needs to push the sperm up the female reproductive tract to the egg (Figure 17-5).

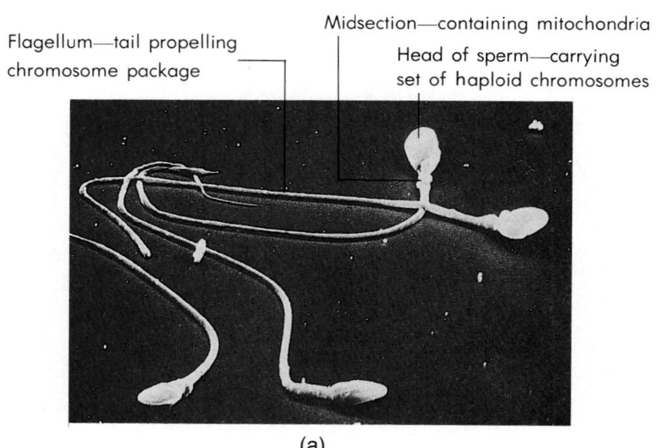

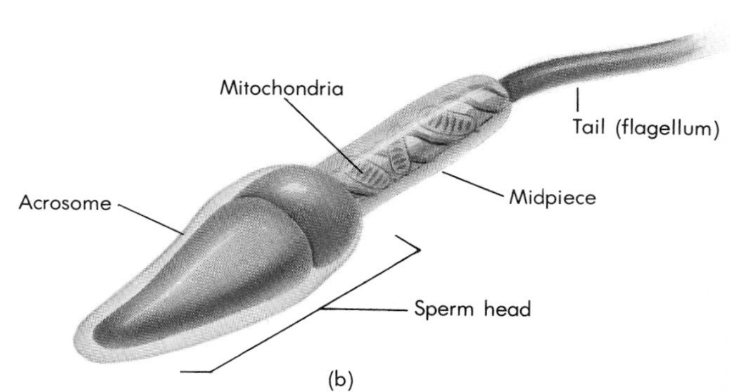

FIGURE 17-5 (a) Human sperm cell (× 4500). (b) Side view of a sperm's head area.

The Testes

Did you know that the words *testes* and *testimony* are related? Both date back to ancient times, when it was a common practice for men to place their hand over the genitals while swearing to tell the truth (*testis* is Latin for witness).

The sperm are produced in the two testes, which are normally suspended in a thin-skinned sac called the **scrotum,** hanging beneath the groin (Figure 17-6a). The testes originate in the embryo's body cavity. However, they leave the pelvic cavity and descend into the scrotum before birth. This is necessary because the testes are temperature sensitive. The testes' position outside the body leaves them somewhat vulnerable, but the scrotum is about 3.1 degrees Celsius (°C) cooler than the interior of the body cavity. If the testes remain inside the abdomen, as in the condition known as **cryptorchidism,** they fail to produce functional sperm because they are too warm. Sperm production is so thermally sensitive in some individuals that taking hot baths or spending too much time in the sauna may lead to a temporary lowering of fertility. Thermal jockstraps have even been proposed as a possible means of birth control. In one study, participants used jockstraps that raised the scrotal temperature by 1.0°C and reduced the sperm count by about 25 percent.

Each testis contains 200–300 compartments, or lobules. In each lobule lies a mass of coiled **seminiferous tubules** (Figure 17-6b), where the sperm are produced by a process called **spermatogenesis.** The outer walls of the tubules are lined by **spermatogonia,** germ cells that divide by mitosis to produce both more spermatogonia and **primary spermatocytes.** The primary spermatocytes divide by meiosis. The first meiotic division of a primary spermatocyte produces two haploid **secondary spermatocytes.** Each secondary spermatocyte undergoes a second meiotic division and generates a pair of tailless **spermatids.** The spermatids embed themselves in surrounding **Sertoli cells,** which support them as they develop into mature sperm cells.

Sperm production is controlled by a negative-feedback system mediated by hormones. Beginning

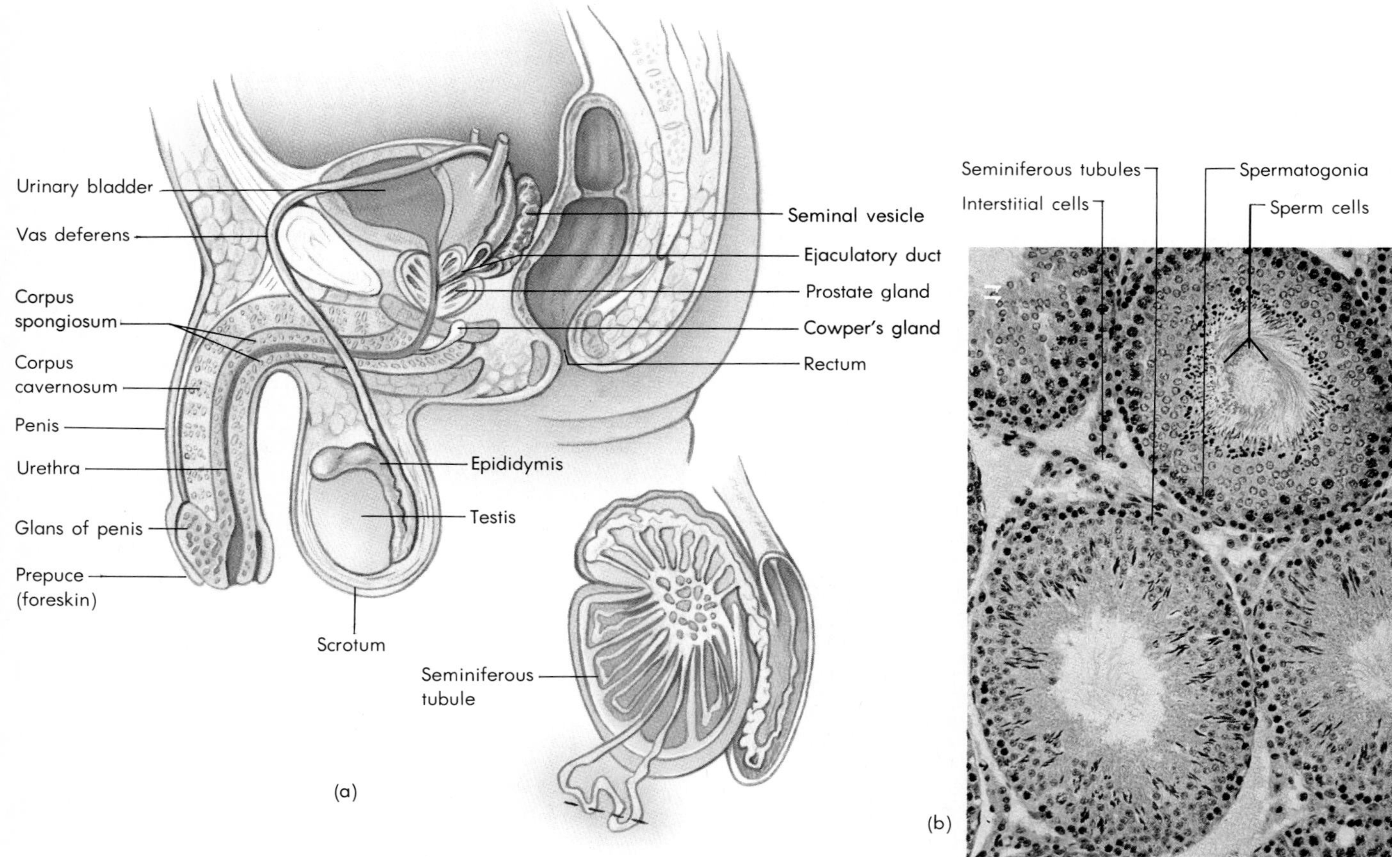

FIGURE 17-6 (a) The male reproductive system. (b) The seminiferous tubules of the testes are the sites of sperm production. Sperm cells are released from the walls of the tubules into the lumens. Clusters of interstitial cells, between the tubules, manufacture and release the male sex hormone, testosterone.

at puberty, the anterior pituitary gland secretes the gonadotropins **luteinizing hormone (LH)** and **follicle-stimulating hormone (FSH)** in response to hypothalamic gonadotropin-releasing hormone (GnRH) (see Chapter 13). Because in the male luteinizing hormone stimulates the **interstitial cells** lying between the seminiferous tubules to secrete the male sex hormone, **testosterone,** LH is also called interstitial-cell-stimulating hormone. Testosterone causes the bodily changes the male undergoes at puberty, including the maturation of the testes and other male sex organs. Follicle-stimulating hormone stimulates spermatogenesis. As mature sperm accumulate in the tubules, at a rate of about 300 million per day, the Sertoli cells secrete **inhibin,** which inhibits GnRH and FSH secretion and slows further sperm production. Inhibin is currently being studied as the potential basis for a male birth control pill.

The Ducts and Glands

Once produced, sperm are stored until they can be transferred to a female by the process of **ejaculation.** Storage and transfer are the tasks of a series of tubes or ducts that link the testes to the penis. Sperm produced in each testis are transferred via collecting ducts to a pair of storage tubules called the **epididymides.** Each epididymis is a coiled tubule about 20 feet long wrapped in a connective tissue sheath and sitting atop each testis. Over a period ranging from 18 hours to 10 days, the sperm finish maturing in the epididymides, becoming capable of fertilizing an ovum. The epididymides can store sperm for up to a month; after that, the aging sperm cells are destroyed and the body reabsorbs them, making way for their replacement by fresh sperm cells. During ejaculation, the epididymides expel sperm by peristalsis, waves of contractions of the smooth muscles within their walls.

The narrow tail of each epididymis merges into a **vas deferens.** This duct also stores sperm and uses peristalsis and the actions of cilia, which line the duct system, to move the sperm through the tract. Each vas deferens extends from the scrotum into the pelvic cavity, where it arches behind and below the urinary bladder. Shortly before it joins the single urethra, each vas deferens is joined by a duct from a small gland [about 2 inches (in.) long], the **seminal vesicle.** The two seminal vesicles contribute secretions rich in the sugar fructose, which partially fuel the activities of the sperm cells and contribute to their motility. The short duct, continuing from where the vas deferens joins with the duct from the seminal vesicle, is called the *ejaculatory duct* (Figure 17-7). Each ejaculatory duct connects with the back wall of the urethra, below the bladder. It is here that the semen flows into the base of the penis for ejaculation.

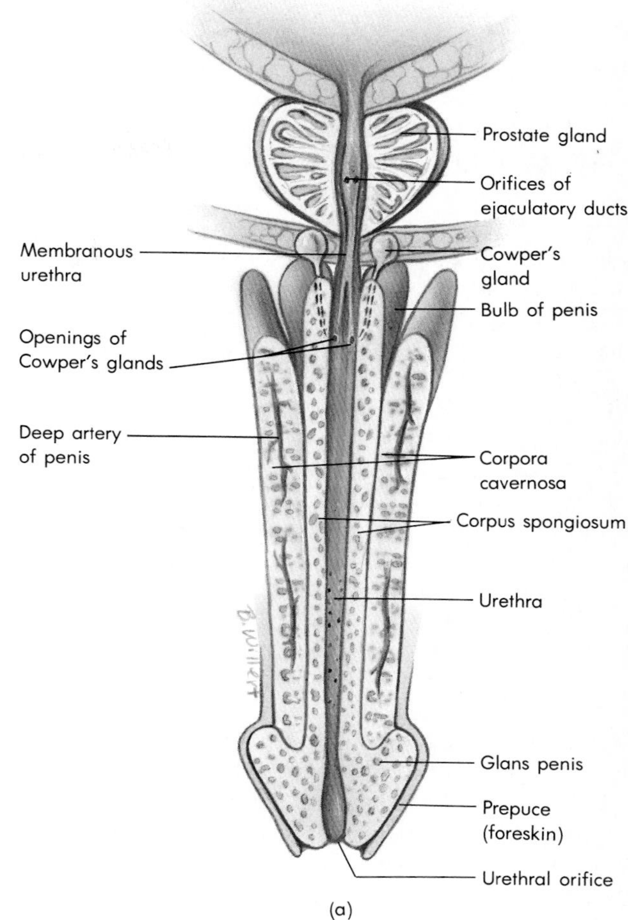

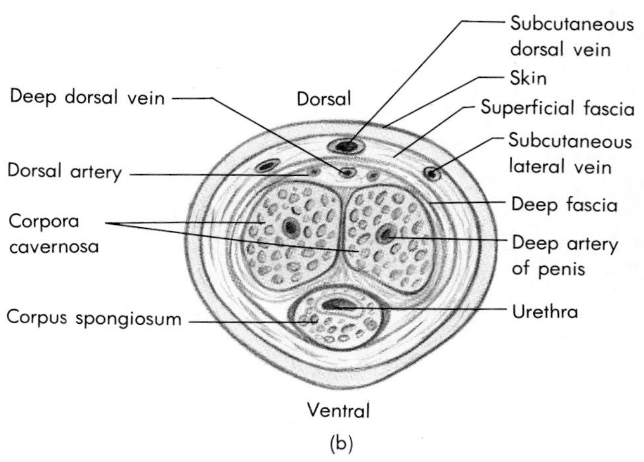

FIGURE 17-7 The structure of the penis. (a) Longitudinal section. (b) Cross section.

At the base of the bladder, surrounding the urethra and the ejaculatory ducts, is a fairly muscular, chestnut-sized organ, the **prostate gland** (note the spelling—it is not a prost*R*ate gland). The prostate's secretion provides the bulk of the semen released during ejaculation. It is also responsible for the characteristic odor of semen. It is a milky secretion, with alkaline components that neutralize and protect the sperm from the acid ingredients found in the female reproductive tract. Semen also contains an antibiotic and enzymes that cause it first to coagulate and then, within a few minutes, to liquefy, freeing the sperm to spread through the female reproductive system.

Two more glands, **Cowper's glands,** sometimes called the bulbourethral glands, connect to the urethra below the prostate. Cowper's glands secrete lubricants for the male urethra during foreplay. These secretions often show as a droplet of clear fluid appearing at the opening of the penis. Such small quantities of Cowper's secretions are released, however, that they are not considered to be one of the principal lubricants for the sex act. Cowper's secretions may have a cleansing action that prepares the male urethra for the arrival of the sperm.

Sperm are best able to fertilize ova when they have been ejaculated into a female reproductive tract. Exposure to the acid fluids found within a vagina activates the sperm so they can travel through the female reproductive system toward the ovum and then actually penetrate and fertilize it. This final ripening or sperm activation process is known as **capacitation.** It includes enabling the sperm's flagella to propel them through the female tract.

The Penis

The **penis** is the delivery system for semen. It is a fleshy cylinder containing three columns of spongy, erectile tissue and a tube, the **urethra,** for the passage of urine and semen from the body. At times of sexual arousal, the autonomic nervous system dilates the arterioles carrying blood into the penis. As blood flows into the columns of spongy tissues, the two **corpora cavernosa** and the single **corpus spongiosum,** the penis enlarges and becomes rigid (producing an erection). The rising hydraulic pressure in the penis constricts the veins that drain the spongy tissues, maintaining the rigidity. After ejaculation, the arterioles constrict once more, the inflow of blood slows, the pressure falls, the veins dilate, and the penis loses its rigidity. A flaccid penis may be only 2 or 3 in. long. The average size for erect penises is 6 in. in length.

The penis has three main sections. The **bulb** lies deep within the body, below the bladder and prostate. The **shaft** is the main length of the organ. The acorn-shaped **glans** caps it (again, be careful of the spelling—there is no *d* in *glans*) and is covered with a **prepuce,** or foreskin (often removed in infancy in the process of **circumcision**). The urethra conducts urine and semen through the penis (though usually not simultaneously), where it exits through a hole in the tip of the glans (Figure 17-7).

The glans is very rich in nerve endings. When stimulated by friction, as during sexual intercourse, these nerves activate the ejaculatory reflex. During orgasm, autonomic nerve signals from the spinal cord initiate peristalsis in the vas deferens and the epididymides and rhythmic contractions in the smooth muscle fibers of the prostate, causing the release of secretions from the glands and the pulsing ejaculation of semen from the penis. The result is a discharge of 2.5–5.0 milliliters of semen containing 50–100 million sperm per milliliter.

---

1. Describe the path of sperm from the testes to the glans.
2. List the components of semen.
3. Many people think that if a man ejaculates before sexual intercourse, he has exhausted his supply of sperm and pregnancy cannot occur. What is wrong with this idea?

THE HUMAN FEMALE REPRODUCTIVE SYSTEM

The function of the female in reproduction is to produce ova, mate, provide a site for fertilization, shelter the growing embryo until the time of birth, and provide milk and care (a responsibility usually shared with her male partner) for the infant after birth. She thus plays a much larger biological role than the male, and her reproductive system is accordingly more elaborate.

The Ova

The human ovum, or egg, is one of the few cells large enough to be visible to the unaided eye. Unlike sperm cells, it lacks a flagellum, its nucleus is not condensed, and it has a large amount of cytoplasm. It is also surrounded by a clear membrane, the **zona pellucida.** The ova of many animals, such as birds, contain large amounts of stored nutrients, or **yolk,** to feed the growing embryo. Human ova contain relatively little stored yolk. They do contain stores of growth regulators and other substances that help to control the initial development of the embryo.

The Ovaries

The ova are produced in the ovaries, two small organs, each about 4 centimeters (cm) long, suspended by bands of connective tissues on either side of the abdomen. The ova arise from the layer of epithelium covering the ovary; the interior is a mass of connective tissue, the **stroma** (Figure 17-8).

The ova begin their development in the ovary even before birth. **Oogonia** in the ovarian epithelium divide by mitosis to produce about 200,000 large **primary oocytes** in each ovary. Each primary oocyte is surrounded by a layer of stroma to form a **primary follicle.** The primary follicles remain dormant until puberty, when pituitary gonadotropins stimulate their further development. Then, each month, one or more follicles begin to enlarge. The follicle's cells secrete the female sex hormones estrogen and progesterone. These hormones stimulate the further development of the follicle and cause the lining of the uterus to thicken. At puberty, they also stimulate the development of the secondary sexual characteristics and support the maintenance of these female characteristics in the adult.

While the follicle enlarges, a first meiotic division turns the primary oocyte into a single haploid **secondary oocyte** and a **polar body.** The polar body is a tiny partial cell consisting of very little cytoplasm and half the primary oocyte's chromosomes; it is essentially a method of casting away the extra chromosomes while retaining the primary oocyte's cytoplasm and its nutrients for the benefit of any eventual embryo (Figure 17-9).

At the same time, the follicle develops a fluid-filled cavity, to one side of which, surrounded by follicular cells, lies the developing ovum. The ripened or mature follicle is now called a **Graafian follicle** (Figure 17-10). **Ovulation** occurs when the follicle bursts, releasing the secondary oocyte from the ovary (Figure 17-11). It remains surrounded by a sheath of follicular cells, the **corona radiata.**

Some women are aware of ovulation as a painful sensation in their lower abdomen. In some cases, the pain occurs over the ovary that is actually doing the ovulating. Since ovulation occurs in the middle of the female cycle, such pains are referred to as *Mittelschmerz* (German for "middle pain"). After ovulation, the secondary oocyte undergoes its second meiotic division only when (and if) triggered by the entry of a sperm cell in fertilization. At this time, it casts off a second polar body and becomes an **ootid.** The ootid then matures to become the ovum. Only then do the sperm and egg nuclei unite.

After the Graafian follicle bursts, its remaining cells fill with a yellowish glandular material, transforming it into the **corpus luteum** (*corpus* = body; *luteum* = yellow). It secretes the hormones progesterone and estrogen, which further stimulate the

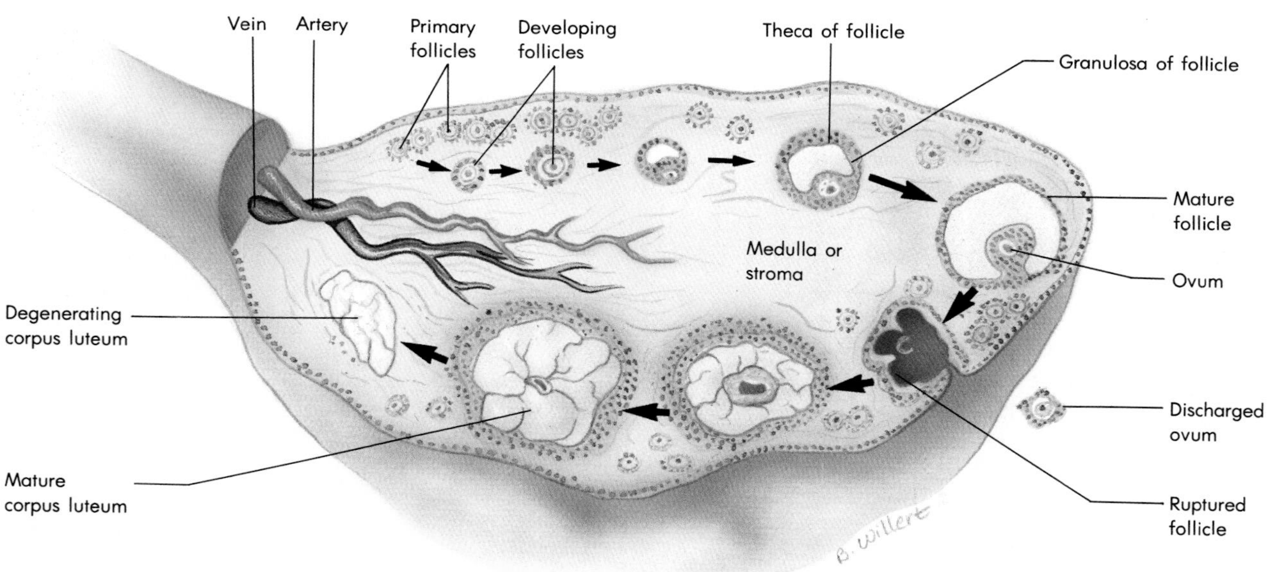

FIGURE 17-8 The structure of the ovary.

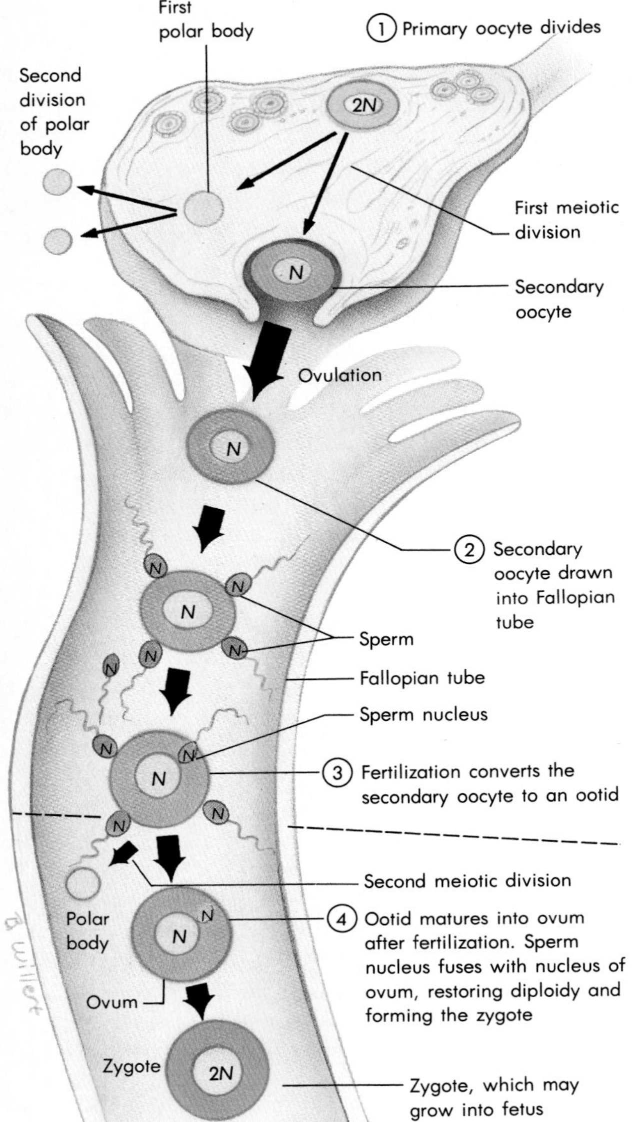

FIGURE 17-9 Oogenesis in the human female. (1) The primary oocyte divides in the ovary to give rise to a polar body and a secondary oocyte. (2) At ovulation, the secondary oocyte escapes the ovary and enters the Fallopian tube. Only after fertilization (3) does it go through the second meiotic division and become an ootid, which matures into an ovum. (4) When the sperm and ovum nuclei finally fuse, the haploid ovum becomes a diploid zygote, which may grow into a fetus.

development of the endometrial lining of the uterus, so that it is ready to receive and support any developing embryo. Estrogen also promotes the development and maintenance of other female sex characteristics. If fertilization does occur, the corpus luteum enlarges and continues its task to support the early stages of pregnancy. If fertilization does not occur, the corpus luteum atrophies, becoming a small mass of scar tissue, a **corpus albicans,** within the ovary.

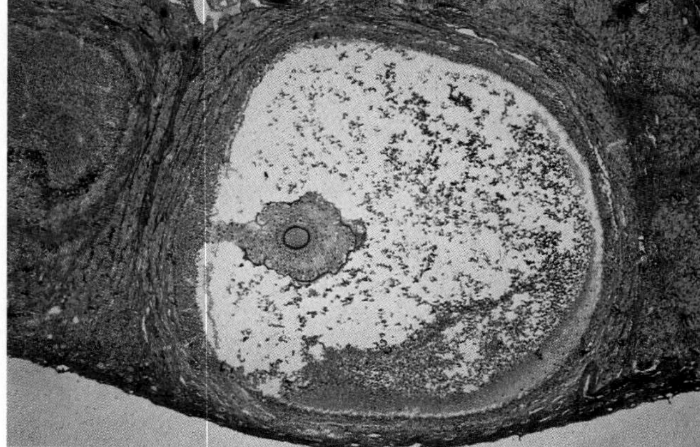

FIGURE 17-10 Photomicrograph of a mature human follicle.

1. Imagine that you have been handed an ovary, together with the tools (such as a microscope) necessary to examine it in detail. What clues would tell you the ovary had come from a prepubertal female?
2. What clues would tell you the ovary had come from a pregnant woman?
3. How might you estimate the age of the woman to whom the ovary had belonged?

The Fallopian Tubes and the Uterus

When the egg escapes the ovary during ovulation in a burst of follicular fluid, it is still a long way from its ultimate destination—the uterus. Its path lies through one of a pair of **Fallopian,** or **uterine, tubes** that arch up from the uterus and over the ovaries, where they end in funnel-shaped openings fringed with fingerlike **fimbriae.** The fimbriae and the lin-

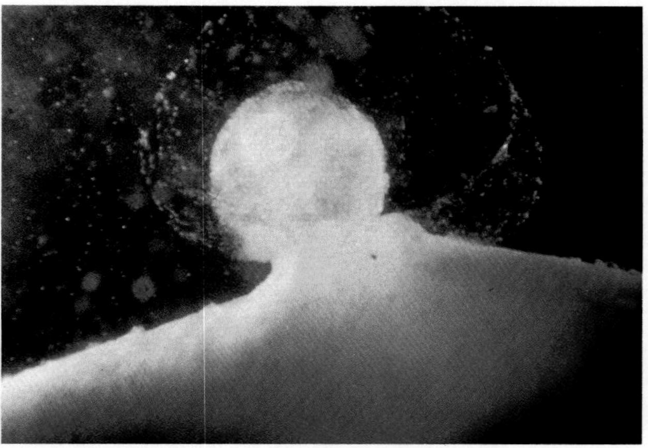

FIGURE 17-11 Ovulation is the moment when the follicle ruptures and releases the ovum.

ing of the tubes bear cilia, whose action sets up currents in the fluid around the ovaries. Such currents draw the ovum into one of the Fallopian tubes, where peristaltic contractions of the smooth muscles in the tube walls help propel the egg toward the uterus. The egg is normally fertilized by sperm while it is still in the Fallopian tubes. However, if the tube is blocked by scar tissue, which sometimes happens after gonorrhea or chlamydial infections, fertilization cannot occur (Figure 17-12). (Sexually transmitted diseases are discussed in Chapter 23.)

The **uterus** (womb), a pear-shaped organ within the pelvic cavity, supports the growth and development of the fetus. The Fallopian tubes connect to the upper side walls of the uterus. The narrow, lower end of the uterus forms the **cervix,** which protrudes into the upper end of the vagina. Normally, the cervical opening is plugged by a thick mucus. About the time of ovulation, this mucus thins, presumably to permit the passage of sperm passing in from the vagina.

The thick wall of the uterus is mostly smooth muscle. Contractions of these uterine muscles propel the baby through the birth canal during childbirth (parturition). However, the wall is not only muscle. The interior of the uterus is lined with a thick, glandular epithelium, the **endometrium.** Each month, under the influence of the hormones released by the developing follicle in the ovary, the endometrium thickens. Its glands secrete a glycogen-rich mucus capable of nourishing a young embryo, but only until the embryo can burrow into the endometrium, form the placenta, and then tap the mother's bloodstream for nutrients.

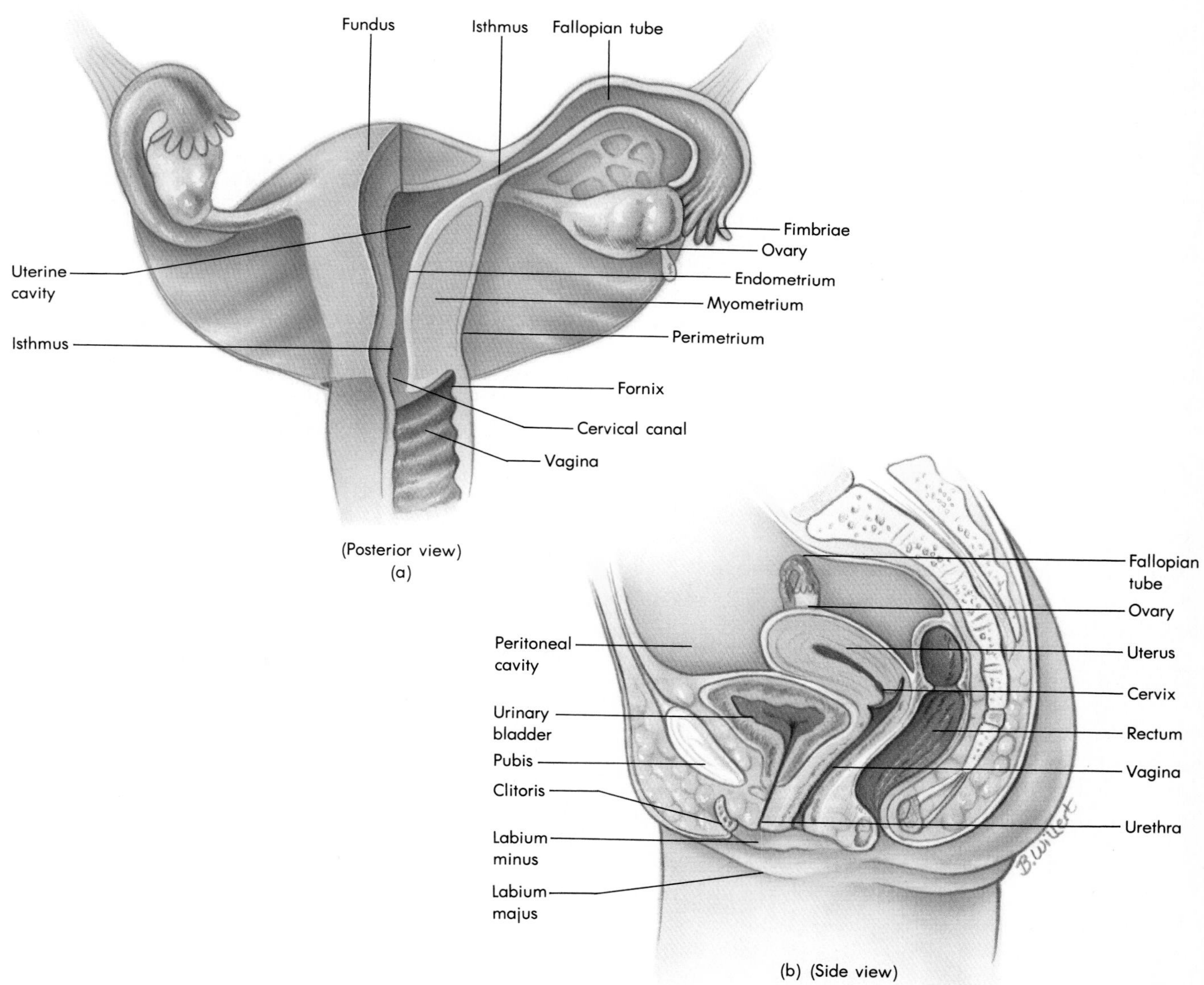

FIGURE 17-12 The relative positions of the ovaries, Fallopian tubes, uterus, cervix, and interior end of vagina: (a) posterior view; (b) side view.

Menstruation

The endometrium has two layers. The basal layer is permanent. The surface layer is periodically sloughed and renewed, about every 28 days in a nonpregnant female. The line separating the two endometrial layers is crossed by a number of spiral arterioles. When an embryo reaches the uterus, it secretes hormones that cause the endometrium to thicken still more. When an embryo does not reach the uterus, as happens in most months, these hormones are not present. The corpus luteum begins to atrophy, and its hormone output declines. The spiral arterioles go into spasm and break. Blood accumulates between the layers of endometrium, causing the surface layer to separate. The blood and scraps of endometrium are then discharged through the cervix and vagina as menstrual flow; the process is called **menstruation.**

The **menstrual cycle** repeats every month from the onset of puberty **(menarche)** until the **menopause,** between the ages of 40 and 50, when the cycle becomes irregular and stops as the ovaries lose their ability to respond to pituitary gonadotropins. With menopause, the ovaries thus reduce their production of estrogen and progesterone. Since these hormones are essential to maintaining the secondary sexual characteristics, menopause is sometimes accompanied and followed by reduction in breast size and a decrease in the natural secretion of sexual lubricants. It is not unusual for a woman to take two years to complete the changes of the menopause. During that time she can experience hot flashes, headaches, joint pain, irritability, and periods of depression. The most important contributor to a smooth transition is a supportive family. Only one woman in four seeks medical help for problems related to menopause.

Some younger women find that menstruation is a painful experience accompanied by uterine cramps, nausea, headaches, diarrhea, and other symptoms. This condition, known as **dysmenorrhea,** appears to be due to the overproduction by the uterus of very active chemicals called prostaglandins (see Chapter 12). Prostaglandins cause uterine contractions, among other things, and have been used to induce childbirth. Severe dysmenorrhea can be effectively treated with drugs that inhibit prostaglandin synthesis.

Premenstrual syndrome (PMS) refers to the breast swelling and tenderness, joint pain, irritability, depression, anxiety, fatigue, headaches, and other symptoms that precede menstruation in many women. The cause is unknown, but women have found relief after treatment with hormones, vitamins, exercise, and stress management techniques.

The Vagina

The **vagina** is a tube about 10 cm (4 in.) long leading from the uterine cervix to the outside of the female body. It receives the male's penis during copulation. The vagina has a muscular wall and a folded inner lining that provides the frictional stimulation that promotes ejaculation in the male. The folds, called **rugae,** also allow the birth canal to expand to accommodate the birth of a baby.

The vaginal mucous membrane, like that of the uterus, secretes a mucus containing glycogen. Bacteria and fungi that normally dwell in the vagina break this carbohydrate down to form acids and create an environment hostile to many other microorganisms. These acids are also hostile to sperm, but the semen counteracts the acidity with its own alkalinity.

The Vulva

In spite of the fact that male and female genitals do not look very similar, many of their structures have the same embryonic origins. Structures that have a common embryologic origin are said to be *homologous*. The external genitalia of females are collectively referred to as the **vulva** or **pudendum.**

There are three openings on the external floor of the female pelvis. From front to back, they include the urethra, the vagina, and the anus. The most conspicuous structures of the vulva include two pairs of skin folds flanking the openings to the urethra and the vagina. The outermost pair of folds are the **labia majora,** which are covered with pubic hair in mature females; they also contain fatty tissues and numerous sebaceous (oil) and sweat glands and are formed from the same skin that becomes the scrotum in the male. The labia majora meet in front, as shown in Figure 17-13, in the **mons pubis** or **mons veneris** (Mount of Venus), a mound of fatty tissue overlying the pubic bone.

Medial to these outer lips are the **labia minora;** they are hairless, are covered with mucous membrane, contain numerous nerve endings that promote part of the pleasurable sensations from sexual activity, and also have many sebaceous glands. The space between the labia minora, just outside the opening to the vagina, is called the **vestibule.** The labia minora meet at the **clitoris,** a small mass of erectile tissue homologous to the male's penis. As

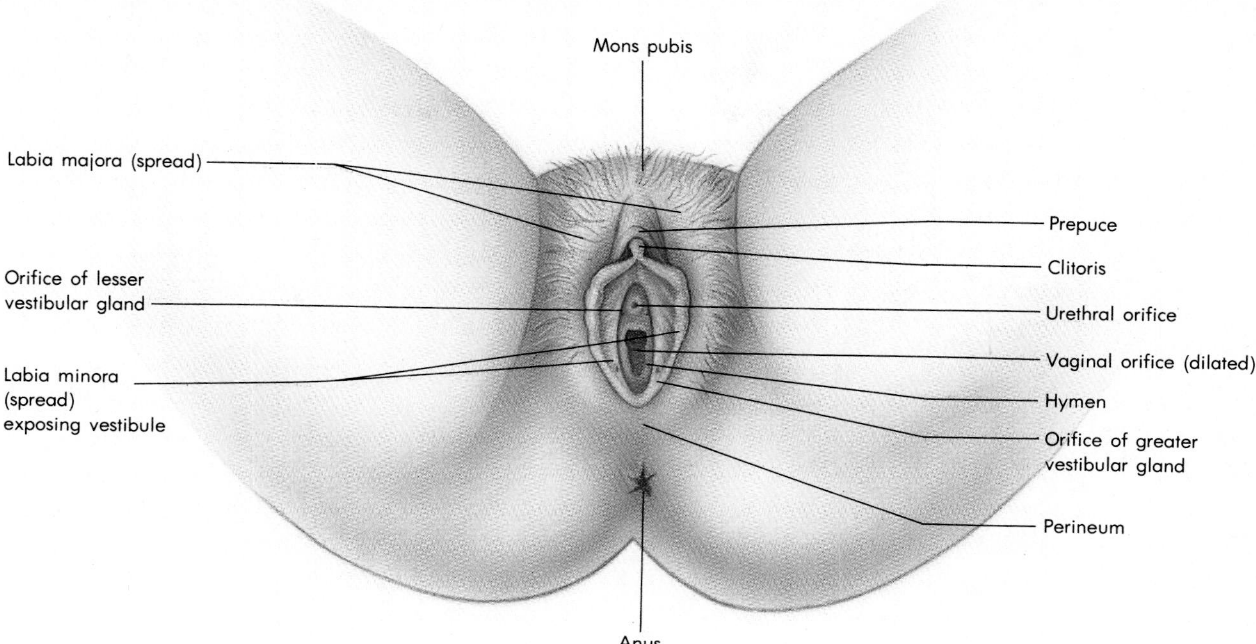

FIGURE 17-13 The vulva, the external genitals of the female.

in the penis, it consists of a shaft made up of columns of erectile tissues; at the end of the shaft, there is a tiny head called the *clitoral glans*. A small fold of skin, the *prepuce*, covers most of the clitoris's length. The clitoris contains many nerve fibers and serves as a focus for many of the pleasurable sexual sensations for females. Unlike the penis, which also serves as part of the delivery system for semen, the clitoris has no function other than to give pleasure.

When a woman is sexually aroused, the mucous membrane of the vagina and vestibule secretes a lubricating fluid. At the same time, the *vestibular glands* on either side of the vestibule supply more mucus.

Just within the vaginal opening of virginal women is often found a ring of membranous tissue, the **hymen.** Many people believe that an intact hymen is a sign of virginity, even though it is easily torn by exercise. In Japan, this belief is so strong that a number of physicians have specialized in surgical restoration, or "recreation," of hymens.

The Breasts

An entire class of animals, the mammals (including humans), derives its name from the **mammary glands** that produce milk for their offspring. In humans the mammary glands are modified sweat glands, called breasts, that sit over the pectoral muscles on the chest. In other mammals, like cows, the mammary glands are located in a large baglike appendage, the udder, found suspended from the belly closer to the hind legs. They mature during pregnancy, when the hormonal levels increase significantly.

The female breasts consist mostly of fatty tissue. In addition, each one contains 15–20 lobes of glandular tissue subdivided into lobules and alveoli (Figure 17-14). After the birth of a child, the alveoli of the mother's breasts manufacture and secrete milk. However, they cannot do so without exposure to prolactin, a hormone secreted by the anterior pituitary during and after pregnancy. The actual release of milk from the alveoli is due to oxytocin, a posterior pituitary hormone. Both prolactin and oxytocin are released when a suckling infant stimulates nerve endings in the nipple and activates the **neuroendocrine reflex.** The oxytocin causes **myoepithelial cells** in the alveoli to contract and squeeze milk into ducts leading to the nipple. The reflex takes only about a minute to act; the resulting release of milk is known as the "milk letdown."

For the first few days after birth, the breasts do not secrete milk. Instead, they produce a yellowish fluid called **colostrum** that is rich in protein and in antibodies that may offer some protection against disease. Colostrum is less nourishing than true milk, but it contains enough nutrients to maintain the baby during its first few days.

---

1. Occasionally, a male newborn will leak "witch's milk" from his nipples, apparently

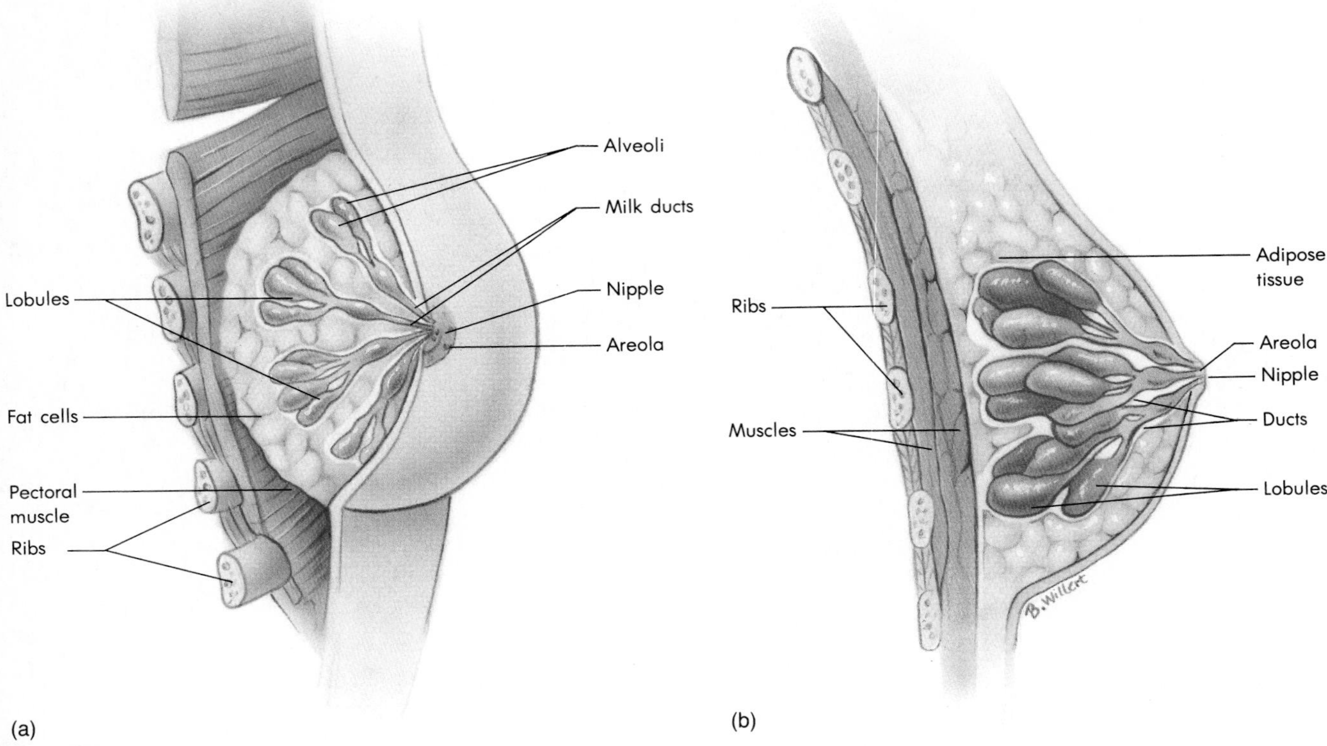

FIGURE 17-14 The female breast: (a) front view; (b) side view.

because his breasts have responded to his mother's high hormone levels. Do you think an adult male could ever nurse a baby?

2. If a man tries to nurse a baby, the baby's stimulation of his nipples will activate a neuroendocrine reflex, just as in a woman. Why would this reflex not cause milk secretion in the male's breasts?
3. With what hormones would you have to pretreat the man in order to make his breasts capable of responding to the neuroendocrine reflex and secreting milk?

HORMONAL CONTROL

We have already seen a number of ways in which hormones affect the male and female reproductive systems. Now we will discuss how hormones regulate the female menstrual cycle. Let us begin with a brief look at the hormones secreted by the ovary.

We have said that the ovary secretes **estrogen.** To be more accurate, we should say that it secretes at least six related *estrogens,* of which the most plentiful are beta-estradiol, estrone, and estriol. All the estrogens stimulate the development of, and maintain, the sex organs and secondary sexual characteristics. In addition, they stimulate the monthly thickening of the endometrium, control fluid balance, and increase protein synthesis throughout the body.

Progesterone strengthens the effect of the estrogens on the endometrium and the breasts. A third ovarian hormone is **relaxin.** Secreted near the end of pregnancy, it causes the pelvic ligaments, including the symphysis pubis and the cervix, to soften and become elastic. It thus helps these structures to stretch or change position to ease the passage of the baby through the birth canal. Part of the discomfort associated with labor and delivery is caused by the shifting around of the various pelvic elements.

Secretion of these hormones is controlled by the anterior pituitary. As we discussed in Chapter 13, the hypothalamus controls this endocrine gland with releasing hormones. Beginning at puberty, it produces GnRH, which stimulates the anterior pituitary to release FSH and LH. None of these hormones is secreted steadily, however; they all rise and fall in step with the menstrual cycle (see Figure 17-15).

The average human female has a 28-day menstrual cycle (Figure 17-15). Day 1 of this menstrual cycle is the first day of the monthly bleeding, or "period." Most females pass through their period

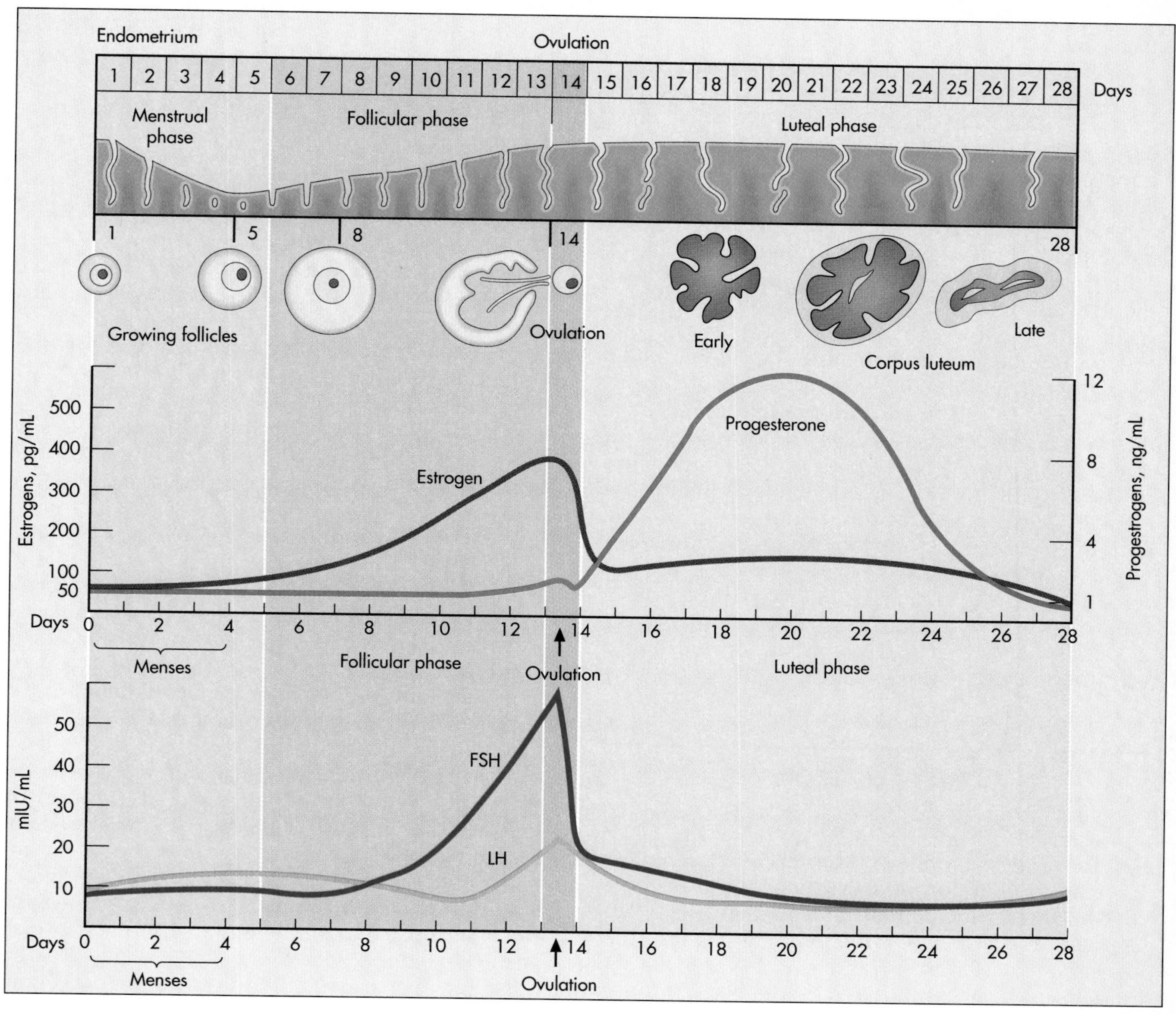

FIGURE 17-15 The menstrual cycle of the human female is marked by cyclic changes in hormone levels, endometrial structure and thickness, and follicular development.

in 3 to 5 days, and it is during this time that the endometrial lining of the uterus is shed. Day 28, of course, is the day before the next period begins. Another "landmark" day in the female reproductive cycle occurs, on average, about day 14—statistically, the day when women are most likely to ovulate.

A cyclic drop in the circulating levels of estrogen and progesterone causes the endometrium to break down, triggering the menstrual flow. By the fourth day of the cycle, the pituitary has resumed its secretion of FSH, provoking the next generation of follicle cells to begin their secretion of estrogen. Beginning about day 6, estrogen levels climb steeply as the follicles enlarge. A spike of LH production immediately precedes—and triggers—ovulation while estrogen declines. After ovulation, as the corpus luteum develops, estrogen production rises slightly again while progesterone reaches its peak. Toward the end of the cycle, estrogen and progesterone both decline, setting the stage for the next cycle.

We can see the effects of the feedback relations between the various hormones in the graph in Figure 17-16. High levels of estrogen inhibit the release of hypothalamic GnRH and hence the release of FSH and LH; we can see that both FSH and LH decline after the initial estrogen peak. With estrogen, the follicle cells also secrete inhibin, which inhibits both FSH and GnRH production. When estrogen and progesterone levels decline, GnRH escapes from its inhibition and FSH and LH begin to rise again, promoting the postovulatory rises in estrogen and progesterone secretion, which once more shut down GnRH production. If pregnancy occurs, the estrogen and progesterone levels stay high, for the developing embryo secretes *human chorionic gonadotropin* (HCG), which keeps the corpus luteum from degenerating (see Chapter 18).

It is interesting to compare the graph of hormone levels with events in the follicle and uterus. In the preovulatory phase of the menstrual cycle, FSH stimulates the follicle to enlarge. Follicular estrogens then stimulate the thickening of the endometrium. In the postovulatory phase, progesterone and estrogens secreted by the corpus luteum under the influence of LH cause the endometrium to thicken still more. The endometrial glands become coiled, the tissue stores fluid and glycogen, and the endometrial surface acquires numerous blood vessels. Menstruation begins when progesterone and estrogen decline and cease their stimulation of the endometrium. Human chorionic gonadotropin prevents menstruation by stimulating a continuing supply of these hormones.

FERTILIZATION

The male reproductive system has as its functions both the production of sperm and the transfer of that sperm to the female during the sexual act. **Copulation** occurs when the male inserts his erect penis into the female's vagina. Friction produced by the sex act eventually elicits the ejaculation of semen containing 200–500 million sperm, more or less. Using their flagella and currents created in the female's system, these sperm then swim through the cervix and uterus and into the Fallopian tubes, where they may meet and fertilize a secondary oocyte, which we generally—and incorrectly—call an ovum or egg.

Sperm usually are capable of fertilizing an ovum for up to 48 hours after their ejaculation—a few remain capable of fertilization even after 72 hours. Ova remain viable for up to 48 hours after ovulation. The timing of intercourse thus does not have to be exact. As long as viable sperm and ova are present at the same time, they can get together. Ovulation usually occurs in the middle (day 14) of each menstrual cycle, and pregnancy can occur if sperm are deposited within two or three days before or up to two days after this event. However, because of uncertainty about the actual time of ovulation, a woman must regard herself as fertile for roughly the middle third of her cycle.

Fertilization normally occurs in the outer third of the Fallopian tube (closest to the ovary). Sperm move through the female system in several ways. Swimming under the power of their flagella, sperm can move only about 1–2 cm (less than 1 in.) per hour. Yet some will be found in the Fallopian tubes within 60–90 minutes after their release. Some feel that uterine contractions accompanying orgasm can speed them through the female's reproductive tract. Of those sperm that make it through the uterus, about half enter the wrong (ovumless) Fallopian tube. Of those sperm cells entering the correct Fallopian tube, only about 2000 are able to swim the final 2 in. through the tube, against currents set up by the cilia lining the tube, to meet with the ovum.

As the sperm gather around the ovum (Figure 17-16), they release enzymes from their acrosomes,

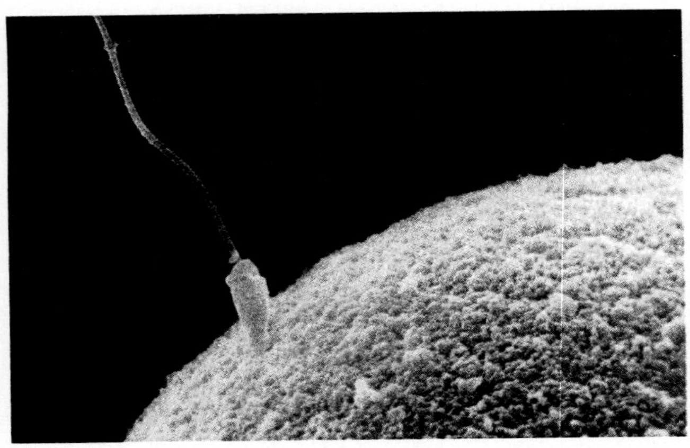
(a)

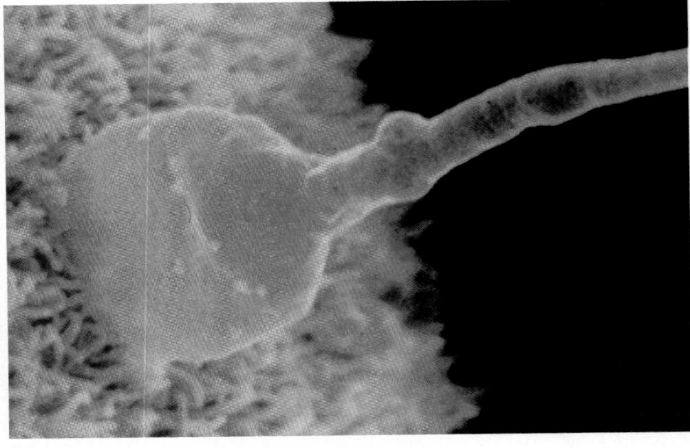
(b)

FIGURE 17-16 Fertilization: (a) sperm contacts the ovum; (b) sperm head in the process of penetrating the ovum.

both to break down the glue that holds the cells of the corona radiata around the ovum and to weaken the zona pellucida. Finally, one sperm cell is able to penetrate the ovum's cell membrane. Immediately afterward, electrical changes in the ovum's cell membrane block the penetration of any other sperm. At the same time, the ovum undergoes its second meiotic division, sheds a second polar body, and becomes a true ovum.

As the sperm enters the ovum, it sheds its flagellum. Soon after entry, its nucleus expands and becomes the male **pronucleus** of the newly fertilized ovum (Figure 17-17). The nucleus of the ovum becomes the female pronucleus, and the two fuse, turning the fertilized ovum into a **zygote.** The sperm and ovum nuclei each contained 23 chromosomes. The zygote has 46, the diploid number. Half the zygote's chromosomes come from each parent.

Test Tube Babies

Fertilization need not occur in the female's body. In recent years, it has become possible to treat women with drugs that induce multiple ovulations. (Spontaneous multiple ovulations are responsible for fraternal or nonidentical twins, triplets, and so on; identical twins result when a fertilized ovum splits into *two* zygotes.) Physicians can then use a suction device, inserted through a small incision in the navel, to remove the ova. They put the ova in a glass dish together with sperm from the woman's husband (or another donor). Such out-of-body procedures are referred to as *in vitro* (in glass) fertilizations. Later, they transfer any successfully fertilized ova back into the woman's uterus (or Fallopian tube) through the cervix, hoping that at least one will continue to develop. In this way they can bypass infertility caused by blocked Fallopian tubes (due to scar tissue). The first child to result from this *in vitro fertilization* was Louise Brown, born in England in 1978. Since then, the procedure has become common, although the success rate remains low.

1. Reproductive technologists have learned how to induce multiple ovulations, fertilize the ova in the body, flush the developing embryos out, and implant the embryos in other females. Embryos produced by in vitro fertilization can also be implanted in "nonsource" females. Such techniques have made it possible for a prize cow to have dozens of calves in a year using less valuable "scrub" cows as host mothers. What circumstances do you think might justify using this technique with humans? Discuss how the interests of biological (genetic) and host mothers might conflict.
2. It is also possible to freeze embryos in liquid nitrogen, much as sperm banks do with semen, and to thaw them later for implantation in a female's womb. Under what circumstances might this technique be used with humans?

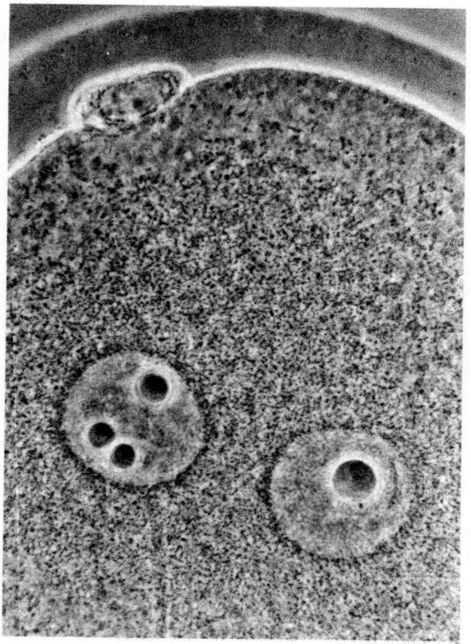
(a)

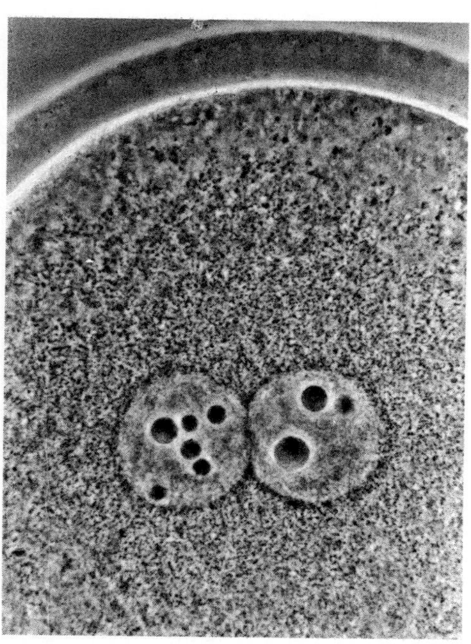
(b)

FIGURE 17-17 (a) A freshly fertilized hamster ovum with two pronuclei. One pronucleus contains the chromosomes contributed by the mother, in the ovum; the other contains those contributed by the father, in the head of the sperm cell. (b) The pronuclei have converged to the center of the cell just prior to the first cleavage.

IMPLANTATION

Once formed, the zygote begins dividing by mitosis. Each division is called a **cleavage** since it looks as though the original cell were being cleaved in two, then four, and so on. By the end of the third day after fertilization, the zygote has become a ball of 16 cells. A few days later, it has become a mass of cells resembling a blackberry or mulberry and is called a **morula.** Not long after this stage, the morula develops an interior hollow region and becomes known as a **blastocyst.** The wall of the blastocyst is only one cell thick, except in one region, where a mass of cells, the **inner cell mass,** develops. This mass will become the actual embryo (Figure 17-18).

While the zygote is developing into the blastocyst, it is traveling through the Fallopian tube toward the uterus. About a week after fertilization, it reaches the uterus. The cells of the blastocyst, which is still no larger than the original ovum, secrete an enzyme that loosens the bonds among the cells of the endometrium. The blastocyst then burrows into the uterine lining, with the inner cell mass usually oriented toward the lining. This is **implantation.** Once it has occurred, the blastocyst's cells can draw nutrients from the endometrial secretions and from the mother's bloodstream and begin to grow in size. It also begins to elaborate the membranes that will become the placenta, an organ that gives the growing embryo access to its mother's bloodstream (see Chapter 18).

Implantation does not always occur in the uterus. Sometimes the blastocyst embeds in the wall of the Fallopian tube. Occasionally the ovum even fails to enter the Fallopian tube in the first place, and the sperm reach and fertilize it in the abdominal cavity. Implantation may then occur on the outer surface of the ovary or Fallopian tube or on the lining of the abdominal cavity, the peritoneum. Such errors lead to **ectopic pregnancies;** they are usually unsuccessful, but when they do go to term, the infant must be delivered surgically, by cesarean section.

BIRTH CONTROL

The various types of birth control operate in several different ways. Some methods prevent the production or release of viable gametes, others block fertilization, some prevent implantation, and the most controversial—abortion—involves terminating pregnancy before birth.

Of course, abstinence (having no sexual contact) is the most certain way to avoid the risks of pregnancy. However, it is also the least enjoyable, and people have long sought more acceptable compromises. One is "the high school boy's promise," or **coitus interruptus,** in which the young man swears an oath to withdraw the penis just before ejaculation. The failure rate, understandably, is high; about 20 percent of the females willing to participate under those terms can expect to become pregnant in less than a year. Another is the **rhythm method,** which depends on refraining from intercourse for a few days before and after ovulation. The difficulty of predicting the time of ovulation explains the very high failure rate, also about 20 percent. Fortunately, there are now ways to im-

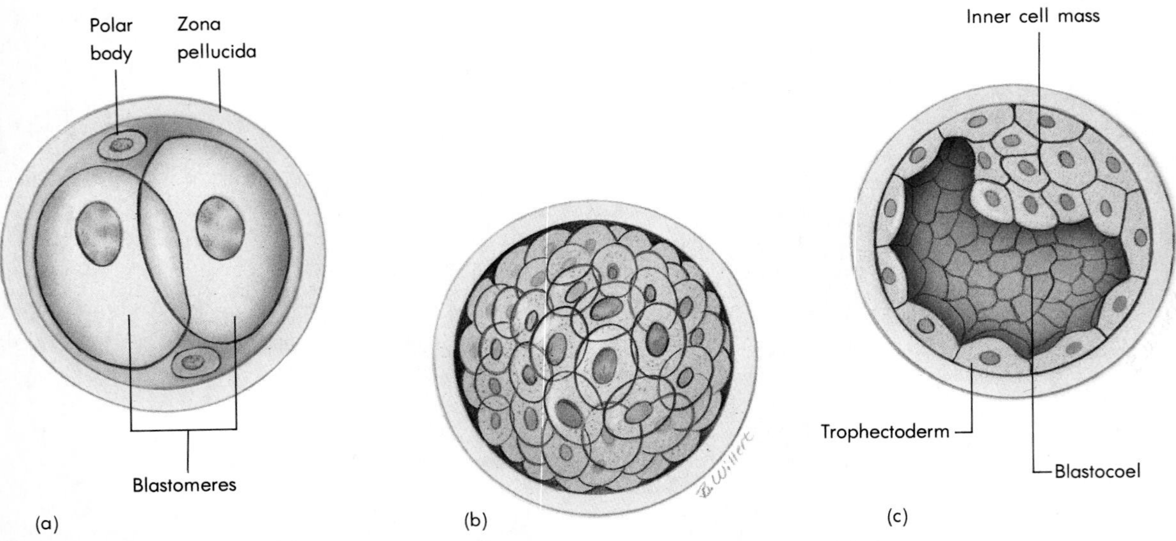

FIGURE 17-18 Early embryo: (a) the two-cell stage; (b) the morula; (c) the blastocyst.

prove the odds. A woman's normal body temperature peaks at the time of ovulation, and careful tracking of her body temperature can guide the rhythm method fairly effectively. Her cervical mucus also thins just before ovulation, and the use of special kits for testing this mucus can also help. Nevertheless, the rhythm method remains untrustworthy, and its practioners are usually called "parents." It is preferred only by those who, usually because of religious beliefs, refuse other methods.

The leading method of birth control in the United States is **sterilization** (see Figure 17-19). Males undergo a **vasectomy,** a simple operation in which both vas deferens are cut and tied off to prevent the movement of sperm from the testes to the penis. It does not affect ejaculation, for the bulk of the semen is composed of secretions from the prostate gland and the seminal vesicles. It is very effective in preventing fertilization and hence pregnancy; only about a tenth of one percent of all couples relying on vasectomies for birth control experience a pregnancy within the first year after the operation.

Female sterilization has a slightly higher failure rate of 0.4 pregnancy per 100 women within a year of the operation. It most often takes the form of **tubal ligation,** an operation in which the Fallopian tubes are cut and tied to prevent movement of ova to the uterus and of sperm from the uterus. It can also take the form of a **hysterectomy,** removal of the uterus, but this is clear overkill. Organs should be removed only when they represent a threat to the patient (as in cases of cancer). Unfortunately, studies have indicated that many hysterectomies are performed "without cause."

Both tubal ligations and hysterectomies are more hazardous than is a vasectomy, which is usually done in the doctor's office with the patient awake. It thus makes more sense for the male, not the female, to be sterilized. The biggest drawback to sterilization is that it is difficult to undo; it is usually chosen only by couples who do not want more (or any) children.

"Barrier" methods of birth control, which interfere with the movement of the sperm toward the ovum, are also widely used; some are shown in Figure 17-20. One common barrier is the **condom,** a rubber sheath that fits over the erect penis and traps semen before it can reach the female reproductive tract. Condoms have not been very popular

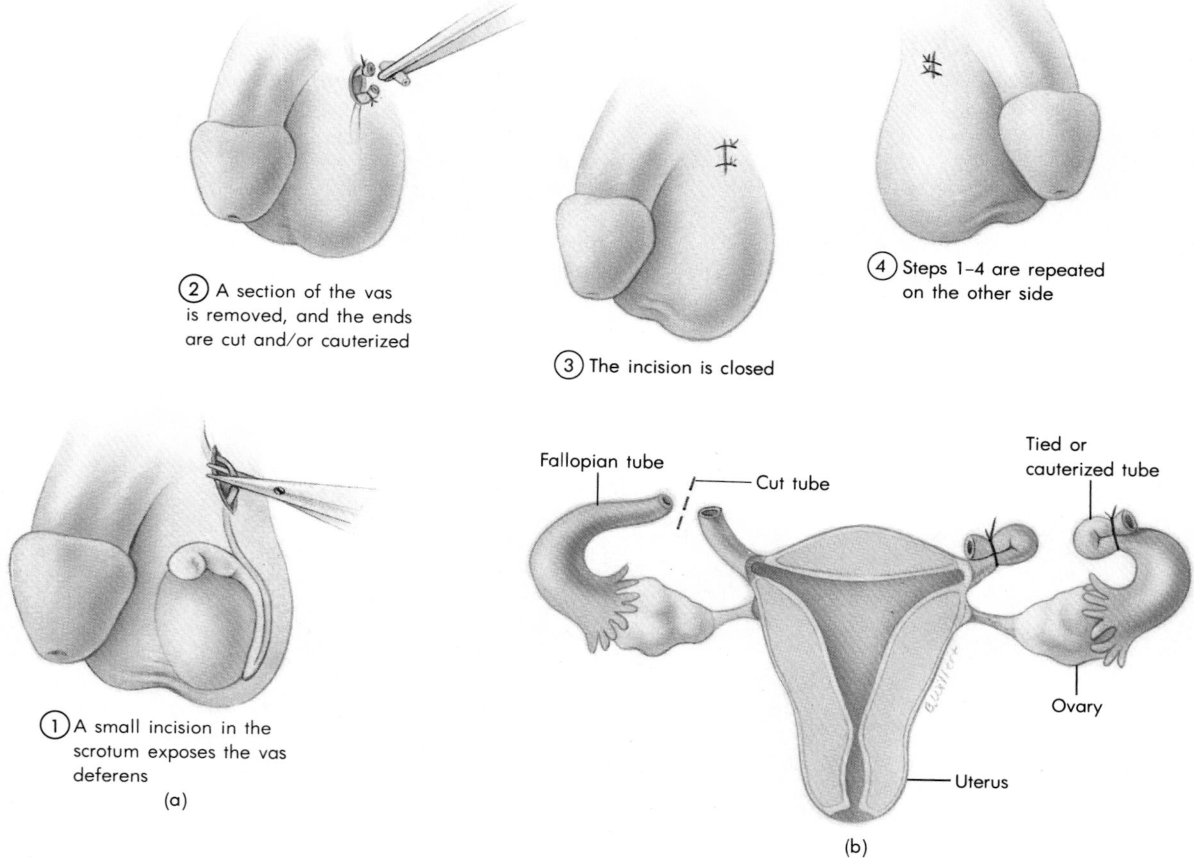

FIGURE 17-19 Sterilization is the most common method of birth control: (a) vasectomy; (b) tubal ligation.

with males since the introduction of alternative forms of birth control because they interfere with some of the pleasurable feelings of sexual activity. However, condom use has recently increased because they seem to be effective barriers for preventing the transmission of the acquired immunodeficiency syndrome (AIDS) and herpes viruses. About 12 percent of the couples using condoms as their only form of birth control can expect to get pregnant in any given year, a significant reduction in risk over coitus interruptus or the rhythm method.

Less efficient (about 18 percent per year failure rate) is the **diaphragm,** a dome-shaped sheet of rubber with a spring-loaded rim that fits around the cervix. Used with a spermicidal (sperm-killing) jelly or foam, it prevents live sperm from entering the uterus. A **cervical cap,** a smaller device that fits snugly over the dome of the cervix, works similarly. Spermicides can also be used alone, sometimes held in place with a small **cervical sponge,** but their failure rate is high, between 18 and 28 percent.

The most common *reversible* method of birth control is the "pill," or **oral contraceptive** (Figure 17-21). The several forms of this drug all use mixtures of synthetic female sex hormones to interfere with the feedback regulation of the female cycle. They load the bloodstream with levels of hormone, usually estrogens and progesterone, that resemble those normally seen during pregnancy. The hypothalamus, pituitary, and ovary respond as they do during pregnancy; GnRH, FSH, and LH are not secreted, follicles do not mature, and ovulation does not occur. The failure rate is low, about 3 percent in the first year of use, but some individuals experience side effects, including headaches, fluid accumulation in tissues, and occasionally blood clots that can cause strokes or heart attacks. Oral contraceptives cause less than two deaths per 100,000 users under age 30; the risk of death goes up for older women and for smokers.

A number of new forms of hormonal regulation for women are being developed and tested for contraceptive use. One is small capsules of hormone implanted under the skin of the leg or arm to provide up to five years of birth control. Another is a rubber ring impregnated with hormone; it fits

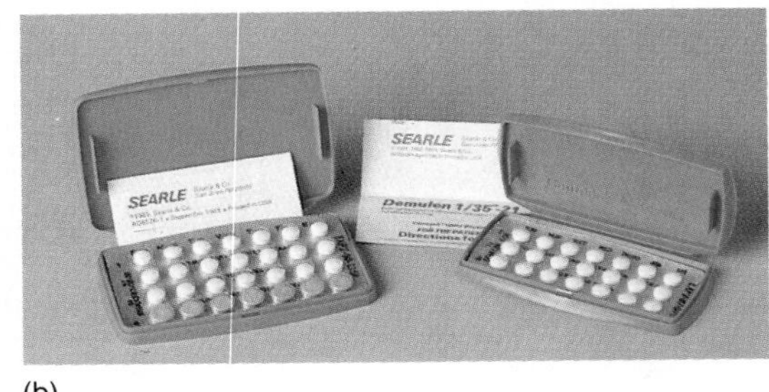

(b)

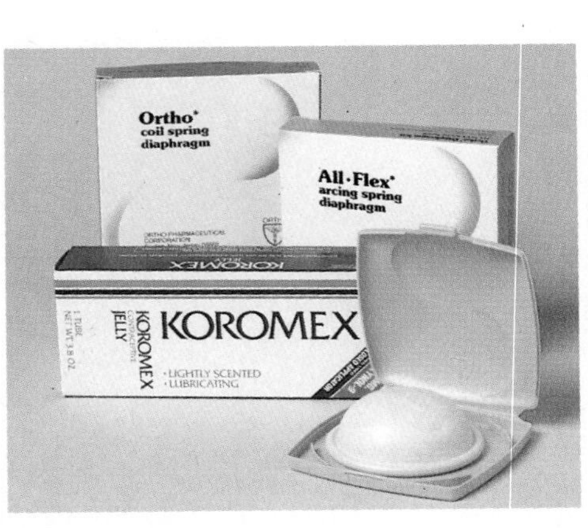

(c)

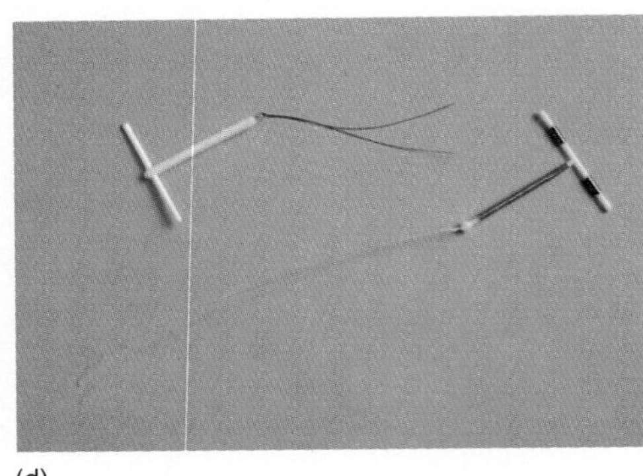

(d)

FIGURE 17-20 Common birth control devices: (a) Condoms. (b) Birth control pills. (c) Diaphragm and spermicides. (d) IUDs.

around the dome of the cervix and releases its hormone slowly where it affects mainly the reproductive organs.

It is also possible to control births by intervening after fertilization. The **intrauterine device (IUD)** is a small plastic device implanted in the uterus. It apparently works by preventing implantation; it has a failure rate of 6 percent in the first year of use. Unfortunately, it can cause heavy menstrual bleeding, cramps, infection, and perforation of the uterine wall. Because these side effects have led to a number of expensive lawsuits, few IUDs are now available on the U.S. market.

The threat of lawsuits as well as the fear of protests and boycotts by those who cannot tolerate the thought of controlling human fertility have kept U.S. pharmaceutical corporations from pursuing research into more advanced forms of birth control. Other, less timid nations have developed the drug RU 486, which blocks endometrial (and other) receptors for progesterone and can thereby prevent or undo implantation (it can also prevent ovulation). Proponents of this drug consider it a "morning-after" pill; its opponents, who belong to the "prolife," or antiabortion, movement, call it an "abortion pill" and are lobbying vigorously to prevent the Food and Drug Administration from approving its use in the United States.

A number of contraceptive techniques are under development around the world. They include a male "pill" that would prevent the release of viable sperm and vaccines. The vaccines would induce the body to make antibodies that destroy the hormones that induce sperm production or support pregnancy.

The ultimate after-the-fact birth control method, chosen by about 1.5 million U.S. women per year, is **abortion.** It involves physically removing a growing embryo from the uterus. It is usually—and preferably—performed during the first three months of pregnancy, but it can also be performed during the second three months.

Box 17-1 Accidental Pregnancies: A Sexually Transmissible Disorder?

The birth rate among adolescent females is higher in the United States than in any other developed country. It is twice as high as the birth rate for the same group in Sweden and 17 times higher than in Japan.

In a recent survey at a California community college, it was found that 25–35 percent of the female freshman and sophomore students taking health education classes had already been pregnant at least once. This statistic translates to more than a million unintended teenage pregnancies per year; that is, every 35 seconds, somewhere in this country, another teenage female gets pregnant.

Most of these young women, and their male partners, are not very happy about what they have "caught." Indeed, they might be happier if they had caught gonorrhea; that, after all, can be cured with antibiotics. It is therefore not very surprising to find that some people view unintended pregnancy as a communicable disease that women catch from men. If this way of looking at the situation is correct, we are in the middle of an epidemic.

If you are a sexually active female and are not practicing birth control, what are your chances of getting pregnant within a year? What would you change if you found out there was a 90 percent chance of your getting pregnant? Who has the most to lose or gain if you get pregnant? Would you have to quit school? Would you consider an abortion? If not, are you—all by yourself—prepared to put in the years necessary to raise a child? Have you ever called the local offices of Planned Parenthood and talked to a female counselor about your birth control options?

If you are a sexually active male, imagine—if you haven't already experienced the situation—that your female partner has just said, "We're going to have a baby!" What would you say in response? Is simply having had sex with someone a good enough reason to get married? If you respond by telling your partner to get an abortion, are you prepared to pay for that procedure? How would you handle a surprise pregnancy if you or she were philosophically opposed to abortion? Are you prepared to pay child support for the next 18 years? Who is the one person on Earth who can best help you to avoid facing that difficult situation?

Learning Focus

HUMAN REPRODUCTION

Use this Learning Focus to summarize what you have learned about human reproduction. Answer the questions as fully as you can.

1. Where does meiosis occur in the testis?

2. List the sites of the four stages of sperm maturation from spermatid to fertilization.

3. What are the components of semen?

4. What is the source of testosterone?

5. Where does meiosis occur in the female?

6. Describe the maturation of the ovarian follicle.

7. What is (are) the source(s) of estrogen and progesterone?

8. Describe the effects of the sex hormones on the body at puberty.

9. Describe the effects of the sex hormones on the body after puberty.

10. What triggers ovulation?

11. Name five events that accompany or immediately follow ovulation.

12. What is the function of the corpus luteum?

13. Describe the mating process. How does it ensure that sperm are able to reach the ovum?

14. Where does fertilization occur?

15. What happens during fertilization?

16. Name two stages at which mitosis is involved in the reproductive process.

17. When does implantation occur? What is implanted and where?

18. What is the function of sexual reproduction?

19. How do meiosis and fertilization help to accomplish this function?

20. List four sites where contraceptive methods interfere with fertilization.

21. List the failure rates for each of the common types of birth control.

SUMMARY

Human sexual reproduction begins with the meiotic division of germ cells in the gonads to produce gametes, sperm in the male and ova (eggs) in the female. Various tubes then transport and support the gametes until they can meet and fuse in the process of fertilization. Fertilization reconstitutes a diploid cell, the first cell of a new individual, the zygote. It contains the unique mixture of genes that will eventually grow into the cells, tissues, and body of a new human being.

Male and female differ in primary and secondary sexual characteristics. The former are those features that have directly to do with reproduction; they include the primary sex organs, the gonads, and the secondary sex organs, the tubes and glands that transport and support gametes and embryos. The latter are superficial traits such as the male's facial hair, heavy musculature, and deep voice, and the female's subcutaneous fat and breasts.

The male produces sperm in the seminiferous tubules of the testes as spermatogonia divide to form primary and secondary spermatocytes and spermatids, which mature into spermatozoa. Interstitial cells between the tubules secrete the male sex hormone testosterone. Once formed, sperm move into the epididymis and vas deferens, where they finish maturing. At the time of ejaculation into the female reproductive tract, the prostate and Cowper's glands and the seminal vesicles add an alkaline fluid containing mucus and nutrients to form the bulk of the semen. Sperm leave the body when peristaltic contractions of the vas deferens and epididymal walls and muscle fibers of the prostate gland propel the sperm through the urethra out of the penis. The penis is stiffened for intercourse by three columns of spongy tissue that become engorged with blood.

The female produces ova by meiosis of epithelial cells (oogonia) on the surface of the ovary. Before birth, these cells become primary oocytes, each one surrounded by connective tissue cells from the ovary's interior, or stroma, to form follicles. Meiosis turns these primary oocytes into secondary oocytes and polar bodies and, later, into ova. After puberty, under the influence of hypothalamic releasing factors and pituitary gonadotropins [follicle-stimulating hormone (FSH) and luteinizing hormone (LH)], the follicles enlarge and their cells secrete the female sex hormones estrogen and progesterone. Each month thereafter, until menopause, one or a few follicles enlarge greatly, develop a fluid-filled cavity, and burst to release the ova within them. This process is ovulation. The released ova are drawn into the Fallopian tubes and transported to the uterus. The empty follicle becomes the corpus luteum.

While the follicle is enlarging each month, the estrogen it secretes stimulates the endometrium lining the uterus to thicken in preparation for pregnancy. If a fertilized ovum fails to reach it, the endometrium sloughs off in the process of menstruation. Otherwise, the endometrium continues to thicken for a time.

The uterus opens via the cervix into the vagina, a muscular, elastic channel leading to the vulva and the outside of the body. The vagina receives the male penis and semen and functions as part of the birth canal. The vulva, or external genitalia of the female, consists of the mons pubis, labia majora, labia minora, the clitoris, and the vaginal opening. Glands and the vaginal lining provide a lubricating fluid during sexual arousal.

The breasts, or mammary glands, of the female have two functions: sexual and supportive. They stimulate sexual interest in her male partner, provide pleasurable sensations for the woman during sexual activity, and provide milk for her offspring. Milk is secreted in response to the hormones prolactin and oxytocin and to the neuroendocrine reflex stimulated by a sucking infant.

Both men and women mature sexually at puberty under the influence of the sex hormones. The male sex hormone, testosterone, stimulates the maturation of the male genitals and the development of the male secondary sexual characteristics. Female sex hormones, especially estrogen, stimulate the development of ovarian follicles and the development of the female secondary sexual characteristics, including the breasts. In the female, however, the role of hormones is more complex. Hypothalamic, pituitary, and ovarian hormones interact to control the menstrual cycle and the readiness of the body for pregnancy. When pregnancy occurs, they interact to prepare the body for birth and the breasts for lactation.

Fertilization usually occurs in the Fallopian tubes, when about 2000 sperm finally locate and surround the ovum and use the enzymes in their acrosomes to break through the corona radiata and zona pellucida surrounding it. When one sperm penetrates the ovum cell membrane, sperm and ovum nuclei fuse to form the zygote. The zygote then divides by mitosis while traveling toward the uterus. When it has formed a hollow blastocyst, it attaches to the endometrium and burrows into it, or implants. At that point, pregnancy has started.

The several forms of birth control work in various ways. The rhythm method relies on restricting intercourse to "safe" times in the woman's monthly cycle and is notoriously ineffective. Oral contraceptives, vasectomies, and tubal ligations all prevent the release of viable gametes; all three are effective, and sterilization is the most common birth control method in the United States. Condoms, diaphragms, cervical caps, and spermicides block the movement of sperm toward the ovum, either physically or by killing the sperm. Intrauterine devices seem to prevent or interfere with implantation. Abortion removes the embryo.

Table 17-1 provides a summary of selected diseases of the reproductive system.

Table 17-1

Clinical Summary: Selected Diseases of the Reproductive System

DISORDER	DESCRIPTION
Breast cancer	One of most often fatal cancers in women; rare in men. Incidence rises rapidly after menopause. Risk factors include family history of breast cancer, having had no children, having a first child after age 34, and previous breast cancer. Best protection lies in regular self-checks for lumps and in periodic mammograms (x-rays). Treatment may require removal of the affected breast and underlying tissues.
Cervical cancer	Common but easily detected early with a Pap test (microscopic examination of cervical cells for abnormalities). Treatment often involves removal of the cervix and uterus (hysterectomy).
Cryptorchidism	Failure of the testes to descend before birth from the abdominal interior to the scrotum, resulting in infertility. Treated by surgically moving the testicles to the scrotum.
Dysmenorrhea	Painful menstruation, apparently due to uterine overproduction of prostaglandins. Symptoms include uterine cramps, vomiting, headache, and diarrhea. Treated with drugs that inhibit prostaglandin synthesis and release.
Endometriosis	Occasionally, scraps of endometrium reach the pelvic cavity through the Fallopian tubes and fasten to the surfaces of pelvic organs, where they follow the normal menstrual growth-slough rhythm. Common among women aged 30–40. Symptoms include premenstrual pain and unusual menstrual pain. Treatment may involve inhibiting endometrial proliferation with drugs (hormones) or removing the excess tissue by surgery.
Impotence	Very common inability of the male to produce or maintain an erection. Causes may be inhibitions or anxieties, reactions to medications, or illnesses such as diabetes. Treatment may involve psychotherapy, change in medication, or attention to the illness.
Premenstrual syndrome (PMS)	The set of symptoms—breast swelling and tenderness, irritability, joint pain, constipation, fatigue, depression or anxiety, etc.—that often precede menstruation. The cause is unknown but may involve hormonal imbalances.
Toxic shock syndrome	High fever, sore throat, fatigue, irritability, diarrhea, vomiting, abdominal pain, and even death resulting from exposure to toxins secreted by bacteria growing in vaginal tampons left in place too long. Changing tampons frequently or using pads prevents it.

STUDY QUESTIONS

1. Discuss the differences between asexual and sexual reproduction.
2. Why must gametes be haploid?
3. Explain the differences between primary and secondary sexual characteristics.
4. Describe how hormones control sperm production.
5. What is capacitation, and why is it essential to normal sperm function?
6. Explain the function of polar bodies.
7. In what ways do the male and female external genitalia resemble (are homologous to) each other?
8. Describe the interaction of hormones that regulates the menstrual cycle.
9. Discuss why a male must ejaculate 200–500 million sperm so that just one may fertilize an ovum.
10. Discuss the advantages and disadvantages of the various methods of birth control.

GLOSSARY

Abortion Removal of embryo from uterus before it has reached the time for birth.

Accessory sex organs Organs that transport and support gametes in the body.

Acrosome Tip of sperm cell; carries enzymes to break through barrier around ovum.

Asexual reproduction Reproduction without gametes, by budding, fission, runners.

Blastocyst Post–morula stage of embryonic development; hollow ball of cells.

Capacitation The final activation of sperm for fertilization that occurs on exposure to vaginal fluids.

Cleavage First mitotic cell divisions of zygote.

Corpus luteum Postovulatory follicle; gland secreting estrogen and progesterone.

Ejaculation Ejection of semen from the penis.

Endometrium Lining of uterus.

Estrogen A female sex hormone; stimulates endometrial development.

Fallopian (uterine) tubes Ducts that lead released ova from ovary to uterus.

Fertilization Fusion of gametes.

Gametes Haploid cells (sperm and ova) that form a diploid cell by fusion.

Gonads Organs that produce gametes and sex hormones.

Graafian follicle Mature follicle, containing fluid-filled cavity and the ovum.

Implantation Attachment to and burrowing into endometrium of blastocyst.

Meiosis A mode of cell division that produces haploid daughter cells.

Menarche First menstruation during puberty; a young woman's first period.

Menopause Time of life when menstruation and reproductive cycling cease.

Menstrual cycle Monthly growth and shedding of endometrium and maturation and release of ova.

Menstruation Monthly discharge of blood and endometrial fragments shed when pregnancy fails to occur.

Morula Berrylike clump of cells formed by first cell divisions of zygote; early embryo.

Neuroendocrine reflex Release of oxytocin in response to suckling.

Ovary Female gonad.

Ovulation Release of ovum from Graafian follicle.

Ovum Female gamete.

Polar body Nonviable, rudimentary daughter cell of meiosis of oocyte; vehicle for discarding surplus chromosomes.

Primary follicle Primary oocyte surrounded by layer of stromal cells.

Primary sex organs Gonads.

Primary sexual characteristics Body features having directly to do with reproduction.

Progesterone A female sex hormone; reinforces effects of estrogen.

Secondary sexual characteristics Body features that distinguish male from female but are not involved directly in reproduction.

Seminiferous tubule Tubule within the testis in which sperm are formed.

Sexual reproduction Reproduction by growth of new individuals from fused, haploid gametes.

Spermatogenesis Process by which sperm are formed.

Spermatozoa (sperm cells) Male gametes.

Sterilization Birth control by cutting and tying off the vas deferens or Fallopian tubes or by removing the uterus, ovaries, or testes.

Testis Male gonad.

Uterus Womb; the organ that supports the developing embryo.

Zygote Fertilized ovum after fusion of male and female pronuclei.

Chapter 18

Human Development

Each human begins as a single cell. That cell, the fertilized ovum or zygote, is produced when ovum and sperm unite in a woman's Fallopian tube. As the zygote travels toward the uterus, it divides by mitosis to become a cluster of cells. Eventually, this cluster of cells implants in the endometrium of the uterus and develops, through the nine months of gestation, into a baby. Over the next 13–15 years, the infant becomes a child and then an adolescent, passing through the somewhat tumultuous processes of maturation. With maturation, the body prepares itself for its own role in the reproductive processes; what started as a single cell is now ready to produce its own offspring.

This chapter will describe the processes of human development from zygote to birth. It will pay relatively little attention to development after birth, though it will say something about the human life cycle.

FROM FERTILIZATION TO PREGNANCY

Before pregnancy, fertilization of the ovum must take place in the distal third of one of the Fallopian tubes, near the ovary. The first divisions of the zygote begin within 24 to 30 hours after fertilization has taken place. The first two cells formed by the division of the zygote are called **blastomeres** (Figure 18-1). The blastomeres continue dividing at a rate of about one division per day for the next several days. Three days after fertilization, the **conceptus** consists of a fairly compact ball of 12–16 cells called the **morula,** which usually enters the uterus by the fourth day. At this point, the conceptus is referred to as a **blastocyst,** which is made up of 50–100 cells surrounding a fluid-filled cavity, the blastocoel.

By the sixth day following fertilization, the blastocyst attaches to the lining of the uterus, usually near the middle, on its inner posterior wall. At this point the blastocyst begins the process of **implantation** by invading the endometrial lining of the uterus (Figure 18-1). The successful completion of implantation marks the official beginning of **pregnancy.** This condition normally lasts 280 days, give or take a few, from the beginning of the last menstruation until childbirth. It can be detected in several ways. During the first 6–12 weeks of pregnancy, the simplest sign is the absence of menstrual periods; a pregnant woman's uterus does not shed its endometrium. Other early signs include weight gain, mood swings, slight fever, morning sickness (nausea), and tender, swollen breasts and eventually abdominal distension. To confirm these clues, a physician tests for the presence in the blood of **human chorionic gonadotropin (HCG),** a hormone secreted by the fetal membranes. There are also versions of this test for use at home, though they are less reliable and must be confirmed by a physician.

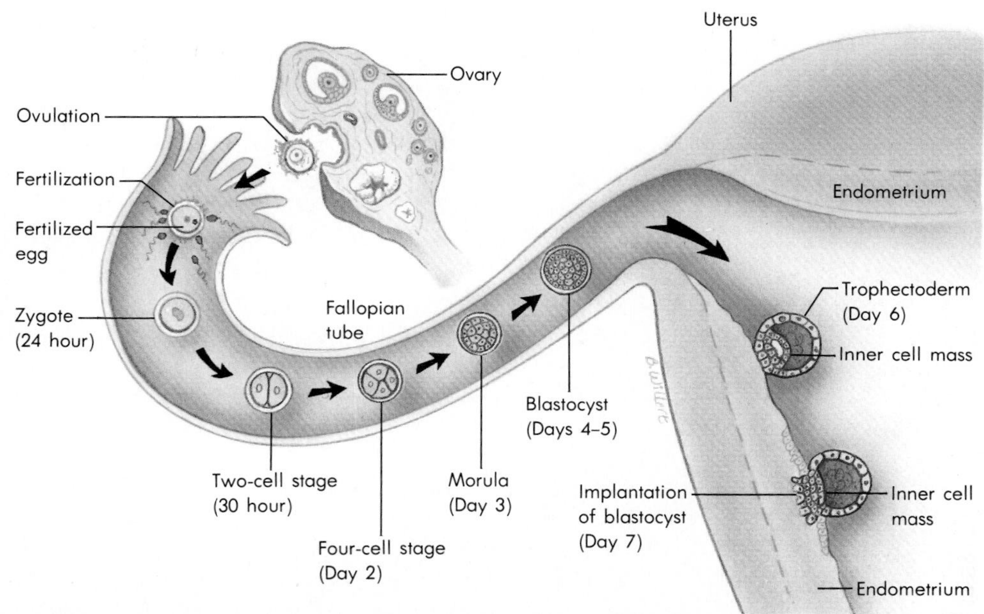

FIGURE 18-1 After fertilization, the zygote moves slowly down the Fallopian tube toward the uterus. As the zygote divides, a growing cluster of cells (or blastomeres) forms. When the cell cluster resembles a berry, it is called the morula. The morula becomes a hollow ball of cells, the blastocyst. About 5 days after fertilization, the blastocyst reaches the uterus, where it will connect with the maternal circulation. On day 6, it sticks to the endometrial lining and begins "digesting" its way into the endometrium. Implantation has occurred by day 7.

The Fetal Membranes

The wall of the blastocyst, the **trophectoderm,** is only one cell thick, except in the region of the **inner cell mass,** a small knot of cells that will become the actual embryo. After implantation, the cells of the inner cell mass begin to develop differences. The cells multiply and form three layers, as shown in Figure 18-2. Between the two layers nearest the trophectoderm, there appears a small **amniotic cavity.** The middle layer, called **ectoderm,** will eventually give rise to the nervous system and the epidermis of the skin. The third layer, the **endoderm,** will eventually become the linings of the digestive organs and glands.

By 12 days after fertilization, the endoderm and ectoderm form an **embryonic disc.** As shown in Figure 18-3, the amniotic cavity is surrounded by a layer of cells, the **amnion,** grown out from the ectoderm. The endoderm has produced an expanding bubble of membrane called the **yolk sac.** In egg-laying animals, such as birds and reptiles, the yolk sac encloses the yolk of the egg and serves as a source of nutrients. Scattered **mesoderm** cells have appeared; they will in time give rise to muscle, bone, blood vessels, and other structures.

Within the next two days, the embryonic disc develops a central groove through which mesodermal cells migrate to form another layer of cells in the middle of the disc, sandwiched between the endoderm and ectoderm. As development continues, the embryonic disc curls upward and inward at the edges, pinching off the yolk sac and forming a cavity, the **primitive gut,** within the disc. The mesodermal cells also multiply and form a layer of cells lining the trophectoderm. This two-layered membrane now becomes the **chorion** and develops fingerlike protrusions, the **chorionic villi,** which

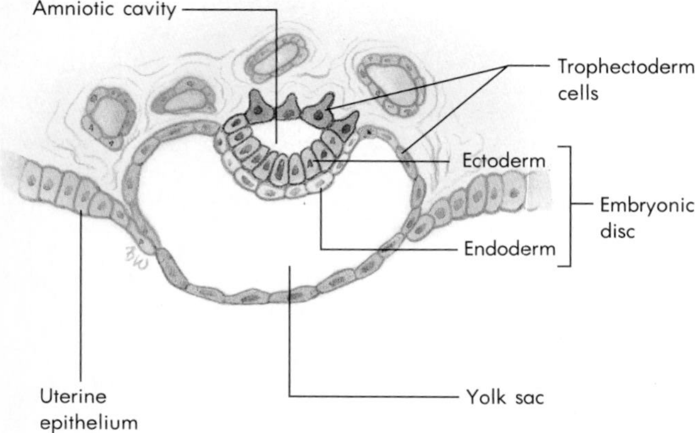

FIGURE 18-2 The primary germ layers are visible in the developing embryo by day 12.

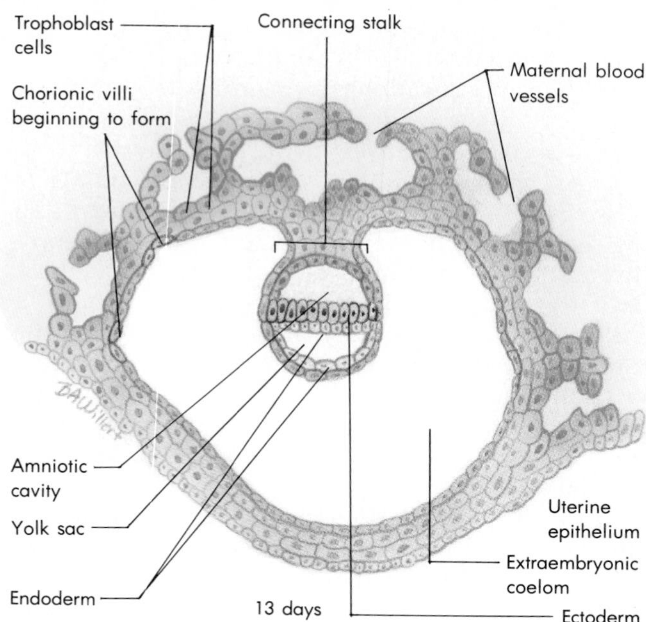

FIGURE 18-3 By day 13, the embryonic disc has begun to curl, and the yolk sac is developing.

project into the endometrium and encourage the formation of chambers filled with maternal blood from which the embryo can draw oxygen and nutrients. The chorionic membrane performs the important task of keeping the blood of the fetus and the mother separated. It also secretes HCG, which stimulates the corpus luteum to continue secreting estrogens and progesterone. These in turn prevent the uterus from sloughing its endometrium and ending the pregnancy with a menstrual period.

By 25 days after fertilization, the embryonic disc has begun to resemble a fishlike animal, complete with tail and gill clefts (see Figure 18-4). The amnion has ballooned to surround the embryo, protecting it within a bag of **amniotic fluid** (the "bag of waters" that breaks just before childbirth). The chorion has thickened and its villi have enlarged. The embryo is now attached to the chorion by a short stalk, which will soon incorporate the yolk sac and become the **umbilical cord.** Blood vessels in the stalk bring nutrients and oxygen to the embryo and carry wastes away.

The Placenta

As the link between the chorion and the endometrium strengthens, the two tissues form a distinct organ, the **placenta.** Its function is to provide a place where the bloodstreams of the developing embryo and its mother can meet, separated only by

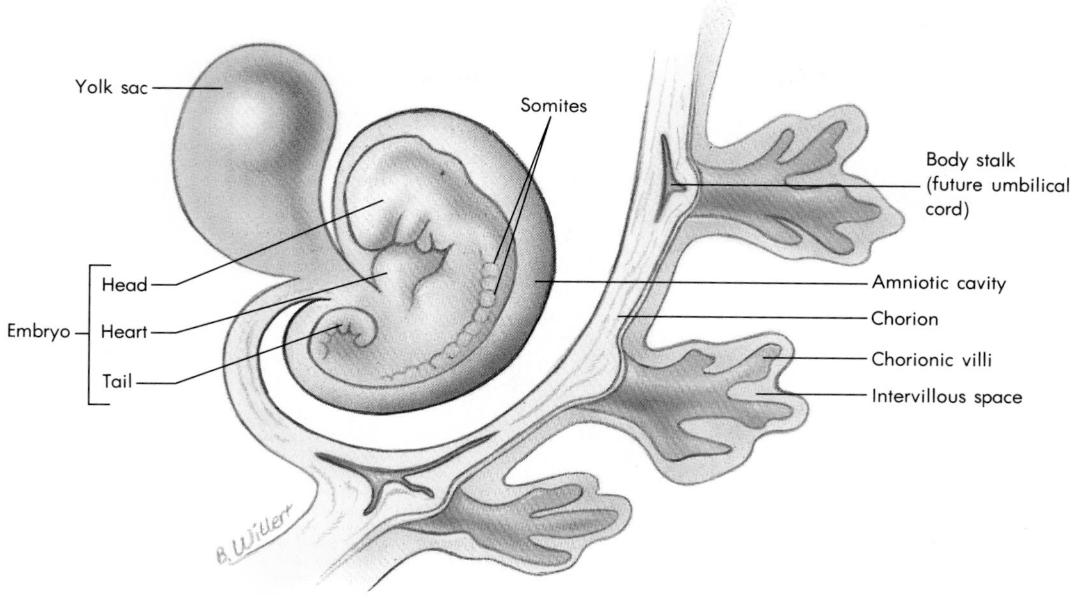

FIGURE 18-4 Embryonic development by day 25.

thin membranes, for the exchange of nutrients, oxygen, and wastes. It blocks the passage into the embryo of some drugs and toxins, but not all. For this reason, pregnant women must be very cautious about medications, tobacco, alcohol, and various other substances (see what follows).

In addition, the placenta secretes several hormones. One is HCG. Its production declines in the fourth month of pregnancy, when the placenta takes over the production of estrogen and progesterone, making the ovaries' contributions of those substances unnecessary. The placenta also secretes a hormone that stimulates placental HCG production. Still others are **human chorionic somatomammotropin,** which stimulates breast development late in pregnancy, and **relaxin,** which softens the links between the bones of the pelvis and helps the cervix dilate for childbirth.

The placenta is linked to the embryo by the umbilical cord, which contains three major blood vessels: two *umbilical arteries* that carry blood from the embryo to the placenta and one *umbilical vein* that returns blood to the embryo. Capillaries proliferate within the chorionic villi, which are bathed in maternal blood in the placenta (see Figure 18-5).

1. Trace the paths of nutrients and oxygen from the mother through the placenta and into the embryo.
2. Trace the path of wastes in the opposite direction.

FETAL DEVELOPMENT

The fertilized ovum is considered to be **totipotent** because it can give rise to the many different cell types that form the various tissues of a complete human being. This capability lasts for only a very short time. When the embryo consists of only two cells—or even four—the cells can be separated and each one will develop into a complete individual. Natural separations of this sort are in fact the source of identical (monozygotic) twins (and triplets and quadruplets), which share all their genes because they are descendants of the same fertilized ovum. Nonidentical (dizygotic) or fraternal twins (etc.) occur when the mother ovulates two (or more) ova at a time and each one is fertilized by a different spermatozoon. Consequently, fraternal twins are no more genetically related to each other than any other brother or sister in the same family.

Differentiation

A cell removed from an eight-cell or larger embryo will develop only into an incomplete embryo and die. It still has all the genes of the zygote, as does any cell of an adult, but it seems to have lost the ability to use some of the genes it needs to develop completely. As the cells of the embryo multiply, the resulting cells eventually become locked into a par-

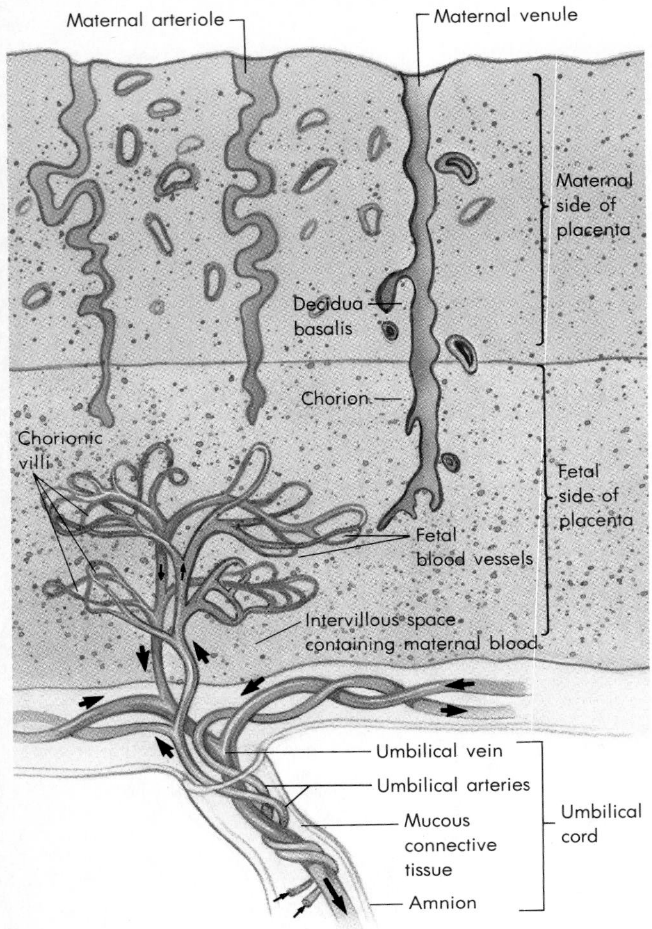

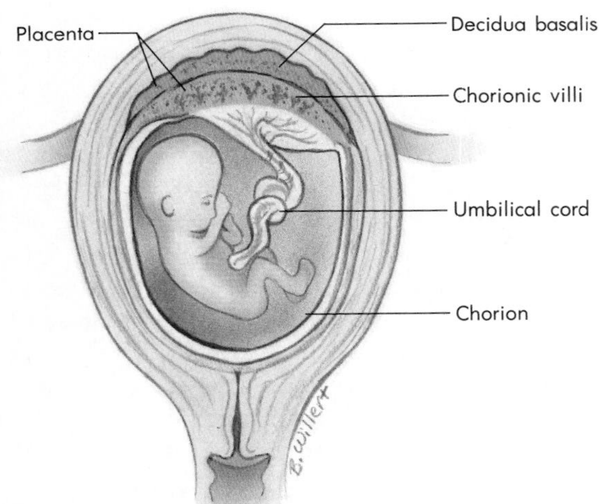

FIGURE 18-5 The fetus, umbilical cord, and placenta. Note that fetal blood never mixes with maternal blood.

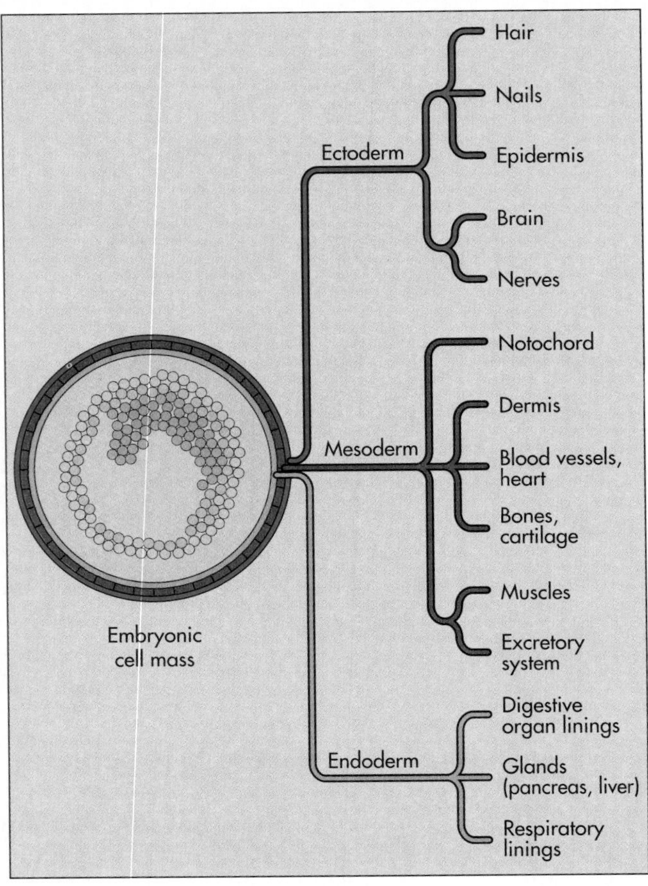

FIGURE 18-6 The embryonic germ layers and their destinies.

ticular, restricted destiny. We can see the broad outlines of these destinies in the fates of ectoderm, mesoderm, and endoderm cells. From these **germ tissues,** as outlined in Figure 18-6, germinate all the various tissues of the finished individual. We see the restrictions on their fates in the simple fact that the germ tissues are not interchangeable. An animal embryo whose mesoderm, for instance, has been removed surgically will not produce muscle, bone, or other mesodermal structures.

The loss of totipotency is part of the process of **differentiation.** The embryo's cells become different from each other apparently because a switch-like mechanism selectively turns on or off certain genes to form the various kinds of cells. The fertilized ovum contains all the genes necessary to specify all the structures and functions of a complete individual. So do all the cells of an embryo or an adult human. However, as the embryo develops, various cells chemically modify some of their genes. Genes not needed for their programmed fates are shut down, made unavailable for later use. Other genes are kept in a form that allows the cells to activate them when they are needed. The turning on and off of genes seems to be controlled in at least two ways. Some cells change their form and function after a set number of cell divisions. Most change in response to chemical signals released by neighboring cells. For instance, a tooth bud in the dermis of the jaw's skin releases substances that **induce** overlying epidermis to form the outer enamel layer of a tooth.

The factors that make all genes of a fertilized ovum available for use seem to lie in the ovum's cytoplasm. Researchers can transplant a nucleus, with all its turned-off genes, from a fully differentiated cell into an ovum whose nucleus has been removed, as shown in Figure 18-7. The cytoplasmic factors then reactivate the turned-off genes and the hybrid cell develops into a normal embryo. The embryo's genes are identical to those of the animal that donated the nucleus; they have nothing in common with the source of the ovum. The transplant process thus offers a way to duplicate—or **clone**—the nucleus donor. It has worked with frogs, fish, and mice; someday, it may allow the genetic duplication of a human or perhaps provide a controversial source of compatible organs for transplants. It will probably not be used to grow armies of identical soldiers or to duplicate famous scientists, artists, or political leaders since heredity is by no means the only thing that determines personality. Environment—experience and education—plays a crucial role.

Some adult animals, such as lizards and salamanders, are able to reverse the differentiation of their cells enough to regrow, or **regenerate,** severed tails and limbs. Humans cannot. Their regenerative abilities are very limited. They can regrow small amounts of lost skin and up to half of a damaged liver, but only because nearby cells can multiply. True regeneration shows only in children up to 12 years of age, who can regrow as much as the lost tip of a finger.

Stages of Embryonic Development

We saw some of the earliest steps in embryonic differentiation when we discussed the development of the fetal membranes, but not all of them. While the embryo is growing toward its fishlike stage, it is

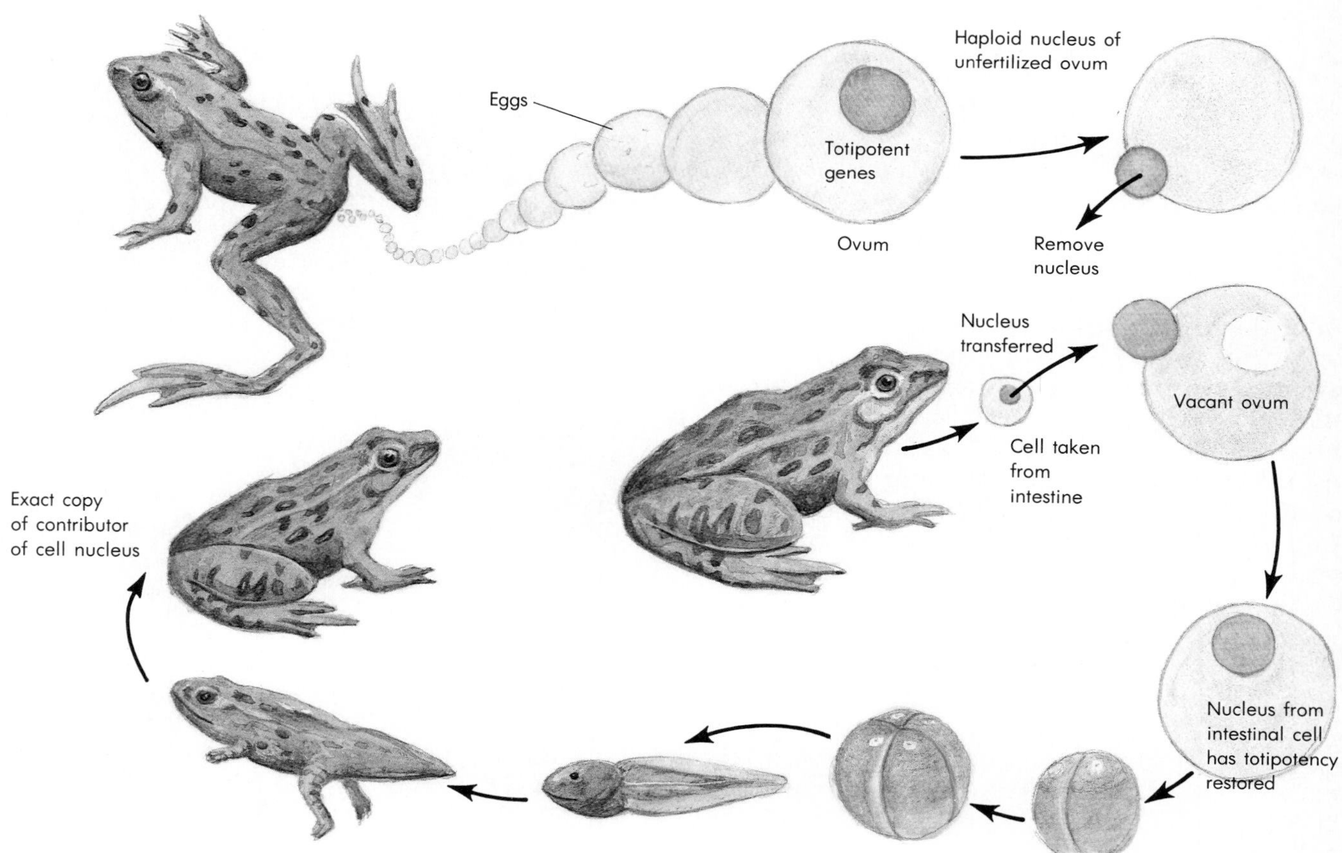

FIGURE 18-7 Cloning occurs when the nucleus of a body cell (here a cell from a frog's intestine) is transplanted to an ovum whose own nucleus has been removed or destroyed. The ovum's totipotency and the transplanted nucleus's genes then combine to produce an embryo and eventually an adult. This adult is the genetic duplicate of the animal from which the transplanted nucleus came.

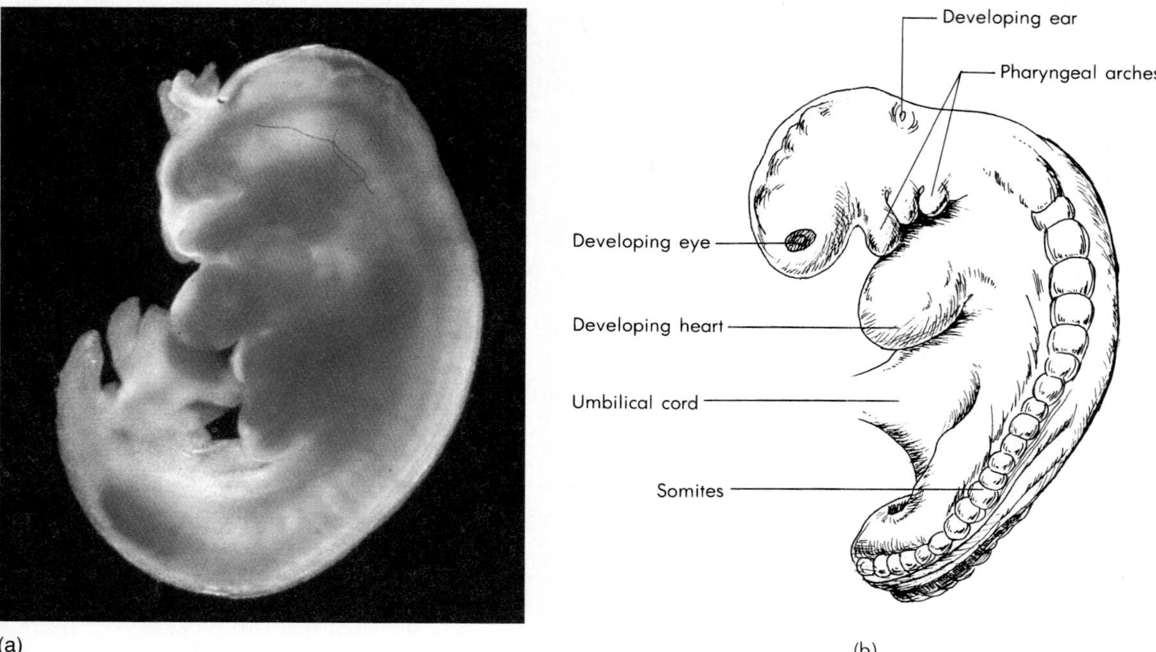

FIGURE 18-8 (a) A four-week embryo, showing somites. (b) Labeled drawing of four-week embryo. Using (b), can you identify the structures in the photo?

developing internally as well. This development can be conveniently divided into three stages, the trimesters (three-month periods) that most often label the phases of pregnancy.

The First Trimester

After the mesodermal layer appears in the embryonic disc, the disc develops the beginnings of the eventual central nervous system (CNS). The groove through which the mesodermal cells have migrated becomes a **neural plate,** which then folds to form a **neural tube.** Mesodermal cells form the blocks, or **somites,** of future muscle and bone visible in Figure 18-8. The first blood vessels form in the fourth week, and the heart begins to beat, making the circulatory system the first organ system to function in the embryo. The eyes begin to take shape as cup-shaped outgrowths from the embryonic brain.

By the fifth week after fertilization, the embryo is still less than a quarter inch long, but it is beginning to lose its resemblance to a fish. It retains the distinct tail and the grooves or gill clefts in its neck, but it has tiny buds where its limbs will later be, and its eyes are visible (Figure 18-9). In several ways, it now resembles the embryo of an amphibian or reptile. It is, so to speak, moving up the evolutionary ladder as it develops, passing through forms that resemble those of its distant ancestors or their embryos (see Chapter 24).

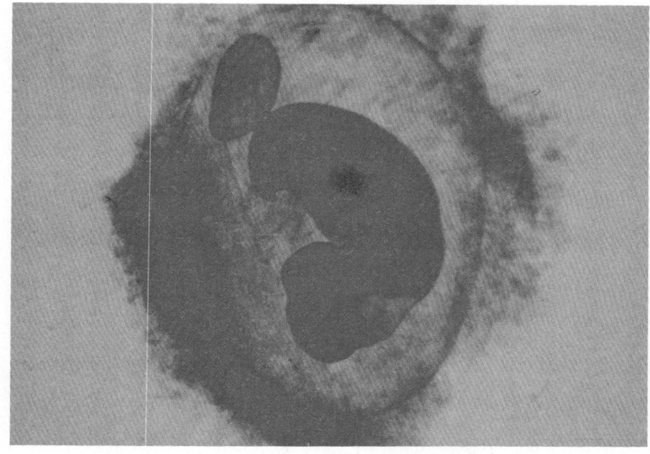

FIGURE 18-9 A five-week embryo.

In weeks 6–8, the limb buds expand into recognizable arms and legs, the gill clefts vanish, and the tail diminishes in size. The embryo begins to look more human, even in its face, though its head is bulbous. The bones begin to take shape within its body. By the third month, the embryo is no longer called an embryo, but a **fetus.** By the end of this month, the embryo's organs are all fairly well developed. It can now move its arms and legs, make facial expressions, and suck its thumb, though none of these movements can be felt by the mother (Figure 18-10).

The seventh week marks the time at which male and female embryos become different. Until then, they are indistinguishable, even though their sex has long been defined by the sex chromosomes they carry (XX for females; XY for males; see Chapter 19). The embryo's genital region is marked only by a swelling, and the gonads within are neither ovary nor testis. Instead, the gonads are **indifferent gonads** with an outer cortex and an inner medulla.

In male embryos, the medulla of the indifferent gonads develops into the testes and the ducts of the indifferent gonads develop into components of the male reproductive tract, including the epididymis, vas deferens, and seminal vesicles. In female embryos, the cortex of the indifferent gonads develops into the ovaries and the ducts develop into the Fallopian tubes, uterus, and vagina (see Figure 18-11). The male ducts develop under the influence of a "regression factor" and testosterone, the male sex hormone, released by the embryonic testes. Testosterone alone stimulates the development of the male external genitalia. It may also create differences in the size and organization of certain brain regions. Various researchers feel that these sex differences in the brain may account for some of the apparent innate differences in the behavioral patterns of men and women.

The female ducts and external genitalia form later than the male reproductive tract, when there is an absence of these chemical signals. Thus, hormone abnormalities in the mother or the embryo can produce sexual abnormalities. If the mother has an androgen-secreting adrenal tumor, her baby can be genetically female yet have what appear to

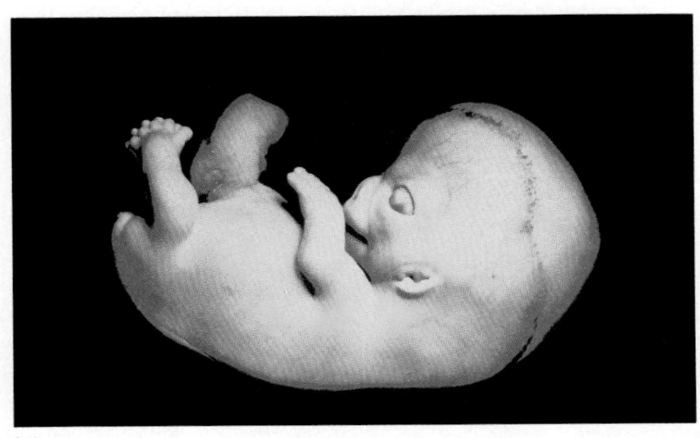
(a)

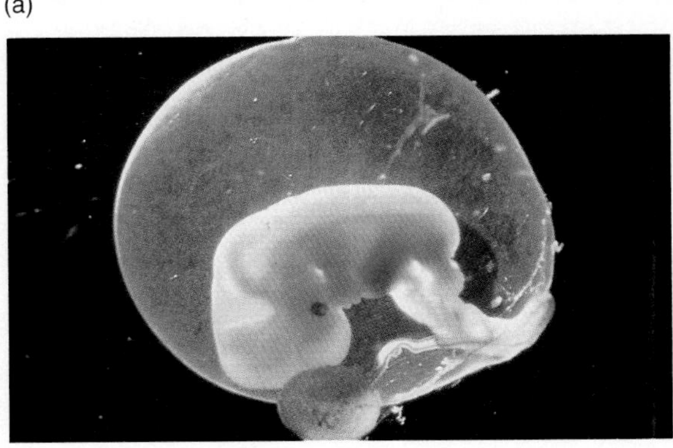
(b)

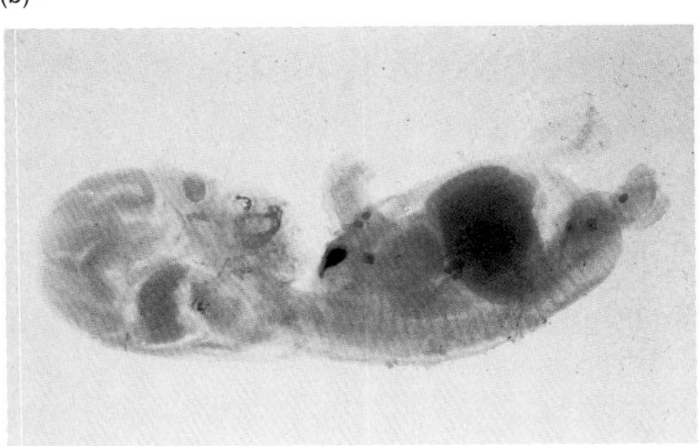
(c)

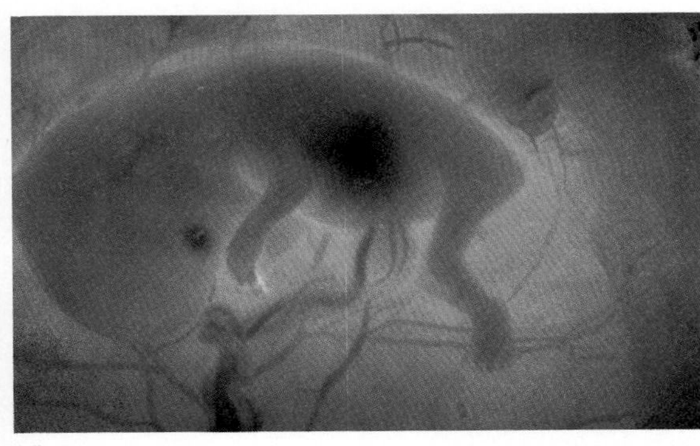
(d)

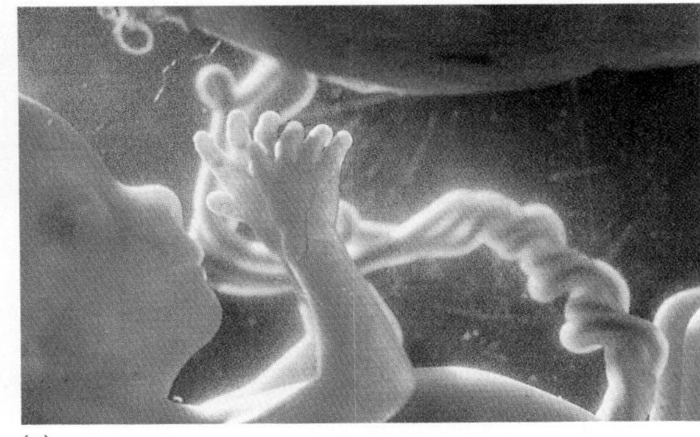
(e)

FIGURE 18-10 (a) Human embryo at 44 days. (b) Seven-week embryo. (c) Eight-week embryo. (d) Twelve-week embryo. (e) Fifteen-week embryo.

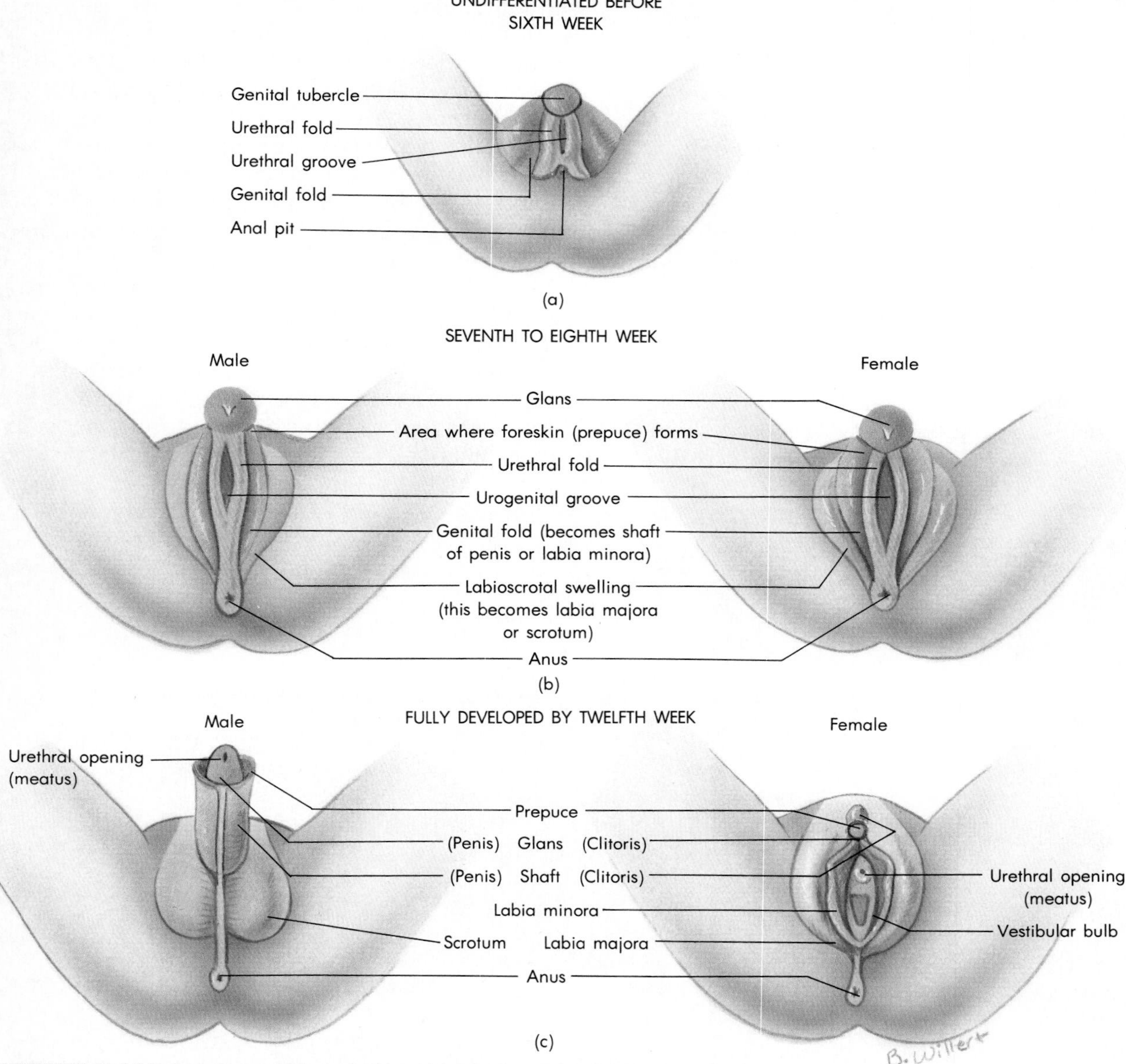

FIGURE 18-11 (a) Before the seventh week, the embryo looks neither female nor male. About that time, the indifferent gonads develop into either testes or ovaries. (b) The embryonic testes secrete testosterone, which provokes the genital region to develop a male configuration. The female genital structures develop in the absence of testosterone. (c) Fully developed male and female genitals. Note how both sexes develop from the same basic set of embryonic structures.

be male gonads and genitalia. If a male embryo fails to produce enough testosterone, it can have testes hooked up to what appear to be female genitalia. Fortunately, such anomalies are rare and can now be corrected surgically.

---

1. Discuss why we could call femaleness a more "basic" aspect of the human condition than maleness.
2. In what ways is the male body plan a modification of the female body plan?

The Second Trimester

The fetus's movements become noticeable to the mother during the fourth month. During this time of "quickening," many people believe the fetus becomes a person in its own right. Some people support abortion in the first trimester, before the movements of the fetus can be felt by the expectant mother. However, after the time of "quickening," the fetus takes on certain human characteristics that make it impossible for these people to continue supporting the concept of abortion in the later stages.

During the fifth month the placenta enlarges to cover about half the uterus. The fetus weighs about half a pound and is beginning to grow hair. The heartbeat is audible through a stethoscope. For the first time, the fetus's organs are mature enough to let it survive outside the womb, but such a premature baby, or "preemie," must be kept warm in an incubator, protected from infections, and supplied with special, easy-to-digest foods (Figure 18-12). Most preemies do not survive unless they have undergone six months (or more) of development and weigh a pound and a half.

The Third Trimester

After six months of development, a fetus has most of what it needs in order to survive if it is born prematurely. It has the necessary organs, and they are all working. Yet there are still three months before its birth. During these three months, its organs enlarge and grow stronger to make survival more likely. Growth is especially vigorous in the brain and nervous system, for it is in this stage that many nerve pathways are formed. A shortage of protein in the mother's diet late in pregnancy can result in stunted brain growth and hence in a lifelong mental deficit. It can also result in a smaller than normal baby, but body weight is relatively easily made up after birth. A stunted brain never makes up its lost development; this is probably the greatest tragedy of famine.

By the end of the third trimester, the fetus has reached its birth weight of 3–4 kilograms (6–9 pounds), which includes a good deal of stored nutrients. These will help support the baby for its first few days of life outside the womb, when babies typically lose some weight. Toward the end of the third trimester, as the placenta begins to deteriorate, antibodies enter the fetus's blood; these proteins will help to protect the baby against infection until its immune system can begin to function and produce its own antibodies.

Perinatal Changes

The fetus—now an infant—undergoes several additional changes just before and after birth. These changes enable the infant to shift from reliance on the mother for oxygen, nutrients, and temperature control to reliance on its own systems. They begin as the placenta starts to detach from the wall of the uterus and the supply of oxygen to the infant drops. Carbon dioxide builds up in its blood, stimulating the breathing centers in the medulla of the brain. As the baby emerges from the birth canal, its skin temperature drops. When the skin of the buttocks (which usually emerge after the head) cools, a reflex triggers the first deep breath. The lungs inflate for the first time, and oxygen begins to enter the baby's body as it does in an adult.

Fetal Circulation

At the same time, the fetal circulatory system undergoes some dramatic changes as it adjusts to the loss of the placenta as a source of oxygen and the need to shunt blood through the lungs. Prior to birth, the pulmonary circulation through the lungs is not operating, since the fetus gets its oxygen and nutrients by way of the umbilical cord. However, within minutes of being born, the baby must shift circulatory gears and get the pulmonary supply system up and running. In a free-living individual, the chambers of the right side of the heart receive deoxygenated blood from the various organ systems of the body. They then pump it out, by way of the right ventricle, to the lungs, where the hemoglobin is reloaded with oxygen and carbon dioxide

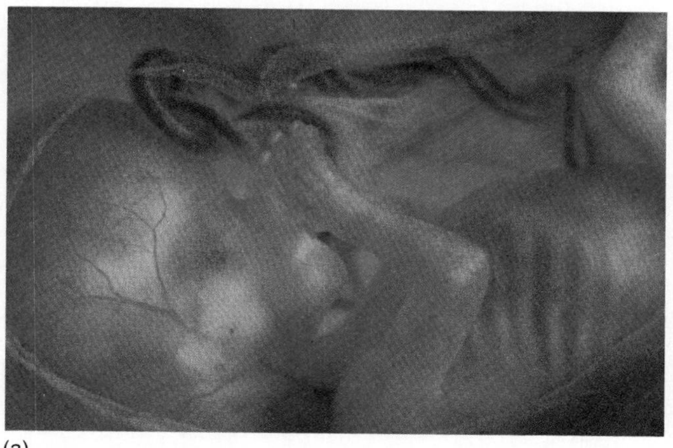

(a)

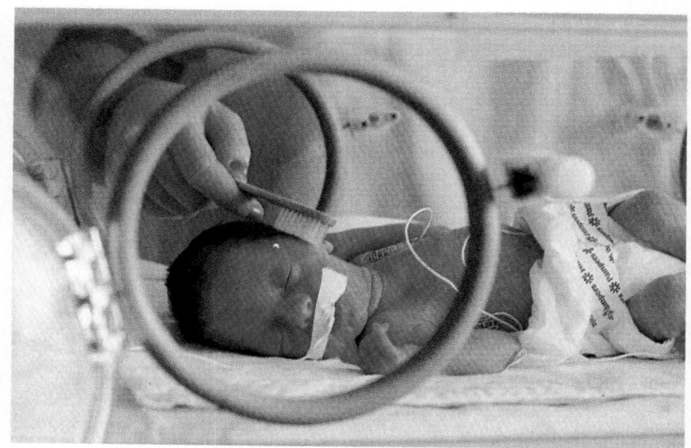

(b)

FIGURE 18-12 (a) Sixteen-week fetus. (b) Premature baby in incubator.

is released. The chambers of the left side of the heart normally receive oxygenated blood from the lungs and then pump it back out to the body's various organ systems. The supply situation in the fetus is very different (see Figure 18-13).

The fetus receives its oxygen and nutrients by way of the umbilical vein coming through the cord from the placenta. Since the pulmonary cardiovascular circuit is not operating, it is necessary to transfer the oxygen and nutrient-rich blood directly to the systemic circulation operating out of the left side of the heart so it can be distributed to the tissues around the body.

The blood enters the fetal body through the umbilical vein by way of the liver. It is transferred from there through a short blood vessel, the *ductus venosus,* to the inferior vena cava, where it joins with the oxygen-poor blood returning to the right atrium from the tissues. Since the lungs are not yet functioning, the oxygen/nutrient-rich blood bypasses the pulmonary circuit and is shunted to the systemic circuit through two openings or channels, the *ductus arteriosus* and the *foramen ovale.* The ductus arteriosus is a short channel, directly above the heart, through which the oxygen-rich blood in the pulmonary artery is transferred directly to the aorta. The foramen ovale is an oval opening in the wall between the atria; through it, blood passes directly from the right atrium to the left atrium. However, with the first breath inhaled by a newborn these fetal channels have to be closed off. At the same time, the pulmonary circuit must be activated and isolated from the blood flowing through the systemic channels of the cardiovascular system. Once these systems are running efficiently, the fetus can live independently, supported by its own respiratory and cardiovascular systems.

At birth, the ductus arteriosus, the vessel connecting the pulmonary artery to the aorta, closes as its muscular wall goes into spasm; later, it grows permanently shut and is replaced by connective tissue. The umbilical vein, where it joins the infe-

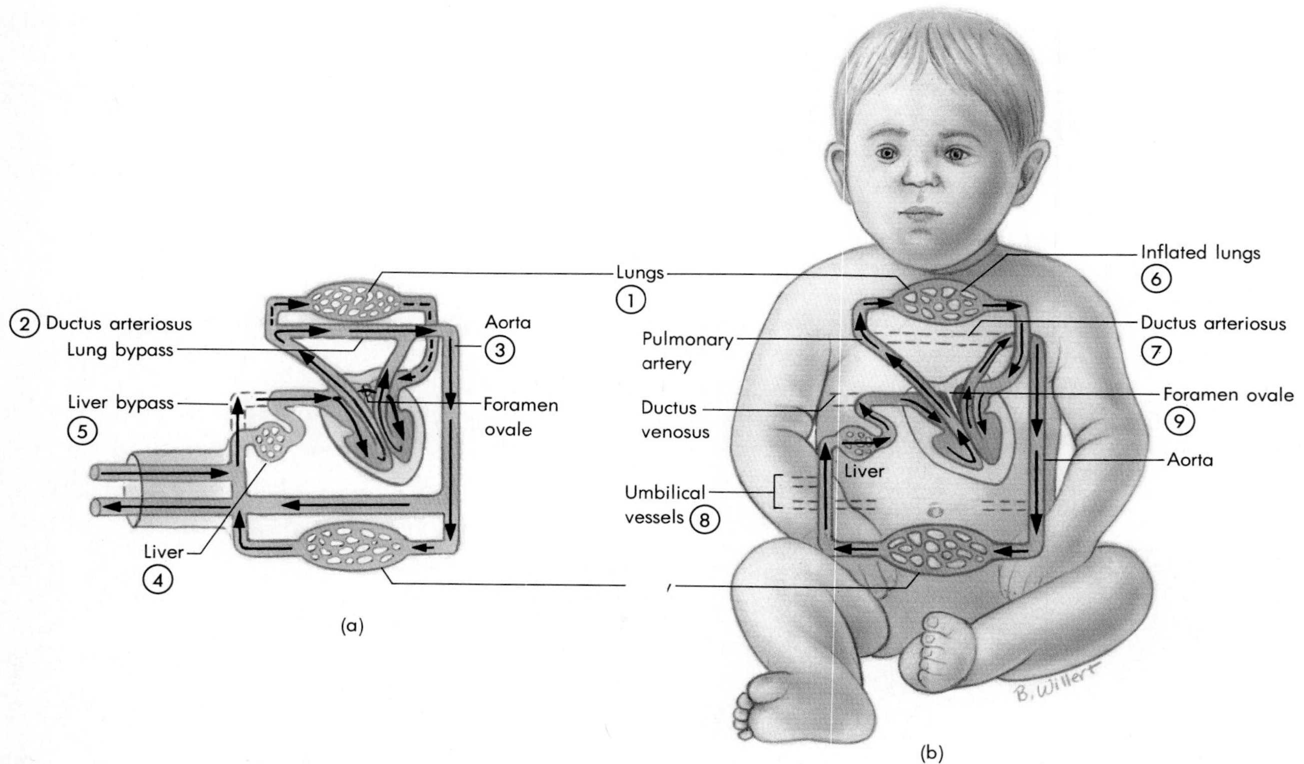

FIGURE 18-13 The fetal circulation changes drastically near the end of the birth process. (a) In the fetus, the lungs are collapsed and little blood flows through them (1). Instead the blood is shunted through the ductus arteriosus and the foramen ovale (2) to the aorta (3). It goes to the placenta through the umbilical artery (4), is recharged with oxygen and nutrients, returns through the umbilical vein, and bypasses the liver in the ductus venosus (5) on its way back to the heart. (b) After birth, the lungs inflate (6) and blood must flow through them instead of through the placenta. Accordingly, the ductus arteriosus (7) and ductus venosus (8) close and a flap of tissue seals off the foramen ovale (9).

rior vena cava as the ductus venosus, also closes off; it too is later replaced by connective tissue (Figure 18-13).

The changes in blood flow patterns that accompany the first breath also produce pressure changes in the heart. As a result, two flaps of tissue come to cover the foramen ovale. Normally, these flaps then fuse with the heart tissue and seal the hole. When this process fails and the hole remains, unoxygenated blood can enter the aorta. This gives the baby's skin a bluish tinge; such "blue babies" used to die young, but today their defect can be repaired by open-heart surgery.

Immediately after birth, the baby's respiration rate and pulse rate are high. The liver, not yet working properly, may be releasing excessive amounts of bile pigments into the blood. (As many as half of all newborns develop *jaundice,* a yellowish cast to the skin, within their first week of independent life; high levels of the pigments responsible for the coloration can cause brain damage and even death.) Within the next few days, the baby's production of red blood cells and hemoglobin increases, and both respiration and pulse rate decline. Liver function improves. Fetal hemoglobin, designed to function well when less oxygen is available, as in the placenta, is replaced by adult hemoglobin. The number of white blood cells in the blood, which is initially several times the adult level, drops.

---

1. In a breech birth, the baby's buttocks emerge from the birth canal before the head. Why might this be hazardous to the baby? How might the hazard be minimized with the aid of cloths soaked in hot water?
2. Do you think "blue babies" would be unlikely to survive without surgery? Why?

BIRTH DEFECTS

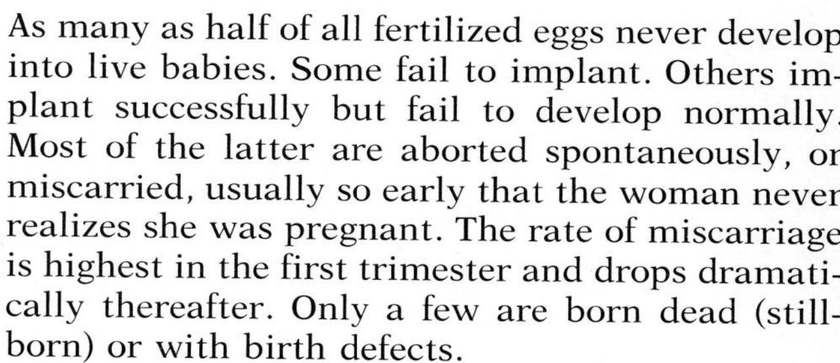

As many as half of all fertilized eggs never develop into live babies. Some fail to implant. Others implant successfully but fail to develop normally. Most of the latter are aborted spontaneously, or miscarried, usually so early that the woman never realizes she was pregnant. The rate of miscarriage is highest in the first trimester and drops dramatically thereafter. Only a few are born dead (stillborn) or with birth defects.

Some birth defects are very dramatic. Failures of the twinning process can produce babies with two heads or Siamese twins whose bodies are joined at the hip, shoulder, breast, or elsewhere. Failures of development can produce babies without arms, legs, eyes, brains, or heads. The esophagus can fail

Learning Focus

HUMAN DEVELOPMENT TIME LINE

For the events of embryonic development listed in the following, determine when they occur in the nine months of pregnancy. What events have been left out of the list and when do they occur? Add details you have gleaned from the text and the lectures you have attended. Repeat the process from memory as you prepare for exams.

Fertilization

Blastocyst

Implantation

Embryonic disc

Chorion

Placenta

First blood vessels

Limb buds

(Continued on p. 418)

Recognizable limbs	Be sure you have inserted the following items in the sequence at the appropriate times:
End of sexually indifferent stage	Germ tissues Mesoderm Neural plate
First movements	HCG Amnion
First noticeable movements	End of totipotency Somites Gill clefts
Brain maturation	Tail Tail absorption Visible eyes
Closure of ductus arteriosus	Audible heartbeat
First breath	Closure of ductus venosus
	Closure of foramen ovale
	Ability to survive outside womb

to connect to the stomach. The heart can lack one or more of its four chambers. The vertebrae can fail to close over the spinal canal, leaving the spinal cord exposed and vulnerable (spina bifida). Blocked drainage of cerebrospinal fluid can allow the buildup of pressure in the cranium, a ballooning of the skull, and damage to the brain.

Other birth defects are less visible but no less serious. We have already mentioned brain damage from maternal malnutrition and the blue baby syndrome. There are also cases of *mental retardation, multiple fingers* or *toes, cleft palate, clubfoot,* and many other defects. About 4 percent of all live births have severe problems of one kind or another.

There are many causes of birth defects. As we will see in Chapter 19, some lie in the genes. Having a sixth finger is an inherited trait, as are *cystic fibrosis* (overproduction of viscous mucus that blocks the respiratory passages and pancreatic duct), *phenylketonuria* (inability to metabolize the amino acid phenylalanine), *color blindness, hemophilia,* and some kinds of mental retardation.

Many birth defects are caused by outside influences or **teratogens** (*terato* = monster). Exposure of the mother to radiation, such as x-rays, can cause *microcephaly* (small head), mental retardation, and misshapen bones. Excessive alcohol consumption—as little as a drink or two a day—can cause low birth weight, slow growth before and after birth, small head, malformed face, arms, and legs, defective heart and other organs, and mental retardation, the signs of **fetal alcohol syndrome (FAS).** A milder version, known as fetal alcohol effects, may not be noticed until later, when the child has difficulties in learning. Cigarette smoking may cause *anencephaly* (absence of a brain) and heart problems, and it may be involved in causing cleft palate. Children of smoking mothers also have more problems of the digestive and respiratory tracts. Insecticides, herbicides, industrial and household chemicals, and drugs can also cause birth defects. One classic example is thalidomide, a tranquilizer once used to ease morning sickness. Taken during the first trimester of pregnancy, it proved to interfere with limb development. "Thalidomide babies" tend to have flipperlike limbs; their needs have stimulated considerable improvement in prosthetic devices (see Figure 18-14). Sometimes the effect is delayed, as in the case of diethylstilbestrol, a drug once used to prevent miscarriages. Years later, the children of mothers who had used it showed an increased tendency to develop cancers of the genitalia.

Fortunately, it is often possible to detect birth defects early and to prepare for their impact or even to treat them. Detection has become possible only in the last decade or two, for it requires access to the fetus or its cells. Because they present their own hazards, x-rays are not appropriate. Less dangerous is **sonography,** which uses high-frequency sound waves (ultrasound) to examine the fetus. A device sends beams of ultrasound into the pregnant

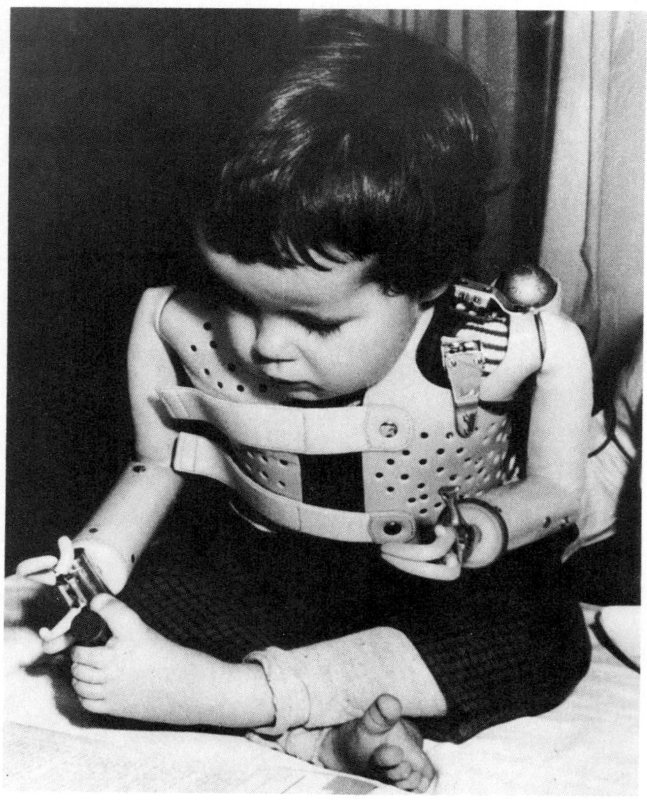

FIGURE 18-14 This child's mother had taken thalidomide during her pregnancy. The prosthesis makes it possible for the child to play in the sand and feed himself.

woman's abdomen and then receives, records, analyzes, and displays the echoes from the structures of the fetus's body. Sonography can detect structural defects in the brain, heart, limbs, and other organs; sometimes it can even produce images that let the physician tell a male fetus from a female (Figure 18-15).

A second important method of detecting birth defects is **amniocentesis.** It relies on the fact that an embryo or fetus sheds some of its cells into the amniotic fluid that surrounds it. During the second trimester, and sometimes earlier, there are enough cells in this fluid to let a physician insert a hypodermic needle through the mother's abdominal and uterine walls to withdraw a sample of amniotic fluid and embryonic cells (see Figure 18-16). The fluid itself can be analyzed for the presence of chemicals that indicate some abnormalities. The cells can be examined for signs of over 50 different genetic disorders, including cystic fibrosis, hemophilia, sickle cell anemia, and Tay-Sachs disease, and over 300 chromosomal defects, including Down (mongolism, due to a third 21st chromosome), Klinefelter (a male with two, not one, X chromosomes), and Turner (a female with one X chromosome, not two) syndromes. The likelihood of some of these chromosomal defect syndromes increases with the mother's (and even the father's) age.

Chorionic villi sampling involves removing a piece of the placenta. Since the chorionic villi in the placenta are derived from the early embryo's tissue, this too can provide fetal cells for analysis. Chorionic villi sampling has two advantages over amniocentesis: it can be done sooner—as early as the eighth week of pregnancy—and it can yield results more promptly.

Similar results can come from fetal blood tests, for it is also possible to draw samples of a fetus's blood through the mother's abdominal and uterine walls. But what can parents do when amniocentesis, chorionic villi sampling, or blood testing re-

(a)

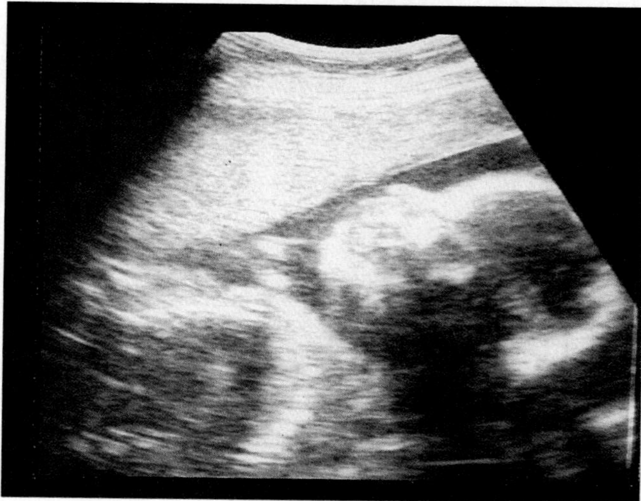

(b)

FIGURE 18-15 (a) Pregnant mother undergoing a sonogram examination. The little girl is anxiously waiting to see her new brother's image appear on the monitor. (b) Sonogram image of fetus in the womb.

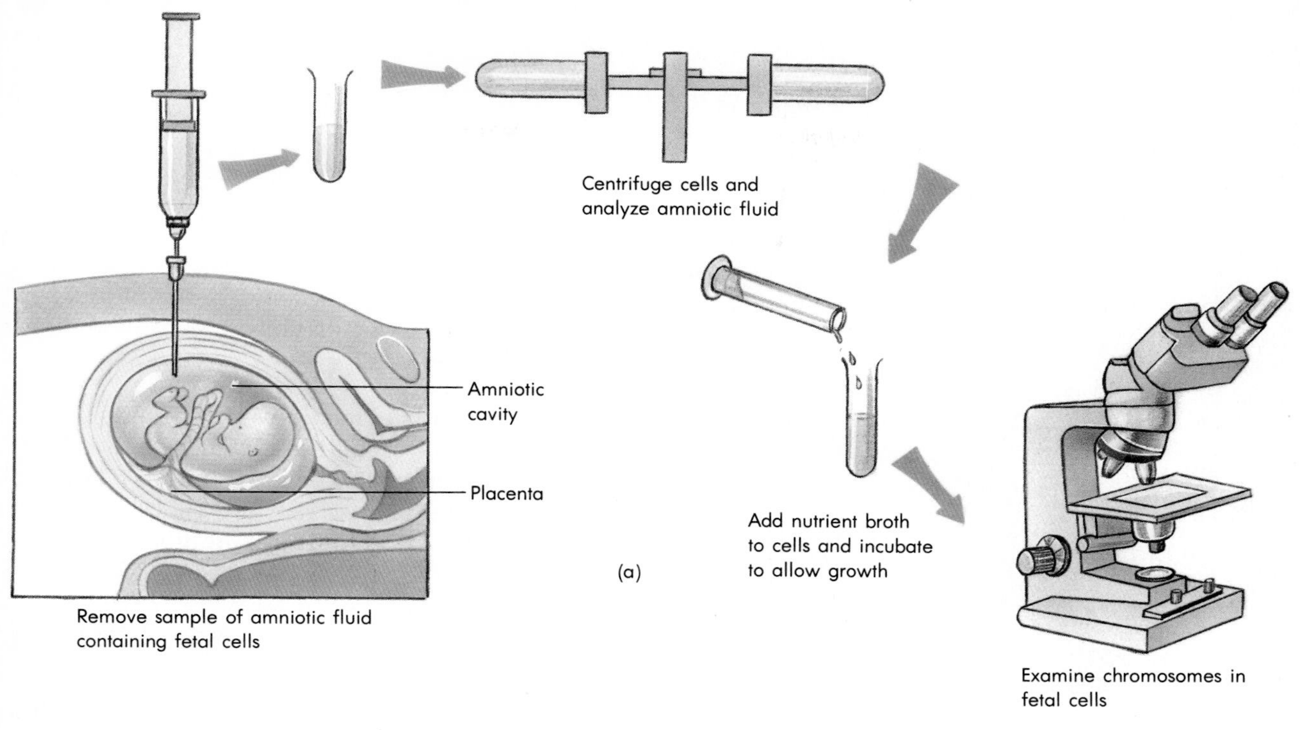

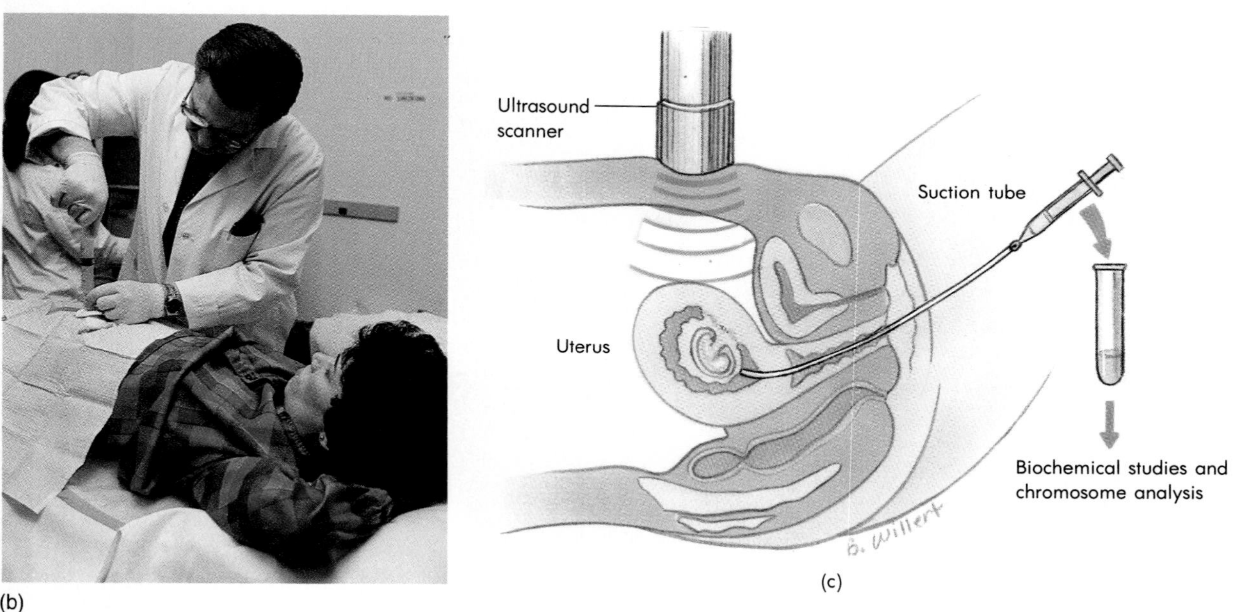

FIGURE 18-16 (a) In amniocentesis, fetal cell samples are drawn with the amniotic fluid through the maternal abdominal and uterine walls, cultured, and then examined for evidence of chromosomal abnormality and genetic disease. (b) Photo of the procedure being done. (c) Another noninvasive technique for fetal examination is performed using computer-enhanced ultrasonic imaging while the fetus is still in the womb. Also shown is another technique for taking fetal cells for biochemical and chromosomal analysis.

veals a serious problem with their unborn child? They often need to do nothing. Most embryos with chromosomal abnormalities never make it to birth. In one study of 15,000 spontaneous abortions (miscarriages), half had some kind of chromosomal abnormality; 1350 of the 7500 had Turner's and 350 had Down syndromes. However, these ratios change among the fetuses that somehow manage to survive—one newborn in 750 has Down; one newborn male in 1000 has Klinefelter's; and one newborn female in 5000 has Turner's. *Some* problems do survive, and they must be coped with.

The most serious birth defects are those, such as Down syndrome, for which there are helpful treatments but no cures. On the other hand, some defects permit very simple cures. For instance, jaundice has long been treated by exposing the newborn to special lights, which break down the bile pigment in the blood; a new treatment exposes the newborn's blood to an enzyme that breaks down the pigment. On a more complex level, surgical reconstruction can cure malformed hearts, cleft palates, and other structural defects. The surgery usually comes after birth, but recently surgeons have developed ways to repair some defects while the fetus is still in the womb. In the future, more such techniques will become available. Genetic engineering may even make it possible to replace malfunctioning genes, supply missing chromosomes, and even remove extra chromosomes. Applied early in pregnancy, genetic engineering may thus someday provide a cure even for Down syndrome.

---

1. Some people believe that abortion is an appropriate choice when sonography, amniocentesis, or chorionic villi sampling reveals that a fetus will become a baby with a serious birth defect. What kinds of birth defects—if any—do you think are serious enough to warrant abortion?
2. Given the chance, many people might choose to have male children instead of female children. (Yes, there is such a bias.) Discuss the hazards to society of using sonography or amniocentesis to identify the sex of a fetus and then using abortion to "choose" the sex of one's child.

CHILDBIRTH

Near the end of pregnancy, a woman's body prepares for childbirth. In the ninth month, placental production of human chorionic somatomammotropin peaks to prepare the breasts for lactation. Together, the placenta and the ovaries release relaxin to soften the cervix and the ligaments that bind the bones of the pelvis. Placental production of progesterone, which inhibits uterine contractions, declines. At the end of the nine-month **gestation** period, uterine production of prostaglandins, which stimulate uterine contractions, seems to increase. The posterior pituitary releases oxytocin, which also stimulates uterine contractions. When all is ready, **labor** begins (Figure 18-17).

Labor consists of waves of contractions of the uterine muscles, sweeping down the uterus. Since the baby is usually positioned with its head down, the contractions press the baby's head against the cervix. The resulting repeated surges of pressure dilate the cervix. They also arouse reflexes that apparently increase the secretion of oxytocin by the pituitary. The increased levels of oxytocin then stimulate more frequent and stronger uterine contractions, which increase the stretching of the cervix and the oxytocin-releasing reflex. Labor is thus one of the rare examples of *positive feedback* in the human body.

Not surprisingly, labor is usually painful. Each wave of contractions produces a labor "pain." At first, the pains come slowly, but they soon speed up and strengthen. Because some women find them excruciating, physicians have long favored the use of anesthetics to ease childbirth. However, the drugs administered often impaired the newborn baby. General anesthetics have been largely replaced by "nerve blocks," local injections of anesthetics to deaden the nerves that supply the pudendal area (female genitals). However, even these drugs may affect a baby for a day or two after birth. Many women today prefer "natural childbirth," by which training in breathing techniques and relaxation helps them control both the process and the pain of labor. The father is often present during both training and delivery. Anesthetics may be essential only in problem births, such as when the placenta blocks the cervix, the baby emerges buttocks first ("breech delivery"), or malformations of the mother's pelvic bones (congenital or acquired, as through fractures) narrow the birth canal. Sometimes, such problems are so severe that birth is possible only by *cesarean section,* surgical removal of the baby through an incision made in the abdominal and uterine walls.

As labor proceeds, the cervix dilates enough to admit the baby's head into the vaginal birth canal. At or before this stage, the amnion bursts, releasing its "waters." Repeated uterine contractions then push the baby into and through the birth canal, as shown in Figure 18-17. With the emergence of its buttocks, the baby takes its first breath. Further uterine contractions then squeeze blood from the placenta through the umbilical cord and into the baby. The umbilical blood vessels constrict. The physician clamps or ties the umbilical cord and cuts it; later the cord will dry up and fall off, leaving the scar we call the navel (or belly-button). Finally, the uterus expels the placenta, sometimes aided by the further release of oxytocin stimulated by the baby's first attempt to suckle at its mother's breast.

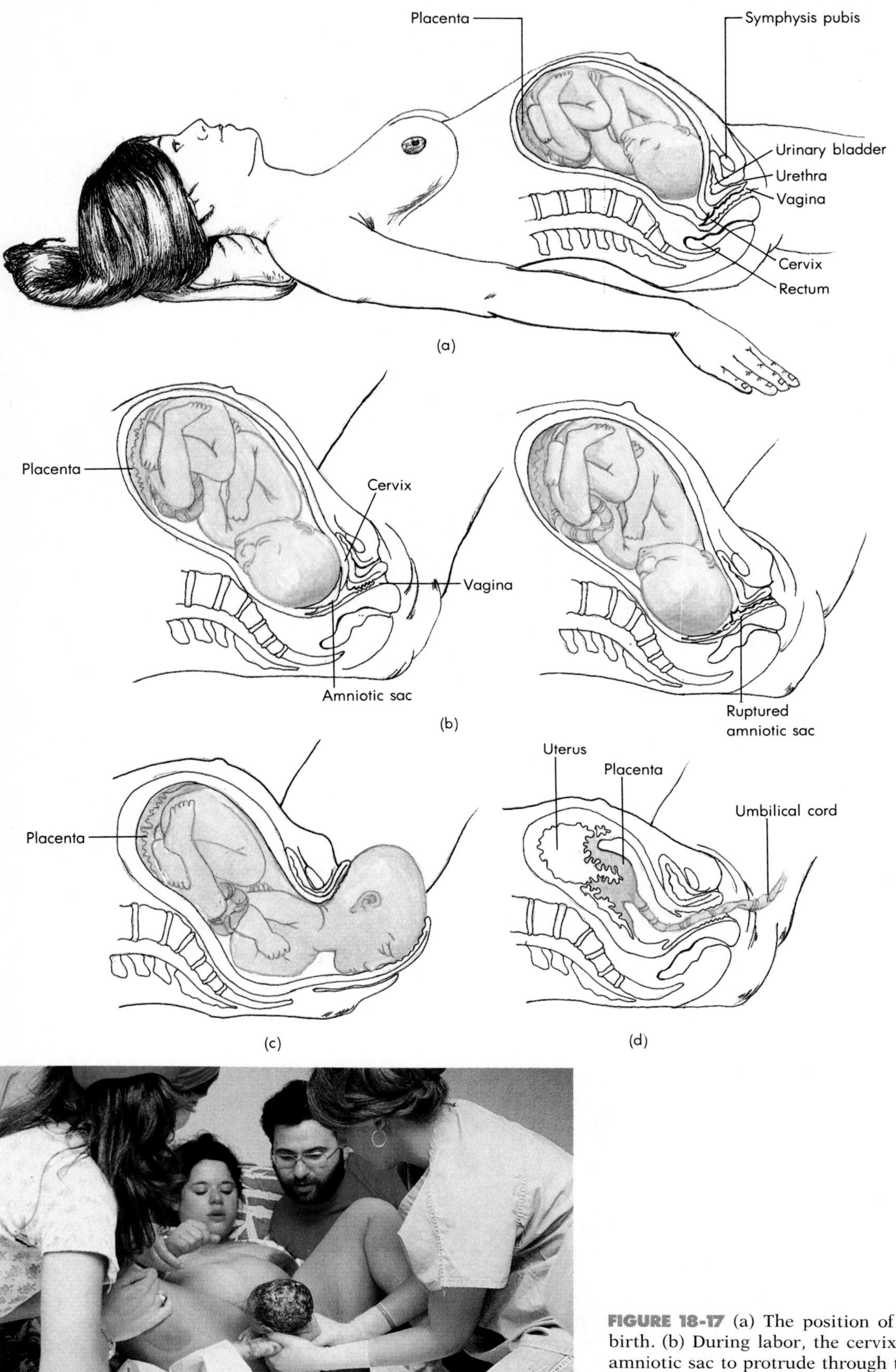

FIGURE 18-17 (a) The position of the fetus just prior to birth. (b) During labor, the cervix dilates and allows the amniotic sac to protrude through the opening. (c) The sac ruptures, spilling the amniotic fluid, and the fetus is expelled through the birth canal. (d) The placenta detaches and is released as the afterbirth. (e) An emerging baby.

POSTNATAL DEVELOPMENT

Human development does not stop at birth. A new infant has a great deal of growing to do, both physically and mentally, and the growth begins immediately. At birth, the bones of the skull are soft and the gaps between them, the **fontanels** (see Figure 18-18), are large. This arrangement gives the head a flexibility necessary to pass through the birth canal; during delivery, a baby's head is often deformed enough to alarm the parents, but it soon returns to the normal shape.

In the next year or so, the fontanels fill in and the bones harden while the brain enlarges rapidly. As it enlarges, the infant acquires the neural equipment to control its body and to learn language and other features of human behavior. The body also enlarges, often doubling in size within the first few months. For several years, however, the head remains much larger in proportion to the body than an adult's, perhaps a reflection of the child's need for rapid brain development and learning. It also has an interesting implication for the design of dolls and cartoon characters intended to appeal to children. Charlie Brown, with his oversized, round head, is an exaggeration of normal child proportions, and it is very easy for children to identify with him.

At first, the baby is protected against disease by the antibodies that crossed the placenta from its mother's blood to its own. Later, it receives more antibodies in its mother's milk (which is one reason why breast feeding is better than bottle feeding). Only about six months after birth does its immune system begin to function well. Then, as the baby is exposed to bacteria, viruses, and other hazards, it begins to generate its own antibodies.

GROWTH AND MATURATION

By the age of 2, a baby is about half as tall as it will be when adult. It has also begun to talk and walk, and it is learning more intensively than it ever will again, even in college. Ten years later, it has completed most of its growth and is ready to enter puberty. At this time, hypothalamic and pituitary hormones awaken the reproductive system, and the individual develops its secondary sexual characteristics (see Chapter 16). As the sex hormones build up in the body, they first accelerate and then end the growth of the skeleton; by the late teens most young people are as tall as they will ever be. Further change involves the maturation of body contours, growth of facial hair, and learning. A 20-year-old has the body of an adult and is rapidly acquiring the social skills, contacts, and educational and job experiences that will allow him or her to take a place in society.

Psychological maturation continues throughout life, so that in a sense people never stop growing. Physical maturation peaks in the twenties, when athletes are in their prime. After that, the body

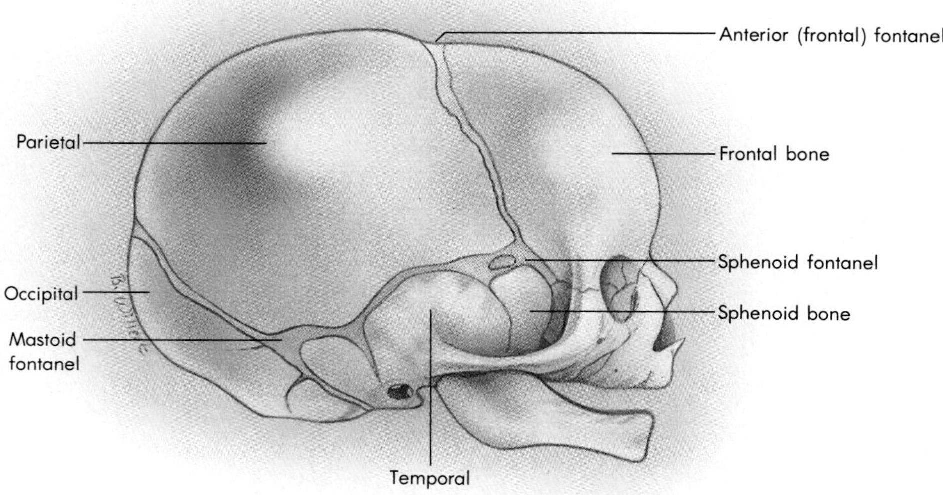

FIGURE 18-18 A newborn's fontanels, or "soft spots," are zones where the skull plates are incomplete. This incompleteness allows the skull to flex in order to fit through the birth canal. It also allows rapid growth of the brain in the months immediately after birth.

loses more cells than it can produce. The brain alone loses some 100,000 cells per day. However, this loss does not necessarily result in problems of mental function. The reason is both that the loss is scattered (not concentrated in a single area of the brain) and that in even a long lifetime, the loss does not amount to a very large percentage of the brain. After all, the brain contains about 10^{12} (a million million) cells; it would take 2500 years for the daily loss to consume even one-tenth of the brain's cells.

A more serious cause of decline with age, or **senescence,** is the common buildup of cholesterol-laden plaques in arteries, which restricts the delivery of oxygen and nutrients to vital tissues. There is also the buildup of waste products of cellular metabolism in cells, of cross-linkages (and hence stiffening) between molecules of connective tissue protein, and of damage to the DNA in cells (which may lead to cancer). As the immune system loses efficiency with age, it loses the ability to fight off infections and cancers. Finally, there seems to be a built-in limit on the number of times connective tissue and blood-forming cells, and others that must divide repeatedly throughout life, can divide.

All these factors together add up to limit the human life expectancy. Single humans have claimed to have lived for 165 years or more, but their claims are usually poorly documented. One hundred and twenty years seems to be more like the true maximum human life span, though very few people can hope to reach such a ripe old age. Life expectancy, the number of years the average person can expect to live, is now about 75 years in the United States; women can expect to live a little longer than men. The citizens of many undeveloped countries, especially in Africa, can expect to live less than 50 years.

It is interesting to look at the way life expectancy has changed in the United States. As recently as 1950, whites could expect at birth to live for 69 years; nonwhites could expect to live for 61 years. The numbers climbed as the United States improved the quality of its diet, medical care, and sanitation, largely because of the way these factors affect infant and childhood mortality. However, our ability to increase the life span of the average American has slowed dramatically. Improvements in the treatment of heart disease, cancer, and other diseases of the older members of the population have had very little effect on the average life span. It is as if we each have within us a preset, ticking life-clock.

Medicine is a long way from learning how to reset the clock, but there is a great deal an individual can do to help him or her last—and enjoy a happy, healthy, vigorous life—until that clock runs down. The most important things are even fairly simple: Avoid smoking and alcohol abuse, exercise moderately but regularly, and eat a varied diet. And it cannot hurt a bit to use seat belts and (your mother was right!) look both ways before crossing the street.

SUMMARY

Human development begins with fertilization and implantation, as discussed in Chapter 17. After implantation, continued division of the blastocyst cells produces the embryo. The inner cell mass differentiates into ectoderm, endoderm, and mesoderm, and from its edges extend the amnion and yolk sac. The third fetal membrane, the chorion, arises as mesodermal cells line the blastocyst's trophectoderm. The chorion develops villi that project into pools of the mother's blood to supply the embryo with oxygen and nutrients. Together, the chorion and endometrium form the placenta, which not only nourishes the fetus but also secretes several hormones needed to maintain pregnancy and prepare the mother's body for childbirth.

Initially, the fertilized egg has the capacity to generate every kind of cell and tissue in an adult body. That is, it is totipotent. It loses its totipotency as its daughter cells differentiate, first into the germ tissues and then into the many other tissues of the body. In humans, because some genes appear to be permanently turned off, differentiation is almost entirely irreversible.

Pregnancy lasts nine months, divided into three three-month stages, or trimesters. During the first trimester, the embryo takes on the appearance of a human being and begins to move. The sexes become different from each other. The internal organs are all formed and beginning to function. In the last month of this period, the embryo is called a fetus.

During the second trimester, the fetus's movements become obvious to the mother. The fetus develops until it is able to survive, with help, if born prematurely. During the third trimester, it grows toward its birth weight of 3–4 kilograms. Its nervous system develops in important ways, and proper nutrition of the mother is crucial for the fetus's future intelligence. The final steps toward independence from the womb occur at and soon after the time of birth. Among other things, the lungs inflate for the first time and the circulatory system adapts to a new source of oxygen.

Development often goes awry. As many as half of all conceptions fail to lead to a live birth, and about 4 percent of all live births have something wrong, a birth defect. Birth defects can be caused by radiation, alcohol, tobacco, industrial and household chemicals, drugs, and sometimes just bad luck. Many birth defects can be detected in the womb with such techniques as sonography (ultrasound) and amniocentesis (examination of cells taken from the amniotic fluid). Knowing a fetus has a birth defect permits parents to choose among their medical options or to prepare to cope with their child's future problems. Such knowledge can also permit attempts to treat the defect, sometimes while the fetus is still in the womb.

After nine months of development in the womb comes the time of childbirth. Placental, uterine, and pituitary hormones set the stage for and initiate labor, waves of contraction of the uterine muscles that expel the newborn. After birth, the baby grows rapidly, both physically and mentally. Physical growth ends in the teen years, after the sex hormones surge. Not long after that, the body begins its long decline toward senescence and death. Human life expectancy is limited to about 75 years, though some individuals may live 50 years more.

STUDY QUESTIONS

1. When does human development begin and end?
2. Describe the development of the amnion.
3. Describe the development of the chorion. How does it become the placenta?
4. Which of the embryo's organ systems is the first to appear? Which is the first to become functional?
5. Describe the functions of the placenta.
6. Discuss the origins and destinies of the germ tissues.
7. What is totipotency? How does it end?
8. Discuss how the two human sexes become different from each other during embryonic development.
9. Why are the various changes in a baby's body that accompany and follow birth necessary to its proper functioning?
10. Discuss the changes in lifestyle that might help a pregnant woman minimize the chances that her baby will have a birth defect.
11. When is it advisable for a woman to choose amniocentesis (or some other technique) to check on whether her baby has a birth defect?
12. Describe the "positive-feedback" system involved in childbirth.
13. In what sense does growth continue beyond the time of puberty?
14. Discuss why human life expectancy seems to have a built-in limit.

GLOSSARY

Amniocentesis The obtaining and examining of fetal cells afloat in samples of amniotic fluid withdrawn from the uterus in order to detect genetic and chromosomal defects.

Amnion Fetal membrane surrounding the amniotic cavity and enclosing the developing embryo.

Amniotic cavity Cavity formed as the inner cell mass forms its first tissue layers; the wall of the cavity next to the trophectoderm will become the amnion.

Blastocyst A hollow ball of 50–100 embryonic cells that enters the uterus from the Fallopian tube and eventually implants. Once implantation has been achieved, pregnancy has officially begun.

Chorion Two-layered membrane formed as mesodermal cells line the trophectoderm.

Chorionic villi sampling The removal of a small fragment of the chorionic villi in the placenta; the fetal cells in the fragment can be analyzed to detect genetic and chromosomal defects.

Differentiation The process by which cells give rise to different types of daughter cells in a fetus.

Ectoderm Germ tissue that will give rise to skin and nervous tissue and their derivatives.

Embryonic disc Circular mass of ectoderm and endoderm that will become the embryo.

Endoderm Germ tissue that will give rise to the linings of the digestive organs and glands.

Fetus An embryo, after movements begin and the organs are formed in the fourth month.

Germ tissues Ectoderm, endoderm, and mesoderm; the embryo's first tissues, which give rise to adult tissues.

Human chorionic gonadotropin (HCG) Hormone secreted by the chorion; it stimulates the corpus luteum to continue secreting estrogens and progesterone.

Implantation The process by which the blastocyst invades the endometrial lining of the uterus.

Indifferent gonad The embryonic gonad, before it has differentiated into ovary or testis.

Induction The causing of differentiation by secretions of neighboring cells.

Inner cell mass Mass of cells within the blastocyst that will become the embryo.

Mesoderm Germ tissue that will give rise to muscle, bone, blood cells, and other tissues.

Morula The compact ball of 12–16 cells formed as the conceptus moves down the Fallopian tube toward the uterus.

Placenta Organ formed by endometrium and chorion; nourishes embryo; supplies hormones of pregnancy.

Pregnancy State of carrying an embryo or fetus in the uterus.

Teratogen An agent causing birth defects.

Totipotency Ability of a fertilized egg to give rise to all the cell types of an adult organism.

Chapter 19

Human Genetics

We expect close resemblances between parents and children, brothers and sisters, and even cousins. We do not expect them between unrelated persons, although we are sometimes surprised by meeting a near duplicate, a *doppelganger*, of someone we know.

We are surprised because we assume that unrelated persons should be different, at least within the limits of similarity that define the human species. Most humans have approximately the same body shape and organs, move their limbs in the same way, act in at least roughly the same fashion, can speak, and are generally more intelligent than chimpanzees, dogs, horses, and other animals. The differences lie in the details—hair color and texture, skin color, size, facial configuration, levels of talent and intelligence, blood types, and even the efficiencies of enzymes. And there are enough possible differences to let every one of the more than five billion humans on Earth be unique. Duplicates and near duplicates are unusual. In fact, they are so unusual that in some cultures people have considered both twins and unrelated doubles supernatural creatures. Sometimes this view has meant crediting them with divine powers; sometimes it has meant calling them cursed. In some cultures, when twins were born, one or both were killed.

THE MENDELIAN REVOLUTION

Before the mid-nineteenth century, people had no idea of the biological mechanisms controlling resemblances and differences. They knew that many characteristics, including such birth defects as extra fingers, "ran in families" and that related persons usually resembled each other more than unrelated persons but that there were always differences even within families. They tried to explain their knowledge by supposing that each individual contained a recipe of some kind that spelled out his or her assortment of characteristics. They thought the recipe resided in the blood, and they spoke of the "blue blood" of aristocrats, with the recipe for nobility, and the "bad blood" of criminals.

When our predecessors tried to explain why children both resembled and differed from their parents, they said that in mating the parents had mixed their bloods. The children thus inherited characteristics from both parents, and often they showed characteristics intermediate between those of their parents—just as if one kind of blood had diluted the other. These ideas seemed supported by the light brown, intermediate skin tones often seen in the mulatto children produced by marriages between whites and blacks.

However, the nature of inheritance is clearly not so simple. Children frequently show characteristics missing in their parents but present in other relatives (see Figure 19-1). Baby Joe might have Uncle Edgar's broad nose, Aunt Ellie's tiny earlobes, and Grandpa Luigi's black hair even though both parents have small noses, large earlobes, and blond hair. The pattern suggested that the "recipes" in the blood can have elements that lie concealed, cropping up in a family only occasionally.

The pattern remained unexplained until 1866, when Gregor Mendel, a monk living in what is now Czechoslovakia, published the results of his work on peas. Over the previous 10 years, he had learned that the recipe for an organism's features was a collection of discrete inheritable elements that came in pairs and were passed from parent to offspring as individual bits of information (we now call them *genes;* see Chapter 4). Each individual gets half its genes from each parent, thanks to the mechanics of fertilization. And the interactions of the genes account for the patterns of inheritance. When Mendel's work caught the attention of the scientific mainstream in 1906, it quickly led to our modern understanding of heredity, **genetics,** and eventually molecular biology and genetic engineering.

FIGURE 19-1 The members of a family typically both resemble and differ from each other.

Mendel had observed that certain varieties of pea plants have several *either-or* sorts of features. For instance, they can be tall or short but not of medium height. They can have yellow or green seeds but not chartreuse seeds. They could also have smooth skinned or very wrinkled seeds but not moderately wrinkled seeds, and so on. Furthermore, the seeds produced by a pure strain of tall plants give rise only to tall offspring. Likewise, seeds produced by a pure-breeding strain of short plants give rise only to short offspring. That is, each type of pea plant breeds true.

Pea plants breed true because they are self-fertilizing. Pea flowers are closed and do not admit pollen from other flowers. They therefore cannot interbreed, and there is no mixing of traits when different types of plants are grown near each other (Figure 19-2).

Mendel learned to open the pea flowers by peeling back the petals and introducing pollen from other plants (Figure 19-2). He could then force specific kinds of cross-breedings, or "crosses," and study the results. His first discovery was that when he crossed a tall plant with a short one, all the offspring were tall. When he then crossed these offspring, the next generation of offspring turned out to include one short plant for every three tall ones (see Figure 19-3). Shortness occurred even though it was missing in the parents. Genetic crosses that focus on a single trait are called **monohybrid crosses.**

Mendel's ability to focus on a single or perhaps a few traits at a time and ignore all the plant's other possible traits was in many ways one of his most important contributions to the study of genetics. His choice of traits and plant species was also serendipitous.

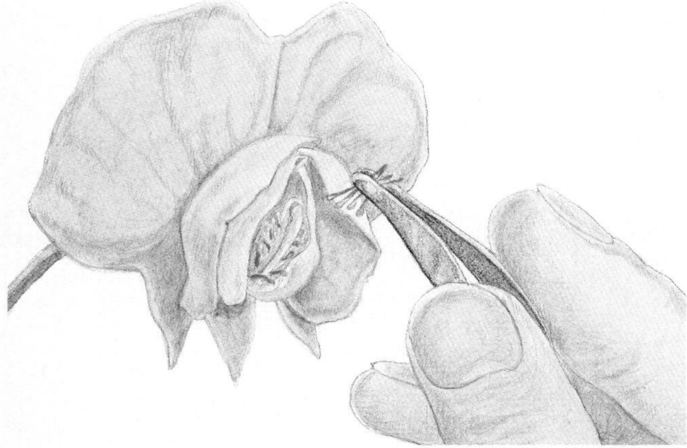

FIGURE 19-2 When genetic researchers cross-fertilize pea flowers, they cut away the flower's petals and deliver pollen by hand. This type of controlled breeding experiment led to the discovery of the mechanisms of inheritance by Gregor Mendel in the last century.

Another major discovery occurred when he examined how two different traits are distributed to the next generation, in what are now called **dihybrid crosses.** When he crossed a *tall plant* with *yellow seeds* with a *short plant* with *green seeds,* all their offspring were tall with yellow seeds. When he crossed plants from this next generation, they produced their offspring in a fairly predictable ratio: nine tall plants with yellow seeds, three tall with green seeds, three short with yellow seeds, and one short with green seeds (see Figure 19-3). Once again, traits missing in the parents (though they were there in the grandparents) showed up, and wholly new combinations of traits appeared. It is here that the genius of Mendel really showed, for from these data he was able to determine how genes and heredity operate.

Mendel did not find any blending of traits in inheritance (though this genetic clarity was partly a matter of luck on his part). He saw either-or inheritance leading to specific ratios of traits in the second generation offspring. An approximate 3:1 ratio was typical for each second-generation monohybrid (one-trait) cross he studied. An approximate 9:3:3:1 ratio was typical for every second-generation dihybrid (two-trait) cross he studied. These ratios and the either-or nature of inheritance led him directly to the basic principles of genetics.

THE PRINCIPLES OF MENDELIAN GENETICS

Mendel realized that he could explain his discoveries very simply:

1. The hereditary recipe in any pea plant had to have blocks of information, the genes.
2. The genes existed in pairs that were separated during the production of gametes and then recombined during fertilization.
3. It was possible to have two different genes, like those for tallness and shortness in pea plants, that governed the same trait (in this case, plant height); we now call such genes **alleles.** One gene can have more than two alleles.
4. In the traits Mendel studied, one allele happened to be clearly **dominant** over the other, which is now usually called the **recessive** allele. Mendel found that if a pea plant inherited a gene for tallness (dominant allele) from one of its parents and another gene for shortness (recessive allele) from the other, only the tallness allele affected the plant's size.

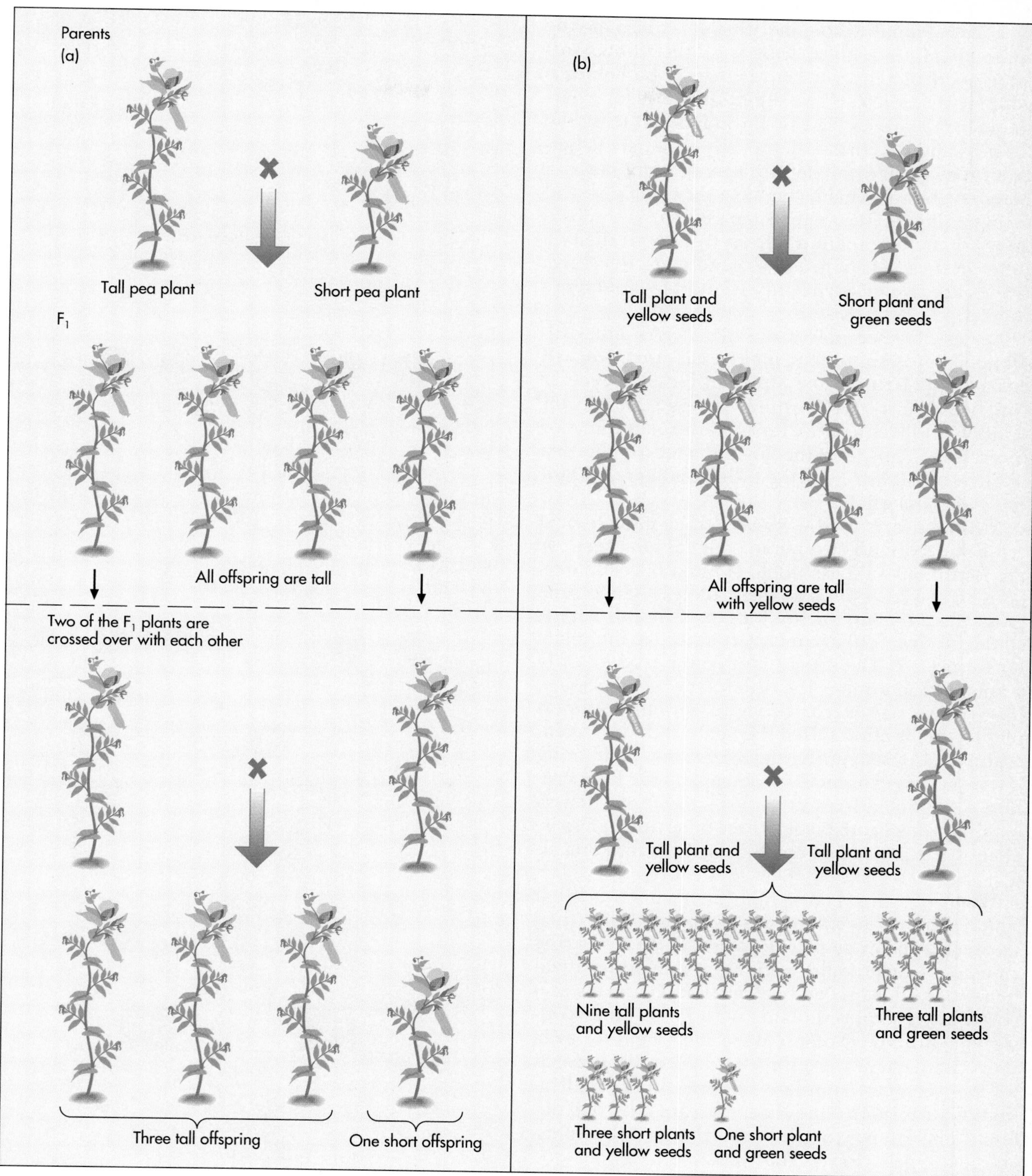

FIGURE 19-3 (a) Monohybrid cross. This is what Mendel saw when he started his breeding experiments. Crossing tall pea plants with short ones produced all tall offspring in the first (F_1) generation. Crossing the tall F_1 plants produced second- (F_2) generation plants with three tall plants for each short plant. How would you interpret these results? (b) Mendel also found some predictable mathematical ratios for offspring produced by dihybrid (two-trait) crosses. First he crossed a tall yellow-seeded pea plant with a short green-seeded one. All their offspring (F_1) were tall with yellow seeds. When he cross-bred two of the tall yellow-seeded F_1's, they produced offspring in the following proportions: nine tall yellow-seeded plants; three tall green-seeded plants; three short yellow-seeded plants; and one short green-seeded plant. From these mathematical relationships, Mendel was able to predict a great deal about the nature of the gene. Using his figures, what can you say about how these hereditary units might behave?

An individual who inherits two copies of the same allele for a given trait (e.g., the tallness allele), one from each parent, is said to be **homozygous** for that allele. If two different alleles for a particular trait are carried by an individual (e.g., the tallness and shortness alleles), the organism is **heterozygous** for that allele. Because of dominance, a heterozygous organism can have the same physical form, or **phenotype,** as an organism that is homozygous for the dominant allele, even though the two differ in their sets of genes, or **genotypes.** Recessive alleles affect phenotype only when they are homozygous.

Let us now see how these ideas work out with Mendel's crosses. As is usual in genetics, we will use a single letter of the alphabet for each gene or trait we discuss, with capital letters standing for dominant alleles and lowercase letters standing for recessives. Thus, for the height gene of the pea plant, we will use *T* for the tallness allele and *t* for the shortness allele.

Mendel's early studies began with a monohybrid cross started with two plants, one homozygous for the dominant tallness allele (*T*) and one homozygous for the recessive shortness allele (*t*). That is, he crossed a *TT* plant with a *tt* plant. The tall parent could produce only gametes containing the allele for tallness, *T*. The short plant could produce only gametes containing the allele for shortness, *t*. (Since the series of crosses begins with these parent plants, this is often called the "parental," or P, generation.) Consequently, fertilizations combining the two types of gametes produced by the P generation plants could result only in the formation of heterozygous *Tt* offspring. These offspring had to be phenotypically tall because *T* is the dominant allele.

When this first generation (F_1 means first filial generation) of heterozygous *Tt* plants was then crossed, each parent produced both *T*- and *t*-bearing gametes, which could recombine during fertilizations to produce second-generation (F_2 means second filial generation) plants with *TT*, *Tt*, and *tt* offspring. Only the *tt* plants were short, and for each short plant there were three tall (*TT* or *Tt*) ones, as we can see if we simulate the process of fertilization (Figure 19-4). Note that there are only two different phenotypes for plant height but three different genotypes. How is that possible?

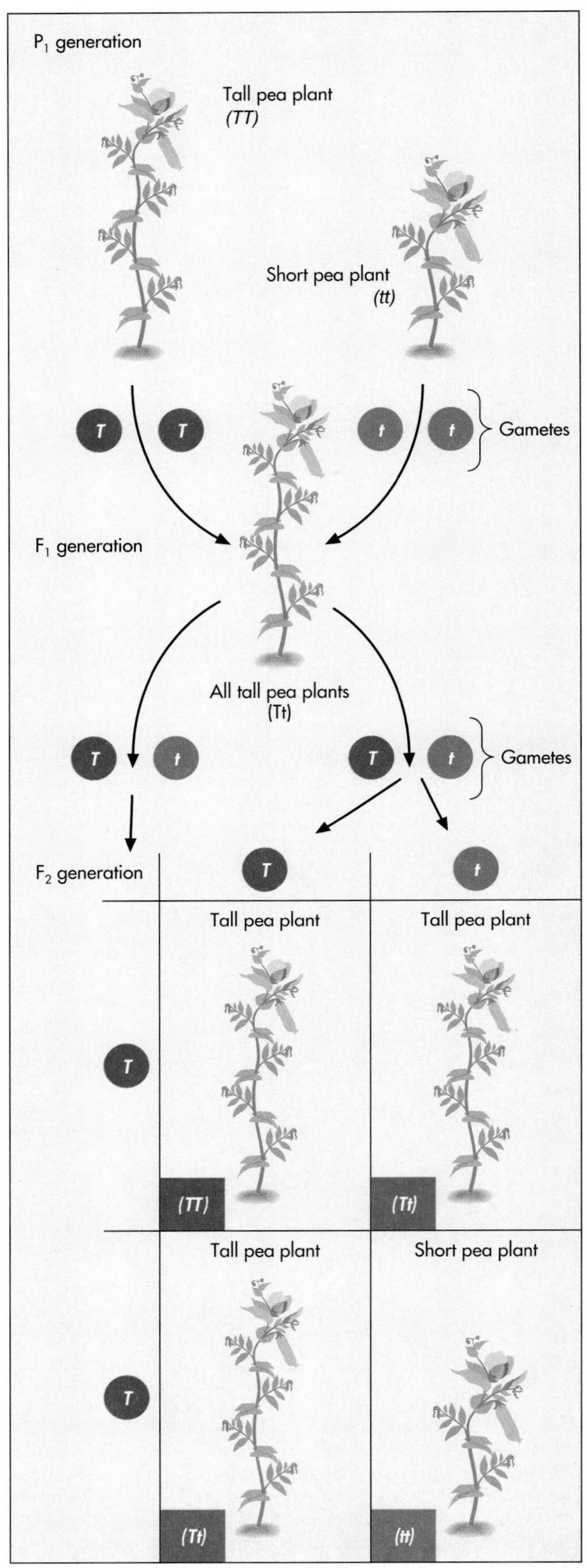

FIGURE 19-4 We can lay out the parental gametes and offspring genotypes on a *Punnett square* grid. When we do so for a monohybrid cross, we immediately see how Mendel thought the paired heredity units, the alleles, were operating.

1. Mendel's genes had to come in pairs that were separated during gamete production and were recombined during fertilization. Compare this idea with what you now know of meiosis and fertilization. With what intracellular structure do Mendel's genes seem to correspond?
2. Tall pea plants can be either homozygous (*TT*) or heterozygous (*Tt*). How could you tell which is the case by crossing a tall plant with some other pea plant? What would you expect the results to be if you crossed a *TT* plant with a *Tt* plant? With a *tt* plant? What would the results be if you crossed a *Tt* plant with a *tt* plant?

The pattern works out similarly for the dihybrid cross. Like tallness, yellowness of seeds (*Y*) is dominant to greenness (*y*). When a *TTYY* (tall, yellow-seeded) plant cross-breeds with a *ttyy* (short, green-seeded) plant, they produce *TY* and *ty* gametes, respectively, which recombine during fertilization to yield *TtYy* F_1 offspring, heterozygous for both genes. When these plants are then bred with other heterozygous F_1's, their offspring are usually produced in the approximate proportions of nine tall yellow (*T_Y_*) to three tall green (*T_yy*) to three short yellow (*ttY_*) to one short green (*ttyy*), as shown in Figure 19-5. There are four phenotypes but nine genotypes. How is that possible? Why are there blank spots left in three of the preceding four genotypes?

It was the monohybrid cross alone that led Mendel to one of the most basic ideas of genetics. The results of this cross are not explainable unless the genes come in pairs and the members of each pair are separated and isolated from each other by the preparation of gametes. We now call this separation Mendel's **law of segregation.** The dihybrid cross confirmed this idea and added clear evidence for the **law of independent assortment.** This law says that different genes segregate entirely independently of each other. That is, the pea alleles for tallness and yellow seeds behave as if they are not attached or linked to each other. Each one is parceled out to the gametes and offspring without regard for the other. As an exercise, work out the phenotype and genotype ratios you should expect if the tallness and yellow-seed alleles *were* linked to each other.

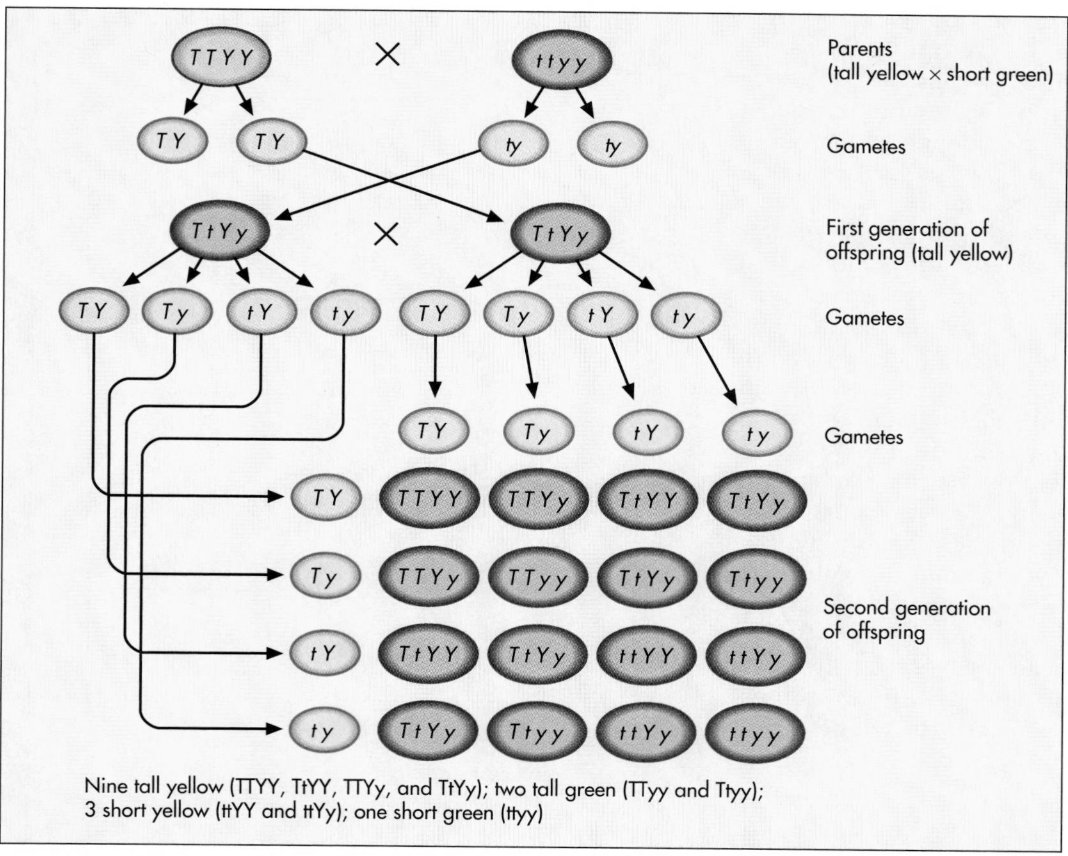

FIGURE 19-5 The same Punnett square method works for the dihybrid cross.

Eventually, other biologists realized that the laws of segregation and independent assortment described the movements of the chromosomes during meiosis. Each person inherits two haploid sets of 23 chromosomes—one set from each parent—during conception. Each chromosome of a homologous pair governs the same traits as the other member of the pair, but the chromosome inherited from the father may contain alleles for those traits different from the alleles on the homologous chromosome inherited from the mother. You learned in Chapter 5 that pairs of homologous chromosomes separate during meiosis, when a diploid cell produces the next generation's haploid gametes. Each gamete receives one copy of each homologous pair, but the chromosomes are sorted out to the gametes independently. Fertilization then restores diploidy.

Are genes chromosomes? At first, people thought they might well be. However, it was not long before genes were found that seemed to violate Mendel's laws. We now know that such genes share the same chromosome and travel together through meiosis and fertilization; they are *linked*. Mendel was fortunate in his choice of genes to study. They each occupied a different chromosome, and he encountered no linkages to confuse his studies.

Learning Focus

PRINCIPLES AND PENNIES

People often have trouble grasping just how the genes and alleles sort themselves out to yield the Mendelian 3:1 and 9:3:3:1 ratios. *You* will avoid this trouble entirely by playing the simple game that follows with a friend or roommate. If you play it as an in-class exercise, you and your fellow students should pair off, two by two.

To understand the 3:1 ratio, each player needs four "pea plants," two homozygous and two heterozygous. Prepare the homozygous plants by glueing pairs of pennies together to make one two-headed (*PP*) and one two-tailed (*pp*) coin (heads are dominant; tails are recessive). You each also need a regular head-tail coin to represent the heterozygote plants (*Pp*) that you will cross in later experiments.

The two sides of each coin represent the two genes in a diploid cell, and the three different kinds of coins give you the three possible pairs of alleles (*PP, Pp,* and *pp*). Flipping one of these coins is the equivalent of making a haploid, one-gene gamete by meiosis. (Why?)

To begin the game, you or your partner should take the *PP* coin. The other should take the *pp* coin. Both of you should flip your coins several times to generate "gametes." Each time you toss your coins, you should create "offspring" by recombining the gametes the coins generate; that is, you should write down the results of each toss as a pair of symbols (one for each player), *PP, Pp,* or *pp*.

What is the result?

How does this compare with the first stage of Mendel's monohybrid cross?

Next, you and your partner should flip your *Pp* coins and mark down the results. Do this repeatedly, 100 times or more. Count the number of times you get *PP, Pp,* and *pp*. Record your results in the square array that follows:

	P	*p*
P	*PP*	*Pp*
p	*Pp*	*pp*

Total number of tosses for each pair of coins ____________

Number of times the pairs came up:

PP __________ } Total __________

Pp __________

pp __________

Calculate the ratio of

*P*____ : *pp* ____:____

Calculate the ratio of *P_* and *pp* "offspring." What is the result?

How does this compare with the second stage of Mendel's monohybrid cross?

To understand the 9:3:3:1 ratio of the dihybrid cross, you and your partner should each prepare three more "pea plants," using nickels to give you *NN, Nn,* and *nn* gene pairs. You or your partner should then flip the *PP* and *NN* coins (to represent a *PPNN* pea plant) to generate two-gene gametes. The other should flip the *pp* and *nn* coins. Write down the results as quadruplets of symbols (*PPNN, Ppnn, ppnn,* etc.).

What is the result?

How does this compare with the first stage of Mendel's dihybrid cross?

Next the players should flip their *Pp* and *Nn* coins many times (100 times or more). Record the various combinations of *P*'s, *p*'s, *N*'s, and *n*'s in the square array that follows:

Calculate the ratio of *P_N* to *P_nn* to *ppN* to *ppnn*. What is the result?

How does this compare with the second stage of Mendel's dihybrid cross?

To understand the idea of linkage, imagine what the results would be if you welded your doubled pennies and nickels together side by side.

	PN	*Pn*	*pN*	*pn*
PN	*PPNN*	*PPNn*	*PpNN*	*PpNn*
Pn	*PPNn*	*PPNn*	*PpNn*	*Ppnn*
pN	*PpNN*	*PpNn*	*ppNN*	*ppNn*
pn	*PpNn*	*Ppnn*	*ppNn*	*ppnn*

Total number of tosses for each pair of coins ______________

Number of times the pairs came up:

	Raw score	Ratio
P__ N__		
P__ nn		
ppN __		
ppN		

Calculate the ratio of

P__N__ : P__ nn : ppN__ : ppN

________ : ________ : ________ : ________

THE MENDELIAN HUMAN

Mendel's laws do not apply only to pea plants. They also apply to every other diploid organism—other plants, animals, and even humans. To see the law of segregation in detail in humans, we need to look at a human trait, such as one of those described in Table 19-1 and illustrated in Figure 19-6. Chin fissures, ear pits, Darwin tubercles, congenital ptosis, epicanthus, camptodactyly, middigital hair, tongue rolling, PTC tasting, and *S*-methyl thioester smelling are all due to dominant alleles. Having dry, granular ear wax instead of brown, sticky ear wax is due to a recessive allele. In each case, the law of segregation shows in the offspring of heterozygous parents. However, it does not always show as a simple 3:1 ratio.

Fortunately, we can see the effects of the law of segregation even without clear ratios. Consider a person who lacks the ability to detect the smell of the *S*-methyl thioester present in urine after eating asparagus. This "nonsmeller" individual has to be homozygous for the recessive allele (*ss*). If his or her spouse has the ability to smell thioester, she or he must be either homozygous (*SS*) or heterozygous (*Ss*) for the dominant allele (*S* confers the ability to

Table 19-1

Common Mendelian Traits in Humans

TRAIT	DESCRIPTION
Chin fissure	Dimple or cleft in chin
Ear pit	Hole or pit at root of external ear
Darwin tubercle	Bump on upper ear cartilage
Congenital ptosis	Drooping organ, usually an eyelid
Epicanthus	Skin fold at edge of eyelid near nose
Camptodactyly	Crooked finger
Middigital hair	Hair on middle segment of fingers
Tongue rolling	Ability to roll tongue into a tube
Phenylthiocarbamide (PTC) tasting	Ability to taste (as bitter) PTC
S-methyl thioester smelling	Ability to detect characteristic odor in urine after eating asparagus
Ear wax	Wet, sticky versus dry, crumbly ear wax

(a)

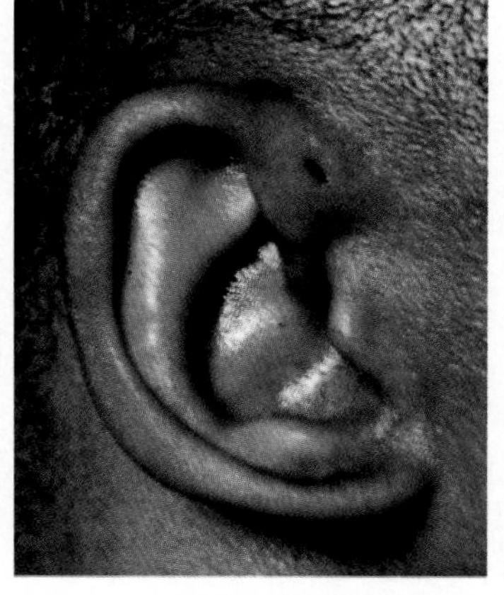
(b)

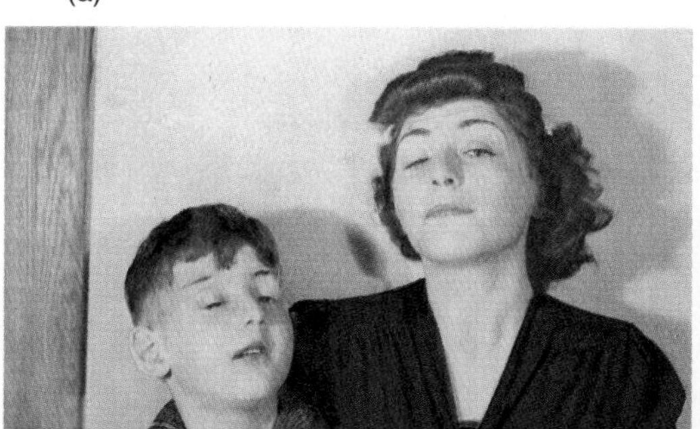

(c)

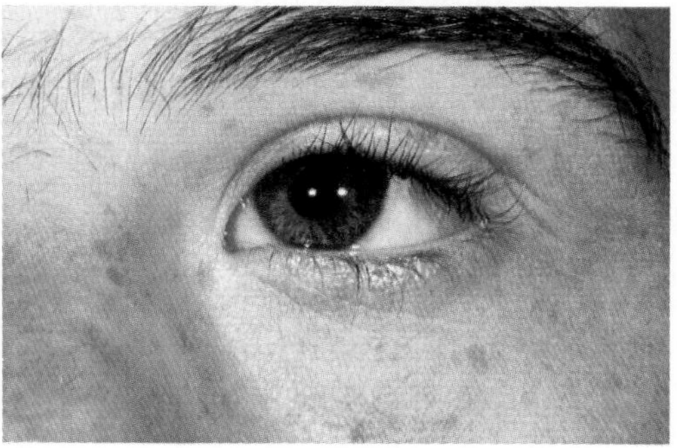
(d)

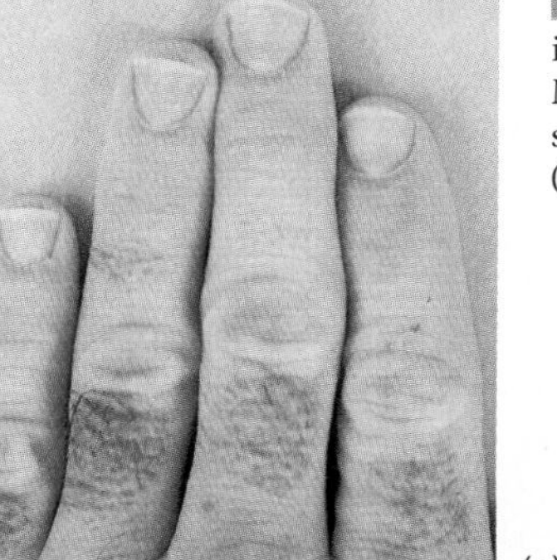
(e)

FIGURE 19-6 Many human traits are inherited according to the same rules Mendel found in pea plants: (a) chin fissure; (b) ear pit; (c) congenital ptosis; (d) epicanthus; (e) middigital hair.

smell it). In the former case, all their children will have the ability. In the latter, only half their children will be able to smell *S*-methyl thioester.

---

1. Consider two parents who can both smell *S*-methyl thioester. One is *SS*; the other is *Ss*. What can you say about their children?
2. What if both of these parents are *Ss*?

The simplest way to see the law of independent assortment is with the classic 9:3:3:1 ratio. However, human families simply do not have enough children to reveal clear 9:3:3:1 ratios. Researchers who study human genetics must therefore group the offspring of many different matings to reveal the pattern. They will find several couples whose members are each heterozygous for two different traits, such as *S*-methyl thioester smelling (*Ss*) and tongue rolling (*Rr*). They can then regard these couples' children as the offspring of the same *SsRr* × *SsRr* mating and count the various possible phenotypes. When they do this, they find the distribution of traits for all the children to be approximately nine smellers and rollers to three smellers but nonrollers to three rollers but nonsmellers for each nonsmeller and nonroller.

EXCEPTIONS TO THE PATTERNS

The Mendelian patterns are products of probability. Chromosomes and genes do not sort themselves out in meiosis and fertilization to give precise 3:1 and 9:3:3:1 ratios. Although the actual numbers never quite match them, when the numbers of offspring studied are large, the numbers can come very close. For instance, a dihybrid cross with 80 offspring might produce an actual ratio of 42:14:20:4 instead of a perfect 45:15:15:5.

When the numbers are small, they may bear little resemblance to the ratios. For instance, two *Ss* people with only two children are quite likely to have one *S*-methyl thioester smeller (*SS* or *Ss*) and a nonsmeller (*ss*) or even two nonsmellers. But the law of segregation remains clear. They can have a nonsmeller child only if they each contain a recessive allele.

Statistics thus accounts for some exceptions to the Mendelian patterns. There are many other causes for such exceptions as well. We can see one in the camptodactyly trait, which shows as a crooked finger due to a too-short tendon. The dominant allele responsible does *not* always show its effect on phenotype; the allele is thus said to have **incomplete penetrance** (an allele with *complete* penetrance always shows in the phenotype). The allele's effect can also vary in severity; it is said to have **variable expressivity.** An allele whose effect is always the same has *constant* expressivity.

We can see another cause for exceptions to the Mendelian patterns in chin fissures. This trait too has variable expressivity, for it can vary from a small dimple to a Y-shaped furrow. It also varies in penetrance. It has almost complete penetrance in males, but not in females, in whom only half of those who have the allele show the trait. It is thus said to be **sex influenced.** Common baldness is a sex-influenced trait with an additional peculiarity. The baldness allele is dominant in males, who begin to show a receding hairline about age 20, but recessive in females, who begin to lose their hair only about age 30. It too varies in penetrance and expressivity.

Additional exceptions are due to **incomplete dominance, codominance, multigenic inheritance, linkage, crossing over,** and **sex linkage.** We will now discuss each of these phenomena in turn.

Incomplete Dominance, Codominance, and Multiple Alleles

Early geneticists doubted Mendel's results because they knew of cases where parental traits did indeed seem to blend in the offspring. One such case, an example of incomplete dominance, was that of the primrose, for when red-flowered plants (*RR*) were crossed with white-flowered plants (*rr*), the offspring had pink flowers. The explanation emerged when the pink-flowered plants were crossed with each other, for their offspring showed a ratio of one red flowered to two pink flowered to one white flowered. The pink-flowered plants, just as Mendel's law of segregation predicted, had a *Rr* genotype, and their offspring were one *RR* to two *Rr* to one *rr*. The difference was caused by the lack of dominance; when both alleles were present in a plant, both affected the phenotype.

Such cases have helped to explain how incomplete dominance works. In Chapter 4, we discussed how genes control the synthesis of enzymes, which in turn make various substances necessary to the cell. In primroses, the *R* (red) allele specifies the structure of the enzyme that makes the red pigment. The *r* (white) allele specifies the same enzyme, but in a defective form. An *rr* plant cannot make the red pigment. An *Rr* plant makes only half as much enzyme and red pigment, and the flowers are only half red, or pink, as shown in Figure 19-7.

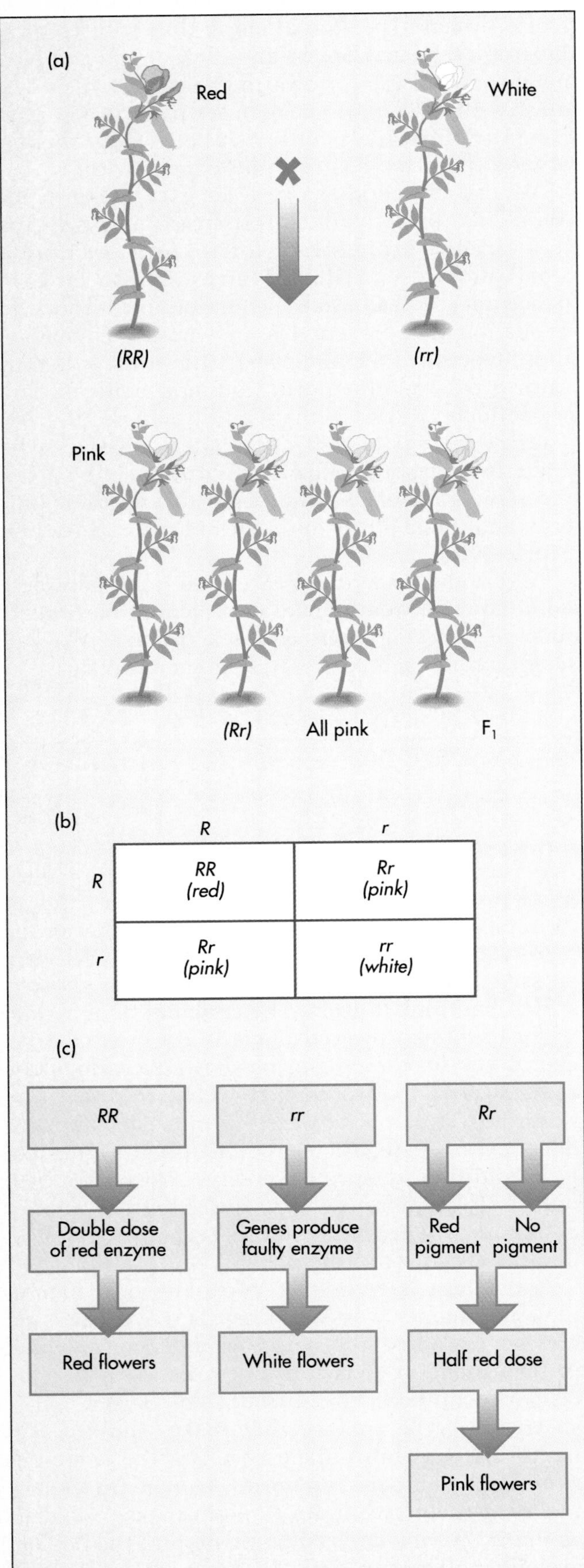

Codominance

We see true dominance when a single working gene is enough to provide an organism with all the enzyme it needs to fill its requirements for the enzyme's product. Then the homozygous dominant and the heterozygote are indistinguishable.

Codominance exists when both alleles in the heterozygote contribute features to the phenotype. One of the best examples in humans is the ABO blood types, which also illustrate that many genes may have more than two alleles.

On their surfaces red blood cells bear "identification factors" (antigens)—chains of sugar groups attached to cell membrane proteins—that let us classify blood as type A, type B, type AB, or type O. Three alleles of a single gene determine the blood type. They are I^A, I^B, and i^O. The i^O allele is recessive to the other two blood type alleles; it specifies no label or antigen on the red blood cell surfaces (the "O" really stands for zero), and only a person homozygous for this allele can have type O blood. The I^A and I^B alleles are codominant with each other, and both are dominant to the i^O allele. Thus, I^AI^A and I^Ai^O are the two possible genotypes for a person with type A blood. A person with type B blood can have either I^BI^B or I^Bi^O as the genotype. Because of codominance, a person with the I^AI^B genotype has both A and B antigens on his or her blood cells and is said to have type AB blood (see Figure 19-8).

Blood types are important to us because they affect the success of blood transfusions. The body's immune system does not normally make antibodies (see Chapter 23) against the antigens found on its own red blood cells. However, a person's plasma does contain antibodies (even though that individual has never been previously exposed to "foreign" blood cells) that will work against foreign blood cells bearing different antigens from those of the host. If a person is accidentally transfused with the wrong type of blood, these antibodies will bind to the foreign antigens on the incoming red blood cells and destroy them. Type O people make antibodies against both A and B antigens. Type A people make antibodies against B antigens. Type B

FIGURE 19-7 Incomplete dominance. (a) Crossing pure-breeding red-flowered plants (*RR*) with white-flowered plants (*rr*) produces all pink-flowered heterozygous offspring (*Rr*). (b) When pink-flowered F_1's are crossed with each other, one-fourth of their offspring have red flowers, one-half have pink flowers, and one-fourth have white flowers. (c) The *RR* flowers produce a double dose of red-pigment-producing enzymes; the *rr* flowers produce no enzyme for producing red pigment and consequently remain white, and the *Rr* produce half the amount of red pigment possessed by a full red flower and are pink.

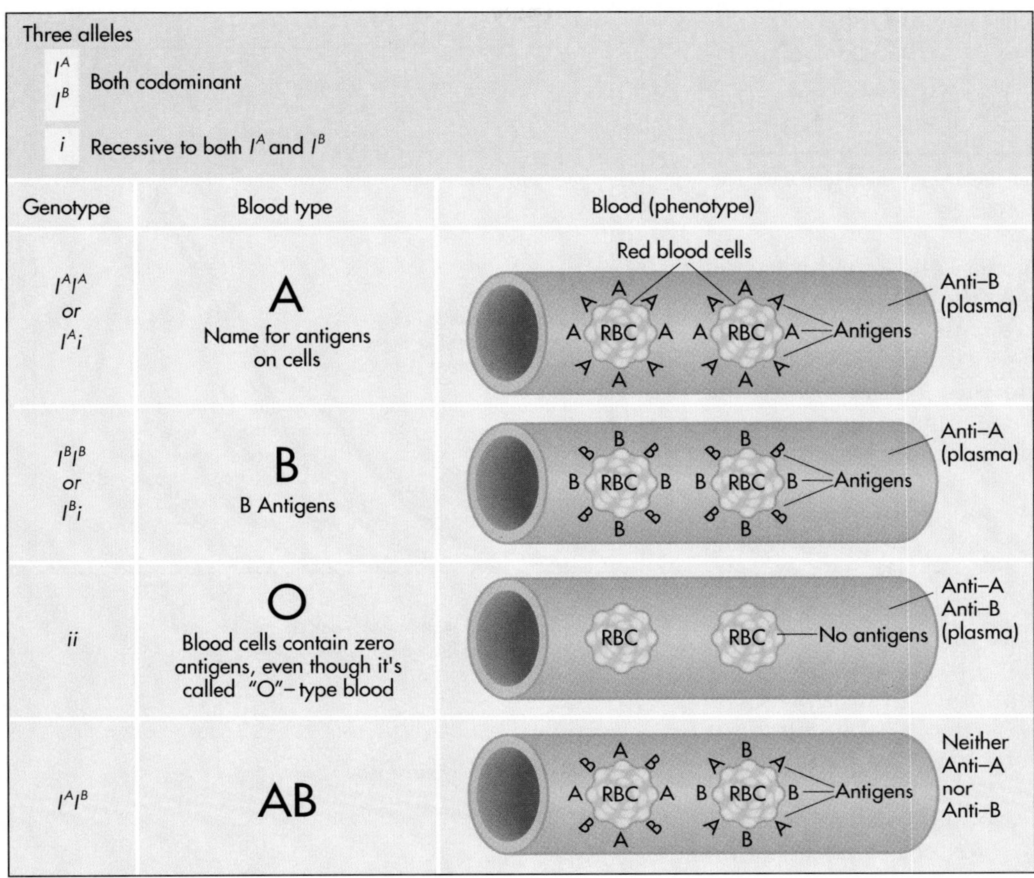

FIGURE 19-8 A, B, O, and AB blood types.

people make antibodies against A antigens. Type AB people make antibodies against neither antigen. As a result, type O people cannot receive blood of types A, B, or AB, but they can supply blood to anyone; they are "universal donors." Type AB people can give blood only to other AB's, but they can receive blood from anyone; they are "universal recipients."

Genetics of the Rh Factor in Blood

There are many other blood type systems, for the red blood cells bear many different antigens, and each system is controlled by a single gene with two or more alleles. One of the best known of these other blood type systems is the Rhesus, or Rh factor. People who are Rh^+ have the Rh antigen on their red blood cells and are either homozygous or heterozygous for the dominant allele. Those who are Rh^- do not have the Rhesus antigen on their red blood cells and are therefore said to be negative for the trait. Since the Rh^- allele is recessive, a person with negative blood must be homozygous for the recessive allele. When accidentally exposed to the Rh antigen, the white blood cells of Rh^- people make antibodies against it.

Physicians must thus take Rh type into account when matching transfusion donors and recipients. They must also take it into account when caring for a pregnant Rh^- woman married to a Rh+ man. The mother is sometimes exposed to the Rh^+ blood of her fetus during childbirth, when the placenta separates from the uterus. She can respond to this by making antibodies to the Rhesus antigens. In the condition known as *erythroblastosis fetalis,* these antibodies may attack and damage a later Rh^+ baby, for the mother passes some of her normally protective antibodies across the placenta shortly before the baby is born. Treatment used to involve giving the second Rh^+ newborn of a Rh^- mother an immediate and total blood transfusion. Today, the physician prevents the problem from arising by injecting into the Rh^- mother of a *first* Rh^+ child antibodies that destroy whatever Rh^+ cells may have reached the mother's system; this keeps the mother from becoming sensitized to those cells and producing antibodies that may attack the blood cells of her subsequent babies. If the mother has not been so treated, the physician can give her antibodies that destroy any anti-Rh antibodies she may have made in response to a previous Rh^+ baby.

Gene Interactions and Multigenic Inheritance

Traits are controlled by genes, and our discussion so far would seem to suggest that one gene controls only a single trait. However, numerous individual genes have effects on several different traits. This phenomenon, called **pleiotropy,** shows with the ABO blood types. For some reason, people with type A blood seem to have a greater susceptibility to stomach cancer than do type O people. Type O people seem to be more prone to duodenal ulcers and less prone to blood clots in the blood vessels than are people of types A, B, or AB. The reasons are not clear.

An additional factor complicating the study of human genetics is that many traits are affected by several different genes. Common baldness is a clear example, for though it is due to a single gene, it behaves as a dominant gene in males and a recessive in females. Its expression therefore depends on the way the hair follicles respond to different levels of the male sex hormone, testosterone, and thus on the genes that specify gender. Skin color, intelligence, and height provide additional examples in humans. Each of these traits is controlled by more than one gene, and their inheritance is not an either-or matter.

The difference in skin color between United States whites and blacks seems to be due to three or four genes whose alleles are codominant, so that heterozygotes have skin shades intermediate between those of homozygotes (Figure 19-9). To understand this, consider how inheritance of skin color would have to function if skin color were controlled by only three genes, each with two alleles. For each gene, one allele (*A, B,* or *C*) specifies an enzyme that makes pigment; the other allele (*a, b,* or *c*) specifies an enzyme that does not make pigment. An individual can thus have zero, one, two, three, four, five, or six pigment alleles and have any of seven shades of skin color. Someone with no pigment alleles is white; someone with six is black; those with intermediate numbers of pigment alleles have intermediate shades of brown.

The mating of a white person (*aabbcc*) with a black (*AABBCC*) produces children with a *AaBbCc* genotype and a midrange skin color (mulatto). This pattern is exactly what people have long observed and what supported the idea of "blending" inheritance that Mendel replaced with either-or genes and alleles.

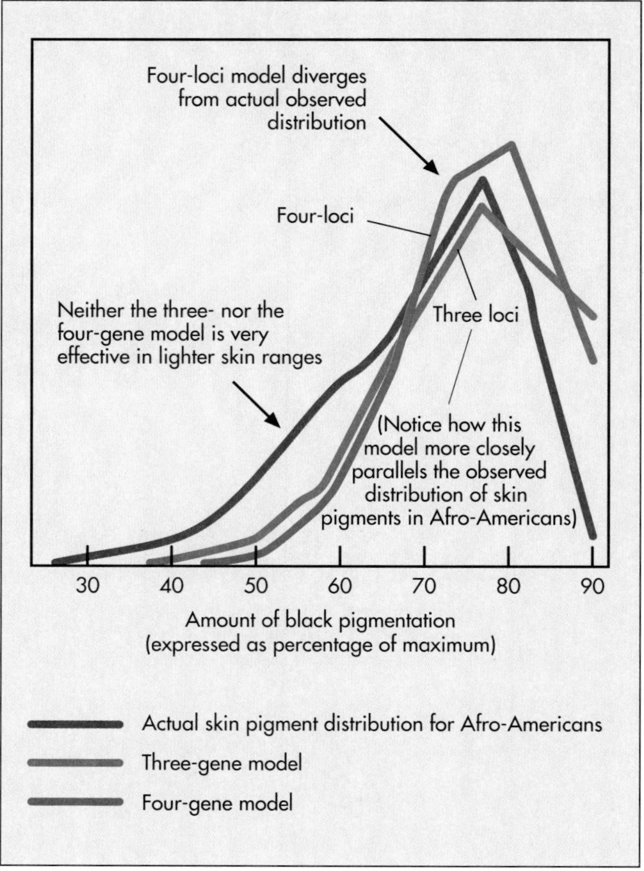

FIGURE 19-9 Distribution of observed dark skin pigments among Afro-Americans compared to theoretical models employing three and four gene loci with codominant alleles to explain the pigmentation patterns.

1. People have long believed that the children of mulattoes, known as quadroons, have skin colors lighter than their parents. However, the truth is not so simple. Work out the genotypes (the numbers of pigment alleles) that the children of two *AaBbCc* mulattoes can have. Can mulattoes have white children? Can they have fully black children?
2. We mentioned that skin color may be controlled by four genes, not three. How would this change your answers to the last question?

Intelligence and height also seem to be controlled by many genes, but nobody knows how many genes are involved. We do know that people seem to be heterozygous for many of the genes involved, so that even though children tend to resemble their parents in height and intelligence, they can vary greatly. People need not worry, as some do, that the greater birth rates of the poor and uneducated will lead to a less intelligent version of the human species. The heterozygosity of the genes for intelligence guarantees that people of below-average intelligence can produce children with a wide range of intelligence.

Perhaps more to the point, it is hardly reasonable to blame poverty and ignorance (which many people equate with low intelligence) on heredity alone. It is clear that genetics is not the only influence on human traits. Environment also plays a strong role. Exposure to the sun affects skin color. Nutrition affects both height and intelligence. Childhood mental stimulation—for example, cuddling, toys that (unlike television) encourage imagination and manipulation, reading aloud, and opportunities to play with other children—can drastically raise intelligence levels. This is why programs such as Head Start have been so successful at improving later performance in school and life. And school helps cure ignorance.

People have long argued the relative importance of heredity and environment. The truth seems to be that with height and intelligence (and other traits), the genes define a potential, a maximum. Then, the better your environment—the better your nutrition and stimulation and education—the closer you can come to reaching your personal potential. Conversely, lack of adequate nutrition, stimulation, and education may doom you to a life of poverty.

(a)

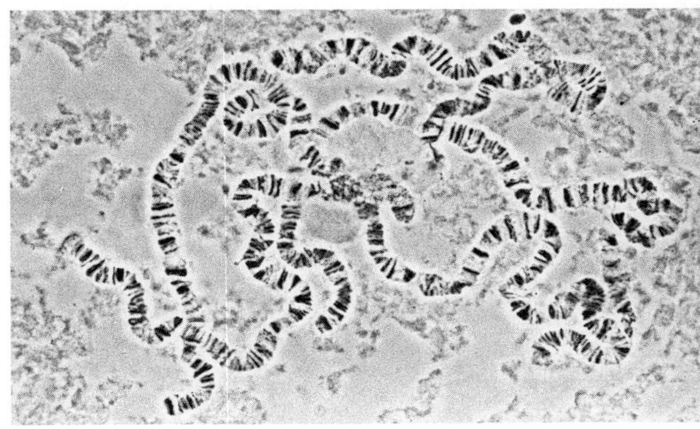

(b)

FIGURE 19-10 (a) The fruit fly, *D. melanogaster*, has only four pairs of chromosomes. (b) The polytene chromosomes of the cells of the larva's salivary glands are very large and thus easy to study.

Linkage and Crossing Over

Early in the twentieth century, Thomas Hunt Morgan found the tool that would make it possible for scientists to work out many of the details of genetics. He introduced the fruit fly, *Drosophila melanogaster*, to genetics labs for two main reasons: It is a small organism that can easily be raised in large numbers in small vials and it has a short life cycle, so that geneticists can see the results of a cross in weeks rather than in months or years (Figure 19-10).

The early work with fruit flies soon showed that these organisms have many more genes than chromosomes and that blocks of genes tended to be inherited together. They did not assort independently but traveled as groups from generation to generation. Breeding experiments revealed that there were four of these *linkage groups*. Since there were also four chromosomes, Morgan and his co-workers promptly concluded that these blocks of genes were on the same chromosomes and that independent assortment worked only for genes on separate chromosomes.

However, linkage was not perfect. Morgan soon noticed that when he crossed flies with two genes, *A* and *B*, that belonged to the same linkage group or chromosome, the linkage would break some of the time. Consider the cross *AABB* × *aabb*. The Mendelian laws predict that the result will be *AaBb* offspring in the F_1 generation which, when crossed, will give a 9:3:3:1 ratio among the offspring in the F_2 generation. In a pure, unbroken linkage the *AaBb* × *AaBb* cross would produce a 3:1 ratio (3 *A_B_*:1 *aabb*), in which the *A* and *B* genes behave as if they were a single gene. There would be no *A_bb* or *aaB_* offspring at all.

Morgan observed that some small percentage of the offspring were in fact *A_bb* and *aaB_*. It was as if the linkage groups were breaking and reforming to make new combinations of alleles. The answer emerged when careful microscope work revealed that during the first meiotic prophase, the four chromatids from each pair of homologous chromosomes gathered together in a bundle, or **tetrad.** In this bundle, neighboring chromatids from different members of the chromosome pair would break and rejoin; the new connections sometimes resulted in the exchange of chromatid segments between the different chromosomes. This exchange of segments is now known as crossing over (Figure 19-11). It can

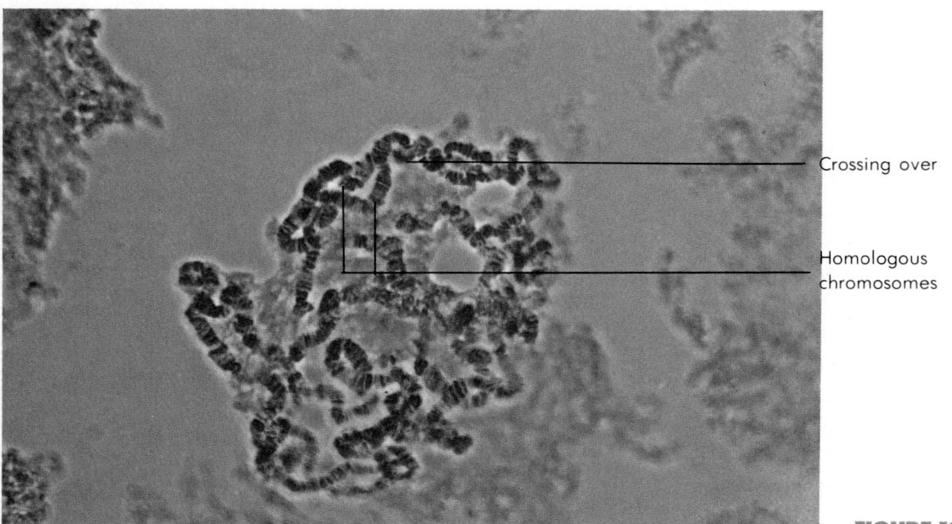

FIGURE 19-11 Crossing over in *Drosophila*.

rearrange sequences of alleles, producing *aB* and *Ab* chromatids from *AB* and *ab* chromatids. It is thus one important way in which sexual reproduction increases genetic variability between generations. When sexual reproduction combines the chromosomes and alleles of two parents in a new individual, crossing over produces new combinations of alleles on single chromosomes.

Linkage is clearly a violation of the Mendelian laws in one sense. All genes do not assort independently. But chromosomes do. Crossing over seems to be a more fundamental violation. But it too broadens our understanding of what is going on in the cell's nucleus. Mendel saw only a small part of the truth. He was right, and his choice of plant subjects and which traits to study was fortunate, but he was not complete. His successors, like those of most scientists, greatly extended his vision.

Chromosome Mapping

In the process, Mendel's successors found a useful tool. Crossing over occurs between different pairs of genes with different frequencies. It has turned out that the crossover frequency depends on how far apart two genes are on a chromosome. Measurements of crossover frequency in terms of "crossover units" (one crossover unit is 1 percent of crossover frequency) have allowed geneticists to arrange the genes on a chromosome in sequence and draw chromosome maps (see Figure 19-12). A major international effort—the Human Genome Project—is currently underway both to prepare a detailed map and to determine the sequence of nucleotides in all the genes on the human chromosomes.

Sex Linkage

So far, we have used "linkage" to refer to the way genes are physically associated with each other on chromosomes. We encounter another meaning for the term in the phenomenon of sex linkage, which refers to the way some traits seem to show up only or mostly in males. These traits include hemophilia, color blindness, and Lesch-Nyhan syndrome (see Table 19-2).

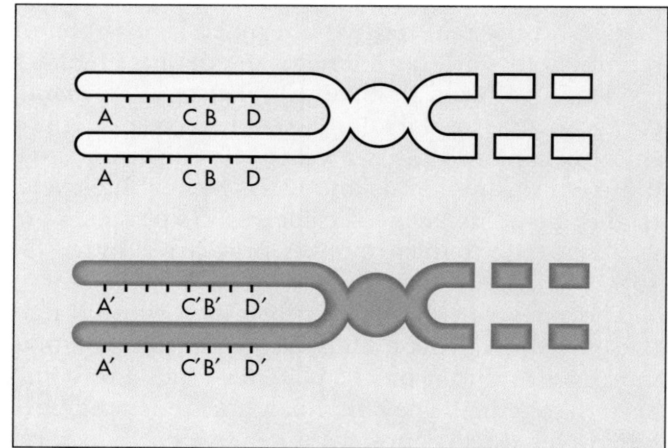

FIGURE 19-12 We can find the relative positions of genes along the length of a chromosome by determining how frequently crossing over occurs between them when homologous chromosomes pair in meiotic prophase. The greater the distance between any two gene loci, the more frequently crossing over is likely to occur. Here, such measurements have shown that genes A and B are five crossover units apart; C and D are three units apart; B and C are one unit apart; and A and C are four units apart. From this, we can deduce the unique sequence: A (4) C (1) B (2) D.

Table 19-2

Clinical Summary: Selected Genetic Disorders in Humans

CONDITION	DESCRIPTION
Albinism	Lack of melanin in the skin, hair, and eyes (recessive)
Chondrodystrophic dwarfism or achondroplasia	Premature closure of areas of long bone growth; produces dwarfs with short, curved limbs and normal-sized trunk (dominant)
Huntington's chorea	Progressive deterioration of the central nervous system, affecting both mental and motor abilities; onset of symptoms is often delayed to after age 35 (dominant)
Sickle cell anemia	Defective gene causing hemoglobin to distort (or sickle) and stiffen red blood cells when oxygen tension is low (codominant)
Tay-Sachs disease	Progressive degeneration of mental abilities starting at about 5 months; visual impairment and physical deterioration follow, finally ending in death by about age 4 (recessive)
Cystic fibrosis	Production of glycoproteins that interfere with normal salt metabolism and cause thick mucus to clog lungs, pancreas, and liver; pneumonia results, fibrous growths develop in the pancreas, and digestion and bile production problems commonly occur (recessive)
Brachydactyly	Extremely short fingers; middle joint of each finger reduced so hand seems all thumbs (codominant; causes lethal skeletal deformities in homozygotes)
Marfan syndrome	Production of abnormal connective tissues leading to various symptoms, including spider fingers and toes, pigeon chest, overly flexible skeleton, scoliosis of the spine, and cardiovascular and eye defects (dominant)
Ehlers-Danlos syndrome	Poorly structured collagen leading to hyperplastic skin and blood vessel walls and hypermobility of the skeletal joints (dominant)
Klinefelter's syndrome	Receipt of XXY chromosomes at conception; children appear normal, but adolescents show undeveloped testes, unusual height, feminine muscular development, some breast development, high voice, sterility, and often mental retardation; sexual behavior can be normal (1 out of 400 male births is Klinefelter's)
Turner's syndrome	Receipt of XO chromosomes at conception (X-bearing sperm fertilizes an egg that lacks an X chromosome); children appear normal until puberty, but skin flaps give neck a webbed appearance; female organs fail to develop, there is no breast development or menstruation (1 out of 2500 female births is Turner's)
XYY syndrome	Produced when a YY sperm fertilizes an egg with a single X chromosome; males carrying the extra Y chromosome are likely to be above average in height and tend to have lower than average intelligence; questionable studies suggest that aggressive behavior in some males may be caused by the extra Y chromosome; more studies needed (1 out of 1000 male births is XYY)
Color blindness (red-green)	Defective gene on the X chromosome leading to a defect in the green-sensing cones, which makes it impossible to distinguish between green and red light; more likely to affect males than females
Hemophilia	Bleeder's disease that affects about 1 male in 10,000; occurs when defective X chromosome gene fails to produce one of the needed clotting factors, substance VIII [or antihemo-

(Continued on p. 442)

CONDITION	DESCRIPTION
	philic globulin (AHG)] without which blood clots only with difficulty even after minor injuries; much of the bleeding is internal, caused by ruptured capillaries
Lesch-Nyhan disease	Caused by a gene that produces errors in purine metabolism; afflicted children produce excess amounts of uric acid causing brain damage, spastic muscles, and tendencies for self-mutilations, which includes biting their lips and hands severely, producing serious injury (caused by a recessive X-linked gene)
Down syndrome (Trisomy-21) (mongolism)	Caused by nondisjunction in the no. 21 pair of homologous chromosomes; effects include a skin fold on the upper eyelid, mental retardation, short, stocky body, round face, and difficulty in speaking; Down's victims are also likely to have heart defects and suffer from frequent respiratory infections (1 Down's baby in 800 births overall: 1 in 1500 for mothers in their 20s; 1 in 70 for mothers in their 40s; 1 in 25 births for mothers over 45)

Humans have 23 pairs of chromosomes. Twenty-two of these pairs are **autosomes.** One pair is the **sex chromosomes,** and it differs between males and females. Normal females have two large X chromosomes. Normal males have one X chromosome and one much smaller Y chromosome. The Y chromosome carries very few genes, but its presence is essential for maleness. Without it, even a person who has only one X chromosome is phenotypically female.

We can tell a person's sex from a single cell. One method is to culture white blood cells until they enter the metaphase stage of mitosis, soak them in low-salt (hypotonic) solution to swell them, and spread their chromosomes out, photograph the results, and finally cut out and arrange the chromosomes. The resulting **karyotype** (Figure 19-13) reveals the X and Y chromosomes as well as all the rest.

Another method is to stain cells obtained, for instance, by scraping the lining of the cheek and examine their nuclei. Normal females show a small knot of inactivated chromosomal material near the edge of the nucleus. This knot is a **Barr body** (Figure 19-14). The cells of a female use only one of their pair of X chromosomes to guide the production of enzymes; they turn off the other X chromosome. (Individual cells are thus essentially haploid for X chromosome genes and consequently behave as if they are homozygous for each allele carried on the active X chromosome; but since the cells "choose" which X chromosome to turn off at random, the body as a whole remains diploid and can show intermediate, heterozygous phenotypes.) Male cells, which have only one X chromosome, do not inactivate it and show no Barr bodies. They are haploid and homozygous (the proper term is *hemizygous*) for X alleles.

Barr bodies can reveal a number of sex chromosome anomalies (Figure 19-14). A woman who shows no Barr bodies has to have only one X chromosome; her femaleness shows that she has no Y chromosome. Such XO people are said to have Turner's syndrome; they are short, fail to mature sexually, and have some intellectual deficits. A man with a Barr body has to be XXY (Klinefelter's syndrome). He is always sterile, often tall, and sometimes mentally retarded; he may show some female secondary sexual characteristics such as breast development. Rarely, people are born with more than three sex chromosomes (up to XXXXXXY); their physical and mental defects grow worse as the number of sex chromosomes increases.

Sex linkage exists because normal males have only one X chromosome and therefore cannot be heterozygous for X chromosome genes. Consider a woman, heterozygous for some X chromosome gene (*Aa*), who mates with a man whose X chromosome carries the dominant allele (*A*). Their daughters will get the man's X chromosome and one of the woman's X chromosomes; half will be *AA* and half will be *Aa*. Their sons will get the father's Y chromosome and one of the mother's X chromosomes; they must be half *A* and half *a*. Therefore, the recessive trait will be seen only in the sons.

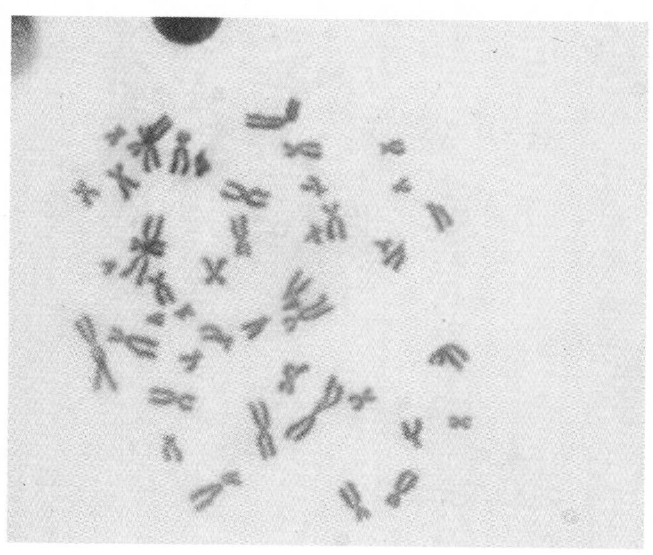

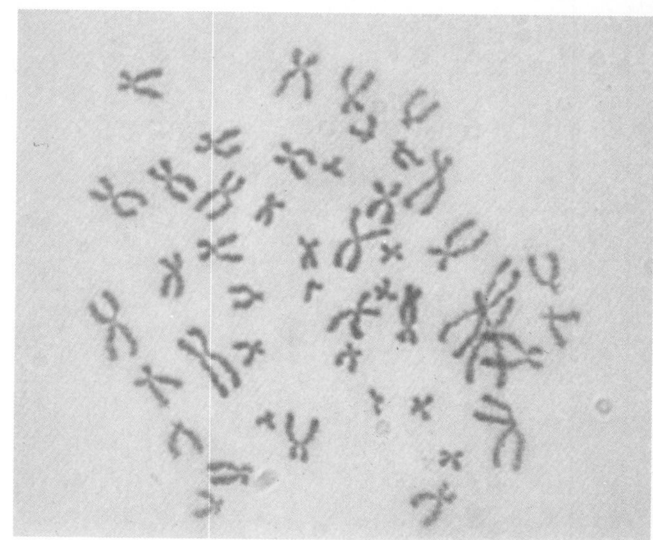

(a)

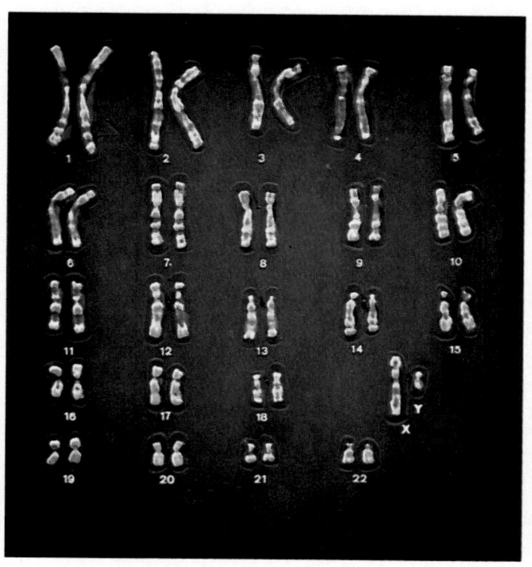

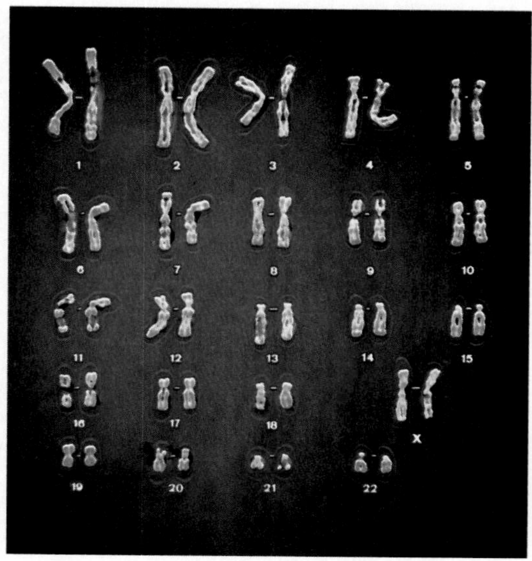

(b)

FIGURE 19-13 (a) Two full metaphase sets of stained chromosomes taken from white blood cells of a male and a female. (b) The photographs of each karyotype are then cut up and arranged in chromosome groups for further study. The female and the male both have 22 pairs of autosomes and one pair of sex chromosomes; the female has XX, the male has XY.

Since half the sons and none of the daughters show the recessive trait, the trait is said to be sex linked. The heterozygous mother is called a **carrier** of the recessive trait.

---

1. Consider a cross between an *Aa* woman and an *a*-bearing man (the *A* gene is on the X chromosome). What will be the genotype and phenotype ratios of their daughters? Of their sons?
2. Consider an *aa* × *A* cross. What will the ratios be in this case for daughters? For sons?

A second kind of sex linkage shows in the inheritance of genes on the Y chromosome, but it is relatively less important. Only a few genes are known to lie on the Y chromosome. One Y chromosome gene is for the hairy-ears trait, which occurs only in males. Another gene is for the H-Y antigen found in fetal males. Still another gene discovered recently lies near the end of the short arm of the "y" chromosome and may specify the testis determining factor.

Hemophilia is a hereditary disorder of the blood-clotting mechanism. Affected individuals cannot make one of the factors essential for normal clot-

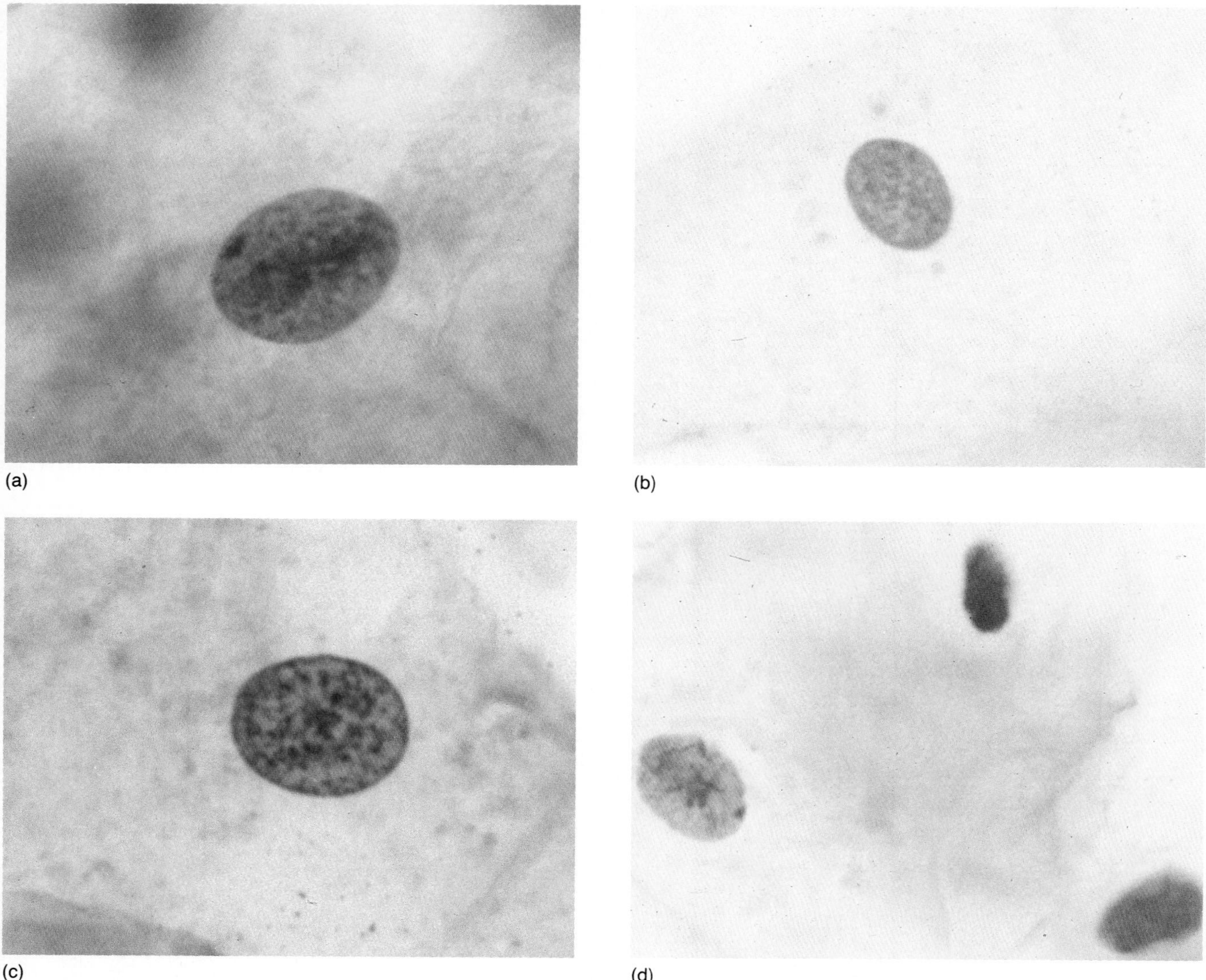

FIGURE 19-14 In the 1940s, Barr and Bertram noted that the nuclei of female somatic cells (e.g., white blood cells or epithelial cells) stain differently from male cells. A pigmented area, the Barr body, is found at the edge of the nucleus. There is one Barr body fewer than there are X chromosomes in the cell. (a) Therefore, a cell from a normal female (XX) has a single Barr body. (b) A female suffering from Turner's syndrome (one X) and (c) a male (one X) have no Barr body. (d) Cells from an abnormal male (XXY) with Klinefelter's syndrome also have a single Barr body. Note that embryos with extra sex chromosomes are formed when the sex chromosomes fail to separate properly in meiosis.

ting. Wounded, they continue to bleed. Simple operations—and even bruises and nicks—can be fatal, and they often were, before physicians learned to provide the necessary clotting factors as injections. With care, male hemophiliacs can survive long enough to breed. They can even produce hemophilic daughters, though the hemophilia allele is rare enough to make the meeting and mating of an affected man and a carrier woman very unlikely. However, hemophilic daughters would surely die at puberty, with their first menstruation, unless they were also medically supplied with the clotting factor.

Hemophilia is one of the best known sex-linked traits both because it is so dramatic and because it affected a major royal family, that of Queen Victoria of England. Study of the royal family chart or pedigree, in fact, revealed for the first time the pattern of sex linkage. This pattern is strikingly clear (see Figure 19-15, p. 447), and it would show as clearly in any other affected family's pedigree.

Red-green color blindness, the most common

Learning Focus

PEDIGREES AND HEREDITY

To a human geneticist, a *pedigree* is a chart that shows you as well as your brothers, sisters, cousins, parents, aunts, uncles, and ancestors as far into the past as possible. It also shows who has or had some trait, such as hemophilia or color blindness. It therefore reveals how that trait is inherited and can help you figure the odds as to whether your children will have the trait.

Pedigrees use several standard symbols, as indicated in the illustration that follows. Circles stand for females, squares for males. An open circle or square indicates someone who lacks the trait in question. A solidly filled-in circle or square indicates someone who has it. A partially shaded circle or square (or one with a large dot in the center) indicates a heterozygous carrier (for recessive traits).

A horizontal line linking a circle and a square marks a marriage (or mating) (see the family tree that follows). A vertical line connects the marriage link to the children of that marriage; the verticals for several children are connected by another horizontal, as in the figure. The symbols of the pedigree are either numbered or named to help us discuss them. Note that it is possible to fill in a pedigree by working backward; if we know that one child of a couple displays a recessive trait but the parents do not, we also know that both parents must be heterozygotes.

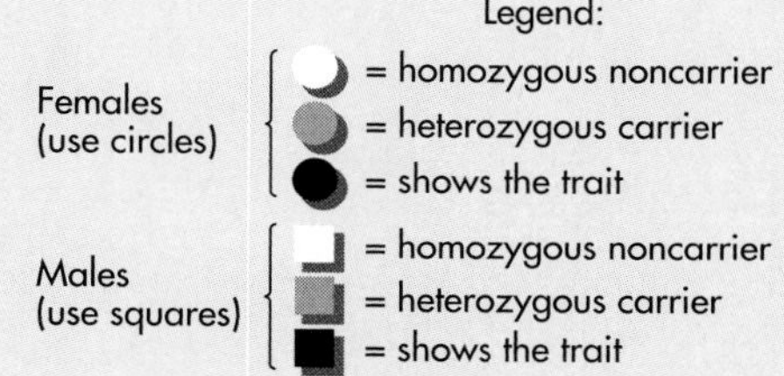

1(a) Fill in the pedigree chart showing the inheritance of a dominant allele by individual number 6; individuals 3, 4, and 5 do not have the trait. Fill in the chart as it might look.

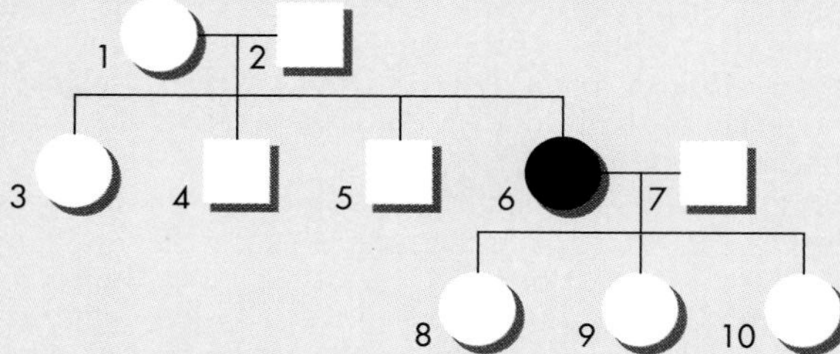

1(b) How would the pedigree for the family in 1(a) change if both individuals 1 and 2 were heterozygotes?

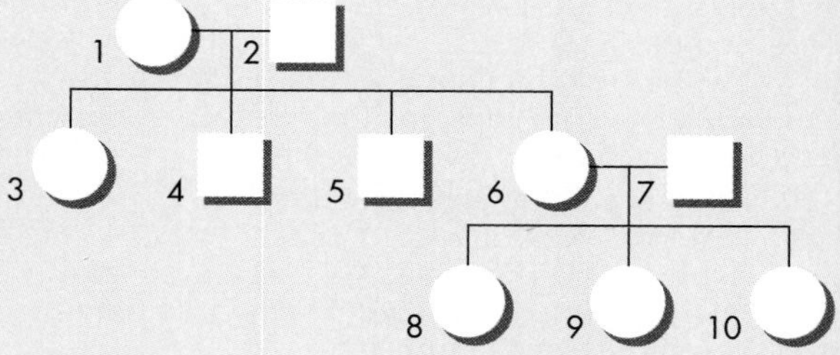

2(a) Inheritance of a dominant allele; individual 1 is homozygous. Fill in the rest of the chart as it might look.

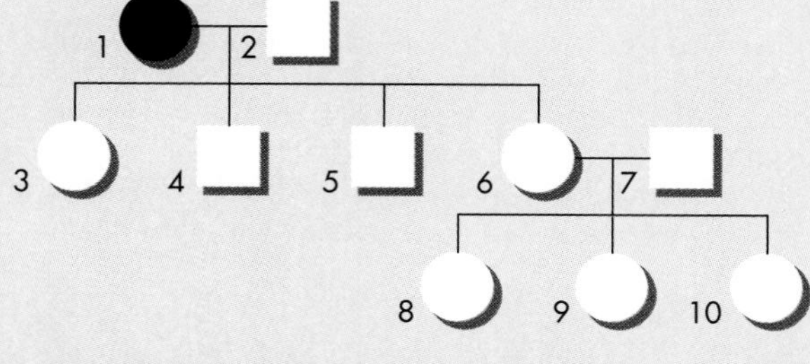

2(b) How would 2(a) change if individual 1 was a heterozygote?

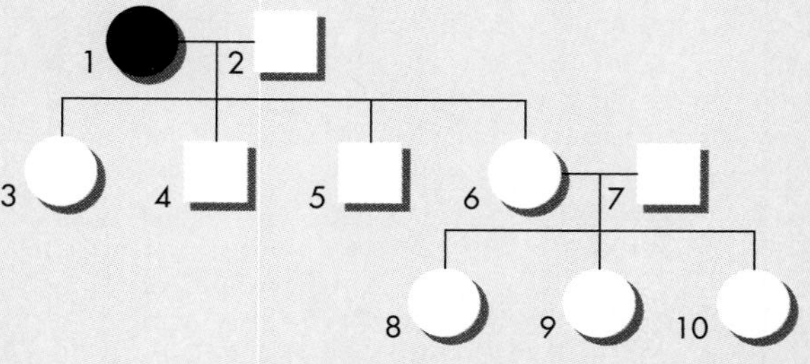

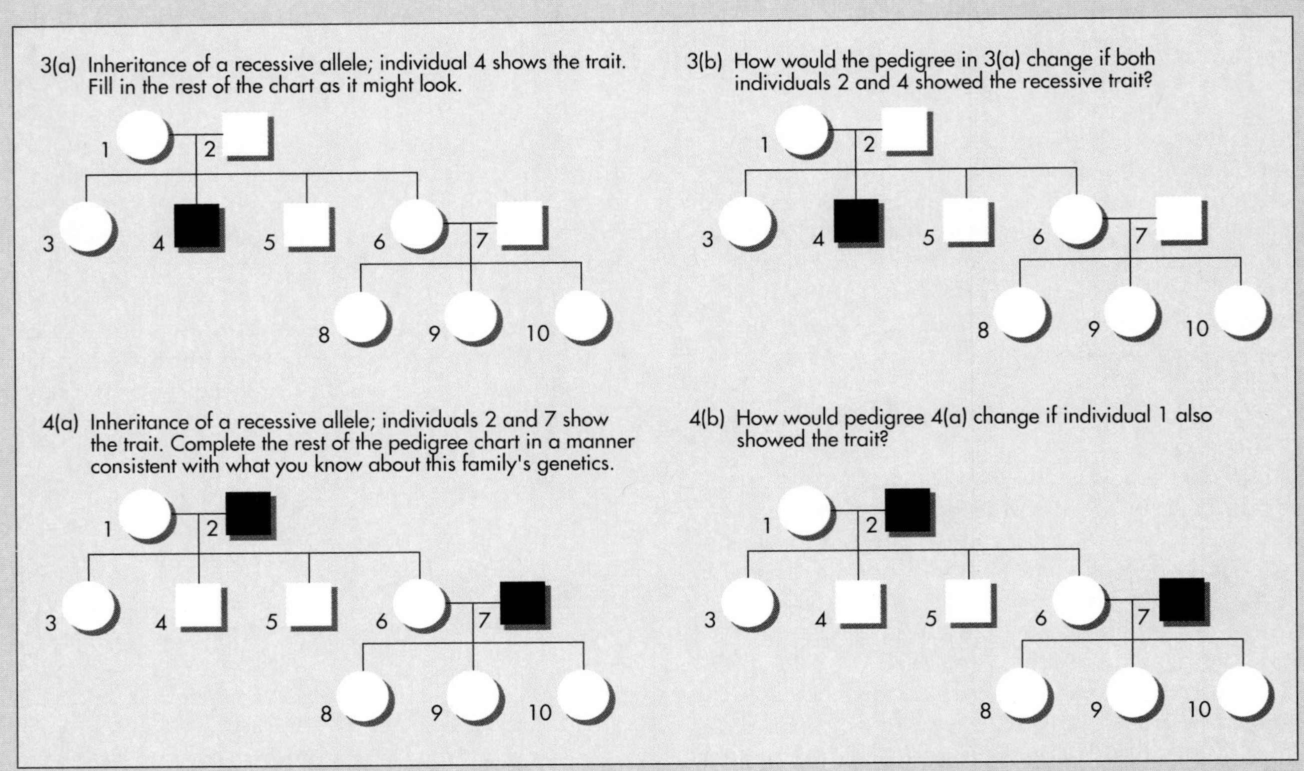

ANSWERS

1(a) Since the trait is dominant, we know that at least one of the parents, individuals 1 or 2, also showed the trait. If the trait is rare, then the other parent had to be homozygous for the recessive gene and does not show the trait. Since individual 6 has the trait, then she is heterozygous and one or two of her children would probably also have the trait.

1(b) Both the parents, individuals 1 and 2, would have the trait. You would also expect three of their four offspring (individuals 3, 4, 5, and 6) to show the trait. One or two of individual 6's children would also probably have the trait.

2(a) Since individual 1 is homozygous for the dominant trait, then all her offspring (3, 4, 5, 6) will also show the trait. It's confusing, because if they are all showing the trait, then their symbols should be darkest. They technically all are also heterozygotes, but because they are actually showing the dominant trait they are not simply carrying the unexpected gene as if they were heterozygous carriers of a recessive trait. Did you get all that?

2(b) If individual 1 was a heterozygous carrier of a dominant gene, then about half her children (3, 4, 5, 6) would have it. If individual 6 happened to inherit the dominant trait, then half of her children (8, 9, 10) would also show it. However, if individual 6 does not show the trait, then none of her children will have it.

3(a) If individual 4 shows a recessive trait, then he had to be homozygous. That means that both his parents (1 and 2) are heterozygous carriers. Mathematically, you would also expect two of individual 4's three siblings to be heterozygous carriers, but you have no concrete data to suggest which ones might be the carriers.

3(b) The father (2) should be darkened to show he has the trait and is homozygous recessive. The mother (1) is still a heterozygous carrier. That means that mathematically you would expect the children (3, 4, 5, 6) also to show the trait by being homozygous recessives. Even if individual 6 was one of them, assuming the trait is rare, chances are her spouse (7) does not carry the gene, and it is unlikely that any of their children will have it.

4(a) You know that individuals 2 and 7 are both homozygous and show the recessive trait. That means that individuals 3, 4, and 5 are all heterozygotes. So mathematically you would expect one or two of individual 6's children (about half) also to show the trait.

4(b) Everyone in the chart would show the trait. Why?

Can you think of any other possible situations that were not covered by our answers?

(a)

FIGURE 19-15 Partial pedigree of hemophilia A, traced back through the family of the Tsar of Russia to Queen Victoria.

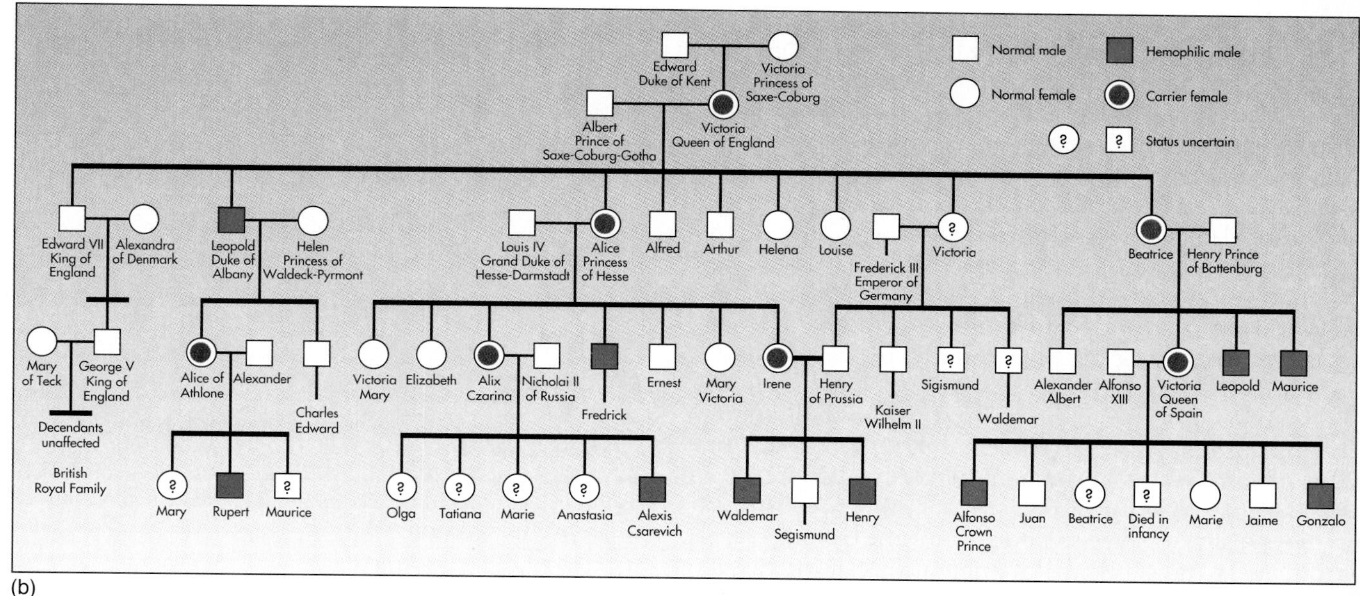

(b)

type, follows the same pattern as hemophilia. Unlike hemophilia, it is not fatal (although it does make spotting traffic signals more difficult); however, like hemophilia, color blindness is due to a defective allele. Males show the trait when they inherit a single copy of the defective allele from their carrier mother. Females can show the trait too, but they must inherit two copies of the allele, one from a carrier mother and one from a color-blind father. Although affected females are rarer than affected males, they do exist.

At present, *Lesch-Nyhan syndrome* is even more likely to be fatal than hemophilia, for we do not know how to treat it. It is due to a defect in the X chromosome gene for the enzyme HGPRT. Males with the defective allele cannot make the enzyme and accumulate uric acid. The symptoms include gout, arthritis, and kidney stones. Central nervous system symptoms include lack of coordination and self-mutilation. Victims usually die before age 5. Since very few reach breeding age—and none apparently breed successfully—all cases result when a carrier mother passes an affected X chromosome to her son. It seems impossible, not just unlikely, for the disorder to affect females.

GENETIC DISEASE

There are many more hereditary diseases than we have mentioned so far. Most are rare, but some occur more frequently than others. All cause a great deal of suffering, in part because we live in an age when nongenetic diseases have become largely treatable while genetic disorders persist in defying our faith in medicine. Fortunately, the day is not far off when we may be able to repair some defective genes.

Some genetic disorders are relatively common. *Familial hypercholesterolemia* is due to a dominant allele for which one person in 500 is heterozygous. Heterozygotes have higher than normal blood levels of cholesterol, develop atherosclerosis early in life, suffer a much higher incidence of heart disease, and develop yellowish nodules in their tendons, especially in the hands. Homozygotes are more severely affected than heterozygotes; they often die of heart attacks before age 20.

Most genetic disorders are rarer than hypercholesterolemia. *Huntington's chorea,* due to another dominant allele, is a degeneration of the central nervous system that strikes in middle age. *Familial emphysema,* due to a recessive inability to make the enzyme alpha-1-antitrypsin, is a breakdown of the lung alveoli, leading to difficulty in obtaining sufficient oxygen; it can drastically shorten the life span. *Cystic fibrosis* is a recessive condition marked by secretion of a thick mucus that blocks the ducts of the pancreas and other glands, malnutrition due to incomplete digestion of food, and cirrhosis of the liver. *Tay-Sachs disease,* also due to a recessive allele, results when a victim's cells are unable to make the enzyme that breaks down a substance (ganglioside GM_2) in brain cells. The substance accumulates, leading to blindness, seizures, loss of sensory and motor abilities, and death by age five.

From where do all these curses upon the human condition come? They result from random changes or **mutations** in the genetic material and hence in the machinery of human metabolism. In fact, genetic disorders are often called "inborn errors of metabolism." The responsible mutations are changes in the sequence of nucleotides in the deoxyribonucleic acid (DNA), as we discussed in Chapter 4. They result in changes in the amino acid sequences of proteins such as enzymes. A single changed nucleotide, which can result in a single changed amino acid, can have severe effects.

One of the best understood examples is *sickle cell anemia.* This condition and all its symptoms result from one small difference in the hemoglobin molecule. Sickle cell hemoglobin differs from normal hemoglobin in just one amino acid because of a change in one nucleotide in the hemoglobin gene. As a result, the hemoglobin tends to crystallize when exposed to low levels of oxygen like those in the tissues during exercise or at high altitudes. Cells containing crystallized hemoglobin take on a crescent or sickle shape, as shown in Figure 19-16. They then are less capable of carrying oxygen and can get stuck in capillary beds, blocking the blood supply to vital tissues in the brain, lungs, and elsewhere. The resulting symptoms include rapid tiring, retarded development in children, fever, pain, and, in pregnant women, miscarriages.

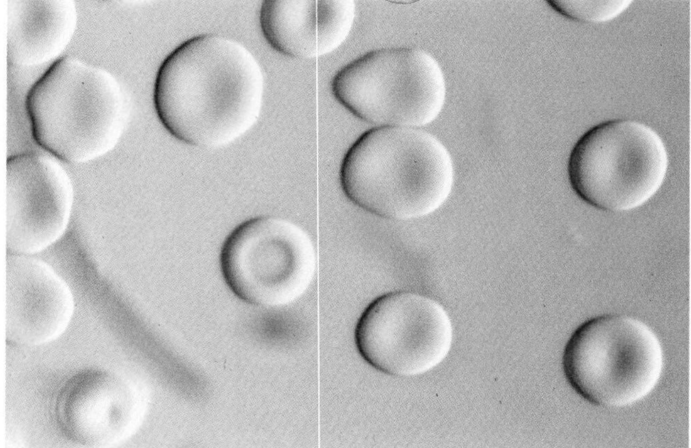

(a)

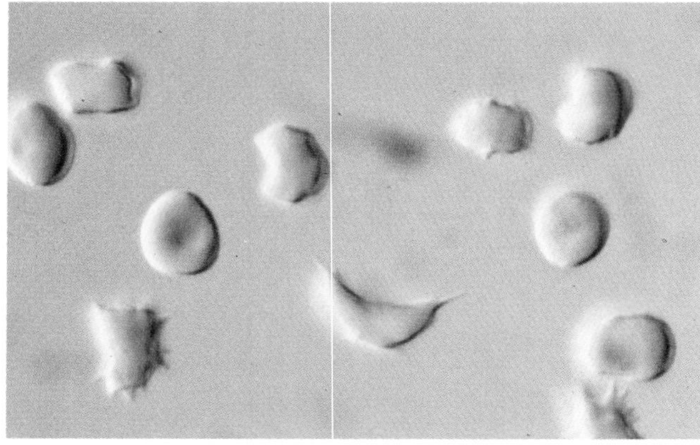

(b)

FIGURE 19-16 (a) With an adequate oxygen level, human erythrocytes assume their normal shapes as biconcave disks. (b) In people carrying the sickling trait (for sickle cell anemia), the erythrocytes twist into crescent shapes when the oxygen level drops below a certain level.

Sickle cell heterozygotes are said to show the "sickle cell trait." Heterozygous carriers for the sickle-cell allele do *not* have trouble with the low-oxygen tensions resulting from vigorous exercise or high altitudes. They show no symptoms except when their oxygen levels drop severely. Homozygotes are said to have "sickle cell disease." Even mild exercise can be enough to provoke symptoms in them.

The sickle cell allele is rare in whites. Among U.S. blacks, one person in 10 is heterozygous for the allele, and one child in 400 has the disease. It is so common because, despite its damaging effects, it also confers a benefit. The ancestors of many U.S. blacks came from parts of Africa where malaria is common, and the sickle cell allele is most common in just such areas, where one person in six is a heterozygote. These heterozygotes are more resistant to malaria than people with only the normal allele, for their red blood cells respond to invasion by the malaria parasite (a protozoan) by sickling and being destroyed. Heterozygotes who mate with heterozygotes may lose one child in four to malaria and one to sickle cell disease, but half their children will also be heterozygotes. People with normal blood easily fall prey to malaria and die. Therefore, carriers of the sickle cell trait breed more successfully than normal people in malaria-ridden areas and are numerous in those populations.

The allele is still carried by people moving out of a malarial region and it is then no longer an adaptive trait. However, it does become less common as these people interbreed with other people who lack the allele. The frequency of the sickle cell allele is thus less in the United States than in malarial regions of Africa. It also helps us see why "outbreeding"—mating with unrelated people—is generally beneficial. Unrelated people are not likely to carry the same recessive alleles, and their children are unlikely to show recessive, detrimental traits. The children of intergroup crosses may be much healthier than those of in-group crosses. This phenomenon, known as **hybrid vigor,** is well known in animals and plants.

The opposite of outbreeding, mating with related people, is **inbreeding.** Since related persons are likely to carry the same recessive alleles, their children are likely to be homozygous for some of these alleles, some of which cause severe disorders. Incest, mating with family members, is an extreme example of inbreeding, and it is most likely to produce genetically defective children. Many people believe that the human species' awareness of this hazard accounts for an almost universal taboo on incest.

Chromosomal Mutations

Not all mutations involve changes in single nucleotides. Some involve rearrangements of the chromosomes. Chemicals, radiation, and even viruses can cause pieces of chromosomes to break loose and reattach in new positions or to other chromosomes. When a piece is lost, the change is a **deletion.** When a piece turns around (switching ends) before it reattaches, the result is an **inversion.** When a piece attaches to a different chromosome, the result is a **translocation.** Some translocations have been found to be associated with certain cancers (Figure 19-17), perhaps because they upset the relations of

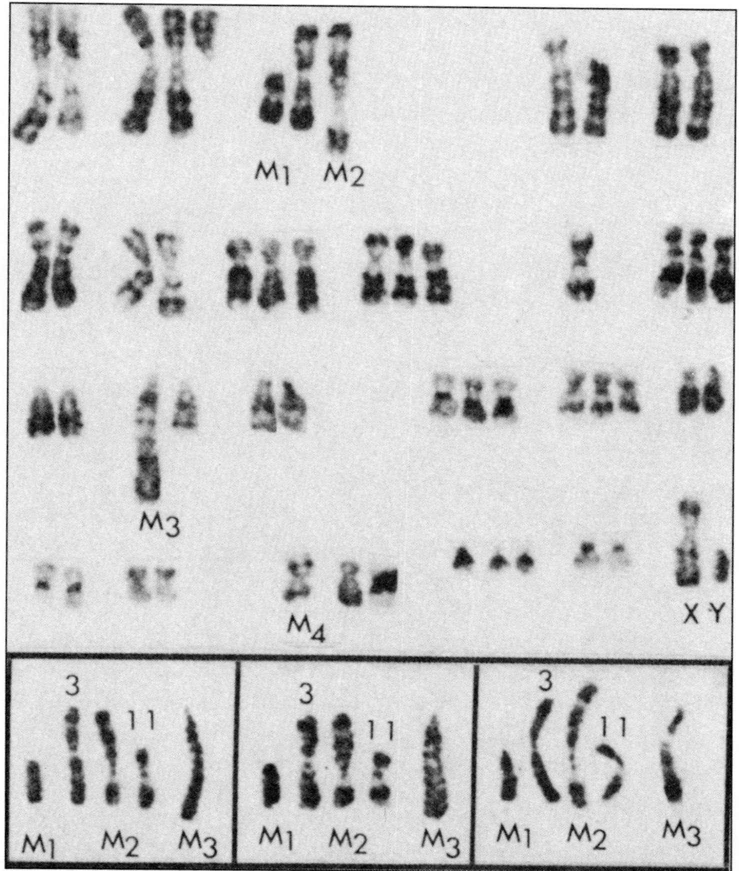

FIGURE 19-17 Cancer cells often show chromosomal abnormalities. In the karyotype seen here, kidney cancer cells have one normal number 3 and one normal number 11 chromosome. However, the other members of both pairs, 3 and 11, have undergone translocations in which a portion of number 3 has transferred to number 11. Some think such genetic rearrangements may precipitate the cell changes that lead to cancer.

certain genes with other genes that serve to regulate them.

Duplications of chromosome pieces as small as a few nucleotides or as large as whole chromosomes can also happen. Such changes appear to provide the raw material for the development of new genes, for they supply surplus DNA that can then accumulate mutations until it becomes a useful gene or genes. However, whole extra chromosomes are also associated with several serious disorders. Failures of meiosis can produce children with extra sex chromosomes (see the preceding) or others. An extra chromosome number 21 is the cause of Down syndrome, or mongolism, marked by mental retardation, delayed skeletal maturation, poor muscle tone, and distinctive facial features. Children with this syndrome often have heart defects and other physical difficulties, and their lives tend to be short (Figure 19-18). Children with extra chromosome numbers 18 (Edwards syndrome) and 13 (Patau syndrome) have even worse symptoms and are unlikely to survive for more than a few months.

Controlling Genetic Disorders

Not much can be done to treat most genetic disorders. Deformities can be corrected surgically. Drugs are being sought, with at least a little success, to control the symptoms of sickle cell anemia. Down syndrome can be helped by surgery and by care.

Unfortunately, there are not yet any cures. So far, treatment is most effective when it is not treatment, but control or prevention—that is, being aware of genetic disorders in the family history and using tests to tell whether one carries alleles for genetic disorders. Such tests are available for many disorders, and *genetic counselors* work in many hospitals and clinics to help people understand what it means to be heterozygous for a recessive defect and then to choose whether or not to take the chance of having an afflicted child. For many disorders, amniocentesis and related tests can then check the fetus in the womb (see Chapter 18). If the fetus proves to have a particular defect, some people choose to abort that fetus and try again. Others use the knowledge to prepare themselves to cope with afflicted children.

Most genetic disorders cannot be detected until the symptoms begin to show. Fortunately, this list is growing shorter every year, and some disorders that cannot be detected directly can nevertheless be spotted with the aid of "marker" genes. A marker gene lies so close on the chromosome to the gene responsible for a disorder that crossing over rarely or never breaks the linkage. In addition, the marker has one allele that almost always accompanies the defective allele and can be detected.

Recently, researchers studying an extended family with a high incidence of Huntington's chorea—caused by a dominant allele that destroys the central nervous system and that usually waits to show symptoms until well after victims have reproduced—found such a marker. They are now using the marker to tell whether the young, childless children of Huntington victims have the Huntington allele. When they find the marker, they can predict with fair certainty that the person will develop Huntington's chorea later in life and that his or her children will have a 50:50 chance of inheriting the Huntington allele. The person can then choose to start a pregnancy, check for the allele with the aid of amniocentesis, abort the fetus if it is found to be carrying the gene, simply have the children and risk the 50:50 chance that they will get the disease, or not have children at all.

There are signs of more promising approaches, even of actual cures. Researchers are using the techniques of genetic engineering to isolate the normal alleles of genes whose defective alleles cause hereditary disorders. They are learning to implant these alleles in viruses and to use the viruses to implant the alleles in defective cells, where they function as replacement genes and cancel out the defect. So far, most of the experiments have been done in cells cultured outside the human body and in laboratory animals. But researchers have already transplanted the first cells whose genetic defects have been repaired in this way back into the child they came from, where they may successfully supply an enzyme whose absence keeps the immune system from functioning.

This approach to treating genetic disorders is called *somatic* (body) cell therapy. It does not repair the genetic defect in the germ cells that produce gametes; it therefore offers no cure for the yet unborn. Researchers and regulators are reluctant to take that step, for it seems to intrude on the rights of future generations by passing on to them possible unforeseen side effects. It also opens the door to redesigning perfectly normal alleles, tailoring children's health, intelligence, facial features, and even body form in ways that may suit the parents but not the children. However, germ cell therapy seems bound to happen. It is such a simple concept—an injection of tailored virus carrying a normal replacement for a defective allele and one's children are safe for all time. It seems too hopeful to suggest there may come a time when genetic disorders will be unknown, except for those that arise from new mutations.

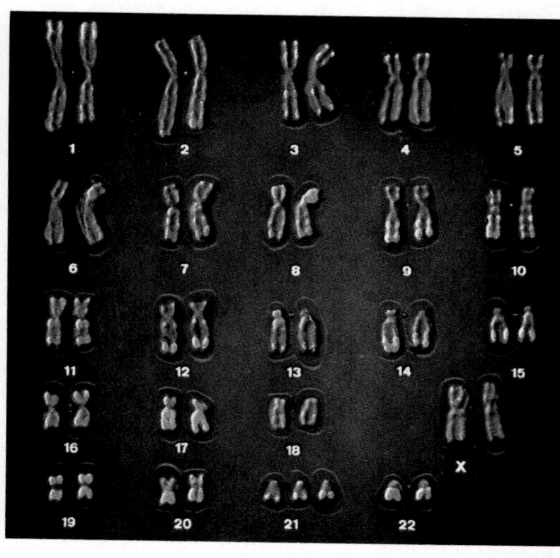

(a)

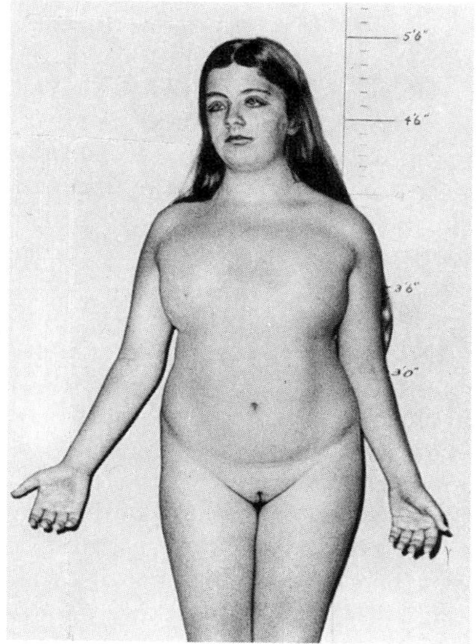

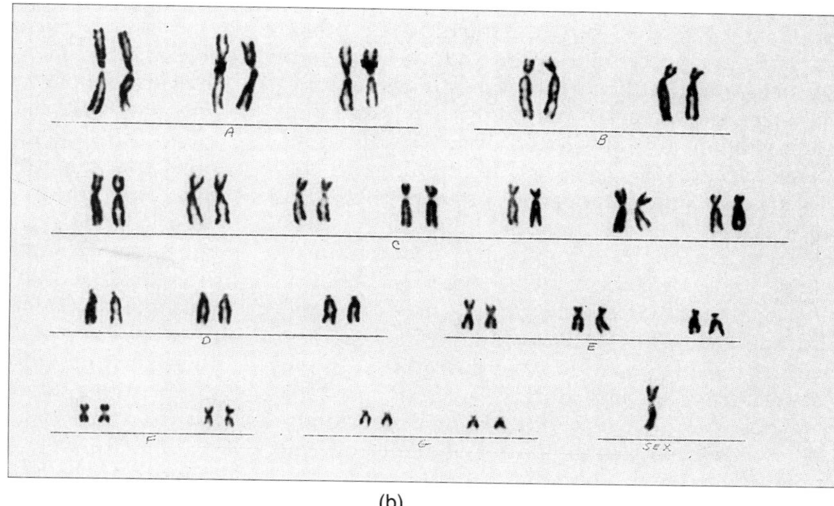

(b)

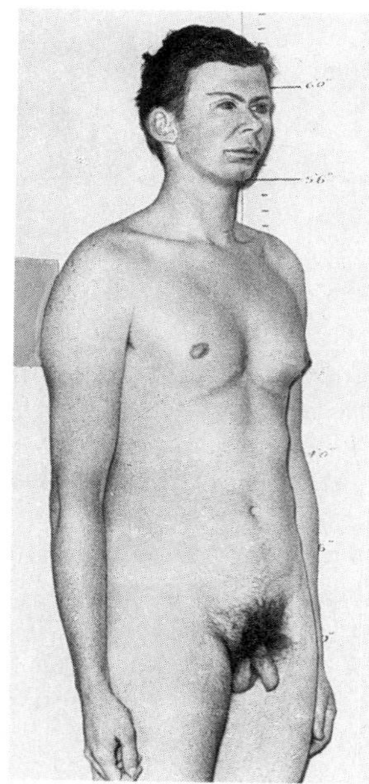

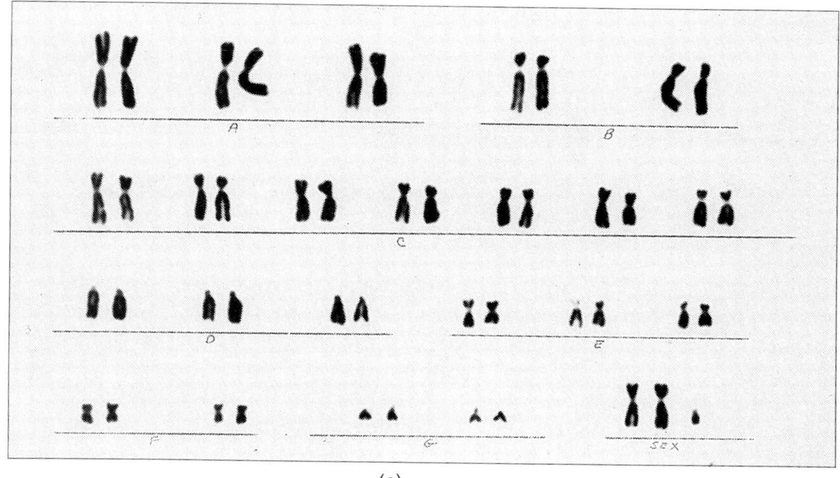

(c)

FIGURE 19-18 Human diseases caused by whole chromosome errors. (a) Twin boys with Down syndrome and their karyotype. (Notice three representatives of pair number 21.) (b) Turner's syndrome plus karyotype showing only one sex chromosome, an X. (c) Klinefelter's syndrome plus karyotype with three sex chromosomes, XXY.

---

1. We invite you to think about the implications of germ cell genetic engineering. How would you redesign your own children? Would you give them the muscles of dancers, the bones of football players, and the intelligence of a Nobel Prize winner? Would they thank you?

SUMMARY

Gregor Mendel was the first person to realize that living things passed their hereditary information on in discrete units, or genes. He also saw that in plants and animals the genes exist as pairs that separate during the production of gametes and recombine when sperm (or pollen) fertilizes an egg and that two genes separate and recombine without regard for each other. Thus each individual represents a random recombination of genes and traits from its two parents. He expressed his findings in the laws of segregation and independent assortment.

Concepts essential to any understanding of genetics are dominance, incomplete dominance, codominance, recessiveness, homozygosity, heterozygosity, alleles, phenotype, and genotype. Early geneticists added the idea that many genes fail to assort independently; they travel together in linkage groups that correspond to the chromosomes. Crossing over during meiotic prophase can shift alleles from one copy of a chromosome to another, thus creating new combinations of alleles on one chromosome. Additional apparent exceptions to the Mendelian patterns result from incomplete dominance and codominance, when heterozygotes have a phenotype distinct from either homozygote; multigenic inheritance, when many genes control a single trait; and sex linkage, when alleles on the X sex chromosome show up in male offspring more often than in female. Phenotype can also vary in response to environmental influences, as in the cases of skin color, intelligence, and height.

Hereditary disorders can be due to dominant or recessive alleles that presumably arose as mutations in the deoxyribonucleic acid (DNA). Some, such as sickle cell anemia, serve adaptive purposes in certain environments. All tend to show up more often in inbred populations and less often in outbred populations, an observation that may partially explain the hybrid vigor observable in crosses between individuals of very different populations.

Some hereditary disorders are due to changes in chromosome structure. These changes can be deletions, duplications, or inversions of chromosomal segments or their transfer to other chromosomes. They can also be duplications of entire chromosomes, as in Down, Edwards, and Patau syndromes.

Unfortunately, there are few real cures for hereditary disorders. The best we can usually do is to calculate the odds that a carrier for a recessive allele for a disorder will have afflicted children. The carrier must then choose whether to have children or not. In the future, some disorders may lend themselves to gene replacement therapy by genetic engineering.

STUDY QUESTIONS

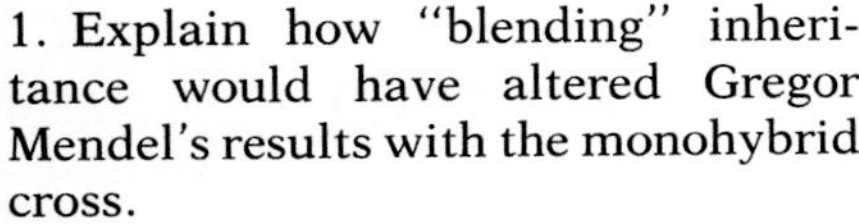

1. Explain how "blending" inheritance would have altered Gregor Mendel's results with the monohybrid cross.
2. How do dominance and recessiveness show in the results of the monohybrid cross?
3. What is the law of segregation?
4. What is the law of independent assortment?
5. How does linkage between the genes for two traits affect the results of a dihybrid cross?
6. Explain the concepts of incomplete dominance and codominance.
7. What is the difference between a sex-influenced and a sex-linked trait?
8. Under what conditions can a woman show a sex-linked trait?
9. Give an example to show how environmental influences can increase the number of phenotypes beyond what the inheritance mechanisms of a trait would seem to allow.
10. How does crossing over increase the number of different possible gene combinations?
11. How can inbreeding increase the chances of producing a genetically defective child?
12. How can marker genes help in the detection and prevention of genetic disorders?

GLOSSARY

Alleles Alternate versions of a gene controlling the same trait.

Autosome A non–sex chromosome.

Codominance When two alleles both affect phenotype in the heterozygote and neither allele is truly dominant nor recessive.

Crossing over The exchange of segments between chromosomes during meiotic prophase.

Dominant allele An allele that affects phenotype whether homozygous or heterozygous.

Genetics The study of inheritance.

Genotype An organism's set of genes or alleles for particular traits.

Heterozygous Possessing two different alleles for a particular trait.

Homozygous Having both alleles of a pair the same.

Inbreeding Mating with related individuals, who share more recessive alleles.

Incomplete dominance A kind of codominance in which both of the alleles contribute features to the phenotype.

Linkage Failure of genes to assort independently because they are on the same chromosome.

Multigenic inheritance Control of a trait by several genes.

Mutation A change in the DNA.

Phenotype An organism's observable features.

Pleiotropy When a single gene or allele affects several traits.

Recessive allele An allele that affects phenotype only when homozygous.

Sex chromosome A chromosome whose presence determines sex; females have two X chromosomes; males have one X and one smaller Y chromosome.

Sex-influenced trait A trait whose expressivity differs in male and female.

Sex linkage The presence of a gene on a sex chromosome; recessive traits controlled by such genes on the X chromosome tend to show more often in males.

Tetrad A bundle of four chromatids formed during meiotic prophase when the members of a homologous pair line up.

Chapter 20

The Digestive System

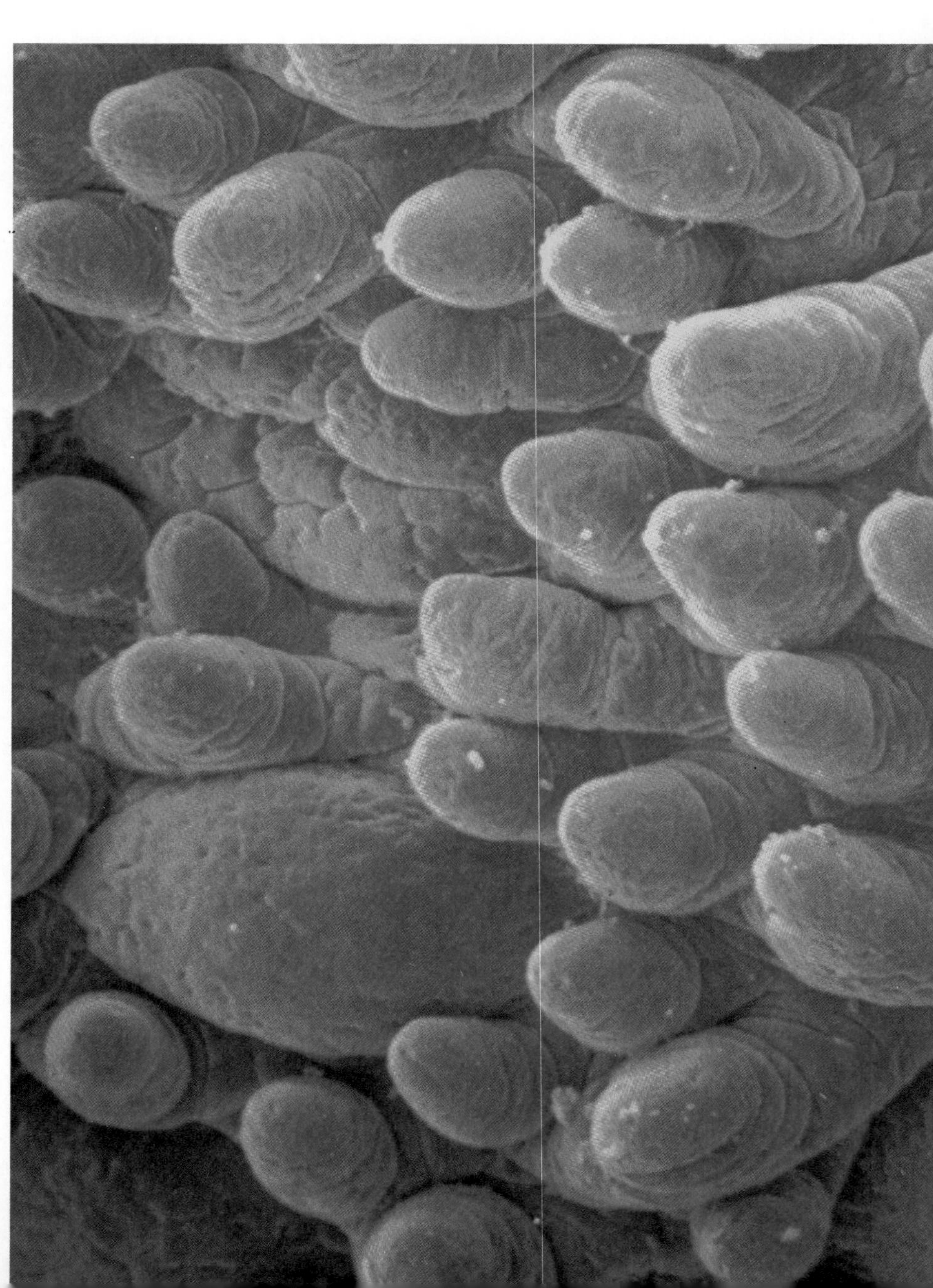

GLOSSARY

Alleles Alternate versions of a gene controlling the same trait.

Autosome A non–sex chromosome.

Codominance When two alleles both affect phenotype in the heterozygote and neither allele is truly dominant nor recessive.

Crossing over The exchange of segments between chromosomes during meiotic prophase.

Dominant allele An allele that affects phenotype whether homozygous or heterozygous.

Genetics The study of inheritance.

Genotype An organism's set of genes or alleles for particular traits.

Heterozygous Possessing two different alleles for a particular trait.

Homozygous Having both alleles of a pair the same.

Inbreeding Mating with related individuals, who share more recessive alleles.

Incomplete dominance A kind of codominance in which both of the alleles contribute features to the phenotype.

Linkage Failure of genes to assort independently because they are on the same chromosome.

Multigenic inheritance Control of a trait by several genes.

Mutation A change in the DNA.

Phenotype An organism's observable features.

Pleiotropy When a single gene or allele affects several traits.

Recessive allele An allele that affects phenotype only when homozygous.

Sex chromosome A chromosome whose presence determines sex; females have two X chromosomes; males have one X and one smaller Y chromosome.

Sex-influenced trait A trait whose expressivity differs in male and female.

Sex linkage The presence of a gene on a sex chromosome; recessive traits controlled by such genes on the X chromosome tend to show more often in males.

Tetrad A bundle of four chromatids formed during meiotic prophase when the members of a homologous pair line up.

Chapter 20

The Digestive System

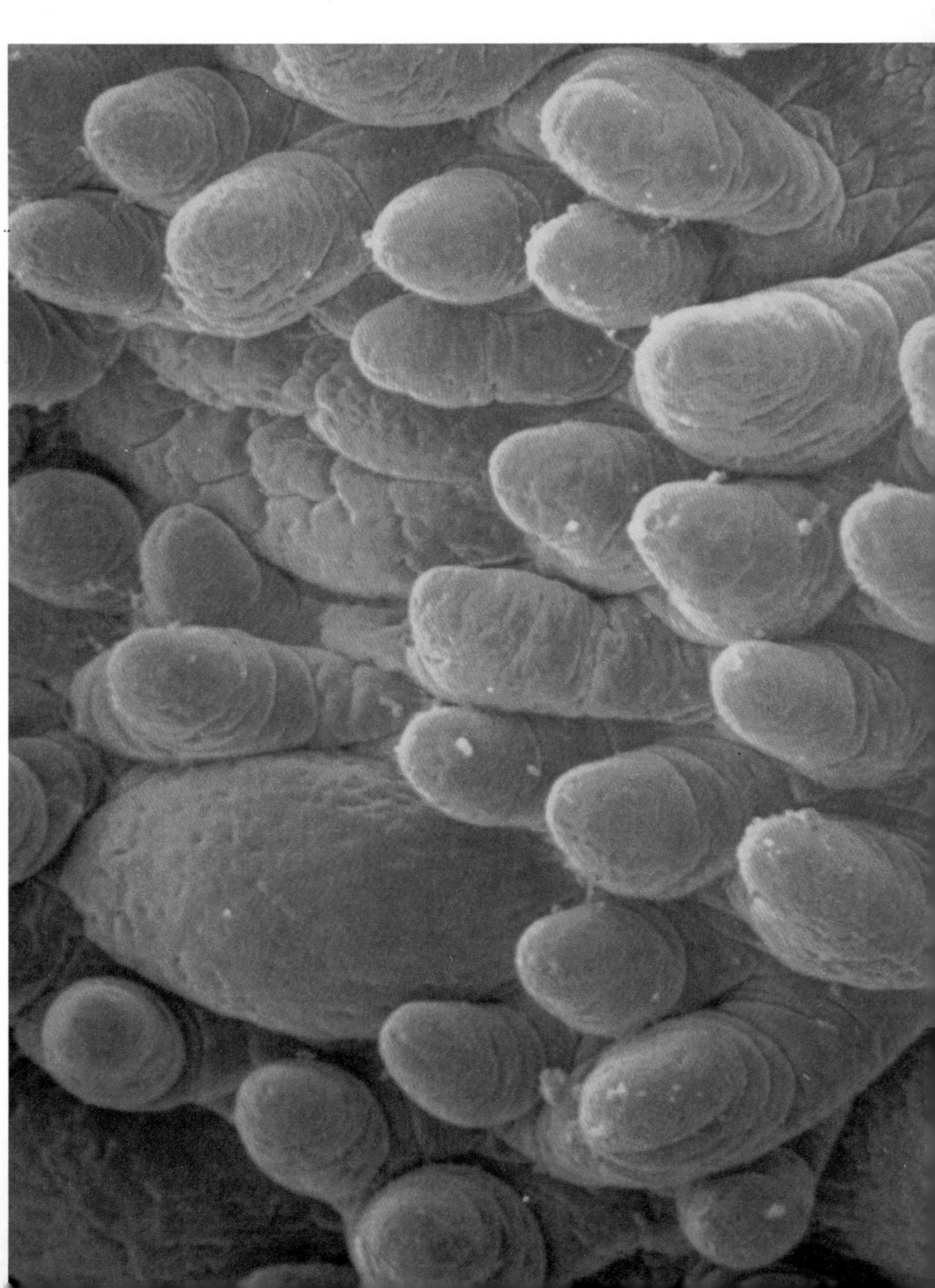

You have spent a long day cleaning your apartment, waxing your car, and playing racketball at the Health Club. You have had no break since breakfast, and your hunger pangs gently remind you it is time to replenish your reserves. You need to restore your energies and replace some of the nutrients your body has consumed during the course of these activities. That large soda, order of fries, taco salad, bean and cheese burrito, and piece of apple pie will certainly help—even though it may be less than an "ideal" diet.

Some simple animals (like tapeworms and other parasites) can soak up their food and other nutrients directly through their skins, but you cannot. Before your body uses the food molecules you ingest for energy or tissue building, you have to process that food. You have to put it in your mouth, grind it, mix it with digestive solutions, and move it from chamber to chamber down the tubeway of the **digestive system,** where a variety of nutrient-releasing enzymes can act upon it. The whole process is called **digestion.** It is followed by **absorption,** which transfers the freed nutrients from the digestive tube into the blood.

Digestion has two phases. The first is **mechanical breakdown** of food. It begins with chewing, which divides large chunks of food into smaller particles. It continues as the muscular walls of the stomach and intestine mix food and liquid into a soupy fluid. The second phase of digestion is **chemical breakdown.** It is accomplished by enzymes secreted by the salivary glands, stomach, pancreas, and small intestine. They break the large molecules of starch, protein, fat, and nucleic acid in the food into their component units—sugars, amino and fatty acids, glycerol, and the nitrogen-containing bases of nucleotides. In the process, minerals, salts, and vitamins are also released from the foods.

The process of absorption then draws these units, together with water, vitamins, and minerals, from the interiors of the stomach and intestines to the bloodstream for distribution to the body's cells. The unabsorbable residue, including "fiber," becomes the bodily wastes or fecal material that is eliminated from the body through the anus.

THE DIGESTIVE ORGANS

As strange as it sounds, the chambers of your digestive organs are technically outside your body. How can that be? Well, imagine a donut. The hole passing through it is really not part of the donut. In fact, the hole is outside the donut. Now stretch that donut into a thick-walled tube open at both ends. The hole is still not part of the material forming its outer walls. Stretch the inner wall more than the outer so the hole bulges, pinches, kinks, and coils. Add arms, legs, eyes, a bellybutton, and so on, to the material surrounding the passageway. In a sense, the human body is built around a passageway that is open at both ends. The inside of your stomach is then as much outside your body as the spaces between your toes.

This passageway through your body is made up of various compartments called the **true digestive organs** (mouth, stomach, intestines). The glands and organs that release various secretions into these chambers or provide other services are called the **accessory digestive organs** (salivary glands, liver, pancreas). The true digestive organs begin with your mouth, which passes food to your **esophagus,** or gullet, when you swallow. The esophagus carries foods and liquids through the chest to an expanded chamber just below your diaphragm, called the stomach. The stomach passes the nutrient mixture, in turn, to the small intestine, where most digestion and absorption occur. The residue passes to the large intestine, then to the rectum, where the wastes are prepared for elimination, and finally out of the body through the anus (Figure 20-1). The walls of the esophagus, stomach, and intestines share a basic structural scheme (Figure 20-2), with regional modifications.

The Mouth

As the entry to the digestive system, the mouth has several important features. The lips manipulate incoming food and liquid, modify the sounds associated with speech, and help control the food during chewing. The teeth are tools used to tear or slice manageable pieces from meat, fruit, and other foods. They are also used to grind the food, increasing its surface area and thereby enhancing its exposure to the digestive enzymes. We use the tongue to manipulate food within the mouth, positioning it between the teeth so it can be broken up. Once chewing is complete, the tongue moves the food to the back of the throat so it can be swallowed. The tongue also bears the taste buds (see Chapter 16).

Easing the task of chewing and swallowing and helping to get the food molecules into solution are three pairs of **salivary glands.** One pair, the *parotids,* is embedded in the cheeks in front of the ears. A second pair, the *sublingual glands,* lies in the floor of the mouth beneath the tongue. The third pair, found just below the angle of the jaw, is called the

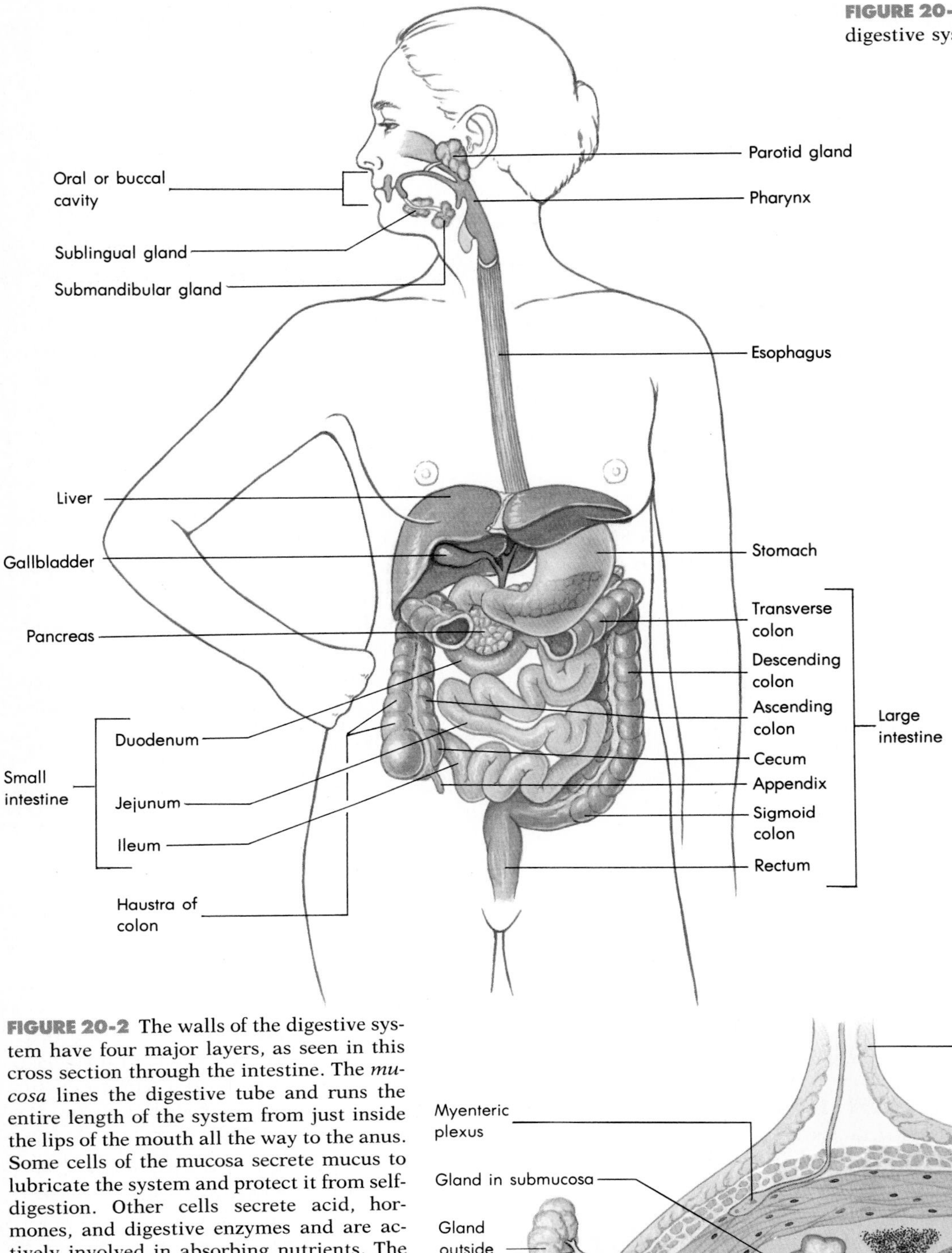

FIGURE 20-1 The major organs of the digestive system.

FIGURE 20-2 The walls of the digestive system have four major layers, as seen in this cross section through the intestine. The *mucosa* lines the digestive tube and runs the entire length of the system from just inside the lips of the mouth all the way to the anus. Some cells of the mucosa secrete mucus to lubricate the system and protect it from self-digestion. Other cells secrete acid, hormones, and digestive enzymes and are actively involved in absorbing nutrients. The *submucosa* is a layer of loose connective tissue containing the blood and lymph vessels, lymph nodes, and nerve fibers that service the digestive system. The *muscularis* contains two or three layers of smooth muscle whose contractions mix food with digestive juices and propel it along the tube. The *serosa* covers the outer surface of the digestive system and merges with the peritoneum. Folds of the peritoneum also form the mesenteries that support or suspend the intestines.

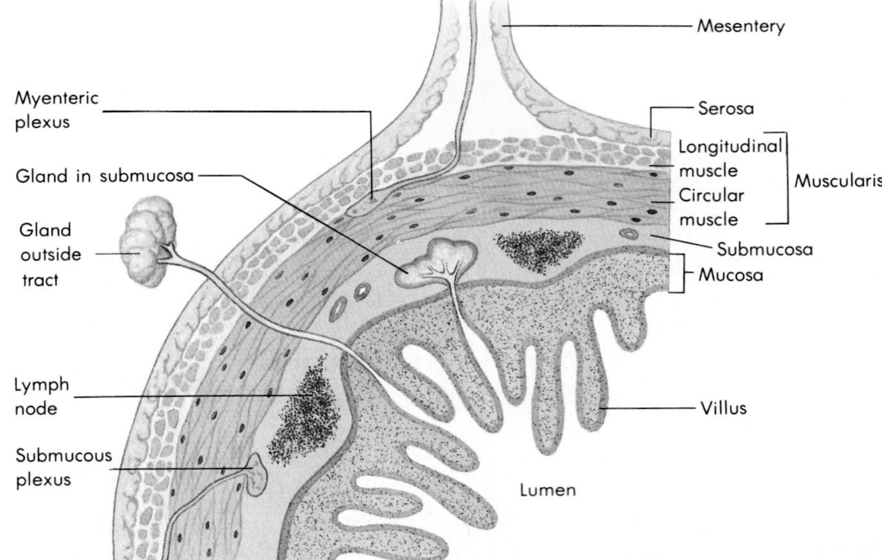

submandibular salivary glands. Together they secrete about 1.2 liters per day of fluid, the **saliva.** Secretion peaks when we smell, taste, or even think of food.

Saliva lubricates the mouth, mixes with dry food to form a mush that can easily be swallowed, and dissolves flavor molecules so the taste buds can sense them. It is mostly water, but it contains mucus, antibacterial agents (see Chapter 24), and the first digestive enzyme the body applies to its food. This enzyme, **salivary amylase,** breaks down starches to sugars; it is salivary amylase that causes bread or other starchy food to turn sweet after it has been chewed for a time. The enzyme is inactivated by the acidity of the stomach juices (Figure 20-3).

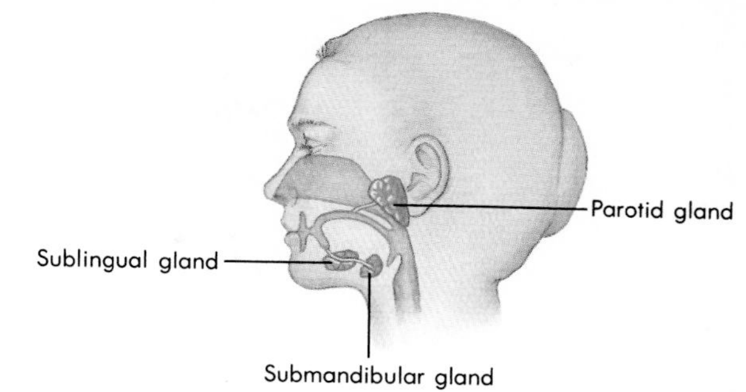

FIGURE 20-3 The three pairs of salivary glands.

The Pharynx and Esophagus

During swallowing, the tongue pushes a ball, or **bolus,** of food past the **uvula,** the flap that hangs

Box 20-1 Teeth and Tooth Decay

The mouth contains several kinds of teeth (Figure 1). The **incisors** in front are bladelike slicers, good for biting off chunks of foods such as apples. There are four incisors each in the upper and lower jaws. Flanking them are the pointed **canines** (the "eye teeth"), used for tearing meat and biting off chunks of food; when they are enlarged, as they are in dogs, baboons, and lions, we call them fangs. Immediately behind the canines, tooth structure becomes much more complex. The first two teeth posterior to each canine are the **premolars.** The last three teeth on each side of both jaws are the **molars.** The complex structure and ridges of the premolars and the molars provide the elaborate surfaces for grinding tough, fibrous foods.

The teeth are anchored in sockets within the jaw bones by a layer of dense connective tissue, the **periodontal ligaments.** Their substance is much like that of bone, consisting of phosphates and carbonates of calcium. The core of each tooth is composed of relatively soft **dentin.** Where the tooth protrudes from the gum and is exposed to the greatest wear, the dentin is covered by a layer of denser, stronger **enamel** (enamel is the hardest material in the human body). Below the gum line, the dentin is covered by **cementum,** which joins the tooth to the periodontal ligaments. Each tooth is built around a hollow core, the **pulp** cavity, which is filled with connective tissue, blood vessels that nourish the tooth, and the nerves that make it so sensitive to heat, cold, pressure, and pain (Figure 2).

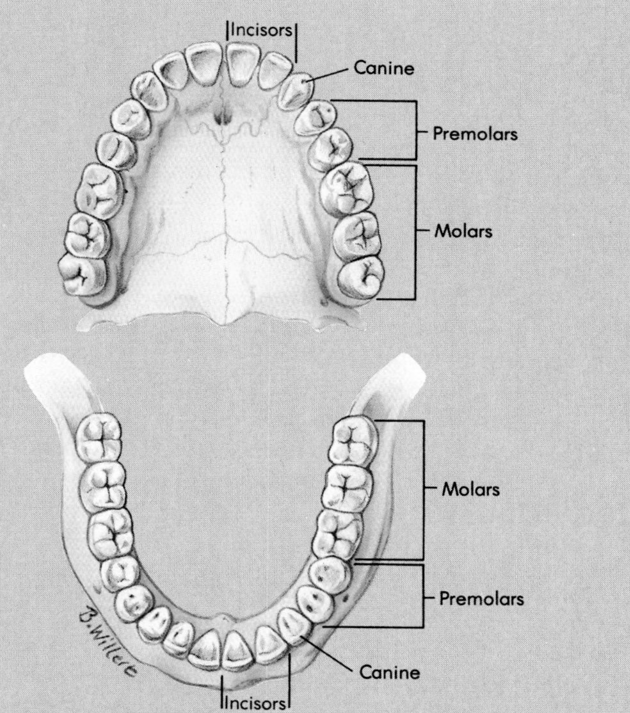

FIGURE 1 Tooth arrangements in the human mouth.

Unfortunately, the mouth is moist and warm, and it is repeatedly resupplied with high-quality nutrients. If it were not for the antibacterial agents in saliva, it would be overrun with bacteria. As it is, some bacteria are able to resist the effects of these defenses. They make their

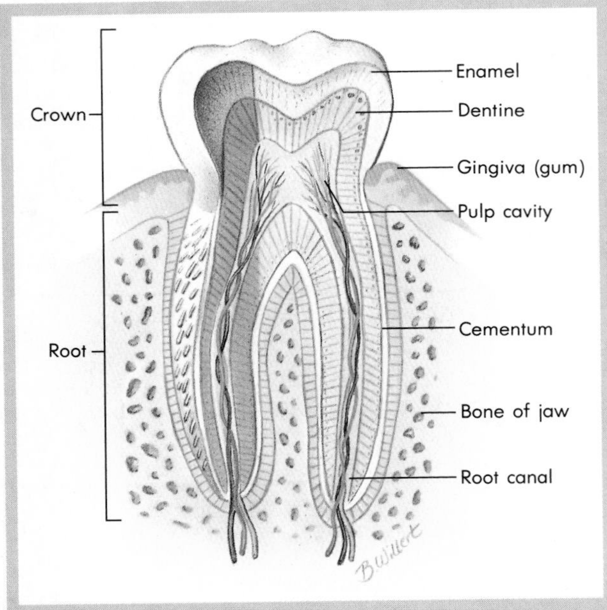

FIGURE 2 Anatomy of a single tooth.

home in the mouth, especially in the deposits of food sludge that build up in the crevices between the teeth and along the gum line. These deposits, knitted together by mats of bacteria and strengthened by calcium salts, form what we call **plaque.** They are a precursor of **caries,** or tooth decay, for many of the bacteria in them secrete acid that attacks the tooth enamel. Acid production is increased when sugars are available to nourish the bacteria. The worst sugars for tooth decay are those that linger near the teeth, such as honey and candy. Sweetened beverages, because they pass by so quickly, are less harmful.

Tooth decay bacteria can also produce noxious odors. We thus floss and brush our teeth to remove plaque and bacteria and prevent both caries and bad breath. The fluoride often added to toothpastes and drinking water enhances the enamel's resistance to acid and protects against tooth decay, but it does nothing to kill the bacteria and prevent bad breath.

from the back of the soft palate, into the **pharynx,** or throat. The larynx rises and the epiglottis folds down over the opening to the trachea, keeping food out of the trachea and forcing the bolus into the esophagus (Figure 20-4). Its presence triggers reflexive, involuntary contractions of the smooth muscles in the esophageal wall. A wave of contraction forms above the bolus and flows down the esophagus to the stomach, pushing the food before it. This one-way process is **peristalsis.** It also occurs in the stomach and intestines, where it pushes food through the digestive tract.

The Stomach

The **stomach** is an expandable sac where the first *major* stage of chemical digestion occurs. Unlike the rest of the digestive tract, the stomach wall has three layers of smooth muscle, one circular, one longitudinal, and one oblique (slantwise). Their contractions knead the stomach's contents and mix food thoroughly with the stomach's secretions for 3.5–4 hours, producing a soupy fluid known as **chyme.** Peristaltic contractions propel the chyme toward the lower end of the stomach, the **pylorus.** The chyme moves out of the stomach into the small intestine through a muscular valve called the **pyloric sphincter** (Figure 20-5).

The mucosa of the stomach wall is marked by prominent folds, or **rugae,** that serve to increase its surface area (Figure 20-6). The mucosa is also studded with glandular **gastric pits** that contain several specialized cells (Table 20-1). **Parietal cells** secrete both hydrochloric acid, which makes the stomach contents strongly acidic and serves as a germicide, and intrinsic factor, which is necessary for the absorption of vitamin B_{12} in the small intestine (see Chapter 22). **Chief cells** secrete **pepsinogen,** which hydrochloric acid converts to **pepsin,** an enzyme that breaks proteins into polypeptides. **Enteroendocrine cells** secrete **gastrin,** a hormone that stimulates parietal and chief cells and relaxes the pylorus. **Mucous cells** secrete a layer of mucus over 1 millimeter (mm) thick that, together with the acid- and pepsin-resistant membranes of the mucous cells, usually shields the stomach wall from the effects of acid and pepsin.

1. Heartburn is the feeling you get when some of the stomach contents enters the lower end of the esophagus, which is normally kept tightly closed by the pressure of the organs outside it. Why do you think taking antacids works to relieve the pain?
2. Stomach ulcers result when hydrochloric acid and pepsin digest the stomach's wall. Which of the stomach's secretory cells do you think might malfunction to cause this?

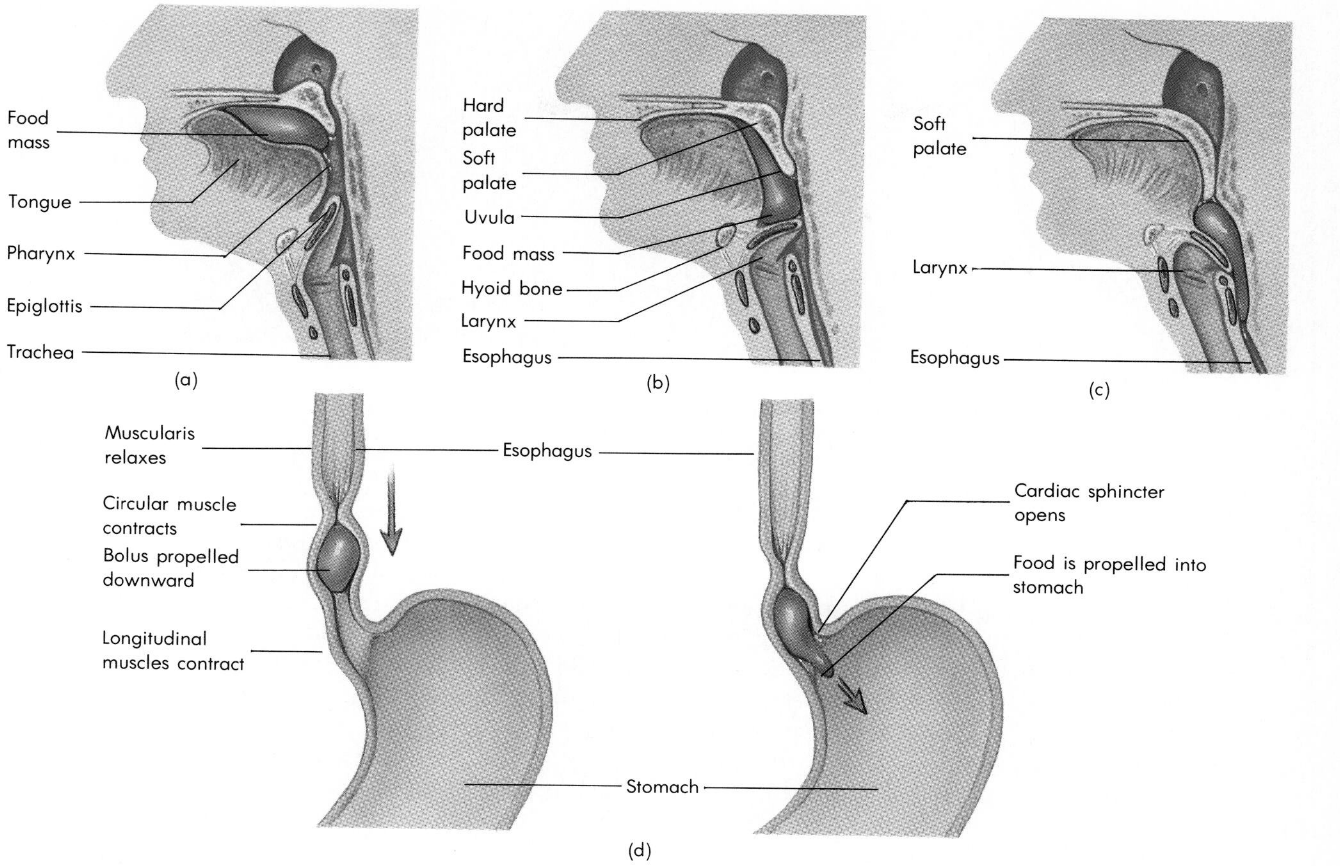

FIGURE 20-4 Swallowing and peristalsis. (a) The tongue pushes food toward the pharynx. (b) The soft palate and larynx rise, the epiglottis folds down to close off the trachea, and the throat muscles relax, allowing the esophagus to open. (c) The throat muscles contract and force the food into the esophagus. (d) Peristalsis then propels the food toward the stomach.

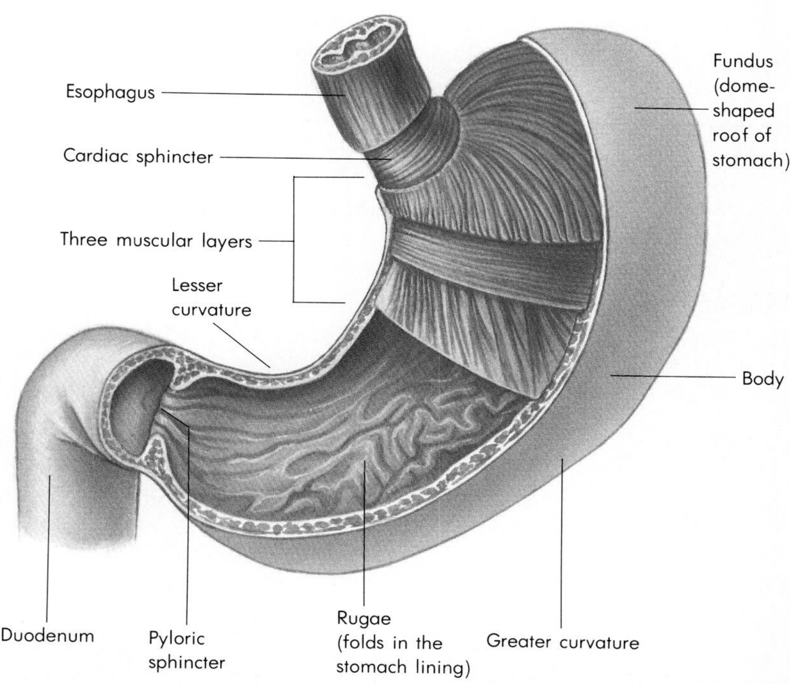

FIGURE 20-5 A dissected view of the stomach.

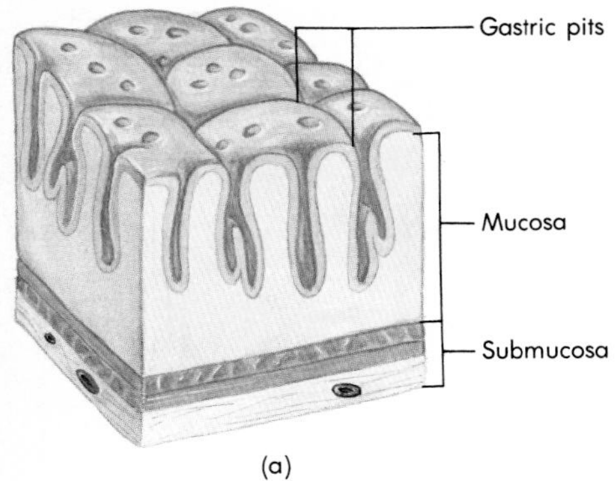

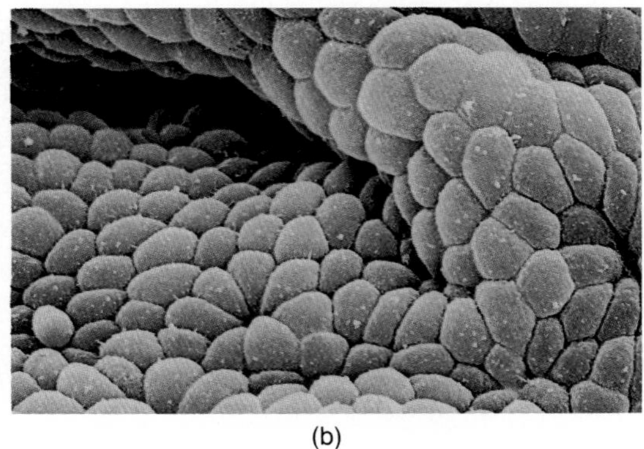

FIGURE 20-6 The stomach lining. (a) Diagrammatic representation of the stomach lining. (b) Scanning electron micrograph of the stomach lining.

The Small Intestine

The **small intestine** is a long tube, about 2.5 centimeters (cm; 1 in.) in diameter, connecting the stomach to the large intestine. It has three segments. The **duodenum** is the 25 cm (10 in.) immediately after the pylorus. Because it receives the highly acid chyme from the stomach, it is the part of the small intestine that most often develops ulcers (duodenal ulcers), usually when the stomach produces excess acid for some reason, such as stress or heredity. (Recent research indicates that bacterial infections may play an important role in causing some ulcers.)

Secretions from the pancreas and liver are released into the duodenum through a duct formed by the **common bile duct** and the **pancreatic duct.** (We discuss how these secretions are controlled later in the chapter in a Learning Focus.) The bile duct delivers **bile,** a mixture of cholesterol, bile pigments (breakdown products of hemoglobin), and bile salts produced by the liver and stored in the **gallbladder.** The bile salts serve as emulsifiers; that is, they break fat in food into tiny droplets, suspended in water, that fat-digesting enzymes can attack efficiently. The pancreatic duct delivers pancreatic juice from the **pancreas.** Because this juice is alkaline, it neutralizes the stomach acid in the chyme and halts the action of pepsin. The pancreatic juice also contains numerous digestive enzymes that work best in an alkaline environment. They include enzymes that break down starches, proteins and peptides, fats, and nucleic acids (Figure 20-7).

The 2.5 m (8 ft) of intestine following the duodenum is the **jejunum.** The final 3.6 m (12 ft) is the **ileum;** it ends at the **ileocecal valve,** where the small intestine joins the large intestine. The mucosa of the entire small intestine is studded with

Table 20-1

Secreting Cells of the Stomach

TYPE OF CELL	SECRETION	FUNCTION
Parietal cells	Hydrochloric acid	Makes stomach acid; acts as germicide; activates pepsin
	Intrinsic factor	Aids the absorption of vitamin B_{12}
Chief cells	Pepsinogen	Converted to pepsin; digests protein
Enteroendocrine cells	Gastrin	Stimulates parietal and chief cells to produce gastric juices
Mucous cells	Mucus	Shields the stomach walls from acid and pepsin

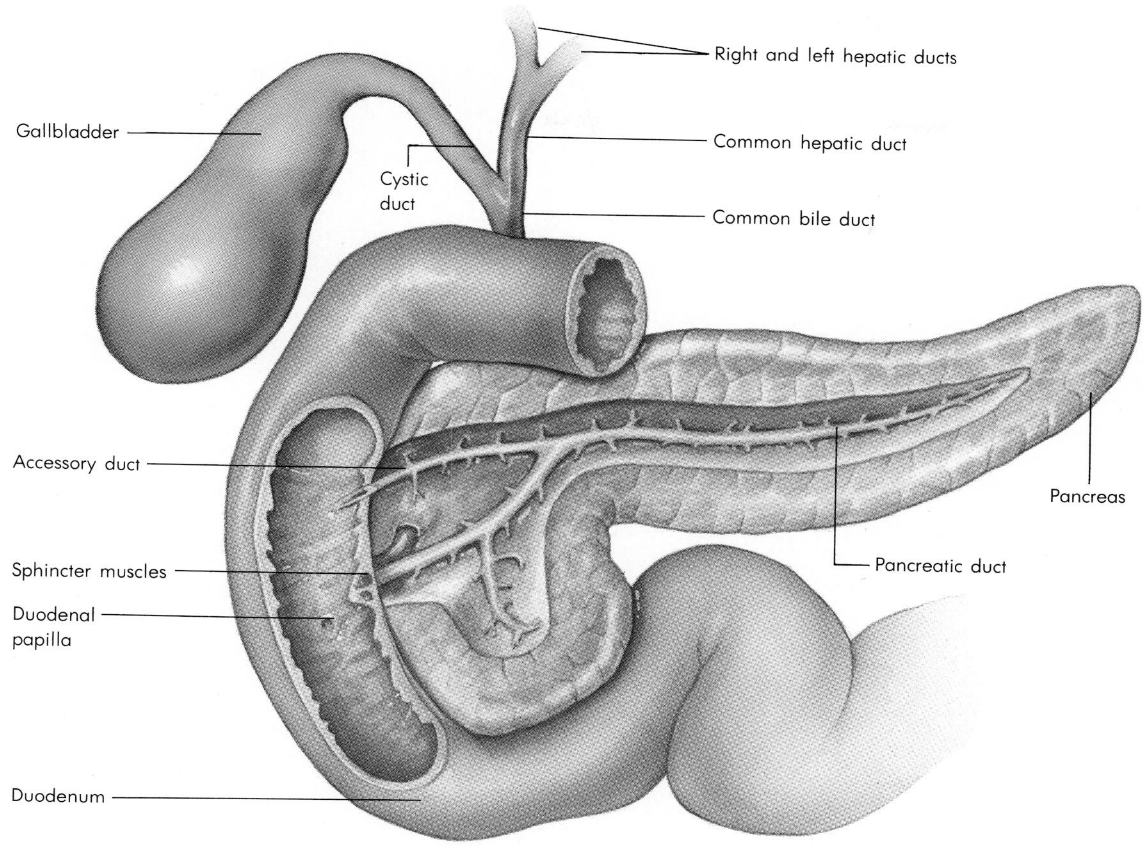

FIGURE 20-7 The duct system connecting the gallbladder and pancreas to the duodenum of the small intestine.

intestinal glands, pits whose cells secrete mucus and digestive enzymes. The glands of the duodenum secrete an alkaline mucus that helps neutralize the acidic chyme and protect against ulcers.

The digestive enzymes secreted by the cells of the intestinal glands complete the work begun by the enzymes in the pancreatic and stomach juices and saliva. The enzymes do not attack fats or starches, for those food elements are adequately digested by the salivary and pancreatic enzymes. Nor do they attack complete proteins. Rather, the intestinal enzymes break disaccharides such as sucrose, maltose, and lactose into monosaccharides, peptides into amino acids, and nucleotides into five-carbon sugars (pentoses) and bases.

Like the stomach, the small intestine has its surface area increased by folds. The largest of these folds are the **plicae circulares,** which resemble the wrinkles in a shirt sleeve that has been pushed up an arm. Superimposed on the plicae are the **villi,** fingerlike projections of the intestinal mucosa. The cells that cover the villi show still more folds, for their surfaces, where they face the intestinal cavity, are subdivided into **microvilli;** under the microscope, each cell seems to bear a brushlike edge, its **brush border.** The total surface area of the small intestine is roughly that of a football field. All of this area is available for secreting enzymes and absorbing nutrients (Figure 20-8).

The villi are the sites where the simple nutrients released by the action of the digestive enzymes are absorbed. Each villus projects about 1 mm from the intestinal mucosa and is covered with a layer of digestive epithelium one cell thick. The core is occupied by loose connective tissue. It holds one arteriole, a venule, and a tiny connecting net of capillaries. There is also a lymphatic capillary into which the epithelial cells shunt absorbed fat, which gives the contents of the lymphatic capillary a milky appearance after a meal and earns it its name of **lacteal.**

Peristalsis moves food through the small intestine, just as it does in the esophagus and stomach. However, the muscles of the small intestine spend much more of their time in a different sort of activity. In peristalsis, rings of circular smooth muscle contract in sequence, so that a wave of contraction moves slowly down the intestine, at a speed of

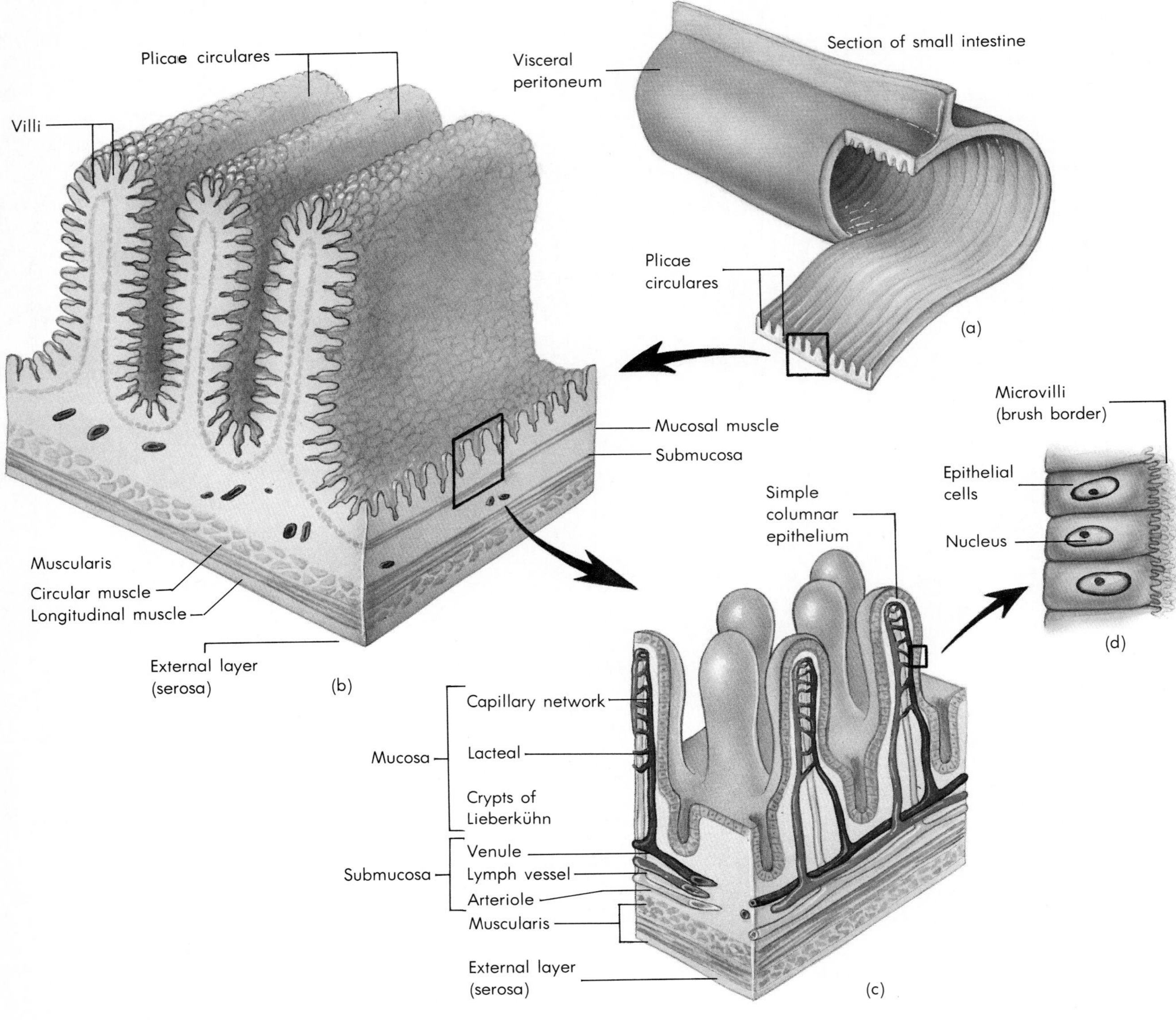

FIGURE 20-8 The wall of the small intestine. (a) The plicae circulares. (b) Detail of plicae circulares showing villi. (c) Close-up of a villus. (d) The epithelium covering the villi, showing the microvilli.

about 1 cm/min; chyme remains in the small intestine for 3–10 hours. In **segmentation** (Figure 20-9), the rings alternate their contractions 12–16 times per minute, so that each segment of the intestine repeatedly kneads its contents, mixing food with enzymes and spreading it over the intestinal surface for absorption. We discuss how these actions are controlled later in the Learning Focus.

The Large Intestine

The large intestine, or **colon,** is the final processor of the materials passing through the digestive system. From the small intestine it receives chyme that has been exhausted of most absorbable nutrients. Over the next few hours, it recovers the small amount of water still remaining and a few nutrients and compacts the residue into feces for elimination. Much of the bulk of the feces consists of bacteria that live in the colon and feed on leftover or unusable nutrients.

These bacteria serve several roles. They break down the bile pigments to simpler pigments, which give feces their characteristic color. They also ferment carbohydrates into gas and break proteins remaining in the chyme into amino acids and simpler compounds, including hydrogen sulfide. Some

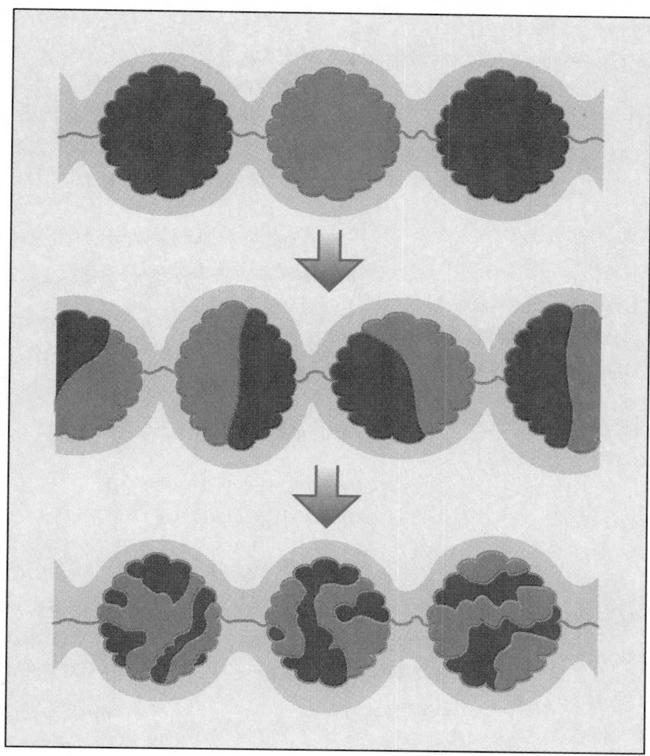

FIGURE 20-9 Segmentation is a process of localized muscle contraction in the smooth muscles of the small intestine that thoroughly mixes the digestive juices.

of these compounds give feces their odor. Others are absorbed, transported to the liver, and converted to less toxic compounds for excretion in the urine. In addition, bacteria manufacture and secrete several B vitamins and vitamin K, which, once absorbed from the colon, contribute to human health (see Chapter 22).

The colon is larger in diameter than the small intestine. It is divided by folds into pocketlike segments, the **haustra.** Its mucosa secretes primarily a lubricating mucus (Figure 20-10). Where the small intestine connects to the colon, it joins the saclike first chamber of the colon, called the **cecum.** Attached to the cecum is the **vermiform appendix,** which seems to serve as a place where bacteria can be trapped and identified by the body's immune system. **Appendicitis** occurs when the appendix becomes infected. It must then be removed before it bursts, spreading bacteria throughout the peritoneal cavity. The resulting larger infection, or **peritonitis,** can be fatal.

For proper function, the colon depends on the delivery of adequate bulk by the rest of the digestive system. Most of this bulk is **fiber,** or roughage, made up of indigestible carbohydrates such as cellulose from plant foods (see Table 20-2). When the diet lacks fiber, the colon takes longer to pass its

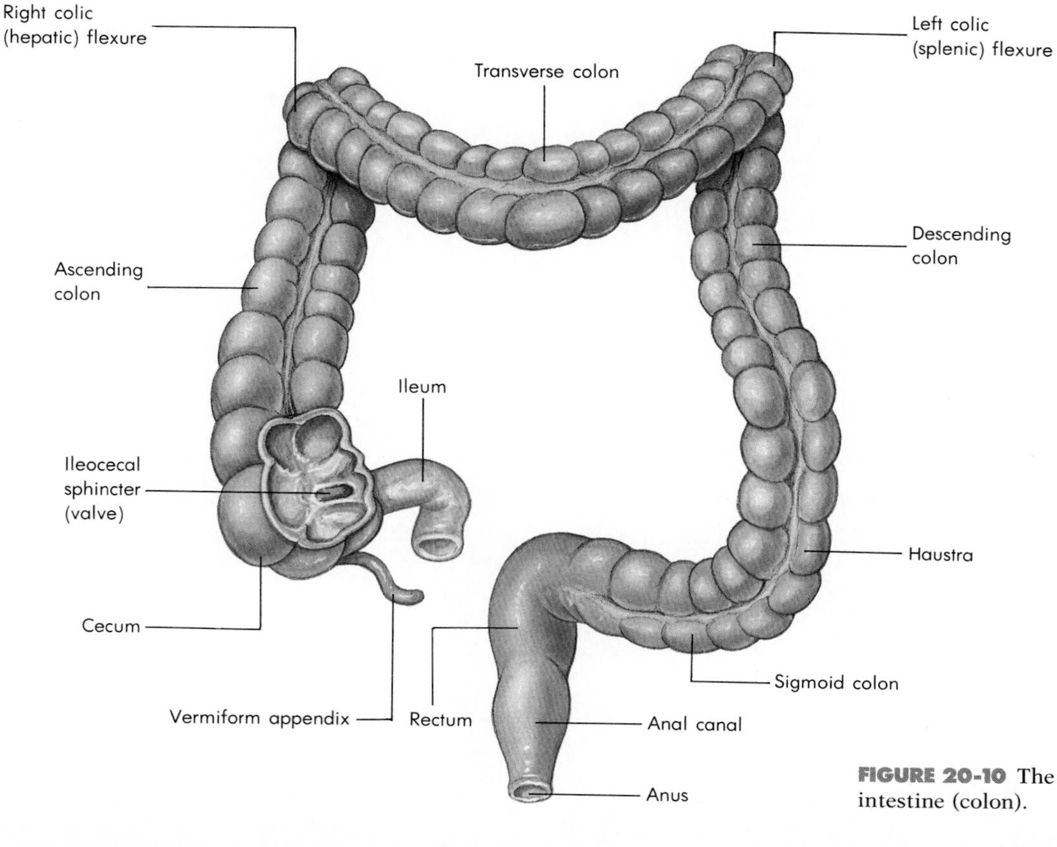

FIGURE 20-10 The gross anatomy of the large intestine (colon).

Table 20-2

Components of Fecal Matter

COMPONENT	PERCENT
Water	75
Solid matter	25
Composition of Solid Matter	
Bacteria	30
Fat	10–20
Inorganic matter	10–20
Protein	2–3
Undigested roughage	30

contents to the rectum and anus for expulsion. The colon then suffers more prolonged exposure to whatever toxic compounds may be present; some of these compounds can cause colon cells to become cancerous. Indeed, people who eat less fibrous foods seem to have a higher incidence of colon cancer.

A shortage of dietary fiber has other effects as well. One is **diverticulosis.** It occurs when the colon wall develops a weak spot and balloons outward as a blind pouch or diverticulum. When a diverticulum becomes infected or inflamed, the condition is known as diverticulitis; it can be very painful. Diverticula seem to appear when the colon contracts against insufficient bulk. The contractions can be so powerful that the high pressures produced may "blow out" the colon wall. Increasing the fiber content of the diet is an effective way to prevent and treat diverticulosis.

Lack of fiber is also associated with **constipation,** and it is easy to see why. The lack of bulk causes it to take a longer time for enough fecal material to accumulate to stimulate the defecation reflex. In addition, the lack of bulk means little distension of the colon, and hence little stimulus to the colon's muscles, which move the colon's contents (see the Learning Focus). In general, anything that reduces the colon's muscle activity increases the residence time of material in the colon and therefore increases the amount of water that is absorbed from it. (See Table 20-3 for the normal amount of time food or chyme spends as it passes through each section of the digestive system.) When the material becomes dry and hard to expel, the result is constipation. Anything that increases muscle activity, such as some bacterial toxins, decreases residence time and water removal. The result is **diarrhea,** which can result in dangerous dehydration and loss of salts from the body.

Curing constipation requires stimulating the colon's muscles. This can be done with **laxatives,** but they have the drawback of encouraging dependence on them. They can thoroughly empty the colon; since it then takes a long time to refill the colon and generate a normal bowel movement, people often believe they are still constipated and take more laxatives. In time, the colon comes to rely on the stimulus of the laxative to maintain its normal activity. A better treatment is to increase the fiber content of the diet with bran (the hulls of grains such as wheat and rice), whole grains, fruits, and vegetables.

Curing diarrhea requires inhibiting the colon's muscles, the function of many antidiarrhea medications. Some are actually derivatives of morphine; they act as sedatives for the colon. Serious cases, which people can develop with diseases such as cholera, require giving the patient fluids and salts either by mouth or intravenously (Figure 20-11).

Table 20-3

Time Spent by Food Moving Through Various Areas of the Digestive System

AREA OF THE DIGESTIVE SYSTEM	TIME
Mouth	Seconds
Esophagus	Seconds
Stomach	3.5–4 hours
Small intestine	5 hours (3–10 hours)
Large intestine	18–24 hours
Rectum	Seconds to minutes

1. Compare the benefits for colon function of diets high in meat (whose "fiber" is the protein of connective tissue) and diets high in plant foods.
2. Most Americans eat diets that are high in plant foods such as flour. Unfortunately, most flour has been "refined" to remove its fiber. Discuss the effects of this practice on the health of Americans and ways to change the practice or avoid the effects.

Learning Focus

REGULATION OF MOTILITY AND SECRETION

Digestion is an involuntary process, but it is exquisitely regulated by the autonomic nervous system, largely by local reflex loops, and by hormones secreted by the digestive tract itself. Let us see how the process works.

SALIVATION

Parasympathetic nerves originating near the junction of the brain's pons and medulla control the salivary glands. They are activated when we think of, smell, see, or touch food. Food in the mouth is most effective, for it stimulates taste receptors whose signals feed directly to the sources of the nerves that control salivation. In addition, the presence of irritants in the stomach or upper small intestine (as with some foods and with nausea) stimulates local receptors that activate the nerves.

SWALLOWING

Swallowing has both voluntary and involuntary stages. The voluntary stage ends when we flex the tongue against the roof of the mouth and push a food bolus toward the throat. As the bolus passes the uvula, the involuntary stage begins. The bolus mechanically stimulates receptors in the mucosa and activates reflexes through the medulla and pons. These reflexes raise the soft palate and uvula, raise the larynx (Adam's apple) to meet the epiglottis and close the trachea, and trigger esophageal peristalsis.

STOMACH SECRETIONS AND MOTILITY

The sight, smell, or taste of food is enough to begin the stomach's secretion of acid, mucus, and pepsinogen via parasympathetic nerve signals from the brain. Contact of food or irritants with mucous cells lining the stomach (and elsewhere) triggers secretion of mucus by those cells. In addition, local reflexes also stimulate mucus secretion by cells in other areas of the digestive system. The presence of food stimulates stretch receptors embedded in the stomach wall to signal both a control center in the brain and the stomach glands themselves. The control center increases both secretion and the contraction of the stomach muscles. Protein, alcohol, or other food in the stomach stimulates the mucosa of the pyloric region to release the hormone gastrin, which stimulates secretion of gastric juices, increases the contractions of the stomach and intestine, and relaxes the pyloric sphincter. Both distension and gastrin encourage the emptying of the stomach.

The stomach's secretion and contractions can also be inhibited. Anger, fear, and anxiety activate the sympathetic branch of the autonomic system and inhibit the digestive control center. In addition, when chyme reaches the small intestine, its acidity, protein, fats, or irritating contents stimulate the release of the hormones **secretin, cholecystokinin (CCK),** and **gastric inhibiting peptide (GIP)** (see the figure). Secretin inhibits stomach secretion and contractions of both the stomach and intestine; it also stimulates the secretion of bile by liver cells, triggers the release of intestinal enzymes, and promotes the release of pancreatric juices rich in alkaline components that will neutralize the acid nature of chyme from the stomach. The CCK has a similar effect on stomach secretion and on muscle contractions; it stimulates the release of pancreatic juice rich in digestive enzymes, and it stimulates the gallbladder to contract its muscles and release bile. Since CCK is released especially in response to fat, its effect on the gallbladder explains why eating fatty foods can lead to gallbladder attacks. Contractions of the muscular wall of the gallbladder can be very painful when gallstones—masses of crystallized cholesterol—are present (see the figures that follow). Gastric inhibiting peptide inhibits stomach secretion and contractions.

The result of all these feedback mechanisms is that the amount of chyme discharged by the stomach is limited and the small intestine is able to adjust the secretion of bile and enzymes to match the amount of chyme to be processed.

THE SMALL INTESTINE

Contractions of the small intestine's muscles are controlled largely by local reflexes mediated by the intrinsic plexus. The hormones gastrin, CCK, secretin, and GIP also have effects, as already mentioned. The production of bile by the liver is regulated by secretion, autonomic nerve signals, blood flow through the liver, and the availability of bile salts reabsorbed from the small intestine.

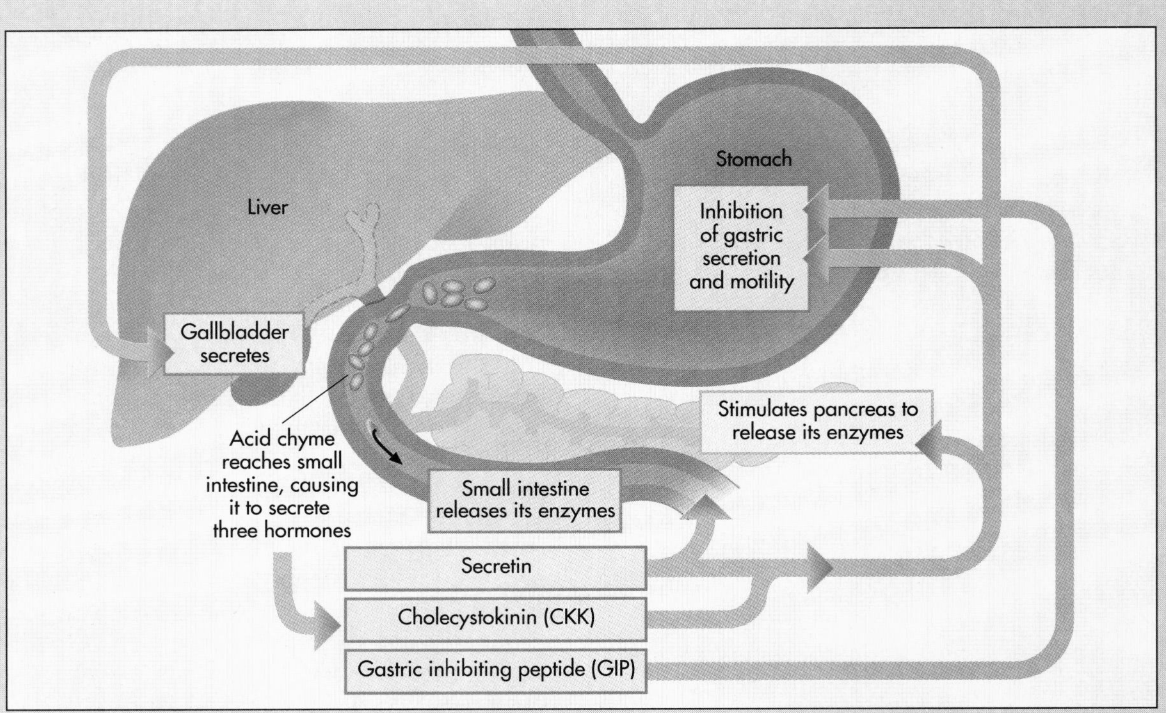

Inhibition of gastric secretion and motility and the effects of food entering the small intestine.

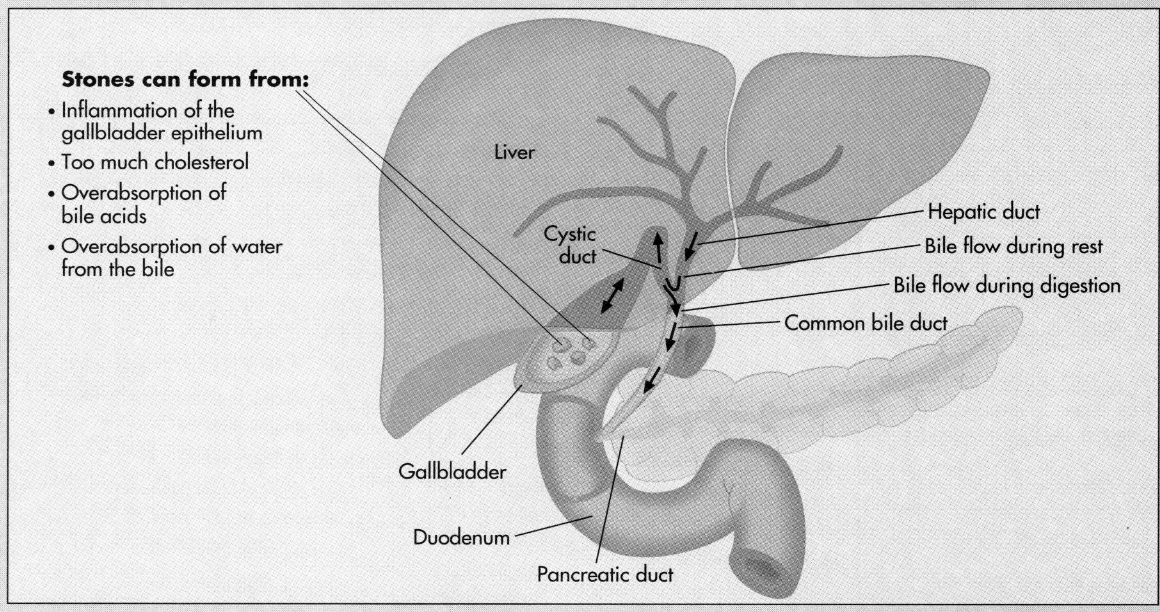

The flow of bile and the formation of gallstones.

THE LARGE INTESTINE

The large intestine secretes mainly mucus, and its movements are under the control of reflexes. When food enters the stomach, a **gastroileal reflex** intensifies peristalsis of the ileum to force chyme into the large intestine. Food in the stomach also triggers a **gastrocolic reflex,** a wave of **mass peristalsis** that sweeps the length of the large intestine and pushes material into the rectum. Distension of the large intestine triggers **haustral churning,** in which full segments of the large intestine (the haustra) contract. Distension of the rectum triggers the defecation reflex, which is usually under voluntary control in adults.

Learning Focus Response

REGULATION OF MOTILITY AND SECRETION

Use the figure that follows to help you summarize and review what you have learned about the regulation of motility and secretion in the digestive system. Make several copies of the drawing. Draw the information from other diagrams and descriptions in the text. Write it directly on this diagram in the appropriate place. Then draw arrows to show the interactions between the various items in the drawing. Add words to describe what is going on. When you are satisfied that you have captured all the important information, repeat the exercise with another of your copies. Do it again when you are preparing for exams.

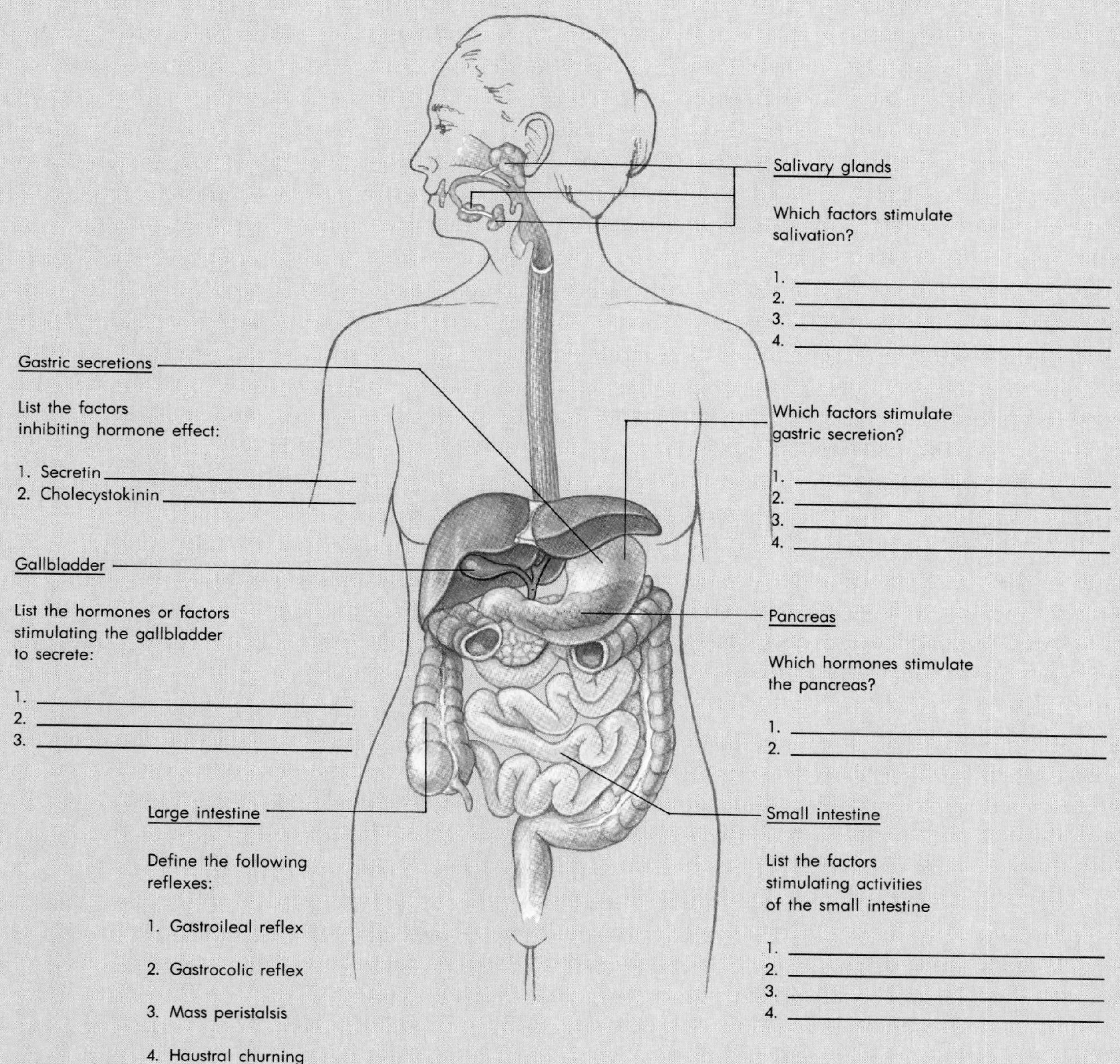

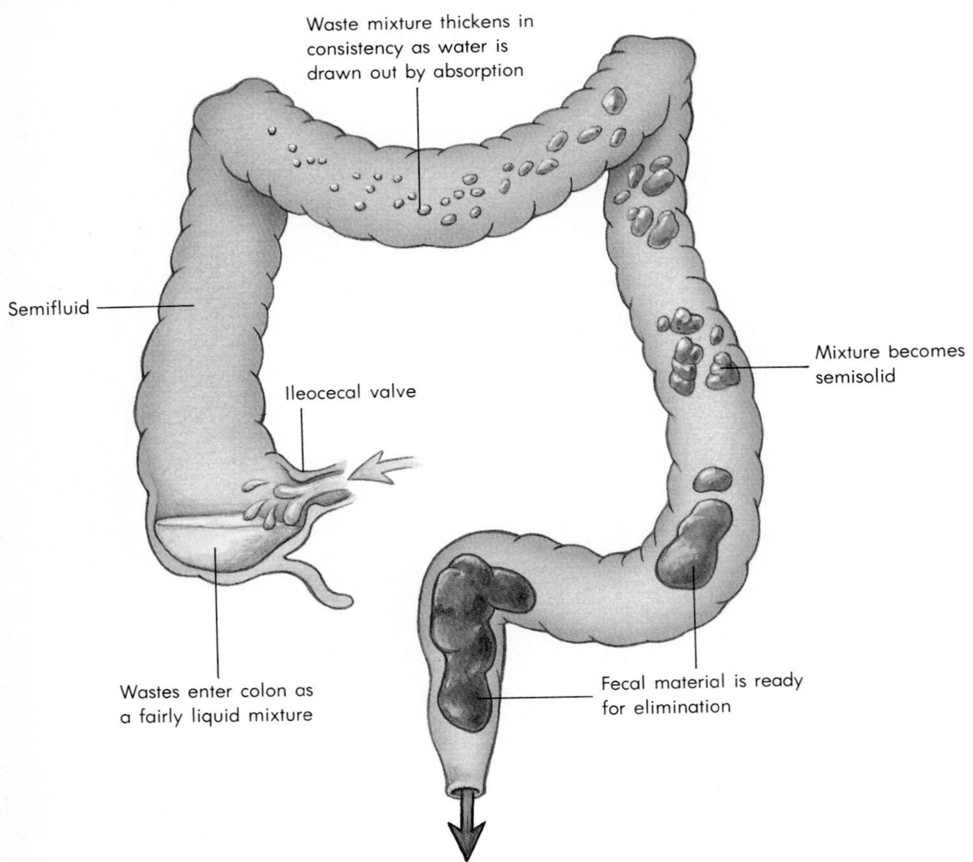

FIGURE 20-11 Preparation of solid wastes by the colon.

ABSORPTION

Digestion achieves its goal when the proteins, starches, fats, and nucleic acids in food have been broken down into their component units. But the digestive process is not yet finished. Next the system makes those nutrients available to the cells of the body. To do this, it must pass the nutrients through its wall into the bloodstream in the process we call absorption.

Nine-tenths of all absorption of nutrients occurs in the small intestine. The mucous membrane of the mouth permits the passage of small amounts of alcohol but very little else. The membrane of the stomach is permeable to water, sugar, salts, aspirin, additional alcohol, and some other drugs. Thus, alcohol reaches the bloodstream very quickly, the reason that drinking on an empty stomach affects the brain so promptly. Food in the stomach dilutes alcohol and slows its absorption.

The nutrients enter the blood and lymph through the villi, passing through the small intestine's mucous membrane by diffusion, facilitated diffusion, osmosis, and active transport (see Chapter 3). Nutrients, water, and salts that are not absorbed in the small intestine are absorbed in the large intestine, along with the vitamins produced by the bacteria that dwell there (Figure 20-12).

Nutrients

The monosaccharide fructose enters the surface cells of the villi by facilitated diffusion. Glucose and galactose enter by active transport; adenosine triphosphate (ATP) powers a molecular "pump" in the cell membrane that carries these sugars, together with sodium ions, into the cell. Once inside the cell, all three sugars diffuse into the capillaries of the villi (Figure 20-12).

Amino acids are also transported actively, along with sodium, into the intestinal cells, thence to diffuse into the blood of the capillaries. Dipeptides and tripeptides (joined pairs and triplets of amino acids) are also transported into the cells; once there they are broken into single amino acids, which diffuse into the blood. Occasionally, whole proteins penetrate the intestinal cells and reach the blood. As a result, the proteins can stimulate the immune system and cause *food allergies* (see Chapter 24).

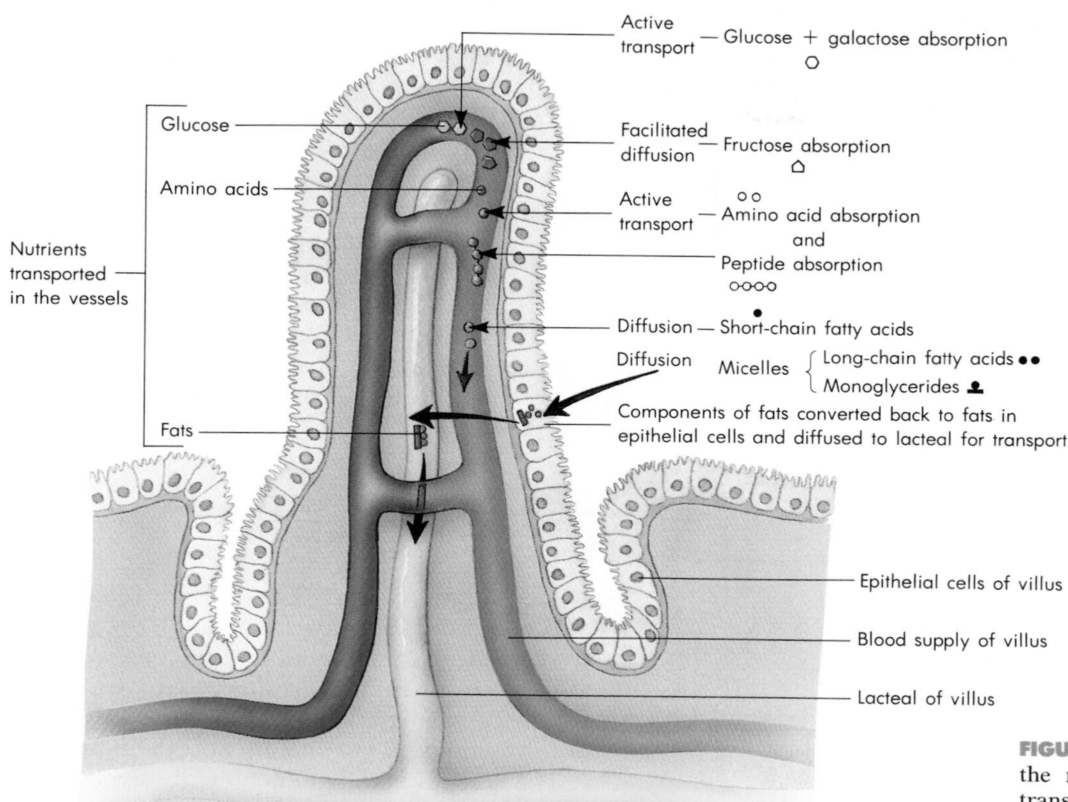

FIGURE 20-12 A summary of the routes of absorption and transport for some common nutrients absorbed through the villi.

Fats exist in food mostly as **triglycerides,** three fatty acids fastened to a glycerol unit. Digestion liberates two of the fatty acids from the glycerol, producing free fatty acids and **monoglycerides.** Short fatty acids diffuse into the cells of the villus and then into the blood, like sugars and amino acids. Longer fatty acids and monoglycerides (as well as the relatively few remaining **diglycerides** and triglycerides) dissolve into clumps of bile salt molecules known as **micelles.** The micelles serve as carriers for the fatty acids and monoglycerides. When they collide with the membranes of villi cells, their cargo diffuses into the villi cells. The micelles then are free to pick up another load and repeat the process. Some of the bile salts of the micelles are themselves absorbed in the ileum and returned to the liver for reuse.

Once within the cells of the villi, the fatty acids and monoglycerides are recombined into triglycerides. These fats are then packaged with phospholipids, cholesterol, and protein into **chylomicrons,** passed to the lacteals of the villi, and delivered by the lymphatic system to the large veins of the circulatory system near the heart.

Table 20-4 compares the usual amounts of each type of nutrient absorbed by the small intestine each day to the potential amount that could be absorbed.

Table 20-4

Nutrient Absorption by the Small Intestine

FOOD CATEGORY	AMOUNT NORMALLY ABSORBED PER DAY	MAXIMUM CAPACITY ABSORBED PER DAY
Carbohydrates	Several 100 grams (g)	Several kilograms (kg)
Fats	100+ g	500–1000 g
Amino acids	50–100 g	500–700 g
Water	8–9 liters	20+ liters

Water

Each day a person secretes about 7.5–8 liters of water in saliva and stomach, intestinal, and pancreatic juices. He or she takes in only about 1.5 liters with food and drink, but the discrepancy is not real, for the secreted water is not lost. It is recycled and absorbed by osmosis, accompanying other nutrients and keeping the contents of the intestinal cells from becoming hypertonic. The small intestine absorbs 8–8.5 liters. The rest passes to the large intestine, where most of it also returns to the body. Only about 100 milliliters (ml) is lost in the feces per day and another 1–1.5 liters is lost as urine, sweat, and exhaled water vapor (Table 20-5).

Salts

Like water, salts (electrolytes) are components of both food and digestive secretions. Sodium ions enter the intestinal cells by diffusion and by active transport. Other ions, such as chloride, can follow sodium passively, drawn by electrostatic attraction. They can also be transported actively. Most ions are absorbed by active transport alone. Calcium requires in addition vitamin D and parathormone. Iron ions form a complex with a protein, **ferritin,** in the intestinal cells and are released as needed to the blood.

NUTRIENT TRANSPORT AND THE HEPATIC PORTAL SYSTEM

Once they reach the bloodstream, the materials absorbed from the digestive system, including nutrients, salts, water, and sometimes poisons, are not simply delivered to cells throughout the body. They are first sent to the liver via the hepatic portal system for processing and detoxification. The venules and veins that drain blood from the various areas of the digestive system merge to form the **hepatic portal vein.** This vein carries blood directly from the capillary beds of the digestive system to the liver, where it branches to form a second capillary bed. The hepatic portal vein is thus unlike most other veins in the body. It does not flow from a capillary bed to the heart. Instead, it links two capillary beds to form the **hepatic portal system** (Figure 20-13).

The Liver

The liver is one of the body's largest organs. It weighs about 1.4 kg (3 lb) and has the unique ability to regenerate up to about half its mass after surgical removal of tumors or damaged tissue. In it, sheets of liver cells are exposed to blood delivered by both the hepatic portal vein and the hepatic artery. The central veins of each of the liver's many lobules then merge to form the hepatic vein, which joins the inferior vena cava. Tiny **canaliculi** between the liver cells collect the bile secreted by the cells and deliver it to bile ducts; the accumulated bile is stored in the gallbladder until needed in the small intestine.

The cells of the liver serve a variety of functions. In addition to making bile, they synthesize most of the proteins found in blood plasma and destroy worn or damaged red and white blood cells. Their enzymes convert the nitrogen released when amino acids are broken down into the less toxic urea, which later is excreted in the urine. Other enzymes detoxify drugs, including alcohol, pesticides, and other harmful compounds present in the blood,

Table 20-5

Water Management in the Digestive System

BODY AREA	AMOUNT SECRETED (ml/day)	AMOUNT ABSORBED (ml/day)
Salivary glands	1200	0
Esophagus	Minimal mucus	0
Stomach	2000	Minimal
Gallbladder	700	0
Small intestine	2700	8000–8500
Large intestine	60–100	400–900

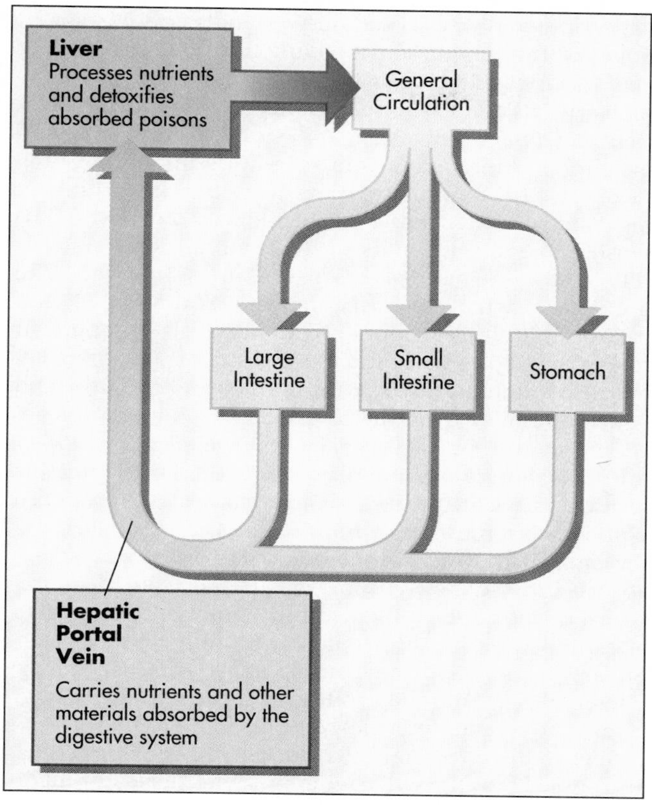

FIGURE 20-13 The role of the hepatic portal system in the circulatory system.

which have been absorbed either with the food or from the air. Those compounds it cannot detoxify, such as dichlorodiphenyltrichloroethane (DDT), it may store. Prolonged exposure to alcohol and some other compounds, including some actually made more toxic by the detoxifying enzymes, can lead to **cirrhosis,** a condition in which dead liver cells are replaced by fat and connective tissue, causing the vital services provided by the liver to be lost to the body.

Most important in the context of this chapter is the fact that the liver processes nutrients. Under the control of the hormones insulin and glucagon (see Chapter 13), the liver collects excess monosaccharides and converts them to glycogen or fat for storage; when the blood sugar level falls, it then releases glucose from glycogen or by conversion of fat or protein. It also stores some vitamins and minerals.

SUMMARY

Digestion is the process of breaking the macromolecules in food (starch, protein, fat, and nucleic acids) into their basic units for absorption and use by the body. It has two components: mechanical and chemical digestion. Mechanical digestion grinds food into small particles accessible to the digestive enzymes that carry out chemical digestion. Mechanical digestion begins in the mouth, with the action of the teeth and the tongue, and continues in the stomach and small intestine, where muscular contractions mix food with fluid secretions. Chemical digestion also begins in the mouth, where food is mixed with saliva, which contains the starch-digesting enzyme salivary amylase. The stomach supplies hydrochloric acid and pepsin, which attack protein. The pancreas and small intestine supply the rest of the necessary enzymes.

The digestive system is a tube running through the body from mouth to anus. It consists of the mouth, pharynx, esophagus, stomach, small intestine, large intestine (or colon), rectum, and anus. It also includes the glands that supply various substances necessary for digestion. These glands are the pits of the stomach and intestinal wall, the salivary glands, the liver, gallbladder, and pancreas.

The lining of the digestive tube has four layers: the mucosa, the submucosa, the muscularis, and the serosa. The mucosa contains or gives rise to the glands; its surface area is increased by numerous folds. The submucosa contains blood vessels and controlling nerve cells. The muscularis contains two or three layers of smooth muscle, generally one circling the digestive tube and one running its length. The serosa covers the outer surface and merges with the peritoneum that lines the abdominal cavity.

The task of the mouth is to pulverize food by chewing and to mix it with saliva, which lubricates the mouth and converts the food into a solution that can easily be processed by the digestive system. In addition, the saliva supplies antibacterial agents and salivary amylase, which begins the breakdown of starch. Chewing is done with the teeth, composed of calcium carbonate and phosphate. The surfaces of the teeth are covered with hard enamel; below the gum line, they are covered with cementum, which joins them to the periodontal ligament that anchors them in their sockets in the jaw bones. Tooth decay (dental caries) is caused by bacteria that live in deposits of food (plaque) left between the teeth and secrete acid that attacks enamel; it can be prevented by flossing and brushing to remove the plaque and by toughening the surface of the tooth with fluoride treatments.

Swallowing propels a bolus of food from the mouth through the pharynx and into the esophagus, where peristaltic muscle contractions push the food into the stomach. In the stomach, muscle contractions mix food with mucus, acid, and pepsin to produce the soupy chyme. Peristalsis then propels the chyme through the pylorus into the small intestine.

The small intestine has three segments: the duodenum, the jejunum, and the ileum. In the relatively short duodenum, the acid chyme is neutralized by the alkaline pancreatic juice. The pancreatic juice also contains numerous digestive enzymes. Bile from the liver serves to emulsify fat and aid in both its digestion and absorption.

Most digestion occurs in the second and third segments, the longer jejunum and ileum, where pancreatic and intestinal enzymes attack the food. The lining of the small intestine has its surface area increased tremendously by folds and fingerlike villi, where absorption occurs. Peristalsis propels food through the small intestine. Segmentation mixes it with the fluids that were added there.

At the end of the small intestine is the ileocecal valve, which opens into the first section of the large intestine, the cecum. The appendix is attached to the cecum as a finger-shaped pouch. The function of the colon is to absorb the few remaining nutrients, excess water, and vitamins produced by bacteria living there and to compact the residue for elimination as feces. To function properly, it requires large quantities of indigestible material, that is, fiber, or roughage. Bulk in the colon stimulates the muscular contractions that move food residues to the rectum. A shortage of bulk—or fiber—can lead to constipation, diverticulosis, and other ills.

The secretory and motor activities of the digestive system are regulated by both nerve signals and hormones. Some of this regulation is voluntary, as with chewing and the first stages of swallowing. Some involves the higher levels of the nervous system, for thought and emotion can affect salivation and stomach secretion and motility via autonomic control centers in and near the medulla. Still other regulation involves reflexes linking mechanical (distension) and chemical (protein, acid, fat) sensors to control centers in the brain and to the nerve cells of the intrinsic plexus in the intestinal wall. The hormones gastrin, secretin, cholecystokinin (CCK), and gastric inhibiting peptide (GIP) also play important roles. They not only affect the digestive tract's muscles but also stimulate the release of bile from the liver and pancreatic juice from the pancreas.

The mouth can absorb alcohol, the stomach water, salts, alcohol, and some drugs, and the large intestine residual nutrients, water, salts, and bacterially produced vitamins, but most absorption occurs through the villi of the small intestine. Amino acids and most sugars are absorbed by active transport, as are most salts (electrolytes). Fructose, fatty acids, and monoglycerides enter intestinal cells by diffusion. Water enters by osmosis. Sugars, salts, water, and amino acids diffuse from the intestinal cells into the capillaries of the villi. Long fatty acids and monoglycerides are assembled into triglycerides in the intestinal cells, packaged as chylomicrons, and passed into the villi's lymphatic capillaries, or lacteals.

Blood travels from the small intestine in the hepatic portal vein, which breaks up into a second capillary net in the liver. The liver processes and stores nutrients and detoxifies many potentially harmful compounds. The lymph from the lacteals is delivered to the bloodstream near the heart; it reaches the liver only as part of the general circulation.

Table 20-6 provides a summary of selected diseases of the digestive system.

Table 20-6

Clinical Summary: Selected Diseases of the Digestive System

DISORDER	DESCRIPTION
Appendicitis	Inflammation (infection) of the appendix. The swollen appendix may burst, releasing fecal material and pus into the abdominal cavity and thus causing peritonitis. Symptoms are abdominal pain and vomiting. Treatment calls for removal of the appendix.
Caries	Tooth decay, erosion of the tooth substance by acid produced by bacteria living in deposits of food and plaque on the teeth. Treatment requires removing the deposits (brushing and flossing) and avoiding acid-promoting foods such as sugar. Fluoride treatments also strengthen the tooth enamel. Sealant treatments plug pores and cracks and thus interfere with both attachment of plaque and penetration of acid.

DISORDER	DESCRIPTION
Cirrhosis	Liver damage caused by exposure to toxins such as alcohol, liver parasites, and infections. Dead liver cells are replaced by fat and connective tissue. In severe cases, liver function is lost. Symptoms include jaundice, swelling of the legs, and poor blood clotting, among other things. Treatment involves curing the infection or halting the exposure to toxins.
Diverticulitis	Inflammation of colon diverticula (out-pouchings); affects about 15 percent of people with diverticulosis. Increasing intake of high-fiber foods often improves symptoms. Severe cases may require surgery.
Peritonitis	Inflammation (infection) of the membranes lining the abdominal cavity and covering the abdominal organs. May result from wounds (e.g., bullet or knife wounds) or perforations of the intestines. If untreated, death is likely. Treatment requires large doses of antibiotics.
Ulcer	A damaged portion of stomach (gastric ulcer) or duodenal (duodenal ulcer) wall exposed to gastric juices. Causes may be excess secretion of gastric acid or pepsin or deficient secretion of gastric mucus, triggered by bacterial infections of the stomach, stress, anxiety, foods, medications, or other factors. Consequences may be internal bleeding or perforation of the stomach or duodenal wall and peritonitis. Treatment requires reducing stress, changing diet or medications, or taking drugs that reduce acid secretion.

STUDY QUESTIONS

1. What is the difference between digestion and absorption?
2. A peptic (stomach) ulcer is a hole or lesion digested by acid and pepsin in the mucous membrane of the stomach. As such a hole grows deeper, what layers of the stomach wall does it penetrate? List them in order, beginning with the mucous membrane.
3. What makes the mouth a favorable environment for bacterial growth? What makes it an unfavorable environment?
4. List the ways in which the digestive system maximizes its internal surface area for the purposes of secretion and absorption.
5. Discuss the differences between the digestive enzymes secreted by the pancreas and those secreted by the small intestine.
6. What is the difference between peristalsis and segmentation?
7. Why should you be sure that your diet contains adequate amounts of fiber?
8. Despite their name, lacteals have no more to do with absorbing the nutrients in milk than with absorbing those in other foods. What function do they serve?
9. What is the difference between chylomicrons and triglycerides?

GLOSSARY

Absorption The process of taking nutrients from the digestive tract into the blood.

Accessory digestive organs The various glands that release secretions into the chambers of the digestive system, including the salivary glands, pancreas, and gallbladder.

Bile A mixture of cholesterol, bile pigments, and bile salts, which serve as emulsifiers for fat digestion.

Chief cells Stomach gland cells that secrete pepsinogen.

Cholecystokinin (CCK) A hormone secreted by the small intestine that inhibits stomach secretion and muscle contractions of the stomach and intestines and stimulates release of bile by the gallbladder and enzyme-rich pancreatic juice.

Colon The large intestine.

Digestion The process of breaking food down into its component nutrients.

Digestive system The array of organs that carries out digestion; mouth, pharynx, esophagus, salivary glands, stomach, small and large intestines, liver, pancreas, appendix, rectum, and anus.

Duodenum The first 25 cm of the small intestine.

Enteroendocrine cells Stomach gland cells that secrete the hormone gastrin.

Esophagus The gullet; the tube from throat to stomach.

Gallbladder The storage sac for bile secreted by the liver.

Gastrin A hormone secreted by stomach gland cells (and by cells in the small intestine) that stimulates stomach secretion and relaxes the pylorus.

Hepatic portal system Carries nutrients absorbed by the capillaries of the digestive system to the liver for processing.

Lacteal The lymphatic capillary within a villus.

Microvilli Fingerlike protrusions of individual intestinal cell membranes.

Pancreas The gland that secretes many of the digestive enzymes.

Pepsin Protein-digesting enzyme secreted in the stomach as pepsinogen.

Peristalsis A wave of muscular contraction that propels food through the digestive tract.

Pharynx The chamber forming the throat region behind the nasal and oral cavities and in the vicinity of the larynx. It serves as a common passageway for liquid, food, and air.

Pylorus The end of the stomach nearest the small intestine.

Salivary amylase A starch-digesting enzyme in saliva.

Salivary glands The three pairs of glands that release saliva into the mouth.

Secretin A hormone secreted by the small intestine that inhibits stomach secretion and stomach and intestinal muscle contractions and stimulates secretion of bile and alkaline pancreatic juice.

Small intestine Main site of digestion and absorption in the digestive system; links stomach and large intestine.

Stomach A distensible sac that receives food and liquid from the esophagus and mixes it with gastric juices, which include water, acid, and the enzyme pepsin, and prepares it for the small intestine.

True digestive organs The various chambers and compartments of the digestive system, including the mouth, esophagus, stomach, small intestine, large intestine, and rectum.

Villi Fingerlike protrusions from the small intestine wall, through which most nutrient absorption occurs.

Chapter 21

Energy Metabolism

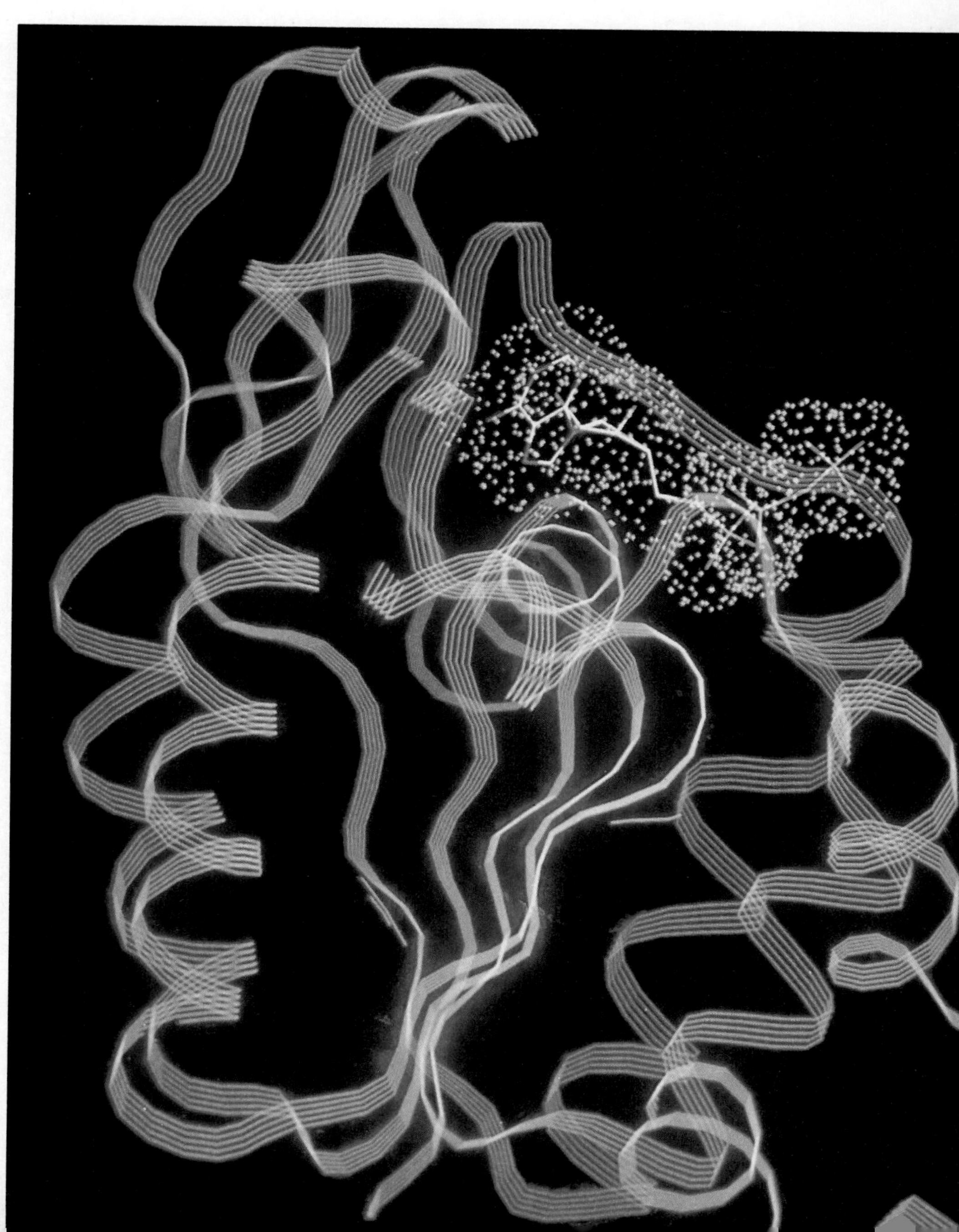

The Environmental Protection Agency requires that every new-car price sticker tell consumers how many miles per gallon of gasoline (city and highway) they can expect the car to get as well as the estimated fuel costs for the first year of operation. Similar "mileage ratings" could be attached to the human body. Just as a car requires **energy** in the form of the fuel it consumes in its engine, so does the body require energy to fuel the metabolic engines of human cells (Figure 21-1).

The fuels powering machines and bodies have a common source, the sun. In both cases, the fuels are organic molecules made by living organisms, ei-

(a)

(b)

(c)

(d)

FIGURE 21-1 Fuel consumption in living and nonliving machines. (a) Gas shortages in the 1970s caused long lines of cars to form around gas stations. (b) People in line to refuel their bodies at a familiar Tokyo fast food vendor. (c) A young male lion is refueling at the expense of a very dangerous Cape buffalo. Note that the buffalo's right horn is broken, making it more vulnerable to such attacks. (d) Plants refuel by capturing a portion of the sun's energy during photosynthesis.

ther now or in the past. For cars, the energy-rich organic molecules of gasoline were originally formed millions of years ago, when the **photosynthesis** of plants living in ancient swamps and seas trapped solar energy in a chemical form that could be preserved by burial. The petroleum industry draws these rich deposits from subterranean pools and refines them into gasoline. For the human body, the fuel also consists of organic molecules. We get them as components of the food we eat in the form of carbohydrates, fats, and proteins. Ultimately, human fuel molecules are also formed by the photosynthetic activities of contemporary plants or by animals that have eaten plant products.

The need for a continuous supply of energy is the common denominator shared by all life forms. It forms the basis of food chains in nature and, indeed, is responsible for the tremendous diversity we see in the living realm. The story of life, to a large extent, is the story of complex energy-manipulating systems. If we are to understand life, including our own lives, we must understand how these energy-processing systems operate.

REFINING OUR FUELS: DIGESTION

Our cells cannot use the organic molecules we ingest in food directly. With the aid of digestion, we break down our foods into simple sugars, amino acids, and fat components that are absorbed through the walls of the digestive system into the blood (see Chapter 20). Once in the blood, these substances are delivered to all the trillions of cells of the body. These nutrient molecules can then be used for many purposes—as building materials for cell growth and repair, as coenzymes (vitamins), and as sources of the energy needed to provide the power for life itself. It is the latter we will focus on in this chapter.

METABOLISM

The sum total of the body's chemical activities is called **metabolism.** It includes reactions such as breaking down organic molecules in order to produce ATP, converting one organic substance to another, constructing proteins by chaining sequences of amino acids together, and thousands of other processes. The ATPs generated during cellular respiration are used to pay the energetic expenses for all the body's chemical activities (more about this later). Consequently, it is not uncommon to measure the rate of cell respiration (or more properly, oxygen consumption) to measure how fast the body's metabolism is functioning (Figure 21-2).

These metabolic processes can be divided into two broad groups, those of anabolism and those of catabolism. Stated simply, **anabolism** includes all those reactions in which complex molecules are synthesized. It includes the processes by which proteins, fats, and nucleic acids are manufactured from simpler building blocks. Usually, anabolic reactions consume energy; that is, they must be powered by a source of ATP (Figure 21-3).

Catabolism, on the other hand, includes all reactions in which substances are broken down into simpler units. For instance, the digestive enzymes that break down ingested proteins into amino acids are said to trigger catabolic reactions. Very often, catabolic reactions release energy as the more complex starting molecules are broken down. Some of the end products of catabolic reactions can be fed into the Krebs cycle to extract the energy they contain. Other catabolic processes continually destroy

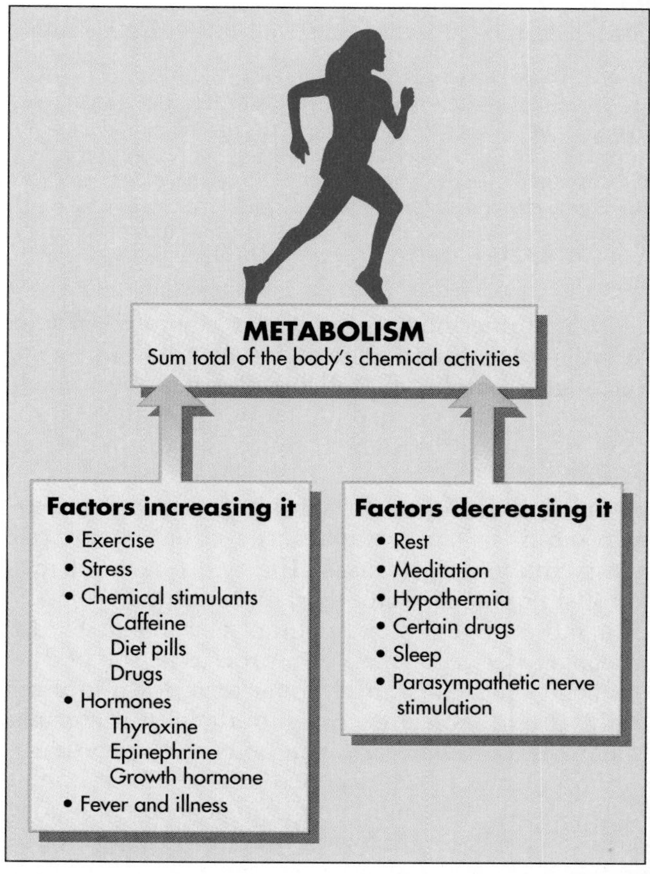

FIGURE 21-2 A few of the factors influencing metabolic rate.

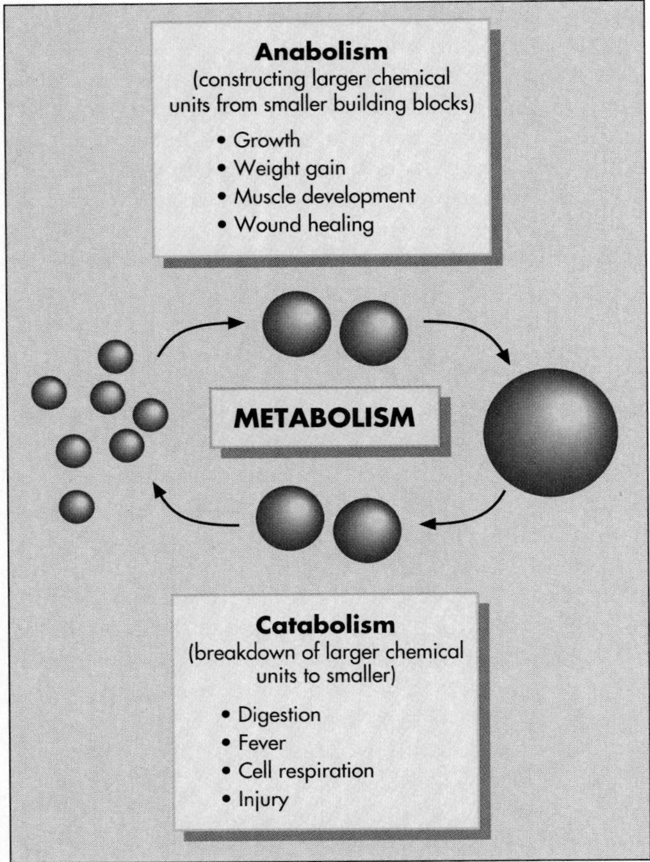

FIGURE 21-3 The processes of metabolism fall into two broad categories: Catabolism is those processes that tear down more complex chemicals into their simpler chemical components. In contrast, anabolism consists of building up more complex materials from simpler building blocks. Both of these processes are ongoing in the body. Normal physiologic systems use a blend of both processes. The examples listed in the diagram are very simple. Give some additional examples of anabolism and catabolism.

aged, nonfunctional cells, digest ribonucleic acid (RNA) molecules that have served their purpose, and digest patches of cell membrane, often using the lysosomes described in Chapter 3. As they do so, they release amino acids, lipids, and other compounds.

Catabolism thus tears the body down. Anabolism builds it back up again, recycling the components catabolism releases. The two phases of metabolism balance each other, maintaining the body in a more or less steady state. As a result of the interplay between these two processes, the body's molecules are constantly being torn down and rebuilt. Often, molecules are rebuilt with materials taken from elsewhere in the body, but eventually every molecule is replaced with materials obtained in the food. Although this change occurs sooner with some tissues than with others, over the course of about seven years every atom in the body is replaced by new atoms taken from food.

1. An interesting puzzle is that of "Achilles's ship." Achilles was an ancient Greek hero. His ship was an old tub that had been repaired so often that one day Achilles realized that not one plank, peg, or bit of rope was the same as it had been when the ship was new. He then asked, "Is my ship indeed the same ship I bought so long ago? Or is it a new ship?" Clearly, you can ask a similar question: Is your body the same as it was seven years ago? Is your "identity" the same? Are you still you?
2. If your answer is yes, you should now ask yourself why you are indeed the same person you always were. What does your present body have in common with your old one? How about your "identity"?

Basal Metabolic Requirements

Food provides both raw materials and energy to the cells of the body. In this chapter, we are most concerned with the energy needed by the body to support its various activities. We need first to ask just how much energy the body requires to survive and, beyond that, how much additional energy is required to pay for all the body's other activities as well. Too little energy leads to illness or even death (through malnutrition), just as surely as too much food leads to a weight gain.

Obesity, defined as being 20 percent or more overweight, is a serious condition affecting about 34 million Americans. It is known to shorten life spans by causing or aggravating many diseases, including high blood pressure, atherosclerosis, diabetes, and even some types of cancer.

To answer our questions, we must recognize that the body takes the energy it needs from its food, converting glucose and amino and fatty acids to ATP through cellular respiration and then using the ATP to power its activities. When the food that is eaten contains more energy than the body needs, the body stores the surplus, largely as fat in adipose tissue. If this excess continues for a long enough time, obesity can result. When the food that is eaten contains less energy than the body needs in a given day, the body makes up the deficit by drawing energy from its own pool of organic chemicals and processing these. Thus, when the diet fails to meet the body's needs, the body loses weight by consuming its own resources as alternative fuels. In extreme cases, it can eventually become desperate enough to consume even the essential proteins of the heart. Of course, unless this trend is reversed, death shortly follows.

Calories: The Measure of Energy

When we eat only enough to pay the energy costs of our activities, our weight stays constant. It is thus possible to estimate how much energy we need simply by measuring how much we eat during periods when we are neither gaining nor losing weight. However, it is simpler to take advantage of the fact that energy of all kinds can be expressed in terms of heat. In fact, whenever energy is used in the body, or elsewhere, it is ultimately converted to heat.

We measure heat as calories. One **Calorie** (Cal) is the amount of heat necessary to warm one liter of water by one degree Celsius. (Be careful to capitalize Calorie—it equals 1000 calories, and it is also called a kilocalorie, or kcal.) If we burn organic substances, like food, in a **calorimeter,** a device that allows all the heat released by burning to warm a quantity of water, we find that 1 g of carbohydrate (starch or glucose) contains 4.1 Cal; a gram of protein contains 5.4 Cal; and a gram of fat contains 9.3 Cal. "Burned" in the body, carbohydrate and protein release about the same amounts of heat; protein, because it is not completely burned, releases only 4.2 Cal (Table 21-1).

Energy Costs and Lifestyle

We could measure the heat released by people and other animals as they use their food by putting them in a calorimeter too. It is usually more practical, however, to measure their oxygen consumption. Because of the close link between the use of oxygen and the production of ATP, each liter of oxygen consumed results in the production of 4.8 Cal of energy. The rate of energy production is known as the **metabolic rate.**

When a person is relaxed and resting quietly, some hours after a meal (so that no excess energy is being used in digestion) and without any drugs in the body, he or she is using energy only to run the heart, lungs, kidneys, and other life support organs and to maintain baseline levels of nervous activity, muscle tone, and body temperature. The rate of energy use for these purposes is the **basal metabolic rate (BMR).** It is the minimum expenditure of energy needed to stay alive and awake (it falls about 10 percent during sleep). It increases with the size of the body and decreases with age, and it averages 1650 Cal per day for males and 1350 Cal per day for females.

However, the actual energy a person expends per day includes not only their basal metabolic rate but also any additional energy needed to support their activities beyond simple survival. Activities like sitting up, talking, going to work or school, and anything else are extra and are paid for by dietary adjustments in the energy intake over the long haul. These other activities are sometimes referred to as **voluntary work.** The relationship is described in the following equation (Figure 21-4):

$$\text{Energy needed} = \text{basal metabolic requirement (BMR)} + \text{energy cost of activities (voluntary work)}$$

There is a relatively easy method of calculating the energy needs of moderately active people. If you are *female,* simply multiply your body weight by *15 Cal/lb per day.* The answer, in Calories needed per day, includes energy costs of both the BMR and the additional activities associated with a moderately active lifestyle (like going to work or school, socializing, etc.). Thus, a 100-lb moderately active female (a fairly small female) probably burns up about 1500 Cal on an average day. On the other hand, a moderately active *male* burns about *18 Cal/lb per day.* Therefore, a moderately active, average-sized male (155 lb) uses about 2790 Cal per day. Of course, the more active a person is, the greater the Caloric costs; construction workers may use 4000 or more. Researchers have also figured the additional energy costs of many single activities, a few of which are listed in Table 21-2.

Table 21-1

Caloric Values of Nutrients (Cal/g)

NUTRIENT	CALORIES RELEASED IN CALORIMETER	CALORIES RELEASED IN BODY
Carbohydrate	4.1	4.1
Protein	5.4	4.2
Fat	9.3	9.3

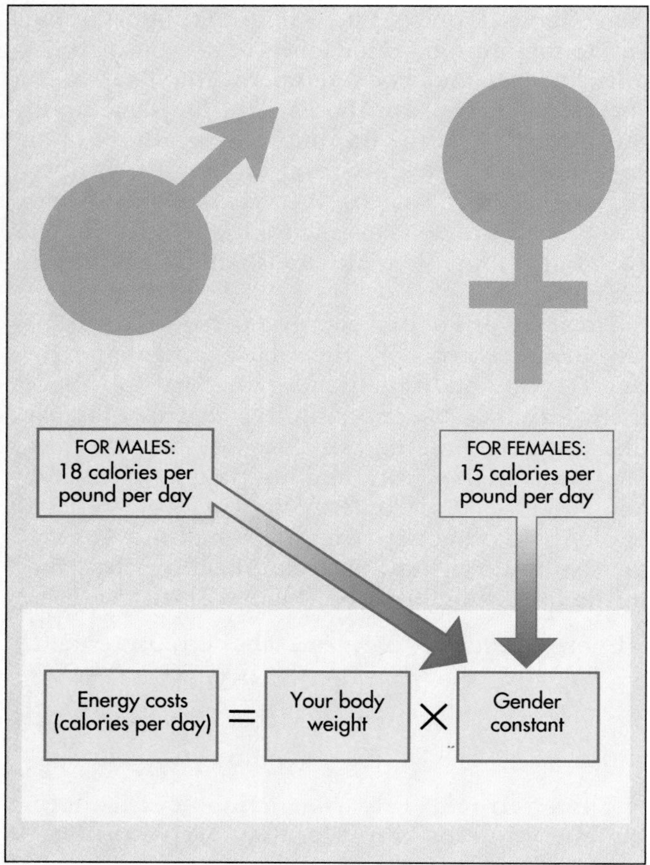

FIGURE 21-4 Calculating energy costs for supporting a moderately active lifestyle.

Metabolic Rate and Body Temperature

Metabolic rate also varies with body temperature. Each one-degree change in temperature changes the metabolic rate by about 10 percent, largely because the chemical reactions of metabolism proceed more quickly at higher temperatures. People whose temperatures are higher than normal—either because of fever or because they simply "run hotter"—thus use more calories. People with lower body temperatures use fewer calories.

Table 21-2

Energy Costs of Selected Activities

ACTIVITY	ENERGY COST, CAL/HR
Lying down	90
Sitting	96
Driving a car	168
Bicycling leisurely	270
Bicycling vigorously	666
Dancing slowly	312
Golfing	300
Tennis	426
Walking rapidly	312
Swimming	300–666
Jogging	560
Scrubbing floors	204

1. Consider two people. One has a body temperature half a degree above the 37°C human average; the other's temperature is half a degree below the average (both are well within the normal range of human variation). They each consume enough calories to meet the needs of an individual of normal body temperature with their level of activity. What can you predict about the shapes of their bodies? That is, which one has excess calories to store as body fat? Which one must draw on body reserves?
2. Does the dependence of metabolic rate on body temperature suggest that steam baths might help a weight loss program? Try a little arithmetic: How many calories per hour does the average college-age male use? How many more would he use if his body temperature were raised by one degree in a steam bath or hot tub? How much good would an hour's soak do him? (We are not, however, advocating the use of a steam bath or a spa as part of a weight-loss program.)

Metabolic Regulation

It is possible to use drugs to alter one's metabolic rate. Caffeine is a metabolic stimulant, raising the rate. Amphetamines are much more powerful stimulants, one of the features that makes them attractive as "diet pills" (they also suppress appetite). There are also suppressants that turn down the rate of metabolic activity. One of the most common ways to slow metabolism is to lower the body temperature, which reduces the body's need for oxygen and nutrients; the cooled state, known as **hypothermia,** is especially helpful during heart surgery, when the circulation may be interrupted for a considerable time.

The body controls its metabolic rate in several ways. During vigorous exercise, when the body's production of energy may increase immensely, most of these control methods are apparent. Body temperature goes up during exercise. As the body warms, the metabolic rate also goes up. In addition, two hormones, thyroid hormone (thyroxine) and epinephrine, both secreted during exercise, increase the release of glucose from glycogen in muscle and the rate of glucose use and ATP production. Growth hormone, also secreted during heavy exercise, increases the release of fatty acids from fat for use by muscle cells; it also increases glucose release but not glucose use. Still another control method involves the sympathetic nervous system; the norepinephrine its cells release can boost the metabolic rate greatly.

Similar reactions occurring during times of stress explain why the metabolic rate goes up when one is anxious or afraid and hence why "nervous" people tend to be thin. Related effects explain why the metabolic rate goes up during pregnancy.

ENERGETICS OF EXERCISE AND WEIGHT CONTROL PROGRAMS

The bookstores are full of diet books promising to slim you down like the people you see on television, improve your sexual and athletic prowess, and reduce your risks of having a heart attack. Every week, several new diet books reach the markets offering their own special approaches for helping you attain personal perfection. Some say the route is most efficiently traveled by eating seaweed; others suggest that the secret lies in drinking large volumes of water each day. Regardless of the gimmick, the basic assumptions are similar for most weight control diets. And here they are:

The target of most weight control programs is fat, and each *pound of fat* contains about *3500 usable Calories* (Figure 21-5). Therefore, for each pound of fat loss, the dieter (or exerciser) must force his or her metabolism to burn off 3500 Cal from the fat reserves of the body. That can be achieved in two main ways. The person can make the body more expensive to operate by increasing the activity level through an exercise program without eating more to compensate. Or the person may keep the same activity level but reduce the food intake through dieting. The most effective approach for many people is to eat less and exercise more.

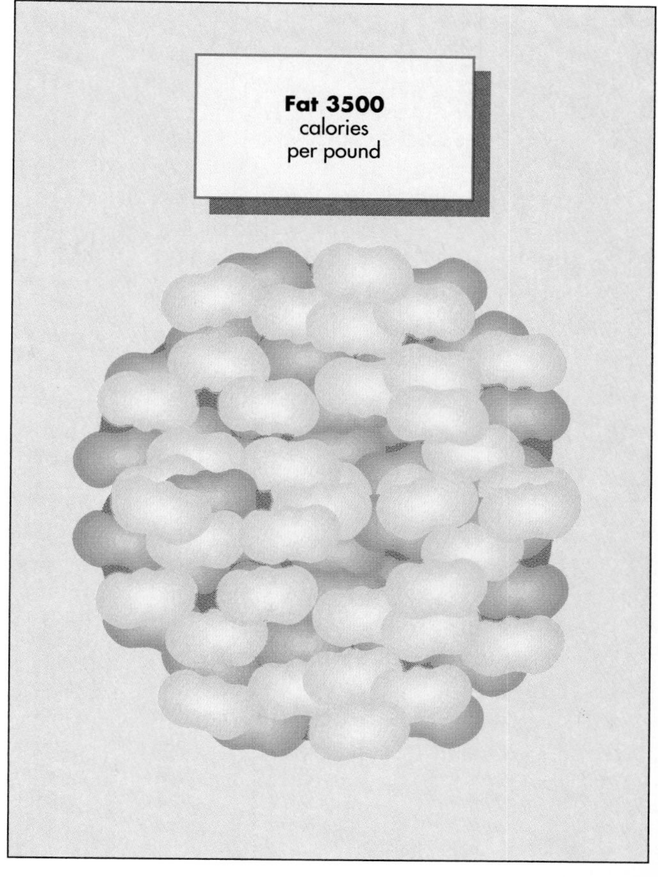

FIGURE 21-5 Each pound of fat contains about 3500 usable Calories. That figure is used in most diet calculations, so if a person's goal is to lose a pound of fat in a week, they must somehow run a 3500 caloric deficit (burn that many more calories than they take in with their diet) for that week. How many calories would you have to cut back on to lose 10 lbs?

(Continued on p. 484)

Box 21-1 Sample Exercise Program

Here is a sample exercise program, so you can see how the numbers work. After you see how simple it is, you can tailor it to your own situation and design your own program.

First, you need a goal. Suppose our subject is a 5-ft-tall, 129-lb, moderately active female. She has weighed that for a couple of years, and this year she decides to go on a weight loss program. She has heard that exercise is a great way to lose unwanted pounds and she decides to start gradually. Her goal: exercising off 1 lb of fat in a week. Nothing else in her life is going to change; she will eat the same diet and maintain the same activity level for everything else.

Second, you need to pick a type of exercise.

Our subject selects jogging. Table 21-2 indicates that jogging burns up about 560 Cal/hr for the average person. Since the target is 3500 Cal of fat, divide the caloric target by the cost of the exercise to find out how long it will take to lose the weight.

Hours of exercise needed =

$$\frac{\text{Calories of fat}}{\text{cost of exercise per hour}}$$

Therefore,

Hours of exercise =

$$\frac{\text{3500 Cal}}{\text{560 Cal/hr (jogging cost)}}$$

$$= \text{6.25 hours of jogging}$$

Surprised? From these figures it should be obvious that though exercise may be great for building strength, endurance, and cardiovascular fitness, it is not a particularly efficient way to lose weight. In order for this woman to achieve her weight loss goal, she would have to run almost an hour a day—seven days a week. Since most joggers run only three or four times a week and few trained athletes run for over two hours in a stint, it seems unlikely that she will be able to lose the pound of fat in that time.

Dieting may be a more realistic approach to attaining her goals. For instance, her goal is to drop 3500 Cal of fat in 7 days. Dividing 7 into 3500 indicates she could achieve her goal by eating 500 fewer Calories than she needs each day for a week. She can do this with hardly any changes in her lifestyle. We know, for instance, that she has weighed 129 lb for some time. By multiplying her body weight times 15 Cal per day (for moderately active females), we find that she has been consuming and using, on average, about 1935 Cal per day. If we reduce her daily intake by 500 Cal, she can still consume over 1435 Cal a day (which is a substantial amount of food—12 baked potatoes without butter or sour cream) and still attain her goal.

Learning Focus

CALORIES, LIFESTYLE, AND EXERCISE

Every year Americans spend millions of dollars in pursuit of the "body beautiful" that seems to lurk inside their skin. Most of these programs are simply variations on the same theme. The calculations that underlie them all are basically the same. Let us review them:

CALCULATING THE NUMBER OF CALORIES YOU NEED EACH DAY

Most people are moderately active—they work, have a social life, go to school, and so on. And all these activities require a certain number of Calories from food to pay for them. The number of Calories that they need varies with their body size, level of activity, and sex. In the text you can find the constants used to calculate the number of Calories needed by a moderately active male and female. Write these constants in the spaces provided:

Males need ________ Cal/lb of body weight per day.

Females need ________ Cal/lb of body weight per day.

Write down your body weight here: ____________ lb.

Calculate the number of Calories you would need per day to be moderately active by multiplying your body weight by the correct factor for your sex. Write your answer here:

Calories you need per day = ____________ Cal

How many Calories would you need per day if you had the same body size but were the opposite sex? Write the answer in the space provided here:

Calories needed per day = ____________ Cal

CALCULATING THE COSTS OF EXERCISE

Suppose, in addition to having a moderately active lifestyle, you also engage in some daily exercise. The costs of that additional activity must be added to the Calories you need for your moderate activities in order to get the *total* amount of food energy you need per day. For example, suppose a 150-lb male lives a moderately active lifestyle but also takes a one-hour leisurely bicycle ride after work each day. How many Calories per day does he regularly use? To get the answer refer to Table 21-2:

Calories needed = 150 lb × 18 Cal/lb per day + additional cost of exercise

= 2700 Cal + 270 Cal (cost of one hour's bicycle riding)

= 2970 Cal

Now do it for yourself. Suppose you decided to add a one-hour bout of high-speed walking to your daily routine. How many Calories do *you* need per day?

Calories needed = total for moderate activity + costs of exercise

= ________ + ________ Cal

YOUR CHOICE OF EXERCISE

Select from the list of exercises in Table 21-2 the one you would prefer to include in your own daily routine and record the information in the following:

My choice of exercise is ________.

The energy cost for that type of exercise is

________ Cal/hr.

Recent studies suggest that excellent health benefits can be derived from an exercise program that consumes about 2000 Cal per week. Using walking as an example, how many hours per week would you have to engage in to reach that level of benefit?

Hours per week = 2000 Cal/cost of exercise per hour

For example, using walking,

$$\frac{2000 \text{ Cal}}{312 \text{ Cal/hr}} = 6.41 \text{ hours of walking per week}$$

= 55 minutes of walking per day

Using your choice of activity,

$$\text{Hours per week} = \frac{2000 \text{ Cal}}{\text{cost of your choice}}$$

= ________ hours per week

If you exercise daily, how many hours a day would you have to practice your preferred type of exercise to reach your 2000 Cal per week goal?

________ hours per day (seven days per week)

How many hours per session would you have to exercise if you could work out only four times a week?

________ hours of exercise per session (four times a week)

EXERCISE AND WEIGHT LOSS

Many people are surprised to learn that exercise is not a particularly efficient way to lose weight. In general, the target of most weight control programs is fat. For that reason we need to know how much energy is contained in each pound of fat. Look that answer up in the text and record it here:

Calories per pound of fat = ________ Cal

Using the energy costs of your favorite exercise, determine how many hours it would take for you to burn up a single pound of fat (assuming that your diet was not increased to compensate):

$$\textit{Hours of exercise} = \frac{3500 \text{ Cal (of fat)}}{\text{exercise cost per hour}}$$

Calculate it for your chosen exercise:

$$\text{Hours} = \frac{3500 \text{ Cal of fat}}{\_\_\_\_\_\_\_\_ \text{ Cal/hr}} = \_\_\_\_\_\_\_\_ \text{ hours}$$

From the figure you just calculated, would you say that exercise is an efficient way to burn Calories for weight loss? Of what benefit *is* exercise?

The bottom line (no pun intended) is that the best thing you can do for yourself is to eat a varied, good quality diet and get a reasonable amount of exercise on a regular basis.

Dieting and Adult Body Weight: A Continuing Controversy

There has always been a great deal of controversy surrounding the concepts of dieting, nutrition, exercise, and adult body weight. In fact, many of our traditional ideas about these subjects are being reexamined. A recent study of identical twins, separated in childhood and raised in different family situations, indicates the single, most important determinant of adult body weight is genetic, not environmental. That, coupled with the idea that diets do not really work (about 90 percent of the people disciplined enough to achieve their goal for a weight change through dieting will regain the weight in less than a year), is causing a revolution in the way we view these subjects.

One line of thought suggests that we each inherit from our parents a "fat set point" that is preset for an acceptable range of body weights. Supposedly, this thermostat is located in the hypothalamus of the brain, where it keeps track of the level of glycerol (a chemical component of fat). In theory, the fatter a person is, the more glycerol there is in circulation, and vice versa. Most people starting a diet find they can lose a few pounds fairly easily. But beyond that, the fat set point resists further weight changes by activating various metabolic systems that make the body much more efficient at converting ingested Calories to fat. Increases in weight above a range acceptable to the set point are also metabolically resisted (Figure 21-6).

It seems that the weight tends to fluctuate, plus or minus 10 percent, around a genetically determined set point for each adult body. A 129-lb woman can thus expect to add or lose a few pounds in either direction fairly easily but can expect to achieve a change of more than 13 lb only with great difficulty and discomfort, as her own physiology can shift in either direction to resist her efforts.

It may be that the gradual increase seen in adult body weight as time passes may be related to upward changes in the setting of the "fat-o-stat" rather than simply being related to a more sedentary lifestyle. At any rate, the concept is interesting and the answers may soon be coming.

Contemporary weight loss programs take the fat set point into consideration. Certain chemicals seem to quiet the set point mechanism, making it easier to control weight comfortably. Some people believe that cigarette smoke contains ingredients that turn the set point to a lower acceptable body weight, accounting in part for the weight gains experienced by many smokers who quit. Others believe that diet pills also work by turning down

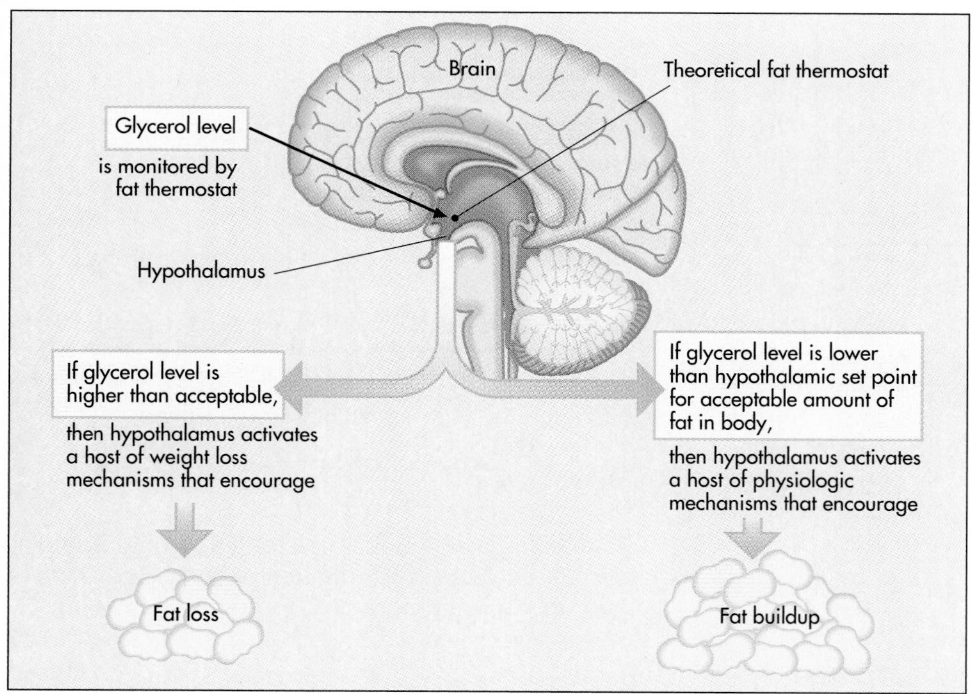

FIGURE 21-6 A proposed hypothalamic "fat thermostat." Some investigators believe that adult body weight is kept within a genetically determined (inherited) "acceptable" range through the actions of the hypothalamus, which may contain cells responsible for monitoring the amount of fat in the body, by assaying the glycerol level in the circulation.

the set point, allowing someone to be comfortably slimmer than their inherited fat thermostat would normally allow.

But you do not need to smoke or become a diet pill addict to keep the weight off. Evidence suggests that exercise can contribute to weight loss in ways other than simply burning excess Calories. It may be that some of the body's natural chemicals, produced during exercise, also turn down the fat thermostat—making thinness an easier goal for those who include regular exercise in their lifestyle.

CONTROLLING BODY TEMPERATURE

Body temperature is generated by the heat released by metabolic reactions. The difference between *warm-blooded* mammals and birds and *cold-blooded* fish, amphibians, and reptiles is that the warm-blooded groups keep their body temperatures nearly constant by controlling the production and loss of heat. Humans control heat loss by regulating the flow of blood to the skin and by sweating. They control heat production by adjusting metabolic rate. In both cases, the control center is in the brain's hypothalamus, where cells continually monitor the temperature of the blood and compare it against a "set point." The set point is normally 37°C, but it can be changed, as it is with fever (see Chapter 23).

Elevated blood temperatures stimulate the hypothalamic "thermostat" to inhibit the sympathetic nervous system's effect on the smooth muscles that constrict the blood vessels to the skin. The vessels open, blood flows into the skin, and heat is lost from the body's surface as from a radiator. The hypothalamus also activates the sweat glands and decreases heat production by decreasing muscle tone and inhibiting the release of epinephrine, norepinephrine, and thyroid hormone (Figure 21-7).

Low blood temperatures stimulate the hypothalamus to activate the sympathetic nervous system, thus shutting down blood flow into the skin and conserving heat, while sympathetic norepinephrine boosts the metabolic rate. At the same time, the hypothalamus stimulates production of thyroid hormone and adrenal epinephrine. When the need for heat is great enough, it can also activate the skeletal muscles, generating heat by enhancing muscle tone and stimulating the repeated contractions of shivering—and even the flailing of arms and stomping of feet when shivering is inadequate.

Various abnormalities can affect the control system for body temperature and metabolic rate. For

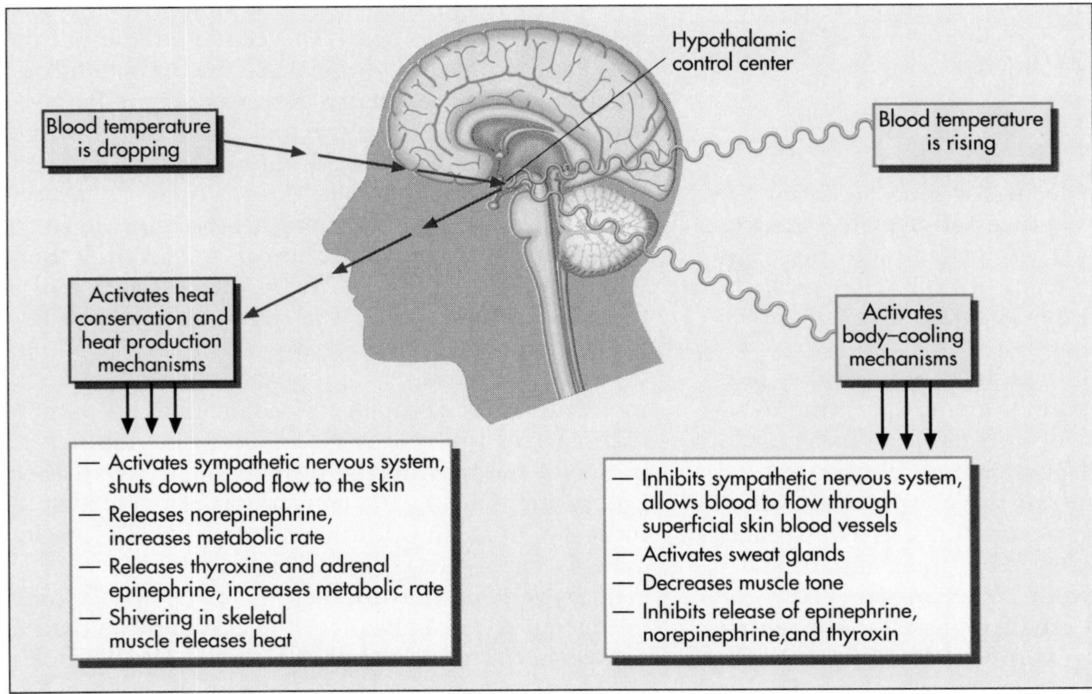

FIGURE 21-7 Body temperature is controlled by feedback effects of blood temperature, via the hypothalamus, on the sympathetic nervous system, thyroid, adrenal medulla, musculature, skin, blood vessels, and sweat glands.

instance, people who secrete excess thyroid hormone tend to have higher than normal metabolic rates. They thus generate more body heat than normal; they often shovel snow in their shirtsleeves or they seem able to eat all they want without gaining weight. Tumors of the adrenal medulla can have a similar effect. People who secrete less thyroid hormone than normal have slower metabolisms and generate less heat; they tend to feel cold when the rest of us are comfortable. They also tend to gain weight easily since, though they may eat no more than others, they need fewer calories and have more excess to store as fat.

CELLULAR RESPIRATION

We will now consider, in some detail, the processes that convert the raw energy contained in foods to a form that can be used by cells.

To release energy from the chemical bonds in which it is stored, cells completely dismantle fuel molecules in a series of chemical reactions collectively called *cellular respiration.* One of the most important fuels for the body's cells is the sugar glucose. As energy is released from the breakdown of this sugar, it is used to convert adenosine diphosphate (ADP) to adenosine triphosphate (ATP), the refined form of chemical energy that drives most cellular activities. The equation summarizing cellular respiration is as follows:

$$C_6H_{12}O_6 + 6O_2 \longrightarrow 6CO_2 + 6H_2O + \text{usable energy (ATP)}$$

It states that one molecule of glucose, using six molecules of oxygen, is processed to yield six molecules of carbon dioxide, six water molecules, and usable energy in the form of ATP. However, the equation is somewhat misleading, for the process is not that simple.

The events of cellular respiration are really those of three separate reaction systems operating in series with each other: glycolysis, the Krebs cycle, and the electron transport chain. Cellular respiration begins with **glycolysis** (sugar splitting), which occurs in the cell's cytoplasm. Since glycolysis does not use the oxygen taken in by breathing, it is also called *anaerobic* (without air) *respiration.* The end products of glycolysis are then passed to the matrix of the mitochondria for complete breakdown in the **Krebs cycle.** Finally, the energized electrons and hydrogens released by the Krebs cycle are gathered up by a carrier molecule, nicotinamide adenine dinucleotide (NADH), and transferred to the electron transport chain in the **cristae** of the mitochondria. There, most of the ATP is actually formed. When the electrons and hydrogens have passed completely through the electron transport chain, they are joined to the oxygen obtained by breathing to make water molecules. This phase of respiration is called *aerobic respiration.* Each of these processes will be described in some detail in what follows (Figure 21-8).

Glycolysis

Glucose enters each cell by active transport. Once it is in the cytoplasm, enzymes attach phosphate groups from two ATPs. The molecule is then split into two two-carbon sugars, which in turn are converted to two molecules of pyruvic acid. During the process, energized electrons and hydrogen ions are removed and transferred to two molecules of the electron carrier NADH. The two NADHs can be processed later to contribute the energy they contain to the total ATP yield of cellular respiration (Figure 21-9).

For each glucose molecule, glycolysis generates a total of four molecules of ATP. Because it takes two ATPs to get glycolysis started, there is a net profit of two ATPs for each glucose molecule processed in the glycolytic mill.

Cellular respiration does not stop with the formation of pyruvic acid. The center of the action now shifts from the cytoplasm to the mitochondria, where most of the energy that was originally contained in glucose is extracted. However, pyruvic acid itself cannot enter the mitochondria. It must first be converted to **acetyl coenzyme A** (acetyl CoA). The conversion begins with the removal from each pyruvic acid molecule of a carbon atom, which is then eliminated from the cell as a molecule of carbon dioxide. The remainder of each pyruvic acid molecule is a two-carbon *acetyl group.* Each acetyl group is temporarily attached to a molecule of coenzyme A to form acetyl CoA, which then carries the energy-rich molecular fragments, or acetyl group, into the mitochondria for further processing. During the conversion of pyruvic acid to acetyl CoA, an additional pair of NADH molecules is formed (see Figure 21-8).

If oxygen is not available, the energy yield stops at the four ATPs of glycolysis. The pyruvic acid does not go to the mitochondria, and the NADH molecules are not processed to yield more energy. Instead, the NADHs are used to convert the pyruvic acid to other compounds such as ethyl alcohol, lac-

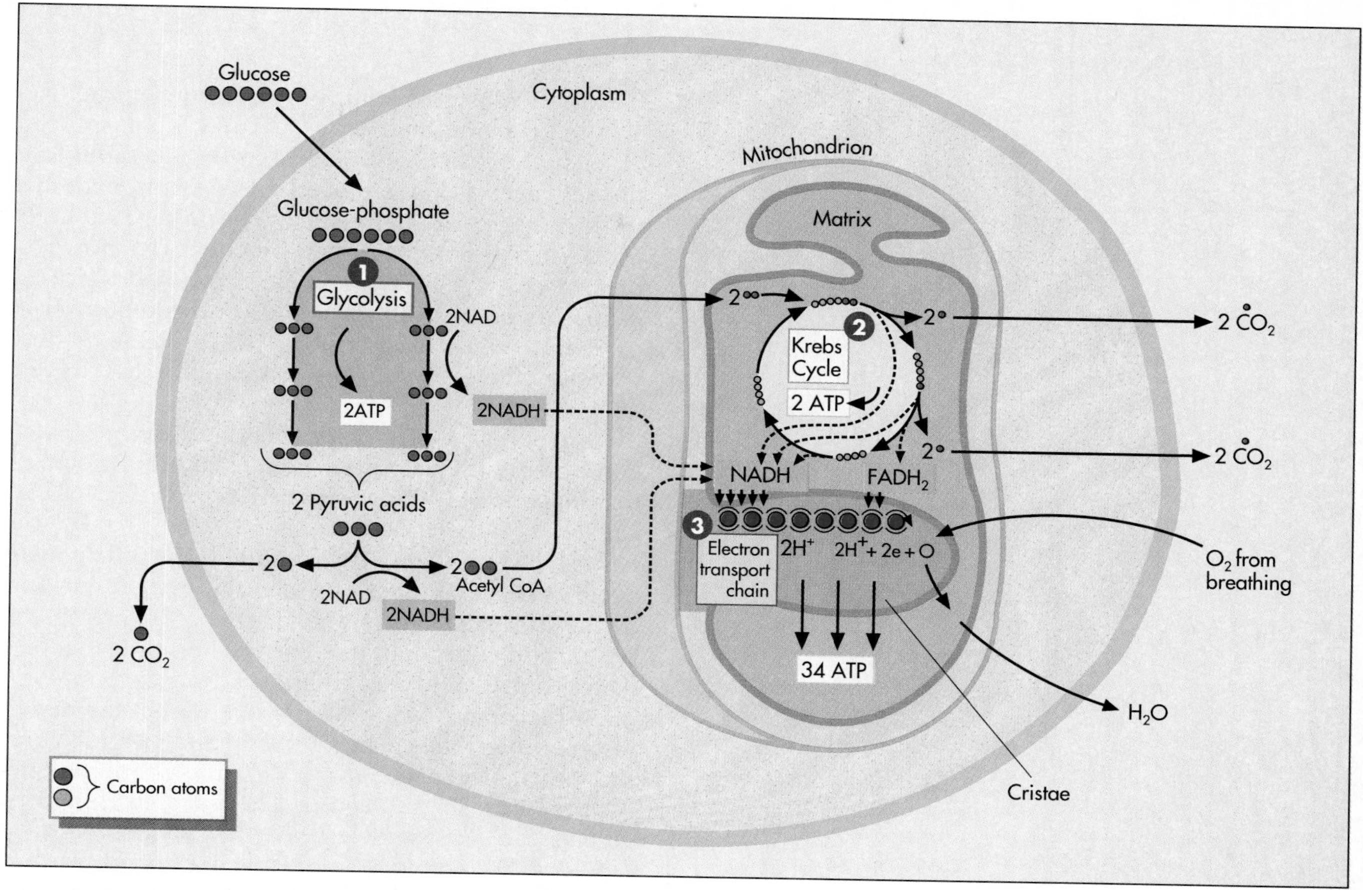

FIGURE 21-8 The three stages of cellular respiration. (1) Cellular respiration begins with the splitting of glucose (a six-carbon molecule) in the cytoplasm during *glycolysis*. The process yields a net profit of two ATPs and an additional two NADHs, which can be processed later to contribute to the total ATP produced during cellular respiration. The end product of glycolysis is two pyruvic acid molecules. (2) Each pyruvic acid molecule loses one of its three carbons as carbon dioxide, generating in the process two more NADHs. The two remaining carbons (an *acetyl group*) combine with coenzyme A (CoA) to form acetyl CoA, which enters the *Krebs cycle* in the mitochondrial matrix. There the acetyl group combines with a four-carbon "pickup" molecule. Further chemical reactions remove the acetyl carbons as more carbon dioxide and generate one ATP. In the process, electrons and hydrogen ions are also released; they are picked up by NADH and $FADH_2$. The Krebs cycle yield from each glucose molecule is two ATPs, six NADHs, and two $FADH_2$s. (3) The energized NADHs and $FADH_2$s are transferred to the electron transport chain mounted in the cristae of the mitochondria. There they release pairs of electrons and H^+ ions. The electrons flow down the chain, from link to link, and the energy released is used to synthesize additional ATPs. A total of 34 additional ATPs come from the electron transport chain. Oxygen (from breathing) accepts the electrons and the H^+ ions, forming water.

tic acid, or other two- or three-carbon organic acids or alcohols. This diversionary reaction is called **fermentation.** Human cells convert pyruvic acid to lactic acid, which builds up in muscle when exercise is so great or prolonged that it exhausts the oxygen supply or exceeds the cardiovascular system's ability to deliver oxygen to the cells. Yeast cells instead convert pyruvic acid to the ethyl alcohol of beer, wine, and whiskey. People exploit bacterial fermentation to make industrial chemicals such as acetone (Figure 21-10, p. 489).

1. Glycolysis produces four ATPs. Why do we say it produces a "profit" of only two ATPs?
2. Under what circumstances can glycolysis produce a profit of more than two ATPs?

Other Fuel Sources

Although glucose may be the easiest fuel molecule for cells to process, it is not the only fuel. During

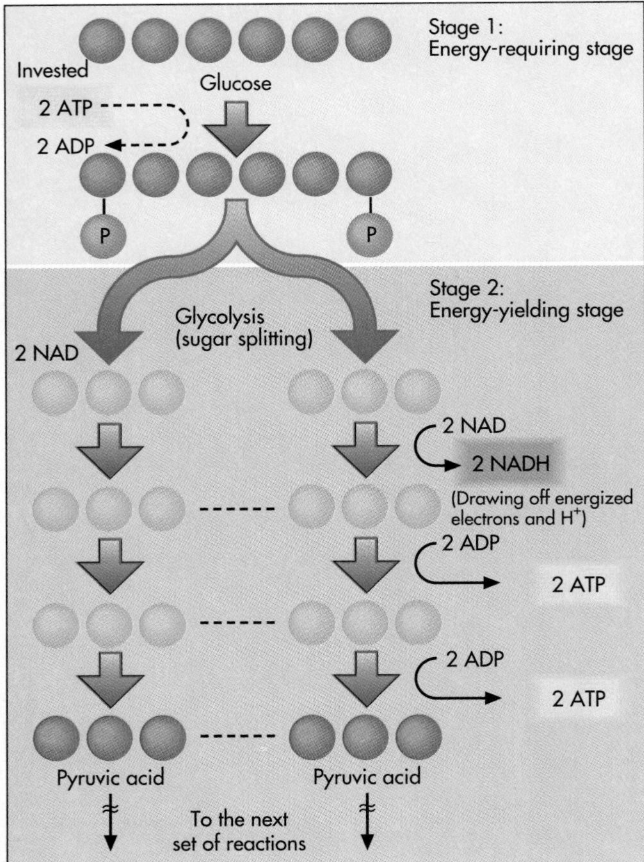

FIGURE 21-9 The events of glycolysis occur in two stages. Stage 1—the "endergonic" (energy-requiring) stage—begins by attaching additional phosphates, from ATP, to the sugars being processed. This is the energetic "pump priming" needed to get things started. After a second phosphate is attached, the sugar is split into two three-carbon intermediates that are sent on to the energy extraction reactions, beginning in stage 2. In stage 2—the exergonic (energy-releasing) stage—the two three-carbon intermediates are rearranged, four ATPs are generated, and energized electrons and hydrogen ions are passed to two NADHs for later harvesting of the energy they contain. The result is two molecules of pyruvic acid, which, if enough oxygen is available, can be sent on to the mitochondria for the extraction of additional energy.

periods of dieting, starvation, or increased energy expenditure, the body's cells can extract energy from other types of organic substances, such as fats and proteins. When fatty and amino acids are to be processed for energy, however, they do not go through glycolysis. These alternative fuels are also broken into two-carbon units, or acetyl groups, which can then be attached to coenzyme A for entry into the mitochondria. Certain amino acids can enter the Krebs cycle at various other steps as well. Once these materials have entered the mitochondria, they can have the energy they contain extracted and made available to the cell in the form of ATP.

The Krebs Cycle

It is the Krebs cycle that processes the energy-rich acetyl CoA obtained from pyruvic acid or from other fuels. When sufficient oxygen is available to the cell, acetyl CoA is moved into the mitochondria so that the energy harvest can continue. The Krebs cycle then functions like a biochemical conveyor belt to carry the energy-rich two-carbon glucose remnants delivered by acetyl CoA through a series of reactions that completes the breakdown and transfers the energy in the remnants to NADH and ATP.

As acetyl CoA enters the mitochondrial matrix, it combines with a four-carbon "pick-up" molecule (oxaloacetic acid) to enter the Krebs cycle as a six-carbon molecule of citric acid (this is why the Krebs cycle is sometimes called the **citric acid cycle**). Chemical reactions then remove the carbons of the acetyl group one at a time, converting the six-carbon citric acid to a five-carbon intermediate (ketoglutaric acid) and then to a four-carbon compound (succinic acid). Succinic acid is rearranged into another four-carbon intermediate (malic acid) that is then converted back to the original four-carbon pick up molecule. The carbons removed during the Krebs cycle leave the cells as carbon dioxides. Each glucose molecule processed, since it gives rise to two acetyl CoA fragments in glycolysis, requires two complete circuits or revolutions of the Krebs cycle (Figure 21-11, p. 490).

Most of the energy harvest from the Krebs cycle is delayed. Only one ATP is produced during the conversion of a five-carbon ketoglutaric acid to a four-carbon succinic acid (two ATPs are thus produced for each glucose processed). However, as the Krebs cycle breaks down the acetyl groups that enter it, several energy-rich electrons and hydrogen ions are released. They are captured in the form of NADH and $FADH_2$. Their energy, extracted later by the electron transport chain, provides most of the ATP yield of cellular respiration.

The Electron Transport Chain

The NADH and $FADH_2$ molecules produced by the Krebs cycle, glycolysis, and the reaction that converts pyruvic acid to acetyl CoA deliver their highly energized electrons and hydrogen ions to the **electron transport chain.** This structure is a series of iron-containing pigments, the cytochromes, that specialize in routing the flow of energized electrons. The iron portion of each cytochrome contains loosely held electrons that facilitate the flow (Figure 21-12, p. 491).

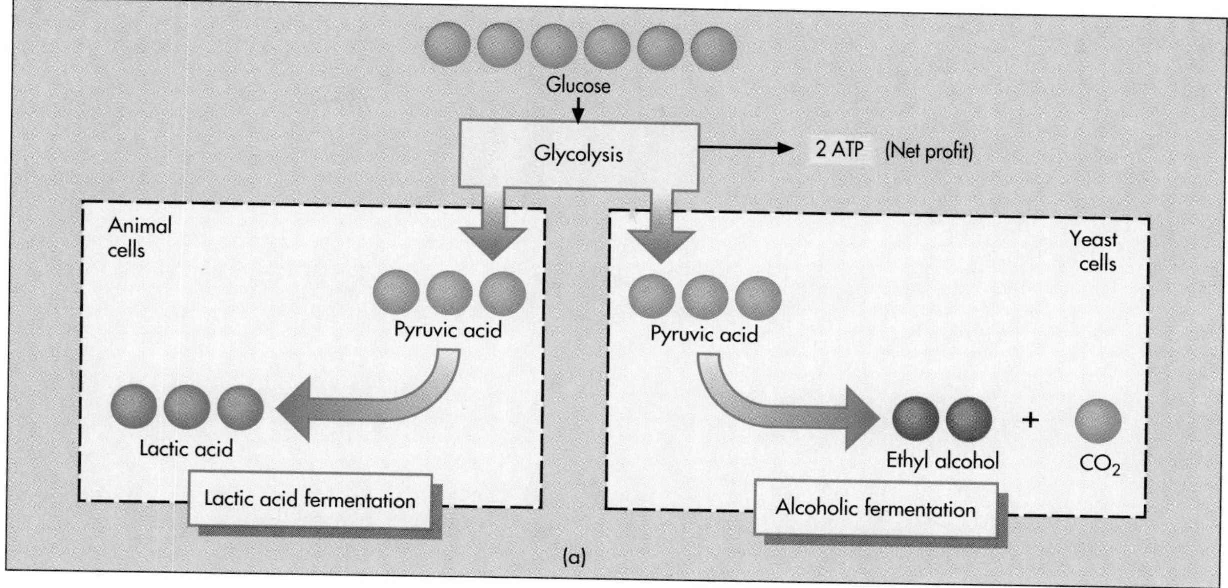

(b)

FIGURE 21-10 (a) The fermentation biochemical pathway. If oxygen is not available, glycolysis can continue to yield two ATPs per glucose by routing pyruvic acid through the fermentation process. Animal cells convert pyruvic acid to lactic acid. Yeast cells convert it to ethyl alcohol and carbon dioxide. Bacteria can make alcohol or other chemicals. (b) Winemakers have used fermentation to produce alcohol from a mixture of grape sugars for thousands of years. In this sequence of photos the grapes are first crushed, fermented, aged, and tested. Usually the yeasts (found on the grape skins) are first cultured aerobically in the grape juice (when oxygen is available, the cells multiply rapidly). Once the juice contains enough yeast cells, the oxygen is cut off, forcing the yeast to rely on fermentation to survive. Alcohol and carbon dioxide are released as by-products of fermentation, converting grape juice to wine. In sparkling wines such as champagne, the carbon dioxide is retained in the wine (it provides the bubbles). In nonsparkling wines, the carbon dioxide is allowed to escape.

The cytochromes are embedded in the inner membranes of the mitochondrial cristae, where they accept the electrons carried by NADH and $FADH_2$. The hydrogen ions are passed into the outer compartment between the inner and outer membranes of the mitochondria. As they accumulate there, they produce a *chemiosmotic* (pH and concentration) gradient across the inner membrane.

The negatively charged electrons flow down the cytochrome chain. At the end of the chain, the electrons and hydrogen ions react with oxygen (from breathing) to form water. To reach the scene, the hydrogen ions move down their chemiosmotic gra-

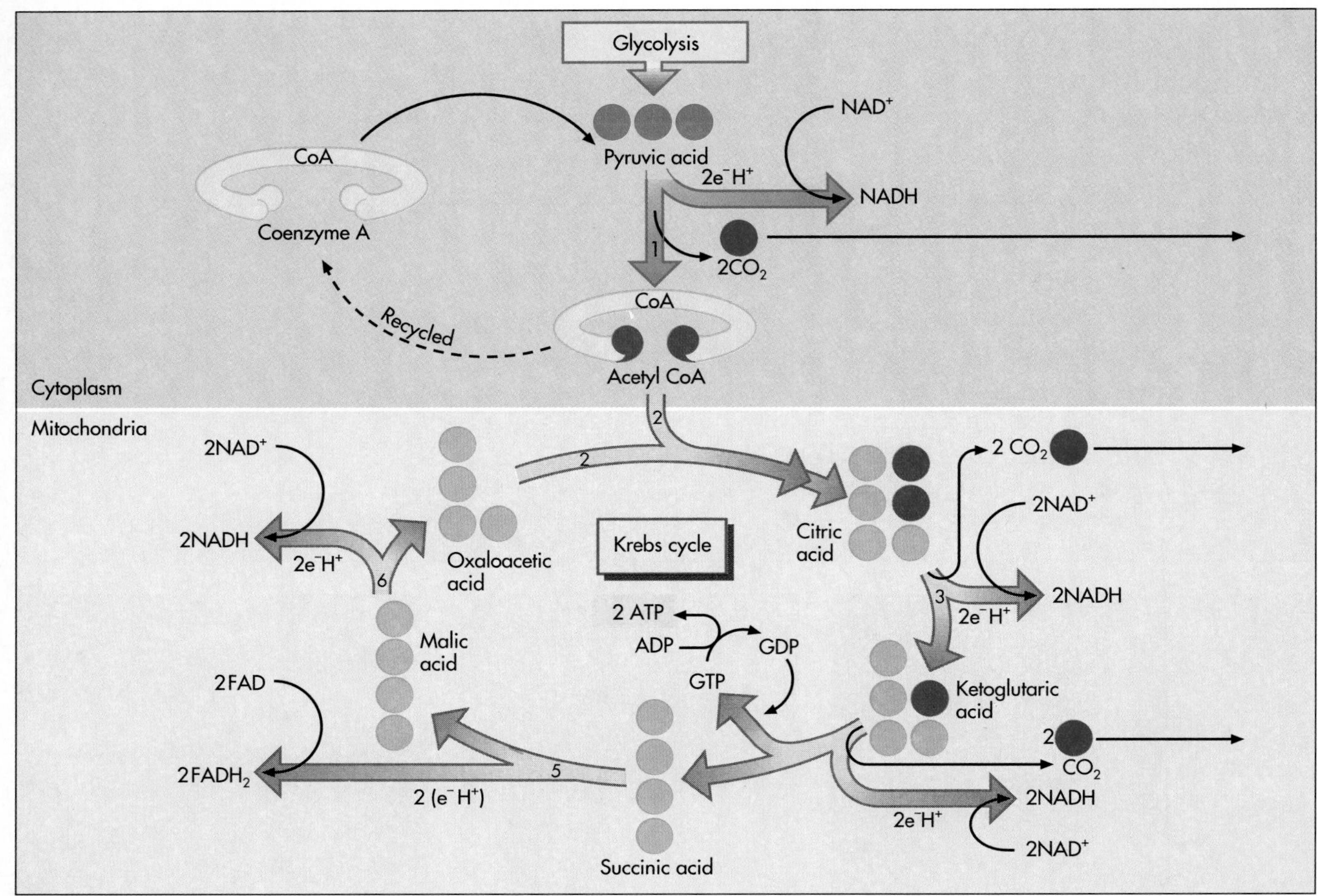

FIGURE 21-11 The Krebs cycle. The two acetyl fragments of glycolysis are transferred to the mitochondrial matrix where they are picked up by the four-carbon oxaloacetic acid, forming the six-carbon citric acid, and inserted into the Krebs cycle. In the course of the cycle, the citric acid molecules pass through a series of chemical reactions that strip away the added carbons, expelling them from the cell as carbon dioxide. Energy-rich electrons and H^+ ions released in the process are gathered up by NADH and $FADH_2$ and passed to the electron transport system for further processing. Two molecules of ATP are also made during the two turns of the Krebs cycle needed to process a single glucose molecule.

dient back to the inner mitochondrial compartment by flowing through the F_1 particles in the inner membrane. In the process, known as **chemiosmotic phosphorylation,** they give up energy to power the attachment of phosphate groups to ADP, making ATP (Figure 21-12).

Each NADH delivered to the cristae produces three molecules of ATP as electrons flow along the electron transport chain. Since the electrons released by $FADH_2$ enter the chain further downstream, each one of these molecules yields only two ATPs. Altogether, chemiosmotic phosphorylation produces 34 ATP molecules for each molecule of glucose processed.

The chemical yields for the various components of cellular respiration are summarized in Figure 21-13. Note that glycolysis yields a profit of only two ATPs without depending on the electron transport chain. Two more are produced by the Krebs cycle. However, by far the bulk of cellular energy production depends on chemiosmotic phosphorylation and on the availability of oxygen.

1. Consider a six-carbon fatty acid. When processed for energy, it is broken into three acetyl groups. How many ATPs will it yield? How does this number compare with the yield from a six-carbon glucose? (Actually, fats yield much more energy than this comparison suggests; as they are broken into acetyl groups, they produce both ATP and NADH.)
2. Why does chemiosmotic phosphorylation require that the aerobic respiration apparatus be confined within a membrane?

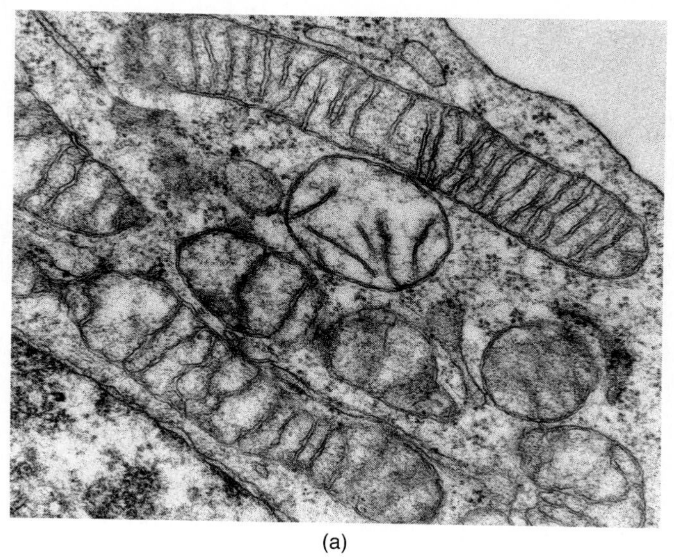

(a)

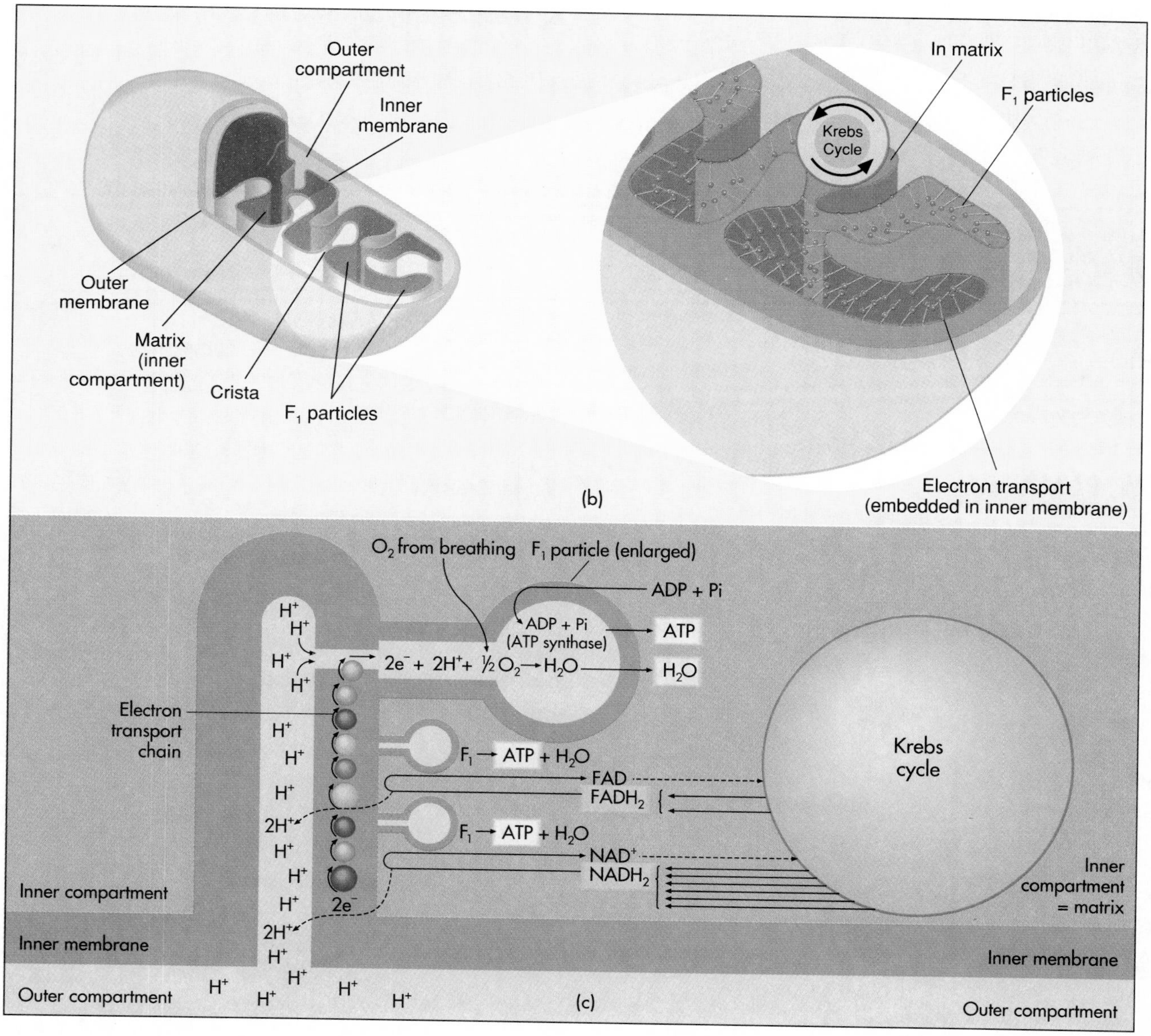

FIGURE 21-12 (a) The shelflike cristae that subdivide mitochondria are visible in electron micrographs. (b) The relationships of the mitochondrion's inner and outer membranes, cristae, matrix, and F_1 particles become clearer in a drawing. (c) The ATP-generating system for the electron transport chain. NADH and $FADH_2$ molecules deliver their high-energy electrons and H^+ ions to the electron transport chain. The energized electrons flow from cytochrome to cytochrome down the chain. The H^+ ions pass through the inner membrane to the outer compartment of the cristae. The electrons and the H^+ ions eventually flow into the F_1 particles where they combine with oxygen to form water. The chemiosmotic (pH and concentration) gradient of H^+ ions across the inner membrane supplies the energy for the formation of ATP (using the enzyme ATP synthase). The NADH and $FADH_2$ deliver their electrons to different parts of the electron transport chain. Consequently, each pair of electrons delivered by NADH produces three ATPs, while each pair from $FADH_2$ yields only two ATPs.

Learning Focus

CELLULAR RESPIRATION SUMMARY

Shortly after the American Revolution, the French scientist Georges Cuvier gave the world a lesson in productivity. The leading comparative anatomist of his day, he helped shape the field of taxonomy (plant and animal classification), worked out a theory of "catastrophism" to explain the presence of fossils without conflicting with his own religious views, and was an astute politician. He was also a scientific philosopher whose ideas are still being talked about today. He used the principle of divide and conquer to help him maximize his productive efforts. He had a different desk for each of his areas of interest—comparative anatomy, taxonomy, philosophy, and so on. He could move rapidly from one project to another simply by going from desk to desk. His approach to handling complexity was to divide it into small manageable units, master those, and then put the small units together into a larger picture.

We suggest you use a similar approach in mastering the events of cellular respiration. Using Figures 21-8, 21-9, 21-10, and 21-11, and the related text, organize the following information.

GLYCOLYSIS

1. Describe in your own words what happens in glycolysis.

2. What is the net production of ATP in glycolysis?

3. Why is the net production of ATP different from the actual number of ATPs formed during glycolysis?

4. How many NADHs and $FADH_2$s are formed during glycolysis?

5. How many ATPs can be produced by running the electrons and hydrogens carried by one NADH through the electron transport system?

KREBS CYCLE

6. Describe in your own words what happens in the Krebs cycle.

7. How many ATPs are produced directly from the Krebs cycle?

8. How many NADHs and $FADH_2$s are formed during each turn of the Krebs cycle (when one acetyl group is processed)?

9. For what are the NADHs and $FADH_2$s used?

10. How many times is the Krebs cycle operated for each molecule of glucose processed by the cell?

11. What are the products of the Krebs cycle and where are they sent?

ELECTRON TRANSPORT SYSTEM

12. Describe the electron transport system and what it does.

13. Into which mitochondrial compartment are the hydrogen ions delivered to the electron transport system by NADH and $FADH_2$ likely to be sent?

14. For what is the energy released by the electrons moving down the electron transport system likely to be used?

F_1 PARTICLES

15. The F_1 particles seem to be the sites for ATP synthesis. Describe how it may happen.

16. Where is the oxygen from breathing consumed, and how?

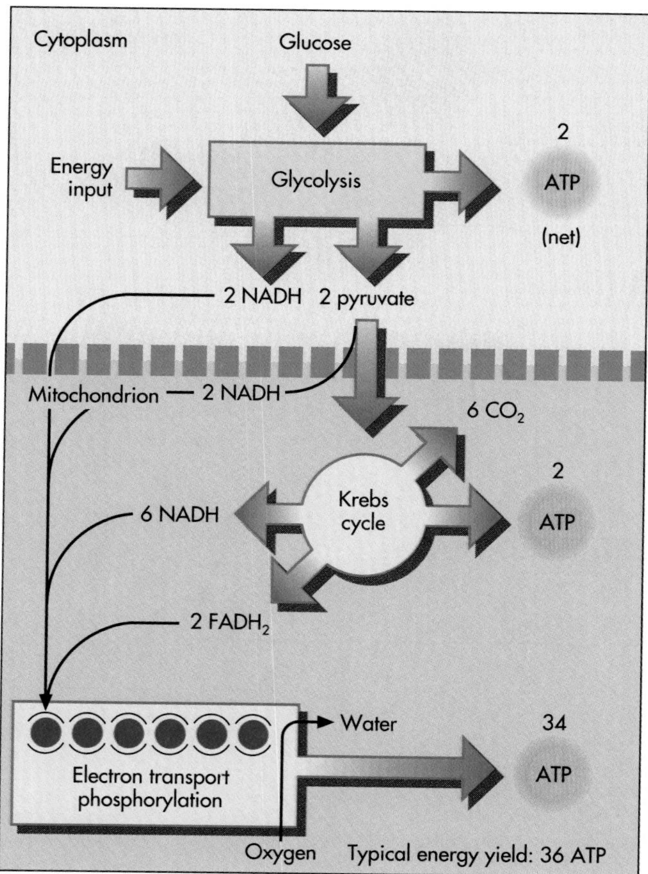

FIGURE 21-13 Summary of energy production and flow through the various components of cellular respiration.

OXYGEN DEBT

When you exercise strenuously, your muscles use ATP more rapidly than it can be regenerated. As the exertions continue, the ATP reserves in muscle cells fall and the rate of cellular respiration is stepped up to meet the increased demand. Just how far the rate of cellular respiration can be increased is limited, in part, by the body's ability to deliver adequate amounts of oxygen to the tissues. In fact, many of the changes produced in the body by athletic conditioning involve improving the blood flow (cardiovascular fitness) through the "trained" tissues. Even then, at some point, the cells attempting to operate the Krebs cycle and the electron transport system at their maximum rates fall behind, and the continuing demand for ATP outstrips the system's ability to produce enough.

However, the cells have an "ace in the hole" they can play to improve their yield of ATP. They can still run the glycolysis portion of the respiratory apparatus to continue producing two ATPs per glucose—even though the aerobic respiratory channels are operating at maximum speed. When the cells are forced to dig this deeply for energy, however, they produce lactic acid as a by-product. This lactic acid is responsible for some of the fatigue that develops, for its acidity impairs the ability of muscle cells to contract.

When you end an exercise session, your body recovers in part by removing the lactic acid built up during your exertions. To do this, liver and muscle cells convert the lactic acid back to pyruvic acid. They then route some of the pyruvic acid to the Krebs cycle, using oxygen to generate ATP. They use some of the ATP to convert the remaining pyruvic acid all the way back to glucose and then to glycogen for storage. Muscle cells also rebuild their depleted ATP reserve. The oxygen needed to "finance" this return to normal, along with the oxygen needed to restore blood levels of oxygen to normal, is called the **oxygen debt** incurred by exercise. Paying the debt calls for a period of heavy breathing—panting—after exercise, which lasts until the body's need for oxygen returns to normal (Figure 21-14).

---

1. Under what circumstances is oxygen debt not repaid but continually increased? What is the end result? Why?
2. Cyanide binds to the enzymes of the electron transport system and keeps them from working. In what way is its effect similar to severe oxygen debt?

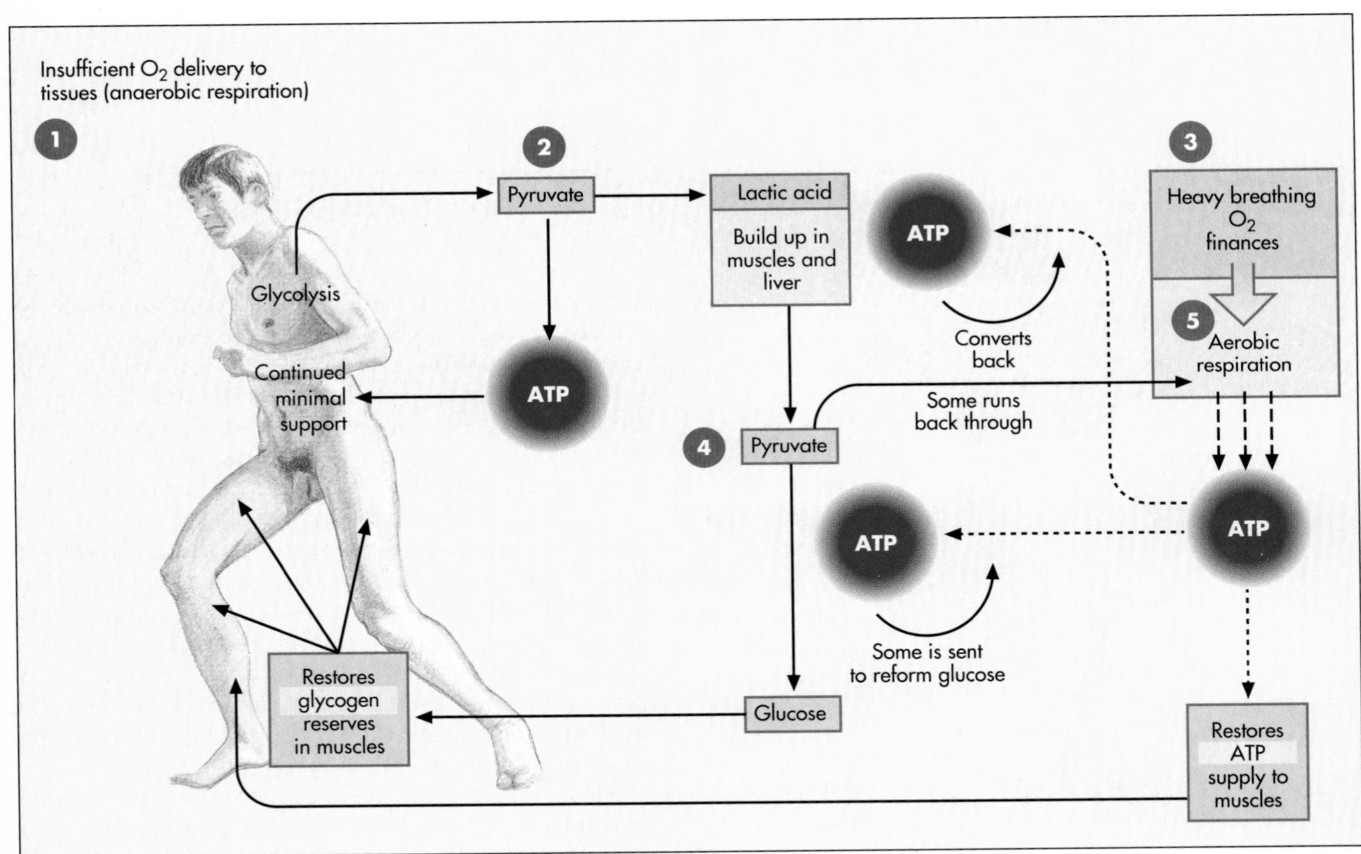

FIGURE 21-14 Oxygen debt. During vigorous exercise (1), the body's demand for oxygen for aerobic respiration can fall behind the circulatory system's ability to deliver it. However, the body can continue to obtain ATP from glycolysis by converting pyruvate to lactic acid (2), which builds up in the cells of the liver and muscles. After exercise, the body goes through a period of heavy breathing (3). The oxygen so supplied is used in aerobic respiration to generate ATP, some of which is used to convert some of the lactic acid back to pyruvate (4). The pyruvate can then be metabolized to generate more ATP, which can be used to remove more of the accumulated lactic acid. In some cases (5), some of the pyruvate can be converted all the way back to glucose (or even glycogen, for long-term storage).

SUMMARY

Many of the activities of living systems are centered around the quest for and the processing of energy. Animals get energy from the foods they eat. But the molecules ingested must undergo a significant amount of processing before the energy they contain can be used in the metabolic machinery of the animal's cells. The food molecules first have to be broken down by the processes of digestion, which produce simple sugars, amino acids, and fat components that can be absorbed into the blood and delivered to all the cells. Once in the cells, energy release and transfer takes place in a complex set of reactions collectively referred to as cellular respiration. Ultimately, the energy contained in the various organic nutrients is transferred to adenosine triphosphate (ATP) for use by the cells.

Metabolism is the term for all the chemical reactions that occur in the body, including those of cellular respiration. Anabolism is that portion of metabolism that builds larger molecules and cellular components from smaller molecules. Anabolic reactions usually require energy. Catabolism is that portion of metabolism that breaks down larger molecules and cellular components. Catabolic reactions often release energy; they include the reactions of cellular respiration. Catabolism and anabolism continually interact, renewing the body's structure by recycling its components and by replacing components with new ones derived from food.

When we take in more energy in our food than we need to run our bodies, we often store the excess as fat. The amount of energy we need depends on our activities, but we do need a certain minimum to stay alive. This minimum, expressed as a rate or speed of energy use—or heat production—is the basal metabolic rate (BMR). The metabolic rate is about 10 percent less during sleep. It rises with activity.

The metabolic rate can be raised by drugs such as caffeine and amphetamines. It also varies with body temperature, rising about 10 percent for each one degree increase in temperature. The body controls its metabolic rate with epinephrine from the adrenal medulla, norepinephrine from sympathetic nerves, growth hormone, and thyroid hormone. Body temperature is controlled by a "thermostat" in the hypothalamus, which continually adjusts metabolic rate and heat production by controlling sympathetic activity, the release of thyroid hormone and epinephrine, and skeletal muscle tone and activity (shivering).

Obesity results from eating more food than one needs. However, losing weight is not simply a matter of reducing one's caloric intake. The body seems to regulate the size of its fat cells, and when one forces those cells to shrink by dieting, regulatory mechanisms lower the metabolic rate and increase the appetite until the cells are back to "normal." Many scientists feel the most productive way to beat this annoying resistance to good health is to combine dieting with exercise, for exercise increases the metabolic rate even as it improves fitness in many ways.

Cellular respiration occurs in several stages. Glycolysis uses no oxygen as it converts one molecule of glucose to two pyruvic acid molecules and generates two ATP and two nicotinamide-adenine dinucleotide (NADH) molecules. Aerobic respiration begins in the mitochondria, with the removal of a carbon dioxide from each pyruvic acid to make a pair of acetyl coenzyme A (CoA) molecules and extract more energy as NADH. (Acetyl CoA can also be produced from fatty and amino acids.) Each acetyl CoA then enters the Krebs cycle. There, two more carbon dioxides are removed and three NADH, one ATP, and one flavin adenine dinucleotide ($FADH_2$) molecules are made for each acetyl CoA processed. The NADH and $FADH_2$ molecules release their hydrogens and electrons into an electron transport chain where their energy is used to generate more ATP. The electrons from each NADH generate three ATPs; those from each $FADH_2$ generate two. At the end of the electron transport chain, oxygen accepts the hydrogens and electrons to form water.

When oxygen is not available, the electron transport chain cannot work. Cellular respiration stops with glycolysis, whose NADHs are used to convert pyruvic acid to lactic acid in human cells and to ethanol in yeast cells. This reduces the energy yield of cellular respiration to just 2 ATPs, much less than the 38 ATPs normally possible with the use of oxygen.

During strenuous exercise, the needs of muscle cells get ahead of the body's ability to supply oxygen. They must then rely on glycolysis, which generates lactic acid. After the exercise stops, liver and muscle cells convert the lactic acid back to pyruvic acid. These cells then use oxygen to generate ATP from some of the pyruvic acid; they use this ATP to convert the rest of the pyruvic acid back to glucose. The oxygen needed to finance this process is referred to as the oxygen debt incurred by exercise.

STUDY QUESTIONS

1. What are the two main uses of nutrients derived from ingested food?
2. Why must the digestive system break proteins, nucleic acids, and starches into their component units before the body can use them?
3. Explain the difference between *basal metabolic rate* and *metabolic rate*.
4. What is the relationship of body temperature to metabolic rate?
5. Why is it difficult for obese people to lose weight—and to keep the weight off—simply by dieting?
6. Why is exercise an important aid to weight control?
7. Under what circumstances is the energy yield of glycolysis only two ATP molecules?
8. Explain why, though NADH is the energy equivalent of three ATPs, $FADH_2$ is the energy equivalent of only two ATPs.
9. Explain the concept of oxygen debt.

GLOSSARY

Acetyl coenzyme A An energy-rich two-carbon molecule that is shunted into the Krebs cycle to have its energy harvested and the hydrogens it contains loaded into NADH and $FADH_2$ for further processing.

Anabolism Those metabolic reactions in which larger molecules are synthesized from smaller building blocks.

Basal metabolic rate The minimum amount of energy use needed to stay alive and awake.

Calorie The amount of energy required to raise the temperature of a liter of water by one degree Celsius (1 Calorie = 1 kilocalorie = 1000 calories).

Catabolism Metabolic reactions in which larger molecules are broken down into smaller ones; often releasing energy in the process.

Chemiosmotic phosphorylation The attachment of phosphate groups to ADP (making ATP) using energy liberated as hydrogen ions travel down a pH and concentration (chemiosmotic) gradient from the outer compartment to the inner compartment of mitochondria.

Cristae The shelflike membranous platforms running at right angles to the long axes of the mitochondria. The electron transport system is embedded in the cristae.

Electron transport chain A sequence of enzymes and coenzymes that channels the flow of energized electrons. The energy released by these electrons is used to pump hydrogen ions across the inner membrane of the mitochondrion and to synthesize ATP.

Energy The capacity to do work (mechanical, chemical, electrical, etc.).

Fermentation The metabolic pathway that, in the absence of oxygen, converts pyruvic acid to lactic acid, ethyl alcohol, or other compounds.

Glycolysis Literally, "sugar splitting"; the extraction of energy from organic molecules without the aid of oxygen.

Krebs cycle The self-regenerating sequence of reactions that extracts the energy and raw materials to form NADH, $FADH_2$, and ATP from acetyl groups.

Metabolic rate The speed of energy use or release (Calories per day) by a living organism.

Metabolism The sum total of all the body's chemical activities.

Oxygen debt The amount of oxygen needed to remove the lactic acid built up during exercise as well as to restore blood levels of oxygen to normal.

Photosynthesis The energy-trapping activities of plants and other autotrophs, in which a portion of the energy in sunlight is trapped and stored in carbohydrate molecules.

Chapter 22

Nutrition

Most of the roughly 65 million Americans involved in some form of dieting at any given time are trying to lose weight—usually pounds of unwanted fat. A very few dieters (usually adolescent boys and body builders) are trying to gain weight. Some dieters are simply trying to improve the nutritional quality of the food they ingest.

Your acquaintances surely include a few strict vegetarians (or perhaps you are a vegetarian). Some speak about their nutritional lifestyles with the zeal of missionaries. You, or someone you know, may think they are buying nutritional insurance by taking a handful of vitamin pills, purchased in "power packs" sold in most college bookstores. And then there are the meat-and-potatoes people who just like to keep things simple. Some speak out passionately against all "artificial" food additives. Others are junk food addicts. Who is right?

Nutrition is a subject about which science has something to say regarding the quality of your life. It is a very technical and rigorous science. But it is also a subject about which a great deal of misinformation is passed around by those who have read only a little or who have heard something from a friend. You should be suspicious about your sources of information. Anyone who is selling you nutritional products (such as vitamins or food supplements) or an exercise or weight control program or who has any vested financial interest in how you spend your nutritional dollars has a conflict of interest. In fact, the worst advice about how to manage your health often comes from people working in health clubs or health food stores.

You probably harbor some of these bits of misinformation. Take the true-false quiz that follows and you may be surprised by the number of things you thought you knew, but do not.

NUTRITIONAL FACTS AND FICTIONS

1. *Statement:* Athletes involved in high-intensity training programs should enhance their diets with protein supplements. (True or false?)

Answer: False. Reputable scientists have produced not one shred of evidence indicating that athletes in training benefit from or have their performances enhanced by taking protein supplements. In fact, there is a substantial body of evidence suggesting that such supplements are more likely to cause problems than be beneficial. Body builders are often victims of this scam.

2. *Statement:* All fats should be eliminated from your diet. (True or false?)

Answer: False. While we tend to eat more fat than we need, we do need about 20 percent of our caloric intake in the form of fats (the average American diet is currently 35–45 percent fat). There are certain types of essential fatty acids we must consume; fats are important components of cell membranes, and they are needed to help transport the fat-soluble vitamins (A, D, K, and E). Moderation, not elimination, is the rule.

3. *Statement:* Everyone should take a one-a-day multiple vitamin pill in order to make sure their nutritional bases are covered. (True or false?)

Answer: False. This is a very popular misconception fostered by the pharmaceutical companies, health food stores and clubs, and virtually everyone else with vitamins to sell. This one costs the American public literally billions of dollars per year—needlessly! It is easy to cover your daily nutritional needs by eating a variety of foods that include various fruits, vegetables, and animal products. Variety in your diet is the key. People in special situations, like pregnant women, rapidly growing children, and people with special metabolic needs might benefit from a vitamin supplement upon advice from their doctors (Figure 22-1).

4. *Statement:* Athletes can replace their bodily fluids faster by drinking commercially prepared "sports drinks" (electrolyte solutions). (True or false?)

Answer: False. Exercise physiologists now agree that nothing beats clear, cool water. In fact, some evidence suggests these electrolyte fluids are actually absorbed more slowly than water. In addition, there is no evidence suggesting that athletic activities lead to substantial depletions in any of the body's electrolytes.

5. *Statement:* Fresh fruits and vegetables are nutritionally better for you than frozen. (True or false?)

Answer: Usually false! The only time when this is true is when you are picking them directly out of the farmer's field or from a roadside stand. In most cases, "fresh" vegetables and fruits are stored or transported around the country for several days or weeks before they are actually consumed. Fruits and vegetables that have been frozen within one or two days of harvesting are actually more nutritious, especially if salt, sugar, or other substances have not been added during processing.

6. *Statement:* Fat people eat more than slender people. (True or false?)

Answer: False. This is often not the case. Many factors besides food intake influence body weight. In many cases, overweight people actually eat less than their slender counterparts.

FIGURE 22-1 Many dietary supplements benefit the seller of the product more than the user. Most people (including body builders) do not get much physiologic benefit from taking protein supplements. With the exception of children and pregnant women, few people in this country even need daily vitamin supplements.

7. *Statement:* Dried fruits and nuts are excellent health-type snacking foods. (True or false?)

Answer: False. Actually they are calorie dense. Since dried fruits have lost their water, they consist mostly of highly concentrated sugars. Nuts contain substantial amounts of fats. Weight-conscious people should eat these foods with a great deal of discretion. Dried fruits also contain preservatives such as sulfites, which some people cannot tolerate.

NUTRIENTS

Good health requires a certain number of calories every day, as we saw in Chapter 20, but it also requires minimum amounts of many specific substances we call nutrients (see Table 22-1). These nutrients provide three basic services to the body:

1. They provide the body with the fuels it requires for all its metabolic activities.
2. They provide the raw building materials that will be used for growth and maintenance.
3. They supply substances that help regulate various body processes.

The ideal, or balanced, diet provides these nutrients in a variety of foods that are not top heavy on calories. By the end of this chapter, you will know what the basic essential nutrients are, what foods are good sources of each, and what some of the consequences are of not getting enough of any one of them. You will also be able to recognize the truth—or lack of truth—behind some of the claims for many nutritional supplements and "health foods."

The body needs nutrients for several purposes. Proteins are needed for building the structural and functional components of cells and for making the body's chemical tools, its enzymes. Carbohydrates and lipids (fatty materials) are important sources of energy. Some lipids are essential for building cell membranes and absorbing fat-soluble vitamins. The body needs vitamins and minerals in much smaller quantities than it does carbohydrates, lipids, and proteins, but these substances

Table 22-1

Classes of Nutrients

NUTRIENT CLASS	FUNCTIONS IN THE BODY
Proteins	Structural materials; enzymes; hormones; a source of amino acids
Lipids	Energy storage in animals; components of cell membranes; bodily insulation and padding; transport of fat-soluble materials
Carbohydrates	Raw fuels; energy storage
Vitamins	Cofactors for enzymes; help enzymes do their jobs
Mineral elements	Electrolytes, salts, skeletal matrix
Water	Solvent medium for the living system

are equally essential; they help enzymes work and help govern the body's internal environment. Some minerals serve structural purposes, as in bones. Others help to regulate the amount of water in the body. Water is the solvent medium for protoplasm. Each of the major nutrient categories is discussed in what follows.

PROTEIN

Cells use proteins in several ways. Protein molecules are the building blocks of the intracellular skeleton, the meshwork of microfibrils and microtubules that gives shape to each cell and supports the organelles, and of the fibers that surround and link the cells in the body (Figure 22-2). Some, found on the surfaces of cells, act as receptors for intercellular messages.

Other proteins have the ability to use the energy of adenosine triphosphate (ATP) to shorten, giving muscles their ability to contract. Still others are the enzymes that control chemical reactions within the body. A few small ones (polypeptides) even serve as hormones and neurotransmitters.

Protein is thus essential to the normal operation of the body. However, it is most crucial in the diet of growing children. Both adults and young people break down unneeded proteins into their constituent amino acids and then use them to build new proteins. Because adults metabolize some of the amino acids to yield energy, they need about 60 grams (2 oz) of protein in their diets per day. Since meat contains about 25 percent protein, the adult requirement is met by 8 oz of hamburger per day—the equivalent of two "quarter pounders." Because children and embryos are growing rapidly, they must have a continuously increasing supply of protein to build new tissue. A one-year-old child needs only 25 g per day, but this is much more per pound of body weight than adults need.

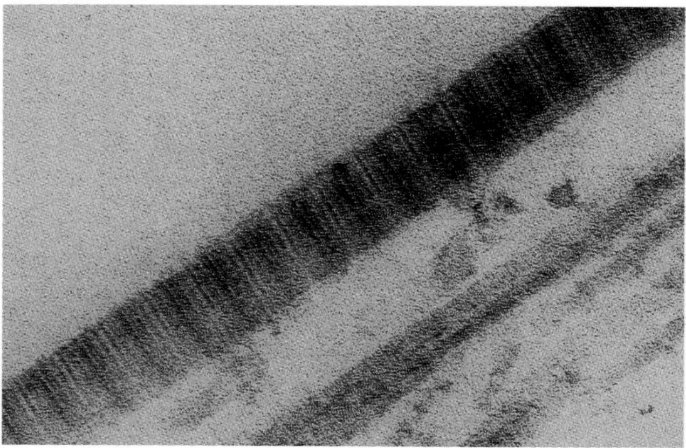

FIGURE 22-2 Collagen, a protein secreted by connective tissue cells, forms a network of fibers that bind cells together. It is also an important part of bone, cartilage, tendon, and ligament, lending strength and resiliency to the body's supports and links.

Essential and Nonessential Amino Acids

As we discussed in Chapter 2, proteins are large molecules, "macromolecules," made of smaller units called amino acids. Only 20 different amino acids are used in human proteins (though many more exist). Of these 20 amino acids, human cells can manufacture only 12. The 8 starred in Table 22-2 are called **essential amino acids** because the body *cannot* manufacture them; therefore, they are an "essential" part of the diet. The body cannot make proteins that require any of these essential amino acids unless the diet supplies them.

Complete and Incomplete Proteins

Proteins that contain all the essential amino acids are called **complete proteins.** The best sources of proteins for the human diet are animal proteins, including meat, poultry, fish, dairy products, and eggs. Virtually all sources of animal protein are complete proteins, because animals are physiologically similar enough to humans that they need basically the same amino acids as we do. The minimum daily requirements for these essential amino acids are indicated in Table 22-3. As you can see, the average American consumes from three to six

Table 22-2

The Twenty Amino Acids Used in Human Proteins

Alanine	*Leucine
Arginine	*Lysine
Asparagine	*Methionine
Aspartic acid	*Phenylalanine
Cysteine	Proline
Glutamic acid	Serine
Glutamine	*Threonine
Glycine	*Tryptophan
Histidine	Tyrosine
*Isoleucine	*Valine

Note: Asterisks indicate essential amino acids.

Table 22-3

Essential Amino Acid Minimum Daily Requirements (g/day)

AMINO ACID	WOMEN	MEN	AVERAGE AMERICAN DIET
Isoleucine	0.45	0.70	4.2
Leucine	0.62	1.10	6.5
Lysine	0.50	0.80	4.0
Methionine	0.55	1.10	3.0
Phenylalanine	1.12	1.10	4.1
Threonine	0.30	0.50	2.8
Tryptophan	0.16	0.25	0.9
Valine	0.65	0.80	4.2

times more of the essential amino acids than the minimum requires.

Plant foods also contain protein and amino acids, but they often do not contain all of the essential amino acids. Proteins lacking one or more of the essential amino acids are known as **incomplete proteins.** Grains such as corn are deficient in lysine, isoleucine, and tryptophan. Legumes such as beans are deficient in methionine and cysteine. Leafy greens lack methionine. Vegetarians must therefore include several different sources of plant proteins in their diets to ensure they receive all the essential amino acids. They must eat several different incomplete proteins in order to provide themselves with complete protein. Many cultures that do not eat much meat follow this practice. For example, rural Mexicans cover their nutritional bases by relying largely on a mixture of corn and beans.

A carnivorous diet also has hazards. People who eat only meat may not get enough of the vitamins present in vegetables (discussed later in this chapter). If they do not eat enough meat to satisfy their need for both protein and energy, they may still suffer from protein deficiency.

Deficiency Diseases

The lack of any one essential amino acid—or other nutrient—leads to a **deficiency disease.** The lack of sufficient protein in the diet leads to the particular deficiency disease known as **kwashiorkor.** Visible symptoms include a belly swollen by accumulated fluid, due to lack of blood proteins; thin limbs, due to lack of muscle mass; and stunted growth. Less visible is the effect on brain development in children. Lacking protein, the brain as well as the body cannot grow as it should. Unfortunately, though the body can make up for some lost growth once protein becomes available, the brain often cannot. The result of childhood protein deficiency can thus be lifelong mental retardation (Figure 22-3).

1. Is kwashiorkor truly a *protein* deficiency disease? Why would it be more accurate to call it an amino acid deficiency disease?
2. Some vegetarians eat only plant foods. Others add dairy products to their diets. Still others add both dairy products and eggs. Still others eat fish as well. Which of these diets would you consider most nutritious?

We can therefore predict one of the saddest consequences of famines. In Africa in the early 1980s, prolonged drought greatly impaired the produc-

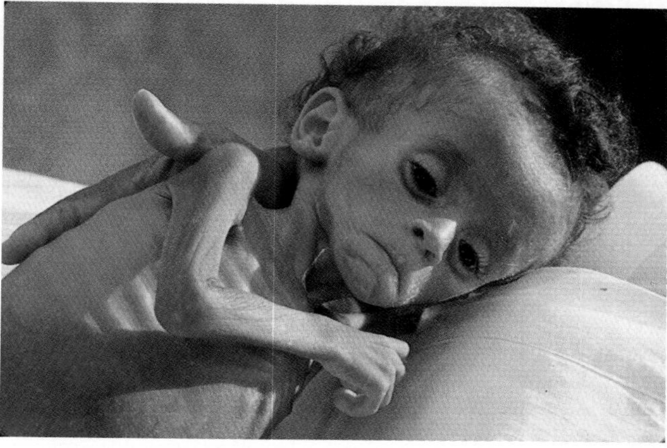

FIGURE 22-3 Children with diets severely lacking in protein, like this child from Brazil, suffer from kwashiorkor. Their bodies eventually are forced to consume their own stored protein simply to stay alive. Carrot-colored hair is produced in children with diets so inadequate that hair cannot even be pigmented properly. The distended abdomens are caused by edema.

tion of food, while unrestrained population growth increased the demand for food and damaged the environment. Many people died of starvation. The survivors subsisted largely on grain donated by more fortunate nations, including the United States, and kwashiorkor was common among the children. The problem eased when the drought ended in the autumn of 1985, but it has returned since then and will return again. Preventing future famines will require improvements in both agricultural techniques and population control. Unfortunately, some experts say, the brain damage done to the present generation of children in protein-deficient populations will make such improvements impossible or extremely difficult to achieve (Figure 22-4). The generations of people experiencing such nutritional deficiencies will be less able to do the planning or solve the problems that will alleviate the crisis.

LIPIDS

Fats are the most abundant nutrient in the American diet, after water and carbohydrates. They are part of a diverse chemical family called **lipids,** which include many different substances that dissolve in organic solvents such as ether and chloroform but not in water. Besides fats, lipids include fatty acids, cholesterol, and phospholipids (Figure 22-5).

FIGURE 22-4 Famines precipitated by drought and aggravated by overpopulation were responsible for millions of deaths during the mid-1980s. Even reaching a refugee camp such as this one in Ethiopia, where some food and water might be available, improved only the chances of survival—and, in some cases, barely.

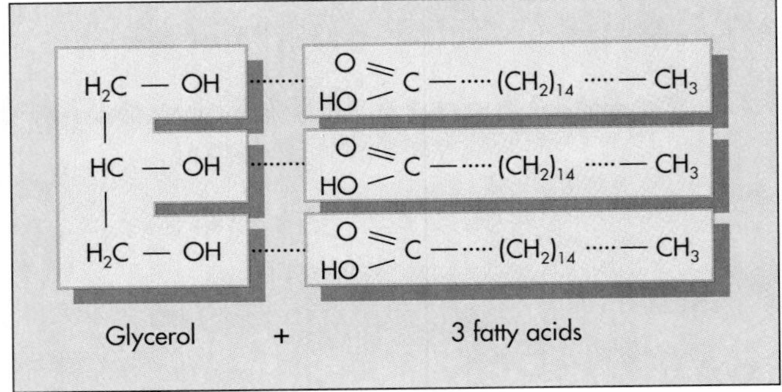

FIGURE 22-5 Structure of fat. Fats are formed by combining three fatty acids with glycerol.

Fats are found in animal foods such as meat, eggs, milk, cheese, and butter as well as in many plant foods. Even lean steak contains about 28 percent fat, which accounts for 70 percent of the calories it contains. Also, because fats are slow to leave the stomach during digestion (they usually remain in the stomach for 3.5 hours), they contribute significantly to feelings of satiety after eating a meal. For that reason, it is often recommended that some fats, in the form of butter (or margarine) on vegetables or whole milk, be included in weight loss diets because they make it easier for a "satisfied" dieter to stay on the diet (Figure 22-6).

Each fat molecule has as a superstructure a **glycerol** molecule to which is attached three fatty acids

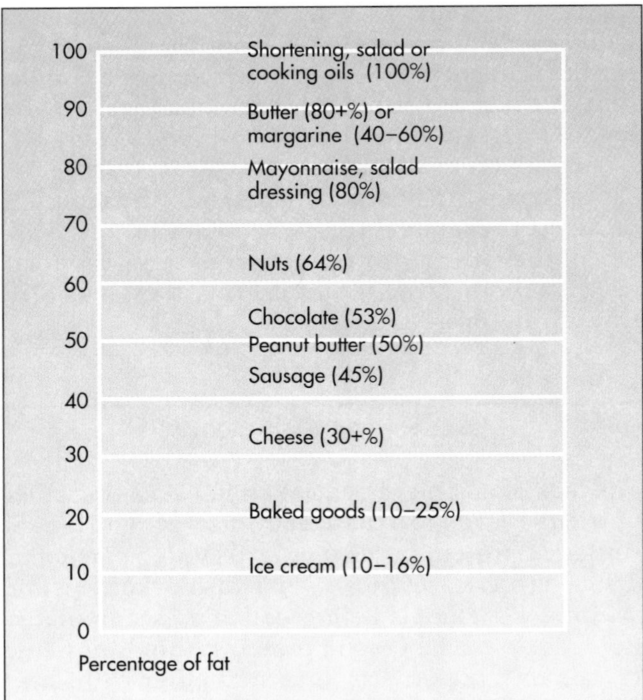

FIGURE 22-6 Fat-rich foods.

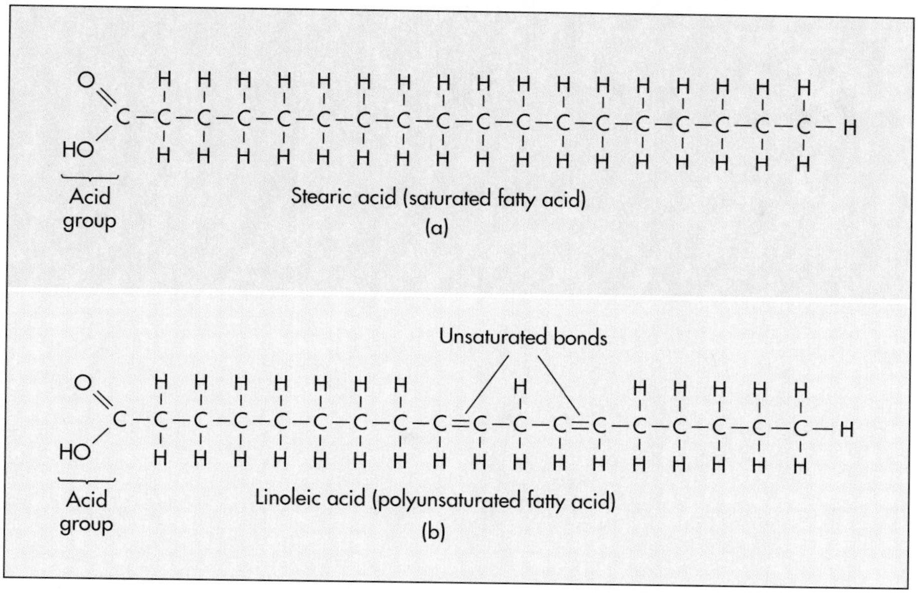

Acid group

Stearic acid (saturated fatty acid)
(a)

Unsaturated bonds

Acid group

Linoleic acid (polyunsaturated fatty acid)
(b)

FIGURE 22-7 (a) A saturated fatty acid. (b) A polyunsaturated fatty acid.

(Figure 22-5); a fat can thus also be called a **triglyceride** (see Chapter 2).

Typically, animal fats are **saturated** fats whose component fatty acids lack carbon-to-carbon double bonds. Oils, which are liquid fats found in plant products, have **unsaturated** fatty acids containing one or more carbon-to-carbon double bonds. **Polyunsaturated** fatty acids contain several double bonds. We call them "unsaturated" because at each double bond, the molecule can accept two more hydrogen atoms; saturated fats have no room for additional hydrogens (Figure 22-7).

Fats have more than double the calories of protein or carbohydrate. Foods high in fat, such as fried foods, are therefore good to avoid if you are trying to lose or control weight. However, fats are a necessary part of the diet. The liver receives absorbed fats via the hepatic portal system (see Chapter 20) and uses them to make the fatty acids the body needs to provide the phospholipids essential to the structure of cell membranes. It can also synthesize fatty acids from fragments of sugar and protein. But it cannot make two particular fatty acids, linolenic acid and linoleic acid, which must be obtained in the diet (they are present in many plant foods). These two fatty acids are **essential fatty acids.** Deficiencies lead to impaired growth, skin problems, and poor kidney function.

Cholesterol is a multiringed molecule needed as raw material for the synthesis of sex and adrenal cortex (steroid) hormones and as a component of cell membranes (Figure 22-8). The human body makes much of the cholesterol (about 1000 mg per day) it needs from saturated fatty acids, even excreting an excess in the form of bile salts. But people also get a great deal of cholesterol in their food, especially in eggs and organ meats such as liver (Table 22-4). Some studies have linked high consumption of cholesterol and saturated fatty acids to high blood pressure and heart attacks caused by **atherosclerosis,** clogging of the arteries by deposits of cholesterol (and other substances). Reducing cholesterol intake to less than 300 mg per day, either by changing the diet or by taking a drug that limits cholesterol absorption, may significantly reduce the risk of heart disease (Figure 22-9).

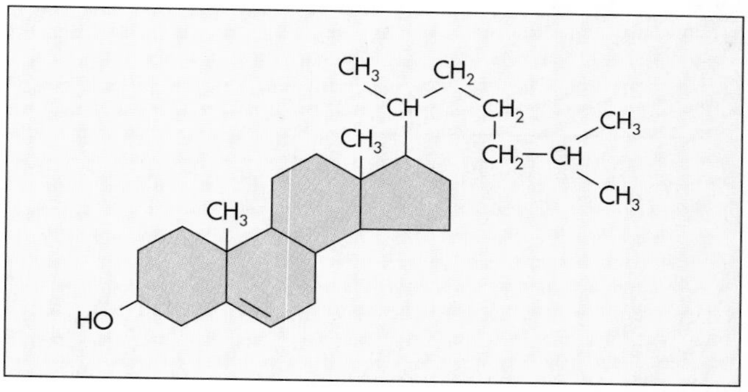

FIGURE 22-8 The cholesterol molecule.

Cultures in which people consume large amounts of cholesterol and saturated fats also show a greater incidence of breast, colon, and prostate cancer. The National Academy of Sciences and various government agencies have therefore recommended that people reduce their fat consump-

Table 22-4

Cholesterol Content of Certain Foods (Ascending Order)

FOOD	SERVING SIZE	CHOLESTEROL CONTENT (MG)
Milk, skimmed	1 cup	5
Cream, light table	1 oz (fluid)	20
Cottage cheese	½ cup	24
Cream, half and half	¼ cup	26
Ice cream	½ cup	27
Cheese, cheddar	1 oz	28
Butter	1 tablespoon	35
Chicken, turkey		
Light meat	3 oz cooked	67
Dark meat	3 oz cooked	75
Beef, pork, lobster	3 oz cooked	75
Lamb, veal, crab	3 oz cooked	85
Egg	1 yolk or egg	250
Liver: beef, calf, hog, or lamb	3 oz cooked	370
Brains	3 oz raw	1700+

Source: "Fats in Foods and Diet," Agricultural Information Bulletin No. 361, U.S. Department of Agriculture, Washington, D.C., 1974.

tion by at least 25 percent by relying on fish, poultry, lean meat, and plant foods for their protein; trimming fat off their meat; going easy on eggs and liver; minimizing consumption of butter, cream, and other sources of saturated fat; and avoiding fried foods.

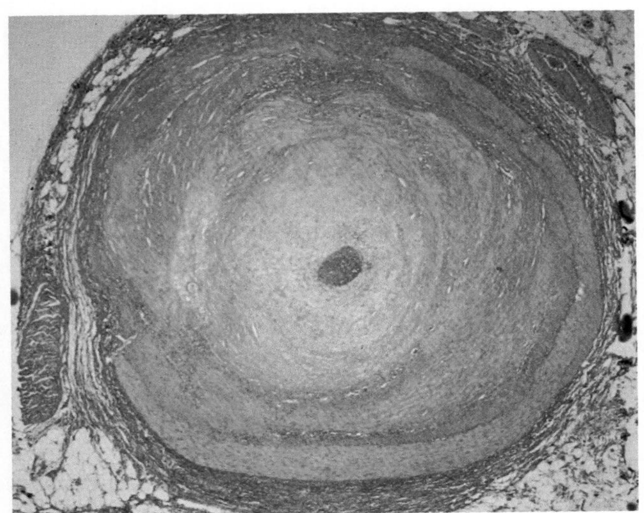

FIGURE 22-9 An artery clogged by atherosclerotic plaque is nearly completely blocked by lipid deposits such as cholesterol.

CARBOHYDRATES

Just as proteins are strings of amino acids and fats are assemblies of fatty acids, carbohydrates are strings of single sugars, or **monosaccharides.** The human body needs carbohydrates for adding sugars to other molecules to make glycoproteins and glycolipids, which serve various roles, notably as receptors for hormones and identification tags embedded in cell membranes. But it only has to obtain a portion of these sugars from the diet. It can make any sugar it needs; there are no essential sugars.

Nevertheless, the average human on Earth consumes a diet that is 65 percent carbohydrate. This percentage can range from about 40 percent to over 80 percent. The higher figure usually applies to poor people, even in the United States, who cannot afford more expensive foods such as meat.

Most of the carbohydrate people eat is starch, but simple sugars make up a large proportion for some people. Americans consume about 60.5 kilograms (kg; 133 lb) of sugars apiece every year. Half of this figure takes the form of the disaccharide **su-**

crose (table sugar); most of the rest is corn sweeteners, mixtures of the monosaccharides fructose and glucose. Three-quarters of these sugars are added to foods—candies, beverages, baked goods, and even canned and frozen vegetables and fruits—by the food processing industry.

Individual consumers thus have relatively little control over how much sugar they eat. This may help explain why obesity is such a serious problem for Americans. Too many of the calories we consume are hidden, and hence hard to avoid. That 60.5 kg of sugars per person per year is the caloric equivalent of 26.7 kg (59 lb) of body fat.

Carbohydrate in the human diet provides all the fuel for the brain, which consumes 140 g (0.31 lb) of glucose every day. Carbohydrates also provide most of the caloric energy the rest of the body needs, without imposing large metabolic costs. People can also use protein as a source of calories, but not so economically as carbohydrates. They must then expend energy to dispose of the nitrogen that is part of every amino acid molecule; this is the source of the 25–30 g of urea excreted daily in the urine. People can also use fat as a source of calories, but they must then cope with the release of acidic ketone bodies to the blood. Carbohydrate converts quickly and thoroughly to acetyl groups, which enter the Krebs cycle and are broken down to carbon dioxide and water (see Chapter 21). These same acetyl groups permit excess carbohydrate to be converted to fat for storage. By-products of the Krebs cycle can also be converted to the nonessential amino acids.

---

1. The caloric value of carbohydrate and protein is 4.1 kilocalories per gram (kcal/g). That of fat is 9.3 kcal/g. How much pure carbohydrate would you need to meet your daily energy requirement of about 2000 kcal?

2. How much pure protein?

3. How much pure fat?

4. Which of these three possibilities does your own diet most closely resemble?

When carbohydrate is digested in the small intestine, it is broken into its component sugars, mostly glucose. The sugar is then absorbed and carried in the hepatic portal system to the liver, where nonglucose sugars are converted to glucose and the glucose is stored as animal starch, or **glycogen.** Thereafter, the liver releases glucose slowly in response to the interplay of the pancreatic hormones insulin and glucagon (see Chapter 12). It normally maintains the level of glucose between 80 and 120 milligrams per 100 milliliters (mg/ml) of circulating blood. When the amount of glycogen in storage is inadequate, the liver can synthesize glucose from amino and fatty acids. It draws the fatty acids from fat stored in adipose tissues elsewhere in the body. When the fat reserves are depleted, it draws upon the body's proteins. To prevent acidification of the blood by ketone bodies and destruction of cells, the diet should contain at least 100 g of carbohydrate per day.

Famine leads to protein deficiency when the available food contains too little protein. However, when the available food also offers too few calories, this causes **protein-calorie malnutrition (PCM),** or **starvation.** The symptoms are those of kwashiorkor, plus general emaciation, loss of energy, and death. The body uses up its stores of glycogen in the liver and muscles, exhausts its fat supplies, and consumes the proteins of its muscles and other tissues. While fat is being used, ketone bodies give the breath a distinctive, acetonelike odor resembling that of nail polish remover. When the body begins to tap the proteins of the heart muscle itself, death is near. Fortunately, most of the damage is reversible. All it takes to restore physical health is food. In areas of severe famine, during war or drought, or as a consequence of crop failure, it is not unusual to have famine victims completely recover after experiencing a 25 percent weight loss. However, people recover rarely or only with great difficulty after a 50 percent weight loss. In such cases, some of the damage is permanent, especially in the young. Their brains may never compensate for lost growth and development.

VITAMINS

One of the most important principles contributing to good nutrition can be summarized in a single word—*variety.* We have already discussed our need for the major categories of foods, namely proteins, carbohydrates, lipids, and the energy they contain in the form of calories. But we also need about 40 other specific kinds of nutrients, including the vitamins and various minerals. Furthermore, we need a continual supply of all of these nutrients, for we are constantly using up what is already in the body. As shown in Table 22-5, the body's reserves for stored nutrients are limited, and they vary according to nutrient.

You should therefore eat a wide assortment of foods. Healthy diets include a variety of fruits, veg-

Table 22-5

Body Reserves for Stored Nutrients

TYPE OF NUTRIENT	TIME UNTIL DEPLETION OF RESERVE
Water	4 days
Foods	
Carbohydrates	About 13 hours
Amino acids	Several hours
Fat	20–40 days
Vitamins	
Thiamin (vitamin B_1)	7–14 days
Ascorbic acid (vitamin C)	60–120 days
Niacin	60–180 days
Riboflavin (vitamin B_2)	60–180 days
Vitamin A	90–365 days
Mineral nutrients and salts	
Sodium	2–3 days
Iron	750 days (men); 125 days (women)
Iodine	1000 days
Calcium	2500 days (almost 7 years)

Source: H. A. Guthrie, *Introductory Nutrition,* 4th ed. (St. Louis: C.V. Mosby).

etables, meat and fish, and dairy products. If you make a conscious effort to include these items in your daily diet, your needs for these special items will probably be satisfied.

Vitamins are organic compounds, needed in very small amounts, that the body cannot make for itself. There are about 14 different vitamins known to be required by the human body. Many serve as essential parts of enzymes, or **cofactors.** Without them the enzyme cannot work and the body's functions fail. Others play roles in vision and blood clotting. A few are synthesized by bacteria that dwell in the large intestine, but most must be obtained entirely from the food you eat.

There are two broad groups of vitamins. The **fat-soluble vitamins** are obtained in fatty foods such as butter, fats of meats and fish, cream, and vegetable oils and can be stored in the body's fatty tissues. To complicate matters, some of these substances, like vitamin K, are found also in green leafy vegetables—which are not considered fatty. As shown in the Learning Focus that follows, they include vitamins A, D, E, and K. The **water-soluble vitamins** do not dissolve in fat, are obtained in nonfatty foods, and cannot be stored in the body. They include the B vitamins and vitamin C.

THE FAT-SOLUBLE VITAMINS

Each of the fat-soluble vitamins is distinct—they do not form an easily identified, discrete chemical family (Figure 22-10). They do have certain physical and chemical properties in common, in spite of their structural diversity:

1. They all dissolve in organic solvents such as fats, oils, and ether.
2. None of them contains nitrogen.
3. They are more stable in heat and less likely to be lost during the cooking and processing of foods than water-soluble vitamins.
4. Since fat-soluble vitamins do not dissolve in water, they are not eliminated in the urine. They are usually stored in the liver and deficiency symptoms are slow in developing.

Vitamin A

On a worldwide basis, vitamin A deficiencies are still among the most serious nutritional problems

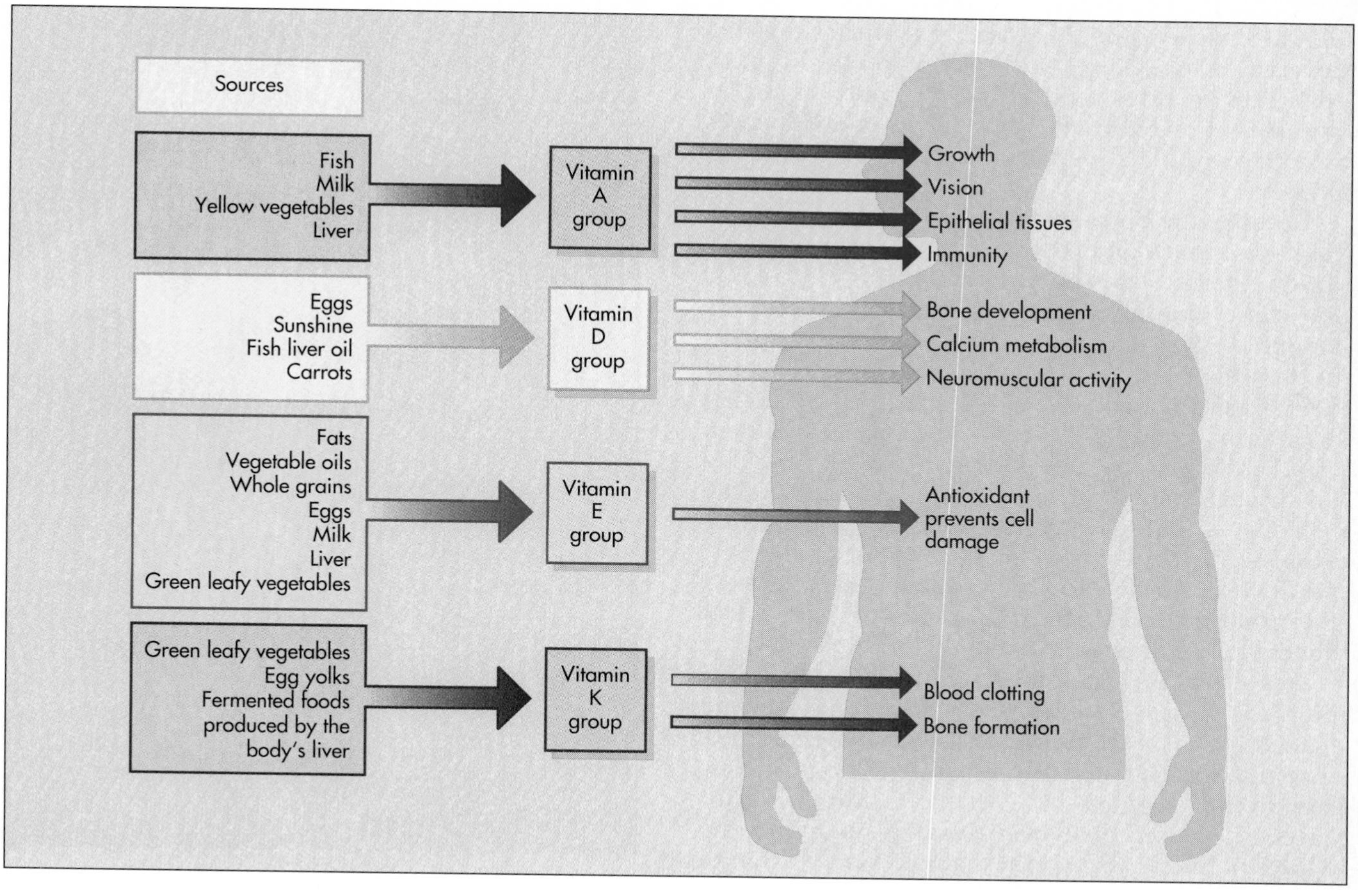

FIGURE 22-10 The fat-soluble vitamins.

facing human populations, especially in developing countries. Such deficiencies are particularly common among the children of these populations. A deficiency of vitamin A leads to night blindness (loss of the ability to adapt to dim light). It also increases the formation of keratin in such nonskin tissues as the cornea and lungs; the results can thus include dry, hard eyes (and possible blindness) (Figure 22-11) and respiratory infections. In addition, vitamin A is essential for growth in young animals. In some poor countries, vitamin A deficiency is a leading cause of childhood mortality and blindness.

We get vitamin A, or **retinol,** in our food. Most of our need for this vitamin, however, we satisfy by making retinol from **carotenes,** the pigments that make carrots and sweet potatoes orange and tomatoes red. Carotenes are also found in green vegetables, apples, apricots, beets, and other foods. One of the major uses for vitamin A is in vision. It is a main component of the pigment **(rhodopsin)** used by the rods of the retina. Because it is necessary for vision in dim light, we often say carrots are good for night vision (see Chapter 16).

Hippocrates, in ancient Greece, prescribed eating liver (an excellent source of vitamin A) for people suffering from night blindness. Near the turn

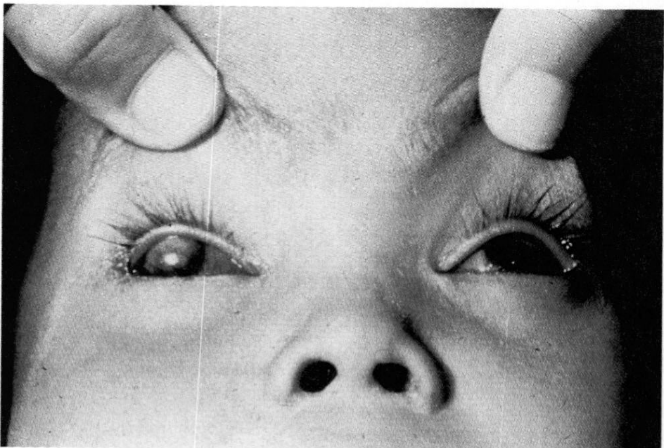

FIGURE 22-11 Vitamin A deficiency is the leading cause of childhood blindness in developing countries because it causes the tear-secreting apparatus to dry up, leading to ulcerations of the cornea. The tragedy of this is that it only takes a few cents per year to supply a child with all the vitamin A required for healthy vision.

of the century, the Japanese scientist Mori discovered, before much was known about vitamins, that certain fatty foods also contained a night-blindness-preventing factor. For vitamin A sources in American diets see the Learning Focus on vitamins.

Because the body stores substantial amounts of both vitamins A and D, vitamin overdosing—a condition referred to as **hypervitaminosis**—is possible for these vitamins. For that reason, these are among the few vitamins controlled by prescription in their higher doses. Under normal conditions the body takes in 5000–10,000 international units (IU) of vitamin A per day. Hypervitaminosis A occurs in small children who have received 75,000 IU per day for over six months (2 oz of liver contains 30,280 IU). The children's symptoms include dry, itching skin, swellings over the long bones, and irritability. Adults experience nausea, headache, and diarrhea. Extreme hypervitaminosis A leads to removal of bone calcium and weakening of the skeleton.

In spite of its potential for causing harm, a few physical fitness faddists in southern California loaded up on great quantities of carotene in hopes that the orange color of the pigment would translate to the "golden tan" in their skin. Unfortunately, it colored them bright orange, including the whites of their eyes and their gums—in addition to making them sick.

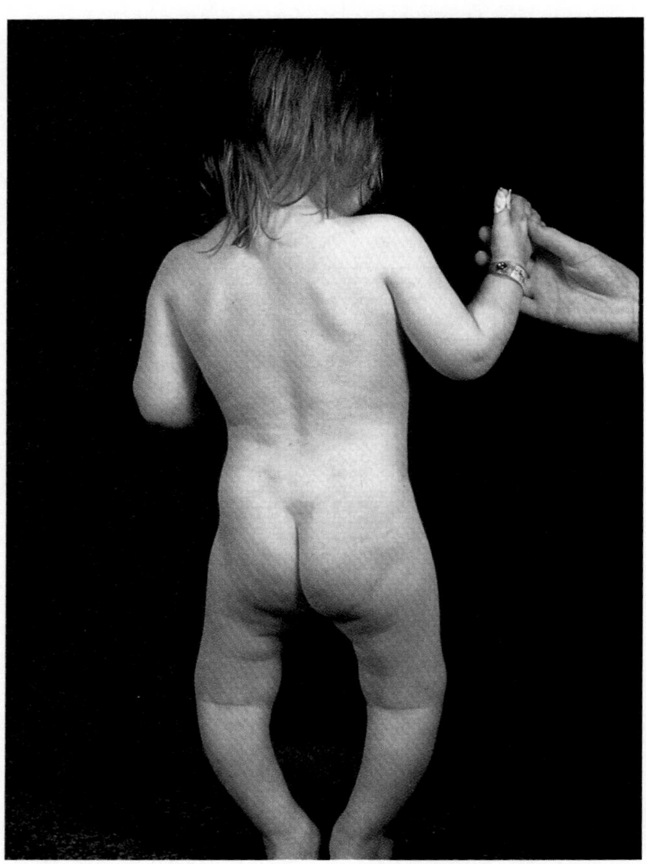

FIGURE 22-12 In children, lack of vitamin D hinders proper calcium absorption and use. The result is rickets, a condition marked by pronounced skeletal deformity, often most visible as bowed legs.

Vitamin D

Vitamin D, or **calciferol,** is actually a family of related compounds and is involved in regulating the absorption of calcium and phosphorus from the small intestine and their deposition in the bones. When it is lacking, the body does not get enough of these minerals, and the bones lack strength. One sign of childhood vitamin D deficiency, in fact, is bowed, rickety-appearing legs, which gives the disease its name of **rickets** (Figure 22-12). The adult version of rickets is called osteomalacia. Frequent broken bones are a side effect of too little vitamin D.

These signs of calcium mismanagement are rare in the tropics, where everyone gets plenty of sunlight. But in more northerly climes, where the sky is often cloudy and people stay indoors much of the winter, rickets used to be common. It was so common among the slum children of industrial England that rickets became known as the "English disease." When people finally made the connection between sunshine and health, the way was paved for the discovery of the vitamin.

Vitamin D is normally synthesized in the skin's dermis upon exposure to the ultraviolet component of sunlight. Once formed, it is released to the blood for later absorption and storage in the liver. The liver converts it to the active "vitamin D hormone," which is released into the bloodstream, as needed, to help in the absorption of calcium and phosphorus. It is also found in fish oils (cod liver oil), egg yolks, butter, and milk (Figure 22-13). Today, some foods are "fortified" with added vitamin D to ensure that children get enough. Significant amounts of vitamin D are passed across the placenta to the fetus and in the breast milk to the infant.

Excess vitamin D can lead to the buildup of calcium and phosphorus salts in normally soft tissues, such as the lungs or kidneys. It can also lead to loss of appetite and poor growth in children. In severe cases of hypervitaminosis D, the subject shows serious symptoms of toxicity that include vomiting, diarrhea, weakness, loss of weight, and kidney damage.

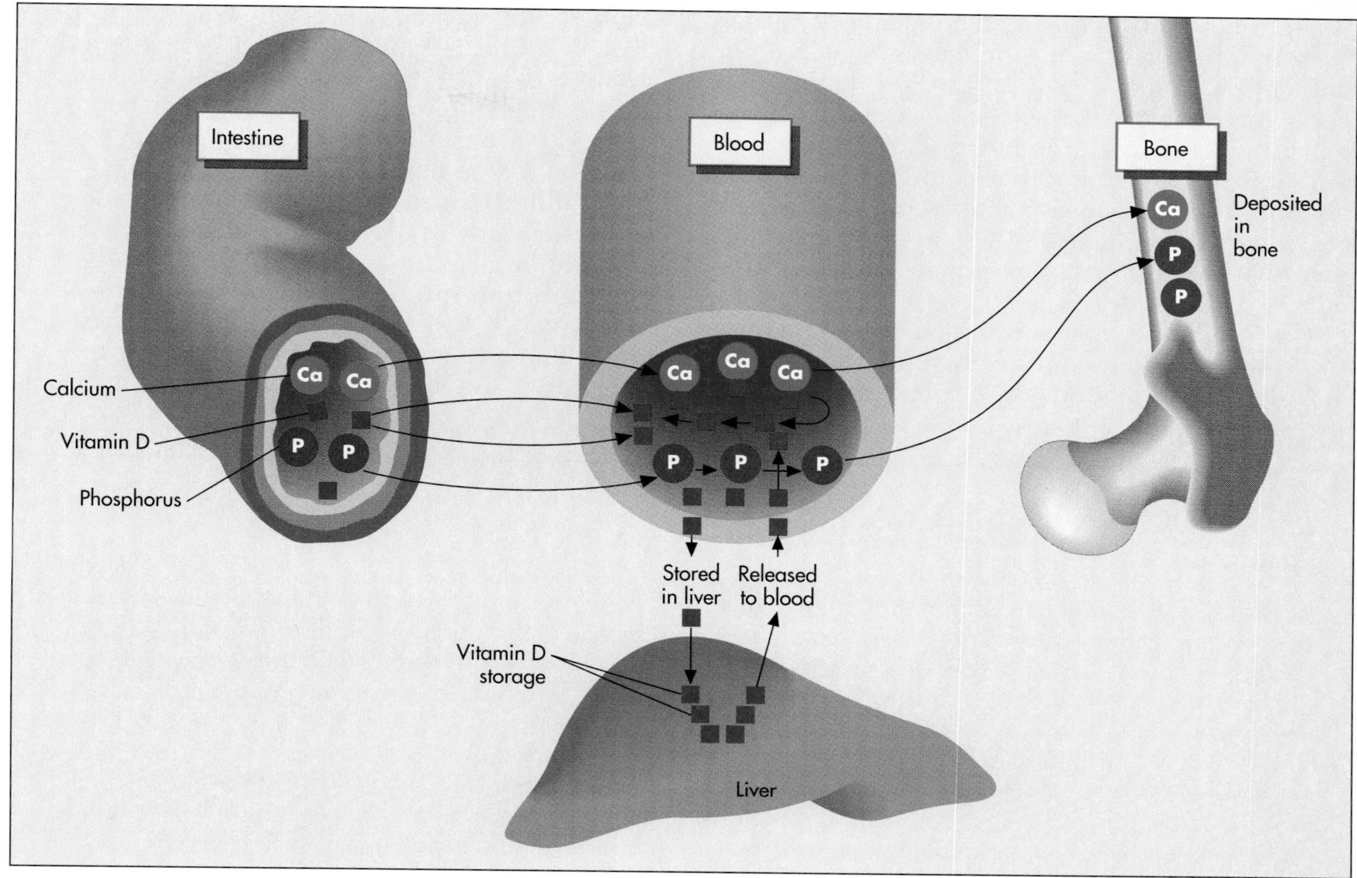

FIGURE 22-13 The action of vitamin D in calcium metabolism.

Vitamin E

Vitamin E, or **tocopherol,** has been called the "immortality vitamin," for it is an antioxidant. That is, it prevents damage to molecules that can occur when they are exposed to oxygen. Such damage builds up in cells as they age, and there is evidence that adding vitamin E to the diet can extend the lifespan of some animals. The vitamin is also involved in the formation of deoxyribonucleic acid (DNA) and ribonucleic acid (RNA) and red blood cells; it aids wound healing and the functioning of the reproductive and nervous systems. Deficiencies can cause muscular dystrophy in monkeys and sterility in rats. However, there is no evidence as yet that vitamin E is essential for any purpose in humans.

Vitamin K

Vitamin K is essential for normal blood clotting. It appears to be a coenzyme needed by the liver to synthesize prothrombin—one of the clotting proteins circulating in the plasma. When there is a vitamin K deficiency, the blood is slow to clot. Vitamin K received its name from the German word for clotting.

People rarely suffer any problem with vitamin K deficiencies. The vitamin is present in many foods, including spinach, cauliflower, and liver, and it is produced by intestinal bacteria. However, heavy doses of antibiotics can destroy the intestinal bacteria. For this reason, vitamin K injections are sometimes given with antibiotic treatments. Vitamin K injections are also the way to treat accidental poisoning with some rat poisons (D-Con, or Warfarin) that act by destroying vitamin K in the blood and causing internal bleeding.

THE WATER-SOLUBLE VITAMINS

The B Vitamins

About the turn of this century, vitamin B was discovered as a component of rice hulls (present in brown rice) that prevented the deficiency disease **beriberi,** an affliction of people who ate white rice without the hulls (see Box 22-1). Over the next few years, researchers learned that "vitamin B" was not just a single vitamin but rather a complex of nine different vitamins that could be chemically separated into various fractions, each of which could cure and prevent different deficiency diseases. The beriberi vitamin became known as vitamin B_1.

The B vitamins are critical members of many of the cells' major coenzyme systems and, consequently, are intimately involved in many of the body's vital chemical processes. For that reason, a dietary source is required for each member of the B complex so that the body can maintain proper nerve and brain functions, growth, and reproduction. They are involved with almost every cellular reaction taking place in the body.

Vitamin B_1, also known as **thiamine,** is a coenzyme for many of the enzymes of the Krebs cycle; in addition, it is also essential for the synthesis of the neurotransmitter acetylcholine. Vitamin B_2 is **riboflavin,** a coenzyme for enzymes of carbohydrate and protein metabolism, especially in the skin, eye, and blood. Riboflavin deficiency can

Box 22-1 The Discovery of Vitamins

Many people have heard that vitamin C was the first vitamin to be discovered. As the story goes, European sailors on long voyages developed joint pain, loose teeth, and other symptoms of the disease called *scurvy.* The symptoms vanished as soon as the sailors could get fresh vegetables and fruits—especially citrus fruits. When this connection was recognized near the end of the eighteenth century, ships began to carry suitable supplies. The British navy relied on lime juice, to be taken with the daily ration of rum; British sailors are still called "limeys."

However, no one realized that what prevented scurvy was a substance present in fruit and vegetables that was essential to human health—what we now call a vitamin. That insight had to wait more than another hundred years, and it depended on a fundamental mistake in the study of a second vitamin deficiency disease, *beriberi.*

Beriberi, from a Sri Lankese word meaning "very weak," was the Oriental sailor's equivalent of scurvy. The Japanese navy fed its sailors white rice, fish, and vegetables. The sailors had plenty of vitamin C, but they still developed weak limbs, poor energy, and nerve damage. They stayed healthy when barley, meat, and evaporated milk were added to their diet, but no one knew why until the Dutch physician Christiaan Eijkman came to Indonesia in 1884 to study beriberi.

The germ theory of disease had just been discovered, and Eijkman was sure that bacteria were the cause of every disease. Accordingly, he devoted his efforts to trying endlessly to infect chickens with whatever bacterium was responsible for beriberi. One day his chickens did develop symptoms remarkably like those of beriberi, but he could find no bacterium that might be responsible. His frustration peaked when, in defiance of all his hopes, the chickens got well.

He found the answer when he learned that the chickens had recovered right after his new cook had started work. The old one, he found, had fed the chickens with white rice left over from the hospital kitchen. The new one, believing that it was a waste to give food fit for people to mere chickens, had replaced the white rice with brown rice. White rice is brown rice with its outer covering, or hull, removed. The hull contains oils that go rancid in storage; white rice keeps better.

Eijkman immediately realized that, like scurvy, beriberi was not a germ disease, but one caused and cured by diet. His mistake came when he decided that white rice must contain a poison and that the hulls must contain the antidote. He was wrong, but his recognition that trace components of food could cause or cure illness prompted much useful research. In 1901, his assistant and successor, Gerrit Grijns, realized the truth—that it was not the presence of some substance that caused beriberi, but the lack of one, what we now call vitamin B_1, or thiamine.

In 1929, Eijkman shared the Nobel Prize in physiology and medicine for his insight, even though he was wrong about important details. He therefore illustrates the fundamental truth that in science the truth of a theory itself can be less important than the truth the theory leads to by stimulating further thought and study.

cause cataracts, dermatitis, anemia, and other problems (Figure 22-14).

Vitamin B_3 is **niacin,** a part of a coenzyme, nicotinamide-adenine dinucleotide (NAD), involved in energy-releasing reactions. It also inhibits cholesterol synthesis and aids fat breakdown. Niacin deficiency disease is **pellagra,** marked by dermatitis, diarrhea, and mental disturbances (Figure 22-15).

Vitamin B_6, or **pyridoxine,** is involved in fat and amino acid metabolism and helps in the production of antibodies (see Chapter 23). Deficiency symptoms include dermatitis, nausea, and slow growth. Vitamin B_{12}, or **cyanocobalamin,** was the last B vitamin to be discovered (in 1948) and is essential for red blood cell formation and synthesis of acetylcholine. It is absorbed from the small intestine as a complex with an **intrinsic factor** secreted by the lining of the stomach. The only vitamin not found in plant foods, it can be obtained from meat, eggs, and dairy products. Although some is produced by intestinal bacteria, vegetarians still heavily risk B_{12} deficiency, which leads to **pernicious anemia** and degeneration of spinal nerve cell axons (Figure 22-16).

Biotin, folic acid, and pantothenic acid are also B vitamins, though they do not have numbers. With vitamins B_2, B_6, and B_{12}, they are produced in part by intestinal bacteria. **Biotin** is involved in the metabolism of pyruvic acid and the synthesis of fatty acids and nucleotides; deficiency can result in mental depression, muscle pain, dermatitis, nausea, and fatigue. **Folic acid** is also involved in nucleotide synthesis, as well as normal production of red and white blood cells; deficiency leads to the production of extra-large red blood cells. **Pantothenic acid** is a necessary part of coenzyme A in the Krebs cycle (and elsewhere). Deficiencies produced in the laboratory have led to muscle spasms, fatigue, nerve and muscle degeneration, and shortages of adrenal steroids. However, deficiencies of pantothenic acid do not seem to occur outside the lab; this vitamin is widely available in foods, and unlike most other water-soluble vitamins, it is stored in the liver and kidneys.

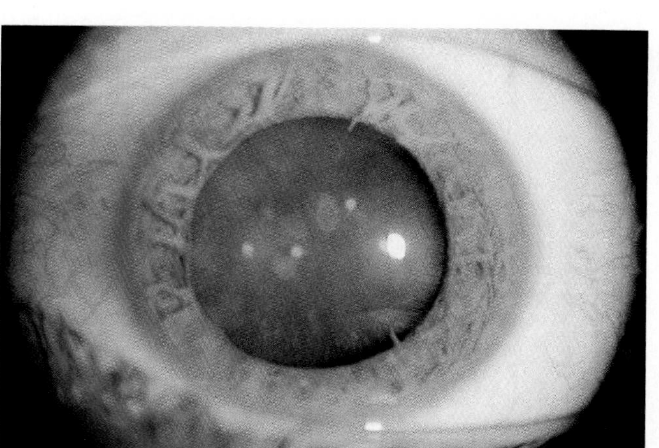

FIGURE 22-14 Cataracts due to riboflavin deficiency.

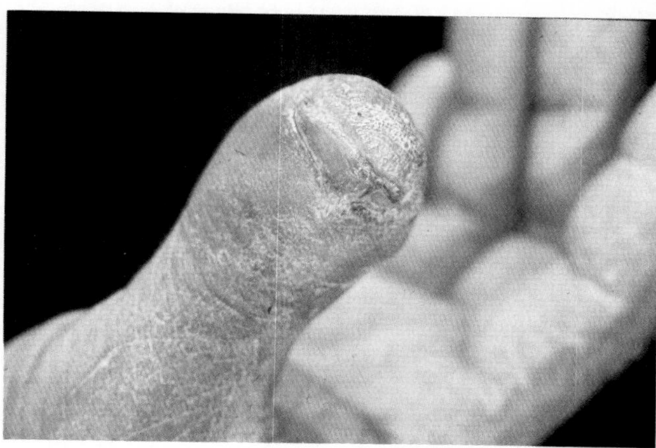

FIGURE 22-15 The symptoms of pellagra include dermatitis.

1. Many people believe that taking vitamins will give them more energy. Why is that line of reasoning incorrect?
2. However, there are at least two ways vitamin B deficiencies can interfere with "pep." Name them.

Vitamin C

Long before anyone knew of vitamin C, the effects of its absence from the diet were well known. A serious deterioration of the connective tissues, called **scurvy,** occurs when vitamin C is not in adequate supply. Descriptions of this disease, marked by gangrene of the gums, loss of teeth, and painful legs, date back to 450 B.C. It was common, too, in fifteenth- and sixteenth-century Europe, when the diets of most people did not include enough fresh fruits and vegetables, our principal sources of this important vitamin. The disease was particularly a problem for men embarked on long ocean voyages. Only in the late 1700s did the English learn enough about the disease to protect their mariners from its ravages.

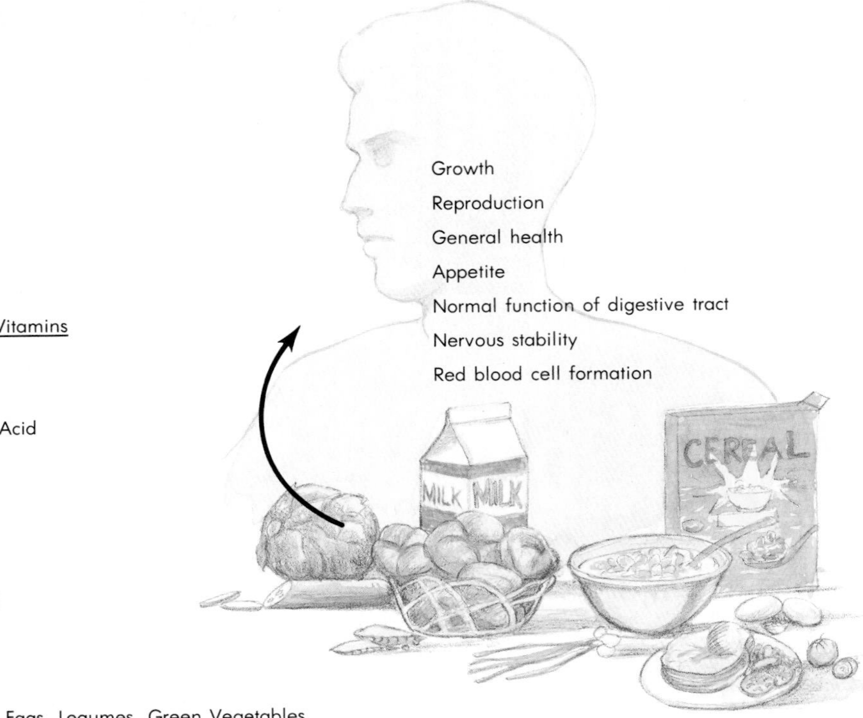

FIGURE 22-16 The B complex vitamins.

It was another 150 years before anyone could identify vitamin C's chemical makeup. We now know that vitamin C, or **ascorbic acid,** is involved in the formation of connective tissue, which explains why scurvy is marked by loosening of teeth, tender gums, retarded growth, and other signs of defective connective tissue (Figure 22-17). Vitamin C is essential for the synthesis of collagen, a critical constituent of most connective tissues. It also plays an important role in the manufacture of the intercellular cements that hold tissues together and is therefore vital to wound healing. Since it also plays a role in antibody functions, vitamin C deficiency is also marked by low resistance to disease.

Some people believe that large doses of vitamin C can help fight off disease, but the evidence is not clear. Despite the claims of Nobel laureate Linus Pauling, other nutritional studies have *not* shown vitamin C to help in the prevention or cure of the common cold. Nor does it seem to help cure cancer. On the other hand, vitamin C may indeed help to make various poisons harmless—a process called detoxification. The controversy about vitamin C is likely to continue for years.

Studies have shown that the bodies of normal adult men use about 13–30 mg of vitamin C per day. They have also shown that the body needs to consume only about 45 mg of vitamin C per day to maintain its normal body pool of 1500 mg. For that reason, the recommended daily allowance of 60 mg is more than enough to cover the nutritional needs of the average person.

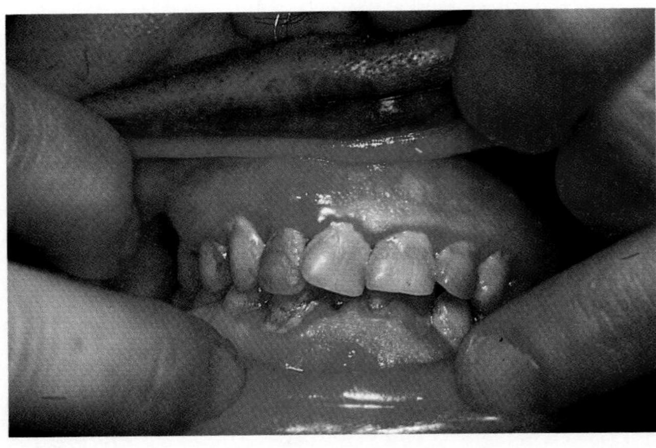

FIGURE 22-17 Lack of vitamin C causes scurvy, a weakening of connective tissue often most visible in loss of teeth and bleeding gums.

Learning Focus

THE VITAMINS

Study the following information carefully. After you have mastered it, go to the Learning Focus Response on p. 514 and focus your learning by answering the questions.

	NAME	NEEDED FOR	SOURCES
Fat-soluble vitamins			
A	Retinol	Dim-light vision, healthy skin, growth, maintenance of epithelial linings of chambers opening to outside; tear secretions	Dairy products, eggs, fish liver oil, yellow and green leafy vegetables
D	Calciferol	Calcium and phosphorus metabolism, skeleton	Fish liver oils, egg yolk, added to milk, made in skin
E	Tocopherol	Not known in humans, antioxidant	Meat, dairy products, wheat germ, leafy greens
K	—	Blood clotting	Intestinal bacteria, leafy greens, pork liver
Water-soluble vitamins			
B_1	Thiamine	Carbohydrate metabolism, growth, nerve function	Whole grains, beans, liver, intestinal bacteria
B_2	Riboflavin	Metabolism	Dairy products, eggs, whole grains
B_3	Niacin	Growth, metabolism	Liver, eggs, whole grains
B_6	Pyridoxine	Fat and amino acid metabolism	Bananas, potatoes, wheat germ, organ meats
B_{12}	Cyanocobalamin	Red blood cell maturation	Meat, fish, intestinal bacteria
	Biotin	Healthy skin, eyes	Liver, eggs, fresh vegetables
	Folic acid	Red blood cell production	Liver, leafy greens
	Pantothenic acid	Nerve and immune function	Most foods
C	Ascorbic acid	Connective tissue formation, cellular metabolism, health of teeth and bones	Citrus fruits, tomatoes, cabbage, broccoli

Learning Focus Response

THE VITAMINS

Using your various resources—including the preceding table and the text—answer the following questions.

1. If your physician tells you that you are suffering from poor visual sensitivity in the dark due to a vitamin deficiency, what foods should you eat to correct the problem?
2. What would you do to correct a vitamin D deficiency?
3. What symptoms would you expect to accompany a deficiency of niacin? What foods would you eat to ease the symptoms?
4. Anemia is most often linked to a deficiency of which vitamin? What foods would you eat to cure anemia due to such a deficiency?
5. What foods might you eat to tighten loose teeth? Why?

PREVENTION OF VITAMIN DEFICIENCY DISEASES

One of the marvelous things about vitamin deficiency diseases (summarized in Table 22-6) is that they can usually be cured quickly and easily. As soon as two days after a missing vitamin is added to a patient's diet, the symptoms of the disease begin to diminish. Recovery is prompt and complete, although destroyed nerve cells cannot be replaced and lost growth cannot always be made up.

Just as marvelous is the fact that only tiny quantities of vitamins—milligrams per day—are needed to maintain health or cure a deficiency disease. It is thus not surprising that after the discovery of vitamins, people tended to see them as miracle drugs, or that people still tend to think that if a little is good, more must be better.

Interestingly, the study of nutrition is still in its early stages. Recommended daily allowances (RDAs) for the various nutrients are adjusted every few years as new findings come in.

However, we do know that some vitamins are toxic in excess. Some never reach toxic levels because, when the body has more of such vitamins as ascorbic acid, thiamine, or riboflavin than it needs, it excretes the excess in the urine. It is thus fruitless—and even dangerous—to take large doses of vitamins. The excess over the body's immediate needs is simply wasted.

MINERALS

In addition to all the organic nutrients we have already discussed, there are some inorganic nutrients. These **minerals** are just as essential to good health as vitamins. Some even serve as part of a vitamin's structure, as cobalt does with vitamin B_{12}. Others help enzymes serve their functions, as iron does with hemoglobin. Still others serve as vital constituents of the body's fluids, as do sodium, potassium, and chlorine. Calcium and phosphorus even play structural roles in the bones.

Table 22-6

Deficiency Diseases: Their Causes and Effects

MISSING NUTRIENT	DISEASE	EFFECTS
Protein	Kwashiorkor	Swollen belly; fluid accumulation; thin limbs; lack of muscle mass; stunted growth; underdeveloped brain
Protein and calories	Protein-calorie malnutrition	Starvation; available protein used for fuel instead of for growth or maintenance
Vitamin A	Night blindness	Inability to produce rhodopsin, the pigment used in dim-light vision
	Xerophthalmia	No tear secretion, ulceration of cornea; leading cause of childhood blindness
Vitamin D	Rickets	Improper calcium and phosphorus absorption and malformed skeleton in children
	Osteomalacia	Adult rickets, weak bones
Vitamin K	Slow blood clotting	Liver unable to make adequate amounts of the blood-clotting protein prothrombin
Vitamin B_1	Beriberi	Weakness; poor energy metabolism; nerve damage
Riboflavin	Problems with carbohydrate and protein metabolism	Cataracts; dermatitis; anemia
Vitamin C	Scurvy	Connective tissue problems: gum disorders; loose teeth; poor wound healing
Niacin	Pellagra	Dermatitis; diarrhea; mental disturbances
Vitamin B_{12}	Pernicious anemia	No formation of red blood cells; degeneration of neuron axons in spinal cord

The body needs different amounts of various minerals. Those minerals it needs in fairly large amounts we call **macronutrients.** Those it needs only in tiny amounts we call **micronutrients,** or **trace elements.** The macronutrients include calcium, phosphorus, magnesium, potassium, sodium, chlorine, and sulfur. The more numerous micronutrients include iron, cobalt, copper, iodine, fluorine, manganese, zinc, chromium, selenium, and more. They may even include such "poisons" as arsenic, which is apparently essential for red blood cell production in rats, although it has not been proved necessary for human health. Most minerals function in the body as ions; thus, chlorine and fluorine do not appear in the body as the elements, which would be highly toxic, but as chloride and fluoride ions. The following Learning Focus lists many minerals and their functions.

Learning Focus

THE MINERALS

Study the following table carefully. After you have mastered the information, go to the Learning Focus Response and focus your learning by answering the questions.

MINERAL	SOURCES	FUNCTIONS
Macronutrients		
Calcium	Dairy products, egg yolk, shellfish	Bones, teeth, blood clotting, muscle and nerve activity, secretion; 99% stored in skeleton
Phosphorus (as phosphate)	Dairy products, meat, fish, nuts	Bones, teeth, DNA, RNA, ATP, blood buffer, nerve and muscle activity; 80% stored in skeleton
Magnesium	Widespread	Part of many coenzymes, cytochromes, needed for nerve and muscle activity
Potassium	Widespread	Main positive ion in intracellular fluids; needed for nerve and muscle activity
Sodium	Widespread	Main positive ion in extracellular fluids, needed for nerve conduction, part of blood's bicarbonate buffer system, helps control water balance by osmosis
Chloride	Widespread	Important in controlling acid-base and water balance in body, main negative ion in extracellular fluids
Sulfur	Meat, eggs, beans	Component of hormones, vitamins, proteins
Micronutrients		
Iron	Meat, egg yolk, legumes, dried fruits, nuts, cereals	Essential for hemoglobin, cytochromes
Cobalt	Meat, dairy products	Part of vitamin B_{12}

MINERAL	SOURCES	FUNCTIONS
Copper	Eggs, whole grains, beans, beets, asparagus, chocolate	Synthesis of hemoglobin and melanin
Fluoride	Widespread, fluoridated water	Strengthens teeth
Manganese	Cereals, legumes	Enzyme activator, needed for hemoglobin synthesis, growth, reproduction
Zinc	Widespread	Part of carbonic anhydrase, needed for growth, prostate function, fertility, digestion
Chromium	Brewer's yeast	Carbohydrate metabolism, DNA transcription
Selenium	Meat, seafood	Liver function, antioxidant
Iodine	Seafood, iodized salt	Part of thyroid hormone

Learning Focus Response

THE MINERALS

1. Name three minerals that help give us strong teeth. What roles do they play?

2. Name three minerals involved in the production of hemoglobin. Name another involved in blood clotting.

3. Name five minerals essential to the proper functioning of nerve and muscle.

4. Which mineral is essential to preventing pernicious anemia? Why?

5. A deficiency of one mineral leads very quickly to lethargy, swollen neck, and other symptoms. Why? Why is this deficiency uncommon in the present-day United States?

1. Review the section of this chapter dealing with vitamins. Notice that vitamin D exerts its effects by controlling the body's use of the minerals calcium and phosphorus. How would the symptoms of a shortage of these minerals in the diet resemble those of a vitamin D deficiency?
2. What would be the symptoms of a cobalt deficiency?
3. Review Chapter 13, on hormones. What would you expect the symptoms of an iodine deficiency to be?

It is possible to have a shortage of any of these minerals in the diet and hence to suffer mineral deficiency diseases. One of the simplest of these diseases is iron deficiency, which leads promptly to anemia. Recent research suggests that copper deficiencies may lead to heart disease and that as many as three-quarters of Americans consume less than the 2–3 mg of copper they need per day.

Chromium deficiency seems to interfere with sugar metabolism, apparently because chromium helps the insulin molecule bind to its target cells. Chromium deficiency may also have more serious effects, for this mineral seems to be involved in DNA transcription. Zinc deficiencies—which may be common—can lead to skin rashes, loss of appetite, poor healing, mental lethargy, and hair loss, among other things. People need 15 mg of zinc per day, but many get less, especially when they are on weight loss diets. Selenium deficiency may weaken the body's resistance to cancer, for peoples whose diets are low in selenium seem more prone to cancer.

Unfortunately, it is not always easy to correct mineral deficiencies, for excesses of many minerals can also lead to problems. The body excretes excesses of some minerals in the urine or feces. However, some minerals remain in the body and do damage. For instance, taking zinc tablets can raise the body's zinc level so much that it interferes with calcium absorption and leads to loss of calcium—and strength—from the bones, a condition called osteoporosis. Too much sodium in the diet (from table salt) seems to be associated with high blood pressure. Recent research, however, suggests the real cause of this problem may be too little calcium. Excess iron can produce constipation. It is thus wise to take mineral supplements only under a physician's supervision and not fall prey to health food fads.

FLUID BALANCE

The amount of water in the human body varies with age. Newborns are about 75 percent water; the level drops to about 63 percent after the first year, and then it varies between 50 and 60 percent for most of your life. In old age, the water level drops to 42–49 percent of the body weight in women over 60 years of age and 50–54 percent in men of the same age group.

In a sense, water is a much more important nutrient than anything else. It has no calories, it is neither a vitamin nor a mineral, and it does not become part of other important molecules. It does, however, serve as a carrier for everything else. Dissolved in it, other nutrients travel from the digestive system to the cells of the body, hormones coordinate the body's activities, and wastes travel to the kidneys and lungs. Suspended in it, enzymes

Box 22-2 Eat Dirt?

For centuries, people have been eating dirt for what ails them. To be specific, they have been eating clay as a treatment for diarrhea and as a nutritional supplement during pregnancy.

Modern Americans use one kind of clay (kaolinite, in Kaopectate) to stop diarrhea, but it was nevertheless a surprise when researchers found that Nigerian villagers use similar clays to treat diarrhea and dysentery as well as some problems of pregnancy. Later reports revealed that clay tablets were used for digestive problems in the Middle Ages in Europe and in ancient Rome. More recently, they have been used in the Middle East, and in Central America clay tablets officially blessed by the Roman Catholic Church are often consumed.

Eating clay is a common practice for pregnant women in developing countries around the world. Since the clay contains many minerals and trace elements, it may be an important source of these essential substances for the developing fetus.

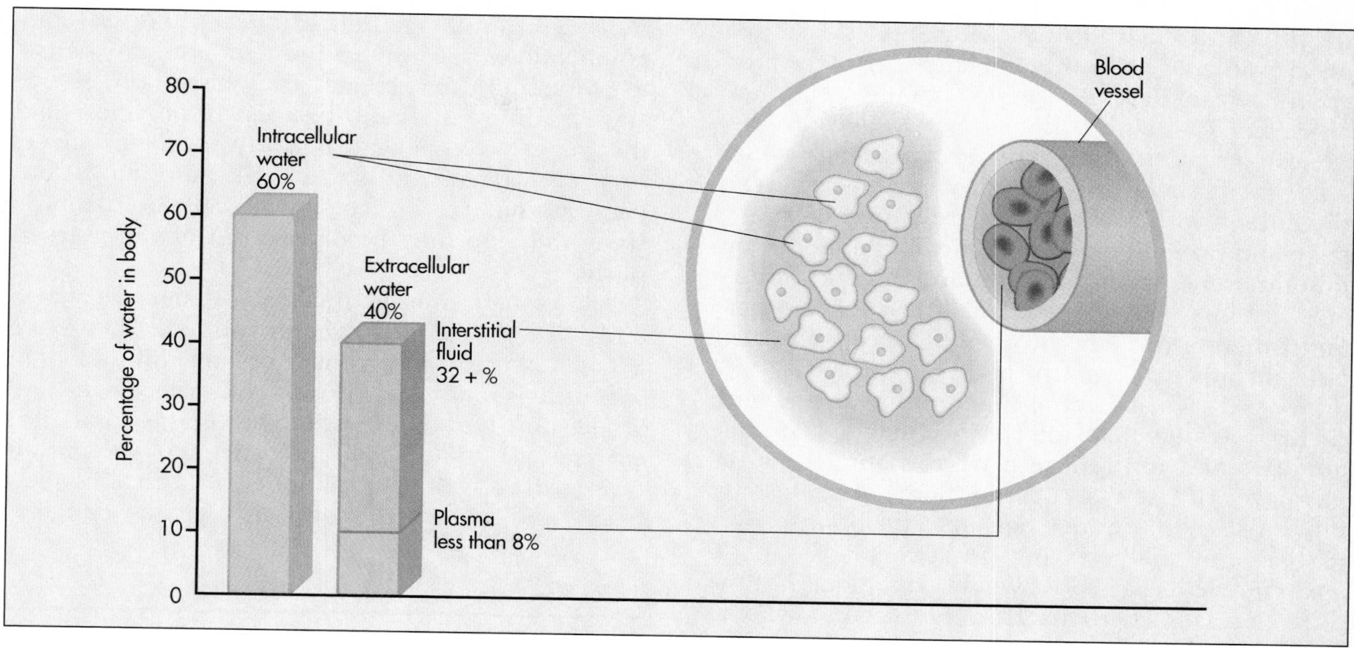

FIGURE 22-18 Fluid compartments in the body.

control chemical reactions, cells course the blood vessels, and the parts of the cell perform their many functions. Without water, life could not exist (Figure 22-18).

---

1. Every day, the average individual takes in 2–2.5 liters of water. Of this total, 600–1200 milliliters (ml) comes in food and 500–1000 ml comes in drink. Review the material on the electron transport chain in Chapter 21 to understand why the rest is called "metabolic" water.
2. Why is it crucial that every day the body lose approximately as much water as it takes in?
3. List the four ways in which the body loses water.

Water deficiency leads promptly to problems with temperature regulation, waste removal, and blood flow. Water excess, because it dilutes the mineral salts in the blood, can draw ions from nerve cells in the brain and cause convulsions. "Waterholics" have been known to drink large amounts of water—gallons per day—deliberately to produce a state of intoxication.

We do not usually need to worry much about getting enough water, for most foods contain water and drinkable water and other beverages are widely available. Nevertheless, our bodies devote a good deal of energy to regulating our content of water. We actually use some of the minerals we need to regulate the water in our bodies. Sodium and chloride especially play crucial roles in the kidney's osmotic control of the concentration of urine and thus in the conservation of water when it is in short supply or the excretion of excess water (see Chapter 12). They are aided by antidiuretic hormone and the renin-angiotensin mechanism.

NUTRITION AND THE CONSUMER

We have already indicated that the miraculous claims for vitamin and mineral supplements are unfounded. They will not restore youth, prolong life, or add pep in the majority of people who even loosely follow a balanced diet and are not suffering from genuine deficiencies. Nor will they cure such diseases as colds and cancer, even though some health food advocates do insist that these diseases are due to nutritional deficiencies. They *are* a great help, even a lifesaver, for those who have not been getting proper amounts of vitamins and minerals, just as adding protein to a protein-deficient diet can return a life-threatened child to health.

Refined Foods

As consumers, you should be aware that many of the things you hear about nutrition are similarly misleading. There is often some basis in fact, but the full truth is both less exciting and less frightening. For instance, a common claim is that the modern American diet is so "refined" that it lacks many nutrients. The truth is that some foods are treated in such a way that nutrients are lost. We have already mentioned white rice, produced by removing the hulls from brown rice in order to improve its taste and prolong its storage life. As the hulls contain thiamine, other vitamins, and minerals, nutrients *are* lost. Nutrients are also lost when wheat and other grains are processed to remove their hulls (bran) and germ (the plant embryo in a whole grain seed). The object of this refining step is both to remove oils that interfere with good storage qualities and to produce a white flour that gives attractive (if less tasty) bread. Some vitamins and minerals are then added back into the flour to "fortify" it, but the resulting product has distinctly fewer nutrients than does whole grain flour. Yet the difference is likely to matter only to people who do not eat a varied diet. Most people get all the vitamins and minerals they need from other foods, without counting those in their bread.

In addition, though the vitamins added to fortify foods such as flour and milk are synthetic—that is, they are produced in a factory, not by an organism—they are precise duplicates of natural vitamins, and their benefit to health is precisely the same.

Food Additives

Another claim is that food additives are dangerous to your health. These additives are preservatives, emulsifiers, and artificial colors and flavors (including artificial sweeteners). Preservatives prevent food from spoiling on its way from the processor to your kitchen, either by the growth of bacteria and molds or by chemical changes in the food (such as the oxidation of fats), which can alter the flavor. In many cases, not using preservatives is more dangerous to the consumer because of the various types of food poisonings that can develop in unprotected foods. Emulsifiers prevent mixtures of oil and water, as in salad dressings, from separating. The functions of coloring and flavoring agents are obvious.

Many of these chemicals are simply synthetic duplicates of "natural" compounds, ones also produced by plants, animals, bacteria, and molds. Most have been in use so long that we can be confident that they are reasonably safe to use; the Food and Drug Administration lists them as "Generally Recognized as Safe" (GRAS). Some have actually been tested; those that have proved capable of causing cancer, such as Red Dye No. 2, have been removed from the market. The exceptions are compounds such as nitrites, used to preserve bacon, wieners, and other meats. They have been shown statistically to increase the chances of cancer, but they seem to compensate partially for their dangers by preventing the growth of still more hazardous bacteria, such as those responsible for botulism. In time, they too may be phased out of the marketplace.

A few, such as the artificial sweeteners saccharin and cyclamate, remain controversial. They have been shown to cause cancer or other ill effects in laboratory animals. However, experience and study show that the hazard to humans is small, if it exists at all. In addition, the benefits of such products to diabetics, who cannot eat "natural" sweeteners, and to dieters may outweigh the dangers.

Organic Foods

Some people claim that "organically grown" foods are more nutritious than "nonorganically grown" foods. This is not so. In this context, "organic" means simply that the farmer has not used chemical fertilizers and pesticides, relying instead on manure and nonchemical pest control techniques. In either case, crop plants synthesize their own carbohydrates, proteins, lipids, and vitamins. They obtain their minerals from the soil and ultimately from whatever fertilizer the farmer has used. But the minerals are the same no matter whether they come from a chemical factory or a cow. Organic and nonorganic foods are precisely the same in nutritive value. They do not even differ when the farmer fails to supply all the necessary minerals, either organically or nonorganically, for plants grow and produce harvestable food only to the extent allowed by their nutrient supply.

On the other hand, the chemicals nonorganic farmers use to control insects and fungi really can make a difference in food. These toxic compounds are often present on or in the harvested crop. They can enter the human diet, and they hold at least the potential for harm. Nevertheless, adverse effects can be minimized by proper timing of chemical use and sometimes by careful washing of fruits and vegetables before eating.

The biggest difference between organic and nonorganic farming is the effect on the soil. Soil fertility depends on the presence of organic matter (humus), and chemical farming tends to deplete this vital substance. Organic farmers, because they constantly renew the soil's organic matter with manure and other organic materials, preserve the soil's fertility far more effectively.

SUMMARY

For good health the human body needs not only calories but also proteins, carbohydrates, lipids, vitamins, minerals, and water. It therefore requires support by a balanced, varied diet that is not dominated by high-calorie "junk foods."

Proteins are the raw materials from which the body builds its structural components, enzymes, some hormones and neurotransmitters, and cell surface receptors. Both adults and children recycle their proteins, but they do require a continuous supply. Adults need about 60 g per day. As children are growing and adding to their tissues, their need is far more crucial. Although they use about 25 g per day, that is much more than the adult requirement on a per-pound basis.

Proteins are made of 20 different amino acids, of which 8 cannot be synthesized in the adult human body. These essential amino acids must be obtained in the diet. They are most easily obtained from protein of animal origin. They can be obtained from plant protein, but vegetarians must be careful to mix their plant foods, since single sources of plant proteins lack some of the essential amino acids. A diet deficient in these amino acids leads to the deficiency disease kwashiorkor, marked by retarded growth, brain damage, and several other problems.

Lipids include fats, oils, fatty acids, and cholesterol. Fats and oils and their component fatty acids are important energy sources. They also provide the raw materials for construction of cell membrane lipids as well as transport and store the fat-soluble vitamins that can neither dissolve in water nor remain in aqueous parts of the cellular environment. The body can synthesize many of these raw materials, but two fatty acids are essential fatty acids. Like essential amino acids, they must be obtained in the diet. When they are missing, the result is impaired growth, skin problems, and poor kidney function.

The body can synthesize cholesterol from saturated fats, which lack carbon-carbon double bonds, but it gets a great deal with its food, especially from animal foods. Cholesterol is used as a raw material for steroid hormones and as a component of cell membranes. Excess cholesterol has been linked to atherosclerosis and heart trouble. Excess cholesterol and saturated fat have also been linked to colon, breast, and prostate cancer. Adults should thus limit their intake of these substances.

The bulk of the human diet is carbohydrates—starches and sugars. These nutrients supply most of the energy the body needs and all the energy the brain requires. The liver, obeying the signals of insulin and glucagon from the pancreas, converts glucose to glycogen and glycogen back to glucose to maintain the blood sugar level between 80 and 120 mg of glucose per 100 ml of blood. When carbohydrate is lacking from the diet, the body can use fat and protein. When they too are lacking, the body converts its own substance into fuel to meet its needs. The persistence of this condition leads to protein-calorie malnutrition, or starvation. As the body devours itself, it grows emaciated. The symptoms of kwashiorkor appear. Eventually, it dies.

The body needs vitamins in small amounts as cofactors for enzymes, as a key component of one visual pigment, and as a blood-clotting agent. Most are obtained from food; some are made by intestinal bacteria. The fat-soluble vitamins (A, D, E, and K) are the only ones the body can store in any quantity; A and D can be toxic in excess. Excess amounts of most of the water-soluble vitamins (the B vitamins and vitamin C) are excreted in the urine or feces.

Vitamin A, or retinol, is the one that functions in vision. Vitamin D, or calciferol, is necessary for the absorption and use of calcium and phosphorus. Vitamin E, or tocopherol, does not seem to be essential to humans. Vitamin K is needed for blood clotting. The B vitamins are cofactors for various metabolic enzymes. Vitamin C is essential to the formation of normal connective tissue. It was the study of various vitamin deficiency diseases that first led to the realization that health could depend on minute quantities of nutrients other than protein, carbohydrate, and lipid and then to the discovery of vitamins. A varied, balanced diet is far more effective than large doses of supplemental vitamins (megavitamins) to a healthy body.

The largely ionic minerals serve as structural materials (calcium and phosphorus in bone), parts of enzyme cofactors (cobalt in vitamin B_{12}), key components of important proteins (iron in hemoglobin) and hormones (iodine in thyroid hormone), and crucial constituents of body fluids (sodium, potassium, and chloride) involved in nerve signal transmission and the regulation of the body's water content. Those the body needs in large amounts we call macronutrients. Those it needs only in tiny amounts are micronutrients. Lack of any mineral leads to a corresponding deficiency disease, just as with vitamins. Deficiencies of some minerals are associated with such diseases as cancer, heart disease, and lack of energy. Some people blame such deficiencies for many diseases, but there is little evidence that they are right. Supplemental mineral doses can be hazardous, because excess amounts of some minerals are also linked to disease.

Since the human body is 60 percent water, water is also an essential nutrient. It is needed as a solvent and vehicle; without it, nutrients, wastes, and messages cannot move within the body. A deficiency of water leads more quickly to death than does a deficiency of anything else except oxygen.

There are a number of misconceptions about nutrients. Refined foods do lack some nutrients, but anyone who eats a balanced, varied diet will suffer from no deficiency diseases. Food additives are largely safe; a few are used despite recognized potential hazards, because they are beneficial in other important ways. Organic foods are no more nutritious than nonorganic foods; they contain precisely the same nutrients. The real difference lies in the possible contamination of nonorganic foods by pesticides and in the better maintenance of soil fertility by organic farming methods.

STUDY QUESTIONS

1. Why is "junk food" junk?
2. Name three functions of proteins in the body.
3. Why is protein most essential for children and embryos?
4. Kwashiorkor is due to a deficiency of which specific amino acids?
5. People never (or rarely) suffer from a deficiency of cholesterol. If they did, what would you expect the symptoms to include?
6. Vitamin A is essential for vision. Why, then, is it also essential to the health of blind people?
7. Thiamine is essential for the synthesis of acetylcholine. Discuss how this could account for the muscular weakness of beriberi.
8. Megavitamins can be dangerous because they can produce toxic doses of which vitamins?
9. Under what circumstances would you expect eating dirt to aid health?
10. In what ways are organic foods better than nonorganic foods?

GLOSSARY

Atherosclerosis Clogging of arteries by deposits of cholesterol.

Beriberi Vitamin B_1 deficiency disease.

Cofactor A chemical compound needed to make an enzyme function properly; often a vitamin.

Complete protein Protein containing all the essential amino acids.

Deficiency disease Illness produced by a lack of some essential nutrient.

Essential amino acids Amino acids the human body cannot synthesize and must obtain in the diet.

Essential fatty acids Fatty acids the body cannot synthesize and must obtain in the diet.

Fat-soluble vitamin A vitamin (Vitamin A, D, E, or K) that dissolves in lipid and that can be stored in the body.

Hypervitaminosis Overdosage with vitamins.

Incomplete protein Protein lacking one or more essential amino acids.

Macronutrient A mineral needed in relatively large quantities.

Micronutrient A mineral needed only in tiny amounts.

Mineral An inorganic substance, usually ionic, essential as parts of vitamins, hormones, and other molecules; some play crucial roles as constituents of body fluids.

Pellagra Niacin deficiency disease.

Pernicious anemia Vitamin B_{12} deficiency disease.

Protein-calorie malnutrition (PCM) Starvation due to lack of both calories and protein.

Rickets Vitamin D deficiency disease.

Scurvy Vitamin C deficiency disease.

Starvation Bodily deterioration due to lack of nutrients and energy.

Trace element A micronutrient.

Vitamin Organic compound needed in the diet in small amounts, often as a cofactor for enzymes.

Water-soluble vitamin A vitamin (the B's, C) that dissolves in water and cannot be stored in the body.

Chapter 23

The Threat: Disease and Infection

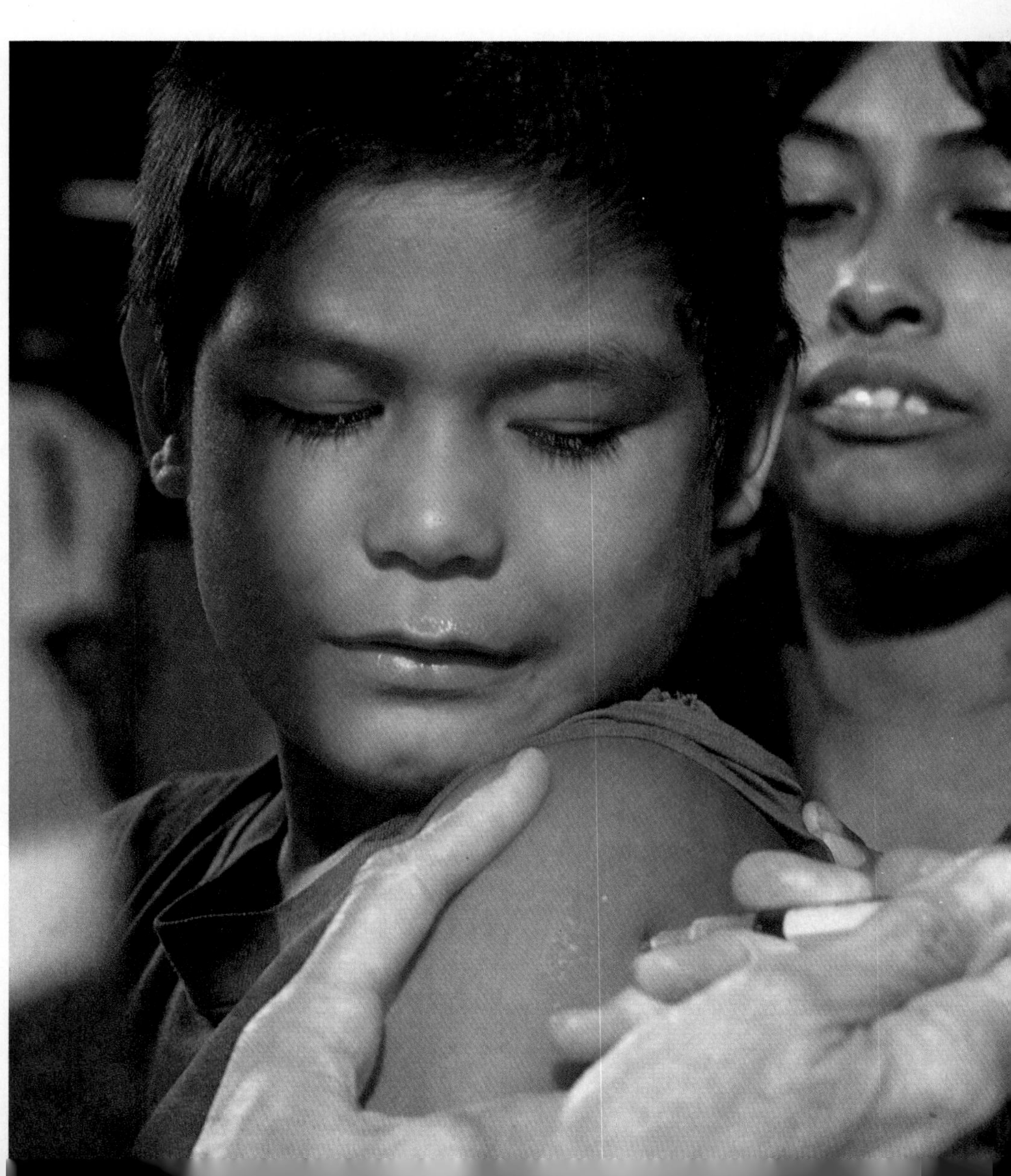

The human body is an organic structure so complex that it makes the largest and most complicated of machines look simple. Its trillions of component cells cooperate—as tissues, organs, and organ systems—to maintain life. They process nutrients, remove wastes, provide movement, coordinate each other's activities, and even reproduce the whole package faithfully from generation to generation. Together, they compose an example of reliability and interdependence that philosophers sometimes hold up as a goal for human society.

Yet, like society, the human body suffers from terrorists when its own cells create havoc as cancer cells; from subversives when viruses take over its machinery from within; and from invaders when bacteria and other parasites establish beachheads and destroy its tissues. In spite of these vulnerabilities, an average human survives between 70 and 80 years. One reason for its continued health is the body's immune system (Chapter 24), which has the job of blocking, detecting, and destroying threats of all kinds—terrorists, subversives, and invaders.

The terrorists—the many kinds of cancer—are described in Box 23-1; we will return to them in Chapter 24. The rest of this chapter will focus on the subversives—the viruses, viroids, and controversial prions that exist at the edge of life—and the invaders—the bacteria, protozoa, fungi, and multicellular parasites.

THE SUBVERSIVES

Those agents of disease and infection we call subversives, the viruses, viroids, and prions, are so simple that many biologists argue over whether they should even be considered alive. They lack many of the features we ordinarily identify with life, including basic cellular structure, organelles, and the enzymes necessary for processing energy. Viruses have only a protective protein outer covering and the genetic equipment necessary to take control of the host cells they attack, subverting their resources into manufacturing additional copies of the virus and often killing the host cell in the process. Viroids and prions have even fewer structural elements.

Viruses

Viruses pose real problems for microbiologists and taxonomists. The problem is that, as disease agents, they seem in some ways to act as living things. Yet they lack cell membranes, cytoplasm, most enzymes, and organelles; they are not cells. Since cells are defined as the fundamental unit of life, viruses are by definition not alive. Of course, some biologists argue they are alive, refer to them

Box 23-1 Cancer

According to the latest statistics, a quarter of the students reading this text will get cancer; one-sixth will die of the disease. Cancer has overtaken heart disease as the primary cause of death for middle-aged Americans.

Is cancer on the rise? Some researchers believe that because people are not dying so often of other diseases, they are now living long enough to develop cancer. Indeed, it seems that everyone who lives long enough will eventually get cancer, for throughout our lives individual cells spontaneously undergo the changes that mean cancer. Fortunately, our immune system normally identifies the changed cells and destroys them before they cause us grief. Unfortunately, as we age, our immune system becomes less efficient and more likely to miss such cells. When it does miss one, a cancer starts.

Other researchers say the techniques of detection and diagnosis are so much better than they used to be that more cancer cases are being counted. Still others attribute the apparent rise in cancer cases to our increasing exposure to industrial chemicals, pesticides, air and water pollutants, and cigarette smoke. There is even good evidence that the incidence of cancer goes up when people eat too much red meat or fat or eat few vegetables. (According to the National Academy of Sciences, cabbage, cauliflower, broccoli, brussels sprouts, and related vegetables, as well as fresh fruits containing vitamins A and C, actually offer some protection against cancer.)

What is cancer? Very simply, it is loss of cellular control. It begins when cells stop properly regulating their own reproduction. They no

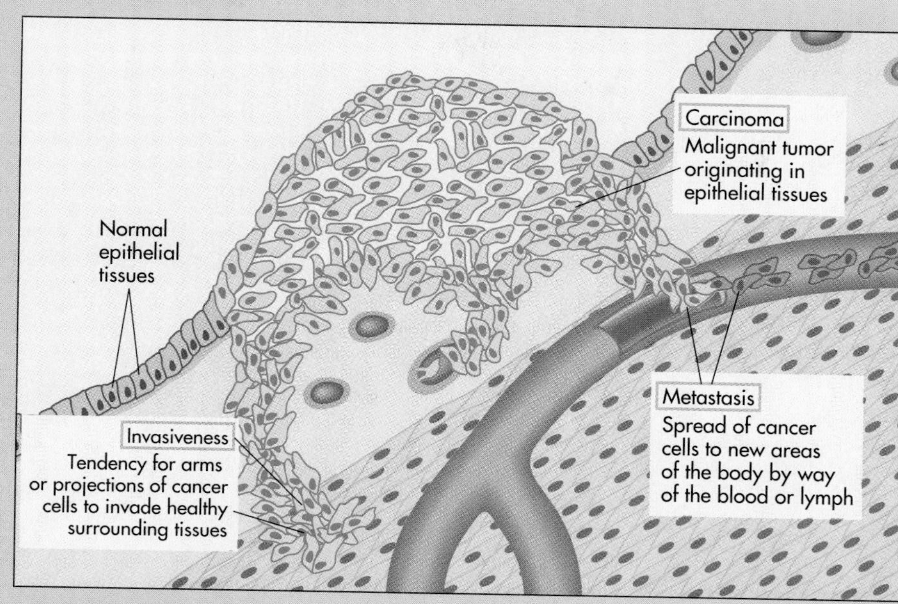

FIGURE 1 A cancerous tumor, showing metastasis.

longer respond to signals from neighboring cells that normally prevent them from multiplying (Figure 1).

CARCINOGENESIS

The process in which a normal cell turns into a cancer cell is called **carcinogenesis.** It has two steps, initiation and promotion, which appear to be triggered by separate mutations in growth-controlling genes, caused by exposure to either chemicals or radiation. Carcinogenesis can also be triggered by *oncogenes,* so-called cancer genes. Sometimes these genes are carried into the cells by viruses. Oncogenes can also appear when chemicals or radiation alter genes with normal functions in noncancerous tissues. Particularly potent oncogenes can repeat within families and account for the way certain cancers are passed from generation to generation. Often, oncogenes (and their normal counterparts) are involved in controlling the cell's responses to growth factors and other intercellular signals.

Initiation "transforms" cells, freeing them from mitotic inhibitors; once transformed, the cells no longer respond to "Stop dividing" signals from neighboring cells. In cancer cell cultures, transformed cells show no sign of contact inhibition; they continue to multiply even after they touch each other, and the bottom of the laboratory dish is covered with mounds of cells. In the body, this freedom from inhibition can lead to **tumors,** unrestrained masses of cells that interfere with blood flow, nerve action, and other bodily functions as they encroach upon neighboring organs. Tumors also interfere with bodily functions by preempting nutrients.

Not all tumors are cancerous, or **malignant.** Some are **benign;** that is, their cells multiply and produce a mass of tissue—sometimes weighing many pounds—but they do not grow endlessly or invade other tissues in the body. They behave as if they had become less sensitive, but not insensitive, to inhibitory signals. Often, they are contained within fibrous casings produced by normal, surrounding connective tissue.

Transformation by itself is not enough to cause a cancer. A second step, *promotion,* is necessary, for transformed cells do not take full advantage of their freedom from control until they have been exposed to a promoter, a virus, a chemical, or some other agent able to produce a second genetic change. A promoter may also be simply a change in the cells' environment that permits or encourages growth.

METASTASIS

If growth were all there were to cancerous tumors, cancer would be relatively easy to treat with surgery. Unfortunately, cancerous cells can undergo further changes. One such change makes them able to detach from the original tumor and travel, usually in the blood or lymph, to other parts of the body, where they take root and establish new tumors. This spreading of the cancer cells is called **metastasis.** Unfortunately, because metastasis is common, surgery often misses some cancerous cells. Curing cancer means killing *all* cancerous cells, wherever they may be in the body, and this is difficult. Physicians use toxic drugs and x-rays to interfere with cell division or kill actively dividing cells, even though such treatments damage normal cells too. They are also developing ways to use the body's immune system to attack cancer cells more selectively (see Chapter 24).

1. Why do you think anticancer drugs might be especially hazardous to use on children?
2. How do you think genetic engineering might eventually offer a way to treat cancer?

TYPES OF CANCER

Cancers are named for the type of tissue in which they arise. Thus, *lymphomas* are cancers of lymphoid tissue such as the lymph nodes. *Leukemias* are cancers of the stem cells in the bone marrow that normally produce the white blood cells. *Sarcomas* are cancers of muscle, bone, and connective tissue. *Carcinomas* are cancers of epithelial tissue; *adenocarcinomas* involve glandular epithelium, as in the liver, salivary glands, and breast. Breast cancer is one of the most common cancers, although lung cancer is more often fatal (Table 1).

SURVIVING CANCER

The best way to survive cancer is not to get it in the first place. Unfortunately, there is no guaranteed way to avoid this disease. The best you can do is to avoid risks such as tobacco smoking (lung and mouth cancer), toxic chemicals (liver cancer, leukemias), and excessive exposure to sunlight (skin cancer) and radiation and eat less fat and more fiber, fruit, and vegetables.

Even then, however, you may get cancer. The early clues that this has happened are as follows:

1. An unusual bleeding or discharge.
2. A lump or thickening in the breast or elsewhere.
3. A sore that is unusually slow to heal.
4. Any change in frequency of or pain during defecation or urination.
5. Persistent hoarseness or cough.
6. Persistent indigestion or difficulty in swallowing.
7. Any change in a wart or a mole.

Of those who get cancer, only about half are still alive five years later. For some cancers, however, the odds are much better. If you get breast cancer, for instance, and treat it before it can spread, you have a 96 percent chance of living. The odds are, of course, much worse if you detect the cancer later in its development. For this reason, you should get regular medical check-ups and see a physician if you notice any of the clues in the preceding list.

Table 1

The Ten Most Common and the Ten Most Deadly Cancers

MOST COMMON	MOST DEADLY
Breast, female	Lung, male
Lung, male	Lung, female
Prostate, male	Breast, female
Colon and rectum, female	Colon and rectum, female
Colon and rectum, male	Colon and rectum, male
Lung, female	Prostate, male
Uterus, female	Leukemias and lymphomas, male
Urinary tract, male	Leukemias and lymphomas, female
Leukemias and lymphomas, male	Urinary tract, male
Leukemias and lymphomas, female	Ovary, female

Source: American Cancer Society.

as microorganisms, and write of their life cycles.

Our position in this book is that they are nonliving disease agents. They are simply bits of genetic material wrapped in protective protein coverings. Like video game cartridges, they can take control of complex machines—the cells—and force the cells to play a new game. This game directs the cells to produce more copies of the virus that invaded them. Viruses are no more alive than is a computer chip. But they can be devastating to the interests of the cells they control.

The smallest **viruses** are about the size of a ribosome and contain only 10 genes. The largest approach the size of a bacterial cell, about 1 micrometer in diameter, and have 500 or more genes (Figure 23-1). All consist of a protective covering, or shell, of protein molecules called a **capsid,** surrounding a core of nucleic acid (Figure 23-2). The protein can be limited to just one or two kinds or it can include dozens, serving as structural components, enzymes, and "receptors" that recognize and attach to the surface molecules of host cells (Figure 23-3). The nucleic acid can be either DNA or RNA—

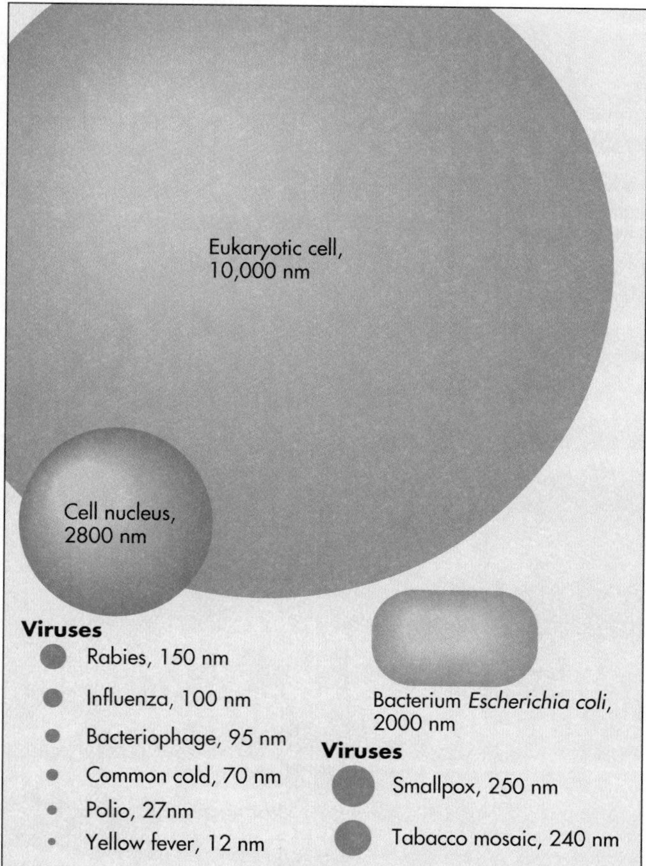

FIGURE 23-1 Relative sizes of eukaryotic cells, prokaryotic bacteria, and assorted viruses.

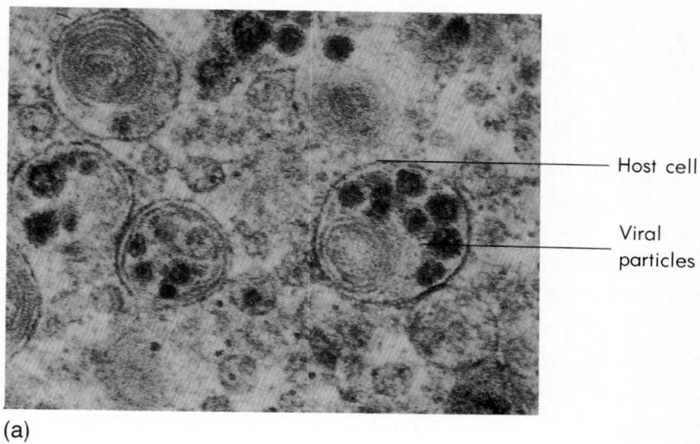

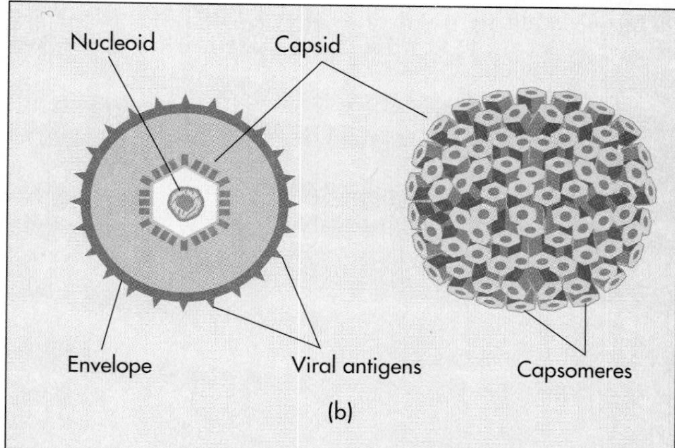

FIGURE 23-2 (a) An electron micrograph of a human cell containing a cluster of Epstein-Barr virus particles. This virus causes infectious mononucleosis and has been linked to some cancers. (b) Many human viruses are encased in a phospholipid envelope produced when the virus budded from its host cell; this envelope is a fragment of the host's cell membrane. Within the envelope are the capsid and the nucleoid, containing the virus's nucleic acid, DNA or RNA (never both). The individual protein molecules forming the capsid are called *capsomeres.*

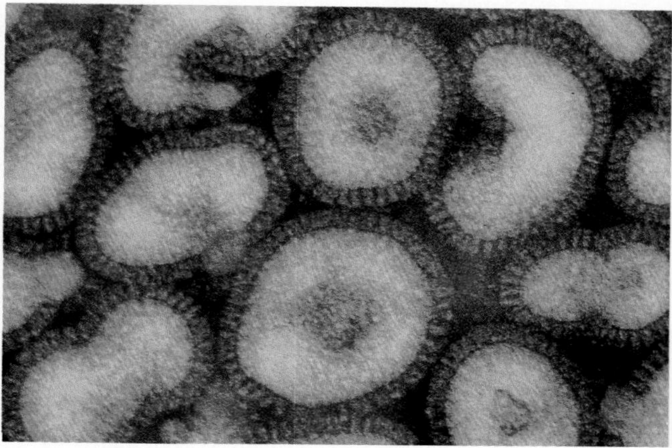

FIGURE 23-3 The capsid of the influenza virus bears spikes.

never both. The core can also contain enzymes needed to replicate the nucleic acid once it is inside the host cell.

Some viruses are encased in a membranous covering, or **envelope,** outside the capsid. The envelope is not made by the virus; it pinches off as a portion of the host cell's cell membrane when the virus leaves. Human viruses not covered by an envelope are known as "naked" viruses.

Viruses are thus simple structures, but they are still responsible for many human diseases. They cause the common cold, mumps, measles, chickenpox, flu, polio, rabies, and even some cancers. In fact, one particular virus is responsible for the great plague of the 1980s and 1990s, acquired immunodeficiency syndrome (AIDS). Box 23-2 describes the AIDS epidemic. The disease's effects on the body are discussed in Chapter 24, with the material on the immune system, because the AIDS virus attacks the cells of that system.

Viruses owe many of their effects to the way their "life" cycle results in the destruction of cells. They owe others to the body's sometimes overly aggressive attempts to fight off viral infections (see Chapter 24). Still others are caused by the way viruses can insinuate some of their own genetic material into the cell's genome (genetic material) and alter cellular function.

Viruses are classified into three groups according to the shape of their capsids. Those with the most complex shapes are called **complex viruses** (Figure 23-4). Herpes viruses, chicken pox, and sev-

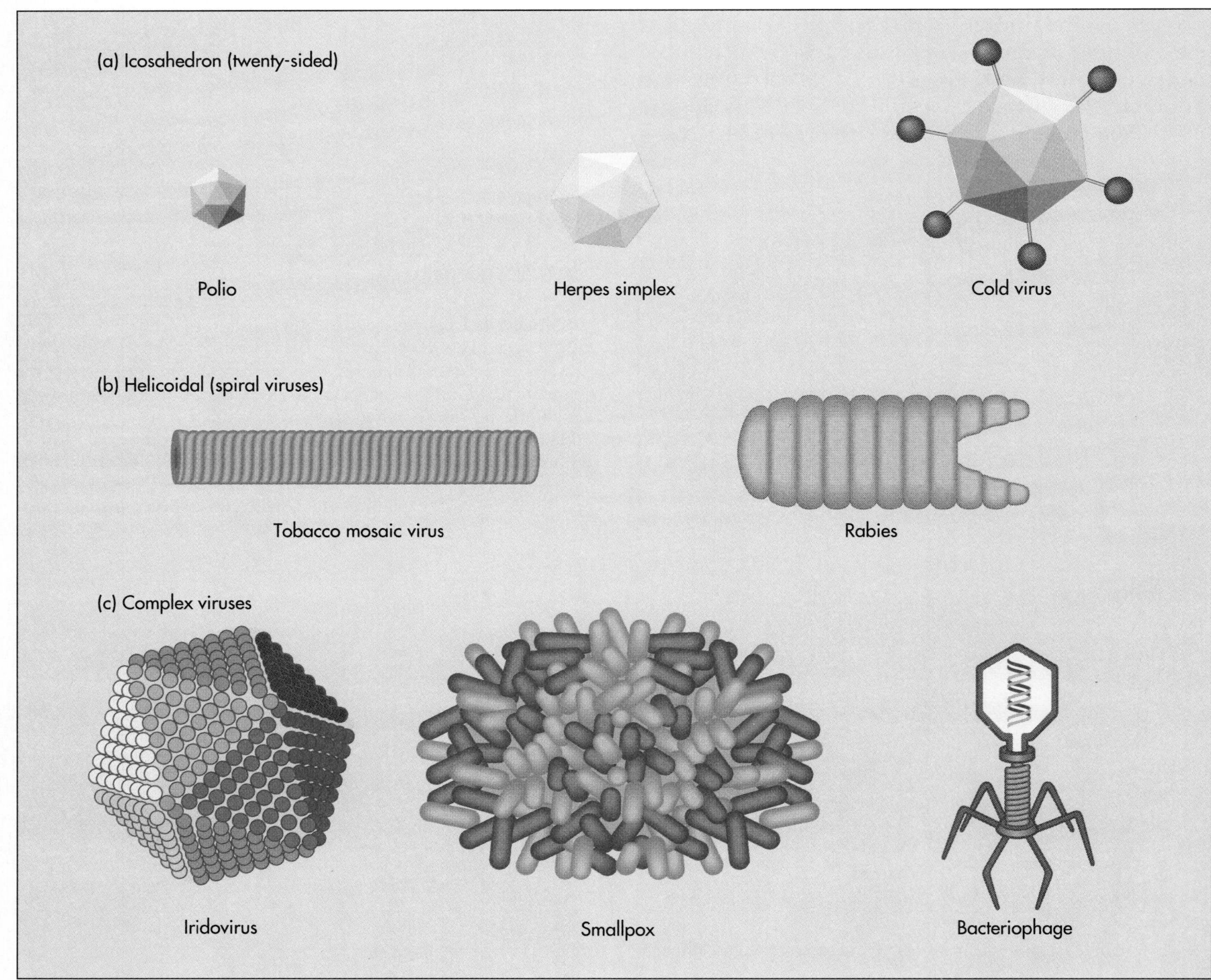

FIGURE 23-4 Representative viral shapes: (a) icosahedron (20-sided); (b) helicoidal (spiral); (c) complex.

eral of the tumor viruses have an icosahedral configuration. The helical viruses include the rabies virus and the tobacco mosaic virus (one of the first viruses studied). The complex viruses show wide variations in their external shapes. Bacteriophages (discussed next) and smallpox and influenza viruses are complex viruses.

Bacteriophages

The first viruses studied in detail were the **bacteriophages** (or phages) that infect bacteria. Many of these viruses are constructed like tiny hypodermic needles. They carry their genes as DNA in a multifaceted capsid. From the capsid extends a tubular tail with several spindly hairlike "legs" projecting from it. The proteins of the legs recognize and bind to surface proteins on the bacterium under attack. The tubular tail then injects the phage's DNA into the bacterium's interior (Figure 23-5).

Once the phage DNA is inside the bacterium, it can preempt the normal functions of the bacterium's enzymes, using them to replicate the DNA and make messenger RNA (mRNA). These bits of mRNA are working copies of the phage genes that will be used as templates for synthesizing the phage proteins (see Chapter 4). As DNA copies and proteins become available, new phages are assembled. When the cell is full of phages, it bursts (a phenomenon called *lysis*), releasing the phages to infect other cells (Figures 23-6 and 23-7).

Yet the phage is not always such a forthright attacker. It is especially subversive when it inserts its DNA into the bacterium's DNA as **prophage.** To do this, it uses bacterial enzymes known as **restriction endonucleases** that can cut and rejoin lengths of DNA; genetic engineers have found these enzymes useful for inserting foreign genes into bacteria, yeast, and other cells. After the insertion, the phage DNA does not preempt the bacterial enzymes. Instead, it lies quiet, reproducing with the bacterium for generations, awaiting some change in conditions that will stimulate it to pull free of the bacterial DNA and resume its more destructive activities. It regulates its own behavior by controlling the synthesis of a few phage proteins that inhibit phage multiplication.

Some bacteriophages carry RNA, not DNA, as their nucleic acid. Their life cycles hold the same two options, but in each case, infection is followed by the making of a DNA copy of the RNA, using another bacterial enzyme, RNA-directed DNA polymerase, that has also proved useful to genetic engineers.

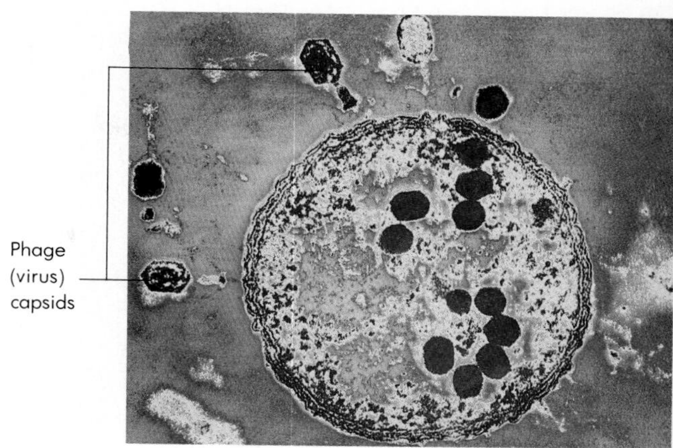

FIGURE 23-6 Electron micrograph showing newly formed T_2 bacteriophage particles within bacterial *(Escherichia coli)* cells. The phages attach themselves to the surface of the host bacterial cell, inject their DNA, and preempt the bacterial cell's resources, using them to synthesize new viral particles. In time, the bacterial cell bursts and releases the new viruses into the surroundings.

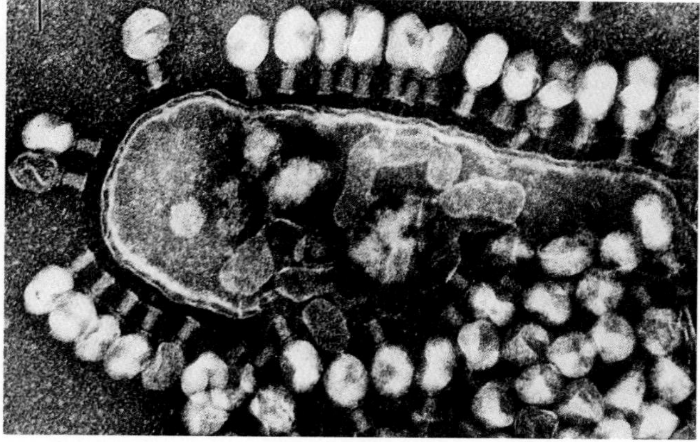

FIGURE 23-5 Electron micrograph of T_4 bacteriophage virus, which infects and destroys bacterial cells that normally reside in the human gut. Inside the viral tail is a tube that penetrates the bacterial cell membrane. As the tail shortens, it injects the DNA contents of the viral head into the host cell.

Animal Viruses

Viruses that attack animal cells are usually larger than phages and contain more genes. They too can have DNA or RNA as their nucleic acid. Most DNA viruses reproduce in the nucleus. The pox viruses (such as smallpox) and RNA viruses reproduce in the cytoplasm. Like phages, some animal viruses rupture their host cells at maturity, emerging as "naked" virus particles. Others synthesize proteins that become embedded in the host cell membrane. When the viruses leave the cell, they emerge coated

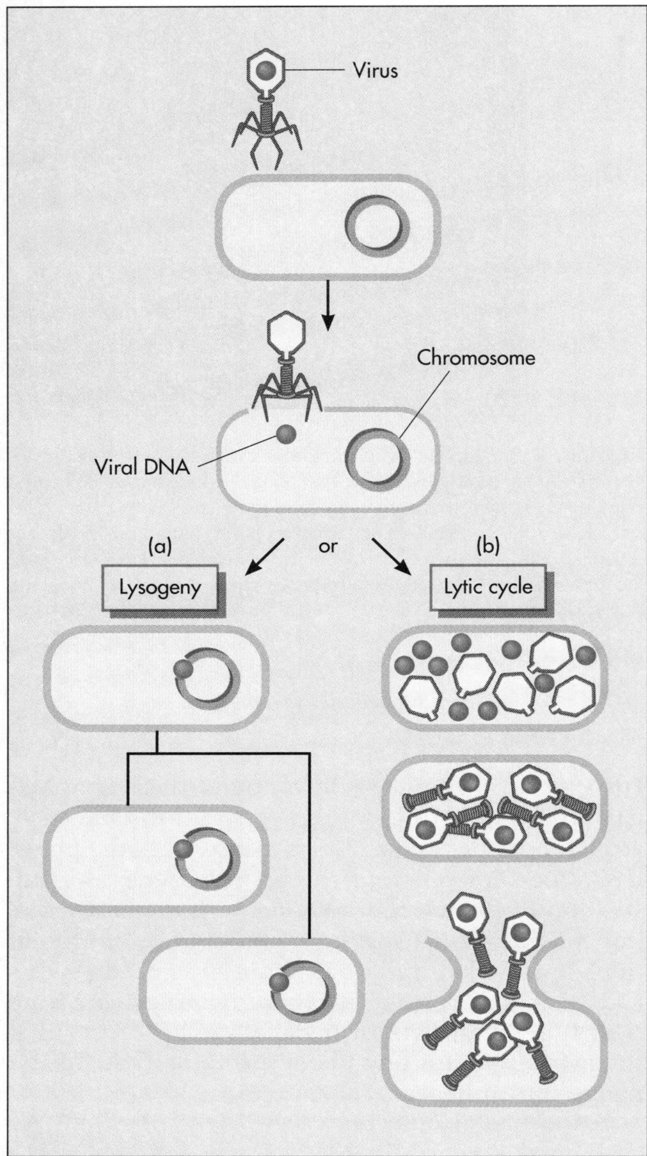

FIGURE 23-7 Viral invasion of a host cell may have two results. (a) In lysogeny, the viral nucleic acid is incorporated into the cell's chromosomes. Every time the cell reproduces, the viral genome is replicated and passed to the host cell's offspring. The virus is quiet during this time and does not cause disease symptoms. (b) The lytic cycle is violent, producing cell destruction and disease. The viral nucleic acid takes over the host cell's resources to make new viral components. These components assemble into new viruses and are released when the host cell membrane ruptures.

by a bubble of host cell membrane studded with viral proteins (Figures 23-8 and 23-9). The flu viruses and the virus causing AIDS (human immunodeficiency virus, or HIV), which contain RNA, behave in this way. The proteins flu viruses supply to the membrane include enzymes that thin mucus and help the viruses reach the cells that line the respiratory tract; these enzymes form the spikes in Figure 23-3.

After attaching to the host cell membrane and being absorbed, animal viruses shed their shell proteins, exposing their nucleic acid core. The RNA of some RNA viruses immediately begins to function as mRNA, commanding the synthesis of viral proteins, including RNA polymerase for multiplying the RNA. Other RNA viruses carry an RNA polymerase in their core. Still others, the **retroviruses,** use a *reverse transcriptase* enzyme they carry in their core to translate their RNA into DNA, which is then processed by the host cell as if it were a set of host genes. The DNA viruses, and the DNA copies of retroviruses, immediately preempt host enzymes to replicate their DNA and to make viral mRNA.

The first viral proteins that are made, the "early proteins," inhibit host cell activities, diverting the host's resources to meet viral needs. The "late proteins" are made only after the viral nucleic acid has been replicated many times. They include the components needed to assemble new virus particles. One of the late proteins is an enzyme that ruptures (or lyses) the cell to release the completed virus particles.

Animal viruses have three additional, more subversive ways to affect their hosts. For instance, the *herpes simplex* virus ordinarily reproduces slowly (Table 23-1, p. 532), generating just enough virus particles to infect daughter cells. The infection becomes obvious as a cold sore or fever blister only when a change in physiological conditions speeds up the viral lytic cycle, causing the destruction of host cells and facilitating the spread of new viruses into the surrounding cells. This virus and its relative, the cause of genital herpes, appear to establish their **latent infections** in the axons of nerve cells, from which they can break free to damage skin cells.

In **persistent infections,** the virus reproduces slowly within host cells and is passed to daughter cells, but it is also extruded from host cells at a rate sufficient to maintain a low level of virus in the blood at all times. The *hepatitis B* virus works in this way. It is transmitted mostly by blood, blood products, and hypodermic needles that have not been promptly and thoroughly cleaned after use. Eight to 22 weeks after it invades liver cells, it goes through a cell destruction (lysis) stage whose symptoms include fever, jaundice, abdominal pain, blood in the urine, and inflammation. After these symptoms subside, 10–12 percent of patients show the continual, slow release of virus into the blood typical of persistent viral infection. There is strong evidence that this virus can cause liver cancer.

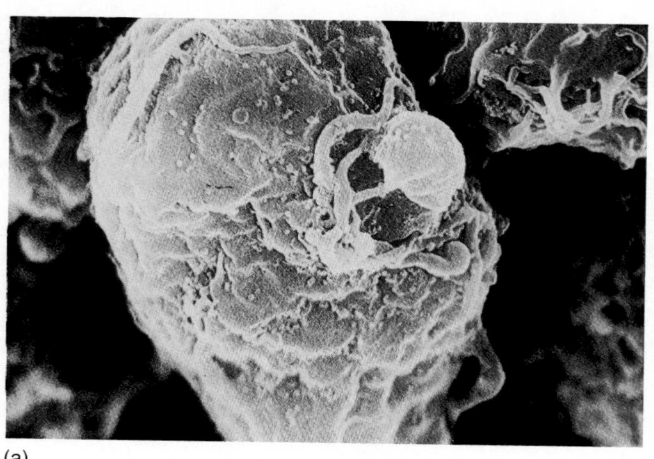

(a)

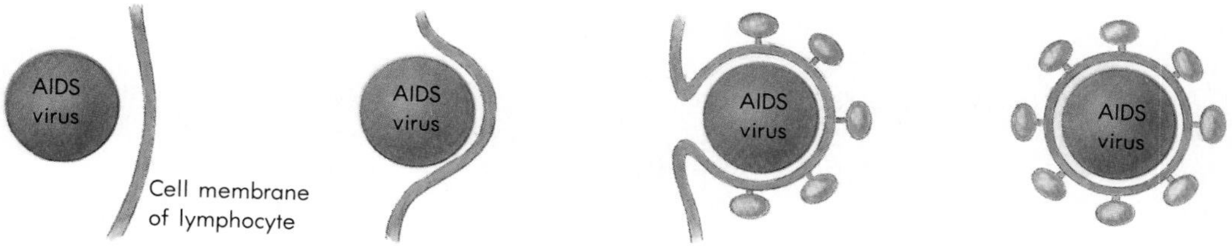

AIDS virus

Cell membrane of lymphocyte

AIDS virus

AIDS virus

AIDS virus

(b)

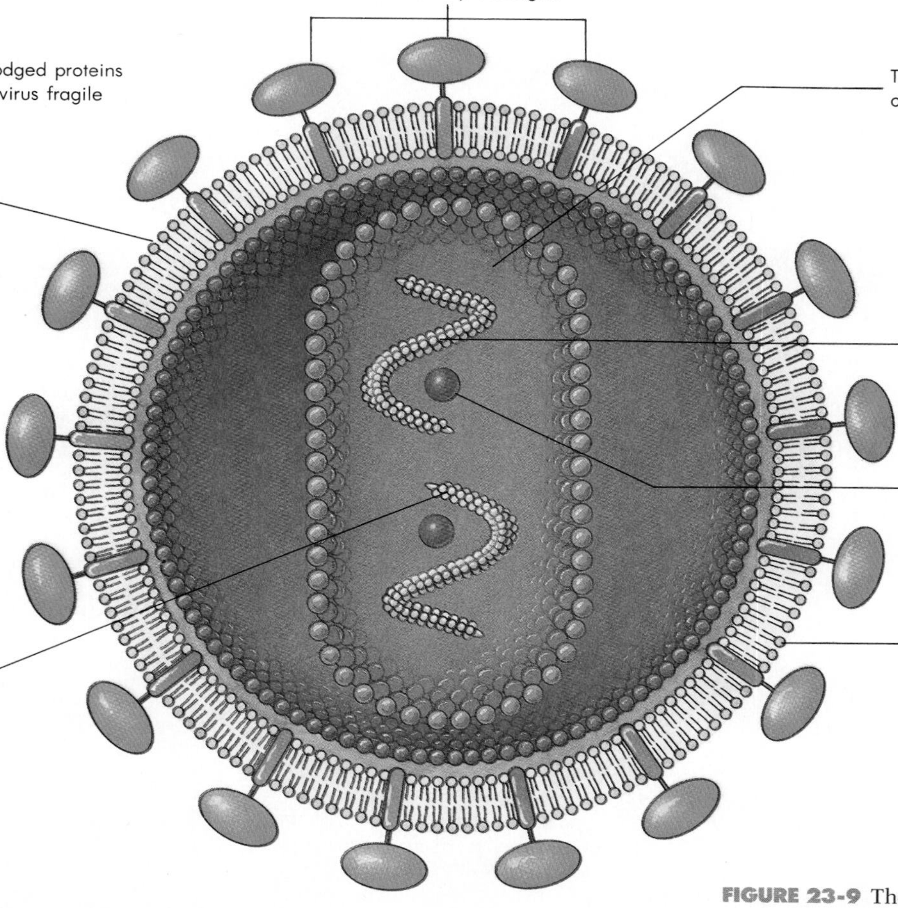

FIGURE 23-8 (a) AIDS viruses (HIV-1) can be seen in this scanning electron micrograph budding their way through the surface of a cultured T-4 lymphocyte. (b) Diagram of HIV-1 virus budding its way through the plasma membrane.

Fragile surface proteins that are easily dislodged

Easily dislodged proteins make the virus fragile

The core is cylindrical; other viruses are spherical

The HIV envelope is formed from host cell's plasma membrane

HIV RNA carries potentially deadly gene combinations that can take over host cell and eventually kill it

The AIDS envelope is thick for a virus of its sort

Reverse transcriptase

The RNA of AIDS has three extra genes. When they are copied in a cell, portions float free—and may kill the cell.

Lipid bilayer

FIGURE 23-9 The AIDS virus.

Table 23-1

Time Required for Viral Replication[a]

TYPE OF VIRUS	HOURS TO REPLICATE	INVADED CELLS
Cold viruses	48	Respiratory cells
Polio viruses	6 to 8	Nerve cells
Herpes viruses	12 to 30	Epidermal cells

[a]Duration of lytic cycle from the host cell's invasion to its destruction, releasing additional viral particles.

The hepatitis A virus, on the other hand, does not produce persistent infections. It is most often spread by fecally contaminated food and water, and it too can damage the liver, but most patients recover promptly.

Finally, some animal viruses, including herpes viruses and the RNA-containing retroviruses, can **transform** their hosts. That is, they can cause infected cells to become cancerous. To do this, a DNA virus must first merge its DNA with its host's. A retrovirus must merge the DNA copy of its genes with those of the host cell. The merging is controlled by some of the virus's early proteins. Once merged, the virus is known as a **provirus.** Like prophage, it continues to command the synthesis of proteins that maintain the merged state (Figure 23-10).

Transforming viruses do more than merge with their hosts. In addition, they carry as part of their genomes **oncogenes,** also known as cancer genes. There are several types of these genes. All bear strong resemblances to genes naturally present in host cells, where they often appear to control the cell's responses to growth factors. In humans, these natural oncogenes are thought to control the rapid growth of fetal development and childhood. After the child reaches a certain age or size, these genes are "turned off." Viral tampering with these oncogenes may cause the host cell to lose its ability to respond to normal controls on its growth and multiplication. It becomes cancerous.

Oncogenes have also been found in host cells that are not infected with virus. Thus many researchers think that viral oncogenes must have originated as altered host cell genes that were

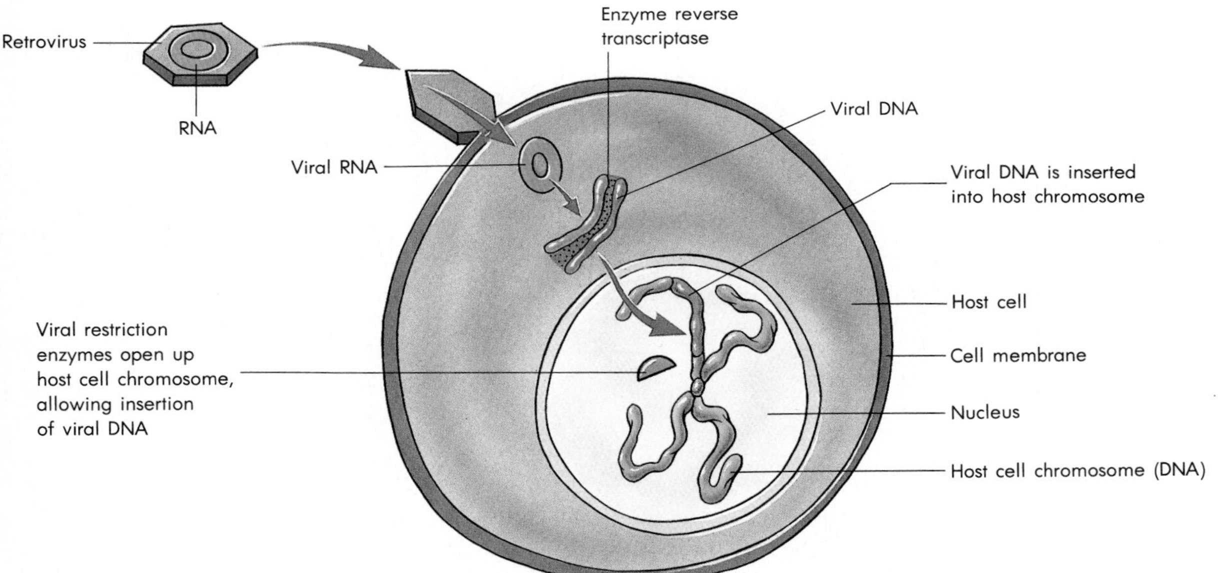

FIGURE 23-10 Retroviruses carry their genetic information in RNA rather than DNA. (1) The virus enters the host cell by shedding its protein coat and injecting the viral RNA. (2) Using the reverse transcriptase enzyme carried by the virus, together with the cell's own resources, the cell makes DNA copies of the viral genes. (3) Viral restriction enzymes snip the host DNA and the viral genes are merged into the host chromosomes. Once merged, the viral genes are indistinguishable from the host's DNA to the cellular equipment. The viral genes may remain quiet for several days, weeks, or even years. Once activated, however, they may induce the cell to produce and release vast quantities of new viral copies or they may induce pathogenic and uncontrolled cell reproduction in the form of cancer.

picked up by viruses. The picking-up process is easy to observe, for when proviruses break loose from host DNA to become active and multiply, as they do periodically, they often carry with them a bit of the host's DNA. Genetic engineers have exploited this phenomenon as a way to pluck genes and parts of genes from cells. They have also used viruses that form proviruses to insert genes and parts of genes, and this procedure promises to become a major tool for future transplants of genes to correct genetic deficiencies.

1. Describe briefly how a genetic engineer might obtain a virus to whose genome has been added a specific human gene.
2. How might the genetic engineer use this virus to transfer the gene to a human who lacks it?

Box 23-2 AIDS: The First 10 Years

The U.S. Department of Health currently estimates that over one million Americans are carrying HIV, which causes "the most devastating public health problem of this century"—AIDS. You cannot glance through a daily newspaper or watch the evening news without encountering the aura of tragedy that shrouds this disease. One day, a friend, or even a family member, may get it. And it is unlikely that it will disappear within your lifetime.

In this box, we look at the trail that HIV has left among the records of the Centers for Disease control (CDC) in Atlanta, Georgia. The records for AIDS statistics reflect two broad time periods: the first before 1981 (which was before anyone knew about the virus and we were not watching for it). The second begins in 1981, when we first realized we were dealing with a viral plague. The statistics illustrated here are cumulative for the period beginning in 1981 up through January 1991 in the United States. A word of caution in interpreting these data: Because AIDS is not a reportable disease (it is not mandatory that a physician report its diagnosis), the figures given here are probably underestimates of the actual numbers.

WHO GETS AIDS?

In the first 10 years that AIDS was recognized as a disease, the CDC registered 164,129 men, women, and children that had been diagnosed with AIDS. So far, about 80 percent of the victims have been men (145,394 cases), about 10 percent have been women (15,894 cases), and 2 percent have been children (2841 cases). Unfortunately, the figures for the numbers of women and children are likely to increase as time passes (Figure 1).

Figure 2 outlines the age of onset for diagnosed AIDS cases. This bar chart shows two disturbing patterns. First, it is apparent and logical that sexually active college-age populations are in the most vulnerable period—the late twenties and early thirties—for coming down with AIDS. However, you must look closely at the chart to see one of the most disturbing trends: The number of male victims vastly outnumbers the females for all age categories but one. In children under five years of age the male-to-female ratio is almost equal, because most of these victims are infected with HIV while still in the uterus. The jump in AIDS cases in this very young category is symptomatic of the widening invasion of HIV into the heterosexual population.

WHAT ARE THE MOST COMMON AVENUES LEADING TO AIDS IN THE UNITED STATES?

There is no question that high-risk male homosexual behavior (such as unprotected anal inter-

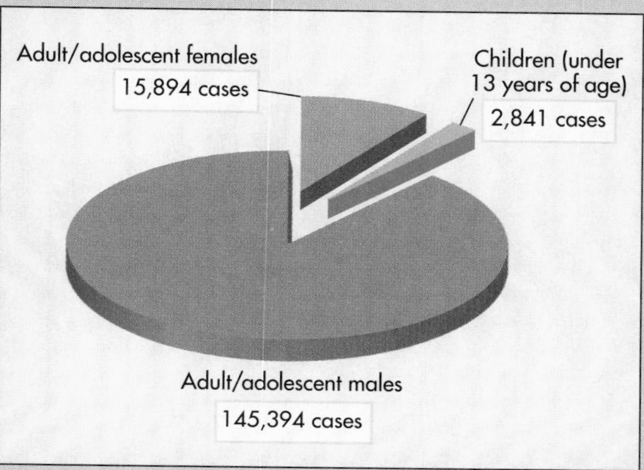

FIGURE 1 Relative numbers of AIDS cases in the United States by sex and age (from 1981 to January 1991).

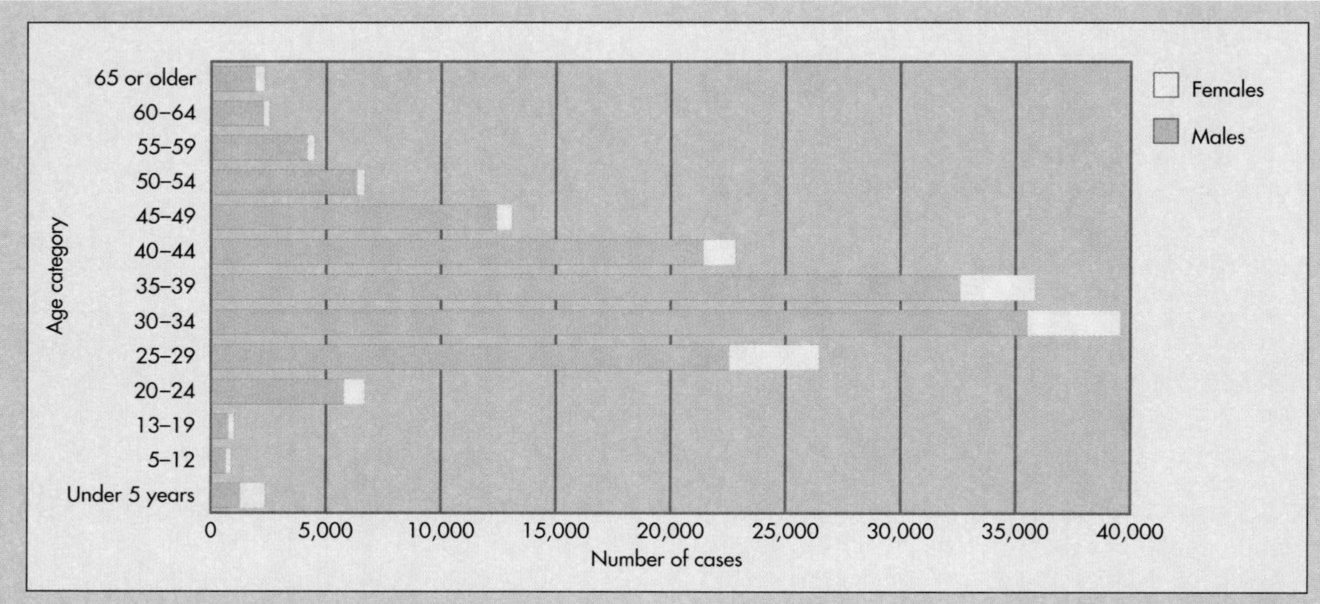

FIGURE 2 AIDS cases in the United States by sex and age of diagnosis (through January 1991).

course) has been responsible for most of the AIDS cases in this country. By itself, it accounts for about 58 percent of the total. However, that statistic is a little low because homosexual activity combined with drug abuse accounts for an additional 7 percent of the cases (Figure 3). Seventy-six percent of the AIDS cases among whites come through homosexual contacts.

The second most common and the fastest growing avenue for the spread of AIDS is direct involvement in intravenous drug use by females and heterosexual males. It currently accounts for about 21 percent of the total AIDS cases.

Both blacks and Hispanics are more likely to contract AIDS through intravenous drug use. Thirty-nine percent of the AIDS cases among blacks and 40 percent of the cases in the Hispanic community are caused by intravenous

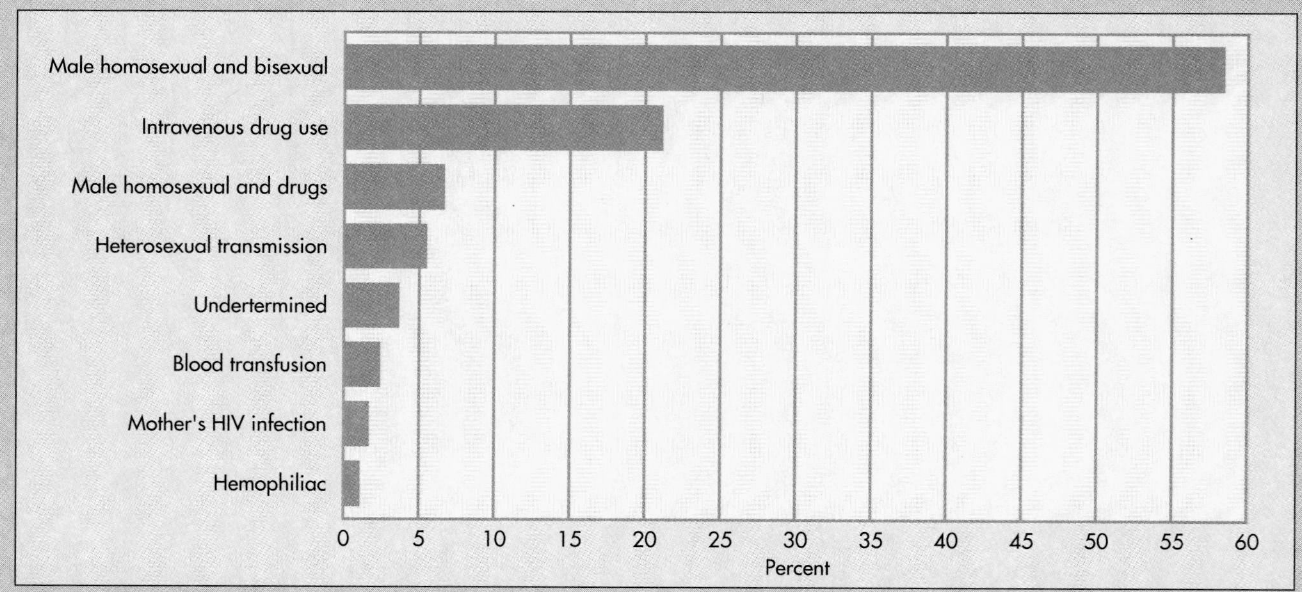

FIGURE 3 Percentage of AIDS cases caused by various modes of transmission (from 1981 through January 1991).

drug use. The sharp increase in the use of drugs during the late 1980s and in the practice of "sex for drugs" created by the crack cocaine epidemic have dramatically accelerated the incidence of most sexually transmitted diseases (STDs) and is most likely to provide the channel for the spread of AIDS to the heterosexual community. Only 8 percent of the AIDS cases among whites are transferred through drug use.

Heterosexual transmission accounts for only about 5 percent of all cases. Transmission due to transfusions of blood or blood products accounts for only a few percent of the cases.

HOW DO WOMEN GET AIDS?

Most AIDS in this country is being transmitted through unsafe homosexual practices by males to other males. However, women get it too—most of them, 8082 cases (51 percent), by sharing needles during intravenous drug use (Figure 4). Needle sharing means using a hypodermic to inject oneself with drugs when that hypodermic has been used by someone else and contains small amounts of that person's blood. Drug addicts are now being advised to make an attempt at sterilizing their needles by rinsing them with dilute chlorine bleach.

The second most common avenue for females to be exposed to HIV is through heterosexual contact, 5206 cases (33 percent). Of these heterosexual contacts, 3259 were with intravenous drug users; only 510 cases involved women having sex with bisexual males.

There are also racial and ethnic implications in the figure. Under current conditions, black and Hispanic women are at much greater risk of getting AIDS than are white women. Seventy-two percent of the women that have contracted AIDS so far are either black or Hispanic. If we adjust the figures for population differences in these three groups, a black woman is 13 times more likely and a Hispanic woman is 8 times more likely to get AIDS than a white woman. In addition, about 72–75 percent of the babies born with AIDS are also from Hispanic or black families. These figures may be reflections of the impact on lifestyle of other issues such as racism, poverty, and drugs.

HOW ARE OTHER STDs RELATED TO AIDS?

Active cases of STDs such as herpes, gonorrhea, chlamydia, and syphilis make catching AIDS easier because they create breaks in the skin or mucous membranes through which HIV readily enters the victim.

Cases of STDs in people who already have AIDS are predictably exacerbated. For instance, normal herpes sufferers usually have active lesions during a flare-up time of 8–10 days. Those same lesions in someone with AIDS are likely to persist for months.

CASES VERSUS PEOPLE

Using the graphs in this box, we have discussed the recent statistics of thousands of AIDS cases. Such discussions gloss over the personal tragedy of individuals and families coping with AIDS. Any AIDS case is a catastrophe of unimaginable proportions. As the disease increases, some of our talented students sicken and die a most terrible death. *Protect yourselves.*

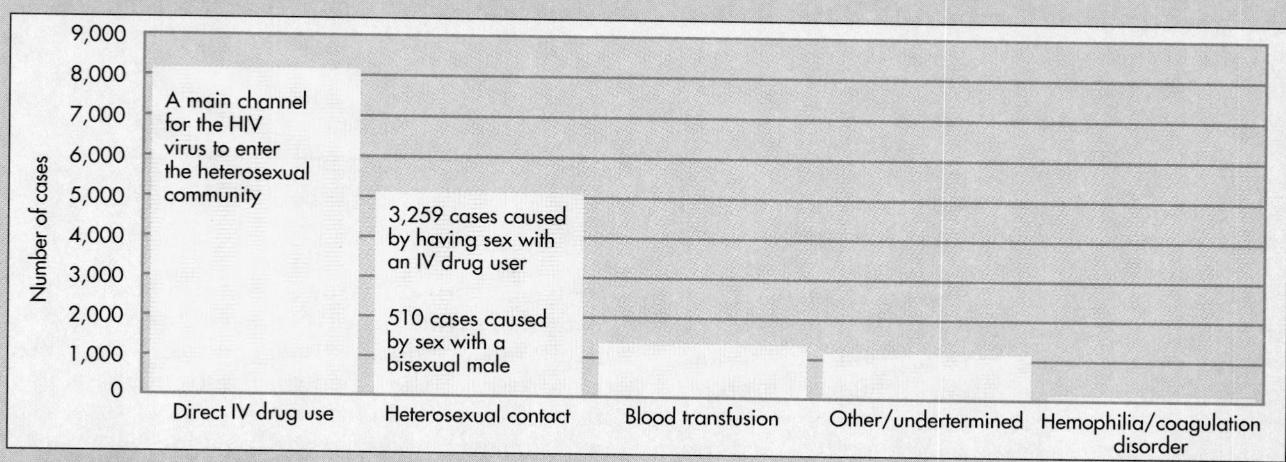

FIGURE 4 Mode of exposure of female AIDS victims (from 1981 to January 1991).

Viroids

Viroids cause a number of diseases of commercially important crops such as potatoes, cucumbers, hops, coconuts, and citrus fruits. They differ from viruses in that they have no protein. They consist only of a short "naked" strand of RNA, just large enough to code for a single relatively small protein. However, their RNA does not act as mRNA. It is copied by cellular enzymes, and the copies can interfere with the production of growth hormones to cause the symptomatic disturbances of growth. The copies can interact with cellular DNA, inappropriately turning on or off the activity of specific genes.

Until fairly recently, it was thought that viroids, or "slow viruses," also caused some slowly developing animal and human diseases. One candidate was scrapie, a brain degeneration of sheep and goats (named because victims scrape off their wool). Another was kuru, a human brain degeneration whose symptoms begin with motor difficulties and progress to dementia; found only in New Guinea, it was communicated by eating the brains of dead family members until authorities stopped the practice. Other possibilities were Creutzfeldt-Jakob disease, a rare human dementia, and Gerstmann-Straussler syndrome, which resembles kuru. However, some researchers now believe these diseases are due to something considerably stranger, prions (pronounced "pree-ons").

Prions

A **prion** is a *protein infectious particle.* It consists only of protein, lacking nucleic acid of any kind. It is found in the brains of victims of scrapie, kuru, and other degenerative brain diseases, where it forms tangled, fibrous masses, or *amyloid plaques.* When this protein is injected into disease-free animals, it appears to multiply, the plaques form again, and the disease develops (Figure 23-11).

Are prions truly infectious? The diseases prions are associated with *do* seem to be transmissible from person to person. But the prions themselves seem to be normal elements of the body. Using DNA synthesized to match the amino acid sequence of the prion protein, researchers have found that animal cells seem to contain the prion gene even when they do not have a prion disease and that the prion protein is closely related to proteins that regulate the production of receptors for neurotransmitters on nerve and muscle cells. Whatever is the true cause of the prion diseases, it seems to make affected cells produce excess amounts of the prion protein or make an altered form of the protein.

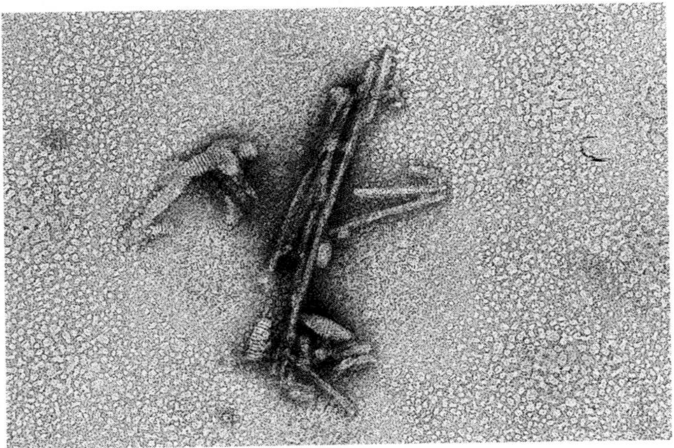

FIGURE 23-11 Scrapie-associated fibrils (once called prions) are infectious pieces of protein.

What lies behind the prions? We do not yet know, although experiments have shown that no DNA or RNA seems to be involved. Prions thus seem to be the most subversive of the subversive agents of disease and infection. They lack even a virus's or viroid's genetic independence. They may, in fact, represent no more than a trigger for a self-perpetuating malfunction.

There is a precedent for such a view, for some biologists believe that viruses and viroids may have begun as escaped genes or gene fragments with the ability to force cells to copy them. Others believe that viruses and viroids may be the degenerate descendants of parasitic bacteria that have lost much of the apparatus we normally consider to be necessary for life.

1. What are the essential, defining features of life?
2. Should we draw the line between life and nonlife to call prions alive? Viroids? Viruses?

THE INVADERS

We called viruses, viroids, and prions subversives instead of invaders because, even though they must invade an organism's body and cells in order to cause disease, they then survive by subverting normal cellular functions. The agents of disease and infection we choose to call invaders differ in that they carry their own machinery of life. They penetrate the body in food, air, and water and through cuts and scrapes. Some colonize the digestive and reproductive tracts and even the skin. Others flourish in the blood and other tissues. All draw nutrient from the body's substance, and many destroy cells, either by consuming them or by secreting toxins. They are bacteria, protozoa, fungi, and multicellular parasites.

Bacteria

Bacteria are an ancient line of cells that originated with the dawn of life on Earth. Structurally, they are much simpler cells than those found in the human body, and because they lack nuclei and other membrane-bound organelles, we call them *prokaryotes* (before nuclei). In addition, they are generally much smaller than the more modern *eukaryotic* (true nuclei) cells such as our own. Bacteria have ribosomes for protein synthesis, cell membranes to separate them from the surrounding environment, and cell walls to give them their characteristic shapes; some have flagella for locomotion—but no other organelles (Figure 23-12).

Many bacteria live independently in the soil or water, but many others have adapted to thrive on or in other life forms. Some bacteria live in the human intestine, where they secrete some of the vitamins we require for health. (In animals such as cows, bacteria actually aid the digestion of food.) Some live on the human skin and in such other areas as the vagina, where their secretions help to maintain an environment inhospitable to other bacteria that cause disease.

Although there are many different kinds of bacteria (Figure 23-12), all are **decomposers** (see Chapter 26). As such, they are crucial to the recycling of nutrients from plant and animal bodies and wastes

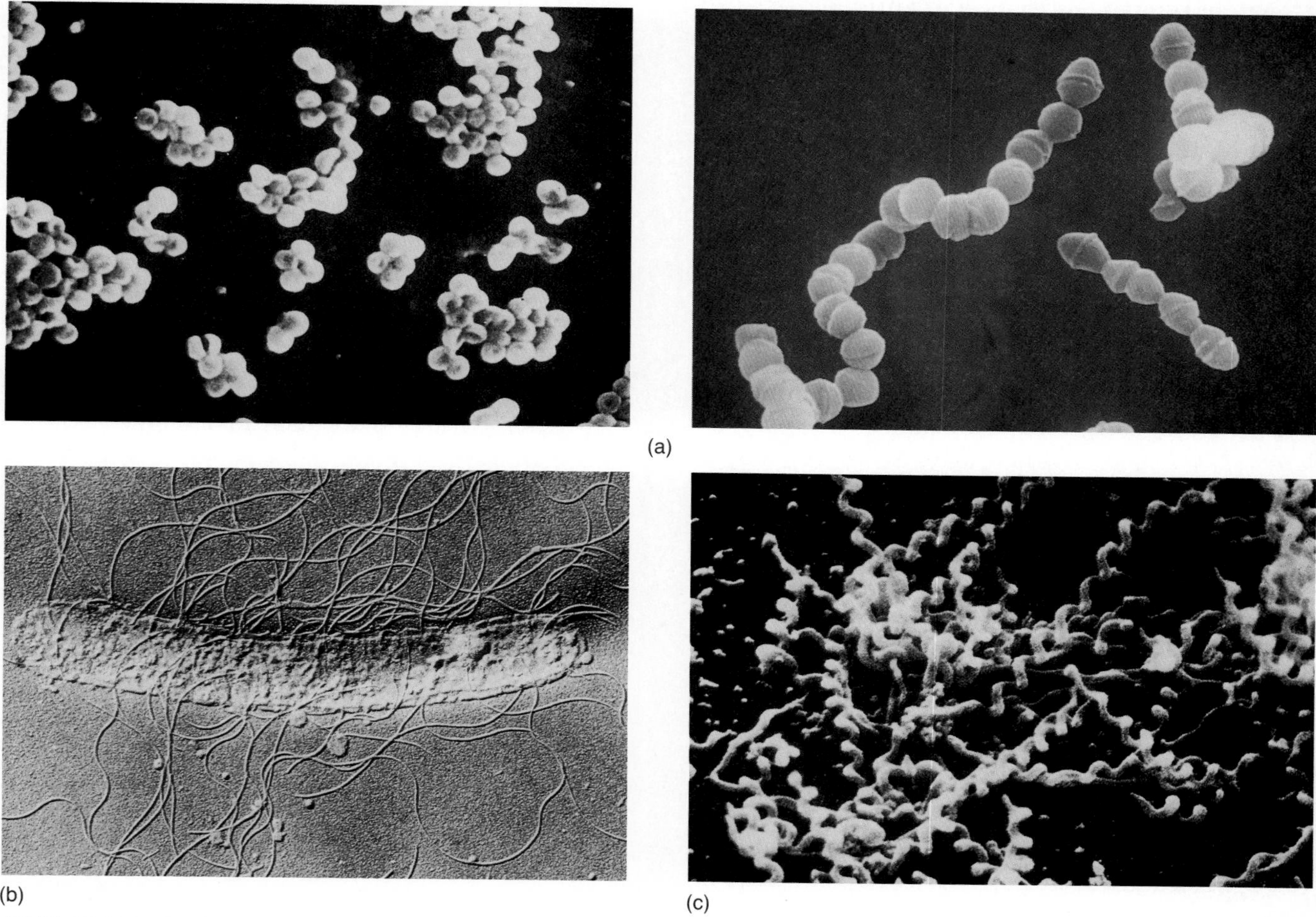

FIGURE 23-12 The three major types of bacteria are defined in terms of cell shape. (a) Coccal forms are essentially spherical. Examples include grapelike clusters of *Staphylococcus* cells and the "string of beads" arrangement characteristic of *Streptococcus*. (b) Bacilli are rod-shaped or cylindrical. Shown is a flagellated form of the bacillus that causes Legionnaire's disease. (c) Helicoidal cells such as this spiroplasma are coiled or spiral; a later figure shows the spirochete that causes syphilis.

back to the soil. They live by breaking down complex organic substances with enzymes they secrete into their environment. People exploit this characteristic of bacteria by using them and their enzymes to release the cellulose fibers of flax to make linen, to change milk sugar into lactic acid to make yogurt, and to alter milk protein to make cheese. Researchers are even finding that some bacteria can make enzymes that will decompose such resistant substances as toxic wastes and petroleum; in fact, such bacteria helped clean up Alaska's Prince William Sound after the 1989 oil spill.

Perhaps the most exciting facet of our relationship with bacteria is developing as we move farther into the age of genetic engineering, for we are now domesticating bacteria, just as we did animals and plants over 10,000 years ago. Although we keep them in vats instead of pastures, we still use the products they produce to support human activities. In this case, we are inducing them to mass produce chemicals, such as hormones (e.g., insulin and growth hormone), antigens, and antibodies, to protect us from disease and to correct metabolic defects (see Chapters 4 and 24).

Pathogens

Relatively few bacteria cause disease. Those that do we call **pathogens** (*pathos*, sadness, pain; *gen*, to generate or produce), but not all pathogens are of equal **virulence.** That is, not all pathogenic bacteria are equally able to make us ill. Weakly virulent bacteria may be able to invade and sicken the body only when given an unusual opportunity. These **opportunistic** bacteria include many that normally live on the skin or in the gut without causing problems. Their opportunities arise when the skin is broken by a cut or scratch or when the body's resistance to infection is low, as when radiation therapy for cancer has weakened the immune system or when the body is debilitated by other disease or malnutrition. The tuberculosis bacterium, often normally present on the skin, causes tuberculosis only when the body is weakened in such ways.

More potent pathogens can attack the body at any time, though opportunity still plays a role. The tetanus bacterium, *Clostridium tetani,* gets into the body when a rusty nail or other object bearing the bacteria or its spores punctures the skin. Once it is deep enough within the body to be shielded from the oxygen that inhibits its growth, it multiplies and secretes the potent toxin that causes the convulsions and severe muscle crampings of tetanus. Cholera bacteria, borne by water contaminated by human wastes, are highly successful at colonizing the digestive tract; their toxins cause severe diarrhea and dehydration and often death.

Great virulence appears to be a temporary feature of pathogens. Diseases often lose some of their seriousness over decades or centuries—partly because people who are resistant enough to survive the disease pass the genetic basis of their resistance on to their children, who in turn form the next generation. Thus, the resistance of the population to that particular disease increases from generation to generation. In addition, extreme virulence is not really an advantage to the pathogen, for when it kills its victim, it too must die out. Over time, pathogens adapt to their hosts and their effects become less severe. The classic example is syphilis, which once caused death within just a few years of infection. Today, people can remain sick with this **sexually transmitted disease (STD)** for decades.

Food Poisoning

Food poisonings are not infections. Rather, they are severe reactions to toxins generated and released by bacteria inhabiting food. The bacteria responsible for food poisoning depend on careless food preparation or lack of refrigeration for their opportunities. The most deadly form of food poisoning, **botulism,** is caused by *Clostridium botulinum,* a rod-shaped soil bacterium whose spores are present on a number of foods (Figure 23-13, p. 543). Trouble is likely to develop when nonacidic foods such as sausage (*botulus* is Latin for sausage), tuna, beans, peas, olives, or asparagus are not canned at high enough temperatures. The spores on these foods may survive and grow in the can. The poison produced by the bacteria is so toxic that simply tasting contaminated food can be fatal. In fact, the botulinum toxin may be the most poisonous chemical we know—it appears to be 10,000 to 100,000 times more toxic than diphtheria toxin or the most potent snake or spider venom. Some studies suggest 1 ounce of pure botulism toxin would be enough to kill every man, woman, and child in the United States—half a pound could wipe out the entire population of the world.

Less severe food poisonings are caused by *Staphylococcus* and *Salmonella* bacteria when food is inadequately refrigerated or food preparers fail to keep their hands and equipment scrupulously clean. These bacteria grow especially well in dairy and egg products such as cream and mayonnaise, and *Salmonella* has been found in unopened eggs, apparently because the bacteria were in the hens that laid the eggs. The toxins these bacteria produce affect the digestive tract, causing stomach aches, vomiting, and diarrhea. One of the problems with food poisoning is that spoiled food often gives no outward signs that something is wrong with it—it smells fine, tastes good, and looks palatable.

Learning Focus

SEXUALLY TRANSMITTED DISEASES

Viral and bacterial diseases can be spread in many ways. Their agents can enter the human body with food and water, on objects—including dirty fingers—that enter the mouth, with droplets put into the air by coughs and sneezes, with the saliva of biting insects, and from the fur—and bites—of wild animals and pets.

Some diseases can also enter the body more treacherously. These diseases were once called the "venereal" diseases, after Venus, goddess of love. Today, we recognize more cynically—or realistically—that much intimate contact has less to do with love than with friendship, lust, loneliness, and other human needs. We say that these diseases are "sexually transmitted" and leave it at that.

Sixty-three percent of all STDs occur in people under 25 years of age. Some of these diseases, such as yeast or protozoan infections of the genital mucous membranes, are fairly innocuous. Others, such as syphilis and AIDS, can be fatal. Still others, such as chlamydia and gonorrhea, cause reproductive problems. All are embarrassing, painful, and dangerous, and all can be passed to one's sexual partners (see the figure below).

CHLAMYDIA

Surprisingly, one of the most serious diseases in the United States—and the most common STD other than yeast and protozoan infections of the vagina—is *not* genital herpes, or gonorrhea, or for that matter, syphilis, but chlamydia! It is caused by a tiny bacterium, *Chlamydia trachomatis,* that takes up residence inside its host's cells. Its painful dysuria (difficulty in urinating) and sometimes sterilizing pelvic infections cause 4 million or more new cases each year. Recent studies suggest that 10 percent of all college students are already infected.

Until the 1980s, there was no simple test for the presence of the chlamydial agent because of the way it hides inside host cells, away from easy scrutiny. A continuing problem is that the symptoms of the disease are often misleading. It is likely to be misdiagnosed as gonorrhea in afflicted males, for it frequently causes inflammation of the urethra and other areas of the excretory system that results in a milky discharge, similar to that seen in the other disease.

It also infects the female reproductive system, moving from the vagina through the uterus and into the Fallopian tubes. It produces lesions in the Fallopian tubes that can eventually seal the tubes as a result of scarring—making the woman infertile in the process. Often, women first discover their problem as a consequence of their inability to become pregnant (see the figure at the top of p. 540).

Chlamydial infections can also increase the incidence of potentially life-threatening ectopic or tubal pregnancies. In such cases, the tubes are open enough to allow fertilization but not enough to allow the young embryo to travel to the uterus for implantation. Infected mothers can also transmit the infection to their babies—

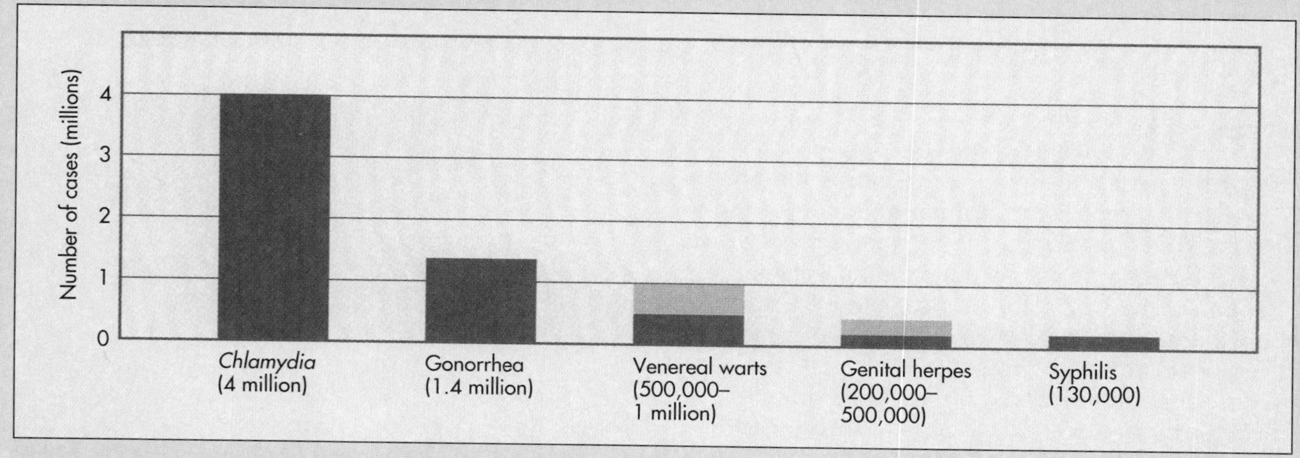

Estimated number of new cases of STDs per year (excluding AIDS).

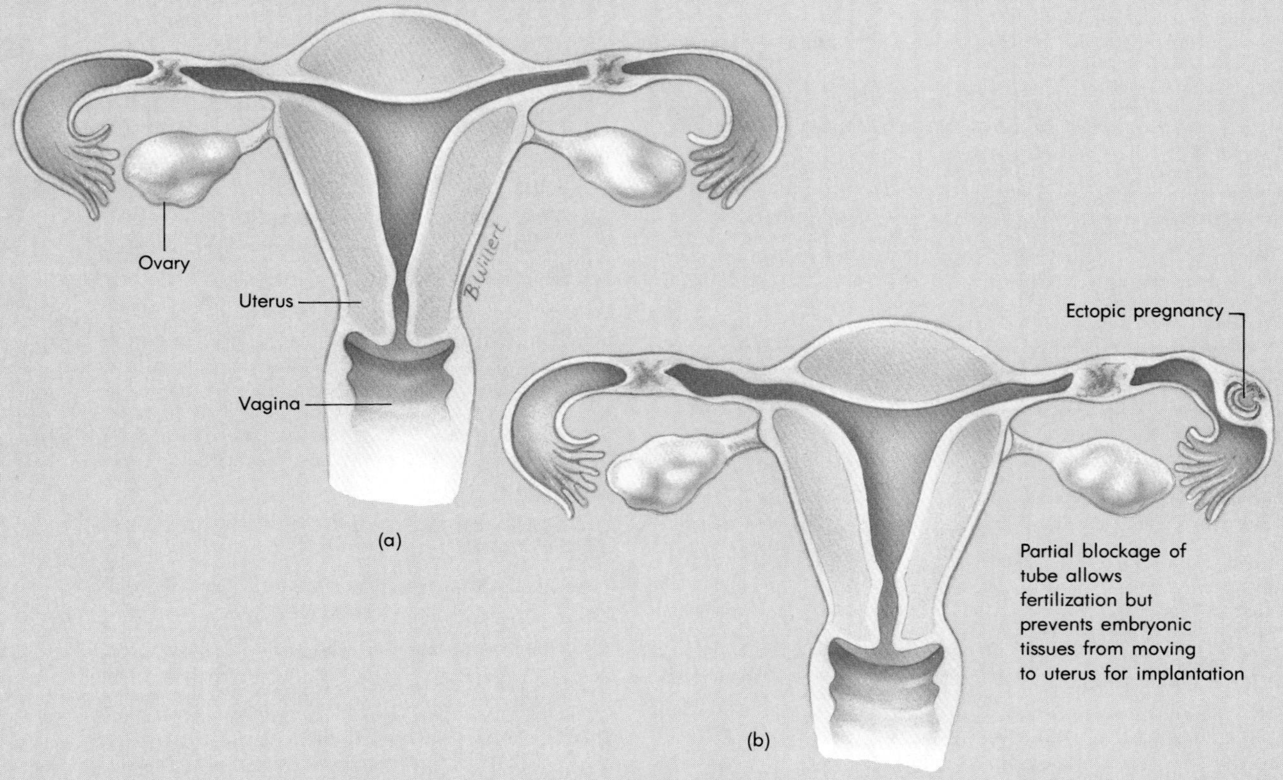

Infertility
Scar tissue blocking Fallopian tubes prevents fertilization

Ovary

Uterus

Vagina

(a)

Ectopic pregnancy

Partial blockage of tube allows fertilization but prevents embryonic tissues from moving to uterus for implantation

(b)

Two of the most serious side effects of chlamydial infections are: (a) infertility caused by damage to, and the formation of scar tissue in, the Fallopian tubes, and (b) ectopic or tubal pregnancies, which can occur when scar tissue blocks movement of the ovum to the uterus but not movement of the sperm to the ovum.

usually in the form of conjunctivitis (inflammation of the eyes) or pneumonia.

It is possible to have more than one STD at once. In fact, studies suggest that 40 percent of the women being treated for gonorrhea, and 20 percent of the men, also have chlamydial infections. Unfortunately, though chlamydia is easily treated with the antibiotic tetracycline, it is not cleared up by the penicillin usually prescribed for gonorrhea.

SYPHILIS

Syphilis, caused by the spiral bacterium, or spirochete, *Treponema pallidum,* is one of the oldest STDs. In Africa, it takes the form of yaws, a skin disease. In cooler climates, it settles deeper in the body, for it is sensitive to cold and dryness. It spreads only by direct contact, usually sexual, with the sores that mark the disease. Once in the body, the disease goes through three stages. Ten days to three months after infection, a sore, or chancre, appears at the site of infection (see the figure at right). The

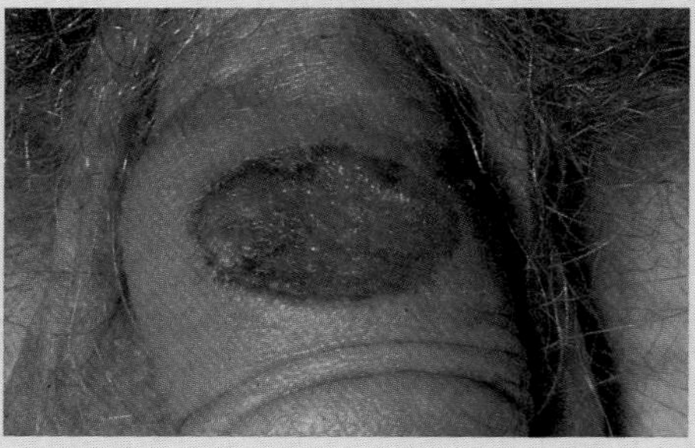

(a)

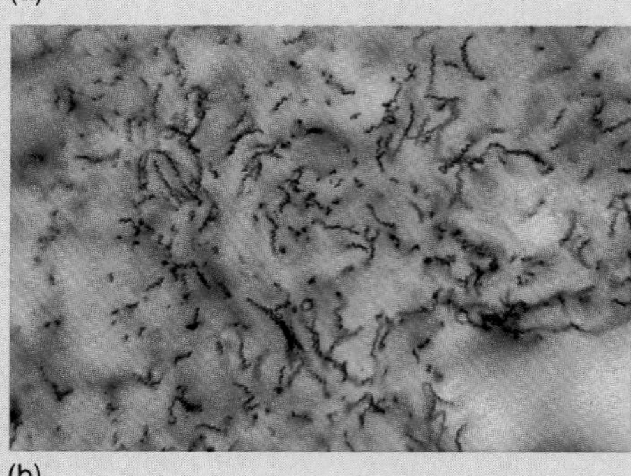

(b)

(a) A syphilitic chancre on the penis. (b) Spirochetes causing syphilis, as seen in human liver.

chancre heals promptly, but two to six months later *secondary lesions* appear as a skin rash or sores on mucous membranes; these sores give syphilis its other name of "the great pox" (see the illustration of the second stage of syphilis below). The secondary lesions also heal by themselves, and in some cases, the victim appears to recover completely. Often, however, the bacteria have spread throughout the body and settled in. Later—even years later—they may attack other organs, including the brain. The result can be strokes, madness, and death.

Congenital syphilis occurs when the bacterium passes from a woman to her unborn child. It can cause miscarriages and birth defects. Fortunately, it is fairly easy to cure syphilis with antibiotics, and in most states of the United States and many other countries a couple cannot get a marriage license until they have had a blood test to see whether they have or have had the disease. The test is also common for pregnant women.

GONORRHEA

Like chlamydia, **gonorrhea** has become increasingly common since the 1960s, as a result in large part of increases in numbers of sexual partners and in the ability of the responsible bacterium—the gonococcus—to resist common antibiotics (see the figure above). Like syphilis, it can develop in the sexual organs, the mouth, or the anus. Most often, the gonococci infect the male urethra and the female urethra, vagina, and cervix. In men, the infection is marked by a thick, white discharge from the penis and pain in urinating. Because women show milder symptoms, they are less likely to seek treatment. They then unknowingly carry and spread the disease,

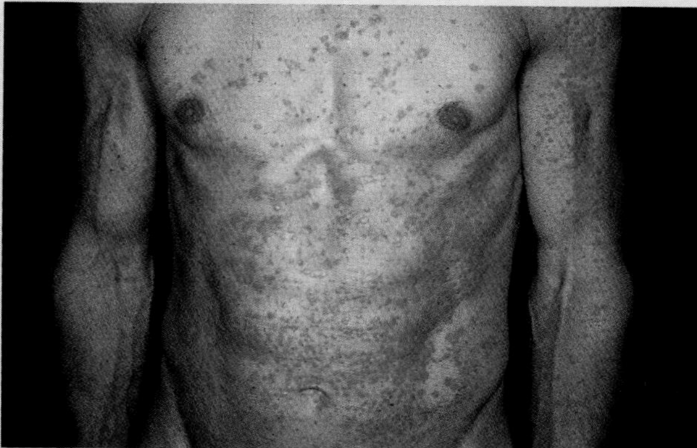

The rash characteristic of the second stage of syphilis.

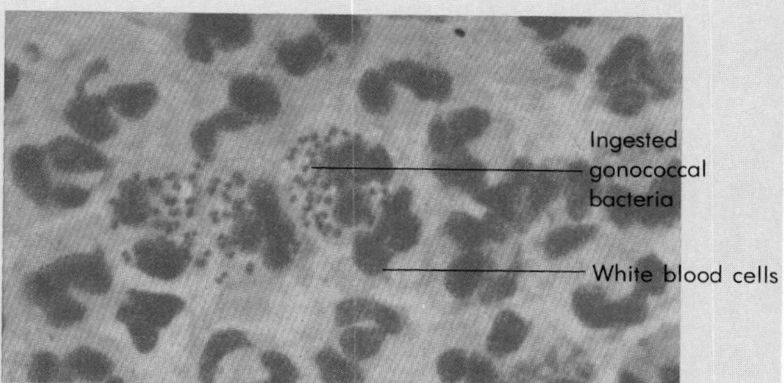

Microscopic section of a positive smear test for gonococcal organisms. The bacteria *Neisseria gonorrheae* have been ingested by white blood cells working in the site of infection. The bacteria appear as groups of tiny dots inside the white blood cells.

and they can suffer severe side effects. As with chlamydia infections, the gonococci can spread to the Fallopian tubes to produce scarring that interferes with fertility. They can also spread to other organs, including the heart. When gonococci enter a baby's eyes at birth, they can cause blindness; to prevent this effect, newborns routinely receive eyedrops of silver nitrate or antibiotics.

OTHER STDs

Some STDs are due to viruses. Genital warts are caused by a virus that has been linked to cervical cancer. Acquired immunodeficiency syndrome (see Box 23-2 and Chapter 24) is also caused by a virus. So is **genital herpes,** caused by the herpes simplex type II virus. This virus produces sores, resembling cold sores, on the genitals or anus; contact with the sores spreads the virus to new victims. A first attack can also produce fever and meningitis. The symptoms last a few weeks and then subside. They may never return, but often the virus becomes latent; it periodically erupts anew, triggered sometimes by stress.

Bacterial STDs can be treated with antibiotics. Viral STDs cannot, though there are drugs that show some effect against herpes when used as ointments on the infective sores, and work is progressing toward vaccines for AIDS. However, it is far wiser to avoid catching these diseases than to have to worry about curing them. Fortunately, there are ways to avoid them. Abstinence can eliminate the chance of infection entirely. Condoms can minimize the chance. Restricting the number of one's sexual partners can also help, for promiscuity can make infection all but inevitable.

Learning Focus Response

SEXUALLY TRANSMITTED DISEASES

Fill in the following table with all the information you have gained about STDs from the Learning Focus and from the rest of the chapter.

The AIDS virus was covered in Box 23-2 and will receive more attention in Chapter 24; do not hesitate to peek ahead in the book.

DISEASE	CAUSE	HOW CAUGHT	SYMPTOMS	TREATMENT
Chlamydia				
Syphilis				
Gonorrhea				
Herpes				
AIDS				

There are a number of other STDs. In the tropics, people can encounter *Granuloma inguinale, Lymphogranuloma venereum,* and *chancroid.* Elsewhere in this chapter, we have said a little about yeast and protozoan (*Trichomonas*) infections. What do you now know about them?

What does your new knowledge of STDs suggest to you about the desirability of various lifestyles? Do you see any risks in having numerous sexual partners? Does til-death-do-us-part monogamy make sense?

Methods of Attack

Bacteria have many ways to produce disease besides their toxins. When large numbers of bacteria are growing in the body, they can preempt the nutrients the body needs for its own well-being. Pockets of pus, formed of bacteria and the white blood cells that fight them, can grow to become painful boils or abscesses. In the wrong places, they can so press upon neighboring organs that they interfere with normal function. Some bacteria live inside white blood cells and kill them, interfering with the body's ability to fight off other infections. To help themselves spread through the body, some secrete enzymes such as collagenase that break down tissues, including blood vessel walls, and cause internal bleeding. Still others wall themselves off from the body's defenses by secreting agents that cause the blood around them to clot; such clots can interfere seriously with blood flow, and if the clots break loose, they can cause strokes, heart attacks, and life-threatening pulmonary emboli.

In many cases, it is the body's response to a bacterial infection that causes the symptoms of disease. The painful, swollen mucus membranes of "strep throat" appear when the body responds to an infection of streptococci with inflammation (see Chapter 24). Capillaries of the region become more permeable, and fluid accumulates in the tissue, causing it to swell; the capillaries dilate, causing

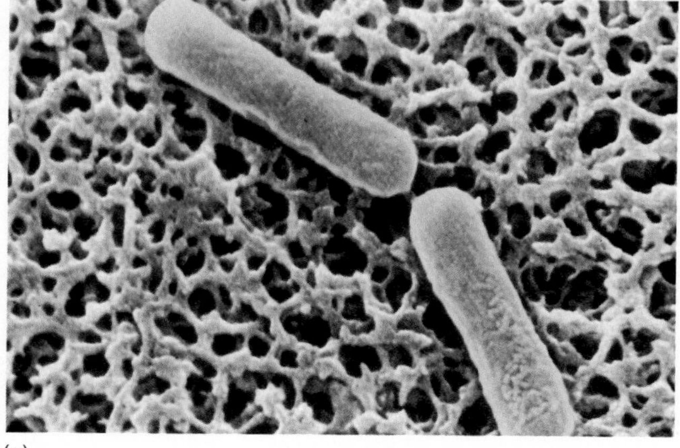
(a)

(b)

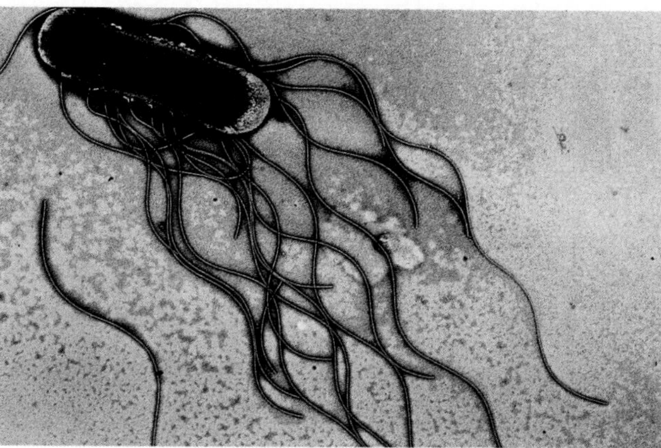
(c)

FIGURE 23-13 The bacteria most involved with food poisonings. (a) One of the most powerful toxins known to science is secreted by the bacteria causing botulism food poisoning, *Clostridium botulini.* (b) Scanning electron micrograph of *Staphylococcus aureus.* (c) The flagellated bacillus *Salmonella* contaminates food and can cause death.

the redness and warmth; and the swelling plus substances released by white blood cells stimulate pain-sensing nerve endings. Similar bodily reactions account for the accumulation of fluid in the lungs of pneumonia patients, which can be severe enough to mechanically prevent the transfer of oxygen to the blood.

Responding to Infection

The best way to deal with bacterial (and viral) infections is to prevent them in the first place. This requires an understanding of how they are carried and how they penetrate the body. It leads directly to our modern belief in the importance of isolating sick people, both because they are more vulnerable to secondary infections and because they can be a source of infection to others. It also leads to our reliance on sanitation or cleanliness to keep bacteria out of our food and water and even our air.

Most bacterial (but not viral) infections respond quickly and completely to treatment with **antibiotics,** substances that interfere with bacterial metabolism in various ways, killing them or inhibiting their reproduction until the body's defenses can do the job. However, since bacteria often become resistant to antibiotics (as we will consider later in this chapter), other answers to infections are also necessary. One of these other answers is the use of vaccines to bolster the body's ability to destroy bacteria, which we will discuss in Chapter 24.

---

1. What diseases have we mentioned that are caught from water contaminated by human wastes?
2. What sanitation practices might be most effective for the prevention of food poisoning?

Protozoans and Other Parasites

Bacteria are by no means the only invaders of the body that cause disease, though they are the most common. Especially in tropical and subtropical areas protozoans can be important pathogens, as can various fungi, worms, and other **parasites.** Many of these organisms are related to free-living forms that can live independently, by themselves. However, because they have adopted a parasitic lifestyle, they have given up many of their chemical and physical abilities, making them utterly dependent on their hosts for many of the processes they need to survive. Apparently, their life in the host has removed their need to keep these abilities.

Protozoans

Protozoans are thought to resemble the cell line that eventually gave rise to the first animals to appear on Earth. They are eukaryotic organisms (their cells have nuclei) and have so many organelles that protozoan cells may be the most complex of all cells found on Earth. Most of the 65,000 different species of protozoans live free in soil and water.

Protozoans are classified according to their modes of locomotion. The **flagellates** use long, whiplike flagella as propulsive organs. The **amoebae** move by extruding portions of themselves as pseudopods (false feet) and then pulling themselves into these extensions as they move forward. Nonparasitic amoebae also use their pseudopods to capture prey items, such as bacteria. The **ciliates** are covered with bristlelike cilia which they use as oars.

Only a few protozoans are parasites and can cause disease. Of these pathogens, none is a ciliate. Some amoebae can infect humans when they are ingested as cysts with contaminated water or food; an example is the amoeba *Entamoeba histolytica*, responsible for amoebic dysentery (Figure 23-14a). The cyst passes through the stomach relatively unchanged. As it continues its journey down the small intestine, its outer covering is digested away, releasing the amoeboid form. The amoeba penetrates the intestinal lining by releasing proteolytic enzymes that irritate these tissues, provoking the dysenteric response—which may include extreme diarrhea and sometimes bloody stools. The amoebae eat the host's red blood cells.

Many pathogenic protozoans are flagellates. They include the organisms responsible for African sleeping sickness (*T. gambiense;* see Figure 23-14b) and some vaginal infections (*Trichomonas*). The **sporozoans** include the organism responsible for malaria (*P. vivax;* see Figure 23-14c). Many protozoan diseases are communicated by insects. Mosquitoes spread the parasites causing malaria, and tsetse flies spread those causing sleeping sickness.

The life cycles of pathogenic protozoans can be complex. For example, when an *Anopheles* mosquito bites a human, it injects an anticlotting agent with its saliva and several thousand "ride along" malarial parasites. These cells quickly invade liver cells, where they develop into multinucleate *schizonts.* The schizonts then subdivide into many *merozoites,* the liver cells burst, and the merozoites are released into the blood. There they invade red blood cells, where they develop into new schizonts to repeat the cycle. The release of merozoites, which follows a fairly rigid rhythm, floods the blood with debris from damaged cells, causing the periodic bouts of fever, and sometimes delerium, seen in malaria.

Some of the malarial cells in the blood are converted to gametes. The life cycle then continues when a mosquito bites a person infected with malaria. With its blood meal, it takes in some of the sex cells. In the mosquito's stomach, the gametes fuse to form zygotes. The zygotes attach to the mosquito's stomach wall, cross it, and form cysts full of new parasites. When a cyst bursts, the parasites migrate in the insect's body fluids to its salivary

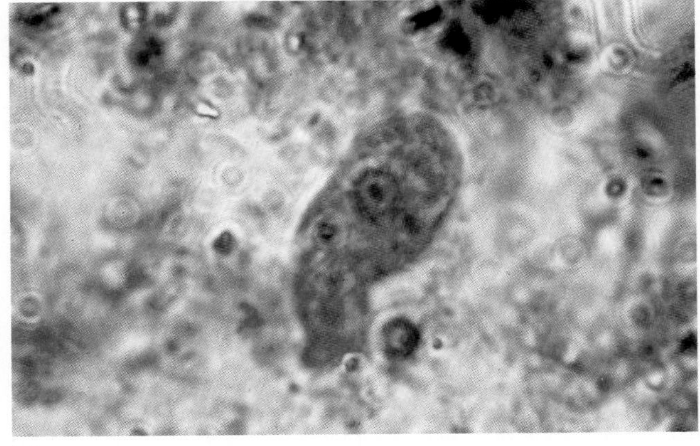
(a)

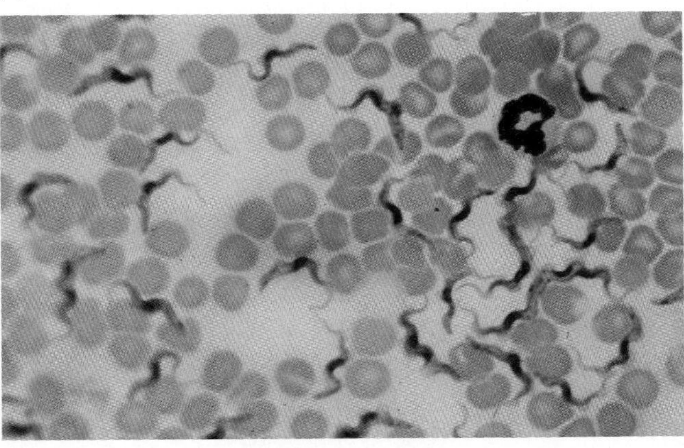
(b)

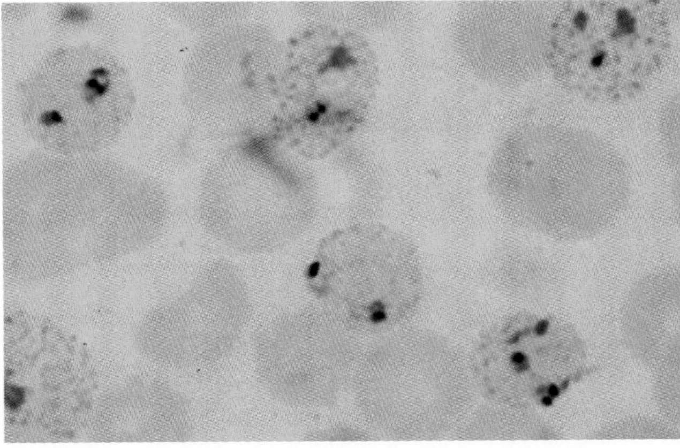
(c)

FIGURE 23-14 The protozoan parasites responsible for three deadly diseases: (a) the cause of amoebic dysentery *(Entamoeba histolytica);* (b) the agent responsible for African sleeping sickness *(Trypanosoma gambiense);* (c) the cause of malaria *(Plasmodium vivax).*

glands, where they await injection into another unsuspecting warm-blooded victim (Figure 23-15).

Malaria is a difficult disease to control because the parasite spends much of its life cycle inside human cells, out of reach of the human immune system. There are prophylactic (disease-preventing) drugs that can be used to make initial infection less likely, but they are not feasible for protecting the large numbers of people who live in malarial regions of the world. There are also drugs, of which quinine is one of the oldest, that relieve symptoms and reduce the number of malaria parasites in the blood, but they are rarely totally effective. "Recovered" malaria victims are therefore not allowed to donate blood. The most effective tactic has been to use insecticides to kill the mosquitoes that spread the disease, and this approach did in fact greatly reduce the incidence of malaria around the world. However, the mosquitoes evolved resistance to the pesticides, and one of the most effective pesticides [dichlorodiphenyltrichloroethane (DDT)] fell into disuse when it proved to persist in the environment and to be toxic to other organisms. The malarial death toll has been rising in recent years.

Fungi

Only about 100 of the 100,000 species of fungi (mushrooms, toadstools, yeasts, and molds) are pathogenic for animals, including humans, and only a dozen can be fatal. Fungal diseases are called **mycoses.** Systemic mycoses are widespread infections that involve various systems in the body. More common are superficial mycoses such as skin infections. The pathogens include a number of tropical organisms, but Americans are far more likely to encounter yeast infections, most often of the vagina, and the various versions of "ringworm." The latter invade the skin of the scalp

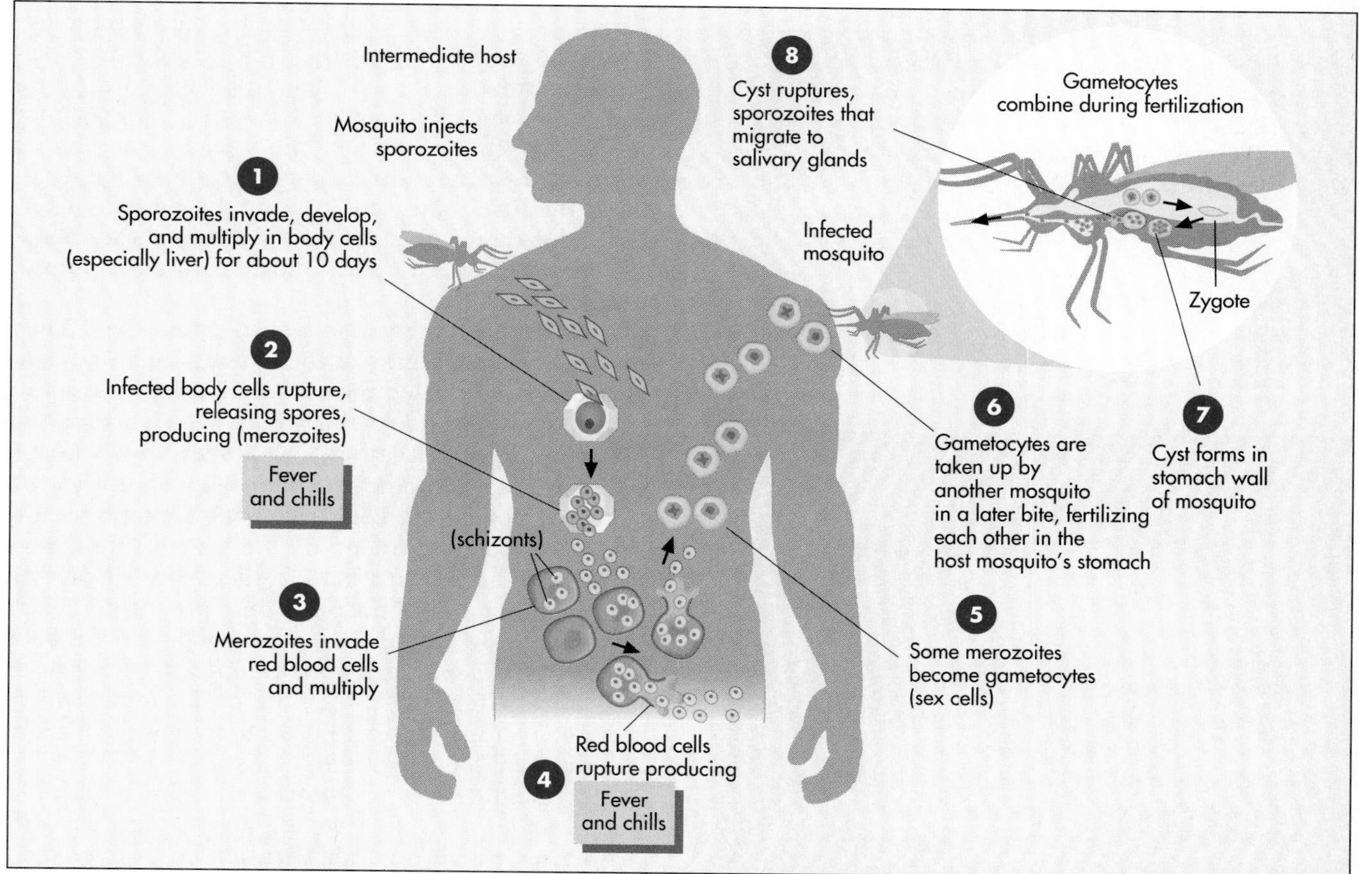

FIGURE 23-15 The malaria cycle is spread over two hosts. When a female mosquito (which needs a blood meal from a warm-blooded host in order to reproduce) bites a human, saliva carries the malaria parasite into the human's body. The parasite invades human cells, multiplies, and in time produces gametocytes that can be taken up when a later mosquito bites the infected human.

(*Tinea capitis*), the groin (as "jock itch," or *T. crusis*), and the feet (as "athlete's foot," or *T. pedis*) and secrete substances that cause the characteristic irritation (Figure 23-16). They affect males more often than females, and they can be passed from pets to their owners.

Yeast infections are common, especially in people undergoing antibiotic therapy to treat a bacterial infection. The bacteria and fungi that dwell on the skin normally balance each other, but when antibiotics destroy both pathogenic and the normal skin bacteria, which produce secretions that keep the fungal populations in check, the yeast cells undergo a population explosion. The result is a secondary infection that must also be treated.

Moist skin is far more hospitable to fungal growth. This is why fungal infections are more common in the humid tropics and in the summer and why superficial mycoses are likely to involve areas of the body that are often sweaty—such as the feet or groin. Many of these fungi are so prevalent that there is little we can do to prevent them, except try to stay dry. Fortunately, a number of medications successfully treat these superficial skin disorders.

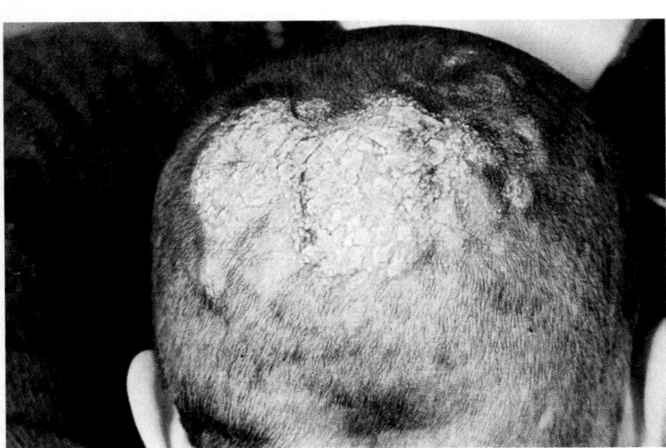
(a)

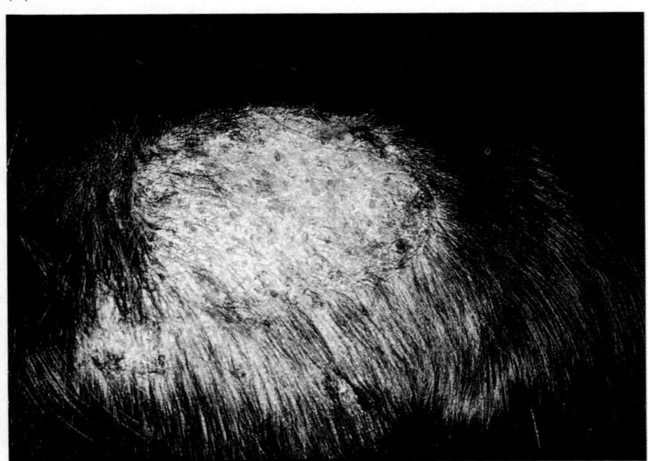
(b)

FIGURE 23-16 Ringworm is a fungal infection of the skin. (a) The fungus responsible for scalp infections is *Tinea capitis*. (b) The appearance of the infection accounts for its name.

Worms and Flukes

In some parts of the world, such as tropical Africa, well over half the people are hosts to multicellular animal parasites. These parasites include nematodes (or roundworms) and platyhelminths (flatworms, tapeworms, and flukes). **Nematodes** include about 50 species that can parasitize humans. Some dwell in the intestine, sharing their host's food and shedding their eggs in the feces. Others, such as the **hookworms,** invade more deeply. The larvae of these tiny worms live in the soil and penetrate the skin of bare feet. Once in the body, they travel in the blood to the lungs. They then move up the trachea to be swallowed. They mature in the intestine and damage the intestinal lining to get to the blood. Their activities can cause severe anemia. They produce eggs that are eliminated in the feces. The eggs hatch in the soil, and the new larvae are then available to repeat the cycle. Hookworms are easily avoided—shoes or boots block the initial infection very effectively. However, subsistence farmers in many parts of the world often cannot afford shoes. They also cannot afford sewage treatment facilities, and they use human wastes as fertilizer for their fields. It is not surprising that hookworms are one of the most common human parasites.

Tapeworms are also intestinal parasites. They lack digestive systems of their own. Each one consists of a head, called the scolex, that bears specialized suckers and hooks that anchor the worm in the intestine, where it can absorb a share of its host's food. The tapeworm body consists of numerous flattened segments, containing the metabolic and reproductive equipment of the worm. New segments are formed just behind the worm's head. As new segments form, the older ones are pushed back toward the terminal end of the worm. The beef tapeworm (*Taenia saginata*) can attain a length of almost 8 meters (m; 25 ft); the pork tapeworm (*T. solium*) averages 2.5 m (8 ft) in length and contains in excess of 2000 segments.

Each segment, or proglottid, contains both male and female reproductive apparatus and is self-fertilizing. Sometimes fertilization occurs between different segments. Egg-filled proglottids break off from the end of the worm and pass out of the body with the wastes. There they await accidental ingestion by another animal. Once ingested, the embryos are released and then migrate to the animal's skeletal muscles, where they form cysts. (Some may encyst in the brain.) They are spread when another animal eats the uncooked infested muscle tissue. Thus, there are two ways to be infected with a tapeworm—eating the eggs or eating infected meat.

Tapeworms do not normally cause great problems for their hosts. Generally, a mild form of diarrhea can occur early in the association. The worm and the host eventually develop a mutual tolerance for each other as their relationship matures. In some cases, however, the worm infestation can become so dense that an intestinal blockage develops.

We might be led to believe that because tapeworms compete for the host's nutrients, an infestation leads to emaciation. In the 1930s, entrepreneurs with this idea marketed dried tapeworm segments as diet pills (of course, they were not advertised as having anything to do with tapeworms!). The logic was that since tapeworms compete for nutrients with their host, their presence should help people lose weight without cutting back on the diet. Unfortunately, the results were less than satisfactory, though they did help pave the way for such organizations as the U.S. Food and Drug Administration.

Flukes are another, more devastating type of parasite with a worldwide distribution. They are a group of flattened, leaf-shaped worms capable of damaging many of the body's systems. They are most common in tropical areas, where they alternate between humans (or other vertebrates) and snails as hosts. The adult, sexually mature flukes are the ones that live in humans. They can infest the blood, liver, or lungs. Blood flukes, also known as **schistosomes,** travel in the blood to the liver, where they mature. Adult flukes establish themselves in the veins of the pelvis and intestine. Their eggs then work through the walls of the intestine and bladder to leave the body with the wastes (Figure 23-17). Masses of eggs also accumulate in the body's organs, including the brain, where they become surrounded by connective tissue and form distinctive nodules. The presence of these nodules damages the liver, bladder, and intestine and accounts for many of the debilitating symptoms of the disease **schistosomiasis.** One symptom, blood

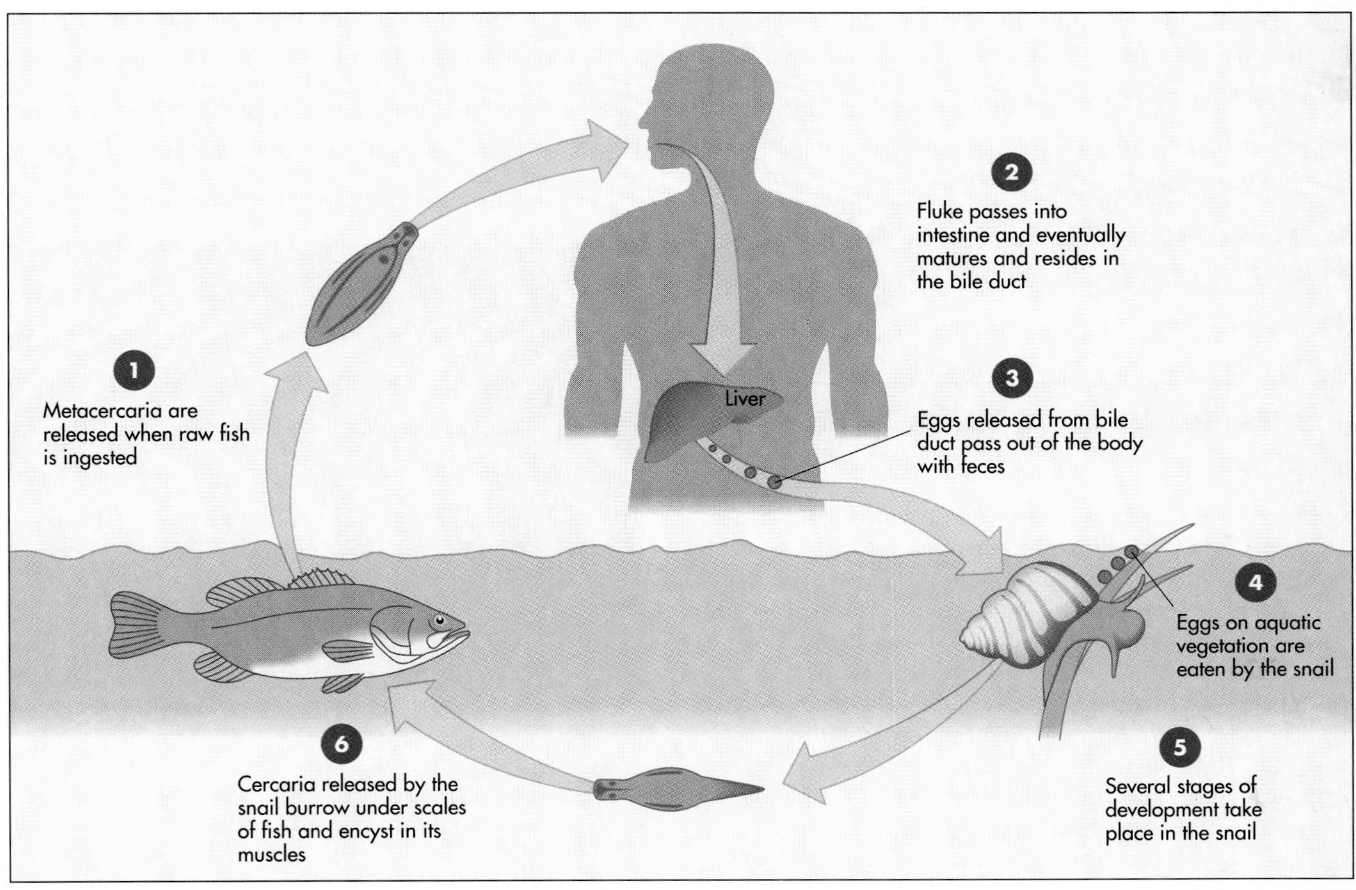

FIGURE 23-17 The human liver fluke life cycle.

dripping from the penis, is so common that it formed the basis for the ancient Egyptian hieroglyphic symbol for the disease.

When excreted fluke eggs reach water, as in an irrigation ditch or rice paddy, they hatch. The larvae then infect freshwater snails, reproduce asexually inside the host, and leave the snail to swim in the water. When they then encounter human skin, they burrow in to renew the cycle.

Schistosomiasis is one of the most widespread health problems in the underdeveloped world, affecting 200–300 million people, 1 person in 20 worldwide. However, it is not easy to either prevent or cure. In principle, a very effective way to prevent the disease would be to break the fluke's life cycle by killing the snails, but currently available snail-killing chemicals are too toxic to add to water supplies, where they would harm both humans and other creatures. Wearing boots when working in a rice paddy or irrigated field is an effective way to avoid infection. So is refraining from eliminating wastes in water in which one will wade, swim, wash, or do laundry or that one will drink. Unfortunately, the tropical lands where the flukes and their host snails thrive are populated by peoples who cannot afford boots, lack alternative sanitary facilities, and depend on irrigation for their food supply.

However, there is now hope for a cure to this pervasive disease. Researchers have begun to find ways to attack blood flukes in the body. Certain new drugs are effective, although they are also expensive. Vaccines are also under development, as well as other treatments made possible by recent progress in molecular biology.

THE PROBLEM OF DRUG RESISTANCE

Only recently have researchers begun to find drugs that promise to fight infections by viruses and some parasites. Other parasites and bacteria have long been treatable with such medications as antibiotics. Unfortunately, although these drugs can be very effective, they can also cause a problem.

The problem arises because not all the members of a species are identical. They vary in their genes and therefore in their characteristics. One of these characteristics is the ability to resist the effects of toxic chemicals such as antibiotics (Figure 23-18).

Consider that a person infected with a bacterial disease contains millions or billions of bacteria. When the patient takes an antibiotic, especially if he or she stops short of taking all the prescribed capsules, a few of these bacteria are bound to survive. As they reproduce in the patient's body, they give rise to millions of new bacteria, all with the same genes and characteristics, including the ability to resist the effects of the antibiotic. We say they have become **drug resistant,** even though they have merely concentrated a resistance that was already there by the selective deaths of the drug sensitive bacterial cells in the reproductive population.

The new population of resistant bacteria (the next generation) will still contain variation; that is, some will be more resistant than others. If the patient later takes a larger dose of antibiotic but still not enough to kill all of the more resistant bacteria, some will again survive and the resulting infection will be still more resistant. For this reason, physicians usually treat bacterial infections with heavy doses of antibiotics the first time, hoping to kill even the most resistant members of the population. Unfortunately, many people stop taking their prescription when the symptoms die down; they thus lower the dose and give resistance a chance to develop.

Bacteria can, of course, develop resistance to more than one antibiotic. When bacteria are exposed to many antibiotics in large doses, only the very resistant can survive. This problem is especially serious with common bacteria, such as those that cause the STDs syphilis and gonorrhea. These infections could once be treated easily and effectively with a shot of the antibiotic penicillin. However, resistance has developed and the resistant

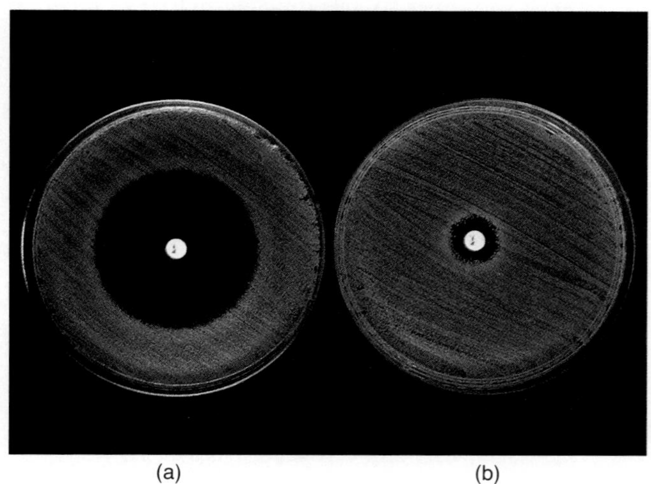

FIGURE 23-18 Developing penicillin resistance in bacterial cultures of *Staphylococcus pyogenes.* In the center of each bacterial culture is placed a disc loaded with penicillin that leaches out into the surrounding medium. The penicillin in the sensitive culture (a) inhibits bacterial growth in a wide radius surrounding the disc. (b) In the penicillin-resistant culture the bacteria can grow right up to the edges of the antibiotic source without being inhibited.

bacteria have spread widely. Now these diseases require treatment with enormous doses of penicillin or other, more expensive antibiotics.

Resistance is an especially serious problem in hospitals. They expose many kinds of bacteria to many antibiotics over long periods of time, and in them are found some astonishingly resistant bacteria, virtually immune to all the common antibiotics. For this reason, among others, there is a continual demand for new antibiotics.

The phenomenon of resistance is hardly unique to bacteria and antibiotics. We can see it in the reaction of malaria parasites to antimalaria drugs and of insects to pesticides. In a broader sense, it is but one example of the process of *natural selection* that is the mechanism of evolution (see Chapter 25).

THE PROBLEM OF RECOGNITION

Chapter 24 deals with how the body recognizes foreign invaders such as viruses, bacteria, and parasites in order to defend itself against them. Here we wish to say only that the key to recognition lies with the proteins and other molecules these invaders bear on their surfaces. They are unlike the molecules that mark the surfaces of the body's own cells, and the body's immune system can sense and react to the difference. We call these surface molecules *antigens;* the immune system responds to them in part by generating *antibodies* that can attach to them, marking the foreign cells bearing them for destruction and removal.

VIRULENCE

As the evolutionary relationship between parasite and host progresses, there is a contest of sorts, played for very high stakes—namely, survival. About as fast as the body's defenses accumulate new ways to protect the body from foreign invaders (see Chapter 24), the parasites accumulate new tricks for avoiding those defenses. As viruses, bacteria, and other parasites accumulate techniques or tricks making them more difficult for the defenses to handle, they are said to become more virulent.

The most simple and common virulence-enhancing strategy is that a parasite multiplies so rapidly the body's defenses cannot stop it before it spreads to new hosts. Bacteria are capable of reproducing (by fission) every 30 minutes under ideal conditions. A population started by a single such bacterium could thus exceed hundreds of millions in less than 24 hours. Many infectious agents also have a variety of ways of avoiding recognition and destruction. Viruses hide inside the body's cells. They can be detected and removed only when they are between cells, free in the blood, or when their proteins appear in cell membranes. In the latter case, their destruction requires destruction of the body's cells as well, a process that actually produces the symptoms of many viral diseases. Some bacteria wear mucuslike coatings that make detection and destruction difficult. Some parasites, including the schistosomes, change their surface molecules frequently enough to evade destruction. These characteristics make difficult the development of vaccines, which stimulate the body to produce antibodies against a disease organism's antigens.

SUMMARY

The human body is vulnerable to attack by viruses, prions, bacteria, protozoa, fungi, and larger parasites. These pathogens, or agents of disease and infection, vary in virulence, the ability to do damage.

Viruses are the simple agents of infection responsible for mumps, flu, chicken pox, colds, and many other diseases. They consist of a protein shell and a core of deoxyribonucleic acid (DNA) or ribonucleic acid (RNA). They live by penetrating cells and using the cells' metabolic machinery to manufacture new viral protein and nucleic acid. When a bacterium is full of bacteriophage (bacterial virus) particles, it bursts to release the virus to infect other cells. Animal viruses can also burst out of their host cells; often, however, they are extruded through the host's cell membrane, emerging wrapped in a bubble of membrane. In addition, both bacterial and animal viruses can merge with their host's DNA, as prophage or provirus, to reproduce with the host until conditions change and stimulate the virus to break free and take over the cell's machinery.

The RNA viruses come in two forms. Some use the enzyme RNA polymerase to replicate their RNA as if it were DNA. Others, the retroviruses, use host enzymes to make a DNA copy of their RNA.

All viruses function by using their nucleic acid as messenger RNA (mRNA) or as the template for mRNA to guide the synthesis of viral proteins. The first proteins made inhibit host cell activities. Later ones are viral components and cell-bursting enzymes. The early proteins also control the type of infection. Latent infections produce virus particles just fast enough to maintain the infection as the host cell reproduces. Persistent infections release virus particles slowly into the blood. Some viruses can also transform their hosts, causing them to lose control of their growth and become cancerous; they carry oncogenes, or cancer genes, often related to genes in the host cell's own genome.

Viroids contain no protein. They consist only of small RNA molecules that apparently affect host cells by interacting with genes for growth hormones. They infect only plants, although for a time it was thought they caused several human diseases now thought to be caused by prions, "protein infectious particles." Prions consist only of protein, and they seem to act by affecting the production of a protein that serves a normal function within the cell.

Many diseases are caused by bacteria, simple prokaryotic (without nuclei) cells. Given the opportunity to invade the body through a break in the skin or to colonize the mucous membranes of the respiratory or digestive system, they cause illness by secreting toxins, damaging tissues, causing blood clots and inflammation, and preempting nutrients. Many are normally present on the skin or in the intestinal contents; they cause disease most often when the body is weakened by malnutrition or other illness or when the immune system is weakened by treatments for cancer such as radiation. Others reach the body through contaminated food, water, or air and by contact with infected persons. Bacterial, but not viral, infections can usually be treated effectively with antibiotics.

Amoebae, flagellates, and sporozoans are one-celled protozoans. Most are free living, but a few species cause disease. The flagellates include the agents responsible for African sleeping sickness (*Trypanosoma gambience*, carried by the tsetse fly) and for one sexually transmitted disease (STD) (*Trichomonas*). The sporozoans include the organism responsible for malaria, spread by mosquitoes.

Of the 100 pathogenic species of fungi, only a dozen are fatal to humans. In the United States, the most common fungal diseases are not so threatening. They are genital yeast infections and ringworm (including athlete's foot).

The larger parasites include roundworms, tapeworms, and flukes. Tapeworms and some roundworms live in the intestine, causing illness by using nutrients the body needs and obstructing the digestive system. Other roundworms, including hookworms, and the flukes invade the body through the skin, consume blood, cause anemia, and cause painful lumps or cysts in tissues. Blood flukes, or schistosomes, require snails as an alternate host; larval flukes escaping the snails infect humans who are wading in irrigation canals or rice paddies. The flukes infect the snails when eggs excreted in human wastes reach the water in which the snails live. An effective way to prevent schistosomiasis is to use proper sanitation and sewage treatment. Other methods, such as killing the snails, are less effective; medications for preventing or curing the human infection are only now becoming available.

The phenomenon of drug resistance is a serious problem in treating bacterial diseases and parasites. Because of genetic variation, a few members of a population of disease organisms are likely to survive treatment with antibiotics or antiparasite drugs. These survivors then give rise to a new generation of more resistant pathogens. Pathogens that are resistant to many drugs are common, and many infections today require large doses or new drugs. Far more effective in most circumstances is the body's immune system, which exploits the proteins and other molecules on the surfaces of viruses and cells to tell which ones are native to the body. It reacts against those that are not, destroying or removing them.

Table 23-2 lists some common infections and diseases.

Table 23-2

Clinical Summary: Causes of Common Infections and Diseases

Viruses	Syphilis
Colds	*Chlamydia*
Flu	Food poisoning
Rabies	Boils (*Staphylococcus*)
Chicken pox	Tetanus
Herpes	Gas gangrene
AIDS	Anthrax
Viral pneumonia	Bacterial dysentery
Viral meningitis	Typhoid
Measles	Diphtheria
Mononucleosis	Tuberculosis
Hepatitis	
Yellow fever	Protozoa
Bacteria	Malaria
Bacterial pneumonia	Amoebic dysentery
Bacterial meningitis	Sleeping sickness
Gonorrhea	*Trichomonas* (vaginitis)

Fungi
- Yeast infections
- Athlete's foot
- Jungle rot
- Ringworm
- Valley fever (coccidiomycosis)

Worms and flukes
- Tapeworms
- Flukes
- Schistosomiasis
- Trichinosis
- Pin worms
- Hookworms
- Filariasis (elephantiasis)

STUDY QUESTIONS

1. Why do we call viruses, viroids, and prions subversives?
2. What is the difference between RNA viruses and retroviruses?
3. Why would it be unreasonable to say that all bacteria are dangerous?
4. Why do parasites often lose abilities they would need to live independently of a host?
5. Why is it difficult to attack malaria parasites within the body?
6. Name two techniques useful for controlling both hookworm and schistosomiasis.
7. How many ways can disease organisms get into the human body?
8. Name at least one way to block each of the invasion routes you just listed.
9. Is there any method that might block more than one of these routes?

GLOSSARY

Antibiotic Substances originally produced by one life form, such as a fungus, that inhibits or kills bacteria. We can now synthesize many of these antibiotics in the laboratory without relying on the original living agent that first synthesized it.

Capsid The protective protein coat surrounding the nucleic acid of a virus.

Drug resistance Ability to resist the toxic effects of antibiotics and other drugs.

Flukes Parasitic worms that live and reproduce in the blood, liver, and lungs; some require snails as alternate hosts.

Hookworms Parasitic roundworms that invade the body through the skin of the feet. They migrate to the lungs and eventually are coughed up into the throat and swallowed. They spend the sexual phase of their life cycle in the digestive system ulcerating the intestine, consuming the host's blood, and laying eggs that pass out of the host's body into the soil.

Latent infection Virus infection that produces just enough virus particles to maintain the infection in the host cell's progeny.

Mycoses Any of the diseases caused by fungi. They include deep diseases involving many of the body's systems (systemic mycoses) and the mycoses of the skin, such as athletes foot (referred to as superficial or dermamycoses).

Nematodes Roundworms; many are parasitic.

Oncogene A gene that causes cancer.

Parasite Any organism that lives on or in its host and consumes the nutrients or substance from that host.

Pathogens Organisms that cause disease.

Persistent infection Virus infection that produces enough virus particles to maintain a low level of virus in the blood.

Prion An apparently infectious protein lacking nucleic acid.

Prophage Bacteriophage whose DNA has joined the DNA of its bacterial host.

Protozoa Extremely complex one-celled organisms, thought to have given rise to the cell line that eventually became the animals. Some are parasites.

Provirus Plant or animal viral nucleic acid incorporated into its host's DNA.

Restriction endonuclease Enzyme that cuts DNA at a specific sequence of nucleotides.

Retrovirus An RNA virus that translates its RNA to DNA in its host cell.

Schistosomes Blood flukes.

Sexually transmitted disease (STD) Any disease transmitted by sexual contact.

Tapeworms Tapelike, segmented intestinal parasites.

Transformation Viral conversion of a host cell to a cancerous cell, without control over its growth and multiplication.

Viroid A viruslike structure consisting of RNA but containing no protein; infects plants only.

Virulence Ability to cause disease.

Virus Tiny intracellular parasite consisting of a shell of protein and a core of RNA or DNA.

Chapter 24

The Defenders: The Immune System

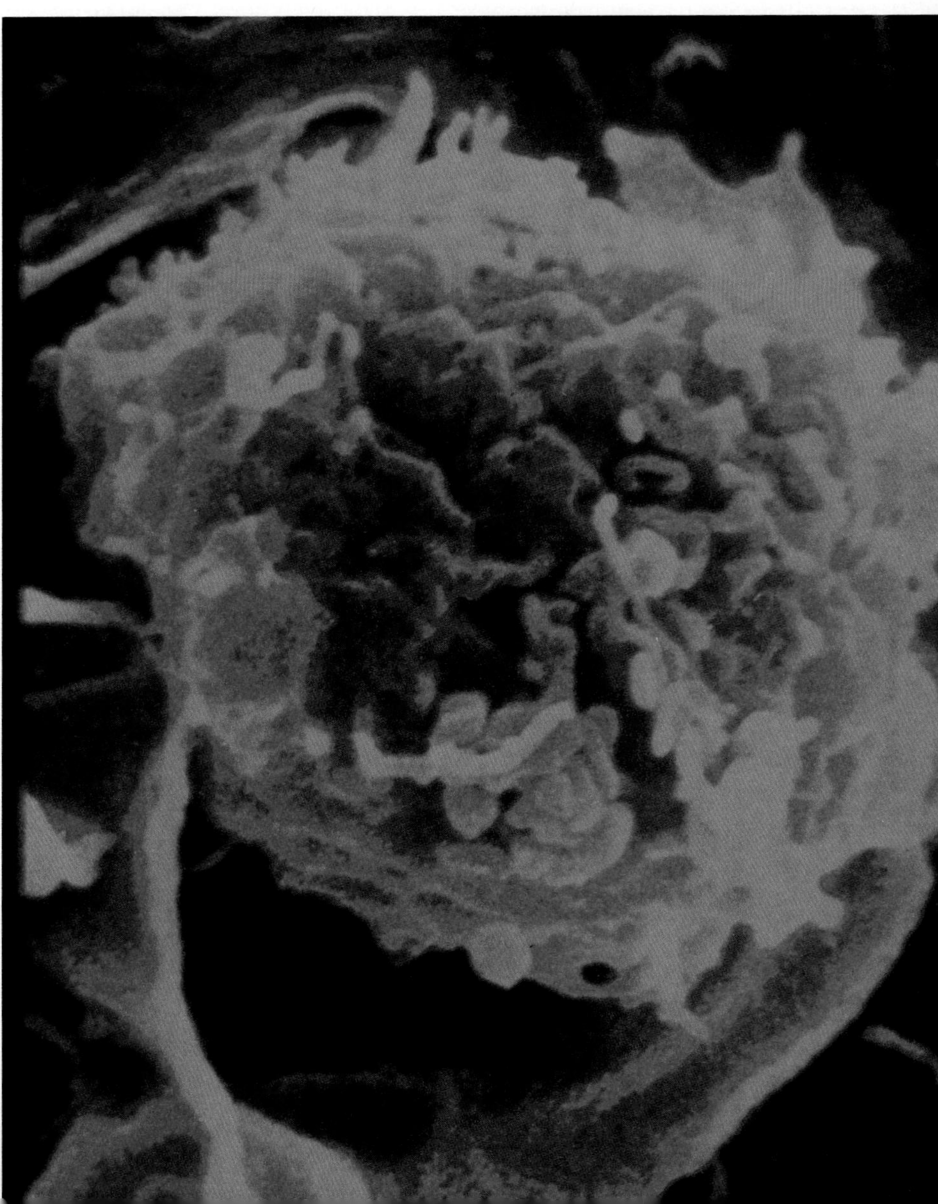

Every living thing is continuously being invaded, damaged, parasitized, or colonized by the various agents of infection we discussed in Chapter 23. Our ability to keep our bodies intact for up to a century (and sometimes more) is a credit to the efficiency of that elaborate array of defenses we call the **immune system.** Even in the "best" of health, we are dealing with so many potential serious problems—cuts, colds, and even cancerous cells—that we might best define "health" as a controlled state of disease.

The study of the immune system is one of the great frontiers of medical and biotechnological research for the 1990s. Studying the complexities of the immune system will help us avoid many of the degenerative diseases of aging and deal more efficiently with serious infection, and may eventually help us combat cancer.

RECONNAISSANCE OF THE BODY'S DEFENSES

The body's defenses can be divided into two broad categories: nonspecific resistance and specific resistance. **Nonspecific resistance** includes a host of physical barriers, filters, mucus traps, and an army of "aggressive" white blood cells that work against a wide variety of harmful substances such as bacteria, viruses, fungi, and harmful chemicals. Each of these protective devices works against many potential intruders. If all fail and somehow the invaders gain a "foothold," then the body defends itself with white blood cells—the macrophages and neutrophils—that take on all comers.

The second level of defense, **specific resistance** (sometimes referred to as *specific immunity*), consists of very efficient chemical and cellular mechanisms tailored to deal with specific threats posed by certain bacteria, fungi, viruses, and toxins. This type of resistance depends on the body's ability to recognize the difference between self (its own body cells) and nonself (foreign cells). Each type of invader carries on its surface unique chemical configurations, called antigens, that make it possible for the white blood cells to recognize them as nonself. We will discuss each of these levels of resistance in detail in the coming sections.

THE DISEASE SCENARIO

Perhaps the most productive way to look at the body's defenses is to describe how the body responds to attack. Four processes have to occur in order for a disease to develop. They are discussed in the following sections.

Exposure or Contact

Bacteria and viruses in the environment can land on the skin. They are harmless as long as they remain on the surface of the skin, where their numbers are kept in check by various mechanisms. In fact, many bacteria, fungi, and other organisms normally dwell on the skin and in its pores and hair follicles. Certain portions of the skin, however, are vulnerable. For instance, moist creases in the skin, such as between the toes, are fertile gardens for the fungus responsible for athlete's foot, the eyes are hospitable to many bacteria, and the genitals provide an easy pathway into the reproductive tract for bacteria, yeasts, and protozoa such as those associated with the sexually transmitted diseases.

Airborne bacteria and viruses and those in food are another matter. Many can cause disease or illness without even crossing the walls of the respiratory or digestive systems. They can produce symptoms of illness by irritating the membranes lining these systems or by producing poisons that are absorbed through the walls of these systems. The body has special mechanisms for removing invaders from these areas.

Invasion or Penetration

Once contact has been made with the disease agent, the next step is invasion. There are numerous channels through which an invasion can be launched. Sexually transmitted disease bacteria migrate along the mucous membranes of the urogenital tract. Human immunodeficiency virus (HIV), which causes acquired immunodeficiency syndrome (AIDS), enters through cuts, abrasions, and sores in the skin or mucous membranes (it can also enter when a blood transfusion is obtained from an infected person or a hypodermic needle is shared by an infected drug user). Common cold viruses invade cells that line the respiratory passages. Pneumonia bacteria take hold in the lungs and conjunctivitis bacteria on the moist lining of the eyes. Some bacteria, including those responsible for cholera and dysentery, colonize the contents of the large intestine.

Actual penetration of the body occurs when fungi, such as those causing athlete's foot and ringworm, insinuate themselves past the surface layers to grow in the deeper layers of the skin. Penetration is also literal when a scratch or a cut actually tears the skin open, allowing opportunistic bacteria to enter the tissues beneath the skin (Figure 24-1).

Multiplication or Incubation

Once the invasion is successfully under way and the invaders have reached their "preferred" location in the body, the next step toward a full-blown infection is an increase in numbers of the disease organisms until they can produce the symptoms of the disease. The period from the time of invasion until enough disease agents are produced through multiplication to cause symptoms is called the **incubation period.**

Illness: Overt Signs of Physiological Warfare

Symptoms of illness do not begin abruptly. There is about a 24-hour transition period, the **prodromal stage** (see Figure 24-2), between the end of the incubation period and the beginning of the acute (or attack) stage. During this prodromal period people suffer fatigue, headache, muscle aches, jitteriness, and general "dis-ease." The **acute stage** is the period of major damage to the body, caused by bacterial toxins, cellular destruction, or the release of damage products from the body's cells. (As we will see later in this chapter, the body's own responses to the invasion can be more damaging than anything the invaders might do.) This stage may last only a day or, in the case of severe or incurable diseases, the rest of one's life. People who survive the acute stage enter the **convalescent stage,** in which their immune systems defeat the pathogen and they regain their strength and vigor. This stage may last only a few days or many years.

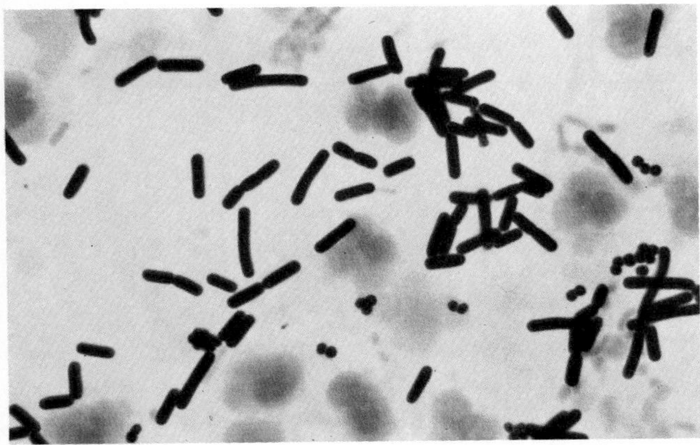

FIGURE 24-1 Section through an infected wound. The blue-black rods are *Clostridium perfringens,* the bacteria that cause gas gangrene. The larger reddish blobs are stained white blood cells. Also visible are smaller spherical cells of *Staphylococcus pyogenes,* the commonest cause of pus-forming infections in humans. Faintly stained pink, smaller rod-shaped bacteria are *Escherichia coli,* a common inhabitant of the intestine.

---

1. List the stages in the disease process and state what happens in each one.
2. Draw your own version of the disease time line, describing what is happening at each stage.

NONSPECIFIC RESISTANCE: EXTERNAL

Nonspecific resistance employs a wide range of protective devices that do not depend on identifying the precise nature of the attacker. They act to prevent invasion, suppress multiplication, and control the spread of disease agents by using physical, chemical, and cellular defensive techniques. These methods fall into several categories.

Mechanical Resistance

The Skin

The most prominent element of nonspecific resistance is the body's skin. The skin supports its own natural external "microflora" that contains many types of bacteria and fungi. Moreover, when intact, the skin offers a thick barrier that can rarely be passed by bacteria. It is also strong and resilient enough to resist many cuts and punctures, which might allow bacterial entry. However, some invaders specialize in attacking the skin. Certain species of fungi, such as athlete's foot, can live in the superficial skin layers.

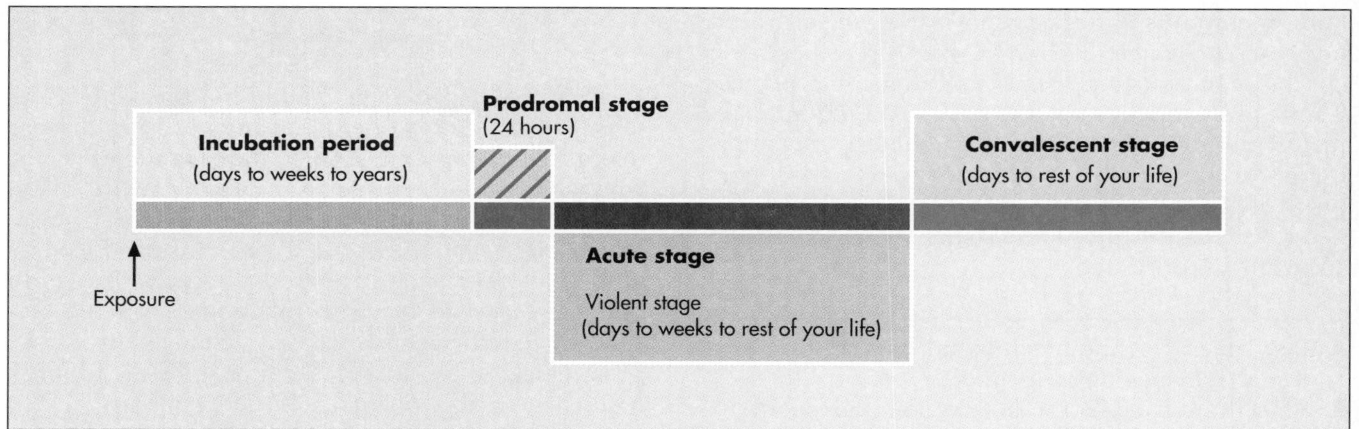

FIGURE 24-2 A time line of disease. The history of a disease is divided into four stages. The incubation period runs from when the person was first exposed to the agent until it has multiplied to the point of causing symptoms. The prodromal phase (usually only 24 hours) is the day when the body is becoming sick. It includes jitteriness, headache, and fatigue. The acute stage is the violent stage, when the damage to the system is actually being done. If the disease is serious enough, the person will not live through the acute stage. The convalescent period begins when the body actually starts to win the battle.

The Mucous Membranes

Mucous membranes are extensions of the skin that line body cavities that open to the outside, such as the mouth, digestive, respiratory, and reproductive tracts, and eye sockets. These membranes are kept moist by a secreted layer of mucus that traps many potential disease organisms (note the spelling change: *mucous* is the adjective describing the type of membrane; *mucus* is the noun naming the secretion). In the nose there are also hairs that filter out dust and bacteria. In addition, the side walls of the nasal cavities are folded to swirl the air as it passes through, making it more likely that particles in the air will strike and be trapped by the mucus.

Mucous membranes are less effective as barriers than skin, because some bacteria thrive on the moisture they provide. Some bacteria can even penetrate the membranes into the body. Whether they succeed seems to depend on how many bacteria are present, on whether they secrete cell-damaging toxins, and on whether there is previous damage from viruses or irritants (such as tobacco smoke).

Cilia

The cells of the mucous membrane lining the trachea have cilia that keep the mucus moving upward, away from the lungs and toward the mouth, at a speed of 1–3 centimeters per hour. Coughing and sneezing also help remove foreign materials from the respiratory tract by catapulting them from the system (Figure 24-3).

Chemical Resistance

Some mucous membranes have additional protective mechanisms. The eye washes its membranes clean with a continual flow of tears. The urethra is periodically flushed by the flow of urine. The mouth is washed clean by saliva, and both saliva and tears contain an enzyme, **lysozyme,** that can break down bacterial cell walls. Lysozyme is also found in nasal mucus and sweat, which serves to wash some microorganisms from the skin.

The skin gains additional protection from the **sebum** secreted by its sebaceous or oil glands.

FIGURE 24-3 Cilia lining the wall of the trachea function as part of the nonspecific resistance system by conveying particulate matter that has been trapped in the mucus up to the throat where it can be ejected from the system. Also visible are nonciliated tracheal cells.

Sebum contains a large proportion of unsaturated fatty acids, which can kill some disease-causing bacteria. In addition, the skin gains some protection from its acidity, or low pH of 3–5, due in part to the metabolic by-products of bacteria that normally dwell on the skin. Acid also plays an important role in the stomach, where hydrochloric acid destroys many microorganisms that enter the stomach with the food.

1. List several of the mechanical barriers for external nonspecific resistance.
2. State how each of the barriers is nonspecific in its action.
3. Give examples of chemical versions of nonspecific resistance.

NONSPECIFIC RESISTANCE: INTERNAL

Once disease-causing microorganisms manage to bypass the external defenses and enter the body, other defenses come into play. Some, like interferon, are purely chemical. Others involve the active responses of living cells.

Phagocytosis

Many immunologists think phagocytosis may be the single most important line of day-to-day immune protection. As we discussed in Chapter 10, phagocytosis is a form of cellular eating performed by phagocytes, various members of the white blood cell community, especially the neutrophils and the macrophages. These cells are particularly effective at ingesting bacteria that may have entered the body through a cut in the skin or by violating one of the body's other protective barriers.

Phagocytosis occurs when phagocytes encounter bacterial invaders, surround them with extensions of their cell membranes (called pseudopods), and then fold in the cell membrane as a pocket or bubble containing the captive pathogens. As discussed in Chapter 3, the phagocyte then uses the enzymes in lysosomes to digest what it has eaten.

Phagocytosis is used not only to keep the numbers of invading bacteria down but also to remove some of the body's own worn-out or damaged cells. Because phagocytosis is used against a wide range of targets, it is considered a form of nonspecific resistance.

Interferon

Viruses pose special problems for the body's defenses because they operate as intracellular disease agents (see Chapter 23). Once they invade the cells, they are less vulnerable to attack by white blood cells or the body's other nonspecific defensive measures. **Interferon** is a group of proteins produced by cells that have been invaded by viruses. It acts to delay the spread of viral infections until the systems of specific resistance can take over. It is part of nonspecific resistance because it seems to work against a variety of intracellular problems.

Viral invasion causes the infected cells to produce interferon. The main stimulus triggering the production of interferon seems to be the double-stranded ribonucleic acid (RNA) molecules injected into the victim cell during attack by an RNA-containing virus. The presence of this brand of nucleic acid stimulates the cell's own genes to produce messenger RNA (see Chapter 4), enabling the invaded cell to manufacture interferon. Interferon leaves the infected cells and binds to surface receptors on the host's neighboring cells. The interferon-receptor complex stimulates these uninfected cells to produce enzymes that prevent viral infections. These enzymes remain inactive, unless these cells are likewise invaded by double-stranded RNA from viral intruders.

Interferon slows the spread of viral infections in other ways, too. It inhibits the release of viruses from host cells that are already infected. It also changes viruses released by infected cells in ways that make it difficult for them to invade new cells.

Because many researchers have thought that interferon holds the promise of effective treatment for viral diseases, including the common cold, it has been one of the first human proteins to be produced by the new technology of genetic engineering (see Chapter 4). Now available in quantity, it is being tested against viral diseases, including herpes infections, the common cold, and some cancers. It has proved effective in a few cases, but it has not so far fulfilled the expectations of its researchers.

Inflammation

Inflammation is a general response to a wide range of insults to the body that cause cell damage. The inflammation response can be triggered by a blow, a cut, various types of infections, allergic responses, burns, poisons, exposure to radiation (sunburn, among others), and chemical agents such as poison ivy, insect stings, pollutants, and acids (Figure 24-4).

The damaged cells release various substances, including **histamine,** that cause local blood vessels to dilate and become more permeable. Increasing the local blood supply aids in the removal of toxins and debris and increases the delivery of nutrients and oxygen to the damaged tissues. This inflow of blood leads to swelling and reddening of the inflamed tissues. It also causes those tissues to feel warm to the touch. The increased permeability of the blood vessels allows white blood cells and antibodies to leave them and reach the damaged area. The phagocytes remove debris and bacteria.

The inflamed cells may also release **prostaglandins,** which cause feelings of pain in the damaged area. Since the pain causes some loss of function, it causes the damaged area to be protected and rested, thus speeding healing. Aspirin is often prescribed for the pain associated with inflamed tissues because it inhibits the synthesis of prostaglandins. For that reason it is often taken by people who suffer from arthritis (inflammation of the joints).

When the damage includes broken blood vessels, a blood clot promptly forms (see Chapter 10). However, even when no blood vessel is broken, fibrinogen leaves the blood through the more permeable vessel walls. Fibrin forms among the cells around the injury site and serves to limit the movement of bacteria and their toxins into the rest of the body. The bacteria, the living and dead phagocytes, and the debris from the damaged tissue cells form a pocket of **pus.**

The fluid that escapes the blood and collects in and around the damaged site accounts for the swelling, or **edema.** Swelling, nerve injury, bacterial toxins, and the by-products of cell damage all contribute to the pain of inflammation.

The function of the inflammatory process is to

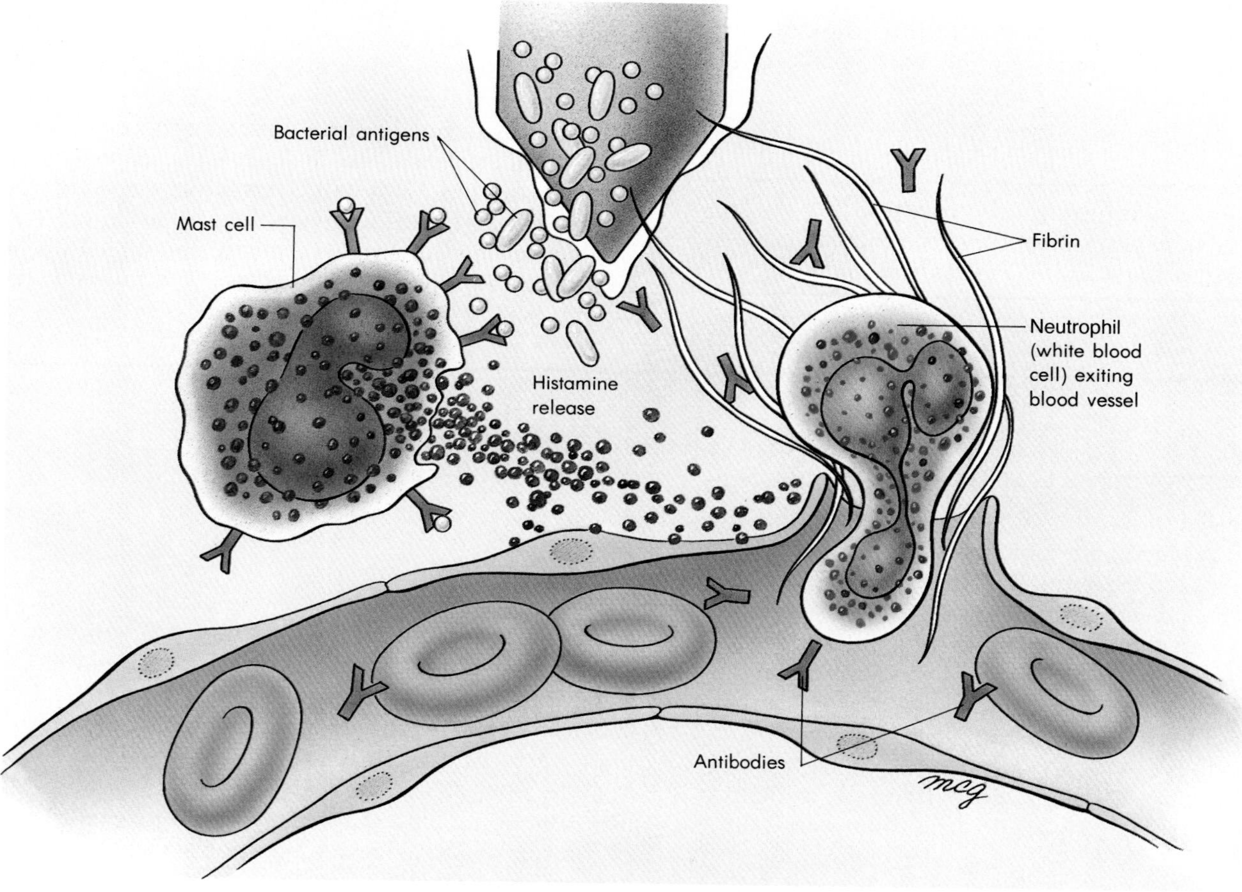

FIGURE 24-4 An inflammation reaction can be set off in numerous ways. In this case, the contaminated nail has injected bacterial contaminants. Proteins (antibodies or receptors: discussed later in the chapter) on the surface of the mast cell react with bacterial components (antigens), setting off a discharge of histamine. The histamine increases the permeability of capillary and venule walls, allowing neutrophils (white blood cells) to leave the circulation and migrate to the site of the infection. Circulating antibodies are also able to leave the blood vessels, as is fibrin, the netlike protein that normally is activated during blood clotting. In the tissues, the fibrin net helps prevent the spread of the infection.

neutralize and remove all harmful substances, including bacteria. Once this is done, the damaged area can be repaired, but the repair process begins before inflammation subsides. Connective tissue cells divide and multiply to fill in damaged areas. If the damage is extensive, as with burns, skin or other tissues may be replaced with **scar tissue.** Unfortunately, scar tissue is sometimes so stiff that it can make movement both painful and difficult (see Box 24-1). If the damage is less extensive, epithelial cells will fill in gaps in skin; epithelial cells can be induced to fill in wider areas, but only with difficulty.

To support the defense and repair processes, the body responds to inflammation and infection just as it does to other stresses. The heartbeat increases to enhance the local blood supply. Hormones command the release of blood sugar from storage and an acceleration of the metabolic rate, making additional energy and metabolic resources available for the repair processes. The increased metabolic rate and the additional warm body fluids moving into the damaged area contribute heat to the site of inflammation.

1. Describe the inflammation reaction.
2. Discuss how the inflammation reaction contributes to nonspecific resistance.

Fever

Another weapon in the body's nonspecific defense arsenal lies in its ability to raise its temperature, as it often does in response to bacterial or viral infections. We call such a temperature rise **fever,** and there is a biological reason for it. It is a response of the body to toxins generated by the infection. In addition, affected cells often enhance the body's fever response by releasing chemicals called **pyrogens,** which reset the body's thermostat, located in the hypothalamus. Besides elevating the body temperature, pyrogens are thought to act against bacterial infections by stimulating the liver to remove iron from the blood. Bacteria need iron to grow properly. Fever apparently makes the body less hospitable to the disease organisms. These organisms have evolved to function best at particular temperatures; they seem to prefer normal human body temperature. The fever also probably speeds up the body's defenses, the same way heat speeds up chemical reactions in a laboratory test tube.

We are currently reexamining the way we think about fever. Until recently, it was popular to break out the aspirin bottle at the first sign of an elevated body temperature. The human body can survive fevers as high as 110°F (43°C) for short times. It was believed that fevers approaching these dangerous levels offered the threat of brain damage. Recent studies have indicated that fevers (even high fevers)

Box 24-1 Skin Replacement for Burns

Extensive burns offer physicians one of their most troublesome problems. When large areas of skin are destroyed, bacteria can invade the body easily, leading to widespread, potentially deadly infections. At the same time, when the body loses much of its skin, it also loses the ability to control the loss of fluids by evaporation and even seepage. Burn victims can thus die of dehydration.

Antibiotics, creams, and bandages can help. But the best help comes from an intact layer of skin, and modern burn specialists spend much of their effort restoring that layer. If a patient has some intact skin, physicians can graft it in strips and dots to the damaged areas. As skin cells multiply, both the graft donor sites and the graft recipient sites fill in. However, skin growth takes time, and the most severely burned patients may have too little skin left to provide grafts for the whole body. In fact, only the skin of their armpits and other sheltered regions may be undamaged.

In such cases, physicians have used skin from cadavers and pigs, artificial skin made from the connective tissue protein collagen, and cultured skin. They make cultured skin by taking some of the patient's few remaining skin cells, dispersing them on a membrane, and inducing them to multiply and spread out. Within days, they can have enough cultured skin to cover all or most of the burn victim's damaged areas. This aids the healing process and minimizes the formation of stiff, unsightly scar tissue.

The process is not yet perfect, but it is bound to improve. As it does so, the day will near when the most difficult task in treating a burn victim is keeping the patient alive until the new skin is ready.

are not likely to produce brain damage unless they are part of a disease that directly attacks the brain. Further, it has been shown that we may be much better off if we do not take aspirin to alleviate a fever. Instead, if we allow the fever to run its course, then the duration of the disease is likely to be much shorter and the chance of life-threatening complications is much lower. Our hypothalamic thermostat resets to normal as soon as the pyrogens vanish from the body.

1. What is fever and how does it contribute to nonspecific resistance?
2. Why should we not be too ready to employ fever-suppressing drugs, such as aspirin?

SPECIFIC RESISTANCE

Infectious agents occasionally overpower the body's systems of nonspecific resistance. When they do, the body has a very sophisticated additional defense system, called *specific resistance.* Specific resistance uses two main lines of defense (Figure 24-5). The first is **cell-mediated immunity.** It depends on the ability of certain white blood cells—the **T-cell** lymphocytes and phagocytes—to identify materials that do not belong in the body and to remove or destroy them. These cells are always present in the blood, tissues, lymph nodes, and other lymphatic tissues (such as the tonsils, spleen, and parts of the intestine), ready to attack any material they might identify as "nonself." Detection of such an invader sets off a wide range of responses that ultimately results in its isolation, attack, and removal.

The second form of specific immunity is **humoral immunity.** It depends on the ability of those lymphocytes known as **B cells** to produce and release specific protein molecules, the antibodies, which operate by complexing with antigens carried on the surfaces of most foreign invaders.

Antigens

The key to identifying nonself material is molecular shape. Every molecule has its own unique

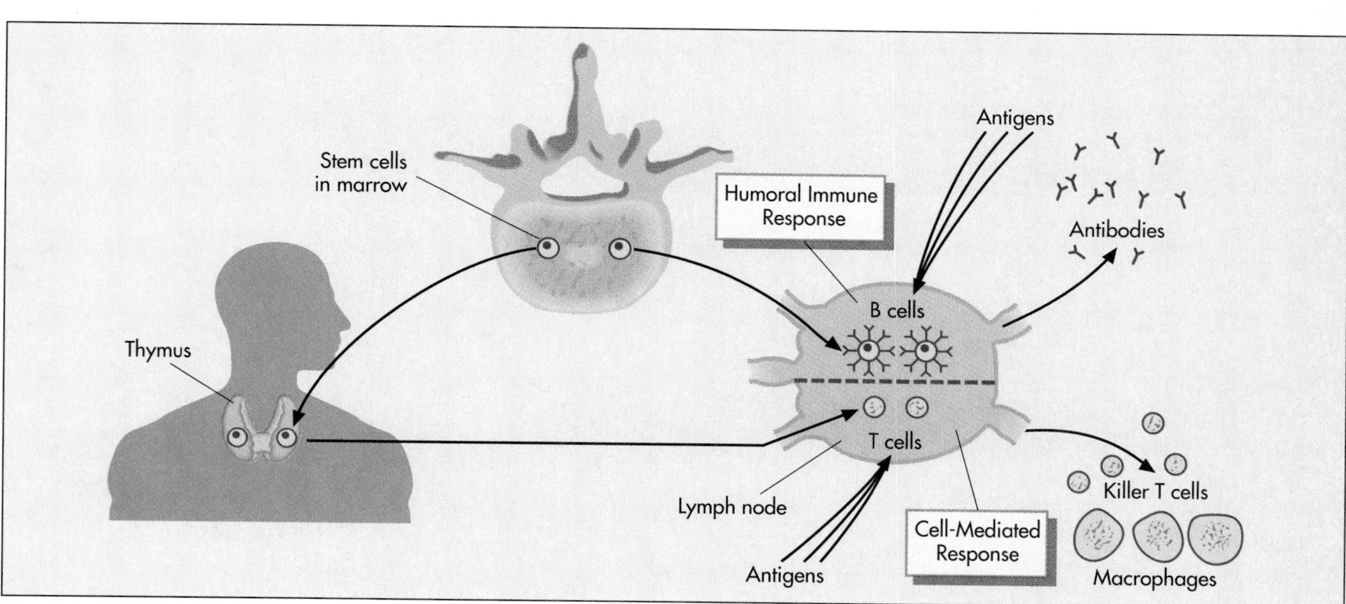

FIGURE 24-5 There are two major divisions in the body's specific resistance systems: The humoral immune response (also called the antibody system) and the cell-mediated response. The principal combatants in these systems are the lymphocytes. The T-cell lymphocytes play the major role in the cell-mediated immune response and B cells in the humoral (antibody) response. Both types of lymphocytes originate from dividing stem cells in the bone marrow. About half of these cells migrate to the thymus gland where they are converted to T cells (*T* for thymus). The other half become the B-cell lymphocytes. Eventually both B and T cells concentrate in the lymph nodes. If they are exposed to the specific antigen they are sensitized to, both cell types begin dividing. Eventually an army of cytotoxic (killer) T cells is produced that destroys the invaders. The B cells respond by producing substantial quantities of specific antibodies and releasing them to locate and inactivate the antigens.

shape, and the shapes of molecules differ according to the molecules' compositions. Proteins, in particular, differ from species to species and even from individual to individual within a species. That is, although two people have similar proteins in their bodies, these proteins are very likely to differ slightly in amino acid sequence and hence in shape.

Since proteins dot the surfaces of all cells, they offer a system of labels that permits cells to be identified as "self" or nonself. Such recognizable proteins are called **antigens.** Antigens are usually cell surface proteins and glycoproteins, but they can be lipoproteins, nucleoproteins, and in rare cases, nucleic acids released from damaged cells or viruses. They can also be bacterial toxins, snake venoms, and other molecules. They are generally large molecules with minimum molecular weights ranging from 8000 to 10,000 daltons. (A dalton is the mass of a single proton or neutron. A molecule of water has a molecular weight of 18 daltons; a molecule of glucose has a molecular weight of 180 daltons.)

Lymphocytes "scan" the proteins on the surfaces of each of the particles or cells they encounter during their chance collisions as they move about the body (Figure 24-6). If they recognize these proteins as nonself, they attack. Lymphocytes can "read" these labels because their cell membranes bear proteins, recognition factors, whose shapes fit the foreign antigens. There are also lymphocytes carrying "reader" proteins on their surfaces that fit the body's own cells, but these are usually suppressed and do not trigger a defensive response when they encounter proteins designated as self.

Antibodies

The chemicals the body actually uses to recognize these foreign antigens are a class of proteins called **antibodies.** These antibodies recognize antigens in much the same way enzymes recognize their substrates, by molecular fit (see Chapter 3). Each one binds to a small portion of an antigen; in the case of protein antigens, this portion is 3–10 amino acids long. Several different antibodies may thus bind to several different portions of an antigen. Because the binding depends on shape and there is an almost limitless number of possible shapes, the body must be able to make antibodies tailored to fit virtually any antigen. We are only beginning to understand how the relatively few genes involved in antibody formation can provide the structural variety

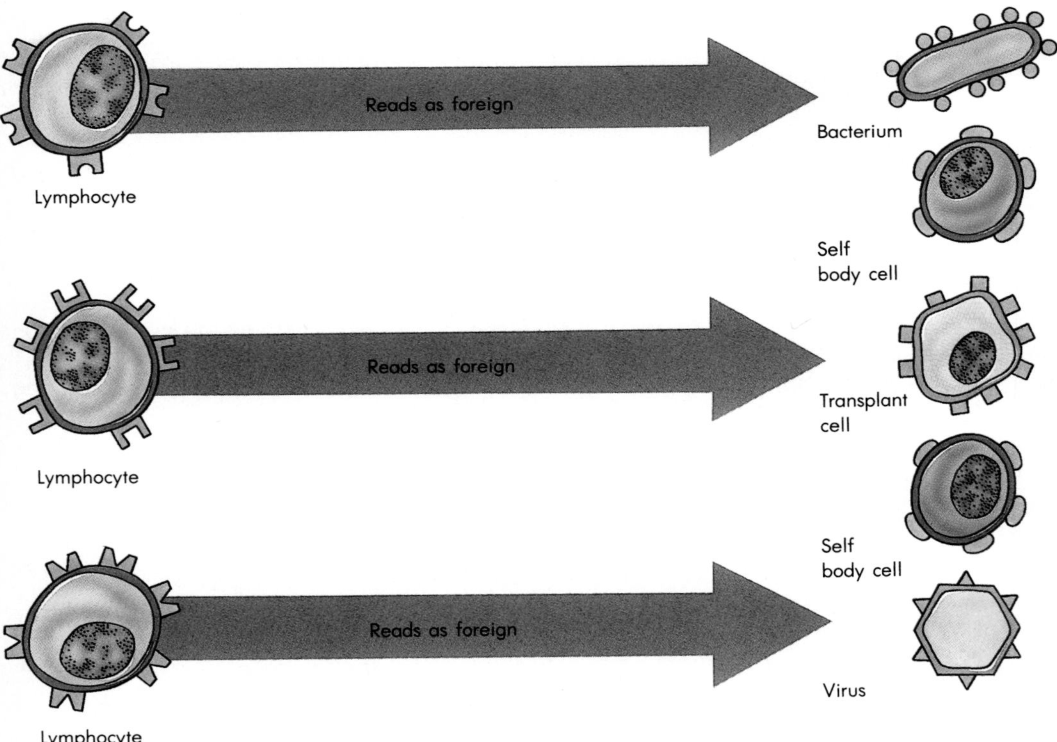

FIGURE 24-6 Lymphocytes are continuously "patting down" the cells (and debris) they encounter. If the surface recognition factors on the cells they are examining fit the receptors on the lymphocyte surfaces, those cells are designated foreign and are attacked by the lymphocytes.

of antibodies needed for the body's defense (more about that later).

Antibody Shape

Antibodies are multiple-unit proteins shaped rather like the letter Y. They are produced by a class of lymphocytes called B cells (see Figure 24-5). Antibodies are constructed of two long chains and two short chains of amino acids, as shown in Figure 24-7. Each chain has a "constant" region that is the same for all antibodies of a given category (there are several categories of antibodies); the constant part of the antibody is usually found in the lower part of the Y. The sequences of amino acids in the spreading arms of the Y are fitted to the specific types of antigen the antibody is responsible for recognizing (Figure 24-8). This portion of the antibody is called the variable region. It is *so* variable that there are separate, distinct antibodies for each of the several million different antigens the body might be exposed to during a lifetime. At birth, the body contains a few B cells capable of responding to each antigen. On later exposure to an antigen, the corresponding B cells multiply by cloning themselves and produce substantial quantities of their particular brand of antibody.

---

1. What is the difference between antigens and antibodies?
2. How does specific resistance differ from nonspecific?

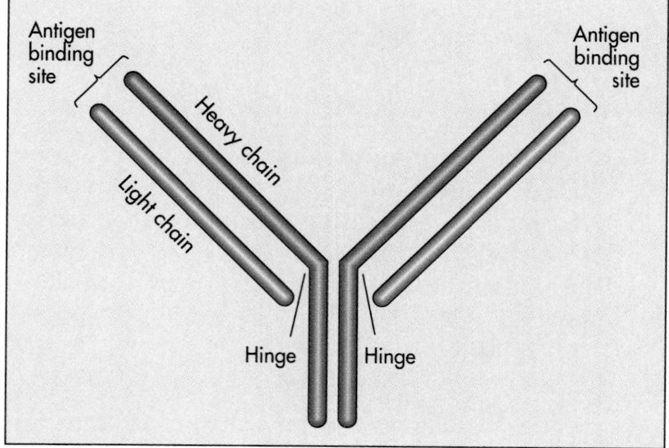

FIGURE 24-7 Most antibodies consist of four chains of amino acids arranged in a Y configuration. The lower parts of the Y are the constant portions of the molecules. The variable upper ends of the spreading arms of the Y are shaped so the antibody specifically complexes with "its" antigen.

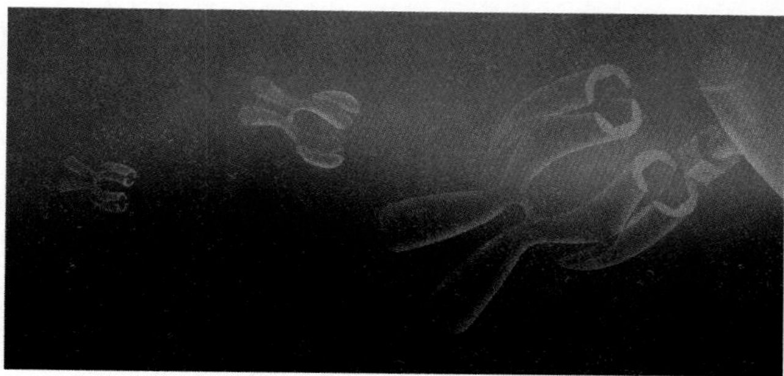

FIGURE 24-8 An antibody complexing with its antigen. Note the fit between the ends of the antibody arms and the antigen.

Gamma Globulins, or Immunoglobulins

Once formed, the antibodies either circulate freely in the plasma and lymph or become attached to the surfaces of lymphocytes and mast cells, where they help these cells recognize troublesome antigens.

Antibodies were first known as the blood proteins called the **gamma globulins,** or **immunoglobulins** (see Chapter 10). Each time the body develops a new type of antibody, it joins the immunoglobulins in the plasma and lymph. Thus, by identifying the particular antibodies in a person's blood, physicians and researchers can often tell what antigens and diseases a person has been exposed to in the past. Tests for antibodies to syphilis bacteria, the AIDS virus (HIV), hepatitis viruses, and other disease organisms help screen blood donors and keep the blood used in transfusions as safe as possible.

Blood-borne antibodies also support the processes of **passive immunity.** Some diseases strike so quickly that victims have no time to "actively" produce the antibodies against them. In such cases, it is possible to take blood from a person or animal (often a horse) that has previously been exposed to that particular disease or antigen. Allowing this blood to clot removes cells and some proteins from the blood, leaving a straw-colored fluid called serum. Serum is rich in antibodies, including large amounts of antibodies to the specific disease or antigen. Plasma is also rich in antibodies, and it can be used in much the same way as serum. Plasma can be prepared by spinning a blood sample at high speed in a centrifuge; centrifugal force presses the heavier blood cells to the bottom of the sample, leaving the plasma on top.

Injected into a person as "gamma globulin shots," the concentrated solutions of "ready-made" antibodies in serum are instantly available to help fight disease. They are frequently used to protect the fetus, as when a woman has been exposed to German measles during the first trimester (three months) of her pregnancy. They are also used to treat snakebite; antivenin is serum prepared from the blood of an animal exposed to snake venom. Unfortunately, since the patient's body is not continually manufacturing these antibodies, they quickly disappear from the blood. Thus, the immunity provided by gamma globulin shots does not last long.

We can see a natural example of passive immunity in the transfer of antibodies across the placenta, from the mother to the fetus, just before birth. Armed with these maternal antibodies, the newborn gains partial protection for the very vulnerable first few months of life. In addition, the mother continues to pass antibodies to her nursing baby through her breast milk.

Antibody Action

Antibodies work by binding to antigens. If the antigen is a toxin or venom, the antibodies interfere with its toxicity. If the antigen is on a cell or virus, the antibodies complex with it, marking it for destruction by the phagocytes. In addition, antibodies can help complement (see below) destroy foreign cells. When the branches of the antibody molecule have attached to a foreign cell's surface, the stem of the molecule exposes a site that can bind complement. When complement binds, it can then begin to form a hole in the foreign cell's membrane.

Many disease agents are difficult for phagocytes to capture and ingest. One class of antibodies complexes with antigens to form handlelike attachments, making it easy for the phagocytes to capture them. Another group reacts with several antigen-bearing cells or particles at once, causing them to form large clusters of antigenic material, again making it easy for the phagocytes to capture them.

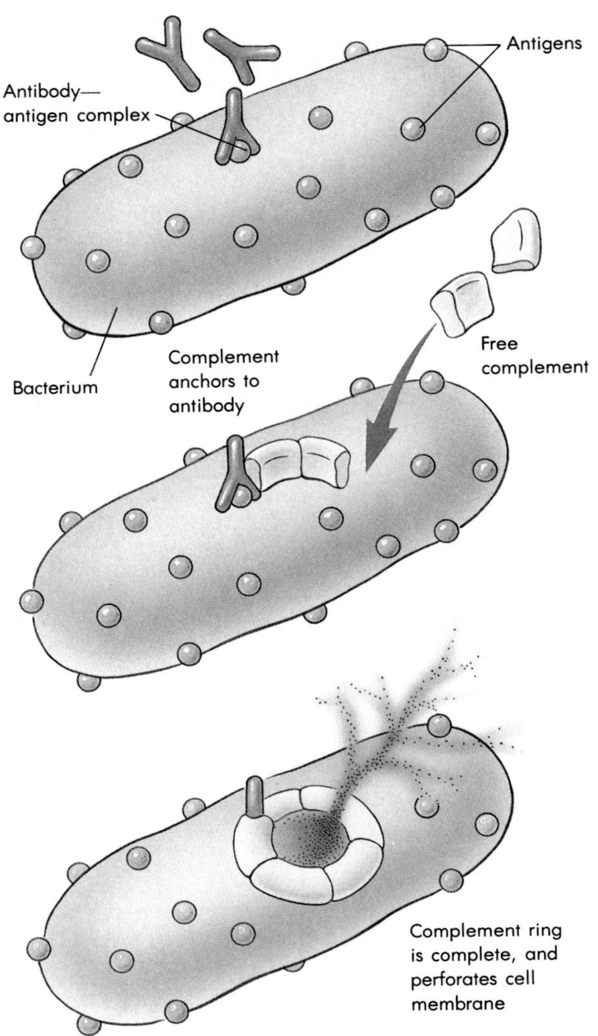

FIGURE 24-9 Complement works in conjunction with circulating antibodies to perforate bacterial cell membranes. The antibody first binds to the foreign antigen. The antibody-antigen complex serves as the first point of attachment for the complement molecules, which attach in sequence until the final complement molecule completes the perforation.

Complement

Complement is a group of 11 blood proteins that interact with antibodies to serve several functions. When activated by attaching to antibodies that have already bound antigen, some of these proteins punch leaky holes in bacterial cell membranes, thus killing the cells (Figures 24-9 and 24-10). Others bind to receptors on the membranes of white blood cells and stimulate phagocytosis of bacteria. Some serve to attract white blood cells to an area of infection. Some stimulate the release of histamine from mast cells, white blood cells, and platelets, thus helping the development of the inflammatory response.

---

1. What are the immunoglobulins? How are they related to the gamma globulins?
2. Describe how the different classes of antibodies work.
3. Distinguish between interferon and complement.

Origins of T Cells and B Cells

At any given time, there are about a trillion lymphocytes (about 2 pounds) in a human body. They

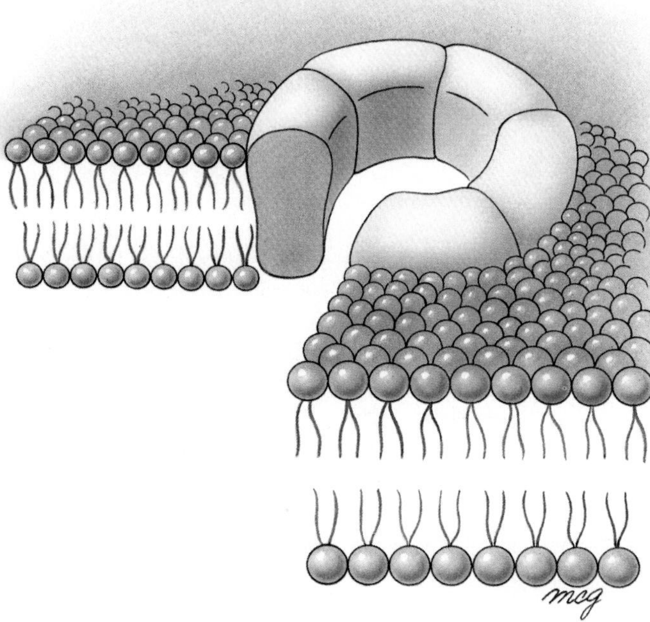

FIGURE 24-10 Magnified version of a complement lesion through a bacterial membrane.

arise from **stem cells** in the red bone marrow at a rate of about 200,000 per second. As they mature, they move to the blood and lymphatic system. About half the lymphocytes travel to the thymus gland, where they are converted to T-cell lymphocytes. The T cells eventually become our principal defense against viruses and fungi.

The other cells become B-cell lymphocytes. It is not sure where they develop, although some people feel they develop in the liver, spleen, or intestinal wall; the prevailing opinion is that they probably complete their maturation in the bone marrow. The B cells are responsible for antibody-mediated, or humoral, immunity.

Before birth and in the years immediately thereafter, the body generates B and T cells that can respond to virtually any conceivable antigen, both self and nonself. For both types of cells, the maturation process involves the suppression or killing of those cells that recognize self (this prevents those cells from attacking one's own tissues) and a proliferation of those cells involved in protecting against nonself antigens. Thus, the body's arsenal is stocked against foreign encroachments.

B Cells

At any one time, the body holds a complete set of B cells that can quickly produce a specific antibody to virtually any specific antigen. Each B cell carries attached to its cell membrane 100,000 or so molecules of the antibody it can produce, but it does not normally secrete antibody. It must first be activated by exposure to the appropriate antigen.

The B cells are most often activated when the macrophages, phagocytic connective tissue cells, consume antigen-bearing foreign cells such as bacteria. These macrophages break down the foreign cells and mount the antigens they contain on their own cell membranes. They then "present" these antigens to the B cells. Only the appropriate B cell's membrane antibodies or receptors will bind to the antigens on the macrophage. When the connection is made, the matching B cell enlarges and begins to multiply. Many of its daughter cells become **plasma cells,** which secrete great quantities of the specific antibody into the blood. A few of its daughter cells become the next generation of **memory cells,** which do not produce antibody but rather remain in circulation to initiate another full-blown response if that antigen is encountered again in the future. Once the body sets up memory cells for a specific antigen, the immunity against that antigen is usually life long, because individual memory cells are thought to be able to live for decades.

Active Immunity

Each system of specific resistance takes several days to several weeks to develop after the first exposure. However, each exposure to an antigen increases the number of B cells that can respond to it (Figure 24-11, p. 566). Thus, protection develops much faster on subsequent exposures to the same antigen and a disease is often defeated before symptoms can develop. If the body manufactures its own antibodies, as a consequence of being exposed to the antigen, the immunity that results is called **active immunity.**

For many diseases, the body can successfully defend itself and set up its own natural immunity before the disease kills the patient. However, some diseases are so deadly that they weaken and kill the patient before these defenses can be established. For these diseases, it is advantageous to set up the antibody systems to deal with them in advance, just in case the body is exposed. We do this by means of vaccination.

Vaccines

Today we prevent many diseases by providing the necessary first exposure to an antigen in the form of

Learning Focus

B-Cell Activation and the Humoral Response

Study the accompanying diagram carefully. (After you have mastered the information, go to the Learning Focus Response and practice what you have learned by filling in the blank version of the diagram.) Beginning in the upper left portion of the diagram, circulating macrophages encounter and phagocytize antigen-bearing infectious agents. After these foreign agents are digested, their antigens are moved to the macrophage's surface and presented to each B cell the macrophage encounters. Each B cell identifies the antibody it can produce by wearing copies of that antibody embedded in its cell membrane. Once the match is found, complexing with the correct antigen sets the B cell off. It enlarges and begins to divide, rapidly producing two types of cells: memory cells and plasma cells. The memory cells are copies of the parent B cell; these cells do not produce antibody, but they do remain in the circulation and go into action later if the same antigen is ever encountered again. The remaining daughter cells become plasma cells, the antibody factories. They produce large quantities of the specific antibody and release it into the circulation where it can meet and bind to the antigen.

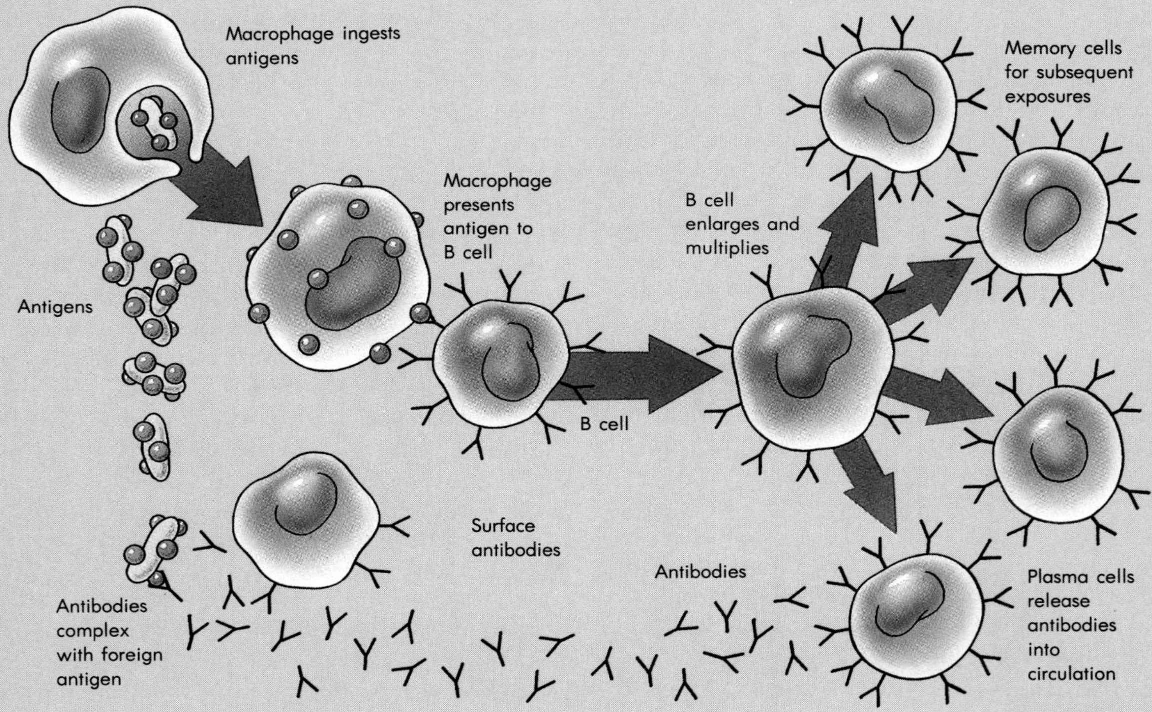

Learning Focus Response

B-Cell Activation and the Humoral Response

Make a few copies of the diagram shown here and use it to summarize and reinforce all that you know about B cells—why they activate, how they relate to foreign antigens, what the macrophages do, how long-term immunity is produced, what are the plasma cells and the memory cells. To see how you are doing, refer to the version of the figure in the Learning Focus. Also, add any interesting tidbits to the diagram that you may uncover from the text or from class lectures. Then repeat the exercise. Go over it again when you are preparing for exams.

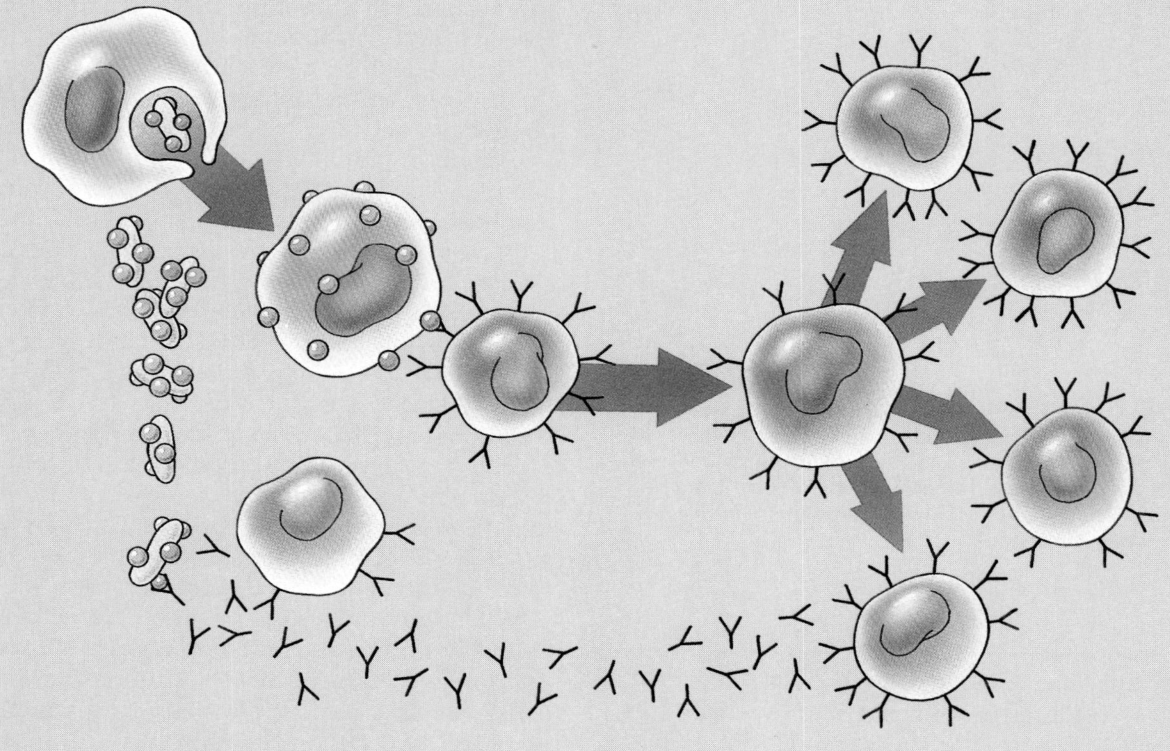

a **vaccine,** a preparation of dead or weakened bacteria, viruses, or toxins, of proteins extracted from them. Traditionally, vaccines are weakened or inactivated, a process referred to as *attenuation,* by drying the infectious agent or exposing it to ultraviolet radiation before injecting it into the patient.

Some vaccines are now being made by genetic engineering. The gene for a surface protein is removed from a disease agent and transplanted into a host bacterium, which produces the surface protein in quantity. The antigen is purified; since it is no longer associated with the disease agent, it makes a much safer vaccine. The immune system then constructs an antibody that protects the individual from a later invasion by the real agent. In addition to being safer, it is also cheaper, since many disease agents are difficult to grow in quantities necessary to produce large amounts of vaccine.

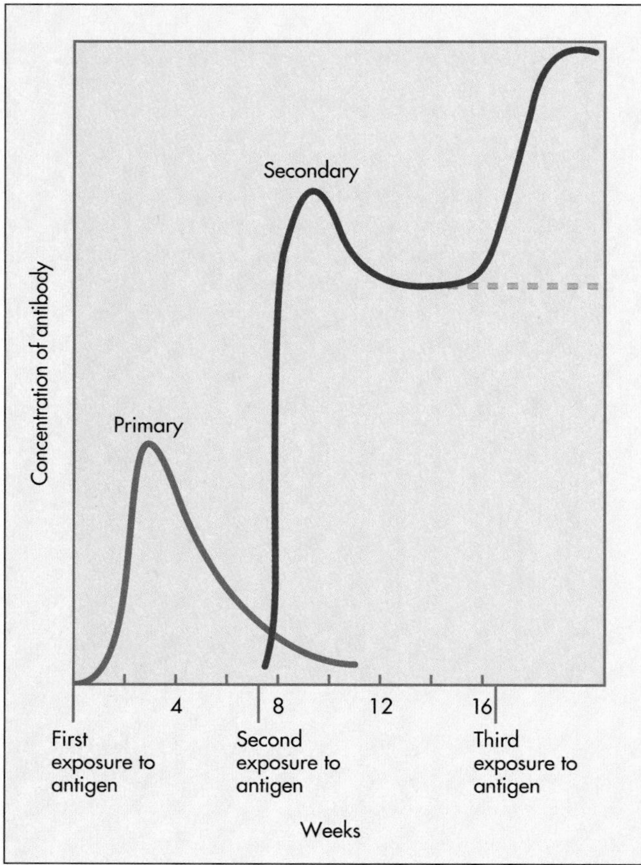

FIGURE 24-11 Each time the body is exposed to a specific antigen, the antibody level for that particular antigen increases significantly. Each exposure also increases the population of memory B cells for that antigen, so that each subsequent response is faster and more substantial. This is the principle behind booster injections for vaccines.

1. Why must vaccines contain antigen?
2. What is the relationship of the antigen in a vaccine to the disease the vaccine is intended to prevent?
3. Why do vaccines offer lasting protection against disease?

T Cells

The T cells compose the other half of the body's lymphocytes. They differ from B cells in several ways. Their surfaces are smoother, their membrane receptors are not antibodies, and they do not secrete antibodies.

The T cells get their name because they mature largely in the thymus gland. However, many may mature partly in the skin, which is related in structure and origin to the thymus. In the skin, T cells seem to become sensitized to antigens that come into contact with the skin. Because ultraviolet light interferes with this process and thus impairs the immune system's defense functions, the skin's ability to tan—and thus to protect the skin's T cells from the sun—is especially important to our continued health.

The T cells are responsible for cell-mediated immunity. Like B cells, once activated by antigens presented by macrophages, they enlarge and multiply. However, the daughter cells do not differentiate into only two types. Instead, as shown in Figure 24-12 (p. 569), they form separate groups of *memory, cytotoxic (killer), helper,* and *suppressor* T cells.

Once again, *memory T cells* serve to enhance future responses to the same antigen. *Cytotoxic (killer) T cells* secrete substances that kill invading cells. They also secrete a *macrophage activating factor* that attracts macrophages and stimulates their phagocytic activity.

Cytotoxic (killer) T cells are most effective against slowly developing bacterial infections, fungi, transplant cells, and cancer cells. In fact, T cells may actually discover and destroy cancerous cells many times during an individual's life. Because many types of cancer cells undergo changes in their surface factors, the T cells are able to identify and kill them before they cause problems. Patrolling T cells thus form part of what is called the **immunosurveillance system.** Some cancers, however, do not develop unique antigens or for another reason are not detected or removed in time. Cytotoxic T cells also destroy cells infected by viruses, since those cells are often marked by viral antigens that have migrated to the cell's surface.

When activated, *helper T cells* release chemicals called lymphokines that stimulate the rapid production and maturation of additional T cells and B cells. Some lymphokines stimulate macrophage phagocytosis. Other lymphokines stimulate stem cells to produce more lymphocytes. Helper T cells thus orchestrate a significant part of the immune system's response to an antigen-based threat.

The *suppressor T cells* have a very different function. Their job is to shut down the immune system after the threat has been defeated and to keep the immune system from accidentally turning on the body's own tissues. When they fail to function properly, the result can be an *autoimmune* disease (see what follows).

The T cell's membrane receptor for antigens is not an antibody. However, the receptor molecule does resemble an antibody in that it has both constant and variable regions. In addition, the many variations in the receptor molecule that allow different receptors to match up with different anti-

Learning Focus

T-CELL ACTIVATION

Macrophages take in foreign antigens, digest them, and move them to their surface membranes. They present the antigens to T cells with the appropriate receptors. The activated T cells enlarge and begin to multiply; they produce four different types of T cells—cytotoxic (killer), suppressor, helper, and memory T cells.

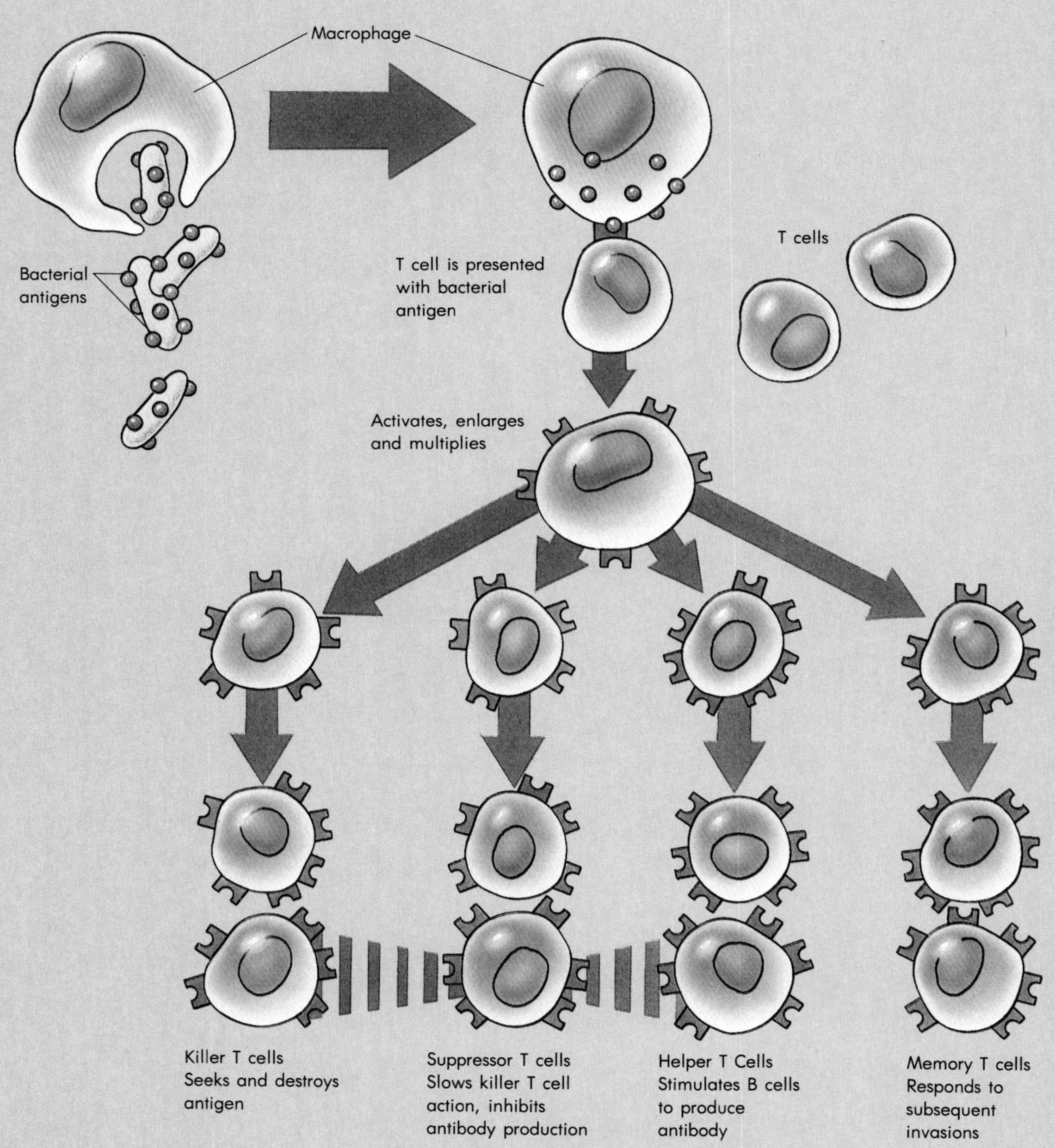

Learning Focus Response

T-Cell Activation

Make a few copies of the diagram shown here and use them to summarize the events of T-cell activation. Indicate the role of each of the four types of T cells. Check your progress by referring to the version of the figure in the Learning Focus. Also add any notes that you might have picked up from the text or class lecture notes. Then repeat the exercise. Go over it again when you are preparing for exams.

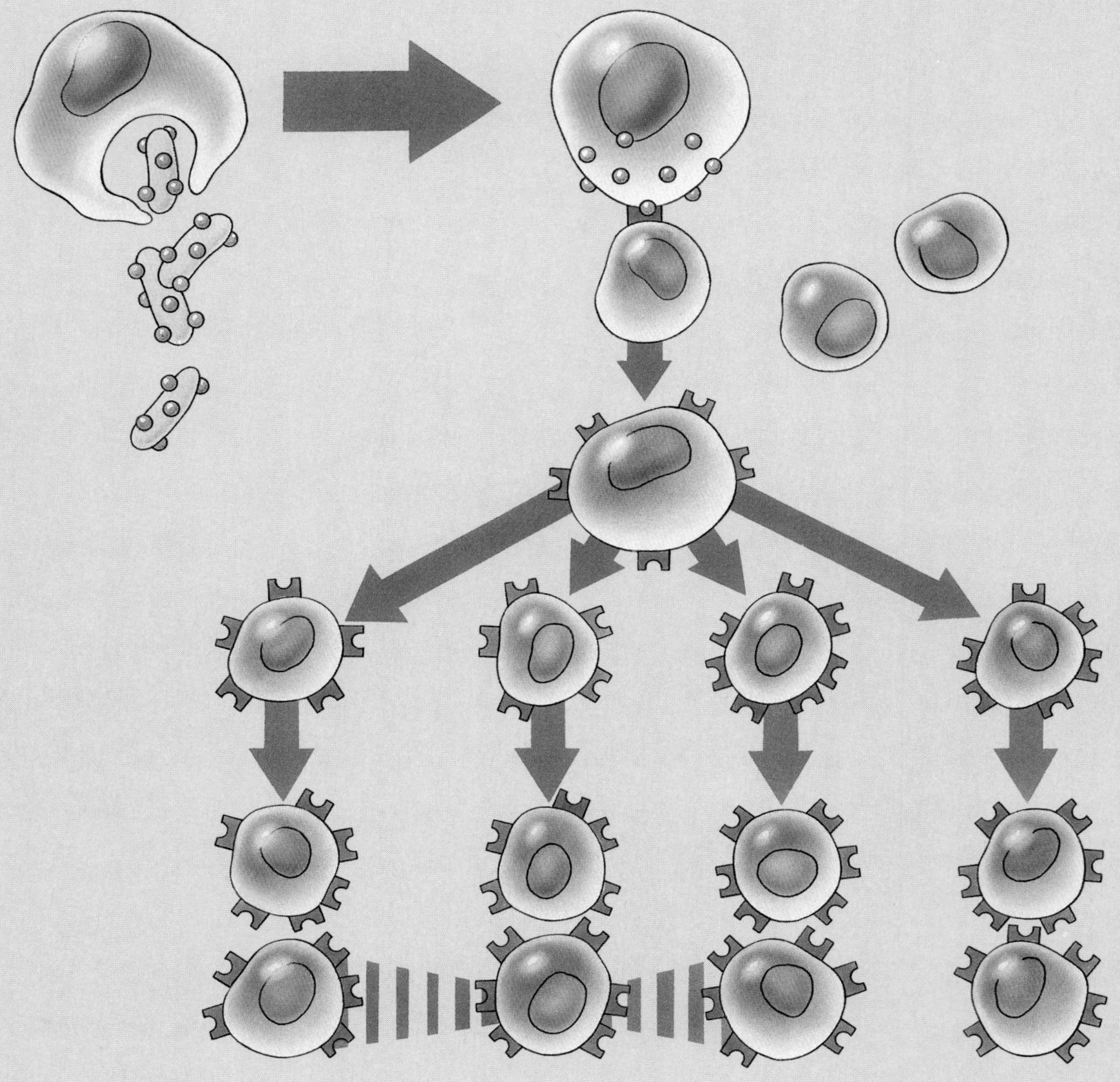

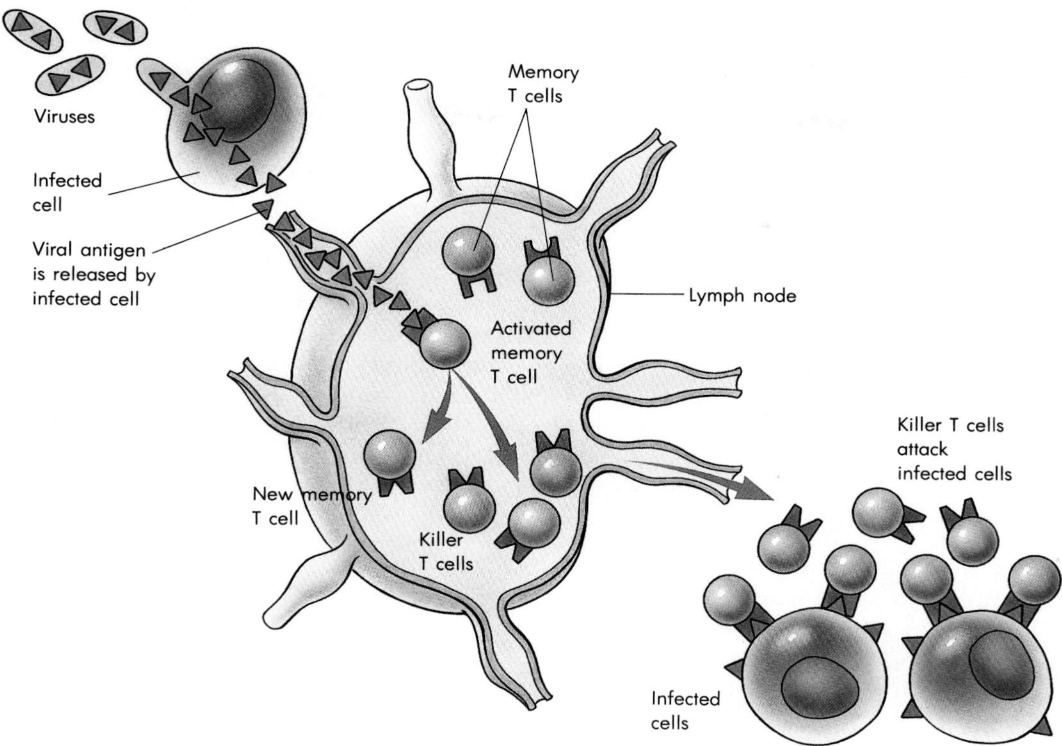

FIGURE 24-12 Cell-mediated immunity. The cytotoxic (killer) T cells are particularly effective at defending the body against viral infections. The viruses invade body cells. Some viral antigens reach the infected cell's surface and are eventually released by the infected cell. In the lymph nodes these viral antigens are intercepted by the appropriate memory T cell. The memory cell begins dividing, producing more similar memory cells and an army of cytotoxic (killer) T cells. The cytotoxic T cells search for other body cells showing the same viral antigens on their surfaces and destroy these infected cells.

gens seem to arise in the same way as they do for antibodies. We might think that the thousands of variations, since they amount to thousands of different protein molecules, would call for thousands of distinct genes, one for each variation. However, there are only a few genes. The variations for both B cells and T cells arise early in their development. The few genes interchange segments in a process similar to the crossing over of meiosis (see Chapter 5), thus generating all the necessary variants.

1. Distinguish between T cells and B cells.
2. How are they formed and what does each cell type contribute to immunity?

IMMUNITY AND SYMPTOMS

At times, a disease's symptoms are due less to the effects of a bacterium or virus than to the immune system's efforts to fight off the disease. In mice, the viral disease lymphocytic choriomeningitis is fatal because of the action of T cells and not because of the direct action of the virus. For several days after infection, the virus multiplies freely in many organs. Once antibodies to the virus begin to appear, however, the mouse develops an inflammation of the membranes surrounding the brain (the meninges) and dies. The meningitis occurs because the T cells attack cells that carry viral antigens in their cell membranes. The attack interferes with viral reproduction, but it also causes the release of cell damage products that trigger an inflammatory reaction, which causes changes in fluid distribution and blood pressure, pain, and even death. The disease's symptoms can be prevented by destroying the mouse's immune system before infection. Injecting the mouse with B cells from an immunized mouse then has no effect, but injecting T cells from an immunized mouse is promptly fatal.

Similar interactions between T cells and virus-infected body cells seem to account for the symptoms of many diseases, including mumps, measles, hepatitis, and even herpes. Antibodies alone can cause symptoms, for they can bind to viral antigens in cell membranes and let complement destroy the infected cells. In addition, they can bind virus particles, toxin molecules, and bacterial cells into clumps that can clog small blood vessels and damage the kidneys, heart, and brain.

ALLERGY

The symptoms of **allergy** are due entirely to the actions of the immune system. Antigens that trigger allergic responses are called **allergens.** They generally do not stimulate the B cells to produce antibodies as strongly as other antigens do. A first exposure to an allergen does not provoke an allergic reaction, but it does lead to the production of antibodies. The stems of these antibodies bind to the surfaces of some of the body's own cells, especially mast cells, leaving the antigen-binding tips of the antibody "Y" exposed. On a later exposure, allergen molecules bind to the antibodies on the mast cells and cause these cells to release large quantities of substances such as histamine. The histamine release stimulated by an injected allergen, such as an insect sting or an antibiotic, may lead to a general inflammation throughout the body—swollen, flushed skin, dilated blood vessels, low blood pressure, increased mucus production, constricted respiratory passages, and hives. When severe, this general inflammation takes the form of **anaphylactic shock;** it can be fatal.

More often, the inflammation is a local disturbance, appropriately called **local anaphylaxis.** It occurs in the skin and the mucous membranes of the nose, sinuses, and respiratory passages where most mast cells are found. It accounts for the sneezing episodes often seen in hay fever victims when the pollens they are sensitive to are in season or in people sensitive to animal dander when they visit friends with pets.

1. What are allergens? How do they differ from typical antigens?
2. Describe the events in an allergic reaction.
3. How does anaphylactic shock differ from local anaphylaxis?

Box 24-2 How the Human Immunodeficiency Virus Attacks the Body

Practically every day, you hear or read something about the deadly disease, AIDS. It is mentioned on television, on the radio, and in the daily newspaper. One day a friend, or worse yet, a family member may get it. And it is unlikely that it will go away within your lifetime.

Although AIDS may have existed for much of this century, it was first recognized only in 1981 among male homosexuals and among Haitian immigrants to the United States. By February 1991, over 164,000 cases of AIDS had been identified in this country. Over 100,000 had already died. Over a million Americans were estimated to harbor the virus. By 1989 it had already become the number 2 killer of men between the ages of 25 and 44, second only to accidental death (including murder).

If those statistics do not seem horrifying enough, consider that the United Nations' World Health Organization reports 340,000 documented AIDS cases worldwide and estimates that the true figure is about 1 million. At the same time, 8–10 million people are infected with the virus, and in some parts of the world one pregnant woman in ten is infected. By early in the twenty-first century, experts say, urban infection rates could be as high as 16 percent, and 40 percent for people in their 30s. Up to a fifth of the African work force could die of AIDS, and over the next decade the disease could produce 10 million orphans on that continent. James Curran, director of the Division of HIV/AIDS at the Centers for Disease Control in Atlanta, Georgia, says, "Things are going to get much, much worse before they get better."

By the mid-1980s, it was believed that people diagnosed with AIDS would die within two years of the diagnosis. Now, however, the lives of patients in the developed nations are being prolonged with drugs. The first drug to help AIDS patients was azidothymidine (AZT), which blocks the reverse transcription of the viral RNA to DNA in the patients' cells. (See Chapter 23 for more on retroviruses.)

The virus attacks both macrophages and helper T cells, readily killing the latter. It thus deprives the body of the immune response that is normally orchestrated by helper T cells. Because the virus less often kills the macrophages, these cells both protect it from immune attack and help it spread throughout the body, even into the brain (where it damages brain cells) and into the bone marrow (where it destroys the stem cells that produce more T cells and macrophages). Figure 1 shows the pathway taken by the AIDS retrovirus.

This frontal assault on the prime players of the immune system leaves the victim susceptible to numerous infections and some cancers. In fact, one of the signs that led to the discovery of AIDS was a sudden increase in the incidence of a rare form of pneumonia caused by the protozoan *Pneumocystis carinii*. The protozoan is very common, but it causes disease only in people whose immune systems are weakened.

Another sign that led to the discovery of AIDS was a sudden increase in the incidence of the once rare Kaposi's sarcoma, a cancer marked by painless bluish or brownish nodules in the skin. It affects 15 percent of AIDS patients and is about 20,000 times as common in this group as it is in the general population. It seems to develop because the HIV virus's protein stimulates the growth of certain cells in the skin; some researchers, however, believe that it is due to a second sexually transmissible disease organism, partly because Kaposi's sarcoma seems to be more common in people who acquired their HIV infection through sexual contact.

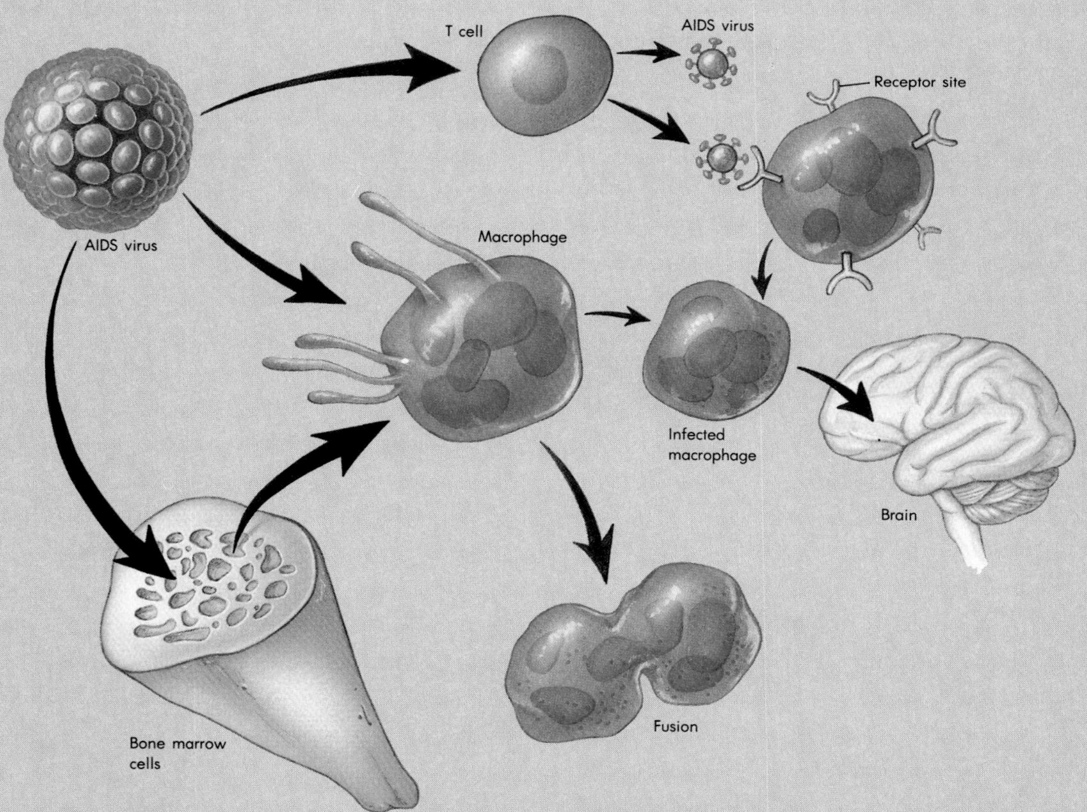

FIGURE 1 The AIDS retrovirus (HIV) enters the body through breaks in the skin. Once inside, it invades T cells, macrophages, and the bone marrow cells that give rise to T cells and macrophages. It enters these cells either by being phagocytized or by attaching to receptors on the cells' cell membranes. Inside macrophages, the HIV particles multiply and are stored in vacuoles. The virus-laden macrophages move about the body, passing the viruses to other macrophages and T cells by fusing with them. As long as the viruses remain inside the macrophages, they are protected from attack by antibodies. Researchers believe that infected macrophages may also migrate into the brain and transfer virus particles to brain cells, accounting for the neurological complications of AIDS. The viruses can also spread from cell to cell by escaping from infected T cells.

IMMUNE SYSTEM PROBLEMS

The immune system seems overzealous in its defense of the body when it recognizes antigens on the surfaces of cells in a transplanted kidney or heart as foreign and attacks them, leading to *rejection* of the organ (Figure 24-13). Patients sometimes need replacement parts in order to live, and we might wish that the immune system could recognize the need and leave transplants alone. Yet the immune system is not overreacting; rather it is just doing its job of removing nonself materials from the body.

Transplant Rejection

Physicians have long sought ways to fight or prevent the rejection of transplants. They have suppressed the immune system by destroying it with radiation and inhibiting it with drugs. Unfortunately, complete suppression of the immune system opens the body to invasion by bacteria and viruses. By shutting down immunosurveillance, it also dramatically increases the risks from undetected cancer.

The "immunosuppressant" drug cyclosporine has proved a highly selective inhibitor of T cells, which are responsible for the bulk of the rejection process. Its use has dramatically improved the success rate in organ transplants; studies have shown that such drugs are equally helpful in preventing transplant rejection whether the donor is a close relative or a stranger to the recipient. However, physicians still seek organ donors that are as closely matched to the recipient as possible; that is, they try to find donors whose cells bear the same **histocompatibility antigens** as those of the recipient. These antigens define tissue types in much the same way red blood cell surface antigens define blood types.

Cancer

Genuine failures in the immune system occur in many cancer cases. A cancer begins as a single mutated, virus-infected, or otherwise changed cell. This cell multiplies wildly, often to form a tumor whose cells bear antigens unlike those on the cells of the rest of the body. Apparently, these antigens are usually detected and the cancer cells are destroyed by the immune system. However, the immune system loses efficiency with age, so that as a person grows older, there is an increased likelihood that cancer cells will escape detection. Similarly, an immune system that has been inhibited, as for

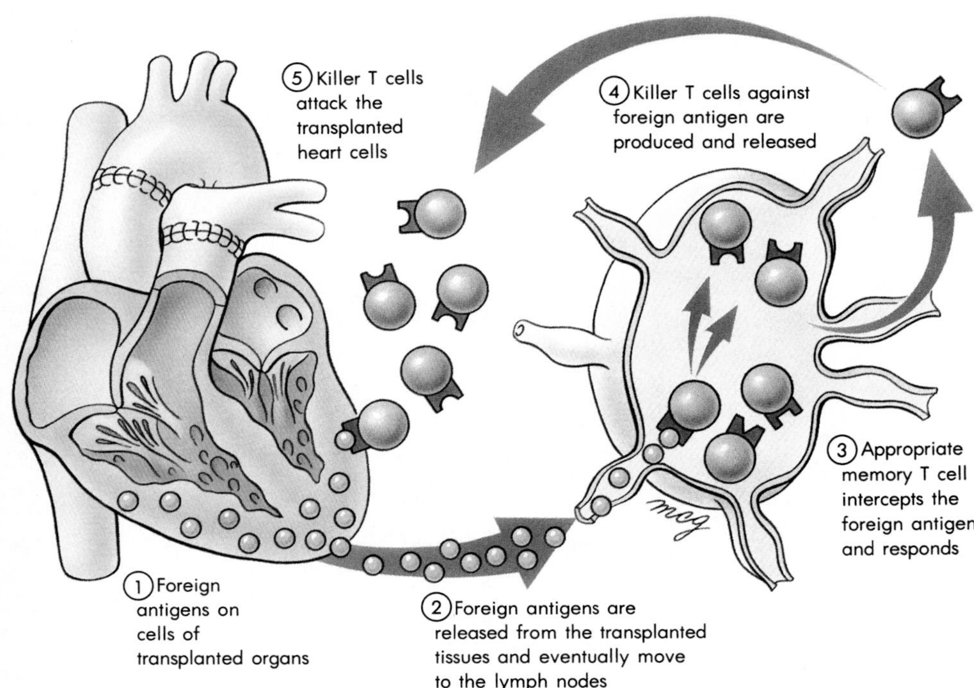

FIGURE 24-13 Transplant rejection.

transplant surgery, may miss cancerous cells. In addition, some cancer cells wear a gelatinous coat that shields their surface antigens against detection, while others change their antigens so quickly and often that the immune system cannot mount an effective attack.

Monoclonal Antibodies and Cancer Treatment

Plasma cells offer another opportunity for biological engineering. It is possible to fuse plasma cells with cancer (myeloma) cells to produce hybrid cells that multiply indefinitely, like cancer cells, and produce large amounts of antibody to a single antigen. The fused cells are known as **hybridoma cells;** their product is **monoclonal antibodies** (Figure 24-14). These pure, one-antigen antibodies, unlike the antibody mixtures extractable from blood, are invaluable in research and promise many opportunities for specific attacks on such diseases as cancer. One approach being developed is to develop monoclonal antibodies to the antigens that mark cancer cells, attach them to toxic drugs, and use them to deliver the drugs directly and precisely to the cancer cells.

Autoimmune Diseases

In autoimmune diseases, the immune system attacks some portion of the body itself. Generally T and B cells carrying antibodies or receptors that would work against self cells are suppressed or removed to prevent attacks on the body's own tissues. For unknown reasons, sometimes the suppression fails and an attack is launched. It could be that a mutation occurs, modifying the T cells and B cells in a way that allows them to escape suppression. Or it might be that a disease agent invades, bearing antigens that are so similar to self antigens that the antibodies produced in response also work against the body's own cells. This might be what happens in those strep throat bacterial infections that lead to rheumatic fever. In this disease, the body's immune cells turn against and damage the valves of the heart, apparently because they have surface antigens similar to those found on streptococcal bacteria.

The autoimmune diseases include severe anemias, where the attacked self antigens are on red blood cells; chronic hepatitis, where they are on liver cells; myasthenia gravis, a wasting muscle disease caused by muscle cell destruction; multiple sclerosis, in which the myelin sheath on the nerve fibers is destroyed; and possibly psoriasis, rheumatoid arthritis, some cases of diabetes and thyroid disease, and others. Treatment is difficult, but the symptoms can sometimes be eased by reducing or suppressing the immune system's inflammatory response with drugs or radiation.

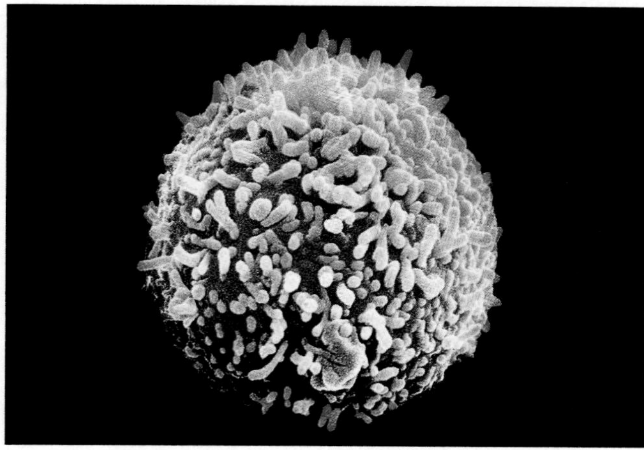

FIGURE 24-14 Hybridoma cells are grown in culture by fusing plasma cells (the antibody producers) with cancer cells (myeloma). The hybrid cells retain their abilities to produce antibodies and acquire the ability to reproduce extremely quickly. Thus, they are an important "living" source for producing vast quantities of specific antibodies that can be used for medical or industrial applications.

1. Describe the role of the immune system in transplant rejection.
2. Why would you expect the drug cyclosporine to have side effects that resemble the symptoms of AIDS?
3. How is cancer related to declining efficiency of the immune system?
4. What are the autoimmune diseases? Give examples.

Failure of the Immune System to Develop

In some individuals, the immune system fails to develop properly, perhaps because of a hereditary defect. The result is the individual's inability to produce T or B cells or both. Such a situation occurred in the case of David, who had to be raised in a germ-free environment—it amounted to a large plastic bubble, and he was called the "Bubble Boy"—unable to touch other people directly, even his parents. He died early in 1984, shortly after an attempt to graft functioning B and T cells into his bone marrow.

Stress and the Immune System

The immune system can also fail in a less severe way. When the body is under psychological or physical stress, it increases its production of natural painkillers, the endorphins and enkephalins. These substances seem to inhibit both B and T cells. They may therefore account for why people under stress are more vulnerable to infections of all kinds and even to cancers. Since the B and T cells also seem to produce their own endorphins and enkephalins, this inhibitory effect of the substances may reflect a natural control mechanism. The mechanism may be designed to prevent excessive activity of the immune system, but in the case of stress it does not act appropriately.

Box 24-3 Avoiding AIDS

Nothing can yet be done to cure AIDS victims. Your best hope of avoiding the disease is avoiding exposure to HIV. You maximize your chances of doing this if you realize that HIV gets into the system through blood-to-blood contact. A few people have been infected by receiving blood or blood products (such as clotting agents) from infected persons. Today blood-collecting agencies such as the Red Cross take great care to test blood obtained from donors, and the risk of getting AIDS in this way is very low. (As of mid-1991, the chance of contracting AIDS from a contaminated transfusion was less than one in 61,000.)

The main way the AIDS virus is transmitted from person to person is by sexual contact. Sexual activities sometimes abrade the sex organs, making small scrapes or tears in the skin that can allow HIV to enter. Since human semen carries numerous white blood cells (which may be carrying the virus), the virus can pass directly from the semen into the blood of the recipient. People with sexually transmitted diseases that produce open lesions (sores), such as herpes, chlamydia, and gonorrhea, are easily infected by HIV.

The second leading route of exposure is by intravenous drug use. People who inject illicit drugs often share their needles with each other, and a used needle carries a bit of the most recent user's blood with it.

There are no guarantees, but if you wish to reduce your risk of being exposed to HIV, you can take some very simple steps:

1. Avoid sexual promiscuity. Long-term monogamy confers a great deal of protection for both heterosexuals and homosexuals. Lacking that, condoms (rubbers), especially when used with a spermicidal foam or jelly, can play a protective role.
2. Do not participate in anal intercourse (homosexual and heterosexual).
3. Avoid the use of injectable "recreational" drugs.

SUMMARY

Disease always begins with exposure to some disease-causing organism. This organism may contact skin or mucous membrane or it may be introduced into the body by a wound. Either way, once the organism has penetrated the body, it must become established there. Only then can it multiply enough to cause the actual disease; symptoms of disease usually begin slowly, in the prodromal stage, get worse in the acute stage, and ease in the convalescent stage.

The body has two major lines of defense against disease. The first, nonspecific resistance, includes mechanical factors such as the skin; cleansing flows of urine, sweat, and tears; dust-trapping wax in the external ear and mucous membranes lining the body's openings; the filters and baffles in the upper respiratory system; hairs in the nose and cilia in the trachea; and armies of phagocytic white blood cells. Their function is to prevent disease agents from entering or becoming established in the body. Nonspecific chemical factors such as lysozyme also help prevent invasion, while chemical factors such as interferon and complement work to prevent multiplication of disease agents within the body or to destroy invaders. Inflammation is a process that restricts the spread of disease agents and summons phagocytes to destroy them. Fever raises the body temperature in an effort to make the body a less hospitable environment for disease agents (although a relatively few diseases, like syphilis, thrive at the higher temperatures seen in fevers).

Specific resistance is the function of the immune system's B and T cells. The B cells secrete antibodies, proteins that bind to specific foreign (nonself) materials, or antigens, marking them for phagocytosis or destruction by complement. Cytotoxic (killer) T cells attack both foreign cells and self cells that may be harboring viruses. Helper T cells stimulate the actions of both B and T cells. Suppressor T cells eventually shut down the immune response. Both B cells and T cells respond to antigens by multiplying to equip the body with enough specialized cells to deal with the immediate problem and "memory" cells that will respond to future exposures to the same antigen.

The immune system's efforts to fight off disease can be responsible for the symptoms of some diseases, and its responses to some weak antigens, called allergens, are responsible for the allergic responses. The immune system fails to work as desired (or perhaps it works too well) when it rejects transplanted organs; it must then be suppressed with drugs or radiation. It also fails when it does not kill a cancer cell and allows a tumor to develop, either because of suppression for transplant surgery or because of loss of efficiency with age. Autoimmune diseases, in which the immune system fails to distinguish self from nonself and attacks the body's own cells, represent another way the system can fail. Finally, the immune system can be impaired, as when the HIV virus destroys T cells and causes the vulnerability to infection and cancer typical of acquired immunodeficiency syndrome, or AIDS.

STUDY QUESTIONS

1. Why do you think the study of immunity promises to make some researchers wealthy?
2. Which stage of the disease process—exposure, penetration, incubation, or the acute phase (attack)—is the most important to prevent? Why?
3. Vitamin A deficiency leads to dry eyes. Why might it also lead to eye infections?
4. Why might it be unwise to try to relieve a mild to moderate fever?
5. Distinguish carefully between antibodies and antigens.
6. What is the chief difference in function between B cells and T cells?
7. Explain how suppressor T cells might be involved in the process of telling "self" from "nonself."
8. Why do we say the immune system causes the symptoms of some diseases?
9. Do you think it will ever be possible to transplant organs on a routine basis? Why or why not?
10. Do you think vaccines can hold any hope for preventing autoimmune diseases? Why or why not?

GLOSSARY

Active immunity Immunity acquired by the body's actually having been exposed to an antigen and then developing memory cells for it.

Allergen A weak antigen capable of triggering an allergic response.

Allergy An overreaction of the immune system to a weak antigen.

Anaphylactic shock The life-threatening, generalized inflammation reaction associated with serious allergic responses.

Antibodies Defense proteins tailored to fit specific antigens. The resulting antibody/antigen complex inactivates or marks for attack many types of disease agents.

Antigens Recognizable proteins used by the immune system to distinguish between "self" and "nonself." They are often surface recognition features on cells.

B cells Antibody-producing lymphocytes not maturing in the thymus. They are major figures in the humoral-mediated immune response.

Cell-mediated immunity A system employing principally the T cells to directly attack the foreign antigens.

Complement A group of 11 blood proteins that help defend the body in both nonspecific and specific ways.

Histamine A chemical released from mast cells (and others, when damaged) that increases the permeability of local capillaries. This encourages the movement of white blood cells, antibodies, and fibrin from the blood into a damaged area.

Humoral immunity Immune response mainly employing antibodies and their ability to complex with antigens.

Immune system The physical, mechanical, and cellular systems the body uses to defend itself.

Immunosurveillance system The continuous checking of all the tissues and debris in the body by the T cells and macrophages. If they detect "nonself" or troublesome changes in the surface recognition factors on any of the cells, they call for a defensive attack.

Inflammation A general body response to a wide range of physical, chemical, and cellular insults. It usually involves an increased flow of blood into the damaged area, which provokes redness, heat, swelling, pain, and loss of function.

Interferon A chemical agent produced by cells that have been invaded by viruses. It helps prevent the spread of the infection in numerous ways.

Macrophages Large cells in the tissue spaces and the cardiovascular and lymphatic systems that phagocytize foreign cells, debris, and potential toxins.

Memory cells Both T-cell and B-cell lymphocytes, bearing specific surface recognition factors that enable them to respond rapidly to future exposures to antigens to which they have been previously sensitized.

Monoclonal antibodies Specific antibodies produced in large quantities by cells cloned for that purpose.

Nonspecific resistance An assortment of different defenses that work against a wide range of disease and infectious agents.

Passive immunity Temporary immunity produced when a person is given someone else's antibodies by injection (as in a gamma globulin shot), by being passed through the placenta, or by being taken in with breast milk from the mother.

Plasma cells B cells that have the responsibility for manufacturing antibodies.

Specific resistance An array of cellular and chemical defenses specifically tailored to attack a certain agent.

T cells Lymphocytes that mature in the thymus gland and are the principal combatants employed in cell-mediated immune responses.

Vaccine An injected solution of antigens used to stimulate antibody development, to protect individuals that might later be exposed to the real agent.

Chapter 25

Human Evolution

"Well, I'll be a monkey's uncle!" This figure of speech first arose during the nineteenth century, as a result of the great controversy that followed Charles Darwin's theory of evolution. Darwin said that all living things are descended from other, different living things. When he added that he did indeed mean that humans are descended from creatures that resembled monkeys and apes, he upset a great many people. The idea jibed neither with what they had learned in church nor with their sense of human dignity, and it soon gave rise to the above greeting for the impossible.

Humans *are* related to monkeys and apes, but they are at best cousins ten thousand times removed. Humans are *not* descended from modern monkeys or apes. Evolution does not work in quite that way. It is much better to say than modern humans, monkeys, and apes shared common ancestors thousands of generations and millions of years ago.

Besides our obvious connections with primates, humans are also related to horses and cows, dogs and cats, mice and birds, even fish and angleworms. The similarities are fewer, and the number of generations that separates them is far, far greater, but a relationship exists. Every living thing descends from its predecessors, and ultimately every living thing descends from a single, first cell that arose on the planet Earth almost four billion years ago. The endlessly varied chain of life that has populated our world since then is the product of evolution, of mutation and recombination, of natural selection and punctuated equilibrium. In this chapter we shall discuss some of the ways evolution works, and then examine how those principles apply to the development of the human species.

HOW EVOLUTION WORKS

The raw material of evolution is change. Cosmic rays, background radiation, environmental chemicals, accidents, and even the processes of cell division all cause mutations in the genes of living things, including humans. These fundamental changes enter the genetic deck of cards for each species. Sexual reproduction—and its component processes of crossing over, independent assortment, and fertilization—then shuffles that deck and deals out new combinations of alleles again and again (see Chapters 4, 5, 17, and 19). The result is both genetic and physical variation within every species, as each individual embodies a unique combination of inherited characteristics. Individuals also vary in noninherited, physical characteristics acquired through their exposure to the environment—such as stunted growth, loss of limbs, damage from injury, muscular development from an exercise program, and poor brain development due to childhood illness or malnutrition. However, because these physical or phenotypic differences cannot be passed on to offspring, they do not count in evolution.

Environmental change also encourages changes in a species through time, for while it does not directly affect the form of individuals, it does change the conditions in which they must survive and reproduce. Because the thrust of evolution is the adaptation of populations to their environments, as environments change, so do the adaptations necessary for survival and reproduction. The link between environment and adaptation, or evolution, is *natural selection.*

Natural Selection: Accumulation of Gradual Changes

As we discussed in Chapter 1, Charles Darwin and Alfred Russell Wallace explained the diversity of life in terms of gradual changes that slowly adjusted the form of a species, continuously fine-tuning it for success in an impermanent environment. Building change upon change meant that new generations looked considerably different from their predecessors. As new environments opened, populations would spread into and colonize these new opportunities, and each new pioneering population would be molded for success in its new surroundings (Figure 25-1).

When Darwin and Wallace outlined the theory of evolution over a century ago, they did not begin with change. Rather, they started with a seeming contradiction. First, they noted, all living things have an immense **reproductive potential.** That is, a codfish or a maple tree produces thousands or millions of eggs or seeds every year. If every one of these "propagules" were to survive and develop into a mature individual, the world would soon be miles deep in codfish or maple trees. Though humans produce many fewer propagules, a handful of children in a lifetime, they too hold the potential for covering the Earth in human flesh in relatively few generations. (We are approaching that condition in some areas of the world right now.)

However—here is the contradiction—this potential is usually not reached. In general, over a

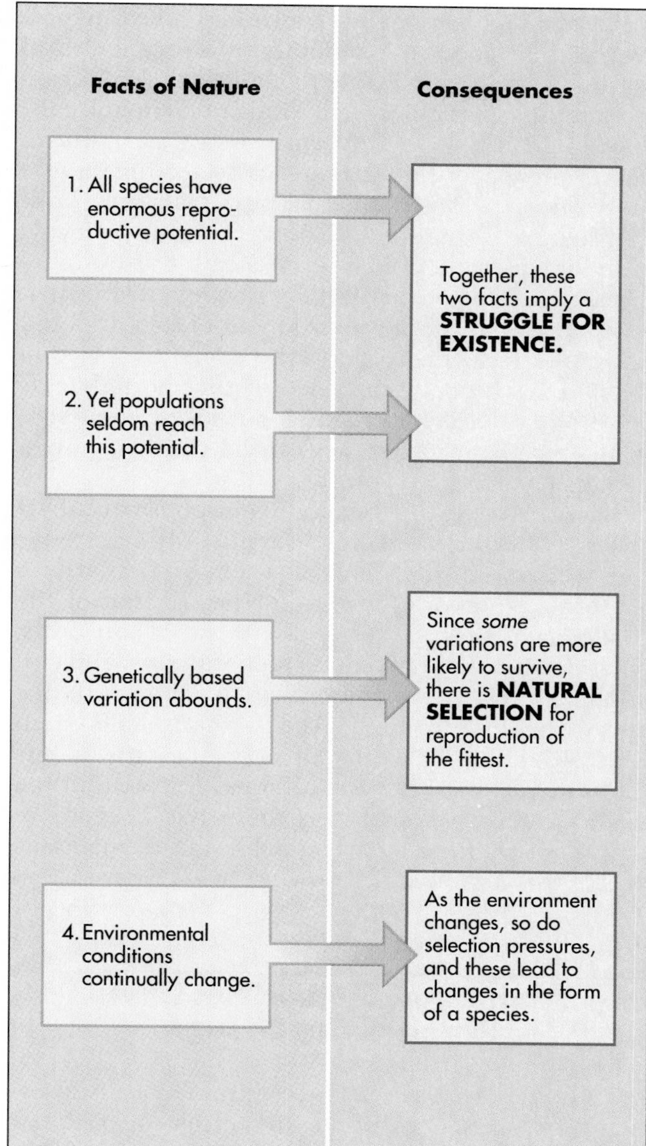

FIGURE 25-1 An outline of Darwinian evolution.

long period, each species on the planet usually has a *stable*, relatively small population. The death rate matches the birth rate, and most propagules by far never make it to maturity.

The necessary conclusion is that since only a limited number survive, there must be a **struggle for existence** as the members of each species compete for those resources they need to reach maturity and to reproduce—space in the sun for plants, nest sites for birds, food for animals, and so on. It is the limits on these resources that constrain the sizes of populations, for when any species becomes more numerous than the resources can support, the surplus must die. Even if the surplus does not die, it still lacks the health and vigor, or the living space or nest sites, to reproduce successfully, and it does not contribute to future generations.

The "struggle for existence" is an unavoidable fact of nature. It follows from the existence of reproduction and the fact that the environment can offer only limited resources, and it leads to the theory of evolution through two questions: Which members of each generation survive? Which ones successfully reproduce and contribute the greatest number of offspring to the next generation? The answers to these questions, seen among the characteristics possessed by the offspring that form the next generation, indicate the direction that evolution necessarily takes, toward ever-better adaptation to an ever-changing environment (Figure 25-1).

The individual members of a species vary from each other because of the differences in their genes. They vary in their physical characteristics—such as tooth size, growth rate, keenness of smell, intelligence, speed of running, their reproductive needs or capabilities, resistance to disease or cold, and so on. These variations mean differences in abilities to get food, tolerate adverse conditions, escape predators, and in other ways survive until the organisms can successfully reproduce. That is, certain individuals in the population possess a blend of genetically based characteristics that enables them to produce a greater number of offspring (with similar advantageous characteristics.) Such differential reproductive success is called **natural selection.**

Those who *do* survive are usually those whose physical characteristics favor survival and the production of offspring that survive and reproduce in their own turn. If those variations are genetic, they are passed on to the generations spawned by the survivors. Therefore, we say that the requirements of survival—or reproduction—*select for* certain genetic variants by giving them an advantage. The carriers of these advantageous traits produce more surviving offspring under the prevailing conditions. Those that have other traits, causing them to produce fewer surviving offspring under the prevailing conditions, are said to be *selected against*.

---

1. Why do you think *nongenetic* variations in members of a species do not contribute to evolution?
2. Can you think of any way nongenetic variations *could* contribute to evolution?
3. Which matters more to the process of natural selection, sheer survival or reproduction of the fittest? Why?

Clearly, in an unchanging environment, the logical consequence of natural selection would be ever-increasing adaptation to that environment. Once the adaptation was perfect, changes in existing species would cease. However, environments are never unchanging, except in the very short term. Continents shift, climates cool or warm, natural disasters occur, and new diseases appear. Species also move into and colonize new environments. And with all these changes come changes in what it takes to be competitive and hence changes in the pressures that define which genetic variants are selected for or against.

As species undergo natural selection in changing or new environments, they necessarily change their forms. Over time, through the processes of natural selection, they *gradually* become new types of living things, as we can see in the sequences of forms preserved in the Earth's rocks as fossils. Just as species can change through time, they can also gradually change across geographic gradients. As populations spread into new environments, away from the parental population, they are adapted to and molded by the pressures existing in their new situations. As time passes, these separated populations eventually diverge from each other in form, physiology, and behavior. Given enough time and isolation, the different forms may become different species. We sometimes see a graded series of related species distributed along various environmental gradients, as across a mountain range, from jungle into desert, from the equator northward, or from the sea to shore. Such changes, such progressions from species to species, are **evolution,** the result of natural selection.

Artificial Selection

Some environmental factors are likely to be more important than others in determining which members of a species will successfully reproduce. These same factors will therefore be more important in determining the direction the evolution of the species will take, for it is individuals with characteristics that make them more successful than others in coping with these important conditions that will have more surviving offspring. They will be selected for, and their next generation will have a significantly greater proportion of individuals with the same, successful characteristics. Characteristics that make their carriers less likely to reproduce in a given situation are selected against, and they become less frequent in the species as time passes.

Natural selection occurs when natural environmental conditions are exerting the pressures that determine the direction the evolution of a species will take. However, some evolution is human directed, particularly in our controlled breeding programs for domestic animals and agricultural plants. We have been able to control the evolution of virtually all the different kinds of farm animals, dogs, pigeons, and many varieties of agricultural plants simply by encouraging, or "selecting," those individuals to breed that possess traits we have decided are valuable and preventing undesirable individuals from breeding. We call this human-directed evolution **artificial selection.** Ironically, some of the earliest practitioners of evolutionary principles were the livestock breeders discussed in the Old Testament, who selected the spotted goats out of their flocks for breeding purposes. Their controlled breeding program ensured that they prospered.

Given enough time, genetic and physical differences—whether naturally or artificially selected—can accumulate to the point where a population's present form bears few similarities to that of the original stock. Dogs are a case in point; pit bulls, mastiffs, poodles, labradors, pomeranians, and chihuahuas are all descendants of the same wolflike ancestors (Figure 25-2). And humans have been controlling the breeding of dogs for only a few thousand years. Given the billions of years that life has existed on this planet, there has been more than enough time to account for the tremendous variety of life forms we now see.

Punctuated Equilibrium: Rapid Evolutionary Change

Darwin and Wallace and their immediate successors explained evolution in terms of gradual changes that gently molded the shapes and physiologies of new species through the processes of natural and artificial selection. According to their model, species evolved as the frequencies of advantageous and deleterious gene combinations gradually varied through time. And long periods of time were essential for explaining the fantastic diversity of life that we now see on our planet.

However, recent and extensive examinations of the fossil record indicate that while the concept of natural selection is alive and well, it may not always be so gradual a process or require so much time. A great deal of evolutionary change seems to occur very rapidly. Many species continue almost unchanged for millions of years. Then, within a few thousand years, or even less, perhaps because of some rapid change in the environment, what was a common species suddenly disappears to be replaced by a substantially different new species or a group of related species. Such rapid and dramatic

FIGURE 25-2 Human-controlled evolution: dog breeding. Virtually all dogs came from breeding the various ancestral strains of the genus *Canis*. Two of the possible ancestral strains are (a) the wolf, *C. lupus,* and (b) the black-backed jackal, *C. mesomelas*. By interbreeding these with domesticated canids ancestral species, humans were able to produce the tremendous variety we now see among the various dog breeds. All modern breeds are considered to be of the same species, *C. familiaris*. Included are (c) the German shepherd, (d) the chihuahua, (e) the Shar-Pei, and (f) the saluki. The diversity seen among various members of *C. familiaris* stretches the concept of species somewhat.

changes, we say, "punctuate" the long-lasting "equilibrium" condition and give the new view of evolution its name of **punctuated equilibrium.**

Founder Effect and Genetic Drift

Small populations sometimes evolve rapidly (i.e., experience rapid changes in the frequencies of certain alleles in the gene pool) as a consequence of a catastrophe, such as an earthquake, a volcanic eruption, an epidemic, a fire, or a flood. Such events sometimes wipe out most of the individuals in a population and with them most of the alleles they carried. The new population, formed when the relatively few survivors breed, often shows a tremendous reduction in genetic variability. The traits of the new population are those of its founders, and the difference in traits between the new population and the original we credit to the **founder effect.**

The reduction of genetic variability of a population by such a catastrophe can threaten the survival of the species. A small breeding pool is more likely than a large one to produce offspring that inherit disadvantageous homozygous recessive gene combinations. In addition, the laws of chance say that in a small breeding population some alleles may not be passed to the next generation. When alleles disappear in this way, we call the process **genetic drift.**

Such problems threaten, for example, the survival of the California condors. The last free-ranging condor was captured in 1987 and put into the San Diego Zoo's condor breeding program. Environmentalists hope that this endangered species can be bred in captivity in sufficient numbers and their offspring can eventually be released back into the wild. However, there are relatively few breeding individuals in the program, and some of the participants are reluctant to breed in captivity. Some people therefore consider the program to be controversial and of questionable value. Certainly, the surviving population will show the signs of the founder effect and be vulnerable to genetic drift.

Populations and Species

Thus far in our discussion, we have used two terms that may seem unfamiliar: species and population. They are not the same thing, and the differences are crucial in any discussion of evolution. A species is a natural reproductive unit. As such, a **species** is a group of physically similar organisms that *can* interbreed, at least potentially, with other members of the same species. That is, they can mate and produce viable, fertile, competitive offspring. They are reproductively isolated from (jargon for "They can't breed with") members of different species.

Members of the same species that are separated from each other by great distance could theoretically interbreed but do not have the opportunity to get together. Thus, a whitetail deer in Maine could probably mate successfully with a Virginia whitetail.

This potential for interbreeding is one of the crucial tests biologists use when they want to tell whether two similar organisms actually belong to the same species. The weakness in this definition of a species is that it makes no mention of asexual species—which are simply groups of physically, chemically, and genetically similar organisms that usually produce offspring that are genetically identical to the parent.

A **population** consists of all the members of the same species occupying a defined region during a specific interval of time. All the humans living in the world in the year 2000 are thus a population. A different, more narrowly defined population might be all the cancer victims being treated in a specific New York hospital in March 1996.

When we speak about natural populations, we usually mean all the members of a species occupying a certain area at a certain time that actually have a chance of breeding with each other. Their members thus share genes. Furthermore, mutations or new gene combinations that arise from crossing over quickly spread through the population, helping to fine-tune the population to be successful in its specific environment. These same new genes and gene combinations may not spread to other populations within the same species because of geographic separation. Thus, various populations may differ somewhat in their genetic constitutions. When such differences between two populations are noticeable (usually to the eye), we call each population a *subspecies.* If two subspecies are geographically isolated from each other long enough for the differences between the two to accumulate to the point where their members can no longer interbreed, we consider them to be *distinct species.*

---

1. In Chapter 1, we said that individuals do not evolve but species or populations do. Which do you think is the actual unit of evolution, species or populations? Why?
2. Clearly, two separated populations may accumulate enough differences to make them in-

capable of interbreeding. They will then be separate species. What could prevent this from happening?

3. We explained genetic drift in terms of populations shrunken by catastrophe. Can you think of other situations that might reduce the genetic variability of a population and allow the founder effect or genetic drift?

Reproductive Isolation

The remains of past organisms left in the rocks beneath our feet are **fossils.** They are usually bones and pieces of bones that have been replaced by minerals. They can also be footprints, casts of leaves and skin surfaces, and even frozen bodies such as the mammoths found in the Siberian permafrost years ago (Figure 25-3). Comparing the re-

(a)

(b)

(c)

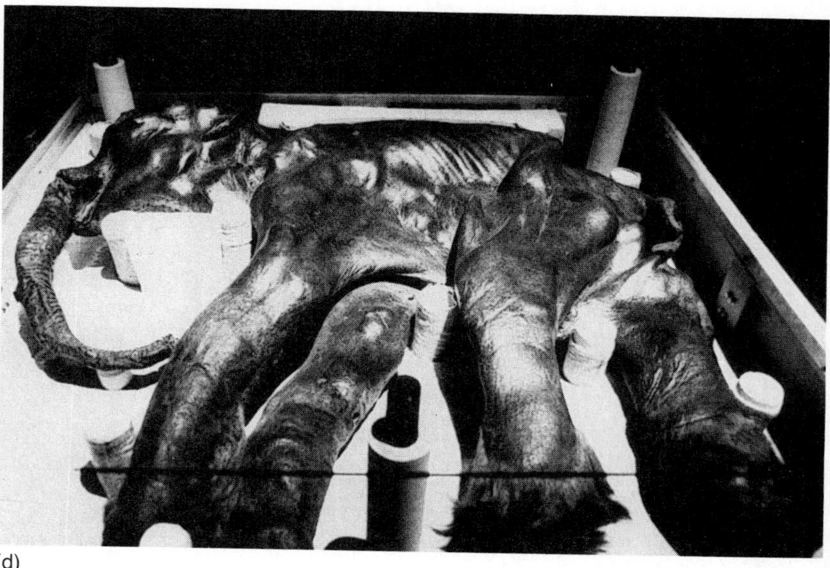

(d)

FIGURE 25-3 Samples of fossils. (a) Paleontologist, working at Dinosaur National Monument in Utah, cleaning sandstone away to reveal the fossil of *Diplodocus*, one of the longest dinosaur species. (b) Fossilized dinosaur tracks found on a Navajo reservation in Arizona. (c) Fossil sassafras leaf from the Cretaceous period (when dinosaurs roamed the Earth). (d) A baby mammoth (about 1 meter tall) found frozen in 1977 in Siberia; estimates suggest that she had been buried for about 40,000 years.

mains of these ancient life forms to similar species alive today, we can see that the differences are substantial and accumulate sequentially through time.

It is possible to date fossils according to the ages of the rocks in which they are embedded, often by measuring the amounts of various radioactive isotopes that have decayed since the formation of the Earth. With such information we can sometimes unravel how species arise from their predecessors over time—those predecessors accumulate, under the guiding pressure of natural selection, new inherited characteristics, and as the number of these characteristics increases, their possessors become increasingly different from their ancestors.

It is not hard to imagine how, as a species changes body size and shape, it becomes physically unable to breed with its predecessors. Since all dogs are technically the same species, *Canis familiaris,* theoretically they should be able to interbreed. Similar subspecies (each breed is a subspecies), like German shepherds and huskies, can and do interbreed. However, as selective breeding programs have emphasized the physical differences between the various breeds, some of the subspecies have been effectively **reproductively isolated** from each other. Physical differences between female chihuahuas and male Great Danes make breeding impossible. Thus, human-controlled dog breeding programs could lead to speciation in a relatively short period of time. It is not hard to imagine how similar differences have developed through time between modern species and their predecessors.

Reproductive isolation can take numerous forms. In general, it usually begins as a period of **geographic isolation,** when two populations come to be separated physically, as when an offshoot of one population colonizes an island or when a river, mountain range, or rift between continents splits a species' territory in two. As genetic differences accumulate, isolated populations may come to differ behaviorally as their breeding seasons or courtship patterns also change. Reproductive isolation can also be anatomical, that is, be due to differences in body size or shape. It can be chemical, as when two closely related species cannot fertilize each other. In general, the mixing of genes in cross-species hybrids goes no further than the one offspring, and the species remain distinct.

In the case of the horse and the donkey, their anatomies and behaviors permit mating and their chemistries permit the one's sperm to fertilize the other's egg, but the resulting offspring, a mule, is almost invariably sterile (Figure 25-4). On the rare occasions when it is not, the explanation seems to

(a)

(b)

(c)

FIGURE 25-4 When a (a) mare (female horse) is bred with a (b) jack (male donkey), the result is a hybrid known as a (c) mule. The mule is often stronger and has more endurance than either of its parents. However, because of the way a mule's chromosomes are likely to line up during metaphase of meiosis, the mule is almost always sterile.

be that even though a horse's and a mule's chromosomes do not match up well enough to be divided neatly in meiosis, occasionally the meiotic lottery can produce a gamete containing a complete set of "horse" or "donkey" chromosomes. In a 1984 case, a female mule apparently produced an egg with nothing but horse chromosomes. When she then mated with a horse, she produced an offspring that was pure horse. The two species—horse and donkey—thus remained distinct.

The Persistence of Similarity

So far we have talked mostly about how evolution accounts for the accumulation of differences between species. But evolution never affects every characteristic of an organism at once. It leaves many traits intact, and even species whose last shared ancestor lived many millions of years ago retain numerous similarities. It is thus possible to learn a great deal about human anatomy—bone, muscle, and organ shape and location—by dissecting a cat or fetal pig or even a frog (Figure 25-5).

These similarities speak eloquently for the kinship of all living things. The anatomical and physiological similarities are greatest within the most closely related groups, such as the primates (apes, monkeys, and humans). They are nearly as great within the **mammals** to which we humans belong (mammals are warm-blooded, hairy animals that suckle their young) and even among the **vertebrates** (animals with backbones). To these similarities, we can add similarities in development. Each animal, as it matures from embryo to adult, goes through stages that resemble the stages through which its predecessors went (Figure 25-6).

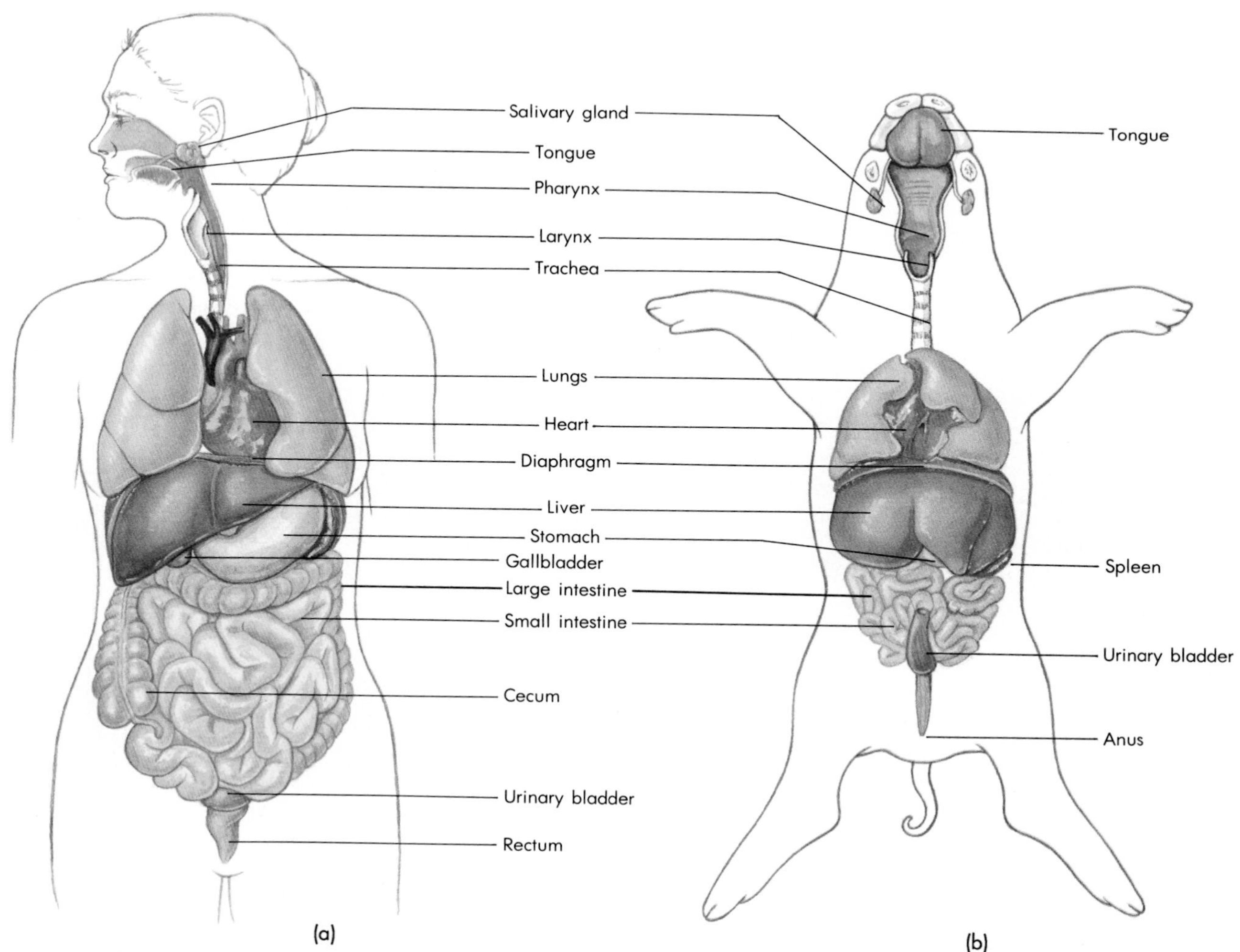

FIGURE 25-5 The similarities in the internal anatomies of the human and the fetal pig reveal similarities in the inherited basic mammalian body plan.

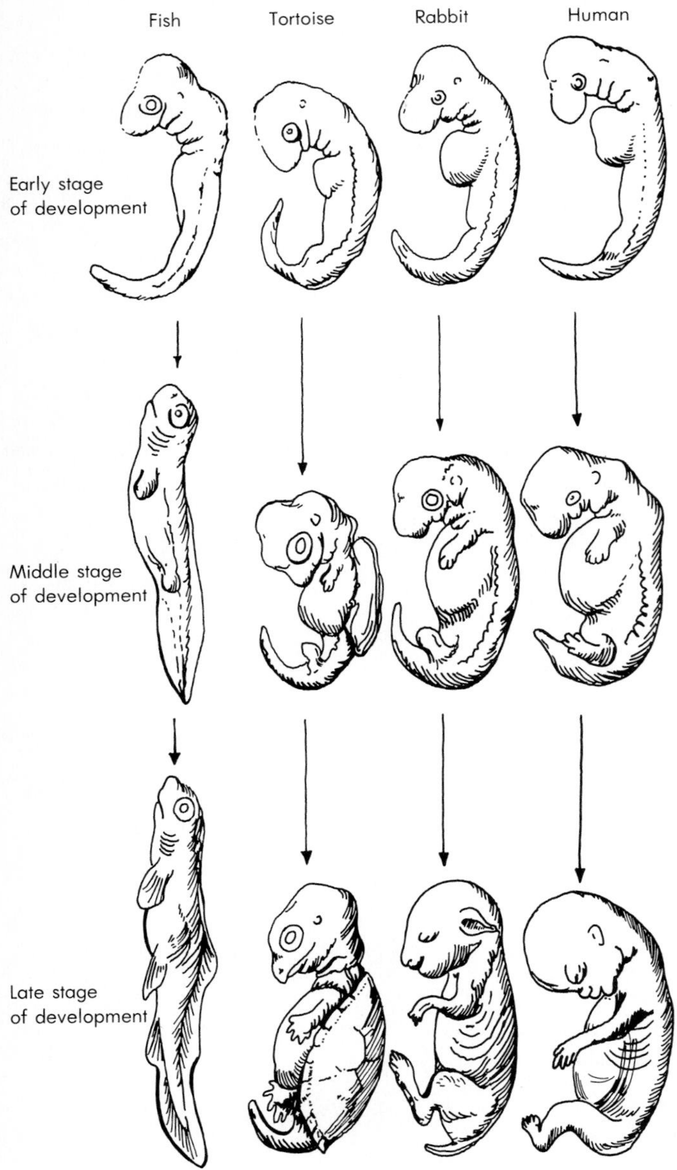

FIGURE 25-6 The early developmental stages for fish, tortoise, rabbit, and human embryos are almost identical. Each displays basic vertebrate characteristics, which include gill slits, arches, tails, and large heads. Continuing development eventually allows each species to express its own unique characteristics.

We can also see similarities in biochemistry, for all living things use the same kinds of macromolecules (nucleic acids and proteins), such as cytochromes. Since mutations are changes in deoxyribonucleic acid (DNA) and hence often in the amino acid sequences in proteins, they show up as small, single-unit differences in these macromolecules, and it is possible to measure the degree of difference between two species by counting the amino acid or nucleotide differences between their proteins or genes. In fact, since mutations are often random events and they happen to a given nucleotide roughly once per billion cell generations, counting mutations in a gene or protein can give us a rough measure of how long it has been since two species shared a single ancestor. The number of differences becomes large only between two organisms that are very distantly related, such as humans and plants. As indicated in Table 25-1, the number of amino acid differences in cytochrome *c* between humans and other organisms increases with the amount of time that has passed since they shared a common ancestor.

Table 25-1

Number of Amino Acid Differences in Cytochrome *c* between Humans and Selected Other Organisms

Organism	Differences
Rhesus monkey	1
Dog	11
Chicken	13
Bullfrog	18
Fruit fly	29
Wheat	43
Baker's yeast	45

Learning Focus

HOMOLOGY AND ANALOGY

Through processes like natural selection and punctuated equilibrium, species acquire new traits and may eventually give rise to new species. However, although different species have different characteristics, they also preserve many similarities. We therefore assume that the more closely related two species are to each other, the more recently they shared a common ancestor and the more characteristics they share. These shared characteristics can be structural (bones, fins, and body shapes), behavioral (courting patterns, cries, and gaits), and even chemical, as we have already seen. However, not all similarities between different species exist because they have been preserved during evolution. Sometimes they arise as independent adaptations to similar ways of life.

Two very different ways to fly. While butterfly and bird wings are both devices for flying, each species has evolved its own anatomical solutions to the problems of flight. The wings of birds and butterflies developed from completely unrelated anatomical features and are thus analogous.

When two species have structures, behaviors, or chemistries that have come to resemble each other because the species have separately adapted to similar ways of life, we say that the characteristics are **analogous.** Thus a bird and a butterfly both have wings, but they did not inherit their wings from some shared ancestor; their evolutionary lines diverged long before the development of wings of any kind. Furthermore, their wings are built along very different lines; they are not at all the same kinds of structures even though they serve similar functions. Finally, their wings do not grow in the same way and they develop from very different parts of the embryos.

Structures that serve similar functions but are built and develop differently we also call analogous. Structures that are anatomically and developmentally related but may serve *different* functions we call **homologous.** We can see an excellent example of homology if we compare a bird's wing with a human arm. The two limbs are built around a framework composed of similar bones. They have many of the same muscles. They develop from similar parts of the embryonic animals. And both the avian wing and the human arm represent the accumulation of diverging naturally selected changes in a single ancestral structure, the front limb of an amphibian ancestor that lived hundreds of millions of years ago.

Now turn to the Learning Focus Response to apply your understanding of these two terms, analogy and homology, to two new examples.

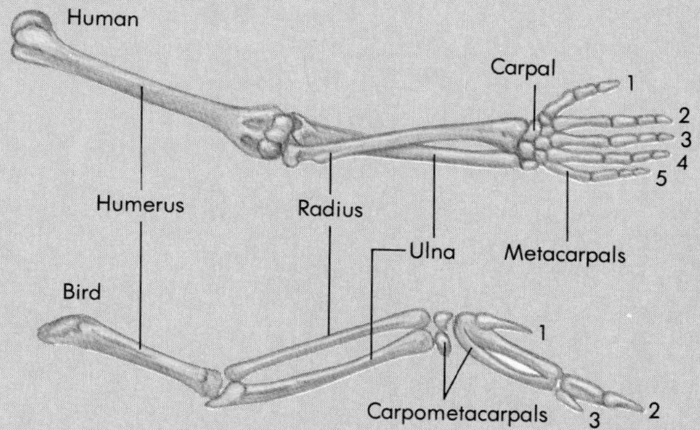

The forelimbs of humans and birds share many of the same structural elements but have been modified along different lines to support different lifestyles. Such anatomically related but functionally different structures are thus homologous.

Learning Focus Response

HOMOLOGY AND ANALOGY

1. Bats probably evolved from shrew-sized ancestors with four limbs much like those of a shrew or mouse. In time, wings evolved from the bones of their "hands." They retained the basic bony structure, exploiting the finger bones as supports for the wing surface. They did not develop feathers, using instead a membrane of skin to achieve the same purpose.

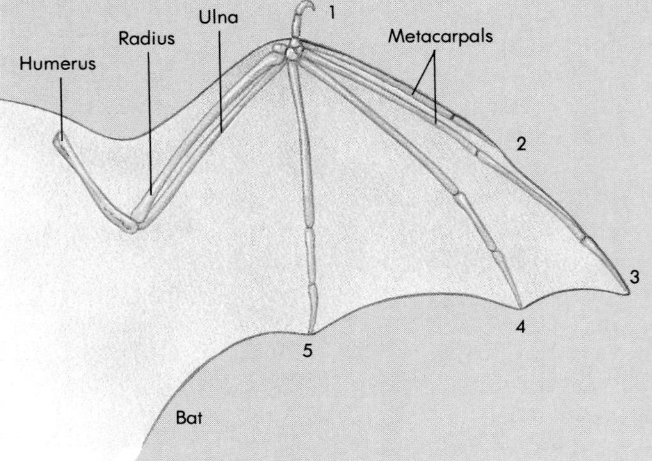

Bat wings are very different from bird wings. While both are modifications of the vertebrate forelimb, the superstructure of a bat wing is really made up of modifications of the bones of the hand.

Is the bat wing homologous or analogous to a bird wing?

Are the structures similar or different?

Did they evolve from similar structures?

Do they develop similarly in the embryo?

Do they serve similar or different functions?

Which of these features seems least crucial to the distinction between homology and analogy?

2. Now let us look at the body forms of two very different animals, a shark and a porpoise. The two resemble each other, for both are smoothly streamlined marine (ocean-dwelling) creatures. However, whereas the shark has always been marine, the porpoise is a mammal descended from four-legged land dwellers. Millions of years ago, its ancestors returned to the sea, lost their legs, and adapted to the demands of a swimming life. (So too did such relatives of the dinosaurs as the ichthyosaur.)

Since mammals are ultimately descended from primitive fish, going back far enough into the history of life sharks and mammals have a common ancestor with an aquatically adapted body form. As one of the successful lines of animals that eventually invaded the terrestrial environment, mammals gave up that shape long ago. The porpoise's legless streamlining is a *secondary* adaptation, and it shows us the meaning of *convergent evolution,* the development of resemblance between unlike species exposed to similar selective pressures.

You should be able to see several reasons why we can call the shapes of sharks and porpoises homologous. Can you see why we might call them *analogous* instead?

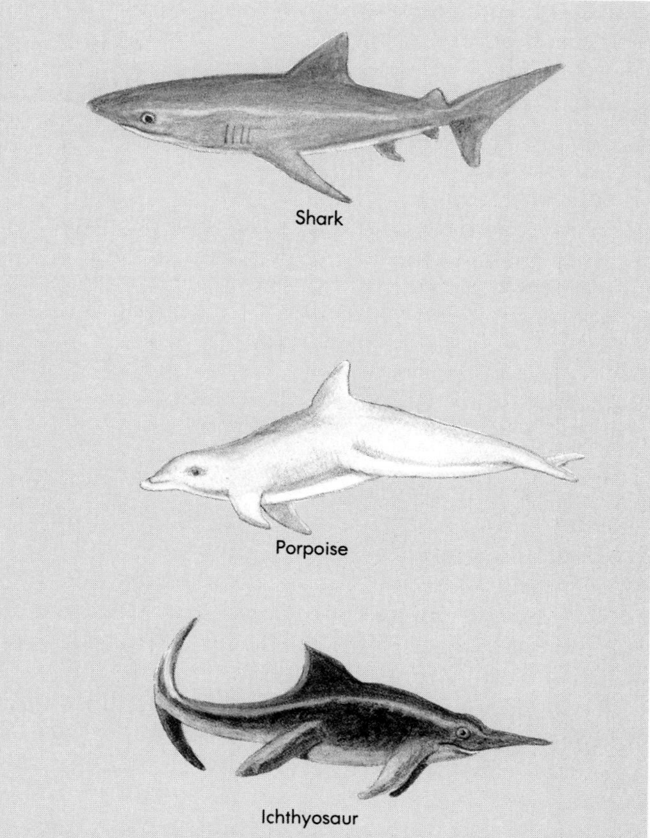

The bodies of living sharks and porpoises and extinct ichthyosaurs are examples of convergent evolution. They all evolved similar shapes, resulting in streamlined forms that pass efficiently through the water.

THE PRIMATES

Humans belong to the group of mammals known as the **primates;** that is, they are related to apes, monkeys, and other creatures that share a number of distinctive characteristics. Each of these features developed as natural selection fit the earliest primates to life in trees, feeding on insects and fruit, and each one of them plays an important part in making humans what they are today. Prehensile (grasping) paws, opposable thumbs (able to meet each finger tip to tip for picking up small items), and nails instead of claws on their fingers and toes help primates manipulate food items as well as tools.

Most primates also have their eyes mounted on the front of their skulls, letting the visual fields overlap to support three-dimensional vision. Such depth perception is advantageous since primates have long been *arboreal*—lived in trees—and often leap from branch to branch. However, depth perception has a cost as well, for it has apparently developed at the expense of the sense of smell, which is most useful for ground-dwelling animals. We might expect smell to remain important for finding food, even in the trees, but primates have applied to that purpose color vision, useful for spotting ripe fruit and other food items at considerable distances, even downwind.

Perhaps most important of all, primates all have relatively well-developed brains. This feature permits rapid tracking and seizing of insect prey and allows primates to learn new ways to cope with changing environments and to interact intensively with each other.

1. Keen, three-dimensional vision would surely aid survival in the treetops, but how would fingernails help?
2. Which primate characteristics do you think were selected for by the demands of diet?

Disastrous Origins

In a sense, the primates owe their existence to a world-shaking catastrophe. The first mammals lived at the same time as the dinosaurs, but they were small, secretive, rat-sized creatures that hid in the underbrush, fearfully avoiding the notice of the monsters that stalked their world. Their opportunities were severely limited.

Then, about 65 million years ago, the dinosaurs all died out, leaving a vacant world and a host of opportunities for the early mammals to seize. At present, the best evidence seems to say that the dinosaurs died out because a comet or asteroid, a chunk of rock about 10 kilometers across, crashed into the Earth. One of the best candidates for the impact site is located in Mexico's Yucatan peninsula. Wherever the rock hit, the result was apparently vast clouds of dust and steam that blocked the sun's heat for several years and chilled the planet, much as current theory says would happen after a nuclear war. Plants died, photosynthesis was suspended, and the dinosaurs, which depended on those plants for food, died too. The early mammals may have survived by eating the decaying remains of the dinosaurs.

Adaptive Radiation

When the sun reemerged from the clouds, it must have shone on a world that seemed dead. But seeds lying dormant in the soil soon sprouted, and the mammals (and other surviving animals) multiplied, free at last from their saurian predators. With time, they accumulated variations that fit them to different parts of the world, underground, in the water, in trees, and so on. The wealth of opportunities for evolution fostered what we call an **adaptive radiation** as mammals adapted to new foods, new homes, and new habits, eventually taking the forms we now know as rodents, carnivores, ungulates, and so on (Figure 25-7). Similar adaptive radiations have happened whenever life has entered a new zone of opportunity, as when life forms first produced shells and, later, internal skeletons; when plants and animals left the water for the land; and when birds and insects first took to the air.

Prosimians

We can see what the earliest primates must have been like if we look at the simplest, most primitive (least monkey like) primates living today. These are the **prosimians** ("first monkeys"). They include the tree shrews, who lack three-dimensional vision but have well-developed brains and grasping paws. They also include the more monkeylike tarsiers, lorises, and lemurs; these have three-dimensional vision, and most are arboreal (Figure 25-8, p. 592).

Monkeys

Monkeys, apes, and humans belong to the group of primates we call the **anthropoids.** They bear a sometimes striking family resemblance, which is one reason why monkeys and apes are so popular in zoos. Most monkeys can walk erect at least some of the time; they have humanlike arms, legs, and hands; and they are quite intelligent. Many also live in humanlike families and form tribes (or troops) led by elders, either male or female. They range in size from that of a small cat to that of a large dog.

There are two main groups of monkeys, the New World and the Old World monkeys. The evolutionary lines that gave rise to these groups seem to have originated in Africa. Over a hundred million years ago, Africa and South America were fused together, along with the other continents, in one giant land mass. The geological processes of continental drift then broke that land mass into the modern continents. What was to become South America slowly pulled away from Africa and the Atlantic Ocean flowed into the widening gap between them. By about 50 million years ago, the two different populations of the common ancestor to both groups were isolated from each other. As the gap widened

FIGURE 25-7 Once the dinosaurs were gone, mammals quickly evolved an immense variety of forms such as these extinct giant mammoths, bison, dire wolves, and saber-toothed cats, which flourished in the Pleistocene era.

and time passed, the New World and the Old World monkeys evolved their distinctive traits. Only the New World monkeys, for instance, can hang by their tails from a branch (Figure 25-9, p. 593).

Apes

The apes are larger than most monkeys, have larger brains, and look and act more like humans. They include the gibbons, orangutans, chimpanzees, and gorillas, and with humans they comprise the **hominoids.** The chimpanzees, whose proteins and DNA are 99 percent identical to our own, seem to be most closely related to humans, though some researchers believe the orangutans are closer cousins (Figure 25-10, p. 594).

The hominoids first appeared about 30 million years ago. Among their distinguishing features was—and is—their ability to swing by their arms from branches, or **brachiate.** The gibbons, which

(a)

(b)

(c)

(d)

FIGURE 25-8 Prosimians: (a) tree shrew *(Tupaia glis)*; (b) tarsier *(Tarsius syrichta)*; (c) a slow loris *(Nycticebus coucang)* eating a grasshopper; (d) ringtailed lemur *(Lemur catta)*.

(a)

(c)

(b)

(d)

FIGURE 25-9 New World monkeys: (a) wooly monkey from the Amazon region, *Lagothrix lagutrichia;* (b) squirrel monkey from Costa Rica, *Saimiri sciureus.* Old World monkeys: (c) Patas monkey from the Sahara, *Cercopithecus patas;* (d) male baboon from Tanzania, East Africa, *Papio anubis.*

(a)

(b)

(c)

(d)

FIGURE 25-10 The great apes: (a) Concolor gibbon, *Hylobates concolor;* (b) an orangutan, *Pongo pygmaeus;* (c) adult chimpanzees, *Pan troglodytes,* engaged in grooming in Gombe National Park, in Tanzania; (d) mountain gorilla, *Gorilla gorilla,* quietly feeding on shrubbery.

split away from the rest of the hominoids some 12 million years ago (Figure 25-11), rely on this ability to travel through the treetops. Other apes (and humans) can do it too, but they prefer different styles of locomotion.

The best evidence for the timing of the various splits in the hominoid lineage comes from studies of differences in proteins (see the preceding). We believe that orangutans became separate about 10 million years ago and that humans had their last shared ancestor with the African apes, the chimps and gorillas, about 6 million years ago. However, there is some controversy over these dates. Fossil bones suggest earlier times for these events. Whatever the truth, it is clear that the human lineage is old. Although it is in fact older than we can easily imagine, given the age of the Earth, it is still but an eyeblink in our world's history.

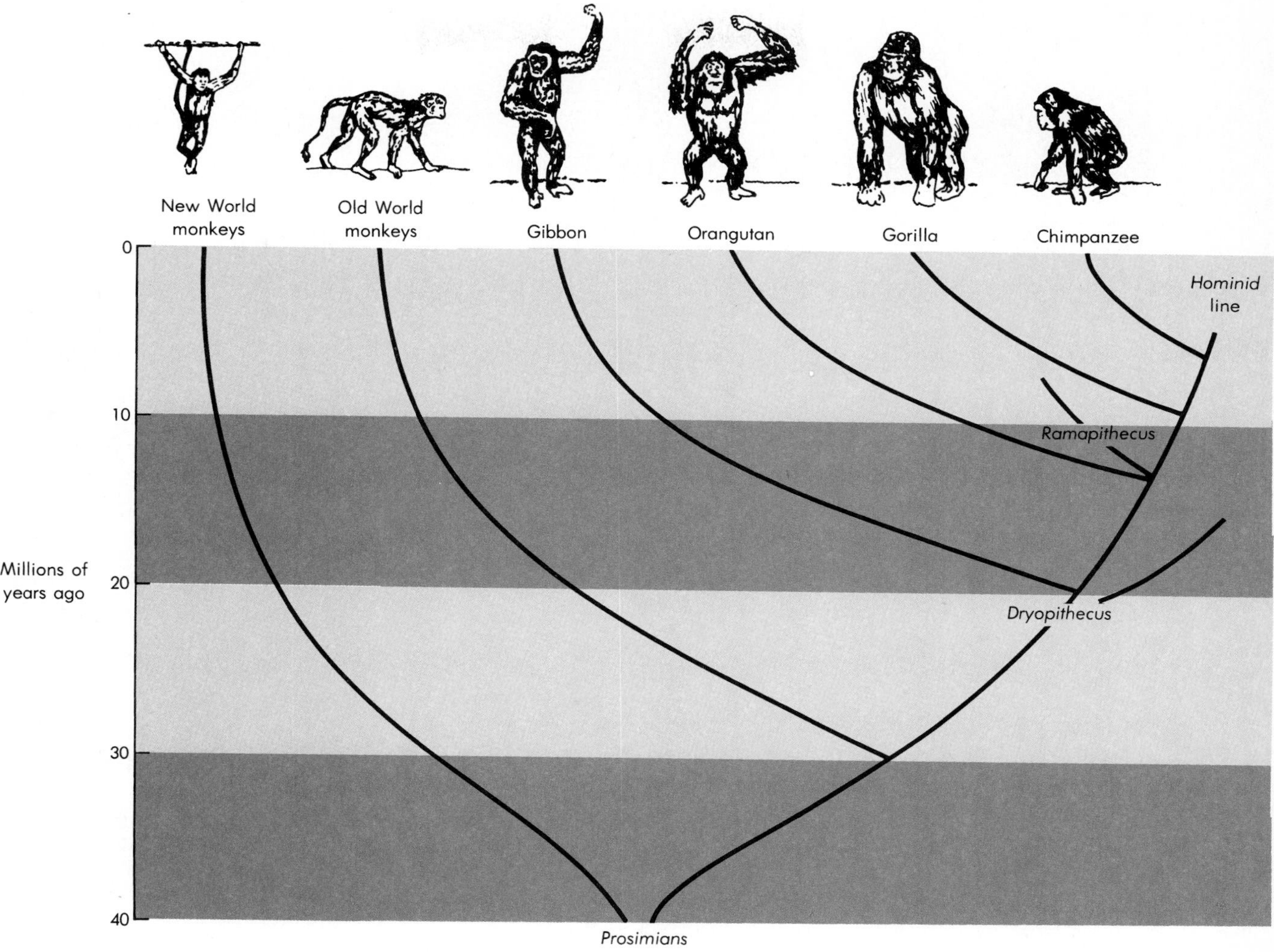

FIGURE 25-11 Evolutionary tree for hominoids.

THE HUMAN ANCESTRY

Humans and their immediate ancestors compose the group of primates known as **hominids** (as opposed to the *hominoids*). Not long ago, it was thought that the first member of this group was the apelike *Ramapithecus,* which lived some 15 million years ago. Now, however, *Ramapithecus* is considered no more than a hominoid, perhaps belonging to the subgroup that gave rise to the orangutans.

The first known hominids were the **Australopithecines,** first discovered in South Africa in 1924 (Australopithecus means "southern ape"). They were typically less than 4 ft tall, weighed less than 50 lb, and had brains only a little larger than those of apes. However, they did walk erect and had recognizably human hands and feet (Figure 25-12). Their erect posture may have been an adaptation to let them see over the high grass of the ground over which they foraged or to free their hands for carrying food back to a home base for other members of their family or tribe (see Table 25-2).

The first Australopithecines lived almost four million years ago, but we have few specimens that are so old. The more numerous younger specimens include skulls, bones, and even some nearly complete skeletons, as shown in Figures 25-12 and 25-13. The skulls do not look much like modern human skulls, but the rest of the skeleton makes the kinship unmistakable. However, there is still a great deal of argument over exactly how the Australopithecines gave rise to modern humans.

So far, it seems best to say that the earliest Australopithecines belonged to the species *Australopithecus afarensis,* named for Africa's Afar Triangle, where many of the fossils have been found. They lived in African grasslands and fed on whatever they could find. They were apparently not tool

(Continued on p. 599)

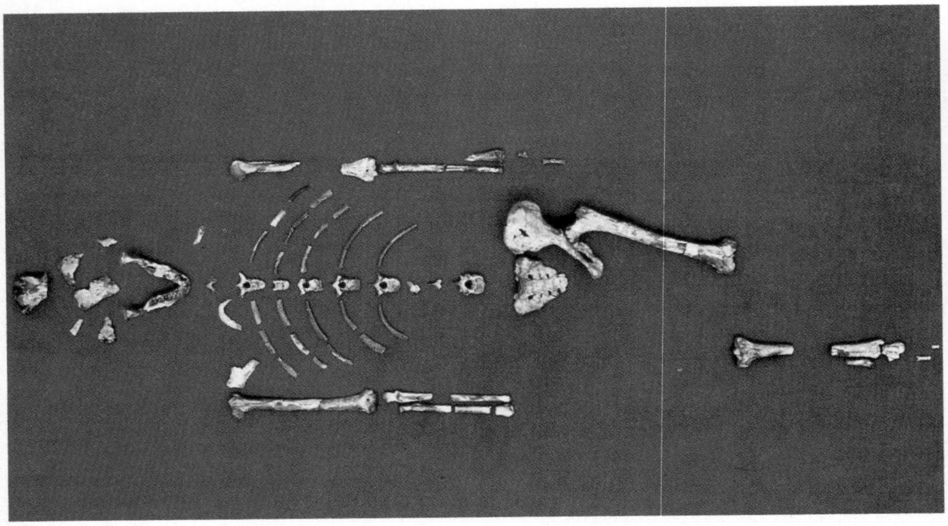

(a)

(c)

(b)

FIGURE 25-12 (a) The Lucy skeleton is one of the earliest fossils classified as truly hominid. She was a 3-ft-tall, adult female who lived 3.7 million years ago. Her pelvis is distinctly humanlike, and the structure of her knees shows she walked erect. (b) Fossil footprints 3.5 million years old provide evidence of erect hominids in northern Tanzania, Africa. The trail on the left is thought to have been made by a single individual. The trail on the right was probably made by two individuals—one walking in the footprints of the other. (c) Close-up of a single footprint of a very humanlike foot that walked Earth millions of years ago.

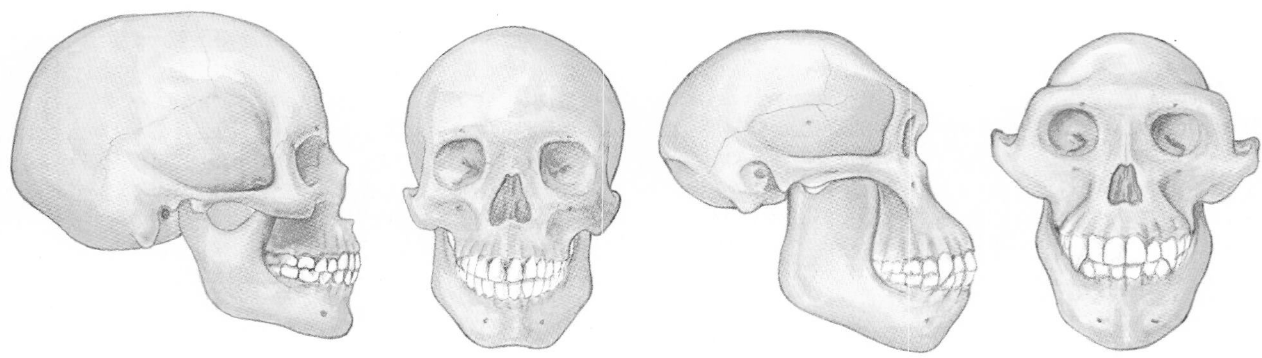

FIGURE 25-13 Side-by-side comparison of a modern skull and an Australopithecine skull.

Table 25-2

Chronological Time Frame for the History of Evolution

AGE (Millions of years ago)	ERA	PERIOD	EPOCH	MAJOR EVENTS
4600	Precambrian			**Earth forms** by a coalescence of local interstellar debris.
3800				**Life begins** with the origin of the prokaryotic cells. Beginning of the Kingdom Monera.
3500				Fossils of well-developed bacterial cells are found in Australian deposits from this time period.
2500				Earliest version of **oxygen-producing photosynthesis develops.** Oxygen build-up in the atmosphere begins.
1500				**First eukaryotic cells** form by chance fusions among bacteria, other prokaryotes, and cyanobacterial cells—giving rise to the complex cell lines that will eventually form the eukaryotic kingdoms.
700				Plant and animal kingdoms begin.
600	Paleozoic	Cambrian		The fossil record undergoes a rapid expansion as the various kingdoms go through adaptive radiation. By this time most of the invertebrate phyla and the algae divisions are well represented in the fossil records.
505		Ordovician		**First vertebrates** form—the jawless fishes. **Marine algae** flourish.
438		Silurian		**Invasion of land** begins—first as vascular plants leave the water, followed by the arthropods (buglike creatures).
408		Devonian		**First amphibians** and **insects** form.
360		Carboniferous		**Extensive land forests form** the rich organic basis of what will later be harvested as coal and oil deposits. **Reptiles originate, seed plants** arise, and the amphibians are the dominant land vertebrates.
286		Permian		Rapid **expansion** of the **reptilian line.** First **mammal-like reptiles** form. Modern insect orders form and mass extinctions of marine invertebrates occur.
248	Mesozoic	Triassic		**First dinosaurs, mammals, and birds. Gymnosperms,** cone-bearing trees, and shrubs with needlelike or scaly leaves become the dominant line of terrestial plants.

AGE (Millions of years ago)	ERA	PERIOD	EPOCH	MAJOR EVENTS
213		Jurassic		**Dinosaurs** are the **dominant** land vertebrates.
144		Cretaceous		**Flowering plants** evolve. Near the end of the Cretaceous—about 65 million years ago—a giant asteroid struck the Earth causing **mass extinctions among dinosaurs. Continental drift** begins tearing and separating the modern continents from what was a single land mass.
65	Cenozoic	Paleogene	Paleocene	Rapid adaptive radiations among the mammals, birds, and some insects. This marks the beginning of the **Age of Mammals.**
54			Eocene	Flowering plants flourish and mammalian radiation continues. Old world and new world monkeys form separate lines as the gaps widen between the drifting continents.
38			Oligocene	Virtually all modern mammalian orders are present, including **apes.** These first hominoids appeared about 30 million years ago.
24		Neogene	Miocene	Mammalian and angiosperm lines continue to diversify and expand.
5			Pliocene	Early hominids appear. **Human** and **ape** lines probably **separated** from a common ancestor about 6 million years ago. ***Australopithecus*** ("Lucy") lived about 4 million years ago. ***Homo habilis*** may also have lived during part of the same time frame as the Australopithecines.
1.8			Pleistocene	Ice ages.
1.7				***H. habilis*** gives rise to ***Homo erectus.***
1.0				All Australopithecines are extinct.
0.4				Earliest versions of ***Homo sapiens*** come into existence.
0.2				*H. erectus* dies out.
0.125				***H. sapiens neanderthalensis*** lived in Europe, the near east, central Asia, and Africa.
0.04–0.30				Evidence suggests that the Neanderthals had been replaced by the modern version of *H. sapiens,* Cro-Magnon man.
0.01			Recent	The **discovery of agriculture** marks the beginnings of a predictable food supply, freeing the human mind for developing the arts, sciences, and technologies.

users, unless they used bits of wood and vine that did not survive the years as well as the stone tools their successors favored; modern chimpanzees are known to use such simple tools.

About three million years ago, this species split into two distinct lineages. One became the later Australopithecines, including the species *A. africanus, A. robustus,* and *A. boisei,* all of which were extinct by roughly one million years ago.

Genus *Homo*

As shown in Figure 25-14, the early Australopithecines also gave rise to the first hominids similar enough to modern humans to be grouped in the same **genus,** or group of related species. This genus is genus ***Homo.*** This early offshoot of the Australopithecines belonged to the species ***Homo habilis.*** We belong to the species ***Homo sapiens.***

Some researchers insist that *H. habilis* was no more than a somewhat advanced variety of *A. africanus* or at best an intermediate between the Australopithecines and genus *Homo*. We prefer the other view, for though the habilines lived at the same time as the Australopithecines, the latter eventually died out. *H. habilis* had a larger brain and apparently used tools (Figure 25-15). These humanlike traits gave it a clear selective advantage, and over 1.5 million years ago it apparently spawned a second species for genus *Homo,* which in turn gave rise to us (see Table 25-2).

The second member of genus *Homo* was ***Homo erectus*** (Figure 25-16). It first appeared 1.7 million years ago, some time before the Australopithecines died out. Its skeleton was much like our own; one specimen, discovered in Kenya in 1984, suggests that some individuals may have been 6 ft tall. In addition, while possessing an average brain size smaller than that for humans, the range of brain sizes for *H. erectus* actually overlapped our own. The major differences are its large jaws and its prominent brow ridges. Its teeth, while smaller than those of the Australopithecines, were larger than our own.

Careful study of *H. erectus* remains in Africa, Europe, and Asia has revealed that they used stone tools, ate the meat of large animals that they may have hunted and killed (they may also have been scavengers), used fire, and lived in semipermanent camps. They may also have talked, for the anatomy of their throats would have permitted them to make all the necessary sounds and the complexity

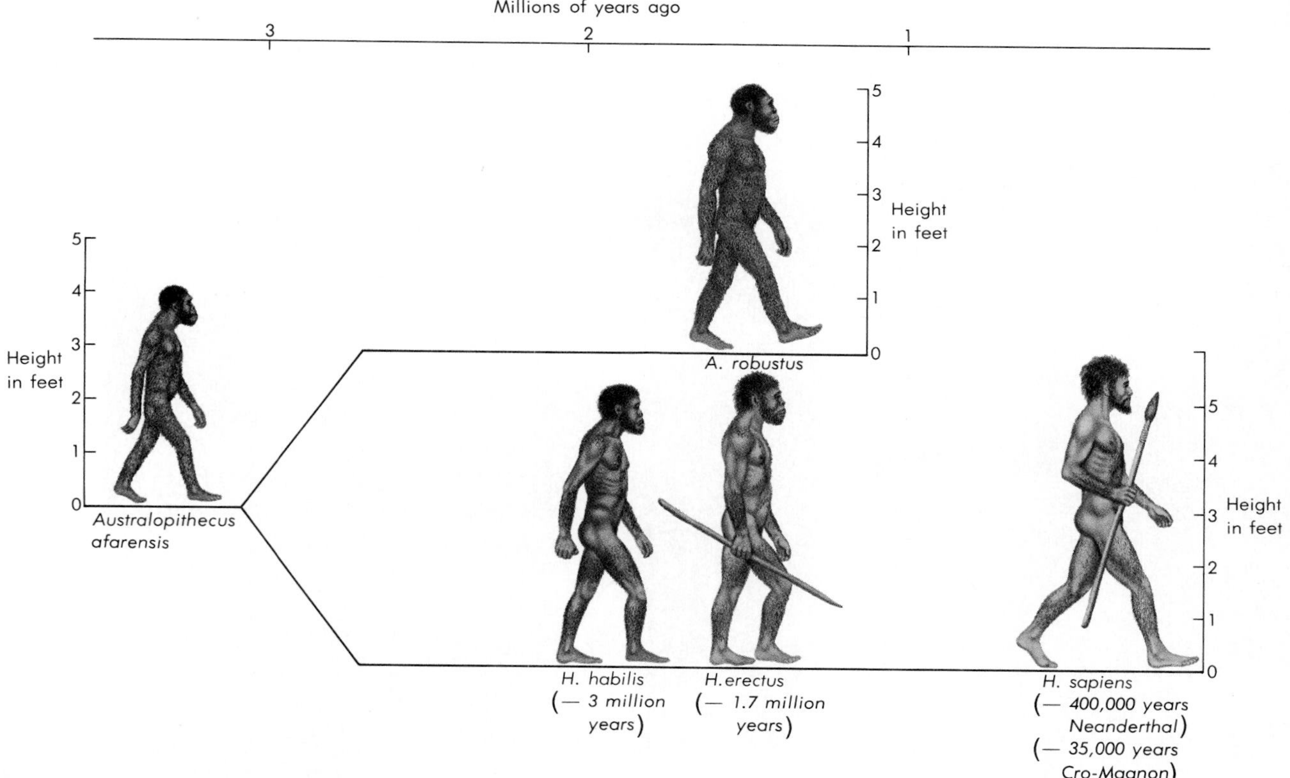

FIGURE 25-14 The hominid family tree.

FIGURE 25-15 Ancient pebble tools used by our early ancestors.

of their lives probably required some oral transmission of ideas and techniques from generation to generation. These early, small-brained, primitive members of our genus, though not of our species, were thus human in many basic respects. If we could meet them, we would surely consider them people, not animals.

Yet if we could meet members of *H. erectus,* we would surely also be impressed by their differences. They were not modern humans. They were more intelligent than any of their predecessors. They had more and better tools. They presumably had language. But their tools stayed simple throughout their million-year history, rarely improving, rarely showing signs that an *erectus* Einstein had ever seen a better way of doing things. And their remains show no sign that their ability to speak—and think—ever lead to a capacity for abstraction, as we can see even in their immediate successors.

1. Which species, *H. habilis* or *H. erectus,* do you think best deserves the name of the first humans? Why?
2. What were the most important ways in which *H. erectus* differed from modern humans?

Homo sapiens

Despite their primitive humanity, *H. erectus* dominated much of the world from their origins until 300,000–400,000 years ago. At that time, there appeared the first members of our species, *H. sapiens,* "thinking man." Presumably, some *erectus* population, reproductively isolated in a valley or on an island, accumulated the genetic variations that gave it a larger brain and greater intelligence. While it is difficult to date the time when the first members of the human species appeared, fossil evidence tells us that by about 125,000 years ago a new human version had shown up in Europe. We call it *H. sapiens Neanderthalensis,* or **Neanderthal**

(a)

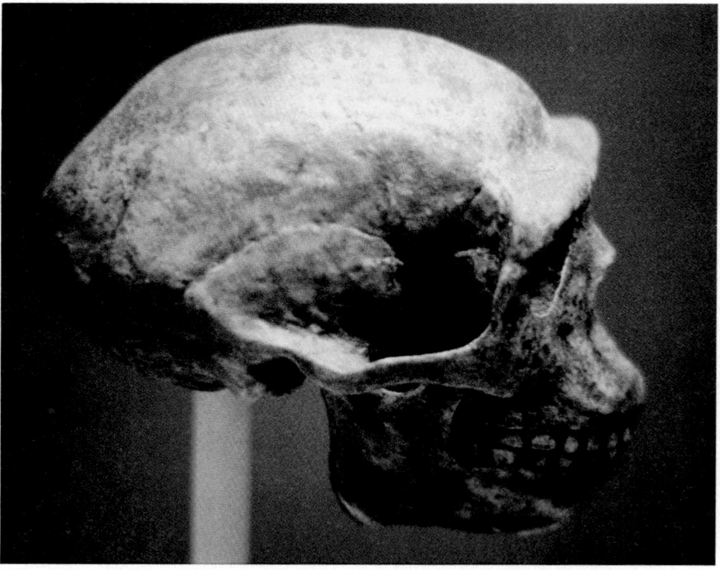

(b)

FIGURE 25-16 *Homo erectus* skull: (a) front view; (b) lateral view.

man, after the place where the first specimen was discovered, the Neander valley (or *thal*) in Germany. Other specimens have been found throughout Europe, the Near East, Central Asia, and Africa.

Once the more modern hominids existed, *H. erectus* could not last long, and in fact the species had vanished by about 200,000 years ago. Presumably, their successors' greater intelligence gave them an edge in the competition for campsites, food, and other resources; there may also have been war.

The Neanderthals differed from modern humans in several ways. They were more heavily boned and muscled and their skulls bore prominent brow ridges. The forehead was relatively low, and the face protruded beyond a relatively weak chin (Figure 25-17). However, these traits were all more pronounced in the older members of the species. Those who lived nearer the end of the Neanderthals' time, when modern humans were beginning to appear, were much more like ourselves. Nevertheless, even the boniest, most brutish seeming Neanderthals were not entirely outside the modern range; it is possible to walk down a city street today and meet people who would not have been physically out of place a quarter of a million years ago.

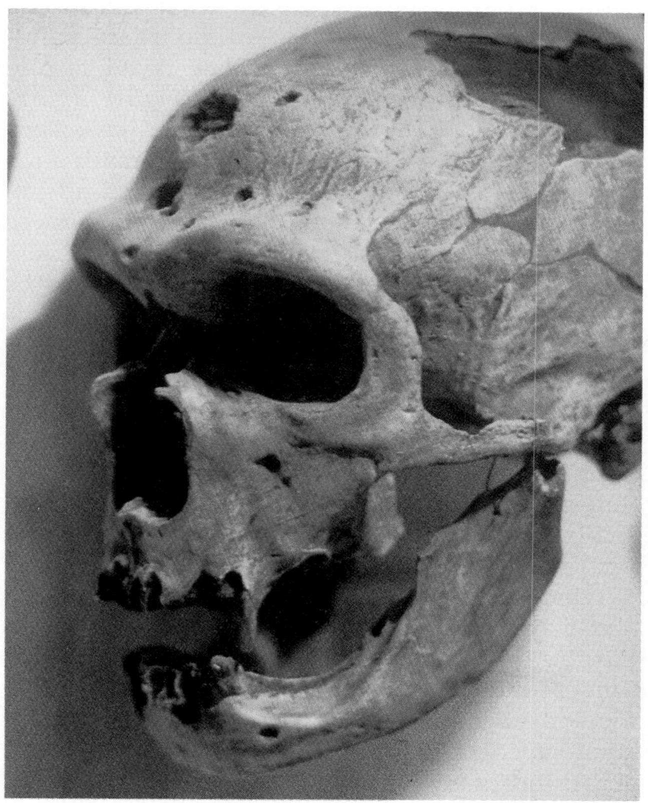

FIGURE 25-17 Neanderthal skull.

The Neanderthals did *not* differ from modern humans in many important ways. Their brains were as large as our own, and the average brain size may have been slightly greater. They used tools that were much more sophisticated than those of *H. erectus,* they improved those tools from generation to generation (showing a taste for progress), and they built shelters. Some of their tools seem designed for processing animal skins for use as clothing (Figure 25-18). They surely talked, in order to pass on cultural traditions, and they buried their dead, sometimes with flowers, animal skulls, and other items that suggest a belief in an afterlife, or religion. Their ability to think had thus advanced to the point where they could handle abstractions. If modern humans had never appeared, they might well have developed much as we have, into an urban, technological species capable of reaching for the stars.

But modern humans did appear. The Neanderthals themselves apparently changed; some populations developed less craggy features, leaner bodies, and presumably keener intelligence, and then—we believe—interbred with other populations. At any rate, the Neanderthal version of humans seems to have disappeared rapidly from the fossil record by about 30,000–40,000 years ago, displaced or replaced by the modern version of *H. sapiens.* The transition is well documented in such places as Iraq's Shanidar cave, which has been continuously inhabited for over 100,000 years. There we can see the oldest skeletons, clearly Neanderthal, overlain by those of more modern humans (Figure 25-19). Another theory suggests that modern humans, the Cro-Magnons, may have developed from the earlier stock in Africa and then migrated north by about 30,000 years ago.

FIGURE 25-18 Neanderthal burial site showing skeletal remains buried with stone tools. These complex burials suggest Neanderthals had religious beliefs that may have included some concept of an afterlife.

FIGURE 25-19 Front view of Shanidar cave, Iraq.

The Birth of Civilization

The Neanderthals were gone (or at least indistinguishable) by about 35,000 years ago, replaced by the modern version of *H. sapiens*. We believe that the transition occurred as the new humans interbred with their predecessors and displaced those populations of the older type that remained by being better able to find and use resources.

The competitive ability of the new humans shows well in a simple comparison: It took the Neanderthals over 300,000 years to spread through Africa, Europe, and Asia. In less than a tenth of that time—by 10,000 years ago—their successors had migrated to the Americas and Australia, covering all the surface of the planet except Antarctica and a few Pacific islands (many of which were reached only in the last thousand years). They were travelers, impelled like us, perhaps, by a driving curiosity about what lay over the horizon and equipped by their intelligence to adapt even to such extreme environments as the Arctic.

We call the first of our kind the **Cro-Magnons.** Their skeletons were indistinguishable from our own. Their brows protruded less, as in modern skulls. Their brains were as large as ours. They made more and better tools than did the Neanderthals, with an emphasis on small, sharp flakes of stone used as knives, scrapers, and spearheads. They also used bone needles, shell fishhooks, and other tools.

The Cro-Magnons were the first to leave behind them definite traces of an intellectual life. Their burials showed more signs of abstract thought than those of the Neanderthals, for they interred their dead with tools and ornaments, and they decorated the bodies with pigments. We have also found statuettes, bits of bone carved with what seem to be records of the phases of the moon, and cave paintings so deftly expressive that they could inspire such a modern master as Picasso (Figure 25-20).

For most of their history, humans have been tribal hunters and gatherers, living in small, often temporary settlements. Hunting provided an important part of their diet, but judging from modern hunters and gatherers, the roots, nuts, fruits, and small animals gathered by the women and children were a large part of the daily menu. This way of life now exists only in a few places, as among some Australian Aborigines and South African Bushmen. Yet even the Cro-Magnons had art, presumably religion, and a simple technology that equipped them adequately to survive in their environment. We can thus say that they had the basic requirements of civilization even if they did not live in cities, drive cars, and watch television.

What we now call civilization did not begin until about 10,000 years ago, probably in the Middle East, as the last ice age was winding down. At that time, people first learned to plant seeds of grain in the soil and grow a more predictable and more abundant food supply. (Agriculture apparently began independently in other human populations as well. The native inhabitants of the New World

(a)

(b)

FIGURE 25-20 Cro-Magnon paintings photographed in Lascaux cave: (a) a wild horse; (b) a bison.

were growing maize long before the arrival of the European explorers.) As people found themselves relying ever more on gathering for their food instead of hunting, someone may have observed that discarded seeds sprouted and supplied more food, closer to home. From that observation, it was a simple step to the first deliberate planting of seeds and harvesting of crops.

---

1. What sort of person do you think must have been the first to deliberately plant a seed?
2. Do you think art, agriculture, cities, or technology is the best sign of the birth of civilization? Why?

One effect of agriculture was that it restricted people to a single area. Temporary settlements became permanent villages. The villages grew into the first cities, and the increased food supply soon led to a leisure class of rulers, priests, and artists. Writing appeared about 5000 years ago. And our form of civilization was on its way.

WHAT MAKES HUMANS HUMAN?

A number of characteristics distinguish human beings from other primates, past and present. The most important may be intelligence, for that is what makes possible all the features of human life that we treasure, from adaptation to virtually any environment to art, from agriculture to science and technology. But that intelligence depends on another uniquely human feature, the largest, most advanced brain of the primate world, and that in turn apparently arose only because of other traits.

The first known hominids, the Australopithecines, had a brain only a little larger than that of their apelike predecessors. However, they already had one feature of modern humans shared by none of the apes: They walked erect all of the time. Chimpanzees and gorillas do this some of the time, but more often they walk hunched over, using the backs of their knuckles like a quadruped's front feet. Their skeletons are not designed for constant uprightness. The evolution of the erect posture was eased by the hominoid adaptation for brachiation, swinging by the arms, and by the common primate ability to sit upright. But it could not occur until the ancestral hominid had left the trees, perhaps in response to a food shortage caused by climatic change or because of competition with other primates. Then erect posture became adaptive, either because it gave the early hominids a higher viewpoint from which to watch for ground predators or food or because it freed the hands for carrying food.

Once the hands were free, the early hominids were able to develop further the tendency to use tools of all kinds, from folded leaves for carrying food to stones for throwing and clubs for bashing. This in turn made increased brain size and intelligence advantageous, and they were selected for until the Australopithecines gave rise to the first members of genus *Homo*.

However, tool using alone is hardly a uniquely human trait. Chimpanzees use simple tools, and so do many other animals. What *is* human about tool using is the tendency to improve tools constantly, to seek new ways of doing things. This is a product of the extra intelligence of the hominid line, and its demands provide selective pressure for constant increase in intelligence. At the same time, it creates a need to teach the young to make the new tools and thus to communicate more subtly than can be done with mere apelike grunts and cries. This need provided the selective pressure that favored the development of those changes in throat and brain structure that would allow speech. Apes seem to have brains that can generate messages, for they can be taught to use the gestures of sign language to convey fairly complex ideas, but they lack the physical ability to speak words.

This ability to speak seems to have appeared first with *H. erectus,* which was also the first hominid to use fire. With speech came greater demands for improved brains, leading to *H. sapiens.* With fire came less obvious changes, for fire makes possible a broader choice of foods. It makes tough roots and leaves edible and tenderizes meat, and we can see its effect in the decrease in tooth size from the Australopithecines to *H. erectus* and then to ourselves.

What is the most distinctive single characteristic of human beings? Speech, tools, curiosity, intelligence, and a fondness for change, all have their precursors in other primates. But no species has refined this fondness for change like humans. Ever since their origins, humans have been improving their tools. They have also been inventing new abstract ideas, religions, languages, and social systems. There are no signs that the process will ever end.

THE ENGINEERED FUTURE

At the moment we can see the prospect of a brand new kind of change. Recent advances in molecular biology have given rise to the technology of genetic engineering (see Chapter 4). So far, this technology has produced bacteria, yeast, and other cells that synthesize various biochemicals that we find valuable. Researchers are implanting genes for disease resistance, herbicide resistance, and other features in crop plants. The first attempt to use genetic engineering to cure a human genetic defect took place in the fall of 1990. Eventually, larger changes will be tried, and within your lifetime it seems likely that the genetic engineers will have the *capability* (if not the wish) to redesign the human body. Futurists are saying that in time there may be a new kind of human being, designed for greater health and longevity, higher intelligence, and less aggression (among other desirable features). There may even be an immense number of different new kinds of human beings, as parents design their children to suit their dreams.

The possibilities are endless, but there are reasons for restraint. The trouble with genetic manipulation is that it affects not just the individual receiving the changes but also the individual's children and grandchildren. And we must ask ourselves whether we have the right to impose our own functional and aesthetic preferences on future generations.

SUMMARY

The keys to evolution are that the members of a population or species vary from each other, that many of the variations are inheritable (genetic), and that each individual has the potential to produce many more offspring than will survive to reproduce in their turn. The offspring that survive and reproduce are those whose variations make them more able to find and use the resources necessary to survival and reproduction. Their special features fit them to their environment and are said to be selected for by that environment; features that interfere with survival and reproduction are selected against. This is the basic recipe for natural selection and hence for evolution. A similar process, under human control, shapes domestic animals and plants and is known as artificial selection. Evolution can progress gradually or in jumps and bursts (punctuated equilibrium).

The unit of evolution is neither the species nor the individual, but the population, a group of similar organisms that actively interbreeds and shares genes. New species arise when old ones are split into reproductively isolated segments, either in time or in space, and accumulate enough genetic and physical differences to make them unable to interbreed. However, even when two populations have become different enough to be called separate species, they retain many similarities of form, development, chemistry, and behavior. These similarities allow us to trace relationships between species that last shared a common ancestor millions of years ago.

With prosimians, monkeys, and apes, humans comprise the primates. Most primates share three-dimensional color vision, prehensile paws, opposable thumbs, finger and toe nails, and advanced brains. With monkeys and apes, humans make up the anthropoids. With the apes, they are the hominoids. By themselves, but including their extinct humanlike ancestors, they comprise the hominids. The earliest known hominids were the Australopithecines, who walked erect and may have used tools. About three million years ago, they gave rise to the first member of the human genus, *H. habilis. H. erectus,* which appeared 1.7 million years ago, was the first *Homo* to use fire and, presumably, speak, but—though its brain was larger—it still used only simple tools. Only with the appearance of *H. sapiens Neanderthalensis* 300,000–400,000 years ago did technological evolution, the only truly unique feature of human beings, begin. The Neanderthals gave way to modern humans, in the person of the Cro-Magnons, roughly 35,000 years ago. From those days, we have the remains of art and signs of religion and abstract thought. Only 10,000 years ago did humans invent agriculture, which soon led to cities and the modern form of civilization.

Human evolution has not stopped, for natural selection still operates. In addition, with such new technologies as genetic engineering will come the capability for humans to control and direct their own genetic and physical modification. Where the future will take our species cannot be known, but we can expect the journey and the destination to be fascinating.

STUDY QUESTIONS

1. Why can we not say that humans are descended from monkeys?
2. Describe the key concepts of evolution.
3. In what way is it true that natural selection does *not* mean "survival of the fittest"?
4. What role in evolution does environmental change play?
5. What is the difference between a species and a population?
6. What is the importance to evolution of reproductive isolation?
7. Of what use are the similarities we can see in structure, development, and biochemistry between different species?
8. Outline the categories of primates, showing which groups contain which others. To which groups do humans belong?
9. How long ago did the first recognizable humans appear? Which hominid would this have been?
10. Which hominid features, if any, are uniquely human?

GLOSSARY

Adaptive radiation The proliferation of new species from existing species exposed to various opportunities to adapt to new ways of life.

Analogy Similarity in function but not in structure and development.

Australopithecines The earliest known hominids; they walked erect and may have used tools.

Cro-Magnons The first humans of modern form.

Evolution The development of species from preexisting species by natural selection or punctuated equilibrium.

Fossils The remains of past organisms; they can be bones, imprints, casts, or even frozen bodies.

Hominids Humans and their immediate, nonape ancestors.

Hominoids The group of primates including apes and humans.

Homo erectus The second known member of genus *Homo;* used fire and may have been able to speak.

Homo habilis The first known member of genus *Homo.*

Homology Similarity in structure and development but not necessarily in function.

Homo sapiens "Thinking man," the species to which modern humans belong.

Natural selection The process of differential reproduction that makes each succeeding generation better adapted to its environment.

Neanderthals An early form of *Homo sapiens.*

Primates The group of mammals including prosimians, monkeys, apes, and humans.

Reproductive isolation The inability to interbreed due to separation in time or space, or to anatomical, biochemical, or fertility obstacles.

Reproductive potential The number of offspring an organism can have.

Struggle for existence The competition for resources necessary for successful reproduction.

Chapter 26

Human Ecology

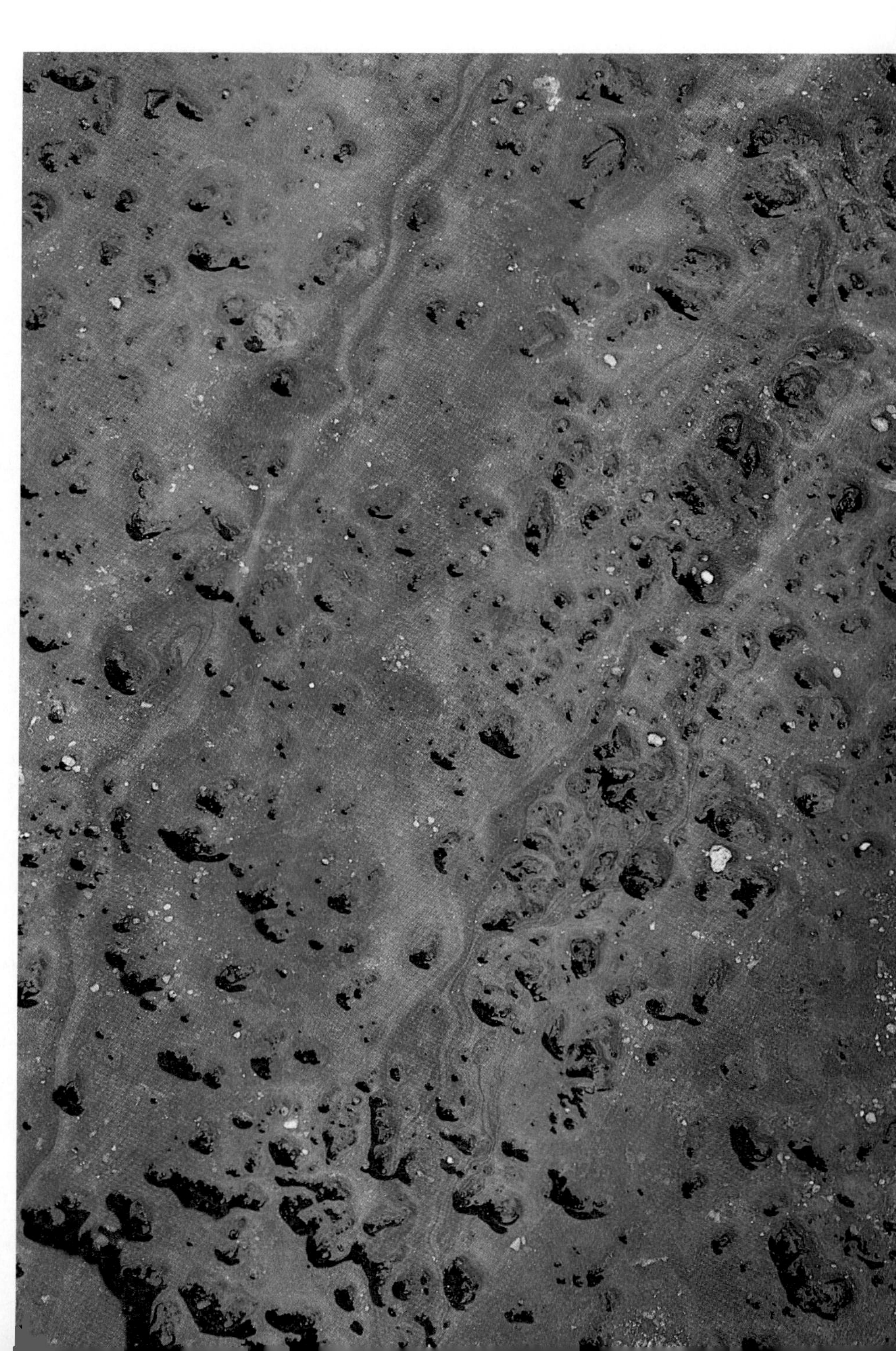

Just 300 years ago the average human's lifespan was only 39 years and women were very likely to die in childbirth. Almost everyone died of infections, such as syphilis, plague, malaria, or smallpox, or of war or starvation. In the seventeenth century, we developed the microscope, discovered microorganisms, and took a major step forward in our battle with infection and disease. We did it again in the nineteenth century with the beginnings of vaccination. In the 1940s, we developed the anti-infection drugs, the antibiotics we now take for granted.

This last century has been filled with astounding improvements in the human condition. Humans have learned to fly and to build automobiles, vast industrial complexes powered by fossil fuels, and of course, computers. At the same time, we developed agricultural systems so efficient they eliminated many of the famines that haunted our ancestors. We have visited the moon and have plans for visiting the nearby planets. Genetic engineering, which may turn out to be the most exciting innovation of all, is just in its infancy.

Yet for all the glory of human achievements and aspirations, there is also the potential for disaster.

HUMAN ECOLOGY: THE UNCOMFORTABLE SCIENCE OF ACCOUNTABILITY

Humans share this small planet with a multitude of other living things, from bacteria to redwoods. We live in a complex environment that includes both living (biotic) and nonliving (abiotic) components, with each interacting intimately and crucially with the other. The study of these intricate interactions is called **ecology.** It may be the ultimate science, since virtually all other branches of science somehow affect the living or nonliving factors on the Earth.

Human ecology focuses specifically on the interactions and impact that human activities have on our environments. It is a useful field of biology, in part because it allows us to step back from our own immediate concerns and take an overview of the world around us.

Our findings show us that the continuation of all life, including humans, ultimately depends on how our activities affect the various life support systems of the biosphere. The issues are so important that human ecology has been described as the "religious science." It goes beyond simply describing mechanism and observation, as do the other branches of science, and focuses our attentions on the "quality of life." It suggests *what we should* be doing and not just *what is.*

Of course, ecological "shoulds" are the fodder of crusades, and certainly you will see many "environmental causes" promoted in your lifetime. We are not talking of issues and decisions that will have to be made and problems that will have to be faced by generations of the distant future. *Yours* is the generation that must show the wisdom and make the sacrifices. (And yes, earlier generations should have been wiser and less greedy.)

If the current mathematical models are correct, there may be nearly ten billion people living, eating, consuming, and polluting by the year A.D. 2030 (twice as many as now). If you are a 20-year-old college student, you will be nearly 60 years old by then. You may have 35-year-old children and 12-year-old grandchildren. The issues will not go away by themselves, in spite of the fact that many people have grown tired of hearing about them.

What can you do? A starting point is to learn something of the structure of nature and of the human place in it. Then you must educate yourself on the issues that are most likely to affect your life. Sooner or later you will have to take a stand on those issues—at the ballot box and in the way you live your own life.

THE BIOSPHERE

Our planet, the Earth, is a tiny ball of rock about 8,000 miles (12,800 km) in diameter, orbiting a nearby star, our sun. Approximately 80 percent of the Earth's surface is covered with water, which, at its deepest, is only about 5 miles (8 km) deep. The combined watery environments, including the lakes, rivers, and oceans, collectively form the **hydrosphere.** It is in this hydrosphere that "life" originally formed. Circulating above the Earth's surface is the mixture of gases and water vapor that collectively form the **atmosphere.** Though the Earth's interior is molten rock, its surface layers have cooled and solidified to form the **lithosphere,** the Earth's crust.

Analogous with these layers is that thin skin of life we have named the **biosphere.** The biosphere occupies the surface of the land, the crevices in the top hundred meters or so of the crust, and, for the most part, only the upper few hundred meters of the seas. Although there are bottom dwellers that live off the steady rain of detritus from the ocean's surface or tap seafloor minerals for energy, they are a distinctly minor component of the biosphere. We also can find a few outliers on the tops of the highest mountains and even—in the form of bacterial and fungal spores—on the air currents high in the atmosphere. But almost all living things dwell in a zone at most several kilometers thick. All the life we know exists in a thin layer, the biosphere, on the outer surface of a single, fairly small planet (Figure 26-1).

ECOSYSTEMS

It is virtually impossible to study meaningfully or contemplate productively something as large and complex as "nature." We deal best with discrete and manageable blocks of information that can be conveniently broken down into their components. Such a definable unit in nature is an **ecosystem.** An ecosystem is a specific area in which there is a dynamic interplay between the living elements of the system and the physical factors (light, temperature, availability of water, weather, and so forth) of the environment.

The living components of an ecosystem—collectively referred to as the **community**—include all the plants, animals, bacteria, fungi, and protistans living inside the system. Other biotic factors are nonliving products of living systems, such as their remains, dead leaves and bodies, and waste materials.

FIGURE 26-1 This disquieting photo of Earth was taken by the astronauts of Apollo 10 from a quarter of a million miles away. The west coast of Africa can be seen through breaks in the clouds that cover most of the continent. In it, we see not just the beauty of our planet but also its frailty. All life as we know it is drifting through space clinging to this single tiny ball of water and rock.

A convenient feature of the ecosystem concept is its definability. We can consider the entire world as an ecosystem or we can contract our focus and study desert or urban ecosystems. Just as conveniently, we can construct our own private ecosystem in an aquarium; we can even consider the human body as a personal ecosystem. This flexibility leads to some vagueness because the boundaries between one ecosystem and another are sometimes not obvious (Figure 26-2).

Large and complex systems can be **complete ecosystems** if they incorporate enough resources, photosynthetic fixation of energy, and recycling of essential chemicals by the processes of decomposition so that the system can persist indefinitely and be self-sustaining. Large areas like the midwestern United States or the state of California may be complete ecosystems. The limiting factors that are most likely to keep a system from being a complete ecosystem are insufficient area for photosynthesis to meet the system's needs for energy or limited availability of water or light.

An **incomplete ecosystem** is one that is not large or complex enough to be self-sustaining. A small town, for instance, would be considered an incomplete ecosystem, since it is likely to need food, energy, and perhaps water from outside areas. For that reason, the human body is also, at best, an incomplete system.

---

1. Clearly, some abiotic ecosystem components are long-term products of living things. How does this comment apply to air?
2. How does the comment in question 1 apply to soil?

Those parts of the world we generally call ecosystems are natural or discrete units named for the plant species that dominate them or for some other conspicuous feature they possess. There are coniferous forests, hardwood forests, old field ecosystems, cave ecosystems, deserts, and rain forests (to mention just a few). In addition, several major ecosystems, called **biomes,** are spread across large geographic areas (see Box 26-1).

The boundaries between different ecosystems are usually distinct enough to be recognizable. A grove of hardwood trees is distinguishable from the cultivated field next to it. However, the boundaries are not like walls. There is always some interaction between neighboring ecosystems, for—even

(a)

(b)

(c)

(d)

FIGURE 26-2 Identify the biotic and abiotic (physical) components of each of the systems depicted in this series of photographs: (a) rain forest; (b) tide pool; (c) a terrarium; (d) a human body.

Box 26-1 Biomes: Major Natural Regions of the World

We use several systems to characterize the various floral (plant) and faunal (animal) associations in the world. One of the most widely accepted is the "biome" system pioneered by an early ecologist, Victor Shelford.

The biome concept is based on the fact that the structure of most natural systems is largely determined by a region's climate, geology, availability of water and light, and vegetation. Each biome is a distinctive association of plants and animals in the "climax" phase of its development. Each biome is also a group of related ecosystems distributed across a fairly broad geographical area. Most biomes get their names from their dominant plant species or from some other easily recognizable feature they possess. For instance, there is a tropical rain forest biome, a temperate grassland biome, and a coniferous forest biome, to mention just a few.

The major biomes are distributed around the hemispheres in east-west bands that in some cases may be hundreds of miles wide. For instance, in the northernmost areas are the arctic ice fields. Just below these lies the tundra biome, followed in the south by the coniferous forest biome, and so on. The distribution of biomes is clearly influenced by latitude; it is also affected by major geographic features, such as the presence of mountain ranges or bodies of water. (See Figure 1a–h.)

(a) Tundra biome: The frozen plain of the tundra extends from the upper reaches of the timberline in the far north to the Arctic ice. It is found as a band across the upper latitudes of North America, Europe, and Asia and also at higher elevations near the tops of mountains in the lower latitudes. Severe cold and dry air characterize a tundra most of the year. A brief summer thaw allows a short (60-day) growing

FIGURE 1 (a) At higher elevations (above the tree line) and in the far north is found the tundra biome, shown here in autumn colors in the Denali National Park in Alaska.

season during which lichens, grasses, sedges, insects, rodents, dwarf woody plants, and migratory animals quickly complete each year's life cycle. It is an area of permafrost (permanently frozen ground) where only the top few inches thaw during each summer. The thaw of the upper layers produces a boggy, wet grassland with numerous ponds, streams, and lakes.

(b) Coniferous forest biome: Extensive bands of coniferous forests extend as broad belts across the northern reaches of North America, Europe, and Asia, beginning at the southern edge of the tundra. The dominant trees (spruces, pines, and firs) are conifers, from which the biome receives its name. All produce their seeds in conelike reproductive structures and possess needles or scalelike leaves that resist desiccation (drying out) in cold, dry winter air. The coniferous forests also extend their southernly ranges in the upper elevations of mountain chains. The coniferous forests produce much of the lumber that is harvested each year. It is also the home of many larger vertebrates including snowshoe hares, lynx, moose, bears, and squirrels (and other rodents), and numerous bird species.

(b) The coniferous forest in the mountains of Alaska; interspersed with the conifers are aspen trees in their yellow autumn colors.

(c) Temperate deciduous forest biome: Abundant and evenly distributed rainfall (30–60 in. annually) and moderate (seasonal) temperatures are usual in areas supporting the temperate deciduous forests. They originally covered most of eastern North America, virtually all of Europe, and parts of Japan, Australia, and the southern end of South America. Most trees and shrubs in this biome drop their leaves during the fall and winter and go into a state of dormancy. Many of these forests have been replaced by cultivated areas and the encroachment of cities and towns.

(c) Temperate deciduous forest of eastern United States.

(d) Savanna biome: Savannas—usually found in warm areas with substantial rainfall (often 40–60 in. annually)—are grasslands with scattered clumps of trees at varying intervals. The largest savanna is found in the plains of Africa (there are similar areas in South America and Australia). Both the grasses and trees in these areas are drought resistant. The African sa-

(d) Savanna biome.

vanna supports the world's largest natural population of hoofed animals (antelopes, giraffes, zebras, etc.) and their predators and scavengers.

(e) Temperate grassland biome: Grasslands occur in areas receiving too little rainfall to support a forest—normally in areas that receive between 10 and 30 in. of annual precipitation. Most temperate grasslands occur in the continental interiors, away from the coasts. The soil is usually rich with an abundance of organic humus. The extensive moist temperate grasslands are the pastures and farmlands that provide much of the food for humans. It is in these areas that the earliest human civilizations probably first flourished. Early conversion of grasslands to farming probably caused few ecological problems since many of the crops are similar in nature to the grasses usually found in those areas. The greatest misuse of grasslands has occurred with overgrazing, which has been responsible for extensive loss of usable land through desertification.

(e) Grassland—short grass prairie.

(f) Chaparral: The chaparral biomes are found in areas of moderate winter rainfall with almost no summer precipitation. In fact, in some chaparral areas the growing season is as short as four months. It is characterized by dense shrubby growth punctuated with occasional small trees. The drought-resistant trees and shrubs often possess small leathery leaves that persist all year. Chaparral is found in the foothills of California, in Mexico, along the shores of the Mediterranean, and on Australia's southern coast. In most regions, fire is an important factor in maintaining the shrubby nature of the community. In fact, some of the most flammable plants in the world grow in the chaparral. Periodic fires (about every 25–50 years) maintain the chaparral community's health by releasing the nutrients locked up in plant tissues and by triggering the sprouting of seeds.

(f) Chaparral of the Santa Monica mountains near the west coast of the United States.

(g) Desert biome: Aridity (usually less than 10 in. of rain per year) is one of the key features characterizing the desert biome. This dryness can be caused by the desert's position on the planet—such as in rainshadows behind mountain ranges, in the subtropics (which are deprived of rain by atmospheric circulation patterns), and at the high altitudes of Tibet and Bolivia—and by climatic changes provoked by human-directed activities such as deforestation or overgrazing. Some areas (northern Chile and central Sahara) are so dry that they receive absolutely no rainfall. Desert plant life must be extremely drought resistant and includes populations of annuals that grow only during periods following rainfall, of succulents that store water in their tissues, and of desert shrubs with elaborate root systems and tough leathery leaves (which may be shed in prolonged periods of drought). Deserts can be classified as hot deserts or cold deserts based on temperatures. Most animal life is insect or nocturnal. Even lizards and snakes that elsewhere are active during the day retire from the heat of the sun.

(h) Tropical rain forest biome: The tropical rain forests include the richest biotic regions of the world. They are broadleafed forests in areas of high rainfall (80–90 in. or more annually), located close to the equator. The rainfall is distributed throughout the year; seasonal "dry" periods still average 5 in. of rain per month. The rain forests occur in three regions of the world—the Amazon and Orinoco basins in South and Central America, central and western Africa, and the Indo-Malay-Borneo and New Guinea areas. In the tropical rain forests, the seasonal variations in temperature are less than the temperature differences occurring between night and day.

(g) Desert biome (Sonoran).

(h) Tropical rain forest, Hawaii.

though ecosystems do not exchange much in the way of material—blowing leaves, flying birds and insects, flowing water, and animal wastes do cross their boundaries.

1. In what biome do you live?
2. What other biomes are within a day's drive of your home or campus?
3. What wild animals share your biome with you? Would you expect to find them in other biomes as well?

TROPHIC STRUCTURE: ENERGY AND NUTRIENT FLOW

There is a great deal of interest in the way energy and nutrients move through the community of an ecosystem. Ecologists have developed several different schemes for describing the patterns of energy and nutrient flow that collectively fall under the heading of *trophic (food) structure*. There are food webs, food networks, and ecological pyramids. Each scheme contributes a slightly different view of the interrelationships in the system. We will discuss only a few of these schemes.

Producers

Since the photosynthetic activities of green plants, algae, and cyanobacteria capture a portion of the energy from the sun and convert it to a form (glucose) that can be used by virtually all other forms of life, they are said to be **producers.** They produce, or at least "package," the energy in a form needed to keep the biosphere in operation. Because in the process they manufacture many of their own nutrients, we also call them **autotrophs** (self-feeders). Thus, they do not need to eat other living things, as do the animals, or **heterotrophs** (other-feeders).

Consumers

All living things that cannot manufacture their own food we call **consumers.** Each different feeding position in the scheme of nutrient flow is called a **trophic level.** The producers form the first step, or trophic level, in virtually every nutritional system. Each time the food is consumed as it passes through a community, it moves through another trophic level. The organisms that eat plant matter (herbivores) are called **primary consumers** because this is the first time the nutrients are consumed after they were first produced or harnessed by photosynthesis.

Animals that eat herbivores (carnivores) are **secondary consumers.** Carnivores that eat other carnivores are **tertiary consumers.** And so on.

Because of the energy lost as heat (see the discussion of the laws of thermodynamics in Chapter 2) each time energy moves through a trophic level, most chains of eater and eaten have only five or six links. Beyond these, there is not enough usable energy left to power further biotic activity.

Most consumers do specialize to some extent in their food habits. Cats and dogs are carnivores for the most part. However, virtually everyone has seen a dog or a cat chewing grass from the lawn. On occasion, they can and do shift their trophic activities from secondary back to primary consumption. Other animals, including humans, are capable of functioning comfortably at many trophic levels. We eat flesh, vegetables, fruits, and grains, qualifying us as omnivores (*omni,* everything; *vore,* eater). It is our dietary flexibility that has contributed greatly to our species' ability to survive in a changing environment.

Detritivores

Some consumers are **detritivores.** They eat only *detritus,* the already dead—fallen leaves, dead bodies, the scraps left by carnivores and herbivores, and the wastes of other consumers. They include scavengers such as the crows that eat the bodies of cats and squirrels killed on the highway.

Decomposers

Decomposers are the bacteria and fungi that break down waste materials and dead remains of other organisms into mineral and organic nutrients that can be used by plants. They are responsible for recycling nutrients and for producing the organic component of soil, **humus,** that is so important to soil fertility. Humans use them in compost piles and in sewage treatment plants, where they reduce noxious sewage to relatively nonpolluting plant nutrients. Some decomposers consume living matter as disease organisms.

One difference between detritivores and decomposers is simple. Detritivores are generally large animals that put their food inside their bodies. Decomposers are small and live inside their food, digesting it outside their bodies with enzymes they secrete.

We have purposely not named many examples of producers, consumers, detritivores, and decomposers. We want *you* to supply the examples by answering the following questions:

1. Think of what you have eaten in the past week and name three producers you have consumed.
2. Was the cow that became your last hamburger a primary or a secondary consumer? Why?
3. What kind of consumer does eating a hamburger make you?
4. Suppose that on a hike in Yellowstone National Park, a friend is killed and eaten by a grizzly bear. What kind of consumer is the bear?
5. What kind of consumer is the mouse that carried off and ate the scraps dropped by the bear?
6. What kind of consumer is the turkey vulture that ate the bear after the ranger shot it?

Food Chains and Webs

Ecology is sometimes defined as the study of who eats whom. This definition is simplistic, for various living things interact with each other in many ways other than as food. But it does have a large component of truth. Life depends on the flow of energy captured from the sun by plants and passed on to consumers, detritivores, and decomposers as living things in turn eat and are eaten.

We can see this truth plainly if we consider a sequence of eater and eaten. In its simplest version, this sequence is that of a **food chain.** It begins with the plants or producers that manufacture organic nutrients—starches, fats, proteins, and vitamins—using the energy of sunlight. Primary consumers eat the plants, secondary consumers eat the primary consumers, and tertiary consumers eat the secondary consumers in turn. All eventually die and become food for detritivores and decomposers (Figure 26-3).

More realistically, however, we should draw a **food web,** for few animals are pure primary or secondary or tertiary consumers. Food webs capture the realities of the way animals vary their diets (Figure 26-4). Most carnivores eat both herbivores and carnivores as well as occasional vegetation. Anyone who has ever kept a dog has seen the variety of the carnivore diet, for dogs will also eat carrion (dead—often long dead—meat), table scraps (including beets, rice, and especially green vegetables), and occasional fresh kills; they are thus detritivores as well as carnivores and herbivores. Similarly, many herbivores will at least occasionally eat meat (a cow has been seen eating a squirrel). Rats and mice eat mostly seeds and berries, but they will also eat insects, and they have been known to gnaw the fingers of children living in slums (Figure 26-5).

However, no food web can be completely realistic. An ideal food web diagram would contain each individual kind of plant and show how many herbivores prefer to dine on only one or a few plant species as well as all the different animals (and plants) the carnivores eat. Of course, almost all ecosystems include far too many different organisms to fit them all—and their eater-eaten relationships—into a single diagram.

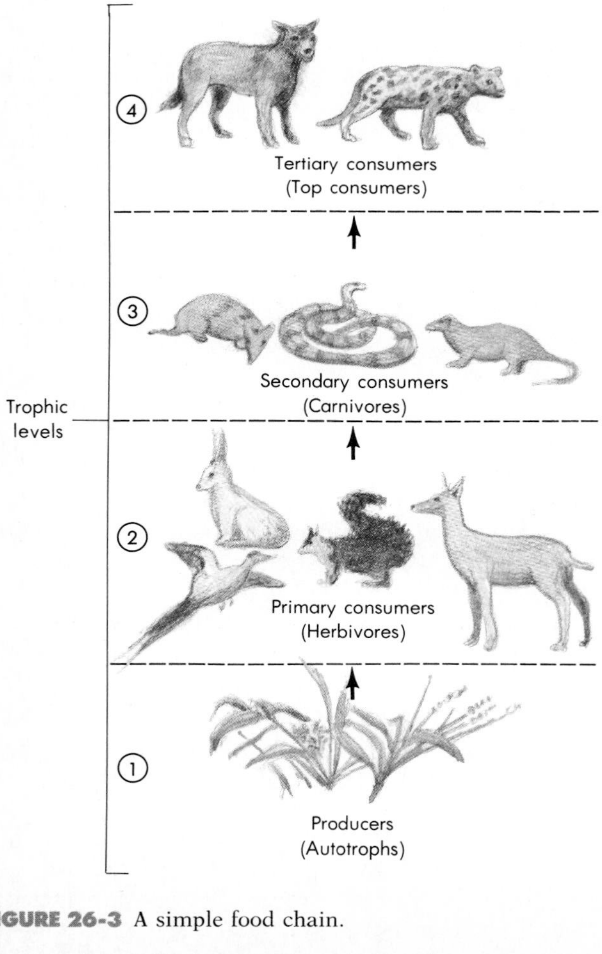

FIGURE 26-3 A simple food chain.

Ecological Pyramids

The second law of thermodynamics says that no energy transaction can be 100 percent efficient (see Chapter 2). Whenever energy is used to do work or is transformed from one form to another, some of it is lost. When we use the electric energy stored in a battery to run a motor, some of that energy is wasted as heat. When we burn gasoline to run a car or oil to heat a house, we do not get the benefit of all the energy in the gas or the oil.

This same principle has an enormous influence on the structure of ecosystems. Plants cannot capture all the energy in sunlight. In fact, they capture in photosynthesis only a little more than 1 percent of all the usable energy in sunlight that reaches the surface of the Earth. The figure is so low because plants can absorb only certain fractions (wavelengths) of sunlight and because, of those fractions plants do absorb, some of the energy they carry is lost as heat. Of the energy that plants actually convert into useful form, some is immediately consumed by the plant to run its own life processes. It is thus unavailable to the rest of the biosphere. A little less than 1 percent of the sunlight that strikes the plants' leaves is converted into plant matter and actually made available as food for consumers.

Efficiency of Trophic Energy Transfers

Animals are also inefficient. When herbivores eat plants, they lose some of the energy in their food as

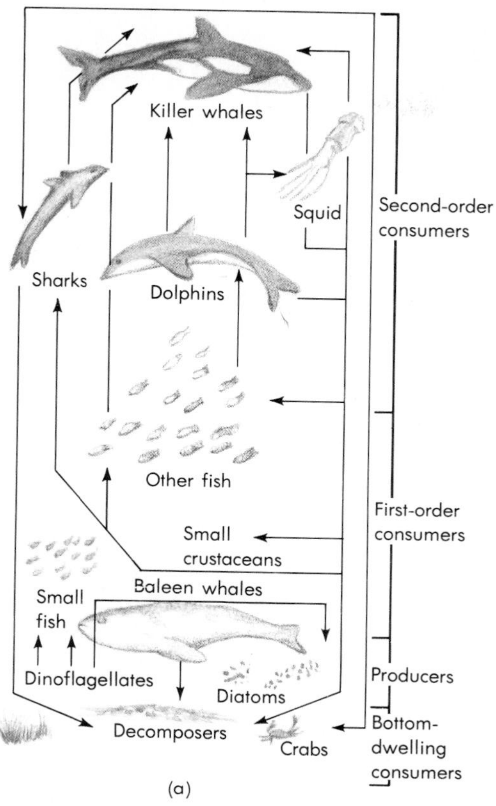

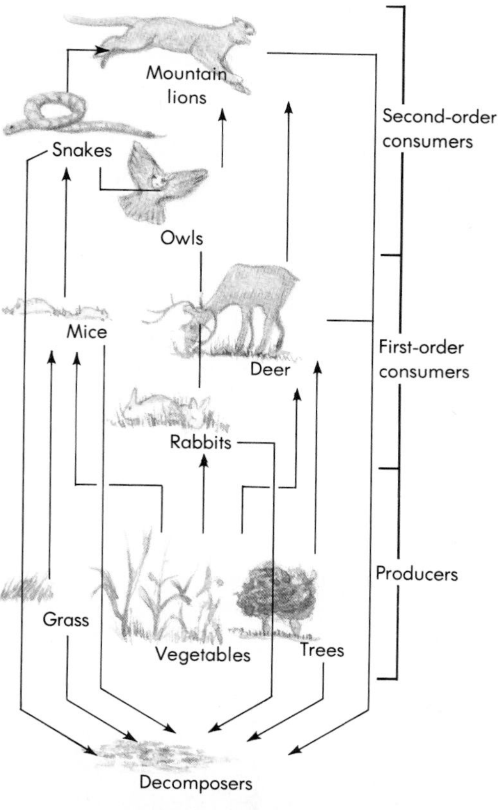

FIGURE 26-4 (a) Marine food web. (b) Terrestrial food web.

FIGURE 26-5 A grasshopper mouse, *Onychomys*, eating a grasshopper.

heat and use some for their life processes. They convert roughly a tenth of the energy in their diet into their own structures. In turn, carnivores are a little more efficient. In converting one form of animal flesh to another, the energy transferred from flesh to flesh ranges from 10 to 20 percent. Detritivores and decomposers also convert roughly a tenth of the energy they consume into their own substance. Since an organism's mass reflects its energy content, we can measure its energy content in terms of its mass. We can then prepare a diagram that shows how the energy contents (or masses) of the various trophic levels depend on each other (Figure 26-6). This diagram, because of its shape, is an example of what we call an **ecological pyramid.** In particular, it is a **biomass pyramid.** It shows that the first trophic level of an ecosystem contains the most mass (and energy) and that each successive level contains about 80–90 percent less mass (and energy).

Ecological pyramids offer us a clear lesson, for they tell us that the lower on a food chain we humans can eat, the more of us the world can support. Think of two human populations. A vegetarian population that subsists on grains and vegetable matter will have 90 percent more transferable energy available to it than will a population that exists by consuming exclusively the flesh of primary consumers, like cattle and sheep. Thus, on the same quantity of arable land, a vegetarian country could support a population roughly 10 times larger than could a carnivorous country. However, no population is exclusively vegetarian or carnivorous.

Survival economics forces many poorer countries (which often have large populations) to tap their nutritional options at the lowest possible level—namely, by consuming producers. For example, the typical European diet receives about 22

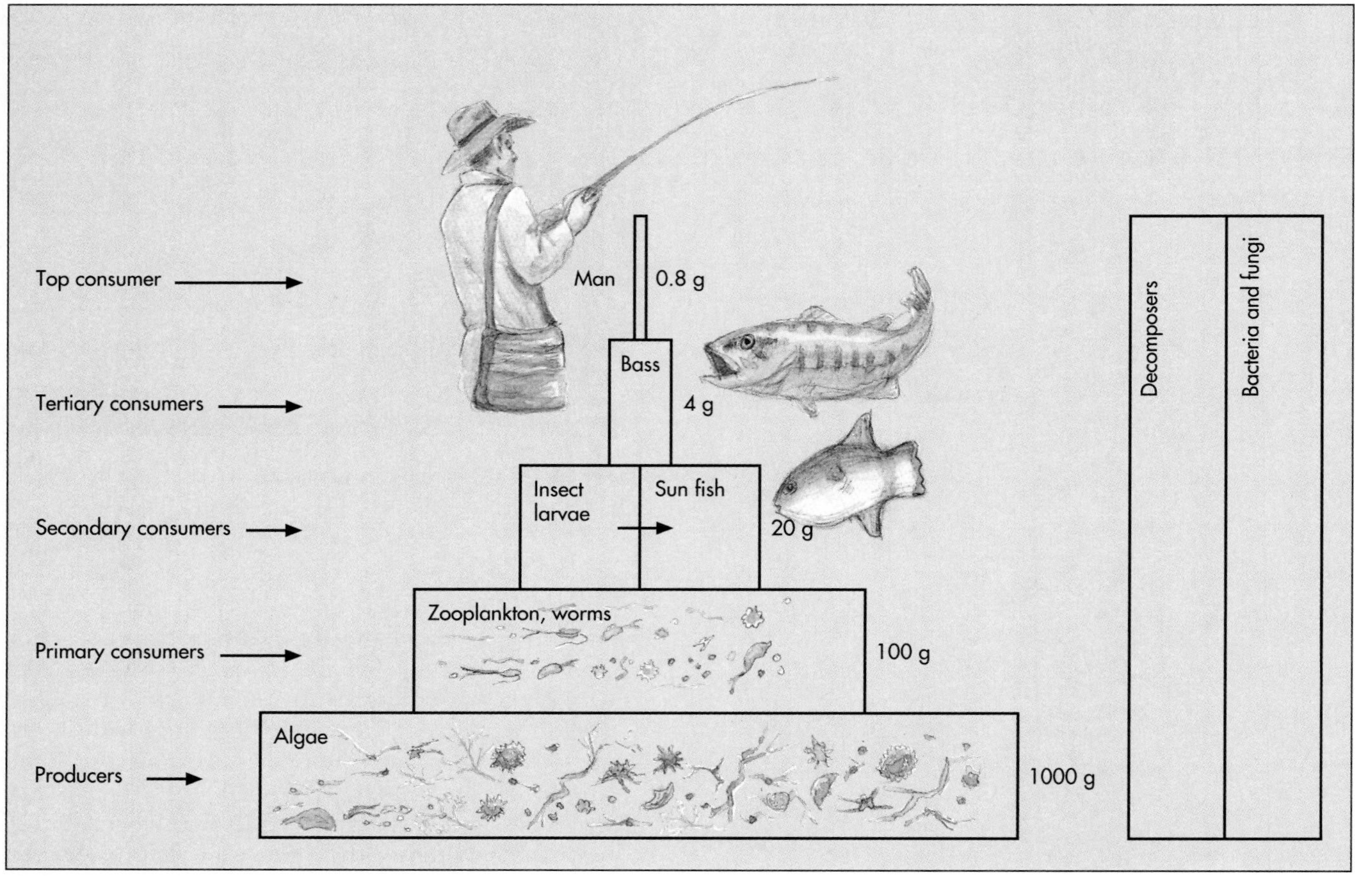

FIGURE 26-6 Biomass pyramid.

percent of its calories from meat, fish, and eggs. For more affluent countries, like the United States and Canada, the figure is over 31 percent. In Asia, only about 3 percent of the calories come from animal sources. We should be aware, however, that even in affluent meat-eating societies like the United States, the majority of caloric intake comes from grains, potatoes, and green vegetables—not meats.

On the other hand, there are many reasons for continuing the production and the consumption of a certain amount of animal protein. Certain kinds of animal protein are produced much more efficiently than our discussion of ecological pyramids suggests. Cows, chickens, rabbits, and some fish convert their feed to milk (and cheese), eggs, and meat with an efficiency of as much as 25 percent. There is considerable variability; chickens, under ideal conditions, convert 2–3 kg of feed into 1 kg of flesh. On the other hand, it is not unusual for cattle to convert 6 kg of feed into 1 kg of beef.

Many animals can be raised on land unsuitable for growing crops. In addition, "integrated systems" produce both plant and animal food on the same land. For example, tilapia fish are raised in the waters of rice paddies, where they consume insect larvae that might harm the rice plants and fertilize the plants with their wastes. Some integrated systems are less deliberate, as in the Philippines, where farmers harvest the rats that flourish in their grain fields.

INTERSPECIFIC INTERACTIONS: SYMBIOSIS

Living things do not interact with each other only as eater and eaten. Burrowing animals aerate and loosen the soil and make conditions more hospitable for plants. Squirrels store seeds by burying them and thus aid plant reproduction. Birds and other animals aid plant dispersal over wide areas by depositing seeds with their droppings. Insects and birds are instrumental in pollinating many plant species. Trees provide shelter and nest sites for many birds. In many cases, the plants and animals that help each other in such ways have evolved together to take advantage of their interactions (Figure 26-7).

(a)

(b)

(c)

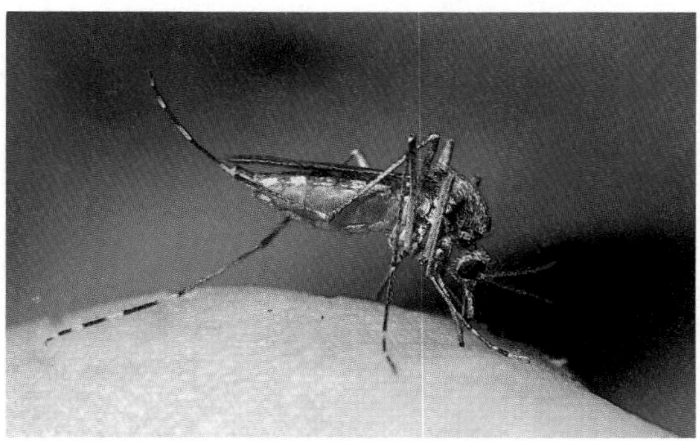

(d)

(e)

(f)

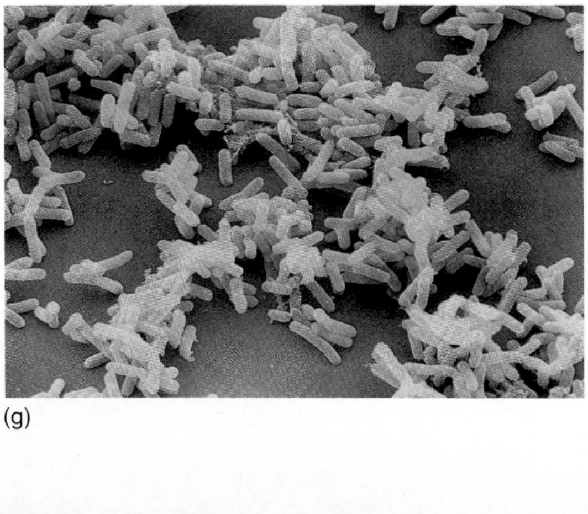

(g)

FIGURE 26-7 Interspecific interactions. (a) Mutualism includes the mutually beneficial relations between honey bees and the flowers they pollinate. (b) Earthworms establish complex interrelationships with many species that depend on their aeration of the soil. (c) Sharks and their ride-along companions, the remora, have a mildly parasitic relationship. We might argue, however, that the impact of the remora on the shark is so minimal that the relationship would qualify as commensal. (d) A parasitic relationship between a female mosquito and its human host. A blood meal is necessary for the female to complete its reproduction cycle. (e) These nesting cardinals have a commensal relationship with their nesting tree. (f) The lioness and her cubs have a predator relationship with their prey. (g) Because both humans and the bacteria *Escherichia coli* in their intestines benefit from the association, the relationship is often called mutualistic.

These **interspecific** (between-species) **interactions** are examples of what we call **symbiosis.** Symbiosis means simply "living together," and there are several ways organisms can live together. When members of two different species benefit each other, like humans and the vitamin-producing bacteria residing in our guts, we call that relationship **mutualism.** It is considered mutually beneficial to both species. When only one species benefits while the other is neither helped nor harmed, we call the relationship **commensalism.** We can see an example of commensalism in the relationship between college students and the birds that feast on the crumbs dropped from their snacks around campus. Another is the small crabs living in the burrows that certain worms make in the ocean bottom; the crabs gain a safe place to live, while the worms seem not to notice their tenants at all.

A less positive kind of symbiosis is **predation,** where one animal species (such as a lion) benefits at the expense of another (such as an antelope). Generally, the predator is larger than its prey, and the predacious act usually results in the quick death of the victim. However, we should not make the mistake of thinking that what the predator eats fails entirely to benefit. Animal predators remove the aged and ill from their prey's populations, enhancing the vigor of the species by guaranteeing that the environment's limited resources are most likely to be shared among the fittest individuals.

Parasitism is a variation on the same theme of one species benefiting at the expense of another. Although the parasite does live at the expense of the host, it is often smaller than its host and usually does not kill it. *Endoparasites* live inside the host, as does a tapeworm or a disease-causing bacterium. The parasite can either compete with its host for its own food or tap the host itself, drawing nutrients from its blood or even its cells. There are also external parasites, *ectoparasites* (like fleas and lice), living on the skin or in the fur; they draw their sustenance by biting through the skin and drawing out body fluids. Some parasites are very harmful to their hosts, robbing them of energy, impairing their growth, and even killing them.

A long-term parasitic relationship may have little effect on the host and, in fact, may even develop into a protective relationship between the parasite and the host. Botfly larvae, for instance, are maggot-like creatures (half the size of your little finger) that develop just below the skin of their hosts, wild rabbits. You might suspect the presence of the larvae would exact a rather severe toll on the host. However, some researchers believe the larvae may actually benefit their host by secreting antibiotics that protect the rabbit from infections while the association exists.

We see the worst effects when a parasite adapted to one host somehow invades another, as when humans eat raw or undercooked pork and are infected with trichinosis. The worm responsible hardly affects pigs (and bears), but it can cause pain, and rarely death, in humans.

1. What selective pressures must be working to reduce the harm a parasite does to its host? That is, how does a parasite that kills its host reduce its own reproductive success?
2. What are the differences between a predatory and a parasitic relationship, since both types of relationship are likely to benefit one species at the expense of the other?

COMPETITION

Still another kind of interspecific interaction is **competition.** The resources—food, space, sunlight, nesting sites, water, and so on—that any species needs are in limited supply, and other species often need the same resources. Many different carnivores seek mice as food, for instance. Many birds wish to nest in the same kinds of trees. Plants all need their space in the sun. There is thus often a race or contest, a competition, between species to seize resources first and to hold them against all comers. The more similar the competing species are in their environmental requirements, the more intense will be the competition between them.

However, every species does not compete with every other species. For one thing, competition can occur only between species that share the same **habitat.** That is, only species that live in the same place, such as a temperate forest or a desert or an abandoned pasture, can compete for local resources.

In addition, not all species seek the same resources. Some birds, for instance, prefer shrubs for their nests, while others favor pine trees or maple trees or even rocky ledges. Plants may grow best in different kinds of soil or different levels of sunlight. Predators may favor different prey. We thus say that species differ in their **ecological niches.**

The best way to think of an organism's habitat may be as its address—where you would go to find it. It is defined by its community, climate, terrain, soil type, and other environmental factors. Its ecological niche, on the other hand, is its profession or way of life. It includes diet, the time of day it prefers to be active, its breeding season, its relationship to other members of the community, and other biological factors. The concept of "niche" is a difficult one because it is an infinite concept; indeed, attempts to describe a niche are somewhat arbitrary and never ending.

Competitive Exclusion

Competition between species almost always concerns niche factors, and it is a truism of ecology that two species cannot occupy exactly the same niche. This **competitive exclusion principle** was first demonstrated by a Russian biologist, Gauss, who put two species of the protozoan *Paramecium* in a dish and found that one species always died out (Figure 26-8).

It seems that in any given environment where two similar species are competing for limited resources, one species will always be slightly more successful in gaining its food or in fulfilling some other critical need. Eventually it will outreproduce the other species and drive it to extinction. If you change the competitive arena ever so slightly, the other species could emerge the victor.

Not every interaction between different species is a form of competition. For example, plants sometimes release into their surroundings chemicals that either stimulate or inhibit the growth of neighboring plants. The primary purpose for some of these chemicals seems to be to defend against insects or to limit the spread of disease in the plant's niche. The effects on the growth of neighboring plants is secondary or accidental.

Realized Versus Potential Niches

Gauss's paramecia are found in the same ponds, but outside the laboratory they do not eat the same food. One is smaller than the other and it prefers

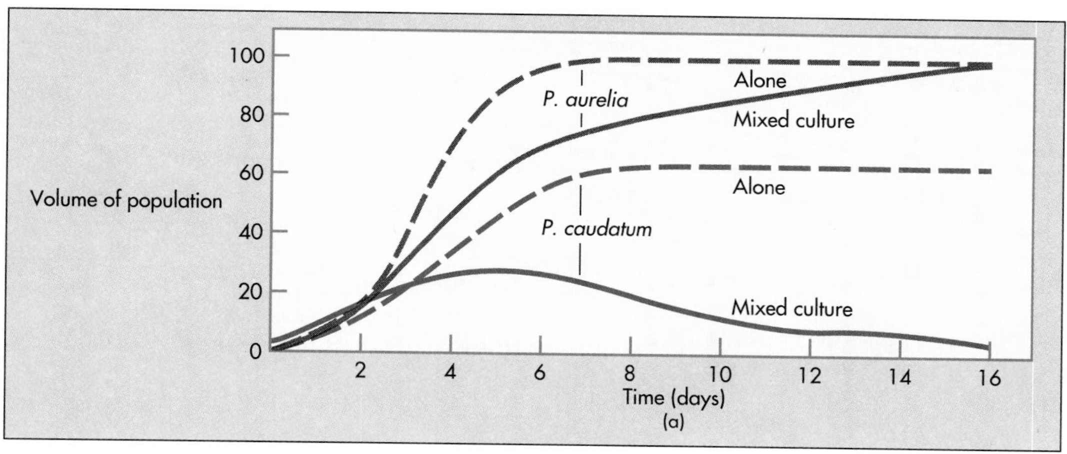

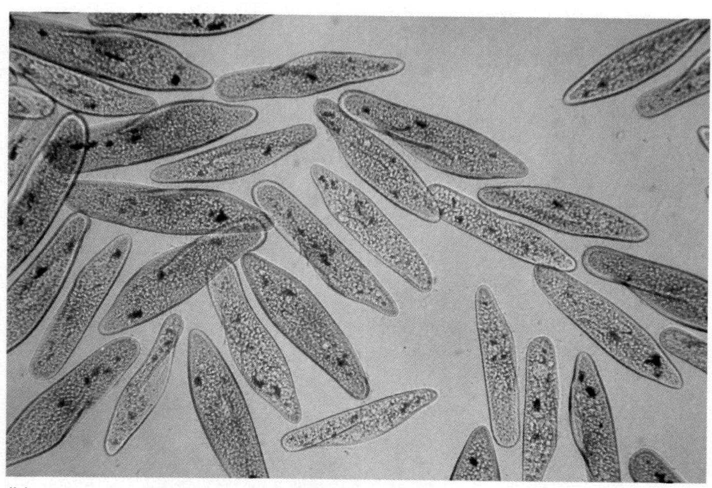

FIGURE 26-8 Gauss's *Paramecium caudatum* vs. *P. aurelia* experiment. (a) When cultures of *P. caudatum* and *P. aurelia* are grown separately, their populations quickly increase and level off. When they are grown together, the population of *P. aurelia* grows slowly, while *P. caudatum* declines. The two similar species cannot share the same niche. (b) Photomicrograph of *P. caudatum*.

smaller prey. There is some overlap, but in nature, where one species is a more able competitor, the other can step aside, concentrate its efforts on some other portion of its niche, and continue to survive. We say that each species has a **potential niche** that it occupies in the absence of competition and a **realized niche** that it occupies when some other species preempts some of the resources it might otherwise use. Because realized niches can also depend on the resources available, manipulating them makes possible a very effective way of controlling some insect pests. For instance, the lygus bug, a cotton pest, prefers alfalfa to cotton and can be kept out of cotton simply by planting alfalfa nearby. Since cotton is the more valuable crop, this tactic lessens the bug's economic impact.

Pesticides and Vacated Niches

Thanks to competition, very few species other than humans occupy their full potential niches. This is one reason why the use of pesticides to kill agricultural pests often leads to outbreaks of crop-eating insects that earlier had not been pests. For example, a farmer sprays only his fields with the pesticide, and he kills only those insects whose niches put them in his fields. Yet other insect species can also eat the crop plants, but they are not in the fields, sometimes because they have been excluded by competitors. When the farmer removes those competitors, the insects that had until then been surviving in the nearby woods and hedgerows promptly invade and begin to devour the crop.

Learning Focus

TEST YOUR UNDERSTANDING OF ECOSYSTEMS

Imagine yourself in a sunny world we will call "Wonderland" (with apologies to Lewis Carroll). Plantlike "momes" grow in the "wabe" or soil, "slithy toves" eat the momes, "gimble gyrers" eat the slithy toves, and "bandersnatches" eat all dead bodies and meal scraps. In the wabe live "snarks" that break down waste materials and those scraps the bandersnatches miss and return their components to the wabe. Now answer questions 1–7.

1. In this scene, identify the producers, primary consumers, secondary consumers, decomposers, and detritivores.

2. What technical term would you apply to that part of Wonderland the scene describes?

3. Sketch the food web of our Wonderland scene.

4. What organism occupies Wonderland's third trophic level?

5. Which Wonderland organism is an autotroph?

6. If there are 100 tons of momes, how many tons would you expect to find (roughly) of slithy toves and of gimble gyrers?

7. What would you expect to happen if a second kind of mome eater appeared in Wonderland?

INTERLOCKING SYSTEMS OF NATURE

We have not yet discussed the "balance of nature" primarily because there is no such thing, at least in the sense of a changeless status quo. Nature is an open system powered by the continuous flow of energy from the sun. As the energy passes through each trophic level, a fraction of it is channeled through the chemical reactions that power each unit of the biosphere. In time, all the captured energy is converted to drive the workings of living systems or is ultimately degraded to heat that flows away from the planet, never to return again.

The various elements of the ecological community interlock in a manner that helps ensure the continued existence of the biosphere. Fueled by the sun, plants provide food and shelter for animals as well as the oxygen released during photosynthesis. Animals in turn release the carbon dioxide the plants need as a source of carbon. They also support plant reproduction by scattering their seeds and loosening the soil. Decomposers keep the chemical constituents of protoplasm moving and available, enabling each generation to draw its share from the recycled reserves to support their own lives. Predators control the numbers of their prey, while the supply of prey itself restricts the numbers of predators.

The balance of nature resembles the dynamic, ever-changing, near steady state of physiology. It is an "ecological homeostasis" maintained by all the processes we have mentioned and many more besides. This homeostasis is so thorough-going that it has been said the Earth's biosphere modulates the supplies of nutrients, water, and oxygen and even affects climate in a manner maintaining its suitability for life. This is the so-called Gaia hypothesis, named for the ancient Greek Earth goddess.

1. Human use of fossil fuels increases the level of carbon dioxide in the atmosphere. Plants increase their use of carbon dioxide in photosynthesis when more carbon dioxide is available. How might these two facts add up to an example of ecological homeostasis?
2. As human population grows, it displaces plant life from the land. How might this process interfere with ecological homeostasis?

We cannot discuss here all aspects of ecological homeostasis. We can discuss only a few of the more obvious examples of how supplies of nutrients are kept in roughly constant supply and then consider, all too briefly, how human activities interfere with the stability of the biosphere. Let us look first at a few basic nutrient cycles.

Ecological Cycles

Plants draw their nutrients from the soil, water, and atmosphere; animals draw theirs from plants and other animals. However, during the decompositional recycling of nutrients back to the soil, some nutrients are bound to be lost as they are washed away to sea and buried in the deep sediments. Yet the land does not seem to become exhausted. It has remained fertile for all the 500 million years that life has been on land, and the seas have remained fertile for much longer. The supply of nutrients must be replenished constantly. To see how this works in the simplest possible way, let us consider the water cycle.

The Water Cycle

Water is among the most essential of all nutrients. Both plants and animals require it as a vehicle in which other nutrients and intercellular messengers can dissolve and be carried throughout their bodies, both within and among cells. It is the medium in which almost all physiological reactions occur. It is also needed for temperature control, both of the planet and of the living body.

Fortunately, the **water cycle** returns water to land as quickly as it runs off. As water warms, it evaporates and enters the air. Then, as the air rises and cools, the water vapor condenses into tiny droplets. Masses of these droplets form clouds that coalesce into larger droplets as temperatures decrease, until they fall as **precipitation,** rain, snow, or hail, some of which falls on land (Figure 26-9).

Most evaporation occurs from the surface of the world's oceans, which cover about 70 percent of the Earth's surface and contain 97.1 percent of the Earth's supply of water. Ice caps and glaciers contain 2.1 percent of the Earth's water. The 0.8 percent of the Earth's water that is both liquid and fresh does not sound like much when you consider that it is all that is available for human needs, but this small fraction is constantly renewed. A quarter of it accounts for the world's lakes, rivers, and streams. The rest is "groundwater," sponged up by soil and running in **aquifers,** beds of gravel, sand, and permeable rock deep below the land's surface. Some of this groundwater is renewed only very slowly, for it can take thousands of years for it to flow from its source (a lake or rainy area where

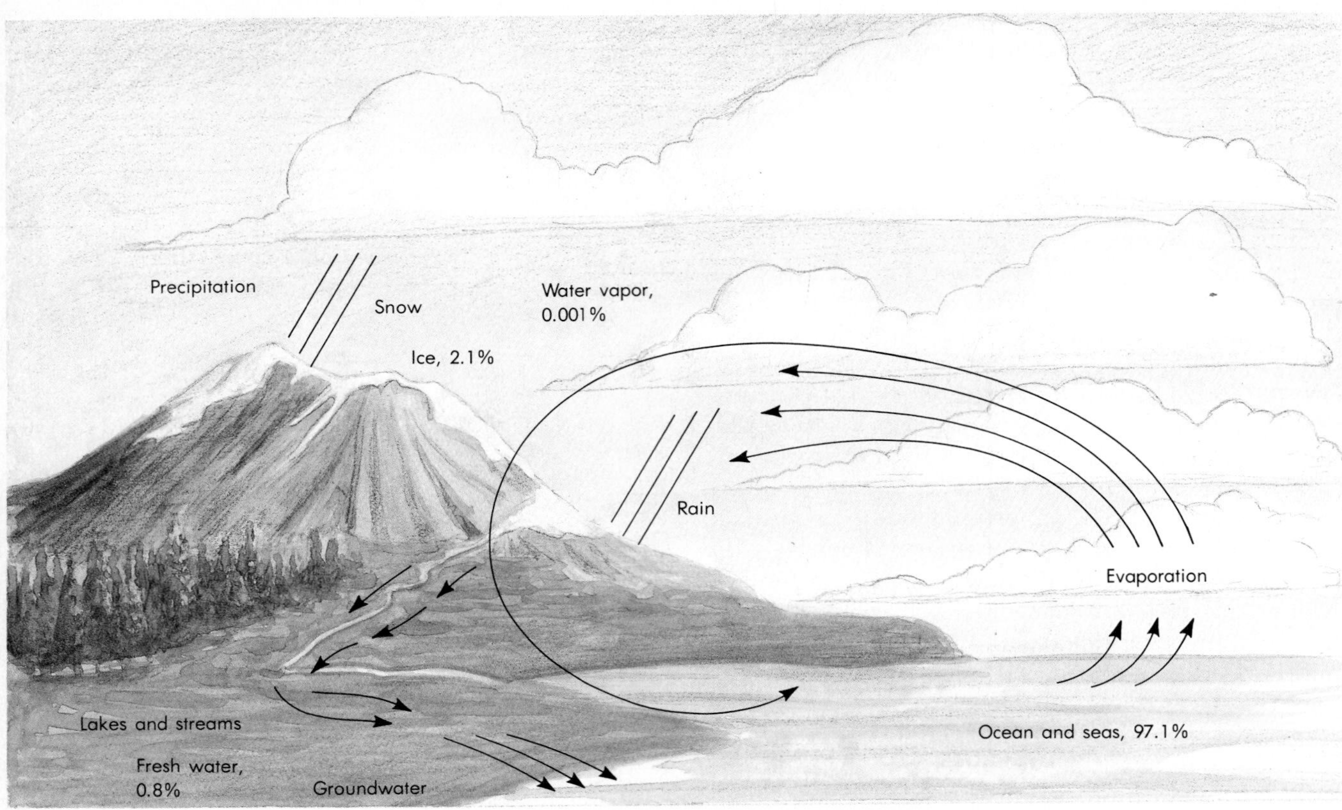

FIGURE 26-9 The water cycle.

water can seep into the aquifer) to the spring where it finally returns to the surface. Some, such as the groundwater beneath the Sahara desert, is not renewed at all, for it has not rained in its source area for millennia.

Humans tap the water in deep aquifers with wells and use it for drinking, washing, and especially irrigation of dry-land crops. However, this practice cannot long continue where humans draw the water from the ground more rapidly than it is being replaced. The water supply can be completely exhausted or, lacking that, its removal can leave open spaces in the rock deep below ground. When these spaces collapse under the weight of the rock, the surface of the Earth settles, opening cracks in the ground and in building foundations. In extreme cases, where the open spaces are actual caverns, huge craters can appear as the ground collapses into sinkholes that swallow cars, roads, and buildings (Figure 26-10).

Evaporation from soil and plant leaves can also be an important source of precipitation. There is concern that the clearing of the rain forest now under way in the Amazon basin of Brazil will so reduce the local rainfall that it will turn what is now jungle into virtual desert.

One of the most serious problems facing the world is the arrival of the time when there will not be enough usable fresh water available to meet all our residential, agricultural, and industrial needs. According to some experts, that time is rapidly approaching, because of increasing demand for water (due to expanding populations) and contamination of water supplies by toxic wastes, sewage, and other pollutants. The impact on human society is only suggested by the water rationing plans implemented in California in 1991, after 5 years of drought. Building permits were suspended, agriculture in the Central Valley was threatened, and homeowners were asked to curtail personal water use by as much as 50 percent.

The Carbon Cycle

Carbon, too, is an essential nutrient. Carbon dioxide is the principal source of organic carbon for the biosphere. It is first absorbed by plants and used during sugar synthesis to store the energy captured

FIGURE 26-10 A Florida sinkhole. The lowering of the water table removes support from the surface and permits the ground to collapse.

during photosynthesis. When animals eat plants (or other animals), they extract this energy and release carbon dioxide. Plants then absorb the carbon dioxide and use it to make more carbohydrates, proteins, and other materials. They also release the oxygen animals need.

Not all carbon dioxide follows this simple **carbon cycle,** however. Much of it dissolves in water as bicarbonate ions and is used by animals such as clams, mussels, oysters, and scallops that make shells of calcium *carbonate* or is incorporated in sediments. It can then become rocks such as limestone. Organic carbon can also be buried in sediments, and this becomes fossil fuels such as coal, oil, and natural gas. The carbon in both stone and fossil fuels is eventually returned to the biosphere by such processes as erosion, weathering, oxidation, heating (as by volcanic action), or retrieval and use by humans (Figure 26-11).

The Greenhouse Effect

The human release of carbon by burning fossil fuels, making cement from limestone, and clearing forests has doubled the amount of carbon dioxide in the Earth's atmosphere over the last century, and it may double it again in the next. Because carbon dioxide absorbs heat emitted from the Earth's surface and reemits some of it back toward the surface, it interferes with the loss of heat from the planet to space, much as does the glass in a greenhouse. Many scientists therefore believe that an increase in carbon dioxide will raise the Earth's temperature. Present projections say this "greenhouse effect" may make the Earth as much as 4.5°C warmer by the year 2100.

As a result, areas near the poles may be much warmer. The polar ice caps will melt, at least partially, and may raise the sea level enough to flood low-lying areas such as Florida. Some estimates suggest that the ocean level could rise by 200 ft if all of the ice caps were to melt. Presently, the ocean level seems to be rising an inch or so every few years because of atmospheric warming. This rate of rising water level would translate to devastation for many coastal cities in the next century. Rising sea levels, coupled with record high tides and storms, have already wreaked havoc on the Pacific coast of the United States, with waves literally breaking over the tops of oil islands just a few miles offshore and destroying virtually every pier in southern California (Figure 26-12).

Unfortunately, because of political pressures, the U.S. government continues to react conservatively to ecological concerns. For example, in 1987, because of pressure from motorists, it raised the 55-mph speed limit back to 65. In environmental terms, this change translated to increased fuel consumption, which accelerates carbon dioxide buildup in the atmosphere, which will exacerbate the greenhouse effect.

The Nitrogen Cycle

Plants and animals both need many more nutrients than water and carbon. They need phosphorus for their nucleic acids and adenosine triphosphate (ATP), sulfur for their proteins, and all the trace minerals that were mentioned in Chapter 22. Each of these substances becomes part of the soil as rocks weather and crumble. They dissolve in the film of water that surrounds soil particles and are collected by plant roots. They enter the metabolisms of plants, and then animals, and they return to the soil when plants, animals, and their wastes decompose. Some of these nutrients wash from the soil into streams, lakes, and seas, where they nourish aquatic organisms. As these organisms die, they become part of the sediments that in time become new rock. They again become available to the bio-

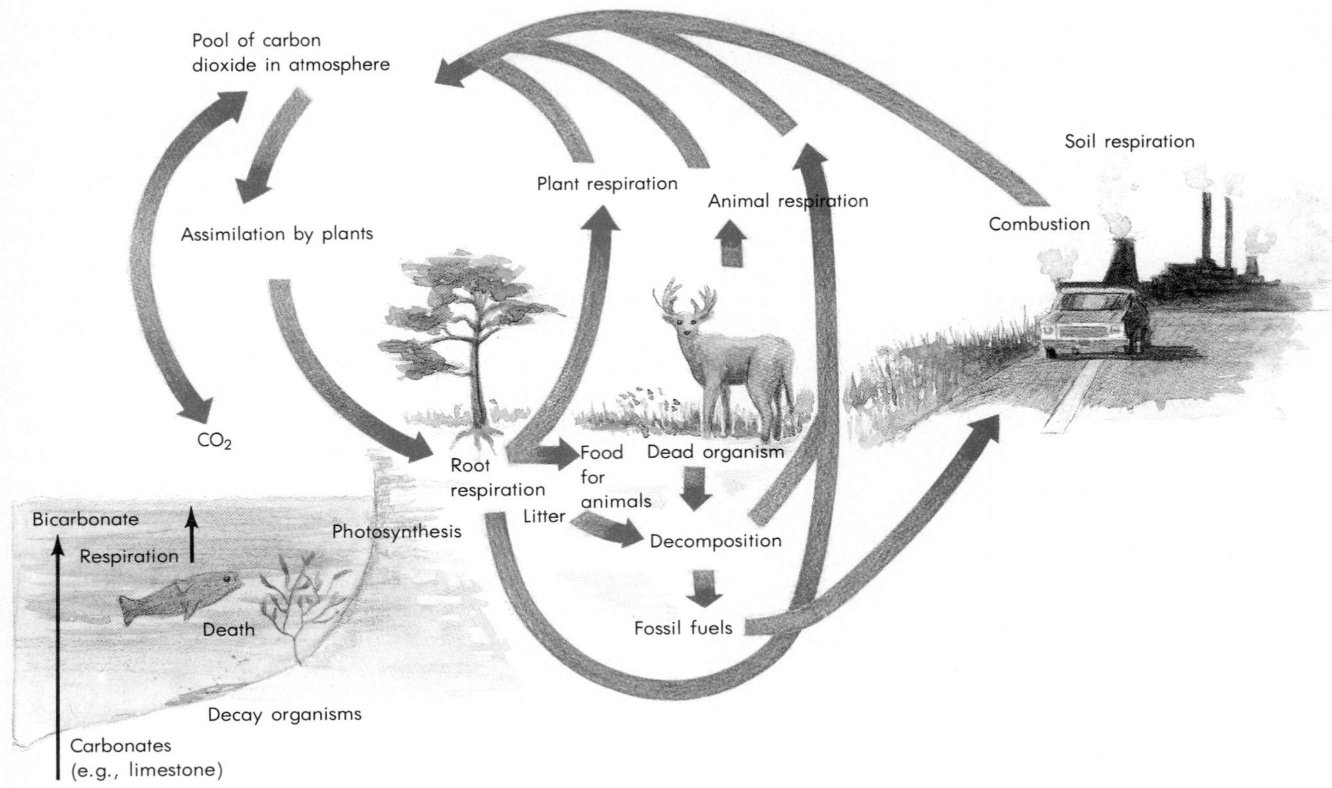

FIGURE 26-11 The carbon cycle.

sphere when geological forces thrust old sea bottom into the air, to weather and crumble once more, renewing the cycle.

We will not discuss these cycles in detail, for the specifics are similar to those of the carbon cycle. However, we will sketch the workings of one very important nutrient cycle, the **nitrogen cycle.** Nitrogen is a crucial nutrient. It is an essential part of every molecule of nucleic acid and protein. It is needed for many vitamins and for the heme groups

FIGURE 26-12 Storm damage to low-lying coastal areas is likely to become both more frequent and more severe as the greenhouse effect raises the sea level.

that let blood carry oxygen. Its cycling depends on living things far more than does the cycling of most other nutrients.

Nitrogen Fixation

The Earth's atmosphere is almost 80 percent nitrogen, but nitrogen gas, N_2, is not usable by most living things. Metabolism requires nitrogen in the form of nitrate (NO_3^-) or ammonia (NH_3), only a little of which is produced by the action of lightning on N_2. Most biologically useful nitrogen is the product of the biological process known as **nitrogen fixation,** carried out mostly by bacteria and cyanobacteria (previously called blue-green algae) that live in soil and water. Many of the nitrogen-fixing bacteria live as mutualistic symbiotes in nodules attached to the roots of plants such as legumes (peas, beans, alfalfa, and so forth). The plants produce carbohydrates to feed the bacteria and a form of hemoglobin to keep oxygen away from the nitrogen-fixing enzymes. The bacteria produce ammonia, which the plants can then use as a nitrogen source, making it almost a self-fertilizing system. They provide the biosphere with some 175 million metric tons of fixed nitrogen every year.

The nitrogen made available by microorganisms becomes incorporated in the tissues of the plant hosts, mostly as protein. From there it moves through the rest of the biosphere. Animals get it when they eat the plants and each other. When decomposers break down organic compounds, they return the nitrogen to the soil to be recycled. However, the recycling is far from perfect. There exist not only nitrogen-fixing bacteria but also **nitrifying bacteria,** which convert ammonia to nitrites (NO_2^-); **nitrate bacteria,** which convert nitrites to nitrates; and **denitrifying bacteria,** which convert nitrates back to nitrogen gas. In addition, some biologically useful nitrogen is always being washed out of the soil and incorporated in sediments. Some organic nitrogen accumulates in vast deposits of dung in bat caves or on islands where seabirds roost and nest (Figure 26-13).

As nitrogen is constantly being removed from the cycle, there is a continuing need for new sources of fixed nitrogen. The need is so great that when humans try to increase the productivity of their farms, they find that natural fixation is not enough. They must mine the deposits of dung, and when those deposits have been exhausted—as they

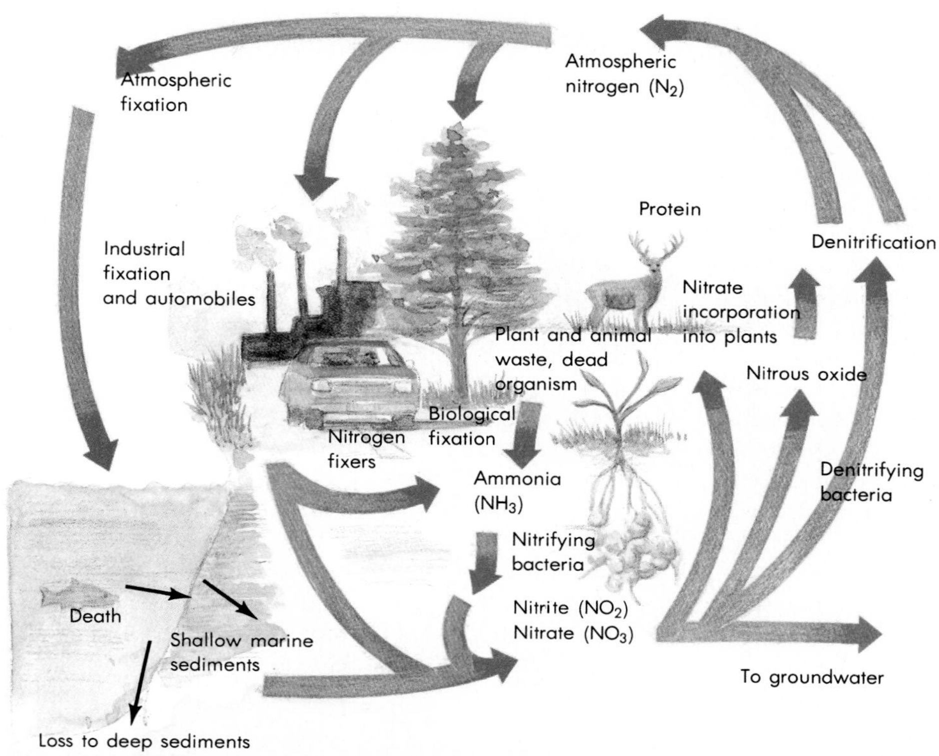

FIGURE 26-13 The nitrogen cycle.

were in fact years ago—they must manufacture nitrogen fertilizers at great expense in energy and money. At the moment, researchers are trying hard to use the new technology of genetic engineering to make plants able to fix their own nitrogen supplies and to make the symbiotic nitrogen-fixing bacteria able to live on the roots of more nonlegume species of crop plants.

HUMAN IMPACTS

The various elements of the biosphere are efficient at keeping essential materials, like energy and key nutrients, in circulation. When some of the critical chemical nutrients are allowed to slip out of this trophic flow, as when carbon or nitrogen or phosphorus escapes to deep sediments and rock, those nutrients are not necessarily lost forever. Over the ages of geological time, even the rocks are recycled. The system resists long-term changes.

However, the system is not immune to damage. In the short term, biological recycling depends on the health of the biosphere. And human activities do not always help. Some major human-linked problems may be so damaging as to cripple the biosphere's self-correcting survival mechanisms by threatening the basic life support systems.

Virtually all the major environmental problems threatening the biosphere have their basis in human overpopulation. There are simply too many humans consuming too many resources, producing too many pollutants, scrambling for too much fuel, and willing to do anything to produce enough food and more money. The results of this overpopulation are hunger, drought, pollution, endangered water supplies, the greenhouse effect, the loss of stratospheric ozone, pesticide and fertilizer contamination of water supplies, deforestation, desertification, and long-term climatic changes affecting vast areas of the globe.

We have already mentioned how humans interfere with interspecies competition with pesticides and thereby create new pests, accelerate the release of carbon and threaten to change world climate, change rainfall patterns, and exhaust groundwater. In addition, oil spills destroy the life in large areas of the sea. Water pollution with sewage, toxic wastes, and agricultural runoff can virtually sterilize lakes and streams and even the estuaries (river mouths) that serve the seas as nurseries for fish, crabs, and a myriad of other organisms, many of them of commercial importance. Air pollution can turn green hills into desert, and rain made acid by sulfur and nitrogen oxides from the human use of coal and oil can kill the fish in lakes as well as erode buildings and statuaries (Figure 26-14).

---

1. We, the authors, are completing this chapter in the spring of 1991, shortly after the Gulf War. As Iraqi troops fled Kuwait, which they had invaded, they set fire to well over 500 oil wells. Those fires are still burning, and the experts are debating the impact of the smoke on the local climate, the environment, and human health. Unfortunately, it is much too soon for us to be able to discuss that impact here. But you are reading these words later in time. We suggest that you go to your campus library and look up what happened, and what is still happening.
2. What did you learn about the impact of all that oil smoke on climate? On weather? On regional agriculture? On human health?
3. Can you think of other situations—other times and places—in which humans put similar amounts of crud into the air? What do you think must be the impacts of those situations?

As we near the end of the twentieth century, another kind of air pollution is also making headlines. This is pollution by chlorofluorocarbons, organic molecules containing chlorine and fluorine. These substances, because they are not toxic, have been widely used as aerosol propellants, refrigerants, and cleaners for electronic components. Unfortunately, when they disperse in the upper atmosphere, they both contribute to the greenhouse effect and, when they break down, stimulate the removal of ozone molecules. Ozone blocks solar ultraviolet from reaching the Earth's surface; without it, we can expect to see increases in the incidence of skin cancer in humans (and animals) and declines in the productivity of crop (and other) plants. So far, the ozone loss is definite only in the Antarctic and Arctic regions, but it seems all too likely that it will become noticeable in more temperate—and more thickly populated—zones. World governments have judged the threat to be severe enough to agree to reductions in the use of chlorofluorocarbons and even outright bans.

Yet even damage by chlorofluorocarbons may prove relatively minor in the long run. The world heals its wounds. Ozone regenerates as solar ultraviolet strikes oxygen molecules. A decade after even a major oil spill there may be few signs of damage. In the United States, federal legislation has mandated control of water pollution, and many bodies of water once more have fish and are safe for swimmers. Air pollution has eased too, though acid rain remains a problem.

(a)

(b)

(d)

(c)

FIGURE 26-14 Examples of the chemical impact of industry on fragile environments. (a) In Sudbury, Ontario, smelters released fumes that killed off the surrounding forest. (b) Acid rain has corroded this bronze plaque at the Gettysburg National Military Park. (c) This helicopter is "liming" a lake left lifeless by acid rains. In theory, the liming may bring the pH of the lake back within a range that allows it to support life. (d) The aftermath of the 1991 Gulf War left over 500 oil wells in flames.

Who Cares If Another Endangered Species Is Lost?

More serious are other kinds of damage. Human beings occupy the broadest of all ecological niches, and they threaten the competitive exclusion of many other species. The human use of land for farms, cities, roads, and dams destroys the habitats of numerous organisms, to the point where today some 280 species of mammals and 350 species of birds are listed as, and 20,000 species of plants are estimated to be, in danger of extinction. Some people feel that the name *Homo sapiens* may join that growing list if we do not make some concerted efforts to turn this around. Without significant changes and a cooperative global response, the problems can become only worse, and the consequences may be dire.

As the biosphere loses species, it also loses interspecies interactions—checks on populations, food supplies, nutrient cycling mechanisms, and other factors essential to the continued health of life on Earth. In human terms, the loss can also be important, for with wild plants and animals we feed our needs for domestic crops, medicines, and beauty. We are also looking to a future when genetic engineering techniques will enable us to transplant genes from one organism to another, building in resistance to diseases and chemicals or providing new desirable features. If we lose the thousands of species that share our world with us, we will thus lose a world of genetic potential.

Still more serious may be what humans do to the

land itself. Careless cultivation of farmland encourages erosion of topsoil by wind and rain; the United States alone loses enough topsoil every year to cover 1.2 million hectares (3 million acres) with a layer 30 cm deep. Strip-mining of coal and metal ores leaves deep, infertile scars in the land (Figure 26-15).

Deforestation

Within a relatively short period of time, vast tracts of forests are being virtually wiped out. Recent estimates suggest that approximately 27 million acres of tropical forests (a combined area the size of Austria) are being destroyed each year. In the Amazon jungles of Brazil and in parts of Central America, much of the devastation is caused by developers destroying the forest to reach the wealth of natural resources they contain, especially timber and minerals (Figure 26-16, p. 630).

Government policies that give the urban poor jungle farmsteads do not help, for each such farmstead eliminates more of the rain forest. Rain forest soils are thin and infertile, and the farms themselves have very short lives, as has long been known by the rain forest's native inhabitants. Clearing a field by cutting (slashing) the trees and burning the logs and stumps, they practice a kind of farming known as "slash-and-burn" agriculture. The ashes provide nutrients to the soil for just two or three years. Then the soil turns hard and unusable, forcing the farmers to move on and destroy more sections of forest.

Deforestation is accelerating in places like India and Pakistan, driven in part by an exploding population's need for wood to burn as fuel. It is difficult to explain to a family that needs to cook and keep warm that it would be better for the ecology of the world if they refrained from stripping every bush, tree, or scrap of wood from their local environment (Figure 26-17, p. 630).

Yet deforestation is not a problem only in the world's less developed nations. The gasoline crises of the 1970s and the dramatic increases in the prices of fuel oil and natural gas that followed spurred the sale of wood-burning stoves in Maine, Idaho, and other forested areas of the United States. Many environmentalists are already concerned about further damage to the fragile forested ecosystems.

---

1. What logical arguments can you offer to a cold or hungry family to stop them from taking whatever wood or plants they need from nearby forests?
2. In what different ways might totalitarian and democratic governments approach the complex problems of deforestation? Which system do you think is more likely to get a workable policy in motion?

Because atmospheric oxygen is a by-product of global photosynthesis, the devastation of extensive tracts of forests may cause measurable changes in its availability. More likely changes, however, will be climatic, agricultural, and of course, economic. Because mature forests hold enormous amounts of carbon in the form of wood, deforestation releases vast amounts of carbon dioxide (due to burning of waste wood and wood used as fuel and to eventual decay of lumber); it therefore contributes significantly to the greenhouse effect. In addition, eliminating large stands of forest has already shifted the patterns of rainfall that affect substantial tracts of agricultural lands. There is a sharp decrease in rainfall in areas, such as Panama, where water was once abundant. Some studies suggest that the agricultural breadbaskets for several continents may find themselves with insufficient water at a time when more food production is needed to cope with a growing world population. Estimates suggest that already over 20 percent of Earth's human population is feeling the detrimental effects directly caused by deforestation.

Some of the most powerful economic and political institutions, including the World Bank, the U.N. Food and Agriculture Organization, the U.N. Development Program, and the Rockefeller Foundation, have called deforestation "the Earth's most serious environmental problem."

---

1. Why must you focus your arguments first on economic issues and then on the environmental issues when attempting to convince developing nations that they should adopt a stringent policy to protect their forests rather than allow them to be fully exploited?
2. If you were an ambassador from the United States charged with the responsibility of selling the "protect-your-forests" policy to a developing country, how would you counter the following response to your arguments? "Our forests are our most important resource. If we do not fully develop them, then we cannot be economically competitive with other nations. We feel there is more to our destiny than simply to become national parks where the rich nations can come to play."

(a)

(b)

(c)

(d)

(e)

(f)

FIGURE 26-15 (a) A strip mine for copper in Butte, Montana. (b) Strip mining for subbituminous coal in Gillette, Wyoming. (c) In-the-pit view of strip coal mining. (d) Severe soil erosion in the Rift Valley, Kenya. (e) Soil erosion along cow paths. (f) Overgrazing by sheep removes vegetation, such as grasses, so completely that it has contributed to major climatic changes in many parts of the world. A considerable amount of the desertification that has occurred in the past few hundred years in the Middle East and Africa has been attributed to overgrazing.

(a)

(b)

FIGURE 26-16 Deforestation. (a) Felling trees and burning them to clear away the rain forest is the basis of the devastating slash-and-burn agriculture in Brazil. (b) The unproductive laterite soil is exposed by clear cutting of an east Malaysian (Borneo) rain forest for the expansion of a plantation.

FIGURE 26-17 Animal dung patties, shown here drying on walls in Nepal, are used as an alternative fuel source because of the dwindling availability of wood.

Desertification

The evidence is strong that extreme environmental damage caused by deforestation, slash-and-burn agricultural practices, and overgrazing by livestock promotes the conversion of productive lands to barren deserts. These practices strip the land of vegetation, expose the soil to erosion, cause water tables to subside, and reduce the amount of moisture in the atmosphere. The result is lowered humidity, less rain, and hotter air. These are, by definition, the ingredients from which deserts are made. As the processes of **desertification** continue, they lead to environmental changes that will take literally thousands of years to reverse.

A significant degree of desertification has occurred rather rapidly in the southwestern United States in the last hundred years or so. The natural vegetation found in the northern areas of the Great Basin desert, which runs from Great Salt Lake through southern Utah, Nevada, and into southern California, has changed dramatically since the arrival of pioneers in the Great Salt Lake valley. Military reservations such as Fort Douglas, which have been continuously protected from grazing of livestock and other agricultural practices, show that the earlier natural plant communities were much lusher than the desert communities that succeeded them after the damage was done.

Similarly, within the relatively recent past, the Middle East was extensively forested. Hence, the phrase "Cedars of Lebanon." At this time, the only cedars to be seen in the deserts of that region are those growing in protected gardens. Environmentalists think that these dramatic changes in productivity and climate resulted primarily from overgrazing by the goats and sheep that once were herded throughout these areas.

Irrigation: At most, a Temporary Solution

The solution to desertification is not simply a matter of irrigation projects. It is much more complicated. Irrigation of arid fields, because small amounts of mineral salts are dissolved in the irrigation water and because these salts remain behind when the water evaporates, leads to the buildup of salt in the soil. This phenomenon has turned valleys in Iraq and Iran that supported some of Earth's first civilizations into infertile wastelands, and it is destroying croplands today in the United States and elsewhere.

Irrigation can lead to other problems as well. In California's San Joaquin Valley, farmers have tried to solve the salinization problems by using excess irrigation water to wash salt out of the soil. This water drains into the Kesterson National Wildlife Refuge, part of a major wintering ground for migratory birds. With it goes the salt and the element selenium, in which the valley's soil is rich. Selenium is an essential nutrient in small amounts, but in large quantities it can cause cancer in laboratory animals. In the wildlife refuge, selenium accumulates as water evaporates, and its concentration in the water there is enough to qualify the refuge as a toxic waste dump. The selenium has caused birth defects and death among water fowl in the refuge, and it appears to be responsible for disappearance of most fish from the refuge and the nearby San Joaquin River. There is even a distinct risk that the selenium will eventually find its way into drinking water.

Droughts, Deserts, and Famines

The effects of desertification were most recently illustrated by the horrible droughts and subsequent famines that took place in Ethiopia and other areas of northern Africa in the mid-1980s. Emaciated, starving, and dying children stared blankly into television cameras that sent their images to people living in North America and Europe. The effect was so touching and dramatic that many members of the entertainment community banded together and recorded a song, "We Are The World," as a massive fund-raising campaign to help the famine victims (Figure 26-18).

Unfortunately, that help was short-lived, for the droughts and famines returned in 1990–1991. They will continue cycling and intensifying in that area, for the Sahara Desert is rapidly expanding across the African continent (Figure 26-19). Estimates suggest it may be spreading south in some areas at over 50 miles per year. A major cause contributing to these problems is overpopulation in a marginally productive environment. Too many people are resorting to overgrazing and destructive agricultural practices merely to eke out a living. And as the desert spreads, the desperate people flee just ahead of its leading edge, priming new areas for its continuing expansion.

---

1. Why do some harsh critics of fund-raising projects, like those for the starving people in Ethiopia, claim that sending money not only will not work, but in actuality will lead to much greater starvation and human suffering later? Discuss the arguments these critics might use to make that point.
2. What better use for the money might the critics of such fund-raising programs be likely to suggest?

FIGURE 26-18 Starving mother and child in food distribution center in Ethiopia, 1984.

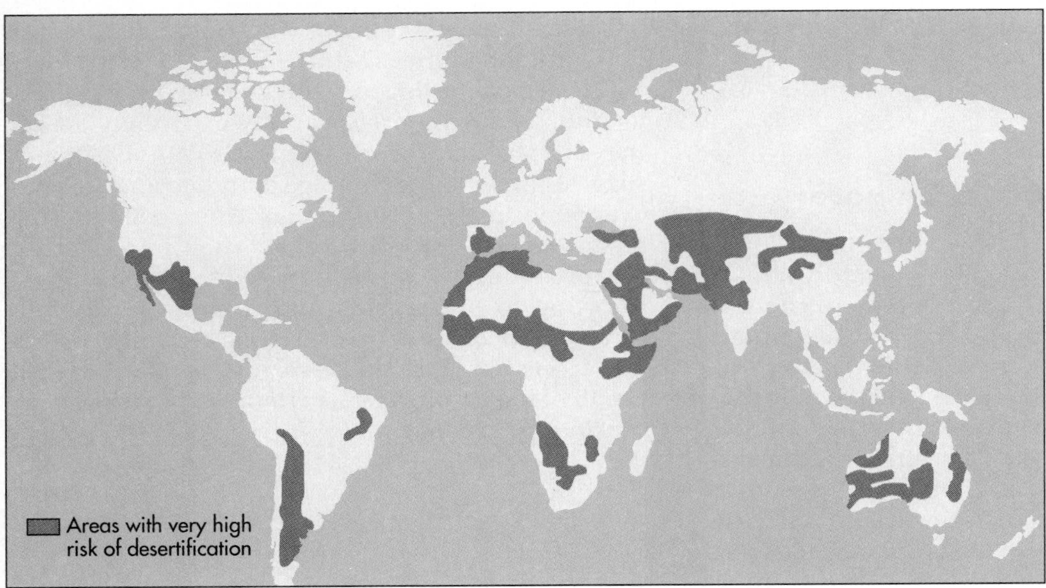

FIGURE 26-19 This world map shows where deserts are growing. The resulting loss of farmland makes it increasingly difficult for many people to thrive—or even survive.

THE POPULATION CRISIS

By 1990 the United States had a little over 250 million people and so much empty, fertile land that it could export food all over the world. Yet the United States is an unusually fortunate land, partly because its people have been using birth control techniques for many decades. Other nations are less fortunate, often because their populations are so large and growing so rapidly that they have outstripped the capacity of their land to support them all, and their lots will worsen greatly unless they get control of their birth rates. We can understand their plight, and the need for birth control, if we look at a few basic concepts of population biology.

The size of any population, now or in the future, depends on its history. Specifically, it depends on its **growth rate,** the number of people it adds per 1000 members per year. The growth rate, in turn, is the difference between the **birth rate,** the number of births per 1000 per year, and the **death rate,** the number of deaths per 1000 per year:

$$\text{Growth rate} = \text{birth rate} - \text{death rate}$$

For the United States, the birth rate is 16 and the death rate is 9. The growth rate is thus 7 persons added per 1000 members of the United States population per year, or 0.7 percent. The United States population will double to 500 million in a century if it continues to grow at the same rate.

For the world as a whole, the growth rate is 18 persons added per 1000 members of the world population per year, or 1.8 percent per year. The doubling time, from the present world population of 5.3 billion to 10.6 billion, is only 39 years. If 1990 growth rates do not change, there will be 10.6 billion people on Earth by the year A.D. 2029.

The difference between the United States and the world as a whole is largely due to the nonindustrialized nations, which have over three-quarters of the world population, an overall growth rate of 2 percent, and the fastest doubling times. Rapidly doubling populations, like those of Asia, include almost 59 percent of the world's people (Figure 26-20). The fastest growing populations of all, in Africa, include 12.42 percent of the world's current population (Figure 26-21). Latin American populations—8.4 percent of the total—are doubling every 33 years. On the other hand, Europe holds 9.42 percent of the world's people, but they are growing so slowly that at current rates it will take 266 years for that population to double.

Many of the world's less developed countries (LDCs) are already experiencing dire problems because of their population sizes and growth rates. Mexico, for instance, with a population of 88.6 million and a growth rate of 2.4 percent, suffers immense poverty and unemployment, and the rapidly growing need for basic services such as roads, sewers, and water supplies for so many people limits efforts of government to find the necessary money for education and economic development. In Sub-Saharan Africa, 15 years of drought aggravated by

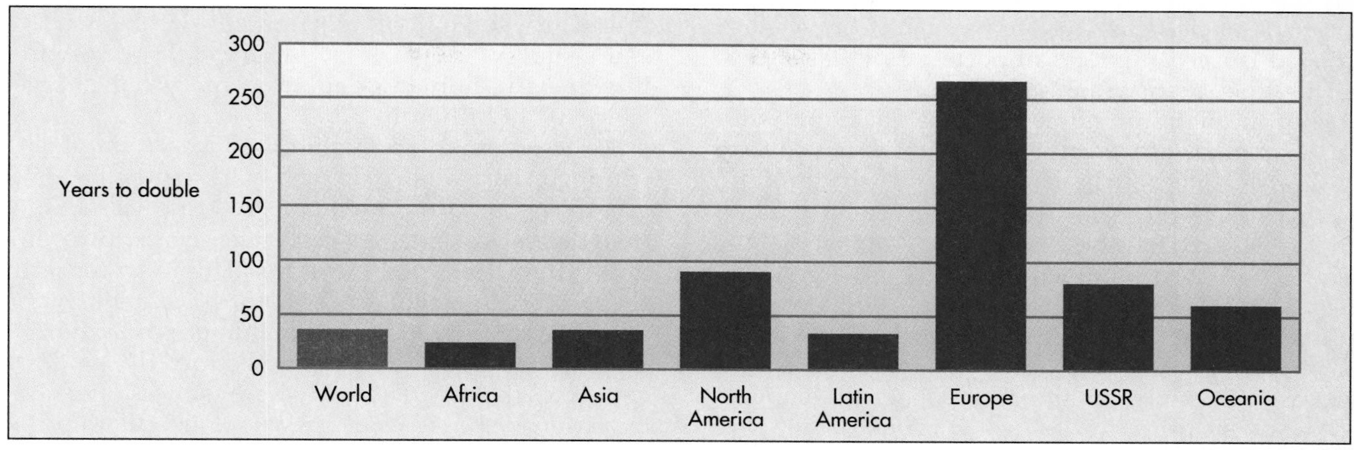

FIGURE 26-20 Population doubling times by region (at 1990 rates of growth).

political difficulties recently crippled agriculture and caused widespread famine. The tragedy is that a smaller population would have had little difficulty in weathering the bad years; a larger one will have much greater difficulty, and bad years—and decades—are inevitable.

High population growth rates are a modern problem. They result from the progress the human species has made in controlling disease with vaccines and sanitation and hence in reducing the death rate. In most parts of the world, the birth rate has also fallen, but not nearly to the same extent. Birth control is essential to prevent disasters such as those that now face much of Africa and to relieve problems such as those in Mexico.

However, while the issues of population growth, birth rates, and death rates may seem like simple extensions of mathematical principles to us, they are not viewed that way by others. For some, birth control is not an option because of religious beliefs that originated long before overpopulation seemed a serious hazard. Others view birth control as a disguised sort of genocide that the developed countries promote to keep populations down in countries that might someday become economic, political, or military threats. Some people object to birth control out of "macho" motivations that interpret "manhood" as the right to have as many babies as they can. Intriguingly, birth control is accepted much more readily where the choice is up to women.

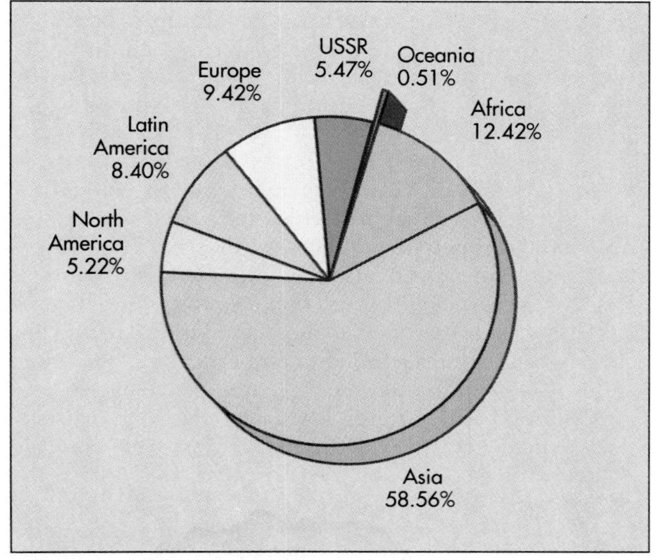

FIGURE 26-21 The 1990 world population distribution by region.

It was once thought that a country that could develop its industry and increase its prosperity to a relatively high level would then naturally reduce its population growth rate. This "demographic transition" was thought to be due to the wish of more prosperous people to spend their new wealth on themselves rather than on their children and on their realization that they no longer needed children to guarantee their survival in old age. Now, however, population researchers are realizing that the process seems to work the other way around. Reduced growth rates must come first, in order for smaller families to have more wealth to invest, and then the national economy will have resources to spare for economic development.

THE NEED FOR RESTRAINT

Most of the Earth, at present, shows relatively few signs of damage. In fact, advances in agricultural technology have given the world a significant excess of food, and some of the world's fastest growing countries—such as India and China—have moved from being food importers in the 1960s to being food exporters in the 1980s.

But the signs are numerous that this happy situation cannot last. World population is now over 5 billion people, and it is growing at a rate of 1.8 percent per year. That is, it will exceed 10 billion before A.D. 2030, and the human impact on the world will then be much greater than it is today. The result will be fewer species, less efficient recycling of nutrients, diminished soil fertility, and a chancier time for the human species, which depends ultimately on the interdependent support systems of nature for its own survival.

Unfortunately, if we continue as we have in the past, sometime in the next century we can expect the human population and its needs to collide head-on with the problem of a deteriorated environment. Food and other resources will be in such short supply that massive famine will be inevitable, and the human population may well plunge from 10–20 billion to 2 billion or less.

Although many people already worry about the coming crisis, we are doing too little about it. We are trying to protect endangered species, shield land from erosion, deforestation, desertification, and salinization, and clean up air and water, but these problems remain. And they seem likely to become worse as long as the human population—and its appetite for resources of all kinds—continues to grow. The solution must thus involve a cessation of population growth and preferably a reversal of it. Unfortunately, most present efforts to control population are notoriously ineffective. Our numbers may not decline until a crisis—such as famine, disease, or war—forces them down.

SUMMARY

Ecology is the study of how living things—including humans—interact with their living and nonliving environments. It can thus help guide human choices and preserve the living world, the biosphere, for the future.

The biosphere can be subdivided into large groupings of similar plants and animals called biomes. Biomes are shaped by temperature and water supply and hence by climate, latitude, and altitude. On land, they include tropical rain forest, savanna, desert, temperate grasslands and forests, northern coniferous forest, tundra, and the polar ice caps. Aquatic biomes are those of still and moving fresh water and of the intertidal zone, the continental shelves, and the deeps at sea.

An ecosystem can be an arbitrarily defined area where the interactions between the abiotic (nonliving) and the biotic (living) factors can be studied. In some cases, ecosystems are natural areas where conditions cause them to be self-sustaining and the major inputs are water and energy, or sunlight. Mineral nutrients are largely recycled between the biotic and abiotic components of the ecosystem. The biotic components include the producers that synthesize organic materials from sunlight, water, and carbon dioxide, the plants; the primary consumers that eat the producers, the herbivores; the secondary and tertiary consumers that eat the primary consumers and each other, the carnivores; the detritivores that eat debris; and the decomposers that return nutrients in dead bodies and wastes to the soil. Their eater-eaten relationships can be portrayed simply in a food chain and more realistically in a food web, which shows, for instance, that a secondary consumer can also eat plants.

The lowest level of a food chain, the producers, is said to occupy the first trophic level. The next level, the primary consumers, occupies the second trophic level. Since energy is lost at each step in the chain, the second trophic level contains roughly about 20 percent of the energy—and a tenth of the mass—found in the first. The third, the secondary consumers, contains a tenth of the energy and mass found in the second. We can therefore see that the lower on a food chain humans find their food, the more human beings the Earth can support.

Living things interact with each other not only as eater and eaten but also in other ways. In mutualistic interactions, two different species each benefit. In commensal interactions, one benefits but the other is not harmed. Eater-eaten interactions are examples of predation. Parasitism is where the eater lives within or on its host; the ideal parasite does minimal harm to its host, for if it kills its host, it too must die and its reproductive success must suffer.

Species with similar requirements for life—that live in the same place or habitat—also interact by competing with each other for the resources they need to survive. They never compete for everything they need, however, for different species always need at least slightly different sets of resources; that is, they occupy different ecological niches. If two species did occupy the same niche, one would be a more efficient competitor and the other would die out. In nature, such "competitive exclusion" usually applies only in those parts of niches that overlap. Species thus occupy "realized niches," portions of their "potential niches" to which they are restricted by competition with other species.

Such processes help to preserve the ecological homeostasis that maintains the world as a place suitable for life. So do the various mechanisms that underlie the recycling of water, carbon, nitrogen, and other nutrients. Unfortunately, human activities can interfere with nature. Humans occupy the broadest ecological niche of all, and their activities threaten to exclude a great many other species. They also damage air, water, and land, preempt or destroy habitats, and disrupt nutrient cycles. The problem exists no matter how many people there are on Earth, but it is serious largely because there are so many people. The world population in 1990 was 5.3 billion, and recent estimates suggest it is growing at a rate of 1.8 percent per year.

In many parts of the modern world, the sheer numbers of people have led to problems of famine, unemployment, and retarded development. The numbers have become excessive because vaccines and sanitation have reduced the death rate much more than the birth rate, leaving population growth rates high. The answer has to begin with some form of birth control.

To preserve the world for the sake of the human future—if not out of respect for the other living things with which we share it—we must control the human impact on it, preferably by reducing human numbers. If we do not, the world will be so damaged that we will have no choice. A struggling biosphere will produce famines, plagues, and other disasters enough to do the job for us.

STUDY QUESTIONS

1. How does "human ecology" differ from "ecology"?
2. What is the difference between a biome and an ecosystem?
3. Some green, photosynthesizing plants, such as pitcher plants, Venus's-flytraps, and sundews, gain many nutrients by trapping, digesting, and absorbing insects. Are such plants producers or consumers?
4. Estimate as best you can the amount of food it takes to grow a child from birth (at 5 kg) to age 10 (at 55 kg). Does your estimate support our generalization that consumers embody roughly a tenth of the energy and mass of the food they eat?
5. Assume that half of the food a child ate was meat from herbivores, which in turn consumed plant food that the child could have eaten. Add up the plant food the child ate and the plant food the herbivores ate. How many children could grow to age 10 (from 5 to 55 kg) on that amount of plant food alone?
6. How would you describe the interspecific relationship between the human and canine species?
7. How might the competitive exclusion principle help explain why there is only one living species descended from the Australopithecines (see Chapter 25)?
8. We can purify polluted water in several ways, including filtering to remove contaminants, treatment with ozone to oxidize contaminants, and distillation (evaporation and condensation). Which is most similar to the natural water cycle?
9. What do you think is the most important reason to protect endangered species?

GLOSSARY

Autotrophs Self-feeders; organisms that synthesize the bulk of their own nutrients; photosynthesizers.

Biome A group of similar ecosystems, defined by climate, and hence by water, temperature, latitude, and altitude.

Biosphere The layer of life that surrounds the planet Earth.

Birth rate Number of births per 1000 members of a population per year.

Commensalism Symbiosis when one species benefits and the other is not affected.

Community The biotic or living elements of an ecosystem. They include all plants, animals, bacteria, fungi, protozoans, and anything else that lives in the defined area.

Competition Struggle between two (or more) species for resources.

Consumers Heterotrophs.

Death rate Number of deaths per 1000 members of a population per year.

Deforestation The loss of forests as a consequence of their exploitation for the resources they contain.

Desertification Conversion of productive land to barren deserts as a consequence of environmental destruction caused by deforestation, overgrazing, and slash-and-burn agriculture.

Ecological pyramid A drawing showing the relative sizes of different trophic levels in terms of energy, mass, or numbers.

Ecological niche How an organism lives. A description

of all the various relationships a species has between itself and other members of its community and the physical factors in its environment.

Ecology The study of the interactions of living things with each other and with their environment.

Ecosystem A defined area in which the interactions between living and nonliving factors produce a fairly stable and somewhat self-sustaining environment.

Food chain Simple, linear sequence of eaters and eaten.

Food web More realistic and complex view of food chain, showing how many organisms occupy more than one trophic level.

Growth rate Speed of growth of a population, measured as number of people added to the population per 1000 members per year; birth rate minus death rate.

Habitat A specific type of environment where a species or a community naturally occurs. For instance, the banks of streams, woodlots, and desert sand dunes are each different habitats.

Heterotrophs Other-feeders; organisms that obtain their nutrients by consuming other organisms.

Mutualism Symbiosis when both species benefit.

Parasitism A version of predation where the parasite lives within or on its prey or host. Parasites usually do not kill the host.

Predation Symbiosis when one species eats the other.

Producers Plants; photosynthesizers; makers of organic materials from sunlight, water, and carbon dioxide.

Trophic level Position in a food chain.

Glossary

Abdominopelvic Cavity Portion of ventral cavity in the belly and pelvis.

Abiotic Nonalive.

Abortion Removal of embryo from uterus before it has reached the time for birth.

Absorption The process of taking nutrients from the digestive tract into the blood.

Accessory Digestive Organs The various glands that release secretions into the chambers of the digestive system, including the salivary glands, pancreas, and gallbladder.

Accessory Sex Organs Organs that transport and support gametes in the body.

Accommodation Changes in the shape of a lens that adapt it for viewing objects at different distances.

Acetabulum The cuplike portion of a pelvic bone that holds the head of the femur.

Acetyl Coenzyme A An energy-rich two-carbon molecule that is shunted into the Krebs cycle to have its energy harvested and the hydrogens it contains loaded into NADH and $FADH_2$ for further processing.

Acids Substances that release hydrogen ions (H^+) in solution. Strong acids release more hydrogen ions than weak acids.

Acne A condition common in puberty, marked by inflammation and infection of the sebaceous glands.

Acrosome Tip of sperm cell; carries enzymes to break through barrier around ovum.

Actin The major protein of the thin myofilaments.

Action Potential The nerve signal; reversal of the membrane potential from -70 to $+30$ mV, and recovery, on disturbance of the membrane.

Activation Energy The energy needed to initiate a chemical reaction.

Active Immunity Immunity acquired by the body's actually having been exposed to an antigen and then developing memory cells for it.

Active Transport Transport methods requiring the cell to expend energy.

Acute Stage The violent phase of a disease, in which the body is undergoing a serious physiological disruption.

Adaptation Fading of a sensation with continued exposure to the stimulus.

Adaptive Radiation The proliferation of new species from existing species exposed to various opportunities to adapt to new ways of life.

Adenosine Triphosphate (ATP) The principal form of chemical energy used by cells.

Adrenal Cortex Outer layer of adrenal gland; source of steroid hormones.

Adrenal Glands Glands above kidneys; secretors of steroid hormones, epinephrine, and norepinephrine.

Adrenal Medulla Inner portion of adrenal gland; source of epinephrine and norepinephrine.

Adrenocorticotropic Hormone (ACTH; Adrenocorticotropin; Corticotropin) Controls adrenal cortex.

Adventitia Tunica externa.

Afferent Arteriole The blood vessel bringing blood to the nephron.

Afferent System The nerves carrying signals to the CNS.

Agonist Of the two or more muscles crossing a joint, the muscle that is contracting during a movement.

Alarm Reaction First stage of general adaptation syndrome.

Albino A person who lacks the ability to synthesize melanin.

Aldosterone An adrenal hormone that stimulates sodium reabsorption in the distal convoluted tubule.

Alleles Alternate versions of a gene controlling the same trait.

Allergen A weak antigen capable of triggering an allergic response.

Allergy An overreaction of the immune system to a weak antigen.

All-or-None Principle The idea that a muscle fiber contracts either fully in response to a stimulus or not at all; the idea that all nerve cell action potentials are the same size; stimuli below a set threshold do not produce action potentials.

Alpha Cells Pancreatic cells secreting glucagon.

Alveolar Sacs Clusters of alveoli; the air sacs of the lungs.

Alveolus Sites of gas exchange in the lungs. Bubblelike swellings in the walls of the alveolar sacs; site of external respiration.

Amino Acids Building blocks from which proteins are made. Human protein employs 20 different amino acids.

Amniocentesis The obtaining and examining of fetal cells afloat in the amniotic fluid in order to detect genetic and chromosomal defects.

Amnion Fetal membrane surrounding the amniotic cavity and enclosing the developing embryo.

Amniotic Cavity Cavity formed as the inner cell mass forms its first tissue layers; the wall of the cavity next to the trophectoderm will become the amnion.
Amniotic Fluid Fluid enclosed with the embryo within the amnion.
Amoeba Protozoan that moves by flowing into pseudopods.
Ampulla Swelling at the base of a semicircular canal; contains the cupula and hair cells.
Anabolism Those metabolic reactions in which larger molecules are synthesized from smaller building blocks.
Analogy Similarity in function but not in structure and development.
Anaphase The third stage of mitosis, in which the spindle fibers draw sister chromatids apart.
Anaphylactic Shock The life-threatening, generalized inflammation reaction associated with serious allergic responses.
Anastomosis A direct link from artery to artery; provides redundancy.
Androgen Adrenal cortex steroid hormone resembling male sex hormone.
Anemia Reduced oxygen-carrying capacity of the blood; usually due to insufficient red blood cells or hemoglobin.
Aneurysm Swelling of artery wall.
Angiotensin I The product of renin's action on angiotensinogen; becomes angiotensin II in the lungs.
Angiotensin II The end product of the renin blood pressure control mechanism; stimulates vasoconstriction and aldosterone secretion.
Angiotensinogen The plasma protein acted upon by renin.
Anisotropic (A) Band The zone spanned by the thick myofilaments, including their overlap with the thin myofilaments.
Antagonist The muscle that because its action opposes an agonist must relax to permit motion.
Anterior Toward the front.
Anterior Cavity The interior of the eye in front of the lens.
Anterior Horn Anterior portion of spinal gray matter; contains motor neurons.
Anterior Lobe of Pituitary Main endocrine portion of pituitary.
Anterior Roots Bundles of motor axons formed from merger of the anterior spinal rootlets.
Anthropoids The group of primates including monkeys, apes, and humans.
Antibiotic Substance originally produced by one life form, such as a fungus, that inhibits or kills bacteria. We can now synthesize many of these antibiotics in the laboratory without relying on the original living agent that first synthesized it.
Antibodies Defense proteins that are tailored to fit specific antigens. The resulting antibody-antigen complex inactivates or marks for attack many types of disease agents.
Anticodon The base triplet on a tRNA that is complementary to an mRNA codon.
Antidiuretic Hormone (ADH; Vasopressin) A pituitary posterior lobe hormone that stimulates water and sodium reabsorption in the distal convoluted tubule and collecting duct.
Antigens Recognizable proteins used by the immune system to distinguish between "self" and "nonself." They are often surface recognition features on cells.
Aorta Largest artery in the body; branches deliver blood to all tissues.
Aplastic Anemia Anemia due to destruction of the red bone marrow.
Apneustic Center Respiratory control center in the brain (pons); stimulates inspiratory center.
Apocrine Referring to sweat glands consisting of several branched tubules; apocrine sweat glands secrete a viscous fluid that bacterial action makes odorous.
Aponeurosis A sheetlike tendon.
Appendicitis Infection of the appendix.
Appendicular Skeleton The bones of the limbs and the pectoral and pelvic girdles.
Applied Research Research designed to solve specific economic, technological, medical, agricultural, or other problems. Applied research usually has practical value.
Aqueous Based in water.
Aquifer Underground bed of sand, gravel, or permeable rock through which water flows.
Arachnoid Delicate, spider-webby middle meninx.
Arrector Pili The smooth muscle attached to the outside of a hair follicle.
Arteriole One of the smallest arteries; subdivides to form capillaries.
Artery A blood vessel carrying blood away from the heart.
Arthritis Irritation or inflammation of the joints.
Articular Capsule A connective tissue sac surrounding a synovial joint.
Articular Cartilage The layer of hyaline cartilage covering the surfaces of bones that meet in synovial joints.
Artificial Selection Differential reproduction of organisms with traits favored by human beings, under human control.
Ascorbic Acid Vitamin C.
Asexual Reproduction Reproduction without gametes, by budding, fission, runners; produces clones.
Asthma Breathing difficulty due to narrowing of the bronchi and bronchioles; often due to allergy.
Astigmatism Distortion of visual image by irregularities in lens or cornea.
Atherosclerosis Clogging of arteries by deposits of cholesterol.
Atherosclerotic Plaque Deposit of cholesterol and smooth muscle on artery wall.
Atmosphere The mixture of gases and water vapor circulating above the Earth's surface.
Atomic Mass The number of neutrons plus protons a given type of atom contains.
Atomic Number The number of protons a specific type of atom contains.
Atoms The tiny fundamental units of matter. They consist of a nucleus made up of positively charged protons

and neutrons orbited by negatively charged clouds of electrons. There are 92 different kinds of atoms in nature.

Atria The chambers of the heart that receive venous blood.

Atrioventricular Bundle The band of specialized muscle fibers that conducts the contraction signal from the AV node to the walls of the ventricles.

Atrioventricular (AV) Node The region of the right ventricle that transfers the contraction signal from the right atrium to the ventricles.

Auditory Association Area Region of temporal lobe cortex devoted to analyzing sound.

Australopithecines The earliest known hominids; they walked erect and may have used tools.

Autonomic Nervous System (ANS) Portion of nervous system controlling smooth and cardiac muscle and glands; not normally consciously controllable.

Autosomes Non–sex chromosomes; homologous autosomes look identical.

Autotrophs Self-feeders; organisms that synthesize the bulk of their own nutrients.

Axial Skeleton The skull, vertebral column, ribs, and sternum.

Axon Nerve fiber carrying signals away from a neuron's cell body.

Bacteriophage A virus that infects bacteria.

Barr Body A condensed X chromosome visible in the nucleus of any cell with more than one X.

Basal Ganglia Nuclei within forebrain devoted to motor control, memory formation, and emotion.

Basal Layer The layer of actively dividing cells at the base of the epidermis.

Basal Metabolic Rate (BMR) The minimum amount of energy use needed to stay alive and awake.

Base Complementarity Base pairing.

Base Pairing The matching, via hydrogen bonds, of adenine to thymine or uracil and of cytosine to guanine along the length of DNA and RNA molecules.

Bases Substances releasing hydroxyl ions (OH^-) in solution. Strong bases release more OH^- than weaker ones.

Basic Research Pure research motivated by curiosity rather than the need to solve a specific problem. Often discoveries made during basic research lead to practical innovations.

Basilar Membrane The wall of the cochlear duct that holds the hearing receptors.

B Cells Antibody-producing lymphocytes maturing in areas of the lymphatic system other than the thymus. They are major figures in the humoral-mediated immune response.

Benign Noncancerous.

Beriberi Vitamin B_1 deficiency disease.

Beta Cells Pancreatic cells secreting insulin.

Bicuspid Valve The two-flapped valve between the left atrium and the left ventricle.

Bile A mixture of cholesterol, bile pigments, and bile salts, which serve as emulsifiers for fat digestion.

Biology The study of life.

Biomass Pyramid An ecological pyramid showing the relative masses of the occupants of successive trophic levels.

Biome A group of similar ecosystems, defined by climate, and hence by water, temperature, latitude, and altitude.

Biosphere Places in the Earth's atmosphere and waters and on the Earth's crust where life can exist.

Biotin A B vitamin essential for pyruvic acid metabolism.

Bipolar Cells Retinal neurons receiving nerve signals from rods and cones.

Birth Rate Number of births per 1000 members of a population per year.

Blastocyst The hollow ball of 50–100 embryonic cells that enters the uterus from the Fallopian tube and eventually implants. Once implantation has been achieved, pregnancy has officially begun.

Blastomeres The first two cells formed by the dividing zygote (or fertilized egg).

Blind Spot Region of retina where ganglion cell axons exit to form the optic nerve; contains no rods or cones.

Blood–Brain Barrier Arrangement of capillary walls, basement membrane, and astrocytes that keeps many molecules carried in the blood from reaching the brain's neurons.

Blood Types (or Groups) Blood classification system based on markers in red blood cell membranes.

Bolus A ball of food for swallowing.

Bones The elements of the skeleton; made of calcium phosphate and carbonate deposited in a collagen matrix.

Botulism A deadly form of food poisoning caused by the bacterium agent *Clostridium botulini.* It is most often caused by improper canning of vegetables, tuna, and olives.

Brachiation Locomotion by swinging by the arms from overhanging branches.

Brainstem Midbrain, pons, and medulla oblongata; contains nuclei for most cranial nerves and control centers for automatic functions such as breathing.

Bronchi Air tubes branching from the trachea and major bronchi; stiffened by cartilage rings and plates.

Bronchioles Branches of the smallest bronchi; not stiffened by cartilage.

Bronchogenic Carcinoma Commonest lung cancer, arising in the bronchi.

Brush Border The edge of a cell bearing microvilli; so named for its appearance under the microscope.

Budding Reproduction by growing offspring from the body of the parent.

Buffer System A combination of a weak acid and its salt that resists changes in pH by the addition of an acid or a base.

Bulb The expansion at the base of the hair follicle; the hair is formed within it; base of penis.

Bundle of His The atrioventricular bundle.

Bursa A sac containing synovial fluid that cushions body parts where they move over each other.

Calciferol Vitamin D; essential for calcium and phosphorus absorption and use.

Calcitonin Thyroid hormone that prevents resorption of calcium from the bones.
Callus A region of thickened, toughened skin formed in response to pressure and friction; the new bone formed at the site of a fracture.
Calmodulin A protein mediating the effect of calcium ions as a second messenger.
Calorie A unit of energy. By definition, it is the amount of energy required to raise the temperature of 1 liter of water by 1°C (e.g., from 14.5°C to 15.5°C) (1 Calorie = 1 kilocalorie = 1000 calories).
Calorimeter A device for measuring energy release. A bomb calorimeter measures the warming of water surrounding a chamber in which the substance is being burned. Another type of calorimeter is occupied by a living animal and the water is warmed by the heat the animal gives off.
Canaliculi The tiny channels that drain bile secreted by liver cells to the bile duct and gallbladder.
Canines The pointed, fanglike teeth flanking the incisors.
Capacitation The final activation of sperm for fertilization that occurs on exposure to vaginal fluids.
Capillary Smallest of blood vessels; wall is endothelium only; site of oxygen and nutrient transfer to tissues and waste transfer to blood.
Capsid The protective protein coat surrounding the nucleic acid of a virus.
Carbon Cycle The path carbon follows from the air to plants and animals and back to the air, with sidetracks in water and sediments; the recycling of carbon.
Carcinogenesis The process of cancer initiation.
Cardiac Cycle The sequence of events marking each contraction of the heart.
Cardiac Muscle The specialized muscle tissue of the heart; striated, but involuntary.
Cardiac Output The amount of blood pumped by either the right or left ventricle of the heart per minute.
Cardiopulmonary Resuscitation (CPR) Technique for restarting stopped heartbeat and/or breathing by external cardiac compression and/or mouth-to-mouth resuscitation.
Cardiovascular System The heart and blood vessels.
Caries Tooth decay.
Carotene The yellow pigment found in the epidermis and dermis of Asian peoples.
Carotenes Plant pigments the body converts to retinol.
Carpals The bones of the wrist.
Carrier Heterozygote for a recessive allele.
Cartilaginous Joint A joint in which the bones are linked by cartilage.
Catabolism Metabolic reactions in which larger molecules are broken down into smaller ones; often releasing energy in the process.
Catalysts Chemical agents that speed up chemical reactions without being consumed in the process.
Cataract Cloudiness of lens or cornea.
Cecum The pouchlike first section of the colon near the ileocecal valve.
Cell Cycle The sequence of events from cell division to cell division.
Cell-mediated Immunity A system employing principally the T cells to directly attack the foreign antigens.
Cell (Plasma) Membrane The active outer covering of the cell; controls which materials enter or leave.
Cells Tiny, modular, fundamental units in which life occurs.
Cellular Respiration The chemical reactions that break down energy-rich sugars and transfer the chemical energy they contain to ATP, consuming oxygen in the process.
Cellulose A structural polysaccharide in plants.
Cementum The surface material of a tooth below the gum line.
Central Canal Extension of brain's fourth ventricle (under cerebellum) down the center of the spinal cord.
Central Nervous System (CNS) Brain and spinal cord.
Centrioles Tiny structures made of microtubules; organizing centers for the spindle apparatus.
Centromere Region linking sister chromatids.
Cerebellum The "little brain" located just behind and below the cerebrum. It controls muscle tone and posture and adjusts the complex movement patterns initiated by the cerebrum in a manner that brings the movement to target in a coordinated manner.
Cerebral Cortex Surface layer of cerebrum.
Cerebral Hemispheres Two halves of the cerebrum; each one controls the opposite side of the body.
Cerebrospinal Fluid Fluid secreted in brain's ventricles; circulates through ventricles and between arachnoid and pia mater.
Ceruminous Glands Modified eccrine sweat glands found in the ear canal; they secrete cerumen, or ear wax.
Cervical Cap A small rubber shield fitting snugly over the cervix.
Cervical Sponge A small sponge, soaked with spermicide, that rests over the cervix.
Cervical Vertebrae The seven vertebrae of the neck.
Cervix Lower end of uterus projecting into vagina.
Chalones Intracellular mitotic inhibitors.
Chancre Sore at site of exposure that is the first sign of syphilis.
Chemical Breakdown The release of sugars, amino and fatty acids, glycerol, etc., by digestive enzymes from food molecules.
Chemical Equation A statement summarizing the events in a chemical reaction. It identifies the reactants, the products, the relative amounts involved, and the direction of the process.
Chemiosmotic Phosphorylation The attachment of phosphate groups to ADP (making ATP) using energy liberated as hydrogen ions travel down a pH and concentration (chemiosmotic) gradient from the outer compartment to the inner compartment of mitochondria.
Chemistry The study of the composition, properties, and structure of matter.
Chief Cells Stomach gland cells that secrete pepsinogen.
Chloroplasts Plant cell plastids containing the apparatus for photosynthesis.
Cholecystokinin (CCK) A hormone secreted by the small

intestine that inhibits stomach secretion and muscle contractions of the stomach and intestines and stimulates release of bile and enzyme-rich pancreatic juice.

Cholesterol A multiringed lipid; raw material for steroid hormones.

Chondroitin Sulfate Intercellular glue.

Chorion Two-layered membrane formed as mesodermal cells line the trophectoderm.

Chorionic Villi Protrusions of the chorion into blood-filled spaces in the endometrium.

Chorionic Villi Sampling The removal of a small fragment of the chorionic villi in the placenta; the fetal cells in the fragment can be analyzed to detect genetic and chromosomal defects.

Choroid The light-absorbing pigmented layer of tissue between the retina and the sclera.

Chromatin The complex of DNA and protein that makes up a eukaryotic chromosome.

Chromosomal Mutation A mutation that affects the structure of a chromosome; deletion, duplication, inversion, or translocation of lengths of DNA, sometimes many genes long.

Chromosomes Threadlike structures of DNA and protein, found in the nucleus; their DNA carries the information of heredity.

Chronic Anemia Long-term anemia, sometimes due to prolonged slow bleeding.

Chylomicron A particle of triglyceride, protein, cholesterol, and phospholipid secreted by an intestinal cell into a lacteal.

Chyme The soupy mixture of food, mucus, hydrochloric acid, and pepsin prepared by the stomach.

Cilia Locomotor organelles built of microtubules.

Ciliary Body The ring of smooth muscle that controls the shape and curvature of the lens in the eye.

Ciliate Protozoan with numerous oarlike propulsive organelles or cilia.

Circulatory Shock Failure of the heart to pump enough blood to meet the body's needs; symptoms include pale, clammy skin; feeble, rapid pulse; shallow, rapid breathing; lowered body temperature; unconsciousness; and often death.

Circulatory System The cardiovascular system.

Circumcision Surgical removal of the prepuce.

Circumferential Lamellae The layers of bone tissue on the surfaces of bones.

Cirrhosis The condition in which liver cells killed by toxic substances (including alcohol) are replaced by fat and connective tissue.

Citric Acid Cycle Another name for the Krebs cycle; so named because citric acid is the first material formed as each cycle begins.

Clavicle The collarbone.

Cleavage First mitotic cell divisions of zygote.

Cleft Palate Birth defect in which the maxillae fail to fuse completely.

Clitoris Erectile organ where the labia minora meet.

Clone An offspring that is a genetic duplicate of the parent.

Cloning The genetic duplication of an organism by transplanting a cell nucleus into an egg cell and stimulating development.

Clotting Formation of a platelet and fibrin plug in broken blood vessels.

Coccyx The tailbone.

Cochlea The coiled tube containing the receptors for hearing.

Cochlear Duct The one of three tubes in the cochlea that contains the hearing receptors.

Codominance When two alleles both affect phenotype in the heterozygote and neither allele is truly dominant nor recessive.

Codon A DNA or mRNA code word, or base triplet, standing for an amino acid.

Cofactor A chemical compound needed to make an enzyme function properly; often a vitamin.

Coitus Interruptus Withdrawal of the penis from the vagina before ejaculation.

Collagen Most abundant human protein; plays an important role in connective tissue.

Collecting Duct A tubule that receives urine from the distal convoluted tubules of several nephrons and delivers it to the renal pelvis.

Colon The large intestine.

Colostrum Clear, antibody-rich fluid secreted by the breasts for the first few days after childbirth.

Commensalism Symbiosis when one species benefits and the other is not affected.

Common Bile Duct The duct from the gallbladder; delivers bile to the small intestine.

Communication As part of the scientific method, it usually means publishing a paper about scientific findings in a journal or in some other way sharing them with the scientific community.

Community The biotic or living elements of an ecosystem. They include all plants, animals, bacteria, fungi, protozoans, and anything else that lives in the defined area.

Compact Bone Dense bone tissue found on the surfaces of bones.

Competition Struggle between two (or more) species for resources.

Competitive Exclusion Principle The idea that when two species occupy the same niche, one must die out, move away, or change its niche.

Complement A group of 11 blood proteins that help defend the body in both nonspecific and specific ways.

Complete Ecosystem An ecosystem that is large enough and complete enough to be self-sustaining indefinitely.

Complete Protein Protein containing all the essential amino acids.

Complex Viruses A variety of viruses possessing capsid configurations more complex than the helix or icosahedron configurations.

Compounds Substances made by chemically combining two or more different kinds of atoms.

Conceptus Product of conception. The conceptus consists of little more than a cluster of cells shortly after fertilization, bearing little resemblance to a baby.

Condom A rubber sheath covering the penis; used to prevent semen from reaching the vagina.

Conduction System That part of the heart specialized for the generation and conduction of electrochemical signals that causes the heart to contract in an orderly and efficient manner.
Cones Retinal light receptors sensitive to color and used for detailed vision and vision in bright light; each cone is sensitive to red, blue, or green light.
Congenital Syphilis Syphilis contracted before birth from the mother.
Constipation The inability to defecate often caused by inadequate amounts of fiber in the diet.
Consumers Heterotrophs.
Contact Inhibition Cessation of cell division when cells touch each other in cell culture.
Continuous Variable A variable that is measured rather than counted. For instance, a person is found to weigh 125.556 lb.
Contractility Ability to shorten.
Control Group The group or population receiving the sham treatment or placebo in an experiment. Results from the control group are compared against those from the experimental group.
Convalescent Stage The recovery period from a disease.
Convergent Circuit A neural circuit in which one neuron receives signals from many others.
Copulation The act of inserting the penis in the vagina.
Cornea The transparent front of the eye.
Coronal Section A section made parallel to the long axis of the body dividing it into front and back portions.
Corona Radiata Layer of follicular cells surrounding the ovum in a Graafian follicle.
Corpora Cavernosa Paired columns of spongy tissue within the penis; responsible for erection.
Corpus Albicans Atrophied remnant of corpus luteum; develops when pregnancy fails to occur.
Corpus Callosum Band of millions of white nerve fibers linking the two cerebral hemispheres.
Corpus Luteum Postovulatory follicle; gland secreting estrogen and progesterone.
Corpus Spongiosum Column of spongy tissue surrounding the urethra; responsible for erection.
Cortisol (Hydrocortisone), Corticosterone, Cortisone Glucocorticoids.
Costal Cartilage The cartilaginous link between a rib and the sternum.
Countercurrent Multiplier Mechanism An arrangement of opposing fluid flows that permits the magnification of concentrations.
Covalent Bonds Chemical bonds based on the sharing of electrons.
Cowper's Glands Small glands that add mucus and a urine-neutralizing compound to the urethra.
Cranial Cavity Portion of dorsal cavity containing the brain.
Cranial Nerves Twelve nerves that enter and leave the brain instead of the spinal cord.
Cranial Vault The major rounded chamber of the skull, formed by fusing its bony plates to provide a form-fitting protective case around the brain.
Cranium The portion of the skull enclosing the brain.
Crenation Shrivelling of blood cells in hypertonic solution.
Cretinism Childhood hypothyroidism.
Cribriform Plate The perforated section of the ethmoid through which nerve fibers for the sense of smell enter the brain.
Cristae The shelflike membranous platforms running at right angles to the long axes of the mitochondria. The electron transport system is embedded in the cristae.
Cro-Magnon The first humans of modern form.
Crossing Over The exchange of segments between homologous chromosomes during meiotic prophase.
Cryptorchidism Failure of one or both testes to descend from the abdomen into the scrotum.
Cupula Gelatinous mass blocking ampulla; it sways when the head rotates, bending the hair cell hairs embedded in it and provoking nerve signals.
Cyanocobalamin Vitamin B_{12}.
Cyclic Adenosine Monophosphate (cAMP) A chief intercellular carrier of hormone signals; a chief second messenger.
Cytochromes Proteins that make ATP in cellular respiration.
Cytokinesis Division of the cell.
Cytoplasm That portion of a eukaryote's protoplasm between the plasma membrane and the nuclear envelope, or all the internal substance of a prokaryote.

Dark Adaptation The increase in visual sensitivity that comes after being in dim light or darkness for awhile.
Datum (plural data) A single measurement.
Dead Air Volume The air of a breath that remains in the airway; 150 ml.
Death Rate Number of deaths per 1000 members of a population per year.
Decomposers Bacteria and fungi that break down organic materials and release the nutrients they contain to the soil.
Deductive Logic A reasoning system that attempts to predict an individual outcome based on a general statement. Starting with the general statement "men are taller than women," you might try to predict (before seeing either of them) that since individual A is a man and individual B is a woman, A will be taller than B.
Deficiency Disease Illness produced by a lack of some essential nutrient.
Deforestation The loss of forests as a consequence of their exploitation for the resources they contain.
Deletion Loss of a chromosomal segment.
Denaturation An irreversible change in the shape of a protein.
Dendrite Nerve fiber carrying signals toward a neuron's cell body.
Denitrifying Bacteria Bacteria that convert nitrates to nitrogen gas.
Dentin The calcium phosphate and carbonate that comprises the core of a tooth.
Deoxyribonucleic Acid (DNA) The substance of heredity; the DNA molecule is a double-stranded helix; each

strand is a chain of alternating deoxyriboses and phosphate groups. From each strand projects a series of nitrogen-containing bases—adenine, guanine, thymine, and cytosine—that encodes the genetic instructions. The bases pair to link the two DNA strands.

Dermal Papillae Upward folds of the dermis.

Dermis The deeper, connective tissue layer of the skin, containing nerve endings, blood vessels, muscles, glands, and hair follicles.

Desertification Conversion of productive land to barren deserts as a consequence of environmental destruction caused by deforestation, overgrazing, and slash-and-burn agriculture.

Detritivores Eaters of debris and wastes; scavengers.

Dextrose Another name for glucose.

Diabetes Insipidus Due to antidiuretic hormone deficiency; marked by great volumes of highly dilute urine.

Diabetes Mellitus Sugar diabetes; high levels of blood glucose caused by lack of insulin (or inability of the body's cells to respond to insulin); symptoms include high urine volume.

Dialysis The exchange of solutes between two solutions separated by a membrane.

Diaphragm Dome-shaped muscle sheet at bottom of chest cavity; expands chest cavity for inhalation on contraction.

Diaphragm A rubber shield covering the cervix; used to prevent sperm from entering the uterus.

Diaphysis The hollow shaft of a long bone.

Diarrhea The production and release of large amounts of watery stool, often as a consequence of intestinal infection or irritation.

Diastole Relaxation of the heart.

Differentiation The process by which cells give rise to different types of daughter cells in a fetus.

Diffusion The movement of molecules from where they are highly concentrated to where they are less concentrated.

Digestion The process of breaking food down into its component nutrients.

Digestive System The array of organs that carries out digestion; mouth, pharynx, esophagus, salivary glands, stomach, small and large intestines, liver, pancreas, appendix, rectum, and anus.

Diglyceride Glycerol attached to two fatty acids.

Dihybrid Cross A cross between parents who differ in their alleles for two genes.

Diploid Number The number of chromosomes in a body cell; an even number.

Dipole A polar molecule.

Disaccharides Double sugars formed by chemically linking two monosaccharides.

Discrete Variable A variable that is easily counted. For instance, the number of people watching a given television program during a particular half hour.

Distal Further from a point of reference, such as a limb's point of attachment.

Distal Convoluted Tubule The coiled segment of the nephron's tubule furthest from the renal corpuscle.

Diuretic A drug that increases urine production and dilution.

Divergent Circuit A neural circuit in which one neuron sends signals to many others.

Diverticulosis Ballooning outward of a "diverticulum" from the wall of the colon.

DNA Gyrase The enzyme that cuts DNA strands and lets them unwind for replication.

DNA Ligase An enzyme that links together short lengths of DNA.

DNA Polymerase An enzyme that links DNA nucleotides together on a DNA template.

Dominant Allele An allele that affects phenotype whether homozygous or heterozygous.

Dorsal Toward the back or spine.

Dorsal Cavity Body cavity containing the central nervous system.

Double Helix The double-stranded spiral of the DNA molecule.

Drug Resistance Ability to resist the toxic effects of antibiotics and other drugs.

Duodenum The first 25 cm of the small intestine.

Duplication Repetition of a chromosomal fragment.

Dura Mater Outermost, toughest meninx.

Dust Cells Lung macrophages that phagocytize foreign material and debris.

Dysmenorrhea Painful menstruation.

Eccrine Referring to sweat glands consisting of a single coiled tubule; eccrine sweat glands secrete a watery fluid.

Ecological Niche How an organism lives. A description of all the various relationships a species has between itself and other members of its community and the physical factors in its environment.

Ecological Pyramid A drawing showing the relative sizes of different trophic levels in terms of energy, mass, or numbers.

Ecologists Individuals studying the relationships between the various living and nonliving parts of the environment and how they interact.

Ecology The study of the interactions of living things with each other and with their environment.

Ecosystem A relatively self-sufficient portion of the world; it receives little more than energy and water from outside itself.

Ectoderm Germ tissue that will give rise to skin and nervous tissue and their derivatives.

Ectopic Pregnancy Pregnancy resulting when implantation occurs in Fallopian tube or on surface of abdominal organs.

Edema Local swelling in the tissues due to an increased flow of fluid into damaged area.

Efferent Arteriole The arteriole leaving the glomerulus.

Efferent System The nerves carrying signals from the CNS.

Ejaculation Ejection of semen from the penis.

Elasticity Ability to return to normal length after stretching or contracting.

Electrocardiogram (EKG) A graphic record of the electrical activity generated in the heart's conduction system; the term is abbreviated *EKG* because in German *cardio* is *kardio*.

Electroencephalogram (EEG) Record of electrical activity of brain neurons, recorded by electrodes on scalp.

Electrolytes Ionic substances dissolved in water that are capable of conducting electricity.

Electromyogram A recording of a muscle's electrical activity.

Electrons The tiny electrically negative particles in orbit around the atomic nucleus.

Electron Transport Chain A sequence of enzymes and coenzymes that channels the flow of energized electrons. The energy released by these electrons is used to pump hydrogen ions across the inner membrane of the mitochondrion and to synthesize ATP.

Elements The various kinds of atoms are called elements.

Embolism An embolus (moving clot) that has blocked a small blood vessel, usually resulting in tissue death.

Embolus A thrombus that has broken loose and is traveling through the bloodstream.

Embryonic Disc Circular mass of ectoderm and endoderm that will become the embryo.

Emphysema Loss of the lungs' internal surface area by destruction of the alveoli.

Enamel The hard material on the chewing surfaces of a tooth above the gum line.

Enamel Organ An embryonic epidermal structure that produces the enamel of a tooth; it also stimulates deeper tissues to produce the rest of the tooth.

Endergonic Reaction An energy-capturing reaction.

Endocardium The inner lining of the heart; endothelium of the heart.

Endochondral Ossification Bone formation in the cartilaginous precursor of a long bone.

Endocrine Glands Organs that secrete hormones into the blood; ductless glands.

Endoderm Germ tissue that will give rise to the linings of the digestive organs and glands.

Endometrium Lining of uterus.

Endomysium The connective tissue sheath surrounding a muscle fiber.

Endoneurium Connective tissue sheath around individual fibers in nerves.

Endoplasmic Reticulum (ER) A series of membranous channels linking the various areas of the cytoplasm; the principal route for molecular traffic around the cell.

End Organs of Ruffini Skin receptors sensitive to heavier, continuous touch.

Endorphins Modulators of pain sensitivity.

Endothelium Lining of blood vessels.

Energy The capacity to do work (mechanical, chemical, or electrical).

Enkephalin A hormone with opiatelike effects in the brain that helps to regulate peristalsis in the small intestine; modulator of pain sensitivity.

Enteroendocrine Cells Stomach gland cells that secrete the hormone gastrin.

Entropy Disorder.

Envelope Outer covering of some animal viruses, derived from host cell membrane.

Enzymes The protein catalysts that speed up virtually all the body's chemical reactions under conditions compatible with life.

Enzyme–Substrate Complex The association formed while the enzyme holds the substrate molecules in position prior to their reaction.

Epicardium Visceral pericardium.

Epidermal Grooves Inward folds of epidermis that mark the skin and provide room for motion at joints.

Epidermal Ridges Raised folds of epidermis that help provide the hands and feet with traction.

Epidermis The surface, epithelial layer of the skin.

Epididymis Tubule coiled atop a testis; serves as pathway from testis and storage and maturation site for sperm.

Epiglottis The cartilaginous flap that covers the larynx during swallowing.

Epimysium The connective tissue sheath surrounding a muscle.

Epinephrine Adrenaline; major hormone of adrenal medulla.

Epineurium Connective tissue sheath around nerves.

Epiphyseal Plate The unossified cartilage between the epiphysis and diaphysis in a growing long bone.

Epiphysis The end of a long bone.

Erythroblastosis Fetalis Destruction of an unborn infant's red blood cells by the mother's immune system; usually caused by Rh factor incompatibilities between the blood of the mother and fetus.

Erythrocytes Red blood cells.

Erythropoiesis Red blood cell formation in the red bone marrow.

Erythropoietin Blood protein stimulating red blood cell production.

Esophagus The gullet; the tube from mouth to stomach.

Essential Amino Acids Amino acids the human body cannot synthesize and must obtain in the diet.

Essential Fatty Acids Fatty acids the body cannot synthesize and must obtain in the diet.

Estrogens Female sex hormones; involved in sexual maturation, menstrual cycle.

Ethmoid The smallest of the cranial bones separating the cranial vault from the nasal cavity.

Euchromatin Uncondensed chromatin, containing active genes.

Eukaryotes Complex modern cells, including those of humans, that have a well-defined nucleus and other membrane-bound organelles.

Eustachian (Auditory) Tube The tube linking the middle ear to the throat; serves for pressure relief.

Evolution The development of species from preexisting species by natural selection.

Excitability Ability to respond to stimuli.

Exergonic Reaction A reaction that releases energy as it proceeds.

Exhaustion Stage Final stage of general adaptation syndrome.

Exocytosis Fusion of a membrane-wrapped bubble of material made in the cell with the cell membrane; mechanism of secretion and membrane growth.
Exon A gene segment that does code for protein.
Exophthalmic Goiter Adult hyperthyroidism.
Experimental Group The group or population in an experiment receiving the treatment being tested by the experiment.
Experimentation Designing and carrying out experiments to test hypotheses.
Expiratory Center Respiratory control center in the brain (medulla); stimulates forced exhalation.
Expiratory Reserve Volume The amount of air exhalable in addition to the tidal volume; 1200 ml.
Extensibility Ability to stretch.
External Auditory Canal (or **meatus**) The tubular channel leading from the pinna to the eardrum.
External Ear The pinna, ear canal, and eardrum.
External Nares The nostrils.
External Respiration The exchange of gases between air in the lungs and the blood.
External Root Sheath The wall of the hair follicle.
External Sphincter A ring of skeletal muscle surrounding the urethra distal to the internal sphincter; controls urine release voluntarily.
Exteroceptor Sense organ for events outside the body.
Extrinsic Pathway The clotting mechanism that begins when damaged cells release tissue thromboplastin.
Eye The cameralike sense organ of vision.

Facilitated Diffusion The transport of materials across a membrane, down a diffusion gradient, assisted by carrier molecules in the membrane.
Fallopian (Uterine) Tubes Ducts that lead released ova from ovary to uterus.
False Rib A rib whose costal cartilage attaches to the costal cartilage of another rib but not to the sternum.
Fascia A sheet of connective tissue between muscle and skin.
Fascicle A bundle of muscle fibers.
Fast Muscle White muscle.
Fat-soluble Vitamin A vitamin (A, D, E, or K) that dissolves in lipid and can be stored in the body.
Fatty Acids Long-chain organic acids that combine with glycerol to form a fat.
Femur The thighbone.
Fermentation The metabolic pathway that, in the absence of oxygen, converts pyruvic acid to lactic acid, ethyl alcohol, or other compounds.
Ferritin A protein in intestinal cells that forms a complex with iron ions; serves as storage mechanism.
Fertilization Fusion of gametes.
Fetal Alcohol Syndrome (FAS) The set of symptoms seen in children born to women who drink excessive amounts of alcohol.
Fetus An embryo, after movements begin and the organs are formed in the fourth month.
Fever An increase in the body temperature above normal, in response to the chemicals released during an infection.
Fiber Roughage; indigestible material (cellulose) in plant foods.
Fibrin Insoluble, fibrous protein formed from fibrinogen when blood clots.
Fibrinogen Plasma protein converted to insoluble fibrin during the clotting process.
Fibrous Joint A joint in which the bones are linked by fibrous connective tissue.
Fibula A bone of the lower leg.
Fimbriae Ciliated fingerlike projections of ovarian end of Fallopian tube; they guide released ova into the tube.
Fission Reproduction by dividing in two.
Flagella Locomotor organelles built of microtubules; longer than cilia.
Flagellate Protozoan with whiplike propulsive organelle.
Flagellum Whiplike, propulsive tail of sperm cell.
Flat Bones Thin plates; the skull bones, ribs, scapula, and pelvis.
Flexion Reflex Wired-in link between skin pain sensors and motor neurons that command leg flexors to contract; involves an interneuron.
Floating Rib A rib without a costal cartilage.
Flukes Parasitic worms that live and reproduce in the blood, liver, and lungs; they require snails as alternate hosts.
Folic Acid A B vitamin essential for nucleotide synthesis.
Follicles Structural and storage units of thyroid gland; a mass of cells surrounding a developing ovum.
Follicle-stimulating Hormone (FSH) Anterior pituitary gonadotropin; stimulates spermatogenesis and development of ovarian follicle.
Follicular Cells Cells that extend all the way through the walls of the thyroid follicle; the cells surrounding a developing ovum.
Fontanels The membranous areas (the "soft spot") between the bony plates of the fetal cranium.
Food Chain Simple, linear sequence of eaters and eaten.
Food Vacuole A membrane-bound chamber in the cytoplasm containing food items consumed during phagocytosis.
Food Web More realistic and complex view of food chain, showing how many organisms occupy more than one trophic level.
Foramen Magnum The opening in the bottom of the cranium through which the spinal cord passes into the vertebral canal.
Forebrain Cerebrum, basal ganglia, thalamus, and hypothalamus; highest level of the brain.
Formed Elements The red and white blood cells and platelets.
Fossils The remains of past organisms; they can be bones, imprints, casts, or even frozen bodies.
Founder Effect Loss of genetic variability when a new population is founded by a small portion of a larger population.
Fovea Centralis Retinal region with greatest density of cones and no rods; used for details, color, and bright-light vision.

Fracture A break in a bone.
Free Nerve Endings Skin receptors sensitive to temperature and pain.
Frontal Bone The single bone forming the front of the cranial vault.
Frontal Lobe Anterior portion of cerebrum; devoted to control.
Fructose Fruit sugar; the sweetest of all sugars. Like glucose, it is a monosaccharide.

G_1 Stage The growth stage following cell division and preceding DNA replication.
G_2 Stage The growth stage following DNA replication and preceding cell division.
Gallbladder The storage sac for bile secreted by the liver.
Gamete Specialized haploid reproductive or sex cells; male and female gametes (sperm and ova) fuse to produce offspring.
Gamma Globulins, or Immunoglobulins Names for the chemical family of plasma proteins consisting of the antibodies.
Ganglia Small collections of nerve cells lying outside the CNS.
Ganglion Cells Retinal neurons receiving nerve signals from bipolar cells; their axons are the fibers of the optic nerve.
Gastric Inhibiting Peptide (GIP) A hormone secreted by the small intestine that inhibits stomach secretion and muscle contractions.
Gastric Pits The stomach glands.
Gastrin A hormone secreted by the cells of the stomach (and the small intestine); stimulates stomach secretion and relaxes the pylorus.
Gastrocolic Reflex The wave of mass peristalsis in the colon that follows the arrival of food in the stomach.
Gastroileal Reflex The intensification of peristalsis in the ileum that follows the arrival of food in the stomach.
General Adaptation Syndrome (GAS) The body's response to stress.
Generalization Identifying a pattern found among many observations.
Genes Segments of chromosomes whose DNA encodes specific characteristics of the cell or organism, usually in the form of structural proteins or enzymes.
Genetic Code The list of codon–amino acid equivalences.
Genetic Drift Loss of alleles in a small population due to chance failure to pass them to offspring.
Genetics The study of inheritance.
Genetic Variability Inherited variations in the physical forms of a species.
Genital Herpes Viral sexually transmitted disease caused by the herpes simplex virus type 2.
Genotype An organism's set of genes or alleles for particular traits.
Genus A group of related species.
Geographic Isolation Reproductive isolation due to separation in space.
Germ Tissues Ectoderm, endoderm, and mesoderm; the embryo's first tissues, which give rise to adult tissues.
Germinal Layer The surface of the nail bed; it produces the cells that become part of the nail.
Gestation The period of pregnancy.
Girdle The bones attaching the arms or legs to the axial skeleton.
Glans Tip of penis.
Glaucoma Excess pressure within the eyeball; untreated, it can lead to blindness.
Globin The protein portion of hemoglobin.
Glomerular (Bowman's) Capsule The double-walled sac enclosing the glomerulus; the input end of the nephron.
Glomerular Filtrate The fluid that moves from the blood, through the glomerulus, and into the nephron; plasma minus most of its proteins.
Glomerulus A tuft or ball of capillaries branching from the afferent arteriole; the source of fluid processed by the nephron.
Glottis The opening between the vocal folds.
Glucagon Pancreatic glucose-releasing hormone.
Glucocorticoids Adrenal cortex steroid hormones controlling energy metabolism and inflammation.
Glucose The most abundant and important of the monosaccharides ($C_6H_{12}O_6$). It is the raw fuel of the biosphere and is used as a repeating unit in most of the more complex sugars and starches.
Glycerol The three-carbon molecular unit to which fatty acids are attached in fats.
Glycogen Animal starch; a polysaccharide formed by branched chains of glucose units. It is found in limited supplies in the liver and muscle cells of animals.
Glycolysis Literally, "sugar splitting"; the extraction of energy from organic molecules without the aid of oxygen.
Glycoproteins Proteins with attached chains of sugar molecules.
Goiter Enlarged thyroid.
Golgi Apparatus A stack of three to seven (or more) flattened chambers that package proteins and lipids made in the cell.
Gomphosis A peg-in-socket fibrous joint.
Gonadocorticoids Adrenal cortex sex hormones.
Gonadotropins Control sex hormone secretion, sex cell production.
Gonads Organs in which gametes and sex hormones are produced.
Gonorrhea Sexually transmitted disease caused by the coccus *Neisseria gonorrheae.*
Gout A disorder marked by high levels of uric acid in the blood and the deposition of uric acid crystals in the joints and kidneys.
Graafian Follicle Mature follicle, containing fluid-filled cavity.
Granular Leukocytes Leukocytes whose cytoplasm contains granules: neutrophils, eosinophils, and basophils.
Growth Factors Substances that stimulate cell growth and division.
Growth Hormone (GH; Somatotropin) Pituitary hormone that controls growth and metabolism.

Growth Rate The speed of growth of a population, measured as number of people added to the population per 1000 members per year; birth rate minus the death rate.

Gustatory (Taste) Cells Sensory cells in taste buds.

Gustatory Hairs Projections of taste cells into pore of taste bud; their membranes bear the protein receptors that bind to flavor molecules.

Gyrus Fold of cerebral cortex.

Habitat A specific type of environment where a species or a community naturally occurs. For instance, the banks of streams, woodlots, and desert sand dunes are each different habitats.

Hair Cells The hearing receptors.

Hair Cortex The part of a hair beneath the cuticle.

Hair Cuticle The layer of scalelike cells covering the surface of a hair.

Hair Follicle The tube of epidermal cells that dips into the dermis and surrounds the hair root.

Hairlip Cleft palate extending through to the nasal cavity.

Hair Medulla The core of a hair.

Hair Root The part of a hair embedded in the skin.

Hair Shaft The part of a hair showing above the skin surface.

Haploid Number The number of chromosomes in a gamete, or the number of homologous pairs of chromosomes in a body cell.

Haustra The pouchlike segments of the colon.

Haustral Churning Muscular activity in the colon that moves material from full segments (haustra) to empty ones.

Haversian Lamellae The concentric layers of bone that make up a Haversian system.

Haversian System The cylindrical structural unit of compact bone.

Heart The muscular organ that pumps the blood.

Heart Murmur The sound made as blood leaks through partially sealed semilunar valves.

Heimlich Maneuver Using sudden chest compression to expel objects in the airway.

Heme Pigment molecule containing iron; the nonprotein portion, and oxygen binding site, of hemoglobin.

Hemoglobin Oxygen-binding protein found in red blood cells.

Hemolysis Bursting of red blood cells as a consequence of exposure to hypotonic solution.

Hemolytic Anemia Anemia due to red blood cell destruction.

Hemorrhagic Anemia Anemia due to blood loss.

Hemostasis Blood clotting.

Hepatic Portal System Nutrients absorbed by the capillaries of the digestive system carry the blood to the liver, through the portal system, where it breaks back down into capillaries for processing by the liver.

Hepatic Portal Vein The vein connecting the capillary bed of the small intestine to the capillaries (sinusoids) of the liver.

Heterochromatin Condensed chromatin; regions of a chromosome containing inactive genes or nonsense DNA.

Heterotrophs Other-feeders; organisms that take their nutrients from other organisms.

Heterozygous Possessing two different alleles for a particular trait.

Hilus The site where the renal arteries and veins and ureter meet the kidney.

Hippocampus Portion of cerebral cortex visible only from within the longitudinal fissure; involved in memory formation.

Histamine A chemical released from mast cells (and others, when damaged) that increases the permeability of local capillaries. This encourages the movement of white blood cells, antibodies, and fibrin from the blood into a damaged area.

Histocompatibility Antigens Those antigens worn as surface factors on cells; often used to distinguish between "self" and "nonself."

Histones Structural proteins of the chromosomes; comprise the nucleosomes.

Homeostasis The process of keeping the chemical and physical environments within the body stable and compatible with life.

Hominids Humans and their immediate, nonape ancestors.

Hominoids The group of primates including apes and humans.

Homo The genus to which humans belong.

Homo erectus The second known member of genus *Homo;* used fire and may have been able to speak.

Homo habilis The first known member of genus *Homo*.

Homologous Chromosomes Matching pairs of chromosomes. Each parent contributes one of the homologues (chromosomes) to each pair. There are 23 pairs of homologous chromosomes in human cells.

Homology Similarity in structure and development but not necessarily in function.

Homo sapiens "Thinking man," the species to which modern humans belong.

Homozygous Having both alleles of a pair the same.

Hookworms Parasitic roundworms that invade the body through the skin of the feet. They migrate to the lungs and eventually are coughed up into the throat and swallowed. They spend the sexual phase of their life cycle in the digestive system, ulcerating the intestine, consuming the host's blood, and laying eggs that pass out of the host's body into the soil.

Hormones Peptides, steroids, and other chemicals released into the blood supply by endocrine glands. They regulate a wide range of bodily processes, such as growth, development, metabolism, and sexual cycling.

Human Chorionic Gonadotropin (HCG) Hormone secreted by the chorion and the placenta; it stimulates the corpus luteum to continue secreting estrogens and progesterone.

Human Chorionic Somatomammotropin Placental hormone stimulating breast development.

Human Ecology The study of the interactions of human beings with their living and nonliving environment.

Humerus The upper arm bone.

Humoral Immunity Immune response mainly employing antibodies and their ability to complex with antigens.
Humus Partially decomposed organic matter; important to soil fertility.
Hyaluronic Acid A common intercellular glue.
Hybridoma Cells Cells engineered for producing large quantities of specific antibodies. These cells are produced by fusing cancer cells (myeloma) (with their ability to divide rapidly and endlessly) with plasma cells (with their ability to produce antibodies).
Hybrid Vigor Increased health resulting when outbreeding increases the number of heterozygous genes in the genotype.
Hydrogen Bonds Weak chemical bonds based on the attraction of opposite small electrical charges (such as those in water dipoles).
Hydrophilic Water loving, or soluble in water.
Hydrophobic Water fearing, or insoluble in water.
Hydrosphere The combined watery environments found on the Earth, including the lakes, rivers, and oceans.
Hymen Ring of tissue partially blocking the vaginal opening in young women.
Hyperglycemia Above-normal levels of blood glucose.
Hyperopia Farsightedness; lens focuses image beyond retina.
Hypertension High blood pressure.
Hyperthyroid An overactive thyroid, releasing too much thyroxine.
Hypertonic Solution Solution containing more solute (less water) than the cells of the body.
Hypervitaminosis Overdosage with vitamins.
Hypoglycemia Below-normal levels of blood glucose.
Hypothalamus Located below thalamus; contains nuclei controlling endocrine system, autonomic system, and emotion.
Hypothermia A condition characterized by a lowering of body temperature. Because chemical processes are slowed by lower temperatures, sometimes a state of hypothermia is medically induced in patients undergoing long, complicated surgical procedures.
Hypothesis A tentative explanation for patterns seen among observations.
Hypothyroid An underactive thyroid, releasing too little thyroxine.
Hypotonic Solution Solution containing less solute (more water) than the cells of the body.
Hysterectomy Removal of the uterus.
H Zone The center of the sarcomere, occupied only by thick myofilaments.

Ileocecal Valve The valve between the small and large intestines.
Ileum The 3.6-m final segment of the small intestine.
Immune System Various physical, mechanical, and cellular systems the body uses collectively to defend itself.
Immunosurveillance System The continuous checking of all the tissues and debris in the body by the T cells and macrophages. If they detect "nonself" or troublesome changes in the surface recognition factors on any of the cells, they call for a defensive attack.
Impermeable Membranes Membranes that allow nothing to pass through.
Implantation The process by which the blastocyst invades the endometrial lining of the uterus.
Inbreeding Mating with related individuals, who share more recessive alleles.
Incisors The chisellike front teeth.
Incomplete Dominance A kind of codominance in which both alleles contribute features to the phenotype.
Incomplete Ecosystem A simple ecosystem that is lacking in resources or possesses an oversimplified trophic structure that prevents it from being self-sustaining over the long haul.
Incomplete Penetrance When a specific genotype does not always produce the same phenotype.
Incomplete Protein Protein lacking one or more essential amino acids.
Incubation Period The period of time between when a disease agent first enters the body and when it has multiplied to the point that it begins to produce disease symptoms.
Incus The anvil; the middle of the three middle-ear bones.
Indifferent Gonad The embryonic gonad, before it has differentiated into ovary or testis.
Induction Activation of genetic transcription of a gene.
Inductive Logic A reasoning system attempting to make general statements based on many individual observations. After measuring the heights of a sample of many men and women, you might be inclined to generalize inductively that men are taller than women.
Inert Stable or nonreactive. Reluctant to participate in chemical reactions.
Infectious Mononucleosis Viral infection marked by increased monocyte and lymphocyte count.
Inferior Below.
Inferior Colliculi Portion of midbrain receiving auditory input.
Inferior Conchae Curved, shelflike plates forming the lateral walls of the nasal cavity.
Inferior Vena Cava The large vein that delivers blood to the heart from below the heart.
Inflammation A general body response to a wide range of physical, chemical, and cellular insults. It usually involves an increased flow of blood into the damaged area, which provokes redness, heat, swelling, pain, and loss of function.
Inhibin Hormone secreted by Sertoli cells and ovarian follicle cells to inhibit GnRH and FSH production; in the male, it inhibits sperm production.
Inhibiting Factors Hypothalamic hormones inhibiting release of anterior lobe hormones.
Inner Cell Mass Mass of cells within the blastocyst that will become the embryo.
Inner Ear The region containing the ear's sensory apparatus; embedded in the temporal bone of the skull.
Inorganic Evolution Natural processes by which nonliving parts of the universe, such as stars, galaxies, mole-

cules, and geological features, were formed and developed.

Insertion The attachment to bone of the more mobile end of a muscle.

Inspiratory Center Respiratory control center in the brain (medulla); stimulates inhalation with a spontaneous rhythm.

Inspiratory Reserve Volume The amount of air inhalable in addition to the tidal volume; 3100 ml.

Insulin Pancreatic hormone lowering blood glucose.

Integument The skin.

Intercalated Disc The dense structure formed by the interdigitating cell membranes of two abutting cardiac muscle fibers.

Interferon A chemical agent produced by cells that have been invaded by viruses. It helps prevent the spread of the infection in numerous ways.

Interleukins Immune system modulators.

Intermediate Pituitary Secretes melanocyte-stimulating hormone; between anterior and posterior lobes.

Internal Nares The pair of openings, at the interior edge of the nasal septum, from the nasal cavity into the pharynx.

Internal Respiration The exchange of gases between blood and tissues.

Internal Root Sheath A layer of cells surrounding the hair root.

Internal Sphincter A ring of smooth muscle surrounding the urethra where it leaves the urinary bladder; controls urine release involuntarily.

Interoceptor Sense organ for events inside the body.

Interphase The stage of the cell cycle between cell divisions.

Interspecific Interaction An exchange of resources between species.

Interstitial Cells Cells lying between the seminiferous tubules; they secrete the male sex hormone.

Interstitial Fluid Liquid in the spaces between the cells of the body's tissues.

Interstitial Lamellae Layers of bone between Haversian systems.

Intervertebral Discs Pads of fibrous cartilage between the vertebrae.

Intervertebral Foramina Openings between the vertebrae for passage of the spinal nerves.

Intestinal Glands Pits in the intestinal mucosa that secrete mucus and digestive enzymes.

Intramembranous Ossification Bone formation in the connective tissue membrane precursor of a flat bone.

Intrauterine Device (IUD) A small object, often plastic, placed in the uterus to prevent implantation.

Intrinsic Factor Substance secreted by the stomach lining essential for vitamin B_{12} absorption.

Intrinsic Pathway The clotting mechanism that begins when platelets release coagulation factors.

Intron A gene segment that does not code for protein.

Inversion Flipping end for end of a chromosomal fragment.

Ion An atom that has lost or gained electrons and hence acquired a net electrical charge.

Ionic Bonds Chemical bonds based on the attraction between oppositely charged ions.

Ionic Compounds Compounds held together through ionic bonds. These compounds often break down into charged ions when they are dissolved in water.

Iris The disk located between the lens and cornea, with an adjustable opening in its center, that controls the amount of light entering the eye. The pigments that give the eyes their characteristic color are also located in the iris.

Irregular Bones Bones that are not flat, long, or short; vertebrae.

Irritability Responsiveness. The ability of living systems to perceive changes in their environment and respond.

Islets of Langerhans Endocrine portions of pancreas.

Isomer One of two or more three-dimensional configurations of a molecule.

Isometric Contraction A muscle contraction in which the muscle does not change length.

Isotonic Contraction A muscle contraction in which the muscle changes length while maintaining a constant tension.

Isotonic Solution Solution containing the same amount of solute as the cells of the body.

Isotopes Atoms of an element that differ in the number of neutrons they contain but not in their number of protons or electrons.

Isotropic (I) Band The clear zone near the Z line, occupied only by thin myofilaments.

Jejunum The 2.5-m second segment of the small intestine, between the duodenum and the ileum.

Joint A junction between two bones.

Juxtaglomerular Apparatus A small organ, composed of cells of the efferent arteriole and the distal convoluted tubule of the nephron, that responds to low blood pressure by secreting renin.

Karyokinesis Division of the nucleus.

Karyotype A photograph made of the full set of chromosomes in a metaphase cell.

Keratin The tough, waterproof protein that fills epidermis, hair, and nail cells.

Ketone Bodies By-products of fat metabolism; notable in diabetics, dieters, and victims of starvation.

Kidneys The blood-filtering organs located between the peritoneum and the back wall of the abdominal cavity; they control blood composition and pH.

Kinins Substances released by white blood cells; mediators of inflammation.

Krebs Cycle The self-regenerating sequence of reactions that extracts the energy and raw materials to form NADH, $FADH_2$, and ATP from acetyl groups.

Kwashiorkor Protein deficiency disease.

Kyphosis Hunchback; exaggeration of the spine's thoracic curve.

Labia Majora Folds of flesh flanking the vaginal opening; they bear pubic hair.

Labia Minora Folds of flesh between the labia majora, without pubic hair.
Labor The process of childbirth, when uterine contractions expel the fetus.
Lacrimal Bones Smallest of the facial bones; form part of eye socket.
Lacteal The lymphatic capillary within a villus.
Lacuna A space between bone lamellae containing osteocytes.
Langerhans Cells Phagocytes found in the skin; they serve to detect foreign materials and alert the immune system to combat them.
Laryngopharynx The part of the pharynx containing the larynx and the opening to the esophagus.
Larynx The voicebox.
Latent Infection Virus infection that produces just enough virus particles to maintain the infection in the host cell's progeny.
Latent Period The delay between a stimulus and a muscle contraction or nerve signal.
Lateral Farther from the midline.
Law of Independent Assortment The statement that alleles of separate genes go to gametes independently of each other.
Law of Segregation The statement that pairs of alleles separate during the production of gametes.
Laxatives Chemical agents that increase intestinal motility.
Lens The mass of transparent protein behind the cornea that focuses light on the retina.
Leukemia Cancer of the leukocyte parent cells in the bone marrow.
Leukocytes White blood cells.
Leukotrienes Immune system modulators.
Ligament A connective tissue strap that links bones across a joint.
Limbic System Areas of the brain devoted to emotion and memory; includes a portion of the cortex and basal ganglia, and some of the hypothalamus and thalamus.
Linkage Failure of genes to assort independently because they are on the same chromosome.
Lipid Organic substance soluble in organic solvents but not water; fats, oils, fatty acids, cholesterol.
Lithosphere The geological components of the Earth. They include rocks, minerals, mountains, and the inorganic components of the soil.
Lobes Subdivisions of lungs that receive air from secondary bronchi.
Lobules Subdivisions of lobes that receive air from terminal bronchioles.
Local Anaphylaxis The local inflammation reaction usually involving specific tissues, like those lining the respiratory system (hay fever).
Long Bones Bones longer than they are wide; limb bones.
Longitudinal Arch The upward curve of the foot from heel to ball.
Longitudinal Fissure Deep cleft between the two cerebral hemispheres.
Loop of Henle A hairpin-shaped loop in the nephron's tubule extending into the renal medulla; builds high salt concentrations in the medulla.
Lordosis Swayback; exaggeration of the spine's lumbar curve.
Lumbar Vertebrae The five large vertebrae located in the small of the back.
Luteinizing Hormone (LH) Anterior pituitary gonadotropin; stimulates ovulation in the female and the secretion of sex hormones from both the ovaries and the testes.
Lymph Tissue fluid that has been collected by the lymph vessels.
Lymphatics Larger lymphatic vessels.
Lymphatic System Drainage and filtering system for excess tissue fluid.
Lymph Capillaries Smallest of lymphatic vessels; they dead-end among tissue cells.
Lymph Nodes Lymph filters containing lymphocytes and their parent cells.
Lymphocytes Leukocytes that secrete antibodies and attack foreign cells.
Lymphoid Organs Organs resembling large lymph nodes; they also contain lymphocytes and their parent cells, but they may serve additional functions as well.
Lymphoma Cancer of lymphoid tissue.
Lysosome A cytoplasmic organelle containing digestive enzymes; used in breaking down the contents of food vacuoles.
Lysozyme An enzyme found in saliva and tears that kills bacteria.

Macronutrient A mineral needed in relatively large quantities.
Macrophages Large cells in the tissue spaces and the cardiovascular and lymphatic systems that phagocytize foreign cells, debris, and potential toxins.
Macula Retinal region surrounding fovea; contains no rods.
Major Calyx Large subdivision of the renal pelvis.
Malignant Cancerous.
Malleus The hammer; the first in the series of ear bones; it is in direct contact with the eardrum.
Mammals All animals that suckle their young; they also share other features, such as hair, a four-chambered heart, and the ability to keep their body temperatures constant by generating heat internally.
Mammary Glands Modified, milk-secreting apocrine sweat glands; the breasts.
Mandible Lower jawbone.
Marrow Cavity The hollow space in the shaft of a long bone.
Mass Peristalsis "Heavy-duty" peristalsis that moves material down the colon and into the rectum.
Matter Anything that occupies space and has mass.
Maxillae Upper jawbones.
Mechanical Breakdown Grinding, mashing, and mixing of food by teeth and stomach and intestinal contractions to produce small particles.
Medial Closer to the midline.

Mediastinum Portion of thoracic cavity containing the major blood vessels, trachea, and esophagus.
Medulla Oblongata Tapering link of brain to spinal cord; contains nuclei of some cranial nerves and nuclei controlling automatic functions such as breathing, heart rate, and blood pressure.
Megakaryocytes Parent cells of platelets.
Meiosis The type of cell division (actually two sequential divisions) employed to produce four haploid daughter cells (often the gametes) from a diploid parent cell.
Meissner's Corpuscles Skin receptors sensitive to light touch and low-frequency vibration.
Melanin The dark (brown or black) pigment of the skin.
Melanocytes Octopuslike cells that synthesize the pigment melanin and transfer it to epidermal cells.
Melanocyte-stimulating Hormone (MSH) Controls pigmentation in lower animals.
Melatonin Pineal hormone; inhibits reproduction, controls breeding season.
Memory Cells Both T-cell and B-cell lymphocytes, bearing specific surface recognition factors that enable them to respond rapidly to future exposures to antigens to which they have been previously sensitized.
Menarche First menstruation during puberty; the young woman's first period.
Meninges (singular *meninx*) Membranes that enclose the brain and spinal cord; dura mater, arachnoid, and pia mater.
Menopause Time of life when menstruation ceases.
Menstrual Cycle Monthly growth and shedding of endometrium and maturation and release of ova.
Menstruation Monthly discharge of blood and endometrial fragments that is shed when pregnancy fails to occur.
Merkel's Discs Skin receptors sensitive to light touch.
Mesoderm Germ tissue that will give rise to muscle, bone, blood cells, and other tissues.
Messenger RNA (mRNA) The form of RNA that carries genetic information from the nucleus to the protein-synthesizing apparatus (ribosomes) in the cytoplasm.
Metabolic Rate The speed of energy use or release by a living organism.
Metabolism The sum total of the body's chemical activities.
Metacarpals The bones of the palm of the hand.
Metaphase The second stage of mitosis, marked by alignment of the chromosomes on the cell's equator.
Metarteriole Capillary linking arteriole and venule directly; has smooth muscle for controlling flow.
Metastasis The movement of cancer cells through the body and the establishment of colony tumors.
Metatarsals The bones of the foot.
Micelle A clump of bile salt molecules that serves to ferry fatty acids and monoglycerides to intestinal cells for absorption.
Microfilaments Thin, nontubular fibrils responsible for changes in cell shape.
Micronutrient A mineral needed only in tiny amounts.
Microtubules Hollow fibers that serve as skeletal and motor structures in the cell.
Microvilli Fingerlike protrusions of individual intestinal cell membranes.
Midbrain Anterior portion of brainstem; contains nuclei for processing visual and auditory input and for some cranial nerves.
Middle Ear The air-filled chamber between the eardrum and the oval window, containing the malleus, incus, and stapes.
Midsagittal Section An anatomical section cut parallel to the long axis of the body down the midline. It divides the specimen into a left and right half.
Mineral An inorganic substance, usually ionic, essential as parts of vitamins, hormones, and other molecules; some play crucial roles as constituents of body fluids.
Mineralocorticoids Adrenal cortex steroid hormones controlling mineral balance.
Minor Calyx Subdivision of a major calyx; embraces a renal papilla.
Mitochondria The sites of cellular respiration in eukaryotes.
Mitosis Cell division that produces two diploid daughter cells from a diploid parent cell.
Mitotic Spindle An array of microtubules, centered on the centrioles, that helps the chromosomes separate in karyokinesis.
Molars The flattened grinding teeth along the sides of the jaws.
Molecular Genetics The study of the gene and its actions on a molecular level.
Molecule The smallest particle of a particular chemical substance, often containing two or more atoms.
Monera The kingdom of prokaryotes.
Monoclonal Antibodies Specific antibodies produced in large quantities by identical cells that have been cloned just for that purpose.
Monocytes Phagocytic leukocytes that give rise to macrophages.
Monoglyceride Glycerol attached to one fatty acid.
Monohybrid Cross A cross between parents who differ in their alleles for one gene.
Monomers Single molecules, like glucose, that are often chained together to form long complex molecules.
Monosaccharide A simple sugar containing three to seven carbons (e.g., glucose, fructose).
Mons Pubis (Mons Veneris) Mound of fatty tissue overlying the female pubic bone.
Morula The compact ball of 12–16 cells formed from the dividing blastomeres as the conceptus moves down the Fallopian tube toward its rendezvous with the uterus.
M Phase The stage of the cell cycle during which cell division ("M"itosis) occurs.
Mucous Cells Stomach gland cells that secrete mucus.
Multigenic Inheritance Control of a trait by several genes.
Muscle A mass of cells specialized for contractility.
Muscle Fiber A muscle cell.
Muscle Spindle Arrangement of modified muscle fibers and nerve endings that responds to muscle stretch.
Muscular System The body's skeletal muscles.
Mutation Change in the DNA.

Mutualism Symbiosis when both species benefit.

Myasthenia Gravis Progressive weakening of the muscles due to destruction of nerve signal receptors in the sarcolemma by the body's immune system.

Mycoses Any of the diseases caused by fungi. They include deep diseases involving many of the body's systems (systemic mycoses) and the mycoses of the skin, such as athlete's foot (referred to as superficial or dermamycoses).

Myelin Lipid-rich sheath around many axons and dendrites; electrical insulator and transmission accelerator.

Myocardium The muscle of the heart wall.

Myoepithelial Cells Contractile cells of mammary gland alveoli; in response to oxytocin, they squeeze milk into the ducts of the breasts.

Myofibrils Bundles of myofilaments within a muscle fiber.

Myofilaments The contractile protein filaments within a muscle fiber.

Myoglobin An oxygen-storing muscle protein; related in structure to hemoglobin.

Myopia Nearsightedness; lens focuses image short of retina.

Myosin The protein of the thick myofilaments.

Myxedema Adult hypothyroidism.

Nail Bed The sheet of epidermis covered by the nail and giving rise to it.

Nail Cuticle The edge of the skin covering the nail root.

Nail Root That portion of a nail buried beneath the skin at its base.

Nails Sheets of hard, keratinized cells covering the tips of the fingers and toes.

Nasal Bones The pair of fused bones that form the bridge of the nose.

Nasal Septum The vertical bony partition separating the nasal cavity into left and right chambers.

Nasopharynx The upper portion of the pharynx, from the internal nares to the border of the soft palate.

Natural Selection The process of differential reproduction that makes each succeeding generation better adapted to its environment.

Neanderthals An early form of *Homo sapiens.*

Negative Feedback Control mechanisms that operate by inhibiting excursions in either direction away from a predetermined set point.

Nematodes Roundworms; many are parasitic.

Nephron The functional, urine-producing unit of the kidney; consists of the renal corpuscle, the proximal convoluted tubule, the loop of Henle, and the distal convoluted tubule.

Nerve A bundle of nerve fibers carrying signals to or from the CNS.

Nervous System The network of nerve cells and fibers that receives and processes electrochemical information and coordinates the body's activities.

Neural Circuit Arrangement of neurons; analogous to an electrical circuit.

Neural Plate Earliest version of the embryonic nervous system.

Neural Tube Second version of the embryonic nervous system; formed as the neural plate curls into a tube.

Neuroendocrine Reflex Release of oxytocin in response to suckling.

Neuroglia Supporting cells of the CNS.

Neuron Nerve cell.

Neurotransmitter Substance released from synaptic vesicles into synaptic clefts; on binding to postsynaptic receptors, it initiates (or inhibits) an action potential in the target cell.

Neutrons The uncharged particles in the nuclei of atoms.

Niacin Vitamin B_3, usually not given a number.

Nitrate Bacteria Bacteria that convert nitrites to nitrates.

Nitrifying Bacteria Bacteria that convert ammonia to nitrites.

Nitrogen Cycle The path nitrogen follows from fixation in the soil to plants and animals and back to the soil, with some return to the air; the recycling of nitrogen from biologically useless to biologically useful and back.

Nitrogen Fixation The conversion by lightning, bacteria, and blue-green algae of atmospheric nitrogen gas to biologically useful ammonia and nitrates.

Node of Ranvier Gap between myelin blocks.

Nonhistones Chromatin proteins that seem to regulate gene activity.

Nonspecific Resistance An assortment of different defenses that work against a wide range of disease and infectious agents.

Norepinephrine Noradrenaline; minor hormone of adrenal medulla.

Nuclear Envelope The double membrane separating the nuclear compartment from the cytoplasm.

Nucleic Acids DNA and RNA.

Nucleoid A DNA-containing region in moneran cells.

Nucleolus The region of the nucleus where ribosomal RNA is manufactured.

Nucleoplasm That portion of the cytoplasm within the nucleus of eukaryotic cells.

Nucleosomes Histone beads around which DNA wraps to shorten its length.

Nucleotides Complex molecules formed by chaining a nitrogenous base to a sugar and a phosphate; the basic building blocks of nucleic acids and ATP.

Nucleus A conspicuous organelle in the center of most eukaryotic cells; contains the chromosomes and nucleolus and serves as the control center for the cell.

Observation Measurements, either formal or informal, of some physical or biological phenomenon.

Occipital Bone The bone that forms the posterior floor of the brain case.

Occipital Condyles Rounded projections by which the skull pivots on the vertebral column.

Occipital Lobe Rear portion of cerebrum; devoted to vision.

Olfactory Cell Neuron embedded in nasal mucosa; its dendrites bear protein receptors for odor molecules.

Olfactory Lobe Portion of brain overlying the nasal

mucosa and receiving axons from the olfactory cells; linked to rest of brain by the olfactory tract.

Oncogene A gene that causes cancer.

One Gene–One Enzyme Theory The idea that each gene specifies the structure of one type of enzyme.

One Gene–One Protein Theory The idea that each gene specifies the structure of one type of protein molecule.

Oogenesis Process by which ova (eggs) are formed.

Oogonia Parent cells of ova in ovarian epithelium.

Ootid Result of second meiotic division (after ovulation) of primary oocyte.

Operator The site to which repressors bind to block transcription of a gene.

Operon A functional genetic unit, regulated as a whole, consisting of one or more structural genes, an operator, and a promoter.

Opportunistic Able to cause disease when given an opportunity.

Optic Chiasma The x-shaped crossing of the optic nerves that occurs just in front of the pituitary; gives rise to the optic tracts.

Optic Tracts The bundles of axons emerging from the optic chiasma; each tract contains axons from one side of both retinas.

Oral Contraceptive The "pill"; a preparation of synthetic female sex hormones that blocks ovulation.

Organ Major structures or compartments of the body made by combining different tissues.

Organelles The tiny structures and compartments that act as the cell's "organs."

Organic Chemistry The branch of chemistry studying and analyzing compounds built on frameworks of carbon atoms.

Organic Compounds Compounds built around frameworks of carbon atoms to which are attached various combinations of hydrogen, oxygen, nitrogen, phosphorus, and sulfur atoms.

Organic Evolution The branch of biological thought focusing on how life originated and then diversified to the many forms we see today.

Organ of Corti The spiral organ.

Organ Systems Groups of organs that cooperatively provide the body with certain services.

Origin The attachment to bone of the relatively immobile end of a muscle.

Oropharynx The middle portion of the pharynx, from the edge of the soft palate to the level of the hyoid bone and epiglottis; that part of the throat visible at the back of the mouth.

Osmosis The diffusion of water through a selectively permeable membrane.

Ossification Bone formation.

Osteoblasts Bone-forming cells; derived from periosteum.

Osteoclasts Bone-destroying cells; derived from monocytes.

Osteocyte A bone cell.

Osteoporosis Bone weakening due to calcium loss.

Otoliths Crystals of calcium carbonate on the surface of the gelatinous masses holding the hairs of the hair cells in the utricle and saccule.

Oval Window Opening in the bone between the middle and inner ears; receives the base of the stapes.

Ovary Female gonad.

Ovulation Release of ovum from Graafian follicle.

Ovum Female gamete.

Oxygen Debt The amount of oxygen needed to remove the lactic acid built up during exercise as well as to restore blood levels of oxygen to normal.

Oxytocin Posterior lobe hormone stimulating uterine contractions, milk release.

Pacemaker The SA node.

Pacinian Corpuscles Skin receptors sensitive to pressure and high-frequency vibration; also found in muscle membranes, joints, and viscera.

Palatine Bones Bones forming the posterior portion of the bony palate.

Pancreas Gland below stomach; as exocrine gland, produces digestive enzymes; as endocrine gland, produces insulin, glucagon, somatostatin.

Pancreatic Duct The duct from the pancreas; delivers digestive enzymes and alkaline fluids to the small intestine.

Pantothenic Acid A B vitamin; part of coenzyme A.

Parafollicular Cells Cells in wall of thyroid follicle that secrete calcitonin.

Parasite Any organism that lives on or within its host and consumes the nutrients or substance from that host.

Parasitism A version of predation where the predator lives within its prey or host.

Parasympathetic Division A branch of the ANS; generally opposes the sympathetic division. The branch of the ANS that encourages physiologic well-being or homeostatis.

Parathormone (PTH) Hormone of parathyroids; stimulates the release of calcium and phosphate from bone.

Parathyroid Glands Small glands on the back of the thyroid; secretors of parathormone.

Parietal Bones The pair of cranial bones that form the posterior roof of the cranial vault.

Parietal Cells Stomach gland cells that secrete hydrochloric acid and intrinsic factor.

Parietal Lobe Upper middle portion of cerebrum; receives and processes sensory data.

Parietal Pericardium The tissue of the pericardial sac; its outer layer is tough and fibrous; its inner layer is serous membrane.

Parietal Pleura Serous membrane attached to the inside of the chest cavity.

Passive Immunity Temporary immunity produced when a person is given someone else's antibodies by injection (as in a gamma globulin shot), by being passed through the placenta, or by being taken in with breast milk from the mother.

Passive Transport Molecular transport mechanisms that do not require a cell to spend energy to make them work.

Patella The kneecap.

Pathogens Organisms that cause disease.

Pectoral Girdle The scapula and clavicle.
Pellagra Niacin deficiency disease.
Pelvic Bones The two halves of the pelvis.
Pelvic Cavity Portion of abdominopelvic cavity below the top of the pelvis.
Pelvic Girdle The pelvis.
Penis Erectile cylinder containing the urethra.
Pepsin Protein-digesting enzyme secreted in the stomach as pepsinogen.
Pepsinogen The precursor of pepsin; activated by hydrochloric acid.
Peptide Bonds The specific chemical bonds employed to chain together amino acids into proteins.
Pericardial Cavity Portion of thoracic cavity containing the heart.
Pericardial Sac The sac enclosing the heart.
Perimysium The connective tissue sheath surrounding a fascicle.
Perineurium Connective tissue sheath around bundles of fibers (fascicles) in nerves.
Periodontal Ligament The connective tissue layer that anchors the teeth in their sockets.
Periosteum The connective tissue membrane covering bone; gives rise to osteoblasts.
Peripheral Nervous System (PNS) The nerves.
Peristalsis A wave of muscular contraction that propels food through the digestive tract.
Peritonitis Infection of the peritoneum.
Permeable Membranes Membranes that freely allow molecules to pass through.
Pernicious Anemia Anemia caused by a lack of intrinsic factor and vitamin B_{12} (necessary for the development of red blood cells).
Persistent Infection Virus infection that produces enough virus particles to maintain a low level of virus in the blood.
pH Scale The system for estimating the relative acidic or alkaline properties of various solutions. A neutral solution is one with a pH of 7. An acid solution has a pH of less than 7 and an alkaline (basic) solution has a pH greater than 7.
Phagocytosis Cellular eating by surrounding an external object with cell membrane.
Phalanges The bones of the fingers and toes.
Pharynx The throat; the rear chamber behind the nasal and oral cavities and in the vicinity of the larynx. It serves as a common passageway for liquid, food, and air.
Phenotype An organism's observable features.
Phospholipids Important components of cell membranes related to fats. They consist of glycerol attached to two fatty acids and a phosphate group.
Photosynthesis The trapping of light energy and its use in the synthesis of glucose in plant cells.
Physiological (Normal) Saline A salt solution isotonic to body fluids.
Pia Mater Innermost meninx, next to brain.
Pineal Gland Secretor of melatonin; located near the middle of the brain.
Pinna (Auricle) The "flap" of the ear attached to the side of the head.
Pinocytosis Cellular drinking by drawing in a bubble of cell membrane.
Pituitary Gland (Hypophysis) Endocrine gland above roof of mouth; many of its hormones control other endocrine glands.
Placenta Organ formed by endometrium and chorion; nourishes embryo; supplies hormones of pregnancy.
Placental Luteotropic Releasing Factor Hormone stimulating HCG production.
Plaque The deposits of food remains, calcium salts, and bacteria that form on the surfaces of teeth and foster tooth decay and gum disease.
Plasma Blood minus its formed elements; a solution of water; proteins, salts, nutrients, and other substances.
Plasma Cells B cells that have the responsibility for manufacturing antibodies.
Plasma (Cell) Membrane The active outer covering of cells. It controls which materials will be allowed to enter or leave cells.
Plastids Organelles containing their own DNA and ribosomes.
Platelets Membrane-enclosed cellular fragments pinched off from megakaryocytes in the bone marrow; essential to blood clotting.
Pleiotropy When a single gene or allele affects several traits.
Pleural Cavity Portion of thoracic cavity containing the lungs.
Pleural Membranes Serous (fluid-secreting) membranes surrounding the lungs.
Plexi (Brachial, Lumbar, Sacral) Zones where many spinal nerves come together; from them emerge nerves to body's organs, each one containing nerve fibers from many spinal nerves.
Plicae Circulares Folds in the wall of the small intestine.
Pneumonia Disease marked by accumulation of fluid and dead white blood cells in the alveoli.
Pneumotaxic Center Respiratory control center in the brain (pons); inhibits inspiratory center.
Point Mutation A mutation that affects a single base by deletion or replacement.
Polar Body Nonviable, rudimentary daughter cell of meiosis of oocyte; vehicle for discarding surplus chromosomes.
Polar Molecule A molecule whose ends have opposite electrical charges.
Polycystic Disease An inherited kidney defect marked by fluid-filled sacs in the nephrons.
Polycythemia Abnormally large numbers of red blood cells in the blood.
Polymers Large molecules formed by chaining together simpler molecular building blocks called monomers.
Polypeptides Chains of one hundred or less amino acids joined together.
Polyribosome The structure formed when several ribosomes attach to and read a single mRNA molecule.
Polysaccharides Complex carbohydrates formed by chemically linking more than two monosaccharides in chains or sheets.
Polytene Chromosomes Specialized chromosomes of the fruit-fly salivary gland; they are bundles of many DNA

strands.

Polyunsaturated Having many double bonds (said of fats and fatty acids).

Polyunsaturated Fats Fats whose fatty acids have more than one double bond. Most plant oils fall into this category.

Pons Band of nerve fibers and cranial nerve and other nuclei anterior to medulla; from it arise stalks of cerebellum.

Population All the individuals of a particular species occupying a defined area during a specific time interval.

Portal System An arrangement of blood vessels consisting of two capillary nets linked by an arteriole or vein.

Posterior Toward the rear.

Posterior Cavity The interior of the eye behind the lens.

Posterior Lobe of Pituitary Release point for two hypothalamic hormones, antidiuretic hormone and oxytocin.

Posterior Roots Bundles of sensory nerve fibers entering spinal cord.

Potential Niche The niche a species occupies in the absence of competition.

Precipitation Rain, snow, sleet, and hail.

Predation Symbiosis when one species eats the other.

Pregnancy State of carrying an embryo or fetus in the uterus.

Premenstrual Syndrome (PMS) Breast swelling and tenderness, joint pain, irritability, depression, anxiety, fatigue, headaches, and other symptoms preceding menstruation in many women.

Premolars The two complex teeth immediately behind the upper and lower canines on either side of the jaw. The premolars are located in front of the more complex molars located in the back of the mouth.

Prepuce Foreskin; fleshy covering of glans; fleshy fold of skin covering the shaft of the clitoris.

Primary Audition Area Region of temporal lobe cortex devoted to analyzing sound.

Primary Consumers Eaters of plants; herbivores.

Primary Follicle Primary oocyte surrounded by layer of stromal cells.

Primary Motor Cortex Strip of cerebral cortex on frontal side of central sulcus; generates voluntary motor commands.

Primary Oocyte Result of mitosis of oogonium before birth.

Primary Ossification Center The area of the diaphysis where endochondral ossification begins.

Primary Sensory Cortex Strip of cerebral cortex on parietal side of central sulcus; receives sensory data from body.

Primary Sex Organs Gonads.

Primary Sexual Characteristics Body features having directly to do with reproduction.

Primary Spermatocytes Results of mitotic division of spermatogonia.

Primary Structure The sequence of amino acids along the length of a protein chain.

Primates The group of mammals including prosimians, monkeys, apes, and humans.

Primitive Gut Endoderm-enclosed cavity formed within the embryonic disc as the yolk sac pinches off.

Principles Fundamental, primary, and general laws or "truths" from which others are derived.

Prion An apparently infectious protein lacking nucleic acid.

Prodromal Stage The 24-hour period just prior to getting sick, when the body is making its transition from the incubation period to the acute stage.

Producers Plants; photosynthesizers; makers of organic materials from sunlight, water, and carbon dioxide.

Products The substances produced by a chemical reaction.

Progesterone A female sex hormone; reinforces effects of estrogen.

Prokaryotes Primitive cells that lack a nucleus and other membrane-bound organelles (bacteria, blue-green algae).

Prolactin Stimulates milk production and release.

Prolactin-inhibiting Factor (PIF) Hypothalamic hormone inhibiting prolactin release.

Promoter The site to which RNA polymerase must bind before it can transcribe a gene.

Pronucleus Nucleus of sperm or ovum after fusion of sperm and ovum but before fusion of the nuclei.

Prophage Bacteriophage whose DNA has joined the DNA of its bacterial host.

Prophase The first stage of mitosis, marked by condensation of the chromosomes, disappearance of the nuclear envelope, separation and duplication of the centrioles, and formation of the spindle.

Proprioceptor Sense organ for muscle stretch or tension or for joint angle.

Prosimians The "first monkeys"; primitive primates similar to the ancestors of monkeys, apes, and humans.

Prostaglandins Fatty-acid-like substances that act as local, short-range hormones. Sometimes described as "tissue hormones."

Prostate Gland Gland surrounding the junction of the vas deferens with the urethra; secretes part of semen.

Protein–Calorie Malnutrition (PCM) Starvation due to lack of both calories and protein.

Proteinoid Proteinlike substance presumably leading to first cells.

Proteins Amino acid polymers with which cells perform many functions.

Protocell First, rudimentary cell.

Protons The positively charged particles found in the nuclei of atoms.

Protoplasm The substance of the living cell.

Protozoa Extremely complex one-celled organisms, thought to have given rise to the cell line that eventually became the animals. Some are parasites.

Provirus Plant or animal virus incorporated into its host's DNA.

Proximal Closer to a point of reference, such as limb's point of attachment.

Proximal Convoluted Tubule The coiled segment of the nephron's tubule nearest the renal corpuscle.

Pseudopods "False feet"; cytoplasmic extensions used by cells to move from place to place.

Pubic Symphysis The pad of fibrous cartilage between the pelvic bones in front.
Pudendum Vulva.
Pulmonary Arteries The blood vessels that carry blood from the heart to the lungs.
Pulmonary Circulation The path of blood from the heart through the lungs and back.
Pulmonary Veins The blood vessels that carry blood from the lungs to the heart.
Pulmonary Ventilation The flow of air in and out of the lungs.
Pulp The soft tissue filling the hollow center of a tooth.
Punctuated Equilibrium The idea, drawn from periods of rapid evolution in the fossil record, showing that some species that existed relatively unchanged for long periods were suddenly replaced by different new species.
Pupil The hole in the center of the iris through which light passes.
Pus Whitish material found in wounds, composed of bacteria, living and dead phagocytes, and debris from damaged tissue cells.
Pyloric Sphincter The muscular valve separating the stomach from the small intestine.
Pylorus The end of the stomach nearest the small intestine.
Pyridoxine Vitamin B_6.
Pyrogens The fever-stimulating chemicals released from damaged cells.

Quaternary Structure Features of protein shape due to interactions with other proteins in multiprotein structures.

Radius A bone of the lower arm.
Reactants The substances consumed or transformed by a chemical reaction.
Realized Niche The niche a species occupies in the presence of competition.
Receptive Field Skin zone in which all the receptors of one kind belong to a single sensory neuron.
Receptors Proteins and glycoproteins in cell membranes, cytoplasm, and nucleus that bind to hormones.
Recessive Allele An allele that affects phenotype only when homozygous.
Recombinant DNA DNA formed by the fusion of genes from two or more organisms; the technique for producing such DNA.
Red Blood Cells Blood cells lacking nuclei and filled with hemoglobin; biconcave disks.
Red Bone Marrow The blood-cell-producing tissue found in the marrow cavities of long bones in the young and in spongy bone in adults.
Red Muscle A muscle with a rich blood and myoglobin supply; specialized for endurance.
Refraction The bending of light rays and images as light passes from one transparent medium to another.
Refractory Period Period during and immediately after an action potential when a nerve cell membrane cannot generate a second action potential.
Regeneration The regrowing of lost parts as cells return to an embryolike state and redifferentiate to form the tissues to be replaced.
Relaxin An ovarian hormone that softens the cervix and symphysis pubis in preparation for childbirth.
Releasing Factors Hormones produced in the hypothalamus that stimulate the release of hormones from the anterior lobe of the pituitary.
Renal Column Extension of renal cortex into the renal medulla between renal pyramids.
Renal Cortex Outer layer (rind) of the kidney; site of renal corpuscles.
Renal Erythropoietic Factor Protein secreted by kidney in response to low oxygen levels; converted to erythropoietin in blood.
Renal Medulla Inner layer (core) of the kidney; site of loops of Henle and the collecting ducts.
Renal Papilla A domelike protrusion of the renal medulla; where collecting ducts drip urine into the renal pelvis.
Renal Pelvis Cavity within the kidney; urine flows through it into the ureter.
Renal Pyramid Subdivision of renal medulla; its tip is a renal papilla.
Renin An enzyme that catalyzes the conversion of angiotensinogen to angiotensin I and hence increases blood pressure and the conservation of water and salt.
Repeatability An important criterion for scientific acceptance, suggesting that experimental findings should be repeatable regardless of who is doing the experimentation.
Replication Repeated measurement of the same treatment or phenomenon; the duplication of a cell's DNA.
Repression Inactivation of genetic transcription of a gene.
Repressor Gene A gene that codes for a repressor protein.
Reproductive Isolation The inability to interbreed due to separation in time or space or to anatomical, biochemical, or fertility obstacles.
Reproductive Potential The number of offspring an organism can have.
Residual Volume Lung volume unavailable in breathing; filled with the first breath at birth; 1200 ml.
Resistance Reaction Second stage of general adaptation syndrome.
Resorption Canal A tunnel driven into compact bone by osteoclasts; the site of a future Haversian system.
Respiratory Bronchioles Branches of terminal bronchioles that deliver air to the alveolar sacs.
Resting Potential The electrical charge (voltage) difference across a neuron's cell membrane; −70 mV (inside of cell is negatively charged with respect to the outside).
Restriction Endonuclease Enzyme that cuts DNA at a specific sequence of nucleotides.
Reticular Formation Network of interconnected neurons in brainstem; acts as a sensory filter and to maintain muscle tone.
Retina The sheet of receptor cells at the back of the eye.
Retinol Vitamin A; essential for rod vision.
Retrovirus An RNA virus that translates its RNA to DNA

in its host cell.

Reverse Transcriptase An enzyme that links DNA nucleotides together on an RNA template.

Rhodopsin The light-sensitive pigment in the retina's rods.

Rhythm Method Birth control by avoiding intercourse around the estimated time of ovulation.

Rib Cage The bony enclosure of the chest formed by the ribs.

Riboflavin Vitamin B_2.

Ribonucleic Acid (RNA) A nucleic acid differing from DNA in that its sugar is ribose, thymine is replaced by uracil, and its structure is single stranded; it is essential to translating the DNA information into protein.

Ribosomal RNA (rRNA) The RNA that forms part of the ribosome.

Ribosomes Tiny granular organelles that are the principal sites of protein synthesis in the cytoplasm.

Rickets Vitamin D deficiency disease.

RNA Polymerase An enzyme that links RNA nucleotides together on a DNA template.

Rods Retinal light receptors responsible for dim-light and night vision. They do not respond to the color of light.

Rough ER Areas of the endoplasmic reticulum covered with ribosomes.

Rugae Folds in vaginal and stomach walls.

Saccharide An organic compound containing sugar or sugars.

Saccule With the utricle, the location of the receptors for head position.

Sacrum The triangular element of the backbone formed of five fused vertebrae; attaches the backbone to the pelvis.

Sagittal Section An anatomical section cut parallel to the long axis of the body that divides the body into left and right portions.

Saliva The fluid added to food in the mouth; contains salts, mucus, antibacterial agents, and salivary amylase.

Salivary Amylase A starch-digesting enzyme in saliva.

Salivary Glands The three pairs of glands that release saliva into the mouth.

Saltatory Conduction Leaping of action potential from node to node of Ranvier.

Sampling Error Failure of a group of experimental subjects (a sample) to reflect the general population in one or more features of interest.

Sarcolemma The cell membrane of a muscle fiber.

Sarcomere The functional unit of muscle fiber contraction; a repeating unit of myofibril structure.

Sarcoplasm The cytoplasm of a muscle fiber; dominated by contractile myofilaments.

Sarcoplasmic Reticulum The endoplasmic reticulum of a muscle fiber; specialized to store and release calcium ions.

Saturated Lacking double bonds (said of fats and fatty acids).

Saturated Fats Fats containing no double bonds in the carbon chains of their fatty acids. Most animal fats fall into this category.

Scala Tympani One of the three tubes in the cochlea.

Scala Vestibuli One of the three tubes in the cochlea; begins at the oval window.

Scapula The shoulder blade.

Scar Tissue New connective tissue replaces tissues damaged by infection or injury.

Schistosomes Blood flukes.

Schistosomiasis Infection by schistosomes.

Schwann Cell Produces myelin for peripheral neurons.

Science Systematic study of physical, material, and biological phenomena and organization of the information discovered about each.

Scientific Method A structured approach to problem solving employed by scientists.

Sclera The tough, opaque, white outer covering of the eye.

Scoliosis Abnormal sideways curvature of the spine.

Scrotum The sac containing the testes.

Scurvy Vitamin C deficiency disease.

Sebaceous Glands The oil glands of the skin; they secrete a lipid-rich waterproofing and softening substance into hair follicles and onto the skin.

Sebum The mixture of fats, cholesterol, proteins, and salts secreted by the sebaceous glands.

Secondary Consumers Eaters of animals; carnivores.

Secondary Oocyte Result of first meiotic division of primary oocyte.

Secondary Ossification Center The area of the epiphysis that starts to ossify some time after primary ossification begins.

Secondary Sexual Characteristics Body features that distinguish male from female but are not involved directly in reproduction.

Secondary Spermatocytes Results of first meiotic division of primary spermatocytes.

Secondary Structure The bending or coiling of a protein strand as a consequence of the position of certain amino acids at specific locations in the sequence.

Secretin A hormone secreted by the small intestine that inhibits stomach secretion and stomach and intestinal muscle contractions and stimulates secretion of bile and alkaline pancreatic juice.

Segmentation Alternating contractions of the intestinal smooth muscles that serve to mix chyme with digestive enzymes.

Selectively Permeable Membranes Membranes that allow certain substances to pass through but prevent, inhibit, or otherwise control the movements of other materials.

Sella Turcica The saddle-shaped depression in the center of the sphenoid that houses the pituitary gland.

Semen The mixture of sperm and glandular fluids discharged from the penis during ejaculation.

Semicircular Canals Hooplike tubes, one lying in each of the three planes of head rotation; when the head rotates, the fluid in the appropriate canal pushes on the cupula to provoke nerve signals.

Semilunar Valves The valves at the entrances to the aorta and pulmonary artery that prevent the backflow of blood into the ventricles; each one is composed of three cups of valvular tissue.

Seminal Vesicles Glands draining into vas deferens near its junction with the urethra; secretes part of semen.

Seminiferous Tubule Tubule within the testis in which sperm are formed.

Senescence The body's loss of vigor and health with age.

Sense Organ Bodily structure that converts an event or stimulus into a form that can activate an action potential in a nerve ending or neuron.

Sensory Receptor Nerve ending or neuron in a sense organ that generates an action potential.

Sertoli (Supporting) Cells Cells that nourish spermatids while they mature to become sperm.

Serum Plasma minus fibrinogen; the fluid left after the blood clots.

Sex Chromosomes X and Y chromosomes; females have two X's; males have an X and a Y.

Sex-influenced Trait A trait whose expressivity differs in male and female.

Sex Linkage The presence of a gene on a sex chromosome; recessive traits controlled by such genes on the X chromosome tend to show more often in males.

Sexually Transmitted Disease (STD) Any disease transmitted by sexual contact.

Sexual Recombination Mixing of genes from two individuals resulting from the fusion of gametes.

Sexual Reproduction Reproduction by growth of new individuals from fused, haploid gametes produced by meiosis.

Shaft Main length of penis.

Short Bones The blocklike bones of the wrist and ankle.

Sickle Cell Anemia Hereditary hemoglobin defect leading to red blood cell destruction at low oxygen levels.

Simple Goiter Enlarged thyroid due to iodine deficiency.

Sinoatrial (SA) Node The pacemaker; the region of the right atrium that initiates each contraction of the heart.

Sinuses Mucous-membrane-lined cavities in the frontal, ethmoid, and sphenoid bones and the maxilla of the skull.

Sister Chromatids Postreplication duplicates of a chromosome, joined at a centromere.

Skeletal Muscle Striated, voluntary muscle; responsible for moving the skeleton.

Skeleton The system of bones that serves the body as a structural framework and a system of levers for movement.

Skull The bones of the head.

Slow Muscle Red muscle.

Small Intestine Main site of digestion and absorption in the digestive system; links stomach and large intestine.

Smooth ER Areas of the endoplasmic reticulum lacking ribosomes.

Smooth Muscle The nonstriated, involuntary muscle of the walls of the body's tubular organs, the iris of the eye, and the arrector pili (hair erectors).

Sodium–Potassium Pump Membrane protein that uses ATP to move sodium ions out of the cell and potassium ions into the cell.

Solute The substance dissolved in a liquid (the solvent) to make a solution.

Solution A chemical mixture formed by dissolving one substance (the solute) in a liquid (the solvent).

Solvent A liquid in which other substances are dissolved.

Somatic Nervous System The voluntary branch of the efferent PNS; serves skeletal muscles.

Somatic Pain Pain felt in skin, muscle, tendon, and joints.

Somatomedins Growth factors aiding growth hormone.

Somite A block of mesoderm that will become muscle and bone.

Sonography The radarlike use of high-frequency sound waves to examine structures, including embryos, within the body.

Species A group of similar organisms able to interbreed.

Specific Resistance A combined version of cellular and chemical warfare (antibodies), in which the defense is specifically tailored to deal with a certain agent.

Spermatids Results of second meiotic division of primary spermatocytes.

Spermatogenesis Process by which sperm are formed.

Spermatogonia Parent cells of sperm lining seminiferous tubules.

Spermatozoa (Sperm Cells) Male gametes.

S Phase The period of the cell cycle when the DNA is replicated or "S"ynthesized.

Sphenoid Bone The butterfly-shaped bone that forms the anterior floor of the cranial cavity.

Spinal Cord Portion of CNS protected by vertebrae; carries signals between brain and body; its neurons handle many reflexes.

Spinal Nerves Thirty-one nerves entering and leaving the spinal cord.

Spinal Rootlets Narrow bundles of nerve fibers entering or leaving the spinal cord; they quickly merge to form the roots of the spinal nerves.

Spine The vertebral column.

Spiral Organ The row of hair cells extending the length of the basilar membrane.

Spleen Lymphoid organ behind the stomach; stores red blood cells and lymphocytes and removes worn-out red blood cells.

Spongy Bone Bone consisting of a network of bony plates and rods.

Sporozoan Protozoan with flagellated gametes and amoeboid adult form; many are pathogenic.

Stapes The stirrup; the ear ossicle that transmits vibrations from the incus to the oval window.

Starch The principal energy storage molecules used by plants. They are formed by chaining glucose (monosaccharide) units together.

Starvation Bodily deterioration due to lack of nutrients and energy.

Stem Cells The bone marrow cells that give rise to the lymphocytes.

Stenosis A narrowing of the passageway through a heart valve that increases the resistance to blood flow.

Sterilization Birth control by cutting and tying off the vas deferens or Fallopian tubes or by removing the uterus, ovaries, or testes.

Sternum The breastbone forming the front of the rib cage.

Steroids A chemical family of lipids whose molecules feature four interconnected rings. Cholesterol is the parent compound for the family; many steroids are powerful hormones.

Stomach A distensible sac that receives food and liquid from the esophagus and mixes them with gastric juices, which include water, acid, and the enzyme pepsin, and prepares them for the small intestine.

Stress Damage or threat of damage to the body (physical or psychological).

Stretch Reflex Wired-in link between sensory neurons that detect muscle stretch and motor neurons; keeps muscle length constant.

Stroke Brain damage due to broken or blocked blood vessels in the brain.

Stroma Connective tissue body of ovary.

Structural Gene A gene that codes for a protein.

Structural Proteins Proteins that make up much of the structural framework of the body. They include fibers, glues, membranes, and a host of other structural materials.

Struggle for Existence The competition for resources necessary for successful reproduction.

Substrates The specific chemicals that a particular enzyme will help to react.

Sucrose Table sugar; disaccharide consisting of joined glucose and fructose monosaccharides.

Sudden Infant Death Syndrome (SIDS) Cause of 10,000 infant deaths per year in the United States; possibly due to failure of the respiratory control centers.

Sudoriferous Glands Sweat glands; they secrete a watery fluid whose evaporation aids heat loss.

Sulcus Groove or cleft between adjacent gyri.

Superior Above.

Superior Colliculi Portion of midbrain receiving visual input; generates eye movement reflexes.

Superior Vena Cava The large vein that delivers blood to the heart from the upper trunk, head, and neck.

Survival of the Fittest One of the basic principles of the theory of evolution. It states that certain individuals in a population possess physical advantages making them more likely to survive, at least long enough to reproduce, than individuals not possessing those traits.

Suspensory Ligament The ligament that attaches the edge of the lens to the ciliary body of the eye.

Sutures The seamlike joints between the bones of the skull.

Symbiosis "Living together"; reliance of one species upon another for one or more resources it needs for survival.

Sympathetic Division Portion of autonomic system arising in thoracic and lumbar cord; coordinates the body for action.

Symphysis A cartilaginous joint marked by an insert of fibrous cartilage between the bones.

Synapse Point at which nerve signals pass from neuron to neuron.

Synaptic Cleft Gap between presynaptic membrane (synaptic knob) and postsynaptic membrane (target cell).

Synaptic Knob Terminal button.

Synaptic Vesicle Membrane-bound organelle in synaptic knob; contains neurotransmitters.

Synchondrosis A cartilaginous joint marked by an insert of hyaline cartilage between the bones.

Syndesmosis A fibrous joint in which the bones are linked by a sheet or membrane of fibrous connective tissue.

Synergist A muscle whose contraction steadies the agonist's origin.

Synovial Fluid The viscous lubricating fluid secreted by the synovial membrane.

Synovial Joint A joint with a fluid-filled joint cavity between the bones.

Synovial Membrane The lining of the articular capsule.

Syphilis Sexually transmitted disease caused by the spirochete *Treponema pallidum*.

Systemic Circulation The path of blood from the heart to the various tissues and organ systems of the body and back.

Systole Contraction of the heart.

Tanning The buildup of melanin in the epidermis in response to exposure to ultraviolet light.

Tapeworms Tapelike, segmented intestinal parasites.

Tarsals The anklebones.

Taste Buds Sense organs of taste; clusters of sensory cells in pores on sides of tongue's papillae.

Taxonomy The branch of biology that attempts to classify all living (or extinct) organisms into evolutionary-related categories.

T Cells Lymphocytes that mature in the thymus gland and are the principal combatants employed in cell-mediated immune responses.

Tectorial Membrane The tissue flap overlying the hair cells.

Telophase The fourth and last stage of mitosis, during which the nuclear envelope re-forms.

Temporal Bones The bones that form the side walls of the cranium. They also contain the apparatus of the inner ear.

Temporal Lobe Lower middle portion of cerebrum; part receives auditory data.

Temporal Summation The merging of successive twitches to produce a stronger, longer contraction.

Tendinitis An inflammation of a tendon sheath.

Tendon A cord or band of connective tissue formed by the merging of epi-, peri-, and endomysium at the end of a muscle; links the muscle to periosteum or fascia.

Tendon Sheath A tube lined by synovial membrane surrounding certain tendons and serving to reduce friction.

Tension Receptor Tension-sensitive sensory receptor in a tendon.

Teratogen An agent causing birth defects.

Terminal Bronchioles Small bronchioles delivering air to lung lobules.

Terminal Button Swelling at tip of terminal branch of axon; presynaptic portion of synapse.
Tertiary Consumers Eaters of carnivores.
Tertiary Structure Shaping of a protein molecule induced as different portions of the protein molecule interact with each other through hydrogen or disulfide bonds.
Testes Male gonads.
Testosterone Male sex hormone.
Tetanus The steady, maximum contraction achieved when stimuli reach a muscle very rapidly.
Tetrad A bundle of four chromatids (two homologous pairs of sister chromatids) formed during prophase I of meiosis; site of crossing over.
Thalamus Portion of brain in wall of third (central) ventricle. It relays incoming sensory signals to the appropriate area of the cerebrum.
Theory A hypothesis that has successfully withstood every test over a long period of time. It is very likely to be a true explanation for the phenomenon.
Thiamine Vitamin B_1.
Thoracic Cavity Portion of ventral cavity in the chest.
Thoracic Vertebrae The 12 vertebrae in the chest region. Each thoracic vertebra is attached to a pair of ribs.
Thrombocytes Platelets.
Thrombus A blood clot in an unbroken blood vessel.
Thymus Gland Lymphoid organ behind breastbone; site of maturation of T cell lymphocytes.
Thyroid Cartilage The large cartilage plate forming the front of the larynx.
Thyroid Gland Secretor of thyroxine; located in the neck below the larynx.
Thyroid-stimulating Hormone (TSH; Thyrotropin) Controls thyroid.
Thyroxine The main thyroid hormone; contains four iodine atoms.
Tibia A bone of the lower leg.
Tidal Volume The air moved in an ordinary, quiet breath; 500 ml.
Tissues Groups of similar cells that have a common function or provide a similar service for the body.
Tocopherol Vitamin E; no known use in humans.
Tonsils Lymphoid organs at back of throat.
Total Lung Capacity All the air the lungs can hold; 6 liters.
Totipotency Ability of a fertilized egg to give rise to all the cell types of an adult organism.
Toxic Poisonous.
Trabeculae The bony plates and rods of spongy bone.
Trace Element A micronutrient; minerals needed only in small amounts.
Trachea The windpipe; the tube carrying air from the pharynx to the lungs; stiffened by U-shaped cartilaginous supports.
Transcription The making of an RNA copy of the information in the DNA.
Transduction Conversion of an event into an interpretable (nervous) signal.
Transfer RNA (tRNA) The RNA that attaches to amino acids and delivers them to the ribosome for use in protein synthesis.
Transformation The conversion of one cell type to another caused by a cell absorbing some of the genetic material from a different cell type; viral conversion of a host cell to a cancerous cell, without control over its growth and multiplication.
Translation The use of the information encoded in RNA to make protein.
Translocation Movement of a chromosomal segment from one chromosome to another.
Transverse Arch The upward curve of the foot from side to side.
Transverse Section A section cut perpendicular to the long axis of the body.
Transverse (T) Tubule An invagination of the sarcolemma that delivers signals to the myofibrils.
Triad A T tubule and its flanking pair of sarcoplasmic reticulum tubules.
Tricuspid Valve The three-flapped valve between the right atrium and the right ventricle.
Triglyceride Fat; glycerol with three fatty acids attached.
Trophectoderm Wall of the blastocyst.
Trophic Level Position in a food chain.
Tropomyosin A control protein in the thin myofilaments.
Troponin A control protein in the thin myofilaments.
True Capillary Branching capillary without muscle (except an initial sphincter).
True Digestive Organs The various chambers and compartments of the digestive system, including the mouth, esophagus, stomach, small intestine, large intestine, and rectum.
True Rib A rib with its own, separate cartilaginous link to the sternum.
Tubal Ligation Cutting and tying off the Fallopian tubes.
Tumor A mass of cells, often cancerous.
Tunica Externa (Adventitia) Outer layer of artery or vein wall; composed of elastic and collagenous connective tissue.
Tunica Interna Inner layer of artery or vein wall; composed of endothelium and elastic connective tissue.
Tunica Media Middle layer of artery or vein wall; composed of smooth muscle and elastic connective tissue (veins have more fibrous connective tissue).
Twitch The contractile response of a muscle to a single, brief stimulus.
Tympanic Membrane The eardrum.

Ulna A bone of the lower arm.
Umbilical Cord Life-supporting cable containing arteries and a vein that attaches the embryo to the chorion (placenta).
Unifying Theory A theory that encompasses and draws together a great deal of information that can be used to explain many things.
Unsaturated Having one or more double bonds (said of fats and fatty acids).
Unsaturated Fats Fats whose fatty acids have one or more double bonds.
Ureter The tube that carries urine from the kidney to the urinary bladder.

Urethra The tube that carries urine from the urinary bladder outside the body; in the male it also carries semen.
Urinary Bladder The storage organ for urine.
Urinary System The organs that produce, transport, store, and expel urine; kidneys, ureters, urinary bladder, and urethra.
Urine The concentrated solution of water, excess salts, and waste materials produced by the action of the nephron on the glomerular filtrate.
Uterus Womb; the organ that receives fertilized ova and holds the developing embryo.
Utricle With the saccule, the location of the receptors for head position.
Uvula The tag of flesh hanging from the posterior edge of the soft palate.

Vaccine An injected solution of antigens used to stimulate antibody development, to protect individuals that might later be exposed to a disease agent.
Vacuole Membrane-bound compartment in a eukaryotic cell; contains food, fluid, or stored material.
Vagina Muscular channel leading from cervix to exterior; receptacle for penis and sperm; birth canal.
Variable The particular thing being measured.
Variable Expressivity When a genotype does not always affect phenotype to the same degree.
Vas Deferens Tube leading from epididymis to penis; pathway and storage site for sperm.
Vasectomy Cutting and tying off the vas deferens.
Vasoactive Intestinal Polypeptide A hormone regulating contractions of the small intestine.
Vein A blood vessel carrying blood toward the heart.
Ventral Toward the belly.
Ventral Cavity Body cavity within the trunk.
Ventricles The chambers of the heart that pump blood into the arteries; cavities within the brain.
Ventricular Folds The false vocal cords; used to close off the airway.
Venule One of the smallest veins; formed when capillaries merge.
Vermiform Appendix A slender organ attached to the cecum; it may serve as a part of the immune system.
Vertebrae The bones of the vertebral column, or backbone.
Vertebral Cavity Portion of dorsal cavity containing the spinal cord.
Vertebral Column The stack of bones (vertebrae) enclosing the spinal cord; the backbone.
Vertebrates All animals with backbones.
Vestibule Cleft between the labia minora.
Villi Fingerlike protrusions from the small intestine wall.
Viroid A viruslike structure consisting of RNA but containing no protein; infects plants only.
Virulence Ability to cause disease.
Virus Tiny intracellular parasite consisting of a shell of protein and a core of RNA or DNA.
Visceral Pain Pain felt in the organs of the body cavities (viscera).
Visceral Pericardium (Epicardium) Serous membrane covering the outside of the heart.
Visceral Pleura Serous membrane attached to the outer surface of the lungs.
Visual Field That portion of the potential field of view whose image falls on the retina.
Vital Capacity Total lung capacity minus residual volume; all the lung volume available for deep breathing.
Vitamin Organic compound needed in the diet in small amounts, often as a cofactor for enzymes.
Vocal Folds The true vocal cords; used to generate sound.
Voluntary Work The amount of energy consumed by the activities of an organism beyond that needed for the basal metabolic requirements. It is the cost of various types of exercise.
Vomer The bony partition that divides the nasal cavity into a left and right chamber (or nostril).
Vulva Female external genitalia.

Warts Small, raised skin thickenings due to viral infection of basal layer cells.
Water Cycle The path water follows from evaporation to condensation to precipitation and back again; the recycling of water.
Water-soluble Vitamin A vitamin (the Bs, C) that dissolves in water and cannot be stored in the body.
White Muscle A muscle with less myoglobin and fewer capillaries than red muscle; specialized for strength and speed.

Yolk Nutrients stored in ova of many animals.
Yolk Sac Bubble of membrane produced by the endoderm.

Z-DNA A variant form of DNA, in which the twist of the double helix is left-handed, not right-handed.
Z Line The border of a sarcomere.
Zona Pellucida Tough, gelatinous layer surrounding ovum.
Zygomatic Bones The cheekbones, which form the lower margins of the eye sockets.
Zygote Fertilized ovum after fusion of male and female pronuclei.

Credits

Unless otherwise acknowledged, all photographs are the property of Scott, Foresman.

1 Chapter 1 Opener: Tropical rain forest in Zaire. James A. Sugar/Black Star. **2** Figure 1.1: Thomas S. England/SS/Photo Researchers. **3** Figure 1.2: Frank Fournier/Contact Press Images/Woodfin Camp & Associates. **3** Figure 1.3: Nathan Benn/Woodfin Camp & Associates. **4** Figure 1.4: The Image Works. **6** Figure 1.5a: Mark Antman/The Image Works. **6** Figure 1.5b: Bob Daemmrich/The Image Works. **13** Figure 1.9: Junebug Clark/Photo Researchers. **17** Figure 1.14: Scala/Art Resource, NY. **18** Figure 1.15a: The Bettmann Archive. **18** Figure 1.15b: The Bettmann Archive. **23** Chapter 2 Opener: Lightning storm. Gary Milburn/Tom Stack & Associates. **26** Figure 2.3: Smithsonian Institution. **37** Figure 2.12: Focus On Sports. **38** Figure 2.13b: Michael Tidwell/Terraphotographics/Biological Photo Service. **38** Figure 2.13c: Tom Hollyman/Photo Researchers. **38** Figure 2.13d: Best Foods Baking Group, Oconomowoc, WI. **41** Figure 2.17: Biophoto Associates/Photo Researchers. **53** Chapter 3 Opener: SEM of nerve cells from cerebral cortex. CNRI/SPL/SS Photo Researchers. **54** Figure 3.1: Lawrence Migdale/Photo Researchers. **55** Figure 3.2a1: Norman Mosallem/Bruce Coleman, Inc. **55** Figure 3.2a2: Biophoto Associates/Photo Researchers. **55** Figure 3.2b1: Richard C. Johnson/Visuals Unlimited. **55** Figure 3.2b2: C. McDaniel/Visuals Unlimited. **55** Figure 3.2c: CNRI/SPL/Photo Researchers. **55** Figure 3.2.d1: Peter Arnold, Inc. **55** Figure 3.2.d2: Jim Solliday/Biological Photo Service. **55** Figure 3.2.e1: John D. Cunningham/Visuals Unlimited. **55** Figure 3.2.e2: Kevin Collins/Visuals Unlimited. **56** Figure 3.3: *Micrographia*, by Robert Hooke, © 1665, published by the Royal Society. **56** Figure 3.4: The Bettmann Archive. **59** Figure 3.7a1: CNRI/SPL/Photo Researchers. **59** Figure 3.7a2: Dwight R. Kuhn. **59** Figure 3.7b: Michael Abbey/Photo Researchers. **60** Figure 3.8a: G. W. Willis, M.D./Biological Photo Service. **60** Figure 3.8b: Runk/Schoenberger/Grant Heilman Photography. **60** Figure 3.8c: Dwight R. Kuhn. **60** Figure 3.8d: M. Abbey/Visuals Unlimited. **60** Figure 3.8e: Runk/Schoenberger/Grant Heilman Photography. **60** Figure 3.8f: G. W. Willis, M.D./Biological Photo Service. **60** Figure 3.8g: Biophoto Associates/Photo Researchers. **62** Figure 3.10b: Dr. T. J. Beveridge, Dept. of Microbiology, University of Guelph/Biological Photo Service. **72** Figure 3.20b: Courtesy, Dr. C. L. Rieder, Biological Microscopy and Image Reconstruction Resource, NIH Biotechnological Resource, Albany, N.Y. From "Correlative Immunofluorescence and Electron Microscopy on the Same Section of Epon-Embedded Material," by C. L. Rieder and S. S. Bowser, *The Journal of Histochemistry and Cytochemistry*, Volume 33, pp. 165–171, 1985. **73** Figure 3.21b: OMIKRON Photo Researchers, Inc. **74** Jim Rosen/Terraphotographics/Biological Photo Service. **76** Science VU/Sidney Fox/Visuals Unlimited. **77** Figure 3.23b: Don W. Fawcett/Visuals Unlimited. **77** Figure 3.23b Dr. Jeremy Burgess/SPL/Photo Researchers. **81** Chapter 4 Opener: Computer-generated double helix of DNA. Dan McCoy/Rainbow. **91** Figure 4.11: Richard Rodewalk, University of Virginia/Biological Photo Service. **96** Figure 4.17b: D. L. Miller/B. R. Beatty/D. W. Fawcett/Visuals Unlimited. **108** Chapter 5 Opener: Human lymphocyte cell undergoing mitotic cell division. CNRI/SPL/Photo Researchers. **109** Figure 5.14: Michael P. Gadomski/Photo Researchers. **109** Figure 5.1b: J. Forsdyke/Photo Researchers. **109** Figure 5.1c: Gregory K. Scott/Photo Researchers. **109** Figure 5.1d: Biophoto Associates/Photo Researchers. **109** Figure 5.1e: David M. Phillips/Visuals Unlimited. **111** Figure 5.2a: Biophoto Associates/Photo Researchers. **111** Figure 5.2b: David M. Phillips/Visuals Unlimited. **111** Figure 5.2c: James R. Fisher/Photo Researchers. **111** Figure 5.2d: Dr. Jeremy Burgess/John Innes Institute/SPL/Photo Researchers. **113** Figure 5.4a: A. Craig-Holmes/Biological Photo Service. **113** Figure 5.5b: "Nucleosomes: The Structural Quantum in Chromosomes," by Donald E. Olins and Ada I. Olins, *American Scientist*, 66:704–711, Fig. 1b and Fig. 4. Reprinted by permission of *American Scientist*, journal of Sigma Xi, The Scientific Research Society, and by permission of Ada L. Olins and Donald E. Olins of the University of Tennessee and the Oak Ridge National Laboratory. **114** Figure 5.7: Jack M. Bostrack/Visuals Unlimited. **118** Figure 5.11: David M. Phillips/Visuals Unlimited. **126** Chapter 6 Opener: Areolar tissue—loose connective tissue. John D. Cunningham/Visuals Unlimited. **130** Figure Table 6.2a: Bruce Iverson/Visuals Unlimited. **130** Figure Table 6.2b: Fred Hossler/Visuals Unlimited. **131** Figure Table 6.2c: Fred Hossler/Visuals Unlimited. **131** Figure Table 6.2d: Bruce Iverson/Visuals Unlimited. **132** Figure Table 6.2e: Bruce Iverson/Visuals Unlimited. **132** Figure Table 6.2f: OMIKRON/SS/Photo Researchers. **132** Figure Table 6.2g: Biophoto Associates/Photo Researchers. **132** Figure Table 6.2h: Biophoto Associates/Photo Researchers. **133** Figure Table 6.2i: Stan Elems/Visuals Unlimited. **134** Figure Table 6.2j: Albert Copley/Visuals Unlimited. **134** Figure Table 6.2k: Stan Elems/Visuals Unlimited. **134** Figure Table 6.2l: Fred Hossler/Visuals Unlimited. **134** Figure Table 6.2m: John D. Cunningham/Visuals Unlimited. **135** Figure Table 6.2n: Dwight R. Kuhn. **135** Figure Table 6.2o: Runk/ Schoenberger/Grant Heilman Photography. **135** Figure Table 6.2p: Manfred Kage/Peter Arnold, Inc. **135** Figure Table 6.2q: Michael Abbey/Photo Researchers. **136** 6a: Biophoto Associates/Photo Researchers. **136** 6b: Fred Hossler/Visuals Unlimited. **136** 6c: Lester V. Bergman & Associates. **136** 6d: John D. Cunningham/Visuals Unlimited. **136** 6e: Ed Reschke/Peter Arnold, Inc. **136** 6f: Eric V. Grave/Photo Researchers. **137** 6g: Fred Hossler/Visuals Unlimited. **137** 6h: Bruce Iverson/Visuals Unlimited. **137** 6i: G. W. Willis, M.D./ Biological Photo Service. **137** 6j: Bruce

Iverson/Visuals Unlimited. **137** 6k: Ed Reschke/Peter Arnold, Inc. **137** 6l: Lester V. Bergman & Associates. **137** 6m: Biophoto Associates/Photo Researchers. **138** Figure 6.4: Custom Medical Stock Photo. All rights reserved. **144** Chapter 7 Opener: Example of many different skin types in our society. SUPERSTOCK. **146** Figure 7.1b: David Scharf/Peter Arnold, Inc. **146** Figure 7.1c: Dr. Tony Brain/SPL/Photo Researchers. **149** Figure 7.2: Runk/Schoenberger/Grant Heilman Photography. **150** Figure 7.3: Runk/Schoenberger/Grant Heilman Photography. **152** Figure 7.6: Dr. Tony Brain/Photo Researchers. **153** Figures 7.8a and b: Lara Harrtley/ Terraphotographics/Biological Photo Service. **156** Figure 7.11a: Richard C. Johnson/Visuals Unlimited. **156** Figure 7.11b: Douglas Richardson/Medical Images, Inc. **156** Figure 7.11c: Ursula Markus/Photo Researchers. **156** Figure 7.11d: SU/SIU/Visuals Unlimited. **159** Chapter 8 Opener: Young boy in skeleton costume. Richard Hutchins/Photo Researchers. **160** Figure 8.1: James M. Mejuto. **166** Figure 8.5b: SPL/Photo Researchers. **175** Figure 8.14b: Biophoto Associates/Photo Researchers. **178** Figure 8.16: Scott Camazine/Photo Researchers. **179** Figure 8.17c: National Osteoporosis Foundation. **184** Figure 8.18a: Biophoto Associates/Photo Researchers. **184** Figure 8.18b: Dr. Gilbert Faure/SPL/ Photo Researchers. **185** Figure 8.20a: James Stevenson/ Photo Researchers. **185** Figure 8.20b: VU/SIU/Visuals Unlimited. **190** Chapter 9 Opener: Gymnast in action. Brian Smith/Black Star. **193** Figure 9.2: R. Calentine/ Visuals Unlimited. **193** Figure 9.3: C. W. Willis, M.D./ Biological Photo Service. **193** Figure 9.4: Biophoto Associates/Photo Researchers. **202** Figure 9.9: Biophoto Associates/Photo Researchers. **207** Figure 9.14: G. W. Willis, M.D./ Biological Photo Service. **207** Figure 9.15: Biophoto Associates/Photo Researchers. **213** Chapter 10 Opener: Red blood cells. David M. Phillips/Visuals Unlimited. **216** Figure 10.1: Grant Heilman Photography. **220** Figure 10.6: Manfred Kage/Peter Arnold, Inc. **221** Figure 10.8b: David M. Phillips/Visuals Unlimited. **224** Figure 10.13a and 10.13b: National Institutes of Health. **228** Figure 10.18: Science VU/Visuals Unlimited. **231** Figure 10.22: VU/SIU/Visuals Unlimited. **236** Figure 10.26c: SIU/Visuals Unlimited. **241** Chapter 11 Opener: Mountain climber on Mt. Rainer wearing oxygen tank. Galen Rowell/Mountain Light Photography, Inc. **246** Figure 11.5b: Fred Hossler/Visuals Unlimited. **248** Figure 11.7b: David M. Phillips/Visuals Unlimited. **249** Figure 11.8a: A. Glaubermann/Photo Researchers. **249** Figure 11.8b: SPL/Photo Researchers. **262** Chapter 12 Opener: Glomerulus, Bowman's capsule of the kidney. Michael Webb/Visuals Unlimited. **265** Figure 12.4: Biophoto Associates/Photo Researchers. **266** Figure 12.6: CNRI/SPL/Photo Researchers. **277** SIU/Peter Arnold, Inc. **281** Chapter 13 Opener: Boy from childhood to young adult. Mark Antman/The Image Works. **285** Figure 13.4b: Biophoto Associates/Photo Researchers. **288** Figure 13.5a: AFIP/SS/Photo Researchers. **288** Figure 13.5b: Lester V. Bergman & Associates. **288** 13.5c: Lester V. Bergman & Associates. **288** 13.5d: Lester V. Bergman & Associates. **289** Figure 13.7: Lester V. Bergman & Associates. **290** Figure 13.8c: Dennis Kunkel/Biological Photo Service. **295** Figure 13.13: Lester V. Bergman & Associates. **301** Figure 13.16a: Bettina Cirone/Photo Researchers. **301** Figure 13.16b: Lester V. Bergman & Associates. **301** Figure 13.16c: SPL/Photo Researchers. **301** Figure 13.16d: Lester V. Bergman & Associates. **309** Chapter 14 Opener: Florence Joyner running. Focus On Sports. **315** Figure 14.5b: C. Raines/Visuals Unlimited. **316** Figure 14.7b: E. R. Lewis/Biological Photo Service. **328** Chapter 15 Opener: Human cerebrum. George Musil/Visuals Unlimited. **330** Figure 15.2b: Fred Hossler/Visuals Unlimited. **331** Figure 15.3c: Fred Hossler/Visuals Unlimited. **335** Figure 15.7a: SIU/Photo Researchers, Inc. **335** Figure 15.7b: SIU/Photo Researchers, Inc. **338** Dr D. Dicksom /Peter Arnold, Inc. **343** Figure 15.12a: SIU/Photo Researchers. **343** Figure 15.12b: Alexander Tsiaras/SS/Photo Researchers. **355** Chapter 16 Opener: Collage of various eyes, ears and noses. **356** Figure 16.1: Michael Grecco/Stock Boston. **359** Figure 16.4: Y. Arthus-Bertrand/Peter Arnold, Inc. **362** Figure 16.5b: SIU/Visuals Unlimited. **362** Figure 16.5c: J & L Weber/Peter Arnold, Inc. **365** Figure 16.8b: Biophoto Associates/Photo Researchers. **365** Figure 16.8c: A. L. Blum/Visuals Unlimited. **365** Figure 16.9: Ralph C. Eagle, M.D./Photo Researchers. **372** Figure 16.15: R. P. Apkarin/Biological Photo Service. **374** Science VU/D. Balkwell/D. Maratea/Visuals Unlimited. **380** Chapter 17 Opener: Forty-day old human embryo. CIW/ Don W. Fawcett/Visuals Unlimited. **382** Figure 17.1a: Runk/Grant Heilman Photography. **382** Figure 17.1b: Dale Jackson/Visuals Unlimited. **382** Figure 17.1c: Stephen Dalton/Photo Researchers. **382** Figure 17.1d: M. I. Walker/SS/Photo Researchers. **382** Figure 17.1e: Biophoto Associates/Photo Researchers. **384** Figures 17.4a and 17.4b: Joel Gordon Photography. **385** Figure 17.5a: David M. Phillips/Visuals Unlimited. **386** Figure 17.6b: Biophoto Associates/Photo Researchers. **390** Figure 17.10: Biophoto Associates/Photo Researchers. **390** Figure 17.11: C. Edleman/La Villette/Photo Researchers. **396** Figure 17.16a: D. W. Fawcett/SS/Photo Researchers. **396** Figure 17.16b: David M. Phillips/Visuals Unlimited. **397** Figure 17.17a and b: Yanagimachi/Biological Photo Service. **406** Chapter 18 Opener: Father and child. The Telegraph Colour Library/FPG. **412** Figure 18.8a: Petit Format/Nestlé/SS/Photo Researchers. **412** Figure 18.9: John D. Cunningham/Visuals Unlimited. **413** Figure 18.10a: John Giannicchi/Photo Researchers. **413** Figure 18.10b: Petit Format/Nestlé/SS/Photo Researchers. **413** Figure 18.10c: Science VU/Visuals Unlimited. **413** Figure 18.10d: Petit Format/Nestlé/SS/Photo Researchers. **413** Figure 18.10e: Petit Format/Nestlé/SS/Photo Researchers. **415** Figure 18.12a: Petit Format/Nestlé/SS/ Photo Researchers. **415** Figure 18.12b: Joseph Nettis/ Photo Researchers. **419** Figure 18.14: OMIKRON/Photo Researchers. **419** Figure 18.15a: Alexander Tsiaras/SS/ Photo Researchers. **419** 18.15b: St. Bartholomew's Hospital/SPL/Photo Researchers. **420** Figure 18.16b: Martha Cooper/Peter Arnold, Inc. **422** Figure 18.17e: Erika Stone/Peter Arnold, Inc. **426** Chapter 19 Opener: Generations of a family. Jon Feingersh/Tom Stack & Associates. **427** Figure 19.1: Bruce Berg/Visuals Unlimited. **434** Figure 19.6a: Custom Medical Stock Photo. All rights reserved. **434** Figure 19.6b: Chet Charles, University of Illinois. **434** Figure 19.6c: Lester V. Bergman & Associates. **434** Figure 19.6d: Lester V. Bergman & Associates. **434** Figure 19.6e: Supplied by Carolina Biological Supply Company. **439** Figure 19.10a: Robert Brons/Biological Photo Service. **439** Figure 19.10b: Biological Photo Service. **440** Figure 19.11: R. Kanuft/ Photo Researchers. **443** Figure 19.13a1: Runk/Schoenberger/Grant Heilman Photography. **443** Figure 19.13a2: Runk/Schoenberger/Grant Heilman Photography. **443** Figure 19.13b1: CNRI/SPL/Photo Researchers. **443** Figure 19.13b2: CNRI/SPL/Photo Researchers. **444** Figure 19.14a: Lester V. Bergman & Associates. **444** Figure 19.14b: Lester V. Bergman & Associates. **444** Figure 19.14c: Lester V. Bergman & Associates. **444** Figure 19.14d: Lester V. Bergman & Associates. **447** Figure 19.15a: Mansell Collection. **448** Figure 19.16a: Makio Murayama/Biological Photo Service. **448** Figure 19.16b: Makio Murayama/Biological Photo Service. **449** Figure 19.17: San Pathak. **451** Figure 19.18a1: Lester V. Bergman & Associates. **451** Figure 19.18a2: CNRI/SPL/ Photo Researchers. **451** Figure 19.18b1: Lester V. Bergman & Associates. **451** Figure 19.18b2: Lester V. Bergman & Associates. **451** Figure 19.18c1: Lester V. Bergman & Associates. **451** Figure 19.18c2: Lester V. Bergman & Associates. **454** Chapter 20 Opener: Intestinal villi. Michael Webb/Visuals Unlimited. **460** Figure 20.6b: Michael Webb/Visuals Unlimited. **475** Chapter 21 Opener: Computer-generated image of ATP transfer. Oxford Molecular Biophysics Laboratory/SPL/Photo Researchers. **476** Figure 21.1a: Tony Korody/Sygma. **476** Figure 21.1b: Robert E. Schwerzel/Stock Boston. **476** Figure 21.1c: Keith Scott/Peter Arnold, Inc. **476** Figure 21.1d: A. Bertrand/Jacana/Photo Researchers. **480** Figure 21.4: Earl Roberge/Photo Researchers. **480** Figure 21.4: Earl Roberge/Photo Researchers **480** Figure 21.4: Earl Roberge/Photo Researchers. **480** Figure 21.4: Catherine Ursillo/Photo Researchers. **484** Figure 21.6a: Paul R. Burton, Cell Biology/University of Kansas/ Biological Photo Service. **489** Figure 21.10b1: Earl Roberge/Photo Researchers. **489** Figure 21.10b2: Earl Roberge/Photo Researchers. **489** Figure 21.10b3: Earl Roberge/Photo Researchers. **489** Figure 21.10b4: Catherine Ursillo/Photo Researchers. **497** Chapter 22 Opener: People enjoying a buffet. Stephanie Maze/Woodfin Camp & Associates. **499** Figure 22.1: Will & Deni McIntyre/Photo Researchers. **500** Figure 22.2: David M. Phillips/Visuals Unlimited. **501** Figure 22.3: John Bryson/Photo Researchers. **502** Figure 22.4: William Campbell/Sygma. **504** Figure 22.9: National Institutes of Health. **507** Figure 22.11: Lester V. Bergman & Associates. **508** Figure 22.12: Biophoto Associates/SS/Photo Researchers. **511** Figure 22.14: Ken Kostuk/Peter Arnold, Inc. **511** Figure 22.15: Lester V. Bergman & Associates. **512** Figure 22.17: Lester V. Bergman & Associates. **523** Chapter 23 Opener: Child being inoculated at health clinic. Bob Daemmrich/The Image Works. **526** Figure 23.2a: G. Musil/Visuals Unlimited. **526** Figure 23.3: K. G. Murti/Visuals Unlimited. **528** Figure 23.5: Lee D. Simon/SS/Photo Researchers. **528** Figure 23.6: Lee D. Simon/SS/Photo Researchers. **530** Figuree 23.8: Science VU/Visuals Unlimited. **535** Figure 23.11: Merz/ Biological Photo Service. **536** 23.12a1: Fred Hossler/ Visuals Unlimited. **536** Figure 23.12a2: David M. Phillips/Visuals Unlimited. **536** Figure 23.12b: Fred Hossler/Visuals Unlimited. **542** Figure 23.13c: David M. Phillips/Visuals Unlimited. **540, center** Biophoto Associates/SS/Photo Researchers. **540, bottom** John D. Cunningham/Visuals Unlimited. **541, top** Raymond R. Otero/Visuals Unlimited. **541, bottom** Biological Photo Service. **542** Figure 23.13a: CNRI/SPL/Photo Researchers. **542** Figure 23.13b: Dr. Tony Brain/Photo Researchers. **542** Figure 23.13c: USDA/SS/Photo Researchers. **543** Figure 23.14a: Larry Jansen/Visuals Unlimited. **543** Figure 23.14b: Arthur M. Siegelman/Visuals Unlimited. **543** Figure 23.14c: Arthur M. Siegelman/Visuals Unlimited. **545** Figure 23.16a: Everett S. Beneke/Visuals Unlimited. **545** Figure 23.16b: Winograd/Biological Photo Service. **546** Figure 23.18: John Durham/SPL/Photo Researchers. **552** Chapter 24 Opener: Cancer cell being swallowed by macrophage. Jay Wasserman/Visuals Unlimited. **553** Figure 24.1: John Durham/SPL/Photo Researchers. **554** Figure 24.3: David M. Phillips/Visuals Unlimited. **560** Figure 24.8: Boehringer Mannheim, Biochemical Products Division, Indianapolis, IN. **572** Figure 24.14: Dr. Jeremy Burgess/SPL/Photo Researchers. **577** Chapter 25 Opener: Dinosaur bones discovered at archaeological dig. David Cupp/Woodfin Camp & Associates. **581** Figure 25.2a: Manfred Danegger/Peter Arnold, Inc. **581** Figure 25.2b: R. S. Virdee/Grant Heilman Photography. **581** Figure 25.2c: Gerard Lacz/Peter Arnold, Inc. **581** Figure 25.2d: Kent & Donna Dannen/ Photo Researchers. **581** Figure 25.2e: Kjell B. Sandved/ Photo Researchers. **581** Figure 25.2f: Jeanne White/ Photo Researchers. **583** Figure 25.3a: Grant Heilman Photography. **583** Figure 25.3b: Runk/Schoenberger/ Grant Heilman Photography. **583** Figure 25.3c: Runk/ Schoenberger/Grant Heilman Photography. **583** Figure 25.3d: AP/Wide World. **584** Figure 25.4a: William J. Weber/Visuals Unlimited. **584** Figure 25.4b: John D. Cunningham/Visuals Unlimited. **584** Figure 25.4c: John D. Cunningham/Visuals Unlimited. **592** Figure 25.8a: Milton H. Tierney, Jr./Visuals Unlimited. **592** Figure 25.8b: Bruce J. Zehnder/Peter Arnold, Inc. **592** Figure 25.8c: Tom McHugh/Photo Researchers. **592** Figure 25.8d: Tom McHugh/Photo Researchers. **593** Figure 25.9a: Tom McHugh/Photo Researchers. **593** Figure 25.9b: F. Gahier/Photo Researchers. **593** Figure 25.9c: Ken Lucas/Biological Photo Service. **593** Figure 25.9d: Michael C. T. Smith/Photo Researchers. **594** Figure 25.10a: Tom McHugh/Photo Researchers. **594** Figure 25.10b: BIOS/Compose/Visage/Peter Arnold, Inc. **594** Figure 25.10c: Biological Photo Service. **594** Figure 25.10d: George Holton/Photo Researchers. **596** Figure 25.12a: Institute of Human Origins. **596** Figure 25.12b: John Reader/SPL/Photo Researchers. **596** Figure 25.12c: John Reader/SPL/Photo Researchers. **600** Figure 25.15: Don W. Fawcett/Visuals Unlimited. **600** Figure 25.16a: John D. Cunningham/Visuals Unlimited. **600** Figure 25.16b: John D. Cunningham/Visuals Unlimited. **601** Figure 25.17: John D. Cunningham/Visuals Unlimited. **601** Figure 25.18: John D. Cunningham/Visuals Unlimited. **602** Figure 25.19: Professor Ralph S. Solecki, Texas A&M University, College Station, Texas. **603** Figure 25.20a: John D. Cunningham/Visuals Unlimited. **603** Figure 25.20b: John D. Cunningham/Visuals Unlimited. **606** Chapter 26 Opener: Iridescent hues of oil slick. Kerry T. Givens/Tom Stack & Associates. **608** Figure 26.1: NASA. **609** Figure 26.2a: J. Fossett/The Image Works. **609** Figure 26.2b: Phil Degginger. **609** Figure 26.2c: E. R. Degginger. **610, top** Stephen J. Krasemann/Photo Researchers. **610, center** William J. Weber/Visuals Unlimited. **610, bottom** Glenn Oliver/ Visuals Unlimited. **611, top** E. R. Degginger. **611, center** John D. Cunningham/Visuals Unlimited. **611, bottom** Grant Heilman Photography. **612, left** Grant Heilman Photography. **612, right** Greg Vaughn/Tom Stack & Associates. **615** Figure 26.5: W. J. Weber/Visuals Unlimited. **617** Figure 26.7a: Hans Pfletschinger/Peter Arnold, Inc. **617** Figure 26.7b: John D. Cunningham/Visuals Unlimited. **617** Figure 26.7c: E. R. Degginger. **617** Figure 26.7d: Hans Pfletschinger/Peter Arnold, Inc. **617** Figure 26.7e: Grant Heilman Photography. **617** Figure 26.7f: R. S. Virdee/Grant Heilman Photography. **617** Figure 26.7g: David M. Phillips/Visuals Unlimited. **619** Figure 26.8b: Tom Adams/Visuals Unlimited. **623** Figure 26.10: Walter Hodge/Peter Arnold, Inc. **624** Figure 26.12: Will & Deni McIntyre/Photo Researchers. **625** Figure 26.13b: George D. Doge & Dale R. Thompson/Tom Stack & Associates. **627** Figure 26.14a: A. J. Copley/ Visuals Unlimited. **627** Figure 26.14b: Runk/Schoenberger/Grant Heilman Photography. **627** Figure 26.14c: R. Wright/Visuals Unlimited. **627** Figure 26.14d: S. Compoint/Sygma. **629** Figure 26.15a: Harvey Lloyd/ Peter Arnold, Inc. **629** Figure 26.15b: William Felger/ Grant Heilman Photography. **629** Figure 26.15c: Theodore Vogel/Photo Researchers. **629** Figure 26.15d: Frank L. Lambrecht/Visuals Unlimited. **629** Figure 26.15e: Grant Heilman Photography. **629** Figure 26.15f: Robert E. Ford/Terraphotographics/Biological Photo Service. **630** Figure 26.16a: G. Prance/Visuals Unlimited. **630** Figure 26.17: Bill O'Connor/Peter Arnold, Inc. **631** Figure 26.18: David Burnett/Contact Press Images/ Woodfin Camp & Associates.

Index

THE METRIC SYSTEM

METRIC PREFIXES

(units: gram, meter, and liter are common suffixes)

Prefix	Multiple	Symbol
(greater than one)		
deka	10	da
hecto	10^2	h
kilo	10^3	k
mega	10^6	M
(less than one)		
deci	10^{-1}	d
centi	10^{-2}	c
milli	10^{-3}	m
micro	10^{-6}	μ
nano	10^{-9}	n
pico	10^{-12}	p

METRIC LENGTH

1 meter (the unit) × 10 = dekameter (10m)
100 = hectometer (10^2 m)
1,000 = kilometer (10^3 m)
1,000,000 = megameter (10^6 m)

1 meter ÷ 10 = decimeter (10^{-1} m)
100 = centimeter (10^{-2} m)
1,000 = millimeter (10^{-3} m)
1,000,000 = micrometer (10^{-6} m)
1,000,000,000 = nanometer (10^{-9} m)
1,000,000,000,000 = picometer (10^{-12} m)
10,000,000,000 = Angstrom (Å) (10^{-10} m) (an older unit of measurement)

METRIC WEIGHTS OR MASSES

1 gram (the unit) × 1,000 = kilogram

1 gram ÷ 1,000 = milligram (mg) (10^{-3} g)
1,000,000 = microgram (μg) (10^{-6} g)
1,000,000,000 = nanogram (ng) (10^{-9} g)
1,000,000,000,000 = picogram (pg) (10^{-12} g)

METRIC-ENGLISH CONVERSIONS

Length

English (USA)	= Metric
inch	= 2.54 cm, 25.4 mm
foot	= 0.30 m, 30.48 cm
yard	= 0.91 m, 91.44 cm
mile (statute) (5,280 ft)	= 1.61 km, 1609 m
mile (nautical) (6077 ft, 1.15 statute mi)	= 1.85 km, 1852 m

Metric	= English (USA)
millimeter	= 0.039 in
centimeter	= 0.39 in
meter	= 3.28 ft, 39.37 in
kilometer	= 0.62 mi, 1,091 yd, 3,274 ft

Weight

English (USA)	= Metric
grain	= 64.80 mg
ounce	= 28.35 g
pound	= 453.60 g, 0.45 kg
ton (short—2000 lb)	= 0.91 metric tons (907 kg)

Metric	= English (USA)
milligram	= 0.02 grains (0.000035 oz)
gram	= 0.03502 oz
kilogram	= 35.27 oz., 2.20 lb
metric ton (1000 kg)	= 1.10 tons

Volume

English (USA)	= Metric
cubic inch	= 16.39 cc
cubic foot	= 0.03 m^3
cubic yard	= 0.765 m^3
ounce	= 0.03 l (30 ml or cc)*
pint	= 0.47 l
quart	= 0.95 l
gallon	= 3.79 l

Metric	= English (USA)
milliliter	= 0.03 oz
liter	= 2.12 pt
liter	= 1.06 qt
liter	= 0.26 gal

1 liter ÷ 1,000 = milliliter or cubic centimeter (10^{-3}l)
1 liter ÷ 1,000,000 = microliter (10^{-6}l)

**Note:* 1 ml = 1 cc